Benthic Habitats and the Effects of Fishing

Benthic Habitats and the Effects of Fishing

Edited by

Peter W. Barnes

U.S. Geological Survey, Menlo Park, California

James P. Thomas

NOAA, Silver Spring, Maryland

American Fisheries Society Symposium 41

Proceedings of Symposium on Effects of Fishing Activities on
Benthic Habitats: Linking Geology, Biology,
Socioeconomics, and Management
Held in Tampa, Florida, USA
12–14 November 2002

American Fisheries Society
Bethesda, Maryland
2005

Suggested citation formats are

Entire book

Barnes, P. W., and J. P. Thomas, editors. 2005. Benthic habitats and the effects of fishing. American Fisheries Society, Symposium 41, Bethesda, Maryland.

Chapter in book

Jennings, S., S. Freeman, R. Parker, D. E. Duplisea, and T. A. Dinmore. 2005. Ecosystem consequences of bottom fishing disturbance. Pages 73–90 *in* P. W. Barnes and J. P. Thomas, editors. Benthic habitats and the effects of fishing. American Fisheries Society, Symposium 41, Bethesda, Maryland.

Front cover images: Top left, Peter Auster and Paul Donaldson, NURC, University of Connnecticut; top right, NOAA/NOS/NCCOS/CCMA Biogeography Program; middle, Kurt Byers, Alaska Sea Grant College Program; bottom left, NOAA; bottom right, Abigail Poray, NOAA and the Florida Keys National Marine Sanctuary.

© Copyright 2005 by the American Fisheries Society

All rights reserved. Photocopying for internal or personal use, or for the internal or personal use of specific clients, is permitted by the American Fisheries Society (AFS) provided that the appropriate fee is paid directly to Copyright Clearance Center (CCC), 222 Rosewood Drive, Danvers, Massachusetts 01923, USA; phone: 978-750-8400. Request authorization to make multiple copies for classroom use from CCC. These permissions do not extend to electronic distribution or long-term storage of multiple articles or to copying for resale, promotion, advertising, general distribution, or creation of new collective works. For such uses, permission or license must be obtained from AFS.

Printed in the United States of America on acid-free paper.

Library of Congress Control Number 2005924550
ISBN 1-888569-60-3   ISSN 0892-2284

American Fisheries Society Web site: *www.fisheries.org*

American Fisheries Society
5410 Grosvenor Lane, Suite 110
Bethesda, Maryland 20814-2199, USA

Contents

Symbols and Abbreviations xix

Introduction

Benthic Habitats and Effects of Fishing: Linking Geology, Biology, Socioeconomics,
and Management in Fisheries—An Introduction 3
James P. Thomas and Peter W. Barnes

Defining the Issues: Status, Management Needs, and Livelihoods

Keeping Our Fisheries Sustainable 11
William T. Hogarth

The Challenge of Protecting Fish Habitat through the Magnuson–Stevens Fishery
Conservation and Management Act 19
Clarence Pautzke

Challenges, Objectives, and Sustainability: Benthic Community, Habitats and Management
Decision-Making 41
Jake Rice

The Role of Scientific Advice in the Management of Benthic Fisheries in Mexico: Present
Status and Perspectives 59
Francisco Arreguín-Sánchez

Ecosystem Consequences of Bottom Fishing Disturbance 73
Simon Jennings, Steve Freeman, Ruth Parker, Daniel E. Duplisea, and Tracy A. Dinmore

National Research Council Study on the Effects of Trawling and Dredging on Seafloor
Habitat 91
*John Steele, Susan J. Roberts, Dayton L. Alverson, Peter Auster, Jeremy Collie,
Joseph T. DeAlteris, Linda Deegan, Elva Escobar Briones, Stephen J. Hall,
Gordon H. Kruse, Caroline Pomeroy, Kathryn M. Scanlon, and Priscilla Weeks*

Destructive Fishing Practices and Evolution of the Marine Ecosystem-Based Management
Paradigm 101
Elliott A. Norse

Symposium Abstract: Approaches to EFH Management for Alaska Groundfish Fisheries
that Fulfill Habitat Management Objectives and Maintain Viable Groundfish Fisheries* 115
J. R. Gauvin

Benthic Habitat Characterization and Advanced Technologies and Their Applications

Fish Habitat Studies: Combining High-Resolution Geological and Biological Data 119
W. Waldo Wakefield, Curt E. Whitmire, Julia E. R. Clemons, and Brian N. Tissot

Benthic Habitat Mapping with Advanced Technologies and Their Application 139
James V. Gardner and Larry A. Mayer

Characterization of Benthic Habitat on Northeastern Georges Bank, Canada 141
Vladimir E. Kostylev, Brian J. Todd, Oddvar Longva, and Page C. Valentine

Mapping and Characterizing Subtidal Oyster Reefs Using Acoustic Techniques,
Underwater Videography, and Quadrat Counts 153
Raymond E. Grizzle, Larry G. Ward, Jamie R. Adams, Semme J. Dijkstra, and Brian Smith

Refining Estimates of Potential White Abalone Habitat at Northern Anacapa Island,
California Using Acoustic Backscatter Data 161
Guy R. Cochrane, John L. Butler, and Gary E. Davis

Using Laser Technology to Characterize Substrate Morphology and Geology of Selected
Lake Trout Spawning Habitat in Northern Lake Michigan 165
Peter W. Barnes, Guy W. Fleischer, James V. Gardner, and Kristen M. Lee

Ground-Truthing Benthic Habitat Characteristics Using Video Mosaic Images 171
George R. Cutter, Jr., Yuri Rzhanov, Larry A. Mayer, and Raymond E. Grizzle

Advances in Processing and Collecting Multibeam Echosounder Data for Seabed
Habitat Mapping 179
Douglas Lockhart, Robert J. Pawlowski, and Edward J. Saade

Classification of Marine Sublittoral Habitats, with Application to the Northeastern
North America Region 183
Page C. Valentine, Brian J. Todd, and Vladimir E. Kostylev

Symposium Abstract: Benthic Habitat Characterization of the Grays Reef National Marine
Sanctuary Using Sidescan, Multibeam, and GIS Techniques* 201
C. Alexander, G. McFall, T. Battista, and R. Bohne

Symposium Abstract: Distribution of Acoustic Backscatter Imagery from NOAA
Hydrographic Surveys* 201
J. K. Brown, D. W. Pritchard, and G. T. Noll

Symposium Abstract: Mapping Seagrass Boundaries with Waveform-Resolving Lidar:
A Preliminary Assessment* 202
T. D. Clayton, J. C. Brock, and C. W. Wright

Symposium Abstract: Fast-Track Methods for Assessing Trawl Impacts* 203
*R. A. Coggan, C. J. Smith, R. J. A. Atkinson, K. N. Papadopoulou, T. D. I. Stevenson,
P. G. Moore, and I. D. Tuck*

Symposium Abstract: Using Lasers to Investigate Deepwater Habitats in the Monterey
Bay National Marine Sanctuary off Central California* 204
C. B. Grimes, M. Yoklavich, W. Wakefield, and H. G. Greene

Symposium Abstract: An Assessment of Fish and Invertebrate Communities along
Trans-Pacific Cable Lines: A Pilot Study with Implications for Marine Reserve Planning* ... 204
T. D. Hart and S. S. Heppell

Symposium Abstract: Benthic Habitat in the Gulf of Alaska: Biological Communities,
Geological Habitat, and Fishing Intensity* 205
J. Heifetz, D. L. Courtney, J. T. Fujioka, H. G. Greene, P. Malecha, and R. P. Stone

CONTENTS

Symposium Abstract: Quantitative Seafloor Habitat Classification Using GIS Terrain
Analysis: Effects of Data Density, Resolution, and Scale* 206
P. Iampietro and R. Kvitek

Symposium Abstract: Sampling Macrozoobenthos from Zebra Mussel Communities
in Lake Erie* 206
G. W. Kennedy, M. C. Fabrizzio, M. A. Blouin, and J. F. Savino

Symposium Abstract: Characterization of Coastal Great Lakes Benthic Habitat* 207
S. J. Lozano, M. Blouin, and N. Wattrus

Symposium Abstract: A System for Classification of Habitats in Estuarine and Marine
Environments: Florida Perspective* 207
K. A. Madley

Symposium Abstract: Sea Floor Mapping on the Scotian Shelf and the Gulf of Maine:
Implications for the Management of Ocean Resources* 208
R. A. Pickrill and B. J. Todd

Symposium Abstract: Seabed Classification with Multibeam Sonars for Mapping
Benthic Habitat* 209
J. M. Preston, A. C. Christney, W. T. Collins, and B. D. Bornhold

Symposium Abstract: Quantitative Measures of Acoustic Diversity to Support Benthic
Habitat Characterization* 209
J. M. Preston, A. C. Christney, W. T. Collins, and R. A. McConnaughey

Symposium Abstract: usSEABED: Towards Unifying Knowledge of Geologic Conrols
on Benthic Habitats* 210
*J. A. Reid, C. J. Jenkins, M. E. Field, M. Zimmermann, S. J. Williams, J. M. Currence,
C. E. Box, and J. V. Gardner*

Symposium Abstract: Fluorescence Imaging Laser Line Scan (FILLS) Imagery for
High-Resolution Benthic Habitat Characterization* 211
M. P. Strand

Symposium Abstract: The Use of Field Calibrated Side-Scan Acoustic Reflectance
Patterns to Quantify and Track Alterations to Benthic Habitat Associated with
Louisiana's Oyster Industry* 211
C. Wilson, H. Roberts, Y. Allen, and J. Supan

Understanding Chronic and Event-Driven Natural Change to Benthic Habitats

Living with Change: Response of the Sea Floor to Natural Events 215
Michael E. Field

Sediment Oxygen Consumption in the Southwestern Gulf of Mexico 219
Elva Escobar and Luisa I. Falcon

Discovery of 100–160-Year-Old Iceberg Gouges and Their Relation to Halibut Habitat
in Glacier Bay, Alaska 235
Paul R. Carlson, Philip N. Hooge, and Guy R. Cochrane

Symposium Abstract: Analyzing Time-Lapse Photographs of the Sea Floor for Changes
in Benthic Community Activity*.. 245
S. E. Beaulieu, H. Singh, and K. L. Smith, Jr.

Symposium Abstract: Fishing Effects on Habitat: The Potential Consequences of
Removing such Habitat Engineers as Red Grouper *Epinephelus morio**............................ 245
F. C. Coleman, C. C. Koenig, M. W. Miller, S. A. Heppell, and K. Scanlon

Symposium Abstract: Geologic Development and Longevity of Continental Shelf
Mudbelt Habitat during the Holocene in the Monterey Bay National Marine Sanctuary,
California*.. 246
E. E. Grossman, M. E. Field, and S. L. Eittreim

Symposium Abstract: Effects of Ice Gouging on Community Structure and the
Abundance of Pacific Halibut *Hippoglossus stenolepis*: Disturbance Does Not Necessarily
Mean Negative Fisheries Effects*.. 247
P. N. Hooge, P. R. Carlson, and G. R. Cochrane

Symposium Abstract: Biodiversity Changes in Space and Time in the Gulf of Alaska:
An Ecosystem Measure of Fishing Effects on Habitat*... 247
R. F. Reuter and S. Gaichas

Linking Fisheries and Supporting Ecosystems to Benthic Habitat Character and Dynamics

Linking Fisheries to Benthic Habitats at Multiple Scales: Eastern Scotian Shelf Haddock 251
John T. Anderson, Jim E. Simon, Don C. Gordon, and Peter C. Hurley

Combining Scientific and Fisher's Knowledge to Identify Possible Groundfish Essential Fish
Habitats... 265
*Melanie Bergmann, Hilmar Hinz, Robert E. Blyth, Michael J. Kaiser, Stuart I. Rogers, and
Mike J. Armstrong*

Delineating Juvenile Red Snapper Habitat on the Northern Gulf of Mexico Continental Shelf 277
*William F. Patterson, Charles A. Wilson, Samuel J. Bentley, James H. Cowan, Tyrrell Henwood,
Yvonne C. Allen, and Triniti A. Dufrene*

Living Substrate in Alaska: Distribution, Abundance, and Species Associations 289
Patrick W. Malecha, Robert P. Stone, and Jonathan Heifetz

Pockmarks on the Outer Shelf in the Northern Gulf of Mexico: Gas-Release Features
or Habitat Modifications by Fish?.. 301
Kathryn M. Scanlon, Felicia C. Coleman, and Christopher C. Koenig

Symposium Abstract: Understanding the Complex Nature of Fish–Seagrass
Associations*.. 313
T. J. Anderson

Symposium Abstract: Geoacoustic and Geological Characterization of Juvenile Red Snapper
Habitat, Northern Gulf of Mexico Continental Shelf*... 313
S. J. Bentley, W.F. Patterson, Y. Allen, W. Vienne, and C. Wilson

Symposium Abstract: Habitat Associations of Upper Slope Rockfishes (*Sebastes* spp.)
and Co-Occurring Demersal Fishes in Ascension Canyon, California*........................... 314
J. J. Bizarro, J. M. Field, H. G. Greene, R. N. Lea, and J. deMarignac

CONTENTS

Symposium Abstract: Data Sets Relevant to Identification of Essential Fish Habitat (EFH) on the Gulf of Mexico Continental Shelf and for Estimation of Effects of Shrimp Trawling Gear* 315
P. Caldwell and P. Sheridan

Symposium Abstract: Habitat and Species Associations of Demersal Fish and Benthic Invertebrates in the New York Bight Apex* 315
S. Chang, J. Vitaliano, and F. Steimle

Symposium Abstract: Fish Landings, Discards, and Benthic Material from Otter Trawling in the Western English Channel* 316
S. P. Cotterell

Symposium Abstract: Decreasing Habitat Disturbance by Improving Fish Stock Assessments: A New Method of Remote Species Identification and Quantification* 316
D. F. Doolittle, M. R. Patterson, Z.-U. Rahman, and R. Mann

Symposium Abstract: Small-Scale Analysis of Subtidal Fish Guilds and Associated Habitat Characteristics along Central California* 317
J. M. Field, M. M. Yoklavich, G. M. Caillet, S. Bros, J. deMarignac, and R. N. Lea

Symposium Abstract: The Sensitivity of Fish and Macro-Epifauna to Habitat Change: An Analytical Approach* 318
S. M. Freeman and S. I. Rogers

Symposium Abstract: Integration of Acoustic Seabed Classification and Fish Census Data for Determining Appropriate Boundaries of Marine Protected Areas* 318
A. C. R. Gleason, A. M. Ecklund, R. P. Reid, D. E. Harper, D. B. McClellan, and J. Schull

Symposium Abstract: Effects of Fishing on the Mid-Atlantic Tilefish Habitat: Restructuring a Structured Habitat* 319
V. G. Guida, P. C. Valentine, and F. Almeida

Symposium Abstract: Development of an Electronic Logbook to Assess Shrimp-Trawl Catch, Effort, and Associated Environmental Data in Areas Fished off Florida and Texas* 320
P. J. Rubec, A. Jackson, C. Ashbaugh, and S. Versaggi

Symposium Abstract: Linking Predator and Prey Species Dynamics in Deep-Water Reefs of the Northeastern Gulf of Mexico* 320
P. E. Thurman, G. Dennis, K. Sulak, and R. S. McBride

Effects of Fishing: Assessment and Recovery

Effects of Fishing on Gravel Habitats: Assessment and Recovery of Benthic Megafauna on Georges Bank 325
Jeremy S. Collie, Jerome M. Hermsen, Page C. Valentine, and Frank P. Almeida

The Effects of Area Closures on Georges Bank 345
Jason Link, Frank Almeida, Page Valentine, Peter Auster, Robert Reid, and Joseph Vitaliano

Effects of Fisheries on Deepwater Gorgonian Corals in the Northeast Channel, Nova Scotia 369
Pal B. Mortensen, Lene Buhl-Mortensen, Donald C. Gordon, Jr., Gordon B. J. Fader, David L. McKeown, and Derek G. Fenton

Susceptibility of the Soft Coral *Gersemia rubiformis* to Capture by Hydraulic Clam
Dredges off Eastern Canada: The Significance of Soft Coral–Shell Associations 383
*Kent D. Gilkinson, Donald C. Gordon, Jr., David McKeown, Dale Roddick, Ellen L. Kenchington,
Kevin G. MacIsaac, Cynthia Bourbonnais, and W. Peter Vass*

Effects of Experimental Otter Trawling on the Feeding of Demersal Fish on Western Bank,
Nova Scotia 391
*Ellen L. Kenchington, Donald C. Gordon, Jr., Cynthia Bourbonnais, Kevin G. MacIsaac,
Kent D. Gilkinison, David L. McKeown, and W. Peter Vass*

Summary of the Grand Banks Otter Trawling Experiment (1993–1995): Effects on
Benthic Habitat and Macrobenthic Communities 411
*Donald C. Gordon, Jr., Kent D. Gilkinson, Ellen L. Kenchington, Cynthia Bourbonnais,
Kevin G. MacIsaac, David L. McKeown, and W. Peter Vass*

Effects of Chronic Bottom Trawling on the Size Structure of Soft-Bottom Benthic
Invertebrates 425
Robert A. McConnaughey, Stephen E. Syrjala, and C. Braxton Dew

Effects of Commercial Otter Trawling on Benthic Communities in the Southeastern Bering Sea 439
Eloise J. Brown, Bruce Finney, Sue Hills, and Michaela Dommisse

Effects of Bottom Trawling on Soft-Sediment Epibenthic Communities in the
Gulf of Alaska 461
Robert P. Stone, Michele M. Masuda, and Patrick W. Malecha

Biological Traits of the North Sea Benthos: Does Fishing Affect Benthic Ecosystem
Function? 477
Julie Bremner, Chris L. J. Frid, and Stuart I. Rogers

The Impact of Trawling on Benthic Nutrient Dynamics in the North Sea: Implications
of Laboratory Experiments 491
Phil Percival, Chris Frid, and Rob Upstill-Goddard

Potential Impacts of Deep-Sea Trawling on the Benthic Ecosystem along the Northern
European Continental Margin: A Review 503
John D. Gage, J. Murray Roberts, John P. Hartley, and John D. Humphery

Immediate Effects of Experimental Otter Trawling on a Sub-Arctic Benthic Assemblage
inside Bear Island Fishery Protection Zone in the Barents Sea 519
*Tina Kutti, Tore Høisæter, Hans Tore Rapp, Odd-Børre Humborstad, Svein Løkkeborg, and
Leif Nøttestad*

Preliminary Results on the Effect of Otter Trawling on Hyperbenthic Communities in
Heraklion Bay, Cretan Sea, Eastern Mediterranean 529
Panayota T. Koulouri, Costas G. Dounas, and Anastasios Eleftheriou

The Effect of Different Types of Otter Trawl Ground Rope on Benthic Nutrient Releases
and Sediment Biogeochemistry 539
*Costas G. Dounas, Ian M. Davies, Peter J. Hayes, Christos D. Arvanitidis, and
Panayota T. Koulouri*

Trawl Fishing Disturbance and Medium-Term Microfaunal Recolonization Dynamics:
A Functional Approach to the Comparison between Sand and Mud Habitats in the
Adriatic Sea (Northern Mediterranean Sea) 545
Fabio Pranovi, Sasa Raicevich, Simone Libralato, Filippo Da Ponte and, Otello Giovanardi

CONTENTS

Short-Term Effects of the Cessation of Shrimp Trawling on Texas Benthic Habitats 571
Peter Sheridan and Jennifer Doerr

Effect of Caribbean Spiny Lobster Traps on Seagrass Beds of the Florida Keys National
Marine Sanctuary: Damage Assessment and Evaluation of Recovery 579
Amy V. Uhrin, Mark S. Fonseca, and Gregory P. DiDomenico

Symposium Abstract: The Effectiveness of Marine Protected Areas on Fish and Benthic
Fauna: The Georges Bank Closed Area II Example* 589
*F. Almeida, P. Valentine, R. Reid, L. Arlen, P. Auster, J. Cross, V. Guida, J. Lindholm,
J. Link, D. Packer, J. Vitaliano, and A. Paulson*

Symposium Abstract: The Impact of Scallop Dredging on American Lobster *Homarus
americanus* in the Baie des Chaleurs, Canada* 590
P. Archambault and L. Gendron

Symposium Abstract: Effect of Shrimp Trawling on Snow Crab Resource in the
Northern Atlantic* 590
G. Brothers and J. J. Foster

Symposium Abstract: The Impact of Oyster Dredging on Blue Cod in New Zealand* 591
G.D. Carbines

Symposium Abstract: Impacts to Coral Reef Benthos from Lobster Trap Gear in the
Florida Keys National Marine Sanctuary* 592
M. Chiappone, D. W. Swanson, and S. L. Miller

Symposium Abstract: Spatial Distribution and Benthic Impacts from Hook-and-Line
Fishing Gear in the Florida Keys National Marine Sanctuary* 592
M. Chiappone, D. W. Swanson, and S. L. Miller

Symposium Abstract: Effects of Fishing on the Benthic Habitat and Fauna of Seamounts
on the Chatham Rise, New Zealand* 593
M. R. Clark, A. A. Rowden, and S. O'Shea

Symposium Abstract: Effects of Smooth Bottom Trawl Gear on Soft Bottom Habitat* 594
*C. L. Cogswell, B. Hecker, A. Michael, F. Mirarchi, J. Ryther, Jr., D. Stevenson,
R. Valente, and C. Wright*

Symposium Abstract: Effects of 135 years of Oyster (*Ostrea chilensis*) Fishing on the
Benthic Habitat, Associated Macrofaunal Assemblages, and Sediments of Foveaux Strait,
Southern New Zealand* 595
*H. J. Cranfield, K. P. Michael, G. Carbines, D. P. Gordon, B. Manighetti, A. Dunn,
and A. A. Rowden*

Symposium Abstract: The Theoretical and Methodological Basis of Estimations of the
Human-Made Influences (Fishing and Constructing) on the Benthic Habitats* 595
V. A. Emelyanov

Symposium Abstract: Impacts of Mobile Fishing Gear on Sponges and Gorgonian Corals
in the Gulf of Alaska* 596
J. L. Freese

Symposium Abstract: Effects of Fishing on Organic Carbon Content of Sand Habitats
on Georges Bank* 596
V. G. Guida, A. Paulson, P. C. Valentine, and L. Arlen

Symposium Abstract: A Before-After-Control-Impact Study of the Sea Scallop Fishing
Grounds of Georges Bank* 597
K. D. E. Stokesbury and B. Harris

Symposium Abstract: Impacts of Scallop Dredging on Marine Bottom Complexity and
Juvenile Fish Habitat* 598
F. Hartog, P. Archambault, and L. Fortier

Symposium Abstract: Community and Life History Divergence of Colonial Hydroids
(Cnidaria, Hydrozoa) from Heavily Trawled Scallop Grounds in the Bay of Fundy,
Eastern Canada* 598
L. M. Henry

Symposium Abstract: Analyzing the Effects of Trap Fishing in Coral Reef Habitats:
Methods and Preliminary Results* 599
R. L. Hill, P. F. Sheridan, R. S. Appledoorn, T. R. Matthews, and K. R. Uwate

Symposium Abstract: A Comparison of Habitat Structure in Fished and Unfished,
Mobile and Immobile Sand Habitats on Georges Bank (Northwest Atlantic)* 600
J. B. Lindholm, P. J. Auster, and P. Valentine

Symposium Abstract: Changes in the Benthic Invertebrate Assemblage following the
Establishment of a Protected Area, the "Plaice Box"* 600
G. J. Piet, J. A. Craeymeersch, and A. D. Rijnsdorp

Symposium Abstract: The Effectiveness of Marine Protected Areas on Fish and Benthic
Fauna: The Georges Bank Closed Area I Example* 601
R. Reid, F. Almeida, P. Valentine, L. Arlen, J. Cross, V. Guida, J. Link, D. McMillan,
S. Muraski, D. Packer, J. Vitaliano, and A. Paulson

Symposium Abstract: Physical and Biological Effects of Shrimp Trawling on Soft
Sediment Habitats in the Gulf of Maine* 602
A. W. Simpson and L. Watling

Symposium Abstract: Ecological Consequences of Lost Habitat Structure for
Commercially Significant Flatfishes: Habitat Choice and Vulnerability to Predators* 602
A. W. Stoner, C. L. Ryer, and R. A. McConnaughey

Symposium Abstract: Bottom Trawling Effects on Cerianthid Burrowing Anemone
Aggregations and Acadian Redfish Habitats in Mud to Muddy Gravel Seabeds of the
Stellwagon Bank National Marine Sanctuary Region, Gilf of Maine (Northwest Atlantic)* 603
P. C. Valentine, J. B. Lindholm, and P. J. Auster

Symposium Abstract: Why Fishing Gear Impact Studies Don't Tell Us What We Need
to Know* 604
L. Watling and C. Skinder

Symposium Abstract: Reduction of Species Diversity in a Cobble Habitat Subject to
Long-Term Fishing Activity* 605
L. Watling and A. Pugh

Symposium Abstract: Ecological Footprints of Scotian Shelf Groundfish Fisheries* 605
K. C. T. Zwanenburg, M. Showell, and S. Wilson

Comparison of Effects of Fishing with Effects of Natural Events and Non-Fishing Anthropogenic Impacts on Benthic Habitat

Comparison of Effects of Fishing with Effects of Natural Events and Non-Fishing Anthropogenic Impacts on Benthic Habitats 609
Hans J. Lindeboom

Extrapolating Extinctions and Extirpations: Searching for a Pre-Fishing State of the Benthos 619
Leonie A. Robinson and Chris L. J. Frid

Symposium Abstract: Using Side-Scan Sonar to Assess the Impact and Persistence of Natural and Anthropogenic Disturbance to Low-Relief Oyster Habitats in Coastal Louisiana* 629
Y. Allen, C. Wilson, H. Roberts, and J. Supan

Symposium Abstract: Survey of Fishing Gear and Fiber Optics Cable Impacts to Benthic Habitats in the Olympic Coast National Marine Sanctuary* 629
M. S. Brancato and C. E. Bowlby

Symposium Abstract: Shrimp and Crab Trawling Impacts on Estuarine Soft-Bottom Organisims* 630
L. B. Cahoon, M. H. Posey, W. H. Daniels, and T. D. Alphin

Symposium Abstract: Did Bottom Trawling in Bristol Bay's Red King Crab Brood-Stock Refuge Contribute to the Collapse of Alaska's Most Valuable Fishery?* 631
C. Braxton Dew and Rorbert A. McConnaughey

Symposium Abstract: Comparative Evaluation of Natural and Trawling Sediment Disturbance via Short-Lived Radionuclides, in situ Monitors and Remote Sensing Techniques in the Pamlico River Estuary, North Carolina* 632
J. E. Frank, D. R. Corbett, T. West, L. Clough, and W. Calfee

Symposium Abstract: Fishing and Environmental Disturbance Indicators in a Shrimp Fishing Ground at the Mexican Central Pacific* 632
E. Godinez-Dominguez, J. Freire, and G. Gonzalez-Sanson

Symposium Abstract: Benthic Perturbations from Walrus Foraging: Are They Similar to Trawling?* 633
C. V. Jay, L. C. Huff, and R. A. McConnaughey

Symposium Abstract: Controversy about Trawling and Santa Maria Key's Causeway Effects on Seagrass* 634
A. Quiros Espinosa, M. E. Perdomo López, and R. Arias Barreto

Symposium Abstract: Scaling of Natural and Anthropogenic Disturbance on the New York Bight Shelf: Implications for Tilefish Communities of the Shallow Continental Slope* 634
M. C. Sullivan, R. K. Cowen, K. W. Able, and M. P. Fahay

Symposium Abstract: Impacts of Trawling and Wind Disturbance on Water Column Processes in the Pamlico River Estuary, North Carolina* 635
T. L. West, D. R. Corbett, L. M. Clough, M. W. Calfee, and J. E. Frank

Extrapolation of Local and Chronic Effects of Fishing and Nonfishing Events to Significant Regions and Time Scales

Spatial and Temporal Scales of Disturbance to the Seafloor: A Generalized Framework for Active Habitat Management 639
Simon F. Thrush, Carolyn J. Lundquist, and Judi E. Hewitt

Muddy Thinking: Ecosystem-Based Management of Marine Benthos 651
Chris L. J. Frid, Leonie A. Robinson, and Julie Bremner

Linking Fine-Scale Groundfish Distributions with Large-Scale Seafloor Maps: Issues and Challenges of Combining Biological and Geological Data 667
Tara J. Anderson, Mary M. Yoklavich, and Stephen L. Eittreim

Spatial and Temporal Distributions of Bottom Trawling off Alaska: Consideration of Overlapping Effort When Evaluating the Effects of Fishing on Habitat 679
Craig S. Rose and Elaina M. Jorgensen

Hydraulic Clam Dredge Effects on Benthic Habitat off the Northeastern United States 691
David H. Wallace and Thomas B. Hoff

Symposium Abstract: The Spatial Extent and Nature of Mobile Bottom Fishing Methods within the New Zealand EEZ, 1989–90 to 1998–99* 695
S. J. Baird, N. W. Bagley, B. A. Wood, A. Dunn, and M. P. Beentjes

Symposium Abstract: Deepwater Trawl Fisheries Modify Benthic Community Structure in Similar Ways to Fisheries in Coastal Ecosystems* 695
M. Cryer, B. Hartill, and S. O'Shea

Symposium Abstract: Detecting the Effects of Fishing on Seabed Community Diversity: Importance of Scale and Sample Size* 696
M. J. Kaiser

Symposium Abstract: Spatial and Temporal Patterns in Trawling Activity in the Canadian Atlantic and Pacific* 696
D. W. Kulka and D. A. Pitcher

Symposium Abstract: A GIS Routine for Assessing Designs that Sample an Area of Fish or Lobster Traps* 697
G. A. Matthews, R. L. Hill, and P. F. Sheridan

Symposium Abstract: Essential Fish Habitat (EFH) in Alaska: Issues in Consistency and Efficiency When Using Geographical Information Systems (GIS) to Evaluate Effects to EFH* 698
R. F. Reuter, C. C. Coon, J. V. Olson, and M. Eagleton

Symposium Abstract: Spatial Distribution of Fishing Activity for Principal Commercial Fishing Gears Used in the Northeast Region of the United States, 1995–2000* 698
D. K. Stevenson

Symposium Abstract: Structure and Use of a Continental Slope Seascape: Insights for the Fishing Industry and Marine Resource Managers* 699
A. Williams, B. Barker, R. J. Kloser, N. J. Bax, and A. J. Butler

Social and Economic Issues and Effects

Perspectives on an Ethic toward the Sea 703
Stephen R. Kellert

Getting to the Bottom of It: Bringing Social Science into Benthic Habitat Management 713
Bonnie J. McCay

Place Matters: Spatial Tools for Assessing the Socioeconomic Implications of Marine
Resource Management Measures on the Pacific Coast of the United States 727
Astrid J. Scholz, Mike Mertens, Debra Sohm, Charles Steinback, and Marlene Bellman

When Do Marine Protected Areas Pay? An Analysis of Stylized Fisheries 745
Harold F. Upton and Jon G. Sutinen

Symposium Abstract: Development of a West Coast Cooperative Research Program,
Working Together toward Better Information* 759
J. Bloesser

Symposium Abstract: Promoting Environmental Awareness and Developing
Conservation Harvesting Technology for the Fishing Industry* 759
G. Brothers

Symposium Abstract: Occupational Endurance and Contested Resources: Managing the
Cultural and Economic Tensions of Lake Michigan's Commercal Fishery* 760
M. J. Chiarappa

Symposium Abstract: Building a Database for Benthic Fisheries Using Tourist Income* 760
G. C. Lane

Symposium Abstract: Impacts of Marine Reserves: How Fishermen Behavior Matters* 761
J. E. Wilen

Determinations of "To the Extent Practicable" Phrase in U.S. Law and Other Legal Issues Concerning Fishing Effects

Symposium Abstract: The Legal Requirement to Address Fishing Effects on Essential
Fish Habitat: Thresholds, Qualifiers, and the Burden of Proof* 765
A. Rieser

Minimizing the Adverse Effects of Fishing on Benthic Habitats: Alternate Fishing Techniques and Policies

Impacts of Fishing Activities on Benthic Habitat and Carrying Capacity: Approaches to
Assessing and Managing Risk 769
Michael J. Fogarty

An Alternative Paradigm for the Conservation of Fish Habitat Based on Vulnerability, Risk,
and Availability Applied to the Continental Shelf of the Northwestern Atlantic 785
Joseph DeAlteris

Habitat and Fish Populations in the Deep-Sea *Oculina* Coral Ecosystem of the Western
Atlantic 795
*Christopher C. Koenig, Andrew N. Shepard, John K. Reed, Felicia C. Coleman,
Sandra D. Brooke, John Brusher, and Kathryn M. Scanlon*

The Impact of Demersal Trawling on Northeast Atlantic Deepwater Coral Habitats:
The Case of the Darwin Mounds, United Kingdom 807
A. J. Wheeler, B. J. Bett, D. S. M. Billett, D. G. Masson, and D. Mayor

Fishing Impacts on Irish Deepwater Coral Reefs: Making a Case for Coral Conservation 819
Anthony J. Grehan, Vikram Unnithan, Karine Olu-Le Roy, and Jan Opderbecke

Symposium Abstract: Biological and Socio-Economic Implications of a Limited-Access
Fishery Management System* 833
R. E. Blyth, M. J. Kaiser, G. Edwards-Jones, and P. J. B. Hart

Symposium Abstract: Results of a Workshop on the Effects of Fishing Gear on Benthic
Habitats off the Northeastern United States* 833
*L. A. Chiarella, D. K. Stevenson, C. D. Stephan, R. N. Reid, J. E. McCarthy,
M. W. Pentony, T. B. Hoff, C. D. Selberg, and K. A. Johnson*

Symposium Abstract: Changes in the Epibenthos Assemblages of the North Sea following
the Establishment of a Protected Area, the "Plaice Box"* 834
J. A. Craeymeersch and G. J. Piet

Symposium Abstract: Fishing for Shellfish in an Internationally Important Nature
Reserve: Do Current Policies Achieve Their Objectives?* 835
B. J. Ens, A. C. Smaal, and J. De Vlas

Symposium Abstract: Developing a Fisheries Ecosystem Plan for the North Sea* 835
*C. L. J. Frid, C. L. Scott, M. F. Borges, N. Daan, T. S. Gray, J. Hatchard, L. Hill,
O. A. L. Paramor, G. J. Piet, S. A. Ragnarsson, W. Silvert, and L. Taylor*

Symposium Abstract: The Ocean Habitat Protection Act: Overdue Protection for
Structurally Complex Seafloor Habitats* 836
H. Gillelan

Symposium Abstract: Reducing Seabed Contact of Bottom Trawls* 837
P. He

Symposium Abstract: Rapid Build-Up of Fish Biomass, but Still Declining Coral Reefs:
Why a Marine Fishery Designation Is Not Enough for the Protection of Reef Epibenthic
Communities* 838
E. A. Hernández-Delgado and A. M. Sabat

Symposium Abstract: Using Ideal Free Distribution Theory to Identify Potential Protected
Areas* 838
H. Hinz, M. J. Kaiser, M. Bergmann, and S. I. Rogers

Symposium Abstract: The Path towards Ecologically Sustainable Fisheries: A Case Study
in the Great Barrier Reef World Heritage Area* 839
D. Huber

Symposium Abstract: Monitoring Changes in the Fully Protected Zones of the Florida Keys
National Marine Sanctuary* 840
B. D. Keller

Symposium Abstract: Identification and Evaluation of Indicators for Environmental
Performance of European Marine Fisheries* 840
L.-H. Larsen, A. Zenetos, and N. Streftaris

Symposium Abstract: Decision Framework for Describing and Identifying EFH,
Mitigating Fishing Impacts, and Designating HAPC in Federal Fishery Management Plans* 841
G. B. Parkes, H. B. Lovett, and R. J. Trumble

Symposium Abstract: The Characteristics and Function of Commercial Fishing Gears:
How These Relate to Their Effects on Seafloor Habitats and the Pursuit of Ways to
Minimize Effects* 841
C. S. Rose

Symposium Abstract: Long-Term, Large-Scale Biological Surveys: A Necessary
Component of Fishery and Ecosystem Management* 842
C. Syms

Symposium Abstract: Approach to Evaluating Fishing Effects on EFH off Alaska* 842
D. Witherell and C. Coon

What Next? What Have We Learned? What More Do We Need to Know? What Should We Act on Right Now?

Moderated Panel and Open Discussion 847

Index 869

---

* Symposium abstracts have been reprinted from the Abstract Volume prepared for the 2002 Symposium on Effects of Fishing Activities on Benthic Habitats: Linking Geology, Biology, Socioeconomics, and Management without further review or editing.

Symbols and Abbreviations

The following symbols and abbreviations may be found in this book without definition. Also undefined are standard mathematical and statistical symbols given in most dictionaries.

A	ampere	K	Kelvin (degrees above absolute zero)
AC	alternating current	k	kilo (10^3, as a prefix)
Bq	becquerel	kg	kilogram
C	coulomb	km	kilometer
°C	degrees Celsius	l	levorotatory
cal	calorie	L	levo (as a prefix)
cd	candela	L	liter (0.264 gal, 1.06 qt)
cm	centimeter	lb	pound (0.454 kg, 454g)
Co.	Company	lm	lumen
Corp.	Corporation	log	logarithm
cov	covariance	Ltd.	Limited
DC	direct current; District of Columbia	M	mega (10^6, as a prefix); molar (as a suffix or by itself)
D	dextro (as a prefix)		
d	day	m	meter (as a suffix or by itself); milli (10^{23}, as a prefix)
d	dextrorotatory		
df	degrees of freedom	mi	mile (1.61 km)
dL	deciliter	min	minute
E	east	mol	mole
E	expected value	N	normal (for chemistry); north (for geography); newton
e	base of natural logarithm (2.71828...)		
e.g.	(exempli gratia) for example	N	sample size
eq	equivalent	NS	not significant
et al.	(et alii) and others	n	ploidy; nanno (10^{29}, as a prefix)
etc.	et cetera	o	ortho (as a chemical prefix)
eV	electron volt	oz	ounce (28.4 g)
F	filial generation; Farad	P	probability
°F	degrees Fahrenheit	p	para (as a chemical prefix)
fc	footcandle (0.0929 lx)	p	pico (10^{212}, as a prefix)
ft	foot (30.5 cm)	Pa	pascal
ft³/s	cubic feet per second (0.0283 m³/s)	pH	negative log of hydrogen ion activity
g	gram	ppm	parts per million
G	giga (10^9, as a prefix)	qt	quart (0.946 L)
gal	gallon (3.79 L)	R	multiple correlation or regression coefficient
Gy	gray		
h	hour	r	simple correlation or regression coefficient
ha	hectare (2.47 acres)		
hp	horsepower (746 W)	rad	radian
Hz	hertz	S	siemens (for electrical conductance); south (for geography)
in	inch (2.54 cm)		
Inc.	Incorporated	SD	standard deviation
i.e.	(id est) that is	SE	standard error
IU	international unit	s	second
J	joule	T	tesla

tris	tris(hydroxymethyl)-aminomethane (a buffer)	α	probability of type I error (false rejection of null hypothesis)
UK	United Kingdom	β	probability of type II error (false acceptance of null hypothesis)
U.S.	United States (adjective)		
USA	United States of America (noun)	Ω	ohm
V	volt	μ	micro (10^{-6}, as a prefix)
V, Var	variance (population)	′	minute (angular)
var	variance (sample)	″	second (angular)
W	watt (for power); west (for geography)	°	degree (temperature as a prefix, angular as a suffix)
Wb	weber		
yd	yard (0.914 m, 91.4 cm)	%	per cent (per hundred)
		‰	per mille (per thousand)

Introduction

Benthic Habitats and Effects of Fishing: Linking Geology, Biology, Socioeconomics, and Management in Fisheries—An Introduction

JAMES P. THOMAS

U.S. Department of Commerce, NOAA Fisheries, Silver Spring, Maryland 20910, USA

PETER W. BARNES

U.S. Department of Interior, U.S. Geological Survey, Menlo Park, California 94025, USA

Abstract. This volume focuses on the effects of fishing activities on benthic habitats and the related science and knowledge needed to understand and quantify those effects, as well as to suggest new ways to address these effects for sustainable fisheries and healthy, diverse ecosystems.

The papers in this volume and the fact that certain topics had limited representation suggest several information and research needs for the future. Foremost among these, in our opinion, are

- Systematic physical and biological characterization (maps) of the U.S. exclusive economic zone seabed in priority areas.

- A national seabed classification scheme.

- Linkage of fisheries and supporting communities to benthic habitat character and dynamics.

- Comparisons of natural change and human induced change caused by fishing and other activities.

- Temporal and spatial distributions of fishing effort by fishery and gear type.

- Extrapolation of site-specific, fishing-caused seabed habitat changes to effects on fishery stocks, the ecosystem and its goods and services, and ultimately the public.

- Understanding socioeconomic effects of fishing, management alternatives, and significance to society, the resource, and the ecosystem.

- Gear modification and development research to minimize adverse effects of fishing on seabed habitat.

- New technology for assessing seabed habitat.

- National research plan to fill the information needs of resource managers and policy makers required to address issue of alterations to seabed habitat caused by fishing.

- Discussion of restrictions or bans on trawling and use of certain other fishing gear in areas where coral, sponge, and other large-structure-forming, seabed organisms exist.

- Including less developed countries in research and management discussions addressing fisheries issues.

- Continued national and international interaction, communication, and development of constructive ways forward.

Background

This volume focuses on the effects of fishing activities on benthic habitats and the related science and knowledge needed to understand and quantify those effects, as well as to suggest new ways to address them for sustainable fisheries and healthy, diverse ecosystems.

Most science organizations are aware that public issues and forefront science can be found at the boundaries between the traditional science disciplines. Joint studies need to evolve into multidisciplinary studies and then on to completely integrated studies. In the marine realm, the benthic boundary between the geosphere and the hydrosphere is biologically rich and exploited by man (as is the atmosphere—geosphere boundary on land). Unlike onshore, however, offshore territory is held and managed as public lands (state and federal) almost exclusively. In the United States, the Department of the Interior (DOI) and the Department of Commerce's National Oceanic and Atmospheric Administration (NOAA) are two primary federal agencies charged with providing the scientific understanding of the character, resources, and processes of the marine benthic boundary ecosystem in the federal public domain while the states focus on ecosystems near shore.

In 1998, NOAA's National Marine Fisheries Service (NMFS) and DOI's U.S. Geological Survey (USGS) held a workshop to review future joint research questions related to fishery habitats and to address research priorities offshore of the United States and in the Great Lakes (Barnes et al. 1999). The workshop emphasized three related issues:

- Fisheries management under the reauthorized Magnuson-Stevens Fishery Conservation and Management Act (i.e., Sustainable Fisheries Act of 1996) addressing (1) description and identification of essential fish habitat (EFH), (2) evaluation of threats to EFH, and (3) measures needed to conserve EFH.

- Great Lakes commercial and recreational fisheries: (1) habitat description and characterization, (2) habitat perturbations, and (3) the potential for change and/or restoration.

- Coral reef resources—relationships between physical factors and loss of biological resources.

The workshop urged NOAA and USGS to combine efforts and developed two recommendations:

- Determine the effects of fishing gear on seabed habitats (EFH).

- Identify and map benthic habitat characteristics and extent of fishing impacts.

These recommendations remain as germane today as they were in 1998 and have yet to be addressed other than in piecemeal fashion.

The 1998 workshop and subsequent regional workshops also urged convening a national/international meeting to address these issues. The papers in this volume result from an international symposium, *Effects of Fishing Activities on Benthic Habitats: Linking Geology, Biology, Socioeconomics, and Management*, held November 12–14, 2002, in Tampa, Florida.

The goals of the symposium, and subsequently this volume, were and are to help ensure sustainable fisheries and healthy, diverse ecosystems. We sought to achieve this by advancing the scientific knowledge available to all stakeholders, but most particularly to resource managers and policy makers to assist them in evaluating and appropriately managing fishing activities that interact and affect benthic habitat. A symposium steering committee and program committee (Appendix) provided invaluable guidance and support for the symposium and this research volume. At the symposium, we encouraged presentations of different perspectives and approaches. Both poster and oral presentations given at the symposium (abstracts) as well as contributed and invited papers derived from the abstracts and meeting discussions are included here. The papers contained in this volume have undergone external and editorial review. The imbalance seen in the number of papers/abstracts per topic, we believe, is reflective not so much of interest in attending the symposium, as it is indicative of the state of the knowledge across these topics.

The symposium and this volume were organized along a path that evolved from defining the issues and information needs, through present knowledge of the benthic ecosystem, to understanding linkages and the effects of natural events, fishing, and other anthropogenic impacts, to assessing the economic and social effects, to existing and alternative management approaches (Figure 1). The issue of effects of fishing activities is extremely complex and controversial and involves commercial and recreational fishers, scientists, environmentalists, lawyers, fishery managers, economists, sociologists, and communities who either provide support for or are supported by fishing and the products of fishing. Good decisions regarding the management of fishing activities are required because such decisions affect people and their livelihoods as well as the ecosystem and the sustainability of fisheries. Good decisions require the best available information, open discussion, and involvement of all stakeholders. We hope that this volume provides some of the best available information for understanding and appropriately managing fishing activities to minimize adverse effects to seabed habitat for the sustainability of fisheries and healthy, diverse ecosystems.

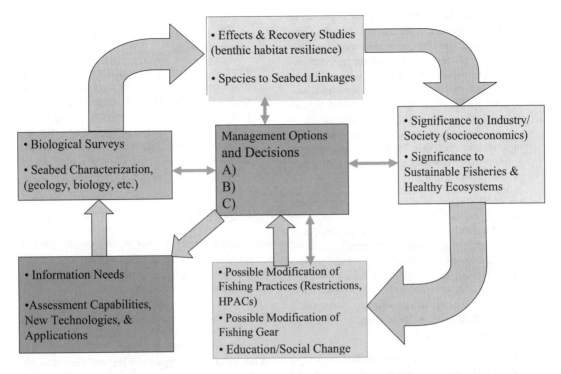

Figure 1. Conceptual diagram illustrating the sequence, interconnectedness, and linkages of the geological, biological, socioeconomic, and management aspects related to benthic habitats and the effects of fishing.

Shortcomings and Needs

The research papers and panel discussion that follows represent a solid cross section of research and analysis related to the physical, biological, socioeconomic, and management aspects of effects of fishing. We would be remiss if we did not point out areas of shortcomings and needs based on our impression from the symposium:

- Less than 5% of the U.S. exclusive economic zone seabed and associated habitat has been characterized and mapped in adequate detail and resolution to provide knowledge of event and chronic changes to habitats as well as extent of habitats. Many symposium participants and papers note the need for a systematic effort to characterize and map benthic habitats in priority areas (e.g., coral reefs, deepwater coral, areas of important fisheries and sanctuaries, primarily on the Continental Shelf and slope) at a sufficient resolution for management use. As part of this effort, a national seabed classification system needs to be developed with the involvement and concurrence of all stakeholders. Such a classification scheme would facilitate management of fisheries that migrate across multiple regions. International agreement on such a classification scheme would help facilitate management of transboundary migrants.

- Few papers herein link fisheries and supporting communities to benthic habitat character and dynamics. However, this appears to be a new, emerging area of research. More site specific work is needed, using new and existing high-resolution remote sensing tools, research submarines, and divers to accelerate progress.

- The need exists for extrapolation of site-specific, fishing-caused habitat alterations to effects on fishery stocks, the ecosystem and its goods and services (i.e., area over which a fishery is managed), and ultimately the public.

- The comparison between natural change (chronic and event driven) and human-induced change from fishing and other activities is poorly represented in this volume. Much effort has focused on natural physical change but little on the change to ecosystems, diversity, and especially the relative impact of human effects. This will require additional difficult and lengthy interdisciplinary, historical, and time series studies of natural and impacted ecosystems to understand the adverse impacts of natural and human events on fish habitat, fishery stocks, and the ecosystem. As part of this effort, precise information is required on the temporal and spatial distribution of fishing effort by fishery and gear type.

- Social and economic effects of fishing, management alternatives, and their significance to society (i.e., the fisher, the industry, the community, the nation, humanity), the resource (i.e., fisheries and protected species), and the ecosystem are among the most important issues to be considered, yet are represented here by few papers. Special efforts are required to more fully engage economists and sociologists on this issue as it affects all stakeholders from individuals seeking to maintain a livelihood to nations concerned with balance of trade payments and quality of life for their people within a healthy, diverse ecosystem.

- Surprisingly, gear modification and development research to minimize adverse effects of fishing on seabed habitat is represented by only one abstract, in spite of an active search and recruitment for the symposium. Increased research involving fishers and gear technologists is suggested that will modify or develop fishing gear that minimizes adverse effects on seabed habitat, protected resources, and bycatch while maintaining catch of targeted species with minimal additional cost to the fisher.

- New technology for assessing seabed habitat using acoustic and optical means, remotely operated vehicles (ROVs), and specialized software was highlighted and will support the research community and managers as this technology becomes funded and readily available. New technology is required particularly for assessing seabed habitat in water less than 30 m. For such shallow water, much of the current technology is either inefficient (time or cost) or does not fully address the needs.

- During the final panel discussion, it was recommended that a national research plan be developed to quantify the benefits and costs to fill the information needs required by resource managers and policy makers to address the issue of alterations to seabed habitat caused by fishing. Such a plan should not be done in isolation, but rather in context with other fisheries management issues.

- Considerable support was expressed during the panel discussion for a global ban on trawling and use of certain other fishing gear in areas where coral, sponge, and other large-structure-forming, seabed organisms exist. However, to move forward with this discussion requires knowledge of the distribution of these types of organisms, particularly deep-sea corals and associated species in areas where our knowledge is incomplete.

- During the panel session, several individuals expressed the need for developed countries to assist less developed countries in addressing fisheries issues by exporting simpler, less expensive ways to manage fisheries.

- Finally, as is often the case when different disciplines communicate, the language of one is often unclear to the other. Thus, for example, the geologist's, fisher's, manager's, and biologists' perceptions of "adverse effects" and need for action are often very different. Continued interaction and interdisciplinary efforts are key to effective communication and development of constructive ways forward.

Acknowledgments

We would like to express our appreciation for the support we have received in developing this publication—support from our agencies, guidance from our steering and program committees, the Ecological Society of America for the symposium, the American Fisheries Society for the publication, and the Homeland Foundation for its financial support. In addition, we could not have accomplished this effort without the willingness of our invited and contributed speakers and presenters, the active participation of panel members and audience, and the generous help of many reviewers of abstracts and papers.

References

Barnes, P., K. Koltes, and J. Thomas. 1999. Relating geology of benthic habitats to biological resources. Unpublished report of a joint USGS and NOAA Scientific Workshop held May 5-7, 1998, Reston, Virginia.

Appendix

Steering Committee

–Emory Anderson (NOAA/NMFS/Office of Science and Technology Liaison to OAR/Sea Grant),
–Jim Balsiger (NOAA/NMFS Regional Administrator, Alaska Region),
–Mary Barber (Ecological Society of America),
–Suzanne Bolton (NOAA/NMFS/Office of Science and Technology; Representative for NRC/OSB),
–Paul Dayton (University of California, San Diego, Scripps Institution of Oceanography),
–Steve Gittings (NOAA/NOS/Sanctuaries),
–Suzette Kimball (USGS/Biological Resources Program, Regional Biologist),
–Justin LeBlanc (Vice President of the National Fisheries Institute),

–Gary Matlock (NOAA/NOS/Acting Director of the National Centers for Coastal Ocean Science),
–Larry Mayer (UNH/Dir. Center For Coastal and Ocean Mapping),
–Barbara Moore (NOAA/OAR/Director of the National Undersea Research Program),
–Rod Moore (Executive Director, West Coast Seafood Processors Association),
–Clarence Pautzke (North Pacific Research Board, formerly NOAA/NMFS/Deputy Assistant Administrator for Fisheries and North Pacific Fishery Management Council),
–Gus Rassam (Executive Director, American Fisheries Society),
–Phil Rigby (Alternate for Jim Balsiger),
–Mike Sissenwine (NOAA/NMFS Science Director, Northeast Fisheries Science Center).

Program Committee

–Chuck Adams (University of Florida),
–Peter Auster (University of Connecticut/NOAA's Undersea Research Program),
–Gary Brewer (U.S. Geological Survey/Biological Resources Discipline),
–Linda Deegan (Marine Biology Laboratory),
–Jon Dodrill (Florida Fish and Wildlife Conservation Commission),
–Don Gordon (Fisheries and Oceans Canada),
–Dave Fluharty (University of Washington and North Pacific Fishery Management Council),
–Jon Heifetz (NOAA/NMFS/Alaska Fisheries Science Center),
–Eric Powell (Rutgers University),
–Page Valentine (U.S. Geological Survey/Woods Hole).

Defining the Issues:
Status, Management Needs, and Livelihoods

Keeping Our Fisheries Sustainable

WILLIAM T. HOGARTH

*Assistant Administrator for Fisheries, National Oceanic and Atmospheric Administration,
1315 East-West Highway, Silver Spring, Maryland 20910, USA*

Abstract. The topic of this symposium, effects of fishing activities on benthic habitats, is a subject charged with controversy, complexity, and socioeconomic significance. Concern about the effects of fishing comes at a time when the sustainability of fisheries and maintenance of healthy ecosystems are being questioned. Are fishing activities detrimental to habitat? When and where might they be more or less damaging? The U.S. Congress, under the Magnuson-Stevens Fishery Conservation and Management Act, has directed us to address the issue. This symposium is one of the means by which we gain new information for future evaluation and appropriate management of fishing activities that affect benthic habitat. We also are addressing the issue through research. With the appropriate information, we will move forward together—fishers, scientists, managers, environmentalists, and other stakeholders.

The Problem

The effects of fishing activities on benthic habitats is a subject charged with controversy, complexity, and socioeconomic significance. Concern about the effects of fishing comes at a time when the sustainability of fisheries and maintenance of healthy ecosystems are under scrutiny. It is the subject of lawsuits, and people's livelihoods are involved. It involves multiple ecosystem components—habitat, target organisms, bycatch, and the fishers themselves. Often, little information is available or is confounded by other factors. The topic is a challenge, yet we must find appropriate ways forward. Congress, in the Magnuson-Stevens Fishery Conservation and Management Act, has directed us to do so.

The problem of the effects of fishing is old, dating back as far as 1376, when a Commons Petition was submitted to the king of England expressing concern regarding a new bottom fishing gear that plowed so heavily "it destroys the flowers of the sea" (DeGroot 1984; Auster et al. 1996; Hall 1999). Yet in 1914, Alexander, Moore, and Kendall (Alexander et al. 1914) reported that the effect of trawling on the seabed was negligible.

Since the late 1800s, when steam power came to the fishing industry, the issue has been intimately related to the development of fisheries with more, larger, and more powerful boats and with more numerous, larger, and heavier gear. These had the potential to affect more habitat in more harmful ways. Fathometers and fish finders along with sophisticated navigational equipment (i.e., global positioning system) have made it possible to precisely target prime fishing areas, habitats, and stocks, thereby not only increasing efficiency but making it possible to affect prime habitats previously protected by their inaccessibility.

Today, the sustainability of fisheries is threatened. The causes are many and the effects are not well-defined. Overcapacity (i.e., too many fishers, too many vessels) combined with social and political pressures have led to overharvesting of stocks. This is especially the case in light of what stocks can sustain under variable and shifting climatic regimes and a variety of anthropogenic insults affecting both habitat and species. Close to shore, the variety of impacts are many, but offshore, the predominant anthropogenic impacts are fishing-related. No other offshore human activity is as widespread, intense, or enduring as fishing. Thus, the accusing finger is pointed at fishing activities and their effects on both habitats and species. When stocks seem to decline, is it an overfishing problem, a regime shift, a habitat problem, or some combination of the three? Frankly, the answers are not simple.

As near-shore fisheries have declined, much of our fishing effort has moved further offshore and downslope to deeper, colder waters. One consequence is that different species and habitats are involved, about which we know little. These areas are sometimes beyond the ranges of typical surveys taken over the past 30–40 years. Deepwater species may be more vulnerable to overfishing because of slower growth rates and lower reproductive potential. We know these habitats are adapted to much less disturbance than occurs in shallow, near-shore habitats, which are subject to frequent storms and anthropogenic activities. Deeper habitats are likely more vulnerable to fishing effects, particularly the high-profile living components. An example is trawling over deepwater coral off southeastern Alaska, which resulted in destruction of the coral. Seven years after the trawling event, there was no evidence of recovery (Krieger 2001).

The National Oceanic and Atmospheric Administration–Fisheries is committed to sustainable fisheries and healthy ecosystems. By healthy ecosystems, we mean "an ecosystem where ecological productive capacity is maintained, diversity of the flora and fauna is preserved, and the ecosystem retains the ability to regulate itself" (NOAA 2002). With industry, the Federal fishery management councils, our state partners through the marine fisheries commissions, and other constituents, we are working actively to recover depleted stocks, maintain healthy stocks, and improve knowledge of stocks about which we know little. We are moving from a species-by-species approach to an ecosystem-based approach in the management of fisheries. Indeed, without healthy ecosystems, sustainable fisheries are not possible. Without sustainable fisheries, the livelihoods of fishers and those in fisheries-dependent occupations are jeopardized, and the contributions of these resources to our food supply, economy, and national health are compromised.

Working with the fishing community is a must. The fishing community has knowledge useful for addressing these problems. Fishers are at sea far more than we, and they regularly use the gear to be assessed. It is also their livelihoods that are at risk if management actions are not effective. We must undertake additional research to fill information gaps, and we must move to ecosystem-based management of fisheries for their long-term sustainability.

The Law and Associated Regulations

In 1996, the U.S. Congress passed the Sustainable Fisheries Act (SFA), which reauthorized the Magnuson Fishery Conservation and Management Act and included a new provision dealing with essential fish habitat. Under the law "essential fish habitat" (EFH) means those waters and substrate necessary to fish for spawning, breeding, feeding, or growth to maturity. According to the Final Rule (NOAA 2002), it is the habitat "necessary to maintain a sustainable fishery and the managed species' contribution to a healthy ecosystem."

Under the SFA, the Secretary of Commerce and the fishery management councils are charged with describing and identifying EFH, threats to that habitat, and actions to encourage the conservation and enhancement of such habitat. It is no accident that the U.S. Geological Survey (USGS) is a co-convener of this symposium. The USGS brings needed elements to the topic—seabed characterization, geological mapping, and geospatial data analysis—all of which are needed for describing and identifying EFH and habitat areas of particular concern (HAPCs). The HAPCs can be ecologically important, sensitive to human-induced environmental degradation, likely to be stressed by development activities, or relatively rare. Seabed habitat characterization data form the underpinning or fundamental data layer upon which biological data may be plotted. It is more than bathymetry. It is backscatter reflectance data along with observations (e.g., by divers, submersibles, and cameras) and samples of substrate that identify the surficial geology of the seabed. Such information forms the backbone of habitat type to which bottom-associated species are linked and EFH and HAPCs are described and identified.

The SFA includes a provision dealing with the potential threat of fishing activities to essential fish habitat and is the principal reason for this symposium. That provision states that any fishery management plan (FMP) prepared by any council or by the Secretary of Commerce shall "minimize to the extent practicable the adverse effects on such [essential fish] habitat caused by fishing." From the Final Rule (NOAA 2002), "Adverse effect means any impact that reduces quality and/or quantity of EFH. ...Councils must act to prevent, mitigate, or minimize any adverse effects from fishing, to the extent practicable, if there is evidence that a fishing activity adversely affects EFH in a manner that is more than minimal and not temporary in nature." Again, it is the changed nature of bottom habitat and the changed nature of bottom-associated communities that matters.

The Final Rule also advises that when considering the practicability of implementing management measures to minimize adverse effects on EFH, "Councils should consider the nature and extent of adverse effect on EFH and the long- and short-term costs and benefits of potential management measures to EFH, associated fisheries, and the nation." Councils also are required to make use of the best available scientific information when developing and implementing management measures; yet we know such information is incomplete and often its application controversial.

Under the law and applicable regulations, councils, through their FMPs, must identify and describe EFH and HAPCs. They also are required to minimize adverse effects on such habitat caused by fishing and do so with concern for practicability and the use of the best available scientific information. This symposium is a significant means for obtaining the latest information relevant to managing fishing activities that affect benthic habitat.

Information and Research Needs

The development of sophisticated technology parallels the dramatic expansion of studies on the effects of fishing (Figure 1). Examination of several bibliographic surveys (Rester 2000; Wion and McConnaughey 2000; Johnson 2002) suggests that research on the effects of fishing began in earnest during the 1970s and blossomed as a major subject in the 1990s. Similarly, the ability to collect highly detailed, digital imagery of the seabed as a base map upon which to plot biological data is equally recent. Lead-lines were used prior to the 1920s, when single-beam echo sounders were invented. Side-scan sonars gradually began replacing single-beam echo-sounders as research instruments in the 1960s. The first commercial multibeam system was developed in the late 1960s; however, it was not until the early 1990s that modern, high-resolution multibeam systems came into use. Additionally, recent advances in computer technology have expanded our ability to handle very large and complex data sets faster and at less cost. As these advances occurred, our understanding evolved and with it our ability to more specifically define informational and research needs and to address these needs. I would like to focus on some of these needs and urge that they be addressed by the research community. Many of these needs have already been presented by Auster and Langton (1999) and the National Research Council (2002) in a recent publication on the effects of trawling and dredging on seafloor habitat (see Steele et al. 2005, this volume).

Habitat Characterization and Mapping

Maps are needed that characterize benthic habitats of the U.S. Continental Shelf and slope. These maps should include surficial geology as well as bathymetry. As already mentioned, these maps are important for identifying and describing EFH and HAPCs, for linking species to habitat type, and for analysis and extrapolation of effects of fishing on benthic habitats to areas over which fisheries are managed. In the 1980s, the USGS and National Oceanic and Atmospheric Administration (NOAA) worked together to accomplish the first coarse survey of the outer portion of the newly designated Exclusive Economic Zone of the United States (Groome et al. 1997). Unfortunately, these surveys of the seabed did not extend into the shallower waters of the upper slope and Continental Shelf (less than 500 to 200 m) where the bulk of U.S. fishing occurs. Today, the need persists for coordinated, systematic surveys characterizing and mapping seabed habitat for most of the U.S. Continental Shelf and slope.

Seabed habitat characterization and mapping for the entire U.S. Continental Shelf and slope is such an immense issue from a time and cost standpoint that efforts must focus on priority areas. Habitat in several areas has priority because of the species involved and the status of our knowledge about these species and their habitats. Deepwater coral is one habitat for priority surveys. The habitat of the economically valuable rockfish *Sebastes* spp. on the continental slope off our U.S. West Coast is another such area. Another is the habitat of the Atlantic red deepsea crab *Chaceon quinquedens* for which a new FMP and Final Rule have been developed and implemented to regulate a growing fishery and prevent potential overharvesting. Other priority areas include the gag *Mycteroperca microlepis* and grouper *Mycteroperca* spp. spawning habitat in the Gulf of Mexico, Atlantic cod *Gadus morhua* habitat in the Gulf of Maine, and many areas of the Gulf of Alaska and Bering Sea. Again, habitat characterization and mapping is needed to elucidate the role habitat structure plays in maintaining sustainable fisheries.

Species Distribution and Abundance

Improved biological data on the distribution and abundance of species over the entire U.S. Continental Shelf and slope is needed to update surveys done over the past 30 to 40 years. The existing information is good, but it should now be based on better knowledge of the distribution of benthic habitats (i.e., seabed characterization). Additionally, new data are needed to provide a better understanding of the biological communities of the continental slope.

Life History Information

Studies are needed of the life history and the relationship or linkage of a species to particular habitats during its life cycle. Such data would provide detailed pictures of the distribution of species and, thereby, an indication

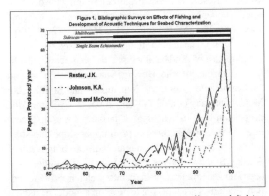

Figure 1. Bibliographic surveys on effects of fishing and development of acoustic techniques for seabed classification.

of how vulnerable that species may be if a particular habitat or portion of that habitat type is impacted or destroyed. Some species may be relatively independent of a particular habitat type; others may be highly dependent and, therefore, highly vulnerable.

Fishing Effects

Even though considerable information exists on effects of fishing activities, additional data are needed to understand effects of each of the gear types on each habitat type and over a gradient of fishing effort. Studies are needed to know how long recovery takes and if a system returns to its previous condition with reduction or cessation of fishing. Such data are needed for identifying gears and fishing efforts that may produce adverse effects on habitat or for determining that no adverse effects occur.

Application and Scaling of Site-Specific Gear Effects

Of particular interest are studies on the application and scaling of site-specific gear effects studies to the much broader areas over which fisheries are managed. It is essential to be able to take effects on a localized set of organisms and extrapolate those effects to populations, communities, and ecosystems or areas over which fisheries are managed. To do this requires characterization and mapping of the habitats of the U.S. Continental Shelf and slope, better understanding of the linkage of species to habitat type, and improved knowledge on the distribution of fishing effort.

Fishing Effects Compared to Natural and Other Anthropogenic Effects

Studies are needed that compare the effects of fishing with other impacts, both natural and anthropogenic. These studies are required to put effects of fishing on benthic habitats into proper perspective with other impacts on the system and warn where and when such effects are potentially out of balance and may need redressing through management measures.

Management Measures

Investigation into methods for minimizing adverse effects of fishing also are needed. These should include tests of modified fishing practices; studies on effects of closures, including rotating closures, on the recovery and maintenance of both species and habitat; and gear modification research to lessen adverse effects while maintaining economic profitability to fishers. Such studies could lead the way to alternative management measures (if needed to prevent or reduce the adverse effects of fishing) and the maintenance of sustainable fisheries.

Where possible, these studies should be carried out jointly with fishers.

Socioeconomic Studies

Also required are socioeconomic studies on the effects of various management actions on fishers, communities, and businesses dependent on fisheries or their products. The spin-off effects of such management measures may not only affect the fishers associated with that fishery but also other fishers, communities, and businesses spatially and temporally distant from that fishery. We need to understand these effects and the ripples they produce through our economy.

Predictive Models

Finally, studies are needed to develop models predicting the likely occurrence of adverse effects with different fishing regimes (e.g., gear, intensity, and geographic coverage), habitats, and species. Such a model should be expanded to include other non-fishing impacts from the comparison studies above and produce susceptibility or vulnerability maps as a product. With such predictive models, we would know what conditions to look for and avoid before a fishery was adversely affected. Such areas would be the focus of management measures.

Relevant NOAA Research

The National Ocean and Atmospheric Administration is carrying out a number of research projects relevant to this symposium. In 1996, the Alaska Fisheries Science Center initiated a number of seafloor habitat studies directed at assessing habitat characteristics as well as evaluating acoustic technology for seabed classification and investigating the impacts of fishing, particularly bottom trawling (von Szalay and McConnaughey 2002; Stone et al. 2005, this volume). Researchers also began determining the distribution, abundance, and life history of deepwater corals and sponges vulnerable to gear impacts (Malecha et al. 2005, this volume). These field studies have focused on specific geographic areas and habitat types: the shallow water soft-bottom areas of the Bering Sea shelf (Smith and McConnaughey 1999; McConnaughey and Smith 2000; McConnaughey et al. 2000, and 2005, this volume), the deeper soft-bottom shelf areas of the Gulf of Alaska (Freese et al. 1999), and the hard-bottom upper slope and shelf habitats of the Aleutian Islands and portions of the Gulf of Alaska (Stone and Wing 2001). Laboratory studies have been initiated to determine the habitat preferences of groundfish. Scientists are working to develop a model of fishing impacts that considers spatial-temporal distributions of bottom trawling (Rose and Jorgensen 2005, this volume), fishing intensity, and habitat recovery rates.

Researchers at the Northwest Fisheries Science Center and Southwest Fisheries Science Center, the NOAA Pacific Marine Environmental Laboratory, and Washington State University are characterizing and quantifying relationships between groundfish populations and seafloor morphology and texture, including the factors controlling these relationships on Heceta Bank (see Wakefield et al. 2005, this volume). This major fishing area is a 50-km-long rocky shoal on the outer shelf of central Oregon.

Researchers from the Southwest Fisheries Science Center, in collaboration with researchers from the University of California-Santa Barbara, Moss Landing Marine Laboratory, California Department of Fish and Game, and NOAA's National Undersea Research Program (NURP) are collecting baseline data to determine the effectiveness of the largest and newest marine protected area (MPA) on the West Coast, the Cowcod Conservation area off Southern California. These researchers, also working in the Big Creek Ecological Reserve off central California, are evaluating the capabilities and effectiveness of laser line-scan technology, an electro-optic imaging technique, as a more efficient way to map and show the relationships between organisms and their seabed habitats. This technology appears capable of surveying relatively large swaths (approximately 10 m × 4 km per hour) at millimeter to centimeter spatial resolutions. And finally, they are characterizing and quantifying relationships between groundfish populations and seafloor morphology and texture (Anderson et al. 2005, this volume).

Researchers at the Pacific Islands Fisheries Science Center, working with NURP scientists, are determining the dependence of fish communities on corals. Also under study are the growth rates of precious corals (>350 m) and black coral (70–150 m) for estimating recovery times from fishing gear-related injuries. The researchers also are documenting previously undescribed fish and habitat assemblages on the moderately deep (30–40 m) summits of the Northwest Hawaiian Islands, where the majority of Hawaii's commercial fishing occurs (Parrish and Boland 2004).

The National Ocean Service of NOAA is mapping corals in the Northwest Hawaiian Islands as well as in the Florida Keys and around Puerto Rico and the U.S. Virgin Islands. They are conducting ecological studies on the relationships between species and seabed habitats (Christensen et al. 2003) and, together with the USGS, they are mapping resources within NOAA's marine sanctuaries. They also are mapping the seabed with multibeam and side-scan systems, particularly in the shallower areas closer to shore, in connection with safe navigation and nautical charting.

Researchers at the Southeast Fisheries Science Center (SEFSC) are studying the effects of various intensities of shrimp trawling on benthic communities and recovery of these communities during a 7-month closure each year (December 1–July 1) in an area off the south Texas coast (see Sheridan and Doerr 2005, this volume). The information will be applied to management of the shrimp fishery and potential use of marine reserves as part of that management.

Together with scientists from the USGS and Florida State University, researchers from the SEFSC also are characterizing and mapping seabed habitat and assessing fishing effects in two protected areas, the coraline Oculina Banks HAPC, off Florida's Atlantic coast, and Madison-Swanson and Steamboat Lumps MPA's, along the shelf-slope break off Florida's Gulf coast. The latter are spawning areas for gag and grouper. Of particular interest is the testing of a variety of methods to restore coral at the Oculina Banks HAPC (see Koenig et al. 2005, this volume).

Finally, in concert with commercial fishers, SEFSC scientists are studying the effects of traps on benthic habitat in open fishing areas and adjoining marine reserves of the Florida Keys, Puerto Rico, and U.S. Virgin Islands. The National Ocean Service has a related study in the Florida Keys National Marine Sanctuary looking at the effects of spiny lobster traps on seagrass beds (see Uhrin et al. 2005, this volume).

Scientists at the Northeast Fisheries Science Center, along with the USGS, University of Rhode Island, University of Connecticut, and NURP Center, are characterizing and mapping benthic habitat over portions of Georges Bank. They also are investigating the effects of bottom trawling and the associated recovery rates of habitats for sea scallop *Placopecten magellanicus*, yellowtail flounder *Limanda ferruginea*, Atlantic cod, and haddock *Melanogrammus aeglefinus* on Georges Bank (see Link et al. 2005, this volume). Similar investigations are underway on habitat for tilefish *Lopholatilus chamaeleonticeps* off New Jersey.

My hope is that those of you who have interest in any of these projects will contact NOAA. We welcome interest and would like to share information, even work jointly with you where feasible, in hopes that cross-fertilization of approaches, ideas, and results would be mutually beneficial and move this entire effort forward in a helpful way.

A Vision for the Future

Only through working together—sharing information, developing new information based on research, and taking full advantage of new technologies—will the effects of fishing and other impacts be identified and understood. This knowledge can then be used to enhance the management of sustainable fisheries, promote healthy ecosystems, and assure the livelihoods of fishers and others dependent on our living marine resources. The task will not be easy, but it is essential that the information learned from this symposium is used and that research is undertaken to accomplish the needed objectives. This will require fishers, scientists, managers, environmentalists, and other stakeholders working together productively.

Acknowledgments

I would like to thank James P. Thomas for assistance in preparing this manuscript. I also would like to thank Jonathan Heifetz, Robert A. McConnaughey, William Collins, Waldo W. Wakefield, Churchill B. Grimes, Cheryl R. Kaine, Jeffrey J. Polovina, Frank A. Parrish, Peter F. Sheridan, Andrew W. David, Amy V. Uhrin, John D. Christensen, Thomas J. Noji, Peter W. Barnes, and Larry A. Mayer for providing material for this paper. Finally, I thank Peter J. Auster, Garry F. Mayer, and Michael P. Sissenwine for review and valuable comments.

References

Alexander, A. B., H. F. Moore, and W. C. Kendall. 1914. Otter-trawl fishery. Report of the U.S. Commission of Fisheries, Appendix VI, Washington, D.C.

Anderson, T. J., M. M. Yoklavich, and S. L. Eittreim. 2005. Linking fine-scale groundfish distributions with large-scale seafloor maps: issues and challenges of combining biological and geological data. Pages 667–678 in P. W. Barnes and J. P. Thomas, editors. Benthic habitats and the effects of fishing. American Fisheries Society, Symposium 41, Bethesda, Maryland.

Auster, P. J., and R. W. Langton. 1999. The effects of fishing on fish habitat. Pages 150–187 in L. Benaka, editor. Fish habitat: essential fish habitat and rehabilitation. American Fisheries Society, Symposium 22, Bethesda, Maryland.

Auster, P. J., R. J. Malatesta, R. W. Langton, L. Watling, P. C. Valentine, C. L. S. Donaldson, E. W. Langton, A. N. Shepard, and I. G. Babb. 1996. The impacts of mobile fishing gear on seafloor habitats in the Gulf of Maine (Northwest Atlantic): implications for conservation of fish populations. Reviews in Fisheries Science 4:185–202.

Christensen, J. D., C. F. G. Jeffrey, C. Caldow, M. E. Monaco, M. S. Kendall, and R. S. Appeldoorn. 2003. Quantifying habitat utilization patterns of reef fishes along a cross-shelf gradient in southwestern Puerto Rico. Gulf and Caribbean Research 14(2):9–27.

De Groot, S. J. 1984. The impact of bottom trawling on benthic fauna of the North Sea. Ocean Management 9:177–190.

Freese, L., P. J. Auster, J. Heifetz, and B. L. Wing. 1999. Effects of trawling on seafloor habitat and associated invertebrate taxa in the Gulf of Alaska. Marine Ecology Progress Series 182:119–126.

Groome, M. G., C. E. Gutmacher, and A. J. Stevenson. 1997. Atlas of Gloria sidescan-sonar imagery of the Exclusive Economic Zone of the United States: EEZview. U.S. Geological Survey Open-File Report 97-540, Reston, Virginia.

Hall, S. J. 1999. The effects of fishing on marine ecosystems and communities. Blackwell Science Ltd., Oxford, UK.

Johnson, K. A. 2002. A review of national and international literature on the effects of fishing on benthic habitats. NOAA Technical Memorandum NMFS-F/SPO-57.

Koenig, C. C., A. N. Shepard, J. K. Reed, F. C. Coleman, S. D. Brooke, J. Brusher, and K. M. Scanlon. 2005. Habitat and fish populations in the deep-sea *Oculina* coral ecosystem of the western Atlantic. Pages 795–805 in P. W. Barnes and J. P. Thomas, editors. Benthic habitats and the effects of fishing. American Fisheries Society, Symposium 41, Bethesda, Maryland.

Krieger, K. 2001. Coral (Primnoa) impacted by fishing gear in the Gulf of Alaska. In J. H. Martin Willison, J. Hall, S. E. Gass, E. L. R. Kenchington, M. Butler, and P. Doherty. Proceedings of the First International Symposium on Deep-Sea Corals. Ecology Action Cener and Nova Scotia Museum, Halifax.

Link, J., F. Almeida, P. Valentine, P. Auster, R. S. Reid, and J. Vitaliano. 2005. The effects of area closures on Georges Bank. Pages 345–368 in P. W. Barnes and J. P. Thomas, editors. Benthic habitats and the effects of fishing. American Fisheries Society, Symposium 41, Bethesda, Maryland.

Malecha, P. W., R. P. Stone, and J. Heifetz. 2005. Living substrate in Alaska: distribution, abundance, and species associations. Pages 289–299 in P. W. Barnes and J. P. Thomas, editors. Benthic habitats and the effects of fishing. American Fisheries Society, Symposium 41, Bethesda, Maryland.

McConnaughey, R. A., K. Mier, and C. B. Dew. 2000. An examination of chronic trawling effects on soft-bottom benthos of the eastern Bering Sea. ICES Journal of Marine Science 57:1377–1388.

McConnaughey, R. A., and K. R. Smith. 2000. Associations between flatfish abundance and surficial sediments in the eastern Bering Sea. Canadian Journal of Fisheries and Aquatic Science 57:2410–2419.

McConnaughey, R. A., S. E. Syrjala, and C. B. Dew. 2005. Effects of chronic bottom trawling on the size structure of soft-bottom benthic invertebrates. Pages 425–437 in P. W. Barnes and J. P. Thomas, editors. Benthic habitats and the effects of fishing. American Fisheries Society, Symposium 41, Bethesda, Maryland.

National Research Council. 2002. Effects of trawling and dredging on seafloor habitat. National Academy Press, Washington, D.C.

NOAA (National Oceanic and Atmospheric Administration). 2002. Magnuson-Stevens Act Provisions; essential fish habitat (EFH). 50 CFR part 600. U.S. Federal Register 67(12):2343-2383.

Parrish, F., and R. Boland. 2004. Habitat and reef-fish assemblages of banks in the Northwestern Hawaiian Islands. Marine Biology 144:1065–1073.

Rester, J. K. 2000. Annotated bibliography of fishing

impacts on habitat. Gulf States Marine Fisheries Commission, Number 73, Ocean Springs, Mississippi.

Rose, C. S., and E. M. Jorgensen. 2005. Spatial and temporal distributions of bottom trawling off Alaska: consideration of overlapping effort when evaluating the effects of fishing on habitat. Pages 679–690 in P. W. Barnes and J. P. Thomas, editors. Benthic habitats and effects of fishing. American Fisheries Society, Symposium 41, Bethesda, Maryland.

Sheridan, P., and J. Doerr. 2005. Short-term effects of the cessation of shrimp trawling on Texas benthic habitats. Pages 571–578 in P. W. Barnes and J. P. Thomas, editors. Benthic habitats and the effects of fishing. American Fisheries Society, Symposium 41, Bethesda, Maryland.

Smith, K. R., and R. A. McConnaughey. 1999. Surficial sediments of the eastern Bering Sea continental shelf: EBSSED database documentation. NOAA Technical Memorandum NMFS-AFSC-104.

Steele, J., S. J. Roberts, D. L. Alverson, P. Auster, J. Collie, J. T. DeAlteris, L. Deegan, E. E. Briones, S. J. Hall, G. H. Kruse, C. Pomeroy, K. M. Scanlon, and P. Weeks. 2005. National Research Council study on effects of trawling and dredging on seafloor habitat. Pages 91–99 in P. W. Barnes and J. P. Thomas, editors. Benthic habitats and the effects of fishing. American Fisheries Society, Symposium 41, Bethesda, Maryland.

Stone, R. P., and B. Wing. 2001. Growth and recruitment of an Alaskan shallow-water gorgonian. In J. H. M. Willison, J. Hall, S. E. Gass, E. L. R. Kenchington, M. Butler, and P. Doherty. Proceedings of the First International Symposium on Deep-sea Corals. Ecology Action Center and Nova Scotia Museum, Halifax.

Stone, R. P., M. M. Masuda, and P. W. Malecha 2005. Effects of bottom trawling on soft-sediment epibenthic communities in the Gulf of Alaska. Pages 461–476 in P. W. Barnes and J. P. Thomas, editors. Benthic habitats and the effects of fishing. American Fisheries Society, Symposium 41, Bethesda, Maryland.

Uhrin, A. V., M. S. Fonseca, and G. P. DiDomenico. 2005. Effect of Caribbean spiny lobster traps on seagrass beds of the Florida Keys National Marine Sanctuary: damage assessment and evaluation of recovery. Pages 579–588 in P. W. Barnes and J. P. Thomas, editors. Benthic habitats and the effects of fishing. American Fisheries Society, Symposium 41, Bethesda, Maryland.

von Szalay, P. G., and R. A. McConnaughey. 2002. The effect of slope and vessel speed on the performance of a single beam acoustic seabed classification system. Fisheries Research (Amsterdam) 56:99–112.

Wakefield, W. W., C. E. Whitmire, J. E. R. Clemons, and B. N. Tissot. 2005. Fish habitat studies: combining high-resolution geological and biological data. Pages 119–138 in P. W. Barnes and J. P. Thomas, editors. Benthic habitats and the effects of fishing. American Fisheries Society, Symposium 41, Bethesda, Maryland.

Wion, D. A., and R. A. McConnaughey. 2000. Mobile fishing gear effects on benthic habitats: a bibliography. NOAA Technical Memorandum NMFS-AFSC-116.

The Challenge of Protecting Fish Habitat through the Magnuson–Stevens Fishery Conservation and Management Act

CLARENCE PAUTZKE[1]

North Pacific Research Board, 1007 West 3rd Avenue, Suite 100, Anchorage, Alaska 99501, USA

Abstract. Eight regional fishery management councils and the National Marine Fisheries Service (NMFS) are charged with managing fisheries outside 3 mi (5 km) in accordance with the Magnuson–Stevens Fishery Conservation and Management Act (MSA) and other federal law. Protecting fish habitat is just one of many management responsibilities. All fishery management plans must describe and identify essential fish habitat using guidelines established by the Secretary of Commerce, and managers must minimize, to the extent practicable, adverse effects caused by fishing and encourage conservation and enhancement of essential fish habitat. This paper summarizes habitat provisions in the MSA, Secretarial guidance, and how they have evolved over time. It reviews the actions of the eight councils to identify and protect habitat and discusses how additional fishing restrictions must be approached within the context of other issues facing managers. Managers are pressured not only to protect habitat from a wide range of fishing gears, but to reduce bycatch, minimize impacts on stressed and endangered species, rebuild and maintain fish stocks, provide optimum yield, protect communities, promote safety, and simultaneously provide for economically feasible fisheries. The North Pacific Fishery Management Council actions to protect habitat are discussed in more detail as an example of the challenges faced in balancing conservation and management needs. Legal issues are discussed, as well as the high costs of conducting the research necessary to identify essential fish habitat and its relation with fish stocks. It is argued that despite the challenges of the council and NMFS fishery management process under the MSA, it still remains the most expedient way to develop habitat protections with higher probability of acceptance by the fishing industry.

Introduction

Fisheries management is under the spotlight, and 2006 could be pivotal in revising and perhaps redirecting its implementation in the United States. Attention is coming from several directions: the U.S. Commission on Ocean Policy and the Pew Oceans Commission are reviewing fisheries within the larger context of a national ocean policy review. The Magnuson–Stevens Fishery Conservation and Management Act (MSA) (Public Law 94-265, as amended) is due for reauthorization, and in response to many lawsuits, the National Marine Fisheries Service (NMFS) is internally reviewing how fisheries regulations are developed and implemented.

The MSA was last reauthorized in 1996 with the Sustainable Fisheries Act (SFA) (Public Law 104-297). The amendments brought sweeping changes to management, including increased restrictions on overfishing and bycatch; a revised definition of optimum yield; three new national standards on protecting communities, reducing bycatch, and promoting safety; and enhanced protection for essential fish habitat from fishing gears. These issues, along with others such as overcapitalization, cooperative research, data deficiencies, and monitoring requirements, again could dominate reauthorization in 2006. An overall trend is toward enhanced ecosystem-based management, and a major component could be increased habitat protection.

Fish habitat protection is receiving attention from other quarters as well. The National Research Council has completed two habitat studies, one on marine protected areas (NRC 2001) and the other on trawling and dredging effects (NRC 2002). In addition, Executive Order 13158, signed in May 2000, is providing momentum for habitat protection via marine protected areas (MPAs). Fishermen normally perceive MPAs as equivalent to prohibitions on fishing, though in reality, the term may apply to any area that has been identified and provided some protections for natural or cultural resources by local, state, federal, or tribal governments. Executive Order 13158 is helping to dispel that misperception. For

---
[1] E-mail: cpautzke@nprb.org

example, the National Oceanic and Atmospheric Administration (NOAA)-maintained inventory at http://mpa.gov includes some 276 federal MPA sites, most of which allow commercial fishing, even though certain gears may be restricted. It is unclear how much additional habitat protection will result from the Executive Order, but any significant new closures most likely will require support of the commercial fishing sector. That sector is accustomed to working through the regional fishery management council system of the MSA, which provides an immediately accessible, and thus preferred, approach to protecting habitat.

Habitat Protection through the Magnuson–Stevens Act

The MSA has been modified many times over the past 26 years since it was first passed in 1976. It established eight councils with voting members representing state and federal fisheries agencies, academia, resource users, and other interested parties. Each council decision is made by recorded vote in public forum after public comment. Final decisions require Secretary of Commerce approval to be implemented as federal regulations. Decisions must comply not only with the MSA but a host of other laws, including the National Environmental Policy Act (U.S. Code, title 42, section 4321 *et seq.*), Endangered Species Act (U.S. Code, title 16, section 1531 *et seq.*), Marine Mammal Protection Act (U.S. Code, title 16, section 1361 *et seq.*), Regulatory Flexibility Act (U.S. Code, title 5, section 601 *et seq.*), and other applicable laws and executive orders. Regulatory changes may take up to a year or even longer to implement, particularly if complex or contentious.

Management is effected through fishery management plans (FMPs) and their revision. The plans contain a variety of management tools that may be revised using plan or regulatory amendments proposed by agencies, user groups, or the public and approved by a council and then the Secretary of Commerce. All regulations must comply with ten national standards, which are indicative of the many competing goals, objectives, and standards of performance envisioned in the MSA. There now are 39 council and three secretarial FMPs covering over 700 species of fish. Habitat protection provisions are contained in those plans and regulations.

Evolution of Habitat Protection Provisions

When first enacted in 1976, the main purpose of the MSA was to displace foreigners from the fishing grounds and Americanize the fisheries. There was little reference to habitat despite the findings of the U.S. Comptroller General that, among other things, many fish stocks were being depleted or threatened with depletion through overfishing by U.S. and foreign fishermen and alteration of habitat (Comptroller General 1976). Nevertheless, habitat was not mentioned directly in the findings or policy sections of the MSA or in the original national standards. Its sole reference was in defining the term "fishery resource" as including any fishery, any stock of fish, any species of fish, and any habitat of fish.

This changed with the introduction of S. 747 into the 99th Congress in October 1985. The bill contained a provision requiring new and existing FMPs to include discussions of habitat issues to the extent practicable. While NMFS already had the option to send comments on federal projects to federal agencies under the Fish and Wildlife Coordination Act, S. 747 required federal and state agencies to respond to the councils within 60 d on any concerns over habitat degradation. A U.S. House of Representatives proposal, HR 1533, stipulated that the councils could comment on any state or federal activity that might affect fishing habitat; federal agencies had 45 d to respond. The final legislation, signed into law in November 1986, amended MSA sections 302(i) and 303(a)(7). Its intent was to give the councils and Secretary of Commerce more leverage over federal agencies in reviewing effects of permits from nonfishing impacts, forestry practices, agriculture, runoff, stream flow, pollution, coastal development, etc., on spawning and other sensitive rearing areas for fish managed by the councils. Fishing impacts were not emphasized.

The MSA was scheduled for another reauthorization in 1990. The councils continued to emphasize nonfishing impacts in testimony before Congress, but protection from fishing gears surfaced as an issue. On 17 May 1989, the environmental organization Greenpeace testified before the Senate Committee on Commerce, Science, and Transportation that habitat provisions had to be strengthened to increase council effectiveness in preserving fish habitat by influencing actions of other agencies. The councils were urged to more thoroughly consider impacts of various fishing gears on habitats in developing fishery management plans and amendments. S. 1025, introduced on 2 August 1990 as the Fishery Conservation Amendments of 1990 and later passed as Public Law 101-627 on 28 November 1990, again emphasized nonfishing impacts. It strengthened the role of councils in commenting on anadromous fish-related habitat but did not address impacts of fishing gear.

Sustainable Fisheries Act of 1996

The next major reauthorization was scheduled for 1993 but dragged on until the Sustainable Fisheries Act was passed in October 1996. During that time, it became increasingly clear that not all was well with U.S. fisheries under council management. Many stocks were overfished, high bycatch and discard characterized many fisheries, and in the North Pacific, the council and fishing industry had just endured a very bitter battle over inshore–offshore allocations of pollock *Pollachius virens*.

In September 1992, a Center for Marine Conservation representative, testifying at a hearing on implementation of the Fishery Conservation Amendments of 1990 held by the National Ocean Policy Study of the Senate Committee on Commerce, Science, and Transportation, urged Congress to require the councils and NMFS to consider habitat effects of fishing practices. Then, in February 1993, NMFS unveiled major reauthorization proposals at a regional council chairmen meeting. One proposal, entitled "Strengthened Protection for Fisheries Habitat," stated that most species utilized in U.S. fisheries were estuarine dependent; fish populations were at or near historic lows in abundance; and coastal, estuarine, and riverine habitat continued to be degraded and lost due to effects of freshwater flow alterations and diversions, physical habitat alteration, contaminants, and nutrient over-enrichment. The NMFS opined that its authority still was insignificant regarding decisions on proposed projects, policies, or programs that could affect habitat for those species. Sidestepping the issue of fishing impacts on habitat, NMFS recommended requiring councils to designate habitat essential for optimum yield and strengthening consultation requirements between NMFS and other agencies.

Emphasis on fishing impacts finally gained momentum in 1995 when NMFS proposed protections for habitat from fishing activities. The U.S. House of Representatives passed HR 39 on 18 October 1995 after much debate (see House Congressional Record, pp. H10213–10247) and also passed a successful amendment requiring councils to include in their fishery management plans measures to minimize, to the extent practicable, fishing impacts on fish habitat. Debate on the floor suggests that insertion of the phrase "to the extent practicable" was a compromise for making habitat provisions mandatory rather than discretionary. It also responded to many comments the U.S. House of Representatives had received from the fishing industry, which was deeply concerned that a strict minimization of fishing impacts on habitat would have very dire consequences for their activities. Similar provisions were incorporated into S. 39, which eventually passed in both the U.S. House of Representatives and U.S. Senate and was signed into law as the Sustainable Fisheries Act (Public Law 104-297) on 11 October 1996. Those requirements remain the basis for addressing fishing impacts on habitat today and are summarized in Table 1.

Secretarial Guidelines on Essential Fish Habitat

The new MSA Section 305(b)(1) directed NMFS to develop guidelines to assist the councils and the Secretary of Commerce in describing and identifying essential fish habitat (EFH) in fishery management plans, identifying adverse effects to EFH, and highlighting those actions required to conserve and enhance EFH. The guidelines also detail procedures the Secretary of Commerce, other federal agencies, and the councils must use to coordinate, consult, or provide recommendations on federal and state actions that may adversely affect EFH. Drafting guidelines that impact the council and fishing public is a difficult process at best. The guidelines give operational meaning to sometimes very brief legislative provisions and, thus, become the battleground for interested parties to try to impose their interpretation as the controlling policy. Nearly every word and nuance are scrutinized and debated. The guidelines are, at the same time, very broad and very detailed.

The difficulty inherent in developing guidelines is reflected in the time required by NMFS to provide final guidelines. The NMFS published an advanced notice of proposed rulemaking (ANPR) on 8 November 1996, less than a month after passage of the SFA. It was intended to solicit comments to assist NMFS in developing an approach for the proposed regulations. A second ANPR was published on 9 January 1997 announcing the availability of a "Framework for the Description, Identification, Conservation, and Enhancement of Essential Fish Habitat," which provided a detailed outline for the proposed regulations. After more public hearings, briefings, and workshops across the nation, NMFS issued a proposed rule on 23 April 1997. More hearings were held before an interim final rule was issued on 19 December 1997, which took effect 20 January 1998. The NMFS had two additional public comment periods on the interim final rule and published a final rule on 17 January 2002 (67 Code of Federal Regulation 2343–2383).

Five general principles were used by NMFS (1998) to guide its treatment of habitat regulations:

(1) Description and identification of EFH must be based on the best information available;

(2) Procedures for describing and identifying EFH should be scientifically defensible;

(3) A risk-averse approach must be used in describing and identifying EFH to ensure adequate habitat protection, because serious or irreversible damage to essential habitats is likely to further reduce the capacity of many managed fish stocks to support sustainable fisheries;

(4) The guidelines must be sufficiently broad to address the habitat needs for many different species in many different regions as well as for varied physical, chemical, and biological processes that affect habitat quantity and quality; and

Table 1. Summary of current essential fish habitat (EFH) provisions in the Magnuson–Stevens Fishery Conservation and Management Act (MSA) after the Sustainable Fisheries Act (SFA) amendments of 1996. Please refer to the MSA for specific language and provisions.

Provision	Description
Section 2(a): Findings	Three of ten findings refer to habitat: (2) Direct and indirect habitat losses have resulted in a diminished capacity to support existing fishing levels. (6) A national program of conservation is needed to facilitate long-term protection of essential fish habitats. (9) One of the greatest long-term threats to the viability of commercial and recreational fisheries is the continuing loss of marine, estuarine, and other aquatic habitats, and habitat considerations should receive increased attention for the conservation and management of fishery resources of the USA
Section 2(b): Purposes	One of seven purposes of Congress is: (7) To promote the protection of essential fish habitat in the review of projects conducted under federal permits, licenses, or other authorities that affect or have the potential to affect such habitat.
Section 3: Definitions	(10) Essential fish habitat means those waters and substrate necessary to fish for spawning, breeding, feeding, or growth to maturity.
Section 305(a): Required provisions of FMPs.	(7) Councils must describe and identify EFH for the fishery based on the guidelines established by the Secretary under section 305(b)(1)(A), minimize to the extent practicable adverse effects on such habitat caused by fishing, and identify other actions to encourage the conservation and enhancement of such habitat.
Section 305(b)(1): Secretarial guidelines and information	Secretary is required within 6 months of enactment of SFA to establish regulatory guidelines for councils to follow in describing and identifying EFH and considering actions to ensure conservation and management of the fisheries and their impacts on habitat. The Secretary also is required to provide each council with recommendations and information to assist in the identification of EFH, the adverse impacts on that habitat, and the actions that should be considered to ensure conservation and management of that habitat.
Section 305(b)(2–4): Council comments on federal or state activities	Federal agencies must consult with NMFS regarding actions that may adversely affect EFH. Councils may comment to any federal or state agency concerning activities or proposed activities authorized, funded, or undertaken that, in the view of the council, may affect the habitat including EFH. For anadromous fish stocks, councils must comment. Federal agencies have 30 days to respond to NMFS and council comments in writing.
Section 404(c)(1) and (2): Areas of research	The Secretary shall initiate and maintain, in cooperation with the Councils, a comprehensive program of fishery research which will include the identification of EFH and gear research to minimize adverse effects on EFH.

(5) Ecological relationships among species and between the species and their habitats require, where possible, that an ecosystem approach be used in assessing and conserving EFH of a managed species or species assemblage.

The guidelines are quite lengthy, and the main components are summarized below as they pertain to fishing impacts.

Identification and Description of Habitat
Text and tables must be used to describe and identify EFH. They must provide information on biological requirements for each life history stage of the species and should summarize available information on environmental and habitat variables that control or limit distribution, abundance, reproduction, growth, survival, and productivity. The EFH definition adopted by Congress in 1996 was very broad: "those waters and substrate necessary to fish for spawning, breeding, feeding or growth to maturity." The guidelines interpreted the definition as follows:

- *Waters.* Aquatic areas and associated physical, chemical, and biological properties, used currently or historically.

- *Substrate.* Sediments, geological features underlying the waters, and associated biological communities, such as coral reefs or submerged aquatic vegetation.

- *Necessary.* Habitat required to support a managed

species or assemblage at a target production level reflecting conscientious stewardship.

- *Spawning, breeding, feeding, or growth to maturity.* Covers a species' full life cycle.

- *Feeding and growth to maturity.* Includes EFH for prey species if the managed species depends on the existence of a specific prey species.

There are four levels of information proposed for defining EFH:

- Level 1: Distribution data are available for some or all portions of the geographic range of the species;

- Level 2: Habitat-related densities of the species are available;

- Level 3: Growth, reproduction, or survival rates within habitats are available; and

- Level 4: Production rates by habitat are available.

Level 1 information identifies the geographic range of the species, while levels 2–4 help identify habitats most highly valued within that range. If habitat degradation may be contributing to an overfished condition for a particular species, EFH may include currently used and historic habitats. Councils should be risk-averse to ensure adequate areas are identified as EFH.

Most councils only had level 1 and 2 information, which resulted in very broad designations of EFH. Off Alaska, for example, rockfishes such as dusky rockfish *Sebastes ciliatus* had large EFH designations even though they lived in relatively narrow bands along the shelf break (Figure 1). Essential fish habitat for walleye pollock *Theragra chalcogramma*, essentially a pan-North Pacific species, covers the whole continental shelf (Figure 2). Other council regions had similar situations, leading the chairmen to comment in June 1999, "If everything is designated as essential, then nothing is essential."

Despite these frustrations, the broad definitions were retained. Curtailing the definitions of essential fish habitat could end up possibly crippling the ability to protect it. It is better to start intensively gathering the types of detailed information required for delineation of the very specific, ecologically significant spawning and nursery areas originally intended in the legislation. Research must be conducted to move from the current, broad level 0–2 EFH descriptions prevalent for most fisheries to define more specific EFH designations based on levels 3–4. The cost of such intense research is discussed further below. The councils have been cautioned by NMFS that they should not designate EFH if no information is available on a given species and habitat usage cannot be inferred by other means. The trend is for councils to continue to identify more valuable habitat areas and designate them as habitat areas of particular concern.

Habitat Areas of Particular Concern

In the proposed rule for EFH, NMFS referred to the identification of "vulnerable habitats" within EFH but did not clarify their meaning. When challenged on the meaning of vulnerable, NMFS urged the councils to consider the ecological function and value of a type or area of EFH, its susceptibility to perturbation, and whether it was stressed or rare. The NMFS renamed vulnerable habitats as habitat areas of particular concern (HAPC) and urged the councils to use them to focus conservation, enhancement, management, and research efforts and to provide insight into relationships between key habitat characteristics and ecological productivity or sustainability and the ways in which human activity adversely affects such habitat and its contribution to population productivity. The guidelines provide the following considerations for identifying HAPCs: (1) importance of ecological function, (2) sensitivity to human-induced degradation, (3) degree to which habitat is stressed or subject to stress, and (4) rarity of habitat type.

Assessment of Fishing Effects

Councils are required to mitigate or minimize adverse effects from fishing, to the extent practicable, if fishing can be shown to be having more than minimal or temporary impact to EFH. Adverse impacts may include direct or indirect physical, chemical, or biological alterations of the waters or substrate and loss of or injury to benthic organisms, prey species and their habitat, and other ecosystem components, if such modifications reduce the quality or quantity of EFH. They could include site-specific or habitat-wide impacts, including individual, cumulative, or synergistic consequences of actions. Councils are urged to rank fishing gears in terms of severity of impact and set priorities for which impacts need to be addressed first. They should consider using marine protected areas to manage gear impacts and other options that include fishing equipment restrictions, season and area restrictions on use of specified equipment, equipment modifications to allow better escapement, prohibitions on use of explosives and chemicals, prohibitions on anchoring or setting equipment in sensitive areas, and prohibitions on fishing activities that cause significant physical damage. Cumulative effects of past actions also need to be assessed, as must the long-term and short-term costs and benefits of potential management measures to protect EFH.

Prey Species

Loss of prey may adversely affect EFH and managed species because the presence of prey makes waters and substrate function as feeding habitat. Because the defini-

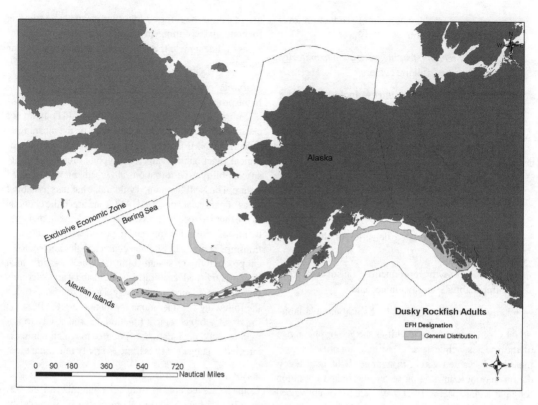

Figure 1. Essential fish habitat designation for Dusky rockfish (C. Coon, North Pacific Fishery Management Council, personal communication).

tion of EFH includes waters and substrate necessary for feeding, actions that reduce availability of prey, either through direct harm or capture or by degrading prey habitat, may be considered adverse effects on EFH.

Regional Fishery Management Council Actions to Protect Habitat

Before enactment of the SFA in 1996 and publication of the secretarial guidelines, the eight councils already had taken numerous steps to protect habitat from fishing. In many cases, the actions were taken to protect a particular species or species complex rather than habitat directly. The councils responded to the SFA by further identifying and describing EFH, designating HAPCs, and considering additional protective measures. The information presented below is based on summaries developed by regional council staffs for this report and was current as of late 2002. The summaries are representative of a much broader array of protective actions and are not intended to describe the full scope of council achievements. Council staffs should be contacted directly for the latest information on their activities to protect habitat.

New England Fishery Management Council

Essential Fish Habitat Designation
The New England Fishery Management Council (NEFMC) developed a single, omnibus EFH amendment for 18 managed species (sea scallops *Placopecten magellanicus*, monkfish (also known as goosefish) *Lophius americanus*, Atlantic herring *Clupea harengus*, Atlantic salmon *Salmo salar*, and 14 species of groundfish). The EFH designations were based on species reports provided by the NMFS Northeast Fisheries Science Center that summarized species life histories and status of stocks and described habitat associations across all life history stages. The reports included summary descriptions and maps of relevant survey data showing relative abundance and range of each species. The NEFMC considered alternatives for identifying EFH ranging from small to large areas based on the unique attributes and status of each species. The baseline alternatives were based on a ranked catch-per-unit-effort index of NMFS survey data for 10-min squares of latitude and longitude corresponding to areas of decreasing relative concentration. To these baseline alternatives, additional information, such as the NOAA Estuarine Living Marine Re-

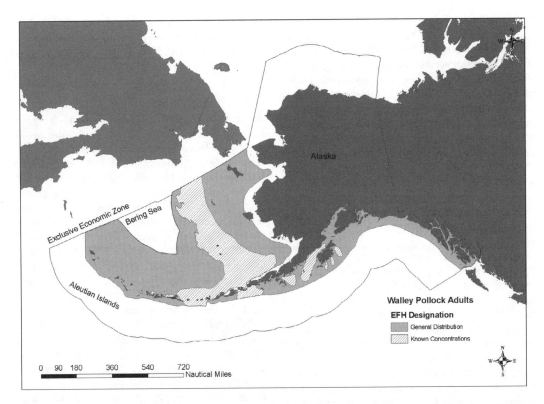

Figure 2. Essential fish habitat designation for walleye pollock (C. Coon, North Pacific Fishery Management Council, personal communication).

sources (ELMR) program reports, state survey information, historical information, and input from members of the fishing industry, was used to build final EFH designations for each species. The ELMR reports were used to help identify bays and estuaries appropriate for EFH designation for most managed species (not all species were included in the reports). Historical information and anecdotal reports from the fishing industry helped in identifying some areas not represented in the survey data.

Habitat Areas of Particular Concern

The HAPCs were designated for juvenile Atlantic cod *Gadus morhua* and Atlantic salmon based on a review of scientific literature describing species–habitat associations and higher levels of scientific information that indicated specific conditions or places where critical habitat existed for a particular life stage of these species. For juvenile Atlantic cod, a small portion of the northeastern Georges Bank was designated within the boundaries of a current closed area. The gravel and cobble substrate provides shelter from predation for juvenile Atlantic cod and has a rich growth of prey items such as bryozoans, hydroids, and worm tubes. This area is vulnerable to bottom fishing with scallop dredges and other mobile gears that reduce habitat structural complexity. The HAPC for Atlantic salmon included 11 rivers in Maine that support the only remaining U.S. populations of naturally spawning Atlantic salmon that have historic river-specific characteristics. These river populations harbor an important genetic legacy that is vital to the persistence of these populations and to the continued existence of the species.

Fishing Restrictions

Fishing-related adverse impacts were identified and assessed, drawing heavily from Auster and Langton (1999). All gear types used in the New England region were described, along with their associated proportion of landings. The potential impacts from aquaculture, fishing-related marine debris and lost fishing gear, and at-sea fish processing also were assessed. In addition, the council evaluated all existing fishery regulations to determine which measures already served to reduce the potential adverse impacts from fishing activities. This evaluation highlighted the fact that the NEFMC's year-round or temporary closures of over 24,000 mi^2 (81,600 km^2) had reduced fishing effort considerably. In 1997, groundfish fishing effort was reduced by 50%. Roughly 30% (6,600 mi^2 [22,440 km^2]) of Georges Bank is locked up within three large closed areas in which all bottom-tending mobile fishing gear is prohibited. On the basis of a habitat risk assessment prepared by the NEFMC, vessels were allowed temporary access to the closed areas to harvest an abundant sea scallop resource but only in places determined to be the least sensitive to the impacts of dredge gear. Though scallop fishing was temporarily allowed in the closed area and potentially could adversely affect the habitat, it reduced the overall scallop fishing pressure region-wide and, thus, should have reduced potential impacts over the entire region. Also, fishing effort is expected to move from some areas of potentially more sensitive habitat (e.g., rocks and gravel) to areas of less sensitive habitat (e.g., flat sand). In areas that are closed to fishing, there has been recovery of emergent corals, bryozoans, hydroids, and sponges.

Roller and rockhopper gear with a diameter larger than 12 in (30 cm) is prohibited in some of the most sensitive habitat in the Gulf of Maine, and streetsweeper trawl gear has been banned since 1998. Nearly 1,200 mi^2 (3,108 km^2) of the Gulf of Maine are closed to most types of bottom-tending mobile fishing gear. The total area closed to trawl and dredge gear in New England is about the size of Massachusetts. Based on this assessment, the NEFMC concluded that no additional measures were practicable at the time. The only management measure included in the EFH omnibus amendment was to change the rationale for the portion of the existing Georges Bank closed area designated as an HAPC to include protection of habitat. Formerly, this area was closed solely for the purpose of protecting spawning stocks for several overfished species.

Mid-Atlantic Fishery Management Council

Essential Fish Habitat Designation
The Mid-Atlantic Fishery Management Council (MAFMC) used Northeast Fishery Science Center bottom trawl survey data for the past 30 or more years, along with ichthyoplankton data from Marine Resources Monitoring Assessment and Prediction (MARMAP) surveys of the 1980s, to identify EFH in the offshore areas. Essential fish habitat was generally based on whether the resource was overfished (usually 90% of distribution) versus under or fully utilized (75%). The EFH in nearshore and estuarine areas was identified using ELMR data.

Habitat Areas of Particular Concern
The HAPCs were identified for summer flounder *Paralichthys dentatus* and tilefish *Lopholatilus chamaeleonticeps*. Summer flounder HAPC included submerged aquatic vegetation and macroalgae beds in nursery habitats for juvenile and larval stages. These habitats provide shelter from predators. For tilefish, the HAPC includes substrate between the 250-ft (75 m) and 1,200-ft (360 m) isobaths within statistical areas 616 and 537 off southern New England and the New York Bight. In recent years, 90% of the tilefish landings have occurred from these two areas.

Fishing Restrictions
The vast majority of the Mid-Atlantic area is fairly shallow, with sandy sediment and not much structure. Because of the lack of information on the extent and permanence of fishing effects on EFH, MAFMC decided to focus its efforts on the nonfishing gear impacts on EFH. No management measures were proposed for gear impacts in any FMP. The MAFMC has, however, imposed several measures that have either reduced the race for fish (Atlantic surfclams *Spisula solidissima*) or reduced the total effort to rebuild the resource. Those capacity reduction measures help reduce the impact on EFH. No gear restrictions have been imposed, however, because the summer flounder HAPC is in state waters (where MAFMC does not have the authority to regulate fishing) and the tilefish HAPC does not have documented gear impacts. If there is any species that appears to be impacted by bottom-tending mobile gear in the mid-Atlantic region, it would most likely be the structure–burrow-oriented tilefish. But the very structure that is used to identify the HAPC is very difficult to fish without losing or damaging gears. In addition, fishermen have expressed concern over possible closures of over 3,000 mi^2 of bottom area without one scientifically documented study directly connecting gear impacts on tilefish habitat. The MAFMC will be considering area closures for egg masses of *Loligo pealei* in inshore New England waters during certain summer months.

South Atlantic Fishery Management Council

Essential Fish Habitat Designation
The South Atlantic Fishery Management Council (SAFMC) embraced an ecosystem approach in addressing the EFH mandate and developed a habitat plan (SAFMC 1998) that supports the designation of EFH and EFH–HAPCs for all managed species or species complexes and also will serve as the foundation for an eventual fishery ecosystem plan for the South Atlantic region. The designations for each major species are comprehensive and detailed.

For peneaid shrimp, for example, EFH includes inshore estuarine nursery areas, offshore marine habitats used for spawning and growth to maturity, and all interconnecting water bodies as described in the habitat plan. Inshore nursery areas include tidal freshwater, estuarine, and marine emergent wetlands (e.g., intertidal marshes); tidal palustrine forested areas; mangroves; tidal freshwater, estuarine, and marine submerged aquatic vegetation (e.g., seagrass); and subtidal and intertidal nonvegetated flats. This applies from North Carolina through the Florida Keys.

For brown rock shrimp *Sicyonia brevirostris*, EFH consists of offshore terrigenous and biogenic sand-bottom habitats from 18 to 182 m in depth with highest concentrations occurring between 34 and 55 m. This applies for all areas from North Carolina through the Florida Keys. Essential fish habitat includes the shelf current systems near Cape Canaveral, Florida, that provide major transport mechanisms affecting planktonic larval rock shrimp. These currents keep larvae on the Florida Shelf and may transport them inshore in spring. In addition, the Gulf Stream is an essential fish habitat because it provides a mechanism to disperse rock shrimp larvae.

Similarly broad and comprehensive designations were made for red drum *Sciaenops ocellatus*; snapper–grouper species; coastal migratory pelagics; golden deepsea crab *Chaceon fenneri*; spotted spiny lobster *Panulirus guttatus*; and coral, coral reefs, and live and hard-bottom habitats. For example, snapper–grouper species EFH includes coral reefs, live and hard bottom, submerged aquatic vegetation, artificial reefs, and medium to high profile outcroppings on and around the shelf break zone from shore to at least 600 ft (but to at least 2,000 ft for wreckfish *Polyprion americanus*) where the annual water temperature is sufficiently warm to maintain adult populations of members of this largely tropical complex. The EFH includes the spawning area in the water column above the adult habitat and the additional pelagic environment, including *Sargassum*, required for larval survival and growth up to and including settlement. In addition, the Gulf Stream is an essential fish habitat because it provides a mechanism to disperse snapper–grouper larvae. Additional inshore areas are designated EFH for specific life stages of estuarine dependent and nearshore snapper–grouper species, areas such as attached macroalgae; submerged rooted vascular plants (seagrasses); estuarine emergent vegetated wetlands (saltmarshes, brackish marsh); tidal creeks; estuarine scrub–shrub (mangrove fringe); oyster reefs and shell banks; unconsolidated bottom (soft sediments); artificial reefs; and coral reefs and live and hard bottom.

Habitat Areas of Particular Concern

The HAPCs were designated in the comprehensive EFH amendment for all species covered under a given FMP rather than for individual species. They include general habitat types such as seagrass habitat and areas of ecological importance such as the Charleston Bump. Many HAPCs were designated. For example, for snapper–grouper, designated HAPCs were medium to high profile offshore hard bottoms where spawning normally occurs, areas of known or likely spawning aggregations, nearshore hard-bottom areas, the Point, the Ten Fathom Ledge, Big Rock, the Charleston Bump, mangrove habitat, seagrass habitat, oyster–shell habitat, all coastal inlets, all state-designated nursery habitats of particular importance to snapper–grouper, pelagic and benthic *Sargassum*, Hoyt Hills for wreckfish, the *Oculina* Bank HAPC, all hermatypic coral habitats and reefs, manganese outcroppings on the Blake Plateau, and SAFMC-designated artificial reef special management zones. The HAPCs are intended to prevent further decreases in biological productivity for each species and long-term economic benefits.

For shrimp, HAPCs include all coastal inlets, all state-designated nursery habitats of particular importance to shrimp, and state-identified overwintering areas. For spiny lobster, HAPCs include Florida Bay, Biscayne Bay, Card Sound, and coral–hard bottom habitat from Jupiter Inlet, Florida, through the Dry Tortugas, Florida. Coral HAPC includes the Ten Fathom Ledge, Big Rock, and The Point (North Carolina); Hurl Rocks and the Charleston Bump (South Carolina); Gray's Reef National Marine Sanctuary (Georgia); the *Phragmatopoma* (worm reefs) reefs off the central eastern coast of Florida; *Oculina* Banks off the eastern coast of Florida from Ft. Pierce to Cape Canaveral; nearshore (0–4 m) hard bottom off the eastern coast of Florida from Cape Canaveral to Broward County; offshore (5–30 m) hard bottom off the eastern coast of Florida from Palm Beach County to Fowey Rocks; Biscayne Bay, Florida; Biscayne National Park, Florida; and the Florida Keys National Marine Sanctuary.

Fishing Restrictions

The SAFMC did not develop specific management measures for the areas designated as HAPCs other than for the *Oculina* Bank HAPC, where the designation and protective measures predated the EFH provisions. It has, however, taken action within the Coral Fishery Management Plan and subsequent amendments to protect coral, coral reefs, and live hard-bottom habitat in the South Atlantic region by establishing an optimum yield of zero and prohibiting all harvest or possession of these resources that serve as essential fish habitat to many managed species. Another measure implemented through the coral plan was the designation of the *Oculina* Bank HAPC, a unique and fragile deepwater coral habitat off southeastern Florida that is protected from all bottom-tending fishing gear damage. The SAFMC also prohibited the following gears in snapper–grouper FMP to protect habitat: bottom longlines inside of 50 fathoms (100 m) or anywhere south of St. Lucie Inlet, Florida; fish traps; -ending (roller-rig) trawls on live bottom habitat; and entanglement gear. Also established under the snapper–grouper plan is an experimental closed area (experimental marine reserve) where the harvest or possession of all species in the snapper–grouper complex is prohibited. Other actions that directly or indirectly protect habitat or ecosystem integrity include the prohibition of rock shrimp trawling in a designated area around the *Oculina* Bank, mandatory use of bycatch reduction devices in the penaeid shrimp fishery, a prohibition of the use of drift gill nets in the coastal migratory pelagic fishery, and a mechanism that provides for the concurrent closure of the Exclusive Economic Zone (EEZ) to penaeid shrimping if environmental conditions in state waters are such that the overwintering spawning stock is severely depleted.

Gulf of Mexico Fishery Management Council

Essential Fish Habitat Designation

The Gulf of Mexico Fishery Management Council (GMFMC) identified EFH as everywhere that each life stage of each managed species commonly occurred. This identification was based largely on the NOAA Gulf of Mexico Data Atlas and ELMR data.

Habitat Areas of Particular Concern

The GMFMC designated HAPCs in the generic EFH amendment, not for individual species, but to benefit all managed species. Certain bays, estuaries, and sanctuaries were identified that had previously been identified in the fishery management plan as important areas for other purposes. They fall under three main habitat types:

(1) Nearshore areas of intertidal and estuarine habitats with emergent and submerged vegetation, sand and mud flats, shell and oyster reefs, and other substrates that may provide food and rearing for juvenile fish and shellfish managed by GMFMC; migration route areas for adult and juvenile fish and shellfish; and areas that are sensitive to natural human-induced environmental degradation, especially in urban areas and in other areas adjacent to intensive human-induced vegetated wetlands, oyster reefs, shellfish beds, and certain intertidal zones.

(2) Offshore areas with substrates of high habitat value and diversity or vertical relief that serve as cover for fish and shellfish. These can be areas with rich epifaunal communities (e.g., coral, anemones, bryozoans, etc.) or various types of live rock and other hard-bottom structures that may be most readily impacted by fishing activates.

(3) Marine and estuarine habitat used for migration, spawning, and rearing of fish and shellfish, especially in urban areas and in other areas adjacent to intensive human-induced developmental activities.

They include designations such as Florida Keys National Marine Sanctuary, Florida Bay, Flower Garden Banks National Marine Sanctuary, three estuarine research reserves, Grand Bay in Mississippi, Florida Middle Grounds, and the Dry Tortugas, a pristine reef area.

Fishing Restrictions

The GMFMC has addressed fishing impacts on habitat since its first fishery management plans. Current management measures such as gear prohibitions and area or time closures minimize fishing effects on EFH. The trawl, trap, and recreational fisheries were reviewed for potential impacts to EFH. Primary information regarding fishing impacts on habitat in the Gulf of Mexico was lacking. New management measures and regulations were not proposed in response to the EFH amendments in 1996. Fishing-related management measures to minimize any identified impacts were deferred to future amendments when GMFMC has the information necessary to decide if the measures are practicable.

Though no special regulations were proposed for the designated HAPCs, GMFMC already has designated a number of sites as marine protected areas, including the Tortugas Shrimp Sanctuary, Cooperative Texas Shrimp Closure, Southwest Florida Seasonal Closure (Shrimp/Stone Crab), Central Florida Shrimp/Stone Crab Separation Zones, Longline/Buoy Gear Area Closure, Florida Middle Grounds HAPC, Madison/Swanson and Steamboat Lumps Marine Reserve, Stressed Area Closure, Flower Garden Banks HAPC, and Tortugas North and South Marine Reserves. In addition to marine protected areas, there are other areas protected to some degree through FMP area closures. Some of these closures are specific gear closures that were established to protect stocks by reducing fishing pressure during certain seasons or year round. These closures also have the effect of protecting the habitat from these gears. Others were established specifically to protect habitat. In total there are about 135,000 mi^2 protected.

Caribbean Fishery Management Council

Essential Fish Habitat Designation

The Caribbean Fishery Management Council (CFMC) used presence or absence of species from a specific habitat, as described in the literature, as the primary basis for identifying EFH in the U.S. Caribbean. Species association with habitat was then used to establish locations of specific habitats (e.g., presence of queen conch *Strombus gigas* eggs represents sandy habitats; presence of spawning red hind *Epinephelus guttatus* might represent reef structures of primarily *Montastrea anularis*). The lack of benthic information and maps only allowed for generalized and broad definitions of EFH. A habitat suitability model was used based on sparse habitat information for St. John, showing that most species could be found almost anywhere around the area. Overlap of species on habitat was significant, and EFH includes all waters and substrates (e.g., mud, sand, shell, rock, and associated biological communities), including coral habitats (coral reefs, coral hardbottoms, and octocoral reefs), sub-tidal vegetation (seagrass and algae), and adjacent intertidal vegetation (wetlands and mangroves). Therefore, EFH includes virtually all marine waters and substrates from the shoreline to the seaward limit of the EEZ. The EFH is identified and described based on areas where various life stages of 17 selected managed species and the coral complex commonly occur.

Habitat Areas of Particular Concern
In the generic EFH amendment, CFMC acknowledged the scarcity of information for life histories, as noted above, and generically designated habitat types, including estuaries in Puerto Rico and the U.S. Virgin Islands, nearshore reefs, and other hard bottom structures, based on ecological function. It also designated an area southwest of St. Thomas known as the Hind Bank, as an HAPC. Amendment 1 to the Coral FMP established a no-take marine conservation district of about 16 mi^2 in this area. The area already was closed seasonally to protect the red hind spawning aggregation. The year-round closure is intended to protect corals and associated flora and fauna. The HAPC was identified on the basis of the specific association between a grouper (red hind) and coral species during a specific life stage (spawning adult).

Fishing Restrictions
The review of Auster and Langton (1999) did not include any information on fishing gears used in the U.S. Caribbean. A review of the literature by CFMC did not find specific scientific evidence to implement management measures dealing with fishing gear impacts. Though no management measures were included in the EFH amendment, Amendment 1 to the Coral FMP establishes a no-take marine reserve in the EEZ southwest of St. Thomas that will allow for comparison among fishing and nonfishing areas and for the baseline description of HAPCs. The CFMC is now working to delineate EFH for queen conch, particularly areas such as seagrass beds associated with clean sandy areas. Seagrass beds found near coral reefs are nursery grounds for the queen conch. Prior to 1996, CFMC already had taken action to designate habitat critical to spiny lobster and reef fish. This led to prohibitions on use of certain gears such as poisons, explosives, drugs, or other chemicals to harvest lobster or reef fish because of their obvious impact on coral reefs and other habitats.

Western Pacific Fishery Management Council

Essential Fish Habitat Designation
Because of large gaps in the scientific knowledge about the life histories and habitat requirements of many managed species, the Western Pacific Fishery Management Council (WPFMC) adopted a precautionary approach in designating EFH. The unavailability of information on geographic variation in the density of managed species or relative productivity of different habitats and, to a lesser degree, species habitat preferences precluded precise designations of EFH. With the exception of the EFH for precious corals, the designations consist of the depth ranges of all FMP species. The preferred depth ranges of specific life stages were used to designate EFH for bottom fish and crustaceans. For crustaceans, the designation was further refined based on productivity data. Water temperature was a useful indicator for the distribution of EFH for pelagic species. Temperature also implies a depth range since many species are confined to mesopelagic waters above a permanent thermocline. It was recognized, however, that certain species make extensive vertical migrations to forage, in some cases below the thermocline. The precious corals designation combined depth and bottom types as indicators but was further refined based on the known distribution of the most productive areas for these organisms. To reduce the complexity and the number of EFH identifications required for individual species and life stages, WPFMC designated EFH for management unit species assemblages. Species were grouped into complexes because available information suggested that many of them occur together and share similar habitat.

Habitat Areas of Particular Concern
The WPFMC designated HAPCs for each of its management unit species. The HAPCs for bottom fish included all escarpments and slopes between 40 and 280 m and three known areas of juvenile opakapaka (*Pristiponoides* sp.). Pelagic species HAPCs included the water column down to 1,000 m that lies above seamounts and banks shallower than 2,000 m. Crustacean HAPCs included all banks with summits less than 30 m in the Northwestern Hawaiian Islands (NWHI; they provide critical recruitment sites for spiny lobster). Precious coral HAPCs included six precious coral beds and Auau Channel, which provides protection for black coral.

Fishing Restrictions
Since 1976, WPFMC has implemented four FMPs for western Pacific fisheries (precious corals, crustaceans, bottom fish, and pelagic) and has recently completed its fifth FMP for coral reef ecosystems. The predominant fishing gear types, hook-and-line, longline, troll, and traps, used in

these fisheries cause few fishing-related impacts to the benthic habitat of bottom fish, crustaceans, and precious corals. The current management regime prohibits the use of bottom trawls, bottom-set nets, explosives, and poisons, and the use of nonselective gear to harvest precious corals is also prohibited. In addition, fishing for crustacean management unit species is prohibited landward of the 10 fathoms in the NWHI to protect habitats utilized by the majority of coral reef species. Further, both the NWHI bottom fish and crustacean fisheries are limited-entry fisheries, allowing only 17 permitted bottom fish fishermen and 15 permitted crustacean fishermen to fish the area.

The WPFMC determined that current management measures to protect fishery habitat were adequate and no additional measures were necessary. However, the council identified the following potential sources of fishery-related impacts to benthic habitat that may occur during normal fishing operations:

- Anchor damage from vessels attempting to maintain position over productive fishing habitat;
- Heavy weights and line entanglement occurring during normal hook-and-line fishing operations;
- Lost gear from lobster fishing operations;
- Illegal fishing for precious corals with tangle nets; and
- Remotely operated vehicle tether damage to precious coral during harvesting operations.

The WPFMC also is concerned that marine debris originating from fishing operations outside its area may have impacts on EFH. The Coral Reef Ecosystem Plan, if approved, would establish a precautionary ecosystem management approach to ensure the sustainable utilization of coral reef resources and designate approximately 12% of the Western Pacific Region coral reef habitat as a no-take marine protected area. These designations represent an important step toward meeting the U.S. Coral Reef Task Force's goal to protect 20% of the regions coral reefs by 2010.

Pacific Fishery Management Council

Essential Fish Habitat Designation

The Pacific Fishery Management Council (PFMC) adopted separate EFH amendments for each fishery management plan. Essential fish habitat for Pacific coast groundfish is defined as the aquatic habitat necessary to allow for groundfish production to support long-term sustainable fisheries for groundfish and for groundfish contributions to a healthy ecosystem. Because of the large number of species and their ranges, the entire EEZ becomes EFH when the more than 400 individual habitats at various life stages are taken together. The FMP groups the various EFH descriptions into seven composite units based on major habitat types These include estuarine habitat (the bays and estuaries along the coast, up to where the saltwater ends upriver); rocky shelf habitat; nonrocky shelf habitat; submarine canyon habitat; the continental slope/basin; the neritic zone (the water column more than 10 m above the continental shelf); and the oceanic zone (the water column more than 20 m above the continental slope).

Essential fish habitat for the Pacific coast salmon includes all streams, lakes, ponds, wetlands, and other currently viable water bodies; most of the habitat historically accessible to salmon in Washington, Oregon, Idaho, and California; and freshwater, including areas above artificial barriers, except for impassable barriers (dams). In estuarine and marine areas, salmon EFH extends from the nearshore and tidal submerged environments within state territorial waters out to the full extent of the EEZ off Washington, Oregon, and California north of Point Conception. Foreign waters off Canada, while still salmon habitat, are not included in salmon EFH because they are outside U.S. jurisdiction, but EFH does include the marine areas off Alaska designated as salmon EFH by the North Pacific Fishery Management Council.

Essential fish habitat for the coastal pelagic species (CPS), including finfish (northern anchovy *Engraulis mordax*, Pacific sardine *Sardinops sagax*, Pacific chub mackerel *Scomber japonicus*, and jack mackerel *Trachurus symmetricus*) and market opalescent inshore squid *Loligo opalescens*, is based upon a thermal range bordered within the geographic area where a CPS species occurs at any life stage, where the CPS species has occurred historically during periods of similar environmental conditions, or where environmental conditions do not preclude colonization by the CPS species. The east–west boundary of EFH includes all marine and estuary waters from the coast to the limits of the EEZ and above the thermocline where sea surface temperatures range between 10°C and 26°C. The southern boundary is the United States–Mexico maritime boundary. The northern boundary is defined as the position of the 10°C isotherm, which varies seasonally and annually.

Defining EFH for highly migratory species such as tuna, swordfish, and sharks is a challenging task. Highly migratory species are usually not associated with the features that are typically considered fish habitat (such as seagrass beds, rocky bottoms, or estuaries). Their habitat may be defined by temperature ranges, salinity, oxygen levels, currents, shelf edges, and sea mounts. Designations of EFH depend on data from fishery and fishery-independent sources, life history information, expert opinion regarding the importance of certain areas, tagging and tracking data, and other information. As with groundfish, EFH is designated for each life stage of stocks listed in the FMP, classified by their oceanographic (e.g., neretic, mesopelagic) and geographic distribution.

Habitat Areas of Particular Concern
The PFMC has not yet designated any HAPCs but is currently working with NMFS in developing the Groundfish Fishery Management Plan EFH Environmental Impact Statement (EIS) to develop a framework for identifying, evaluating, and designating HAPCs.

Fishing Restrictions
Fishing gear restrictions are the primary management measures used to minimize adverse effects of fishing gear on EFH and to protect important habitat areas. Before 2000, large footrope trawls with roller gear or chafing gear were used to bounce over tough rock piles in search of rockfish. Use of such gear to land shelf rockfish has been prohibited since 2000. Although the regulation was primarily intended to reduce catch of shelf rockfish species, it has effectively eliminated bottom contact trawling in most of the rocky reef areas on the West Coast (Hannah and Freeman 2000). The need to reduce harvest of overfished groundfish stocks resulted in extensive time and area closures for most West Coast commercial groundfish fisheries in 2003. Depth-based restrictions for groundfish fisheries in 2003, depending on sector and region, will constrain most bottom contact fishing activities out to 250 fathoms. As a result, fishing gear effects on benthic habitat will likely be substantially less than in recent years.

North Pacific Fishery Management Council

Essential Fish Habitat Designation
North Pacific Fishery Management Council (NPFMC) scientists added a fifth level, level 0, indicating no systematic sampling, to the four levels of information in the Secretarial guidelines. The NPFMC's technical teams determined that only levels 0 and 1 (presence–absence) information was available for most life stages. Level 2 (habitat-related densities) information was generally available for adult life stages. Only salmon had higher 3 and 4 information levels. Though skimpy, this information enabled the teams to describe general distributions and known concentrations of adult and juvenile stages for some species. General distribution is the area containing about 95% of the individuals across all seasons but not a species' entire current and historic range. Known concentrations are subsets of the general distribution and were identified only for life stages that had at least level 2 data. The delineation of known concentrations was based on an examination of survey and fishery hauls and a catch-per-unit-effort measure of density. Boundaries for known concentrations were drawn around areas with an upper third of the densities for a species' life stage. Based on NMFS recommendations, NPFMC used the relatively larger general distributions to identify EFH, believing they would more adequately address unpredictable annual differences in spatial distributions of a life stage and changes due to long-term shifts in oceanographic regimes. The NPFMC also believed that all habitats occupied by a species contribute to production at some level and that long-term productivity is based on both high and low levels of abundance. A species' entire general distribution may be required during times of high abundance.

Habitat Areas of Particular Concern
Habitat areas of particular concern were identified as habitat types (rather than discrete areas) using the criteria specified in the guidelines. Three HAPC habitat types were identified, though for many HAPCs, information is too incomplete to delineate specific boundaries:

(1) Shallow-water living substrates (kelp, mussels, eelgrass, rockweed, etc.) important for feeding and rearing of groundfish species and spawning for Atka mackerel *Pleurogrammus monopterygius* and yellowfin sole *Limanda aspera*. These areas are vulnerable to shore-based activities and relatively rare. They are also important to king crab and herring reproduction and to Pacific salmon migration.

(2) Deepwater living substrates (coral, anemones, bryozoans, etc.), which provide vertical structure for protection and shelter of groundfish.

(3) Freshwater areas used by anadromous fish. These areas include streams, lakes, and other freshwater areas, and also urban sites where human disturbance may be high.

Fishing Restrictions
An independent literature review was used to assess the impacts of all fishing gears used under existing FMPs. The Auster and Langton (1999) analysis was incorporated into the environmental assessment by reference, and current management measures were evaluated relative to EFH and the literature reviews. Prior to the SFA, NPFMC had implemented many conservation measures to protect various managed species. These included a ban on directed fishing for forage fish to protect the prey field for managed stocks; year-round area closures to groundfish trawling and scallop dredging to protect crab habitat; gear restrictions on scallop dredge size and design of crab pots; and various closures to reduce interactions with marine mammals, particularly Steller sea lions *Eumetopias jubatus*. All in all, there are about 30,000 mi^2 in the Bering Sea closed to bottom trawling to protect habitat, an area that roughly approximates the size of Indiana or Maine and is more than twice the size of

Georges Bank. There are even larger closures in the Gulf of Alaska to protect habitat (60,000 mi^2).

The NPFMC viewed those actions as sufficient to comply with the provisions of SFA but did move ahead with protecting a 3.1-mi^2 pinnacle area off Sitka in southeastern Alaska from fishing and anchoring (essentially creating a marine reserve). The area was found to be extremely productive yet vulnerable to human impacts and met all criteria for establishment as an HAPC. However, in reviewing the amendment package, NPFMC decided to separate the pinnacle closure from the EFH provisions and adopt it as a separate amendment. There was concern that such a management measure could set a precedent for all future actions taken to protect EFH.

Legal Challenges

Seven environmental groups and two fishermen's associations did not agree that five councils had taken sufficient action to protect habitat from fishing effects in response to SFA. They filed a complaint in April 1999 in the U.S. District Court for the D.C. Circuit (*American Oceans Campaign v. Daley*, Civ. No. 99-982 D.D.C. [2000]). The plaintiffs included American Oceans Campaign (AOC), Center for Marine Conservation, Florida Wildlife Federation, Reefkeeper International, Natural Resources Defense Council, National Audubon Society, Pacific Coast Federation of Fishermen's Association, and Cape Cod Commercial Hook Fishermen's Association. The original complaint addressed only the Gulf of Mexico Fishery Management Council but was later expanded to include other plaintiffs and councils. It argued that NMFS and five regional councils (New England, Gulf of Mexico, Caribbean, Pacific, and North Pacific) developed and approved amendments that violated MSA and were arbitrary, capricious, and contrary to law in violation of the Administrative Procedures Act.

The complaint alleged that the council EFH amendments:

(1) Did not adequately assess the effects of fishing and fishing gear on EFH;

(2) Did not adequately identify and assess potential measures to minimize adverse effects of fishing on EFH; and

(3) Failed to impose practicable measures to minimize fishing impacts on EFH.

The AOC claimed that bottom trawling and other fishing activities harm EFH and that various fishery management measures could be used to protect EFH from the effects of such fishing activities. The AOC claimed further that the councils failed to investigate whether certain additional restrictive measures would be practicable and ultimately failed to identify, include, and implement practicable measures to protect EFH. The AOC argued that those failures violated nondiscretionary duties imposed upon the councils by the plain language of MSA and implementing regulations.

The AOC also claimed that the environmental assessments performed by the councils inadequately evaluated the environmental effects of the proposed actions and, therefore, lacked justification for the NMFS finding of no significant impact. More specifically, the assessments failed to:

(1) Evaluate long-term or cumulative impacts of approving the amendments on EFH affected by ongoing fishing activities, including but not limited to bottom trawling;

(2) Adequately evaluate practicable methods to minimize the effects of fishing on EFH; and

(3) Address an adequate range of alternatives.

The AOC asked the court to remand the EFH amendments back to the councils with instructions to revise them and bring them into compliance with the MSA by a certain date. They also wanted a whole new analysis, including an assessment of the long-term and cumulative environmental impacts of minimizing the adverse effects of fishing on EFH along with detailed assessments of alternative methods for protecting EFH.

The U.S. District Court responded in a memorandum opinion on 14 September 2000, as follows:

(1) Concluded that the EFH amendments did not violate MSA but did violate the National Environmental Policy Act (NEPA). Defendants had to perform new and thorough environmental assessments (EA) or EIS under NEPA.

(2) Issued a permanent injunction, enjoining federal defendants from enforcing EFH amendments until the Secretary of Commerce performed a new, thorough, and legally adequate EA or EIS for each EFH amendment.

(3) Concluded that it was reasonable for the Secretary of Commerce to approve the amendments after considering whether the amendments complied with MSA, given how little scientific information was available to the council at the time.

(4) Determined that it was reasonable for the Secretary of Commerce to conclude that the amendments did not need to include additional protective measures, given the lack of scientific evidence available to the councils and the Secretary of Commerce and existing protective measures already in place.

(5) Each of the environmental assessments were insufficient for compliance with the requirements of NEPA for a variety of reasons:

- Insufficient analysis and alternatives.

- No hard look at the problem (e.g. the actual environmental consequences and impacts of fishing on the designated EFH).

- Failed to identify relevant areas of environmental concern (the court found that the environmental assessments only discuss fish habitats in general terms, describing the types of EFH that should be protected but not specifying which EFH needed protection and why.

- Failed to make a convincing case that the impacts from the action were insignificant.

- Finally, the environmental assessments failed to demonstrate that any significant impacts were mitigated by the alternative selected.

The focus of the lawsuit was fishing effects on habitat. The court order included EFH amendments developed by all five councils. However, for the Pacific Fishery Management Council, the order affected only the groundfish plan since that was the only one challenged and it relied on its own environmental assessment. The other four councils developed omnibus plan amendments and environmental analyses covering multiple fishery management plans. Although for certain councils AOC only challenged some of their plans, all of the plans covered under the environmental assessments were affected by the court order because the analyses share the same deficiencies with respect to all the plans. Thus, the EFH provisions of the following 22 FMPs must be addressed in new environmental impact statements being developed by the councils:

- New England Fishery Management Council: multispecies, scallops, salmon, monkfish, herring.

- Caribbean Fishery Management Council: reef fish, spiny lobster, queen conch, coral.

- Gulf of Mexico Fishery Management Council: shrimp, red drum, reef fish, coastal migratory pelagics, stone crab, spiny lobster, coral and coral reefs.

- Pacific Fishery Management Council: groundfish.

- North Pacific Fishery Management Council: Bering Sea and Aleutian Islands groundfish, Gulf of Alaska groundfish, king and Tanner crab *Chionoectes bairdi*, scallops, salmon.

The analyses are in various stages of development. The North Pacific Fishery Management Council is discussed in more detail below as an example of the challenge of providing additional habitat protection against a backdrop of an already highly regulated fishery.

North Pacific Fishery Management Council: Next Steps for Habitat Protection

Responding to the lawsuit, NPFMC has been developing and analyzing alternatives to further protect habitat from fishing gears in the Gulf of Alaska (GOA), Bering Sea, and Aleutian Islands. The analysis should be available for final action in 2004 or 2005, and any implementing regulations would take effect a year or two later. The alternatives as of February 2003 were as follows:

Alternative 1: Status quo.

Alternative 2: Gulf Slope Bottom Trawl Closures
Prohibit use of bottom trawls for rockfish in 13 designated areas of the GOA slope (200–1,000 m) but allow vessels endorsed for trawl gear to fish for rockfish in these areas with fixed gear or pelagic trawl gear.

Alternative 3: Bottom Trawl Gear Prohibition for GOA Slope Rockfish on Upper Slope (200–1,000 m)
Prohibit use of bottom trawl gear for targeting GOA slope rockfish species on upper slope area (200–1,000 m) but allow vessels endorsed for trawl gear to fish for slope rockfish with fixed gear or pelagic trawl gear.

Alternative 4: Bottom Trawl Closures in All Management Areas
Prohibit use of bottom trawl gear in designated areas of the Bering Sea and Aleutian Islands for all species and bottom trawling for rockfish in designated areas of Gulf of Alaska. Bottom trawl gear used in the remaining open areas of the Bering Sea would be required to have disks/bobbins on trawl sweeps and footropes to reduce contact with the seafloor.

Alternative 5A: Expanded Bottom Trawl Closures in All Management Areas
Prohibit use of bottom trawl gear in designated areas of the Bering Sea, Aleutian Islands, and GOA for all target species. Bottom trawl gear used in the remaining open areas of the Bering Sea would be required to have disks/bobbins on trawl sweeps and footropes to reduce contact with the seafloor.

Suboption 5B: Expanded Bottom Trawl Closures in All Management Areas with Sponge and Coral Area Closures in the Aleutian Islands
In addition to the measures in Alternative 5A, all untrawled or lightly trawled areas of the Aleutian Islands, as well as areas with historically high coral and sponge bycatch rates,

would be closed to bottom trawling. Total allowable catches for target species would be reduced based on the amount historically taken from the Aleutian Island closure areas. Bycatch limits for corals and sponges in the Aleutian Islands would be established, and a comprehensive plan for research and monitoring would be required.

Alternative 6: Closures to All Bottom Tending Gear
Prohibit use of all bottom-tending gear (dredges, bottom trawls, pelagic trawls that contact the bottom, longlines, dinglebars, and pots) for commercial fishing within approximately 20% of the fishable waters (i.e., 20% of the waters shallower than 1,000 m) in each region.

Any new closures must be viewed in the context of an already highly regulated fishery. They will add to the cumulative burdens of existing area restrictions and other management measures. Area closures have been used off Alaska since the beginning of the groundfish fisheries in the early 1960s when foreign fishermen had area restrictions negotiated in early bilateral agreements prior to U.S. extension of fisheries jurisdiction in 1976. These were incorporated, and often expanded upon, in preliminary management plans shortly after the MSA was passed. Though the council was reticent to impose area restrictions on the developing domestic trawl fleet and potentially slow progress in Americanizing the fisheries in the late 1970s and early 1980s, U.S. fishermen targeting other, more "traditional" species such as Pacific halibut *Hippoglossu stenolepis*, crab, herring (*Clupea* spp.), and Pacific salmon, were not so magnanimous. They demanded protection of their fishing grounds and nursery areas from the groundfish trawlers and were prepared to go toe to toe with domestic and foreign trawl fishermen alike. They ultimately were very successful in securing many significant closures and restrictions on bottom trawling beginning in the mid-1980s. Table 2 presents a general chronology of area restrictions. Table 3 and Figure 3 show size and locations of some of the areas.

Though the main goal of most closures was to protect valuable species targeted by other domestic fishermen and not to protect habitat per se, they significantly protected habitat. Even the many 10 and 20 nautical mile closures around rookeries and haulouts, designed to protect the foraging base for endangered Steller sea lions, provided extensive habitat protection. Those closures illustrate the challenges that inevitably will arise as the council considers additional measures to protect corals and sponges.

Steller sea lion-related closures resulted from regulations developed during the 1990s and biological opinions developed in 2000 and 2001. The biological opinion issued on 30 November 2000 (NMFS 2000) concluded that fisheries for pollock, Pacific cod *Gadus macrocephalus*, and Atka mackerel, as managed by regulations effective in 2000, jeopardized the survival and recovery of Steller sea lions and adversely modified their critical habitat. The biological opinion included reasonable and prudent alternatives that would have placed substantial restrictions on trawling and seriously disrupted the economies of the fishing industry and fishing communities.

In December 2000, the council rejected the 2000 biological opinion developed by NMFS and the accompanying fisheries restrictions to protect Steller sea lions and began developing alternative measures (NMFS 2001b). By June 2001, the council had agreed to analyze five alternatives, ranging from no action to broad restrictions on fishing for Pacific cod, pollock, and Atka mackerel in large areas of critical habitat. A draft supplemental environmental impact statement was released in August 2001, and in October 2001, the council adopted fishery-specific closed areas around rookeries and haulouts, together with seasons and catch apportionments. The council added restrictions such as a total closure of the Aleutian Islands pollock fishery, a reduction in the critical habitat limit for Atka mackerel, and a vessel monitoring system for most vessels fishing pollock, cod, or Atka mackerel.

The analysis demonstrated that the groundfish fishery is a very significant economic engine in the North Pacific. According to statistics for 1999 in the final EIS (NMFS 2001b), over 12,000 people were employed in the fishing and processing sectors, and the total wholesale value exceeded $1 billion. There were about 145 shore and sea-based processors, and over 900 catcher vessels that depended on the fishery. Not only did the fisheries provide employment income that supported the local and regional economies, they also provided significant tax revenues. For example, the Kodiak region's share of fisheries business tax and fishery resource landing taxes exceeded $1.3 million in 1999. In the Alaska Peninsula and Aleutian Islands region, fisheries-related shared taxes accounted for 99.7% of all the shared taxes and fees coming to the region from the state in 1999, and total fisheries-related tax revenues exceeded $7 million (based on all fisheries, not just groundfish).

The effect of the proposed closures to protect Steller sea lions was to move fishermen away from their traditional fishing grounds. Some of the more stringent alternative closure schemes were predicted to cause between 31% and 55% reductions in catcher vessel total harvest of pollock, Pacific cod and Atka mackerel across all regions. In specific areas such as the Alaska Peninsula and Aleutian Island region, there was a predicted 54% to 80% reduction in regionally owned catcher vessel harvests. Even the alternative adopted by the council was predicted to cause a 5–9% reduction in catcher vessel harvest of pollock, Pacific cod, and Atka mackerel across all regions and a 7–17% reduction for specific regions such as the Alaska Peninsula and Aleutian Islands. Spe-

Table 2. General chronology of cumulative restrictions off Alaska, summarized from the Alaska Groundfish Fisheries Draft Programmatic Supplemental Environmental Impact Statement of January 2001 (NMFS 2001a).

Years of restriction	Types of restriction
Foreign Bilaterals: pre-1977 (foreign fisheries only)	• Time/area closures to reduce halibut and crab bycatch and conflict with those fisheries • Foreign fishing closures to protect Pribilof fur seals
Preliminary Management Plans: 1977 (foreign fisheries only)	• Trawl closures to protect spawning pollock and flounders • Time and area closures expanded • Bristol Bay Pot Sanctuary closed to trawling all year • Extensive trawl closures in GOA to protect halibut
First Fishery Management Plans: 1979–1982 (mainly foreign fisheries)	• GOA bottom trawl restrictions to protect halibut • GOA and Bering Sea Aleutian Islands) BSAI expanded time and area closures • Year-round closures to foreign fishing inside 12 mi • Three closures off southeast Alaska • Davidson Bank and Petrel Bank closures
1983–1985 (mainly foreign and joint venture fisheries)	• Southeast GOA closed east of 140 W to all foreign fishing to protect halibut • Foreign fisheries in GOA restricted to off-bottom trawls all year • Kodiak Gear Area closed to foreign trawls to protect crab fishermen and gear • Prohibition on discard of net and debris to protect habitat
1986–1990 (mainly joint venture and domestic fisheries)	• GOA type I–III closures off Kodiak to protect crab • First total closures to U.S. trawlers in BSAI • Bycatch-triggered closures • Walrus Island closure
1991–1995 (fully domestic fisheries)	• Expanded BSAI closures • Pribilof Island closure to trawling • Chum salmon bycatch-triggered area closures in BSAI • Rookery and haul-out closures to protect sea lions
1996–2000+ (fully domestic fisheries)	• Nearshore closures to protect crab • Bottom trawl ban for pollock • Chinook salmon bycatch-triggered savings areas • Commercial fisheries prohibited for forage fish, corals and sponges • More sea lion-related closures • Sitka Pinnacle closure

cific fisheries would take an even greater hit. For example, the alternative would cause a 17–26% reduction in Pacific cod harvest in the Alaska Peninsula and Aleutian Island region for regionally owned catcher vessels. The alternatives also could jeopardize the safety of the fishing fleet to the extent that smaller fishing vessels were required to go further offshore for their catch and the extent that occurrences of accidents and injuries are highly correlated with fishing distance offshore, weather and sea conditions, and vessel size. In determining which alternative to choose, the council had to weigh the impacts of the proposed restrictions on Steller sea lions, on the fishing

Table 3. Examples of groundfish trawl closure areas in Bering Sea and Aleutian Islands groundfish management area. Some closures are triggered by a specified level of bycatch of salmon or crab. Based on Table 2.7-8b in NMFS (2001a).

Closure	Year	Area (mi²)	Gears	Duration
Herring Savings Areas (3)	1991	19,125	Trawl	Partial
Steller Sea Lion Protection	1992	10,900	Trawl	Year-round
Pribilof Islands Habitat Conservation Area	1995	7,000	Trawl	Year-round
Chinook Salmon Savings Area	1995	9,000	Trawl	Partial with trigger
Chum Salmon Savings Area	1995	5,000	Trawl	Partial with trigger
Bristol Bay Red King Crab Savings	1996	4,000	Trawl	Year-round
Near-shore Bristol Bay Closure	1996	15,000	Trawl	Year-round
Opilio crab bycatch limitation zone	1996	90,000	Trawl	Partial with trigger

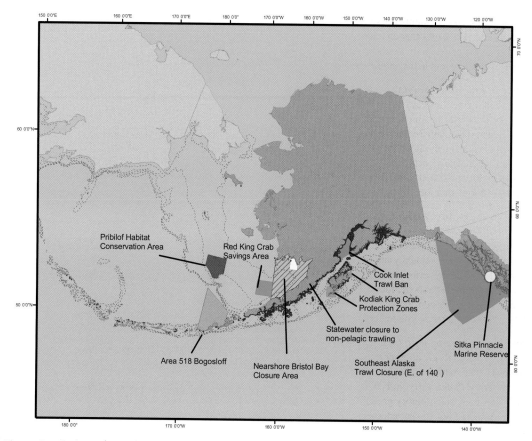

Figure 3. Alaska groundfish fishery areas with year-round or partial closures. (C. Coon, North Pacific Fishery Management Council, personal communication).

fleet and its economic viability, on coastal communities, on target fish species and bycatch, and on a host of other components of the ecosystem including habitat, seabirds, forage fish, cetaceans, and other marine mammals.

These Steller sea lion protective measures created a patchwork of many small closed areas from the Bering Sea down through southeastern Alaska not too dissimilar to what might be needed to protect corals and sponges. New closures will add to the burden of an already highly regulated fishery, and it will be incumbent on the council and NMFS to provide the best information available for pinpointing exactly which specific seafloor habitats need to be given protection. That will require very expensive research.

Habitat Research Costs

Habitat research must address a problem shared by all councils: insufficient information to specify the types and extent of habitat critical to managed fish species. The National Research Council (2002) found that EFH mainly was designated at level 2 (habitat-related densities) using frequency-dependent distributions of fishes as a proxy for habitat. It also noted that levels 3 (growth, reproduction, and survival rates within habitat) and 4 (production rates by habitat) require significant information about the relationship between the managed species and the type of habitat. And while the bathymetry of virtually the entire shelf of the USA has been mapped at a scale of 1:250,000 with 2-m contours, that scale is insufficient for fishery management purposes because some features (e.g., ledges, boulders, depressions) with relief much less than 2 m can be significant habitats for fishes. The report lists three types of mapping systems most often applied to seafloor habitats: side-scan sonar, multibeam, and automatic bottom classification systems. These systems are expensive. Example costs below are based on past experience off Alaska (R. McConnaughey and J. Heifetz, NOAA Fisheries, personal communication). Alaska, of course, presents one of the most difficult working environments for ship-based operations. Similar operations off other coasts nearer to population centers and support facilities should prove to be much more cost efficient than the examples given here.

Basic Bottom Trawl Survey
Basic bottom trawl surveys typically collect biomass and numerical data for benthic invertebrates as part of standard catch processing. Benthic invertebrates sampled are mostly megafauna living on the seafloor. Some infauna also are taken. Daily cost for the Eastern Bering Sea bottom trawl survey is approximately US$6,700/d to characterize 0.25 km^2, or about $27,000/km^2.

Bottom Trawl Survey with Grab Samples for Infauna Identification
Benthic invertebrates are considered an important constituent of seafloor complexity and heterogeneity and must be considered in any serious habitat mapping that will be used to assess the effects of mobile fishing gear on EFH. Daily costs of collection and sample processing are about $38,000 to characterize infauna in 5 m^2 of seabed.

Sediment Mapping with Grab Samples
Surface sediments are another important characteristic of groundfish habitat and can be collected using a van Veen grab sampler. Fifty grab samples collected daily and submitted to laboratory granulometric processing, costs about $12,860 to characterize sediments over 5 m^2 of seabed.

Sediment Mapping Using Single or Split beam Echosounder
Acoustical methods are much more efficient for mapping surface sediments than grab samples, and one option is to use single or split-beam seabed classification using the QTC View system (Quester Tangent Corporation, Sidney, British Columbia). Statistical methods are used to classify (acoustically) distinct seabed types that are then groundtruthed with grab sampling or video. The daily cost would be about $7,700 to characterize 15.5 km^2 of surface sediments and produce a map, or about $500/km^2.

Side-Scan Sonar with Video/Grab Sampling Verification
Interferometric side-scan sonar can provide both backscatter and swath bathymetry and is groundtruthed with video or grab sampling. Total estimated cost is about $13,000 for excellent backscatter imagery and good swath bathymetry over 62 km^2, or $210/km^2.

Multibeam Bathymetry and Pseudo-Side-Scan Sonar
Good quality hydrographic surveys are being conducted in the northern GOA using multibeam bathymetry with pseudo-side scan sonar traces for backscatter in 150–750-m-depth waters. Approximately 1,000 km^2 on Portlock Bank northeast of Kodiak and 500 km^2 off Yakutat have been surveyed. The costs were about $29,000/d to cover about 500 km^2 during a 7-d cruise, or about $400/km^2. Geological interpretation of the maps adds about $10/km^2, and if a Delta manned submersible is used to groundtruth the sediment types and biological communities present, three dives per day run about $10,000 and covers about 100 km^2 of mapping, which adds about $100/km^2. Total costs then are about $510/km^2.

These examples demonstrate that delineating habitat will come at a very high cost. Obviously these cost estimates represent the high end of the cost range, given the remoteness of many areas sampled off Alaska. Survey costs also will be influenced by other factors such as beam width, time to cover an area of interest, required resolution, and whether the survey requires a dedicated cruise or can be combined with other research. Completing the surveys, however, is just the first step in defining EFH. Research must move beyond habitat mapping using bathymetrics and infauna identification to actually determining the biology and ecosystems dynamics of a particular area. Increasing the level of information availability to level 4 fish production estimates and their dependence on specific habitat, as envisioned in the Secretarial guidelines, will cost much more and require many years of research.

Discussion

Many significant issues come into play in the highly complex fishery management regimes that typify most areas under council purview. The manager's dilemma is how to address and balance those competing issues. In most cases, each side of the issue has very ardent and vocal supporters and opponents, different goals and objectives, and potentially very different outcomes, all of which must somehow satisfy the diverse national standards and provisions of the Magnuson–Stevens Fishery Conservation and Management Act and all other applicable law. Issues such as bycatch reduction, overfishing, community protection, endangered species protection, safety considerations, habitat protection, overcapitalization, and many others collide in the management process. Each incremental restriction proposed must be weighed in terms of cumulative impacts of all restrictions on any given fishery. Each solution comes with a cost, and regardless of council region, striking the right balance is a very onerous task.

Habitat protection often is achieved by modifying bottom-tending gear or restricting its use in specific management areas. This paper has focused on the latter approach and argued that such protection is best achieved through the regional council system to which fishermen are accustomed. Without buy-in, proposed restrictions likely will not be implemented as intended or may be watered down to the point of being ineffective in protecting habitat. The reality is that most commercial fishing groups have many avenues to protest a new restriction if they feel it is unfair. First, there is the council itself which has to develop and approve new restrictions through a lengthy public process. If the battle is lost there, fishermen can argue their case, and economic injury, to NMFS during Secretarial review of council-proposed regulations, and they often try to make their case with politically appointed personnel above the NMFS level. They also may initiate legal action or appeal to their congressional representatives for support. Depending on the seniority of their delegation and its clout, regulations may be avoided or at least delayed while compromises are sought. All this happened in the case of proposed regulations to protect Steller sea lions off Alaska and could happen again if new areas are proposed for closure to protect habitat without sufficient justification and buy-in from commercial fishing groups.

A proponent offering up a proposed change in fishery management is by far the easiest part of the management process. The hard part is deciding exactly what to do with it, which areas to close, which gear types to restrict, and for what reason. As noted earlier, advocates for protecting coldwater corals off Alaska have proposed protecting areas of high coral occurrence from commercial fisheries. Corals inhabit many areas off Alaska (Heifetz 2002) and cannot be fished commercially. Some areas already have been impacted by commercial fisheries for other species, some have not. In many instances, coral locations are identified because coral is hauled up incidentally in fishing gear. There has been some underwater exploration with remote cameras and submersibles, but overall, there has been no comprehensive mapping of coral beds, which will require many years and dollars. This will provide level 1 and 2 information, but much more intense research will be needed to describe the actual relationships of fish and coral in the quantitative sense envisioned in level 4 information in the Secretarial guidelines. If and when the NPFMC acts to protect coral habitat in the coming 12–24 months, it will be using relatively sparse information.

The NPFMC also will find itself in an immediate controversy over which gears to restrict. Bottom trawling always receives the most attention in terms of restricting gear types, but other gears such as heavy crab and groundfish pots or longlines dragged over coral areas also have impacts. This will complicate the decision process. At one point in 1997 and 1998, the NPFMC was moving ahead relatively quickly in defining several areas in which to protect corals. The initiative targeted trawl impacts and was advanced by fixed gear fishermen. Then it became widely recognized that fixed gears also brought up corals or knocked them down and that some proposed closure areas outside major fishing communities such as Yakutat and Kodiak were prime fishing grounds for fixed gear. The process quickly came to a halt as the NPFMC regrouped and designated a committee to further flesh out alternatives for possible closures. The current efforts to protect coral habitat are building on that earlier initiative.

This type of cooperative effort between stakeholders and the NPFMC must be nurtured if restrictions are to be implemented successfully. New regulations will impact the economic viability of the fleet. Closing productive areas may displace fishermen into less productive areas and require them to fish longer and harder to take the biologically allowable quota. Safety also could be a big issue if smaller vessels are moved further out to sea and have to run further to the fishing grounds in exposed waters. And often these smaller vessels come from the smaller communities along the Alaska coast that depend heavily on the revenues and taxes generated from the fisheries. Processors in local communities also need to maintain their production base and access to the fishery to continue to support local employment. All these factors make the management process and finding the right balance in habitat protection very difficult.

When we consider restrictions on the fleet, we need to ensure that habitat protection is really needed. We need to carefully define the purpose of habitat protection. Is it to reduce bycatch on nontarget species? Or is it to protect one stock of fish and ensure its sustainability? Can we predict the incremental increase of fish production that would result from the additional protection of habitat? What information level are we working at in defining essential fish habitat? Is it meritorious to close down large areas of highly trawlable fishing grounds that are fished year after year and remain productive? Is the purpose to protect the larger marine ecosystem? Is it to protect biodiversity? Or leave a particularly unique part of the ecosystem such as coral or sponges untouched in pristine condition for future generations? These are not easy questions to answer, and there is no magical formula for determining how much habitat to place off limits to specific gears. If one believes that closed areas are the ultimate solution to bungled or potentially bungled fishery management, then larger areas will be advocated. If fisheries have proven to be managed in a sustainable way, as is the case for the groundfish fisheries off Alaska for the past 25 years, then we need a very clear statement of purpose for additional closures.

Sensitive, vulnerable habitat needs to be pinpointed through research. Research over large areas and many years needs to be funded, and when areas are closed, they need to be monitored and compared against fished areas to determine whether the protection achieved the stated goals. That does not mean we should stand still until all the information comes in. But in being precautionary, we need to strike the right balance and aspire to intelligent regulation that has broad support, provides economic viability, protects the habitat, and ensures the long-term sustainability of the fishery resources. After all, this is not a faceless fishery or ecosystem. The people in small coastal communities depend on the resources, not only commercially but for subsistence. The resources need to be kept abundant, and the users need to be provided with safe fishing opportunities so long as the fish stocks will allow it. We need to make a convincing case that what we are doing as managers, over the long haul, will benefit these people, their children, and future generations to come. As one will find inscribed on the great granite rockworks of the Franklin D. Roosevelt memorial in Washington, D.C.:

"Men and nature must work hand in hand. The throwing out of balance of the resources of nature, throws out of balance also the lives of men."

Acknowledgments

The author wishes to acknowledge the help provided by all eight regional fishery management councils in describing their activities in protecting essential fish habitat. Special thanks go to Cathy Coon and David Witherell at the North Pacific Fishery Management Council for maps of closed areas off Alaska, Jon Kurland of NOAA Fisheries for his review of the legal section, and Robert McConnaughey and Jon Heifetz for the cost estimates for research. Last, I owe a great debt of gratitude to Penny Dalton and Bill Hogarth for the opportunity to work at NMFS Headquarters from September 2000 through June 2001 and develop a true appreciation for their very difficult task in carrying out protection of essential fish habitat and fisheries management in general.

References

Auster, P. J., and R. W. Langton. 1999. The effects of fishing on fish habitat. Pages 150–187 *in* L. Benaka, editor. Fish habitat: essential fish habitat and rehabilitation. American Fisheries Society, Symposium 22, Bethesda, Maryland.

Comptroller General. 1976. Report to Congress: action is needed now to protect our fishery resources. February 18, 1976. Comptroller General, Washington, D.C.

Hannah, R. W., and M. Freeman. 2000. A preliminary assessment of the impact of the "small footrope" regulation on the spatial distribution of Oregon bottom trawl effort in 2000. Oregon Department of Fish and Wildlife, Marine Resources Program, Newport, Oregon.

Heifetz, J. 2002. Coral in Alaska: distribution, abundance, and species associations. Hydrobiologia 47(1–3):19–28.

NMFS (National Marine Fisheries Service). 1998. Draft technical guidance to NMFS for implementing the essential fish habitat requirements for the Magnuson–Stevens Act. 9 January 1998. NMFS, Office of Habitat Conservation, Silver Spring, Maryland.

NMFS (National Marine Fisheries Service). 2000. Endangered Species Act section 7 consultation: biological opinion and incidental take statement for listed species in the BSAI Groundfish FMP and the GOA groundfish FMP, November 2000. NMFS, Protected Resources Division, Juneau Alaska.

NMFS (National Marine Fisheries Service). 2001a. Alaska groundfish fisheries: draft programmatic supplemental environmental impact statement. NMFS, Alaska Region, Juneau, Alaska.

NMFS (National Marine Fisheries Service). 2001b. Steller sea lion protection measures: final supplemental environmental impact statement. NMFS, Alaska Region, Juneau, Alaska.

NRC (National Research Council). 2001. Marine protected areas: tools for sustaining ocean ecosystems. National Academy Press, Washington, D.C.

NRC (National Research Council). 2002. Effects of trawling and dredging on seafloor habitat. National Academy Press, Washington, D.C.

SAFMC (South Atlantic Fishery Management Council). 1998. Habitat plan for the South Atlantic region: essential fish habitat requirements for fishery management plans of the SAFMC, Charleston, South Carolina.

Challenges, Objectives, and Sustainability: Benthic Community, Habitats and Management Decision-Making

JAKE RICE[1]

Department of Fisheries and Oceans, 200 Kent Street, Ottawa, Ontario K1A 0E6, Canada

Abstract. This paper first considers why it has been so difficult to make progress on moving fisheries toward ecological sustainability even in the narrow context of the target species. It then reviews the additional challenges that must be confronted when addressing the ecosystem effects of fishing, particularly impacts on benthic communities and habitats. Several impediments to progress are identified, including excess fishing capacity, the differential time courses of costs and benefits in reductions in fishing, myths and preconceptions regarding precaution and the relationship of sustainability to fishery characteristics, and above all, the complexity of the concept of sustainability itself, which has ecological, economic, and social dimensions. It has proven nearly impossible to find management options that do not lose ground on some dimensions in exchange for positive change on others. The paper evaluates the main tools available for reducing the effects of fishing on benthic communities and habitats, with regard to sustainability on all three axes. Four main classes of tools, including changing the cost–benefit accounting to include ecosystem goods and services, marine protected areas, gear modification and fleet substitution, and eco-certification, were all found to incur significant social or economic costs in order to make significant contributions to reducing impacts of fisheries on benthos. Because of the inescapability of trade-offs in decision making, a structured framework is needed for the decision-making process. Objectives-based fisheries management provides such a framework, but including benthos (and other ecosystem properties) in the list of objectives presents real challenges in keeping the list tractably short and the individual objectives usefully explicit. Work done by expert groups on approaches to meet these challenges is summarized. Overall, although there are many ecological questions about fishing effects on the benthos that have not yet been fully answered, the more urgent challenges are to find ways to use the knowledge we do have more effectively in decision making.

Introduction

Concern about detrimental ecosystem effects of fishing, including effects on benthic habitats, is producing increasingly frequent calls for fisheries managers to take a broad view of their responsibilities. However, commitments to an ecosystem approach to management have already been made at the policy level (e.g., Reykjavik Declaration, Bergen Declaration, and Koge Conference). To the extent that progress in reducing undesirable ecosystem effects of fishing is unsatisfactorily slow, the shortcomings are in practice, not policy. This paper will try to identify the major impediments to adopting practices that are effective at reducing the ecosystem effects of fishing and the shortcomings of the management tools available to support management and decision making at the ecosystem scale. Some inadequacies arise because the ecological problems being addressed are scientifically very difficult. Many papers in this volume address aspects of this issue. However, not all the management problems and impediments to progress are consequences of incomplete ecological knowledge. This paper will focus on clarifying the problems, other than lack of adequate ecological knowledge, that impede progress toward sustainable fisheries in ecosystem contexts compatible with healthy benthic communities.

There is a sound basis to claim decision makers and managers have already embraced, conceptually, an ecosystem approach to fisheries management. The World Summit on Sustainable Development (UN 2002) has plentiful ecosystem language, although as with Agenda 21 (UN 1992), it was negotiated by politicians and advocacy groups from diverse backgrounds and reflects commitments at a very high and abstract level. However, a plethora of agreements and instruments at the next level reflect the approaches and opinions of those working in the management and policy ranks of marine fisheries (Kimball

---

[1] E-mail: ricej@dfo-mpo.gc.ca

2001). Internationally, the United Nations Fisheries Agreement, the Kyoto Declaration, the Reykjavik Declaration, the Bergen Declaration, the Copenhagen Declaration, and the Koge Stakeholders Conference all commit to adopting an ecosystem approach to fisheries (Table 1). Institutional changes also reflect this commitment. In the European Commission, DG Fisheries has created the Directorate on Environment and Ecosystem Health, and DG Environment has established a directorate dealing with marine environmental quality. In the Canadian Department of Fisheries and Oceans, not only is there a full sector on oceans, but the old Fisheries Research Branch has been given equal status to become the Fisheries Research and Biodiversity Science Directorates. The institutional commitments have been made, so it is time to put the tools into the hands of the managers.

Many other papers in this volume document the ecological complexity of the management problems that must be addressed in making fisheries sustainable in an ecosystem context. However, analyzing other issues, such as the clarity with which the management problems are posed, requires a specialized perspective. With 20 years of research on multispecies fisheries management and ecosystem effects of fishing, my primary responsibility since the mid-1990s has been to coordinate scientific advice to fisheries and oceans managers and policy setters in Canada, Europe, and to some extent, the United States. As a full-time science advisor, I routinely facilitate communication between scientists and managers or policy setters, in the search for better decisions and actions toward sustainability. Such communication is often poor despite the numerous good sources documenting the ways fishing can alter

Table 1. International instruments committing governments to an ecosystem approach to fisheries management.

Agreement	Statement in favor of ecosystem approach	Reference
United Nations Fish Stock Agreement (1995)	"the need to avoid adverse impacts on the marine environment, preserve biodiversity, maintain the integrity of marine ecosystems, and minimize the risk of long-term or irreversible effects of fishing operations."	www.un.org/Depts/los/convention_agreements/convention_overview_fish_stocks.htm
Kyoto Declaration on the Sustainable Contribution of Fisheries to Food Security (1995)	"take immediate action to...conduct...integrated assessments of fisheries in order to evaluate opportunities and strengthen the scientific basis for multi-species and ecosystem management...and to minimize post-harvest losses."	www.fao.org/DOCREP/
Reykjavik Declaration (2001)	Specific provisions for (1) Immediate introduction of management plans with incentives for sustainable uses of ecosystems; (4) Advancement of the scientific basis for incorporating ecosystem considerations in management; (9) Collection of information on the sustainable performance of management regimes; and (10) Development of guidelines for incorporation of ecosystem considerations in fisheries management.	http://government.is/
Bergen Declaration (2002)	"The Ministers therefore agree to implement an ecosystem approach by identifying and taking action on influences which are critical to the health of the North Sea ecosystem ... [2.] The Ministers agree that fisheries policies and management should move towards the incorporation of ecosystem considerations in a holistic, multiannual and strategic context [19]"	http://odin.dep.no/md/nsc/declaration/022001-990330/
Copenhagen Declaration (2002)	"ensure that ecosystem considerations, including the effects of human activities and climate and oceanographic conditions, are taken into account [in advice], and frame advice in relation to fisheries management, giving full consideration to the ecosystem context."	www.ices.dk
Koge Stateholders Conference (2002)	"Action 1: The Commission will make proposals for developing an ecosystem-based approach, including ecosystem benchmarks and targets, to ensure conservation and sustainable use of biodiversity."	EC (2002)

ecosystems in general, and benthic habitats in particular (Jennings and Kaiser 1998; Hall 1999; ICES 2000; Jennings et al. 2001). The reasons for poor communication shed light on some of the impediments to progress in reducing or eliminating the undesirable effects of fishing on benthic habitats despite having knowledge of the undesirable effects and the institutional commitments to an ecosystem approach.

The Impediments to Progress

The first impediment is that there is simply too much fishing capacity on the scales of global to very local and industrialized to artisanal (Food and Agriculture Organization of the United Nations [FAO] International Plan of Action on Management of Fishing Capacity, FAO 2001a), and ways have not been found to reduce it effectively. The litany of problems that follow from excessive fishing capacity is well documented, including the vicious cycle of excessive capacity exerting ever-increasing effort levels to try to obtain catches from depleted stocks, depleting them further and requiring more effort (Hilborn and Walters 1992; FAO 1997). This cycle of ever-increasing effort means ever-increasing impact on benthic habitats, and management actions that redistribute excess effort rather than eliminating it can increase those impacts further (Jennings et al. 2005, this volume)

If the damage and inefficiency of excessive capacity have been obvious for so long, why is there so little progress on permanent reductions? One reason is that there is an absence of short-term incentives in the face of large and immediate social and economic impacts of displacing fishers (see below). A second reason is that technology is constantly working against the objective of capacity reduction. Between 1992 and 1997, Canada spent approximately Can$3.9 billion in the Atlantic Fishery Adjustment Package, with explicit goals of reducing the dependence of coastal Atlantic Canada on fisheries and decreasing excessive capacity. When Canada's Auditor General reviewed the value obtained for the money (Government of Canada 1997), the conclusion was that many fishers had indeed undertaken training in new technologies and many vessels had been retired. Unfortunately, the new technologies in which the fishers chose to receive training were very often technologies that they applied to their fishing operations. The net result, after Can$1.9 billion in expenditures, was a 60% increase in functional fishing capacity of the fleet as a whole. Although the offshore trawler fleet had been redirected to other species or dispersed to other parts of the world (another problem needing attention but not the subject of this paper), the smaller-vessel inshore fleet obtained global positioning systems, bottom profilers, and a variety of other instruments, all of which increased the per-vessel fishing capacity. The third reason is that there is actually a global crisis in food security, with over 800,000,000 people in the world under nourished (Watson 2002). There are lucrative markets in the developed world and undernourished people in the underdeveloped world, both of which guarantee fishing is going to be hard to turn off. When the needs for food security and economic development are weighed against the undesirable impacts of fishing on benthic diversity and productivity, the management choice is not guaranteed to favor reducing benthic impacts.

A second impediment is the temporal context in which the management choices are made. Responsible fisheries (resource) management has long struggled with serious disadvantages in decision making. Social and economic costs of major reductions in fisheries are immediate, and there is little uncertainty about them (FAO 1997). Benefits of yield from rebuilt stocks will be higher than the costs of effort reductions but only in a much longer term. How much longer is highly uncertain, as many depleted stocks rebuild very slowly (several stocks of Atlantic cod *Gadus morhua* off Atlantic Canada and the northeast Atlantic, and the North Sea stock of Atlantic mackerel *Scomber scombrus*; ICES 2001a; CSAS 2002a, 2002b). The time lag in benefits makes it hard to achieve equity between those costs and benefits. Those bearing the costs today often doubt that they will be the ones to receive the benefits years into the future. This skepticism produces the common expectations that governments will compensate those bearing the costs, making governments themselves reluctant to act decisively (Dommen 1993; Cochrane 2000).

Moving fisheries management into an ecosystem context is not going to make the management choices easier. Short-term costs are likely to be even higher as reductions and displacements of fisheries have to be sufficiently large to address more types of damage than just excessive harvest of the fisheries' target species (ICES 2000, 2001c). It is widely argued that populations and yields of target species will improve as the result of measures intended to benefit other parts of the ecosystem, including benthos (Sladek Nowlis and Roberts 1999; Roberts et al. 2002). However, there are several reasons to be skeptical that such arguments will necessarily fix the cost–benefit temporal imbalance. First, it is hard to argue that the future yield benefits of measures directed primarily at benthic ecosystem components that are not harvested will necessarily be greater than the future benefits of measures directed specifically at increasing yield of the target species, and the latter have often not carried the day in decision making. Even if such benefits do accrue, they are still longer term than the costs and still do not address the problem of transition costs. Moreover, the benefits only persist as long as fisheries are kept at effort levels reduced greatly enough to exclude many initial participants (ICES 2000), underscoring fears that many will be excluded from sharing the benefits when they are realized. Many of the other benefits that are given "value" by an ecosystem approach are benefits that will not accrue to those who will bear the greatest portion of the cost (Turner et al. 2001; Ledoux and Turner 2002). Mechanisms to address these perceived inequities are few, often expensive to groups with leverage to resist imposed costs, and, consequently, rarely applied (FAO 2002b).

As with overcapacity, these problems with ecological cost accounting systems are not new. Through the 1990s, interest has grown in the concept of "ecosystem goods and services" as a method to make the accounting system more balanced (Costanza et al. 1997). Such accounting methods may be a great conceptual tool, but the core problem in fisheries is the imbalance in the *immediacy* of the costs (short term) versus benefits (long term) because the long-term economic benefits of yields from rebuilt stocks are not disputed. "Ecosystem goods and services" is little help because, in the short term, even industrialized societies do not yet pay directly for those goods and services. Internationally (but anecdotally), the concept is no more helpful; in separate conversations with international colleagues from the Baltic and West Africa, fisheries advisors and managers expressed appropriate interest in the creativity and soundness of the concepts and their total irrelevancy to how decision making occurs in their parts of the world. The risk that sometime in the future their societies will have to pay for goods and services that their marine ecosystems now provide without outlay of currency is a risk that they have to take in exchange for food and trade commodities today. Even governments that are not struggling with debt and poverty show little evidence they are embracing proposals to account explicitly for services that are being obtained for free in the short term. We need tools to change the accounting for ecosystem degradation in the short term, not just accounting methods that will be effective only after the natural capacity of marine ecosystems to provide these goods and services has been degraded to the point where payments are required.

A third impediment to progress on sustainability is that there are a number of myths and preconceptions that impede confronting sustainability issues directly. Two are of particular relevance to the slow progress toward addressing the impacts of fishing on benthic habitats. One preconception is that "small is beautiful" and "big is bad" when it comes to fishing. Neither is categorically true. Small beam trawl vessels can cause substantial perturbation to benthic habitats (Lindeboom and deGroot 1998; ICES 2000), and even smaller artisanal fisheries can cause significant habitat damage with poisons and dynamite (Thomas 1999; McManus et al. 2000). Moreover, fleets of many small fixed-gear vessels present nearly insurmountable problems for on-board vessel monitoring, such that ecosystem gains in the narrow context of protecting benthic habitats from mobile gears may be more than offset by losses due to discarding and underreporting of catches (FAO 1999, 2002c). On the other hand, some large commercial vessels with mobile gears have invested heavily in technology and gear modifications that greatly reduce their habitat impacts while maintaining efficient, closely monitored harvesting (Linnane et al. 2000; ICES 2001c; Kostylev et al. 2002). If we are to have effective tools to reduce impacts of fishing on benthic habitats, we have to deal with the real options available for fisheries, not naïve over simplifications.

The other preconception is that the precautionary approach justifies stopping everything causing unwanted impacts on habitats. Over the last quarter-century, the literature on the precautionary approach and precautionary principle has exploded in the areas of resource conservation, health and safety, and environmental protection. Even in the narrow context of natural resource conservation, various agreements and publications have used a variety of nuances of wording that have major applied consequences, and even formal instrumental and legislation can be unsettlingly vague (Table 2). Any international principle with diverse interpretations risks being ineffective in practice, and currently, there are many efforts to bring consistency to what precaution does and does not mean in terms of binding governments to act in particular ways (EC 2000; Government of Canada 2001). It may be of concern, though, that these efforts are being driven by world trade organizations who are concerned that "precaution" is being used as a nontariff trade barrier and not by resource managers and conservationists interested in resource management decisions that give more protection to benthic habitats. It is too soon to know where these negotiations will end up. However, the government-wide initiatives cited above give no reason for optimism that decisions on what precaution means, for example, in marine habitat protection, will be made by technical experts in that field rather than trade experts. Science needs to take back some of the initiative on this issue.

There is a final barrier to progress toward sustainability of uses of marine ecosystems and their habitats, and that is simply the complexity of the concept of sustainability. From the perspective of benthic habitats, this complexity is less apparent because sustainability is measured almost exclusively on an ecological dimension. It is only necessary to ask, "Are the perturbations of benthic populations and habitats reversible?" The International Council for the Exploration of the Sea (ICES) Working Group on Ecosystem Effects of Fishing (WGECO) conducted such an evaluation 3 years ago (ICES 2000). At the extremes, if complex physical structure of the habitat

Table 2. Wording differences in references to the Precautionary Approach among international instruments and national legislation dealing with applications for environmental conservation and safety.

Instrument	Harm	Uncertainty	Anticipatory action	Capacity	Science basis	Cost-effective
London Declaration (1987)[a]	Damaging effects... most dangerous substances	Possibly damaging ... before a causal link [of] absolute clear scientific evidence	Before a causal link... action to control inputs	No	Yes	No
Bergen Ministerial Declaration (1990)[b]	Serious or irreversible damage... prevent enventual degradation	Lack of full scientific certainty	Anticipate, prevent, and attack the causes	No	Yes	No
Rio Declaration (1992) (Agenda 21)[c]	Threats of serious or irreversible damage	Lack of full scientific certainty	Postponing cost-effective measures to prevent environmental degradation	States according to their capability	Yes	Cost-effective measures
Convention on Biological Diversity (1992)[d]	Threat of significant reduction or loss of biological diversity	Lack of full scientific certainty	Postponing measures to avoid or minimize such a threat	No	Yes	No
Helsinki Convention (1992)[e]	Harm might be caused	When there is reason to assume...no conclusive evidence of a causal relationship	Preventative measures are to be taken	No	No	No
OSPAR Convention (1992)[f]	Hazards to human health, harm living resources and marine ecosystems, damage amenities or interfere with other legitimate uses of the sea	Reasonable grounds for concern...no conclusive evidence of a causal relationship	Preventive measures are to be taken	No	Implied	No
Maastricht Treaty on European Union (1993)[g]	Environmental damage should be rectified as a priority	No	Preventative action should be taken...at source	Taking into account the diversity of situations in the various regions	No	Polluter should pay
Framework Convention on Climate Change (1993)[h]	Threats of serious or irreversible damage	Lack of full scientific certainty	Measures to anticipate, prevent or minimize the causes...and mitigate ...adverse effects	Take into account different socio-economic contexts		Cost-effective so as to ensure global benefits at the lowest possible cost

Table 2. Continued.

Instrument	Harm	Uncertainty	Anticipatory action	Capacity	Science basis	Cost-effective
Cartagena Protocol on Biosafety (1994)[i]	Extent of the potential adverse effects	Lack of scientific certainty due to insufficient relevant scientific information and knowledge	Taking a decision, as appropriate …to avoid or minimize such potential adverse effects	No	Yes	No
Code of Conduct for Responsible Fishing (1995)[j]	Protect [living aquatic resources] and preserve the aquatic environment	Accounting for the best scientific evidence available. The absence of adequate scientific information	Postponing or failing to take measures to conserve…	No	Yes	No
Straddling Stocks Treaty (1996)[k]	Widely to conservation, management and exploitation … to protect the living marine resources and preserve the marine environment.	Information is uncertain, unreliable or inadequate. The absence of adequate scientific information … dealing with risk and uncertainty	States shall be more cautious … postponing or failing to take conservation and management measures	No	Yes	No
Canadian Legislation						
Canadian Environment Assessment Act	Threats of serious or irreversible damage	Lack of full scientific certainty	Postponing cost-effective measures to prevent environmental degradation	No	Yes	Cost-effective measures
Canada's Oceans Act	Applied widely	Err of the side of caution	Unclear	No	No	No
Species at Risk Act	Threats of serious or irreversible damage	Lack of full scientific certainty	Cost-effective measures to prevent the reduction or loss … should not be postponed	No	Yes	Cost-effective measures

[a] Reference: *http://odin.dep.no/md/nsc/declaration/022001-990245.*
[b] Reference: *www.sehn.org/internatneg.*
[c] Reference: *www.un.org/esa/sustdev/agenda21.htm.*
[d] Reference: *www.biodiv.org/default.aspx.*
[e] Reference: *www.helcom.fi/convention/conven92.html.*
[f] Reference: *www.ospar.org/eng/html/convention/welcome.html.*
[g] Reference: *http://europa.eu.int/en/record/mt/top.html.*
[h] Reference: *www.doc.mmu.ac.uk/aric/eae/Global_Warming/Older/FCCC.html.*
[i] Reference: *www.icgeb.org/~bsafesrv/bsfprot.htm.*
[j] Reference: *www.fao.org/fi/agreem/codecond/codecon.asp.*
[k] Reference: *www.un.org/Depts/los/convention_agreements/convention_overview_fish_stocks.htm.*

is damaged, the perturbations may be essentially irreversible, whereas alterations of some fine-scale features in high-energy habitats may be undetectable after a matter of hours to a few days. Likewise, perturbations of biotic components of the benthos can last from decades for long-lived, late-maturing species with low dispersal ability but may be readily reversible for highly opportunistic species that disperse propagules widely.

Finding sustainable fishing strategies on the ecological dimension alone is merely a task of quantifying magnitudes of perturbations caused by different options (for example, for different gears or a single gear used with differing spatial and temporal distributions) and rates of recovery from the perturbations expected to result from each option. This is a scientifically demanding but tractable problem. That, however, is only the ecological problem and not the management problem. The management problem is broader: to balance sustainability on the ecological dimension with sustainability on at least two other dimensions, the economic and the social (Charles 2001; FAO 2002b). Even when managers are not formally mandated with the responsibility to solve all ecological, social, and economic problems associated with a fishery, failure on any of the criteria creates strong pressure of the political decision-making system to rectify the failures (Charles 2001). Except possibly in the presence of very strong top-down governance, these pressures make real-world progress on sustainability on the ecological dimension ephemeral until social and economic dimensions are addressed as well (FAO 2002b).

Bioeconomic modeling and analysis has been an active research field since the 1980s (Clark 1990) or even earlier. On these two axes, solution sets are often hard to find and even harder to reach (Charles 2001). However, solutions to simultaneous sustainability do exist, often through use of property rights to create economic incentives toward responsible fishing (FAO 2000). Although rights generally are allocated because of first-order economic benefits to those receiving shares and a reduction in risk of overfishing of the target stocks, they almost always result in major reductions in effort as well (FAO 2000, 2002b). This reduced effort produces direct reductions in detrimental effects on benthic habitats and communities (ICES 2001d), although the programs rarely include tools that intentionally focus the remaining effort in ways that minimize such impacts.

Social sustainability is a comparatively newer field, with much less documentation (Ommer 1995; Hanna et al. 2000). Even the measures of what comprises the goals of "sustainable communities" may be abstract and difficult to quantify. It is, after all, a matter of preserving coastal cultures, and cultural vitality is not readily tractable to the types of modeling that characterizes the search for sustainable bioeconomic strategies. However, there is no doubt about the importance of social sustainability in decision making about fisheries. Social assistance payments, "emergency" bridge funding, modernization payments, and a host of other government programs are used to maintain coastal communities whose fisheries are failing in an attempt to keep the communities alive (Schrank et al. 1995; Hatcher and Pascoe 1998; Schrank 2001). In less affluent parts of the world, the decision leverage of social sustainability may not be manifest as financial subsidies, but the power is still there. Unrest over continued right to fishing despite resource depletion can lead to violence and even threaten the stability of governments (Cochrane and Payne 1998).

The real challenge to management is not to reduce the ecosystem effects of fishing but to find strategies that are simultaneously sustainable on all three dimensions: ecological, economic, and social (Charles 2001; Rice 2002). A group of international experts from fisheries science, management, economics, and social sciences recently addressed the issue of what drives fisheries so inexorably toward unsustainability (FAO 2002b). The results of the deliberations were not encouraging. The key drivers toward unsustainability could be itemized readily, as could management measures to address each of them (Table 3). Unfortunately, each measure that could move a fishery in the direction of sustainability on one axis where problems were present *always* increased the vulnerability of that fishery to unsustainability on other dimensions of sustainability. No solution sets were found that suggested management could move fisheries toward increased sustainability on all three axes at once..

This is a very important conclusion. The remainder of this paper reviews the most commonly cited tools for managing fisheries in ways that reduce their impacts on benthic habitats. When proposed, these are usually evaluated on the ecological dimension. Only occasionally are the effects of these tools evaluated on the economic dimensions and even more rarely on the social dimension. By addressing the impacts of these management tools on social and economic sustainability, as well as their benefits to benthic habitats, this paper may build a better understanding of what impediments must be overcome to actually improve practices in fisheries that impact benthic habitats.

Evaluating Tools on the Axes of Sustainability

Accounting for Ecosystem Goods and Services

The intent of this economic tool is to adjust improper incentives by giving greater value to uses of ecosystems

Table 3. Factors driving fisheries towards unsustainability (FAO 2002b).

Unsustainability driver	Commentary
Inappropriate incentives	Incentives structures promote short-term profit taking and/or damaging activities and do not reward actions that forego short-term gains for potentially greater benefits in the longer term.
High demand for limited resources	Demand for fish is increasing, both as a contribution to food security and as major trade item for obtaining foreign currencies.
Poverty and lack of alternatives	In many parts of the world, fisheries are an employer of last resort, and even when fisheries are economically unsustainable, particpants have no viable alternatives to remaining in the fisheries, depressing stocks, and profits, further.
Complexity and inadequate knowledge	True of social, economic, and bio-ecological systems and their dynamics: Inadequate or unreliable information and incomplete understanding how systems will respond to natural and anthropogenic forcing makes it hard to identify sustainable courses of action.
Lack of governance	Occurs in several forms. Fisheries may have a low priority in government agendas, governments may be ineffective in influencing behaviour of its citizens, or legitimate authorities or appropriate institutions to manage fisheries may be absent.
Externalities	Factors usually beyond the control of the fisheries sector, such as pollution, environmental variability, and competing demands for use of the areas supporting the fisheries make activities become unsustainable, even when the commitment to sustainable practice was present.

other than the short-term market value of fish products. By increasing the costs resulting from perturbing ecosystems and increasing the benefits from leaving them in their natural states, the expectation is that there will be economic incentives to perturb the ecosystems less. Functioning this way, such accounting would increase ecological sustainability.

Its impact on economic and social dimensions of sustainability can be either positive or negative when the axes are considered individually but must be negative on at least one of them. Accounting for ecosystem goods and services necessarily increases the demand for ecosystem resources; accounting for more goods and services necessarily adds large numbers of new participants to the competition for the resources. For the wealthy in the developed world, the debate about how to account for greenhouse gasses on corporate balance sheets (Pachauri 1999; Bergman 2002) is a simple task compared to accounting for marine ecosystem goods and services. However, the Kyoto Accords mean that such accounting systems have to be developed, and there may be ways to make corporate operations look even more profitable. Moreover, ecologically, many of the new uses that would be accounted for may leave the ecosystem less perturbed. However, the increased competition for ecosystem resources may victimize those driven to fishing through poverty or lack of alternative employment, those fishing for sustenance, and those residents culturally committed to coastal communities (Symes 1999; Cochrane 2000). These groups are not prepared to bid for resources against those drawing most from ecosystem goods and services, and they will simply have fewer opportunities to find food or work (Armstrong and Clark 1997). If these individuals were actually compensated for the value of the ecosystem goods and services provided by the resources they are not harvesting, their burdens of poverty and unemployment could be lightened. However, then economic costs imposed on the comparatively wealthy residents of the global village would look very different. Depending on choices made by the local, national, or global governance systems, accounting for ecosystem goods and services necessarily results in either greater social unsustainability (those fishing due to poverty or for sustenance are further marginalized) or greater economic unsustainability (society pays more for fish, and possibly other benefits currently enjoyed without payment).

The importance of the governance system in determining the consequences of ecosystem goods and services accounting should not be downplayed. Were the accounting for ecosystem goods and services a practical path to better decision making, there would be no debate about taking action on the emissions reductions embodied in the Kyoto Accords (c.f. Bergman 2002). Here, the science basis for both the long-term benefits and the costs is more fully developed than for damage to benthic habitats by fishing practices, yet action has been slow in the most developed global economies (Watson 2002).

This situation also brings into focus the third factor of unsustainability to which ecosystem goods and services accounting is vulnerable: complexity and inadequate knowledge. The contribution of uncertainty in assessments of the status of harvested fish stocks to poor decision making about harvesting is well documented (Walters and Maguire 1996; Alverson 2002). When the accounting becomes more complex, the uncertainty about the status of the much larger number of ecosystem components that provide goods and services, the values of those services, and the impacts of alternative uses on those components will make the debates on fisheries issues simple by comparison.

Marine Protected Areas

Proponents of marine protected areas (MPAs) argue that they have potential to provide benefits on both the ecological and economic dimensions of sustainability, and if there are comments on social impacts, the proponents are also positive about those impacts (Hannesson 1998; Roberts et al. 2002). Without question, MPAs are an effective, possibly indispensable, tool to protect fragile and vulnerable habitat features from damage by fishing. This promotes ecological sustainability. Even on that dimension, however, the accounting needs to be complete (ICES 2000). If an MPA large enough to protect substantial tracts of benthic habitat is implemented, it is necessary to consider where the fishing effort that used to be exerted within the area now protected is displaced. Has effort that had been focused in a preferred area for decades been exported to areas previously lightly impacted by fishing, causing perhaps greater damage to benthos than did the fishing on heavily perturbed areas that are now closed? Even in heavily fished ecosystems, fishing effort is aggregated so much that many sites are rarely fished (Figure 1). Such areas might be altered greatly by effort forced into it by closing preferred fishing grounds (Jennings et al. 2005, this volume). Has effort been exported to areas where catch rates are poorer so actually more fishing effort is required to take the total allowable catch, and on the ecosystem scale, impacts of fishing on benthos are increased? Certainly these considerations do not necessarily negate the ecological gains of MPAs, but managers have to think broadly and include them as part of the equation (Armstrong and Reithe 2000; ICES 2000).

With regard to social and economic sustainability, evaluating costs and benefits is similarly complex. Advocates argue that MPAs increase yield in the medium term (Roberts et al. 2002), although evidence for this claim is equivocal and case specific (Hilborn et al. 2004). In the short term though, compliance with a meaningfully large MPA means that costs of fishing can only increase, and at least some previous users either lose access or must travel further to maintain access to fishing opportunities. These consequences then reveal the factors of unsustainability to which MPAs are vulnerable. The high demand for limited resources is accentuated by making the resources even more limited. The poor and those lacking alternative employment opportunities see their income from fishing diminished further, at least in the short term. Those social effects can be dealt with but only at increased cost on the economic dimension. These consequences of increased demand (which attracts the wealthy fishing sectors) and loss of access (which affects the poorest sectors) both may be incentives to violate the rules of an MPA, which in turn requires increased costs for surveillance and enforcement of the MPA once established. The need for enforcement, in turn, creates vulnerability of the entire system to ineffective governance, either as coastal communities resisting external imposition of MPAs that they do not support or, in areas of the world with unstable central governance systems, increasing regional instability. Co-management of reserves (or fisheries) is not an alternative that reduces the importance of effective governance (Felt et al. 1997; Lee 2004). Rather it is a very sophisticated form of governance and requires effective central governance to prevent interest groups from outside the co-management community from trespassing into the co-management area or otherwise undoing agreements.

None of these factors necessarily negate the value of MPAs as a tool to make progress on the ecological and possibly economic dimensions of sustainability. However, they require thinking broadly about the accounting of costs and benefits and effects on all three axes of sustainability. Fisheries scientists have had very little success in reducing directed fishing effort despite strong arguments about higher yields at lower costs if stocks were allowed to rebuild. There is a lesson in that experience for accounting systems which try to build support for MPAs based on long-term gains on the economic and social dimensions. Arguing for short-term pain for long-term gain often has not been successful when the gains were direct, first-order effects of fishing on fish populations. It is again naïve to be confident that such arguments will succeed when the benefits are less direct, higher-order effects of MPAs on future fishery yields.

Gear Modifications and Fleet Substitutions

There are many ways that gears can be modified to reduce their impact on benthic habitats. Such modifications can increase sustainability on the ecological axis, but improvements are not guaranteed. Rather, even well-designed measures can achieve little unless the gears are used as

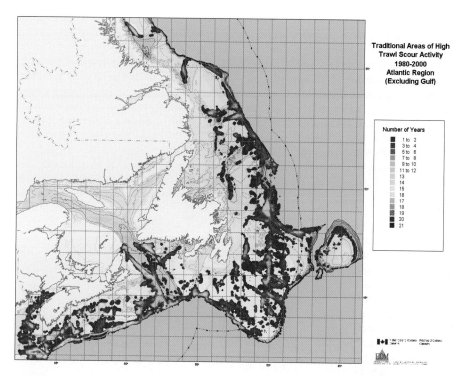

Figure 1. Traditional areas of high scour activity, 1980–2000, in the Atlantic region, Gulf excluded. The multi-year composite map was created to show where high-intensity trawling occurred persistently from year to year by combining all areas of high trawl activity and counting the number of years over the period of study that each location was trawled intensely. For each annual map of effort, areas where the intensity of fishing exceeded 30% were selected. These intesely trawled areas were overlaid in a method analogous to stacking transparencies, one for each year. The desner areas in the stack represent areas that were trawled during more years. The overlay procedure in SPANS (INTERA TYDAC Technologies) represented this as a count of the number of years where the high-intensity trawl areas overlapped. The use of 30% as a threshold for high intensity of tarwling was arbitrary. By varying this proportion, the areas depicted on the map would change in size (the higher the intensity level chosen, the small the area) but not location. Figure from Kulka and Pitcher (2001).

intended. Studies suggest that gear modifications that achieve much of their reduced benthic impact through reduced efficiency of gear performance will not be deployed as intended (ICES 2000, 2003; Linnane et al. 2000). Moreover, reducing the efficiency of gears means that effort will increase to take the same amount of fish unless the gear modifications are accompanied by concomitant reductions in quotas as well. Either misuse of gears or increased effort could dissipate the expected benefits of gear modifications (ICES 2000). The effects on social and economic axes could be positive, neutral, or negative, depending on how the modified gear performs relative to the higher-impact gears. However, because many modifications do reduce efficiency in harvesting the target species, short-term negative economic consequences are likely. In any case, the adoption of such gears is likely to require provision of incentives to offset any increased costs of gear purchase or operation as well as effects of reduced gear efficiency. This limits the promise of these tools for areas where poverty is tying people to the fisheries and where high demand already creates incentives for illegal or unreported fishing. Additionally, use of such gear, if mandatory, requires surveillance and enforcement costs.

Fleet substitution is also promoted as a means to reduce undesirable effects of fishing on benthos (ICES 2000). The advocated changes are usually from mobile gears (trawls) to static gears (gill nets, long-lines, etc.), or from large vessels to smaller vessels (e.g., Lindeboom and deGroot 1999; Kaiser et al. 2002; Valdemarsen and Suuronen 2003). Both changes may not even be considered movement toward increased sustainability if they are viewed in ecosystem contexts larger than just benthic habitats and populations. Hook-and-line and gill-net fisheries present serious bycatch problems for species that cannot support high bycatch mortality; for example, small cetaceans in gill-net fisheries (Kock and Benke 1996; ICES 2001d) and seabirds in net and long-line fisheries (Hunt and Furness 1996; Ashford and Croxall 1998). The latter problems are so serious that FAO has developed an International Plan of Action for Reduction of Seabird Bycatch. Moving to smaller vessels might reduce impacts on benthos if smaller vessels tow lighter gears more slowly. However, such relationships are not guaranteed; rather, the speed–weight relationship is usually reciprocal for towed gears (ICES 2000, 2001b), so impacts may not be reduced by much. More importantly, smaller vessels are less likely to be able to carry or afford independent observers, greatly reducing the effectiveness of surveillance and enforcement of all the fishing regulations (FAO 2002c). Loss of independent observers is always a major setback to conservation interests (FAO 1999). With regard to factors of unsustainability, the increasing demand for fish always works against fleet changes that lower efficiency, and strong governance is always necessary for these changes to be implemented across entire fisheries. On the other hand, in cases when the fleet substitutions lead to increased efficiency, they have the risk of being socially destabilizing (Armstrong and Clark 1997; Fowler 2002).

In all these cases, if one is thinking only of protecting benthic communities, there will be unquestioned gains from gear modifications and fleet substitutions (ICES 2000, 2001b). However, taking a sufficiently broad ecological perspective, each case must be considered on individual merits. The ecological benefits of such changes to fisheries may outweigh the ecological, social, and economic costs, but the outcome of an appropriately broad cost–benefit analysis cannot be presupposed. This class of tools should always be considered as a possible way to reduce impacts of fishing on benthos, but options need careful analysis in each case.

Ecocertification

Major independent certification systems give substantial weight to knowing what the impacts of the fishing gears are on benthic habitats and having evidence that the impacts are sustainable (Marine Stewardship Council 2002). These tools bring strong economic pressure to bear on fisheries, moving them in the direction of greater sustainability on both the ecological and economic dimensions. The measures necessary to obtain and keep certification may increase costs of fishing (Gudmundsson and Wessells 2000; Peterman 2002), but they may increase the value of the catch disproportionately. For those reasons, it warrants serious consideration as a management tool to reduce undesirable effects of fishing on benthos. However, the use of economic instruments to alter fishing practices is argued to be socially destabilizing because the instruments are likely to discriminate against fisheries in under-developed countries that cannot bear the up-front costs of certification (FAO 1998, 2001b). High demand for fish also limits the effectiveness of market-based tools by providing consumers willing to buy non-certified catches. Certification programs also put governance systems under pressure, requiring significant governance efforts to maintain the chain of documentation that prevents abuse and misrepresentations of certification (FAO 1998; Marine Stewardship Council 2002).

Management Objectives for Benthos

The tools discussed above have value only if they are adopted and used, and because each tool has been shown to have associated short-term costs, adoption is not guaranteed. Hence, a new framework, not just an inventory of new management tools, is needed to address the systemic problems with unsustainable fisheries. In the past decade, fisheries management and science have developed such a framework in the process of implementing the precautionary approach. The framework is built upon explicit objectives for management, for which indicators are chosen and around which quantitative targets, limits, and risk management strategies are structured (FAO 1996a, 1996b). This framework has been endorsed by policy makers and managers internationally and forms a core portion of the Code of Conduct for Responsible Fishing (FAO 1997). Recent policy instruments for placing fisheries management into an ecosystem context, such as the Bergen Declaration and the Reykjavik Declaration, have followed this lead and embraced explicit objectives, indicators, and reference points as cornerstones for implementing the approach (Table 4).

By themselves, explicit management objectives are neutral with regard to sustainability. However, the process by which they are selected, the mix of properties of the stocks and ecosystems for which objectives are set, and the methods by which diverse objectives are reconciled into a functional suite all give them the potential to move fisheries toward sustainability on any or all of the dimensions. For all its potential power, there are two main challenges to making the framework effective. One is making the list of objectives functional. The list has to be workably short, each objective must be quantitatively explicit, and the suite has to be inter-compatible. The second challenge is linking the objectives to indicators that are actually informative about status of ecosystem components and reference points that avoid "serious or irreversible harm" without preventing sustainable activities that produce economic and social benefits.

With regard to developing a functional suite of ob-

Table 4. Ecological quality elements and objectives agreed by North Sea Ministers at Bergen, March 2002 (from Annex 3 Table B, Bergen Declaration). Further background information on each element/objective may be found in the Bergan Declaration itself.

Ecological quality element	Ecological quality objective
(a) Spawning stock biomass of commercial fish species	Above precautionary reference points[a] for commercial fish species where these have been agreed by the competent authority for fisheries management
(c) Seal population trends in the North Sea	No decline in population size or pup production of 10% over a period of up to 10 years
(e) By-catch of harbour porpoises	Annual by-catch levels should be reduced to levels below 1.7% of the best population estimate
(f) Proportion of oiled common guillemots among those found dead or dying on beaches	The proportion of such birds should be 10% or less of the total found dead or dying, in all areas of the North Sea
(m) Changes/kills in zoobenthos in relation to eutrophication[b]	There should be no kills in benthic animal species as a result of oxygen deficiency and/or toxic phytoplankton species.
(n) Imposex in dog whelks *Nucella lapillus*	A low (<2) level of imposex in female dog whelks, as measured by the Vas Deferens Sequence Index
(q) Phytoplankton chlorophyll a[b]	Maximum and mean chlorophyll a concentrations during the growing season should remain below elevated levels, defined as concentrations > 50% above the spatial (offshore) and/or historical background concentration
(r) Phytoplankton indicator species for eutrophication[b]	Region/area - specific phytoplankton eutrophication indicator species should remain below respective nuisance and/or toxic elevated levels (and increased duration)
(t) Winter nutrient concentrations (Dissolved inorganic nitrogen [DIN] and dissolved inorganic phosphate [DIP]) [b]	Winter DIN and/or DIP should remain below elevated levels, defined as concentrations > 50% above salinity related and/or region-specific natural background concentrations
(u) Oxygen[b]	Oxygen concentration, decreased as an indirect effect of nutrient enrichment, should remain above region-specific oxygen deficiency levels, ranging from 4–6 mg oxygen per liter

[a] In this context, reference points are those for the spawning stock biomass, also taking into account fishing mortality, used in advice given by ICES in relation to fisheries management.
[b] The ecological quality objectives for elements (m), (q), (r), (t), and (u) are an integrated set and cannot be considered in isolation. ICES will give its further advice during the implementation phase.

jectives, making the objectives quantitatively explicit is simply a matter of self discipline by those setting the objectives. The unpacking exercise (CSAS 2001a) from conceptual objectives of the nature "preserve benthic biodiversity" to an operational set of objectives to either maintain viable populations of specified sensitive and vulnerable species or maintain some preferred biodiversity index requires only patience and good skills at running effective and informed consultations (FAO 2002a).

The task of keeping the list of objectives workably short is harder. Once a jurisdiction starts down the path of objectives-based fisheries management, it is natural for interest groups to want their favorite properties on the list of objectives that management must achieve. This is understandable, and there is some advocacy for assessments that provide qualitative information on many stock and ecosystem properties rather than rigorously quantitative information on fewer (Parsons et al. 2000; CSAS 2001b; Caddy 2002). Unfortunately, this approach hamstrings management and makes it impossible for objectives-based management to employ clear harvest control rules (Rosenburg and Restrepo 1996; Butterworth and Punt 1999; Butterworth 2000) or even quantitative indicators and reference points (Rochet and Trenkel 2003; Trenkel and Rochet 2003). To illustrate the problem, ICES structures fisheries advice around just two objectives: keeping reproductive potential above a conservation limit and keeping exploitation below an overfishing limit. Each objective is supported by a single indicator—spawning stock biomass (SSB) and fishing mortality (F), respectively—each with a conservation limit reference point and an explicit method for dealing with analytical uncertainty. Even in this relatively simple situation, managers have almost always met resistance if they try to reduce harvesting of a stock where F exceeds its reference point but SSB is not at risk of falling below its reference point. Arguments are made that as long as SSB is healthy, it is a societal right to choose to harvest temporarily high biomass "now" rather than "later." As the list of objectives grows longer, the difficulty in guiding decision making will multiply. There will always be some objectives being met and others not being achieved. A long list of ecosystem objectives risks simply becoming a playing field where each special interest uses advocacy arguments for the importance of its objective relative to others, a situation little different from the present one.

How can the list of objectives be large enough to ensure maintenance of healthy ecosystems, including benthos, but small enough to provide unambiguous guides to sustainable harvesting? This is territory just beginning to be explored. One part is collective decision making, getting the classes of ecosystem properties for which objectives are necessary as illustrated by WGECO for the ecological component of sustainability (ICES 2001c) and FAO for an ecosystem approach to fisheries (FAO 2002a). Another part is analytical, testing potential indicators associated with the individual objectives for orthogonality and calibrating their axes across objectives. Little attention has been given to this redundancy and calibration problem, although it may be possible to borrow experience and analytical tools from personality testing in psychology, where researchers have been grappling with this problem for decades (Rice and Rochet, in press).

Making the suite of objectives inter-compatible is also challenging. Part of the challenge is just getting stakeholders who may question the legitimacy of other participants to work together. This is a core challenge of co-management (Felt at al. 1997; Jentoft 2000) and seems to depend partly on the skills of the meeting chair and greatly on the past history of dynamics among the stakeholders. New quantitative tools for multi-criterion optimization have promise (Fernandes et al. 1999; Mardle and Pascoe 1999) for finding inter-compatible objectives once the diverse interest groups have found ways to work together constructively.

Even the most explicit objectives require indicators and reference points to be made operational. There is no shortage of candidate indicators, but selecting ones of high and known information content is a step that is often overlooked. The ICES WGECO has devoted substantial attention to this problem. They developed a series of criteria that could be used in screening candidate indicators for use in decision making (Table 5) and applied them to screening a series of indicators. In a suite of over 60 commonly proposed indicators, the working group concluded that only 12 were even appropriate for in-depth investigation as components of an ecosystem approach to fisheries management. Several important general properties of ecosystems had no indicators that passed even this first screening; only one passed for benthic ecosystem components.

Even the ICES WGECO screening approach does not go too deeply into determining the actual information content of these indicators. It was proposed that formal signal detection theory, a core tool of human-factors research in psychology, might be applied to ecosystem indicators (Helstrom 1968) to quantify the hit, miss, and false alarm rate for decisions based on various indicators (Rice 2001). The Scientific Committee on Oceanographic Research (SCOR) Working Group on Ecosystem Indicators for Fisheries (Working Group 119) is developing the WGECO and signal detection theory framework more completely. A screening sequence covering 15 factors of an indicator has been developed as well as a scoring system. This work has promise to make the selection of indicators to operationalize the objectives a matter of science rather than a matter of taste.

The final step of making the objectives framework operational is choosing reference points on each indicator. Within the precautionary approach, the conservation reference points have a clear meaning in some national law and many international agreements (Government of Canada 2001); they are the point on the indicator beyond which the risk of serious or irreversible harm is unacceptable. Strategies are available to deal with situations where there is insufficient knowledge to determine the point of serious or irreversible harm on candidate indicators (Caddy 2002; ICES 2002; Rochet and Trenkel 2003). One is to use another indicator where there is some basis in science and traditional ecological knowledge as a measure of status relative to one of the management objectives. In fact, the screening criteria proposed by WCECO and under development by SCOR Working Group 119 include the ability to identify conservation reference points as a screening criterion. The other strategy is to use the lowest (or highest, depending on the direction in which harm is measured) value in the historic series as the conservation limit, arguing that beyond that value, system dynamics are unknown (ICES 2001d; CSAS 2002c). This is scientifically sound but has been a weak tool in influencing decision making.

A key part of the precautionary approach is that it must take full account of uncertainty. Almost every agency using conservation limit reference points has these points complemented by a second reference point whose distance from the conservation limit is a function of the magnitude of uncertainty in the assessments and the implementation system. These points go by different names, B_{buf} and F_{buf} in the Northwest Atlantic Fisheries Organization (Anonymous 2002) and B_{pa} and F_{pa} in ICES (ICES 2001a), but they serve the same important function. A risk management decision-making system that is risk neutral with regard to these precautionary reference points will have a high likelihood of preventing conditions from falling outside the corresponding conservation limit. Together, these reference points and the harvest control rules that direct decision making us-

Table 5. Working Group on Ecosystem Effects of Fishing criteria for screening candidate reference points and performance of candidate indicators. In each cell of table, the three values are number of indicators that (a) fail to address important aspects of property, (b) provide limited information, or (c) are fully informative about the general ecological property. In many cases, single indicators might actually represent a large number of related metrics (e.g., species diversity indice—of which scores have been proposed). In such cases, the indicator could receive more than one score if some metrics were thought more promising that others in the class. Note that for spatial integrity, the Working Group was unable to come up with a single indicator considered worthwhile to screen.

Property screening criteria	Biological diversity		Ecologicalfunctionality (5 sets tested)	Spatial integrity (none possible)
	Fish community (7 sets tested)	Benthic community (5 sets tested)		
Comprehensive and communicable	4 – 6 – 4	4 – 1 – 1	4 – 1 – 1	
Sensitive to manageable human activity	1 – 6 – 6	4 – 1 – 1	4 – 1 – 1	
Tight linkage in time to that activity	6 – 5 – 1	5 – 1 – 1	4 – 1 – 1	
Easily and accurately measured	1 – 2 – 6	4 – 1 – 1	4 – 1 – 1	
High signal (human effect) to noise (natural) ratio	5 – 5 – 2	4 – 1 – 1	4 – 1 – 1	
Measurable over mostof area of interest	1 – 2 – 6	4 – 1 – 1	4 – 1 – 1	
Historic baseline long enough for setting objective	1 – 5 – 5	4 – 1 – 0	5 – 1 – 1	
Representative and generalizable	1 – 1 – 6	4 – 0 – 1	4 – 0 – 1	

ing them are a powerful tool for achieving conservation objectives.

There is an obvious shortcoming in conservation and precautionary reference points. They guide decisions away from ecological risks, but they do not do not lead to "healthy" or "desirable" states. That is supposed to be achieved by the third type of reference points: target reference points. Remembering that reference points are on indicators that are intended to measure status relative to objectives, it should be quite straightforward to set target reference points. They are the position on each indicator associated with positive objectives—objectives expressing the benefits society wishes to receive from the ecosystem, not harm that we wish to avoid (FAO 1996a). Some of those benefits may be ecological, and some groups argue that such targets can be set by the technical experts (scientists and holders of traditional ecological knowledge [TEK]) alone. This is flawed logic, however, because workable suites of objectives must balance responsibly benefits on all three dimensions of sustainability: economic, social, and ecological (Lanters et al. 1999). It has proven very difficult to get the policy setters to specify what their economic objectives are and almost impossible to specify what their social objectives are. Without these two pieces, the technical experts cannot identify states of the ecosystem with a maximum likelihood of producing that suite of benefits.

Therefore, even though everyone working within the systems of objectives-based management wants to focus their efforts on achieving sustainable benefits rather than avoiding conservation crises, policy setters are lagging behind. Nonetheless, the framework of explicit objectives, reliable indicators, and quantitative reference points provides a complete framework for improving the care of all the components of marine ecosystems and particularly benthic communities and habitats.

Concluding Messages

Managers have many responsibilities other than protecting benthic communities and habitats. Advocates of improved conservation of those communities and habitats can be maximally constructive when they place their concerns in the broad context that commonly challenges fisheries managers.

Sustainability has social, economic, and ecological dimensions. Almost all choices available to managers require losing ground on at least one dimension in exchange for gains on others. In practice, the promise of long-term benefits is rarely enough of an incentive to empower managers to make choices with high short-term economic and, particularly, social costs.

There are a number of management tools that individually can play a constructive role in improving the conservation of benthic communities and habitats and the sustainability of uses. Key ones include better accounting for ecosystem goods and services, marine protected areas, modifications or substitutions of fishing gears, and eco-certification.

Value of each of these tools is limited without an effective framework in which to exercise them. Objectives-based management provides such a framework. This framework requires explicit objectives as well as conceptual ones, reliable information-rich indicators for each objective, and trios of reference points (limit, uncertainty buffer, and target) for each indicator. There is ample room for benthic objectives within this framework.

Progress toward objectives-based management is advanced for ecological, conservation-based objectives, weaker for economic objectives, and rarely visible for social objectives.

References

Alverson, D. L. 2002. Factors influencing the scope and quality of science and management decisions (the good, the bad, and the ugly). Fish and Fisheries 3:3–19.

Anonymous. 2002. Collection of working papers for the scientific council precautionary approach workshop. Northwest Atlantic Fisheries Organization SCS Doc 02/21, Halifax, Nova Scotia.

Armstrong, C. W., and D. J. Clark. 1997. Just fishing? Equity and efficiency in fisheries management regimes. Marine Resource Economics 12:203–220.

Armstrong, C. W., and S. Reithe. 2000. Comment: Marine reserves: will they accomplish more with management costs? Marine Resource Economics 16:165–175.

Ashford, J. R., and J. P. Croxall. 1998. An assessment of CCAMLR measures employed to mitigate seabird mortality in long-lining operations for *Dissosthichus eleginoides* around south Georgia. Commission for the Conservation of Antarctic Living Marine Resources Science 5:217–230.

Bergman, B. 2002. A gathering storm over Kyoto. McLean's Magazine 115(41):20–21.

Butterworth, D. S. 2000. Science and fisheries management entering the new millennium. Pages 37–51 in M. H. Nordquist and J. N. Moore, editors. Current fisheries issues and the Food and Agriculture Organization of the United Nations. Brill Academic Publishers, Leiden, The Netherlands.

Butterworth, D. S., and A. E. Punt. 1999. Experiences in the evaluation and implementation of management procedures. ICES Journal of Marine Science 56:985–998.

Caddy, J. F. 2002. Limit reference points, traffic lights, and holistic approaches to fisheries management with minimal stock assessment input. Fisheries Research 56:133–137.

Charles, A. T. 2001. Sustainable fishery systems. Blackwell Scientific Publications, Fish and Aquatic Resources Series 5, Oxford, UK.

Clark, C. W. 1990. Mathematical bioeconomics—the optimal management of renewable resources. Wiley, New York.

Cochrane, K. L. 2000. Reconciling sustainability, economic efficiency and equity in fisheries: the one that got away. Fish and Fisheries 1:3–21.

Cochrane, K. L., and A. I. L. Payne. 1998. People, purses and power: feveloping fisheries policy for the new South Africa. Pages 57–71 in A. T. Pitcher, P. J. B. Hart, and D. Pauly, editors. Reinventing fisheries management. Kluwer Academic Publishers, Dordrecht, The Netherlands.

Costanza, R., R. d'Arge, R. de Groot, S. Farber, M. Grasso, B. Hannon, K. Limburg, S. Naeem, R. V. O'Neill, J. Paruelo, R. G. Raskin, P. Sutton, and M. van den Belt. 1997. The value of the world's ecosystem services and natural capital. Nature (London) 387:253–260.

CSAS (Canadian Science Advisory Secretariat). 2001a. Proceedings of the National Workshop on Objectives and Indicators For Ecosystem-based Management. CSAS Proceedings Series 2001/09, Ottawa.

CSAS (Canadian Science Advisory Secretariat). 2001b. Proceedings of the Maritime Region Fisheries Management Studies Working Group. CSAS Proceedings Series 2001/08, Ottawa.

CSAS (Canadian Science Advisory Secretariat). 2002a. Northern (2J3KL) cod status update. DFO Canadian Science Advisory Secretariat, Stock Status Report A2/01(2002), Ottawa.

CSAS (Canadian Science Advisory Secretariat). 2002b. Northern Gulf of St. Lawrence cod (3Pn4RS) in 2001. DFO Canadian Science Advisory Secretariat, Stock Status Report A4/01 (2002), Ottawa.

CSAS (Canadian Science Advisory Secretariat). 2002c. National Workshop on Reference Points for Gadoids. In D. Rivard, and J. Rice, editors. DFO Canadian Science Advisory Secretariat Proceedings 2002/033, Ottawa.

Dommen, E. 1993. Fair principles for sustainable development: essays on environmental policy and developing countries. Edward Elgar Publishing, New Horizons in Environmental Economics, London.

EC (European Commission). 2000. Communication of the commission on the precautionary principle. Commission of the European Communities. Available: *http://europa.eu.int/comm/fisheries/doc_et_publ/factsheets/legal_texts/docscom/en/com_00_1_en.pdf* (September 2004).

EC (European Commission). 2002. Communication from the Commission to the Council and European Parliament: towards a strategy to protect and conserve the marine environment. EC, COM 2002, 529, Brussels.

FAO (Food and Agriculture Organization of the United Nations). 1996a. The precautionary approach to marine capture fisheries and species introductions. FAO, FAO

Technical Guidelines for Responsible Fishing 2, Rome.

FAO (Food and Agriculture Organization of the United Nations). 1996b. Precautionary approach to fisheries, part 2. FAO Technical Paper 350.

FAO (Food and Agriculture Organization of the United Nations). 1997. Fisheries management. FAO, FAO Technical Guidelines for Responsible Fishing 4, Rome.

FAO (Food and Agriculture Organization of the United Nations). 1998. Report of the technical consultation on the feasibility of developing non-discriminatory technical guidelines for eco-labelling of products from marine capture fisheries. FAO, Rome.

FAO (Food and Agriculture Organization of the United Nations). 1999. An introduction to monitoring, control, and surveillance systems for capture fisheries. FAO Fisheries Technical Paper 338.

FAO (Food and Agriculture Organization of the United Nations). 2000. Current property rights systems in fisheries management; proceedings of a conference, Freemantle, WA. FAO Fisheries Technical Paper 404.

FAO (Food and Agriculture Organization of the United Nations). 2001a. Managing fisheries capacity: a review of policy and technical issues. FAO Fisheries Technical Paper 409.

FAO (Food and Agriculture Organization of the United Nations). 2001b. Product certification and ecolabelling for sustainability. FAO Fisheries Technical Paper 442.

FAO (Food and Agriculture Organization of the United Nations). 2002a. Guidelines on the Ecosystem Approach to Fisheries. FAO, FAO Technical Guidelines for Responsible Fishing 9, Rome.

FAO (Food and Agriculture Organization of the United Nations). 2002b. Report and documentation of the International Workshop on Factors of Unsustainability and Overexploitation in Fisheries (Bangkok, Thailand, 4–8 February 2002). FAO Fisheries Report R672.

FAO (Food and Agriculture Organization of the United Nations). 2002c. The costs of monitoring, control, and surveillance of fisheries in developing countries. FAO Fisheries Circular 967.

Felt, L., B. Neis, and B. McCay. 1997. Comanagement. Pages 185–194 in J. Boreman, B. Nakashima, J. A. Wilson, and R. L. Kendall, editors. Northwest Atlantic groundfish: perspectives on a fishery collapse. American Fisheries Society, Bethesda, Maryland.

Fernandes, L., M. A. Ridgley, and T. vant Hof. 1999. Multiple criteria analysis integrates economic, ecological and social objectives for coral reef managers. Coral Reefs 4:393–402.

Fowler, T. F. 2002. Community reaction to a social disaster: a case study. Doctoral dissertation. Memorial University of Newfoundland, nstitute for Social and Economic Research, St. John's.

Government of Canada. 1997. Office of the Auditor General. Chapters 14–16 of the Report of the Auditor General to the House of Commons. Supply and Services Canada, Ottawa.

Government of Canada. 2001. A Canadian perspective on the precautionary approach—framework and discussion paper. Privy Council Office Working Group Document. Available: *www.pco-bcp.gc.ca/raoics-srdc/docs/precaution/Discussion/* (July 2003).

Gudmundsson, E., and C. R. Wessells. 2000. Ecolabeling seafood for sustainable production: implications for fisheries management. Marine Resource Economics 15:97–113.

Hall, S. J. 1999. The effects of fishing on marine ecosystems and communities. Blackwell Scientific Publications, Oxford, UK.

Hanna, S. C., M. Hall-Arber, and S. C. Ridlington. 2000. Change and resilience in fishing. Oregon Sea Grant, Corvallis.

Hannesson, R. 1998. Marine reserves: what would they accomplish? Marine Resource Economics 13:159–170.

Hatcher, A., and S. Pascoe. 1998. Charging the UK fishing industry: a report to the Ministry of Agriculture, Fisheries, and Food. Centre for the Economics and Management of Aquatic Resources Report 49, University of Portsmouth, Portsmouth, Hampshire, UK.

Helstrom, C. W. 1968. Statistical theory of signal detection. Pergamon, Oxford, UK.

Hilborn, R., and C. J. Walters 1992. Quantitative fisheries stock assessment: choice, dynamics, and uncertainty. Chapman and Hall, New York.

Hilborn, R., K. Stokes, J.-J. Maguire, T. Smith, L. W. Botsford, M. Mangel, J. Orensanz, A. Parma, J. C. Rice, and J. Bell. 2004. When can marine reserves improve fisheries management. Ocean and Coastal Managment 47:197–205.

Hunt, G. L., Jr., and R. W. Furness. 1996. Seabird fish interactions, with particular reference to seabirds in the North Sea. ICES Cooperative Research Report 216, Copenhagen.

ICES (International Council for Exploration of the Sea). 2000. Report of the Ecosystem Effects of Fishing Working Group. ICES, CM 2000/ACME:02, Copenhagen.

ICES (International Council for Exploration of the Sea). 2001a. Report of the advisory committee on fisheries management 2001. ICES, Cooperative Research Report 246 (3 vol), Copenhagen.

ICES (International Council for Exploration of the Sea). 2001b. Report of the Working Group on Fishing Technology and Fish Behaviour. ICES, CM 2001/ B:05, Copenhagen.

ICES (International Council for Exploration of the Sea). 2001c. Report of the Working Group on Ecosystem Effects of Fishing. ICES, CM:2000/ACE:01, Copenhagen.

ICES (International Council for Exploration of the Sea). 2001d. Report of the ICES Advisory Committee on Ecosystems. ICES, Cooperative Research Report 246, Copenhagen.

ICES (International Council for Exploration of the Sea). 2002. Report of the Ecosystem Effects of Fishing Working Group. ICES, CM 2002/ ACE:03, Copenhagen.

ICES (International Council for Exploration of the Sea). 2003 Extracts of the Report of the ICES Advisory Committee on Fisheries Management. Available: *www.ices.dk/ committe/acfm/comwork/report* (June 2004).

Jennings, D., and M. J. Kaiser. 1998. The effects of fishing on marine ecosystems. Advances in Marine Biology 34:201–352.

Jennings, S., S. Freeman, R. Parker, D. E. Duplisea, and T. A. Dinmore. 2005. Ecosystem consequences of bottom fishing disturbances. Pages 73–90 *in* P. W. Barnes and J. P. Thomas, editors. Benthic habitats and the effects of fishing. American Fisheries Society, Symposium 41, Bethesda, Maryland.

Jennings, S., M. J. Kaiser, and J. D. Reynolds. 2001. Marine fisheries ecology. Blackwell Scientific Publications, Oxford, UK.

Jentoft, S. 2000. Legitimacy and disappointment in fisheries management. Marine Policy 24:141–148.

Kaiser, M. J., J. S. Collie, S. J. Hall, S. Jennings, and I. R. Poiner. 2002. Modification of marine habitats by trawling activities: prognosis and solutions. Fish and Fisheries 3:114–136.

Kimball, L. A. 2001. International ocean governance: using international law and organizations to manage marine resources sustainably. IUCN: The World Conservation Union, Gland, Switzerland.

Kock, K. H., and H. Benke. 1996. On the bycatch of harbour porpoise (*Phocoena phocoena*) in German fisheries in the Baltic and North Sea. Archive of Fisheries and Marine Research 44:95–114.

Kostylev, V. E., B. J. Todd, O. Longva, and P. C. Valentine. 2005. Characterization of benthic habitat on northeastern Georges Bank, Canada. Pages 141–152 *in* P. W. Barnes and J. P. Thomas, editors. Benthic habitats and the effects of fishing. American Fisheries Society, Symposium 41, Bethesda, Maryland.

Kulka, D. W., and D. A. Pitcher. 2001. Spatial and temporal patterns in trawling activity in the Canadian Atlantic and Pacific. Internationl Council for the Exploration of the Sea, ICES, CM 2001/R:02, Copenhagen.

Lanters, R. L. P., Skjoldal H. R., and T. T. Noji, editors. 1999. Ecological quality objectives for the North Sea. Basic document for the Workshop on the Ecological Quality Objectives for the North Sea. Fisken og Havet 10-1999. RIKZ Report 99.015, Amsterdam.

Ledoux, L., and R. K. Turner. 2002. Valuing ocean and coastal resources: a review of practical examples and issues for further action. Ocean and Coastal Management 45:582–616.

Lee, T. 2004. Working together to achieve long term ecological integrity—governance and Yellowstone Park. In N. Monro, editor. Proceedings of the 5th Science and Management of Protected Areas Association Conference. Science and Management of Protected Areas Association, Wolfville, Nova Scotia.

Lindeboom, H. J., and S. J. deGroot, editors. 1998. IMPACT II—the effects of different types of fisheries on the North Sea and Irish Sea benthic ecosystems. NIOZ-Rapport 1998-1, Texel, The Netherlands.

Linnane, A., B. Ball, B. Munday, B. vanMarlen, M. Bergman, and R. Fonteyne. 2000. A review of potential techniques to reduce the environmental impact of demersal trawls. Irish Fishing Investigations, Number 7, Marine Institute, Dublin.

Mardle, S., and S. Pascoe. 1999. A review of applications of multi-criteria decision-making techniques in fisheries. Marine Resource Economics 14:41–63.

Marine Stewardship Council. 2002. Fisheries certification. Marine Stewardship Council. Available: *www.msc.org* (June 2004).

McManus, J. W., L. A. Menez, K. N. Kesner-Reyes, S. G. Vergara, and S. M. Ablan. 2000. Coral reef fishing and coral-algal phase shifts: implications for global reef status. ICES Journal of Marine Science 57:572–578.

Ommer, R. E. 1995. Fisheries policy and the survival of fishing communities in eastern Canada. Pages 307–322 *in* Deep-water fisheries of the North Atlantic oceanic slope. Kluwer Academic Publishers, Dordrecht, The Netherlands.

Pachauri, R. K. 1999. Protecting the global environment: towards effective governance and equitable solutions. Paper for World Bank. Available: *www.worldbank.org/ research/abcde/eu_99/eu/pachauri.pdf* (June 2004).

Parsons, D. G., L. Savard, and C. Lu. 2000. The traffic light: a colourful but ugly approach to precautionary shrimp stock management. Journal of Shellfish Research 19:222–228.

Peterman, R. M. 2002. Ecocertification: an incentive for dealing effectively with uncertainty, risk, and burden of proof in fisheries. Bulletin of Marine Science 70:669–681.

Rice, J. C. 2001. From science to advice—how to find ecosystem metrics to support management. ICES, CM 2001: T/12, Copenhagen.

Rice, J. C. 2002. Sustainable uses of the ocean's living resources. ISUMA—The Canadian Policy Journal 3:80–87.

Roberts, C. M., J. H. Bohnsack. F. Gell, J. P. Hawkins, and R. Goodridge. 2002. Effects of marine reserves on adjacent fisheries. Science 294:1921–1923.

Rochet, M.-J., and J. C. Rice. In press. Criteria for selecting indicators for measuring ecosystem effects of fishing: a framework and tests. *In* P. Cury and V. Christensen, editors. Ecosystem indicators for fisheries management. ICES Science Symposium series, Copenhagen.

Rochet, M.-J., and V. M. Trenkel. 2003. Which community indicators can measure the impact of fishing? A review and proposals. Canadian Journal of Fisheries and Aquatic Sciences 60:86–99.

Rosenburg, A. A., and V. S. Restrepo. 1996. Precautionary management reference points and management strategies. FAO Technical Paper 350:129–140.

Schrank, W. E. 2001. Subsidies for fisheries: a review of concepts. FAO Fisheries Report 638(supplement):11–39.

Schrank, W. E., B. Skoda, P. Parsons, and N. Roy. 1995. The cost to government of maintaining a commercially unviable fishery: the case for Newfoundland 1981/82 to 1990/91. Ocean Development and International Law 26:357–390.

Sladek Nowlis, J., and C. M. Roberts. 1999. Fisheries benefits and optimal design of marine reserves. U.S. National Marine Fisheries Service Fishery Bulletin 97:604–616.

Symes, D. editor. 1999. Fishery dependent regions. Blackwell Scientific Publications, Oxford, UK.

Thomas, J. D., editor. 1999. Proceedings of the International Conference on Scientific Aspects of Coral Reef Assessment, Monitoring, and Restoration. Bulletin of Marine Science 69(2).

Trenkel, V. R., and M.-J. Rochet. 2003. Performance of indicators derived from abundance estimates for detecting the impact of fishing on a fish community. Canadian Journal of Fisheries and Aquatic Sciences 60:67–85.

Turner, R. K., I. J. Bateman, and N. Adgers, editors. 2001. Economics of coastal and water resources: valuing environmental functions. Kluwer Academic Publishers, Dordrecht, The Netherlands.

UN (United Nations). 2002. Plan of implementation—World Summit on Sustainable Development. UN, New York.

UN (United Nations). 1992. United Nations Convention on the Environment and Development. Available: *www.unep.org/unep/partners/un/unced/home.htm* (June 2004).

Valdemarsen, J. W., and P. Suuronen. 2003. Modifying fishing gears to achieve ecosystem objectives. Pages 321–341 *in* M. J. Sinclair and J. W. Valdemarsen, editors. Proceedings of the Reykjavik Conference on Responsible Fisheries in the Marine Ecosystem. Fao Fisheries Technical Papers, Rome.

Walters, C. J., and J.-J. Maguire. 1996. Lessons for stock assessment from the northern cod collapse. Reviews in Fish Biology and Fisheries 6: 125–137.

Watson, R. 2002 Climate change, a time for action. Talk given to ICES Centenary Meeting. Available: *www.ices.dk/aboutus/robwatson.asp* (June 2004).

The Role of Scientific Advice in the Management of Benthic Fisheries in Mexico: Present Status and Perspectives

FRANCISCO ARREGUÍN-SÁNCHEZ[1]

*Centro Interdisciplinario de Ciencias Marinas del IPN, Apartado Postal 592,
La Paz, 23000, Baja California Sur, México*

Abstract. In many countries, traditional scientific guidance for fisheries management has been based on the population dynamics of the target stocks. Most research developed for management occurs in response to the necessity of protecting stocks from evident overfishing and loss of recruitment. To accomplish this, different strategies have been implemented through specific management measures. The complexity of analytical tools varies according to the degree of knowledge required, from general, simple population-dynamics studies to formal and complex simulation experiments, including risk and uncertainty analyses. Other management initiatives requiring scientific advice are stock conservation strategies such as fishery closures, use of excluder devices, and natural reserves and protected areas. Recently, an ecosystem-based approach has been instituted as an alternative available for management strategies. However, even though this new approach offers more information upon which to base decisions, ecosystem-based management requires participation of all ecosystem users, including fishing fleets, which sometimes results in negative benefits for some stakeholders in order to improve ecosystem health, fishery yields, or stock recovery. The ecosystem approach also is used to evaluate the impact of fishing on the dynamics and structure of ecosystems that experience strong fishing effort. In this paper, I discuss scientific work that seeks to promote ecosystem health, conservation, and sustainable exploitation as common goals of research and management. I illustrate this approach by discussing fisheries from the littoral zone of Mexico. Several misleading scientific concepts presently form the basis for fisheries management in Mexico, and these have precipitated serious consequences for the stocks to which they are applied. These concepts include (a) the idea that fish produce eggs in excess, (b) the belief that allowing fish to reproduce at least once during their lifetime is sufficient to maintain recruitment levels, and (c) the assumption of constant natural mortality in fish stocks. Discussion of these concepts yields the conclusion that marine protected areas are probably the best choice when the goal is to manage Mexico's benthic fisheries for sustainability.

Introduction

The Exclusive Economic Zone of Mexico encompasses almost 3 million square kilometers, 10% of which is situated over the continental shelf. The fishery catch has averaged about 1.33 million metric tons per year for the last two decades. Tunas (mostly yellowfin tuna *Thunnus albacares*), sardines (mostly Pacific sardine *Sardinops sagax caerulea* and some *Opistonema*, *Etrumeus*, *Scomber*, and *Oligoplites* spp.), and squids (mostly giant squid *Dosidiscus gigas*) constitute about half the catch at 0.67 million metric tons per year, and the remaining target species make up the other half at 0.66 million metric tons per year (49% of the total value)

(CONAPESCA 2001). These latter fisheries, with the exception of the industrialized shrimp fishery on the eastern and western Mexican continental shelf and the red grouper *Epinephelus morio* fishery off the northern coast of the Yucatan Peninsula, are classified as small-scale or artisanal fisheries.

Recent studies (DOF 2000; INP 2000) indicate that most fisheries in Mexico are exploited at maximum sustainable levels or are overfished. Projections suggest that no substantial increases in catch can be expected. However, Casas-Valdez and Ponce-Díaz (1996a, 1996b) reported that some potentially unexploited resources off the eastern coast of the Baja California peninsula could increase the current annual national catch by at least 40%.

A number of fisheries are especially important in

---

[1] E-mail: farregui@ipn.mx

Mexico because of their economic value, such as shrimp (*Farfantepenaeus* spp. and *Litopenaeus* spp.), lobster (*Panulirus* spp.), and abalone (*Haliotis* spp.) taken from the continental shelf, and shrimp, oysters (*Crassostrea* spp.), and clams landed in coastal waters (DOF 2000; INP 2000; CONAPESCA 2001).

This paper presents (1) a brief review of the current state of fish stock exploitation in Mexico, (2) a discussion of the impact of fishing on Mexican ecosystems, (3) examples of scientific advice developed for benthic fishery resource management in Mexico, and (4) an outline of the major scientific challenges faced in achieving fishery sustainability in Mexico.

Current Status of the Main Benthic Fisheries in Mexico

Most fisheries in Mexico have been managed on a single-species basis by controlling fishing mortality with fishery closures (season and area), limits on the numbers of licenses and boats, establishment of minimum legal sizes, implementation of catch quotas, and gear prohibitions/restrictions in some areas. Table 1 lists the primary benthic fisheries in littoral zones of the Gulf of Mexico, Caribbean Sea, and Pacific Ocean. For most of these fisheries, the goals of management are well defined and depend on the specific state of exploitation. For overfished and collapsed fisheries, the goal is stock recovery based on specific biological reference points. For fully exploited resources, goals vary depending on the resource and include optimizing captures, increasing biomass, avoiding stock depletion and overfishing, maintaining spawning stock, and establishing a minimum legal catch size that permits reproduction at least once. For fisheries that are not yet fully exploited, the goal is the monitoring of fishing activities and stock size. Some additional goals are aimed at reducing the impact of fishing; for example, management of the shrimp trawl fishery seeks to reduce bycatch and minimize deterioration of sea-floor habitat. There are also controls on the use of gill nets in some areas to reduce bycatch of nontarget species, such as marine mammals.

A number of species are not mentioned in Table 1 because they are locally targeted. As in other tropical and subtropical regions, species diversity in Mexico is high, and many species are captured secondarily to the commercial harvest. Examples are the artisanal fisheries in Bahía de La Paz (Gulf of California) (Ramírez-Rodríguez 1991, 1997; Ramírez-Rodríguez and Rodríguez 1991) and the Campeche Bank red grouper fishery off the Yucatan Peninsula (Fuentes 1991; Arreguín-Sánchez et al. 1996). Overall, about 50 species are targeted in Mexico's diverse fisheries (CONAPESCA 2001).

Impact of Fishing on Mexican Ecosystems

There is much evidence in the literature about the impacts of fishing on ecosystems (see Kaiser and de Groot 2000; Jackson et al. 2001), and the government of Mexico is aware that such impacts exist. Fishing impacts can be assessed in various ways, including measurement of habitat deterioration, biomass depletion, and the extent to which "fishing down the food web" has occurred (i.e., the process by which fisheries target increasingly lower trophic levels as upper levels collapse). Additionally, impacts can be indicated by the strength of public opinion urging conservation or recovery of some endangered or charismatic species.

Two issues of primary importance due to their large scale are (1) the effects of shrimp trawling on the ecosystem and on nontarget species and (2) the use of gill nets, which are nonselective and therefore also cause mortality of nontarget species.

The two main problems caused by trawling are (1) the physical impact of the gear on the sea bottom, and (2) the high biomass of dead bycatch returned to the sea (Chávez and Arvizu 1972; Chapa-Saldaña 1976; Hendrickx 1985; Perez-Mellado and Findley 1985; Christensen 1998). The first has not been evaluated properly, and its effect is quantitatively unknown. However, evaluation of bottom impacts is urgently required, because the recovery of sea-bottom communities may be slower than the rate of mortality induced by trawling, thus condemning these ecosystems to degrade and their fauna to disappear. Additionally, we must recognize that the fishing industry has modified its shrimp trawling gear over time, changing from large, weighted gear, in the 1950s and 1960s, to smaller and lighter gear. The early gear had a greater impact on the sea bottom, and though the newer gear is ecologically more friendly, it also represents an increase in fishing efficiency. Recent technical developments by the National Institute of Fisheries in Mexico have provided new designs that continue to contribute to the reduction in both sea-bottom impact and bycatch (Flores-Santillán and Martínez-Meza 1993). Such developments are still experimental, and their impacts have not been quantified at the commercial fleet level.

Arreguín-Sánchez et al. (2002) evaluated the structural and functional attributes of the central Gulf of California ecosystem, where the shrimp-trawl bycatch was reduced by 30%. The most important effects associated with bycatch reduction (i.e., the increase in living biomass) were changes in structural attributes such as (1) the connectance index (Odum 1971) of the food web (the ratio of the number of actual links to the number of possible links), which decreases 6%; (2) the system omnivory index (Pauly et al. 1993) (a measure of how feeding inter-

Table 1. Status of the most important benthic fisheries in Mexico, for the Gulf of Mexico, Caribbean Sea (top), and the Pacific Ocean (bottom); (source CONAPESCA 2001).

Target group	No. species	Status	Goal of management	Management measures
Shrimp	5	White, overfished Pink, collapsed	Recover stocks, maintain stocks over Limit Reference Point, reduce bycatch and damage on bottoms caused by trawling.	Closed seasons and areas to avoid and protect growth and recruitment overfishing.
Red grouper	1	Overfished	Recover stock over Limit Reference Point.	Limitation of licenses, catch quota, and minimum legal size. Three fleets, small and medium (Mexican), large (Cuban).
Octopus	2	Fully exploited	Optimize catch.	Limitation of licenses, minimum legal size, and catch quota.
Queen conch	1	collapsed	Stock recovery.	Minimum legal size, catch quota and closed seasons.
Red snapper	1	Fully exploited	Avoid reduction of biomass below 0.5 Bo maintenance of spawning stock and recruitment within the maximum stock production level.	Limitation of licenses, maintenance of catch
Spiny lobster	1	Fully exploited		Number of licenses (concessions), closed season, minimum legal size. Catching ovigerous females is prohibited.
Mullets	2	overfished	Minimum legal size must be adults that have reproduced at least once.	Minimum legal size, closed season (currently it was suggested to increase the size of first capture and not increase fleet size).
Abalone	3	Collapsed	Stock recovery. Biomass to its high production level year to year.	Minimum legal size, closed seasons and areas, reduction of catch quotas.
Shrimp	2 (white and blue)brown rock	Overfished Fully exploited potential	Avoid growth and recruitment overfishing, and protect reproductive season. Increase biomass at the end of the fishing seasons.	Closed seasons and areas, control fishing effort and fishing gears, minimum legal size, trawl prohibition within bays. Upper Gulf of California closed to trawling.
Red urchin	1	Overexploited	Reach 0.5 Bo.	Closed season, catch quota per area, minimum legal size, number of licenses.
Red snapper	1	Potential	Resources usage under vigilance.	Fishery access through licenses.
Crab	3	Fully exploited	Recover biomass, maintain an fishing mortality of $F = 2.35$ and an exploitation rate of $E = 0.6$ per year for all recruited ages.	Control of fishing effort through number of pots and boats, minimum legal size.
Mullets	2	Overexploited	Maintain catch level per zone.	Minimum legal size, closed season (currently suggested an increase in size at first capture and no increase of fishing effort).

actions are distributed between trophic levels), which decreased 11%; and (3) the Finn cycling index (the fraction of an ecosystem's throughput—sum of all flows in the system according to Ulanowicz 1986— that is recycled; Finn 1976), which decreases 5%. Odum (1969) suggested that food-chain structure changes from linear to weblike as ecosystems mature. If the structural indices decreased because of the reduction of bycatch, the result could be a less-mature ecosystem, which is also reflected by a decreased production/respiration (P/R) ratio.

Ecosystem information content increases with bycatch reduction, as reflected by ascendancy (the measure of the average mutual information in a system, scaled by system throughput), overhead (the limit on how much the ascendency can increase, reflecting the system's strength in reserve), and developmental capacity (the upper limit for the size of the ascendancy (Ulanowicz 1986). Such increase in these indices are logical because of the reduced removal of living biomass from the system. However, if we assume that ecosystem resilience can be expressed by the overhead/ascendancy (O/A) ratio, the calculated O/A ratio decreases by only 0.25% when bycatch is reduced by 30% from existing levels, and the ratio decreases by only 1.2% when bycatch is completely eliminated.

In general, the above indices illustrate changes in ecosystem structure as lower amounts of living biomass are removed from the system as bycatch. Particularly, some structural indices are reduced because more living biomass is accumulated within the ecosystem groups. These changes are also reflected by the three information indices (Ulanowicz 1986), which increase because greater retention of living biomass (i.e., less biomass removal) also means that more energy flows into the system.

Another way to illustrate the impact of fishing on ecosystems is through temporal changes in the mean trophic level (MTL) of catch and the fishing-in-balance (FIB) index (Pauly et al. 2000). The MTL is the average of the trophic levels (TLs) of all species appearing in the catch; the TL for each species is weighted by the species' contribution to catch biomass. The FIB index enables us to assess how fisheries are using tropic levels and considers transfer efficiencies between them. Figure 1 shows the trends in both indices for continental-shelf ecosystems off Veracruz, Campeche, and Yucatan in the Gulf of Mexico, and for the entire Gulf of California, from 1956 to 1999. Horizontal trends in both indices suggest no changes in the ecosystem structure compared to the TLs exploited by fisheries in the Veracruz and Yucatan shelf ecosystems. For the Gulf of California, the MTL shows an increasing trend, whereas the FIB index remains stable (Figure 1). This can be explained by the historical development of the Gulf of California fisheries. In the 1950s and 1960s, shrimp supported the highest-valued fishery and constituted the highest proportion of the catch, whereas more

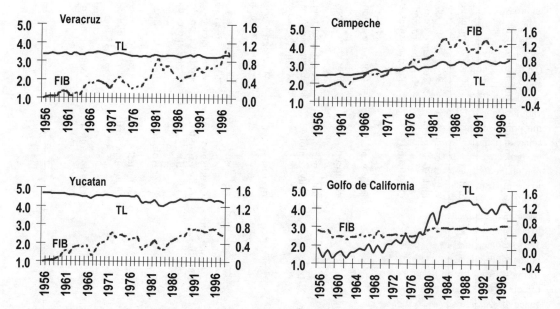

Figure 1. Impact of fishing on the continental-shelf ecosystems off Veracruz, Campeche, and Yucatan in the Gulf of Mexico, and the entire Gulf of California, Mexico, from 1956 to 1999. The primary y-axis illustrates mean trophic level (TL, solid line) and the secondary y-axis illustrates the fishery-in-balance index (FIB, dotted line).

recently, sardine have constituted the highest proportion of the catch (by volume). Both shrimp and sardine occupy a similar TL and have been recently exploited in lower amounts compared to other species. The changes observed reflect the historical access of fishing fleets to the target fish resources.

For the Campeche shelf ecosystem, I observed increases in the MTL and FIB indices as a consequence of the collapse of the shrimp fishery (Figure 1). At the beginning of the 1970s, the shrimp fishery yielded around 25,000 metric tons per year; currently, the fishery yields closer to 1,000 metric tons per year. Whether this collapse was caused by fishing or not (Arreguín-Sánchez et al. 1997a, 1997b; Ramírez-Rodríguez et al. 2000, 2001; Ramírez-Rodríguez and Arreguín-Sánchez 2003), the increase in the MTL and FIB indices reflect changes in ecosystem structure that are consistent with the process of fishing up the food web (Arreguín-Sánchez 2002; Arreguín-Sánchez et al. 2004).

The use of gill nets is extensive along both coasts of Mexico. Even when this fishing gear is highly selective for size, it is not selective for species, thereby causing mortality via entanglement of nontarget taxa. Nontarget mortalities are a problem when there is a need for biodiversity conservation or when nontarget species are endangered or charismatic. Two cases are discussed here: (1) the use of gill nets in the Laguna de Términos (southern Gulf of Mexico), a protected area aimed at the conservation of fauna and flora, and (2) the Gulf of California, an area where a high diversity of marine mammals are being entangled and killed in large numbers (Harcourt et al. 1994; Zavala-González and Mellink 2000).

In early October 2002, the Mexican government (DOF 2002) imposed restrictions on the use of gill nets and emphasized prohibition of shrimp trawling in the upper Gulf of California. The effects of gill nets on the vaquita *Phocoena sinus* (also called the Gulf of California porpoise) and the totoaba *Totoaba macdonaldi*, a sciaenid fish, are of special concern because both species are protected. Protection of the two species is the impetus behind the recent closure of the upper Gulf of California to shrimp trawling and the strong restrictions on the use of gill nets there (Morales-Zárate et al. 2004). The closure is meant to eliminate trawl damage to the ecosystem and to establish a shrimp-to-bycatch proportion (biomass) of 1:1 to replace the present proportion of 1:9. Fishermen have requested the re-opening of the upper Gulf of California for the shrimp fishery, arguing that the negative effects of trawling have not been quantified. In the absence of a specific evaluation, a question emerged: can fishing (shrimp trawls and gill nets) theoretically be maintained in a way that is compatible with conservation of the upper Gulf of California ecosystem, even with the vaquita and totoaba as key protected species?

Based on an ecosystem model constructed by Morales-Zárate et al. (2004), my co-authors and I (Arreguín-Sánchez et al. 2004) approached the question of compatibility between fishing and conservation by use of an Ecosim model (Walters et al. 1997). Table 2 briefly summarizes the possible scenarios explored for making vaquita and totoaba conservation (or recovery) compatible with fishing (i.e., no ecosystem deterioration was permitted). The initial goals were to maintain totoaba and vaquita stocks at their current sizes. Fishing was modeled to remain at least at the current levels.

More than 50 scenarios were explored, but only two of them resulted in maintaining or increasing stock abundances while also maintaining the fisheries within a realistic framework. Realistic scenarios arose from combinations of weighting factors that allowed for changes that the fishing fleet could implement within a short time. For example, a small change in a fishery operating close to the maximum sustainable yield could be selected as realistic, but a fivefold change in fishing mortality for a fully exploited stock was an unrealistic scenario. In the realistic cases, both offshore shrimp-trawl and gill-net fishing quotas remained unchanged or decreased slightly, whereas the coastal shrimp fleet efforts remained constant or increased slightly. Despite this, the low number of feasible outputs suggests a substantial barrier to management of fishing activity in this region, because fleets must operate under strict controls, which usually are not implemented. Deviation from such controls could result in stock depletion. The measures recently imposed seem appropriate to protect the vaquita and totoaba, but at the expense of other fisheries, suggesting that the possible socioeconomic impacts must be evaluated, and alternatives to fishing activity may be required.

Scientific Advice for Management

Mexico's National Institute of Fisheries is the federal institution responsible for providing scientific recommendations for fisheries management. Its strategy during the last decade has been to incorporate uncertainty and risk analyses in the testing of different scenarios, which are then discussed with stakeholders, including fishermen. This strategy has been accompanied by specific recognition of management goals that are established by use of ecosystem, habitat, stock, and socioeconomic reference points.

The ecosystem approach is an additional tool that supports alternative inputs for management. An example is the collapsed fishery for pink shrimp *Farfantepenaeus duorarum* in the southern Gulf of Mexico. In this fishery, recruitment and stock abundance have decreased for three decades (Figure 2). Fishing pressure has been identified

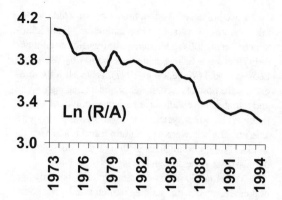

Figure 2. Time series for the recruitment rate index ($\log_e[R/A]$) of pink shrimp in the southern Gulf of Mexico, 1973–1995. R = recruits, A = adults.

as an important contributor to the habitat destruction associated with these declines (Arreguín-Sánchez et al. 1997a, 1997b, 1999; Ramírez-Rodríguez and Arreguín-Sánchez 2003; Ramírez-Rodríguez et al., in press). Simulations based on a single-species, age-structured model were used to explore strategies for pink shrimp stock recovery (Ramírez-Rodríguez et al. 2000). Results indicated that if fishing mortality is reduced by 50% compared to current levels, pink shrimp recruitment will be reduced by 12% at the end of 10 years. If fishing mortality is increased by 50%, pink shrimp recruitment will be reduced by 96% at the end of 10 years. If the fishery is closed for 10 years (100% reduction in fishing mortality), pink shrimp recruitment will increase by only 21%. Such scenarios were discouraging to fishermen, stakeholders, managers, and the residents of the region in general.

By use of the Ecosim model discussed earlier, simulations were also made by varying fleet operations, emphasizing shrimp-stock recovery as a more important goal than the preservation of economic and social criteria, and always maintaining present or enhanced ecosystem structure. Model simulations of shrimp-stock recovery that included no constraints on the fishing fleet resulted in a perturbed ecosystem in which some stocks (dolphins, catfish, porgies, croakers, and grunts) were depleted (Figure 3). In searching for realistic outputs (in terms of fishing fleet activity), I found a combination of assumptions under which shrimp abundance could increase by about 40%. To achieve this increase, the shrimp fleet would need to reduce fishing effort to 64% of current (2002) levels, the artisanal hook-and-line fisheries based on demersal fish and middle-sized pelagic fish would need to decrease effort by 27% and 4%, respectively, and the beach-seine and octopus fleets would need to increase fishing effort by 14% and 56%, respectively. This combined strategy yields a stable, sustainable ecosystem over time (Figure 4).

The scenarios that would produce a stable ecosystem were discussed with fishermen, fleet owners, managers, and scientists. The main concern came from non-shrimp fishers, who questioned the need for them to reduce their fishing activity despite the fact that they do not target shrimp. Scientific demonstration of the benefits of sustainable ecosystems for fishers is an ongoing necessity for gaining their willing participation in stock recovery strategies.

The Paradigm of Sustainability and Its Biological Basis

The sustainability concept has been frequently discussed. Scientists, managers, and fishermen in Mexico recognize that depleted stocks and deteriorated ecosystems must be allowed to recover before specific policies aimed at sustainability can be initiated. Recovery strategies should be implemented first because sustainability must be based on highly productive, rather than depleted, ecosystems. However, experiences with single-species depletion suggest some basic fisheries science concepts must be revised to support scientifically sustainable policies.

Fecundity as A Key Factor

In general, when a fish stock is exploited, the immediate effect is the reduction in life expectancy. Exploitation reduces the probability that fish reach the highest attainable age-classes. Particularly relevant is the age distribution of females in a given population. Analysis of the relationship between age or size and fecundity clearly demonstrates the effect of exploitation on egg production or reproductive potential. Table 3 illustrates this effect with several species; from the table, we can conclude that a population of young females must be one to two orders of magnitude more numerous than a population of older females to maintain equivalent reproductive potential.

A concept that is frequently mentioned by managers, scientists, fishermen, and the general public is that aquatic organisms produce more eggs (offspring) than are required for stock maintenance. This viewpoint assumes that such reproductive "excess" is a mistake of nature or at least an unnecessary loss of energy. These assumptions are probably not based on fact, even though we do not yet understand the functional nature of high egg production. Pauly et al. (2002) suggested that a possible consequence of a drastic reduction in egg production is that recruitment would be more sensitive or vulnerable to environmental changes. Therefore, an apparent excess number of eggs probably acts as a buffer for maintaining recruitment levels, and consequently allows for sustainable stocks.

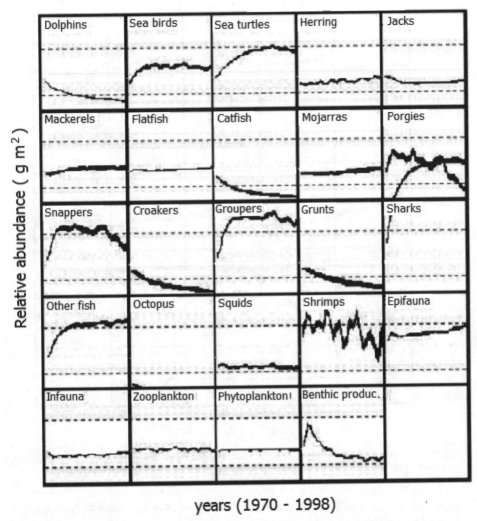

Figure 3. Model output of an ecosystem-based, simulated strategy aimed at recovering the pink shrimp stock in the southern Gulf of Mexico without the use of fishing fleet controls. An unstable ecosystem results from the absence of controls, with depletion of some groups.

A second concept strongly rooted in fishery management is that exploitation must allow each fish to reproduce at least once in its lifetime. The negative impact of this concept is clear, as demonstrated in terms of the number of eggs produced by a fully exploited stock (reproductive potential is greatly reduced) and the high risk of creating unsustainable stocks if use of this concept continues to remain a basis for management.

The arguments above suggest that the concepts of reproductive excess and that a fish must reproduce at least once in its lifetime are misleading, especially for intensively exploited stocks. Recognition of the shortcomings of these concepts is particularly important for management purposes, and even more so if management is aimed at sustainability. Allowing some fish to reach old age, thereby maintaining a relatively high reproductive potential, would be a preferable strategy. However, the obvious question would be how best to implement it.

The Estimation of Natural Mortality

A widely recognized problem of scientific uncertainty is the assumption of constant natural mortality (Caddy 1990, 1996), which is commonly used in fishery stock assessment. However, the consequences of this assumption on stock assessment and on fishery management goals have been poorly evaluated. The essence of the problem is our inability to obtain realistic estimates of natural mor-

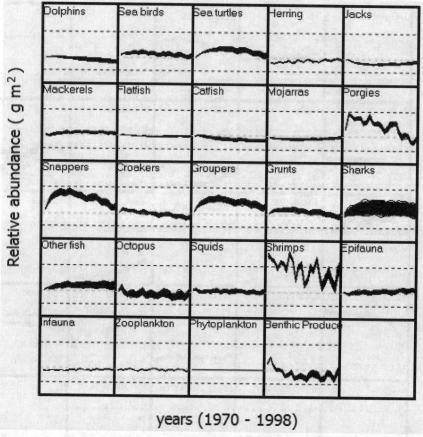

Figure 4. Model output of an ecosystem-based, simulated strategy aimed at recovering the pink shrimp stock in the southern Gulf of Mexico by use of fishing fleet controls.

tality during the life cycle from either models or field data. Figure 5 illustrates this problem with the Pacific sardine *Sardinops sagax caerulea*; in the figure, a constant natural mortality plot is compared to a life-cycle natural mortality plot (Martínez-Aguilar et al. 2005). Assuming an average fecundity of 1 million eggs per female (MacGregor 1957; Torres-Villegas et al. 1985; Lo et al. 1986, 1995; Torres-Villegas 1986; Claramunt et al. 1993), the number of fish from a single brood surviving after 7 years is two individuals, based on the life-cycle estimate of natural mortality. In contrast, based on traditional calculations (cohort analysis), if we compute backwards (starting with two individuals) and reconstruct the stock by substituting a constant value for natural mortality, the initial number of eggs is estimated as only 776. The number of 1-year-old individuals is overestimated by a factor of three compared to the number of 1-year-olds computed under a life-cycle natural mortality scenario. These numbers suggest the urgent need for a proper evaluation of natural mortality estimates for stock assessment work.

The Stock–Recruitment Relationship

In the absence of realistic estimates of natural mortality, the impact of errors in mortality estimates on stock assessments is unknown. Early life stages are clearly underestimated, but young adults (usually the recruits to the fishery) are overestimated.

Such biases enhance the uncertainty of the stock–recruitment relationship and contribute to our inability to predict stock dynamics, yields, and fishing effort. Consequently, there is loss of confidence in scientific recommendations because of the high levels of uncertainty in recruitment predictions and the risks involved in management measures.

Conclusions

Most benthic target fisheries in Mexico are fully exploited, overfished, or even collapsed. Single-species management actions are well defined in terms of goals and spe-

Table 2. Ecosystem based simulations using Ecosim (Walters et al. 1997) to evaluate how fishing can be compatible with conservation (particularly totoaba and vaquita stocks) in the upper Gulf of California. Inputs in the upper panel correspond to weighting factors applied to each criteria and stock level desired. A weighting factor of two indicates a doubling of relevance with respect to another with value of one. A value of two in group columns indicates that relative abundance will be duplicated with respect to the present value. Outputs on the bottom panel represent proportional changes for each criteria and fleet operation magnitude resulting after simulations. Gray rows on the bottom panel represent the only two realistic situations after running 50 simulations.

Simulation inputs

Scenario	Weighing factors for criteria				Weighing factors for stock recovery			
	Economic	Social	Recovery	Ecological	Totoaba	Vaquita	Sea Lion	Whales
1	1	1	5	1	2	2	1	1
2	1	1	5	1	1	2	1	1
3	3	6	5	1	1	2	1	1
4	1	1	4	1	1	1	1	1
5	2	2	8	2	1	1	1	1
6	2	2	8	1	1	1	1	1
7	2	2	8	1	1	2	1	1
8	1	0	0	0	1	1	1	1
9	0	1	0	0	1	1	1	1
10	0	0	1	0	1	1	1	1
11	0	0	0	1	1	1	1	1

Simulation outputs

Scenario	Criteria				Fleets		
	Economic	Social	Recovery	Ecological	Trawl Shrimp	Artisanal Shrimp	Artisanal Fish
1	0.15	0.2	1.41	1.64	0.11	0.9	0.66
2	0.04	0.1	1.26	1.80	0.01	0.8	0.12
3	0.53	0.7	1.00	1.09	0.43	11.0	2.55
4	0.62	0.6	0.99	1.01	0.93	1.7	1.09
5	0.58	0.6	1.00	1.06	0.76	2.3	1.69
6	0.63	0.6	1.00	1.00	1.00	1.3	0.86
7	0.25	0.6	1.14	1.51	0.00	20.0	0.01
8	0.71	0.6	0.78	0.82	1.46	1.0	1.12
9	0.57	0.9	0.63	0.87	0.07	12.0	20.09
10	0.63	0.6	1.00	1.00	0.99	1.0	1.00
11	0.02	0.0	1.00	1.83	0.01	0.2	0.06

cific exploitation strategies, but are not always successful. The incorporation of statistical tools, such as uncertainty and risk analyses, has allowed scientists to provide more detailed advice to managers for use in the decision-making process. However, the improved sophistication of analyses and subsequent recommendations has not been enough to achieve sustainability. Single-species-based management within a given ecosystem frequently results in incompatible strategies or policies that benefit one stock or fishery at the expense of another. The ecosystem approach provides a new perspective, in that different and broader criteria can be explored together to make exploitation compatible with conservation and sustainability.

Of particular relevance to this discussion are the situations in the southern Gulf of Mexico, where recovery of the collapsed pink shrimp fishery is sought, and in the upper Gulf of California, where compatibility between fishing and conservation is needed. These two cases exemplify the types of problems that can be addressed by use of the ecosystem approach.

Another aspect is the use of ecosystem modeling tools to evaluate the impact of fishing on ecosystems and to explore possibilities for mitigating or recovering ecosystem structure and function, including specific stocks. Modeling allows exploration of the ecosystem responses to particular management actions, conservation policies, or changes in habitat conditions.

Analysis of trophic structure and ecosystem dynamics provides knowledge relevant to the evaluation of interdependencies between species, fishing fleets, and ecosystem components (including those components unrelated to fishing). The ecosystem approach is an analytical

Table 3. Comparison between fecundities of young and old adults of different species.

Species	Old female and fecundity	Young females needed for the same number of eggs	Proportion old:young adults[a]	Reference
Lutjanus campechanus[b]	One female of 12.5 kg	212 of 1.1 kg	1:212	Bohnsack (1990)
L. campechanus	80 cm TL produce 49 million eggs	93 females of 40 cm TL	1:93	Collins et al. (1996)
Sardinops caeruleus	24 cm TL produce 1.4 million eggs	122 females of Lm = 14 cm TL	1:122	Computed from Martínez-Aguilar et al. (2005)
Scomberomorus cavalla[b]	One female of 25.6 kg (148.9 cm FL)	176 females of 0.68 kg (44.6 cm FL)	1:176	Funicane et al. (1986)
S. cavalla[b]	One female of 110 cm TL	68 females of 40 cm TL	1:68	Funicane et al. (1986)
Lopholatilus chamaelonticeps[b]	One female of 75 cm TL	106 females of 30 cm TL	1:106	Grimes et al. (1988)
Epinephelus guttatus[b]	One female of 45.7 cm TL	38 females of 24.8 cm TL	1:38	Burnett-Herkes (1975)
Caranx crysos[b]	One female of 1.08 kg (38.5 cm TL)	38 females 0.29 kg (24.3 cm FL)	1:38	Goodwin (1985)
Morone saxatilis[b]	One female of 22.73 kg	230 females of 1.36 kg	1:230	Hugg (1996)
Scomberomorus regalis[b]	One female of 4.94 kg (80 cm FL)	14 females of 0.56 kg (38 cm FL)	1:14	Collette and Nauen (1983)

[a] To produce the same number of eggs as older adult.
[b] Taken from FishBase (Froese and Pauly 2000).

tool that provides a convenient framework for management directed towards a goal of sustainability.

As we strive for sustainability, scientists and managers must recognize that the negative effects of some concepts have been underestimated, or that our understanding has sometimes been wrong. For example, the importance of fish fecundity has been clouded by the misleading concepts used as the basis for management, such as the reproductive excess and one lifetime reproduction concepts. Another problem is the assumption that natural mortality is constant, which yields serious errors in estimating stock size and structure. A better approach is to vary natural mortality over the life cycle to obtain realistic estimates. To move fisheries towards sustainability, some proportion of older females must be allowed to survive, as their higher fecundity contributes to a higher reproductive potential.

Ideally, we would incorporate the concept of maximum legal size to protect higher age-classes, but this cannot be implemented through gear selection. Therefore, the use of marine protected areas as a fishery management tool is probably the best choice for achieving stock sustainability. In addition, marine protected areas are beneficial in that they can redistribute and allocate fishing effort, restore benthic communities, stimulate ecosystems and biological production, and conserve biodiversity.

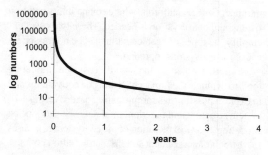

Figure 5. Natural mortality estimates for the Pacific sardine in the Gulf of California. Natural mortality experienced during the life cycle at various stages of development (solid line) is compared to a constant natural mortality value. Data are from Martínez-Aguilar et al. (2005).

Acknowledgments

I would like to gratefully acknowledge support given by the National Polytechnic Institute (project CGPI 20040641) and the National Council of Science and Technology (projects 34865-B and 2002-C01-1231-A1).

Thanks to Dr. Ellis Glazier (La Paz, Baja California Sur, Mexico) for editing the English-language text.

References

Arreguín-Sánchez, F. 2002. Impact of harvesting strategies on fisheries and community structure on the Continental Shelf of the Campeche Sound, southern Gulf of Mexico. Pages 127–134 *in* T. J. Pitcher and K. Cochrane, editors. The use of ecosystem models to investigate multispecies management strategies or capture fisheries. University of British Columbia, Fisheries Centre Research Report 10(2), Vancouver.

Arreguín-Sánchez, F., E. Arcos, and E. A. Chávez. 2002. Flows of biomass and structure in an exploited benthic ecosystem in the Gulf of California, Mexico. Ecological Modelling 156:167–183.

Arreguín-Sánchez, F., M. Contreras, V. Moreno, R. Burgos, and D. Valdés. 1996. Population dynamics and stock assessment of the red grouper (*Epinephelus morio*) fishery on the Campeche Bank. Pages 210–225 *in* F. Arreguín-Sánchez, J. L. Munro, M. Balgos, and D. Pauly, editors. Biology, fisheries and culture of tropical groupers and snappers. ICLARM (International Center for Living Aquatic Resources Management) Conference Proceedings 48, Manila, Philippines.

Arreguín-Sánchez, F., J. A. Sánchez, D. Flores-Hernández, J. Ramos-Miranda, P. Sánchez-Gil, and A. Yáñez-Arancibia. 1999. Stock–recruitment relationships (SRRs): a scientific challenge to support fisheries management in the Campeche Bank, Mexico. Pages 225–235 *in* H. Kumpf, K. Steidinger, and K. Sherman, editors. The Gulf of Mexico large marine ecosystem: assessment, sustainability, and management. Blackwell Science, Inc., Malden, Massachusetts.

Arreguín-Sánchez, F., L. E. Schultz-Ruíz, A. Gracia, J. A. Sánchez, and T. Alarcón. 1997a. Estado actual y perspectiva de las pesquerías de camarón. Pages 185–203 *in* D. Flores-Hernández, P. Sánchez-Gil, J. C. Seijo, and F. Arreguín-Sánchez, editors. Análisis y diagnóstico de los recursos pesqueros críticos del Golfo de México. EPOMEX Serie Científica, Universidad Autónomica, Campeche, México.

Arreguín-Sánchez, F., L. E. Schultz-Ruíz, A. Gracia, J. A., Sánchez, and T. Alarcón. 1997b. Las pesquerías de camarón de altamar: explotación, dinámica y explotación. Pages 145–172 *in* D. Flores-Hernández, P. Sánchez-Gil, J. C. Seijo, and F. Arreguín-Sánchez, editors. Análisis y diagnóstico de los recursos pesqueros críticos del Golfo de México. EPOMEX Serie Científica, Universidad Autónomica, Campeche, México.

Arreguín-Sánchez, F. M. J. Zetina-Rejón, S. Manickchand-Heileman, M. Ramírez-Rodríguez, and L. Vidal. 2004. Simulated response to harvesting strategies in an exploited ecosystem in the southwestern Gulf of Mexico. Ecological Modelling 172:421–432.

Bohnsack, J. A. 1990. The potential of marine fisheries reserves for reef fish management in the U.S. southern Atlantic. Snapper–grouper development team report to the South Atlantic Fishery Management Council. NOAA Technical Memorandum NMFS-SEFC-261.

Burnett-Herkes, J. 1975. Contribution to the biology of the red hind, *Epinephelus guttatus*, a commercially important serranid fish from the tropical western Atlantic. Doctoral dissertation. University of Miami, Coral Gables, Florida.

Caddy, F. J. 1990. Death rates and time intervals: is there an alternative to the constant natural mortality axiom? Reviews in Fish Biology and Fisheries 1:109–138.

Caddy, J. F. 1996. Modelling natural mortality with age in short-lived invertebrate populations: definition of a strategy of gnomonic time division. Aquatic Living Resources 9:197–207.

Casas-Valdez, M., and G. Ponce-Diaz, editors. 1996a. Estudio del potencial pesquero y acuícola de Baja California Sur, volume 1. Centro de Investigaciones Biológicas del Noroeste y Centro Interdisciplinario de Ciencias Marinas del IPN, México.

Casas-Valdez, M., and G. Ponce-Diaz, editors. 1996b. Estudio del Potencial Pesquero y Acuícola de Baja California Sur, volume 2. Centro de Investigaciones Biológicas del Noroeste y Centro Interdisciplinario de Ciencias Marinas del IPN, México.

Chapa-Saldaña, H. 1976. La fauna acompañante de camarón como índice de monopesca. Memoria del Simposio sobre Biología y Dinámica Poblacional de Camarones. Instituto Nacional de Pesca, México D.F. 2:447–450.

Chávez, H., and J. Arvizu. 1972. Estudio de los recursos pesqueros demersales del Golfo de California, 1968-1969: III. Fauna de acompañamiento del camarón. Pages 361–378 *in* J. Carranza, editor. Memorias del IV Congreso Nacional de Oceanografía. México D.F.

Christensen, V. 1998. Fishery-induced changes in a marine ecosystem: insight from models of the Gulf of Thailand. Journal of Fish Biology 53(Supplement A):128–142.

Claramunt, G., G. Herrera, and P. Pizarro. 1993. Fluctuaciones de la fecundidad parcial en sardina española (*Sardinops sagax*) durante la época principal de desove de 1990, en la zona norte de Chile. Scientia Marina 57(1):8–14.

Collette, B. B., and C. E. Nauen. 1983. FAO species catalogue, volume 2. Scombrids of the world. An annotated and illustrated catalogue of tunas, mackerels, bonitos and related species known to date. FAO Fisheries Synopsis 125.

Collins, L. A., A. G. Johnson, and C. P. Keim. 1996. Spawning and annual fecundity for the red snapper (*Lutjanus campechanus*) from the northeastern Gulf of California. Pages 174–188 *in* F. Arreguín-Sánchez, J. L. Munro, M. Balgos, and D. Pauly, editors. Biology, fisheries and culture of tropical groupers and snappers. ICLARM Conference Proceedings 48, Manila, Philippines.

CONAPESCA. 2001. Anuario estadístico de pesca 2001.

Comisión Nacional de Acuacultura y Pesca. Secretaría de Agricultura, Ganadería, Desarrollo Rural, Pesca y Alimentación, México.

DOF (Diario Oficial de la Federación). 2000. Carta nacional pesquera. Diario Oficial de la Federación, México.

DOF (Diario Oficial de la Federación). 2002. Norma oficial de emergencia NOM-EM-139-ECOL-2002. Medidas de protección de los ecosistemas marinos y costeros y de las especies sujetas a protección especial en aguas de la Reserva de la Biosfera del Alto Golfo de California y Delta del Río Colorado. Diario Oficial de la Federación, México.

Finn, J. T. 1976. Measures of ecosystem structure and function derived from analysis of flows. Journal of Theoretical Biology 56:262–380.

Flores-Santillán, A., and J. A. Martínez-Meza. 1993. Evaluación biotecnológica de las redes de arrastre comerciales de Bahía Magdalena. Informe de Investigación. Documentos Internos Instituto Nacional de la Pesca, Centro Regional de Investigatión Pesquera de La Paz, Baja California Sur, México.

Froese, R., and D. Pauly, editors. 2000. FishBase 2000: concepts, design and data sources. ICLARM, Los Baños, Laguna, Philippines.

Fuentes, D. 1991. La pesquería de mero en el Banco de Campeche. Memoria del VII Congreso Nacional de Oceanología, Ensenada B.C., México. 1:361–374.

Funicane, J. H., L. A. Collins, H. A. Brusher, and C. H. Saloman. 1986. Reproductive biology of king mackerel, *Scomberomorus cavalla*, from the southeastern United States. Fishery Bulletin 84(4):841–850.

Goodwin, J. M., IV. 1985. Reproductive biology of blue runner (*Caranx crysos*) from the eastern Gulf of Mexico. Northeast Gulf Science 7(2):139–146.

Grimes, C. B., C. F. Idelberger, K. W. Able, and S. C. Turner. 1988. The reproductive biology of tilefish, *Lopholatilus chamaeleonticeps* Goode & Bean, from the United States Mid-Atlantic Bight, and the effects of fishing on the breeding system. Fishery Bulletin 86(4):745–762.

Harcourt, R., D. Aurioles G., and J. Sánchez. 1994. Entanglement of California sea lions at Los Islotes, Baja California Sur, México. Marine Mammal Science 10(1):122–125.

Hendrickx, M. 1985. Diversidad de los macroinvertebrados bentónicos acompañantes del camarón en el área de California y su importancia como recurso potencial. Pages 95–145 *in* A. Yáñez-Arancibia, editor. Recursos pesqueros potenciales de México: la pesca acompañante del camarón. Instituto de Ciencias del Mar y Limnología, Universidad Nacional Autónoma de México, Progrrama Universitario de Alimentos, Instituto Nacional de la Pesca, México, D.F.

Hugg, D. O. 1996. MAPFISH georeferenced mapping database. Freshwater and estuarine fishes of North America. Life Science Software, Edgewater, Maryland. (listed in FishBase 2000)

INP (Instituto Nacional de la Pesca). 2000. Sustentabilidad y pesca responsible en Mexico: evaluación y manejo 1999–2000. Instituto Nacional de la Pesca, México.

Available: http://inp.semarnat.gob.mx/sustentabilidad/default.htm.

Jackson, J. B. C., M. X. Kirby, W. H. Berger, K. A. Bjorndal, L. W. Botsford, B. J. Bourque, R. H. Bradbury, R. Cooke, J. Erlandson, J. A. Estes, T. P. Hughes, S. Kidwell, C. B. Lange, H. S. Lenihan, J. M. Pandolfi, C. H. Peterson, R. S. Steneck, M. J. Tegner, and R. R. Warner. 2001. Historical overfishing and the recent collapse of coastal ecosystems. Science 293:629–638.

Kaiser, M. J., and S. J. de Groot, editors. 2000. Effects of fishing on non-target species and habitats. Blackwell Science, Oxford, UK.

Lo, N. C. H., J. Alheit, and B. Alegre. 1986. Fecundidad parcial de la sardina peruana (*Sardinops sagax*). Boletín del Instituo del Mar 10(2):45–60.

Lo, N. C. H., P. Smith, and J. L. Butler. 1995. Population of northern anchovy and Pacific sardine using stage-specific matrix models. Marine Ecology Progress Series 127:15–26.

MacGregor, J. S. 1957. Fecundity of the Pacific sardine (*Sardinops caerulea*). Fisheries Bulletin 57:427–449.

Martínez-Aguilar, S., F. Arreguín-Sánchez, and E. Morales-Bojórquez. 2005. Natural mortality and gnomonic time divisions of the life cycle for the sardine (*Sardinops caeruleus*) in the Gulf of California, Mexico. Fisheries Research 71:103–114.

Morales-Zárate M. V., F. Arreguín-Sánchez, S. E. Lluch-Cota, and J. López-Martínez. 2004. Trophic structure and energy fluxes in the ecosystem of the upper Gulf of California, Mexico. Ecological Modelling 174:331–345.

Odum, E. P. 1969. The strategy of ecosystem development. Science 164:262–270.

Odum, E. P. 1971. Fundamentals of ecology. W. B. Saunders Co, Philadelphia.

Pauly, D., V. Christensen, S. Guénette, T. J. Pitcher, U. R. Sumaila, C. J. Walters, R. Watson, and D. Zeller. 2002. Towards sustainability in world fisheries. Nature 418:689–695.

Pauly, D., V. Christensen, and C. Walters. 2000. Ecopath, Ecosim, and Ecospace as tools for evaluating ecosystem impact of fisheries. ICES Journal of Marine Science 57:697–706.

Pauly, D., M. Soriano-Bartz, and M. L. Palomares. 1993. Improved construction, parametrization and interpretation of steady-state ecosystem models. Pages 1–13 *in* V. Christensen and D. Pauly, editors. Trophic models of aquatic ecosystems. ICLARM Conference Proceedings No. 26.

Perez-Mellado, J. L., and L. T. Findley. 1985. Evaluacion de la Ictiofauna del camarón capturado en las costas de Sonora y norte de Sinaloa, México. Pages 201–254 *in* A. Yañez-Arancibia, editor. Recursos Pesqueros Potenciales de México: La pesca acompañante del camarón. Instituto de Ciencias del Mar y Limnología, Universidad Nacional Autónoma de México, Programa Universitario de Alimentos, Instituto Nacional de la Pesca, México, D.F.

Ramírez-Rodríguez, M. 1991. Redes agalleras en Bahía de La Paz, B.C.S. Revista de Investigaciones

Científiques, Serie Ciencias Marinas, México, 2(2), 65–74.

Ramírez-Rodríguez, M. 1997. La producción pesquera en Bahía de La Paz, B.C.S. Pages 273–282 in J. Urbán-Ramírez and M. Ramírez-Rodríguez, editors. La Bahía de La Paz, Investigación y Conservación. Universidad Aunónoma de Baja California Sur, Centro Interdisciplinario de Ciencas Marinas y Acuario Birch del Instituo Oceanográfico Scripps, La Paz, Baja California Sur, México.

Ramírez-Rodríguez, M., and F. Arreguín-Sánchez. 2003. Spawning stock recruitment relationship of pink shrimp *Farfantepenaeus duorarum* in the southern Gulf of México. Bulletin of Marine Science 71:123–133.

Ramírez-Rodríguez, M., F. Arreguín-Sánchez, and D. Lluch-Belda. In press. Collapse of the pink shrimp *Farfantepenaeus duorarum* fishery and sea surface temperature in the southern Gulf of Mexico. Revista Biología Tropical.

Ramírez-Rodríguez, M., E. A. Chávez, and F. Arreguín-Sánchez. 2000. Situación actual y perspectiva de la pesquería de camarón rosado (*Penaeus duorarum*) en la Sonda de Campeche, México. Ciencias Marinas 26(1):97–112.

Ramírez-Rodríguez, M., and M. C. Rodríguez. 1991. Composición específica de la captura artesanal de peces en Isla Cerralvo, B.C.S., México. Investestigaciones Marinas CICIMAR, México. 5(2), 137–141.

Torres-Villegas, J. R. 1986. Evaluación de *Sardinops sagax* por el método de producción de huevos, en Bahía Magdalena, B.C.S. México. Tesis de Maestría en Ciencas. Centro Interdisciplinario de Ciencias Marinas, La Paz, Baja California Sur, México.

Torres-Villegas, J. R., G. García, A. Levy, and R. I. Ochoa. 1985. Madurez sexual, peso promedio, proporcion sexual, frecuencia de desovantes de *Sardinops sagax* en el Golfo de California para noviembre de 1984. Pages 536–549 in CIBCASIO Transactions, Volume 10, 10th annual meeting, La Jolla, California.

Ulanowicz, R. E. 1986. Growth and development: ecosystem phenomenology. Springer-Verlag, New York.

Walters, C. J., V. Christensen, and D. Pauly. 1997. Structuring dynamic models of exploited ecosystems from trophic mass-balance assessments. Reviews in Fish Biology and Fisheries 7:139–172.

Zavala-González, A., and E. Mellink. 2000. Historical exploitation of the California sea lion, *Zalophus californianus*, in México. Marine Fisheries Review 62(1):35–40.

Ecosystem Consequences of Bottom Fishing Disturbance

SIMON JENNINGS[1]

*Centre for Environment, Fisheries, and Aquaculture Science, Lowestoft Laboratory,
Lowestoft NR33 OHT, UK*

STEVE FREEMAN

*ABP Marine Environmental Research Limited, Pathfinder House,
Maritime Way, Southampton SO14 3AE, UK*

RUTH PARKER

*Centre for Environment, Fisheries, and Aquaculture Science, Lowestoft Laboratory,
Lowestoft NR33 OHT, UK*

DANIEL E. DUPLISEA

Institut Maurice Lamontagne, Post Office Box 1000, Mont-Joli, Québec, G5H 3Z4, Canada

TRACY A. DINMORE

*Centre for Environment, Fisheries, and Aquaculture Science, Lowestoft Laboratory,
Lowestoft NR33 OHT, UK*

Abstract. We review the impacts of towed gears on benthic habitats and communities and predict the consequences of these impacts for ecosystem processes. Our emphasis is on the additive and synergistic large-scale effects of fishing, and we assess how changes in the distribution of fishing activity following management action are likely to affect production, turnover time, and nutrient fluxes in ecosystems. Analyses of the large-scale effects of fishing disturbance show that the initial effects of fishing on a habitat have greater ecosystem consequences than repeated fishing in fished areas. As a result, patchy fishing effort distributions have lower total impacts on the ecosystem than random or uniform effort distributions. In most fisheries, the distribution of annual fishing effort within habitats is more patchy than random, and patterns of effort are maintained from year to year. Our analyses suggest that many vulnerable species and habitats have only persisted in heavily fished ecosystems because effort is patchy. Ecosystem-based fisheries management involves taking account of the ecosystem effects of fishing when setting management objectives. One step that can be taken toward ecosystem-based fisheries management is to make an a priori assessment of the ecosystem effects of proposed management actions such as catch controls, effort controls, and technical measures. We suggest a process for predicting the ecosystem consequences of management action. This requires information on habitat distributions, models to predict changes in the spatial distribution of fleets following management action, and models of the impacts of trawling disturbance on ecosystem processes. For each proposed management action, the change in disturbance affecting different habitat types would be predicted and used to forecast the consequences for the ecosystem. These simulations would be used to produce a decision table, quantifying the consequences of alternative management actions. Actions that minimize the ecosystem effects of fishing could then be identified. In data-poor situations, we suggest that management strategies that maintain or maximize the patchiness of effort within habitat types are more consistent with the precautionary approach than those that lead to more uniform fishing effort distributions.

---

[1] Corresponding author: S.Jennings@cefas.co.uk

Introduction

Many towed gears have impacts on benthic communities and habitats. The consequences of these impacts on abundance, diversity, habitat, and community structure are well known, but the large-scale ecosystem consequences of the impacts are not (Dayton et al. 1995; Jennings and Kaiser 1998; Lindeboom and de Groot 1998; Auster and Langton 1999; Hall 1999; Kaiser and de Groot 2000; National Research Council 2002). In part, this lack of information is due to the emphasis of current research on pattern rather than process, a lack of reliable information on the spatial distribution and frequency of fishing disturbance and the difficulties associated with extrapolating the results of small-scale studies (Duplisea et al. 2001; Jennings et al. 2001a).

Towed gears cause reductions in habitat heterogeneity, biomass, production, and diversity, although the relative impacts and recovery times depend on the balance between fishing and natural disturbance (Kaiser et al. 2002). Not surprisingly, in areas where natural disturbance is very low, such as the deep sea, communities and habitats are poorly adapted to disturbance and the effects of towed gears are profound (Koslow et al. 2001; Hall-Spencer et al. 2002). Conversely, in shallow regions with mobile sediments, the effects of disturbance tend to be more difficult to detect and recovery is faster (Kaiser and Spencer 1996; Collie et al. 2000). Since the disturbance caused by towed gears is so widespread, the effects of a single fishing event must be considered in relation to the effects of other fishing events (Bergman and van Santbrink 2000b).

There is a growing recognition of the requirement to take account of the wider impacts of fishing in fisheries management decisions (Rice 2000; Link 2002). This approach is loosely termed the "Ecosystem Approach to Fisheries Management" (Ecosystem Principles Advisory Panel 1999). A key part of this process is to understand how fisheries management actions that result in temporal and spatial changes in the distribution and intensity of fishing effort affect the ecosystem. The development of methods to predict the effects of management actions on benthic communities and habitats and methods to allow managers to mitigate the effects will be an important contribution to the development of an operational "Ecosystem Approach."

We begin this paper by reviewing the impacts of towed gears on benthic habitats and communities and predict the consequences of these impacts for ecosystem processes. Our emphasis is on the additive large-scale effects of fishing. Then we assess how changes in the distribution of fishing activity following management action are likely to affect production, turnover time, and nutrient fluxes in ecosystems. We complete the paper by suggesting how studies of the ecosystem consequences of fishing activities might be used to inform the development of an "ecosystem approach to fisheries management." This is not intended to be a comprehensive review of towed gear impacts; many such reviews have already been published (Dayton et al. 1995; Jennings and Kaiser 1998; Auster and Langton 1999; Hall 1999; Kaiser et al. 2002; National Research Council 2002).

Effects of Bottom Fishing Disturbance

Biomass, Production, and Turnover Time

Many studies of the impacts of towed bottom fishing gears have focused on larger macrofauna and habitat-forming species, primarily because reductions in their abundance and diversity are an important conservation issue and because they provide habitat for bottom-dwelling fishes (Collie et al. 1997, 2000; Auster and Langton 1999; Koslow et al. 2001). Bottom fishing causes mortality of many species because they are crushed directly by the gear or are caught and die by the time they are taken on deck and returned to the sea (Lindeboom and de Groot 1998). Within and among species, mortality is generally size dependent (Bergman and van Santbrink 2000a, 2000b). Thus, larger bivalves and attached epifauna suffer very high mortality while smaller bivalves and polychaetes may suffer lower mortality, possibly because lighter animals are pushed aside by the pressure wave in front of the gear (Gilkinson et al. 1998). Not only are larger species more likely to suffer high mortality, but the mortality rates they can withstand will be lower because they tend to have lower intrinsic rates of increase and slower growth (Banse and Mosher 1980). As a result, habitat complexity is reduced and smaller, free-living species and individuals are relatively more abundant in heavily fished habitats (Jennings et al. 2001a; Duplisea et al. 2002).

The initial effects of bottom fishing in unfished areas have greater impacts on benthic habitats and species than repeated fishing in fished areas. In unfished areas, biomass is dominated by large, slow-growing species that are more vulnerable to fishing disturbance than the smaller, fast-growing species that dominate biomass in fished areas (Duplisea et al. 2002). Since most biogenic habitats are composed of large and relatively slow growing species, these habitats are easily modified by bottom fishing. Biogenic habitats are often particularly complex and slow growing in areas of low natural disturbance, and the initial impacts of bottom trawls can dramatically change the habitat architecture and the abundance and diversity of associated species (Koslow et al. 2001;

Hall-Spencer et al. 2002). However, even in shallow shelf seas where natural disturbance is more frequent and more intense, the first impacts of bottom fishing led to substantial habitat modification. In the Dutch Wadden Sea, for example, dredging led to the loss of reefs of the calcareous tube-building worm *Sabellaria spinulosa* and its replacement by communities of small polychaetes (Riesen and Riese 1982). Tube-building worms have an important role in stabilizing the sediment and providing sites for the establishment of beds of blue mussels *Mytilus edulis*. Similarly, the abundance of another habitat-structuring sabellid worm *Myxicola infundibulum* decreased following commercial scallop dredging in the Gulf of Maine (Langton and Robinson 1990).

In heavily fished shelf seas there have been reductions in sessile habitat-forming species and increases in the relative abundance of small, free-living fauna. For example, on Georges Bank, Collie et al. (1997) compared areas that were subject to different intensities of scallop dredging and showed that infrequently fished areas were characterized by abundant bryozoans, hydroids, and worm tubes which increased the three-dimensional complexity of the habitat. Conversely, in the more intensively dredged areas, the total biomass of the fauna was lower, and hard-shelled bivalves, echinoderms, and scavenging decapods were the dominant species. The total diversity of the community also decreased in the more intensively dredged areas because many of the organisms associated with biogenic fauna were absent. Similar changes in benthic communities due to fishing were described by Auster et al. (1996) in the Gulf of Maine. Here, sponge cover was much reduced in intensively fished areas and sponge-associated species were depleted or absent.

There are many other examples of shifts in community structure following bottom fishing disturbance, but what are the consequences for benthic production? Smaller species have higher intrinsic rates of increase and body growth, so their production per unit biomass is higher and their turnover times are faster (Brey 1999). As a result, the production to biomass (P:B) ratios of benthic assemblages will increase with the frequency of fishing disturbance. However, the increase in relative production of the macrofauna that remain after trawling is rarely sufficient to compensate for the loss of total production associated with the loss in biomass. Indeed, one study of the effects of trawling on production in a soft-sediment community showed that there was an order of magnitude reduction in total infaunal production across a gradient in trawling disturbance, and while production per unit biomass rose, it did not compensate for the loss of total production that resulted from the depletion of large individuals (Jennings et al. 2001a).

Large, long-lived, and slow-growing species have slow turnover times. Since the turnover time of a community is effectively the inverse of the P:B ratio, the shift toward small individuals and species with fast life histories in trawled areas will reduce turnover time. This is expected to increase the variability of total biomass and lead to greater fluctuations in the production that can be consumed by higher trophic levels. In heavily trawled areas, the smallest macrofauna appear to tolerate trawling disturbance but do not benefit from it. Thus, Jennings et al. (2002) investigated the effects of beam trawling disturbance on the production of small benthic infauna (ash free dry mass > 0.78–62.5 mg) at nine sites that were subject to a 17.5-fold range in annual beam trawling disturbance. They developed a generalized additive model to test for relationships between trawling disturbance and infaunal production, after accounting for differences in sediment characteristics and depth (Figure 1). Statistical power analyses showed that they were likely to detect linear and nonlinear relationships between production and disturbance if they existed (Figures 2, 3), and yet the analyses showed that trawling frequencies of 0.3–6 times per year did not have a negative or positive effect on the production of small infauna or polychaetes. Since these small infaunal polychaetes are a key source of food for flatfishes, they concluded that beam trawling disturbance did not have a positive or negative effect on flatfish food supply. In general, beam trawling appears to have created a system where small fish feed on small food items. This may have minimal effects on the growth and production of flatfishes, but the size structure, biomass, and total production of the infaunal and fish communities are fundamentally different from those in the unfished state.

If the smallest macrofauna cannot benefit from reduced competition and predation following trawling disturbance, then the macrofauna community as a whole must assimilate much lower levels of production. This process suggests an increased role for meiofauna or bacteria in processing the detritus that reaches the seabed. Given that bivalves and many sessile biogenic species are phytoplankton feeders in unstratified seas, the shift to small, free-living species may also reduce the direct uptake of phytoplankton production by the benthic community.

Meiofauna may be impacted directly or indirectly by fishing disturbance since the passage of trawls causes immediate mortality or displacement, changes sediment structure and geochemistry, and affects the abundance of predators or competitors. However, meiofauna also have very fast life cycles and high P:B ratios (Schwinghamer et al. 1986), and some species may benefit from reduced competition and predation while

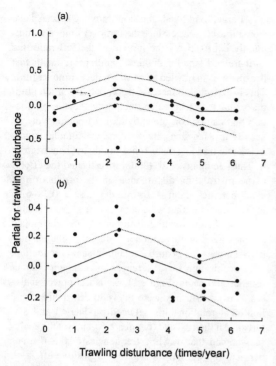

Figure 1. Partial regressions for the production of (a) all small infauna (0.78–62.5 mg ash free dry mass) and (b) polychaetes (0.78–62.5 mg ash free dry mass), having accounted for the effects of sediment particle size, sorting coefficient, and depth. Production was estimated across a 17.5-fold range of trawling disturbance in the central North Sea. From Jennings et al. (2002).

tolerating this disturbance (Schratzberger et al. 2002). In one study of changes in meiofauna community structure across a gradient of trawling disturbance, nematode communities were compared in three beam trawling areas of the central North Sea that were subject to low (trawled one time each year), medium (four times), and high (six times) levels of trawling disturbance. After accounting for the effects of season, sediment type, and depth, the analysis showed that trawling disturbance had a significant impact on the composition of nematode assemblages (Schratzberger and Jennings 2002). In two sampling seasons, the number of species, diversity, and species richness of the community were significantly lower in the area subject to high levels of trawling disturbance than in the areas subject to low or medium levels of disturbance. However, levels of disturbance at the low and medium sites may have been insufficient to cause marked long-term changes in community structure. The extent to which the observed changes in community structure reflect changes in the production of the nematode community remains unknown, but significant increases in production were not expected when there were no clear changes in

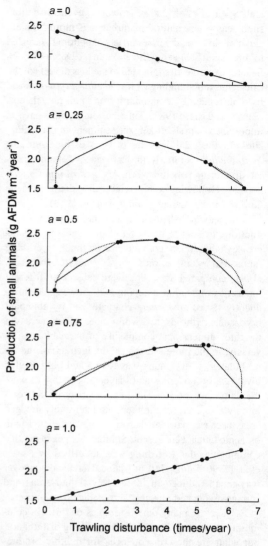

Figure 2. Hypothetical responses of the production of small animals to trawling disturbance for different values of parameter a. Parameter a defines the shape of the response: $a = 0$ suggests that trawling impacts on the production of small animals are entirely negative, $a = 0.25$–0.75 suggests that there is increased production at intermediate levels of disturbance, and $a = 1$ suggests that the production of small animals increases with disturbance because the abundance of larger competitors and predators has been reduced. The observed frequencies of trawling disturbance are shown as circles and the solid lines show how well the response curve can be tracked. Data from a statistical method developed by Mike Nicholson (Cenre for Environment, Fisheries and Aquaculture Science, UK) and described in Jennings et al. (2002). AFDM = ash free dry mass.

abundance or size structure (Schratzberger and Jennings 2002).

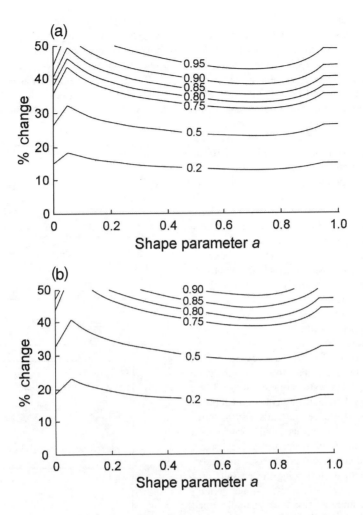

Figure 3. The power of an experiment in the central North Sea to detect proportional changes in the production of (a) all small infauna (0.78–62.5 mg ash free dry mass) and (b) polychaetes (0.78–62.5 mg ash free dry mass) with trawling disturbance. Power is predicted for different responses, as defined by the shape parameter a (see Figure 2).

If the production of macrofauna and meiofauna falls or remains stable in response to trawling disturbance, then bacteria must process inputs of organic carbon and nitrogen to the seafloor. Otherwise, these inputs will accumulate in the sediments or be returned to the water column by resuspension. One study of the effects of dredging on microbial abundance showed that abundance was reduced by around 50% in the top 1 cm of sediment immediately after the passage of gear but that there was little effect on microbial biomass at depths of 4–5 cm or more (Watling et al. 2001). In the surface layer, however, much of the reduction in biomass may have been attributable to the resuspension of surface sediments, and the bacteria may have been deposited elsewhere. Microbial biomass in the surface sediments had recovered a few months after dredging. The impacts of bottom fishing disturbance on bacterial production and the rate of accumulation and breakdown of organic matter in sediments remain key to understanding the effects of disturbance on ecosystem processes.

Bioturbation and Biogeochemistry

Despite the ubiquity of bottom fishing disturbance in shelf seas, relatively little is known of the effects of this disturbance on sediment community function, carbon mineralization, and biogeochemical fluxes (Parker et al. 1999). The physical impact of trawling or dredging, such as scraping of the substrate, sediment resuspension, and mortality of benthos, disrupt the redox status of the sediment and, hence, microbial activity which controls biogeochemical cycling. This may shift the balance of aerobic to anaerobic mineralization, which may, in turn, result in a change in the end products of mineralization. For example, nitrogen cycling and the resulting fluxes of nitrate, nitrite, and ammonium to and from the sediment are a function of anaerobic carbon mineralization (ammonium generation) coupled with aerobic oxidation of ammonium (nitrification) and anaerobic denitrification of nitrate (coupled or uncoupled). Changes in these pathways induced by increased carbon or changes in oxygen regime can lead to changes in fluxes across the sediment–water interface. Similarly, processes such as sulfate reduction can be stimulated at depth by the input of organic carbon and physical mixing of organic material deeper into the sediment. As the rate and balance between such sediment processes are mediated by microbial metabolic pathways, carbon input to sediments, and oxygen state, trawling activities are expected to affect bulk sediment biogeochemistry (Duplisea et al. 2001).

Bottom fishing also resuspends surface sediments (Churchill 1989; Pilskaln et al. 1998; Palanques et al. 2001) and may result in immediate nutrient fluxes into the water column (Krost 1990; Pilskaln et al. 1998; Percival and Frid 2000; Duplisea et al. 2001; Figure 4). Hence, nutrients may be introduced into the water column as a pulse rather than by the usual, slower efflux mechanisms. Duplisea et al. (2001) calculated that pulsed releases of nitrate, ammonium, and silicate following the passage of a beam trawl could be 20, 45, and 26 times greater, respectively, than ambient fluxes from undisturbed sediments. Krost (1990) examined the release of nutrients and other species following trawling-induced sediment resuspension and calculated associated fluxes and changes in nutrient profiles. For example, he calculated an additional annual mobilization of 98–435 metric tons/km^2 nitrogen and 34–167 metric tons/km^2 phosphorus from sediments disturbed by otter boards. Oxygen depletion in the water column as a result of sediment resuspension or chemical oxygen demand (from oxygenation of reduced species) has also been observed. Krost (1990) calculated an annual oxygen demand of 491–2,656 metric tons/km^2 O$_2$ due to the release of hydrogen sulfide by sediment resuspension. Blackburn (1997) proposed that a minimum of 2.4 cm of sediment would need to be resuspended to liberate pore water nitrogen and exchangeable ammonium into the water column. This is possible during trawling as the penetration of most gears is at least 2.4 cm (Lindeboom and de Groot 1998). However, Blackburn (1997) also suggested that the main effect of acute resuspension events was to change the timing of nitrogen release, since nitrogen would have been effluxed naturally from the sediment over longer periods.

Trawling disturbance may increase or alter sediment nutrient flux rates during periods when natural disturbance is relatively low (e.g., after the spring phytoplank-

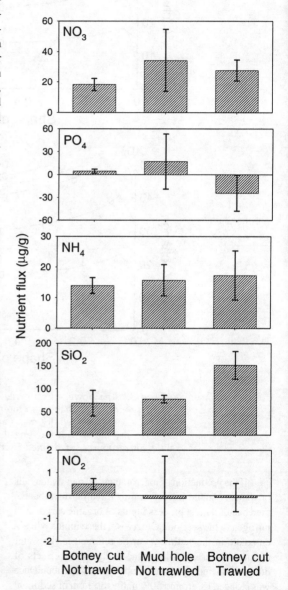

Figure 4. Examples of pulsed nutrient fluxes from intact sediment cores that were resuspended under experimental conditions. Measurements are immediate nutrient releases into the overlying water. Sediment cores were taken from three sites in the North Sea, one of which was known to be heavily trawled. Data from Duplisea et al. (2001).

ton bloom and when wind stress is low in the northern hemisphere). These fluxes may provide nutrients to fuel pelagic production. Work by Fanning et al. (1982) on continental shelf sediments indicated natural or anthropogenic resuspension as a source for silica, nitrate, and nitrite release. They estimated that resuspension of only 1 mm could double or triple the nutrient flux into the photic zone. Increased ambient fluxes of nutrients (e.g., ammonium) may have a similar effect. The productivity stimulated by this injection may be determined by the timing of the release (seasonal), and effects may be harder to detect in coastal areas where the background nutrient concentrations are higher. Resuspension may accelerate nutrient recycling and result in an overall increase in primary productivity and organic carbon export rates (Chavez 1996; Pilskaln et al. 1996).

Sediment disturbance by scallop dredges has also been shown to increase the depth of chlorophyll and protein maxima, while anaerobic bacteria, which normally reside deeper in sediments, were mixed up to the surface layers (Mayer et al. 1991). Thus, Mayer et al. (1991) hypothesized that sediments disturbed by fishing activities for extended periods would show shifts to anaerobic microbial carbon mineralization pathways. These shifts would affect the storage of carbon in sediments and may sequester organic carbon in microbial food webs which are inaccessible to larger sediment fauna. A related study subsequently confirmed that larger sediment fauna did not tend to recover in numbers until fishing activities had ceased, and high-quality carbon sources, such as chlorophyll and proteins, built up in the surface layers (Watling et al. 2001).

Macrofauna are responsible for disturbing and restructuring sediments (Asmus 1986). Open-burrow structures can increase the sediment movement and sediment–water interfacial area, but they can also increase sediment heterogeneity by increasing ventilation, oxygen penetration, and the export of regenerated nutrients (Marinelli 1992, 1994; Clavero et al. 1994). Bioturbation will also have a significant effect on the rate of denitrification (Pelegri et al. 1994). The behavioral patterns of different species and size-classes of macrofauna will influence sediment geochemistry. For example, many benthic macroinvertebrates will pump water through their burrows by active ventilation, and the burrows act as channels for advective hydraulic flow into or out of the sediment. In addition, bioroughness (the surface relief created by benthic organisms) has been observed to cause strong pore water replacement down to 5-cm depths and a link between sediment layers as deep as 15 cm and the water column (Huettel and Gust 1992). In general, such bioirrigation increases oxygen penetration into a sediment and can change the relative dominance and distribution of nitrogen cycle oxidation–reduction reactions and other redox species such as Mn^{2+}, Fe^{2+}, and HS^-. Ziebis et al. (1996) found that under a sediment mound (~4 cm) constructed by the mud shrimp Callianassa truncata, the oxygen penetration depth was increased twofold (to ~1 cm) in a North Sea intertidal flat. Enhanced sediment oxygenation produced by burrows and bioturbation can increase the overall rate of nitrification (Huettel 1990), although the influence of burrows on the flux of NO_3^- into or out of the sediment is dependent on burrow radius, oxidized wall thickness, burrow spacing, and the rate of denitrification (Aller 1988).

Rates of denitrification are also increased by the presence of burrows, and the coupling between nitrification and denitrification can be altered by changes in burrow spacing and type (i.e., it is species-specific and size-specific). Several studies have observed that Corophium spp. (Pelegri et al. 1994), Bivalvia, Polychaeta, and Crustacea (Henriksen et al. 1983) all increase the rate of ammonium production, nitrification, and denitrification in their burrow environment. Species with high irrigation capacities also increase the flux of nitrate to the overlying water column. Aller (1982, 1988) demonstrated how the density and size of macrofauna burrows in sediments can affect the direction and magnitude of nitrogen flux across the sediment–water interface.

Given the important role of benthic fauna in controlling sediment structure and biochemical processes, the major changes in the structure of benthic communities following trawling disturbance (see *Biomass, Production, and Turnover Time*) will have a profound impact on sediment biogeochemistry. In particular, trawling disturbance rapidly depletes populations of the most active bioturbators (Figure 5). Duplisea et al. (2001) used an existing model of biogeochemical processes in a soft-sediment system (Duplisea 1998) to investigate the potential effects of fishing disturbance on carbon mineralization. In areas with low natural disturbance, their simulations suggested that the presence of bioturbating macrofauna in unfished sediments allowed sediment chemical storage and fluxes to reach equilibrium. This is because the macrofauna consume carbon and reduce the magnitude of carbon fluxes. Conversely, in fished sediments, the biomass of macrofauna was markedly reduced (see *Biomass, Production, and Turnover Time*) and larger fluctuations in benthic carbon fluxes and storage are expected.

There remain relatively few studies of the effects of fishing disturbance on nutrient fluxes in soft-sediment communities, and yet all the evidence suggests that such effects may have a profound influence on the marine ecosystem. Further work on the biogeochemical impacts of fishing activities should focus on the large-scale experimental work that is still needed to assess the biogeochemical impacts of entire fisheries.

Food Webs

It is clear that bottom fishing disturbance leads to large reductions in the total biomass and production of free-living and sessile macrofauna (see *Biomass, Production, and Turnover Time*). The reductions in production will clearly decrease the total energy flux in macrobenthic food webs, but do they influence the function of those food webs and the mean trophic level of the benthic community? For complex biogenic communities and their associated fauna in areas of low natural disturbance, such as the deepwater reefs and seamount communities studied by Hall-Spencer et al. (2002) and Koslow et al. (2001), the effects of fishing disturbance on trophic structure are not known. However, they are expected to be highly significant since these communities are lost or fundamentally modified following fishing disturbance and the dominant pathways of energy transfer must also change. Many biogenic habitats, in particular, are dominated by filter feeders, and when these habitats are changed by fishing and increasingly dominated by small, free-living species (see *Biomass, Production, and Turnover Time*), an increasing proportion of benthic production is likely to be supported by detrital pathways.

In mobile sediments in areas of high natural disturbance, sessile biogenic species may be rare or absent. Here, fishing disturbance will cause large reductions in biomass and production, but the remaining animals in the community may reassemble into food webs that have similar structural characteristics. In one study of the effects of fishing disturbance on trophic structure, there were order-of-magnitude decreases in biomass across a gradient of fishing disturbance, but the mean trophic levels of the infaunal and epifaunal communities did not change (Jennings et al. 2001b). Thus the trophic structure of these benthic invertebrate communities may be an emergent property of the community that remains after trawling and maximizes the transfer efficiency of energy among those individuals that have sufficiently high turnover times to withstand the levels of mortality imposed by trawling. Indeed, in the benthic epifaunal community where predator–prey relationships are strongly size based, there was no evidence for a change in the mean predator–prey body mass ratio (Jennings et al. 2001b).

In general terms, the architecture of food webs in areas disturbed by fishing is determined by the animals that are left alive. Even in communities where biomass and production are greatly reduced, the remaining animals may reassemble into food chains.

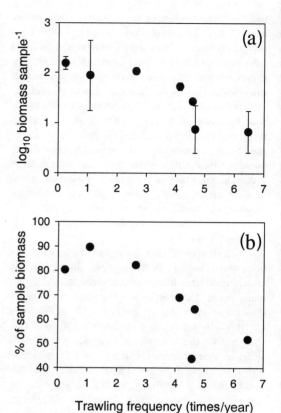

Figure 5. Relationship between the biomass of (a) bivalves/spatangoids or (b) polychaetes and the frequency of beam trawling disturbance in the central North Sea. The lower panel gives the biomass of bivalves/spatangoids or polychaetes as a proportion of total biomass at the different frequencies of trawling disturbance. The bivalves/spatangoids are important bioturbators. Data from Jennings et al. (2001b).

Thus, the greatest effects of fishing on marine food webs are to reduce total production and to change the relative roles of macrobenthic and other pathways in the ecosystem and to force a greater proportion of energy through "faster" pathways, leading to greater instability of biomass and production.

Patterns of Trawling Disturbance

Patterns of Effort in Real Fisheries

Bottom fishing always has a local impact, as any heavy object dragged across the seafloor will crush animals and change habitats. However, at the scale of the ecosystem, the significance of impacts is determined by (1) the balance between cumulative levels of fishing and natural disturbance and (2) the spatial and temporal distribution of fishing impacts in relation to the distribution of species and habitats. The study of the spatial and temporal distribution of fishing impacts in relation to the distribution of species and habitats has assumed increasing prominence of late, and there has been a growing focus on mapping habitats and spatial patterns of fishing effort.

Historically, the resolution of fishing effort data has been quite poor, and when fine-scale effort data were available, they could only be obtained for a small proportion of fishing vessels or for a small sea area (Rijnsdorp et al. 1998). However, even at large spatial scales, such as the International Council for the Exploration of the Sea (ICES) rectangles (211 rectangles of 0.5° latitude by 1° longitude; area of rectangle = 3,720 km^2 at 53°N) used to collate fisheries and environmental data in the northeastern Atlantic, fishing effort was clearly patchy. Thus in the North Sea from 1990 to 1995, where total annual beam trawling effort was approximately 1.2 million hours, the majority of rectangles were beam trawled for less than 2,000 h per year. Since a typical beam trawler will impact 534.5 km^2 of seabed in 2,000 h, it would have taken at least 7 years to trawl any rectangle that was fished for less than 2,000 h per year. Patterns of disturbance for the combined otter and beam trawl fleets were also very patchy (Figure 6).

Large-scale patchiness in effort is the result of many factors, including the distribution of target fish species, the distribution of habitats that cannot be fished with certain gears, fishery regulations, steaming costs, and the avoidance of habitats and obstructions that could damage gears. As a result, patchiness of effort will not ensure that representative areas of different habitat types are unaffected by fishing, unless there is patchiness of effort within habitat type. This is likely to result from trawlers returning to fish tows that are stored on their navigation systems, that are known to be clear of obstructions, and where the highest catch rates have been made in previous years. However, with the advent of new technology, trawlers have much more accurate information on the seabed and may be able to target specific habitats more effectively. This may increase patchiness among habitats but decrease patchiness within habitats, as they can be identified and exploited more easily.

Patterns of fishing effort within the ICES rectangles in the North Sea are also very uneven, as first described quantitatively by Rijnsdorp et al. (1998). They attached position loggers to a proportion of the Dutch beam trawler fleet, and this allowed them to track individual vessels as they fished. The degree of patchiness in fishing effort decreased as the resolution of the analyses was increased from 30 × 30 to 10 × 10 to 3 × 3 to 1 × 1 nautical mi (1 nautical mi = 1.86 km). In 1 × 1 nautical-mi boxes, the distribution of effort was considered random in 90% of cases. In eight of the most heavily trawled ICES rectangles in the southern North Sea (total area of 7,200 nautical mi^2), 5% of the area was trawled less than once in 5 years and 70% trawled more than once a year.

An alternative method for quantifying trawling disturbance on relatively small scales is to use records of vessels sighted by aircraft that patrol fishing grounds. The crew on these aircraft record the description and location of all vessels they see fishing. Relative trawling disturbance can be estimated as the number of actively fishing trawlers sighted per unit of search effort per unit area. While the spatial resolution of fishery protection data are not as good as that collected from position loggers on vessels, the patrol aircraft do record details of all vessels, thereby avoiding bias due to scaling up data for a fleet to the level of the whole fishery (Jennings et al. 2000).

In European waters, the advent of the European Community Satellite Vessel Monitoring System (VMS), which began operating on 1 January 2000, has made it possible to track the activities of the whole North Sea beam trawl fleet greater than 24 m at 2-h intervals (Dann et al. 2001). This has allowed the study of fishing impacts on previously impossible scales (Figure 7). Such fine-scale information on the spatial distribution of effort is increasingly available in other fisheries (Murawski et al. 2000), although access to these data may be restricted. In the case of the European VMS, however, the position data do not indicate whether the vessels are actually fishing or not, and this has to be predicted from their speed.

Implications of Patchy Effort

In the absence of changes to management strategies that

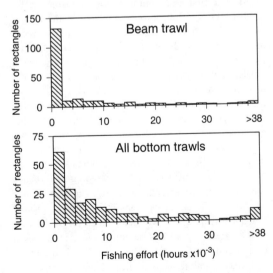

Figure 6. Frequency distribution of mean annual beam trawling effort (top panel) and total trawling effort (bottom panel) (1990–1995) by the International Council for the Exploration of the Sea statistical rectangles in the North Sea. Data from Jennings et al. (1999).

result in the relocation of effort, some consistency in spatial effort distribution is expected from year to year. This is because trawlers tend to return to tows that are stored on their navigation systems and known to be clear of obstructions. Management measures that modify the patchiness in trawling disturbance, or affect the persistence of spatial patterns in effort over time, may have different effects from those that encourage small areas to be fished repeatedly. To demonstrate the potential effects of changes in the spatial distribution of fishing effort, we have simulated the effects of various effort distributions on the production, mean P:B ratios, and mean turnover time of a benthic community in the central North Sea.

The analyses are based on observed relationships between trawling disturbance and production, P:B ratios, and turnover times in the central North Sea (Jennings et al. 2001a). Production was estimated in small areas subject to different levels of trawling disturbance. Despite the geographical proximity of these areas (typically within a few nautical mi of each other), the communities were very different, and the differences were attributable to the effects of trawling disturbance rather than sediment characteristics or depth (Duplisea et al. 2002). In these analyses, we make the simplifying assumptions that the habitat type in all patches is equal, that the communities in patches respond in the same way to trawling disturbance, that the effects of fishing on production in one area do not have secondary effects on another area, and that there are sufficient animals in other areas to sustain recruitment to the benthic community at all levels of fishing effort. The first assumption is not unreasonable for small areas of fishing ground. Indeed, in the central North Sea, we have recorded a 17.5-fold range in mean annual trawling disturbance on the same habitat (Jennings et al. 2002), and Figure 7 shows how intensively trawled areas are often located next to areas that are virtually unfished. The latter assumptions are unlikely to be valid, as ongoing population depletion will inevitably reduce recruitment success at some point, but for the purposes of this analysis, the assumption is conservative and we have probably underestimated the effects of fishing at the highest fishing intensities.

The effects of patchiness in trawling disturbance were assessed for different levels of mean disturbance and when different proportions of the fishing ground remain unfished. We assumed that effort within the trawled areas was always randomly distributed (based on Poisson distributions; Figure 8). This is a conservative assumption since studies of the patchiness of effort on real fishing grounds show that effort is significantly more patchy than random (Dinmore et al. 2003). We compared the effects of patchiness by simulating the underlying spatial distribution of trawling effort with a Poisson distribution or by assuming that the effort distribution was uniform. Using the Poisson is analogous to situations where most areas are trawled relatively infrequently while only a small proportion of the area is trawled often. The uniform distribution assumes that the same amount of total trawling effort is distributed evenly over the whole area.

When trawling effort is spread uniformly across a habitat, it causes greater decreases in P:B ratios and increases in turnover time than when the same mean effort is randomly (Poisson in our examples) distributed (Figure 9). It also causes greater decreases in production (Figure 10). This reflects the greater sensitivity of previously unfished habitats to disturbance than those that have already been disturbed. Indeed, the simulations suggest that if 20% of an area is accessible to fishing, a mean trawling disturbance of 10 times per year across the whole area will have lower cumulative impacts on production than when the whole area is accessible to trawling but fished only two times per year (Figure 10).

Since unfished areas sustain a high biomass and production of benthic fauna, they may have an important role in providing recruits to recovering areas. The simulations show that aggregate production in a system will be better sustained in a system where some unfished areas remain than in a system that is evenly trawled. Since trawling frequencies of greater than once per year would exterminate some of the largest macrofauna species (e.g., see the mortality rates in Bergman and van Santbrink [2000b] and potential populations growth rates in Brey [1999]), we can only assume that their persistence is due to their presence in infrequently trawled areas. It is also important for the persistence of intermediate-sized macrofauna that similar patterns of spatial heterogeneity in trawling effort are maintained from year to year. Indeed, if typical mean trawling frequencies (often 3–4 times per year) were applied uniformly in many shelf habitats, then large species with specific habitat requirements would be at risk of extinction.

The Effects of Management Actions on Effort Distributions

Our simplistic analyses have shown that changes in the spatial distribution and mean intensity of bottom fishing disturbance will determine the impacts of fishing. Since catch controls, effort controls, and technical measures (e.g., closed areas and mesh size restrictions) are likely to modify the spatial distribution of fishing activities, they will have consequences for the ecosystem as well as for target fish stocks.

Area closures on fishing grounds have a major impact on the distribution of fishing activity (Murawski et

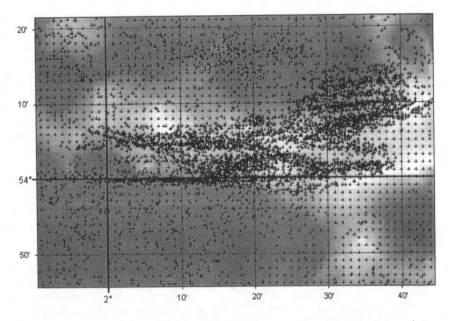

Figure 7. Spatial patterns of fishing effort in the Silver Pit, central North Sea, during 2001 as illustrated by satellite position records (crosses) and sightings per unit search effort data from fishery protection overflights (shading; darker shading = lower effort). The regular grids of satellite positions (shown as crosses) result when positions are reported at low resolution (nearest 1 min latitude and longitude) rather than high resolution (at least 0.1 min latitude and longitude). Analysis by Craig Mills and Richard Stocks from Parker et al. (unpublished).

al. 2000; Rijnsdorp et al. 2001). In the North Sea, the closure of the cod box came into force on 14 February 2001. An area of more than 103,600 km^2 (40,000 mi^2) of the North Sea, almost a fifth of its entire area, was closed to all fisheries likely to catch cod for a period of 12 weeks. Pelagic and sand eel fisheries were allowed to continue under observation. The areas closed included some of the main fishing grounds for the North Sea otter and beam trawl fleets. Most beam trawl vessels did not stop fishing but relocated to other fishing grounds (Rijnsdorp et al. 2001). This had two consequences. First, over the course of the whole year, the closing and then the opening of the cod box led to a more homogeneous spread of total effort (Dinmore et al. 2003). Second, beam trawlers fished in a number of areas that had rarely, if ever, been fished in recent years (Rijnsdorp et al. 2001). Some of these areas were known to harbor dense populations of large bivalves such as *Arctica islandica* and may have been important sources of recruitment to other areas. In one of the areas that was known to us before the cod box closure, displaced beam trawling effort led to marked reductions in the abundance of *A. islandica*, consistent with the high rates of mortality suffered by large bivalves when they are impacted by beam trawls (Bergman and van Santbrink 2000b). The response of the international fleet to closure of one part of the cod box is shown in Figure 11. Outside the closure period, the distribution of effort was similar in 2001 and 2002, but during the cod box period, effort was displaced to the west of the cod box onto grounds that are not normally impacted by fishing (Rijnsdorp et al. 2001; Dinmore et al. 2003).

Dinmore et al. (2003) used an existing size-based

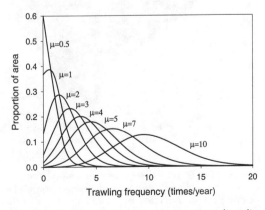

Figure 8. Examples of random (Poisson) trawling disturbance distributions for different values of mean effort.

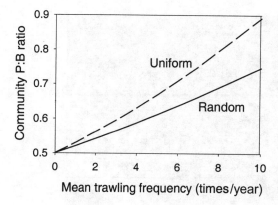

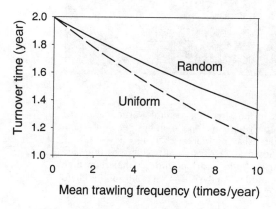

Figure 9. Predicted effect of different frequencies of trawling disturbance on infaunal community production:biomass ratios (top panel) and community turnover times (bottom panel) when disturbance is distributed uniformly or randomly.

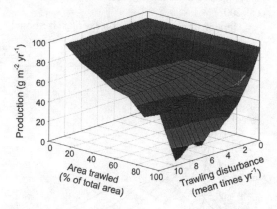

Figure 10. Predicted relationship among the frequency of trawling disturbance, the proportion of the fishing ground impacted by trawling, and the production of the infaunal community. Trawling disturbance is assumed to be randomly (Poisson) distributed within the area impacted by trawling. yr = year.

model of the impacts of fishing on benthic production (Duplisea et al. 2002) to estimate the mean annual production of benthic infauna greater than 0.04 g wet weight in a 30,000-nautical-mi^2 area of the North Sea, based on the distribution of effort observed with and without an annual cod box closure. Their results showed that mean production was reduced if the seasonal closure was implemented because effort would be distributed more widely and a smaller proportion of the area would remain unfished. The analysis was based on the assumption that the model of Duplisea et al. (2002) adequately described the response of the benthic community to trawling throughout the study area. This is unlikely to be true, but the approach provides the first large-scale test of the effects of a management action on benthic production. In the longer term, large-scale analyses of this type could be improved by developing a model that also accounts for the effects of primary production, sediment type, and depth on production.

Both with and without the temporary closure of the North Sea cod box, the distribution of trawling effort was still much patchier than even or random. Indeed, a uniform distribution of effort, akin to trawlers systematically lowing the seabed like a field, would have far greater consequences than the present distributions of effort. Our analyses show that fishery management strategies that lead to patchy but temporally stable patterns of trawling will often have lower total impacts on benthic communities than strategies that lead to more homogeneous and temporally unstable distribution of the same effort.

Managing the Ecosystem Consequences of Bottom Fishing

Toward Ecosystem-Based Management

Ecosystem-based management of fisheries means taking into account the ecosystem effects of fishing when setting management objectives. One way to achieve this is by maintaining aspects of ecosystem structure or function within boundaries that are deemed to be appropriate by society. In order to determine the success of management and to provide management targets, indicators can be used to track changes in ecosystem structure (Rice 2000). The selection of appropriate indicators is driven by science and the preferences and concerns of society. At present, the lists of indicators that have been proposed to assess the effects of bottom fishing disturbance are not comprehensive. Moreover, there is no science to show that keeping the indicators within target levels is sufficient to ensure that human activities are sustainable.

In the short term, it is unlikely that we will be able to quantify the intensities and distributions of bottom fishing disturbance that will lead to the large-scale failure of specific ecosystem functions such as nutrient recycling. However, we will be able to identify the intensities and distributions of disturbance that minimize the ecosystem consequences of bottom fishing. When setting management actions, it is consistent with the precautionary approach to select those actions that minimize the ecosystem consequences of disturbance. So, when alternative management actions meet the management objectives in relation to other issues, the ecosystem consequences of the alternate actions should be assessed and the action that minimizes consequences should be favored.

One step that can be made toward ecosystem-based fisheries management is the a priori assessment of the effects of fisheries management actions such as catch controls, effort controls, and technical measures on the ecosystem. For the ecosystem impacts of fishing we have considered here (benthic production, P:B ratios, and turnover time), we can propose an approach for predicting the effects of management actions (Table 1).

This approach requires information on the spatial distribution of habitats and fishing effort. First, a model of fleet responses to management measures would be developed and used to predict the spatial and temporal redistribution of effort following management action. For bottom impacts, the model would account for redistribution of effort in relation to the expected costs and profits from fishing and the distribution of habitat types that could be fished. Second, the predicted levels of bottom

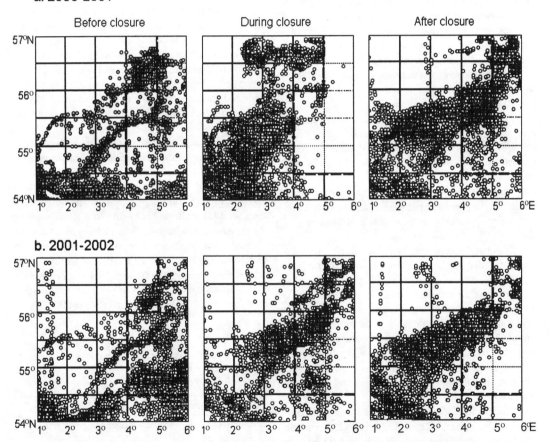

Figure 11. Relocation of effort following the imposition of a temporarily closed area, the cod box, in the North Sea. The upper panels (a) show the location and frequency of satellite records of beam trawlers in a period of 75 d prior to cod box closure (30 November–13 February 2001), during closure (14 February–30 April 2001), and over 75 d after closure (1 May–15 July 2001). The lower panels (b) show the location and frequency of satellite records of beam trawlers in the corresponding periods 1 year later, when the cod box was not closed. The dashed lines indicate the western boundary of the cod box. The grid lines show International Council for the Exploration of the Sea rectangles. Data from Dinmore et al. (2003).

fishing disturbance would be used as inputs to models that predict the effects of disturbance on the ecosystem processes of interest. Finally, the ecosystem effects of a range of potential management options would be compared. In the case of assessing the effects of area closure, such a procedure would be relatively straightforward to implement if habitat mapping had been completed.

An appropriate structure for models that predict responses to area closure already exists (e.g., Holland and Sutinen 1999), but the models need to be examined in new ways to determine how catch and effort controls, and access to improved data on seabed habitats, can influence the spatial distribution of fishing vessels.

Data-Poor Management

We have proposed an approach for predicting the ecosystem consequences of management actions (Table 1). This approach relies on habitat maps, models of fleet responses to management action, and models of the ecosystem consequences of bottom trawling disturbance. Model simulations would be used to produce a decision table, quantifying the consequences of alternate management actions, so that actions that minimize the ecosystem effects of fishing could be identified.

Complex information on fleet dynamics and the ecosystem consequences of fishing may not be available in many fisheries. However, if habitat has been mapped, it is still possible to give useful management advice. Thus, we suggest that management strategies that maintain or maximize the patchiness of effort within habitat types are more consistent with the precautionary approach than those that lead to more uniform fishing effort distributions.

In general terms, managers should aim to manage spatial distributions of effort as much as total effort. There will still need to be decisions as to what society wants the ecosystem to look like, but because the initial impacts of fishing on most aspects of ecosystem structure and function are usually the most severe, the largest changes in ecosystem structure and function are minimized by patchy effort distributions. Patchy distributions of effort could be formalized by permanent area closures, but may be reduced by rotating and temporary area closures.

The methods of comparative risk assessment proposed by the National Research Council (2002) would also be appropriate for analyzing the risk to benthic habitats when detailed habitat maps and effort data are not available. Comparative risk assessment is a qualitative method that compares different types of risk and ranks them. It can be used when scientific knowledge is incomplete because it relies on a combination of available data, scientific inference, and public values (National Research Council 2002).

Role of Habitat Mapping

Mapping the spatial extent of marine habitats is fundamental to understanding the large-scale impacts of bottom fishing and identifying management strategies to minimize them. The habitats of many shelf areas still have not been mapped at a sufficiently high resolution for this purpose (typically classified in cells of 1 km^2 or less).

Many habitat studies have used conventional survey techniques such as grabs, dredges, and underwater cameras to describe the extent and composition of benthic fauna associated with the seabed (Basford et al. 1990; Kaiser et al. 1999; Rees et al. 1999; Ellis et al. 2000). These techniques provide information that is site specific, and extrapolation of these data to previously unsampled areas is subjective without more detailed sampling. In particular, the use of spatial statistics (e.g., interpolation techniques such as kriging) to make inferences about habitat distribution may be statistically unreliable because biological sample stations are often widely spaced on heterogeneous seabeds. If, however, the links between benthic species and their physical habitat components (e.g., sediment grain size) could be established, then extrapolation of data into unsampled areas could be based on the presence or absence of these physical components.

A number of acoustic systems have been used to map seabed properties. These include (1) broad-acoustic beam systems (e.g. side-scan sonar), (2) single-beam acoustic ground-discrimination systems (e.g. QTC VIEW and RoxAnn), (3) multiple-beam swath bathymetry systems, and (4) multiple-beam side-scan sonar systems. Recent advances in the power and reliability of these acoustic mapping technologies have provided an effective tool to delineate and characterize the habitat features of the seabed that are important to demersal fish and benthic faunas (Magorrian et al. 1995; Greenstreet et al. 1997; Kostylev et al. 2001; Freeman and Rogers 2003).

Table 1. An assessment and management process to minimize the ecosystem consequences of bottom fishing disturbance.

Stage	Action
1	Describe the spatial distribution of habitat types.
2	Describe the spatial distribution of fishing effort.
3	Model ecosystem consequences of different effort distributions.
4	Model fleet responses to management action.
5	Produce decision table listing ecosystem consequences of alternative management actions.

Although acoustics provide high-resolution information on seabed structure, they have a number of limitations. Multi-beam sonar, for example, can provide detailed bathymetric profiles of the seabed that can be used to identify gravel veneers, sand waves, and pot marks, but it will only provide limited seabed classification. Moreover, large-scale application of multi-beam sonar may be too costly in time and money because data acquisition and processing are intensive. Single-beam acoustic systems, conversely, provide a higher degree of seabed classification, and the reduced deployment time and more rapid data processing capabilities make them more suitable for the large-scale surveys that would be required to support a move toward ecosystem-based fishery management. However, single-beam acoustic systems lack 100% coverage of the seafloor. All acoustic systems require ground-truthing, and when acoustic classification information is extrapolated to a new geographical location, further ground-truthing is required locally to confirm seabed type. This is partly due to the effects of bathymetry, geological deposits/bedforms, and the presence of structurally complex sessile organisms.

Conclusions

The initial effects of fishing have greater ecosystem consequences than repeated fishing in fished areas. As a result, consistently patchy fishing effort distributions have lower total impacts on the ecosystem than random or uniform effort distributions. In most fisheries, the distribution of annual fishing effort within habitats is much patchier than random, and patterns of effort are maintained from year to year. This is why some vulnerable, free-living, and habitat-forming species have persisted in heavily fished shelf seas.

The adverse impacts of fishing on benthic habitats have to be considered as part of the ecosystem approach to fisheries management. One step that can be taken toward ecosystem-based fisheries management is to make an a priori assessment of the effects of proposed management actions on benthic habitat. This requires information on habitat distributions, models to predict changes in the spatial distribution of fleets following management action, and models of the impacts of trawling disturbance on ecosystem processes. The information would be used to run simulations and produce a decision table, quantifying the consequences of alternate management actions. This would allow actions that minimize the ecosystem effects of fishing to be identified. In data-poor situations, management strategies that maintain or maximize the patchiness of effort within habitat types are more consistent with the precautionary approach. Patchy distributions of effort could be formalized by area closure.

Acknowledgments

We thank the U.S. Geological Survey, National Oceanic and Atmospheric Administration, American Fisheries Society, and Ecological Society of America for their invitation to attend the Symposium on the Effects of Fishing Activities on Benthic Habitats, Craig Mills and Richard Stocks for producing Figure 7, John Crawford and the anonymous referees for helpful comments on the manuscript, and the Department for Environment, Food, and Rural Affairs, UK, for funding this research (MF0731, AE1224).

References

Aller, R. C. 1982. The effects of macrobenthos on chemical properties of marine sediment and overlying water. Pages 53–102 in P. L. McCall and M. J. S. Tevesz, editors. Animal–sediment relations. Plenum, New York.

Aller, R. C. 1988. Benthic fauna and biogeochemical processes in marine sediments: the role of burrow structures. Pages 301–340 in T. H. Blackburn and J. Sorensen, editors. Nitrogen cycling in coastal marine environments. Wiley, New York.

Asmus, R. 1986. Nutrient flux in short-term enclosures of intertidal sand communities. Ophelia 26:1–18.

Auster, P. J., and R. W. Langton. 1999. The effects of fishing on fish habitat. Pages 150–187 in L. R. Benaka, editor. Fish habitat: essential fish habitat and rehabilitation. American Fisheries Society, Symposium 22, Bethesda, Maryland.

Auster, P. J., R. J. Malatesta, R. W. Langton, L. Watling, P. C. Valentine, C. L. Donaldson, E. W. Langton, A. N. Shepard, and I. G. Babb. 1996. The impacts of mobile fishing gear on seafloor habitats in the Gulf of Maine (northwest Atlantic): implications for conservation of fish populations. Reviews in Fisheries Science 4:185–202.

Banse, K., and S. Mosher. 1980. Adult body mass and annual production/biomass relationships of field populations. Ecological Monographs 50:355–379.

Basford, D., A. Eleftheriou, and D. Raffaelli. 1990. The infauna and epifauna of the northern North Sea. Netherlands Journal of Sea Research 25:165–173.

Bergman, M. J. N., and J. W. van Santbrink. 2000a. Fishing mortality of populations of megafauna in sandy sediments. Pages 49–68 in M. J. Kaiser and S. J. de Groot, editors. Effects of fishing on non-target species and habitats: biological, conservation and socio-economic issues. Blackwell Scientific Publications, Oxford, UK.

Bergman, M. J. N., and J. W. van Santbrink. 2000b. Mortality in megafaunal benthic populations caused by trawl fisheries on the Dutch continental shelf in the North Sea in 1994. ICES Journal of Marine Science 57:1321–1331.

Blackburn, T. H. 1997. Release of nitrogen compounds following resuspension of sediment: model predictions. Journal of Marine Systems 11:343–352.

Brey, T. 1999. Growth performance and mortality in aquatic macrobenthic invertebrates. Advances in Marine Biology 35:153–223.

Chavez, F. P. 1996. Forcing and biological impact of the 1992 El Niño in central California. Geophysical Research Letters 23:265–268.

Churchill, J. H. 1989. The effect of commercial trawling on sediment resuspension and transport over the Middle Atlantic Bight continental shelf. Continental Shelf Research 9:841–864.

Clavero, V., F. X. Niell, and J. A. Fernandez. 1994. A laboratory study to quantify the influence of Neris diversicolor O. F. Mueller in the exchange of phosphate between sediment and water. Journal of Experimental Marine Biology and Ecology 176:257–267.

Collie, J. S., G. A. Escanero, and P. C. Valentine. 1997. Effects of bottom fishing on the benthic megafauna of Georges Bank. Marine Ecology Progress Series 155:159–172.

Collie, J. S., S. J. Hall, M. J. Kaiser, and I. R. Poiner. 2000. A quantitative analysis of fishing impacts on shelf sea benthos. Journal of Animal Ecology 69:785–798.

Dann, J., R. Millner, and R. De Clerck. 2002. Alternative uses of data from satellite monitoring of fishing vessel activity in fisheries management: II. Extending cover to areas fished by UK beamers. Centre for Environment, Fisheries, and Aquaculture Science, Report of EC Project 99/002, Lowestoft, UK.

Dayton, P. K., S. F. Thrush, M. T. Agardy, and R. J. Hofman. 1995. Environmental effects of marine fishing. Aquatic Conservation 5:205–232.

Dinmore, T. A., D. E. Duplisea, B. D. Rackham, D. L. Maxwell, and S. Jennings. 2003. Impact of a large-scale area closure on patterns of fishing disturbance and the consequences for benthic production. ICES Journal of Marine Science 60:371–380.

Duplisea, D. E. 1998. Feedbacks between benthic carbon mineralization and community structure: a simulation-model analysis. Ecological Modelling 110:19–44.

Duplisea, D. E., S. Jennings, S. J. Malcolm, R. Parker, and D. Sivyer. 2001. Modelling the potential impacts of bottom trawl fisheries on soft sediment biochemistry in the North Sea. Geochemical Transactions 14:1–6.

Duplisea, D. E., S. Jennings, K. J. Warr, and T. A. Dinmore. 2002. A size-based model to predict the impacts of bottom trawling on benthic community structure. Canadian Journal of Fisheries and Aquatic Sciences 59:1785–1795.

Ecosystem Principles Advisory Panel. 1999. Ecosystem-based fishery management. National Marine Fisheries Service, Silver Spring, Maryland.

Ellis, J. R., S. I. Rogers, and S. M. Freeman. 2000. Demersal assemblages in the Irish Sea, St George's Channel and Bristol Channel. Estuarine, Coastal and Shelf Science 51:299–315.

Fanning, K. A., K. L. Carder, and P. R. Betzer. 1982. Sediment resuspension by coastal waters: a potential mechanism for nutrient re-cycling on the oceans margins. Deep-Sea Research 29:953–965.

Freeman, S. M., and S. I. Rogers. 2003. A new analytical approach to the characterisation of macro-epibenthic habitats: linking species to the environment. Estuarine, Coastal and Shelf Science 56:749–764.

Gilkinson, K., M. Paulin, S. Hurley, and P. Schwinghamer. 1998. Impacts of trawl door scouring on infaunal bivalves: results of a physical trawl door model/dense sand interaction. Journal of Experimental Marine Biology and Ecology 224:291–312.

Greenstreet, S. P. R., I. D. Tuck, G. N. Grewar, E. Armstrong, D. G. Reid, and P. J. Wright. 1997. An assessment of the acoustic survey technique, RoxAnn, as a means of sampling seabed habitats. ICES Journal of Marine Science 54:939–959.

Hall, S. J. 1999. The effects of fishing on marine ecosystems and communities. Blackwell Scientific Publications, Oxford, UK.

Hall-Spencer, J., V. Allain, and J. H. Fossa. 2002. Trawling damage to northeast Atlantic ancient coral reefs. Proceedings of the Royal Society of London 269B:507–511.

Henriksen, K., M. B. Rasmussen, and A. Jensen. 1983. Effect of bioturbation on microbial nitrogen transformations in the sediment and fluxes of ammonium and nitrate to the overlying water. Ecological Bulletin 35:193–205.

Holland, D. S., and J. G. Sutinen. 1999. An empirical model of fleet dynamics in New England trawl fisheries. Canadian Journal of Fisheries and Aquatic Sciences 56:253–264.

Huettel, M. 1990. Influence of the lugworm Arenicola marina on porewater nutrient profiles of sand flat sediments. Marine Ecology Progress Series 62:241–248.

Huettel, M., and G. Gust. 1992. Impact of bioroughness on interfacial solute exchange in permeable sediments. Marine Ecology Progress Series 89:253–267.

Jennings, S., J. Alvsvåg, A. J. Cotter, S. Ehrich, S. P. R. Greenstreet, A. Jarre-Teichmann, N. Mergardt, A. D. Rijnsdorp, and O. Smedstad. 1999. Fishing effects in northeast Atlantic shelf seas: patterns in fishing effort, diversity and community structure. III. International fishing effort in the North Sea: an analysis of temporal and spatial trends. Fisheries Research 40:125–134.

Jennings, S., T. A. Dinmore, D. E. Duplisea, K. J. Warr, and J. E. Lancaster. 2001a. Trawling disturbance can modify benthic production processes. Journal of Animal Ecology 70:459–475.

Jennings, S., and M. J. Kaiser. 1998. The effects of fishing on marine ecosystems. Advances in Marine Biology 34:201–352.

Jennings, S., M. D. Nicholson, T. A. Dinmore, and J. E. Lancaster. 2002. Effects of chronic trawling disturbance on the production of infaunal communities. Marine Ecology Progress Series 243:251–260.

Jennings, S., J. K. Pinnegar, N. V. C. Polunin, and K. J.

Warr. 2001b. Impacts of trawling disturbance on the trophic structure of benthic invertebrate communities. Marine Ecology Progress Series 213:127–142.

Jennings, S., K. J. Warr, S. P. R. Greenstreet, and A. J. Cotter. 2000. Spatial and temporal patterns in North Sea fishing effort. Pages 3–14 in M. J. Kaiser and S. J. de Groot, editors. Effects of fishing on non-target species and habitats: biological conservation and socio-economic issues. Blackwell Scientific Publications, Oxford, UK.

Kaiser, M. J., J. S. Collie, S. J. Hall, S. Jennings, and I. R. Poiner. 2002. Modification of marine habitats by trawling activities: prognosis and solutions. Fish and Fisheries 3:114–136.

Kaiser, M. J., and S. J. de Groot. 2000. The effects of fishing on non-target species and habitats: biological, conservation and socio-economic issues. Blackwell Scientific Publications, Oxford, UK.

Kaiser, M. J., S. I. Rogers, and J. R. Ellis. 1999. Importance of benthic habitat complexity for demersal fish assemblages. Pages 212–223 in L. R. Benaka, editor. Fish habitat: essential fish habitat and rehabilitation. American Fisheries Society, Symposium 22, Bethesda, Maryland.

Kaiser, M. J., and B. E. Spencer. 1996. The effects of beam-trawl disturbance on infaunal communities in different habitats. Journal of Animal Ecology 65:348–358.

Koslow, J. A., K. Gowlett-Holmes, J. K. Lowry, T. O'Hara, G. C. B. Poore, and A. Williams. 2001. Seamount benthic macrofauna off southern Tasmania: community structure and impacts of trawling. Marine Ecology Progress Series 213:111–125.

Kostylev, V. E., B. J. Todd, G. B. Fader, R. C. Courtney, G. D. Cameron, and R. A. Pickrill. 2001. Benthic habitat mapping on the Scotian Shelf based on multibeam bathymetry, surficial geology and sea floor photographs. Marine Ecology Progress Series 219:121–137.

Krost, P. 1990. The impact of otter-trawl fishery on nutrient release from the sediment and macrofauna of Kiel Bay (western Baltic). Institut fur Meereskunde, Report 200, Germany.

Langton, R. W., and W. E. Robinson. 1990. Faunal associations on scallop grounds in the western Gulf of Maine. Journal of Experimental Marine Biology and Ecology 144:157–171.

Lindeboom, H. J., and S. J. de Groot. 1998. The effects of different types of fisheries on the North Sea and Irsh Sea benthic ecosystems. Netherlands Institute of Sea Research, Texel, The Netherlands.

Link, J. S. 2002. What does ecosystem-based fisheries management mean? Fisheries 27(4):18–21.

Magorrian, B. H., M. Service, and W. Clarke. 1995. An acoustic bottom classification survey of Strangford Lough, Northern Ireland. Journal of the Marine Biological Association of the United Kingdom 75:987–992.

Marinelli, R. L. 1992. Effects of polychaetes on silicate dynamics and fluxes in sediments: importance of species, animal activity, and polychaete effects on benthic diatoms. Journal of Marine Research 50:745–779.

Marinelli, R. L. 1994. Effects of burrow ventilation on activities of a terebellid polychaete and silicate removal from silicate pore waters. Limnology and Oceanography 34:559–577.

Mayer, L. M., D. F. Schick, R. H. Findlay, and D. L. Rice. 1991. Effects of commercial dragging on sedimentary organic matter. Marine Environmental Research 31:249–261.

Murawski, S. A., R. Brown, H. L. Lai, P. J. Rago, and L. Hendrickson. 2000. Large-scale closed areas as a fishery management tool in temperate marine systems: the Georges Bank experience. Bulletin of Marine Science 66:775–798.

National Research Council. 2002. Effects of trawling and dredging on seafloor habitat. National Academy Press, Washington, D.C.

Palanques, A., J. Guillen, and P. Puig. 2001. Impact of bottom trawling on water turbidity and muddy sediment of an unfished continental shelf. Limnology and Oceanography 46:1100–1110.

Parker, R., S. J. Malcolm, and S. Jennings. 1999. The impact of disturbance on sediment ecosystem function and the quality of the coastal environment. Contract report to the Ministry of Agriculture, Fisheries and Food, UK, London.

Pelegri, S. P., L. P. Nielson, and T. H. Blackburn. 1994. Denitrification in estuarine sediment stimulated by the irrigation activity of the amphipod *Corophium volutator*. Marine Ecology Progress Series 105:285–290.

Percival, P., and C. Frid. 2000. The impact of fishing disturbance on benthic nutrient regeneration and flux rate. International Council for the Exploration of the Sea, C.M. 2000/Z:07, Copenhagen.

Pilskaln, C. H., J. H. Churchill, and L. M. Mayer. 1998. Resuspension of sediment by bottom trawling in the Gulf of Maine and potential geochemical consequences. Conservation Biology 12:1223–1229.

Pilskaln, C. H., J. B. Paduan, F. P. Chavez, R. Y. Anderson, and W. M. Berelson. 1996. Carbon export and regeneration in the coastal upwelling system of Monterey Bay, central California. Journal of Marine Research 54:1149–1178.

Rees, H. L., M. A. Pendle, R. Waldock, D. S. Limpenny, and S. E. Boyd. 1999. A comparison of benthic biodiversity in the North Sea, English Channel and Celtic Seas. ICES Journal of Marine Science 56:228–246.

Rice, J. C. 2000. Evaluating fishery impacts using metrics of community structure. ICES Journal of Marine Science 57:682–688.

Riesen, W., and K. Riese. 1982. Macrobenthos of the subtidal Wadden Sea: revisited after 55 years. Helgolander Meeresuntersuchungen 35:409–423.

Rijnsdorp, A. D., A. M. Bujis, F. Storbeck, and E. Visser. 1998. Micro-scale distribution of beam trawl effort in the southern North Sea between 1993 and 1996 in relation to the trawling frequency of the sea bed and the distribution of benthic organisms. ICES Journal of Marine Science 55:403–419.

Rijnsdorp, A. D., G. J. Piet, and J. J. Poos. 2001. Effort allocation of the Dutch beam trawl fleet in response to a temporary closed area in the North Sea. International Council for the Exploration of the Sea, C.M. 2001/N:01, Copenhagen.

Schratzberger, M., T. A. Dinmore, and S. Jennings. 2002. Impacts of trawling disturbance on the biomass and community structure of meiofauna. Marine Biology 14:83–93.

Schratzberger, M., and S. Jennings. 2002. Impacts of chronic trawling disturbance on meiofaunal communities. Marine Biology 141:991–1000.

Schwinghamer, P., B. Hargrave, D. Peer, and C. M. Hawkins. 1986. Partitioning of production and respiration among size groups of organisms in an intertidal benthic community. Marine Ecology Progress Series 31:131–142.

Watling, L., R. H. Findlay, L. M. Mayer, and D. F. Schick. 2001. Impact of a scallop dredge on the sediment chemistry, microbiota, and faunal assemblages of a shallow subtidal marine benthic community. Journal of Sea Research 46:309–324.

Ziebis, W., S. Forster, M. Huettel, and B. B. Jorgensen. 1996. Complex burrows of the mud shrimp *Callianassa truncata* and their geochemical impact in the sea bed. Nature 382:619–622.

National Research Council Study on the Effects of Trawling and Dredging on Seafloor Habitat

JOHN STEELE[1]

*Woods Hole Oceanographic Institution, Marine Policy Center, MS 41,
Woods Hole, Massachusetts 02543-1138, USA*

SUSAN J. ROBERTS[2]

*Ocean Studies Board, NA-752, National Research Council, The National Academies,
500 5th Street, Northwest, Washington, DC 20001, USA*

DAYTON L. ALVERSON

Natural Resource Consultants, 1900 West Nickerson Street, Suite 207, Seattle, Washington 98119, USA

PETER AUSTER

*National Undersea Research Center and Department of Marine Sciences, University of Connecticut,
1084 Shennecossett Road, Groton, Connecticut 06340, USA*

JEREMY COLLIE

*University of Rhode Island, Graduate School of Oceanography, Narragansett Bay Campus,
South Ferry Road, Narragansett, Rhode Island 02882, USA*

JOSEPH T. DEALTERIS

University of Rhode Island, Department of Fisheries and Aquaculture, Kingston, Rhode Island 02881, USA

LINDA DEEGAN

*Marine Biological Laboratory, The Ecosystems Center, 7 MBL Street,
Woods Hole, Massachusetts 02543, USA*

ELVA ESCOBAR BRIONES

*Universidad Nacional Autonoma de Mexico, Instituto de Ciencias del Mar y Limnologia,
A.P. 70-305 Ciudad Universitaria 04510, Mexico*

STEPHEN J. HALL

Australian Institute of Marine Science, PMB No 3, Townsville MC, Queensland 4810, Australia

GORDON H. KRUSE

*University of Alaska, Fairbanks, School of Fisheries and Ocean Sciences, 11120 Glacier Highway, Juneau,
Alaska 99801-8677, USA*

---

[1] E-mail: jsteele@whoi.edu; chaired the committee.
[2] E-mail: sroberts@nas.edu; study director for the committee.

CAROLINE POMEROY

*University of California, Santa Cruz, Institute of Marine Science, 1156 High Street,
A316 Earth and Marine Science Building, Santa Cruz, California 95064, USA*

KATHRYN M. SCANLON

*U.S. Geological Survey, Coastal and Marine Geology Program, 384 Woods Hole Road,
Woods Hole, Massachusetts 02543, USA*

PRISCILLA WEEKS

Houston Advanced Research Center, 4800 Research Forest Drive, The Woodlands, Texas 77381, USA

Abstract. This paper summarizes the results of the National Research Council Study on the Effects of Bottom Trawling and Dredging on Seafloor Habitat (National Research Council 2002). The report concludes that integration of existing data on the effects of trawls and dredges, level of fishing effort, and distribution of seafloor habitats would facilitate development of habitat management plans. Current and new management measures should be assessed regularly to provide a better understanding of how various restrictions affect fish habitat and to determine the socioeconomic impacts on the fishing industry and local communities. Resolution of the different, and at times conflicting, ecological and socioeconomic goals will require not only a better understanding of the relevant ecosystems and fisheries but also more effective interaction among stakeholders.

Introduction

Fishing has a variety of effects on marine habitats and ecosystems, depending on the type of gear, the level of fishing effort, and the spatial extent of fishing. After passage of the Sustainable Fisheries Act in 1996, interest in the impacts of fishing on the seafloor increased because fishery management plans were now required to address the effects of fishing on habitat.

At the request of the National Marine Fisheries Service, the Ocean Studies Board of the National Research Council (NRC) convened a committee to undertake a review of the scale of trawl and dredge fisheries and their effects on various types of bottom habitat. Specifically, the committee was given the following statement of task: (1) summarize and evaluate existing knowledge on the effects of bottom trawling on the structure of seafloor habitats and on the abundance, productivity, and diversity of bottom-dwelling species in relation to gear type and trawling method, frequency of trawling, bottom type, species, and other important characteristics; (2) summarize and evaluate knowledge about changes in seafloor habitats associated with trawling and with the cessation of trawling; (3) summarize and evaluate research on the indirect effects of bottom trawling on non-seafloor species; (4) recommend how existing information could be used more effectively in managing trawl fisheries; and (5) recommend research to improve understanding of the effects of bottom trawling on seafloor habitats.

The NRC study examined only gear that is towed across the seabed. Stationary bottom gear may also affect seafloor ecosystems, but consideration of those types of gear impacts were beyond the scope of this study. The committee met with scientists, managers, fishermen, and environmentalists in three regions around the United States (Boston, Massachusetts; Galveston, Texas; and Anchorage, Alaska) to gather information about the local fisheries. The committee's report (National Research Council 2002) contains a review of the literature on trawling and dredging effects, an inventory of trawl and dredge fishing activities, suggestions for future research, and recommendations for using this information to manage the effects of bottom gear on seafloor habitat.

The policy context for addressing the effects of fishing on habitat is found in Essential Fish Habitat (EFH) provisions specified in the Sustainable Fisheries Act. The Sustainable Fisheries Act requires regional fishery management councils to (1) describe and identify EFH for each fish stock managed under a fishery management plan, (2) minimize, to the extent practicable, adverse effects on such habitat caused by fishing, and (3) identify other actions to encourage the conservation and enhancement of such habitat.

The regional councils found it difficult to develop criteria for designating EFH due to gaps in existing knowledge on the distribution of benthic life stages of fishes and other species and the physical and biological characteristics of the seafloor. Similarly, the councils struggled with the requirement to assess the effects of bottom trawling and dredging because they had insufficient data on the spatial scale and extent of bottom fishing effort and lacked guidelines for generalizing the results of research on specific gears and habitats. These problems relate to the committee's task to recommend ways for using existing information in the management of the habitat effects of trawl and dredge fisheries. This paper briefly summarizes the report's review of the effects of trawling and dredging, then discusses a framework for decision making to facilitate management, and ends with a description of the committee's recommendations. More detailed information is available in the full report (National Research Council 2002).

Review of Trawling and Dredging Effects

A complete assessment of the ecosystem effects of trawling and dredging requires three types of information: (1) gear-specific effects on different habitat types (obtained experimentally); (2) the frequency and geographic distribution of bottom tows (trawl and dredge fishing effort data); and (3) the physical and biological characteristics of seafloor habitats in the fishing grounds (seafloor mapping).

Most research programs have examined changes in seafloor structure and biological communities after experimental disturbance by various types of mobile fishing gear. The results document acute effects that may be categorized by gear type, habitat characteristics, composition of the benthic community, and frequency of disturbance. However, these studies analyze a relatively small affected area. To convert these results into an assessment of the ecosystem-level effects on seafloor habitats requires analysis of the frequency of bottom trawl and dredge activity and fine-scale mapping of this effort relative to the geography of seafloor habitats in the fishing grounds. Few studies have been of sufficient duration to document the effects of chronic disturbance by mobile bottom gear that would provide a more accurate assessment of the effects of habitat disturbance on the productivity of commercial and recreational fisheries.

Gear-Specific Effects

Many studies on gear effects provide evidence in support of the ecological theory that stable communities of low-mobility, long-lived species are more vulnerable to acute and chronic physical disturbance than are short-lived species in changeable environments. Habitat complexity is reduced by towed bottom gear that removes or damages biological and physical structures. The extent of the initial effects depends on the type of towed gear and the stability of the habitat. Dredging typically causes more initial damage than otter or beam trawling. The more stable biogenic, gravel, and mud habitats experience the greatest changes and slowest recovery rates. In contrast, less consolidated coarse sediments in areas of high natural disturbance show fewer initial effects. Because these habitats tend to be populated by opportunistic species that recolonize more rapidly, recovery is also faster.

Frequency and Geographic Distribution of Trawl and Dredge Effort

Bottom trawl or dredge gears are commonly used in fisheries around the United States, but the frequency varies widely (Table 1). Data on the geographic distribution and frequency of trawling and dredging suffer from limitations in the spatial resolution of the data, regional variation in reporting methods, and inconsistency in recording methods used from year to year in some regions. In some regions, trawling effort data (defined as the number and duration of tows made in a particular area over a specified period) are averaged over reporting areas as small as 25 km^2, while in other regions, the reporting areas may be 100 times larger. Even in heavily trawled regions, effort is not evenly distributed. As a consequence, some areas may be trawled several times per year while other areas are trawled infrequently, if at all. Over the past decade, there has been a decrease in the level of seabed disturbance by trawl and dredge fishing as reductions in fishing effort, area closures, and gear restrictions have been instituted by managers to rebuild fish stocks, reduce bycatch, and minimize

Table 1. Summary data on fishing effort in three regions: Gulf of Mexico offshore shrimp fishery; New England bottom trawl fisheries (groundfish); and Alaska bottom trawl fisheries (groundfish). Details and references are given in the report (NRC 2002). mt = metric tons.

Region	Crew/vessel size (Number of vessels)	Landed catch (Value)	% fishing grounds trawled per year
Gulf of Mexico	1–3/7.6–26 m (3,600)	94,800 mt ($440 million)	255%
New England	2–7/14–30 m (750)	81,200 mt ($132 million)	115%
Alaska	3–16/17–47 m (200)	39,700 mt ($104 million)	53%

interactions with endangered species (Figure 1, see case studies below).

Habitat Mapping

The largest information gap for evaluating the effects of fishing on habitat is the lack of high-resolution seafloor mapping data in trawled or dredged areas. For most areas, only coarse maps are available on habitat distribution. This mismatch in the spatial scales of experimental results, habitat maps, and trawl effort reporting data makes it difficult to accurately assess the ecosystem level effects of trawling and dredging. Also, the lack of a standardized classification system for seabed habitat makes it difficult to assess the status of habitats in different regions. Because habitat disturbance has been studied primarily at small spatial scales using observations of short-term, acute disturbance, there are few analyses of the landscape-scale impacts of trawling and dredging.

Framework for Decision Making

Risk Evaluation in Decision Making

The decision-making process used to identify and evaluate risks to seafloor habitats and prioritize management actions may be described in terms of risk assessment, risk management, and management actions (Figure 2). Ecological risk assessment represents the first step in the decision-making process and is fundamentally a scientific undertaking based on research and analysis. Risk management applies the results of both the ecological assessment and a social science assessment of the social, economic, and institutional factors that define the issue. The social science assessment characterizes fisheries in terms of gear use, vessel and crew size, landing value, and existing restrictions on fishing. In addition to the ecological and social science assessments, risk management incorporates social values and legal mandates.

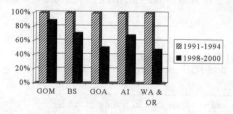

Figure 1. Effort reduction during the 1990s. The percentage decline in fishing effort (number of tows or fishing days) were calculated for the Gulf of Mexico (GOM), Bering Sea (BS), Gulf of Alaska (GOA), Aleutian Islands (AI), and Washington and Oregon (WA and OR). The 1998–2000 effort is depicted relative to the effort level in the 1991–1994 period.

In the final stage of decision making, a management strategy is chosen and regulations are enacted. At this stage, ecological and social sciences once again become important for developing new approaches to solve the problem or in monitoring the chosen measures. If the effects are considered large enough, new regulations may be adopted and research programs may be initiated to clarify uncertainties, especially questions about indirect effects on the resource, on other parts of the ecosystem, and on the response of the fisheries to new regulations. Decision making reflects societal valuation, the domain of policymakers and stakeholders, as well as scientific evaluation.

Ecological Risk Assessment

Ecological risk assessment is "the characterization of the adverse ecological effects of environmental exposures to hazards imposed by human activities" (National Research Council 1993). Two approaches to ecological risk assessment were examined for their utility in the management of benthic habitats. The exposure assessment model is a quantitative approach that has been borrowed from the fields of human health and toxicology and focuses on each risk individually. In this case, trawling represents the risk, and the dose corresponds to the amount of fishing effort in a given area. Response is assessed as the damage to benthic communities, such as a reduction in biomass or species diversity. The second method is comparative risk assessment. This method compares multiple types of risks and addresses the larger question of how best to minimize habitat degradation in its entirety (versus focusing on only one type of stressor). Mobile bottom gear is only one of many factors contributing to the degradation of benthic habitats; the comparative approach provides a method for simultaneous consideration of a wide range of risks to benthic habitats. Other factors might include pollution, drilling for mineral resources, and natural disturbance. In addition to data, comparative risk analysis relies on expert judgment, scientific inference, and deliberation. This method does not provide quantitative information on relative risk from different stressors.

Comparative risk assessment has particular value when there are gaps in the information necessary to complete an exposure assessment. Although fisheries managers continually collect data to improve decision making, limitations in resources and time require managers to assess the effects of fishing in the absence of complete information. In this context, comparative risk assessment provides an approach for evaluating the effects of bottom trawling and dredging. This method brings together the various stakeholders to identify risks to seafloor habitats and prioritize management actions within the context of current statutes. Stakeholder involvement improves the sharing of information among different groups and aids in the development of solutions that have broad societal support.

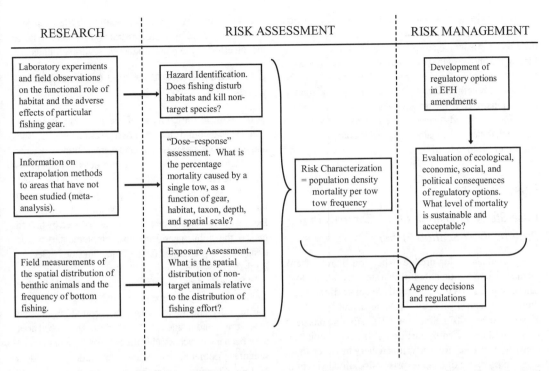

Figure 2. Elements of risk assessment and risk management (modified from National Research Council 1993).

Risk Management

Risk management requires an appreciation of the characteristics of the fishery and the social and legal contexts. Fisheries in different regions and for different species vary dramatically in their use of gear, size of vessel, number of crew, and value of catch (Table 1). These characteristics will affect what types of restrictions will be effective for a particular fishery. For example, increased observer coverage is not a practical option for small-boat fisheries that may employ as few as one or two crew members. Monitoring of these fisheries may be improved by satellite-based remote vessel monitoring systems. Similarly, a high-value fishery may be more amenable to absorbing the costs of changing gear configurations than a fishery with a small profit margin.

Management Actions

The last decision-making step is the institution of new management practices. The committee identified three primary management approaches for addressing effects of fisheries on seafloor habitat: (1) effort reduction, (2) closure of vulnerable habitats to bottom trawl and dredge fisheries, and (3) gear modification.

Fishing Effort Reductions (See Case Study 1)

Effort reduction is the cornerstone of managing the ecological effects of fishing, including, but not limited to, effects on habitat. Other management tools (gear restrictions or modifications and closed areas) may also require effort reduction to achieve maximum benefit. However, fishing effort reduction measures may not be sufficient to reduce effects in highly structured habitats with low recovery potentials.

Establishment of Areas Closed to Fishing (See Case Study 2)

Closed areas have proved useful for protecting biogenic habitats (e.g., corals, bryozoans, hydroids, sponges, seagrass beds) that are disturbed by even low levels of fishing effort.

Modifications of Gear Design or Restrictions in Gear Type (see Case Study 3)

Disturbance depends on the extent of contact of the gear with the seafloor; hence, gear designs that minimize bottom contact can reduce habitat disturbance. In addition, shifts to a different gear type or operational mode may be considered, but the social, economic, and ecological consequences of gear reallocation should be recognized and addressed.

It is unlikely that any one management measure alone can resolve seafloor habitat impacts. The choice, utility, and limitations of a particular combination of these three measures to control fishing effects on seafloor habitats in a specific situation depends upon the current regulatory setting, social and economic characteristics of the existing fishery and its participants, available habitat types, and the specific fishery management goals and objectives. Ideally, the choice of the particular mix of these three tools for any one case should be informed by analyses of the full suite of benefits and costs over reasonable time frames. As demonstrated in the three case studies, creative solutions can be found to lessen the effects of fishing on seafloor habitats while maintaining viable commercial fisheries over the long term.

Case Studies

Case Study 1: Fishing Effort Controls in the Browns Bank Scallop Fishery

The Canadian scallop dredge fishery on Browns Bank on the western Scotian Shelf illustrates how technology can be applied to reduce the total area of seafloor swept through de facto effort controls (Manson and Todd 2000; Kostylev et al. 2003). Since the 1970s, fishing has been prosecuted on different portions of the bank, with inconsistent success. Recently, this fishery has been managed on the basis of an enterprise allocation system in which each of seven companies receives a share of the annual total allowable catch (Department of Fisheries and Oceans 2000).

A recent collaboration among the Department of Fisheries and Oceans, the Geological Survey of Canada Atlantic, and the fishing industry is documenting the relationships among giant scallops *Placopecten magellanicus* and substrate using data from multibeam bathymetry, high-resolution seismic reflection, side-scan sonar, extensive bottom sampling, video, photographic surveys, and a scallop-catch sampling program. The research demonstrates that scallops are strongly associated with gravel lag deposits, readily distinguished from sandy bottom by multibeam bathymetry.

Although the industry's prime motivation, initially, was to improve efficiency, other benefits have accrued, as evidenced by the following tabular comparison of fishery attributes before (1998) and after (1999) the application of the bathymetry data (Table 2).

Table 2. Increased efficiency of the Browns Bank scallop fishery after bathmetric mapping.

Value	1998	1999
Scallop quota (kg)	13,640	13,640
Time on bottom (h)	162	43
Distance towed (km)	1,176	311
Time lost (h)	15	0
Value of lost gear	$10,000	$0
Fuel used (L)	72,697	17,545

The area disturbed by the scallop dredge was substantially reduced by using substrate information to target fishing effort to the most productive scallop grounds.

Case Study 2: Closed Areas on Georges Bank

In response to the collapse of the principal groundfish species, year-round closure of two areas on Georges Bank and one in southern New England were instituted by emergency action in December 1994. These areas, totaling 17,000 km^2, were closed to all bottom-fishing gear and remained closed except for partial and temporary openings for scallop dredging in 1999 and 2000 (Murawski et al. 2000). During the closures, fishing effort was reduced by half for most of the mobile gear fleets, and complementary regulations were implemented on the Canadian side of Georges Bank.

Since the implementation of closures and effort reduction, scallop and some groundfish stocks have rebuilt substantially, and higher densities of groundfish and scallops were found inside the closed areas. The closed areas have been most successful in the conservation of the more sedentary demersal fishes and sea scallops. Haddock *Melanogrammus aeglefinus* and yellowtail flounder *Limanda ferruginea* have recovered to an abundance last observed in the 1970s, and scallop biomass increased 14-fold (Murawski et al. 2000).

The area closures, combined with effort reductions in the fishery, have reduced fishing mortality on the principal groundfish stocks and have protected the seafloor habitat from the physical effects of bottom fishing. Particularly in the northern part of Closed Area II, there has been a rapid increase in epifauna on gravel sediments. In 1998, the New England Fishery Management Council designated part of the closed area as habitat areas of particular concern (HAPC) on the basis of the occurrence of juvenile groundfish on gravel–cobble sediment.

The success is largely attributable to the closure of areas with the highest groundfish and scallop catch rates. Simultaneous effort reduction measures (fewer days fished) helped reduce the consequences of displaced effort. The current fishing effort is concentrated around the edges of the closed areas, which suggests that they are acting as sources for surrounding areas. In the future, the boundaries of the closed areas could be adjusted to enhance larval production and protect nursery areas, spawning concentrations, and migration corridors (Murawski et al. 2000).

Case Study 3: Gear Modifications in the Alaskan Pollock Fishery

The fishery for walleye pollock *Theragra chalcogramma* of the eastern Bering Sea is one of the largest in the world.

Walleye pollock occur on the sea bottom and in midwater up to the surface, with most catches taken at 50–300 m. The fishery is managed by total allowable catch (TAC) but has been constrained by bycatch of pelagic and demersal species.

In 1990, concerns about bycatch led the North Pacific Fishery Management Council (NPFMC) to apportion 88% of TAC to the pelagic trawl fishery and 12% to the non-pelagic trawl fishery (North Pacific Fishery Management Council 1999). For practical purposes, non-pelagic is defined as trawl gear that results in the vessel having 20 or more crabs of any species larger than 38 mm (1.5 in) at the widest dimension onboard at any time. Crabs were chosen as the standard because they live only on the seabed, thereby indicating that the trawl contacted the bottom.

By the mid-1990s, most vessels participating in the walleye pollock fishery had voluntarily switched to pelagic trawls. If bycatch limits were exceeded, premature fishery closures would take effect before the walleye pollock TAC was taken. Even though non-pelagic trawls accounted for only 2% of the walleye pollock catch in 1996, they were nearly one-third of the halibut bycatch and about one-half of the crab bycatch. In November 1999, with broad industry and public support, the NPFMC banned bottom trawl gear use in the Bering Sea walleye pollock fishery. The fishery now has modest bycatch rates and is able to harvest the full TAC.

Although the trawl gear was modified to reduce bycatch, it is postulated to have had the secondary effect of diminishing the impact on seafloor habitat. However, these trawls may contact the seafloor, especially in shallow water (<50 fathoms). If the trawls never touched the bottom, the pelagic trawl definition could be set at zero crab tolerance. Because typical pelagic trawls have large mesh webbing in the lower section of the net and are affixed to chain footropes, many demersal fish and invertebrates will fall through the large mesh panels and reduce the amount of bycatch recorded by onboard observers. The actual extent of bottom contact and seabed disturbance associated with the modified gear has not been studied and quantified.

Recommendations

The following section summarizes the consensus recommendations of the NRC committee.

Although there are still habitats, gears, and geographic regions that have not been adequately studied and characterized, there is an extensive literature on the effects of fishing on the seafloor. It is both possible and necessary to use this existing information to more effectively manage the effects of fishing on habitat. The following recommendations fall into three categories: (1) interpretation and use of existing data, (2) integration of management options, and (3) policy issues raised by existing legislation. These recommendations are intended to build upon the strengths of existing approaches to management.

(1) Interpretation and Use of Existing Data

Recommendation
Fishery managers should evaluate the effects of trawling based on the known responses of specific habitat types and species to disturbance by different fishing gears and levels of fishing effort, even when region-specific studies are not available.

The lack of area-specific studies on the effects of trawling and dredging gear is insufficient justification to postpone management of fishing effects on seafloor habitat. Extrapolations from common trends observed in other areas provide useful first-order approximations of fishing effects for use in habitat management. Management actions should be adjusted as more site-specific information becomes available.

Recommendation
The National Marine Fisheries Service and its partner agencies should integrate existing data on seabed characteristics, fishing effort, and catch to provide geographic databases for major fishing grounds. Effort data should be collected on a regular basis and reporting methods should be standardized across regions and include gear type, location, and length of tow.

Available datasets have been collected by different agencies and currently exist in different formats, at variable levels of resolution, in separate archives. Integration of these databases into a single, geographic information system will assist managers in evaluating regional needs for habitat conservation.

(2) Integration of Management Options

Recommendation
Management of the effects of trawling and dredging should be tailored to the specific requirements of the habitat and the fishery through a balanced combination of effort reduction, gear use and modification, and closed areas.

The optimal combination of these management approaches will depend on the characteristics of the ecosystem and the fishery—habitat type, resident seafloor species, frequency and distribution of fishing effort, gear type and usage, and the socioeconomics of the fishery.

Recommendation
The regional fishery management councils should use comparative risk assessment to identify and evaluate risks to seafloor habitats and to prioritize management actions within the context of current statutes and regulations.

Comparative risk assessment provides multiple advantages for the task of benthic habitat protection because it: (1) relies on a combination of available data, scientific inference, and public values that facilitates management when there is substantial scientific uncertainty, (2) provides simultaneous analysis of a wide range of risks to benthic habitats such as pollution, drilling, and natural disturbance, and (3) enables stakeholder involvement in the decision-making process.

(3) Policy Issues Raised by Existing Legislation

Recommendation
Guidelines for designating EFH and HAPC should be established based on standardized ecological criteria.

Although EFH places appropriate emphasis on managing the total ecosystem in addition to regulating catch of exploited species, designation of EFH currently lacks consistent criteria for designating habitat for species covered by fishery management plans. Typically, EFH designations are too extensive to form a practical basis for managing fisheries. Although this approach may assist in mitigating some habitat threats, it provides little guidance for evaluating the impacts of trawling and dredging. The EFH designations need to be based on a clear understanding of the population biology and spatial distribution of each species.

Habitat areas of particular concern form a subset of EFH based on the ecological value of the area, its susceptibility to perturbation, and whether it is rare or currently stressed (National Marine Fisheries Service 1997). Although these areas are known to play a vital role in the life cycle of exploited fish populations, no HAPC safeguards are specified in existing legislation and regulations. Specific guidelines should be established for regulating fishing activities in HAPC and the effectiveness of the designation should be regularly evaluated.

Recommendation
A national habitat classification system should be developed to support EFH and HAPC designations.

Efforts to inventory and construct regional or national habitat maps require a classification system with common designations. Such a system would facilitate tracking of changes over time and would provide the basis for determining functional links between seafloor ecosystems and fisheries production. A classification system would assist in: (1) ranking different habitats according to the resilience of their biological communities and associated fisheries; (2) estimating the vulnerability of the habitat to disturbance; and (3) managing habitat impacts based on the generalized results of research conducted in other geographic areas.

Research Needs

Additional studies on gear effects will be required to develop a predictive capability to link gear type and effort to bottom disturbance, fish production, and recovery times in particular habitats. High-resolution mapping of the continental shelf and slopes in heavily fished areas should be used to provide a landscape-scale perspective of the effects of trawling and dredging on the seabed.

Studies on the dose–response relationship as a function of gear, recovery time, and habitat type will be needed to evaluate effects of repeated disturbances by fishing gear. In heavily trawled fish habitats, long-term trend data for benthic production versus fisheries production should be collected. Research should be designed to determine the rates and magnitude of sediment resuspension, nutrient regeneration, and responses of the plankton community in relation to gear-induced disturbance. Active engagement of resource users in research will help ensure that mitigation strategies are practical, enforceable, and acceptable to the fishing community.

Acknowledgments

The National Research Council study upon which this paper was based was sponsored by the National Marine Fisheries Service.

References

Department of Fisheries and Oceans. 2000. 2000 Scotia–Fundy offshore scallop integrated fisheries management plan: maritimes region. Department of Fisheries and Oceans, Dartmouth, Nova Scotia.

Kostylev, V. E., R. C. Courtney, G. Robert, and B. J. Todd. 2003. Stock evaluation of giant scallop (*Placopecten magellanicus*) using high-resolution acoustics. Fisheries Research 60:479–492.

Manson, G., and B. J. Todd. 2000. Revolution in the Nova Scotia scallop fishery: seabed maps turn hunting into harvesting. Fishing News International 39(2):20–22.

Murawski, S. A., R. Brown, H. L. Lai, P. J. Rago, and L. Hendrickson. 2000. Large-scale closed areas as a fishery-management tool in temperate marine systems: the Georges Bank experience. Bulletin of Marine Science 66(3):775–798.

National Marine Fisheries Service. 1997. Magnuson-Stevens Act provisions: essential fish habitat. National Oceanic and Atmospheric Administration, Interim Final Rule. Federal Register 62(244):66531–66559.

National Research Council. 1993. Issues in risk assessment. National Academy Press, Washington, D.C.

National Research Council. 2002. Effects of trawling and dredging on seafloor habitat. National Academy Press, Washington, D.C.

North Pacific Fishery Management Council. 1999. Environmental assessment/regulatory impact review/initial regulatory flexibility analysis for amendment 57 to the FMP for the groundfish fishery of the Bering Sea and Aleutian Islands area to prohibit the use of nonpelagic trawl gear in directed pollock fisheries. North Pacific Fishery Management Council, Draft for Secretarial Review, Anchorage, Alaska.

Destructive Fishing Practices and Evolution of the Marine Ecosystem-Based Management Paradigm

ELLIOTT A. NORSE[1]

*Marine Conservation Biology Institute, 15806 Northeast 47th Court,
Redmond, Washington 98052-5208, USA*

"The greatest and noblest pleasure which man can have in this world is to discover new truths; and the next is to shake off old prejudices." Prussian King Frederick II (The Great) (1712–1786)

"Doubtless, it is because of the youthfulness of fishery biology itself that the methodology of fishery conversation has so little developed." Elmer Higgins (1934)

Abstract. The recent increase in interest about effects of trawling and dredging on seafloor ecosystems and their fisheries can be understood by examining three phases in the history of conservation thinking. The primary focus in nonmarine conservation thinking and management worldwide is on maintaining biodiversity, while marine managers are still focused mainly on use of marine life. Marine conservation lags behind nonmarine conservation, as shown using key measures of scientific publication, species protection, and ecosystem protection. Because fishing is the human activity that most affects marine biodiversity, marine fisheries biology has a particularly large role in determining the fate of the sea's biodiversity. Unlike management-oriented nonmarine fields including wildlife biology and forest biology, marine fisheries biology has yet to incorporate key insights from the science of ecology, including the importance of maintaining abundance and diversity of predators and structure-forming species. Growing concern about the loss of structure-forming species, such as corals and sponges, and their role in providing fish habitat cannot be addressed using traditional stock assessment techniques. This creates the need for the evolution of a new marine ecosystem-based management paradigm that incorporates modern understanding of ecology and conservation biology.

Introduction

Conservation evolves in response to new understanding about the workings of nature and ways that individuals and institutions affect it. The pace of conservation evolution has increased rapidly in recent decades, paralleling rapidly increasing awareness of the damage to nature caused by human activities. New scientific fields have been spawned and gained thousands of recruits, including conservation biology and restoration ecology. Conservation's theoretical framework, public awareness, laws (see Bean and Rowland 1997 for a useful overview of U.S. wildlife laws), and government programs evolved first in nonmarine realms, but the increasing harm to the marine realm now compels marine scientists, legislators, managers, users, and conservation advocates to devise new ways to conserve what people value in the sea.

Having been trained as a marine ecologist but having received "battlefield commissions" in marine and forest conservation biology, I have observed how both terrestrial and marine conservation thinking and practice have evolved. Here, I summarize the evolution of conservation as a backdrop for examining the marine management paradigm that most impacts life in U.S. waters, focusing specifically on its effectiveness in dealing with fishing practices that disturb the seafloor. This is a topic that transcends mere academic interest because the greatest challenge humans face in the 21st century is finding ways to live on our planet without destroying its living systems (including marine systems) and their essential functions, on which we depend. Both personal experience with fisheries controversies and examination of the scientific literature and news stories indicate that there is ample reason to be concerned.

Three Stages in Conservation Thinking

To understand the present state in marine conservation, it is useful to compare it to the much longer and better-

---

[1] E-mail: Elliott@mcbi.org

documented history of conservation in nonmarine systems. An astute reader will probably note intriguing parallels with conservation in marine systems as well as some significant differences.

Although the first humans who walked across the Bering Sea profoundly reduced the diversity of North America's terrestrial megafauna and altered its ecosystems (Diamond 1997; Flannery 2001), Europeans who left their biologically impoverished continent were stunned by the abundance of North America's wildlife (Kimball and Johnson 1978). They set out to subdue the new land by making full use of North America's species, including its pines *Pinus* spp. for ship masts, oaks *Quercus* spp. for ship hulls, beaver *Castor canadensis* for hats, and white-tailed deer *Odocoileus virginianus* for meat. The land itself—once bent to our use—could provide an abundance of pasturage, crops, minerals, water, and living space to fuel westward expansion. The attitudes that drove the dominant society's effects on wildlife populations were primarily ones that Kellert (1993, 2005, this volume) classifies as "dominionistic" and utilitarian: We were fulfilling a divine purpose by using everything useful that the land provided.

By the start of the 20th century, the dominant species and ecosystems of the United States had undergone profound changes. The land animal with the greatest biomass—American bison *Bison americanus*—had gone from an estimated 30–60 million (Kimball and Johnson 1978) a few decades earlier to just a few hundred individuals. The most abundant bird—the passenger pigeon *Ectopistes migratorius*—had gone from an estimated 5 billion to the edge of extinction (Blockstein 2002). American bison survived by the narrowest of margins, but the last passenger pigeon died in 1914. The three key attributes—species composition, structure, and functioning—of "virgin" forest and prairie ecosystems that had covered most of the country were almost completely altered by settlers colonizing the land. The few institutions responsible for managing our use of the land were woefully ill-equipped to deal with conservation.

America's conservation movement arose in the 19th century as the impending closing of the frontier led to growing concern about the loss of wildlife and land resources. Reflecting public attitudes, the first stage of conservation was largely utilitarian. Species conservation focused on preventing loss of species that were considered "good" because they were useful (e.g., deer and ducks) and eliminating species (e.g., wolves *Canis lupus* and mountain lions *Felis concolor*) that were considered "bad" because they ate "good" species or otherwise interfered with resource use. High trophic level species—the large carnivores—disappeared throughout large parts of their ranges, as did the biggest grazers, eastern bison and eastern elk *Cervus canadensis*. Many smaller, once-common species, such as white-tailed deer, raccoons *Procyon lotor*, and wild turkeys *Meleagris gallopavo*, were serially hunted to rarity.

At the same time, place-based conservation was driven by concern about loss of scenic and timber values needed for future recreational and industrial use, an ethic exemplified by the thinking of Gifford Pinchot (e.g., Pinchot 1947), the first head of what became the U.S. Forest Service. The U.S. federal government began protecting lands that became our National Parks and National Forests starting in the 1872 and 1891, respectively.

Early species conservation efforts focused mainly on controlling hunting mortality through bag limits (limiting the number of animals a hunter could possess), seasonal prohibitions (e.g., prohibiting hunting of deer until fawns were independent), or restricting hunting methods that were deemed too efficient or wasteful (e.g., a 1910 law prohibiting use of boat-mounted punt guns, which were large bore shotguns that could kill tens of sitting ducks with one shot (see Harvesting the River, Illinois State Museum, http://www.museum.state.il.us/RiverWeb/harvesting/harvest/waterfowl/tools_techniques/guns/punt_gun.html).

Long ago it became clear to nonmarine wildlife managers that controlling hunting mortality was essential but not sufficient. A landmark event in U.S. conservation was the designation of the federal government's National Wildlife Refuge system in 1903 (Lee 1986), the first of more than 500 protected areas dedicated to maintaining and recovering America's disappearing wildlife and their habitat. Another was the founding of a nonprofit organization, Ducks Unlimited, in 1937 for the purpose of protecting prairie pothole breeding and nursery habitat essential to waterfowl in the USA and Canada. Protecting habitat crucial to highly migratory species has a longstanding history in terrestrial and freshwater wildlife management.

Conserving wildlife and wildlands for their own sake, the view espoused by John Muir (Muir 1901) and Aldo Leopold (Leopold 1949), became far more prevalent as the environmental movement flowered in the years following publication of *Silent Spring* (Carson 1962), leading to the passage of laws including the U.S. Endangered Species Act (ESA) of 1973. The ESA clearly represented a major advance in conservation thinking, embracing the idea that all living things are important to conserve for their own right as well as for their utilitarian value. It embodied truly enlightened concepts, as indicated by these excerpts:

- …these species of fish, wildlife, and plants are of aesthetic, ecological, educational, historical, recreational, and scientific value to the Nation and its people.…

- The term "endangered species" means any species which is in danger of extinction throughout all or a significant portion of its range…

- The Secretary shall … determine whether any species is an endangered species or a threatened species because of …(A) the present or threatened destruction, modification, or curtailment of its habitat or range…

- The purposes of this Act are to provide a means whereby the ecosystems upon which endangered species and threatened species depend may be conserved…

The drafters of the ESA clearly considered it crucial to protect species throughout their range whether or not they were demonstrably useful and fully recognized the importance of species' critical habitats and (still more broadly) the importance of protecting ecosystems as a whole. The ESA reflects and institutionalizes conservation ethics including those that Kellert (1993, 2005, this volume) calls scientific, aesthetic, humanistic, naturalistic, and moralistic.

However, there are three important weaknesses with the endangered species approach. The ESA applies only to species that are in serious trouble, and the burden of proof that they need protection falls on those who propose their conservation. This is analogous to having a health-care system consisting of hospital intensive care units that have dauntingly high admission standards. Thus, protection of species and habitat critical to their continued existence under ESA does not take effect until species clearly require heroic—in other words, expensive, painful, and frequently unsuccessful—conservation measures. However essential it is to keep species from "falling off the edge," preventing extinction is hardly a comprehensive, sustainable strategy. One reason is that we seldom have adequate assurance that we can save endangered species before it is too late. Another is that focusing on species that are imperiled ignores the benefits that are roughly proportional to the numbers of organisms, not merely their continued existence. These include the values of species for recreation; few people are willing to spend time searching for wildlife so rare that encountering them is vanishingly unlikely. Many benefits from living things come not only from the numbers of species but their abundance, including ecosystem services (Daily 1997), the natural functions on which humans utterly depend. These include pollination and seed dispersal, preventing proliferation of pest species, amelioration of flooding, building and maintaining soil fertility, decomposing dead plants and animals, cleansing water and air, and maintaining a climate suitable for human life. A focus on conserving endangered species fails to provide the quantitative benefits that ecologists and biogeochemists have demonstrated to result from the multiple roles of living things in the biosphere. A third weakness of the ESA approach is that it functions primarily (although not exclusively) at the species level, giving less emphasis to higher and lower hierarchical levels of biological assets.

The inadequacies of the utilitarian and endangered species stages of conservation created a need for further evolution in conservation thinking that recognizes the importance of living things as resources, and in their own right, but that neither waits until species are about to disappear nor places the burden of proof on those who want to conserve life. This third stage is a focus on maintaining biological diversity, presaged conceptually by Leopold (1949) and Myers (1979); used (in a narrow sense, clearly to mean species diversity) by Risk (1972), National Research Council (1978), Lovejoy (1980a, 1980b), and Wilson (1980); explicitly defined by Norse and McManus (1980); redefined as the term is widely used in conservation today by Norse et al. (1986); and subsequently examined from various perspectives by Wilson (1988), Reid and Miller (1989), and countless others. For those who want to delve into this topic, the evolution of biological diversity as a conservation goal is examined in Norse (1996) and Farnham (2002).

Just as the endangered species ethic does not invalidate the earlier utilitarian ethic but creates a richer, more complex and more encompassing understanding of conservation needs and strategies, the biodiversity ethic broadens and deepens both of these ethics. Conserving biodiversity is a more robust conservation ethic for four reasons:

- It is hierarchical and recognizes the importance of all three major levels of biological organization, the diversity of genes within species, the diversity among species, and the diversity among ecosystems (communities of organisms in their physical settings).

- It transcends mere presence–absence, recognizing that quantities matter.

- It deals with composition, structure, and function, both ecosystem parts and processes, focusing on maintaining the biological integrity that the public intuitively understands as "the health of nature," not merely what is immediately useful, beautiful, or otherwise popular.

- It is precautionary, focusing on maintaining and recovering diversity whether or not there is demonstrable, imminent threat.

In short, focusing on biodiversity conservation is a conceptual framework that makes it possible to conserve life for its own sake and for human benefit whether or not we can gauge the worth of a gene, species, or ecosystem to us at present. For these reasons, maintaining biological diversity has quickly become the primary focus of conservation efforts throughout the world on land and in freshwaters. A recent study (Norse and Carlton 2003) found that biodiversity is mentioned on more Internet sites than other scientific concepts or sciences including relativity, molecular biology, or oceanography. In just 24 years, biodiversity conservation has become the driving force in conservation around the world, except, ironically, in the sea.

Nonmarine and Marine Conservation

In the United States, as in much of the world, the science and practice of conservation are far more advanced on land and in freshwaters than in the sea. Consider the following disparities:

- Market hunting has been illegal for terrestrial animals and quite limited for freshwater animals for many decades; the venison, buffalo, or duck in markets and restaurants is not wild caught. In contrast, most marine wildlife in markets are wild caught; the only way most marine species are managed is by attempting to control adult mortality rather than selectively breeding them, feeding them, controlling their environment, and preventing their diseases. Even marine species (e.g., oysters, shrimps, bluefin tunas *Thunnus thynnus*) that are "farmed" are far less domesticated—that is, they differ far less anatomically, physiologically, and behaviorally from their wild relatives—than their terrestrial and freshwater counterparts.

- Terrestrial wildlife managers and conservation biologists generally estimate populations in terms of numbers of individuals, reflecting their concern about maintaining healthy populations; marine fisheries managers generally measure populations in tons, reflecting their concern about maintaining production of biomass.

- Conservation biologists have tended to overlook the sea. In the first 37 issues (1987 to1995) of *Conservation Biology*, nonmarine papers outnumber marine papers 565 to 37 (Irish and Norse 1996). In subsequent issues (December 1996 to December 2000), nonmarine papers outnumber marine papers 701 to 82.

- Large terrestrial predators are generally protected by laws, regulations, and social peer pressure. Terrestrial apex predators or other high trophic level species including wolves, grizzly bears *Ursus arctos*, big cats, and eagles that were once killed in large numbers are now revered symbols of wildness and the targets of intensive recovery programs. In contrast, many of their marine counterparts—apex predators or other high trophic level species including sharks, billfishes, tunas, and groupers (family Serranidae)—are still considered choice targets for commercial and sportfishing, activities that are subsidized by federal and state governments. Indeed, some nations (Japan, Iceland, Norway, and Canada) support commercial killing of apex predators or high trophic level species including sperm whales *Physeter macrocephalus*, baleen whales, dolphins, and pinnipeds.

- The first terrestrial plants to be protected under the U.S. Endangered Species Act, including the San Clemente Island Indian paintbrush *Castilleja grisea*, were listed in 1977; in contrast, the first marine plant, Johnson's seagrass *Halophila johnsonii*, was listed only in 1998. The first freshwater invertebrates, including the Alabama lampmussel *Lampsilis virescens*, were listed in 1976; the first marine invertebrate, white abalone *Haliotis sorenseni*, was not listed until 2001. The first freshwater U.S. fishes, including Apache trout *Oncorhynchus apache*, were listed in 1967; the first truly marine U.S. fish, the smalltooth sawfish *Pristis pectinata*, was finally listed in 2003. Indeed, the ESA currently offers protection to 746 nonmarine plants, 186 nonmarine invertebrates, and 115 nonmarine fishes but only 1 marine plant, 1 marine invertebrate, and 2 truly marine fishes.

- Laws governing consumptive uses on lands under U.S. jurisdiction are markedly different in their focus on maintaining the full spectrum of species and the resulting benefits they provide. The National Forest Management Act of 1976 (U.S. Code, title 16, sections 1600–1614, August 17, 1974, as amended 1976, 1978, 1980, 1981, 1983, 1985, 1988, and 1990), a law governing use of resources on most federally owned forestlands, requires the U.S. Forest Service to "provide for diversity of plant and animal communities based on the suitability and capability of the specific land area in order to meet overall multiple-use objectives." In contrast, its marine analogue, the Magnuson Fishery Conservation and Management Act of 1976 (U.S. Code, title 16, sections 1801–1882, 90 Stat. 331; as amended by numerous subsequent public laws listed and identified in the U.S. Code), even its 1996 reauthorized and substantially improved version, still has no such explicit mandate. Those who framed our laws on forests were thinking about diversity when those who framed our laws on oceans were not, and that has not changed even in the decades since the concept of biological diversity became the driving force in terrestrial conservation.

- The United States began designating lands that became National Parks, where most terrestrial wildlife species are protected from directed taking, in 1872 and began designating multiple-use lands, which became National Forests, in 1891. There are now 88 National Parks and 125 National Forests. In contrast, the United States began designating National Marine Sanctuaries only in 1975 and has designated only 13 to date. Moreover, few of them have strong provisions barring directed take of marine wildlife. The United States still has no special federal program that focuses on establishing fully protected marine reserves.

- National Parks (from Glacier Bay to Everglades) that fully protect terrestrial wildlife commonly allow fishing, even commercial fishing, in their marine waters.

- In the 2002 federal budget, the U.S. government is spending 67 times as much money for our National Parks and 93 times as much money for our National Forests as for our National Marine Sanctuaries despite the fact that the marine area under federal jurisdiction is not only vastly larger than the federal land base but is actually larger than the entire U.S. land area.

- Habitat restoration (e.g., restoring stream flow regimes, deliberate placement of coarse woody debris, soil amendment, and revegetation with native species) is becoming a commonplace tool for private and public management efforts of terrestrial and freshwater ecosystems. In contrast, there are few effective or large-scale efforts to restore key marine ecosystem features, the closest approximation being the dumping of unwanted materials such as oil production platforms, ships, tanks, subway cars, and tires into the sea to become "artificial reefs." As one illustration, on June 30, 2004, Googlefight.com listed 202 web sites mentioning "anadromous fish habitat restoration" versus only nine web sites mentioning "marine fish habitat restoration."

- Restoration ecologists have tended to overlook the sea. In the eight 2002 and 2003 issues of *Restoration Ecology*, nonmarine papers outnumbered marine papers 102 to 26, even when a special issue, 10(3), having 14 marine papers is included.

- In the last several decades, the U.S. Forest Service has progressively restricted clear-cutting. It is compelled to conduct area-specific environmental reviews before proceeding with each timber "sale" to be logged and requires loggers to pay the federal treasury (that is, the taxpayers who own these lands) for the privilege of extracting resources from public forestlands. In contrast, trawlers, whose impacts on structurally complex bottoms are similar to those of loggers, trawl almost anywhere they wish on federal undersea lands and pay no royalties to the treasury for the public resources they take.

It is understandable why the science and practice of conservation came later to the sea than to the land and freshwaters. Human physiology and senses are much better equipped to see and understand what is happening in those realms. With suitable shelter and clothing, terrestrial biologists can visit and study biodiversity anywhere on a permanent basis. Freshwaters are somewhat more difficult to study but are so shallow and so limited in extent that scientists can readily observe changes to all but the remotest of them (e.g., the depths of the Great Lakes and Lake Baikal). In contrast, the sea's vastness makes research inherently expensive, and the pressure in its depths makes in situ marine observation far more difficult. Indeed, many marine scientists have found that commercial interests that exploit subsea oil deposits or marine fish populations are far more numerous and better equipped than we are. In consequence, the public, lawmakers, and managers know far less about the sea's biodiversity than they know about that of the land or freshwaters. This unfamiliarity has fostered the assumption that the sea is both a never-ending cornucopia and convenient receptacle for humankind's wastes.

Scientific disciplines and management approaches that focus on maintaining biodiversity have not yet become prevailing conservation paradigms in marine ecosystems as they are on land and in freshwaters. As a result, while nonprofits and governments are putting substantial resources to arresting the decline of biodiversity on land, we protect the sea far less effectively. Marine conservation is decades behind its terrestrial and freshwater counterparts. Failure to eliminate this gap may well be the greatest weakness in U.S. conservation.

While the United States exerts management authority over a variety of economic activities within its exclusive economic zone, the fact that fishing is clearly the leading threat to the sea's biological diversity and integrity (Jackson et al. 2001; Pauly et al. 2002; Myers and Worm 2003) means that fisheries management has by far the greatest effect on marine conservation. Although all modern methods of fishing have the potential to overfish target species, the plurality of targeted catch, the majority of bycatch (Figure 1), and the majority of habitat damage is caused by fishing with mobile gear, including bottom trawls and dredges (Chuenpagdee et al. 2003; Morgan and Chuenpagdee 2003). Indeed, trawling and dredging are the most important disturbances on the world's seafloor, especially in waters below 80 m (Watling and Norse 1998). Trawling now occurs on the remotest seamounts and on continental slopes down to 2,000 m (Merrett and Haedrich 1997; Gordon 2001). Since (a) the sea is by far the largest of the Earth's three biotic realms, (b) the vast majority of marine species are benthic, (c) fishing is the greatest threat to marine biodiversity, and (d) trawling is the seafloor's most important fishing method and largest source of disturbance, trawling must have profound implications for the world's biodiversity and the sustainability of humankind's food supply. Yet, despite this, trawling impacts have received remarkably little attention from the public, lawmakers, and managers.

Figure 1. Commercial shrimp trawl catch, northern Gulf of California, Mexico. The bycatch:shrimp biomass ratio in this and other hauls that the author photographed was approximately 20:1 (Norse, Marine Conservation Biology Institute).

Much of the reason why the destructive effects of trawling are largely overlooked and why biodiversity considerations and the ecosystem approach have failed to become integral to marine conservation is the dominance of the stock assessment paradigm that drives marine fisheries management in the United States and some other industrialized countries. Stock assessment focuses on estimating a small number of population parameters within targeted fish "stocks" such as number of individuals at age of first reproduction, spawning stock biomass, growth rate, etc. Assuming an inverse correlation between population density and per capita reproductive success, these estimates are used to calculate a stock's "total allowable catch" (TAC). As Spurgeon (1997), Wilson and Degnbol (2002), and Weber (2002) have pointed out, there can be strong pressure on fisheries biologists to provide population estimates that produce the largest possible TACs, which are then used or further increased by fishery managers who are themselves pressured to increase the TAC. The net result of these intense pressures is that the TACs are often unsustainable, as illustrated by the repeated failure of fisheries managed with this process. An illustrative example described by Rieser et al. (2005) concerned summer flounder *Paralichthys dentatus*, an overfished species caught in U.S. Middle Atlantic states that was subject to a rebuilding plan. In 1999, the National Marine Fisheries Service recommended a summer flounder commercial fishing quota so high that the population had only an 18% probability of achieving established population rebuilding goals. Conservation groups sued the federal government for failing to "ensure" population rebuilding. The federal court's decision said, "only in Superman Comics' Bizarro world, where reality is turned upside down, could the Service reasonably conclude that a measure that is at least four times as likely to fail as to succeed" offers confidence that the population will rebuild. Yet, year after year, the National Marine Fisheries Service admits that a sizeable fraction of exploited fish "stocks" are overfished (and that there aren't enough data to determine the status of a far greater number), a success rate orders of magnitude lower than those considered acceptable in other economic sectors subject to federal regulation. Based on statistics for 2003, the National Marine Fisheries Service reported to Congress that of 909 stocks, 138 are not overfished, 76 are overfished, 1 is approaching overfished condition, and 694 are of unknown status or have no defined fishing mortality theshold (NMFS 2004). Thus, 138/855 = 16 % of U.S. fish "stocks" are demonstrably not overfished. In contrast, the established goal of the Federal Aviation Administration for 2004 is to "reduce airline fatal accident rate to 0.028 per 100,000 departures," which means 99.99997% demonstrably will not have fatal accidents (Federal Aviation Administration 2003). An observer might well ask, is the health of our oceans and their fisheries so unimportant that this disparity in standards is acceptable?

Even more interesting than what stock assessments include is what they exclude. Stock assessments treat each fish species as if it exists in isolation from other species, including non-human predators, pathogens and parasites, alternate hosts, prey, competitors, amensals, and mutualists. Moreover, stock assessments do not take into account the intricacies of populations' habitat relations. Yet, early marine fisheries biologists thought about and examined these topics. Mitchell (1918) noted the importance of parasite outbreaks in some fish species. Coker (1938) stated, "Whatever shelter occurs in the sea it is availed of. Clams, worms and crustacea burrow in the bottom, and even in rock, or find concealment among the shells of the bottom.... Oyster beds and thickets of eelgrass harbor a rich and varied population of small plants and animals." And Herrington (1947) voiced concern about effects of bottom trawling on benthic organisms eaten by Georges Bank haddock *Melanogrammus aeglefinus*. It is difficult to fathom how scientific knowledge of such fundamental significance to fisheries could have been downplayed as marine fisheries biology and management narrowed their focus to stock assessment.

In contrast, scientific understanding of ecology in general, and of the ecology of fishes specifically, has broadened and deepened enormously in the last half century. Ecologists, including fish ecologists, have long known that interactions among species are crucial determinants of population levels, age structure, sex ratios, geographic and local distributions, population genetics, reproduction, physiology, behavior, and relationships to habitat in various life history stages. This ecological understanding has become integral to resource management, including wildlife and forest management and even management of freshwater fishes (such as trout). However, ecological understanding appears to have little influence on marine fisheries biology. To test this, I examined *Science Citation Index Expanded* for papers that cite one of the most influential papers in the history of ecology, a study of the keystone role of a predator in shaping a rocky intertidal community (Paine 1966). Of the 500 most recent papers in peer-reviewed journals (published between 1998 and 2004) that cite this seminal paper, only five were in fishery journals. To someone not trained in the canons of marine fisheries biology, the reductionism of its approach, particularly its failure to incorporate the central ideas in ecology, is difficult to comprehend. That would not be a major problem if marine biodiversity and fisheries were in good shape. But serial depletion of targeted species; population crashes of non-targeted species; increasingly convoluted and inconsistent command-and-control regulation of fish "stocks;" management by lawsuits with outcomes dictated by the courts; the unceasing "state of war" between conservationists, fishermen, and fishery managers; and perpetual socioeconomic crises in fishing communities are not signs that the dominant marine fisheries management paradigm is working. An observer of fisheries science and management is forced to the uncomfortable conclusion that the stock assessment paradigm, while potentially useful in very restricted circumstances, is fundamentally flawed. It is time for what Kuhn (1970) called a paradigm shift.

Lessons for Fisheries Biologists from Terrestrial and Freshwater Realms

Some concepts more readily evolve in one realm than another, with good reason. Anyone who has tried to light a fire with wet wood can understand why the field of fire ecology developed in the terrestrial realm, not in aquatic ones. It is scarcely less obvious why disturbance theory developed on land: Humans are better equipped to see the effects of disturbance on land than in the sea. Even those who were not trained as forest biologists (e.g., Norse 1990) readily see how disturbances that change the structure of a terrestrial ecosystem affect species composition and functioning. Seeing these effects is much more difficult in the sea, but absence of evidence is not necessarily evidence of absence. Not having looked, and therefore not having seen these effects in the sea, may well have lulled many to assume that they are not happening.

In truth, very few people—including fishermen or fishery biologists—have spent much time watching what happens on the seabed below depths that scuba divers can reach. Except for the shallowest tens of meters in clear, warm, nearshore seas, the sea is so hostile to humans and so opaque to the unaided human eye that undersea lands are terra incognita. Most of what people know about the seafloor has been learned by blindly dropping sampling gear—such as otter trawls and box cores—that remove organisms from their habitat context and do not tell much about the biology of the specimens we bring up. Imagine for a moment how little we would know about terrestrial ecology if scientists flying in airships above the clouds had to learn about deer and songbirds by extracting their bodies from sampling gear dragged through forests and grasslands!

Understandably, fisheries biology focuses on things its standard sampling tools allow it to estimate—such as the number of eggs—rather than things its tools cannot estimate (including shelter-seeking behaviors of juveniles and social learning of migratory routes). This recalls the wisdom in the saying, "When the only tool you have is a hammer, you tend to treat everything as if it were a nail."

Fortunately, new research tools such as research submersibles, remotely operated vehicles, and side-scan sonar have allowed marine scientists to gain a much better picture of what is happening on the seabed (Figure 2) than was previously available. But improved technology is a double-edged sword. The ability to see the seabed more clearly has also let fishermen locate and catch fish in their last refuges. In effect, new technologies have "turned the sea transparent."

A terrestrial story from the past provides a faultless analogy with the present-day situation in marine fisheries. From the mid-19th century, when it was the most abundant bird in the world, North America's passenger pigeon was eliminated in just one human lifetime. It disappeared because new technologies—particularly telegraphs and railroads—allowed people to reach and kill large numbers and fell the big beech, oak and chestnut trees that were their essential breeding and nursery habitat, thereby disrupting their nesting aggregations (Blockstein 2002). Management efforts seem doomed to fail when officials do not appreciate the effects of improving technologies. The same is true for failure to appreciate basic ecological principles.

One robust principle in ecology concerns the relationship between the spatial complexity of ecosystems and their diversity. More than four decades ago, a terrestrial ecologist and his physicist brother (MacArthur and MacArthur 1961) published a "must-read" study for anyone interested in habitat. It began, "It is common experience that more species of birds breed in a mixed wood than in a field of comparable size." In deciduous forests around the United States, the MacArthurs found that bird species diversity increases with the structural complexity of their habitats. Stimulated by their work, marine geologist Michael Risk (Risk 1972) showed that fish species diversity increases with benthic structural complexity in coral reefs, so marine scientists have known that structural complexity is crucial for species diversity for decades, well before the 1996 revisions of the Magnuson Fishery Conservation and Management Act (FCMA) of 1976. Since the MacArthurs and Risk published these landmark studies, hundreds of scientific papers have examined the relationship between habitat complexity and species diversity on land, in freshwaters, and in the sea.

As is often the case, such scientific studies merely quantified longstanding common knowledge. In an old African-American folk story called "The Tar Baby," Brer Rabbit is captured by Brer Fox and Brer Bear but tricks them into throwing him into a structurally complex briar patch, allowing Brer Rabbit to escape predation. Most anybody who snorkels or dives has seen fishes avoid approaching humans by hiding in the interstices of a reef. And most freshwater recreational fisherman know that the best fishing spots for trout and black bass *Micropteris* spp. are among tree roots beneath undercut banks, among branches of sunken trees, and in submerged vegetation. The importance of structural complexity is widely understood—by managers and the public alike—to be vital on land and in freshwaters. There is an abundance of scientific studies in fisheries journals such as the *North American Journal of Fisheries Management* on the importance of complex structures in streams and lakes. Yet, even a diligent search through the leading fisheries journals will show that fisheries management has largely overlooked the importance of structural complexity in the sea.

Seafloor Structures

People in vessels on the sea's wavy, reflective surface can be deceived into thinking mainly about the water column, the pelagic realm, which appears almost structureless to the naked eye. The dominant marine fishery management paradigm seems to overlook the fact that more than 98% of marine species live in, on, or immediately above the seabed (Thurman and Burton 2001). Their morphological, physiological, and behavioral adaptations are for living in, on, or just above the seabed. They settle there, extract oxygen from its waters, feed on other members of benthic communities, find shelter from predators, grow and reproduce in, on, or just above the seabed. Hence, the relationship of marine species to their benthic habitat is of paramount importance to understanding the biology of marine life. Failure to apply this understanding

Figure 2. Roller or rockhopper trawl tracks, Stellwagen Bank National Marine Sanctuary, Massachusetts (P. Auster, National Undersea Research Laboratory, University of Connecticut).

contributes to problems associated with the present fisheries management paradigm.

It is important to note that not all benthic species need seafloor structures. Hence, some larger fish and invertebrate species occur on flat, "seemingly" featureless bottoms that may have naturally short disturbance return intervals. By and large, these species—including surf clams *Spisula solidissima*; many penaeid shrimps; raninid crabs; portunid crabs in the genera *Ovalipes*, *Arenaeus*, and *Callinectes*; angel sharks *Squatina* spp.; most dasyatid stingrays; lizardfishes (synodontids); and soles (soleids)—are cryptically colored and/or can quickly bury themselves in the sediment, attributes that allow them to reduce risk of predation.

However, a disproportionately large number of benthic fish and shellfish species are closely associated with geological or biological seafloor structures. A key experimental study by Lindholm et al. (1999) illustrates one reason this seems to be so. It shows that juvenile Atlantic cod *Gadus morhua* are more successful at escaping predators in more complex habitats. But because benthic diversity and biomass are so much greater than in the water column, the seafloor also offers many more feeding opportunities for predators.

As Norse and Watling (1999), Auster and Langton (1999), the National Research Council (2002), and others have shown, dragging heavy trawls and dredges across the seafloor rolls boulders; planes high spots; homogenizes sediments; and crushes, buries, and exposes to scavengers wildlife living in and on the seafloor. Figures 3–8 show the dramatic contrast between untrawled and trawled seafloors in three widely separated areas of the world. Until the 1980s in the United States, fishermen considered rough bottoms too dangerous to trawl, but introduction of roller and rockhopper groundgears have allowed fishermen to trawl on bottoms they had previously avoided.

Figure 4. Trawled glass sponge reef, Hecate Strait, British Columbia. Note trawl track (M. Krautter, Institut fuer Geologie und Palaeontologie, Universitaet Hannover).

Figure 5. Untrawled gorgonian-sponge forest, Northwest Australian shelf (K. Sainsbury, Commonwealth Science and Industry Research Organization)

Figure 3. Untrawled glass sponge reef, Hecate Strait, British Columbia (M. Krautter, Institut fuer Geologie und Palaeontologie, Universitaet Hannover).

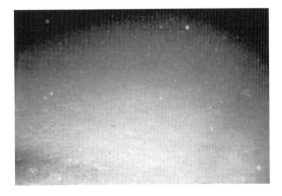

Figure 6. Trawled gorgonian-sponge forest, Northwest Australian shelf (K. Sainsbury, Commonwealth Science and Industry Research Organization)

Figure 7. Untrawled *Oculina* coral reef with gag *Mycteroperca microlepis* and scamp *M. phenax*, Oculina Bank, Florida Atlantic Coast (R. Grant Gilmore, Dynamac Corporation).

The impacts of trawling and dredging are remarkably similar to more familiar terrestrial forms of disturbance such as clear-cutting (Watling and Norse 1998) and chaining, a process for clearing terrestrial vegetation in which a chain is stretched between two bulldozers. Differences, however, include the vastly greater area of the Earth that is trawled and dredged (often repeatedly), the fact that fishing with mobile gear happens out of sight, the generally smaller size of seafloor structure-forming species, the greater vulnerability to disturbance of deepsea ecosystems compared with most forests, and the near absence of effective fish habitat protection even 9 years after the FCMA was amended to require protection of essential fish habitat.

Many fish species of special conservation concern (e.g., Atlantic cod, Nassau grouper *Epinephelus striatus*, gag *Mycteroperca microlepis*, lingcod *Ophiodon elongatus*, bocaccio *Sebastes paucispinis*, yelloweye rockfish *S. ruberrimus*, and darkblotched rockfish *S. crameri*) live in close association with structurally complex seafloor habitats. Undoubtedly, overfishing has contributed substantially to their decline, but both logic and evidence indicate that human-caused disturbance of their habitat from trawling has also contributed substantially to their population decline and failure to recover. The disappearance of these demersal species leads to the conclusion that marine fisheries management needs to advance conceptually and operationally, as have other living resource management paradigms, beyond simply attempting to control human-caused mortality to protecting and managing species within their habitats or—still more broadly—to doing so while maintaining key ecosystem processes.

There is growing concern in the scientific community about loss of benthic structure-formers, including seagrasses, ascidians, bryozoans, tubiculous amphipods, tubiculous polychaetes, tubiculous anemones, pennatulaceans, and—most prominently—corals and sponges. On February 15, 2004, at the annual meeting of the American Association for the Advancement of Science in Seattle, Washington, 1,136 conservation biologists and marine scientists from 69 countries issued an unprecedented Scientists' Statement on Protecting the World's Deep-sea

Figure 8. Trawled *Oculina* coral reef, Oculina Bank Habitat Area of Particular Concern, Florida Atlantic Coast (L. Horn, National Undersea Research Center, University of North Carolina at Wilmington).

Coral and Sponge Ecosystems (Marine Conservation Biology Institute 2004). The statement says that these structure-forming species need to be protected, identifies bottom trawling as the greatest threat to them, and calls for a moratorium on bottom trawling on the High Seas, a ban on bottom trawling within nations' exclusive economic zones where these structurally complex communities are known to occur, prohibition of roller and rockhopper trawling groundgear, research and mapping of deep-sea coral and sponge communities, and establishment of protected areas to conserve these communities.

Areas of seafloor that have been trawled heavily undergo major shifts in species composition, just as clear-cut areas do on land. Reports from a seminal study on the northwestern Australian shelf before and after trawling was initiated (Sainsbury 1987, 1988) showed that removing structure-forming gorgonians and sponges from soft bottoms (Figures 7, 8) changes fish species composition from disturbance-intolerant, commercially high-value species for which benthic structure is important, such as groupers (serranids), snappers (lutjanids) and emperors (lethrinids), to low-value lizardfishes (synodontids) and threadfin bream (nemipterids), which apparently do well on disturbed, flat, seemingly featureless bottoms. There is an intriguing suggestion in the scientific literature (e.g., Engel and Kvitek 1998) that some flatfishes do well on heavily trawled bottoms, perhaps because frequent disturbance increases their preferred foods. Similar patterns have been seen in Alaska and the North Sea. For instance, the Groundfish Forum, an Alaska-focused trawling industry group, says, "flatfish populations are known to be growing very large in proportion to other groundfish and shellfish populations" (see Issues and Initiatives section of www.groundfishforum.org).

In forests, the dramatic change in an ecosystem's dominant species composition is called "type conversion." Under the National Forest Management Act, the U.S. Forest Service must "provide, where appropriate, to the degree practicable, for steps to be taken to preserve the diversity of tree species similar to that existing in the region controlled by the plan." Thus, the U.S. Forest Service—an agency that has long been criticized for favoring resource exploitation over biodiversity protection—nonetheless strongly discourages type conversion on our public lands. In contrast, the U.S. National Marine Fisheries Service allows the commercial fishing industry to cause profound ecosystem shifts on undersea public lands. The FCMA as amended in 1996 requires the National Marine Fisheries Service to consider cumulative effects of fishing. The growing number of lawsuits by conservation organizations (which is approximately the same as the number filed by fishing interests, according to public comments by Eric Bilsky (Oceana, Washington, D.C.) reflects the belief that the National Marine Fisheries Service is not meeting its public trust responsibilities, including those concerning cumulative impacts of fishing. Fisheries managers cannot succeed in meeting public trust responsibilities without restricting or eliminating type conversion of seafloor ecosystems.

Shifting to an Ecosystem-Based Management Paradigm

Leading thinkers in forestry, most notably Jerry Forest Franklin, catalyzed a management paradigm shift in the 1980s by showing the need to maintain structural complexity in the ecosystems with which forest managers were entrusted. Ecosystem management—or, more appropriately, ecosystem-based management (because what we manage is human activities that affect ecosystems)—has rapidly become the new paradigm in forest management (Agee and Johnson 1988; Kohm and Franklin 1997). Franklin (1989) called upon his colleagues to embrace a new, more encompassing way of thinking in saying, "Forestry needs to expand its focus beyond wood production to the perpetuation of diverse forest ecosystems." Growing numbers of managers now accept that forests are about more than producing wood. But, for fisheries managers wedded to the stock-assessment paradigm, marine ecosystems have been about producing meat. At least one marine management regime (Convention on the Conservation of Antarctic Marine Living Resources) has embodied ecosystem-based management since the 1970s, and the most forward-thinking elements of the marine community are increasingly calling for ecosystem-based management (National Research Council 1999; Pew Oceans Commission 2003), but thinking and practice in marine resource management lag behind forest resource management.

It is not too late for fishery biologists and managers to embrace this paradigm shift. A growing number of fisheries biologists are ready to follow their terrestrial counterparts and the urgings of other marine scientists and policy experts in adopting ecosystem-based management. Note that I am not using the term "fisheries ecosystem management" here. Fisheries are just one subset of humans' broader interest in maintaining and recovering the sea's biodiversity. They are a crucial part of the mix, no doubt, but they are not everything.

I would like to offer legislators, managers, and my fellow scientists a new goal and five robust principles that I believe can benefit virtually everyone who catches fish, eats them, watches them, or merely wants to know that they exist. The goal of ecosystem-based management in the sea is to maintain biodiversity and fisheries by protecting and recovering ecosystem composition, structure, and function. From this goal comes five principles:

(1) Stop destructive fishing methods, overfishing, and all other preventable threats. It is essential to stop treating fish populations as if failing to meet management objectives is unimportant and as if quantity and quality of their habitats don't matter.

(2) Allow populations and ecosystems to recover or, where necessary (e.g., Lenihan and Peterson 1998), intervene actively to recover the geological and biological habitat structure that undergirds their species composition and functioning.

(3) Establish comprehensive ocean zoning that fully protects a sizeable fraction of all marine ecosystems and allows sustainable fishing in other zones.

(4) Err on the side of caution; put the resource first. We must use the precautionary approach. As a wise fish biologist and friend says, "No fish, no fishing."

(5) Fully involve fishermen and conservationists in shaping sustainable solutions. We need the wisdom of fishermen as well as the objectivity of public-interest conservationists who do not profit financially from catching fish but who have long-term interests in sustainable ecosystems for their own sake and the benefit of all people, including fishermen.

As the 146 years since the publication of Darwin's The Origin of Species have shown, the idea of evolution worries people who desperately want to cling to what is familiar and constant. Change is frightening. As Samuel Clemens (Mark Twain) noted, "Denial ain't just a river in Egypt." We cannot deny that our evolving technologies have increased our fishing capacity to the point where increasing numbers of fish species are facing ecological, commercial, and perhaps even biological extinction. Therefore, it is incumbent upon fisheries biologists, other marine scientists, conservation advocates, and people entrusted with making decisions to evolve how we manage marine ecosystems to maintain their marine biological diversity and fisheries for our lifetimes and for future generations of both marine life and people.

Acknowledgments

For their visionary ideas of a more rewarding and sustainable relationship with life on Earth, I am indebted to Aldo Leopold, Rachel Carson, and Raymond Dasmann, who, regrettably, I never met, and to Marine Conservation Biology Institute (MCBI) board members John Twiss, Alison Rieser, Jim Carlton, and Michael Soulé, who have blazed new trails in conservation. I thank present and former MCBI staff members Beth Kantrowitz, Jocelyn Garovoy, Fan Tsao, Sara Maxwell, Lance Morgan, Susannah Gill, Hannah Gillelan, and Bill Chandler for their critical input and information. I am indebted to the people who have funded MCBI's work on conserving marine ecosystems, particularly the David and Lucile Packard Foundation, the Pew Charitable Trusts, the Homeland Foundation, the Moore Family Foundation, the Russell Family Foundation, the Edwards Mother Earth Foundation, the Henry Foundation, the Curtis and Edith Munson Foundation, the Sandler Family Supporting Foundation, the Park Foundation, the Richard and Rhoda Goldman Fund, the J.M. Kaplan Fund, the Oak Foundation, the Bullitt Foundation, and the Vidda Foundation, among others. I thank Jerry Franklin and Dave Perry, who were the near equivalents of my dissertation supervisors in forest biology, and Mike Risk, Jim Carlton, Jane Lubchenco, Les Watling, Peter Auster, Ellie Dorsey, Simon Thrush, Callum Roberts, Carl Safina, Dick Haedrich, Larry Crowder, Kristina Gjerde, Phil Kline, Daniel Pauly, Julia Parrish, Mike Hirshfield, Ransom Myers, Boris Worm, Tony Koslow, Martin Willison, Bob George, Don McAllister, Jeremy Jackson, and the conservation biologists, oceanographers, fish biologists, fisheries biologists, fishermen, and others around the world who are helping people become humbler and gentler members of our world ecosystem.

References

Agee, J. K., and D. R. Johnson. 1988. Ecosystem management for parks and wilderness. University of Washington Press, Seattle.

Auster, P. J., and R. W. Langton. 1999. The effects of fishing on fish habitat. Pages 150–187 in L. R. Benaka, editor. Fish habitat: essential fish habitat and rehabilitation. American Fisheries Society, Symposium 22, Bethesda, Maryland.

Bean, M. J., and M. J. Rowland. 1997. Evolution of national wildlife law, 3rd edition. Greenwood Publishing Group, Westport, Connecticut.

Blockstein, D. E. 2002. Passenger pigeon (*Ectopistes migratorius*). In A. Poole and F. Gill, editors. The birds of North America, number 611. The Birds of North America Inc., Philadelphia.

Carson, R. 1962. Silent spring. Houghton Mifflin, Boston.

Chuenpagdee, R., L. E. Morgan, S. Maxwell, E. A. Norse, and D. Pauly. 2003. Shifting gears: assessing collateral impacts of fishing methods in U.S. waters. Frontiers in Ecology and the Environment 1(10):517–524.

Coker, R. E. 1938. Life in the sea II. Scientific Monthly 46(5):416–432.

Daily, G. C., editor. 1997. Nature's services: societal dependence on natural ecosystems. Island Press, Washington, D.C.

Diamond, J. 1997. Guns, germs, and steel: the fates of human societies. W.W. Norton and Company, New York.

Engel, J., and R. Kvitek. 1998. Effects of otter trawling on a benthic community in Monterey Bay National Marine Sanctuary. Conservation Biology 12:1204–1214.

Farnham, T. J. 2002. The concept of biological diversity: the evolution of a conservation paradigm. Doctoral dissertation. Yale University, New Haven, Connecticut.

Federal Aviation Administration. 2003. Federal Aviation Administration National Airspace System Capital Investment Plan Appendix A for Fiscal Years 2004-2008. Available: *www.faa.gov/AboutFAA/PerformanceTargets/PerfTargetDetail/Airline_Fatal_Accident_Rate.htm* (March 2005).

Flannery, T. 2001. The eternal frontier: an ecological history of North America and its peoples. Text Publishing, Melbourne, Australia.

Franklin, J. F. 1989. Toward a new forestry. American Forests 95(11–12):37–44.

Gordon, J. D. M. 2001. Deep-water fisheries at the Atlantic Frontier. Continental Shelf Research 21:987–1003.

Herrington, W. C. 1947. The role of intraspecific competition and other factors in determining the population level of a major marine species. Ecological Monographs 17(3):317–323.

Higgins, E. 1934. Fishery biology, its scope, development, and applications. Quarterly Review of Biology 9(3):275–291.

Irish K., and E. A. Norse. 1996. Scant emphasis on marine biodiversity. Conservation Biology 10(2):680.

Jackson, J. B. C., M. X. Kirby, W. H. Berger, K. A. Bjorndal, L. W. Botsford, B. J. Bourque, R. H. Bradbury, R. Cooke, J. Erlandson, J. A. Estes, T. P. Hughes, S. Kidwell, C. B. Lange, H. S. Lenihan, J. M. Pandolfi, C. H. Peterson, R. S. Steneck, M. J. Tegner, and R. R. Warner. 2001. Historical overfishing and the recent collapse of coastal ecosystems. Science 293:629–638.

Kellert, S. R. 1993. The biological basis for human values of nature. Pages 42–69 in S. R. Kellert and E. O. Wilson, editors. The Biophilia hypothesis. Island Press/Shearwater Books, Washington, D.C.

Kellert, S. R. 2005. Perspectives on an ethic toward the sea. Pages 703–711 in P. W. Barnes and J. P. Thomas, editors. Benthic habitats and the effects of fishing. American Fisheries Society, Symposium 41, Bethesda, Maryland.

Kimball, T. L., and R. E. Johnson. 1978. The richness of American wildlife. Pages 3–17 in H. P. Brokaw, editor. Wildlife and America. Council on Environmental Quality, Washington, D.C.

Kohm, K. A., and J. F. Franklin, editors. 1997. Creating a forestry for the 21st century: the science of ecosystem management. Island Press, Washington, D.C.

Kuhn, T. S. 1970. The structure of scientific revolutions. University of Chicago Press, Chicago.

Lee, W. S. 1986. The National Wildlife Refuge System. Pages 413–460 in R. L. De Silvestro, editor. Audubon wildlife report 1986. National Audubon Society, New York.

Lenihan, H. S., and C. H. Peterson. 1998. How habitat degradation through fishery disturbance enhances impacts of hypoxia on oyster reefs. Ecological Applications 8:128–140.

Leopold, A. 1949. A Sand County almanac, and sketches here and there. Oxford University Press, New York.

Lindholm, J., P. J. Auster, and L. Kaufman. 1999. Habitat-mediated survivorship of juvenile (0-year) Atlantic cod (*Gadus morhua*). Marine Ecology Progress Series 180: 247–255.

Lovejoy, T. 1980a. Changes in biological diversity. Pages 327–332 in The global 2000 report to the President, volume 2, the technical report. Council on Environmental Quality and U.S. Department of State, Washington, D.C.

Lovejoy, T. 1980b. Forward. In M. E. Soulé and B. A. Wilcox, editors. Conservation biology: an evolutionary-ecological perspective. Sinauer Associates, Sunderland, Massachusetts.

MacArthur, R. H., and J. W. MacArthur. 1961. On bird species diversity. Ecology 42:594–598.

Marine Conservation Biology Institute. 2004. 1,136 scientists call for protection of deep-sea corals. Available: *www.mcbi.org/DSC_statement/sign.htm* (March 2005)

Merrett, N. R., and R. L. Haedrich. 1997. Deep-sea demersal fish and fisheries. Chapman and Hall, London.

Mitchell, P. H. 1918. Results and expectations of research on fishery problems. Scientific Monthly 6(1):76–83.

Morgan, L. E., and R. Chuenpagdee. 2003. Shifting gears: addressing the collateral impacts of fishing methods in U.S. waters. Island Press, Washington, D.C.

Muir, J. 1901. Our national parks. Houghton Mifflin, Boston.

Myers, N. 1979. The sinking ark: a new look at the problem of disappearing species. Pergamon Press, New York.

Myers, R. A., and B. Worm. 2003. Rapid worldwide depletion of predatory fish communities. Nature 423:280–283.

National Research Council. 1978. Conservation of germplasm resources: an imperative. National Academy of Sciences, Washington, D.C.

National Research Council. 1999. Sustaining marine fisheries. National Academy Press, Washington, D.C.

National Research Council. 2002. Effects of trawling and dredging on seafloor habitat. National Academy Press, Washington, D.C.

NMFS (National Marine Fisheries Service). 2004. Annual Report to Congress on the Status of U.S. Fisheries - 2003. U.S. Department of Commerce, National Oceanic and Atmospheric Administration, National Marine Fisheries Service, Silver Spring, Maryland. Available: *http://www.nmfs.noaa.gov/sfa/statusoffisheries/statusostocks03/Report_Text.pdf* (March 2005).

Norse, E. A. 1990. Ancient forests of the Pacific Northwest. Island Press, Washington, D.C.

Norse, E. A. 1996. A river that flows to the sea: the marine biological diversity movement. Oceanography 9(1):5–9.

Norse, E. A., and J. T. Carlton. 2003. World Wide Web buzz about biodiversity. Conservation Biology 17(6):1475–1476.

Norse, E. A., and R. E. McManus. 1980. Ecology and living resources: biological diversity. Pages 31–80 in Environmental quality 1980: the eleventh annual report of the Council on Environmental Quality. Council on Environmental Quality, Washington, D.C.

Norse, E. A., K. L. Rosenbaum, D. S. Wilcove, B. A. Wilcox, W. H. Romme, D. W. Johnston, and M. L. Stout. 1986. Conserving biological diversity in our national forests. The Wilderness Society, Washington, D.C.

Norse, E. A., and L. Watling. 1999. Impacts of mobile fishing gear: the biodiversity perspective. Pages 31–40 in L. R. Benaka, editor. Fish habitat: essential fish habitat and rehabilitation. American Fisheries Society, Symposium 22, Bethesda, Maryland.

Paine, R. T. 1966. Food web complexity and species diversity. American Naturalist 100:65–75.

Pauly, D., V. Christensen, S. Guénette, T. J. Pitcher, U. R. Sumaila, C. J. Walters, R. Watson, and D. Zeller. 2002. Towards sustainability in world fisheries. Nature 418:689–695.

Pew Oceans Commission. 2003. America's living oceans: charting a course for sea change. Pew Oceans Commission, Arlington, Virginia.

Pinchot, G. 1947. Breaking new ground. Harcourt, Brace, and Co., New York.

Reid, W. V., and K. R. Miller. 1989. Keeping options alive: the scientific basis for conserving biodiversity. World Resources Institute, Washington, D.C.

Rieser, A., C. G. Hudson, and S. E. Roady. 2005. The role of legal regimes in marine conservation. Pages 362–374 in E. A. Norse and L. B. Crowder, editors. Marine conservation biology: the science of maintaining the sea's biodiversity. Island Press, Washington, D.C.

Risk, M. J. 1972. Fish diversity on a coral reef in the Virgin Islnds. Atoll Research Bulletin 153:1–6.

Sainsbury, K. J. 1987. Assessment and management of the demersal fishery on the continental shelf of northwestern Australia. Pages 465–503 in J. J. Polovina and S. Ralston, editors. Tropical snappers and groupers: biology and fisheries management. Westview Press, Boulder, Colorado.

Sainsbury, K. J. 1988. The ecological basis of multispecies fisheries and management of a demersal fishery in tropical Australia. Pages 349–382 in J. A. Gulland, editor. Fish population dynamics, 2nd edition. Wiley, New York.

Spurgeon, D. 1997. Political interference skewed scientific advice on fish stocks. Nature 388:106.

Thurman, H. V., and E. A. Burton. 2001. Introductory oceanography, 9th edition. Prentice-Hall, Upper Saddle River, New Jersey.

Watling, L., and E. A. Norse. 1998. Disturbance of the seabed by mobile fishing gear: a comparison with forest clear-cutting. Conservation Biology 12:1189–1197.

Weber, M. L. 2002. From abundance to scarcity: a history of U.S. marine fisheries policy. Island Press, Washington, D.C.

Wilson, D. C., and P. Degnbol. 2002. The effects of legal mandates on fisheries science deliberations: the case of Atlantic bluefish in the United States. Fisheries Research 58:1–14.

Wilson, E. O. 1980. Resolutions for the 80s. Harvard Magazine January–February 1980: 22–26.

Wilson, E. O., editor. 1988. Biodiversity. National Academy Press, Washington, D.C.

Symposium Abstract

Symposium abstracts have been reprinted from the Abstract Volume prepared for the Symposium on Effects of Fishing Activities on Benthic Habitats: Linking Geology, Biology, Socioeconomics, and Management without any further review or editing.

Approaches to EFH Management for Alaska Groundfish Fisheries that Fulfill Habitat Management Objectives and Maintain Viable Groundfish Fisheries

J. R. Gauvin[1]

Groundfish Forum, Inc.

While there are still many unknowns and contradictory evidence regarding the effects of on-bottom trawling on benthic habitat and fish populations, there is relatively more agreement among scientists and some stakeholders regarding the need to enact additional measures to manage trawling on hard bottom substrates, particularly those inhabited by concentrations of long-lived and vulnerable invertebrates such as sponges and corals. This is particularly true where these fragile and sessile epifauna occur in waters too deep to be appreciably affected by natural disturbance, and thus where benthic animals and structure would not be expected to be adapted to disturbance events. My paper will present a set of what I believe are non-traditional approaches to management of trawl effects in deep water areas. These alternative measures are designed to meet habitat protection objectives of the M-S Act while allowing the other mandates of the Act to be attained as well. My paper will focus on the hard bottom fisheries off Alaska where, I believe, a set of conditions exist that allow for more flexible approaches to management of trawl fisheries. Fishery managers and stakeholders in Alaska are currently reviewing existing protections for EFH in the context of the court ruling that NMFS had failed to meet NEPA requirements in its earlier analysis of management options for EFH. Some members of the fishing industry in Alaska are interested in consideration of an alternative set of measures from the simplistic time/area closures. We believe that additional sweeping closures could actually compress and intensify fishing effects, possibly triggering a negative outcome for the areas left open to fishing. Hence an alternative approach merits consideration. It is hoped that at least some of the alternatives presented in my paper will have been accepted for analysis in the Environmental Impact Statement currently being developed for our region by the time of Symposium.

---

[1] E-mail: gauvin@seanet.com

Benthic Habitat Characterization and Advanced Technologies and Their Applications

Fish Habitat Studies: Combining High-Resolution Geological and Biological Data

W. Waldo Wakefield[1]

*NOAA Fisheries, Northwest Fisheries Science Center,
2032 Southeast OSU Drive, Newport, Oregon 97365, USA*

Curt E. Whitmire[2]

*NOAA Fisheries, Northwest Fisheries Science Center,
2725 Montlake Boulevard East, Seattle, Washington 98112-2097, USA*

Julia E. R. Clemons[3]

*NOAA Fisheries, Northwest Fisheries Science Center,
2032 Southeast OSU Drive, Newport, Oregon 97365, USA*

Brian N. Tissot[4]

*Program in Environmental Science and Regional Planning,
Washington State University, Vancouver, Washington 98686, USA*

Abstract. Traditionally, estimates of the distribution and abundance of exploited groundfish species and their associated habitats are based on fishery-dependent sampling of catch and fishery-independent survey data using fishing gears such as trawls and a variety of fixed gears. Survey data are often collected as individual samples integrated over a scale of kilometers, compiled at a larger geographic scale (100 km), and extrapolated to an overall estimate of stock size. Considerations of the non-extractive effects of fishing on habitat are extremely limited. Within the past 15 years, a number of collaborations have developed among marine ecologists, fisheries scientists, and marine geologists hallmarked by an integration of sonar mapping of the seafloor with ground-truthing (verification of type of substratum) and direct observation and enumeration of fish and invertebrate populations in the context of their seafloor habitat. An example of such work, targeting a 725-km^2, deepwater, rocky bank from the Oregon continental margin, Heceta Bank, is chronicled in this review. The approaches that have been applied to characterize groundfish–habitat relationships in this region have evolved from stand-alone, human-occupied submersible observations to fully interdisciplinary programs employing the most advanced technologies available to marine research. The combination of multibeam swath mapping sonars and accurate geographic positioning systems has enhanced mapping the seafloor and benthic habitats. The challenge now is to efficiently relate small-scale observations and assessments of animal–habitat associations to the large geographic scales on which fisheries operate. Large-scale benthic habitat characterization at appropriate scales is critical to the accurate assessment of fish stocks on a spatial scale pertinent to fisheries and those natural physical and biological processes and anthropogenic disturbances (e.g., fishing gear impacts) that influence them.

Introduction

The Pacific Fishery Management Council's (PFMC) Fishery Management Plan for the West Coast of the United States currently covers more than 80 species of commercially important West Coast groundfish. Although a number of stocks of West Coast groundfish are in good condition, dramatic declines in several populations have occurred over the past several decades (Ralston 1998; Bloeser 1999; PFMC 2002, 2003). Currently, eight species have been declared "overfished" (current biomass < 25% of estimated maximum exploitable biomass), including seven species of rockfish (family Scorpaenidae; bocac-

---
[1] E-mail: waldo.wakefield@noaa.gov
[2] E-mail: curt.whitmire@noaa.gov
[3] E-mail: julia.clemons@noaa.gov
[4] E-mail: tissot@vancouver.wsu.edu

cio *Sebastes paucispinis*; Pacific ocean perch *S. alutus*; canary rockfish *S. pinniger*; widow rockfish *S. entomelas*; darkblotched rockfish *S. crameri*; cowcod *S. levis*; and yelloweye rockfish *S. ruberrimus*) and lingcod *Ophiodon elongatus* (family Hexagrammidae). The decline in these stocks is shown in Figure 1 as the proportion of the estimated spawning biomass for the period 1970 to 2003. These estimates were derived for the PFMC from recent stock assessment models that incorporate fishery-dependent sampling of catch and fishery-independent survey data. Both sources of data are dependent on fishing gears such as trawls and, thus, represent samples drawn from more homogenous areas of low relief extrapolated to larger and more heterogeneous geographic areas.

The potential bias and uncertainty in groundfish biomass estimates that are limited to trawlable habitats have intensified the need to relate the abundance, distribution, and diversity of groundfish to seafloor habitat type and to study the degree to which fishing activities have affected different habitats (Larson 1980; Richards 1986; O'Connell and Carlile 1993; Yoklavich et al. 2000). Individual pictures taken during remotely operated vehicle (ROV) and human-occupied submersible dives on Heceta Bank, one of the largest and arguably the most important in terms of fisheries of the deepwater rocky banks on the outer continental shelf off Oregon, illustrate the range of habitats for common commercially important demersal fishes (Figure 2). For example, within the collage, Dover sole *Microstomus pacificus* and greenstriped rockfish *S. elongatus* are depicted over mud deposits at the seaward edge of Heceta Bank, and rosethorn rockfish *S. helvomaculatus* and yelloweye rockfish are shown inhabiting differing rock outcrops of high relief within central regions of the bank.

To quantify fish and their associated invertebrate populations in the context of their seafloor habitat, direct observations of the kind presented in Figure 2 are being integrated with geophysical mapping and ground-truthing of seafloor microhabitat types (features of the seafloor on a scale of meters) by a number of North American fisheries biologists, marine ecologists, and marine geologists working in collaboration (e.g., O'Connell and Wakefield 1994; Yoklavich et al. 1997, 2000; Greene et al. 1999; Nasby-Lucas et al. 2002; O'Connell et al. 2002; and review by Reynolds et al. 2001; Martin and Yamanaka 2004). Recently, an expanding number of these regional research programs along the U.S. West Coast (Figure 3) have formed the basis for a coast-wide network of sites where seafloor mapping and direct observation are supporting ongoing habitat-based groundfish research. A systematic approach is emerging for the classification of marine habitats in both shallow and deep water (e.g., see Greene et al. 1999 for one approach), and increasing attention is being given to the inclusion of megafaunal

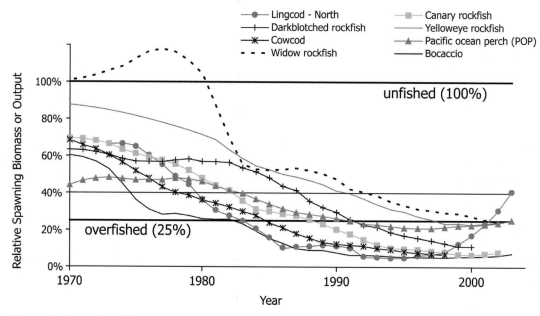

Figure 1. Proportion of the estimated maximum exploitable biomass for eight West Coast groundfish stocks for the period 1970 to 2003. The three horizontal lines represent the overfished threshold (25%), the target level for rebuilding the stock (40%), and the unfished level. Data from stock assessment and fishery evaluations for West Coast groundfish from the Pacific Fisheries Management Council, Portland, Oregon. Data compiled by S. Miller, NOAA Fisheries, Northwest Fisheries Science Center, Seattle.

Figure 2. Representation of the diversity of demersal fishes and benthic habitats of Heceta Bank. The images, acquired during the period 2000–2002, are derived from frame grabs from the *ROPOS* ROV's broadcast-quality video camera and from imaging systems aboard the *Delta* submersible. These images illustrate the range of structural (and to some extent biogenic) habitat for demersal fishes inhabiting the bank. The seven species highlighted with yellow borders were estimated in 2003 to have exploitable biomasses below the overfished threshold.

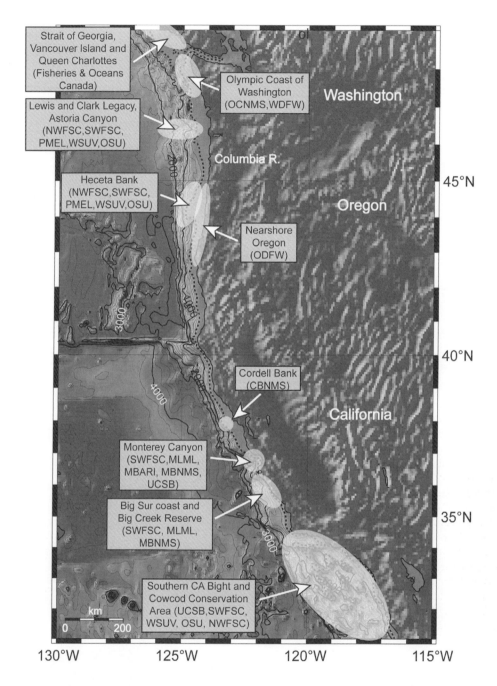

Figure 3. Map of the West Coast of the United States and southern British Columbia, Canada, highlighting growing network of study areas and some of the participating organizations conducting interdisciplinary studies of fish habitat (WDFW, Washington Department of Fish and Wildlife; OCNMS, Olympic Coast National Marine Sanctuary; NWFSC, Northwest Fisheries Science Center; SWFSC, Southwest Fisheries Science Center; PMEL, Pacific Marine Environmental Laboratory; WSUV, Washington State University, Vancouver; OSU, Oregon State University; ODFW, Oregon Department of Fish and Wildlife; CBNMS, Cordell Bank National Marine Sanctuary; MLML, Moss Landing Marine Laboratories; MBNMS, Monterey Bay National Marine Sanctuary; and UCSB, University of California, Santa Barbara).

invertebrates as significant biological components of continental shelf and slope ecosystems. For the remainder of this paper, the history and current focus of work off Oregon will serve as an example of efforts to relate small-scale observations and assessments of animal–habitat associations to the large geographic scales on which demersal fisheries operate. Heceta Bank is an instructive example because it captures the history and progress to date in the study of fish habitat as applied to fisheries research over the past 18 years. It is critical to develop techniques of seafloor habitat characterization that are derived from a spatial scale pertinent to animal distributions, fisheries, and those physical, biological, and anthropogenic (e.g., fishing gear impacts) processes that influence them.

Case Study: Heceta Bank, Oregon

Setting and Historical Work

The following case study from the Oregon margin illustrates a progression in approaches and technologies applied to characterize habitat and the relationships of demersal fishes to habitat. Direct observations of fish and invertebrate assemblages on Heceta Bank (Figure 4) were first carried out in 1987 with the research submersible *Mermaid II*, when a total of 16 dives were completed (Pearcy et al. 1989). On the basis of the discoveries made during this initial exploration, a team of researchers from Oregon State University and the Oregon Department of Fish and Wildlife returned to the bank for 3 consecutive years (1988–1990). They used the human-occupied submersible *Delta* to investigate relationships between the abundance of groundfish (by species and by size-class) and megafaunal invertebrates and the morphology and texture of the seafloor (Hixon et al. 1991; Hixon and Tissot 1992; Stein et al. 1992). A total of 42 dive transects were completed during the 3-year project. The investigators concluded that (1) there were significant statistical relationships between microhabitat variability and fish distribution and abundance by species, and (2) the shallower areas of the bank probably acted as an important nursery ground for juvenile rockfishes (*Sebastes* spp.). At the time of its completion, this work was one of only a few comprehensive habitat–groundfish studies anywhere, and it remains an important baseline for characterizations of Heceta Bank. Even though these studies provided some intriguing initial conclusions and hypotheses about relationships between small-scale seafloor variability and fish populations, the application of the study to the broader setting of the bank (total area > 700 km^2) was problematic because of the limited spatial sampling provided by the submersibles (individual submersible transects were on average 2.3 m wide × 2.8 km long, with patches of contiguous habitat tens of meters in length). The bathymetric maps available at the time of these surveys were based on soundings collected with hydrographic techniques (Figure 4) that are outdated when compared to today's high-resolution swath mapping multibeam sonar (Hughes Clarke et al. 1996). In addition, the maps had a spatial resolution significantly less than the viewing range from a submersible or the length of a transect, and there were no remotely sensed textural data (e.g., acoustic backscatter) to facilitate estimates of habitat coverage beyond the observations made within the submersible viewing range.

Advances in Seafloor Mapping

Side-scan sonars have been used to image seafloor texture for several decades (Clay et al. 1964). The combination of the availability of multibeam swath mapping sonars and accurate geographic positioning systems has enhanced mapping the seafloor and benthic habitats (Hughes Clarke et al. 1996). A survey of Heceta Bank using a Kongsberg Maritime AS EM309 (Horten, Norway) multibeam sonar (30 kHz) was conducted in 1998 (Figure 5) as part of a program to map a larger portion of the Oregon margin (MBARI 2001). The resulting seafloor map, covering an area of 725 km^2, provides a striking contrast to the picture of Heceta Bank available to the investigative team a decade before (Figures 4, 5). The data can be usefully gridded to a horizontal resolution of 5 m on the shallowest portions of the bank (depths of 70 to 150 m) and to 10 m at greater depths down to about 500 m. In 1998, for the first time, investigators could view how the complex terrain creates diverse structural habitat for fish and invertebrate populations. Co-registered imagery of seabed texture (i.e., backscatter) (Figure 5, left panel) and relief (Figure 5, right panel) reveal variations in sediment cover in relation to underlying structures, providing the overall context for establishing groundfish–seafloor relationships. The excellent navigation makes the data essentially seamless so that structures are continuous across the swath boundaries.

Tools for Direct Observation and Sampling

In 2000, a field program was initiated to capitalize on the newly acquired high-resolution seafloor map. Prior to the start of the field program, considerable effort was devoted to an integration of the Hixon et al. (1991) submersible data with the multibeam data in planning for the field program (Nasby 2000; Nasby-Lucas et al. 2002). This field program used the submersible *Delta*, used in Hixon et al. (1991), and an ROV *ROPOS*.

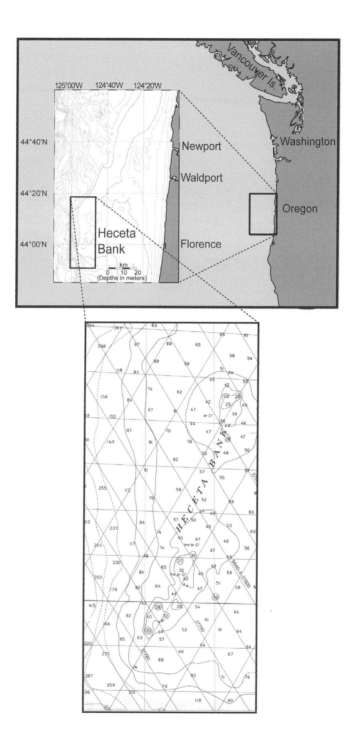

Figure 4. Location of Heceta Bank (top panel) and area sampled during the Hixon et al. (1991) study (bottom panel) as represented on a NOAA National Ocean Service "standard nautical chart" (scale = 1:191,730 and soundings in fathoms).

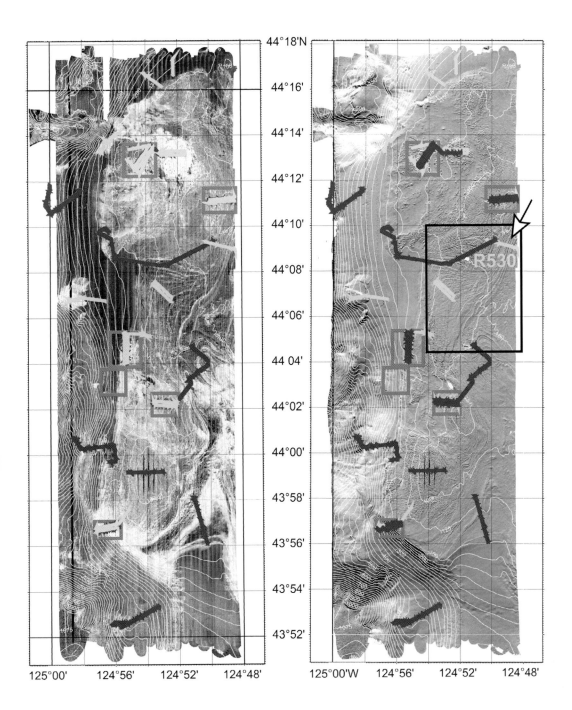

Figure 5. Heceta Bank study areas and bathymetric contours underlain by multibeam backscatter (left panel) where decibel values have been converted to 255 gray-scale values (Kongsberg Maritime AS EM300, Horten, Norway). White depicts high reflectivity and grays depict low reflectivity. The right panel is underlain by sun-illuminated multibeam bathymetry. *ROPOS* ROV dive transects are overlaid in yellow for 2000 and purple for 2001, with orange polygons outlining the historical stations for Hixon et al. (1991) (10-m grid cell size and contour interval = 25 m).

ROPOS (Remotely Operated Platform for Ocean Science) is a powerful, tethered ROV capable of diving to 5,000 m for more than 24 h at a time, with an array of imaging and sampling tools to view and sample both geological structures and marine invertebrates (Shepherd and Wallace 2001; Wallace and Shepherd 2003). The vehicle was operated from a relatively large, dynamically positioned ship (R/V *Ronald H. Brown*) with accommodations for a large, interdisciplinary scientific party. Navigation and visualization tools allowed investigators to locate submersible and ROV transects and associated samples within the high-resolution bathymetric maps. The topographic terrain associated with the data from an ROV dive (Figure 6) shows the sampling resolution and the close match between the *ROPOS* and EM300 bathymetry.

Survey of Marine Benthic Invertebrates

Studies of deepwater fish habitat (i.e., below SCUBA depth) have typically considered geological features and largely ignored benthic invertebrates. This is a serious shortcoming because benthic invertebrates are an important component of the biodiversity, are energeti-

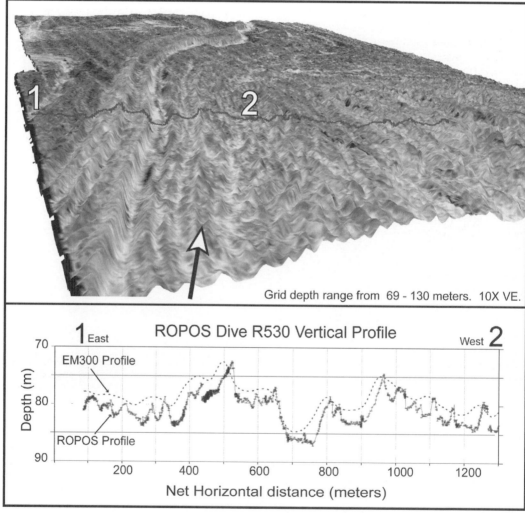

Figure 6. Perspective view of the northern shoal portion of Heceta Bank in side-lit bathymetry looking from the northeast to the southwest, showing details of outcropping geology and *ROPOS* dive R530. The box in Figure 5 shows the context of the perspective view. Cross-sectional profile of the seafloor depth measured from the *ROPOS* ROV overlaid on bathymetry from the EM300 sonar (R. Embley, NOAA Pacific Marine Environmental Laboratory, unpublished data).

cally significant ecological components of continental shelf and slope ecosystems, and provide biogenic structure that may be vital for groundfish. There has also been recent concern that fishing activities may be directly damaging to structure-forming invertebrates such as coldwater corals, sponges, and other large invertebrates (Watling and Norse 1998; Freese et al. 1999; Krieger 2001; Dayton et al. 2002; NRC 2002). To address these issues, invertebrate studies conducted at Heceta Bank consisted of two major components: (1) collection, curation, and identification of voucher specimens, and (2) quantitative ecological analysis of invertebrate distributions and their relationships to habitat and fishes.

Collaborations with curators at the Natural History Museum of Los Angeles County (NHMLAC) proved the value of combining interactive geological and taxonomic fieldwork and the usefulness of ROVs such as *ROPOS* for the collection of intact, identifiable marine invertebrate specimens. Video recordings from ROV transects of the seafloor provided a visual record of the diversity and distribution of megafaunal invertebrates (size > 5 cm) in a range of habitats. At selected sites, video recordings documented the appearance and behavior in situ. Samples of a range of lithologies from boulders and cobbles to finer sediments were sampled using both manipulators and suction samplers. These furnished specimens of smaller, cryptic fauna that are not visible in the video record. Manipulator arms were used to collect specimens of larger organisms. A combination of sampling techniques yielded a diversity of taxa, including ascidians (sea squirts), brachiopods (lamp shells), bryozoans (lace animals), cnidarians (gorgonian corals, sea anemones, sea pens, sea fans), crustaceans (crabs, shrimp, isopods), echinoderms (feather stars or crinoids, sea urchins, sea stars, sea cucumbers, brittle stars), mollusks (bivalves, coiled gastropods, chitons, nudibranchs), poriferans (sponges), sipunculans and polychaetes (worm-like animals), and other groups. The excellent condition of the animals recovered with *ROPOS* permitted the identification of symbiotic relationships such as the association between polychaete worms and deepwater sea stars (G. Hendler, NHMLAC, personal communication). Also, high-quality tissue samples were obtained, which will be critical for conducting future molecular investigations. Specimens have been archived at the NHMLAC.

Building on an emerging database of invertebrate taxa on the continental shelf of the eastern North Pacific, investigators at Washington State University, Vancouver (B. Tissot, unpublished data), have conducted counts of megafaunal invertebrates along video transects and estimated densities of megafauna within patches of contiguous seafloor meso-macrohabitat types (Greene et al. 1999). These data have been used to establish quantitative relationships between invertebrates, fishes, and lithology on Heceta Bank (Figure 7, left panel) and have documented the abundance and distributions of 82 taxa between 2000 and 2002 using methods identical to those used in the earlier 1988–1990 studies (Hixon et al. 1991). Results have confirmed that there are strong relationships between benthic invertebrates, seafloor habitat, and groundfish assemblages (Stein et al. 1992) and that some abundant invertebrates, such as the crinoid *Florometra serratissima*, may also provide structural habitat for juvenile rockfishes (*Sebastes* spp.) (Puniwai 2002). Using geostatistical interpolative tools for spatial analysis (e.g., kriging), invertebrate abundances can be extrapolated across wide areas of habitat, providing visualizations of predicted invertebrate communities and their associated groundfish assemblages (Figure 7, right panel).

Level of Sampling Effort

For the time period of 2000 to 2002, sampling with *ROPOS* (26 transects, Figure 5) and *Delta* (23 transects) was distributed between the six historical sites (Hixon et al. 1991) and new exploratory sites located to maximize the overall coverage of the bank across a wide range of backscatter intensity. For each submersible and ROV transect, data were compiled from reviews in the laboratory of recorded audio and video tapes to determine fish and megafaunal invertebrate abundance, substrata, and geology.

Classification of Substratum (Seafloor)

There are many classification schemes for marine benthic habitats (Greene et al. 1999). The term substratum, as used in historical and ongoing investigations of Heceta Bank, refers to lithology and geomorphology of the seafloor. For consistency, Heceta Bank substratum types were categorized using the Hixon et al. (1991) classification system and two-letter code (also see Nasby-Lucas et al. 2002), variations of which are being increasingly applied along the U.S. West Coast (e.g., Yoklavich et al. 2000). The classification system uses seven classes of substratum across a range of increasing sediment grain size and relief: mud (M), sand (S), pebble (P, diameter < 6.5 cm), cobble (C, 25.5 cm > diameter > 6.5 cm), boulder (B, diameter > 25.5 cm), flat rock (outcrop, F, low vertical relief), and rock outcrop with vertical relief forming a scarp or ridge (R, high vertical relief). The two-letter code denotes primary substratum (>50% of the field of view)

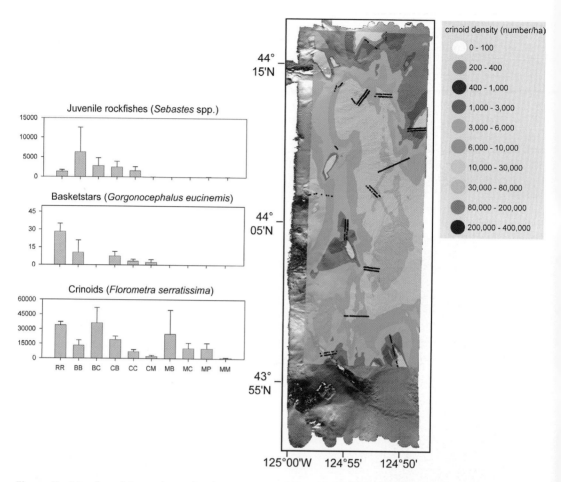

Figure 7. Megafaunal invertebrate distributions on Heceta Bank. Juvenile rockfish distributions (left panel) across major habitats in relation to the distributions of two invertebrates that form physical structure, basketstars, and crinoids (error bars are 1 SE; two-letter code for substrata: M = mud, S = sand, P = pebble, C = cobble, B = boulder, F = flat rock, R = rock ridge; see section on classification of seafloor and Figure 8). Geospatial kriging analysis (right panel) of crinoid abundance across Heceta Bank overlain on sun-illuminated topography, showing areas that range through a spectrum of colors from high (red) to low (white) relative abundance overlain by ROPOS tracklines in black (after Puniwai 2002).

and secondary substratum (>20% of the field of view). If only one substratum was visible, or the primary substratum covered more than 70% of the field of view of the seafloor, then the primary substratum was indicated by both letters of the code (e.g., RR for a high relief rock outcrop). Ultimately, 10 dominant types of substratum were identified on Heceta Bank with the two-letter scheme. Examples of the seven seafloor substrata are shown in Figure 8 as video frame grabs from the *ROPOS* broadcast-quality video camera. For all historical and ongoing investigations at Heceta Bank, the occurrence of both fishes and benthic invertebrates is georeferenced to type of substratum.

Submersible dives conducted in 2002 on Heceta Bank sought to replicate the 1988–1990 dives (Figure 9). The right panel of Figure 9 shows greater detail of the northwestern portion of the bank (gridded at a resolution of 5 m), with areas of both high-relief rock and low-relief sediment and illustrates the resolution of the biological and geological data obtained from the *Delta* submersible. At the end of one of the dive transects (d5702) at station 4, the submersible crossed over a 15-m high rock pinnacle measuring approximately 100 m × 300 m (pinnacles are defined as high-relief mesohabitats with a scale of tens of meters to a kilometer by Greene et al. [1999]). The in situ observer, a veteran of many dives on Heceta Bank, made remarks about the strikingly high concentrations of rockfishes associated with this pinnacle (yelloweye

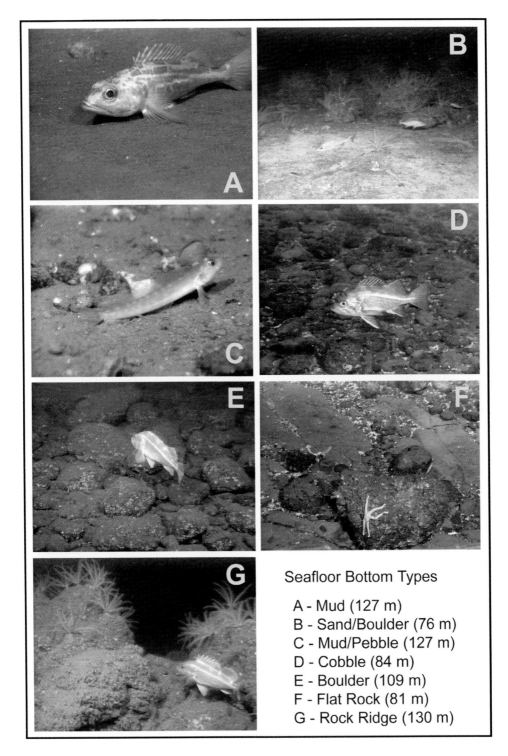

Figure 8. Examples of the seven seafloor types or classes. Fish within these images are: (A) greenstriped rockfish *Sebastes elongatus*, (B) redstripe rockfish *S. proriger* in foreground, (C) northern ronquil *Ronquilus jordani*, (D) canary rockfish *S. pinniger*, (E) yelloweye rockfish *S. ruberrimus*, and (G) yelloweye rockfish.

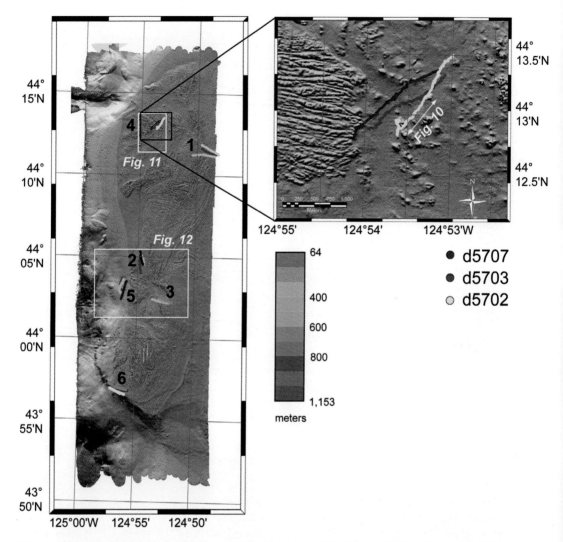

Figure 9. Sun-illuminated topography of Heceta Bank overlaid with 2002 *Delta* submersible dive transects and the historical stations (1–6 numbered in black; left panel), and station 4 topography showing details of *Delta* transects (right panel). See text for description of dive d5702 (yellow line). Locations are shown for Figures 10, 11, and 12.

rockfish, bocaccio, and widow rockfish). Observer and video data were used to quantify the relationship between fish abundance and associated benthic habitats on this transect (Figure 10). Rockfish were found concentrated in the rock ridge and cobble areas and were exceptionally abundant in the area of one isolated rock pinnacle, confirming the observer's impressions. The relative importance of isolated features such as rock pinnacles has not been studied adequately; however, an area of volcanic pinnacles in the eastern Gulf of Alaska is the focus of a recently established marine reserve (O'Connell et al. 1998). This pinnacle area (with pinnacles ranging in size from 300 to 0.8 m at their widest dimension and rising 66 to 92 m above the surrounding seafloor) provides habitat for spawning, breeding, growth, and maturity for a number of commercially important groundfish species and is extremely productive.

Whitmire (2003) used multibeam bathymetry data to identify areas of high topographic relief such as pinnacles and other high-relief mesohabitats that are known to be significant to groundfish. To differentiate areas of high relief (e.g., ridges, pinnacles) from those of little or no relief (e.g., gullies, flat seabed), a topographic position index (TPI) was derived for the entire multibeam survey area (Figure 11). Topographic position index is a

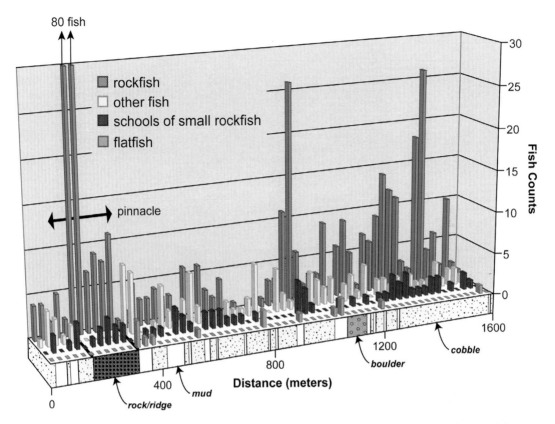

Figure 10. Fish abundance relative to type of substratum along *Delta* dive transect d5702. Fish were grouped into four categories: rockfish, small schooling rockfish (<10 cm), flatfish, and other fish not in previous three categories. Fish abundance is plotted as total numbers per 1-min interval along the transect.

measure of relative elevation, comparing the elevation of each pixel in a bathymetric grid to the mean elevation of a surrounding neighborhood. Positive TPI values correspond to pixels that are higher than the mean elevation, while negative TPI values represent locations that are lower than the mean elevation within the neighborhood. Dive transects, overlaid on sun-illuminated topography (Figure 11) where TPI values have been used to locate pinnacles, ridges, and gullies, illustrate the distribution and abundance of these geomorphologic features, including the pinnacle observed during the *Delta* dive. This illustrates that the depth and dimensions of potentially important features for fish, such as pinnacles, can be quantitatively determined from multibeam bathymetry.

Spatial Analysis

To facilitate spatial analysis of habitat and biological data along transects on Heceta Bank, the dynamic segmentation geographic information system (GIS) data structure developed by Environmental Systems Research Institute (ESRI, Redlands, California) was used (Nasby 2000). Dynamic segmentation permits the visualization of changing "events" along a linear feature (e.g., submersible transect). This spatial analytical tool was employed to show changes in type of seafloor and fish abundance along historical submersible transects (Nasby 2000; Nasby-Lucas et al. 2002) and more recently to depict changes in seafloor type along recent ROV transects at Heceta Bank (Whitmire 2003). Figure 12 illustrates the utility of dynamic segmentation when used in the context of co-registered topography and acoustic backscatter (reflected signal amplitude) imagery. The overlay of dive transects, depicting classified in situ observations (Figure 12, panel A), shows the match between changes in seafloor habitat type observed in the submersible videos to the boundaries between rock outcrop (ridges and pinnacles), mixed substratum (boulder, cobble, and sand), and low relief muds interpreted in the multibeam imagery. For the western-most dives, the backscatter is relatively low (darker gray), corresponding to mud observed in the

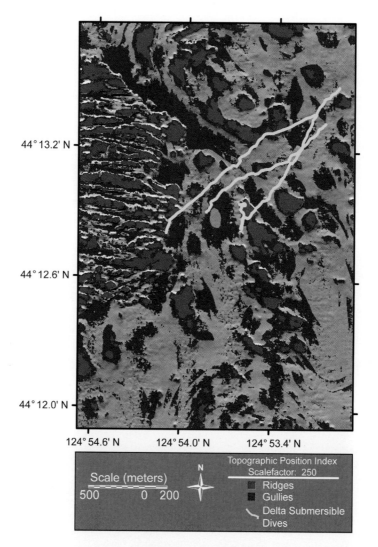

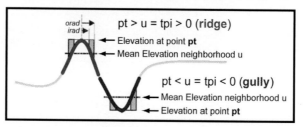

Figure 11. Sun-illuminated topography for a portion of historical station 4 (see Figure 9) with 2002 *Delta* submersible transects d5702, d5703, and d5707 (highlighted in yellow). Topographic position index (tpi) was used to identify ridge and gully features by comparing the depth of each pixel (pt) to the mean depth (u) of all pixels within a specific neighborhood (see Weiss 2001). Scale factor = outer radius (m), dem = digital elevation model, annulus = the shape of the analysis window, focalmean = a function that calculates the mean depth of all pixels that have centers within the annulus, irad = inner radius of annulus (number of pixels), and orad = outer radius of annulus (number of pixels).

video data (yellow segments). In contrast, within the more complex terrains (variable backscatter and structure) in the eastern portion of this region, substratum types observed were rock outcrop (pink segments) and mixed (blue segments). These habitat trends correspond to varying abundances of Dover sole and rosethorn rockfish, two species commonly associated with mud and rockier habitats, respectively (Hixon et al. 1991). The greatest abundance of Dover sole encountered (Figure 12, panel B, purple segments) correlates to mud habitats, whereas rosethorn rockfish (Figure 12, panel C) were found in lowest abundance in these same areas. Note, however, that along one isolated transect, rosethorn rockfish were found in high abundance (purple segment). This suggests that the resolution of the multibeam data is insufficient in resolving isolated boulder, cobble, and pebble patches in

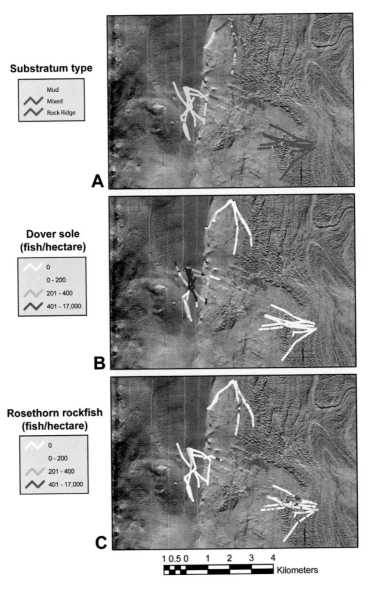

Figure 12. Application of dynamic segmentation data structure: submersible transects segmented by (A) substratum type and abundance of (B) Dover sole and (C) rosethorn rockfish. Transects are overlaid on co-registered topography and backscatter imagery of the central portion of Heceta Bank where backscatter imagery has been draped over bathymetry (see Figure 9 for location). Modified from Nasby-Lucas et al. (2002).

predominantly mud terrain. This is supported by the historical *Delta* submersible video data that show a small percentage of MP, MC, and MB in this region (Nasby 2000).

In addition to visualization, Nasby-Lucas et al. (2002) used dynamic segmentation to identify areas of homogeneous terrain and texture adjacent to submersible transects, thereby facilitating the extrapolation of fish density measurements. In another example, Whitmire (2003) used dynamic segmentation to compile a database of quantitative parameters derived from the multibeam data (e.g., slope, backscatter) that describe substratum types on a pixel-by-pixel basis.

Classifying and Mapping Habitats Over Large Areas

Habitat-specific abundances of fishes and invertebrates observed and enumerated on the scale of isolated transects can be expanded to larger geographic areas by a number of methods of seafloor classification ranging from expert visual interpretation of acoustic imagery to more automated and quantitative interpretative techniques. Commercial products and published algorithms are available to automate the classification of seafloor characteristics from sonar data (e.g., Quester Tangent QTC View [Sidney, BC] and Triton-Elics SeaClass [Watsonville, California]). In addition, several investigators have developed new analytical techniques that utilize quantitative parameters from high-resolution multibeam bathymetry and backscatter imagery (Dartnell 2000; Diaz 2000; Whitmire 2003). Whitmire (2003) applied spatial analytical methods to the 1998 Heceta Bank multibeam sonar data set in an effort to model the geographic extent of habitat classes that have been statistically shown to correlate with a variety of resident demersal fish distributions (Hixon et al. 1991). Whitmire's model was based on four quantitative parameters (e.g., depth, slope, TPI, backscatter) derived from topographic and textural patterns in the multibeam data. By identifying correlations between qualitative seabed characteristics (i.e., substratum types) observed in situ and quantitative parameters derived from the sonar data, Whitmire (2003) predicted the coverage of four habitat classes: mud (unconsolidated), sand (unconsolidated), boulder/cobble (highly reflective), and rock outcrop (high vertical relief) (Figure 13). With the availability of a complete map of habitat structure for the bank, and upon completion of the analysis of all of the video transects, it will be possible to extrapolate habitat-specific estimates of density for selected groundfish species from submersible and ROV transects for the entirety of Heceta Bank.

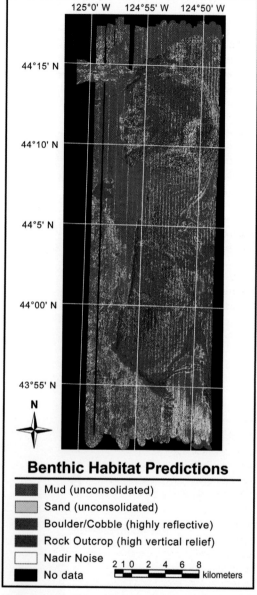

Figure 13. Predicted benthic habitats for Heceta Bank, Oregon. The map shows the output of the decision tree created in Erdas Imagine Knowledge Engineer (Leica Geosystems GIS & Mapping, Atlanta, Georgia; a raster-based GIS software for extracting information from images). An algorithm was employed to remove some of the nadir noise.

Future Directions

The final stage of habitat-based groundfish research on Heceta Bank and in other regions will need to relate information on habitat associations of commercially important fishes to other subdisciplines in fisheries, including design of surveys, stock assessments, identification of essential fish habitat, assessment of risks to habitat from human activities, and input to siting of marine reserves. In the eastern Gulf of Alaska, O'Connell and Carlile (1993) and O'Connell et al. (2002, 2003) have based stock assessments of yelloweye rockfish on a combination of habitat-specific abundances from human-occupied submersible dives and estimates of the distribution of rocky habitat, the preferred habitat for this species of rockfish. In the early stages of their work, O'Connell et al. based the extent of rocky habitat on information gleaned from National Ocean Service "standard nautical charts," a network of submersible dive transects, and commercial logbook data from the directed demersal shelf rockfish fishery. Subsequently, they allocated resources to new side-scan and multibeam sonar surveys in high-priority areas and have been able to identify areas of rocky habitat with a higher degree of accuracy and update and improve their areal estimate of preferred (rocky) habitat, using the modified classification scheme of Greene et al. (1999). However, for the purposes of commercial fisheries management, O'Connell et al. have collapsed their detailed seafloor classification into two substrata: rocky and non-rocky. This is because in areas without geophysical survey data, there is no way to quantify detailed habitat categories. They, therefore, rely on an estimate of yelloweye rockfish abundance for all rocky habitats combined and the estimate of total rocky habitat to estimate biomass. Fish abundance changes between different types of rocky substrata such that the complexity and the presence of suitably sized refuge spaces are key to the occurrence of certain species and assemblages (O'Connell and Carlile 1993; Wakefield et al. 1998). A more detailed understanding of the associations of fishes with habitat and the distribution of habitat types may provide better predictive capability for the occurrence of groundfish species that could be used in stock assessments and management. As the regional coverage of multibeam sonar survey data and the resulting geomorphologic parameters become more complete, there will be increasing need for automated methods for classifying and mapping habitat into more discrete classes.

Off the West Coast of the United States (California, Oregon, and Washington), there are currently several programs underway to create GIS databases for benthic habitats (e.g., Goldfinger et al. 2003; Greene et al. 2003). An overarching goal of these efforts is to create a comprehensive and easily accessible, multi-layered GIS of the geological and geophysical data for the West Coast. An interim seafloor lithology map for Oregon and Washington, version 1.1, has recently been completed (Goldfinger et al. 2003; Romsos 2004). This GIS database capitalizes on many years of data collection, mapping, and numerous data sources. The construction of data–density and data–quality layers provides a guide to users of this database (Romsos 2004). The Oregon–Washington database has been linked to a parallel database for California compiled by Gary Greene and Joe Bizzarro at the Center for Habitat Studies at Moss Landing Marine Laboratories (Greene et al. 2003). The Oregon portion of the resulting West Coast-wide map is shown in Figure 14. The linkage was facilitated by the development of a unifying seafloor classification system (based on Greene et al. 1999). The combined databases represent the first coast-wide delineation of rocky and unconsolidated substrata. The lithology map and underlying GIS database are critical elements in a current project to identify essential fish habitat for West Coast groundfish and conduct a comparative risk assessment from anthropogenic impacts (e.g., commercial fishing gears; PSMFC 2004). It is envisioned that evolving versions of this habitat GIS, when combined with additional layers of information (e.g., distribution and abundance of commercial fishes and benthic invertebrates, fishing effort, oceanographic conditions), will become a tool for both researchers and resource managers working along the West Coast.

The challenge now is to efficiently relate small-scale observations and assessments of animal–habitat associations to the large geographic scales on which fisheries operate. Significant progress has been made in mapping the seafloor of the continental shelf and slope within regions of the United States EEZ (exclusive economic zone), but large areas remain uncharacterized and unmapped at scales necessary for managing marine resources. Although expensive, coast-wide seafloor habitat characterization is critical to the accurate assessment of fish stocks on a spatial scale pertinent to fisheries and those physical, biological, and anthropological (e.g., fishing gear impacts) processes that influence them.

Acknowledgments

We would especially like to thank the following colleagues who are part of the Heceta Bank research group: Bob Embley, Mary Yoklavich, Bill Barss, Chris Goldfinger, Gordon Hendler, Mark Hixon, Susan Merle, Noelani Puniwai, David Stein, Ángel Valdés, and Keri York. Clare Reimers, Ric Brodeur, Phil Levin, three anonymous reviewers, and the book's editors commented on drafts of this manuscript. Chris Goldfinger, Chris Romsos, and Rondi Robison have been responsible for the development of the interim lithological map for Oregon and Washington and Gary Greene and Joseph Bizzarro for California. Susan Merle contributed to a number of the figures for this paper. Steve Copps from the NOAA Fisheries northwest

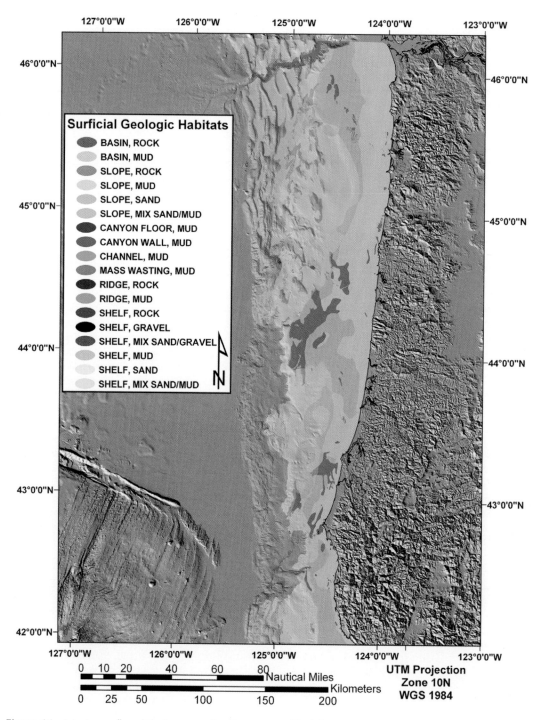

Figure 14. Interim seafloor lithology map for Oregon (modified from Goldfinger et al. 2003; Romsos 2004).

region provided constructive input. The Heceta Bank project was funded by the West Coast and Polar Programs Undersea Research Center of NOAA's National Undersea Research Program, NOAA's Office of Ocean Exploration, the Northwest and Southwest Fisheries Science Centers, and NOAA's Pacific Marine Environmental Laboratory.

Several individuals involved in various aspects of this research are funded by NOAA through the Cooperative Institute of Marine Resource Studies at Oregon State University. We would like to thank the professional personnel who operate the submersibles, ROVs, and ships involved in the Heceta Bank program (*Mermaid II*, *Delta*, *ROPOS*, *Ocean Alert*, *Ronald Brown*, *Pirateer*, and *Velero IV*).

References

Bloeser, J. A. 1999. Diminishing returns: the status of West Coast rockfish. Pacific Marine Conservation Council, Astoria, Oregon.

Clay, C. S., J. Ess, and I. Weisman. 1964. Lateral echo sounding of the ocean bottom on the continental rise. Journal of Geophysical Research 69:3823–3835.

Dartnell, P. 2000. Applying remote sensing techniques to map seafloor geology/habitat relationships. Masters thesis. San Francisco State University, San Francisco.

Dayton, P. K., S. Thrush, and F. C. Coleman. 2002. Ecological effects of fishing in marine ecosystems of the United States. Pew Oceans Commission, Arlington, Virginia.

Diaz, J. V. M. 2000. Analysis of multibeam sonar data for the characterization of seafloor habitats. Masters thesis. University of New Brunswick, Fredericton.

Freese, L., P. J. Auster, J. Heifetz, and B. L. Wing. 1999. Effects of trawling on seafloor habitat and associated invertebrate taxa in the Gulf of Alaska. Marine Ecology Progress Series 182:119–126.

Goldfinger, C. C. Romsos, R. Robison, R. Milstein, and B. Myers. 2003. Interim seafloor lithology maps for Oregon and Washington, version 1.1. Active Tectonics and Seafloor Mapping Laboratory, Publication 03-01, CD-ROM, Oregon State University, Corvallis.

Greene, H. G., M. M. Yoklavich, R. M. Starr. V. M. O'Connell, W. W. Wakefield, D. E. Sullivan, J. E. McCrea, and G. M. Cailliet. 1999. A classification scheme for deep seafloor habitats. Oceanologica Acta 22:663–678.

Greene, H. G., J. J. Bizzarro, M. D. Erdey, H. Lopez, L. Murai, S. Watt, and J. Tilden. 2003. Essential fish habitat characterization and mapping of California Continental Margin. Moss Landing Marine Laboratories, Technical Publication Series Number 2003-01, Moss Landing, California.

Hixon, M. A., B. N. Tissot, and W. G. Pearcy. 1991. Fish assemblages of rocky banks of the Pacific Northwest (Heceta, Coquille, and Daisy Banks). U.S. Minerals Management Service, OCS Study 91-0052, Final Report, Camarillo, California.

Hixon, M. A., and B. N. Tissot. 1992. Fish assemblages of rocky banks of the Pacific Northwest. U.S. Minerals Management Service, OCS Study 91-0025, Final Report Supplement, Camarillo, California.

Hughes Clarke, J. E., L. A. Mayer, and D. E. Wells. 1996. Shallow-water imaging multibeam sonars: a new tool for investigating seafloor processes in the coastal zone and on the continental shelf. Marine Geophysical Research 18:607–629.

Krieger, K. J. 2001. Coral (*Primnoa*) impacted by fishing gear in the Gulf of Alaska. Pages 106–116 in J. H. M. Willison, J. Hall, S. E. Gass, E. L. R. Kenchington, M. Butler, and P. Doherty, editors. Proceedings of the First International Symposium on Deep-Sea Corals. Ecology Action Centre and Nova Scotia Museum, Halifax.

Larson, R. J. 1980. Competition, habitat selection, and the bathymetric segregation of two rockfish (*Sebastes*) species. Ecological Monographs 50:221–239.

Martin, J. C., and K. L. Yamanaka. 2004. A visual survey of inshore rockfish abundance and habitat in the southern Strait of Georgia using a shallow-water towed video system. Canadian Technical Report of Fisheries and Aquatic Sciences 2566.

MBARI (Monterey Bay Aquarium Research Institute). 2001. MBARI, northern California and Oregon margin multibeam survey. MBARI, Digital Data Series Number 5, CD ROM, Moss Landing, California.

Nasby, N. M. 2000. Integration of submersible transect data and high-resolution sonar imagery for a habitat-based groundfish assessment of Heceta Bank, Oregon. Masters thesis. Oregon State University, Corvallis.

Nasby-Lucas, N. M., B. W. Embley, M. A. Hixon, S. G. Merle, B. N. Tissot, and D. J. Wright. 2002. Integration of submersible transect data and high-resolution multibeam sonar imagery for a habitat-based groundfish assessment of Heceta Bank, Oregon. Fishery Bulletin 100:739–751.

NRC (National Research Council). 2002. Effects of trawling and dredging on seafloor habitat. Phase 1: effects of bottom trawling on seafloor habitats. NRC, Committee on Ecosystem Effects of Fishing, National Academy Press, Washington, D.C.

O'Connell, V. M., and C. W. Carlile. 1993. Habitat-specific density of adult yelloweye rockfish *Sebastes ruberrimus* in the eastern Gulf of Mexico. Fishery Bulletin 91:304–309.

O'Connell, V. M., and W. W. Wakefield. 1994. Workshop proceedings: applications of side-scan sonar and laser line systems in fisheries research. Alaska Department of Fish and Game, Special Publication 9, Juneau.

O'Connell, V. M., W. W. Wakefield, and H. G. Greene. 1998. The use of a no-take marine reserve in the eastern Gulf of Alaska to protect essential fish habitat. Pages 127–134 in M. M. Yoklavich, editor. Marine harvest refugia for West Coast rockfish: a workshop. NOAA Technical Memorandum NMFS-SWFSC-255.

O'Connell, V., C. Brylinsky, and D. Carlile. 2002. Demersal shelf rockfish stock assessment for 2003. Alaska Department of Fish and Game, Division of Commercial Fisheries, Regional Information Report 1J02-44, Juneau.

O'Connell, V., C. Brylinsky, and D. Carlile. 2003. Demersal shelf rockfish stock assessment and fishery evaluation report 2004. Alaska Department of Fish and Game, Division of Commercial Fisheries, Regional Information Report 1J03-39, Juneau.

Pearcy, W. G., D. L. Stein, M. A. Hixon, E. K. Pikitch, W. H. Barss, and R. M. Starr. 1989. Submersible observations of deep-reef fishes of Heceta Bank, Oregon. Fishery Bulletin 87:955–965.

PFMC (Pacific Fisheries Management Council). 2002. Status of the Pacific Coast groundfish fishery through 2001 and acceptable biological catches for 2002. Pacific Fisheries Management Council, Portland, Oregon.

PFMC (Pacific Fisheries Managament Council). 2003. Status of the Pacific Coast groundfish fishery through 2003 and stock assessment and fishery evaluation, volume 1. Pacific Fisheries Management Council, Portland, Oregon.

PSMFC (Pacific States Marine Fisheries Commission). 2004. Pacific Coast groundfish EFH, analytical framework, version 4 (10 February 2004). Prepared for the PSMFC by MRAG Americas, Inc., Tampa, Florida; Terra Logic GIS, Inc., Stanwood, Washington; NMFS Northwest Fisheries Science Center, FRAM Division, Seattle; and NOAA Fisheries Northwest Region, Seattle.

Puniwai, N. P.-F. 2002. Spatial and temporal distribution of the crinoid *Florometra serratissma* on the Oregon continental shelf. Masters thesis. Washington State University, Vancouver.

Ralston, S. 1998. The status of federally managed rockfish on the U.S. West Coast. Pages 6–16. *in* M Yoklavich, editor. Marine harvest refugia for West Coast rockfish: a workshop. NOAA Technical Memorandum NMFS-SWFSC-255.

Reynolds, J. R., R. C. Highsmith, B. Konar, C. G. Wheat, and D. Doudna. 2001. Fisheries and fisheries habitat investigations using undersea technology. Pages 812–820 *in* Proceedings of Oceans 2001, MTS/IEEE Conference, volume 2. Marine Technology Society, Columbia.

Richards, L. J. 1986. Depth and habitat distributions of three species of rockfish (*Sebastes*) in British Columbia: observations from the submersible PISCES IV. Environmental Biology of Fishes 17:13–21.

Romsos, C. G. 2004. Mapping surficial geological habitats of the Oregon continental margin using integrated interpretive and GIS techniques. Masters thesis. Oregon State University, Corvallis.

Shepherd, K., and K. Wallace. 2001. Deep precision deployment and heavy package recoveries by the ROPOS R.O.V. system. Pages 1100–1105 *in* Proceedings of Oceans 2001, MTS/IEEE Conference, volume 2. Marine Technology Society, Columbia.

Stein, D. L., B. N. Tissot, M. A. Hixon, and W. Barss. 1992. Fish-habitat associations on a deep reef at the edge of the Oregon continental shelf. Fishery Bulletin 90:540–551.

Wakefield, W. W., V. M. O'Connell, H. G. Greene, D. W. Carlile, and J. E. McRea. 1998. The role of sidescan sonar in seafloor classification with a direct application to commercial fisheries management. International Council for the Exploration of the Sea, CM1998/O:36, Copenhagen.

Wallace, K., and K. Shepherd. 2003. The streaming data management challenge: integrating and logging multiple channels of real-time data from various sources in a flexible and familiar environment. Sea Technology 44:37–44.

Watling, L., and E. A. Norse. 1998. Disturbance of the seabed by mobile fishing gear: a comparison to forest clearcutting. Conservation Biology 12:1180–1197.

Weiss, A. D. 2001. Topographic position and landform analysis. Indus Corporation, Vienna, Virginia.

Whitmire, C. E. 2003. Using remote sensing, in situ observations, and geographic information systems to map benthic habitats at Heceta Bank, Oregon. Masters thesis. Oregon State University, Corvallis.

Yoklavich, M., R. Starr, J. Steger, H. G. Greene, F. Schwing, and C. Malzone. 1997. Mapping benthic habitats and ocean currents in the vicinity of central California's Big Creek Ecological Reserve. NOAA Technical Memorandum NMFS-SWFSC-245.

Yoklavich, M. M., H. G. Greene, G. M. Cailliet, D. E. Sullivan, R. N. Lea, and M. S. Love. 2000. Habitat associations of deep-water rockfishes in a submarine canyon: an example of a natural refuge. Fishery Bulletin 98:625–641.

Benthic Habitat Mapping with Advanced Technologies and Their Application

JAMES V. GARDNER[1]

U.S. Geological Survey, Menlo Park, California 94025, USA

LARRY A. MAYER

University of New Hampshire, Durham, New Hampshire 03824, USA

The ability of today's scientists to map the seafloor was unheard of two decades ago. Navigational accuracies, as well as elevation and spatial resolutions, have now reached the decimeter scale. But are today's resolutions fine enough for biologists trying to characterize specific benthic habitats? Do biologists know what resolutions are necessary to define their benthic habitat of interest? Once biologists have high-resolution data, do they have the technologies to visualize and analyze this newly acquired data? Do agencies have the budgets required to use 21st century technology? High-resolution seafloor mapping technologies come in a variety of flavors with a variety of resolutions, from airborne lidar to underwater photography. Each system has its own pros and cons relative to the particular goal of the seafloor mapper (see Figure 1). For instance, a living platform coral reef can be efficiently mapped with an airborne lidar, but the spatial resolution is 2 m × 2 m at best, with a depth resolution of a few centimeters. The data are spectacular, albeit costly. But are these resolutions good enough to characterize the platform coral reef habitat for biological or management purposes? If not, then maybe underwater video and/or still photography is required. It is very labor intensive to acquire and process underwater video and/or still photography into useful imagery. Although the spatial resolution can be mm scale, there is poor vertical resolution unless stereo photography is collected.

---

[1] Corresponding author: jimgardner@unh.edu; present address: Center for Coastal and Ocean Mapping, 24 Colovos Road, University of New Hampshire, Durham, New Hampshire 03824, USA.

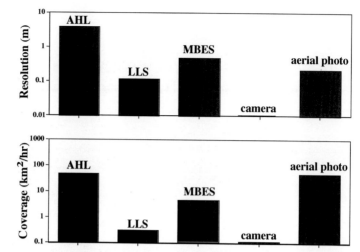

Figure 1. Comparisons of resolution and coverage for airborne hydrographic lidar (AHL), laser linescanner (LLS), multibeam echo sounder (MBES), underwater camera (camera), and aerial photography (aerial photo). The plotted values are based on manufacturers' specifications modified by typical expectations from our experience.

Characterization of Benthic Habitat on Northeastern Georges Bank, Canada

VLADIMIR E. KOSTYLEV[1] AND BRIAN J. TODD

Geological Survey of Canada (Atlantic), Bedford Institute of Oceanography, 1 Challenger Drive, Post Office Box 1006, Dartmouth, Nova Scotia B2Y 4A2, Canada

ODDVAR LONGVA

Geological Survey of Norway, N-749, Trondheim, Norway

PAGE C. VALENTINE

U.S. Geological Survey, 384 Woods Hole Road, Woods Hole, Massachusetts 02543, USA

Abstract.— Seafloor habitats of the Canadian part of Georges Bank were assessed and mapped following the habitat template theory (Southwood 1988). The approach considers the primary selective forces (habitat disturbance and adversity of the environment) that have shaped the existing communities of benthic species and that have defined the life history traits of species found in different habitats. The disturbance axis of the template is modeled based on the information on sediment, currents, and bathymetry. The adversity axis is modeled based on chlorophyll concentration, bottom water temperature, salinity, and seasonal variability in temperature. A preliminary sediment map needed for assessment of the natural disturbance rate was developed from high-resolution multibeam backscatter groundtruthed with archive and current sediment sample data. The distribution of megabenthos assemblages identified from underwater photography was found to follow gradients in disturbance and adversity on the bank. We suggest that application of the habitat template theory is useful for ocean managers in defining areas that are more or less likely to suffer from adverse human impacts. If natural rates of habitat disturbance are high, then risk of harmful habitat alteration and degradation is lower than in naturally stable areas. Similarly, if the natural adversity of the environment is high, then adding additional stressors will further reduce the scope for growth of organisms, which makes natural populations in adverse environments less likely to recover than populations in benign environments.

Introduction

Mapping seafloor habitat is the fundamental step necessary for scientific fisheries management, for monitoring environmental change, and for assessing the impact of anthropogenic disturbance such as fishing and seabed engineering on benthic organisms. The purpose of the study reported here was to map seafloor habitats based on habitat template theory (Southwood 1988) using the Canadian part of Georges Bank as an example.

The variety of approaches to benthic habitat mapping follows closely the number of researchers involved in studies of the seafloor. Thus, for some, a habitat map would be a bathymetric chart while for others, it would be a map of surficial sediments, and yet for others, it would be a map of an area occupied by a certain species. The strict definition of habitat is a place where an organism lives (Begon et al. 1996). Because seafloor is populated by various living organisms, we are interested in distinguishing places that are somehow different and where the differences are meaningful to inhabitants. Therefore, our definition of habitat, formulated for the purpose of habitat mapping is: a spatially defined area where the physical, chemical, and biological environment is distinctly different from the surrounding environment (Kostylev et al. 2001). With this definition, we seek to distinguish distinct areas that share an ensemble of chemical, physical, and biological characteristics, not just geological settings or a single characteristic species. A set of descriptors that can be used in classifying benthic habitats is presented by Valentine et al. (2005, this volume).

---

[1] E-mail: vkostyle@nrcan.gc.ca

Habitat is an organism-centric term. In our approach to habitat mapping, the assemblages of species rather than individual species are considered because distributions of groups of organisms best reflect the distribution of physical factors. We analyzed relationships of benthic assemblages, species, and selected physical variables on Browns Bank (Todd 2000, 2001; Kostylev et al. 2001, 2003) and in the Sable Island gully (Kostylev 2002). These studies showed that the interaction of several factors strongly affected the distribution of benthic assemblages. Typically, substrate type and water depth serve as proxies for a number of co-varying physical factors (e.g., current strength, light penetration, water density). The explanation of this relationship is based on an ecological interpretation of interactions of organisms and their environment and is focused on the current state of benthic assemblages, regardless of history of succession, colonization, and anthropogenic disturbance.

In this paper, we propose an approach to habitat mapping that is based on the evolutionary perspective, namely, on selective forces that have shaped the existing communities of benthic species and that have defined life history traits of species we find in different habitats. The evolutionary perspective on habitat was strongly encouraged by Southwood (1988) who promoted the idea of a habitat template, a classification based on defining habitats in terms of two main characteristics: adversity and stability of the environment. The disturbance axis reflects the rate of habitat destruction or alteration or durational stability of physical habitat in general. In this context, mobile sediments stirred by currents are considered disturbed, and bedrock surface is considered stable habitat. The axis interprets relative degrees of disturbance, and natural disturbance agents may be different in different environments (e.g., ice scouring being the major natural disturbance factor in Arctic ecosystems). Adversity is related to the severity of environmental factors that entail additional physiological costs for growth, survival, and reproduction of benthic organisms and may be thought of as physiological stress that limits an organism's scope for growth. For example, freshwater environment is considered adverse for marine species. In the evolutionary perspective, an adverse environment will select species for their tolerance to extremes of physical factors. A naturally disturbed environment will favor short-lived species that can quickly colonize an area and produce live offspring. Adversity–disturbance domain also defines biotic interactions between species or biotic axis (Southwood 1988), which describes nonlinear change in interspecific interactions and diversity along an imaginary path from adverse disturbed to undisturbed benign conditions. One may expect, for example, stronger competition for space in stable and benign habitats than in adverse and disturbed ones. The habitat template approach appears economical in terms of habitat mapping because it allows us to reduce the number of visualized environmental variables to two (adversity and disturbance) and, thus, alleviates difficulties of interpreting a large set of independent data layers (e.g., separate maps of sediment grain size, bathymetry, water temperatures, salinities, etc.). In a sense, this methodology allows us to pool a number of environmental factors together and graphically display our theoretical understanding of habitats on a map. We do not attempt to give a complete description of physics of the environment and ecology of the bank, which are exhaustively described elsewhere (e.g., Backus and Bourne 1987). Rather, we aim to show the usefulness of the habitat template approach in conjunction with new geological interpretations based on high-resolution acoustic mapping and optical sampling using the Canadian part of Georges Bank as an example.

Methods and Results

In contrast to other banks in the Gulf of Maine, evidence suggests that Georges Bank (Figure 1) was not overridden by glacial ice (Todd et al. 2003). Glacial ice occupied the Northeast Channel to the north of the bank (Figure 1), and outwash sediments were deposited southward onto Georges Bank as the ice melted. Following glaciation, a low stand of relative sea level occurred from 13,000 years before present (BP) until ca 8,000 years BP (as determined by a ^{14}C technique), subaerially exposing large areas of the continental shelf (Shaw et al. 2002). The glacial sediments were eroded, winnowed, and reworked by the subsequent sea-level transgression. The remnant sediment continues to be reworked by wave and current action, building the complex pattern of present-day bedforms. The multibeam bathymetric image (Figure 2A) shows the Canadian portion of Georges Bank.

Two major types of surficial sediment deposits have been identified on Georges Bank. Following Wentworth's (1922) grain size terminology, mobile sand (0.0625–2 mm) dominates the shallowest part of the bank, and less mobile gravel (>2 mm) dominates the remainder of the bank. The sand comprises sheets and superimposed sand wave fields oriented perpendicular to the predominant semidiurnal tidal flow (Todd et al. 2003). The maximum measured sand wave height is 14 m, and bedforms of several meters high are ubiquitous.

Geophysical, Geological, and Biological Information

Multibeam bathymetric data were collected on Georges Bank in 1999 and 2000 by the Canadian Hydrographic Service using a Simrad EM1002 (Kongsberg Simrad, Norway) multibeam bathymetric system. This 95-kHz system produces 111 beams arrayed over an arc of 150 degrees and operates by ensonifying a strip of seafloor

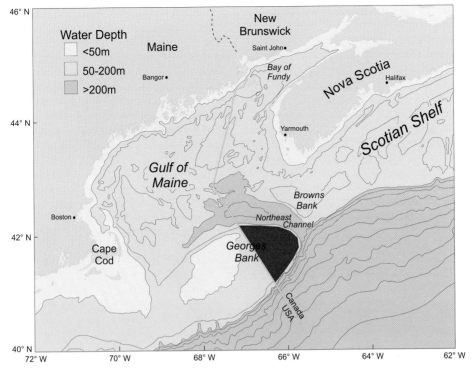

Figure 1. Location map of the study area. The darkest shaded area is the Canadian section of eastern Georges Bank.

across track and detecting the bottom echo. The swath of seafloor imaged on each survey line was five to six times the water depth. Line spacing was about three to four times water depth to provide ensonification overlap between adjacent lines. Navigation was by differential global positioning system, providing positional accuracy of ± 3 m.

During the survey, water depth values were inspected and erroneous values were removed using CARIS/HIPS (Hydrographic Information Processing System) software (Universal Systems Limited, Fredericton, New Brunswick). Within HIPS, the data were adjusted for tidal variation using tidal predictions from the Canadian Hydrographic Service. The data were archived in the form of raw datagrams on magnetic tapes. Multibeam bathymetric data were extracted from the datagrams and were gridded in 10-m (horizontal) bins and shaded with artificial illumination using software developed by the Ocean Mapping Group at the Geological Survey of Canada (Atlantic). Relief maps, color coded to depth, were developed and displayed on a Hewlett-Packard workstation using GRASS (Geographic Resources Analysis Support System; U.S. Army Construction Engineering Research Laboratories; this software is in the public domain). In addition to the bathymetric data, backscatter strengths ranging from 0 to –128 decibels (dB) were logged by the Simrad EM1002 system. Backscatter strengths are computed using calibration values for the electronics and transducers at the time of instrument manufacture.

Interpretation of surficial sediment type was based on the Geological Survey of Canada and the U.S. Geological Survey data, which was composed of 166 sediment samples collected by grab, dredge, and submersible for grain size analysis and 116 photographic stations (1,884 seafloor photographs). Grain size-classes were described following Wentworth (1922). Sampling locations are shown on Figure 3.

Although the relationship between backscatter and sediment texture is not simple and direct, it can be used in simple two-phase systems for distinguishing coarse and fine sediments. A multibeam backscatter intensity map (Figure 2B) was used as a proxy for sediment type distribution because it commonly relates to the mean grain size of surficial sediment (Kostylev et al. 2003). Backscatter intensity is usually depicted in gray scale with strong echo returns shown as dark tones and weaker returns as lighter tones. High backscatter values (dark tones) are typically –10 to –30 dB for gravel, and low backscatter values (light tones) range from –30 to –60 dB for fine-grained sand (Mitchell and Hughes Clarke 1994; Shaw et al. 1997). Analysis of the relationship between

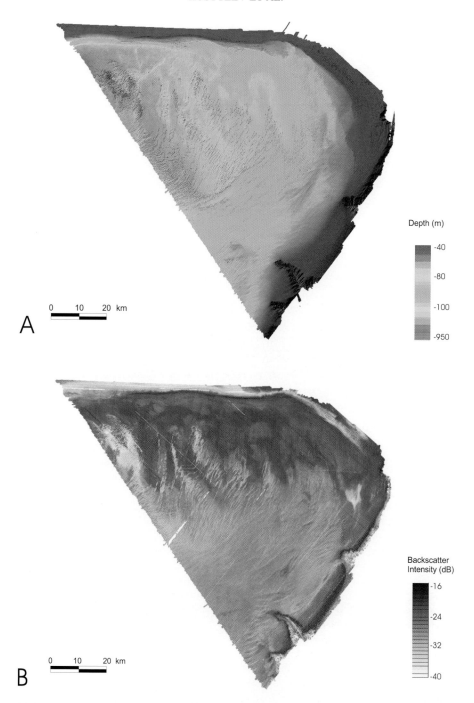

Figure 2. (A) Multibeam sonar bathymetry of Georges Bank. Depths are color coded to the hypsometric curve of the area: shallow water is red-orange (37–73 m) and deep water is blue-violet (>100 m). (B) Multibeam sonar backscatter strength of Georges Bank. Dark (strong) backscatter areas signify coarser-grained substrates then light-toned areas.

pixel brightness on a backscatter map (linear function of acoustic return strength) and average grain size from sediment samples (Figure 4A) revealed significant log–linear correlation ($r = 0.670$, $P < 0.0001$, $n = 166$, $F = 133.8$),

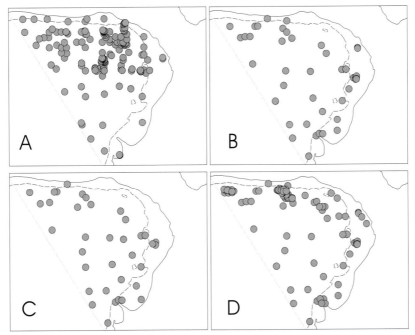

Figure 3. Locations of sampling stations on Eastern Georges Bank with 100-m and 200-m water depth contours. (A) Grain-size samples from U.S. Geological Survey (USGS) and Geological Survey of Canada (GSCA) data archives; (B) Photo and video stations used in the analysis of megafaunal distribution; (C) Grab samples for the analysis of benthic fauna; (D) USGS and GSCA photographic stations used for identification of sediment type.

and the obtained relationship was used to convert the backscatter map into an average grain size map. Standard deviation of grain sizes in a sample generally decreases with backscatter brightness ($r = 0.335$, $P < 0.001$, $n = 166$, $F = 20.8$), with well-sorted fine sediment producing less backscatter. This makes predictions of areas occupied by fine-grained sediment more reliable than the prediction of areas of mixed sediments. Observations on the average fraction of sand and gravel in grab samples (Figure 4B) and average surface cover of sediment types at each sampling station estimated from underwater imagery (Figure 4C) corroborate this relationship and allow reclassification of the backscatter information into areas of sand, gravel, and mixed sediment. Backscatter map areas with pixel brightness below 90 were classified as gravel, 110–190 as sand, and 90–110 as mixed sediment. These subdivisions correspond to areas with more than 70% gravel, areas with more than 70% sand, and areas where both gravel and sand are represented equally (Figure 4B). This preliminary map of sediment type (Figure 5) will be superseded by a geology map based on geophysical and geological groundtruth.

Underwater photographs from 42 recent photo transects collected in years 2000 and 2002 (Figure 3C) were analyzed for presence or absence of megabenthos (organisms generally larger then 1 cm), and organisms were identified to the highest possible taxonomic resolution. Frequency of occurrence of different taxa was calculated for each station and used in further classification and statistical analyses. Continuous video recording with two cameras (oblique and vertical) from each station revealed larger mobile species (e.g., fishes) and provided insight into morphology of the seafloor.

Oceanographic Information

Seasonal data on water temperature, salinity, and near-bottom currents were obtained from the Gulf of Maine circulation model developed by the Department of Fisheries and Oceans (Hannah et al. 2001) and augmented by a database of oceanographic observations (Yashayaev 1998). The data were stored in MapInfo (Toronto, Canada) geographic information system, and continuous coverage for the study area was produced by inverse distance weighting interpolation.

All data used in the analyses were gridded with Vertical Mapper (MapInfo, Toronto, Canada) to a resolution of 200 m, which was optimal for the reduction of noise in empirical data and produced a grid file of manageable size for the subsequent computations (Figure 6).

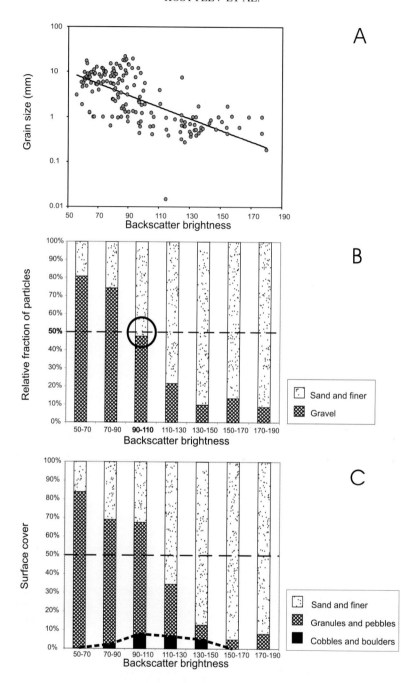

Figure 4. Relationship between sediment grain size and backscatter brightness (assessed as gray scale pixel value as discussed in text). (A) Mean grain size from grab samples and archive data; (B) Percent sand/gravel from archive data, backscatter map areas with pixel brightness below 90 were classified as gravel, 90–110 as mixed sediment, and 110–190 as sand. Note, equal proportion of sand and gravel fractions in the mixed sediment class is marked by an open circle. (C) percent surface cover of sediment types identified from image analysis. Note that gravel is subdivided into two categories: (1) granules and pebbles, and (2) cobbles and boulders.

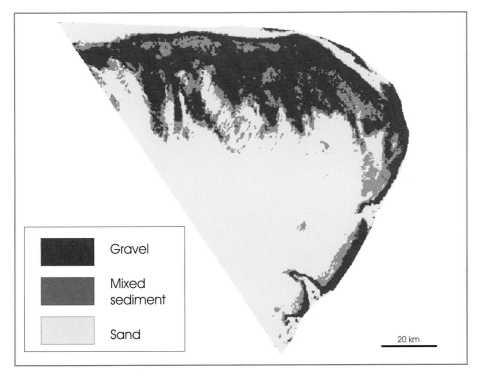

Figure 5. Reclassified backscatter map showing dominant sediment type based on the relationship of gain size and backscatter. Areas with greater than 70% gravel are classified as gravel, areas with greater than 70% sand and finer sediments are classified as sand. The area classified as mixed sediments is where both gravel and sand are represented equally.

Mapping of Habitat Disturbance

In this study, we modeled the natural disturbance of seafloor sediments on the bank, based on the knowledge of grain size distribution and strengths of the M2 tidal currents from Hannah et al. (2001). Difference between the major tidal current and critical current velocity needed to initiate movement of particles of a given size was calculated following Butman (1987) and served as a guide for defining areas where sediment is disturbed only by tides (Figure 7). These areas correspond to the regions where sedimentary bedforms are observed on a multibeam image (compare with Figure 2B). The mean circulation current comprises about 10% of the major tidal currents and was not taken into account. Storm waves and swells can initiate sediment mobility down to considerable depths, confirmed by our observations of moderately active sand waves on the bank at depths of 100 m. Because information on storms for the study area was not available, water depth was used as a proxy for the storm-related disturbance. A simplified approach to modeling disturbance was taken by standardizing (0–1) the following ratios for water depth, grain size, and current strength and using them in an additive model:

$$\text{Disturbance} = \frac{1}{3}\left(\frac{G_{\text{range}}}{G - G_{\text{min}}} + \frac{D_{\text{range}}}{D - D_{\text{min}}} + \frac{U - U_{\text{min}}}{U_{\text{range}}}\right)$$

where G is grain size (mm), D is water depth (m), and U is a lunar semidiurnal (M2) component of mean spring tides within a 10-m layer above seafloor (m/s). This index is calculated for each map cell and reflects a notion that naturally disturbed areas are likely to have strong currents, be shallow, and be dominated by finer-grained sediment. Within the study area, where sand and gravel are the main sedimentary facies, we assumed a linear relationship between sediment mobility and grain size.

The disturbance index was reclassified into high and low values by splitting the range of observed values at the median, based on frequencies of map cell values (Figure 7). The blue isoline on the stability map delimits the area with tide-related disturbance. The red isoline separates the area characterized as mostly disturbed from mostly undisturbed. The tide-generated sediment mobility occurs within the model-defined high disturbance area, which covers an area twice the size of the former. Sediment features observed below tide-disturbed depths are possibly created by low-frequency events, such as

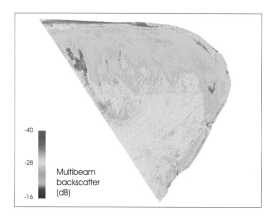
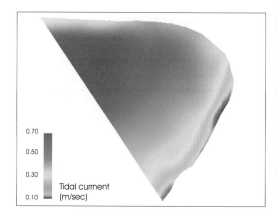
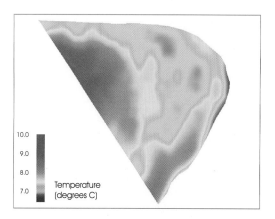
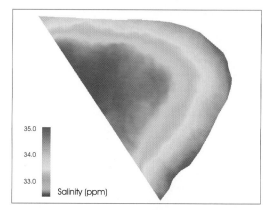
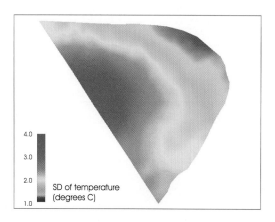
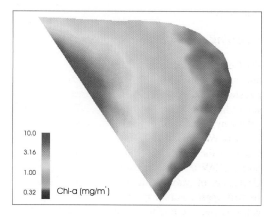

Figure 6. Oceanographic variables used in the habitat classification: backscatter (upper left); major tidal component (m/s; upper right); average bottom water temperature (°C; middle left); bottom water salinity (ppm; middle right); variability in water temperature, standard deviation (°C; bottom left); and chlorophyll concentration (mg/m^3; bottom right).

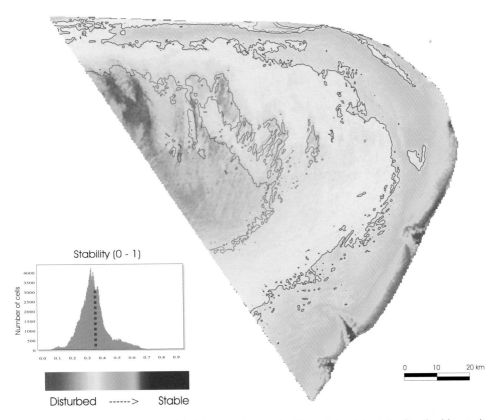

Figure 7. Stability component of habitat template scaled from 0 (low) to 1 (high). The blue isoline on this map delimits the area with tide-related disturbance, the red isoline separates the area characterized as disturbed from undisturbed and corresponds to the value shown with the vertical red line on cell frequency histogram.

large storms.

Mapping of Habitat Adversity

For Georges Bank, annual averages of water temperature, salinity, and chlorophyll concentration (as defined by satellite imagery) were used as proxies for the adversity of habitats. Hargrave and Peer (1973) showed that chlorophyll concentration is a useful indicator for benthic biomass. We estimated an average concentration of chlorophyll in the surface waters in 2001 from monthly composite SEAWIFS images (School of Marine Sciences, University of Maine, http://wavy.umeoce.maine.edu/seawifs_jpgs/), and the reciprocal of the resulting variable was used as one of the indicators of habitat adversity. We also assumed that, for the benthic fauna of the outer shelf within this biogeographic zone, lower average temperature, lower salinity, and higher seasonal variability in water temperature would all increase the adversity of the environment. Physiological adjustments caused by each of these factors will require additional energy expenses that will limit organisms' scope for growth and reproduction. Some factors, such as temperature, directly limit growth rates of living organisms through a decrease in metabolic rate. A composite adversity index (Figure 8) was calculated for every map cell as a maximum value of any of the standardized (0–1) ratios at a given location in the following manner:

$$\text{Adversity} = \text{Max}\left(\frac{T_{\text{range}}}{T - T_{\text{min}}}; \frac{P_{\text{range}}}{P - P_{\text{min}}}; \frac{DT - DT_{\text{min}}}{DT_{\text{range}}}; \frac{S_{\text{range}}}{S - S_{\text{min}}}\right)$$

where T is average yearly bottom temperature (°C), P is chlorophyll concentration (mg/mL), DT is seasonal variance of water temperatures (°C), and S is water salinity (ppm). The resulting variable is dimensionless. All of the initial variables were given equal weight, and all of them were allowed to be limiting for benthos. Because the re-

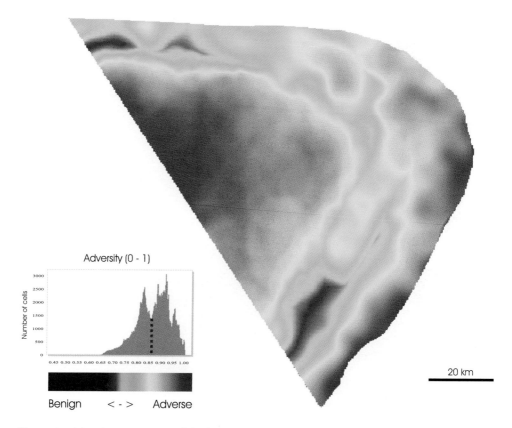

Figure 8. Adversity component of the habitat template scaled from 0 (low) to 1 (high). Red is high adversity and blue is low.

sulting index is expressing adversity on a relative scale, within the study area, adverse and benign environments were defined by subdividing the adversity variable by the median based on the map cell counts. Frequency distribution of adversity values exhibits two distinct peaks (Figure 8), with the median separating them into high-adversity and low-adversity classes.

The two resulting grids were queried for adverse disturbed, adverse undisturbed, benign disturbed, and benign undisturbed combinations of factors and mapped. The resulting map (Figure 9) shows these four classes of environment.

Benthic Assemblages and Habitat Template

Analysis of frequencies of occurrence of benthic megafauna from optical sampling serves as a good indicator of habitat preference. Stations were classified using the Bray-Curtis similarity coefficient. The resulting distribution of different clusters (Figure 9) corresponds well to the habitat template. Stations belonging to the same clusters show strong fidelity to a single habitat type (Pearson chi-square = 97.196, df = 27, $P < 0.0001$). For example, stations of the disturbed adverse part of the bank (blue area) are grouped in one cluster (brown circles). A single habitat type, however, may contain several clusters of stations, representing several types of benthic assemblages. Stations of disturbed benign area (yellow) are represented by two clusters: typical hard substrate fauna (blue circles) and sand dwelling fauna (light blue circles). Mostly mobile or burrowing fauna is dominant in more disturbed habitat of the bank (whelks, hermit crabs, scallops) while various sessile species (e.g., anthozoans and brachiopods) represent stable areas. Suspension feeders (scallops, horse mussels, and serpullid polychaetes) are extremely common in benign areas while scavengers or predatory animals (whelks, hermit crabs, anemones) dominate adverse environments. Detailed description of benthic assemblages and habitat associations of the eastern part of Georges Bank will be presented and discussed elsewhere. There is a much weaker relationship between the clustering of stations based on species identified from grab samples and the habitat template. We assume that

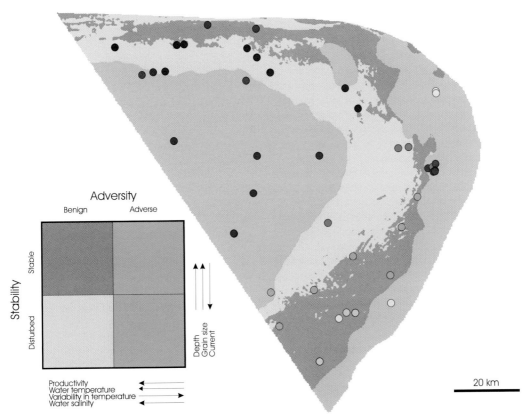

Figure 9. Habitat template mapped onto study area. Blue is adverse and disturbed, yellow is benign and disturbed, brown is benign and stable, green is adverse and stable. Clustering of stations based on the Bray Curtis similarity of fauna is shown with circles of different color, based on optical data only.

the disagreement in matching the habitat template in the clustering of stations based on grabs (points) is based on the nature of grab sampling, which is localized, while optical sampling (transects) characterizes an order of magnitude larger extent of the seafloor. Disturbance and adversity axes are continuous and show a relative scale of these factors. By dividing the template into four distinct regions, we approximate a factorial design with two arbitrary levels. It is likely, therefore, that the degree of correspondence between assemblages and habitat template depend on the exact definition of levels for both adversity and disturbance. Such definition is achievable through physical modeling of mechanical disturbance and of the scope for growth.

Conclusions

Habitat template is a novel concept for summarizing habitat properties, which are relevant to benthic organisms. The main benefit of the classification is in finding commonalities between different parts of the seabed based on assumed life history traits of benthic species selected through their response to natural disturbance and adversity of benthic environment. The approach incorporates evolutionary and ecological considerations in spatial mapping of habitat types and allows the number of variables required for interpretations of observed patterns to be minimized. Such classification may be particularly useful for managers as a guide for planning various seabed activities. For example, benthic communities adapted to naturally adverse and disturbed environment are more likely to be resistant to anthropogenic disturbances than communities of undisturbed benign areas. This broad-scale classification may be further subdivided based on dominant sediment type or geomorphology for better description of species composition of benthic assemblages. For example, undisturbed areas, such as gravel habitats on the northern edge of the bank and stable sands in the south of the bank, have faunas which should carry similar sets of life history traits but that are composed of different sets of species.

The clear quantitative definition of adversity and disturbance has not been agreed upon yet (Southwood 1988), and modeling of the habitat template axes requires additional theoretical work. The work is underway on incorporating more detailed shear stress and grain size data in a physical model of sediment dynamics (Li and Amos 2001) which would also incorporate wave climatology and allow more precise definition of the disturbance axis. In its current state, the approach can be described as transferring our understanding of relative degrees of disturbance and adversity in a study area onto a map. Despite some limitations, the model can be applied locally in order to discover general environmental trends and highlight the differences between habitats and benthos of different areas. The habitat template approach in conjunction with geographic information system mapping allows overcoming subjective differentiation of habitats based on a single physical characteristic of environment, and the resulting map of habitat types is repeatable and testable.

Additionally, habitat template can be considered as a guide for management practices. One can assume, for example, that if natural rates of habitat disturbance are high, then risk of harmful habitat alteration and degradation is lower than in naturally stable areas. Similarly, if the natural adversity of environment is high, then adding additional stressors will further reduce the scope for growth of organisms, which makes natural populations in adverse environments less likely to recover than populations in benign environments.

References

Backus, R. H., and D. W. Bourne. 1987. Georges Bank. MIT Press, Cambridge, Massachusetts.

Begon M., J. L. Harper, and C. R. Townsend. 1996. Ecology: individuals, populations and communities, 3rd edition. Blackwell Scientific Publications, Cambridge, UK.

Butman, B. 1987. Physical processes causing surficial sediment movement. Pages 147–162 in R. H. Backus and D. W. Bourne, editors. Georges Bank. MIT Press, Cambridge, Massachusetts.

Hannah, C. G., J. Shore, J. W. Loder, and C. E. Naimie. 2001. Seasonal circulation on the western and central Scotian Shelf. Journal of Physical Oceanography 31:591–615.

Hargrave, B. T., and D. L. Peer. 1973. Comparison of benthic biomass with depth and primary production in some Canadian east coast inshore waters. ICES, CM 1973/E: shellfish and benthos committee, Copenhagen.

Kostylev, V. E. 2002. Benthic assemblages and habitats of the Sable Island Gully. Canadian Technical Report of Fisheries and Aquatic Sciences 2377:22–35.

Kostylev, V. E., R. C. Courtney, G. Robert, and B. J. Todd. 2003. Stock evaluation of giant scallop (*Placopecten magellanicus*) using high-resolution acoustics for seabed mapping. Fisheries Research 60:479–492.

Kostylev, V. E., B. J.Todd, G. B. J. Fader, R. C. Courtney, G. D. M. Cameron, and R. A. Pickrill. 2001. Benthic habitat mapping on the Scotian Shelf based on multibeam bathymetry, surficial geology and sea floor photographs. Marine Ecology Progress Series 219:121–137.

Li, M. Z., and C. L. Amos. 2001. The upgraded and better calibrated sediment-transport model for continental shelves (SEDTRANS96). Computers and Geosciences 27(6):619–645.

Mitchell, N. C., and J. E. Hughes Clarke. 1994. Classification of seafloor geology using multibeam sonar data from the Scotian Shelf. Marine Geology 121:143–160.

Shaw, J., R. C. Courtney, and J. R. Currie. 1997: The marine geology of St. George's Bay, Newfoundland, as interpreted from multibeam bathymetry and back-scatter data. Geomarine Letters 17:188–194.

Shaw, J., P. Gareau, and R. C. Courtney. 2002. Paleogeography of Atlantic Canada 13–0 kyr. Quaternary Science Reviews 21:1861–1878.

Southwood, T. R. E. 1988. Tactics, strategies and templets. Oikos 52: 3–18.

Todd, B. J., K. W. Asprey, A. S. Atkinson, R. Blasco, S. Fromm, P. R. Girouard, V. E. Kostylev, O. Longva, T. Lynds, W. A. Rainey, P. L. Spencer, M. S. Uyesugi, and P. C. Valentine. 2003. Expedition report CCGS Hudson 2002–026: Gulf of Maine. Geological Survey of Canada Open File Report 1468, Ottawa.

Todd, B. J., V. E. Kostylev, G. B. J. Fader, R. C. Courtney, and R. A. Pickrill. 2000. New approaches to benthic habitat mapping integrating multibeam bathymetry and backscatter, surficial geology and sea floor photographs: a case study from the Scotian Shelf, Atlantic Canada. ICES CM 2000/T:16. ICES 2000 Annual Science Conference. 27–30 September. Bruges, Belgium.

Todd, B. J., P. C. Valentine, V. E. Kostylev, and R. A. Pickrill. 2001. Habitat mapping in the Gulf of Maine. Geological Survey of Canada, Open File Report 3995, Ottawa.

Valentine, P. C., B. J. Todd, and V. E. Kostylev. 2005. Classification of marine sublittoral habitats, with application to the northwestern North America region. Pages 183–200 in P. W. Barnes and J. P. Thomas, editors. Benthic habitats and the effects of fishing. American Fisheries Society, Symposium 41, Bethesda, Maryland.

Wentworth, C. K. 1922. A scale of grade and class for clastic sediments. Journal of Geology 30:377–392.

Yashayaev I. 1998. Computer atlas of the northwest Atlantic. Atlas of Ocean Sections, v2, CD-ROM. Department of Fisheries and Oceans, Dartmouth, Nova Scotia.

Mapping and Characterizing Subtidal Oyster Reefs Using Acoustic Techniques, Underwater Videography, and Quadrat Counts

RAYMOND E. GRIZZLE[1]

Jackson Estuarine Laboratory, 85 Adams Point Road, Durham, New Hampshire 03824, USA; and Department of Zoology, University of New Hampshire, Durham, New Hampshire 03824, USA

LARRY G. WARD[2]

Jackson Estuarine Laboratory, 85 Adams Point Road, Durham, New Hampshire 03824, USA; and Department of Earth Sciences, University of New Hampshire, Durham, New Hampshire 03824, USA

JAMIE R. ADAMS[3]

Jackson Estuarine Laboratory, 85 Adams Point Road, Durham, New Hampshire 03824, USA; and Center for Coastal and Ocean Mapping, University of New Hampshire, Durham, New Hampshire 03824, USA

SEMME J. DIJKSTRA[4]

Center for Coastal and Ocean Mapping, University of New Hampshire, Durham, New Hampshire 03824, USA

BRIAN SMITH[5]

New Hampshire Fish and Game Department, Marine Fisheries Division, 225 Main Street, Durham, New Hampshire 03824, USA

Abstract. Populations of the eastern oyster *Crassostrea virginica* have been in long-term decline in most areas. A major hindrance to effective oyster management has been lack of a methodology for accurately and economically obtaining data on their distribution and abundance patterns. Here, we describe early results from studies aimed at development of a mapping and monitoring protocol involving acoustic techniques, underwater videography, and destructive sampling (excavated quadrats). Two subtidal reefs in Great Bay, New Hampshire, were mapped with side-scan sonar and with videography by systematically imaging multiple sampling cells in a grid covering the same areas. A single deployment was made in each cell, and a 5–10-s recording was made of a 0.25-m^2 area; the location of each image was determined using a differential global position system. A still image was produced for each of the cells and all (n = 40 or 44) were combined into a single photomontage overlaid onto a geo-referenced base map for each reef using ArcView geographic information system. Quadrat (0.25 m^2) samples were excavated from 9 or 10 of the imaged areas on each reef, and all live oysters were counted and measured. Intercomparisons of the acoustic, video, and quadrat data suggest: (1) acoustic techniques and systematic videography can readily delimit the boundaries of oyster reefs; (2) systematic videography can yield quantitative data on shell densities and information on reef structure; and (3) some combination of acoustics, systematic videography, and destructive sampling can provide spatially detailed information on oyster reef characteristics.

[1] E-mail: ray.grizzle@unh.edu
[2] E-mail: lgward@cisunix.unh.edu
[3] E-mail: jamie.adams@unh.edu
[4] E-mail: s.dijkstra@unh.edu
[5] E-mail: bmsmith@starband.net

Introduction

Overharvesting, disease, pollutants, and other factors have resulted in long-term declines in populations of the eastern oyster *Crassostrea virginica* in many areas

(Rothschild et al. 1994; MacKenzie 1996; Hargis and Haven 1999), including the present study area in New Hampshire (Langan 1997, 2000; Smith 2002; Trowbridge 2002). Hence, oysters are a major concern of coastal managers, and in most areas, they are regularly monitored. In the present study area, New Hampshire, oyster distributions and abundances have been monitored using various methods, including quadrat sampling by divers, tonging, and dredging (Ayer et al. 1970; Nelson 1982; Banner and Hayes 1996; Langan 1997, 2000). Shellfish managers in other areas use similar approaches (e.g., Jordan et al. 2002). Typically, these "traditional" methods yield distribution maps that are useful with respect to general location and average abundances in selected areas, but they rarely provide spatially detailed data because of costs and other constraints.

Recent research has explored remote sensing techniques as supplements to traditional methods for characterizing and mapping oyster reefs. Aerial photography has been used effectively for intertidal oyster reefs (Grizzle 1990; Finkbeiner et al. 2001; Grizzle et al. 2002). Subtidal reefs, however, usually require techniques such as acoustic sounders and underwater videography. Acoustic techniques can differentiate between oyster bottom and other substrate types, particularly soft sediments (Powell et al. 1995; Mayer et al. 1999; Wilson et al. 2000). Hence, they can provide high-resolution maps of reef location and spatial extent, but their potential for determining reef characteristics such as densities of living oysters versus nonliving shell has not been demonstrated. Underwater videography only recently has been explored as a routine monitoring tool for oysters (Paynter and Knoles 1999; J. R. Adams, R. E. Grizzle, L. G. Ward, S. Dijkstra, and J. Nelson, abstract from Benthic Ecology Meeting, 2002).

The objective of this research note is to provide a preliminary assessment of a comprehensive mapping–monitoring protocol involving acoustic techniques, underwater videography, and destructive sampling (quadrat counts).

Methods

Two oyster reefs (Nannie Island and Adams Point) in Great Bay, New Hampshire, were mapped in fall 2001 (Figure 1); the mapping techniques included acoustic remote sensing by multichannel vertical incidence and side-scan sonar, underwater videography, and quadrat sampling by divers. Both reefs are worked regularly and extensively by recreational harvesters (mostly with tongs) and have low vertical relief (see Discussion below). Water depths over both reefs range from 1 to 3 m at mean low water. The surficial sediments in Great Bay range from muds to silty sands, with some sand deposits (Armstrong 1974). The substrates near the Nannie Island study site are primarily silts to sandy silts. The surficial sediments at the Adams Point site are largely muds to silts.

All acoustic mapping work was done using standard hydrographic surveying techniques by laying out a series of grid lines. Side-scan sonar (a developmental version of the system 5000 MKII loaned to us by the its manufacturer Klein Associates, Inc., Salem, New Hampshire) was used for both the Adams Point and the Nannie Island reefs. This system has a dynamically focused multibeam transducer array with five simultaneous digitally formed beams per side. To enable work in the very shallow water covering the reefs, the sonar was hull mounted on the R/V *Little Bay*, a pontoon boat that was specially adapted for acoustic mapping in extremely shallow water. The operating frequency was 255 kHz, and the pulse length was 50 ms, resulting in an across-track resolution of approximately 3 cm. The range scale was set to 50 m, leading to an along-track resolution of better than 20 cm. A regular grid with 40-m line spacing was used on both reefs. This protocol allowed us to make better radiometric corrections than normally possible. A PosMV system was used for motion sensing and differential global positioning system (DGPS) was used for positioning.

Multichannel vertical incidence data were obtained using a Navitronic Seadig 21 system only at the Adams Point reef, for bottom characterization. The Navitronic system was installed on the Canadian Department of Public Works vessel R/V *Miramichi Surveyor* that was on location as part of a different project. As installed, the Seadig 21 system had 12 channels and used a 50-ms pulse length, logging a single depth value for each ping on each channel. A DGPS was used for positioning, so no motion sensor was required. For bottom characterization, the signal coming out after the rectification stage (before any variable gains are applied) was fed to a Quester Tangent ISAH-S system that performed an analog to digital conversion (Collins et al. 1996). This procedure allowed identification of the bottom and extraction of over 160 features from this return, both from the time and frequency domains using the Quester Tangent Impact software. The number of features was then reduced to three using principal component analysis, followed by a cluster analysis in a three-dimensional feature space that provided characterization of the data (Quester Tangent 2002).

Video imagery was obtained on both reefs using a custom-made camera system consisting of an underwater black and white camera (Aqua-Vu model IR) with integral infrared lighting (not used in present study) mounted on steel frame, Garmin DGPS unit (model GPS 76), and Sony digital video camera (model DCR-TRV103) for

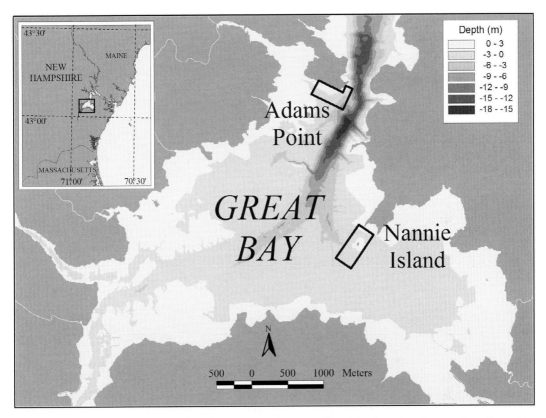

Figure 1. Location of two study reefs, Adams Point and Nannie Island, in Great Bay Estuary, New Hampshire. Note that polygon shapes of reefs approximate shapes and orientations of areas video imaged as shown in Figure 3.

recording (Figure 2). The approximate area of each reef was overlaid with a systematic sampling grid consisting of 40–44 sampling cells. A 5–10-s recording was made of a single position in each cell. Each recording was reduced to a still image using a combination of Enivronmental Systems Research Institute's ArcInfo and Adobe Photoshop, and all the stills (40–44) from each reef were combined into a geo-referenced photomontage. At 9 or 10 of the video-imaged cells on each reef, divers excavated a 0.25-m² quadrat by hand, removing only the surface layer of shell. All living oysters were counted and measured (shell height to nearest mm) using calipers. Quadrats were taken from the exact area that was video imaged, thereby allowing a direct comparison of data derived from video imagery with quadrat counts.

Three individuals examined each of the 19 video images from the two reefs independently. In each image, all objects that could be identified as an oyster shell were counted. This count was further refined by counting all obviously dead shells, usually identified by observation of a light-colored shell interior with dark adductor muscle scar. This yielded three numbers for each image: total

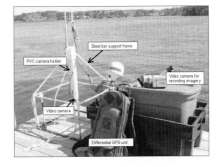

Figure 2. Custom-made videographic camera system consisting of underwater black and white camera (Aqua-Vu model IR) with integral infrared lighting (not used in present study) mounted on steel frame, video camera for recording imagery, and differential global positioning system.

shells, dead oysters, and possibly live oysters (equal to number of total shells minus number of dead oysters). For the present study, the number of possibly live oysters was emphasized and used to compare with the quadrat data that consisted only of live oysters excavated by divers.

Results and Discussion

Side-scan sonar produced easily interpretable imagery data at the Nannie Island location, clearly showing reef boundaries (Figure 3). At the Adams Point reef, the acoustic data identified reefs boundaries but required an experienced analyst to interpret the data because of differences in topography. Vertical incidence data at Adams Point showed a number of distinctly different areas that were not depth dependent. Hence, these data indicate substantial potential for single-beam sonar as a low-cost tool for mapping reefs.

Comparison of maps produced by acoustic techniques and videography indicated that both approaches were capable of delimiting reef boundaries (Figure 3). A major difference between the two is that much higher resolution of reef boundary shape was obtained acoustically. It should be noted, however, that the number of images obtained determines boundary resolution in video maps. Although it is possible to approach the resolution of acoustics using video imaging, this would be practical for only small areas due to the number of images that would have to be taken and processed. Another difference between the two techniques is that data on shell densities (and potentially size distribution) can be obtained from video images. Hence, video maps (photomontages) directly provide information on reef characteristics potentially useful to managers.

The present study corroborates previous research by demonstrating that acoustic techniques can effectively differentiate oyster bottom from surrounding substrate types. For example, research in the Chesapeake Bay, Maryland (DeAlteris 1988); Galveston Bay, Texas

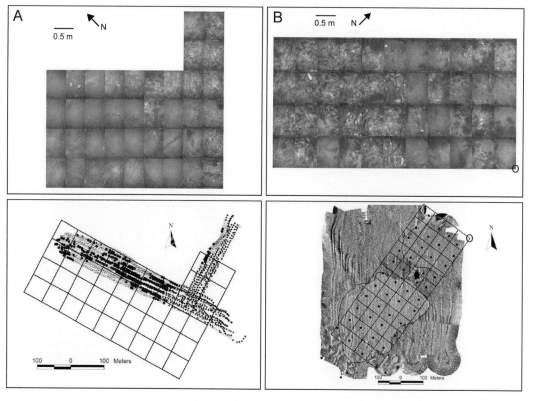

Figure 3. Videographic photomontage (top) and acoustics-derived map (bottom) from study reefs shown in Figure 1. (A) At Adams Point reef, multichannel vertical incidence data showed five different sediment classes; black dots represent shell bottom and approximate reef area. Note that sampling grid chosen for this reef was too coarse to give adequate video coverage of the actual reef area. (B) Side-scan sonar map of Nannie Island reef; dashed lines indicate outline of southern and northern portions of reef. Note, match circles in corners for proper orientation of images.

(Simons et al. 1992; Powell et al. 1995); and Louisiana (Roberts et al. 1999; Wilson et al. 2000) has demonstrated the utility of single-beam sounders and side-scan sonar in mapping subtidal reefs and discriminating between oyster bottom and several other bottom types. Ongoing programs in several areas continue to refine the use of side-scan sonar and single-beam techniques (e.g., Roberts et al. 1999; Smith et al. 2001). Multibeam approaches show considerable promise for reef mapping, but they need to be fully tested (Mayer et al. 1999).

To our knowledge, very little research has been done on videography for mapping and characterizing oyster reefs. Paynter and Knoles (1999) used video to characterize the general conditions of constructed oyster reefs in the Chesapeake Bay but did not rely on videography for mapping. The photomontage approach described here is a new technique we are developing to make maximum use of video imagery in bottom habitat mapping generally. At a minimum, it provides a "picture" consisting of geo-referenced photographs of the mapped bottom area. In the case of oyster reefs, the picture shows relative shell densities, orientation, and potentially other features. Each image, however, is exaggerated in two dimensions because each of the stills represents only a small portion of the actual area occupied by that image on the overall map. In other words, the overall boundaries of the reef are spatially accurate and geo-referenced, but each individual still image is at a much larger scale. For example, if the still images in Figure 3 were at the same scale as the overall map, each would represent only about 1/16,000 of the cell it fills. In a photomontage, the amount of exaggeration decreases as the number of cells imaged increases.

As mentioned above, both reefs have low vertical relief probably because they are heavily worked by harvesters using tongs. Most oysters on both reefs occurred as singles or in small clumps. These characteristics are particularly evident in the images from Nannie Island (Figure 3b). Other reef characteristics potentially inferable from videography include the level of sediment accumulation, presence and extent of shell fouling, and presence of larger reef-associated organisms (Paynter and Knoles 1999; Smith et al. 2001).

One important question that has not been assessed for videography is its potential for counting live oysters. Counts of possibly live oysters made directly from our video imagery from both reefs were only weakly correlated with quadrat data (live oysters extracted from each quadrat by divers) for the entire 19-sample data set (Figure 4a). However, when counts exceeding 25 oysters per quadrat (for either live oysters or video counts) were omitted, there was a strong correlation between the two (Figure 4b). This suggests that videography might only be useful when oyster densities are low, perhaps less than 100 individuals/m^2. The explanation, however, is a bit more complicated for our data set. Two of the omitted data pairs had high numbers of small oysters, many less than 40 mm in shell height. Oysters of this size would be more easily missed in video counts than would larger individuals. A third data pair had large numbers of dead shell, suggesting that some empty valves were among the oysters counted as possibly live. Overall, these data suggest that videography potentially can be used to infer density and other reef characteristics, but limitations exist. This will be an important area of research in future studies.

At least three conclusions can be drawn from the present study: (1) acoustic techniques and systematic videography can readily delimit the boundaries of oyster reefs; (2) systematic videography can provide data on shell (whether live or dead) densities, reef characteristics such as vertical relief, and potentially data on densities of live oysters; and (3) reef characteristics such as shell den-

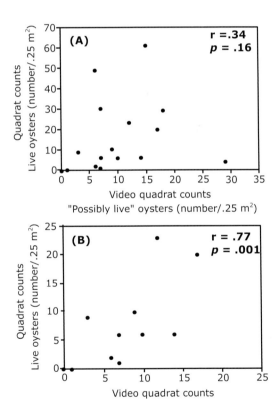

Figure 4. (A) Possibly live oysters counted (mean of three different individuals independently inspecting each image) from video images of quadrats versus corresponding quadrat counts of all live oysters extracted by divers from same 0.25-m^2 area. $n = 19$, $r = 0.34$, $P = 0.16$. (B) Same data set except all counts (video or live counts) exceeding 25 oysters were omitted. $n = 14$, $r = 0.77$, $P = 0.001$.

sity may be extracted from acoustic data, but its full potential remains to be tested. Taken together, these findings suggest that some combination of acoustics, systematic videography, and destructive sampling can improve upon traditional methods by providing more spatially detailed information on oyster reef characteristics.

Acknowledgments

Funding for this study was provided by the National Oceanic and Atmospheric Administration, Office of Sea Grant, state of New Hampshire, and the Cooperative Institute for Coastal and Estuarine Environmental Technology (CICEET). The project was also supported by the Center for Coastal and Ocean Mapping (NOAA Grant NA97OG2041) and Jackson Estuarine Laboratory at the University of New Hampshire, and the New Hampshire Fish and Game Department. The manuscript was improved by the comments of three anonymous reviewers. This is the Center for Marine Biology and Jackson Estuarine Laboratory Contribution Series Number 403.

References

Armstrong, P. 1974. Copper, zinc, chromium, lead, and cadium in the unconsolidated sediments of Great Bay Estuary, New Hampshire. Master's thesis. University of New Hampshire, Durham.

Ayer, W. C., B. Smith, and R. D. Acheson. 1970. An investigation of the possibility of seed oyster production in Great Bay, New Hampshire. New Hampshire Fish and Game, Marine Survey Report 2, Concord.

Banner, A., and G. Hayes. 1996. Important habitats of coastal New Hampshire. U.S. Fish and Wildlife Service, Gulf of Maine Project, Falmouth, Maine.

Collins, W., R. Gregory, and J. Anderson. 1996. A digital approach to seabed classification. Sea Technology 37:3–14.

DeAlteris, J. T. 1988. The application of hydroacoustics to the mapping of subtidal oyster reefs. Journal of Shellfish Research 7:41–45.

Finkbeiner, M., B. Stevenson, and R. Seaman. 2001. Guidance for benthic habitat mapping: an aerial photographic approach. National Oceanic and Atmospheric Administration Coastal Services Center, Charleston, South Carolina.

Grizzle, R. E. 1990. Distribution and abundance of *Crassostrea virginica* (Gmelin, 1791) (eastern oyster) and *Mercenaria* spp. (quahogs) in a coastal lagoon. Journal of Shellfish Research 9:347–358.

Grizzle, R. E., Adams, J. R., and L. J. Walters. 2002. Historical changes in intertidal oyster (*Crassostrea virginica*) reefs in a Florida lagoon potentially related to boating activities. Journal of Shellfish Research 21:749–756.

Hargis, W. J., Jr., and D. S. Haven. 1999. Chesapeake oyster reefs, their importance, destruction and guidelines for restoring them. Pages 329–358 *in* M. W. Luckenbach, R. Mann, and J. Wesson, editors. Oyster reef habitat restoration: a synopsis and synthesis of approaches. Virginia Institute of Marine Science, Gloucester Point.

Jordan, S. J., K. N. Greenhawk, C. B. McCollough, J. Vanisko, and M. L. Homer. 2002. Oyster biomass, abundance, and harvest in northern Chesapeake Bay: trends and forecasts. Journal of Shellfish Research 21:733–741.

Langan, R. 1997. Assessment of shellfish populations in the Great Bay Estuary. Office of State Planning, New Hampshire Estuaries Project, Final Report, Concord.

Langan, R. 2000. Shellfish habitat restoration strategies for New Hampshire's estuaries. Office of State Planning, New Hampshire Estuaries Project, Final Report, Concord.

MacKenzie, C. L., Jr. 1996. Management of natural populations. Pages 707–721 *in* V. S. Kennedy, R. I. E. Newell, and A. F. Eble, editors. The eastern oyster, *Crassostrea virginica*. Maryland Sea Grant, College Park.

Mayer, L., J. Hughes-Clarke, and S. Dijkstra. 1999. Multibeam sonar: potential applications for fisheries research. Journal of Shellfish Research 17:1463–1467.

Nelson, J. I., Jr. 1982. Great Bay Estuary monitoring survey. New Hampshire Fish and Game Department and Office of State Planning, Concord.

Paynter, K. T., and T. E. Knoles. 1999. Use of videography to assess differences between restored and non-restored areas in the Chesapeake Bay. Journal of Shellfish Research 18:725.

Powell, E. N., J. Song, M. S. Ellis, and E. A. Wilson-Ormond. 1995. The status and long-term trends of oyster reefs in Galveston Bay, Texas. Journal of Shellfish Research 14:439–457.

Quester Tangent. 2002. Using Impact. Quester Tangent, Sidney, New Brunswick.

Roberts, H. H., J. Supan, and W. Winans. 1999. The acquisition and interpretation of digital acoustics for characterizing Louisiana's shallow water oyster habitat. Journal of Shellfish Research 18:730–731.

Rothschild, B. J., J. S. Ault, P. Goulletquer, and M. Héral. 1994. Decline of the Chesapeake Bay oyster population: a century of habitat destruction and overfishing. Marine Ecology Progress Series 111:29–39.

Simons, J. D., T. M. Soniat, E. N. Powell, J. Song, M. S. Ellis, S. A. Boyles, E. A. Wilson, and W. R. Callender. 1992. An improved method for mapping oyster bottom using a global positioning system and an acoustic profiler. Journal of Shellfish Research 11:431–436.

Smith, B. 2002. Shellfish population and bed dimension assessment in the Great Bay Estuary. New Hampshire Estuaries Project, Final Report, Concord.

Smith, G. F., D. G. Bruce, and E. B. Roach. 2001. Remote acoustic habitat assessment techniques used to characterize the quality and extent of oyster bottom in the Chesapeake Bay. Marine Geodesy 24:171–189.

Trowbridge, P. 2002. Environmental indicators report, shellfish. New Hampshire Estuaries Project, Final Report, Concord.

Wilson, C. A., H. H. Roberts, and J. Supan. 2000. MHACS: marine habitat acoustic characterization systems, a program for the acquisition and interpretation of digital acoustics to characterize marine habitat. Journal of Shellfish Research 19:627.

Refining Estimates of Potential White Abalone Habitat at Northern Anacapa Island, California Using Acoustic Backscatter Data

GUY R. COCHRANE[1]

U.S. Geological Survey, MS-999, 345 Middlefield Road, Menlo Park, California 94025, USA

JOHN L. BUTLER[2]

National Marine Fisheries Service, Southwest Fisheries Science Center, Post Office Box 271, La Jolla, California 92037, USA

GARY E. DAVIS[3]

Channel Islands National Park, Ventura, California 93001, USA

Abstract.—Acoustic backscatter data provides exceptional detail of rocky seafloor habitat on the continental shelf north of Anacapa Island off Southern California. In the depth range of 25–65 m, there are approximately 60 ha of rocky bottom which may be suitable for abalone *Haliotis* spp. The distribution of rocky bottom is divided into two separate areas on the northeastern and northwestern sides of the island, which are separated by a central area of predominantly sandy bottom. The northeastern area has a far greater area of rocky bottom. When habitat estimation is restricted to the perimeter of rocky areas, the difference in habitat area between the east and west lessens. The northeastern area may be more attractive for seeding of white abalone *Haliotis sorenseni* because it has more rocky bottom and recently has been designated as a marine reserve with no fishing, whereas the northwestern area is now a marine conservation area with limited fishing. This methodology will help locate sites for planting of captive-bred white abalone as part of the white abalone restoration project if numerous other requirements for the abalone, such as food sources, are mapped also.

Introduction

The white abalone *Haliotis sorenseni* (Figure 1) is the only marine mollusk currently listed under the Endangered Species Act by the National Marine Fisheries Service (NMFS). The NMFS published a final rule listing the white abalone as an endangered species on May 29, 2001 (NOAA 2001). A short-lived commercial white abalone fishery began in the early 1970s, peaked mid-decade, and collapsed in the 1980s. Only occasional landings occurred after that time. White abalone were taken by recreational divers also, but actual landings are unknown. Within the last 30 years, the population of white abalone off California and Mexico has collapsed to less than 0.1% of its estimated preexploitation size (Hobday et al. 2001). Recent studies suggest that this species has likely suffered reproductive failure resulting from the severe overharvest (NOAA 2001). The fishery was closed in 1996.

Unlike more mobile animals, abalone are slow-moving creatures, confined to a small area for their entire postlarval life. They reproduce by broadcasting their eggs and sperm into the seawater. For fertilization to occur, the spawning abalone need to be within 1 m of a member of the opposite sex. No neighbors mean that the remaining animals are effectively sterile (Lafferty 2001). The Abalone Restoration Consortium, a team of biologists from the University of California, Santa Barbara (UCSB), U.S. Geological Survey (USGS), Channel Islands Marine Resource Institute (CIMRI), National Park Service, California Department of Fish and Game, and NMFS, has accumulated 15 white abalone captured from the wild, which they are nurturing at UCSB and at CIMRI in Port Hueneme, California. The consortium plans to find and collect 200 white abalone to launch a large-scale captive breeding program.

The final step in restoring white abalone is to estab-

---

[1] Corresponding author: gcochrane@usgs.gov
[2] E-mail: John.Butler@noaa.gov
[3] E-mail: gary_davis@nps.gov

Figure 1. Image of white abalone (in white box) situated in low-relief, mixed hard and soft bottom.

lish self-replacing populations on a variety of reefs through its historic range (Davis et al. 1998). The center of their distribution in the 1970s was the California Channel Islands (Cox 1960; Tutschulte 1976). Anacapa Island is the most easterly island in the Northern Channel Islands chain, located at the northern extreme of the San Diego biogeographic province and in the historic range of the white abalone. The northern side of the island has been recently designated part of a network of marine protected areas (MAP; CDFG 2002), which makes this an attractive option for reseeding abalone reared in captivity. Davis et al. (1998) estimated the area of rocky reef in various areas within the historic range of the white abalone as a proxy for white abalone habitat. We combine recent sonar mapping of the seafloor around Anacapa Island, video observations of white abalone habitat, and geographic information systems statistics to refine this estimate of habitat north of the island to assist in planning for the recovery of the species (Figure 2). Additional areas have been mapped and will be similarly analyzed using the methods discussed in this paper.

Methods

Recent video surveys by Davis et al. (1998) and Butler et al. (personal communication) suggest that the substrata preferred by white abalone are near the bases of edges of high rock areas with hard bottoms or mixed hard and sandy bottoms (Figure 1). The hard bottom consists of low-relief bedrock and large clastics such as boulders and cobbles. A dusting of sand or shell fragments is often observed, which is protected from currents by the adjacent high-standing rocks. We use statistical analysis of side-scan backscatter data to extrapolate known habitat from video observations into a continuous seafloor map of bottom type (Cochrane and Lafferty 2002).

In side-scan sonar, data the amount of acoustic energy reflected back from the seafloor is recorded as a grid of pixels with gray levels. A higher gray level value indicates greater reflected energy due to harder seafloor and other bottom characteristics such as slope of the seafloor relative to the sonar device. The use of gray level to assign a classification to a pixel has proven inadequate to classify natural images (Shokr 1991; Blondel 1996). Because high-relief rocky areas are hard but also have greatly varying slope relative to the sonar device, this is especially true in our case, where one side of a rock facing the sonar will have the highest gray level while the opposite side of the same rock will have the lowest gray level. We use a statistical approach to classifying the sonar called textural analysis, described in Cochrane and Lafferty (2002). We classify the sonar image into the bottom induration types of Greene et al. (1999): hard bottom, mixed hard and soft bottom, and soft bottom.

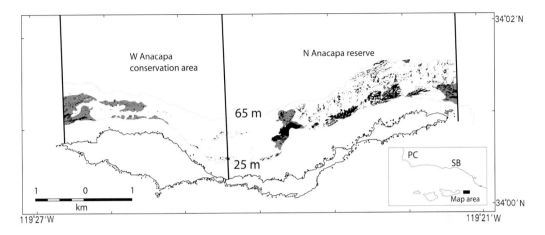

Figure 2. Map showing distribution of hard bottom (black areas), mixed hard and soft bottom (gray areas), and soft bottom (white areas) on the seafloor north of Anacapa Island between the 25-m-depth and 65-m-depth contours. Index map shows location of map area relative to Santa Barbara (SB) and Point Conception (PC). Data used in this publication from Cochrane et al. (2003).

We use Arc/Info to convert classified sonar images to grids and then polygons. Arc statistics can then be used to sum areas and perimeters of rock areas. Assuming that white abalone will be restricted to a 1-m-wide perimeter of rocky areas where the area is greater than the perimeter, we sum the perimeter of those polygons and add it to the sum of the area of small polygons whose perimeter is greater than their area. A further refinement can be achieved using polygon-arc within Arc/Info to restrict the statistics to those polygon arc segments where a rock polygon borders a mixed hard and soft bottom polygon. This eliminates portions of the perimeter of the rocky area that border soft bottom.

Results and Discussion

Davis et al. (1998) estimated that 2×10^5 square meters (20 ha) of rocky substrata in a depth range of 25–65 m, which would be suitable habitat for the abalone, exist around the island. We use recently published USGS side-scan sonar data (Cochrane et al. 2003) to quantify more accurately the area and spatial distribution of habitat in the nearshore waters north of Anacapa Island. Our estimate of white abalone habitat (6.1×10^5 m^2; Table 1) for just the north side of the island exceeds that of Davis et al. (1998) for the entire island about three fold if the white abalone will occupy all rock area in the 25–65-m-depth range. If abalone are more selective and occupy only the perimeter of large rock areas the estimate is reduced by a factor of four (1.4×10^5 m^2; Table 1), and if the habitat is only mixed rock and sand areas adjacent to rock, the estimated available habitat is reduced by over an order of magnitude (0.4×10^5 m^2; Table 1). Further research into the preferences of the abalone is needed for accurate estimates; however, choosing one habitat definition as a proxy will allow comparison of this area to others.

Mapping of the bottom reveals another interesting facet of the problem of repopulating the Anacapa Island area with white abalone. The suitable rocky habitat is divided into two general areas, each within one MPA and separated by approximately 3 km of predominantly sandy bottom in the center of the northern shelf (Figure 2). This is likely caused by currents sweeping sediment off the shelf on either side of the island, whereas sediment accumulates in the relatively calm central island shelf. Since close proximity is required for the abalone to reproduce, it may be necessary to choose one of the rocky areas over the other for the repopulation effort. Consideration of other factors, such as direction of bottom currents, food sources, and perhaps future predation, will need to be made before either of these rocky areas is selected for placement of captive, reared abalone. The North Anacapa Reserve will be closed to commercial and recreational fishing, which might make poaching more obvious than it would be in the West Anacapa conservation area, which will be open to commercial and recreational lobster fishing.

Distribution of exposed rocky bottom will vary greatly from one area to another depending on lithology and sediment supply; no conclusions can be drawn for other areas in the historic range of the white abalone without obtaining data of sufficient resolution in those areas and performing this type of analysis to delimit rocky areas. Existing geologic maps that are based primarily on data collected for petroleum resource estimates do not have the required resolution

Table 1. Estmates of white abalone habitat broken down into areas for each of the two marine protected areas (MPA) and increasingly restricted from total area of rock to only the 1-m-wide perimeter of rock areas adjacent to mixed rock and sand bottom.

MPA	Rock area (m^2)	% of area from 25 to 65 m	Rock perimeter (m^2)	%	Rock bordering mixed perimeter (m^2)	%
North Anacapa	5.4×10^5	15%	1.2×10^5	3%	0.19×10^5	0.5%
West Anacapa	0.7×10^5	3%	0.2×10^5	1%	0.20×10^5	0.9%
Total	6.1×10^5	10%	1.4×10^5	2%	0.39×10^5	0.7%

and should not be converted to habitat maps for a management problem of this type.

References

Blondel, P. 1996. Segmentation of the Mid-Atlantic Ridge south of the Azores, based on acoustic classification of TOBI data. *In* C. J. MacLeod, P. A. Tyler, and C. L. Walker, editors. Tectonic, magmatic, hydrothermal and biological segmentation of mid-ocean ridges. Geological Society, Special Publication 118, Boulder, Colorado.

CDFG (California Department of Fish and Game). 2002. Marine protected areas in NOAA's Channel Islands National Marine Sanctuary. CDFG. Available: *www.dfg.ca.gov/mrd/channel_islands/* (June 2003).

Cochrane, G. R., and K. D. Lafferty. 2002. Use of acoustic classification of sidescan sonar data for mapping benthic habitat in the Northern Channel Islands, California. Continental Shelf Research 22:683–690.

Cochrane, G. R., N. M. Nasby, J. A. Reid, B. Waltenberger, and K. M. Lee. 2003. Nearshore benthic habitat GIS for the Channel Islands National Marine Sanctuary and Southern California state fisheries reserves, volume 1. U.S. Geological Survey Open-File Report 03-85. Available: *http://geopubs.wr.usgs.gov/open-file/of03-85/* (June 2003).

Cox, K. W. 1960. Review of the abalone of California. California Fish and Game 46:381–406.

Davis, G. E., P. L. Haaker, and D. V. Richards. 1998. The perilous condition of white abalone *Haliotis sorenseni*, Bartsch, 1940. Journal of Shellfish Research 17:871–875.

Greene, G. H., M. M. Yoklavich, R. M. Star, V. M. O'Connel, W. W. Wakefield, D. E. Sullivan, J. E. McRea, and G. M. Cailliet. 1999. A classification scheme for deep seafloor habitats. Oceanologica Acta 22:663–678.

Hobday, A., M. J. Tegner, and P. L. Haaker. 2001. Overexploitation of a broadcast spawning marine invertebrate: decline of the white abalone. Reviews in Fish Biology and Fisheries 10:493–514.

Lafferty, K. D. 2001. White abalone restoration. USGS Western Ecological Research Center. Available: *www.werc.usgs.gov/coastal/abalone.html* (June 2003).

NOAA (National Oceanic and Atmospheric Administration). 2001. White abalone (*Haliotis sorenseni*). NOAA, Office of Protected Resources. Available: *www.nmfs.noaa.gov/prot_res/species/inverts/White_AB.html* (June 2003).

Shokr, M. E. 1991. Evaluation of second-order texture parameters for sea ice classification from radar images. Journal of Geophysical Research 96:10625–10640.

Tutschulte, T. C. 1976. The comparative ecology of three sympatric abalones. Doctoral dissertation. University of California, San Diego.

Using Laser Technology to Characterize Substrate Morphology and Geology of Selected Lake Trout Spawning Habitat in Northern Lake Michigan

Peter W. Barnes[1]

U.S. Geological Survey, 345 Middlefield Road, Menlo Park, California 94025, USA

Guy W. Fleischer[2]

U.S. Geological Survey, 1451 Green Road, Ann Arbor, Michigan 48105, USA

James V. Gardner[3] and Kristen M. Lee[4]

U.S. Geological Survey, 345 Middlefield Road, Menlo Park, California 94025, USA

Abstract. As part of a strategy to reestablish native stocks of lake trout *Salvelinus namaycush*, six areas of offshore and coastal Lake Michigan benthic habitat were mapped with a bathymetric laser system. This allowed us to visualize and map morphologic detail by at least an order of magnitude over existing data. Decimeter elevation–bathymetric data, referenced to the International Great Lakes datum of 1985, were obtained on a 4-m grid over a total area of about 200 km^2 in water depths from 0 to 30 m. Based on the laser-derived morphology, regional geology and sparse samples, three geologic regimes were used for substrate–habitat classification: (1) bedrock carbonates of Silurian through Devonian age, indicated by bedding scarps and lineations at or near the surface at all of the mapped areas; (2) glacial deposits that appear as compacted clay till lineations and cobble and boulder moraines with outwash features; and (3) modern sand deposits seen in thin down-drift (to the east) bedforms, sand sheets, and depositional lobes. Preferred spawning substrate—clean cobble and gravel deposits with adjacent deep water—is present in sections of all mapped areas. However, laser data cannot discern cleanliness, and video data indicate the cleanliness on this substrate may be compromised by recent algal and mussel growth.

Background

Lake Michigan supported the largest and most valued commercial and recreational fishery for lake trout *Salvelinus namaycush* in all the Great Lakes before it was driven to collapse in the 1950s by overfishing, predation by the exotic sea lamprey *Petromyzon marinus*, and habitat degradation (Holey et al. 1995). With sea lampreys now under control, the current restocking strategy seeks to recolonize lake trout (Hansen and Peck 1995). The reefs in northern Lake Michigan, one of two historical areas that produced the majority of spawning activity in the early 1930s (Dawson et al. 1997), are prime targets for restocking and also targets for benthic morphologic–habitat assessment in this survey. We seek to identify the presence of ideal offshore and coastal spawning substrate—shallow, clean, gravel–cobble substrate adjacent to deeper water—and focus the interpretation of laser mapping on Boulder Reef as a segment of a larger survey (Barnes et al 2003).

To date, stocked lake trout have become abundant enough to support sport and commercial fisheries and have developed significant populations on some targeted sites on rocky reefs in northern Lake Michigan (Holey et al. 1995). However, many of these fish have strayed to varying degrees into other areas during spawning. A study was initiated to document these movements, to understand the habitat attraction of these other areas, and to quantify the distribution of spawning lake trout and associated physical habitat. An important part of this study is gaining an understanding of the geologic attributes of successful spawning substrate and the potential for those attributes to change the regional distribution of that habitat.

---

[1] E-mail: pbarnes@usgs.gov
[2] E-mail: Guy.Fleischer@noaa.gov; present address: NOAA Northwest Fisheries Science Center, 2725 Montlake Boulevard East, Seattle, Washington 98112, USA.
[3] E-mail: jvgardner@usgs.gov
[4] E-mail: kmlee@usgs.gov

Southwest–northeast striking Paleozoic carbonate rocks underlie most of the northwestern rim of the Michigan Basin and crop out at the coast on offshore islands (Farrand et al. 1984; Anonymous 1987) and at Hog Island Reef (Sommers 1968). Sommers's (1968) reconnaissance diver and submersible studies indicate the presence of glacial outwash, glacial till, and modern lake sediment as surface deposits in addition to the Paleozoic outcrops. A 1989 benthic study using side-scan sonar and video mapped patchy sand, gravel, cobble, rubble, and boulder units on the south end of Boulder Reef (Edsall et al. 1989). The coast and upland areas of the islands and adjacent mainland are also draped with a wide variety of glacial features (Farrand et al. 1984). All these deposits might occur in the offshore study area.

Methods

Recent detailed morphologic assessments of the Great Lakes basins (Holcombe et al. 1996) are based on historic depth measurements, with data densities derived on a 2-km to 90-m grid spacing. We use this grid (National Geophysical Data Center 2004) as the regional setting for our more detailed study.

This study is a collaborative effort with the U.S. Army Corps of Engineers SHOALS (Scanning Hydrographic Operational Airborne Lidar Survey) program, which mapped six areas of offshore and coastal lake bed in late summer 2001 using a bathymetric laser system. These systems acquire decimeter depth accuracy and rapid sample rates but are limited by water clarity, typically 2–3 times the depth that can be seen with the eye. Decimeter elevation–bathymetric data were obtained on a 4-m grid covering a total area of about 200 km^2. Data was obtained from water depths of 0–30 m. Data and images are available at *http://geopubs.wr.usgs.gov/open-file/of03-120/* (Barnes et al. 2003). I (J.V.G.) am also working to develop and apply new techniques to extract spectral reflection and amplitude information from the laser data to allow additional classification of lake bed character.

Elevation data of detailed substrate morphologies were obtained from six coastal and offshore study sites (Figure 1). The relatively good water clarity allowed the SHOALS system to record bottom reflections from 25–30 m. Differential global positioning system navigation and elevation data were referenced to the International Great Lakes Datum. The long-term (1900–1990) lake level is close to 176.5 m (Great Lakes Research Lab 2003). Lake level at the at the time of our survey was about 176 m (National Ocean Service 2004). This is the elevation (176 m) to which we referenced water depth. We have chosen Boulder Reef (Figure 2) to highlight in this note, as there is substantial additional data to support interpretation of the morphology.

The SHOALS decimeter elevation–location data were merged with coarser National Oceanic and Atmospheric Administration gridded bathymetry (Holcombe et al. 1996) and a 30-m U.S. Geological Survey digital elevation model topography to generate the map images that underpin this report. Sun illumination angle, color coding depth values, and vertical exaggeration were subjectively combined to best enhance viewing the substrate morphology with vertical and oblique perspectives.

Observations

Although laser data cannot discern biologic growth on the substrate, a comparison of 1989 (Edsall et al. 1989) and 2001 videos taken of the same habitat morphologies indicates that the relatively clean cobble and boulder substrate seen in 1989 had been altered by extensive mussel and algal growth on the upper surfaces of the larger exposed cobble and boulder substrate in 2001.

Three geologic features can tentatively be interpreted from the morphology in the absence of extensive ground truth information: (1) subsurface bedrock trends, (2) glacial outwash of sand, cobbles, and boulders, and (3) modern lake bed sediments. Interpretation is based primarily on morphologic character, video data, and previous local and regional offshore mapping and sampling (Edsall et al. 1989; Barnes et al. 2003).

Bedrock

Southwest–northeast striking Paleozoic carbonate rocks underlie most of the northwestern rim of the Michigan Basin (Figure 1). Several distinct but subdued morphologic bedding lineations that are aligned southwest–northeast are tentatively interpreted as bedrock at or near the surface on the flat northern half of Boulder Reef (Figure 2).

Glacial

Overlying the Paleozoic bedrock are glacial deposits that are morphologically expressed most clearly on the southern part of Boulder Reef (Figure 2) as rough relief and ridges that the videos show as gravel to boulder rubble (Edsall et al. 1989; Barnes et al. 2003). These features have lineations and push features in the morphology that suggest northwest to southeast movement of glacial ice. Dive observations on the south flank of Boulder Reef (deeper than the laser data) reported compacted clay tills and outwash gravel and sand (Sommers 1968) indicating glacial material. A 3-km-diameter lobe-like cobble and boulder moraine dominates the southern

CHARACTERIZATION OF LAKE TROUT SUBSTRATE MORPHOLOGY AND GEOLOGY 167

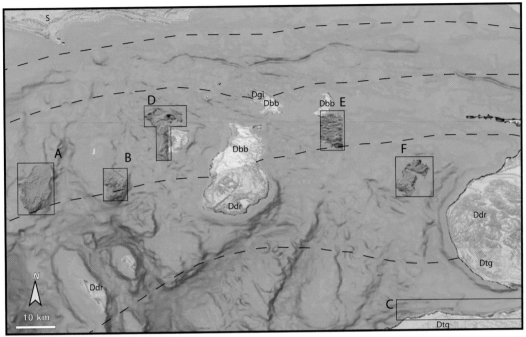

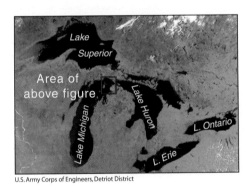

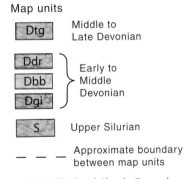

EXPLANATION

Color coded shaded relief map (available at: <http:www.ngdc.noaa.gov/mgg/greatlakes/lakemich_cdrom/html/>)

Map units

Dtg — Middle to Late Devonian

Ddr
Dbb } Early to Middle Devonian
Dgi

S — Upper Silurian

– – – Approximate boundary between map units

Figure 1. Location map of (A) Boulder Reef and the five other areas (B, Gull Island Shoal; C, southern coast of Little Traverse Bay; D, Trout Island and western coast of High Island; E, Hog Island Shoal; F, Dahlia Shoal) mapped (Barnes et al. 2003) along with regional geology (Farrand et al. 1984; Soller 1998).

part of the reef. Ridges radiate outward to the east and south from the shallowest (3 m) portion of this feature, which may reflect outwash deposition. Post-glacial reworking of the glacial deposits (seen as an overprint of smoother morphology of modern–hydraulic sediments) appears minimal in depths greater than 10 m. The coarse textured glacial material extends as a surface deposit to the west and south beyond the depths of the laser survey in the video data (to more than 40 m; T. Edsall and G. Kennedy, U.S. Geological Survey, personal communications).

Modern Lake Bed Sediments

Modern sedimentary deposits appear as overprints on the glacial and bedrock morphologies. A sand sheet drapes what we are interpreting as bedrock lineations on the northern part

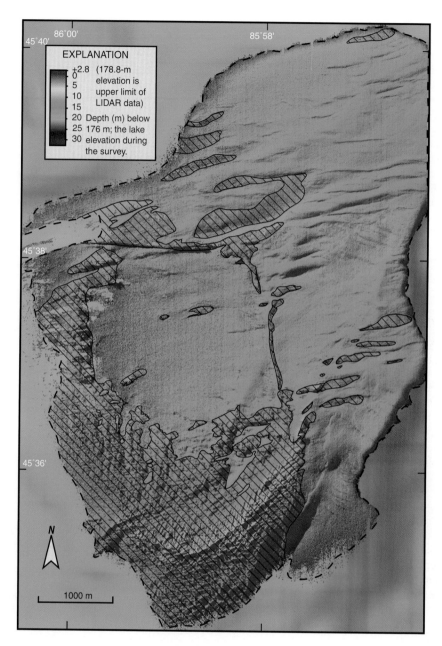

Figure 2. Shaded relief map showing bathymetry and morphology of Boulder Reef from laser hydrographic data. Dashed line is limit of survey data. Sun illumination from the northwest. Areas of rough relief are interpreted as coarse cobble and gravel (thought to be preferred lake trout habitat) are shown with red hatchers.

of Boulder Reef (Figure 2). These deposits also commonly occur as thin down-drift (to the east) bedforms, sand sheets, and depositional lobes such as that seen at the southeastern corner of the image.

Discussion

Relevance to Lake Trout Spawning Habitat

Three criteria were used to define ideal lake trout spawning habitat: (1) coarse (boulder or cobble) substrate with voids;

(2) clean substrate (devoid of biologic growth and fine sediment); and (3) adjacency to steep slopes with access to deeper water. In terms of these ideal characteristics, we believe that the areas where glacial deposits are not being buried by modern sediment could be important habitat. These two deposits are tentatively distinguished in Figure 2 based primarily on the remotely sensed morphology and video data, although more sample data are needed. The substrate and morphology of the gravel to boulder ridges and lobate glacial forms at south end of Boulder Reef seem to meet the three criteria sought for ideal lake trout spawning habitat. However, the cleanliness criteria of these substrates may be compromised by algae and zebra mussel growth. Laser waveform data is being analyzed for benthic albedo information (J.V.G.), which we hope will lead to an ability to assess both substrate texture and also substrate cleanliness and facilitate remote assessment of benthic habitat. No additional studies are currently planned.

Conclusions and Summary

Laser data allowed us to visualize and map substrate detail by at least one order of magnitude over existing data. The lake trout spawning reefs in northern Lake Michigan typified by Boulder reef show complex morphology and habitat variability related to patchiness of bedrock and glacial deposit exposures and to an overprint of modern geologic deposits. The spawning criteria of coarse substrate with adjacency to deeper lake water are common in the areas mapped. The presence of clean spawning substrate was not determined by mapping but by video from pre- and post-zebra mussel infestation, which suggested growth of these mussels and also algae on cobble and gravel substrate since 1989. The laser data gathered in this study will serve as a long-term baseline for monitoring the morphologic character and extent of lake trout habitat, and the technique could have mapping applications elsewhere in clear water regions of the Great Lakes.

References

Anonymous. 1987. Bedrock geology of Michigan. Michigan Department of Environmental Quality, Geological Survey Division. Map at 1:500,000. Available: *www.dnr.state.mi.us/spatialdatalibrary/pdf_maps/geology/Bedrock_Geology_pp.pdf* (April 2004).

Barnes, P. W., G. W. Fleischer, J. V. Gardner, and K. M. Lee. 2003. Bathymetry and selected perspective views of 6 reef and coastal areas in northern Lake Michigan: preliminary geologic interpretation. U.S. Geological Survey Open-File Report 03-120, Menlo Park, California. Available: *http://geopubs.wr.usgs.gov/open-file/of03-120/* (April 2004).

Dawson, K. A., R. L. Eshenroder, M. E. Holey, and C. Ward. 1997. Quantification of historic lake trout (*Salvelinus namaycush*) spawning aggregations in Lake Michigan. Canadian Journal of Fisheries and Aquatic Sciences 54(10):2290–2302.

Edsall, T. A., T. P. Poe, R. T. Nester, and C. L. Brown. 1989. Side-scan sonar mapping of lake trout spawning habitat in Northern Lake Michigan. North American Journal of Fisheries Management 9:269–279.

Farrand, W. R., D. M. Mickelson, W. R. Cowan, and J. E. Goebel. 1984. Quaternary geologic map of the Lake Superious 4° x 8° quadrangle, United States and Canada. U.S. Geological Survey, Miscellaneous Geological Investigation Map I-1420(NL-16). 1:1,000,000.

Great Lakes Research Lab. 2003. Lake Michigan-Huron water levels in meters (IGLD1985). Available: *http://www.glerl.noaa.gov/data/now/wlevels/lowlevels/plot/Michigan-Huron.gif* (May 2003).

Hansen, M. J, and J. W. Peck. 1995. Lake trout in the Great Lakes. Pages 244–247 in E. T. LaRoe, G. S. Farris, C. E. Puckett, P. D. Doran, and M. J. Mac, editors. Our living resources: a report to the nation on the distribution, abundance, and health of U.S. plants, animals, and ecosystems. U.S. Department of the Interior, National Biological Service, Washington, D.C.

Holcombe, T. L., D. F. Reid, W. T. Virden, T. C. Niemeyer, R. De la Sierra, and D. L. Divins. 1996. Bathymetry of Lake Michigan. A color poster with descriptive text and digital data available on CD-ROM. National Oceanic and Atmospheric Administration, Nat. Geophysical Data Center. Report MGG-11.

Holey, M. E., R. R. Rybicki, G. W. Eck, E. H. Brown, Jr., J. E. Marsden, D. S. Lavis, M. L.Toneys, T. N. Trudeau, and R. M. Horrall. 1995. Progress toward lake trout restoration in Lake Michigan. Journal of Great Lakes Research 21(Supplement 1):128–151.

National Geophysical Data Center. 2004. Bathymetry of Lake Michigan. National Oceanic and Atmospheric Administration, National Geophysical Data Center, Boulder, Colorado. Available: *http://www.ngdc.noaa.gov/mgg/greatlakes/michigan.html* (December 2004).

National Ocean Service. 2004. Preliminary water level data. National Oceanic and Atmospheric Administration, National Ocean Service, Port Inland, Michigan. Available: *http://co-ops.nos.noaa.gov/cgi-bin/* (September 2004).

Soller, D. R. 1998. Sheet B of map showing the thickness and character of Quaternary sediments in the glaciated United States east of the Rocky Mountians. U.S. Geological Survey, Miscellaneous Investigations Series Map I-1970, scale 1:1,000,000.

Sommers, L. H. 1968. Preliminary report on geological studies in northern Lake Michigan using underwater observation techniques. International Association for Great Lakes Research, Proceedings, 11th conference on Great Lakes Research 239–244.

Ground-Truthing Benthic Habitat Characteristics Using Video Mosaic Images

George R. Cutter, Jr.,[1] Yuri Rzhanov,[2] and Larry A. Mayer[3]

*University of New Hampshire, Center for Coastal and Ocean Mapping,
Durham, New Hampshire 03824, USA*

Raymond E. Grizzle[4]

*University of New Hampshire, Jackson Estuarine Laboratory,
Durham, New Hampshire 03824, USA*

Abstract. Subtidal benthic habitats from the Piscataqua River, New Hampshire and Maine, have been delineated by an automated segmentation technique using bathymetry derived from multibeam echo sounder data. The map, produced by segmentation of seafloor textures, represents a hypothetical benthic habitat map that requires ground-truthing. Video mosaics are being used to ground-truth substrate composition and transitions apparent in the bathymetry data map and to describe biological features and organism occurrences and densities. Here, we describe the utility of video mosaics for ground-truthing benthic habitat characteristics and present two examples of their use. Video mosaics acquired along two transects in the Piscataqua River were used to detect substrate transitions apparent in the bathymetry that were identified as distinct hypothetical habitat types and to quantitatively assess coverages of distinct sediment conditions, density of megafaunal organisms (lobsters), and bioturbational features (crab feeding pits).

Introduction

Benthic habitat mapping efforts have benefited from high resolution, full coverage data from multibeam and side-scan sonar systems (Kostylev et al. 2001). However, because of the complex interactions among seafloor composition, geometry, acoustic reflection, and backscatter (see Urick 1983), multibeam and side-scan maps must be ground-truthed to confirm sedimentological and biological characteristics. Video imagery can be used for ground-truthing and to provide continuous image data along large distances; however, quantitative analysis of video can be difficult or tedious. Mosaics can be constructed from overlapping video image sequences to convert the many individual video frames to a single still image representing the entire imaged tract. Underwater image mosaics have also proven valuable to ground-truth microbathymetric maps for marine archeological studies (Singh et al. 2000), been used to navigate underwater vehicles (Gracias and Santos-Victor 1998), to process remotely operated vehicle video in real time (Marks et al. 1995) and to map deep-sea vent clam beds (Grehan and Juniper 1996), and suggested as potentially valuable to benthic habitat mapping (Rzhanov et al. 2000; 2001).

We have constructed mosaics representing tens of meters or more of seafloor from which it was possible to detect, identify, and measure large epifauna, large bioturbational features, substrate transitions, and seafloor attributes important to acoustic data. More importantly, mosaics directly represent spatial scales inherent to beams and range cell sizes of common multibeam echo sounders and side-scan sonars. Mosaic construction does not require positioning data; however, positioning data are necessary for placement of the mosaics within sonar data maps and for interpretation with respect to sonar data maps.

To map benthic habitats of the Piscataqua River, New Hampshire and Maine, multibeam bathymetry data were analyzed by Cutter et al. (2003) using an automated segmentation and classification technique (Figure 1) involving the local Fourier histogram (LFH) texture feature classification described by Zhou et al. (2001). Effectively, the LFH texture feature analysis represents distributions of spatial frequency components comprising the series defined by the data values surrounding

[1] Corresponding author: gcutter@cisunix.unh.edu
[2] E-mail: yuri.rzhanov@unh.edu
[3] E-mail: larry.mayer@unh.edu
[4] E-mail: ray.grizzle@unh.edu

each data point, thus providing a multiscale roughness measurement for every point. The technical details of the LFH technique are described in Zhou et al. (2001), and details concerning application to multibeam echo sounder data for seafloor habitat mapping are described in Cutter et al. (2003). Seafloor texture feature regions resulting from the analysis of Piscataqua River data suggested good correspondence to gravel, sand, and rock facies identified through sampling in the study area (Ward 1995) and to multibeam and side-scan sonar acoustic backscatter intensity data. We suggest that the automatically classified texture feature map presented in Cutter et al. (2003) represents a hypothetical physical attributes map or what many would consider a hypothetical benthic habitat map of the seafloor. The hypothetical habitat map presents the opportunity to test hypotheses about seafloor characteristics and associated benthic fauna and, if properly ground-truthed, can be considered a true habitat map. The ability to go from a hypothetical habitat map to a true map of the spatial extent and distribution of benthic habitats will be strengthened by the use of other data such as acoustic backscatter intensity, substrate composition, energy regime, and primary solute concentrations (i.e., consideration of water column and benthic boundary layer conditions). The better these key environmental factors can be described, the more likely it is that the hypothetical habitat map will be an accurate representation of biological habitats. However, even if other environmental data are lacking, a hypothetical habitat map can be developed to predict spatial distributions of habitat and seafloor characteristics.

A habitat map should be specific to a target species or groups of species that associate similarly. Ground-truthing is needed to determine how the species occur, how they are distributed, and how they utilize the seafloor. We can assess the occurrence and distribution of some species using remote sensing techniques. For sessile or limited-motility epibenthic megafauna, video mosaics can be used to assess presence and density of organisms as a function of seafloor region (i.e., by hypothetical benthic habitats derived from segmentation of gridded seafloor bathymetry [Cutter et al. 2003]), thus providing descriptions of essential fish habitat (EFH) levels 1 and 2 (Able 1999). Essential fish habitat level 1 pertains to occurrence of a species in a habitat, and EFH level 2 pertains to abundance (Able 1999). The purpose of this work is to demonstrate the utility of image mosaics as a ground-truthing technique and for habitat characterization by providing examples of mosaics and analyses of mosaics to (1) delineate and describe transitions of sub-

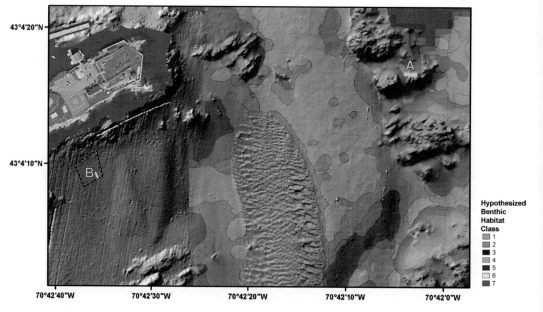

Figure 1. Section of the seafloor (980 × 630 m) in the mouth of the Piscataqua River, part of which was delineated using automated segmentation techniques involving a modified implementation of local Fourier histogram texture feature classification applied to gridded bathymetry data from a Reson 8125 (455 kHz) multibeam echo sounder (Cutter et al. 2003). Seven hypothesized seafloor habitat classes resulting from the segmentation are colored in the figure. The letter A marks the location of the transect where video for the rocky region mosaic was acquired (Figure 2a), and the letter B marks the location of the lobster experimental enclosure mosaic (Figure 2b).

strates; (2) assess occurrence and density of an important megafaunal species, the northern lobster *Homarus americanus* (also known as American lobster); and (3) estimate coverages of substrate characteristics including large bioturbational feeding pits.

Methods

Hypothetical Habitat Map

Bathymetry data in the study area were collected aboard the R/V *Coastal Surveyor* (University of New Hampshire) by Science Applications International Corporation (SAIC) using a Reson 8125 (455 kHz) dynamically focused multibeam echo sounder. The data were collected for the Shallow Water Survey 2001 Common Dataset (Mayer and Baldwin 2001). Data were cleaned using standard hydrographic processing approaches and then gridded using a weighted mean grid with 1-m spatial resolution. Delineations of seafloor configurations representing apparent benthic habitats, constituting a hypothetical habitat map, were produced by Cutter et al. (2003) using an automated segmentation procedure involving LFH texture feature classification (Zhou et al. 2000) applied to the gridded multibeam echo sounder bathymetry data.

Video Mosaic Imagery

Mosaic images were generated from underwater video footage of the seafloor collected by diver and by towed camera using a featureless coregistration technique involving frequency domain processing of images to solve automatically for rigid affine motion parameters (translation, rotation, and zoom; Rzhanov et al. 2000). Each frame is coregistered with the previous frame in the sequence. The magnification of each frame is adjusted to the initial zoom level or to a specified global zoom. Using that approach, the mosaic has a uniform distance scale throughout. After the assembly of mosaics, colors were manually adjusted to compensate for ambient lighting effects by adjusting independently the ranges of the red, green, and blue color channels. Automated frequency domain mosaicking (Rzhanov et al. 2000) overcomes some of the problems experienced by Grehan and Juniper (1996) due to uneven illumination and does not require user intervention or selection of tie points associated with manual, feature-based coregistration techniques. As long as overlap is sufficient (>50%) and projective distortions (due to camera tilt with respect to the target surface) are minimal, the automated frequency domain coregistration technique is robust and produces high-quality, seamless mosaics. When projective distortions occur or the seafloor is not flat, more complicated models can be applied to accomplish mosaicking (Mann and Picard 1997).

Positioning data for the towed camera deployments were provided by differential global positioning system (DGPS) determination of the vessel position where the antenna was offset from the camera by a maximum of about 5 m. For the diver deployments, an acoustic transponder positioning system was used. Mosaic construction does not require positioning data, however positioning data are necessary for placement of the mosaics within sonar data maps and for interpretation with respect to features in the maps. We used the vessel positioning data to place the mosaic from the towed camera into the map and acknowledge positioning errors. We lacked positioning capable of providing feature-for-feature recognition between the sonar data and video. However, habitat regions depicted by the hypothetical habitat map typically had dimensions of 50–100 m across; therefore, even with positioning uncertainty typically associated with DGPS and a towed device, we could determine which region was being imaged. If apparent habitat regions had dimensions of 10 m or less, more precise positioning methods would be required. Our intentions here are to present examples of the utility of mosaics for assessing habitat attributes and, therefore, georeferencing details are not addressed. However, to provide accurate ground-truthing of seafloor attributes (features and transitions), accurate positioning and georeferencing would be required. If the need were to determine that a delineation drawn from sonar data maps existed in a specific location as confirmed by optical imagery, knowledge about the position of imaging devices would be critical. Here, we simply present what can be gained from the mosaics: sedimentary features, transitions between habitats, bioturbational features, and fauna.

Results

The mosaic from the rocky region (Figure 2a) reveals a transition from shelly gravel sediments to boulders and bedrock. The transition observed in the mosaic was marked in the figure using colors corresponding to the nearest texture feature class regions from the segmented bathymetry map. On the rocks, several species of epifaunal sponges, bryozoans, and tunicates were evident.

The mosaic from an experimental enclosure for lobsters (Figure 2b) reveals fine sediment (determined by divers to be silty, fine sand) occupied by benthic megafaunal lobsters (northern lobsters) and crabs (*Cancer irroratus* [known as Atlantic rock crab] and/or *C. borealis* [known as Jonah crab]) and large infaunal razor clams (*Ensis directus* [known as Atlantic jackknife and possibly others), evident from their siphons and empty shells and identified by divers. Large bioturbational pits were apparent, some occupied by crabs and lobsters. Diver observations suggest that the pits were crab feeding pits, excavated by crabs in pursuit of prey of *E. directus*. The bathymetry in the region of the lobster enclosure has not yet been classified by texture feature analysis.

Analysis of Mosaic Images

Substrate Transitions in Piscataqua River Rocky Region Mosaic

Substrate transitions are visible and delineations easily drawn for the mosaic from a rocky region on the eastern side of the river mouth channel (Figure 2a). Despite the fact that the mosaic was acquired from what appeared to be a rocky outcrop in the bathymetry image, various substrates existed. Patches of gravel, shell hash, shell valves, and boulders existed in addition to bedrock. It was evident that a variety of sediments with a wide grain-size distribution cover portions of the outcrops. Texture feature classification similarly suggests that at the scales of the analysis (1 m to 8 m) at least two seafloor textures exist there. The positioning data accompanying the video used to construct the rocky region mosaic were not precise enough to determine whether the transitions delineated in the mosaic corresponded to bathymetric features apparent in the bathymetry map. Our goal for future deployments is to collect positioning data suitable for accurate georeferencing of the video imagery and mosaics. Using video from several transects in the delineated study area, we will construct mosaics that will allow us to describe substrates and biology for all regions and to assess the accuracy of the texture feature classification technique used to delineate apparent habitats. We can then use that accuracy assessment for adjusting and understanding the impact of the texture feature classification parameters.

Lobster Enclosure Video Mosaic

The two primary goals of the analysis of the lobster enclosure mosaic (Figure 2b) were to detect and enumerate occurrences of specific fauna, in this case, lobsters (specifically northern lobsters), and delineate substrate surface conditions according to three apparent classes: (1) sediment with a micro-algal layer coverage; (2) bare sediment, without algal coverage, although perhaps shallow and bioturbated; and (3) deep bioturbated excavations (bio-excavations). In general, for the region from which the mosaic imagery was ac-

Figure 2. (a) Rocky region video mosaic image. The transition marked in the mosaic corresponds to the nearest boundary for the texture feature class regions along the transect and represents transition from shelly, gravelly sediments to bedrock with boulders. On the rocks, several species of sponges, bryozoans, and tunicates are evident. (b) Shallow shoal video mosaic image from an experimental enclosure for tracking lobster movements (lobster enclosure). Lobsters (northern lobsters) are labeled with the letter L, and the three substrate conditions are labeled by number: 1 = sediment with a micro-algal layer coverage; 2 = bare sediment, without algal coverage, although perhaps shallow and bioturbated; and 3 = deep, bioturbated excavations (bio-excavations), apparently crab feeding pits.

quired, the sediment surface conditions were easily distinguishable by coloration such that, consistent with diver observations, the adjusted color imagery showed that (1) sediment apparently covered by algae or microbial veneer was reddish-brown; (2) bare sediment was olive to olive-gray; and (3) deep, bioturbated sediment was bluish to bluish-gray due to digging and burrowing of crabs (*C. irroratus* or *C. borealis*) and lobsters that exposed the anoxic subsurface sediments. The leftmost part of the image was not included in analysis because of uneven illumination. That excluded portion represented approximately 70 cm along the long axis of the mosaic.

Abundance of Megafauna in Lobster Enclosure Video Mosaic

Simple visual analysis of the mosaic accomplishes our first goal, to detect and enumerate lobsters. In the mosaic, seven lobsters were present with more than 50% of the body in the image. Partial lobsters crossing edges were not counted.

Density of Megafauna in Lobster Enclosure Video Mosaic

Since all images were adjusted to the same zoom level during automated coregistration and the beginning of the transect contains a physical target for which dimensions are known (fence with 3.8-cm mesh), the coverage area could be estimated. For the portion of the mosaic analyzed, the coverage area was 5.87 m^2. That allowed estimation of lobster density as 1.2/m^2. That high estimate reflected the higher frequency of occurrence of lobsters in that part of the enclosure. Subsequent parts of the video sequence suggest a much reduced density.

To accomplish a density estimate by analysis of a video sequence would require knowledge of precise distances traveled between any frame as well as height from the target surface. Such information could be provided by a surface positioning system but is unlikely to produce accurate estimates. Accurate estimates of the bottom area imaged require either coregistration of the image series to reproduce the entire tract or detailed instantaneous data from precise motion and position sensors incorporated with the camera. The latter option requires much more sophistication and expense than available to many studies.

Sediment Surface Condition in Lobster Enclosure Video Mosaic

Although sediment conditions were distinguishable by coloration, sediment surface condition was difficult to delineate into continuous regions. To aid visual determination of condition, the histogram from each color channel of the image (red, green, and blue) was range adjusted, eliminating the highest and lowest 1% of values, approximately. Seldom was a particular area of the seafloor totally covered by one class. Some generalizations were made and, therefore, surely some errors in the estimates exist. Three primary conditions were distinguished and regions were delineated according to predominant condition. Transitions between bare sediment and sediments with algal or microbial veneer were not always distinct, even after color adjustment. Deep bioturbational pits and recently exposed bare sediments were easily delineated. The coverage areas with the prespecified conditions were determined to be algal or microbial covered = 3.46 m^2; bare or shallow bioturbated = 2.27 m^2; and deeply bio-excavated = 0.14 m^2.

Discussion

Mosaic image measurements can be made at specific intervals of distance or area covered to provide substrate conditions and organism density estimates for any part of the seafloor imaged. In our mosaic from the lobster enclosure, large biogenic features from megafauna were apparent, such as the razor clam (specifically *E. directus*) siphons. These siphons were just large enough to resolve in the mosaic because of the height of the camera off bottom. For the assessment of infauna or small epifauna, deployment requirements will differ. Images must be collected close to the bottom and lighting must be good and uniform. One disadvantage with that type of deployment is that coverage is reduced. Given a fixed camera field of view, the smaller the range to the target, the smaller the imaged area. However, close-range imaging allows smaller features to be resolved and larger features to be seen with more clarity. The scale and resolution of the imagery ultimately determine the level of detail at which habitat characteristics can be quantified. In this case, we were able to describe megafaunal occurrences and densities as well as megafaunal bioturbational features, adequate for assessing EFH levels 1 and 2. However, we would not be able to use the same imagery to assess even occurrence (EFH level 1) for macrofauna or smaller organisms.

Issues Related to Optical Ground-Truthing

One fundamental issue is that standard optical images, such as point sample seafloor photos that typically image 1 m^2 or less, do not directly represent acoustical features because of resolution disparities. How well do the photos represent what is sensed by the sonar? That cannot be directly determined (without assumptions about variability between points) until the optical images span the same spatial scales as the acoustic data. Videography and video mosaicking are cost-effective means (versus other optical imaging techniques such as laser line scan) to provide optical seafloor imagery that can represent spatial scales corresponding to those of acoustic footprints (though perhaps not in all directions). It is unlikely that features will have acoustic responses and optical representation clearly correlated, feature-to-feature correspondence between acoustic and optical sensing will be difficult, and therefore, we must consider examining covariability over distances or areas imaged. Video mosaics provide optical image data spanning spatial scales at which that may be accomplished.

Why not simply acquire video or photos at greater heights above the seafloor since greater camera-target distance would produce a larger imaged area? Underwater optical imagery is limited by water column conditions, primarily related to attenuation of light by suspended particles. That effect is most pronounced in coastal waters but exists throughout most of the oceans. Even when particle concentrations are low, light attenuation through clear seawater limits the distance and, therefore, the area that can be imaged reliably. In the deep sea or other waters where benthic boundary layer turbidity can be minimal, the limitation is the intensity of artificial light sources and camera image receptors. Video mosaics provide a cost-effective way of providing large areal coverage with high resolution and, thus, an appropriate tool to ground-truth acoustic data.

Conclusions

Mosaicked imagery is beneficial for ground-truthing sonar data and assessment of habitat characteristics. As demonstrated by our case studies, video mosaics are useful for determining seafloor characteristics, detecting seafloor transitions, and measuring biological attributes including habitat-specific occurrence and abundance of organisms. Mosaics can represent spatial scales from mm to tens of m and are important to overcoming spatial scale mismatches between typical sonar data and standard imagery. Mosaics are particularly useful for expanding effective spatial coverage of seafloor imagery in coastal and estuarine waters where low visibility can compromise the spatial coverage of individual images. Since mosaic images overcome data–spatial scale mismatches, mosaics are suitable for ground-truthing hypotheses made about habitat from maps derived from multibeam echo sounder and side-scan sonar data.

Acknowledgments

We thank Lt. Shep Smith, National Oceanic and Atmospheric Administration (NOAA), for cleaning bathymetry data and for grid generation and SAIC for collecting the Reson 8125 multibeam echo sounder data. This work was supported by NOAA Grant NA970G0241.

References

Able, K. W. 1999. Measures of juvenile fish habitat quality: examples from a National Estuarine Research reserve. Pages 134–147 in L. R. Benaka, editor. Fish habitat: essential fish habitat and rehabilitation. American Fisheries Society, Symposium 22, Bethesda, Maryland.

Cutter, G. R., Y. Rzhanov, and L. A. Mayer. 2003. Automated segmentation of seafloor bathymetry from multibeam echosounder data using local Fourier histogram texture features. Journal of Experimental Marine Biology and Ecology 285/286:355–370.

Gracias, N., and J. Santos-Victor. 1998. Automatic mosaic creation of the ocean floor. Pages 257–262 in Oceans '98 conference proceedings. Institute of Electrical and Electronics Engineers, Piscataway, New Jersey.

Grehan, A. J., and S. K. Juniper. 1996. Clam distribution and subsurface hydrothermal processes at Chowder Hill (Middle Valley), Juan de Fuca Ridge. Marine Ecology Progress Series 130(1–3):105–115.

Kostylev, V. E., B. J. Todd, G. B. Fader, R. C. Courtney, G. D. Cameron, and R. A. Pickrill. 2001. Benthic habitat mapping on the Scotian Shelf based on multibeam bathymetry, surficial geology and sea floor photographs. Marine Ecology Progress Series 219:121–137.

Mann, S., and R. W. Picard. 1997. Video orbits of the projective group: a simple approach to featureless estimation of parameters. IEEE Transactions of Image Processing 6:1281–1295.

Marks, R. L., S. M. Rock, and M. J. Lee. 1995. Real-time video mosaicking of the ocean floor. IEEE Journal of Oceanic Engineering 20:229–241.

Mayer, L. A., and K. Baldwin. 2001. Shallow water survey 2001: papers based on selected presentations from the second international conference on high resolution surveys in shallow water. Marine Technology Society Journal 35:3–4.

Rzhanov, Y., G. R. Cutter, and L. Huff, L. 2001. Sensor-assisted video mosaicing for seafloor mapping. Pages 411–414 in ICIP 2001 international conference on image processing. Institute of Electrical and Electronics Engineers, Piscataway, New Jersey.

Rzhanov, Y., L. M. Linnett, and R. Forbes. 2000. Underwater video mosaicing for seabed mapping. In 2000 IEEE international conference on image processing. Institute of Electrical and Electronics Engineers, Piscataway, New Jersey.

Singh, H., L. Whitcomb, D. Yoerger, and O. Pizarro. 2000.

Microbathymetric mapping from underwater vehicles in the deep ocean. Computer Vision and Image Understanding 79:143–161.

Urick, R. J., 1983. Principles of underwater sound for engineers. McGraw-Hill, New York.

Ward, L. G. 1995. Sedimentology of the lower Great Bay/Piscataqua River estuary. Department of the Navy, NCCOSC RDTE division report, San Diego, California.

Zhou, F., J. Feng, and Q. Shi. 2001. Texture feature based on local Fourier transform. Pages 610–613 *in* ICIP 2001 international conference on image processing. Institute of Electrical and Electronics Engineers, Piscataway, New Jersey.

Advances in Processing and Collecting Multibeam Echosounder Data for Seabed Habitat Mapping

Douglas Lockhart[1]

*Fugro Pelagos, Inc., formerly Thales GeoSolutions (Pacific), Inc.,
3738 Ruffin Road, San Diego, California 92123, USA*

Robert J. Pawlowski[2]

*Fugro Pelagos, Inc., formerly Thales GeoSolutions (Pacific), Inc.,
615 E. 82nd Avenue, Suite 304, Anchorage, Alaska 99518, USA*

Edward J. Saade[3]

*Fugro Pelagos, Inc., formerly Thales GeoSolutions (Pacific), Inc.,
3738 Ruffin Road, San Diego, California 92123, USA*

Abstract.—Backscatter data from Reson multibeam echo sounders (MBES) can be captured as a single time series for each beam footprint. Referred to as snippets, this data has a few advantages over MBES pseudo sidescan and, in some cases, true sidescan data. Snippets can be precisely co-registered to the bathymetric surface using the fact that the snippet and sounding are from the same place. This process implicitly corrects the snippet position for water column refraction. The resulting mosaic has improved signal to noise qualities as a result of precise positioning. Other advantages include increased resolution, automated mosaic assembly and potential for automated image classification. Data products generated from snippet processing are useful for habitat classification.

Introduction

Advances in the collection and processing of backscatter data from multibeam echosounder survey technology is providing information about seabed habitat that previously had been available only through a comparison of echosounder and side-scan sonar surveys or from direct visual surveys. By collecting backscatter data in conjunction with routine multibeam echo-sounder data collection, a coregistered data set is built that can be used to classify seabed types. A backscatter data set is derived through the analysis of the reflectance signal strength on a beam-by-beam basis across each swath of the multibeam echosounder survey. Backscatter data can be processed into pseudo-side-scan seabed mosaics that depict bottom hardness and sediment characteristics. Mosaics are then compared with seabed digital terrain models to characterize seabed structure and marine habitats to a new level of detail and accuracy. With this approach, quantitative, precisely georeferenced, pseudo-side-scan imagery from automated processing is replacing qualitative, labor-intensive, side-scan imagery positioned in a quasi-accurate georeferenced system. This approach allows increased data collection rates with improved signal processing and leads to increased area coverage for each multibeam system. The precise, quantitative, georeferenced data set can be used to develop substrate baselines and to quantify marine habitat change over a period of time. The processing tools and techniques for backscatter analysis are described, with examples of recent habitat characterization in Alaska provided. This technology is commercially exportable.

These advances are based on the current capability to record the raw backscatter data of each beam for every ping of the Reson 8000 series multibeam echosounder (MBES; Reson, Goleta, California) system data. The backscatter for any individual beam is referred to as a "snippet" (the rawest form of data available from a MBES, representing the reflectance energy acquired from a single acoustic beam reflection) or "footprint time series." For each beam and each ping, the MBES collects a backscatter record of signal intensity. The backscatter data is geographically registered

---
[1] E-mail: Doug.Lockhart@thales-geosolutions.com
[2] E-mail: Bob.Pawlowski@thales-geosolutions.com
[3] E-mail: Edward.Saade@thales-geosolutions.com

by draping the snippet over the footprint ensonified by the given beam. This procedure ensures accurate three-dimensional registration of each pixel. Data points are mosaicked on the seafloor terrain using a pixel size no greater than 0.1% of water depth per pixel, which produces 5-cm pixels in 50-m water depths. This methodology yields highly accurate placement of the acoustic data and depth resolution to the sub-meter level. This represents a significant improvement in georeferencing, resolution of the seabed, and accuracy over traditional towed side-scan sonar systems.

Acoustic backscatter content collected and processed as an integral part of the MBES data offers advantages of (1) precise coregistration of the backscatter data with the multibeam bathymetry data; (2) improved signal-to-noise ratios compared to conventional imaging sonar; (3) increased resolution for the final product from smaller pixel size; and (4) automated analysis of quantitative reflectance data within a position depth framework.

These advantages can be seen when simultaneously collected bathymetry and backscatter data sets are overlaid for seabed classification. Being hull mounted, the system enables a higher operational speed for more cost-effective data collection with less risk to equipment.

Multibeam Backscatter Sampling

On August 13–14, 2001, Thales GeoSolutions (Pacific), Inc. (TGPI) surveyed a 16-km^2 test site off Kodiak Alaska, with depths between 13 and 140 m. The site was surveyed with the Reson 8111 MBES system, with snippet capability, to International Hydrographic Organization 1st order accuracy and resulted in 1-m^3 resolution in the bathymetry data and 0.25-m resolution in the backscatter data. Figures 1 and 2 show the high-resolution bathymetry as a digital terrain model and the backscatter mosaic, respectively. These two data products provided the tools for classifying seabed types at Moss Landing Marine Laboratory Center for Habitat Study (MLML) (H. G. Greene, Center for Habitat Studies, Moss Landing Marine Laboratory, personal communication) as depicted in Figure 3. Project results are posted at the joint MLML and TGPI fisheries habitat mapping Web site (*www.fisherieshabitat.net/*). Further processing with ER Mapper software produced the bathymetry and backscatter drape depicted in Figure 4, enabling visual enhancement of coregistered data.

Sampling of raw snippet data on a beam-by-beam basis, in addition to depth measurement, allows the collection of an individual time series for each beam. These shorter time series are the acoustic footprint of the beam

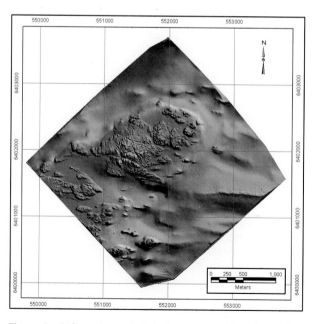

Figure 1. Color-enhanced digital terrain model of the Kodiak test site survey. Image is registered to World Geodetic System WGS-84 and displayed with a sun angle of 45° and sun azimuth of 45°. Depth range is 13–140 m.

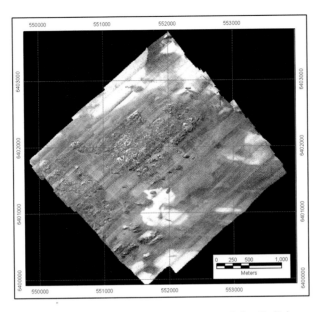

Figure 2. Multibeam backscatter mosaic of the Kodiak test site survey based on multibeam echosounder acoustic swaths.

as it reflects off the seabed and results in the snippet sampling only in the region of bottom detection versus across the water column. When processed, snippets mosaics are crisp, detailed, and accurately registered with the georeferenced bathymetry.

The key to this innovation has been the collection of the raw snippet data at the beam former in the Reson 8000 MBES series instead of processed intensity. This results in a large increase in data and a significant improvement in image resolution.

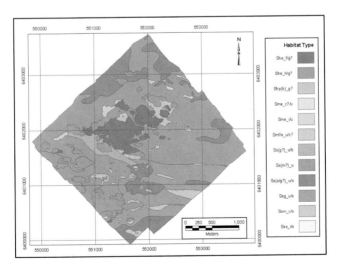

Figure 3. Seabed characterization of Kodiak test site survey through analysis of digital terrain model features and backscatter acoustic reflectance. Classification scheme developed by H. G. Greene (Center for Habitat Studies, Moss Landing Marine Laboratory).

Figure 4. Multibeam backscatter image as draped on the digital terrain model to provide visual enhancement of coregistered data for detailed substrate analysis.

Conclusion

Advances in the processing of snippets of backscatter data from the Reson MBES system is providing habitat managers with improved high-resolution mosaics of the seabed. These mosaics are registered to the georeferenced bathymetry, which is collected simultaneously. As is shown from the results of the Kodiak test site survey, the merging of these two types of data provided state of the art imagery for classifying seabed types and may lead to an improved characterization of benthic fishery habitats.

Classification of Marine Sublittoral Habitats, with Application to the Northeastern North America Region

PAGE C. VALENTINE[1]

U.S. Geological Survey, 384 Woods Hole Road, Woods Hole, Massachusetts 02543, USA

BRIAN J. TODD[2] AND VLADIMIR E. KOSTYLEV[3]

*Geological Survey of Canada (Atlantic), Post Office Box 1006,
Dartmouth, Nova Scotia B2Y 4A2, Canada*

Abstract. Habitats are defined as spatially recognizable areas where the physical, chemical, and biological environment is distinctly different from surrounding environments. A habitat can be delimited as narrowly or as broadly as the data and purpose permit, and this flexibility of scale influences the development of habitat classification schemes. Recent habitat classifications focus on a wide range of habitats that occur in European, American, and worldwide seafloor environments. The proposed classification of marine sublittoral habitats is based on recent studies in the American and Canadian parts of northeastern North America using multibeam and side-scan sonar surveys, video and photographic transects, and sediment and biological sampling. A guiding principle in this approach to habitat classification is that it will be useful to scientists and managers of fisheries and the environment. The goal is to develop a practical method to characterize the marine sublittoral (chiefly the subtidal continental shelf and shelf basin) habitats in terms of (1) their topographical, geological, biological, and oceanographical attributes and (2) the natural and anthropogenic processes that affect the habitats. The classification recognizes eight seabed themes (informal units) as the major subject elements of the classification. They are seabed topography, dynamics, texture, grain size, roughness, fauna and flora, habitat association and usage, and habitat recovery from disturbance. Themes include one or many classes of habitat characteristics related to seabed features, fauna and flora, and processes that we view as fundamental for recognizing and analyzing habitats. Within the classes, a sequence of subclasses, categories, and attributes addresses habitat characteristics with increasing detail. Much of the classification is broadly applicable worldwide (excluding some low-latitude environments), but faunal and floral examples are representative of the northeastern North America region. In naming habitats, the classification emphasizes seabed substrate dynamics, substrate type, and seabed physical and biological complexity. The classification can accommodate new classes, subclasses, categories, and attributes, and it can easily be modified or expanded to address habitats of other regions. It serves as a template for a database that will provide a basis for organizing and comparing habitat information and for recognizing regional habitat types.

Introduction

Recently, increased interest in the management and conservation of marine environments and species has stimulated efforts to produce large-scale maps of the seafloor and, by extension, habitats. Seafloor surveys using modern multibeam sonar and laser technology can generate topographical imagery of seabed features with vertical and horizontal resolutions of centimeters to tens of centimeters, respectively. Seafloor reflectance (backscatter) imagery collected at the same time shows the distribution of seabed materials and reveals patterns that are related to the origin of the materials and their responses to modern processes of erosion, transport, and deposition. These images depict broad geological environments well, and they provide an excellent framework for identifying and classifying habitats at a scale that is useful for scientific research and environmental management.

Habitats are defined as spatially recognizable areas where the physical, chemical, and biological environment

---
[1] Corresponding author: pvalentine@usgs.gov
[2] E-mail: brian.todd@nrcan.gc.ca
[3] E-mail: vkostyle@nrcan.gc.ca

is distinctly different from surrounding environments. Although modern seafloor imagery is a solid foundation for the study of habitats, the classification of habitats requires additional information in the form of video and photographical imagery and geological and biological samples of the seabed. Habitat characterization produces descriptions of habitats based on geological, biological, chemical, and oceanographical observations. Habitat classification produces a set of habitat types based on a suite of standard descriptors of topographical, geological, biological, natural, and anthropogenic features and processes. Habitat mapping is the spatial representation of described and classified habitat units.

The goal of the habitat classification scheme proposed here is to develop a practical method to define marine sublittoral (chiefly subtidal continental shelf and shelf basin) habitats in terms of (1) their geological, biological, and oceanographical attributes and (2) the natural and anthropogenic processes that affect the habitats. The classification is based on recent observations in the American and Canadian parts of northeastern North America using multibeam and side-scan sonar surveys, video and photographic transects, and sediment and biological sampling (Figure 1). The structure of the classification serves as a template for a database. It is designed so that it can be modified easily to classify habitats of other regions.

Recent Habitat Classification Schemes

Several recent classifications have focused on a range of habitats that occur in European land and water environments (European Environment Agency 2004; Connor et al. 2003), in American marine, estuarine, and wetland environments (Cowardin et al. 1979; Allee et al. 2000), and worldwide in deep (subtidal) marine environments (Greene et al. 1999). These five schemes are accessible online (web sites in References section). In 1992, Dethier (1992), who is a coauthor of the Allee et al. (2000) classification, proposed an alternative to the Cowardin et al. (1979) system. It is not the purpose of this paper to exhaustively compare and contrast all of the structural details of these classifications with the scheme proposed here (see Allee et al. [2000] for a review of recent clas-

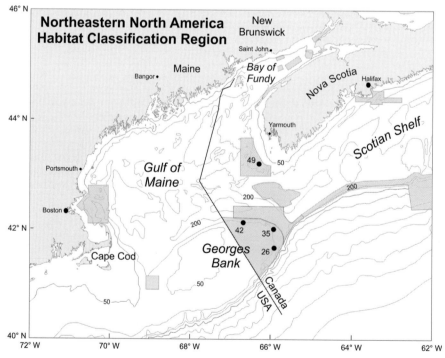

Figure 1. Map showing the geographic region that is the focus of the northeastern North America marine sublittoral habitat classification. The region is characterized by a wide variety of environments in continental shelf, bank and basin, and submarine canyon head settings. Shaded offshore areas have been surveyed using multibeam sonar mapping technology to produce high-resolution topographical and backscatter imagery of the seafloor. Numbered dots indicate locations of stations 26, 35, 42, and 49 that are described as examples 1–4, respectively, in Appendix 2 and shown in Figures 2–5. Depth contours in m.

sifications and a comparison of that classification with the European Nature Information System [EUNIS] classification). All habitat classifications are, for the most part, based on observations of similar kinds of habitat characteristics (e.g., topography, bottom type, and fauna, among others). Classifications are designed by ranking kinds of characteristics to coherently describe habitat structure and function. However, it is somewhat difficult to compare the various schemes in detail because some kinds of habitat characteristics are placed at different levels in the classification hierarchies. Probably, there is not a right or wrong method for classifying habitats. The differences in classification structure reflect the ways the designers have chosen to organize, understand, and rank the structures and functions of natural systems.

The Cowardin et al. (1979) classification is a hierarchy of five levels that range from systems at the highest level (marine, estuarine, and others) to dominance types at the lowest levels (plant or animal forms). The marine sublittoral habitat is treated sparingly in the classification. The Allee et al. (2000) classification is also a hierarchy that ranges from coarse-scale life zones at the top to fine-scale eco-units at the bottom; marine sublittoral environments are addressed in levels 5–13. The EUNIS (European Environment Agency 2004) classification is also a hierarchy that addresses an exceptionally wide range of land and water environments, including the marine sublittoral environment (A3 and A4). The Connor et al. (2003) classification treats the littoral and sublittoral environments of Britain and Ireland. It is a hierarchy of six levels, of which habitat levels 2 and 3 are comparable to the same levels of the EUNIS classification.

In a somewhat different approach, the Greene et al. (1999) classification uses the concept of area as a major criterion for describing habitats, and it recognizes four habitat sizes that include megahabitats (kilometers to tens of kilometers), mesohabitats (tens of meters to 1 km), macrohabitats (1 to 10 m), and microhabitats (centimeters to 1 m). The top level of this partly hierarchical classification is a system (marine benthic), followed by a subsystem (for megahabitats and mesohabitats), a class (for mesohabitats and macrohabitats), two subclasses (for macrohabitats and microhabitats), and modifiers that describe seabed characteristics and processes found in the various habitats. It is a notable improvement on other classifications in that it emphasizes the importance of geological characteristics of seabed habitats.

Regional Approach to Habitat Classification

The classifications discussed above address a broad range of habitats over geographic areas of continental size or larger, except for the Connor et al. (2003) scheme that focuses on the Britain and Ireland region. Our focus is on the marine sublittoral environments of the American and Canadian parts of North America that extend northward from the continental shelf off New Jersey (Figure 1). We found it most practical to develop a habitat classification scheme that is tailored to the region under study and the level of detail we required and is based on extensive seabed observations. However, the classification incorporates concepts that are applicable worldwide.

The marine sublittoral zone, as defined in most classifications, lies below the intertidal zone and extends to the continental shelf edge at a water depth of approximately 200 m. We include in the marine sublittoral zone, the continental shelf basins of the Gulf of Maine region that reach depths of approximately 400 m and the submarine canyon heads that incise the continental shelf and reach depths of up to 800 m. The marine sublittoral zone encompasses by far the major portion of the offshore area of northeastern North America that is valued and managed for its fisheries resources, marine mammal habitats, sensitive environments, and industrial uses such as transportation, petroleum extraction, and cable and pipeline routes.

By necessity, habitat classification schemes utilize many common concepts and terminologies, and in this regard we have drawn freely from the schemes described above. Our classification addresses characteristics and processes that are key components of habitats in general while focusing specifically on marine sublittoral environments (Appendix 1). It recognizes eight seabed themes (informal units) that are the major subject elements of the classification and that emphasize the geological characteristics of habitats. Themes include one to many classes of habitat characteristics related to seabed features, fauna and flora, and processes that we view as fundamental for recognizing and analyzing habitats (Appendix 1). The classes of this scheme are unique formal units, and all reside at the top level (are not hierarchical) and are applied to the classification of each site. Below the classes, a sequence of subclasses, categories, and attributes addresses habitat characteristics with increasing detail. The observations that are the basis for classifying habitats are collected at sites that generally fall into the size range of the mesohabitats (tens of m to 1 km) of Greene et al. (1999). The process of classifying habitats involves the documentation of suites of habitat characteristics at individual sites and the identification of standard descriptors that together represent habitat types at an appropriate scale.

The classification is designed to be a template for a database that will allow the habitat characteristics of a site to be entered easily by selecting terms from lists. The database will (1) be an archive of habitat observations, (2) produce a summary report of the habitat characteristics of individual sites, and (3) export characteristics in spreadsheet format suitable for multivariate statistical analysis, which will aid in the recognition of the basic habitat types, biological communities, and functional groups of the region. The habitat database can be searched for any habitat type or characteristic, and it can provide habitat information for areas of interest to scientists and managers (see Appendix 2 and Figures 2–5 for four examples of habitat site descriptions based on the proposed scheme).

We expect that a well-designed regional habitat classification can be expanded to incorporate new kinds of observations (e.g., classes for seabed chemistry and water column structure and productivity). The classification can be expanded into other environments simply by incorporating applicable terminology. For example, the classification as applied to northeastern North America now does not address low-latitude carbonate environments but could easily be modified for that purpose. As regional classifications of this nature develop, it will be possible to compare habitats of different regions and, ultimately, to merge them if that proves useful.

Habitat Classification Structure

The classification proposed for marine sublittoral habitats has a four-level (class, subclass, category, and attribute), partly hierarchical structure in which many of the levels have a broad geographic application (Appendix 1). However, categories that address fauna and flora in Classes 14–17 are regional in nature. Classes are grouped into eight seabed themes that address major geological, biological, and oceanographical characteristics of habitats and the natural and human processes that create and modify the seabed. The themes, which contain one to many classes, are: (1) topographical setting; (2) seabed dynamics and currents; (3) seabed texture, hardness, and layering in the upper 5–10 cm; (4) seabed grain size analysis; (5) seabed roughness; (6) fauna and flora; (7) habitat association and usage; and (8) habitat recovery from disturbance.

Theme 1, Class 1: Topographical Setting

This class addresses the location of the habitat in terms of seabed slope and major seafloor features and industrial structures. The marine sublittoral zone is subdivided into two depth subclasses: Subclass 1, shallow photic; and Subclass 2, deep aphotic. The shallow photic zone encompasses depths in which epifaunal macrophyte algae occur. Categories and attributes are descriptors of features and structures (physiographic, biogenic, and anthropogenic).

Theme 2, Class 2: Seabed Dynamics and Currents

This class addresses the stability and mobility of seabed materials in response to current types and their strength and frequency of flow. This class recognizes the fundamental role of seabed dynamics in determining the structure and function of habitats. Three subclasses are recognized at present: Subclass 1, mobile substrate; Subclass

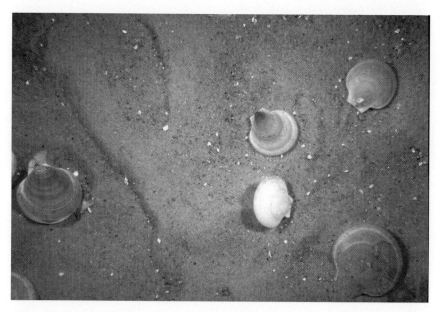

Figure 2. Mobile, coarse-grained, sand habitat (Example 1, Station 26) with sea scallops *Placopecten magellanicus* at eastern Georges Bank. Water depth = 96 m; image size = approximately 51 × 76 cm (20 × 30 in).

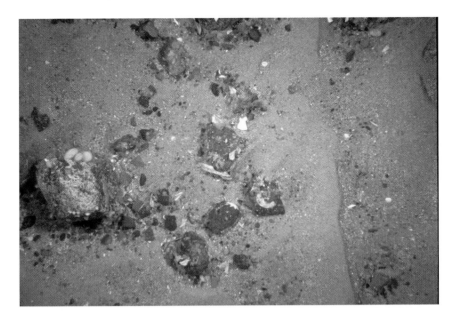

Figure 3. Intermixed mobile sand and immobile gravel habitat (Example 2, Station 35) with epifauna (brachiopods, calcareous worm tubes) attached to gravel at eastern Georges Bank. Water depth = 152 m; image size = approximately 51 × 76 cm (20 × 30 in).

Figure 4. Immobile, pebble, gravel habitat (Example 3, Station 42) with broken mollusk shells, epifauna (calcareous worm tubes *Filograna implexa*) attached to cobbles, and habitat disturbed by scallop dredging. This habitat lies adjacent to an undisturbed, mussel-encrusted, pebble gravel habitat with abundant epifauna at eastern Georges Bank. Water depth = 70 m; image size = approximately 51 × 76 cm (20 × 30 in).

Figure 5. Immobile rock outcrop habitat (Example 4, Station 49) covered by attached epifauna (sponges *Suberites* sp. and *Isodictya* sp., bryozoa, hydrozoa, calcareous worm tubes) at German Bank. Water depth = 75–80 m; image size = approximately 51 × 76 cm (20 × 30 in).

2, immobile substrate; and Subclass 3, intermixed mobile and immobile substrates (e.g., boulders on mobile sand). The mobility (or immobility) of the seabed is a major characteristic that influences habitats in terms of sediment texture, the presence and kinds of bedforms and shell deposits, the erosion and transport of sediment, the presence of biogenic structures, and the presence of many benthic species.

Theme 3, Class 3: Seabed Texture, Hardness, and Layering in the Upper 5–10 Centimeters
This class describes the sediment texture and relative hardness of the seabed by using visual observations of seabed character. Four subclasses include Subclass 1, fine-grained sediment composed of mud, very fine (4 phi) sand, and fine (3 phi) sand; Subclass 2, coarse-grained sediment composed of medium (2 phi), coarse (1 phi), and very coarse (0 phi) sand and gravel; Subclass 3, mixed fine-grained and coarse-grained sediment composed of mud, sand, and gravel mixtures; and Subclass 4, rock or other hard seabed (with or without mud, sand, or gravel). Each of these four subclasses also incorporates terminology that describes visible sediment layering in the upper 5–10 cm (e.g., sand partial veneer on pebbles, mud partial veneer on clay, pebble and gravel on rock outcrop).

Theme 4, Classes 4, 22, 23: Grain Size Analysis
This classification provides the results of sediment texture analysis as follows: Class 4, a general sediment description that includes texture classes (e.g., silty sand, gravelly mud), sorting, skewness, kurtosis, and particle shape; Class 22, major Wentworth size-classes (e.g., sand, gravel, silt, clay, mud) in weight percent; and Class 23, Phi and all Wentworth size-classes (e.g., fine sand, coarse silt) in weight percent.

Theme 5, Classes 5–13, 24: Seabed Roughness
This classification describes the three-dimensionality of the seabed surface in terms of physical and biological structures and the percent of the seabed covered by them. Seven classes of physical structures are: Class 5, bedforms; Class 6, shell materials; Class 7, rough sediments and hard seabeds (features composed variously of pebbles, cobbles, boulders, and rock outcrops that are smaller than the topographical features treated in Class 1); Class 8, biogenic structures (burrows, depressions, mounds); Class 12, anthropogenic marks (trawl and dredge tracks); Class 13, anthropogenic structures (minor structures such as lost fishing gear); and Class 24, physical structures combined. Three classes of biological structures are: Class 9, attached epifauna; Class 10, emergent epifauna; and Class 11, biological structures combined. Seabed roughness classes address a wide range of characteristics that are produced by mobile sediments, immobile hard materials, and biogenic structures and epifauna.

Theme 6, Classes 14–17: Fauna and Flora
This classification enumerates the dominant and typical biological elements that characterize habitats. Four classes are: Class 14, faunal groups (e.g., attached anemones,

erect sponges, polychaetes, flounder); Class 15, faunal species (e.g., Atlantic cod *Gadus morhua*, red hake *Urophycis chuss*, sea scallop *Placopecten magellanicus*); Class 16, floral groups (e.g., calcareous algae); and Class 17, floral species. Fauna and flora are subdivided into groups and species to accommodate and keep separate the varying levels of identification that can be achieved when analyzing habitats with video and photographic imagery and collected specimens. Many important faunal and floral groups can be identified from visual observations. Some faunal and floral species also can be identified using visual methods, but others need to be collected in order to obtain an accurate identification.

Theme 7, Classes 18, 19: Habitat Association and Usage

This classification describes habitats in terms of faunal association, human usage, and state of disturbance. Two classes are: Class 18, faunal-habitat association that documents spawning, juvenile, and adult habitats; and Class 19, human usage of habitat in terms of activities that impact the seabed, such as fishing, waste disposal, construction, and extraction of minerals, among others.

Theme 8, Classes 20, 21: Habitat Recovery from Disturbance

This classification describes the time required for the recovery of physical and biological structures from fishing disturbance (Class 20) and from natural disturbance (Class 21). These classes acknowledge the role of habitats in the life cycles of important fisheries species, the alteration of the seabed that might hinder the habitats' natural function, and the resilience and ability of habitats to resume their natural function. Classes 20 and 21 are included in the classification to address the growing importance of the effects of habitat disturbance to the management of fisheries and environments.

The Question of Scale

A habitat can be delimited as narrowly or as broadly in a geographic sense as the data and the purpose permit. This flexibility of scale necessarily influences the design and development of habitat classification schemes. Scale is particularly important when classifying subaqueous environments where observations are generally far more limited in number than they are in terrestrial environments. Several important questions related to scale should be addressed. What kinds and quantities of data are needed to achieve the required level of detail? What is the appropriate size and complexity of easily recognizable marine sublittoral habitats? Is it possible to define habitats too narrowly for practical application? A guiding principle in this approach to habitat classification is that it will be useful to scientists and managers of fisheries and the environment.

Habitat characteristics used in the proposed scheme are based on seabed geomorphology, video and photographic imagery, and geological and biological sampling. For the most part, physical and biological structures, major faunal and floral groups, and many dominant species that are exposed on the seabed can be adequately observed using video imagery, and it is probably the most common form of seabed data collected. Photographs and biological sampling add a higher level of detail to habitat analysis and are especially helpful for the identification of species that are difficult to identify in video images. Sampling also can provide information on infaunal species that cannot be observed with visual methods. Geological sampling provides materials for grain size analysis that cannot be matched by estimates from video imagery, but sampling with conventional small volume samplers is a poor estimator of the presence of large gravel particles and other materials that are difficult to sample. A practical protocol for "standardizing" seabed observations and therefore habitat classification might be to rely chiefly on video imagery. As often it is not feasible to determine the abundances of individual structures and organisms by counting, this classification scheme allows for the inclusion of visual estimates of the percent of the seabed covered by habitat characteristics (e.g., percent of seabed covered by epifauna, sea scallops, cobbles and boulders on rippled sand, among others).

Our experience in the marine sublittoral zone of northeastern North America has shown that short video transects of 100–200 m are often adequate to characterize areas where habitats are relatively large and homogeneous; whereas transects of 500–1,000 m in length or longer are required to evaluate areas where habitats are relatively small and/or complex. The appropriate transect length will vary from region to region and should be determined by the variability of seabed characteristics, the level of detail required, and a habitat mapping scale that is suitable for practical applications.

The proposed classification need not be based on sonar or laser imagery of the seabed, although imagery has become an almost indispensable basis for the sampling and high-resolution mapping of habitats. The availability of seabed imagery allows video and sampling transects to be conducted across environmental gradients that are expressed in topographical and backscatter images.

We suggest that an appropriate scale for the mapping of sublittoral habitats lies in the range of 1:25,000–1:100,000 (1 cm on the map represents 250–1,000 m on the seabed, respectively) and depends in large part on the data and imagery available. Mappable habitats, therefore, lie in the size range of the mesohabitats and megahabitats of Greene et al. (1999).

Habitat Names

Classified habitats require names in order to facilitate communication among the management, science, and industry communities. However, it is exceedingly difficult to design habitat names (and codes) that are concise, highly descriptive, and decipherable. Cowardin et al. (1979) use a list of five or more descriptors to name habitats. Allee et al. (2000) name habitats by combining an ecotype term (e.g., salt marsh habitat) with a list of descriptors. The EUNIS (European Environment Agency 2004) classification uses short descriptive names for habitats that include faunal, sediment, and oceanographical terms supplemented by a numerical code. For example, a sequence of habitats and codes is: sublittoral muds A4.3; shallow fully marine mud communities A4.31; semi-permanent tube-building amphipods and polychaetes in sublittoral mud or muddy sand, A4.311. Connor et al. (2003) employ habitat names and codes that are based on habitat factors (physical characteristics of habitats), taxon groups (e.g., brittle stars), community features (e.g., tube-building), and species names. For example, the habitat code LS.LSa.MoSa.AmSco describes a habitat characterized by littoral sediment (LS); littoral sand (LSa); mobile sand (MoSa); amphipods (Am); *Scolelepsis* sp. (Sco). Greene et al. (1999) describe habitats by combining a code with an areal size term (e.g., megahabitat, mesohabitat) and a list of descriptors. For example, the large-scale habitat with the code Shp_d1D is named: continental shelf megahabitat (S); flat (1); highly complex (D); hard seafloor (h); with pinnacles (p); differentially eroded (d). There is a separate code for small-scale habitats (Greene et al. 1999). Although habitat name codes can be cumbersome and require some effort to decipher, they are necessary shorthand for communication in situations where long text descriptions are inappropriate.

We have developed habitat-naming conventions for the marine sublittoral classification proposed here. Habitat names (and codes) are descriptive and can incorporate up to three kinds of information including (1) seabed substrate type from Class 3, (2) seabed substrate dynamics from Class 2, and (3) the degree of physical and biological complexity of the seabed from Classes 24 and 11, respectively (Appendix 1). The least descriptive habitat name is based on seabed substrate type only (e.g., mud; M). A more descriptive name links a seabed dynamics term with the substrate type (e.g., immobile; mud; I_M). Seabed dynamics are expressed as mobile (M), immobile (I), and intermixed mobile and immobile (IMI) substrates.

The most descriptive habitat name has three components that include a seabed dynamics term, the seabed substrate type, and seabed structural complexity terms that describe both physical structure and biological structure of the seabed. Structural complexity of habitats is based on the percent of the seabed covered by physical structures and biological structures. The percent of the seabed covered by structures is determined by visual observations (Appendix 3) and is estimated in nine semi-quantitative intervals that range from 0 to 100% (0%, <1%, 1–5%, 5–10%, 10–25%, 25–50%, >50%, >90%, and 100%). These percent intervals are reasonably easy to estimate from video and photographic imagery. The percent of the seabed covered is determined separately for physical and biological structures, and it is possible that the total percent of the seabed covered by physical and biological structures combined could be greater than 100% (e.g., in the case of abundant attached epifauna on a pebble, cobble, boulder substrate). The degree of structural complexity of the seabed is described in terms of the percent of the seabed that is covered separately by physical and biological structures in seven qualitative intervals that range from none to very high (Appendix 3). Wherever more than 50% of the seabed is covered by structures, the structural complexity is considered to be very high. An example of a three-component habitat name and code is: immobile; mud; physical structural complexity low; biological structural complexity very very low = I_M_ps5–10L_bs < 1VVL. In this example, physical structures (ps) cover 5–10% of the seabed and complexity is low, and biological structures (bs) cover < 1% of the seabed and complexity is very very low (see Appendices 2 and 4 for other examples of habitat name construction).

It must be emphasized that names and codes, by definition, do not fully describe a habitat. For example, a habitat name can indicate the degree of structural complexity present, but to gain a full understanding of the nature of a habitat's complexity, it is necessary to examine the supporting observational data to determine the kinds and functions of the physical and biological structures present. Habitat names can be made more descriptive by appending a list of major habitat characteristics such as dominant physical and biological structures, faunal and floral groups and species, biological communities, and functional groups to the kinds of habitat names discussed above (Appendix 4).

Probable Marine Sublittoral Habitat Types

This classification characterizes habitats by emphasizing seabed dynamics (Class 2), seabed texture (Class 3), and seabed structural complexity (Classes 11 and 24). The number of probable habitat types that can occur in the marine sublittoral zone can be determined by combining the characteristics of these classes. For purposes of illustration, we will combine the seabed dynamics characteristics of Class 2 with some of the seabed texture characteristics of Class 3 to identify a representative suite of habitats (Appendix 5). We are leaving out the structural complexity characteristics of Classes 11 and 24 for simplicity.

Among the representative suite of 90 habitats (Appendix 5), 45 are likely to occur in marine sublittoral environments. Habitats that are not likely to occur are those that combine seabed dynamics and substrate types that generally are not compatible in nature (although some of these might occur under certain conditions such as in high energy coastal environments). Examples would include: mobile mud (mud is not likely to be mobile in sublittoral environments); mobile sandy gravel (the gravel component, depending on size, is not likely to be mobile); mobile muddy gravel (mud and gravel components are not likely to be mobile); intermixed mobile and immobile gravel on rock (rock is immobile and the gravel component is not likely to be mobile). The number of habitats that will be recognized and classified in a region depends on the dynamism and physical and biological heterogeneity of the seabed, the kinds and distribution of data available, and the scale chosen for habitat analysis.

Summary

The marine sublittoral habitat classification proposed here is designed to describe and classify habitats in terms of geological, biological, and oceanographical attributes and in terms of the effects of natural and anthropogenic processes on them. The classification emphasizes the importance of seabed substrate type, substrate dynamics, and seabed physical and biological complexity in characterizing and naming sublittoral habitats. The classification is applied to the northeastern North America region. The purpose of the classification is to provide a foundation for scientific research and environmental management of seafloor habitats in a relatively large region. A database will allow data from individual study sites to be summarized, archived, and used to perform statistical analyses for identifying the major habitat types of the region. The information in the database, together with seabed imagery, will be a basis for habitat mapping. The proposed habitat classification serves as a template both for a database and for the development of classifications for other regions. The merging of separate regional classifications that follow the same principles could lead to an integrated national habitat classification for the marine sublittoral zone and other seabed environments.

Acknowledgments

We thank the officers, crews, and scientists who have collaborated with us on many cruises over the years to study the seabed of northeastern North America. We especially thank our shipmates aboard the C.C.G.S. *Hudson* for their support on the June 2002 cruise to Georges and German Banks. Observations and discussions with colleagues on that cruise played a major role in the development of the classification scheme presented here. We also benefited greatly from insightful reviews of the manuscript by H. Josenhans and E. King of the Geological Survey of Canada, K. Scanlon of the U.S. Geological Survey, B. Barr of the National Marine Sanctuary Program, T. Noji of the National Marine Fisheries Service, and G. Greene of Moss Landing Marine Laboratories.

References

Allee, R. J., M. Dethier, D. Brown, L. Deegan, R. G. Ford, T. F. Hourigan, J. Maragos, C. Schoch, K. Sealey, R. Twilley, M. P. Weinstein, and M. Yoklavich. 2000. Marine and estuarine ecosystem and habitat classification. NOAA Technical Memorandum NMFS-F/SPO-43. Available: *www.nmfs.noaa.gov/habitat/ecosystem/ habitatdocs/habitatclassdoc.pdf* (February 2005).

Connor, D. W., J. H. Allen, N. Golding, L. M. Lieberknecht, K. O. Northen, and J. B. Reker. 2003. The national marine habitat classification for Britain and Ireland, version 04.05. Joint Nature Conservation Committee (JNCC), UK. Available: *www.jncc.gov.uk/pdf/ 04_05_introduction.pdf* (February 2005).

Cowardin, L. M., V. Carter, F. C. Golet, and E. T. LaRoe. 1979. Classification of wetlands and deepwater habitats of the United States. U.S. Fish and Wildlife Service FWS/OBS-79/31. Available: *http://wetlands.fws.gov/ Pubs_Reports/Class_Manual/class_titlepg.htm* (February 2005).

Dethier, M. N. 1992. Classifying marine and estuarine natural communities: an alternative to the Cowardin system. Natural Areas Journal 12(2):90–100.

European Environment Agency. 2004. EUNIS (European Nature Information System) habitat classification. Available: *http://eunis.eea.eu.int/habitats.jsp* (February 2005).

Greene H. G., M. M. Yoklavich, R. M. Starr, V. M. O'Connell, W. W. Wakefield, D. E. Sullivan, J. E. McRea, Jr., and G. M. Cailliet. 1999. A classification scheme for deep seafloor habitats. Oceanologica Acta 22:663–678. Available: *www.sciencedirect.com/science/ journals* (February 2005).

Appendix 1. Outline of the Marine Sublittoral Habitats Classification Structure

The classification includes classes, subclasses, categories, and attributes. It is designed as a template for a database. Class numbers are unique. Themes (in italics) are the major subject elements of the classification and can include one to many classes. Themes are informal units and are not incorporated into the habitats database. Observations from Classes 2, 3, 11, and 24 are used to compile habitat names (see Appendix 4 for examples). Category and attribute terms in parentheses are not a complete list. The complete classification structure with all category and attribute terms will be available after publication online at *http://woodshole.er.usgs.gov/project-pages/stellwagen/*.

Theme 1, Topographical Setting
Class 1   Topographical setting: major seabed features and industrial structures
   Subclass 1   Shallow photic (presence of macrophyte algae)
   Subclass 2   Deep aphotic (absence of macrophyte algae)
     Categories   Seabed slope, major physiographic and biogenic features and industrial structures
     Attributes   Angle of seabed slope, types of seabed features (e.g., basin, ridge, shelf edge reef), and industrial structures (e.g., cable, oil platform)
All categories and attributes apply to Subclasses 1 and 2.

Theme 2, Seabed Dynamics and Currents
Class 2   Seabed dynamics and currents
   Subclass 1   Mobile substrate
   Subclass 2   Immobile substrate
   Subclass 3   Intermixed mobile and immobile substrates
     Categories   Types of currents (e.g., tidal, storm wave) and types of events (e.g., storms) causing sediment mobility
     Attributes   Strength of currents and frequency of events (e.g., daily, monthly) causing sediment mobility
All categories and attributes apply to Subclasses 1–3.

Theme 3, Seabed Texture, Hardness, and Layering in the Upper 5–10 cm
Class 3   Seabed texture, hardness, and layering in the upper 5–10 cm
   Subclass 1   Fine-grained sediment: mud, very fine (4 phi) sand, and fine (3 phi) sand
   Subclass 2   Coarse-grained sediment: medium (2 phi) sand, coarse (1 phi) sand, very coarse (0 phi) sand, and gravel (gravel is composed of granules > 1 mm and < 2 mm; pebbles < 64 mm; cobbles < 256 mm; and boulders > 256 mm).
   Subclass 3   Mixed fine-grained and coarse-grained sediment: mud, sand, and gravel mixtures
   Subclass 4   Rock or other hard seabed (with or without mud, sand, gravel)
     Categories   Descriptive sediment and hard seabed types (e.g., mud veneer on clay, gravel pavement, cobbles in muddy sand, sand veneer on rock outcrop)
     Attributes   Percentage of seabed covered by sediment and hard seabed types
Categories apply to appropriate subclasses; all attributes apply to Subclasses 1–4.

Theme 4, Seabed Grain-Size Analysis
Class 4   Seabed grain sizes: general description
   Subclass 1   General sediment description
     Categories   Descriptive texture classification, sorting, grain size distribution, and particle shape
     Attributes   Major descriptive texture classes (e.g., silty sand, gravelly mud), degree of sorting (e.g., well sorted), skewness (e.g., symmetrical), kurtosis (e.g., mesokurtic), and particle shape (e.g., rounded)
Class 22   Seabed grain sizes: major Wentworth size-classes
   Subclass 1   Major Wentworth grain size-classes, weight percent
     Categories   Major Wentworth grain size-classes (e.g., sand, gravel, silt, clay, mud)
     Attributes   Weight percent of major Wentworth grain size-classes
Class 23   Seabed grain sizes: phi and all Wentworth size-classes
   Subclass 1   Phi and all Wentworth grain size-classes, weight percent
     Categories   Phi and all Wentworth size-classes (e.g., fine sand, coarse silt)
     Attributes   Weight percent of phi and all Wentworth size-classes

Theme 5, Seabed Roughness
Class 5   Seabed roughness: bedforms
   Subclass 1   Bedforms (physical structures)
     Categories   Bedform types (e.g., ripples, sand dunes)
     Attributes   Percentage of seabed covered by bedform types

Class 6	Seabed roughness: shell materials
Subclass 1	Shell materials (physical structures)
Categories	Types of shell materials and deposits (e.g., shell fragments, shell deposits)
Attributes	Percentage of seabed covered by shell material and deposit types
Class 7	Seabed roughness: rough sediments and hard seabeds
Subclass 1	Rough sediments and hard seabeds (physical structures)
Categories	Associations of sediment particles, sediment type, seabed structures, and rock outcrops (e.g., cobbles in patches, piled boulders, pebbles in sand dune troughs, irregular rock outcrop)
Attributes	Percentage of seabed covered by rough sediment and hard seabed types
Class 8	Seabed roughness: biogenic structures
Subclass 1	Biogenic structures (physical structures)
Categories	Types of biogenic modifications of the seabed (e.g., crab depressions, fish burrows)
Attributes	Percentage of seabed covered by types of biogenic structures
Class 12	Seabed roughness: anthropogenic marks
Subclass 1	Anthropogenic marks (physical structures)
Categories	Types of marks made on the seabed by human activities (e.g., trawl marks, anchor marks)
Attributes	Percentage of seabed covered by types of anthropogenic marks
Class 13	Seabed roughness: anthropogenic structures
Subclass 1	Anthropogenic structures (physical structures)
Categories	Types of minor man-made structures and equipment on the seabed (e.g., types of fishing gear)
Attributes	Percentage of seabed covered by types of anthropogenic structures
Class 24	Seabed roughness: physical structures combined
Subclass 1	Extent of physical structures
Categories	Types of physical structures
Attributes	Percentage of seabed covered by physical structures by type and all combined

Class 24 summarizes observations for Classes 5–8, 12, and 13.

Class 9	Seabed roughness: attached epifauna
Subclass 1	Attached epifauna (biological structures)
Categories	Epifaunal groups attached to the seabed surface (e.g., erect sponges, tunicates, brachiopods)
Attributes	Percentage of seabed covered by types of attached epifauna
Class 10	Seabed roughness: emergent epifauna
Subclass 1	Emergent epifauna (biological structures)
Categories	Epifaunal groups emergent from below the seabed surface (e.g., burrowing anemones, sea pens)
Attributes	Percentage of seabed covered by types of emergent epifauna
Class 11	Seabed roughness: biological structures combined
Subclass 1	Extent of biological structures
Categories	Types of biological structures
Attributes	Percentage of seabed covered by biological structures by type and all combined

Class 11 summarizes observations for Classes 9 and 10.

Theme 6, Fauna and Flora

Class 14	Faunal groups
Subclasses 1–6	Faunal groups (in several subclasses based on different methods of data collection; e.g., visual observations and/or specimens from various sampler types)
Categories	Faunal groups (e.g., erect sponges, burrowing anemones, sea stars, attached anemones)
Attributes	Presence/absence or percentage of seabed covered by individual faunal groups

All categories and attributes apply to Subclasses 1–6.

Class 15	Faunal species
Subclasses 1–6	Faunal species (in several subclasses based on different methods of data collection; e.g., visual observations and/or specimens from various sampler types)
Categories	Faunal species (e.g., Atlantic cod *Gadus morhua*)
Attributes	Presence/absence or percentage of seabed covered by individual faunal species

All categories and attributes apply to Subclasses 1–6.

Class 16	Floral groups
Subclasses 1–6	Floral groups (in several subclasses based on different methods of data collection; e.g., visual observations and/or specimens from various sampler types)
Categories	Floral groups (e.g., calcareous algae, kelp)
Attributes	Presence/absence or percentage of seabed covered by individual floral groups

All categories and attributes apply to Subclasses 1–6.

Class 17 Floral species
Subclasses 1–6 Floral species (in several subclasses based on different methods of data collection; e.g., visual observations and/or specimens from various sampler types)
Categories Floral species
Attributes Presence/absence or percentage of seabed covered by individual floral species
All categories and attributes apply to Subclasses 1–6.

Theme 7, Habitat Association and Usage
Class 18 Fauna-habitat association: essential fish habitat (EFH)
Subclasses 1–6 Fauna-habitat association (in several subclasses based on different methods of data collection; e.g., visual observations and/or specimens from various sampler types)
Categories Faunal species (e.g., Atlantic cod, haddock *Melanogrammus aeglefinus*, yellowtail flounder *Pleuronectes ferrugineus*)
Attributes Types of fauna-habitat association by species (e.g. adult, spawning, juvenile habitat)
All categories and attributes apply to Subclasses 1–6.

Class 19 Human usage of habitat
Subclass 1 Human usage of habitat
Categories Disturbed, undisturbed, or recovering habitat; kinds of disturbance (e.g., fishing, extraction)
Attributes Types of disturbance activities (e.g., otter trawling, minerals mining)

Theme 8, Habitat Recovery from Disturbance
Class 20 Habitat recovery from fishing disturbance
Subclass 1 Fishing disturbance
Categories Recovery of physical structures and biological structures (e.g., bedforms, attached epifauna)
Attributes Time required for recovery (e.g., months, year, decades)
Class 21 Habitat recovery from natural disturbance
Subclass 1 Natural disturbance
Categories Recovery of physical structures and biological structures (e.g., fish burrows, emergent epifauna)
Attributes Time required for recovery (e.g., months, year, decades)

Appendix 2. Habitat Descriptions

Example 1

Habitat name: mobile; coarse-grained sand; physical structural complexity very high; biological structural complexity none; M_cgS_ps100VH_bs0N
Further habitat characteristics: sand ripples and dunes; scallop depressions; sea scallops; flounder; skate; monkfish; silver hake; red hake; sculpin; sea stars (Figures 1, 2).
Station information: Eastern Georges Bank; C.C.G.S. *Hudson* cruise 2002–026; station 26; water depth 96 m; June 10, 2002; JD 161; visual observations; Campod video/photo station.
Marine sublittoral environment:
Class 1, Topographical setting: deep aphotic; shelf; flat (0–5 degrees)
Class 2, Seabed dynamics and currents: mobile substrate; tidal and storm wave currents
Class 3, Seabed texture, hardness and layering in the upper 5–10 cm: coarse-grained sediment; coarse-grained sand
Classes 4, 22, 23, Seabed grain sizes: no analysis
Class 5, Seabed roughness; bedforms: sand ripples; sand ripple crests fresh; sand dunes low; sand dune asymmetry toward northwest; ripples on dune crests and in troughs; mobile sand dune troughs and crests
Class 6, Seabed roughness; shell materials: shell fragments; shell materials in sand dune troughs; fine shell fragments in sand ripple troughs
Class 7, Seabed roughness; rough sediments and hard seabeds: none
Class 8, Seabed roughness; biogenic structures: scallop depressions
Class 9, Seabed roughness; attached epifauna: none
Class 10, Seabed roughness; emergent epifauna: none
Class 11, Seabed roughness; biological structures combined: none
Class 12, Seabed roughness; anthropogenic marks: none
Class 13, Seabed roughness; anthropogenic structures: none

Class 24, Seabed roughness; physical structures combined: sand ripples, sand dunes, and scallop depressions cover 100% of the seabed
Class 14, Faunal groups: flounder; sea stars; skate; sculpin; sea scallops; no sand dollars
Class 15, Faunal species: red hake *Urophycis chuss*; monkfish *Lophius americanus*; silver hake *Merluccius bilinearis*; sea star *Asterias* sp.; sea scallops *Placopecten magellanicus* cover 10–25% of the seabed
Class 16, Floral groups and Class 17, Floral species: none
Class 18, Fauna-habitat association: habitat for adult monkfish, silver hake, red hake, skate, flounder, and sea scallops
Class 19, Human usage of habitat: unknown
Class 20, Habitat recovery from fishing disturbance: habitat is undisturbed by fishing
Class 21, Habitat recovery from natural disturbance: unknown

Example 2

Habitat name: intermixed mobile and immobile; pebbles and cobbles in sand; physical structural complexity very high; biological structural complexity low; IMI_pciS_ps100VH_bs5–10 L.
Further habitat characteristics: sand ripples; pebbles; cobbles; boulders; attached epifauna; colonial tube worms; sea stars; sponges; brachiopods; anemones; bryozoans (Figures 1, 3).
Station information: Eastern Georges Bank; CCGS Hudson cruise 2002–026; station 35; water depth, 152 m; June 11, 2002; JD 162; visual observations; Campod video/photo station.
Marine sublittoral environment:
Class 1, Topographical setting: deep aphotic; shelf edge; sloping (5–30 degrees)
Class 2, Seabed dynamics and currents: intermixed mobile and immobile substrates; tidal currents
Class 3, Seabed texture, hardness and layering in the upper 5–10 cm: coarse-grained sediment; rippled sand partial veneer on pebbles and cobbles; pebbles and cobbles in rippled sand. Comment: rippled sand migrating through pebble and cobble pavement
Classes 4, 22, and 23, Seabed grain sizes: no analysis
Class 5, Seabed roughness; bedforms: sand ripples; sand ripple crests fresh
Class 6, Seabed roughness; shell materials: shell fragments
Class 7, Seabed roughness; rough sediments and hard seabeds: pebbles and cobbles in patches; boulders in depressions with pebbles
Class 8, Seabed roughness; biogenic structures: none
Class 9, Seabed roughness; attached epifauna: brachiopods; erect sponges; anemones; erect bryozoans; calcareous worm tubes. Comment: epifauna attached to pebbles, cobbles, and boulders
Class 10, Seabed roughness; emergent epifauna: none
Class 11, Seabed roughness; biological structures combined: attached epifauna covers 10–25% of hard substrates and 5–10% of the seabed in total
Class 12, Seabed roughness; anthropogenic marks: none
Class 13, Seabed roughness; anthropogenic structures: none
Class 24, Seabed roughness; physical structures combined: sand ripples, pebbles, cobbles, boulders, and boulder depressions cover 100% of the seabed
Class 14, Faunal groups: sea stars; brachiopods; colonial tube worms; erect sponges; juvenile fish; attached anemones; erect bryozoans; no sand dollars
Class 15, Faunal species: calcareous colonial tube worms *Filograna implexa*; yellow erect sponge *Myxilla* spp.; yellow erect sponge *Suberites* spp.; sea star *Hippasterias* sp.; no sea scallops *Placopecten magellanicus*.
Class 16, Floral groups and Class 17, Floral species: none
Class 18, Fauna-habitat association: habitat for unidentified juvenile fish species
Class 19, Human usage of habitat: unknown
Class 20, Habitat recovery from fishing disturbance: habitat is undisturbed by fishing
Class 21, Habitat recovery from natural disturbance: unknown

Example 3

Habitat name: immobile; pebble gravel; physical structural complexity very high; biological structural complexity very very low; I_pG_ps100VH_bs <1VVL
Further habitat characteristics: pebble gravel; broken shells; disturbed; scallop dredge tracks (Figures 1, 4)
Station information: Eastern Georges Bank; CCGS Hudson cruise 2002–026; station 42; water depth, 70 m; June 12, 2002; JD 163; visual observations; Campod video/photo station
Marine sublittoral environment:
Class 1, Topographical setting: deep aphotic; shelf; flat (0–5 degrees)
Class 2, Seabed dynamics and currents: immobile substrate; tidal currents
Class 3, Seabed texture, hardness and layering in the upper 5–10 cm: coarse-grained sediment; pebble gravel; cobbles on pebble gravel

Classes 4, 22, 23, Seabed grain sizes: no analysis
Class 5, Seabed roughness; bedforms: none
Class 6, Seabed roughness; shell materials: shells partly broken; sea scallop shell debris
Class 7, Seabed roughness; rough sediments and hard seabeds: pebble pavement
Class 8, Seabed roughness; biogenic structures: none
Class 9, Seabed roughness; attached epifauna: colonial calcareous tube worms
Class 10, Seabed roughness; emergent epifauna: none
Class 11, Seabed roughness; biological structures combined: attached epifauna covers <1% of the seabed.
Class 12, Seabed roughness; anthropogenic marks: scallop dredge tracks.
Class 13, Seabed roughness; anthropogenic structures: none.
Class 24, Seabed roughness; physical structures combined: pebble gravel and broken shells cover 100% of the seabed
Class 14, Faunal groups: sea stars; colonial tube worms; no sand dollars.
Class 15, Faunal species: lobster *Homarus americanus*; sea star *Crossaster* sp.; sea star *Solaster* sp.; calcareous colonial tube worms *Filograna implexa*, no sea scallops *Placopecten megallanicus*
Class 16, Floral groups and Class 17, Floral species: none
Class 18, Fauna-habitat association: habitat for adult lobster
Class 19, Human usage of habitat: scallop dredging
Class 20, Habitat recovery from fishing disturbance: habitat is disturbed by fishing. Comment: this habitat lies adjacent to an undisturbed, mussel-encrusted, pebble gravel habitat with abundant epifauna, including calcareous tube worms
Class 21, Habitat recovery from natural disturbance: unknown

Example 4

Habitat name: immobile; pebbles, cobbles, boulders on rock outcrop; physical structural complexity veryhigh; biological structural complexity very high; I_pcboR_ps > 90VH_bs > 50VH
Further habitat characteristics: Rock outcrop; pebbles; cobbles; boulders; attached epifauna; sponges; tunicates; hydrozoans; bryozoans; soft corals; anemones; brachiopods (Figures 1, 5)
Station information: German Bank; CCGS Hudson cruise 2002–026; station 49; water depth, 73–103 m; June 13, 2002; JD 164; visual observations; Campod video/photo station
Marine sublittoral environment:
Class 1, Topographical setting: deep aphotic; bank edge; steep (30–45 degrees)
Class 2, Seabed dynamics and currents: immobile substrate; tidal currents
Class 3, Seabed texture, hardness and layering in the upper 5–10 cm: rocks or other hard seabed (with or without mud, sand, gravel); bedrock outcrop; pebble gravel on bedrock outcrop; pebbles and cobbles and boulders on bedrock outcrop; cobbles and boulders in sand on bedrock outcrop
Classes 4, 22, 23, Seabed grain sizes: no analysis
Class 5, Seabed roughness; bedforms: none
Class 6, Seabed roughness; shell materials: shell fragments; fine shell fragments on sand
Class 7, Seabed roughness; rough sediments and hard seabeds: pebbles in patches; pebbles and cobbles in patches; cobbles and boulders in patches; piled cobbles and boulders with voids between; piled cobbles and boulders with sand between; smooth bedrock; irregular bedrock; stepped bedrock.
Class 8, Seabed roughness; biogenic structures: none
Class 9, Seabed roughness; attached epifauna: anemones; erect bryozoans; erect hydrozoans; brachiopods; soft corals; erect sponges; encrusting sponges; stalked tunicates; encrusting tunicates. Comment: epifauna attached to pebbles, epifauna attached to cobbles and boulders, epifauna attached to bedrock
Class 10, Seabed roughness; emergent epifauna: none
Class 11, Seabed roughness; biological structures combined: attached epifauna covers > 50% of the seabed
Class 12, Seabed roughness; anthropogenic marks: none
Class 13, Seabed roughness; anthropogenic structures: none
Class 24, Seabed roughness; physical structures combined: pebbles, cobbles, boulders, stepped and irregular bedrock cover > 90% of the seabed
Class 14, Faunal groups: attached anemones; soft corals; brachiopods; erect bryozoans; erect hydrozoans; hermit crabs; flounder; sculpin; eel; sea stars; encrusting sponges; erect sponges; palmate sponges; stalked tunicates; encrusting tunicates; no sand dollars
Class 15, Faunal species: redfish *Sebastes fasciatus*; monkfish *Lophius americanus*; yellow erect sponge *Myxilla* sp.; yellow erect sponge *Suberites* sp.; tunicate *Mogula* sp.; stalked tunicate *Boltenia* sp.; soft coral *Gersemia* sp.; erect sponge *Isodictya palmata*; breadcrumb sponge *Halichondria* sp.; sea star *Leptasterias* sp.); attached anemone cf. *Tealia* sp.; attached anemone *Metridium* sp.; sea star *Asterias* sp.; sea star *Hippasterias* sp.; no sea scallops *Placopecten magellanicus*.
Class 16, Floral groups and Class 17, Floral species: none.
Class 18, Fauna-habitat association: habitat for adult redfish, monkfish, and flounder.

Class 19, Human usage of habitat: unknown.
Class 20, Habitat recovery from fishing disturbance: undisturbed habitat.
Class 21, Habitat recovery from natural disturbance: undisturbed habitat.

Appendix 3. **Structural complexity of habitats.** The percent of the seabed covered by physical structures (Class 24) and by biological structures (Class 11) is based on visual observations and is estimated in nine semi-quantitative intervals that range from 0% to 100%. The degree of structural complexity of habitats is described in terms of the percent of seabed covered separately by physical and biological structures in seven qualitative intervals that range from none to very high. Wherever more than 50% of the seabed is covered by structures, the structural complexity is considered to be very high. The percent of the seabed covered by physical and biological structures combined can be greater than 100% (e.g., in the case of abundant epifauna on a rough seabed). See Appendices 2 and 4 for examples showing how structural complexity is incorporated into habitat names.

Estimate of percent of seabed covered by physical (or biological) structures	Degree of structural complexity	Code
0	None	N
<1	Very very low	VVL
1–5	Very low	VL
5–10	Low	L
10–25	Medium	M
25–50	High	H
>50	Very high	VH
>90	Very high	VH
100	Very high	VH

Appendix 4. **Three levels of descriptive habitat names.** Three levels of descriptive habitat names (and codes) are based on seabed substrate dynamics from Class 2, seabed substrate type from Class 3, and the degree of physical and biological structural complexity of the seabed from Classes 24 and 11, respectively (Appendices 1, 3). The least descriptive habitat name is based on seabed substrate type only (e.g., mud). More descriptive names incorporate a seabed dynamics term (e.g., immobile; mud) and a seabed structural complexity term (e.g., immobile; mud; physical structural complexity low; biological structural complexity very low). For further habitat description, a list of major habitat characteristics (dominant physical and biological structures, faunal and floral species, biological communities, functional groups) can be appended to a habitat name. Three examples are given below. Structural complexity of habitats is described in terms of the percent of the seabed covered by physical structures (Class 24) and by biological structures (Class 11). Based on visual observations, the percent of seabed cover is estimated in nine intervals (0%, <1%, 1–5%, 5–10%, 10–25%, 25–50%, >50%, >90%, and 100%). No structural complexity (N) = 0% of seabed covered; Very very low (VVL) = <1% of seabed covered; Very low (VL) = 1–5% of seabed covered; Low (L) = 5–10% of seabed covered; Medium (M) = 10–25% of seabed covered; High (H) = 25–50% of seabed covered; and Very high (VH) = >50% of seabed covered.

Example 1

Habitat name (code) based on seabed type: Mud (M)
Habitat name (code) based on dynamic seabed type: Immobile; mud (I_M)
Habitat name (code) based on dynamic seabed type with physical structure (ps) and biological structure (bs):
  Immobile; mud; physical structural complexity low; biological structural complexity very, very low (I_M_ps5–10L_bs<1VVL)
Explanation: Habitat name is based on seabed dynamics, seabed substrate type, and physical and biological structural complexity. The code is created as follows: Immobile (I); mud (M); physical structures cover 5–10% of the seabed (ps5–10), physical complexity is low (L); biological structures cover less than 1% of the seabed (bs<1), and biological complexity is very very low (VVL)
Further habitat characteristics: fish and crab burrows, flounder, ocean pout, cancer crabs

Example 2

Habitat name (code) based on seabed type: Coarse-grained sand (cgS)
Habitat name (code) based on dynamic seabed type: Immobile; coarse-grained sand (I_cgS)

Habitat name (code) based on dynamic seabed type with physical structure (ps) and biological structure (bs):
   Immobile; coarse-grained sand; physical structural complexity very high; biological structural complexity medium (I_cgS_ps>50VH_bs10–25M)
Explanation: Habitat name is based on seabed dynamics, on seabed substrate type, and on physical and biological structural complexity. Immobile (I); coarse-grained sand (cgS); physical structures cover more than 50% of the seabed (ps>50), physical complexity is very high (VH); biological structures cover 10–25% of the seabed (bs10–25), and biological complexity is medium (M).
Further habitat characteristics: fish and scallop depressions, amphipod tubes, sea scallops, erect sponges, red hake

Example 3

Habitat name (code) based on seabed type: Sandy gravel (sG)
Habitat name (code) based on dynamic seabed type: Intermixed mobile/immobile; sandy gravel (IMI_sG)
Habitat name (code) based on dynamic seabed type with physical structure (ps) and biological structure (bs):
   Intermixed mobile/immobile; sandy gravel; physical structural complexity very high; biological structural complexity low (IMI_sG_ps>90VH_bs5–10L)
Explanation: Habitat name is based on seabed dynamics, on seabed substrate type, and on physical and biological structural complexity. Intermixed mobile/immobile (IMI); sandy gravel (sG); physical structures cover more than 90% of the seabed (ps>90), physical structure complexity is very high (VH); biological structures cover 5–10% of the seabed (bs5–10), and biological structural complexity is low (L).
Further habitat characteristics: pebble and cobble gravel, sand ripples, erect sponges, hydrozoa, skate

Appendix 5. **Probable marine sublittoral habitat types.** The dynamic aspects of seabed substrates (mobile substrate, immobile substrate, and intermixed mobile and immobile substrates) are found in Class 2, Subclasses 1–3. Seabed substrate types are found in Class 3, Subclasses 1–4 (Appendix 1). The number of probable habitats is determined by combining the three dynamic aspects of substrates from Class 2 with a number of seabed types selected from Class 3. Among the representative suite of 90 habitats shown here (more are possible), 45 are likely to occur in marine sublittoral environments. Habitats shown in italics are not likely to occur (although some might occur under certain conditions). The structural complexity component of habitats is not considered here, but it would increase markedly the number of probable habitat types (see Appendix 3). The four examples from Georges Bank (see Appendix 2 and Figures 2–5) are listed below in Subclass 2, numbers 11, 26, and 6, and in Subclass 4, number 7. The following descriptions apply to superscripts: a = mud not likely mobile; b = sandy mud not likely mobile; c = muddy sand not likely mobile; d = can occur in environments with large bedforms; e = Gravel particles >1 cm not likely mobile; f = cobbles and boulders not likely mobile; g = sand component is mobile; h = gravel and fine-grained sand usually do not co-occur; i = mud and gravel not likely mobile; j = rock not mobile. Gravel is composed of granules (>1, <2 mm), pebbles (<64 mm), cobbles (<256 mm), and boulders (>256 mm).

Class 3. Seabed Texture, Hardness, and Layering in the Upper 5–10 cm
Subclass 1. Fine-grained sediment: mud, very fine (4 phi) sand, and fine (3 phi) sand.
  1. Immobile; mud (I_M)
  2. Immobile; sandy mud (I_sM)
  3. Immobile; fine-grained sandy mud (I_fgsM)
  4. Immobile; sand (I_S); also occurs in Subclass 2
  5. Immobile; fine-grained sand (I_fgS)
  6. Immobile; muddy sand (I_mS)
  7. Immobile; muddy, fine-grained sand (I_mfgS)
  8. *Mobile; mud (M_M)*[a]
  9. *Mobile; sandy mud (M_sM)*[b]
  10. *Mobile; fine-grained, sandy mud (M_fgsM)*[b]
  11. Mobile; sand (M_S); also occurs in Subclass 2
  12. Mobile; fine-grained sand (M_fgS)
  13. *Mobile; muddy sand (M_mS)*[c]
  14. *Mobile; muddy, fine-grained sand (M_mfgS)*[c]
  15. *Intermixed mobile/immobile; mud (IMI_M)*[a]
  16. *Intermixed mobile/immobile; sandy mud (IMI_sM)*[b]
  17. *Intermixed mobile/immobile; fine-grained, sandy mud (IMI_fgsM)*[b]
  18. Intermixed mobile/immobile; sand (IMI_S); also occurs in Subclass 2[d]
  19. Intermixed mobile/immobile; fine-grained sand (IMI_fgS)[d]

20. *Intermixed mobile/immobile; muddy sand (IMI_mS)*[c]
21. *Intermixed mobile/immobile; muddy, fine-grained sand (IMI_mfgS)*[c]

Subclass 2. Coarse-grained sediment: medium (2 phi) sand, coarse (1 phi) sand, and very coarse (0 phi) sand, and gravel.
1. Immobile; sand (I_S), also occurs in Subclass 1
2. Immobile; coarse-grained sand (I_cgS)
3. Immobile; gravelly sand (I_gS)
4. Immobile; gravelly coarse-grained sand (I_gcgS)
5. Immobile; gravel (I_G)
6. Immobile; pebble gravel (I_pG)
7. Immobile; cobble and boulder gravel (I_cbG)
8. Immobile; sandy gravel (I_sG)
9. Immobile; coarse-grained, sandy gravel (I_cgsG)
10. Mobile; sand (M_S); also occurs in Subclass 1
11. Mobile; coarse-grained sand (M_cgS)
12. *Mobile; gravelly sand (M_gS)*[e]
13. *Mobile; gravelly, coarse-grained sand (M_gcgS)*[e]
14. *Mobile; gravel (M_G)*[e]
15. *Mobile; cobble and boulder gravel (M_cbG)*[f]
16. *Mobile; sandy gravel (M_sG)*[e]
17. *Mobile; coarse-grained, sandy gravel (M_cgsG)*[e]
18. Intermixed mobile/immobile; sand (IMI_S); also occurs in Subclass 1[d]
19. Intermixed mobile/immobile; coarse-grained sand (IMI_cgS)[d]
20. Intermixed mobile/immobile; gravelly sand (IMI_gS)[g]
21. Intermixed mobile/immobile; gravelly coarse-grained sand (IMI_gcgS)[g]
22. *Intermixed mobile/immobile; gravel (IMI_G)*[e]
23. *Intermixed mobile/immobile; cobble and boulder gravel (IMI_cbG)*[f]
24. Intermixed mobile/immobile; sandy gravel (IMI_sG)[g]
25. Intermixed mobile/immobile; coarse-grained sandy gravel (IMI_cgsG)[g]
26. Intermixed mobile/immobile; pebbles and cobbles in sand (IMI_pciS)[g]

Subclass 3. Mixed fine-grained and coarse-grained sediment: mud, sand, and gravel mixtures.
1. Immobile; mud, sand, gravel (I_MSG)
2. Immobile; mud, gravel (I_MG)
3. Immobile; muddy gravel (I_mG)
4. Immobile; gravelly mud (I_gM)
5. *Immobile; fine-grained, sandy gravel (I_fgsG)*[h]
6. *Immobile; gravelly, fine-grained sand (I_gfgS)*[h]
7. Immobile; muddy, coarse-grained sand (I_mcgS)
8. Immobile; coarse-grained, sandy mud (I_cgsM)
9. *Mobile; mud, sand, gravel (M_MSG)*[i]
10. *Mobile; mud, gravel (M_MG)*[i]
11. *Mobile; muddy gravel (M_mG)*[i]
12. *Mobile; gravelly mud (M_gM)*[i]
13. *Mobile; fine-grained, sandy gravel (M_fgsG)*[e, h]
14. *Mobile; gravelly, fine-grained sand (M_gfgS)*[e, h]
15. *Mobile; muddy, coarse-grained sand (M_mcgS)*
16. *Mobile; coarse-grained, sandy mud (M_cgsM)*[b]
17. *Intermixed mobile/immobile; mud, sand, gravel (IMI_MSG)*[a]
18. *Intermixed mobile/immobile; mud, gravel (IMI_MG)*[a]
19. *Intermixed mobile/immobile; muddy gravel (IMI_mG)*[a]
20. *Intermixed mobile/immobile; gravelly mud (IMI_gM)*[a]
21. *Intermixed mobile/immobile; fine-grained, sandy gravel (IMI_fgsG)*[h]
22. *Intermixed mobile/immobile; gravelly, fine-grained sand (IMI_gfgS)*[h]
23. *Intermixed mobile/immobile; muddy, coarse-grained sand (IMI_mcgS)*[c]
24. *Intermixed mobile/immobile; coarse-grained, sandy mud (IMI_cgsM)*[b]

Subclass 4. Rock or other hard seabed (with or without mud, sand, gravel).
1. Immobile; rock (I_R)
2. Immobile; mud on rock (I_MoR)
3. Immobile; sand on rock (I_SoR)
4. Immobile; fine-grained sand on rock (I_fgSoR)

5. Immobile; coarse-grained sand on rock (I_cgSoR)
6. Immobile; sand and gravel on rock (I_SGoR)
7. Immobile; pebbles, cobbles, boulders on rock (I_pcboR)
8. Immobile; gravel on rock (I_GoR)
9. *Mobile; rock (M_R)*[i]
10. *Mobile; mud on rock (M_MoR)*[i]
11. *Mobile; sand on rock (M_SoR)*[i]
12. *Mobile; fine-grained sand on rock (M_fgSoR)*[i]
13. *Mobile; coarse-grained sand on rock (M_cgSoR)*[i]
14. *Mobile; sand and gravel on rock (M_SGoR)*[i]
15. *Mobile; gravel on rock (M_GoR)*[i]
16. *Intermixed mobile/immobile; rock (IMI_R)*[i]
17. *Intermixed mobile/immobile; mud on rock (IMI_MoR)*[a, i]
18. Intermixed mobile/immobile; sand on rock (IMI_SoR)[g]
19. Intermixed mobile/immobile; fine-grained sand on rock (IMI_fgSoR)[g]
20. Intermixed mobile/immobile; coarse-grained sand on rock (IMI_cgSoR)[g]
21. Intermixed mobile/immobile; sand and gravel on rock (IMI_SGoR)[g]
22. *Intermixed mobile/immobile; gravel on rock (IMI_GoR)*[e, i]

Symposium Abstracts

Symposium abstracts have been reprinted from the Abstract Volume prepared for the Symposium on Effects of Fishing Activities on Benthic Habitats: Linking Geology, Biology, Socioeconomics, and Management without any further review or editing.

Benthic Habitat Characterization of the Grays Reef National Marine Sanctuary Using Sidescan, Multibeam, and GIS Techniques

C. ALEXANDER[1]

Skidaway Institute of Oceanography, Savannah, Georgia

G. MCFALL

Grays Reef National Marine Sanctuary, Savannah, Georgia

T. BATTISTA

National Ocean Service, Silver Spring, Maryland

R. BOHNE

Grays Reef National Marine Sanctuary, Savannah, Georgia

NOAA's Grays Reef National Marine Sanctuary has been completely mapped by the NOAA ship Whiting using multibeam and sidescan sonar techniques. The resulting data mosaics portray the geologic controls on the character of reef habitat as well as the signatures of modern processes affecting the reef. The multibeam data differentiate between the rocky and sandy habitats and illustrate the influence of pre-existing geologic structure on the general morphology of the reef. Sidescan data highlight the fine-scale detail of the rugged reef surface, the influence of bioerosion on the reef surface and the dynamics of mobile, unconsolidated sediments, which periodically alter reef benthic habitats by covering and exposing rocky substrate. GIS techniques, coupled with the high-resolution sidescan data, are being employed to automatically resolve and classify benthic habitats. Diver observations of fish distributions will be compared to benthic habitat distribution determined from the sidescan and multibeam datasets.

---

[1] E-mail: clark@skio.peachnet.edu

Distribution of Acoustic Backscatter Imagery from NOAA Hydrographic Surveys

J. K. BROWN[1], D. W. PRITCHARD, AND G. T. NOLL

*NOAA National Ocean Service, Office of Coast Survey,
Hydrographic Systems and Technology Programs, Silver Spring, Maryland*

Congress appropriated $6.2 million in Fiscal Year 2002 spending for NOAA to upgrade its hydrographic surveying equipment on board the four NOAA survey vessels operating in support of safe navigation. This is NOAA's first fleet-wide hydrographic equipment purchase since 1992. The NOAA Office of Marine and Aviation Operations, together with the National Ocean Service's Office of Coast Survey, specified and procured five Klein 5500 high-speed high-resolution side scan sonarsÑone each for the NOAA ships RAINIER and RUDE, two for the WHITING launches, and one for the BAY HYDROGRAPHER, which NOAA also uses as a systems test platform. In addition, NOAA procured a Reson 8125 high-resolution multibeam echosounder for each vessel, and installed a hull-mounted Kongsberg-Simrad EM1002 multibeam echosounder aboard the WHITING. Ancillary sensors, software, and data storage management purchased to facilitate the use of these systems will help NOAA speed the hydrographic data to the nautical chart. In determining equipment needs, NOAA placed primary focus on meeting Homeland Security requirements in collaboration with the Naval Oceanographic Office. This purchase upgrades NOAA's systems to 21st century technology so that it can continue its mission to produce the navigation products essential to safe and efficient maritime commerce. NOAA will also continue to develop optimized algorithms and work processes, using these Commercial Off the Shelf products, to share with and transfer to the entire hydrographic industry. One of the secondary benefits to this upgrade plan is the increased ability to apply these same technologies to the production of large area maps of benthic habitat. The high-resolution echosounders and side scan sonars will create exciting high-resolution digital terrain models of the underwater environment and provide good estimates of large-scale variations in acoustic backscatter. Demonstration of sample products will be the focus of the poster session.

---

[1] E-mail: Jeff.K.Brown@noaa.gov

Mapping Seagrass Boundaries with Waveform-Resolving Lidar: A Preliminary Assessment

T. D. CLAYTON[1] AND J. C. BROCK

U.S. Geological Survey, Center for Coastal and Regional Marine Studies, St. Petersburg, Florida

C. W. WRIGHT

Laboratory for Hydrospheric Processes, NASA, Goddard Space Flight Center, Wallops Island, Virginia

For ecologists and managers of seagrass systems, the spatial context provided by remote sensing has proven to be an important complement to in situ assessments and measurements. The spatial extent of seagrass beds has been mapped most commonly with conventional aerial photography. Additional remote mapping and monitoring tools applied to seagrass studies include optical satellite sensors, airborne multispectral scanners, underwater video cameras, and towed sonar systems. An additional tool that shows much promise is airborne, waveform-resolving lidar (light detection and ranging). Now used routinely for high-resolution bathymetric and topographic surveys, lidar systems operate by emitting a laser pulse, then measuring its two-way travel time from the plane to reflecting surface(s) below, then back to the detector co-located with the laser transmitter. Using a novel, waveform-resolving lidar system developed at NASA — the Experimental Advanced Airborne Research Lidar (EAARL) — we are investigating the possibility of using the additional information contained in the returned laser pulse (waveform) for the purposes of benthic habitat mapping. Preliminary analyses indicate that seagrass beds can potentially be delineated on the basis of apparent bathymetry, returned waveform shape and amplitude, and (horizontal) spatial texture. A complete set of georectified digital camera imagery is also collected during each EAARL overflight and can aid in mapping efforts. Illustrative examples are shown from seagrass beds in the turbid waters of Tampa Bay and the relatively clear waters of the Florida Keys.

---

[1] E-mail: tclayton@usgs.gov

Fast-Track Methods for Assessing Trawl Impacts

R. A. Coggan[1]

CEFAS Laboratory, Essex, United Kingdom University Marine Biological Station, Millport, Isle of Cumbrae, Scotland, United Kingdom;
Fisheries Research Services Marine Laboratory, Aberdeen, Scotland, United Kingdom

C. J. Smith

Institute of Marine Biology of Crete, Iraklio, Crete, Greece

R. J. A. Atkinson

University Marine Biological Station, Millport, Isle of Cumbrae, Scotland, United Kingdom

K.-N. Papadopoulou

Institute of Marine Biology of Crete, Iraklio, Crete, Greece

T. D. I. Stevenson and P. G. Moore

University Marine Biological Station, Millport, Isle of Cumbrae, Scotland, United Kingdom

I. D. Tuck

CEFAS Laboratory, Essex, United Kingdom

Traditional methods for assessing the impact of towed demersal fishing gear are notoriously slow, taking years to report and imposing undesirable delays in the provision of scientific advice on which fisheries and environmental managers can act. There is a need to develop rapid methods for assessing trawl impacts. We evaluate and compare a suite of rapid methodologies covering a range of readily accessible technologies including:

(1) Acoustic methods: sidescan sonar and bottom discriminating sonar (RoxAnn);

(2) Visual methods: towed video sledge and ROV;

(3) Faunal sampling (epibenthic megafauna): tissue damage, community analysis, population density, functional group composition; and

(4) Sedimentology: granulometry, geotechnical properties and sediment profile imagery.

These methods were applied to otter trawl fisheries in the Clyde Sea, Scotland and the Aegean Sea, Mediterranean, at sites representing a range of trawl impacts. Novel methods of analysis were developed for quantitative interpretation of sidescan and video records. The scientific effectiveness, cost effectiveness and operational constraints of the various methodologies are reviewed. We recommend suitable approaches to the rapid assessment of trawl impacts taking into consideration the variety of resources (such as time, equipment and budget) which may be available. Assessments should employ complementary methods that operate on different scales of resolution (eg. sidescan sonar with either faunal sampling or ROV). Site-specific factors, such as topography and substratum type, will influence choice of methods and survey design. These rapid methodologies can provide results in a matter of days or weeks rather than the months or years associated with traditional assessment methods.

---

[1] E-mail: r.a.coggan@cefas.co.uk

Using Lasers to Investigate Deepwater Habitats in the Monterey Bay National Marine Sanctuary off Central California

C. B. Grimes and M. Yoklavich
National Marine Fisheries Service, Southwest Fisheries Science Center, Santa Cruz Laboratory, Santa Cruz, California

W. Wakefield
National Marine Fisheries Service, Northwest Fisheries Science Center, Newport, Oregon

H. G. Greene
Moss Landing Marine Laboratories, Moss Landing, California

We conducted a 9-day field test of laser line-scan imaging technology (LLS) to investigate benthic marine habitats in and around the Big Creek Ecological Reserve (BCER) off the central California coast. We determined the utility of LLS for determining the distribution and abundance of fish and megafaunal invertebrates, and identifying habitats and species associations by comparing LLS images with those acquired from side-scan sonar and a remotely-operated vehicle. We also evaluated the ability of LLS to detect seafloor disturbance caused by fishing trawl gear. We surveyed a 2.6 km long x 0.4 km wide area inside and directly outside BCER. With the laser we imaged isolated rock outcrops with patches of large Metridium sp., dense schools of fishes, drift kelp, sea pens, salp chains, and sedentary benthic fishes (possibly California halibut, Pacific electric ray, ratfish and juvenile lingcod.). The LLS system offers the advantage of imaging both the biogenic and abiotic components of habitat, and depicts their spatial relationships with detail that currently is not possible using acoustic imaging techniques such as side-scan and multibeam sonar. LLS imagery also provided fine detail of low relief shelf geology such as sand waves and ripples; evaluating these features in a broader context from a post-processed mosaic of the study area could help us understand coastal physical processes that influence dynamic benthic habitats.

An Assessment of Fish and Invertebrate Communities along Trans-Pacific Cable Lines: A Pilot Study with Implications for Marine Reserve Planning

T. D. Hart[1] and S. S. Heppell
Oregon State University, Department of Fish and Wildlife, Corvallis, Oregon

Marine protected areas are currently being considered along the Oregon coast with the intention of rebuilding stocks. But stakeholders have questioned the effectiveness of reserves and little data exist with regard to marine reserves or the effects of fishing activities on benthic habitat along the Oregon coast. A unique opportunity exists to begin to collect such information because of *de facto* no-trawl reserves that exist along submerged coastal corridors where trans-oceanic communication cables have been laid across the seafloor. I intend to investigate a cluster of unburied cables, which extend off the Oregon coast just north of Bandon (summer, 2002). The corridor of reduced fishing impact, according to the trawl log book data, is approximately 2 miles wide and extends out from shore approximately 14 miles to about 70 fathoms. With the use of an ROV (remotely operated vehicle), commercial trawl data, and bathymetry data, I will be able to analyze the impact, if any, these *de facto* refuges have had on bottom-dwelling invertebrates and fish species of commercial importance. Specifically, I will analyze individual groundfish species associations with different substrate types and invertebrates within and outside of the cable corridor. This collaborative research will establish a credible baseline study on which to build further investigation regarding possible design of a successful marine reserve for groundfish and invertebrate species along the Oregon coast.

---

[1] E-mail: Hartt@ucs.orst.edu

Benthic Habitat in the Gulf of Alaska: Biological Communities, Geological Habitat, and Fishing Intensity

J. Heifetz,[1] D. L. Courtney, and J. T. Fujioka
National Marine Fisheries Service, Auke Bay Lab, Juneau, Alaska

H. G. Greene
Moss Landing Marine Lab, Moss Landing, Calilfornia

P. Malecha and R. P. Stone
National Marine Fisheries Service, Auke Bay Lab, Juneau, Alaska

Multibeam, backscatter, and video data were collected on Portlock Bank near Kodiak, Alaska in the vicinity of groundfish fisheries. The objective was to characterize habitat in heavily fished grounds to understand whether habitats in current fishing grounds are vulnerable to ongoing fishing activities. The multibeam and backscatter data indicated at least a dozen macro- or meso-habitats. The megahabitats are the result of past glaciation and are presently being reworked into moderate (cm-m) relief features. Submarine canyons notch the upper slope and provide steep relief with alternating mud-covered and consolidated sediment exposures. The video data from the submersible Delta, indicated little evidence of trawling on the low relief grounds of the continental shelf where perhaps the level bottom did not induce door gouging and there was a lack of boulders to be turned over or dragged. The most common epifauna were crinoids, small non-burrowing sea anemones, glass sponges, stylasterid corals and brittlestars. Occasional large boulders were located in depressions were the only anomaly in the otherwise flat seafloor. These depressions may have afforded some protection to fishing gear, as the glass sponges and stylasterid corals attached to these boulders were larger than were typically observed. In contrast, there was evidence of boulders turned over or dragged by trawling in the areas of the upper slope. The uneven bottom perhaps induced gouging by the trawl doors. The substrate was mostly small boulders, cobble, and gravel. Presently there does not appear to be much habitat in this area that can be damaged by trawling. No large corals and very few large sponges were seen. Whether this is the result of past trawl activity is unclear.

---

[1] E-mail: jon.heifetz@noaa.gov

Quantitative Seafloor Habitat Classification Using GIS Terrain Analysis: Effects of Data Density, Resolution, and Scale

P. IAMPIETRO AND R. KVITEK[1]

California State University Monterey Bay, ESSP, Seafloor Mapping Lab, Seaside, California

There is a great need for accurate, comprehensive maps of seafloor habitat for use in fish stock assessment, marine protected area design, and other resource management pursuits. Recent advances in acoustic remote sensing technology have made it possible to obtain high-resolution (meter to sub-meter) digital elevation models (DEMs) of seafloor bathymetry that can rival or surpass those available for the terrestrial environment. This study attempts to use an algorithmic terrain analysis approach to efficiently, non-subjectively classify seafloor habitats according to quantifiable parameters such as slope, rugosity, and topographic position index (TPI). In addition, we explore the effects of original x,y,z and gridded data density on the results of these analyses, in order to provide insight into how inherent depth-dependent decreases in data density may affect this approach, and to assess the appropriateness of using historical, lower density bathymetric data. Finally, issues of scale with regard to rugosity and TPI are explored and their potential biological relevance discussed.

---

[1] E-mail: rikk_kvitek@csumb.edu

Sampling Macrozoobenthos from Zebra Mussel Communities in Lake Erie

G. W. KENNEDY[1]

U.S. Geological Survey, Great Lakes Science Center, Ann Arbor, Michigan

M. C. FABRIZIO

NOAA Fisheries, Northeast Fisheries Science Center, James J. Howard Marine Sciences Laboratory, Highlands, New Jersey

M. A. BLOUIN AND J. F. SAVINO

U.S. Geological Survey, Great Lakes Science Center, Ann Arbor, Michigan

Macrozoobenthos found on hard substrate areas in central Lake Erie were examined as part of a larger study to understand the use of nearshore areas by larval and juvenile fishes. Large portions of nearshore habitats were colonized in the 1990's by dreissenid mussels, *Dreissena polymorpha* (zebra mussel) and *Dreissena bugensis* (quagga mussel), which makes standard sampling methods for benthos largely ineffective. To provide quantitative collection of macrozoobenthic organisms inhabiting these hard substrate areas, we developed a method involving the use of SCUBA and underwater suction sampling. One-quarter m² quadrats were sampled using a submersible suction sampler to collect free-ranging macrozoobenthos, followed by collection of dreissenid mussels within the quadrat immediately afterward. Comparisons were made of the benthos collected by suction with the benthos remaining among the mussels to determine the effectiveness of the suction sampling method. Preliminary estimates show that over 90% of the total number of organisms and over 95% of the total taxa in each 0.25 m² area were collected by suction sampling. Relationships between the benthos taxa and the dreissenid mussels and density of mussel populations are currently being examined. These data are critical for determining the relationship and linkages between juvenile fishes and invertebrate prey found in these altered Great Lakes habitats.

---

[1] E-mail: gregory_kennedy@usgs.gov

Characterization of Coastal Great Lakes Benthic Habitat

S. J. Lozano[1]

NOAA, Oceanic and Atmospheric Research, Great Lakes Environmental Research Laboratory,

M. Blouin

U.S. Geological Survey

N. Wattrus

University of Minnesota

Colonization by dreissenid mussels, *Dreissena polymorpha* (Zebra) and *Dreissena bugensis* (Quagga), is one of the more ecologically important events to occur in the Great Lakes during the last decade. Since their introduction into the Great Lakes, dreissenids have colonized both soft and hard substrates to depths of 80 m and reached average densities of 40,000 mussels m^{-2} in the littoral zone. Because of high densities and widespread distribution, they have modified habitats for benthos and fishes and fostered growth and proliferation of non-indigenous species, such as the round goby and a Black Sea amphipod, *Echinogammarus*. In 2001, we used multibeam sonar to characterize the benthic habitat in Lake Michigan and acoustic remote sensing technology (sidescan sonar and acoustic bottom classification) to extend and extrapolate information on dreissenid distributions from spatially limited observations. These observations were combined with discrete in situ (video and SCUBA) observations in a geographic information system (GIS). We used this system to link dreissenid distribution with substrate type, morphology and depth. These data are critical inputs to modeling the ecological implications of dreissenid filtering on algal biomass and composition in the Great Lakes over changing environmental conditions.

---

[1] E-mail: Stephen.lozano@noaa.gov

A System for Classification of Habitats in Estuarine and Marine Environments: Florida Perspective

K.A. Madley[1]

Florida Fish and Wildlife Conservation Commission, Florida Marine Research Institute, St. Petersburg, Florida

A standard, benthic habitat classification system for Florida does not exist. Over fourteen different classification systems have been used with Florida mapping projects to date. This is problematic for efforts to compile statewide habitat area estimates, produce habitat maps for the entire state, or compare habitats across regions. Implementation of a standardized classification system will be a large step toward more reliable characterization of Florida seafloor habitats. The Florida Marine Research Institute has studied the classification systems used throughout Florida and the tropics and subtropics as well as successful efforts in terrestrial habitat characterization. The goal has been to combine appropriate components of a variety of systems to form a hierarchical classification system to propose as a strawman for further testing in Florida. We have formed this scheme with guidance from the Allee et al. 2000 NOAA Technical Memorandum for the purpose of creating a habitat characterization system compatible with the forthcoming national classification system. The Gulf of Mexico program has interest in eventually expanding the Florida classification system to encompass habitats for all of the Gulf states. The goal would then be to coordinate adoption of this classification system to be used by all mapping agencies involved with Gulf of Mexico habitat classification. This would enhance fishery habitat comparisons among Gulf states thus assisting fishery and habitat resource managers.

[1] E-mail: kevin.madley@fwc.state.fl.us

Sea Floor Mapping on the Scotian Shelf and the Gulf of Maine: Implications for the Management of Ocean Resources

R. A. Pickrill[1] and B.J. Todd

Geological Survey of Canada (Atlantic), Dartmouth, Nova Scotia, Canada

Multibeam sea floor mapping technologies have provided the capability to accurately, and cost effectively, image large areas of the seabed. Imagery provides base maps of sea floor topography from which targeted surveys can be planned to map sea floor sediments and associated benthic communities. Over the last five years extensive multi-disciplinary surveys have been carried out on Browns, German and Georges Banks. The government of Canada entered into a partnership with the scallop industry to map bathymetry, surficial sediments and benthic communities. The new knowledge has been used by industry, and has implications for fisheries management. Associations between substrate type and benthic community composition have enabled precise maps of scallop habitat to be produced and links between scallop abundance and substrate to be established. The environmental and economic benefits have been immediate, with reduced effort to catch set quota, less bottom disturbance, and containment of fishing activity to known scallop grounds. Stock assessments and management practices are improved. Other pilot projects in Atlantic Canada and the northeastern USA have demonstrated the value of integrated sea floor mapping in designating marine protected areas (The Gully, Stellwagen Bank), in identifying offshore hazards such as landslides, in siting offshore structures, cables and pipelines, and in addressing environmental issues such as the routing of outfalls and disposal of dredge materials. In recognition of the power of these new tools and digital map products, Canada is considering development of a national mapping strategy to provide the foundation for sustainable ocean management in the 21st century.

[1] E-mail: Dick.Pickrill@NRCan.gc.ca

Seabed Classification with Multibeam Sonars for Mapping Benthic Habitat

J. M. Preston,[1] A. C. Christney, and W. T. Collins

Quester Tangent Corp., Sidney, British Columbia, Canada

B. D. Bornhold

Coastal and Ocean Resources Inc., Sidney, British Columbia, Canada.

Seabed images, from multibeam systems or sidescans, convey a lot of information about seabed type. Large-scale rocky relief often gives dramatic images, and morphology such as sand waves can be very evident. Fine-grained sediments affect images in less obvious ways. Statistical processing of the backscatter amplitudes generate features adequate for seabed classification that agree with both large-scale interpretation and fine-grained details. Before calculating features, it is essential to precondition the image by compensating for artifacts due to range and grazing angle. Useful features include ratios of integrals of the power spectrum over various frequency bands, descriptors of grey-level co-occurrence matrices and histograms, means and higher order moments, and fractal dimension. Generating many features and then using multivariate statistical techniques to select the linear combinations that capture most of the variance in the dataset improves the quality and usefulness of the resulting classifications by adapting the classification to each set of images. To complete the classification process, records are assigned to classes by the same clustering process used in the existing Quester Tangent classification products. Maps of these acoustic classes show regions of distinct acoustic character, thus of distinct sediment type in some sense. To make all this useful for benthic studies, one must understand how this acoustic diversity correlates with the distribution of species of interest. Various spatial analysis techniques are available to accomplish this, and several examples of the integration of acoustic and benthic information will be presented.

---

[1] E-mail: jpreston@questertangent.com

Quantitative Measures of Acoustic Diversity to Support Benthic Habitat Characterization

J. M. Preston,[1] A. C. Christney, and W. T. Collins

Quester Tangent Corp., Sidney, British Columbia, Canada

R. A. McConnaughey

National Marine Fisheries Service, Alaska Fisheries Science Center, Seattle, Washington

The fundamental dataset produced by an acoustic classification system is a representation of the acoustic diversity of the sediments in the survey area. Each acoustic record, from a ping, a stack of pings, or a section of a sonar image, is transformed to a feature vector, typically in two or three dimensions. Features may be from spectral analysis or from integration of parts of an echo envelope. Rather than classifying sediments with just these few features, it is often more useful and adaptable to generate many features and use multivariate statistical techniques to select the linear combinations that capture most of the variance in the dataset. Classification can then be done by dividing the records into groups based on the values of the most important, typically three, principal components. A difficult step in this classification process is estimation of the appropriate number of clusters. Motivated by the need for an automated seabed classification process that is both objective and adaptable to a wide variety of survey applications, this paper describes objective methods for choosing the number of clusters, based on information theory. Actual classifications provide insights into acoustic diversity, which can be used as a proxy for change in sediment characteristics including the influence of benthos. QTC IMPACT™ calculated 166 features from each stack of a very large set of echoes from the Bering Sea. An optimum classification scheme, using the three most important principal components, was identified, based on K-means clustering guided by finding minima using information theory techniques.

---

[1] E-mail: jpreston@questertangent.com

usSEABED: Toward Unifying Knowledge of Geologic Controls on Benthic Habitats

J. A. REID[1]

U.S. Geological Survey, Pacific Science Center, Santa Cruz, California

C. J. JENKINS

INSTAAR, University of Colorado, Boulder, Colorado

M. E. FIELD

U.S. Geological Survey, Pacific Science Center, Santa Cruz, California

M. ZIMMERMANN

National Marine Fisheries Service, Alaska Fisheries, Seattle, Washington

S. J. WILLIAMS AND J. M. CURRENCE

U.S. Geological Survey, Coastal and Marine Geology, Woods Hole, Massachusetts

C. E. BOX

U.S. Geological Survey, Pacific Science Center, Santa Cruz, California

J. V. GARDNER

U.S. Geological Survey, Coastal and Marine Geology, Menlo Park, California

The identification of benthic habitat is based, in part, on its underlying geologic character. While some geologic characteristics can be inferred through bathymetry and remotely sensed imagery, knowledge based on actual sampling of the seabed, either through cores or through photography, can be expensive to collect, and in the latter case, difficult to quantify and assess. Hundreds of thousands of sediment cores, photographs, and videos have been collected along the continental shelves of the United States, in very large and very small research efforts, for various purposes and using a variety of equipment. We unify these sets, both numerical lab-based data and word-based data from from core logs, photos, and videos, where we apply fuzzy set theory to parse values from graphed meanings. These quantified combined data are mined for information useful for geologists, biologists, and ecologists into an linked information system, usSEABED, mappable in GISs and queriable in RDBs. We present a ever-growing integrated look at the character of the surficial seabed of the United States (to about 100m depth, where available) that includes textural information, degree of hardness, presence of biota, basic chemistry, and critical shear strength, to name a few. While these data are useful in their own right, most data held within usSEABED are available as baselines for habitat identification and assessment, or in combination with oceanographic, biologic, or geophysical data to a more complete understanding of a variety of critical processes necessary for effective resource management.

---

[1] E-mail: jareid@usgs.gov

Fluorescence Imaging Laser Line Scan (FILLS) Imagery for High-Resolution Benthic Habitat Characterization

M. P. STRAND[1]

Naval Surface Warfare Center, Coastal Systems Station, Panama City, Florida

Laser-based underwater imaging sensors have been developed and matured in the last decade that provide high resolution optical imagery of the sea floor. Laser Line Scan (LLS) and Streak Tube Imaging Lidar (STIL) have been particularly successful. A prototype Fluorescence Imaging Laser Line Scan (FILLS) sensor has been deployed in several underwater environments, yielding high-resolution (~1 cm pixel size) imagery of the associated benthic habitats. The prototype FILLS sensor illuminates the sea floor with 488 nm laser light, and constructs four independent images from light collected at 488 nm, 520 nm, 580 nm, and 685 nm, respectively. The 488 nm image is formed from elastically scattered light (i.e., light scattered with no change in photon energy), while the other images are formed by inelastically scattered light. (The FILLS sensor is routinely operated during nighttime hours so that ambient illumination is negligible). Fluorescence is the primary physical mechanism giving rise to the inelastically scattered light sensed by FILLS. Coral reef environments produce particularly strong (and spectacular!) fluorescence imagery. FILLS was developed primarily for the detection, classification, and identification of man-made objects in underwater environments. In addition it can serve admirably for the characterization of underwater habitats. Examples of FILLS imagery relevant to fish habitat evaluation will be presented.

---

[1] E-mail: strandmp@ncsc.navy.mil

The Use of Field Calibrated Side-Scan Acoustic Reflectance Patterns to Quantify and Track Alterations to Benthic Habitat Associated with Louisiana's Oyster Industry

C. WILSON,[1] H. ROBERTS, Y. ALLEN, AND J. SUPAN

Oyster Geophysics Program, Department of Oceanography and Coastal Sciences, SC&E, Louisiana State University, Baton Rouge, Louisiana

Coastal Louisiana, like many deltaic land masses, faces continued landscape alteration from natural processes and anthropogenic impacts that affect estuarine habitat. The most promising steps to slow/mitigate these changes are river diversions that introduce freshwater and sediment to river-flanking environments and to help establish ideal salinities over historic oyster grounds. Critical to the success of these programs is a rapid and accurate means to qualify and quantify changes in oyster habitat. Digital high-resolution acoustic instrumentation linked to modern data acquisition and processing software was used to build baseline of information for evaluating future changes in shallow water bottoms, with special emphasis on oyster habitats. Application of digital side-scan sonar (100 and 500 kHz), a broad-spectrum sub-bottom profiler (4-24 kHz) for rapidly acquiring water column, surficial and shallow subsurface was used to map over 10,000 ha of water bottom. Geo-referenced side scan sonar mosaics were incorporated into a GIS data base. These data sets, "calibrated" with surface sampling, coring, and other "ground truthing" have established that numerically indexed acoustic reflectance intensities correlate closely with surface shell and oyster reef density. With image processing techniques to analyze mosaic reflectance patterns, we estimated the percent and total acreage of several bottom types.

---

[1] E-mail: cwilson@lsu.edu

Understanding Chronic and Event-Driven Natural Change to Benthic Habitats

Extended Abstract

Living With Change: Response of the Seafloor to Natural Events

MICHAEL E. FIELD[1]

*U.S. Geological Survey, Pacific Science Center, 400 Natural Bridges Drive,
Santa Cruz, California 95060, USA*

Abstract. Natural events are critical factors in determining the character of benthic habitats. All of the characteristics that we identify in habitats are shaped by continuous, chronic processes or by intermittent, more intense processes ("events"), and to a lesser degree, by rare, extreme events. Particle size and sorting, chemical alterations, oxygen and organic carbon content, and many more properties are a time-averaged response to floods, waves, currents, seepage, and numerous other processes. Disturbances that occur on an intermediate scale (in frequency or relative intensity) are recognized by some as a major contributor to community diversity. Disturbances that occur on extreme scales generally do not have a lasting deleterious effect. The greatest potential for negative impacts to habitats occurs when there is an increase in the intensity or frequency of a process, and this includes many human-induced activities.

The Role of Natural Processes

Natural processes are a vital, inseparable component of the biosphere. Their causes, scales, and interactions are complex, and all habitats, such as shallow marine benthic habitats, evolve in accord with the prevailing natural processes. Long-term community structure and trophic patterns develop in response to both chronic processes, such as dominant currents and wave stresses, and intermediate-scale, event-driven processes. As Denny (1995) points out, even the size of organisms is controlled in part by the rigors of the physical environment. All processes, including individual biologic functions (e.g., foraging or burrowing) locally disturb the seafloor but do not necessarily impact or change the seafloor.

Disturbances are essential for maintenance of the habitat through such diverse functions as delivery of nutrients, oxygenation, and introduction of food. The suite of natural processes ultimately controls all of the physical parameters, such as particle size and organic content, and many biologic ones, such as survivability of sessile organisms and continuity of substrate type. In short, every habitat represents a time-averaged response to the dominant physical processes, which is as important in defining the habitat as geologic setting and community structure. For example, many outer continental shelves (approximately 80-m to 120-m water depth) are mantled by sand and gravel (Shepard 1932; Emery 1968). These deposits are remnant from shallow water, near shore settings when sea level was lower than today (Field and Trincardi 1991). Mixed with the sand and gravel are organic-rich mud and shell material derived from modern sediment deposition as well as from biota (the entire spectrum of infaunal, epifaunal, and pelagic organisms from one-celled foraminifera and worms to shelled invertebrates, to fish and mammals). The entire deposit is disturbed by burrowing and feeding organisms, occasional long-period waves, currents, chemical alterations, gas migration, and many more processes (see Knebel 1981). The intensity and frequency of the processes ultimately determine the exact character of the shelf deposits. Thus, on continental shelves commonly disturbed by long-period waves, deep deposits are well-sorted and wave-rippled; on less energetic margins, deposits at similar depths retain the fine-grained organic component until infrequent disturbances re-suspend and remove the fine component (Larsen et al. 1981; Cacchione and Drake 1990).

Do natural processes negatively impact habitats? Only, it appears, at the largest scale, and then only locally. Disruptions from landslides, volcanoes, and other such extreme events do indeed alter habitats extensively, and sometimes permanently. The effect, however, is usually local, in that adjacent areas survive undisturbed to seed the affected area, and the recurrence interval is usually quite large.

---

[1] E-mail: mfield@usgs.gov

Since habitats are adjusted to the prevailing conditions, it is not the process that affects a habitat; rather, it is a change in the process (magnitude or frequency) that results in a negative impact. Assuming that marine habitats are time-averaged responses to the myriad processes that shape them, then it is only a significant change in one of the significant processes that will alter the habitat. Examples of such change are abundant today in coastal areas where discharge of sediment, or nutrients, from the adjacent watershed has increased markedly due to changes in land-use practices (i.e., increased ranching, housing construction, etc.). It is important to note that changes in magnitude or frequency of a process does not necessarily imply an increase in either. A decrease in the repetitions or intensity of a process, such a water mixing by waves, can also lead to negative impacts.

Influence of "Events" on Habitat Character

"Events" are defined here as processes that exceed the typical or average conditions. They can be physical in nature, such as large storm waves, or biologic, such as predator blooms, or chemical, such as anoxia. Because such processes occur at many scales of frequency and intensity, their effect varies accordingly. Events such as large storm waves that occur infrequently (i.e., decadal scales) are termed "intermediate disturbances" (Connell 1978; Figure 1A). As Connell (1978) pointed out, intermediate events open up new niches for colonizing species and increase habitat diversity. This is verified for coral reefs and forests, two very complex communities, and is presumed to be valid for other major habitats, including the open continental shelf. Connell's diagram (Figure 1A) illustrates that if no, or very few, disturbances occur, natural competition will lead to a climax succession marked by low diversity. When natural disturbances occur on a time scale (or conversely, an intensity) that is intermediate in scale, the effect is to maximize diversity (Figure 1B). After each disturbance, the community is dominated by a few hardy, pioneer species. As food and niches become more abundant, diversity increases to a maximum. If unchecked by natural disturbances, the more successful organisms dominate, and diversity begins to decrease. If disturbances become frequent (Figure 1C), the community diversity will theoretically be held at low levels dominated by the opportunistic and robust organisms.

Natural events are of major importance for "resetting the clock" on individual habitats. In contrast, stable habitats (those undisturbed by intermediate events) tend to evolve toward a climax succession whereby the most successful competitor dominates the community. The community structure of many habitats reflects the occurrence of natural intermediate events. In 1976, shellfish and associated benthic fauna on the continental shelf off New Jersey were devastated by oxygen depletion (hypoxia) (Swanson and Sindermann 1979). The event apparently was caused by a combination of factors (including possibly human-induced ones). Although mortality rates of some species were high, repopulation from adjacent unaffected areas, coupled with no recurrences, eventually led to full recovery of the habitat.

Extreme or rare events, on the other hand, have a pronounced impact on a habitat, but typically the devastation is localized. Examples of extreme events include the 1972 freshwater flooding of Chesapeake Bay; covering of Hawaiian coral reefs by lava; and removal of entire habitats by landslides. Yet, since extreme events tend to occur as local phenomena, they are not all that important to habitat survival because most of the habitat remains preserved. Repopulation or replacement is probable because of the abundance of undisturbed genetic stock and the length of time before occurrence of another event of similar magnitude.

The relative balance between different types of natural disturbances and their frequency seems to be a controlling factor in whether a habitat will be adversely affected. Different processes have markedly different frequencies of occurrence and, consequentially, different relative impacts on benthic habitats. The relation between disturbance type and frequency, and the relative impact to the habitat, is conceptualized in Figure 2. Some processes, such as floods and hurricanes, occur somewhat frequently (annual to decadal time scales), but their impact is relatively low because the marine substrate typically is little affected by the rapid passage of a hurricane or the typical plume deposition of floods. Conversely, large-scale processes, such as submarine landslides and volcanic eruptions, have pronounced effects, but only locally and very infrequently. The overall effect of such events is not a threat to the stability of he habitat unless the frequency changes dramatically. Generally, only human-induced processes both occur frequently and have a large impact on marine benthic habitats, marking them as potentially important disturbances. Underscoring this concept is evidence accumulating from coral reefs around the world indicating that those habitats are more susceptible to ecological damage when stresses are repeated or compounded with other stresses than if they are simply one-time events (Connell 1997).

References

Cacchione, D. A., and D. E. Drake. 1990. Shelf sediment transport: an overview with applications to the North-

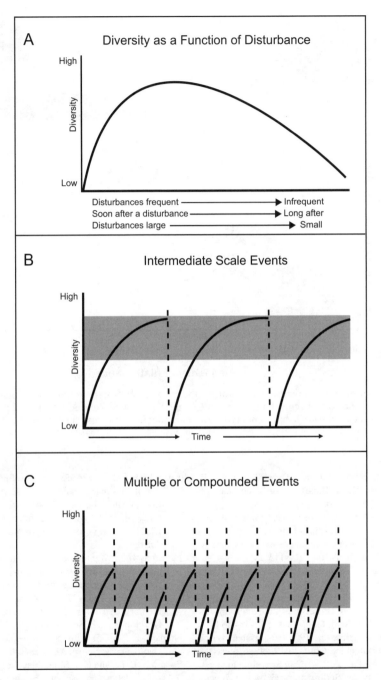

Figure 1. Conceptual diagram illustrating the impact that the frequency of disturbances to the seafloor has on the biologic diversity of a community. (A) Representation of Connell's (1978) diagram showing that species diversity is a function of disturbance, with maximum diversity achieved when disturbances occur that are intermediate in scale (in frequency or magnitude). (B) Redrawing of Connell's diagram showing the influence that large natural events occurring on an intermediate time scale have on maximizing community diversity. (C) If disturbance events are frequent, the community is represented chiefly by opportunistic and robust species, and the overall diversity is held at a low level.

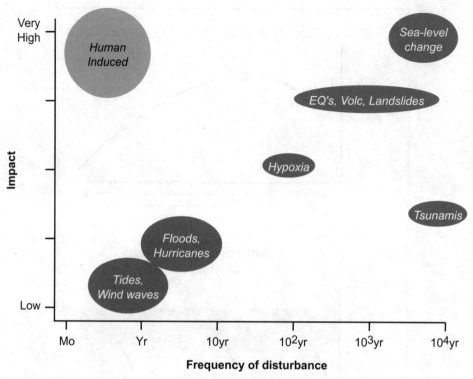

Figure 2. Conceptual diagram illustrating the balance between frequency of disturbance and impact to the habitat. Events that are extreme in terms of impact tend to occur infrequently, making them inconsequential to habitat character. Human-induced activities, such as increased siltation or herbivore removal, can be both frequent and intense, ultimately leading to a major impact to the habitat.

ern California continental shelf. Pages 729–773 *in* B. LeMehaute and D. M. Hanes, editors. The Sea, volume 9. Wiley, New York.

Connell, J. H. 1978. Diversity in tropical rainforests and coral reefs. Science 199:1302–1310.

Connell, J. H. 1997. Disturbance and recovery of coral assemblages. Coral Reefs 16:s101–s113.

Denny, M. 1995. Survival in the surf zone. American Scientist 83:166-173.

Emery, K. O. 1968. Relict sediments on continental shelves of the world. Bulletin of the American Association of Petroleum Geologists 52:445–464.

Field, M. E., and F. Trincardi. 1991. Regressive coastal deposits on Quaternary continental shelves' preservation and legacy. Pages 107–122 *in* Shoreline to abyss. SEPM (Society for Sedimentary Geology) Special Publication 46.

Knebel, H. J. 1981.Processes controlling the characteristics of the surficial sand sheet, U.S. Atlantic outer continental shelf. Pages 349–368 *in* C. A. Nittrouer, editor. Sedimentary dynamics of continental shelves. Elsevier, Developments in Sedimentology 32, Amsterdam.

Larsen, L. H., R. W. Sternberg, N. C. Shi, M. A. H. Marsden, and L. Thomas. 1981. Field investigations of the threshold of grain motion by ocean waves and currents. Pages 105–132 *in* C. A. Nittrouer, editor. Sedimentary dynamics of continental shelves. Elsevier, Developments in Sedimentology 32, Amsterdam.

Shepard, F. P. 1932. Sediments of the continental shelves. Geological Society of America Bulletin 43:1017–1040.

Swanson, L. R., and C. J. Sindermann, editors. 1979. Oxygen depletion and associated benthic mortalities in New York Bight, 1976. NOAA Professional Paper 11.

Sediment Oxygen Consumption in the Southwestern Gulf of Mexico

ELVA ESCOBAR[1]

Universidad Nacional Autónoma de México, Instituto de Ciencias del Mar y Limnología, Apdo. Postal 70-305, México D.F. 04510, México

LUISA I. FALCON[2]

Universidad Nacional Autónoma de México, Instituto de Ecología, México D.F. 04510, México

Abstract. Sediment oxygen consumption (SOC) is metabolically related to the benthic community productivity of continental shelves and margins. These areas have economical relevance for regional fisheries. The aim of this study is to evaluate SOC of the soft-bottom benthic communities in the southwestern Gulf of Mexico and provide rates in areas of varied trawling incidence in the region. Box-cored sediment was incubated at in situ temperature in benthic microcosm incubation chambers and sub-sampled for bacteria, meiofauna, and macrofauna. Environmental factors were measured in bottom water and sediment. The SOC values ranged from 6.12 to 79.9 mL $O_2 \cdot m^{-2} \cdot d^{-1}$ and correlated significantly to depth, grain size, and nitrogen contents in the sediment. A significant correlation was also found between SOC and the infaunal biomass. The proportion of bacteria to meiofauna and bacteria to macrofauna could be related to SOC in different regions. The range of SOC values from the continental shelves in the southwestern Gulf of Mexico were within the range of values recorded in seasonally pulsed sites and highly organic matter enriched areas within the region. We found an order of magnitude difference between the metabolic rates derived from SOC in this study and those reported for the northern Gulf of Mexico. The SOC rates can indicate bottom conditions in a region and should be considered as a potential tool in evaluating natural change and the effects of fishing activities on seabed habitats.

Introduction

Sediment oxygen consumption (SOC), the respiratory metabolism of the entire community of sediment-living organisms (or the measure of the benthic infaunal metabolic rate), has been considered an important oxygen loss to the benthic ecosystems (Officer et al. 1984). Oxygen-based or food chain carbon-based models consider the SOC a zero-order flow across the sediment boundary (Kemp et al. 1994).

Sediment oxygen consumption is measured either in situ with benthic incubation chambers (Pomeroy 1959; Rowe et al. 1997) or aboard ship by incubating undisturbed recovered cores (Pamatmat 1971; Miller-Way et al. 1994). In both methods, incubation chambers contain a known volume of water that is sampled on a periodic basis. The changes in dissolved oxygen reflect the consumption through time by organisms breaking down organic matter or by reactions of the superficial water with dissolved solutes that diffuse through the sediment (Smith and Hinga 1983; Archer and Devol 1992).

Benthic communities are controlled by the abiotic factors that characterize a region, having temperature affect growth in a nonlinear way (van der Have and de Jong 1996). Hence, temperature in tropical marine ecosystems determines that benthic communities are dominated by respiration over production (Rowe et al. 1998). On the shelf of tropical latitudes, photosynthesis is of potential importance, contributing both to the oxygen production during the day and oxygen utilization at night, affecting benthic biomass distribution (Rowe 1971). Phytoplankton cells and debris and zooplankton feces sink to the bottom (Redalje et al. 1994), accumulating and enhancing respiration (Eadie et al. 1994). Additionally, the presence of upwelling, outwelling, and perturbation of the seafloor will affect the nature and amount of organic matter in the sediment (Smith et al. 1974). The river outflow to the shelf and the mesos-

---

[1] E-mail: escobri@mar.icmyl.unam.mx
[2] E-mail: falcon@miranda.ecologia.unam.mx

cale features on the continental margin will contribute to the regulation of primary productivity in the water column and its export to deep water (Fucik and El-Sayed 1979). Depth versus the time span required for the organic matter to reach the bottom will determine the oxygen consumption of benthic communities (Smith and Hinga 1983). This pelagic–benthic coupling on the shelf is exemplified by the bottom waters of the inner continental shelf of the northwestern Gulf of Mexico. These waters become hypoxic on an annual basis during the summer months (Rabalais et al. 1994) and have deleterious effects on the bottom fauna composition and abundance (Renaud 1986). The hypoxic or anoxic conditions of continental shelf habitats result from an imbalance between oxygen consumption and production and physical sources and sinks. Hypoxia has not been recognized as a problem in the study area yet.

The knowledge obtained from SOC will be related to the benthic community productivity of continental shelves and margins where it has economic relevance in regional fisheries. This study will enhance our understanding on how benthic metabolism acts as a component of the global ocean carbon equilibrium in the tropics.

The southwestern Gulf of Mexico is characterized by a complex topography and high variability of habitats. Previous studies in the southwestern Gulf of Mexico have identified sub-regions that maintain enhanced benthic biomass associated with mesoscale features that promote the export of organic matter (Soto and Escobar-Briones 1995). There have been no previous studies of benthic SOC rates that study areas where trawling occurs at different levels. It is our goal to present data on SOC of the soft-bottom benthic communities from the southwestern Gulf of Mexico and then relate this data to locations with different degrees of trawling intensity (Tamaulipas continental shelf, Bay of Campeche continental shelf and margin, and Bank of Campeche in Yucatan). These results will help us evaluate whether SOC rates can be used as proxies of the effect of trawling activity on the seafloor.

Area of Study

The Tamaulipas shelf and the Bay of Campeche are located in the terrigenous province of the western Gulf of Mexico (Figure 1). The former is characterized by the presence of the Mexican Ridges, which were developed from sediment input during the recent transgression. The Tamaulipas carbonate platform forms the underlying foundation beneath the sediment. Trawling activity on this shelf are intermediate.

The Bay of Campeche is influenced by the outflows of the rivers Coatzacoalcos, Grijalva–Usumacinta, and Champoton. The export of biogenic carbon of phototrophic origin from the bay supports the existence of benthic communities and a large shrimp fishery in the southern Gulf of Mexico (Soto and Escobar-Briones 1995). A smooth transition zone marks the transition from the terrigenous province to the carbonate zone off Terminos Lagoon (Lin and Morse 1991) and harbors the major oil production area in the southwestern gulf.

The Campeche Bank, located off Yucatan, is a calcareous platform that has no terrestrial runoff to the sea. Its bottom is a structurally complex mixture of patchy coral reefs and sand bottoms, in contrast to the Bay of Campeche and the Tamaulipas shelf. This region is enriched by the persistent presence of upwelled water generated in the Yucatan Straits. Trawling activities are almost nonexistent on the Campeche Bank due to the rough and complex seafloor morphology.

Methods

Sediment community oxygen consumption rates were measured in replicate undisturbed sediments cores recovered from soft bottoms in the Tamaulipas continental shelf, the Campeche Bay continental shelf and margin, and the Campeche Bank in the southwestern Gulf of Mexico (Figure 1). A total of five cruises were carried out, three of which revisited the Bay of Campeche in three different years, one during the winter storm season and twice during the rainy summer season. The sampling sites included the continental shelf and the continental margin. The former was amply recorded in contrast with the latter (Table 1). The shelf sites were selected based on the geomorphology and commercial trawling incidence. Sediments were collected with box cores and incubated onboard the R/V *Justo Sierra* for a period of up to 12 h. Experiments were terminated earlier if oxygen values reached oxygen concentration levels below 2 mg O_2/L.

The approach for measuring oxygen demand was a ship-board incubation of recovered soft-bottom sediment with overlying water at in situ temperature (Pamatmat 1971). We used a YSI oxygen meter electrode (YSI Environmental, Baton Rouge, Louisiana) placed close to the sediment water interface, yet avoiding resuspension of sediment. Measures were carried on Benthic Incubation Microcosms Chambers (BMICs) that had 0.015 m^2 sediment and that were maintained in the dark and at constant temperature. The oxygen values (mg/L) were transformed to mL/L. The BMICs (Miller-Way et al. 1994) are made of plexiglas and contain two-thirds undisturbed sediment collected directly

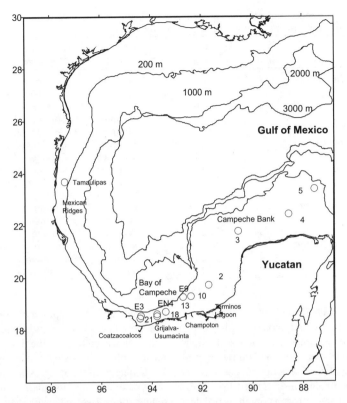

Figure 1. Area of study showing the stations used in the text and tables. E3, E4, and E9 are locations studied in cruises ENOS-1 and ENOS-2 to the Bay of Campeche in winter and summer.

from the box core and one-third seawater collected above the sediment as described in Pamatmat (1978). The chambers were static and were incubated for a period of 8–12 h. Oxygen concentration readings were recovered on a 2-h basis. The oxygen demand estimate is based on the data logged from the single oxygen electrode in each chamber. The electrode was calibrated to the dissolved oxygen content in a saturated oxygen water sample onboard and to the Winkler titration method before and after the cruise. Oxygen consumption adjustments to temperatures measured in situ were obtained with the relationship proposed by Riley (1947). Units were obtained as mL $O_2 \cdot m^{-2} \cdot h^{-1}$ and were converted to mL $O_2 \cdot m^{-2} \cdot d^{-1}$. The data were expressed as metabolic rates (mg $C \cdot m^{-2} \cdot d^{-1}$) using the biomass–metabolism conversion constant in Rowe et al. (1997).

At the end of each incubation, the BMICs were sub-sampled for bacteria, meiofauna, and macrofauna, which were preserved as described in Rowe et al. (1997). The top 1 cm bacterial abundance (cells/cm³ sediment) was estimated by direct count with an epifluorescence microscope as described in Falcon (1998). Bacterial biomass was estimated using the average cell volume and assuming a conversion factor of 220 femtograms C/mm³ (Bratbak and Dundas 1984; Falcon 1998). Sorting the organisms through 54-mm-size sieves and identifying major taxa, aided by a dissecting microscope, in the laboratory was used to obtain meiofauna and macrofauna abundance. Preserved wet weights for meiofauna and macrofauna were obtained from values reported for each taxon or directly obtained on a microbalance. The preserved wet weights were transformed to organic carbon using the conversion factors proposed by Rowe (1983). All biomass values were transformed to organic carbon per unit area (m²). The proportion of each infaunal component was determined for the oxygen consumption at each location.

At every site, the depth, temperature, salinity, and dissolved oxygen were measured from conductivity, temperature, and depth (CTD) profiling; organic carbon and nitrogen, grain size, organic matter, and chloroplastic pigments were determined in the top 1 cm of the sediments from sub-samples obtained from the box core. Dissolved oxygen concentrations in near-bottom water was determined by Winkler titration in the labo-

Table 1. Depth, depth zone, and number of replicates of sediment samples used to determine the oxygen consumption measurements in the southwestern gulf of Mexico

Cruise	Region	Station	Depth (m)	Season	Depth zone	Number of replicates
OG-12 June	Bay of Campeche	2	52	Summer	Continental shelf	3
		10	52	Summer	Continental shelf	3
		13	61	Summer	Continental shelf	3
		18	103	Summer	Continental shelf	3
		21	72	Summer	Continental shelf	3
OG-14 June	Campeche Bank	3	50	Summer	Continental shelf	3
		4	48	Summer	Continental shelf	3
		5	110	Summer	Continental shelf	3
SIG-1 June	Tamaulipas	1	170	Summer	Continental shelf	4
ENOS-1 March	Bay of Campeche	3	142	Winter	Continental shelf	4
		4	203	Winter	Continental margin	4
		9	254	Winter	Continental margin	4
ENOS-2 June	Bay of Campeche	3	169	Summer	Continental shelf	4
		4	203	Summer	Continental margin	4
		9	256	Summer	Continental margin	4

ratory on water samples collected with Niskin bottles set on a rosette and also by continuous recording using a CTD oxygen sensor. Sediment particle size distribution was determined by means of standard sieving and pipette analysis techniques. Mean grain size was expressed as M_z or the graphic mean (Folk 1968). Sediment organic carbon and nitrogen content was determined with a Perkin Elmer CNH elemental analyzer (Perkin Elmer Corporate Headquarters, Wellesley, Massachusetts). Organic matter content was calculated by applying the conversion factor for nitrogen. Chloroplastic pigments in sediment were extracted with acetone in the dark and left for 12 h under refrigeration. After centrifugation, the chloroplastic pigments in the overlying suspension were measured with a Turner bench fluorometer at the extraction bands of 660 and 430 nm for chlorophyl a in 1-cm^3 sediment samples. Three locations were selected and coded 1–3 for the statistical analysis according to the trawling intensity as (1) minimally trawled (2) moderately trawled, and (3) highly trawled based on the fishery reports (Anuario de Pesca) from 1973 to 1999.

Analysis of variance (ANOVA) assumptions of symmetry and heteroscedasticity were checked for untrans-formed physical, chemical, and biological data by examining the residual plots. When the assumptions were not met, log $n + 1$ transformations were applied. Sediment oxygen consumption, biomass, and sediment factors were analyzed for the three regions (Tamaulipas, Bay of Campeche, and Campeche Bank in Yucatan). Seasonality (winter storms and summer rains) was evaluated in samples from the Campeche Bay. The analyses used Statistica, version 5 ANOVA procedures (StatWoft, Inc., Tulsa, Oklahoma), where the model included terms for depth zone, season, and regions. The post hoc comparisons of means were analyzed with a Newman-Keuls critical ranges test on adjusted means following a stepwise analysis ($P = 0.05$). A least-squares multiple regression analysis was applied to SOC to examine its relationships to water depth, temperature, dissolved oxygen concentration, sediment character, and the biomass in a multiple-regression analysis where factors were added one at a time. Trawling incidence on the sediment oxygen consumption among the three areas was analyzed with a multiple analyis of variance (MANOVA) test of post hoc comparisons of the means with a Tukey Honestly, Significantly Different (HSD) test for unequal N (Spjotvoll–Stoline test). A multidimensional scaling analysis displayed the factors closely related to trawling incidence.

Results

Sediment Oxygen Consumption

The SOC ranged from 6.12 to 79.9 mL O_2 m^{-2} · h^{-1} with regional significant differences (ANOVA, Newman-Keuls, $P < 0.05$, HSD for unequal N, Figure 2). SOC rates were four times higher on the Campeche Bank off Yucatan (mean = 69.93 ± 9.45 mL O_2 m^{-2} · h^{-1}; $N = 9$) in contrast to Tamaulipas (14.22 ± 3.65 mL O_2 m^{-2} · h^{-1}; $N = 4$) and the Bay of Campeche (8.74 ± 2.61 mL O_2 m^{-2} · h^{-1}; $N = 20$), which were similar.

Significant differences were recorded (Newman-Keuls, $P < 0.05$, HSD for unequal N at different depths in the Bay of Campeche during the summer rain season. The smallest SOC rates during this season occurred on the shelf of the Bay of Campeche (6.1–12.1

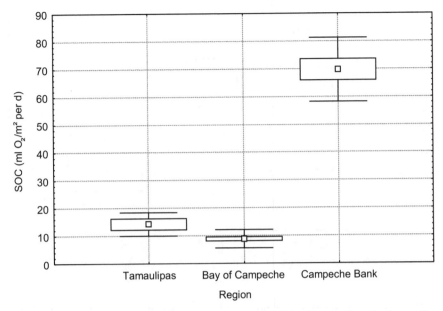

Figure 2. Rates of sediment oxygen consumption (SOC) in three regions in the southwestern Gulf of Mexico. Small squares = mean, large boxes = SE, and bars = SD.

mL O_2 $m^{-2} \cdot h^{-1}$; $N = 15$). The largest SOC rates during this season, eight times larger, were recorded on the continental margin (16.8–44.4 mL O_2 $m^{-2} \cdot h^{-1}$; $N = 9$).

Local differences were recorded between stations, both on the continental shelf and the continental margin, within every region. For example, the two easternmost stations (stations 4 and 5) on the Campeche Bank off Yucatan showed significant differences between each other (Newman-Keuls, $P < 0.05$, $N = 9$) but neither differed significantly from the westernmost station (station 3) while significant differences were recorded between the other westernmost stations on the shelf (Figure 3). (Newman-Keuls, $P < 0.05$; $N = 15$; Figure 4). The SOC rates on the continental margin of Campeche Bay diminished from the west during both the winter storm and the summer rainy seasons (Figure 5). The differences of SOC between sites on the continental shelf were also significant (Newman-Keuls, $P < 0.05$; $N = 8$), as were seasonal differences (Type III SS F 60.67, $P < 0.05$, df = 18; Newman-Keuls, $P < 0.05$). The largest SOC rates recorded during the summer rainy season were approximately twice as high as the rates recorded during the winter stormy season (Figure 5).

Environmental Factors

Bottom-water temperature in the study area ranged from 11–16°C (mean = 14.25 ± 2.22°C) on the cooler continental margin and from 15–24°C on the warmer continental shelf (mean = 19.46 ± 2.81°C; Table 2). These values, although different on the shelf and margin, showed no significant differences for different seasons and regions within the gulf. The highest temperature values were recorded in the Campeche Bank off Yucatan with a mean value of 23.33 ± 1.2°C, followed by Tamaulipas shelf (19.5°C) and the Bay of Campech shelf (18.07 ± 3.5°C). Differences between the two seasons were recorded in the Bay of Campeche both on the continental shelf and on the continental margin. Differences with increasing depth were significant.

The dissolved oxygen concentration in bottom water ranged from 3 to 7.6 mg/L O_2 or 2.1–5.3 mL/L O_2 (Table 2). The values were similar among stations with the exception of the Campeche Bank off Yucatan where values were higher (mean = 7 ± 0.38 mg/L O_2 or 4.9 ± 0.3 mL/L O_2). The lowest values were recorded off Tamaulipas (3 mg/L O_2 or 2.1 mL/L O_2). Seasonal differences were recorded both on the continental shelf and the continental margin in the Bay of Campeche, with lower values recorded in March (4.8 mg/L O_2 or 3.4 mL/L O_2 recorded on the shelf and 4.1 ± 0.1 mg/L O_2 or 2.9 ± 0.1 mL/L O_2 on the continental margin) than in June when values were almost 1.5 times higher. The seasonal variability seen in the Bay of Campeche correlates with the total biomass during the two El Niño-Oscilación del Sur (ENOS) cruises (Pearson $r = -0.60$; $r^2 = 0.36$; $P = 0.002$) and also with the bacteria biomass (Pearson $r = -0.71$; $r^2 = 0.50$; $P = 0.009$) in the ENOS cruise of June (Tables 2, 3).

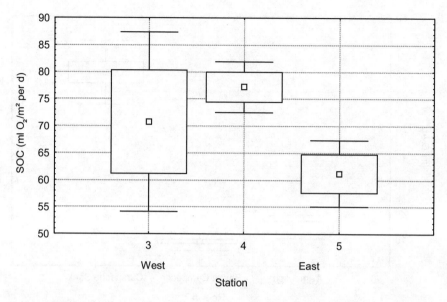

Figure 3. Sediment oxygen consumption (SOC) geographical differences in continental shelf stations from the Campeche Bank off Yucatan. Small squares = mean, large boxes = SE, and bars = SD.

The silt and clay fraction made up 90% of the moderately and poorly sorted sediments at most stations, with the exception of the Bank of Campeche off Yucatan, where sand contributed up to 50% of the sediment. Fragments of coralline algae were found at all three stations of the Campeche Bank. In contrast, nearly all stations on the continental margin of the Bay of Campeche were dominated by the clay fraction and sand provided in average 3.6 ± 2.1% (Table 2). Fine sediments dominated most of the Tamaulipas shelf

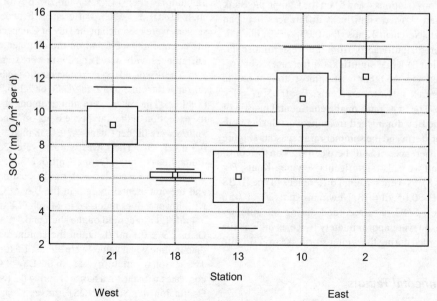

Figure 4. Sediment oxygen consumption (SOC) geographical differences in continental shelf stations from the Bay of Campeche. Small squares = mean, large boxes = SE, and bars = SD.

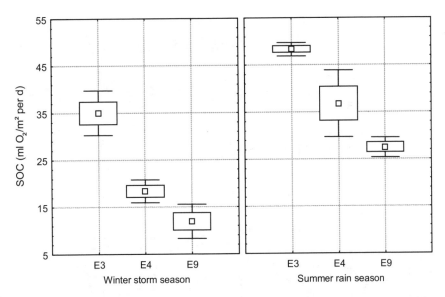

Figure 5. Sediment oxygen consumption (SOC) seasonal differences recorded in different locations from the continental margin in the Bay of Campeche. Small squares = mean, large boxes = SE, and bars = SD.

(Table 2). Chloroplastic pigments were found in sediments of the Tamaulipas shelf and on Campeche Bank off Yucatan (Table 2).

Carbon content of the sediment ranged from 0.41% to 8.02% on the Bay of Campeche shelf. A few carbon values as high as 13.04% occurred on Campeche Bank in Yucatan. The mean total organic carbon content of the sediment in the Campeche Bay stations was twice as high as the values recorded on the Tamaulipas shelf (Table 2). The total nitrogen ranged from 0.01% to 0.2% (Table 2). Nitrogen concentrations were similar among stations with the exception of the Campeche Bank off Yucatan, which had significantly lower values, 0.03% ± 0.02% dry weight (Table 2). The sediment C/N ratios were normal for sediments, ranging from 7.67 to more than 10.5. The C/N values varied significantly within the Bay of Campeche and between regions (Table 2).

The trawling activity in the Tamaulipas shelf is based on the dominant presence of brown shrimp *Farfantepenaeus aztecus* and was coded 2 (moderately trawled). In the Bay of Campeche, the dominant shrimp species is the pink shrimp *P. duorarum*. Trawling activity is notably higher (coded as 3) than over the Tamaulipas shelf (Table 2). Trawl activity is higher at stations 3, 13, and 18, less at stations 4, 9, and 21, and nonexistent at stations 2 and 10 due to a 26-year ban due to the presence of oil rigs. Thus, trawling codes as high (3), and a gradient exists. The trawling incidence in the Campeche Bank is seasonally sustained by diverse shrimp species including the pink shrimp and was coded as minimally trawled (value of 1). In all three regions, the effort has increased, with the number of trips rising threefold compared to trips made prior to 1990.

SOC and Factor Correlations

A least-squares multiple regression analysis was applied to SOC to determine correlations with water depth, temperature, dissolved oxygen concentration, and sediment factors. A regression analysis of the SOC and water temperature after logarithmic transformation was highly significant. The SOC rates varied inversely with water column depth ($r = 0.96$), inversely with the mean sediment grain size ($r = 0.61$), and directly with nitrogen content in sediment ($r = 0.58$). The principal components that explained 73.53% of the total variance were depth–temperature and the sediment grain size (as M_z in Table 2)–nitrogen–organic matter content with loadings more than 0.84. The principle components grouped Campeche Bay and Tamaulipas shelf as one western region. The Campeche Bank in Yucatan sorted out as a separate region in the southwestern Gulf of Mexico.

Benthic Biomass

The standing stocks of the biota, in terms of biomass per m^2, are compared by size categories in Table 3.

Bacteria

Mean biomass values (Table 3) were integrated over 8 cm of the core. Bacterial mean biomass values ranged

Table 2. Descriptive characteristics of the bottom water and sediment of Tamaulipas continental shelf, Bay of Campeche continental shelf and margin, and Campeche Bank shelf off Yucatan. M_z = graphic mean as a graphic measurement of the mean grain size (Folk 1968). CS = continental shelf, CM = continental margin. Mod = moderate.

Cruise/month	Region	Station	Depth (m)	Depth zone	Temp (°C)	mg/L O_2	mL/L O_2	% sand	M_z	Sediment	Pigment	% N	% C	C/N	% m.o.	Trawling
OG-12 June	Bay of Campeche	2	52	CS	19	6.0	4.2	20.3	6.1	Medium silt	Low	0.15	8.02	53.47	2.3	No
		10	52	CS	18	5.5	3.9	6.9	8.7	Clay	Low	0.18	4.13	22.94	2.75	No
		13	61	CS	18.6	6.0	4.2	1.5	8.2	Clay	Low	0.12	2.06	17.17	1.84	High
		18	103	CS	19	5.0	3.5	3.0	6.6	Fine silt	Low	0.13	2.37	18.23	1.99	High
		21	72	CS	18	5.4	3.8	1.3	6.5	Fine silt	Low	0.04	0.41	10.5	0.61	High
OG-14 June	Campeche Bank	3	50	CS	24	7.4	5.2	90.4	2.9	Fine sand	High	0.01	13.04	1,499	0.15	Minimal
		4	48	CS	24	6.8	4.8	54.0	2.9	Fine sand	High	0.02	12.69	920.9	0.31	Minimal
		5	110	CS	22	6.7	4.7	97.4	2.2	Fine sand	High	0.05	11.19	284.3	0.77	Minimal
SIG-1 June	Tamaulipas	1	170	CS	19.5	3.0	2.1	1.4	6.5	Fine silt	Mod	0.11	0.99	9	1.68	Mod
ENOS-1 March	Bay of Campeche	E3	142	CS	17	4.8	3.4	7.8	8.8	Clay	High	0.14	1.13	8.07	2.14	High
		E4	203	CM	16	4.0	2.8	4.9	8.7	Clay	High	0.15	1.34	8.93	2.29	Mod
		E9	254	CM	11	4.2	2.9	2.1	8.2	Clay	Mod	0.2	1.71	8.55	3.01	Mod
ENOS-2 June	Bay of Campeche	E3	169	CM	15	7.6	5.3	7.7	8.8	Clay	Mod	0.15	1.34	8.87	2.29	High
		E4	203	CM	15	5.0	3.5	6.0	8.7	Clay	Mod	0.16	1.38	8.63	2.45	Mod
		E9	256	CM	15	6.2	4.3	1.6	8.2	Clay	Mod	0.13	1.03	7.92	1.99	Mod

Table 3. Infaunal biomass of the continental shelf and continental margin habitats in the three regions of the southwestern Gulf of Mexico. CS = continental shelf, CM = continental margin. Lat = latitude, Lon = longitude.

Cruise/month	Region	Station	Lat (°N)	Lon (°W)	Depth (m)	Depth zone	n	Bacteria mg C/m Average	SD	Meiofauna mg C/m Average	SD	Macrofauna mg C/m Average	SD
OG-12 (Jun)	Bay of Campeche	2	19°57	91°52	52	CS	4	1,039.50	111.88	46.08	54.23	324.00	139.70
		10	19°14	92°24	52	CS	4	1,331.00	280.56	58.23	34.18	504.25	414.80
		13	18°47	93°14	61	CS	4	690.50	30.23	32.28	26.23	590.25	231.91
		18	18°39	93°57	103	CS	4	573.75	56.35	3.90	2.74	377.50	121.92
		21	18°30	94°23	72	CS	4	600.25	142.36	5.75	5.45	475.25	46.59
OG-14 (Jun)	Campeche Bank	3	21°59	90°31	50	CS	3	149.33	24.70	13.20	4.57	280.00	130.22
		4	22°29	88°33	48	CS	3	131.67	96.91	20.50	17.34	225.00	85.98
		5	23°25	87°32	110	CS	3	131.33	21.39	13.20	11.81	137.33	51.79
SIG-1 (Jun)	Tamaulipas	1	23°57	97°13	170	CS	4	475.75	95.96	101.50	31.35	80.00	39.97
ENOS-1 (Mar)	Bay of Campeche	E3	18°39	94°20	142	CM	4	1,432.75	322.61	254.25	189.51	318.25	36.26
		E4	18°56	93°23	203	CM	4	1,748.00	253.13	241.75	100.22	229.25	20.19
		E9	19°16	92°55	254	CM	4	972.25	123.89	36.50	23.06	486.50	236.23
ENOS-2 (Jun)	Bay of Campeche	E3	18°40	94°20	169	CS	4	737.50	71.57	243.33	98.31	277.25	161.33
		E4	18°56	93°23	203	CM	4	919.00	48.26	173.95	147.96	406.50	160.54
		E9	19°17	92°54	256	CM	4	777.25	103.02	82.80	33.40	300.25	131.73

from 131.33 ± 21.39 to 1,748 ± 253.13 mg C/m². In Campeche Bay, biomass declined with depth, increasing again on the margin. The largest biomass values were recorded on the continental margin (737.50 ± 71.57 to 1,748 ± 253.13 mg C/m²). Higher biomass also occurred on the shallower stations of the shelf (690.50 ± 30.23 to 1,331.00 ± 280.56 mg C/m². Bacterial biomass on the continental margin recorded seasonal differences. The Campeche Bank had bacterial mean biomass values that were ten times smaller (131.33 ± 21.39 to 149.33 ± 24.70 mg C/m²) than those observed on the Bay of Campeche shelf (Table 3). The continental shelf off Tamaulipas had intermediate bacterial biomass values (475.75 ± 95.96 mg C/m²; $N = 4$).

Meiofauna

The mean density of meiofauna-sized organisms was $2.87 \times 10^5/m^2$ ($\sigma = \pm 8 \times 10^4$, $N = 25$). The nematodes and harpacticoid copepods were the dominant components. Wet preserved weights per individual, based on volume, had means of 6.8 micrograms for the nematodes and 3.4 micrograms for the harpacticoid copepods. While it is usual for the nematodes to be numerically dominant, it is unusual for them to be bigger than the crustaceans. Meiofaunal mean biomass ranged from 3.90 ± 2.74 to 254.25 ± 189.51 mg C/m² (Table 3). The lowest biomass values were recorded on the Campeche Bank (13.2 ± 4.57 to 20.50 ± 17.34 mg C/m²), except for stations 18 and 21 in the Bay of Campeche. The largest meiofaunal biomass values (36.50 ± 23.06 to 254.25 ± 189.51 mg C/m²) occurred in the Bay of Campeche during the winter and summer on different cruises (Table 3).

Macrofauna

The mean macrofauna density was 318 individuals/m² ($F = \pm 159$, $N = 5$). Polychaetes (65%), mollusks (9%), and crustaceans (27%) were the dominant macrofaunal components. In addition, there were 490 individuals/m² ($F = \pm 147$, $N = 5$) belonging to larger harpacticoid copepods, nematodes, and ostracods. The macrofaunal biomass ranged from 80.00 ± 39.97 to 590.25 ± 231.91 mg C/m² (Table 3). The lowest values (80 ± 39.97 mg C/m²; $N = 4$) were recorded on the Tamaulipas continental shelf (Table 3). The largest values were recorded in the Bay of Campeche continental shelf and were similar to those recorded on the continental margin, displaying no seasonal variation (Table 3). The macrofaunal biomass values on Campeche Bank were half that recorded on the Bay of Campeche shelf (137.33 ± 51.79 to 280.00 ± 130.22 mg C/m² and 324.00 ± 139.70 to 590.25 ± 231.91 mg C/m², respectively).

SOC and Benthic Biomass

The highest SOC values were found over the Campeche Bank, the area with the least trawling activity (Figure 2). In contrast, the SOC values were significantly lower in the shelf off Tamaulipas and the Campeche Bay (MANOVA [Zone; catch per unit effort; SOC] $F = 21.46$; $P = 0.0002$). Significant correlation ($r = 0.71$, $F[3,29] = 9.69$, $P < 0.00014$) was found between SOC and the total infaunal biomass on the continental shelf in the study area. The meiofaunal and bacterial biomass showed a significant correlation ($r = 0.61$, $F[3,40] = 7.91$, $P < 0.00029$) with the SOC in the Bay of Campeche shelf ($P = 0.00047$) and continental margin ($P = 0.0001$). The Tamaulipas region showed a correlation of 0.66 ($N = 4$) between SOC and meiofaunal biomass.

Discussion

The oxygen consumption rate in sediment is determined by local photosynthesis plus the physical process that imports new oxygen-rich waters minus respiration of all living organisms minus chemical oxidation of exported or locally produced reduced metabolic end products (Falkowski et al. 1988). Chemical oxidation of reduced metabolic end products is also important in sediments and is mainly exemplified by the oxidation of sulfide. Biological respiration remineralizes organic carbon to metabolic CO_2. Each process is important to the concentration of oxygen in the sediment and overlying water. Chemical oxidation is the largest consumer of organics in any typical benthic community, and it can be estimated from oxygen fluxes. In general, the sum of most of the aerobic respiration by sediment-dwelling organisms (bacteria, meiofauna, and macrofauna) is equal to the value measured with the incubation chambers (Rowe et al. 1998). The range of SOC values from the continental shelves in the southwestern Gulf of Mexico are within the range of values recorded in seasonally pulsed sites from higher latitudes (e.g., the northwestern Atlantic shelf; Smith et al. 1976; Rowe et al. 1994) (Table 4) and highly enriched areas within the region (Rowe et al. 1994). However, the metabolic rates from the Tamaulipas shelf are not average compared to other sites in the same depth range (Table 4). The respiration rates generally are a function of temperature and oxygen concentration and follow the typical Q_{10} relationship in which respiration doubles for every 10°C rise in temperature. The temperature effect on the sediment–oxygen consumption rate in the southwestern Gulf of Mexico was not examined because differences in temperature (18–19°C on the continental shelf of the Bay of Campeche or 22–24°C on the Campeche Bay in Yucatan) were not enough to modify respiration rates. In contrast, the organic enrichment of the sediment had a relevant role on the benthic biomass and the SOC rates.

The organic matter content of sediments on terrigenous shelves is linked to river outflow and to pelagic–

Table 4. Comparison of environmental conditions of benthos in terrigenous and carbonated continental shelves and their sediment oxygen consumption rates (SOC). D.O. = dissolved oxygen; C_{org} = organic carbon; nd = no data; SEEP = shelf edge exchange processes of the Mid-Atlantic Bight Program.

Method	Depth (m)	Temperature (°C)	D.O. (mL/L)	SOC mL O_2 m^2/h Average	SD	%C_{org}	Study area	Reference
Shipboard core incubation	26	20.5	3.79	51.2	nd	nd	Long Island Sound, western Atlantic	Carey 1967
	35	nd	nd	32.6	nd	nd	Off Galveston Bay, Gulf of Mexico	Rowe et al. 1993
	51	nd	nd	7.8	3.7	nd	SEEP NW Atlantic	Rowe et al. 1994
	68	18.5	5.6	8.7	2.6	3.4	Bay of Campeche, Gulf of Mexico	This study
	69	23.3	7	69.9	9.5	12.3	Yucatan shelf, Gulf of Mexico (carbonate)	This study
	70	10.3	3.96	10	nd	1.8	Puget Sound, Washington, eastern Pacific	Pamatmat 1971
	170	19.5	3	12.6	nd	0.99	Tamaulipas shelf, Gulf of Mexico	This study
In situ benthic chamber	14	nd	3	61.2	2.7	3	Great Harbour, Bermuda, western Atlantic (carbonate)	Smith et al. 1973
	25	27.2	2.7	20	10.47	nd	W of Mississippi river, Gulf of Mexico	Rowe et al. 2002
	29	nd	nd	44.4	nd	nd	W Flower Garden, Gulf of Mexico (carbonate)	Rowe et al. 1993
	30	16	6.1	21	1.95	nd	Off Spanish Sahara, eastern Atlantic	Smith et al. 1976
	40	2	7.19	14.6	1.1	nd	Gay Head-Bermuda transect, western Atlantic	Smith et al. 1976
	50	26	Oxic	32	nd	Sand	Mosquitia shelf, Honduras, Caribbean Sea (carbonate)	Cruz Kaegi 1992
	50	20	3.76	11.04	4.32	1.8	Mississippi river plume, Gulf of Mexico	Rowe et al. 1998
	130	nd	nd	1.43	nd	0.05	SEEP transect, western Atlantic	Rowe et al. 1994

benthic coupling (Carey 1967). In the area of the study, the rivers could influence the organic and sediment loading of the inner shelf sediment samples in the Bay of Campeche. At the Tamaulipas margin, low sediment input to the shelf and sample sites further from the coast indicate this region is not affected by the river input. No rivers deliver sediment to the ocean at Campeche Bank. Photosynthesis on the seabed of the inner continental shelf can almost equal that in the water column when light levels are adequate (Cahoon and Cooke 1992). The dominant biological sources of oxygen at the seafloor in shallow ecosystems are photosynthesis by bottom-dwelling microalgae and phytoplankton in the water column. Some studies have assumed that photosynthesis will be nearly negligible below a river plume (D'Avanzo and Kremer 1994). The remaining sources and sinks of oxygen are products of physical mixing and circulation of water column oxygen. On the seafloor, intrinsic processes either produce or consume oxygen in response to forcing functions from both the water column and the sediment. The water column overlying the bottom in continental shelves is a competing sink for oxygen by heterotrophic bacteria (Williams 1981). Large algal concentrations have been presumed to be responsible for low oxygen when the bulk of the bloom population is located below its compensation depth (Falkowski et al. 1981). Our SOC are less than expected for highly enriched sediments off river outflows but far greater than what might be expected on the continental shelf in tropical environments. On the shelf, oxygen availability can be affected throughout the water column and in the sediment by temporary sediment resuspension caused by trawling activities (Heifetz 1997; Freese et al. 1999). If the trawling activity is sustained, it can affect nutrient input to the water column and enhance of water column photosynthesis (Lohrer and Wetz 2003). The concentration of oxygen in the ecosystem is being controlled by the above processes, but it might be expected that the steady state of SOC is altered for bottoms that are persistently disturbed. Different scenarios have been simulated mathematically in an attempt to estimate how increases in organic input could alter fluxes in the sediment (Rowe et al. 1997). The SOC of the Bay of Campeche continental margin stations resembled that from the continental shelf where bacteria may have a major role in the sediments (Boetius and Damm 1998; Falcon 1998). The values from the continental margin were also similar to those from the upwelling sites (Smith et al. 1974; Rowe 1985), providing an excellent substrate for bacteria and enhancing the presence of metazoans (Smith et al. 1972; Rowe and McNichol 1991). We found differences of an order of magnitude between this study and studies in the northern Gulf of Mexico (Tables 4, 5). Bacteria have been recognized by other authors as playing a major role in benthic oxygen consumption and, thus, in remineralization, where energy would flow through the food web from bacteria to meiofauna and macrofauna (Rowe et al. 1991). The respiration of the individual metazoan taxa is related to size and temperature (Cruz-Kaegi 1992). A knowledge of the relative partitioning of the SOC by infaunal components is required to understand how SOC is altered by geochemical conditions (Lin and Morse 1991). Other carbonate shelves (e.g., Mosquitia shelf in Honduras in the western Caribbean Sea, Cruz-Kaegi 1992; the western Flower Garden, Rowe et al. 2002; and Great Harbor in Bermuda, Smith et al. 1973) have similarly high SOC rates to Campeche Bank (Table 4). The decline in rates and biomass with distance from land and by depth manifested in the SOC values recorded in the southwestern Gulf of Mexico are similar to those cited by other authors (Smith and Hinga 1983; Rowe et al. 1991).

Marine sediments interact with overlying waters in coastal areas and link water column processes to the benthic communities (Aller 1982). Trawling activities are known to modify the seafloor in ways similar to natural biological and physical perturbations that disturb soft sediments at various intensities, frequencies and spatial scales (Hall et al. 1994). These disturbances promote high concentrations of nutrients derived from organic matter in the water column. River outflow and rare but powerful storms can have a similar effect with the input of dissolved organic matter and other dissolved compounds (Lopez-Veneroni and Cifuentes 1994; Trefry et al. 1994). The differences in SOC rates recorded off Tamaulipas, the Bay of Campeche, and the Campeche Bank in Yucatan suggest that the benthic community might be responding to regionally different trawling activities. Our results suggest that sediment and benthos are tightly coupled to processes occurring within the sediment and in the water column and that SOC responds to changes in community structure. Because dredging, spoil disposal, and release of toxic compounds are disturbances that have been shown to have an effect on sediments and their communities (Harrel and Hall 1991; Zajac and Whitlatch 2001), we suggest that the SOC rates could be considered as a potential approach to evaluate disturbance generated on the seafloor by natural events and trawling activities.

Acknowledgments

Support was provided by grants CONACyT G-27777B, 050PÑ-1297, G35442-T; SEP-CONACyT 40158 and DGAPA IN213197, IN217298; IN211200. We acknowledge the invaluable assistance of the captain, crew, and scientific participants of the cruises on board RV

Table 5. Comparison of environmental conditions of benthos from shelf and coastal organic enriched sites (upwelling, outwelling, aquaculture ponds, and bays), from continental margins and their sediment oxygen consumption (SOC) rates. ENSO = El Nino-Southern Oscillation; NEWP = Northeast Water Polynya.

Method	Depth (m)	Temp (°C)	D.O. (mL/L)	SOC (mL O$_2$ m^2/h) Average	SD	% corg	Study area	Reference
				Organic enriched sediments				
In situ benthic chamber	0.5	nd	nd	220	nd	nd	Shrimp ponds	Madenjian et al. 1987
	1	22	nd	115	nd	2.5	Potter pond, RI, western Atlantic	Nowicki and Nixon 1985
	2	21	5.1	22.84	0.7	0.7	N Bermuda, western Atlantic	Smith et al. 1972
	7	18	5.9	73.2	18	2.2	Sapelo Island, GA, western Atlantic	Smith 1973
	20	12.9	nd	22.21	nd	nd	Buzzards Bay, western Atlantic	Rowe and McNichol 1991
	20	24.5	3.3	23.4	1.6	nd	Louisiana shelf, Gulf of Mexico	Dortch et al. 1994
	20	22	2.6	41.67	4.2	nd	Louisiana shelf, Gulf of Mexico	Miller-Way et al. 1994
	30	28.5	Hypoxic	3.7	2.87	nd	W of Mississippi river outflow, Gulf of Mexico	Rowe et al. 2002
	31	12	5.3	33.2	1.8	2.8	New York Bight, western Atlantic (sludge outwelling)	Smith et al. 1974
	23	14	3.14	36.7	1.6	0.2	San Hipolito, eastern Pacific (upwelling)	Smith et al. 1974
	29	13	4.27	28.9	1.2	0.1	Punta Colnett, eastern Pacific (upwelling)	Smith et al. 1974
	50	13	3.96	39.1	1.8	0.2	Punta Asuncion, eastern Pacific (upwelling)	Rowe 1985
	150	nd	4.5	19	nd	2	JOINT NW Africa (upwelling)	Rowe 1985
	300	14	Hypoxic	54	nd	nd	Peru, upwelling, eastern Pacific (upwelling)	Rowe 1985
Shipboard core incubation	200	14.7	4.3	22.7	9.4	1.4	Campeche Bay, Gulf of Mexico ENSO 1997	This study
	210	15	6.3	36	9	1.25	Campeche Bay, Gulf of Mexico ENSO 1997	This study
				Continental margin				
Shipboard core incubation	230	0.04	3.84	1.305	0.08	0.72	Laptev Sea, Arctic	Boetius and Damm 1998
	239	9.59	4.8	8.75	nd	1.95	Puget Sound, Washington, eastern Pacific	Pamatmat 1971
	252	0.26	nd	2.13	1.1	nd	NEWP, Greenland Sea	Rowe et al. 1997
In situ benthic chamber	298	0.5	nd	1.92	nd	nd	NEWP, Greenland Sea	Rowe et al. 1997
	428	nd	nd	1.57	nd	0.05	SEEP transect, western Atlantic	Rowe et al. 1994

Justo Sierra. We also acknowledge the input and creative comments of three anonymous reviewers.

References

Aller, R. C. 1982. The effects of benthos on chemical properties of marine sediment and overlying water. Pages 53–102 *in* P. L. McCall and M. J. S. Tevesz, editors. Animal–sediment relations. The biogenic alteration of sediments. Plenum, New York.

Archer, D., and A. Devol. 1992. Benthic oxygen fluxes on the Washington shelf and slope: a comparison of in situ microelectrode and chamber flux measurements. Limnology and Oceanography 37(3):614–629.

Boetius, A., and E. Damm. 1998. Benthic oxygen uptake, hydrolytic potentials and microbial biomass at the Arctic continental slope. Deep Sea Research 45:239–275.

Bratbak, G., and I. Dundas. 1984. Bacterial dry matter content and biomass estimations. Applied Environmental Microbiology 48 (4), 755–757.

Cahoon, L., and J. Cooke. 1992. Benthic microalgal production in Onslow Bay, North Carolina, USA. Marine Ecology Progress Series 84:185–196.

Carey, A. G. 1967. Energetics of the benthos of Long Island Sound. I. Oxygen utilization of sediment. Bulletin of the Bingham Oceanographic Collection 19(2):136–144.

Cruz-Kaegi, M. E. 1992. Microbial abundance and biomass in the sediments of the Texas-Louisiana shelf. Master's thesis. Texas A&M University, College Station.

D'Avanzo, C., and J. Kremer. 1994. Diel oxygen dynamics and anoxic events in an eutrophic estuary of Waquoit Bay, Massachusetts. Estuaries 17:131–139.

Dortch, Q., N. N. Rabalais, R. E. Turner, and G. T. Rowe. 1994. Respiration rates and hypoxia on the Louisiana shelf. Estuaries 17(4):862–872.

Eadie, B., B. McKee, M. Lansing, J. Robbins, S. Metz, and J. Trefry. 1994. Records of nutrient-enhanced coastal ocean productivity in sediments from the Louisiana continental shelf. Estuaries 17:754–765.

Falcon, L. I. 1998. Consumo de oxigeno y biomasa de la infauna del ambiente de plataforma continental del suroeste del Golfo de Mexico y Peninsula de Yucatan. Bachelor's thesis. Universidad Nacional Autonoma de Mexico, Mexico City, Mexico.

Falkowski, P., C. Flagg, G. Rowe, S. Smith, T. Whitledge, and C. Wirick. 1988. The fate of a spring phytoplankton bloom: export or oxidation? Continental Shelf Research 8:457–484.

Falkowski, P., T. Hopkins, and J. Walsh. 1981. An analysis of factors affecting oxygen depletion in the New York Bight. Journal of Marine Research 38:479–506.

Folk, R. L. 1968. Petrology of sedimentary rocks. Hemphills, Austin, Texas.

Freese, L., P. J. Auster, J. Heifetz, and B. L. Wing. 1999. Effects of trawling on seafloor habitat and associated invertebrate taxa in the Gulf of Alaska. Marine Ecology Progress Series 182:119–126.

Fucik, K., and S. El-Sayed. 1979. Effect of oil production and drilling operations on the ecology of phytoplankton in the OEI study area. Pages 325–353 in C. H. Ward, M. E. Bender, and R. J. Reish, editors. The offshore ecology investigation, effects of oil drilling and production in a coastal environment. Rice University, Rice University Studies 65, Houston, Texas.

Hall, S. J., D. Raffaelli, and S. F. Thrush. 1994. Patchiness and disturbance in shallow water benthic assemblages. Pages 333–375 *in* P. S. Giller, A. G. Hildrew, and D. G. Raffaelli, editors. Aquatic ecology: scale patterns and process. The 34th symposium of the British Ecological Society with the American Society of Limnology and Oceanography. Blackwell Scientific Publications, London.

Harrel, R. C., and M. A. Hall. 1991. Macrobenthic community structure before and after pollution abatement in the Neches River Estuary (Texas). Hydrobiologia 221(3):241–252.

Heifetz, J., editor. 1997. Workshop on the potential effects of fishing gear on benthic habitat. National Marine Fisheries Service, Alaska Fisheries Science Center Processed Report 97-04, Seattle.

Kemp, P., P. Falkowski, C. Flagg, W. Phoel, S. Smith, D. Wallace, and C. Wirick. 1994. Modeling vertical oxygen and carbon flux during stratified spring and summer conditions on the continental shelf, Middle Atlantic Bight, eastern U.S.A. Deep-Sea Research II 41:629–655.

Lin, S., and J. W. Morse. 1991. Sulfate reduction and iron sulfide mineral formation in Gulf of Mexico anoxic sediments. American Journal of Science 291:55–89.

Lohrer, A. M., and J. J. Wetz. 2003. Dredging-induced nutrient release from sediments to the water column in a southeastern saltmarsh tidal creek. Marine Pollution Bulletin 46:1156–1163.

Lopez-Veneroni, D., and L. Cifuentes. 1994. Transport of dissolved organic nitrogen in Mississippi River plume and Texas–Louisiana continental shelf near-surface waters. Estuaries 17:796–808.

Madenjian, C. P., G. L. Rogers, and A. W. Fast. 1987. Predicting night time DO loss in prawn ponds of Hawaii: Part 1. Evaluation of traditional methods. Aquacultural Engineering 6:191–208.

Miller-Way, T., G. S. Boland, G. T. Rowe, and R. R. Twilley. 1994. Sediment oxygen consumption and benthic nutrient fluxes on the Louisiana Continental Shelf: a methodological comparison. Estuaries 17(4):89–815.

Norwicki, B. L., and S. W. Nixon. 1985. Benthic community metabolism in a coastal lagoon ecosystem. Marine Ecology Progress Series 22:21–30.

Officer, C., R. Biggs, J. Taft, L. Eugene Cronin, M. Tyler, and W. Boynton. 1984. Chesapeake Bay anoxia: origin, development, and significance. Science 223:22–27.

Pamatmat, M. M. 1971. Oxygen consumption by the seabed. VI. Seasonal cycle of chemical oxydation and respiration in Puget Sound. Limnology and Oceanography 16:536–550.

Pamatmat, M. M. 1978. Oxygen uptake and heat production in a metabolic conformer (*Littorina irrorate*) and a metabolic regulator (*Uca pugnax*). Marine Biology 48:317–325.

Pomeroy, L. R. 1959. Algal productivity in salt marshes of Georgia. Limnology and Oceanography 4:386–397.

Rabalais, N., W. Wiseman, Jr., and R. E. Turner. 1994. Comparison of continuous records of near-bottom dissolved oxygen from the hypoxia zone along the Louisiana coast. Estuaries 17:850–861.

Redalje, D., S. Lohrenz, and G. Fahnenstiel. 1994. The relationship between primary production and the vertical export of particulate organic matter in a river-impacted coastal ecosystem. Estuaries 17:829–838.

Renaud, M. 1986. Hypoxia in Louisiana coastal waters during 1983: implications for fisheries. U.S. National Marine Fisheries Service Fishery Bulletin 84:19–26.

Riley, G. A. 1947. A theoretical analysis of the zooplankton population of Georges Bank. Journal of Marine Research 6:104–113.

Rowe, G. T. 1971. Benthic biomass and surface productivity. Pages 441–454 in J. D. Costlow, editor. Fertility of the sea. Volume 2. Gordon and Breach Science, New York.

Rowe, G. T. 1983. Biomass and production of the deep-sea macrobenthos. In G. T. Rowe, editor. Deep-sea biology, the sea, volume 8. Wiley, New York.

Rowe, G. T. 1985. Benthic production and processes off Baja California, Northwest Africa and Peru: a classification of benthic subsystems in upwelling ecosystems. Simposio internacional del Afloramiento occidental de Africa, Instituto de Investigaciones Pesqueras de Barcelona 2:589–612.

Rowe, G. T., G. S. Boland, E. Escobar, M. L. Cruz-Kaegi, A. Newton, I. Walsh, and J. Deming. 1997. Sediment community biomass and respiration in the Northeast Water Polynya, Greenland: a numerical simulation of benthic lander and spade core data. Journal of Marine Systems 10:497–515.

Rowe, G. T., G. S. Boland, W. C. Phoel, R. F. Anderson, and P. E. Biscaye. 1994. Deep-sea floor respiration as an indication of lateral input of biogenic detritus from continental margins. Deep-Sea Research. 41(2/3):657–668.

Rowe, G. T., M. E. Cruz-Kaegi, J. W. Morse, G. S. Boland, and E. G. Escobar-Briones. 2002. Sediment community metabolism associated with continental shelf hypoxia, northern Gulf of Mexico. Estuaries 25(6):1097–1106.

Rowe, G. T., and A. P. McNichol. 1991. Carbon cycling in coastal sediments: estimating remineralization in Buzzards Bay, Massachusetts. Geochimica et Cosmochimica Acta 55:2989–2991.

Rowe, G. T., J. Morse, G. S. Boland, and M. L. Cruz-Kaegi. 1998. Sediment metabolism and heterotrophic biomass associated with the Mississippi river plume. Pages 102–105 in Proceedings of the synthesis workshop on nutrient enhanced coastal ocean productivity. National Oceanic and Atmospheric Administration, Silver Spring, Maryland.

Rowe, G. T., M. Sibuet, J. Deming, A. Khripounoff, J. Tietjen, S. Macko, and R. Theroux. 1991. "Total" sediment biomass and preliminary estimates of organic carbon residence time in the deep-sea benthos. Marine Ecology Progress Series 79:99–114.

Smith, K. L. 1973. Respiration of a sublittoral community. Ecology 54(5):1065–1075.

Smith, K. L., K. A. Burns, and J. M. Teal. 1972. In situ respiration of benthic communities in Castle Harbor, Bermuda. Marine Biology 12(3):196–199.

Smith, K. L., C. H. Clifford, A. E. Eliason, B. Walden, G. T. Rowe, and J. M. Teal. 1976. A free vehicle for measuring benthic community metabolism. Limnology and Oceanography 21(1):164–170.

Smith K. L., Jr., and K. R. Hinga. 1983. Sediment community respiration in the deep sea. Pages 331–370 in G. T. Rowe, editor. Deep-sea biology, the sea, volume 8. The sea: ideas and observations on progress in the study of the seas. John Wiley & Sons, New York.

Smith, K. L., G. T. Rowe, and C. H. Clifford. 1974. Sediment oxygen demand in an outwelling and upwelling area. Tethys 6(1–2):223–230.

Smith, K. L., G. T. Rowe, and J. A. Nichols. 1973. Benthic community respiration near the Woods Hole sewage outfall. Estuarine and Coastal Marine Science 1:65–70.

Soto, L. A., and E. Escobar-Briones. 1995. Coupling mechanisms related to benthic production in the SW Gulf of Mexico. Pages 233–242 in A. Eleftheriou, A. D. Ansell, and C. J. Smith, editors. Biology and ecology of shallow coastal waters. Proceedings of the 28th European Marine Biology Symposium, Crete. Olsen & Olsen, Fredensborg, Denmark.

Trefry, J. S., Metz, T. Nelsen, R. Trocine, and B. Eadie 1994. Transport of particulate organic carbon by the Mississippi River and its fate in the Gulf of Mexico. Estuaries 17:839–849.

Van der Have, T., and G. de Jong. 1996. Adult size in ectotherms: temperature effects on growth and differentiation. Journal of Theoretical Biology 183:329–340.

Williams, P. J. Le B. 1981. Microbial contribution to overall marine plankton metabolism: direct measurements of respiration. Oceanologica Acta 4:359–364.

Zajac, R. N., and R. B. Whitlatch. 2001. Response of macrobenthic communities to restoration efforts in a New England estuary. Estuaries 24(2):167–183.

Discovery of 100–160-Year-Old Iceberg Gouges and Their Relation to Halibut Habitat in Glacier Bay, Alaska

PAUL R. CARLSON[1]

U.S. Geological Survey, 345 Middlefield Road, Menlo Park, California 94025, USA

PHILIP N. HOOGE[2]

U.S. Geological Survey, Glacier Bay Field Station, Glacier Bay, Alaska 99826, USA

GUY R. COCHRANE[3]

U.S. Geological Survey, 345 Middlefield Road, Menlo Park, California 94025, USA

Abstract.—Side-scan sonar and multibeam imagery of Glacier Bay, Alaska, revealed complex iceberg gouge patterns at water depths to 135 m on the floor of Whidbey Passage and south to the bay entrance. These previously undiscovered gouges likely formed more than 100 years ago as the glacier retreated rapidly up Glacier Bay. Gouged areas free of fine sediment supported greater biodiversity of Pacific halibut *Hippoglossus stenolepis* than nearby sediment-filled gouges, probably due to increased habitat complexity. Small Pacific halibut were found more frequently in sediment-free gouged areas, presumably due to higher prey abundance. In contrast, large Pacific halibut were found more frequently on soft substrates such as sediment-filled gouges, where they could bury themselves and ambush prey.

Introduction

Glacier Bay, in southeastern Alaska (Figure 1), was formed by multiple glacial advances and retreats throughout much of the Pleistocene epoch (Goldthwait 1987). In 1794, members of Captain George Vancouver's crew reported the presence of a massive wall of ice blocking what is now the entrance to Glacier Bay (Vancouver 1798). Since then, the glacier has retreated about 100 km up the bay, exposing a magnificent fjord system (Figure 1). As the ice front retreated, it left remnants of end moraines which were dated at 1845, 1857, and 1860 by tree-ring cores (Figure 1; Cooper 1937; Lawrence 1958). The 1845 and 1857 tree-ring-dated moraines provided dates of the ice terminus position nearest to the study area for Whidbey Passage Pacific halibut *Hippoglossus stenolepis* (Figure 1). Since 1879, when John Muir first visited Glacier Bay, the ice front positions have been systematically and accurately mapped (Figure 1), first by boat by numerous scientists including Muir (1895) and Field (1964), by aerial photography (Post and LaChapelle 1971), and eventually by satellite imagery (Hall et al. 1995).

Following the ice front retreat, ecological successions of plants, soil, and terrestrial animals have been observed in this spectacular natural laboratory (Cooper 1923; Lawrence 1951; Dinneford 1990). In the past two decades, biologists have turned their attention to the marine realm (Sharman 1990; Bishop et al. 1995) and recently have joined forces with marine geologists to study the biological and physical characteristics of bayfloor habitats in Glacier Bay (Carlson et al. 1998b, 2002; Cochrane et al. 1998; Hooge and Carlson 2001). This paper reports the discovery of some large, complex gouges in a deepwater habitat of Pacific halibut within Whidbey Passage, located in the western-central part of the lower bay and even longer gouges in shallower water depths 20 km south of Whidbey Passage in the southernmost part of Glacier Bay (Figure 1). We discuss the probable age of the gouges, their physical characteristics, how they were formed, and how they have been modified, and we make some preliminary associations

---

[1] Corresponding author: pcarlson@usgs.gov
[2] Present address: Denali National Park, Post Office Box 9, Denali, Alaska 99755, USA; e-mail: philip_hooge@nps.gov
[3] E-mail: gcochrane@usgs.gov

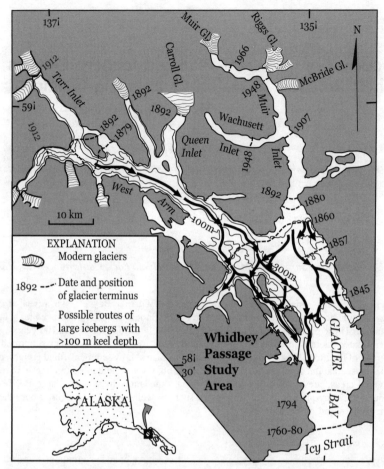

Figure 1. Location map of Alaska and Glacier Bay National Park showing terminus positions and dates during retreat of glacier. Polygon outlines Whidbey Passage study area. Bathymetry measured in meters. Arrows show possible routes of travel of large icebergs with deep keel depths (>100 m) that likely excavated complex and numerous gouge patterns in Whidbey Passage.

of size and age of Pacific halibut occurrences to the variations in benthic substrate.

Effects of Glacial Retreat on the Seafloor

The glacier that filled Glacier Bay began its retreat from the mouth of the bay about 200 years ago (Goldthwait 1963). The massive glacier retreated past the Whidbey Passage study area by about 160 years ago and reached the upper end of the main bay by 1860 (140 years ago), where the bay-filling glacier bifurcated (Figure 1). As the glacier retreated from Whidbey Passage to the head of the lower bay (~1845–1860), calving from the terminus of the massive glacier likely generated huge bergs. Some of the bergs, as they were channeled down Whidbey Passage, had deep enough keel depths to impact the bay floor and form gouges (Figure 2). Subsequently, the West Arm glacier retreated rapidly up the fjord (~2 km/year) until 1879, whereas, in Muir Arm, the glacier was pinned on a shallow entrance moraine from sometime after 1860 until at least 1892 (Seramur et al. 1997) and then began its rapid retreat (Figure 1). Massive icebergs from both West Arm and Muir Inlet may have contributed to the gouging, but the West Arm bergs had the most direct and deeper-water route (up to 400-m depth) into Whidbey Passage (Figure 1). In contrast, the deepest keeled iceberg to come from the Muir terminus soon after 1860 appears to be limited due to the 60-m depth of the moraine at the mouth of the inlet. Additional evidence providing support for abundant ice transiting from West Arm into the main bay

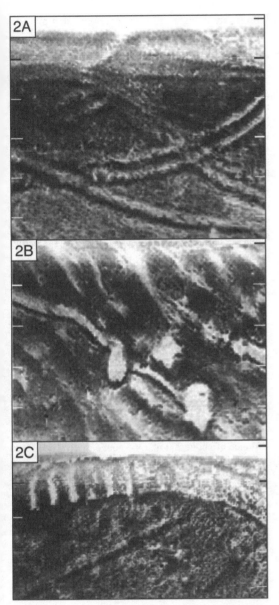

Figure 2. Examples of side-scan sonar images showing variety of iceberg gouges in study area. Scale lines are 25 m apart. Panel (2A) is a portside view of a side-scan sonar image of bottom sediment of Whidbey Passage showing the complex nature of iceberg gouges. Dark shading indicates high backscatter indicative of hard bottom consisting of coarse gravel to boulder-sized sediment. Light shading indicates low backscatter indicative of fine sediment (silt and clay). Note how soft sediment is beginning to obliterate outline of gouges near top of image. Panel (2B) is a side-scan image of iceberg wallow marks, and panel (2C) shows the chatter marks evident on a side-scan image in the northern part of Whidbey Passage.

was reported by Ovenshine (1967). He found many glacial erratics that had mineralogy typical of the West Arm geology (such as staurolite, chiastolite, and biotite-quartz schists) on the beaches of the lower part of the bay.

Water depths of 100 m throughout much of Whidbey Passage and as much as 50 m in the lower bay requires large icebergs, in order for their keels to plow the bottom. Considering that about 85% of a berg's mass is submerged, the total thickness of the berg must be about 120 m in order to scour the bottom in most of Whidbey Passage (Figure 1). Glacial scour, visible as glacial polish, and striations high on the rock walls of adjacent Willoughby Island (elevation 494 m above sea level; USGS 1990) plus 100 m to the floor of the passage, suggests ice thickness of near 600 m; thus, the calving of icebergs less than one-quarter of that thickness is quite reasonable to assume. Iceberg lowing or scouring has been reported from considerably deeper water in other areas in the world. Two examples are the northern Barents Sea, north of Norway and Russia, where Solheim et al. (1988) have imaged intense iceberg flow marks in water depths of 210–220 m, and Scoresby Sund, Greenland, where Dowdeswell et al. (1993) have collected acoustic records of iceberg scours most prevalent at depths of 300–400 m.

Field Methods and Observations

Pacific halibut have been studied in Glacier Bay for several years (Carlson et al. 1998a). In this Whidbey Passage study, Pacific halibut were caught and measured. More than 1,500 have been marked with wire tags. An additional 97 Pacific halibut had 3.5-kHz transmitters surgically implanted. These fish were tracked using a bow-mounted, dual hydrophone that was capable of tracking the fish at distances up to 2 km and at depths to 500 m. Searches for these sonically tagged fish were conducted every 2–3 months for 4 years to assess location and movement of the fish.

In 1998, we used a Klein towed side-scan sonar system (SSS) and an attached 3.5-kHz acoustic profiler to map habitats in Whidbey Passage for comparison to locations of Pacific halibut caught by longline in 1996–1997 (Carlson et al. 1998a). Navigation was by differential global positioning system (DGPS) that provides vessel location to an accuracy of about 1–5 m. Images revealed some spectacular gouges (Figure 2) on the 100-m-deep floor of Whidbey Passage, a U-shaped, bedrock-walled, 2.5-km-wide by 15-km-long valley (Figure 1). Some of the SSS images consist primarily of high backscatter (HBS). The HBS indicates a hard surface where little fine sediment filled the gouges and adjacent area. In some places, the SSS image consists mainly of low backscatter (LBS), thus, a softer surface with some faint gouge outlines that suggested the gouges were nearly filled with fine sediment.

Two video camera drift transects were occupied in the Whidbey Passage area at the end of the 1998 SSS cruise. One camera site was located where little fine sediment filled the gouges and the adjacent area. At the second site, the gouges were nearly filled in with fine sediment. We chose the camera sites based on the variation in backscatter on the SSS images. In the area of little fine sediment, there was HBS. At this video station, we observed numerous cobbles and boulders of varying sizes. As our boat drifted, we visually observed a seafloor relief of a couple meters, usually the result of large amounts of cobbles and boulders, in the form of a ridge, likely one wall of a gouge. At the second site, with LBS, the video captured imagery of billowing clouds of fine sediment being stirred up when the video sled contacted the passage floor. There were also noticeably fewer boulders and cobbles, probably because many had been covered by a blanket of fine sediment. Many of the boulders, particularly in the HBS area, were very large (up to several meters in diameter). In the area of HBS, many of the boulders and cobbles had sessile organisms, such as basket stars *Gorgonocephalus* sp., attached.

On subsequent cruises, in 1999 and 2000, we ground-truthed some of the SSS images using scuba dive transects. Scuba lines transected areas with and without surface expressions of ice gouging, which we will refer to as gouged and ungouged areas. Scuba dive transects were conducted at water depths between 25 and 60 m. On four dive transects across areas imaged on SSS as having little fine sediment (i.e., HBS), one of us (Hooge) observed parallel ridge and trough features with numerous pebbles, cobbles, and boulders and an estimated relief from trough to ridge of 1–3 m. These features were interpreted to be gouges. The central portions of the gouges were covered by sediment, and the troughs of two of the gouges were excavated to greater depth than the surrounding seafloor. Nearby gouge-filled areas were dominated by fine sediments with little or no pebbles or cobbles and only occasional boulders.

In June 2001, a hull-mounted RESON SeaBat 8111 multibeam echo sounding system (MB) was used to collect imagery throughout the main bay to supplement the side-scan coverage of benthic habitats and to determine the broader distribution of gouge features. On this cruise, navigation was also by DGPS. The MB imagery revealed additional seabed features, including bedrock knobs and even longer gouges up to 5 km in length, near the bay entrance. The preservation of these presumably older gouges in the lower part of the bay was even more startling in this shallower water region, previously thought to be an area dominated by sediment deposition.

Morphologic Features and Likely Modes of Formation

Iceberg gouges imaged by SSS and MB systems in Whidbey Passage and the lower part of the bay are quite variable in linear appearance, ranging from single and straight to crisscrossing to sinuous to simple curves and, in some cases, to double gouges (Figure 2a). The gouges most likely were created by deep-draft keels of large icebergs being transported through the bay waters by the tidal currents and perhaps slightly affected by wind acting upon the relatively small part of the iceberg projecting above the water. In several places, we discovered impact pits or wallows about 20 m in diameter, sometimes as a single feature, and once, as many as three pits along a single gouge (Figure 2b). These features form where the berg temporarily comes to rest on the bottom and then lifts off, perhaps due to a flood tide that causes the berg to rise. Similar features were caused by large pieces of sea ice coming to rest in nearshore waters of the Beaufort Sea (Reimnitz and Kempema 1982). Along one gouge track (~20 m wide) in Whidbey Passage, we observed chatter marks (Figure 2c). Apparently the keel was very close to the bay floor and in some rhythmic way bumped along, touching the bottom in a fairly regular manner over a distance of about 500 m. One gouge, several km long, was imaged by MB 20 km south of Whidbey Passage (Carlson et al. 2002). It had a pronounced zigzag pattern probably caused by several reversals of the tide during the time the berg was in intermittent contact with the bay-floor sediment.

The gouges ranged in width from 5 to 20 m and had an estimated relief of 1–2 m. The longest ice gouges that we have imaged on our side-scan sonar records were about 1 km long. However, in the southernmost part of the bay, several gouges imaged by multibeam were several km long (Figure 3), and one gouge measured 5 km long. For comparison, Syvitski et al. (1983) observed iceberg scour marks from a submersible in the Canadian Arctic that varied in width from 10 to 30 m and relief from 0.5 to 6 m.

Various seabed features, from large to small, are present in the passage. Gouges and attendant ridges consisting of boulders (up to 3 m in diameter) to sand-size material built up on sides of gouges (also called berms) are often present. In addition, grounded or drifting bergs in shallower water overturn and dump sediment on the bay floor, sometimes creating mounds of boulders, gravel, and finer sediment. Small boulders to cobbles (often with attached sessile organisms such as sea pens

Ptilosarcus gurneyi and basket stars), small gouges, and sand waves often were observed. The smallest features include pebbles, shells, small pits, and mounds, as well as burrow openings, mud volcanoes, piles of fecal debris, ripple marks, fecal coils, protrusions of infauna such as polychaete worm tubes, siphon expulsion holes, and trails from organisms such as green sea urchins *Strongylocentrotus droebachiensis*, Oregon Triton snails *Fusitriton oregonensis*, hermit crabs (family Paguridae), and Tanner crabs *Chionoecetes bairdi*.

Overlying these bottom features is sediment deposited from suspension in the water column. Suspended particulate matter, including inorganic particles of silt and clay, and organic matter, produced by diatoms and other microscopic plant and animal matter, is constantly raining through the water column in various concentrations. Fine-grained sediment sources include freshwater streams and glacial melt water issuing from glaciers and the surrounding shores and the fine sediment released by melting of the icebergs. Muddy sediment that issues from the glacier terminus as suspended sediment can be carried far down bay before it settles out. However, much of the settling occurs near the active glacial terminus where the concentration of suspended sediment can exceed 500 mg/L (Cowan and Powell 1990). In Whidbey Passage, some of the gouges are comparatively free of the very fine sediment, whereas others have been partially filled in by it. In other places, the suspended sediment has nearly to completely covered the gouges to the extent that only a faint outline of the gouge remains. In the lowermost bay (Figure 3), the ice gouges appear to be relatively free of fine sediment. This is likely due to the strong flushing action of the currents that attain speeds of up to 14.6 km/h (8 knots) through the narrows located about 12 km south of Whidbey Passage (Hooge et al. 2001).

Based on the above, seabed features of Whidbey Passage can be characterized by four different substrates based on the SSS imagery (Figure 4): (1) bedrock (high backscatter, irregular but unpatterned); (2) gouges nearly free of fine sediment (linear gouges that have mostly high backscatter; it is not likely that any gouge areas are completely free of fine-grained sediment deposited from the overflow plume that issued from the glacier terminus); (3) gouges partly filled with fine sediment (a mix of high and low backscatter indicating that the suspended sediment has been deposited in sufficient quantities to partially fill in the gouge areas); and (4) areas of low backscatter (the gouge outlines are nearly to completely obliterated by the blanket of fine suspended sediment).

Effects of Ice Gouging on Pacific Halibut Community

The Pacific halibut catch locations were superposed on an SSS-derived substrate map (Carlson et al., Geo-

Figure 3. Multibeam image of lower Glacier Bay. Extensive iceberg gouges from just above Icy Strait to Willoughby Island (W.I.) through Whidbey Passage (W.P.) are visible beyond the narrows.

logical Society of America abstract, 1999). The effects of ice gouging on the benthic community were examined by both direct observations of the number of sessile species and by the distribution of Pacific halibut. The number of species observed in gouged areas by the drop camera and on scuba transects was significantly higher than in nearby gouge-filled or ungouged areas.

Four scuba transects ($N = 4$) were combined with two video transects from the drop camera ($N = 2$). Presence and absence of all identifiable sessile fauna were recorded (Wilcoxon matched pairs signed rank test, $N = 6$, $Z = -2.201$, $P = 0.027$; Hooge, unpublished data). Differences in species numbers between the substrate types were large; a total of 55 species from 9 phyla

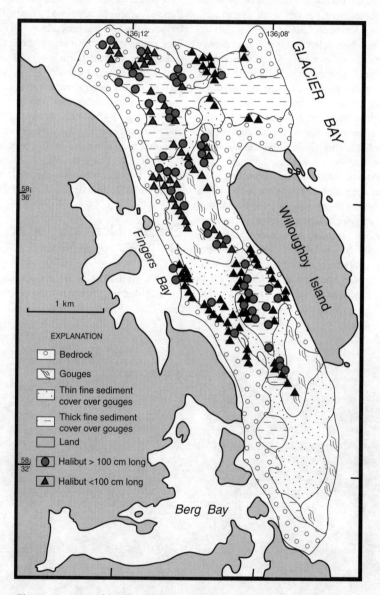

Figure 4. Map of bayfloor habitats based on SSS imagery in Whidbey Passage area and catch locations of large (>100 cm long) and small (<100 cm long) halibut. Types of bay floor habitat in Whidbey Passage are: bedrock; ice gouges essentially free of fine sediment cover; gouges partially filled with fine sediment; and gouges barely perceptible to completely covered by fine suspended sediment (clayey silt) deposited from meltwater runoff plumes.

were present in gouged areas, while 24 species from 4 phyla were found in gouge-filled areas. Gouged areas displayed a mix of species, including all 24 of those from the soft-bottomed areas, as well as additional species associated with harder substrates. The species composition observed in gouged areas was similar to that of other areas in Glacier Bay with a mix of both hard and soft substrates and similar vertical structure from rocks and boulders.

Pacific halibut locations were correlated with the four categories of physical characteristics of the floor of Whidbey Passage derived from the SSS imagery (Figure 4). Of 304 Pacific halibut captured on research longlines in Whidbey Passage, there was a highly significant tendency for smaller halibut (<100 cm fork length) to be caught both on bedrock and on substrate with detectable gouges. In contrast, large Pacific halibut were found more frequently on soft substrates. Small Pacific halibut (>30 cm and <100 cm fork length) were found more frequently on bedrock and exposed gouges (categories 1 and 2) than on soft-bottomed habitats (categories 3 and 4; Fisher's exact test, $P < 0.02$). Removing the high association between small Pacific halibut and bedrock habitats (category 1), there was still a significant tendency for small Pacific halibut (<100 cm fork length) to be captured on exposed gouge habitats (category 2) as compared to soft-bottomed habitats (categories 3 and 4; Fisher's exact test, $P < 0.05$). After adjusting the expected Pacific halibut numbers for the proportions of habitat types found within the area fished, there was a highly significant difference between expected and actual habitat use (chi-square = 14.32, df = 3, and $P < 0.003$). Areas with bedrock and unfilled gouges (categories 1 and 2) were selected more frequently than expected by small Pacific halibut, and soft-sediment areas (category 3 and 4) were selected more frequently than expected by large Pacific halibut (Figure 4). These trends correspond to ontogenetic diet differences that we have observed in Glacier Bay, where small Pacific halibut appear to forage by active predation and large Pacific halibut by sit-and-wait tactics (Chilton et al. 1995; Carlson et al. 1998a). We hypothesize that active foraging should be more productive in rocky habitats, where preferred or more abundant prey may be available due to both the increased sessile species diversity and to the enhanced physical structure of the environment. Likewise, ambush foraging should be more successful in soft-bottomed habitats where the larger Pacific halibut could bury themselves. Rocky iceberg-gouged zones, therefore, represent unrecognized productive benthic habitat.

These results demonstrate that extensive gouging observed in the seafloor of central and lower Glacier Bay is most likely a product of historical ice scour from large bergs calved during the catastrophic retreat of the glacier. These gouges, with little or no soft-sediment fill, are associated with significant differences in benthic habitat and community structure compared with sediment-filled gouges or areas lacking evidence of gouging. Whereas ice scour has detrimental effects on community structure on short time scales (Conlan et al.1998), over a longer time period, it may increase species diversity by providing a variety of interspersed habitat types.

Conclusions

(1) Ice gouges are plentiful on the floor of much of the lower portion of Glacier Bay, as observed first by side-scan sonar collected in Whidbey Passage and then by multibeam imagery of the lower bay.

(2) Gouges observed in Whidbey Passage require large icebergs with keel depths more than 100 m. These icebergs probably traveled through the area shortly after the glacier retreated (between 1845 and 1860) when the lower bay was being deglaciated and until about 1879 when the West Arm glacier bifurcated and began retreating into Johns Hopkins and Tarr inlets. West Arm was a major contributor of large icebergs, because the deeper waters of this arm as compared to Muir Inlet allowed deeper draft bergs to enter Whidbey Passage.

(3) Four types of seafloor geologic habitats were identified—(1) bedrock, (2) gouges with sparse fine–sediment cover; (3) gouges partly filled with fine sediment; and (4) gouges nearly to totally covered by the fine glacial flour (clayey silt).

(4) Pacific halibut caught in the study area were divided into two size-groups. Large halibut, more than 100 cm in length, preferred an unstructured seafloor of soft, fine sediment, where they likely burrowed into the substrate to wait for prey. Small Pacific halibut, less than 100 cm in length, that are much more active pursuing their prey, preferred the harder substrate of bedrock and coarse sediment prevalent in the unfilled ice-gouge complexes.

Acknowledgments

We thank Elizabeth Hooge, Erk Reimnitz, and three unidentified AFS referees for their insightful reviews.

References

Bishop, G. H., P. N. Hooge, and S. J. Taggart. 1995. Habitat correlates of Pacific halibut and other groundfish species in Glacier Bay National Park. Pages 215–220 *in* D. R. Engstom, editor. Proceedings of the third Glacier Bay science symposium, 1993. U.S. National Park Service, Anchorage, Alaska.

Carlson, P. R., T. R. Bruns, G. R. Cochrane, K. R. Evans, P. N. Hooge, and S. J. Taggart. 1998a. Marine geology of benthic biohabitats in Glacier Bay, Alaska: 10th western groundfish conference. National Marine Fisheries Service, Pacific Grove, California.

Carlson, P. R., T. R. Bruns, K. R. Evans, J. T. Gann, D. J., Hogg, S. J. Taggart, and P. N. Hooge. 1998b. Cruise report of M/V *QUILLBACK* in Glacier Bay, Alaska: physical characteristics of Dungeness crab and halibut habitats. U.S. Geological Survey Open-File Report 98-134.

Carlson, P. R., P. N. Hooge, G. R. Cochrane, A. J. Stevenson, P. Dartnell, and K. Lee. 2002. Multibeam bathymetry and selected perspective views of main part of Glacier Bay, Alaska. U.S. Geological Survey Open-File Report 02-391. Available: *http://geopubs.wr.usgs.gov/open-file/of02-391* (November 2003).

Chilton, L., P. N. Hooge, and S. J. Taggart. 1995. Prey preference of Pacific halibut (*Hippoglossus stenolepsis*) in Glacier Bay National Park. Pages 209–214 *in* D. R. Engstrom, editors Proceedings of the third Glacier Bay science symposium, 1993. U.S. National Park Service, Anchorage, Alaska.

Cochrane, G. R., P. R. Carlson, J. F. Denny, M. E. Boyle, S. J. Taggart, and P. N. Hooge. 1998. Cruise report M/V Quilback cruise Q-1-97-GB. Physical characteristics of Dungeness crab and halibut habitats in Glacier Bay, Alaska. USGS Open-File Report 98-791. Available: *http://geopubs.wr.usgs.gov/open-file/of98-791/ofr98-791.html* (November 2003).

Conlan, K. E., H. S. Lenihan, R. G. Kvitek, and J. S. Oliver. 1998. Ice scour disturbance to benthic communities in the Canadian High Arctic. Marine Ecology Progress Series 166: 1–16.

Cooper, W. S. 1923. The recent ecological history of Glacier Bay, Alaska. Ecology 4:223–246.

Cooper, W. S. 1937. The problem of Glacier Bay, Alaska: a study of glacier variations. The Geographical Review 27(1):37.

Cowan, E. A., and R. D. Powell. 1990. Suspended sediment transport and deposition of cyclically laminated sediment in a temperate glacial fjord. Pages 75–89 *in* J. A. Dowdeswell and J. D. Scourse, editors. Glacimarine environments: processes and sediments. The Geological Society, Special Publication 53, London.

Dinneford, B. 1990. Moose colonization of post-glacial sites in southeastern Alaska. Pages 83–85 *in* A. M. Milner and J. D. Wood, Jr., editors. Proceedings of the second Glacier Bay science symposium. U.S. National Park Service, Anchorage, Alaska.

Dowdeswell, J. A., H. Villinger, R. J. Wittington, and P. Marienfeld. 1993. Iceberg scouring in Scoresby Sund and on the East Greenland continental shelf. Marine Geology 111:37–53.

Field, W. O. 1964. Observations of glacier variations in Glacier Bay, southeastern Alaska, 1958 and 1961. American Geographical Society, New York.

Goldthwait, R. P. 1963. Dating the little ice age in Glacier Bay, Alaska. Pages 37–46 *in* International geological congress, 21st session Norden, part 27. International Geological Congress, Copenhagen.

Goldthwait, R. P. 1987. The glacial history of Glacier Bay park area. Pages 5–16 *in* P. J. Anderson, R. P. Goldthwait, and G. D. McKenzie, editors. Guidebook for INQUA Commission II field conference, June 1986. Byrd Polar Research Center, Miscelaneous Publication 236, Columbus, Ohio.

Grove, J. M. 1988. The little ice age. Cambridge University Press, Cambridge, UK.

Hall, D. K., C. S. Benson, and W. O. Field. 1995. Changes of glaciers in Glacier Bay, Alaska, using ground and satellite measurements. Physical Geography 16(1):27–41.

Hooge, P. N., and P. R. Carlson. 2001. Benthic mapping at Glacier Bay, Alaska: integrating physical structure and biohabitats. Available: *www.absc.usgs.gov/glba/benthic_mapping.htm* (November 2003).

Hooge, P. N., P. R. Carlson, and A. J. Stevenson. 2001. Mapping benthic habitat using geological and oceanographic tools—Glacier Bay, Alaska. U.S. Geological Survey, Information sheet, October 2001, Menlo Park. California.

Lawrence, D. B. 1951. Recent glacier history of Glacier Bay, Alaska, and development of vegetation on deglaciated terrain with special reference to the importance of alder in the succession. Yearbook of the American Philosophical Society 1951:175–178.

Lawrence, D. B. 1958. Glaciers and vegetation history in southeastern Alaska. American Scientist 46:89–122.

Muir, J. 1895. The discovery of Glacier Bay. Century Magazine. New Series 28:234–247.

Ovenshine, A. T., 1967. Provenance of recent glacial ice in lower Glacier Bay, southeastern Alaska. U.S. Geological Survey Professional Paper 575D:198–202.

Post, A., and E. R. LaChapelle. 1971. Glacier ice. University of Washington Press, Seattle.

Reimnitz, E., and E. W. Kempema. 1982. Dynamic ice-wallow relief of northern Alaska's nearshore. Journal of Sedimentary Petrology 52:451–461.

Seramur, K. C., R. D. Powell, P. R. Carlson, and E. C. Cowan. 1997. Muir Inlet morainal bank complex, Glacier Bay, southeast Alaska. Pages 92–93 *in* T. A. Davies, T. Bell, A. H. Cooper, H. Josehans, L. Polyak, A. Solheim, M. S. Stoker, and J. A. Stravers, editors. Glaciated continental margins: an atlas of acoustic images. Chapman and Hall, London.

Sharman, L. 1990. Marine intertidal community; development following glacial recession in Glacier Bay, Alaska. Pages 108–115 *in* A. M. Milner and J. D. Wood, Jr., editors. Proceedings of the second Glacier Bay science symposium. U.S. National Park Service, Anchorage, Alaska.

Solheim, A., J. D. Milliman, and A. Elverhoi. 1988. Sediment distribution and sea-floor morphology of Storbanken: implications for the glacial history of the northern Barents Sea. Canadian Journal of Earth Science 25:547–556.

Syvitski, J. P. M., G. B. Fader, H. W. Josenhans, B. MacLean, and D. J. W. Piper. 1983. Seabed investigation with Pisces IV. Geoscience Canada 10:59–68.

USGS (U.S. Geological Survey). 1990. Glacier Bay National Park and Preserve 1:250,000 scale topographic map. USGS, Gustavus, Alaska.

Vancouver, G. 1798. A voyage of discovery in the North Pacific Ocean and round the world in which the coast of Northwest America has been carefully and accurately surveyed...performed in the years 1790, 1791, 1792, 1793, 1794, and 1795, etc. G. G. and J. Robinson and J. Edwards, London. (Not seen; cited in Grove 1988.)

Symposium Abstracts

Symposium abstracts have been reprinted from the Abstract Volume prepared for the Symposium on Effects of Fishing Activities on Benthic Habitats: Linking Geology, Biology, Socioeconomics, and Management without any further review or editing.

Analyzing Time-Lapse Photographs of the Sea Floor for Changes in Benthic Community Activity

S. E. BEAULIEU[1] AND H. SINGH

Applied Ocean Physics and Engineering Department, Woods Hole Oceanographic Institution, Woods Hole, Massachusetts 02543, USA

K. L. SMITH, JR.

Marine Biology Research Division, Scripps Institution of Oceanography, La Jolla, California

Time-lapse photographs or repeated photographic surveys of the sea floor can be used to study the response of benthic fauna to a natural or anthropogenic disturbance. We are interested in the responses of epibenthic megafauna to a temporally varying food supply, or flux of particulate matter to the sea floor. At a deep-sea study site, we have amassed ~10 years of time-lapse photographs, taken once per hour, of ~20 m^2 of the sea floor. We would like to analyze this 10-yr time series for seasonal changes as well as long-term trends in the benthic community. In addition to species composition, abundance, size, and activity of megafauna (with an activity index based on area traversed per unit time), we would like to trace sediment features such as mounds and tracks. Because manual analysis of the large number of photographs is very labor-intensive, we developed image-processing routines that make it easier to analyze oblique photographs, such as detecting organisms and their tracks. Our methods include: 1) digitizing the film, 2) adjusting light on the images (histogram equalization), 3) converting oblique photographs to plan view, and 4) automated image processing, with routines based on edge detectors and morphological operators. We will present results for a 4-mo time series at the deep-sea site, with natural disturbance from a massive accumulation of phytodetritus on the sea floor. We plan to use these algorithms for photographs taken in other soft-bottom habitats, including images transmitted in real-time from the Hawaii-2 Observatory in the abyssal Pacific.

---

[1] E-mail: stace@whoi.edu

Fishing Effects on Habitat: The Potential Consequences of Removing such Habitat Engineers as Red Grouper *Epinephelus morio*

F. C. COLEMAN[1] AND C. C. KOENIG

Department of Biological Science, Florida State University, Tallahassee, Florida

M. W. MILLER

National Marine Fisheries Service, Miami, Florida

S.A. Heppell and S. S. Heppell
Oregon State University, Department of Fisheries and Wildlife, Corvallis, Oregon

K. Scanlon
U.S. Geological Survey, Woods Hole, Massachusetts

Mass removal of species that restructure the architecture of habitat and thus increase its complexity can have multiple effects on ecosystems, including loss of biodiversity and altered biogeochemical pathways. In this paper, we report on the contributions made to habitat heterogeneity by the engineering capabilities of red grouper, Epinephelus morio, throughout its life. We demonstrate that this fish starts excavating habitat at first settlement, provides important structure and enhances biodiversity in nearshore communities of the west Florida shelf as juveniles, and contributes significantly to the structure of low-relief continental shelf edge areas as adults. We discuss the potential benefits of using side-scan sonar imagery to track grouper-induced changes in habitat over time (developing a time-series of images both within marine reserves and in nearby reference sites). We also discuss the implications of red grouper fishery removals to overall productivity of the continental shelf of the northeastern Gulf of Mexico and the particular management problems presented by knowledge of this behavior. Current management decisions to move the longline grouper fishery further offshore may increase pressure on red grouper and other excavating species, such as tilefish, have a significant negative influence on habitat heterogeneity, with potential to cause cascading problems throughout shelf-edge communities.

[1] E-mail: coleman@bio.fsu.edu

Geologic Development and Longevity of Continental Shelf Mudbelt Habitat during the Holocene in the Monterey Bay National Marine Sanctuary, California

E. E. Grossman[1], M. E. Field, and S. L. Eittreim
U.S. Geological Survey, Pacific Science Center, Santa Cruz, California

Recent degradation of benthic habitat and fish stocks is related to both anthropogenic and natural causes. Subsurface geological investigations augment seafloor and habitat mapping to provide constraints on habitat development, longevity, and variability due to natural geophysical processes. A principal geologic feature of the Monterey Bay National Marine Sanctuary is a 421 km^2 mudbelt that extends across a vast proportion of the continental shelf and reaches a maximum thickness of ~32 m. Basal ^{14}C ages of ~14 ka indicate the mudbelt is Holocene and ^{210}Pb accumulation rates show it is presently accreting at 0.24-0.39 cm/yr. Lithologic variations within cores show that the accumulation of this deposit occurred episodically under significantly different depositional energy. Seismic reflection profiles show that mudbelt development on the underlying fossil terrace was governed by complex interactions between fine sediment input (primarily from three major rivers) and transport (cross-shelf and along-shelf) during the Holocene sea-level transgression. Lateral variability in accumulation would have profound impacts on surrounding habitats as muds and sands were partitioned and deposited where they exist today. The composition and age of sediment within mudbelt cores help to define the nature of seabed sediment through time, its longevity as potential essential fish habitat, and its vulnerability to forces acting on the seafloor. Understanding the evolution and rates of sediment transport and accumulation of this and similar mudbelts that occur within important and threatened groundfish habitat along most modern coasts will provide a context for interpreting modern changes to essential fish habitat.

[1] E-mail: egrossman@usgs.gov

Effects of Ice Gouging on Community Structure and the Abundance of Pacific Halibut *Hippoglossus stenolepis*: Disturbance Does Not Necessarily Mean Negative Fisheries Effects

P. N. Hooge[1]
U.S. Geological Survey, Gustavus, Alaska

P. R. Carlson and G.R. Cochrane
U.S. Geological Survey, Menlo Park, California

Side-scan sonar and multbeam imagery of Glacier Bay, Alaska, revealed complex iceberg gouge patterns at water depths to 135 m on the floor of Whidbey Passage and south to the Bay's entrance. These previously undiscovered gouges formed more than 100 years ago as the Little Ice Age glacier retreated rapidly up Glacier Bay. Gouged areas supported greater biodiversity than nearby ungouged areas or sediment-filled gouges, probably due to increased habitat complexity. Small Pacific halibut *Hippoglossus stenolepis* were found more frequently in gouged areas, presumably due to higher prey abundance. These results contrast with the disturbance effects of recent, shallow ice gouging on community composition observed in the Arctic.

---

[1] E-mail: philip_hooge@usgs.gov

Biodiversity Changes in Space and Time in the Gulf of Alaska: An Ecosystem Measure of Fishing Effects on Habitat

R.F. Reuter[1] and S. Gaichas
National Marine Fisheries Service, Alaska Fisheries Science Center, Seattle Washington

Ecosystem resilience, in theory, is related to species and habitat diversity. It is believed that high diversity in a system may act as 'insurance' against any type of disturbance. Disturbance from fishing activities may result in diversity changes through time by the removal of select species and by gear effects on the bottom habitats. Historical bottom trawl survey data can be used to assess changes in diversity in the Gulf of Alaska. Survey data collected between 1960 and 2001 was used to map and biologically classify benthic habitats in terms of species diversity for different classes of marine animals. Although identification of benthic invertebrate species was limited by changing survey priorities, groundfish species such as rockfish and flatfish that occupy distinct habitat types were also used to classify areas. A wide variety of diversity indices are available and in this study several were explored. Maps of species diversity from pre-fishing trawl surveys were created using a geographical information system (GIS) and used to indicate historically important habitats. Inclusion of fishery observer data, when analyzing these maps, may indicate possible fishery effects in heavily fished areas, and suggest natural rates of change in less fished areas. This study will complement more direct experimental approaches for assessing fishing effects on benthic habitat by establishing the historical context of variability in an ecosystem.

---

[1] E-mail: rebecca.reuter@noaa.gov

Linking Fisheries and Supporting Ecosystems to Benthic Habitat Character and Dynamics

Linking Fisheries to Benthic Habitats at Multiple Scales: Eastern Scotian Shelf Haddock

JOHN T. ANDERSON[1]

*Northwest Atlantic Fisheries Centre, Department of Fisheries and Oceans,
Post Office Box 5667, St. John's, Newfoundland A1C 5X1, Canada*

JIM E. SIMON

*Department of Fisheries and Oceans, Marine Fish Division, Bedford Institute of Oceanography,
Post Office Box 1006, Dartmouth, Nova Scotia B2Y 4A2, Canada*

DON C. GORDON

*Department of Fisheries and Oceans, Environmental Sciences Division, Bedford Institute of Oceanography,
Post Office Box 1006, Dartmouth, Nova Scotia B2Y 4A2, Canada*

PETER C. HURLEY

*Department of Fisheries and Oceans, Marine Fish Division, Bedford Institute of Oceanography,
Post Office Box 1006, Dartmouth, Nova Scotia B2Y 4A2, Canada*

Abstract. Historical distributions ($n = 32y$) of age-1 juvenile haddock *Melanogrammus aeglefinus* from the eastern Scotian Shelf population varied linearly with year-class strength, indicating a dependence on demersal habitats. Distributions of haddock ages 2–5 were weakly density dependent, indicating weaker associations with benthic habitats. The preferred areas ($\geq 75\%$) of occurrence for age-1 haddock changed with the spatial scale of analysis. As bin size was reduced from 1,342 km^2 (400 nm^2) to 755 km^2 (225 nm^2) to 336 km^2 (100 nm^2) to 84 km^2 (25 nm^2), the boundaries of preferred areas shifted in location and the total area increased in size. As the spatial scale of bin size was reduced, the frequency of missing data increased, making it difficult to determine the true nature and extent of high-preference areas. The historical data indicated that preferred areas occur at the smallest scale analyzed of approximately 100 km^2. Therefore, preferred areas ($\geq 75\%$ occurrence) and nonpreferred ($\leq 25\%$ occurrence) areas 10 km by 10 km were selected on three banks of differing size for directed studies. Acoustic surveys were carried out in the selected study areas using a normal incidence echosounder to determine fine-scale (16–18 m) bathymetric structure. These areas ranged in mean depths from 42 to 84 m. Bathymetric relief (m/km) was always greater in preferred areas within each bank. Spatial auto-correlation of bathymetric relief had smaller decorrelation scales for preferred areas within banks. Preferred areas for age-1 haddock were always more rugged at finer spatial scales than nonpreferred areas, indicating that preferred habitats may be more complex. There was a bank-scale dependency in surface structure where smaller banks were less rugged at finer spatial scales. We hypothesize that there may be a bank-dependent scaling of habitats where larger banks have a greater variety of habitats that span a greater range of spatial scales.

Introduction

Fish stocks in Atlantic Canada have declined significantly since the mid-1980s, with many stocks of Atlantic cod *Gadus morhua* and haddock *Melanogrammus aeglefinus* closed to fishing due to historically low levels of abundance and poor recruitment (FRCC 2002). In addition to closing entire fisheries, seasonal and area closures have been used to protect spawning fish and enhance juvenile survival. A closed area (Figure 1) was established on the eastern Scotian Shelf in 1987 to protect juvenile haddock (Frank et al. 2000). It was assumed that this area encompassed important juvenile habitat and that closure of this area to fishing would enhance juvenile survival. How-

---

[1] E-mail: andersonjt@dfo-mpo.gc.ca

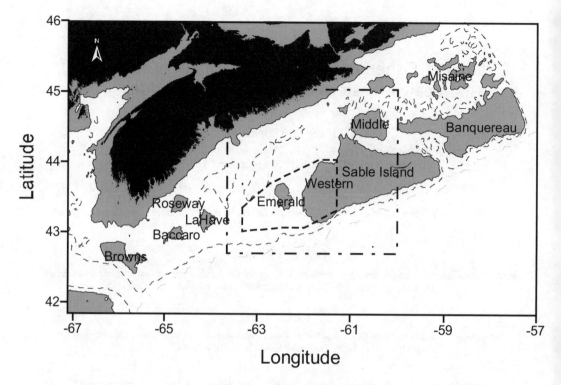

Figure 1. Banks on the Scotian Shelf bounded by the 91-m depth contour (gray shading). The haddock closed area is the enclosed polygon (heavy dashed line) surrounding Emerald and Western banks. The rectangle (heavy dashed-dot line) outlines the study area. Black represents land, and the light dashed and dashed-dot lines are 200-m and 1,000-m bathymetric contours.

ever, initial results demonstrated that simply closing this area to fishing did not increase recruitment to the population (Frank et al. 2000).

Haddock on the Scotian Shelf are managed as two separate stocks, the southwestern stock that includes Browns, Baccaro, Roseway, and La Have banks and the eastern stock that includes Emerald, Western, Sable Island, Middle, Missaine, and Banquereau banks (Figure 1). Previous work demonstrated that haddock preferred depths less than 100 m, temperatures 3–10°C, and salinities from 32 to 34 practical salinity units during summer (Smith et al. 1994). From the available ranges of each variable within each area, haddock did not select for specific depths less than 100 m but did select for warmer, more saline water. This study was based on an analysis of research vessel survey strata data for all ages at the scale of the stock units, which approximated 50,200 km² and 97,600 km² for each stock, respectively.

Theoretically, local fish abundance (number/m²) is highest within optimal habitats that have abundant resources, and abundance is lowest in marginal habitats where resources are scarce (MacCall 1990). As population abundance increases, optimal habitats become saturated, causing the distributional area to expand and fish abundance to increase rapidly in marginal habitats. Long-term survey data can be used to infer home ranges and optimal habitats for fish populations (MacCall 1990). Analyses of these data can identify life history stages that directly depend on benthic habitats and where such habitats occur.

On the southwestern Scotian Shelf, density-dependent habitat selection was demonstrated for age-1 and age-2 juvenile haddock based on an analysis of research vessel survey strata (Marshall and Frank 1994, 1995). Juvenile fish are defined as those fish that have metamorphosed from the larval stage into juveniles with a calcified endoskeleton but are immature sexually. On the Scotian Shelf, survey strata are divided into shallow (<91 m), intermediate (91–183 m), and deep (183–366 m) regions (Halliday and Koeller 1981). Marshall and Frank (1994, 1995) reported that abundance of juvenile haddock was always highest within the shallow strata that defines Browns Bank (Figure 1). When year-class strength was higher, juvenile haddock expanded into the deeper strata east of Browns Bank. Browns Bank is a single shallow strata encompassing an area of 2,249 km². Strata occupied to the east of Browns Bank were intermediate in depth and ranged in size from 536 km^2 to 5,076 km^2, for a combined area of 7,325 km^2. We believe analyses carried out at these spatial scales are too coarse to accurately define habitat for juvenile haddock. Recently, young-of-the-year settling flounder were found to occur primarily at spatial scales of 10 km to 25 km on the New York Bight region (Sullivan et al. 2000), which approximate 100 km^2 to 625 km^2.

The purpose of this study was to utilize the long-term research vessel trawl survey data to delineate the occurrence of preferred and nonpreferred areas for juvenile haddock on the eastern Scotian Shelf. In particular, we wanted to know how precisely these historical data could determine the spatial scale of preferred and nonpreferred habitats, where these areas occurred, and to what degree the seabed differed between such areas. For our study, we chose the middle Scotian Shelf (Figure 1). This area comprises the primary haddock spawning area and distribution of juveniles for the eastern Scotian Shelf management unit (Frank et al. 2001). The area includes 4 of 10 banks on the Scotian Shelf as well as the Haddock Closed Area (HCA; Figure 1). Starting in 1987, the HCA was closed only to mobile fishing gears, but this was extended to fixed-gear fisheries in 1993 (Frank et al. 2000). We based our analysis on 32 years of research trawl survey data (1970–2001) carried out each July by the Department of Fisheries and Oceans, Canada. These data were used to define the likelihood of finding juvenile haddock and, ultimately, to determine preferred and nonpreferred areas. We defined preferred and nonpreferred areas as the likelihood of finding haddock at different ages based on their presence and absence each year as a proportion of the entire time series. We carried out our analysis at different spatial scales that ranged from approximately 1,300 km^2 to 80 km^2 in bin size. Based on the historical data, we selected six 10-km × 10-km areas that represented preferred and nonpreferred areas for juvenile haddock. In 2002, high-resolution acoustic surveys were carried out using a normal incidence echosounder to determine the fine-scale bathymetric structure within the six areas as an initial step in characterizing juvenile haddock habitats.

Methods

Bottom trawl surveys have been carried out on the Scotian Shelf each summer since 1970. These surveys use a depth-stratified random design with depth as the major stratifying variable, where depth intervals were less than 91 m, 91–183 m, and 184–366 m (Halliday and Koeller 1981). Trawling was conducted using a Yankee 36 (1970–1981; Bedford Institute of Oceanography, Dartmouth, Nova Scotia) or a Western IIA (1982–2001; Bedford Institute of Oceanography, Dartmouth, Nova Scotia) trawl equipped with a 19-mm cod end liner towed a distance of 3.24 km at a constant speed of 6.5 km/h (3.5 knots) for 30 min with an estimated wing spread of 12.5 m, resulting in a bottom swept area approximating 0.041 km^2. Fork length of individual haddock was measured to the nearest centimeter for all surveys. Aging was based on otolith analysis following international haddock aging techniques (Hurley et al. 1996). All fish were assigned an age based on age–length keys.

We defined four levels of spatial resolution, where bin sizes were 37 km (20 nm), 27 km (15 nm), 18 km (10 nm), and 9 km (5 nm) on each side. This corresponded to bins ranging from 86 km^2 to 1,373 km^2. To determine the likelihood that haddock occurred within a particular bin, we analyzed the number of times (y) haddock of a given age were caught within a bin for all years sampled and then expressed the result as a quartile (%). The analysis was carried out for ages 1 to 10+ for 32 years of trawl surveys, 1970–2001. For 20-nm bin size, the number of tows ranged from 15 to 85 per bin. For 15-nm, 10-nm, and 5-nm bin sizes, the number of tows ranged from 5 to 58, 5–28, and 3–13, respectively. We dropped bins with less than 5 tows for the 15-nm and 10-nm data sets and bins with less than 3 tows for the 5-nm data set.

Year-class strength of juvenile haddock was estimated as the mean catch per tow within the study area for ages 1–5. We compared our estimates to population estimates for the eastern Scotian Shelf haddock stock estimated by sequential population analysis (SPA) (Frank et al. 2001) to confirm that our estimate of year-class strength was representative of the population. Regression of abundance-at-age estimated by SPA (dependent variable) versus catch-per-tow for the study area (independent variable) was done using SAS (SAS 2000).

We used the results of the spatial analysis by bin size together with a detailed examination of the historical catch distributions of age-1 haddock to select areas where the likelihood of finding haddock was both high (preferred areas) and low (nonpreferred). We defined these areas to be 10 km square and selected one preferred and one nonpreferred area on each of Emerald Bank, Western Bank, and Sable Island Bank (Figure 2). The number and size of study areas were limited by the operational constraints imposed in designing subsequent field surveys. Based on the historical distributions of age-1 haddock, we expected that the preferred areas would represent suitable habitats for juvenile haddock whereas the nonpreferred areas would represent nonsuitable or marginal habitats.

High-resolution acoustic surveys were carried out in the three preferred and three nonpreferred areas in September 2002. Parallel 10-km north–south transects were run at 800-m spacing. For the Sable Island Bank preferred area, the acoustic survey was run at 1,600-m spacing

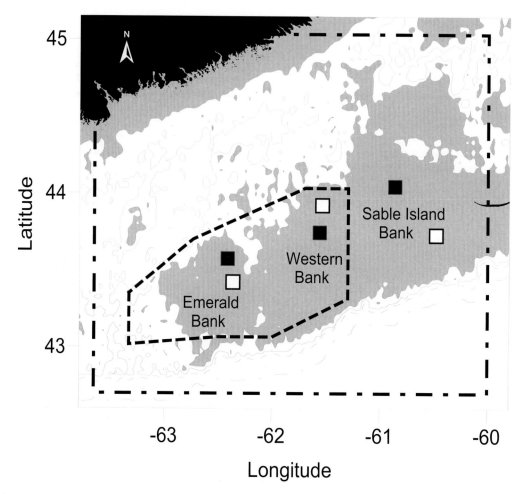

Figure 2. Juvenile haddock preferred areas (black squares) and nonpreferred areas (open squares) are shown on Emerald Bank, Western Bank, and Sable Island Bank. The haddock closed area is the enclosed polygon (dashed line) surrounding Emerald and Western banks. The rectangle (heavy dashed-dot line) outlines the study area. Black represents land and the light dashed and dashed-dot lines are 200-m and 1,000-m bathymetric contours.

due to time constraints. Acoustic data were collected using a dual frequency (120–38 kHz) digital BioSonics DT6000 (BioSonics Inc., Seattle, Washington) echo-sounder. Data were collected at 1 pings per second with a pulse width of 0.4 milliseconds and a ship speed held constant at approximately 3 m/s (6 knots). Bottom depth data were generated using the bottom pick algorithm of QTC IMPACT (IMPACT 2001). To minimize noise associated with ship motion and DGPS (Differential Global Positioning System) error (±3 m), five consecutive pings were normalized using QTC IMPACT to generate a bottom pick, smoothed using an 11-point running average, and offset by 5 m to account for the transducer depth relative to the sea surface. Latitude and longitude were converted into UTM (Universal Transverse Mercator) units from which distance (m) was calculated. Typical horizontal distances between depth observations were 16 to 18 m for all transect lines. Relief was defined as the change in depth (m) between adjacent observations along transects standardized to 1 km horizontal distance (m/km).

We calculated the auto-correlation of absolute relief along transect lines for each area. We interpreted high auto-correlation as a measure of similarity in small-scale bottom relief, whether this be a smooth or rough surface. We interpret the rate at which correlation declined at increasing lags as a measure of the scale over which surface roughness changed. When auto-correlation declined slowly, then relief was similar over larger distances. We expect such seabeds to be characterized as relatively low relief at coarser scales. Conversely, a rapid decline in auto-correlation indicates that relief was changing more rapidly over smaller distances. Such seabeds should be characterized as rough and more complex at finer spatial scales.

Results

Haddock Distributions

The results at 20-nm bin size demonstrated that age-1 juvenile haddock occurred most frequently in two areas. The highest likelihood (≥75%) of encountering age-1 haddock occurred on Emerald Bank, extending to the western edge of Western Bank (Figure 3a). A second high area occurred on the northwestern corner of Sable Island Bank. Depths for these two areas ranged from approximately 45 m to 80 m for the Emerald–Western bank area and from 30 m to 70 m for the Sable Island area. The two areas of high likelihood were linked by an area of moderate likelihood of occurrence (50–75%). The high and moderate areas were bounded by areas where juvenile haddock were less likely to occur, categorized as low (25–50%) and lowest (≤25%) areas.

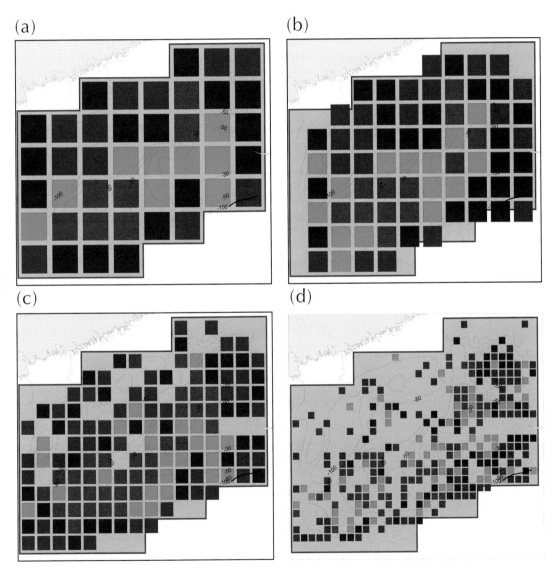

Figure 3. Distributions of age-1 juvenile haddock within the study area for different likelihoods of occurrence (%) at four spatial scales of analysis: (a) 20 nm, (b) 15 nm, (c) 10 nm, (d) 5 nm. Red = ≥75%, green = 50–74%, blue = 25–49%, purple = ≤25%. Depth contours are in meters. Land is shaded yellow. The red-line, gray-shaded polygon outlines the historical data analyzed at 20-nm bin size.

Using 15-nm bin size, the area of highest likelihood (≥75%) on Emerald and Western Banks shifted to the east. On Sable Island Bank, the highest likelihood area became fragmented and extended to the east and south (Figure 3b). Using 10-nm bin size, the highest likelihood area on Emerald and Western banks shifted to the north and expanded in both east and west directions (Figure 3c). On Sable Island Bank, the highest likelihood area expanded again to the east and south (Figure 3c). At 10-nm bin size, the highest likelihood areas were now almost continuous through the study area. Using 5-nm bin size, there were so many missing cells throughout the study area that it was difficult to determine the spatial structure of haddock occurrence (Figure 3d).

At 20-nm bin size, the high likelihood areas (≥75%) accounted for 16.3% of the study area, and this increased to 24.6% of the study area at 10-nm bin size. The total area for the high likelihood areas increased from 9,604 km^2 at 20-nm bin size to 11,662 km^2 at 10-nm bin size. This increase in area occurred even though 19% of the study areas was not sampled at 10-nm bin size. At 5-nm bin size, the proportion of high-likelihood areas was 24.2% of all sampled areas compared to 16.3% at 20-nm bin size. However, at 5-nm bin size, 61% of the area had missing observations, making it impossible to determine what the true distributions were. The problem of missing cells with decreasing bin size confounds across scale comparisons.

The boundaries of the high-likelihood areas changed both position and dimension with the observational scale. At finer spatial scales of analysis, the boundaries became more complex. Spatial complexity, estimated as the perimeter (km) divided by the area (km^2), increased from 0.062 at 20-nm bin size to 0.084 at 15-nm bin size to 0.097 at 10-nm bin size. It was not possible to calculate the spatial complexity at 5-nm bin size, but the data suggests that the trend of increasing complexity continued. Taken together, these observations indicate that an analysis scale of 5 nm or less would be required to determine the true spatial distribution of the high-likelihood areas for age-1 haddock and, therefore, the spatial distributions of optimal habitats.

The spatial extent of high-likelihood areas (≥75% occurrence) for haddock increased from approximately 25% of the entire study area for age-1 haddock to 46% for age-4 haddock based on a 10-nm bin size (Figure 4). For ages 1–4, the moderate-likelihood areas (50–75% occurrence) remained constant around 18% while the low-likelihood areas (<50% occurrence) decreased from 57% at age 1 to 35% at age 4. Beginning at age 5, the high-likelihood areas decreased in extent while the moderate-likelihood and low-likelihood areas increased up to

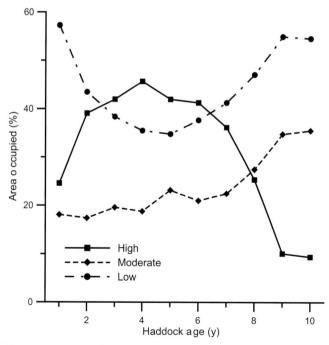

Figure 4. Percent of the study area occupied by haddock at 10 nm bin size within the study area as a function of age for high (≥75%), moderate (50–74%) and low (<50%) likelihood of occurrence.

age 9 and older. The increasing area occupied by age-2–4 haddock was directly related to catchability in the trawl surveys, which we estimated as the average catch rate (numbers/tow) ratios for year-classes at successive ages. The catch rate increased by 2.4 times from age 1 to age 2 and then 1.7 times from age 2 to age 3. The catch rates for ages 3 and 4 were approximately equal, indicating that catchability was still increasing enough to age 4 to offset natural mortality. Only by age 5 did the catch rate decrease to 0.83, indicating that haddock were fully available to the trawl.

Comparison of haddock catch rates from our study area (Figure 1) to population abundance estimated by SPA of the eastern Scotian Shelf stock for age 1–5 demonstrated statistically positive relationships in all cases ($P < 0.0005$, df = 30, R^2 = 35.8–55.4%). Therefore, we conclude that our estimates of abundance reflects the population year-class strength of haddock for these ages. We used the proportion of non-zero sets observed each year for each age to determine the degree to which area occupied was a density-dependent function of year-class strength. There was a highly significant relationship between the proportion of non-zero trawl sets and year-class abundance for age-1 haddock, where the explained variation was 60% (Table 1, Figure 5a). For ages 2–5, there were positive statistical relationships, but the degree of statistical power was much less than for age 1, where explained variance only ranged from 11% to 32% (Table 1; Figure 5b–e). Examination of the data demonstrated that the positive relationships for ages 2 to 5 were dependent on four survey years, 1973–1976. If years prior to 1977 were removed from the analysis, then the only statistical relationship was for age-1 haddock, where the explained variation increased to 70.5% (Table 1). Together, these analyses demonstrate that the spatial distribution of age-1 haddock was strongly density dependent, whereas for ages 2 to 5, the relationship was weak to nonexistent. We speculate that whatever conditions produced the measured responses in 1973–1976 occurred early in the time series and was not representative of conditions since the mid-1970s.

Bathymetry

Average depth in the six 10-km-square study sites decreased from west to east from Emerald Bank to Sable Island Bank (Table 2). The range of depths within each area varied from 12 m to 25 m. The higher variance (CV%) in depth for the Sable Island Bank preferred site resulted, in part, from the lower number of observations due to sampling time constraints.

We examined along-transect depth variations for each study site. As an example, we present here the middle transects surveyed within the Sable Island preferred and nonpreferred sites (Figure 6). At the scale of 10 km, the depth gradient was greater for the Sable preferred site. At finer spatial scales, there was much more variation within the Sable preferred site, indicative of a more rugged topography. Bathymetric relief (see Methods) averaged for each study site varied from 2.65 m/km in the Sable Island Bank nonpreferred site to 5.27 m/km in the Sable Island Bank preferred site (Table 2). For each bank, average relief was greater for the preferred sites compared to nonpreferred sites. We lagged the along-transect relief data to determine the degree of spatial auto-correlation within each study site. The rate at which correlations decline as distance between observations increases is a measure of the relative scale of seabed bathymetric structure.

The rate of decline in lagged correlations was always greater for preferred sites within each bank (Figure 7). The Western preferred site had the highest rate of decrease in lagged correlations of relief, indicating that this study site would have the most rugged surface structure. In contrast, the Sable Island nonpreferred site had the smallest rate of decrease. The different areas were ranked as Western preferred (–0.103); Sable preferred (–0.083); Western nonpreferred (–0.051); Emerald preferred (–0.048); Emerald nonpreferred (–0.033); and Sable nonpreferred (–0.014). The largest contrast in the rates of decrease between preferred and nonpreferred sites occurred on Sable Island Bank (5.9:1), followed by Western Bank (2.1:1), with the smallest difference occurring on Emerald Bank (1.5:1).

We chose a de-correlation scale of lagged relief for correlation values less than 0.5 to compare the scale of bathymetric structure for each study site (Figure 7). Emerald preferred had the smallest de-correlation scale, which occurred at two lags, approximately 34 m, while Sable

Table 1. Summary of linear regression analyses of the relationship between non-zero trawl sets and the abundance of juvenile haddock (number/ tow) for ages 1–5. df = degrees of freedom; *P*-level = the probability level; R^2 = the coefficient of determination.

Age	df	P-level	R^2 (%)	Slope	Intercept
		All years, 1970–2001			
1	30	<0.0001	59.7	0.341	2.829
2	30	0.0749	10.5	0.077	3.801
3	30	0.0188	14.8	0.103	3.794
4	30	0.0033	26.1	0.128	3.767
5	30	0.0005	32.0	0.173	3.699
		1977–2001			
1	23	<0.0001	70.5	0.314	2.956
2	23	0.4045	3.2	0.028	4.030
3	23	0.3851	0.0	–0.025	4.274
4	23	0.4348	0.0	0.017	4.176
5	23	0.2907	0.7	0.024	4.163

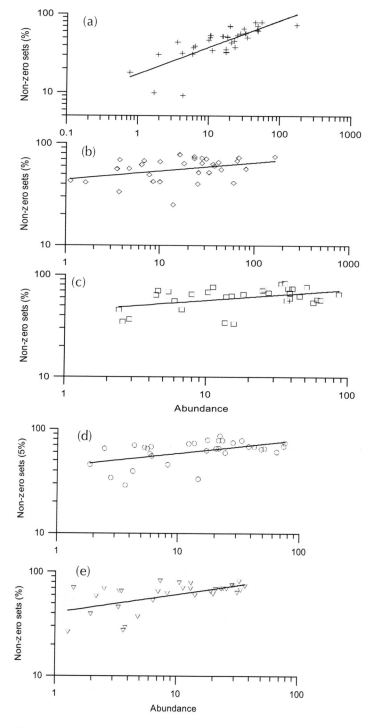

Figure 5. Proportion of non-zero sets (%) and year-class abundance (number/tow) for (a) age 1, (b) age 2, (c) age 3, (d) age 4, and (e) age 5 in the historical data set (1970–2001).

Table 2. Summary of depth (m) and bathymetric relief (m/km) for the preferred (PA; ≥75% occurrence) and nonpreferred (NPA; ≤25% occurrence) areas selected on Emerald Bank, Western Bank and Sable Island Bank. CV = coefficient of variation.

Area	n	Depth or relief	SD	Min	Max	CV (%)
Emerald NPA	7041	83.9	4.15	77	95	5.0
		2.68	2.276	0	18.6	84.9
Emerald PA	7076	78.7	4.26	71	90	5.4
		3.04	2.550	0	32.8	83.8
Western NPA	6827	53.4	4.53	47	72	8.5
		3.24	2.675	0	24.5	82.3
Western PA	6626	58.9	2.52	52	64	4.3
		4.74	4.267	0	33.5	90.1
Sable Island NPA	6810	57.5	2.69	51	64	4.7
		2.65	2.058	0	30.9	77.6
Sable Island PA	2792	42.4	4.34	33	50	10.3
		5.27	4.127	0	25.7	78.3

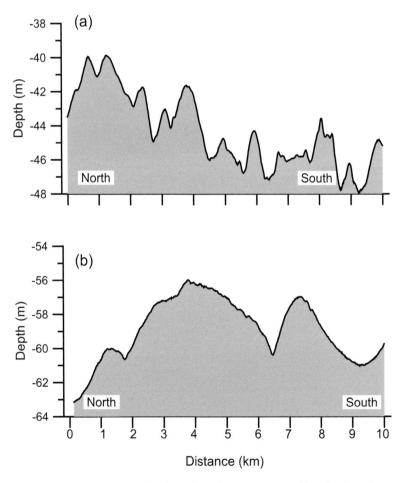

Figure 6. Depth (m) profiles for selected transects on Sable Island Bank (a) preferred and (b) nonpreferred areas from north to south in each area.

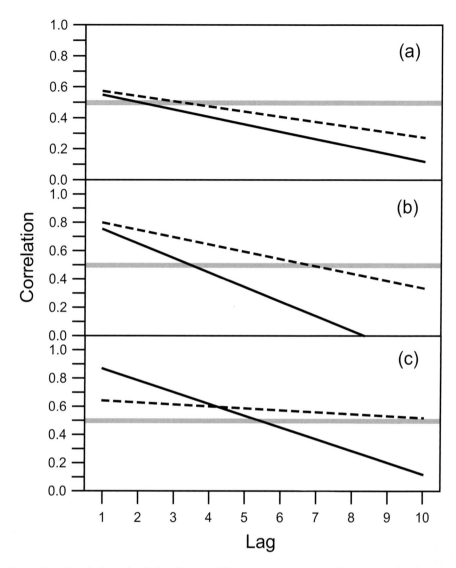

Figure 7. Correlation of relief (m/km) at different lags comparing observations for the preferred (solid lines) and nonpreferred (dashed lines) sites on (a) Emerald, (b) Western, and (c) Sable Island banks. The shaded horizontal line corresponds to a de-correlation limit of $r = 0.5$.

nonpreferred had the greatest de-correlation at 10 lags, which approximates 170 m. Within each bank, the preferred sites always had smaller de-correlation scales than did the nonpreferred areas. When we compared these results to the size of each bank, we found that larger banks had both coarser scales of relief and a greater magnitude of difference between preferred and nonpreferred sites (Figure 8). This suggests there are bank-dependent differences in surface structure where larger banks have coarser scales in surface ruggedness. This result does not appear to be a simple function of separation distances between study sites on each bank.

The preferred and nonpreferred sites were approximately 30 nm apart on Sable Island Bank, whereas they were approximately 10 nm apart on Western Bank and Emerald Bank (Figure 2).

We quantified the distances over which slope was either increasing or decreasing for each area as a direct measure of the spatial surface structure. In all cases, the dominant scale for slopes occurred at 20 m, which approximates our observational scale. However, there were consistent differences between preferred and nonpreferred areas where the proportion of observations at 20 m was always greater for nonpreferred areas, averaging 62%

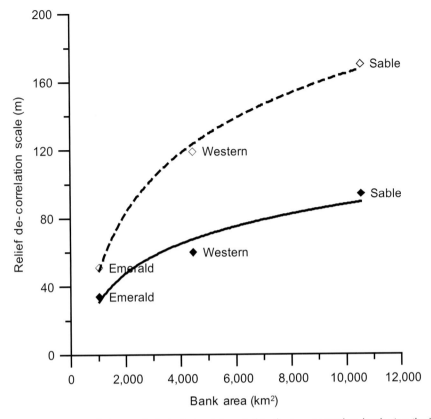

Figure 8. De-correlation scale (m) of absolute relief (m/km) compared to bank size (km²) for preferred (closed diamonds, solid line) and nonpreferred (open diamonds, dashed line) areas. The lines are fitted lognormal relationships.

of all observations, compared to preferred areas, which averaged 41%. This indicates that fine-scale features (i.e., ≤20 m) were more dominant in the nonpreferred areas. At the other extreme, the coarse-scale features dominated in nonpreferred areas. The largest distances measured were 1,400 m, 2,400 m, and 1,800 m for Emerald, Western, and Sable nonpreferred areas, respectively. The number of times that distances of common slopes exceed 500 m was 15, 29, and 34 times, respectively. This result is consistent, with larger banks having coarser-scaled features of bathymetric structure.

Discussion

Our analysis of the 32-year time series of research vessel survey data demonstrated that a bin size of 100 km² approximates the spatial scale of distribution that reflects habitat preferences for age-1 juvenile haddock. This scale was smaller than the resolution of the historical trawl data set based on a random stratified design. We note that there was a high degree of contagion in the spatial distribution of trawl tows. Examination of the distribution of unsuccessful tows did not indicate that areas were untrawlable. It appears that the randomization process used to allocate trawl sets within strata each year limits the utility of these long-term trawl surveys to fully determine the spatial distributions of haddock. We believe that directed studies will be necessary to quantify the distributions of juvenile fish in relation to habitats. Sullivan et al. (2000) demonstrated that settling juvenile fish occurred predominantly at spatial scales of 10 km to 25 km over a 3-year period. These results were consistent for habitats that varied significantly in depth and seabed structure.

Boudreau (1992) demonstrated that variability in haddock distribution occurred at the scale of 100 m and that the ability to detect this variability disappeared at the scale of a trawl tow (≈3.2 km). Distribution of benthic megafauna on the Grand Banks, including fishes, was associated with seabed features at scales of 100 to 500 m (Schneider et al. 1987). Theoretical studies have indicated that spatial heterogeneity at spatial scales of 100 to

1,000 m is important for the maintenance of diversity and stability at population, community, and ecosystem levels (reviewed in Thrush et al. 2001). Making scale-independent observations of marine benthic habitats and associated juvenile fish distributions will require the use of acoustic and optic technologies that can sample continuously from meter to kilometer scales.

Range expansion of haddock was strongly density dependent for age-1 juvenile haddock, where the number of non-zero sets ranged from minima of 10–20% to a maximum of approximately 80%. This range implies that there are large interannual differences in the distribution of age-1 haddock as a function of year-class abundance. By age 2, the distributions were weakly density dependent for the entire data series, and there was no statistical relationship for the period 1977–2001. Our result contrasts with that of age-2 haddock on the southwestern Scotian Shelf, where density-dependent distributions were detected for both age-1 and age-2 haddock, and the relationship was more pronounced for age-2 juveniles (Marshall and Frank 1994, 1995).

The increasing catch rates for age 1–4 is usually interpreted as small fish escape beneath the footrope, whereas there is increasing vulnerability to the trawl at older ages and larger sizes. However, when combined with the increasing area occupied for age 1–4, the results suggest that there is a dynamic between spatial use of habitats and vulnerability to the research trawl. This observation has important implications for establishing closed fishing areas to protect juvenile haddock. We hypothesize that juvenile haddock initially settle and survive in spatially restricted habitats with low vulnerability to the research trawl. As juveniles grow, there is an ontogenetic migration to habitats of increasing vulnerability to the trawl.

Juvenile fish use habitat to reduce the risk of predation (Gotceitas and Brown 1993, 1995; Gregory and Anderson 1997). The movements of juvenile Atlantic cod are limited, and nonmigratory, within areas of settlement during their first 2–3 years of life (Pihl and Ulmestrand 1993; Anderson and Gregory 2000). As predation is size dependent, a more appropriate comparison among populations should be based on fish size. Haddock have varied significantly in size at age on the eastern Scotian Shelf during the time period we analyzed. In the 1970s, an age-1 haddock in summer was approximately 24 cm in length, compared to the 1990s when it was only 18 cm in length, a decrease of 25%. Similarly, an age-2 haddock was 32 cm long in the 1970s, compared to 26 cm in the 1990s. Therefore, an age-1 haddock in the 1970s was approximately the same size as an age-2 haddock in the 1990s. If density-dependent distributions are size dependent, then the relationship between non-zero sets and year-class abundance may have varied over time. We re-ran the regressions comparing non-zero sets (%) to year-class abundance (number/tow) for different time periods corresponding to each decade. There was a weaker relationship for age-1 haddock in the 1970s ($R^2 = 45.8\%$), which was stronger in the 1980s ($R^2 = 69.2\%$) and the 1990s ($R^2 = 64.7\%$). These results are consistent with smaller haddock demonstrating a stronger density dependence in habitat use. For age-2 haddock, the relationship in the 1970s ($R^2 = 10.4\%$) was similar to the long-term relationship (Table 1). It increased in the 1980s ($R^2 = 26.0\%$) and disappeared in the 1990s, where the slope was negative ($R^2 = 9.7\%$). This is not consistent with size dependency in the relationship. Together, these results suggest that the strength of density-dependent relationships with habitat may vary over time, but this was not a simple function of fish size.

The consistent differences in bathymetric structure between preferred and nonpreferred areas for three different banks demonstrates a degree of commonality in habitats for preferred versus nonpreferred sites. Preferred habitat areas were characterized as being more rugged at finer scales, which we interpret as more complex habitats. Habitat complexity is often positively associated with biodiversity, and this can even be true in soft-sediment marine areas typically regarded as habitats with a low degree of structure (Thrush et al. 2001). On the Grand Banks of Newfoundland, densities of benthic invertebrates were greater when relief was greater than 1.0 m/km, measured over 10-km distances (Schneider et al. 1987). On the Grand Banks, increasing degree of habitat roughness was related to sand, shell, gravel, and cobble. On the Scotian Shelf, we do not know yet what features characterize the substrate structure within the preferred and nonpreferred areas nor the scales over which these exist, but these are currently being investigated.

Our initial results examining fine-scale bathymetric structure are encouraging. This relatively simple measure of habitats consistently characterized habitats between preferred and nonpreferred sites and may serve as a proxy variable for fish habitats. We have demonstrated that significant spatial structure in surface relief occurs at scales from tens to hundreds of meters within our 10-km study areas. Ultimately, high-resolution surveys of entire banks will be required to properly measure and map critical fish habitats. On the southwestern Scotian Shelf, multibeam acoustic surveys have been carried out over Browns Bank at a measurement scale of 10 m (Todd et al. 1999). Interpretation of these data, supported by various sources of other observational data, revealed previously unknown structures and features that ranged from meters (e.g., barchan dunes) to 10^3 meters (e.g., Browns Bank moraine).

The relationships between the scale of relief with bank size indicates smaller banks may be less rugged, with finer-scaled features. A recent study demonstrated that finfish species diversity and abundance scaled positively with bank size on the Scotian Shelf (Frank and Shackell 2001). Together, these results suggest that on larger banks, seabed habitat is more complex, occurs at coarser spatial scales, and may contain a greater variety of habitats which support higher biodiversity. We hypothesize that larger banks may be more important in terms of their habitat for juvenile haddock. It is notable that a significant area of high likelihood of occurrence for juvenile haddock occurred on Sable Island Bank outside the HCA. Inclusion of this area within the HCA should increase juvenile haddock survival and recruitment to this stock. Management of coastal fisheries by area closures will require understanding of where critical habitats occur and at what scales in order to quantify their importance to fish production.

Acknowledgments

Denise Davis assisted in the statistical and spatial analysis of the historical fisheries data. Bob Gregory, Edgar Dalley, Gordon Fader, and Ken Frank assisted in the interpretation of the historical data and selection of the preferred and nonpreferred areas. Vanessa Sutton processed the acoustic data collected by Arnold Murphy and Chris Lang. The captain and crew of the CCGS Hudson were instrumental in the successful completion of the 2002 research cruise to the Scotian Shelf. Funding for this research was provided by the Environmental Sciences Strategic Research Fund and the Strategic Science Fund of the Department of Fisheries and Oceans.

References

Anderson, J. T., and R. S. Gregory. 2000. Factors regulating survival of northern cod (NAFO 2J3KL) during their first three years of life. ICES Journal of Marine Science 57:349–359.

Boudreau, P. R. 1992. Acoustic observations of patterns of aggregation in haddock (*Melanogrammus aeglefinus*) and their significance to production and catch. Canadian Journal of Fisheries and Aquatic Sciences 49:23–31.

Frank, K. T., R. K. Mohn, and J. E. Simon. 2001. Assessment of the status of Div. 4TVW haddock: 2000. Canadian Stock Assessment Secretariat, Research Document 2001/100, Canadian Science Advisory Secretariat, Fisheries and Oceans (Station 1256), Ottawa.

Frank, K. T., and N. L. Shackell. 2001. Area-dependent patterns of finfish diversity in a large marine ecosystem. Canadian Journal of Fisheries and Aquatic Sciences 58:1703–1707.

Frank, K. T., N. L. Shackell, and J. E. Simon. 2000. An evaluation of the Emerald/Western Bank juvenile haddock closed area. ICES Journal of Marine Science 57:1023–1034.

FRCC (Fisheries Resource Conservation Council). 2002. 2002/2003 conservation requirements for groundfish stocks on the Scotian Shelf and in the Bay of Fundy (4VWX), in sub-areas 0, 2 + 3 and redfish stocks. FRCC, 2002.R.1., Ottawa.

Gotceitas, V., and J. A. Brown. 1993. Substrate selection by juvenile Atlantic cod (*Gadus morhua*): effects of predation risk. Oecologia 93:31–37.

Gotceitas, V., and J. A. Brown. 1995. Habitat use by juvenile Atlantic cod (*Gadus morhua*) in the presence of an actively foraging and non-foraging predator. Marine Biology 123:421–430.

Gregory, R. S., and J. T. Anderson. 1997. Substrate selection and use of protective cover by juvenile Atlantic cod *Gadus morhua* in inshore waters of Newfoundland. Marine Ecology Progress Series 146:9–20.

Halliday, R. G., and P. A. Koeller. 1981. A history of Canadian groundfish trawling surveys and data usage in ICNAF divisions 4TVWX. Canadian Special Publication of Fisheries and Aquatic Sciences 58:27–41.

Hurley, P. C. F., G. A. P. Black, R. K. Mohn, and P. Comeau. 1996. Assessment of 4X haddock in 1995. DFO, Atlantic Fisheries Research Document 96/30, Canadian Science Advisory, Ottawa,

IMPACT. 2001. QTC IMPACT acoustic seabed classification user guide, version 3. Quester Tangent Corp., Sidney, British Columbia.

MacCall, A. D. 1990. Dynamic geography of marine fish populations. University of Washington Press, Seattle.

Marshall, C. T., and K. T. Frank. 1994. Geographic responses of groundfish to variation in abundance: methods of detection and their interpretation. Canadian Journal of Fisheries and Aquatic Sciences 51:808–816.

Marshall, C. T., and K. T. Frank. 1995. Density-dependent habitat selection by juvenile haddock (*Melanogrammus aeglefinus*) on the southwestern Scotian Shelf. Canadian Journal of Fisheries and Aquatic Sciences 52:1007–1017.

Pihl, L., and M. Ulmestrand. 1993. Migration pattern of juvenile cod (*Gadus morhua*) on the Swedish west coast. ICES Journal of Marine Science 50:63–70.

SAS. 2000. Statistical analysis system, version 8.01. SAS Institute Inc., Cary, North Carolina.

Schneider, D. C., J. M. Gagnon, and K. D. Gilkinson. 1987. Patchiness of epibenthic megafauna on the outer Grand Banks of Newfoundland. Marine Ecology Progress Series 39:1–13.

Smith, S., R. Losier, F. Page, and K. Hatt. 1994. Associations between haddock, and temperature, salinity and depth within the Canadian groundfish bottom trawl surveys (1970–93) conducted within NAFO Divisions

4VWX and 5Z. Canadian Technical Report of Fisheries and Aquatic Sciences 1959.

Sullivan, M. C., R. K. Cowen, K. W. Able, and M. P. Fahey. 2000. Spatial scaling of recruitment in four continental shelf fishes. Marine Ecology Progress Series 207:141–154.

Thrush, S. F., J. E. Hewitt, G. A. Funnell, V. J. Cummings, J. Ellis, D. Schultz, D. Talley, and A. Norkko. 2001. Fishing disturbance and marine biodiversity: the role of habitat structure in simple soft-sediment systems. Marine Ecology Progress Series 223:277–286.

Todd, B. J., G. B. J. Fader, R. C. Courtney, and R. A. Pickrill. 1999. Quaternary geology and surficial sediment processes, Browns Bank, Scotian Shelf, based on multibeam bathymetry. Marine Geology 162:165–214.

Combining Scientific and Fishers' Knowledge to Identify Possible Groundfish Essential Fish Habitats

MELANIE BERGMANN,[1,2] HILMAR HINZ, ROBERT E. BLYTH, AND
MICHAEL J. KAISER

School of Ocean Sciences, University of Wales-Bangor, Menai Bridge LL59 5AB, UK

STUART I. ROGERS

Centre for Environment, Fisheries and Aquaculture, Lowestoft, UK

MIKE J. ARMSTRONG

Department of Agriculture and Rural Development, Belfast, UK

Abstract. Fishers have often complained that standard United Kingdom groundfish survey data do not adequately reflect the grounds targeted by commercial fishers, and hence, scientists tend to make overcautious estimates of fish abundance. Such criticisms are of particular importance if we are to make a creditable attempt to classify potential essential fish habitat (EFH) using existing data from groundfish surveys. Nevertheless, these data sets provide a powerful tool to examine temporal abundance of fish on a large spatial scale. Here, we report a questionnaire-type survey of fishers (2001–2002) that invited them to plot the location of grounds of key importance in the Irish Sea and to comment on key habitat features that might constitute EFH for Atlantic cod *Gadus morhua*, haddock *Melanogrammus aeglefinus*, and European whiting *Merlangius merlangus*. Plotted grounds were cross-checked using records of vessel sightings by fishery protection aircraft (1985–1999). A comparison of the areas of seabed highlighted by fishers and the observations made on groundfish surveys were broadly compatible for all three species of gadoids examined. Both methods indicated important grounds for cod and European whiting off northern Wales, the Ribble estuary, Solway Firth, north of Dublin, and Belfast Lough. The majority of vessel sightings by aircraft did not match the areas plotted by fishers. However, fishing restrictions, adverse weather conditions, and seasonal variation of fish stocks may have forced fishers to operate outside their favored areas on the (few) occasions that they had been recorded by aircraft. Fishers provided biological observations that were consistent among several independent sources (e.g., the occurrence of haddock over brittle star [ophiuroid] beds). We conclude that fishers' knowledge is a useful supplement to existing data sets that can better focus more detailed EFH studies.

Introduction

Subtidal marine habitats are less accessible and, therefore, have received less attention from scientists than have terrestrial habitats (Koehn 1993). As with terrestrial species, the populations of some species of fish may be dependent upon the availability of certain habitat types. Degradation of fish habitat quality may be partially responsible for recent declines in world fisheries (FAO 1995), and the importance of habitat quality needs to be addressed in fisheries science and management (Benaka 1999). Despite centuries of intensive exploitation of fish in European waters, relatively little is known about the small-scale distribution and habitat requirements of commercially exploited marine fish species. Freshwater ecologists, by contrast, have extensively researched the habitat requirements of fish (e.g., Keast et al. 1978; Ebert and Filipek 1988; Koehn 1993). Since the 1980s, the ecological effects of fishing have become a worldwide environmental concern (e.g., Dayton et al. 1995; Jennings and Kaiser 1998; Collie et al. 2000). For example, a consideration (and mitigation) of the effects of fishing on marine habitat that is critical for certain life stages of commercially important fish species became a legal requirement in the United States with the reauthorization of the Magnuson–Stevens Fisheries Conservation and Man-

---

[1] Corresponding author: mbergmann@awi-bremerhaven.de
[2] Present address: Alfred Wegener Institute for Polar and Marine Reesearch, Am handelshafen 12, 27570 Bremerhaven, Germany

agement Act in 1996. These habitats have been termed essential fish habitats (EFH) and include areas that are spawning and nursery grounds, provide prey resources and protection from predators, and form part of a migration route (Benaka 1999). This recent emphasis on EFH has resulted in a number of studies based in North America (see Benaka 1999; Coleman et al. 2000). The present study is the first in Europe that specifically aims to identify key habitats (EFH) for Atlantic cod *Gadus morhua*, haddock *Melanogrammus aeglefinus*, and European whiting *Merlangius merlangus* in the Irish Sea (northeastern Atlantic).

Haddock, Atlantic cod, European whiting, and plaice *Pleuronectes platessa* accounted for 52% of the demersal species landed by UK vessels in 2000 (DEFRA 2000). National landings of haddock and Atlantic cod have decreased from approxiimately 90,000 metric tons (mt) and 75,000 mt to 53,000 mt and 42,000 mt, respectively, between 1996 and 2000. Landings of European whiting and plaice decreased between 1996 and 1998 but have remained constant between 1998 and 2000. Fishing effort for these species remains very high while spawning stocks have fallen below their precautionary level, and the numbers of young fish have generally declined since 1990, raising concerns about the risk of stock collapse (DEFRA 2000).

It is well known that certain fish species are associated with specific habitat features (e.g., reefs, sandbanks), a fact used by fishers to target particular species. Demersal fishers observe samples from the seabed with every haul of their nets, which far exceeds the sampling schemes that scientists can sustain (Maynou and Sardà 2001). Furthermore, experienced fishers may have knowledge accumulated over decades through the knowledge of ancestors (Sardà and Maynou 1998; Freire and García-Allut 1999). In addition, they often maintain detailed records of the location and time when they fished and how much they caught. Nowadays, ship-based electronic instrumentation enables fishers to make links between different seabed types and textures and the fish they seek to catch. Although the ultimate goal of fishers is to provide income from the catch rather than to test scientific hypotheses, many fishers are motivated to understand the links between the marine habitat and the distribution of fish. Despite this obvious wealth of experience, few studies have sought to consider or integrate fishers' views and knowledge on EFH (but see Pederson and Hall-Arber 1999; Williams and Bax 2003). The need to improve the collaboration between scientists and the fishing industry is widely recognized by scientists and fishers alike (e.g., Taylor 1998; Freire and García-Allut 1999; Baelde 2001; Mackinson 2001; Maynou and Sardà 2001; Marrs et al. 2002; Moore 2003). The involvement of the fishing industry in fisheries science might not only improve the credibility of fisheries science but also enhance the support for any regulations upon which it is based.

In the present paper, we assessed the use of two complementary approaches to identify possible locations of EFHs for Atlantic cod, haddock, and European whiting in the Irish Sea for future comprehensive habitat survey. We used existing data from annual national groundfish surveys of fish abundance and compared them with fishing grounds outlined by fishers in response to a questionnaire-type survey.

Methods

Identification of Potential Essential Fish Habitat Using National Groundfish Surveys

Annual groundfish surveys have been undertaken since 1988 to assess the state of marine resources in the Irish Sea across a wide grid of sampling stations (Ellis et al. 2002). Therefore, data from these surveys may provide a unique tool to study the distribution of fish over large areas and longer-term time scales. Areas of the seabed that consistently harbor the highest densities of Atlantic cod, haddock, and European whiting in the Irish Sea (International Council for the Exploration of the Sea [ICES] division VIIa) were identified using these two databases spanning a decade of fishery-independent data from national groundfish surveys. Such grounds may have features that attract fish and, thus, might be candidates to be considered as EFH. The Centre for Environment, Fisheries and Aquaculture Science (CEFAS, Lowestoft, UK) holds a complete data set from 1990 to 1998. Fish were sampled using a 4-m beam trawl at fixed stations every autumn (Symonds and Rogers 1995). The Department of Agriculture and Rural Development (Belfast, UK) database spans a period from 1991 to 2000. Fish were caught by otter trawling at fixed stations every summer or autumn (Ellis et al. 2002). The two sampling gears have a different selectivity for gadoids (Ellis et al. 2002), and therefore, the data could not be combined.

In our analysis, the abundance of each of the three species was separately ranked (based on the populations of stations sampled) for every station and year, and a mean rank over time (per station) was calculated to identify potential EFH for further habitat surveys (reported elsewhere). Plots of mean abundance or total abundance over time were not considered useful to identify habitats that were used consistently from one year to the next, as a strong year-class could skew the results. We converted abundance to ranks within each year. Our rationale for using a rank score was that it is most relevant to know which habitat is consistently attractive to a particular species of fish. The mean rank values for each station were plotted using ArcView geographic information system 3.2 software.

Fishers' Knowledge

The project was first introduced to the fishing community by publishing an article that described the background and purpose of the study in the main national industry paper *Fishing News*. We then liaised with the fishing industry to compare our broad-scale fish distribution maps (from groundfish surveys) with fishers' locations of fishing grounds in terms of the seasonal and spatial distribution of fish. It is often not practical to consult directly with individual fishers who spend most of their time at sea, often for more than a week at the time. Information was gathered in a pilot study through questionnaire-based face-to-face interviews with maps at an annual national fishing exhibition in Glasgow, UK (respondents were selected at random).

Sample size ($n = 19$) was limited by the time available to undertake interviews and the willingness of potential participants. The questionnaire was designed to study fishers' perceptions of the relationship among commercially important fish, habitat features, and changes in abundance and to gain information about the location of potential EFH. It consisted of 16 questions in total (see Pederson and Hall-Arber 1999), which were variously dichotomous, multiple choice, and open ended. Only six questions are analyzed here due to constraints of space (Figure 1). Furthermore, fishers were asked to plot grounds that they perceived as important for their target species on standard maps. Such grounds may be characterized by particular features or the presence of prey organisms and, therefore, harbor high abundances of fish. Thus, they could be indicative of EFH. The hand-drawn plots on the standardized maps were digitized as a chart (ArcView 3.2, ESRI, UK) suitable for comparison with maps that showed mean ranks of fish abundance generated from groundfish surveys.

1. What do you regard as important ground features for your target species ? Please identify seabed structures (e.g., mud, gravel, boulders) or other characteristics of the grounds (e.g., seaweed, sponges) that you associate with your target species. _____

2. What do you regard as the most important factors that affect the grounds that you fish? _____

3. Do you think fishing gear has altered the grounds that you usually fish?   yes ☐   no ☐
If yes, how has it affected the grounds? Please explain. _____

4. Have you noticed any changes over the time that you have been fishing?   ☐ target species
☐ bottom animals and plants   ☐ habitat structure   ☐ fish health   ☐ bycatch   ☐ other changes
☐ other changes. Specify. _____

5. Which of the following have you observed over time for the species that you target?   ☐ no change   ☐ increase   ☐ decrease   ☐ moved to other areas   ☐ replaced by another species
☐ decrease in size. Please describe your observations. _____

6. If you noticed a change to the grounds or species that you fish, please indicate what you think may be the cause(s).   ☐ climate   ☐ pollution   ☐ changes in fishing gear habitat loss
☐ changes in prey abundance   ☐ overfishing   ☐ other. Please explain. _____

Figure 1. Questionnaire format used in face-to-face interviews and mail questionnaires.

More information was then gathered by mailing out revised questionnaires with maps and more detailed information about the project to Sea Fisheries Committees and other relevant fishers' organizations and requesting them to circulate these among their members. Additional interviews were conducted at a fishing exhibition in Newcastle (Northern Ireland; $n = 5$). The responses to questionnaires were analyzed by calculating the frequency of categories ticked and the frequency of key statements made in response to open-ended questions. The fraction of respondents who did not answer a question was excluded when percentage frequencies were calculated.

To compare the fishers' verbal habitat descriptions (in terms of seabed types) with the occurrence of seabed types in the areas they plotted on charts, we overlaid these plots with data from the British Geological Survey (2002). For this purpose, the sediment classification used by the British Geological Survey had to be regrouped so as to match the terminology used by fishers (Table 1). We then calculated the percentage area covered by mud, sand, hard grounds, and gravel or shingle in areas that had been plotted as important fishing ground for each species of fish (MapInfo Professional 7.0, MapInfo Corporation, UK).

Fishers were invited to give either their name or the name of their vessel, which enabled us to cross-check the areas plotted by individual fishers using records of named vessel sightings collated from fishery protection overflights. These aircraft patrolled the fishing grounds around the United Kingdom from 1985 to 1999 and recorded a description and location of all vessels that were observed fishing. The aircraft overflew most ICES subrectangles *c.* 100 times per year (Jennings et al. 2001). It should be noted that the fisheries protection aircraft predominantly flew over UK waters, resulting in very few sightings off the Irish coast. Using these data, we calculated the number of sightings for each fishing vessel (whose identity had been disclosed) that corresponded with the areas plotted by the respective owner. For reasons of confidentiality, the identification of vessels and the corresponding respondents to questionnaires were anonymous.

Results

Fishing Ground Locations and Distribution of Mean Ranks of Fish Abundance

Most fishers were responsive and helpful during face-to-face interviews. We collected a total of 39 questionnaires and 28 maps. Following contacts with the Irish Sea Sea Fisheries Committees, the Fleetwood Fish Forum provided a high-resolution chart detailing the seasonal distribution of commercial fish species in the eastern Irish Sea (Figure 2). This map represents the aggregated knowledge of 50 fishers gathered over a period of ca. 20 years. More responses were obtained from contacts with Sea Fisheries Committees and Fisheries Producer Organizations, but many of these questionnaires were answered by fishers who worked outside the study area or targeted other species. These questionnaires could not be included in this analysis, so only 18 of these maps are included in Figure 3 (includes the Fleetwood chart counted as $n = 1$).

A total area of 13,695 km^2 of fishing grounds were plotted for Atlantic cod, 5,173 km^2 for haddock, and 11,446 km^2 for European whiting. The locations of fishing grounds for Atlantic cod and European whiting were similar (Figure 3A, 3B). The main fishing grounds were located between the Isle of Man and Scotland, around the Solway Firth, north of England and Wales. Similarly, groundfish survey data indicated that the highest mean ranks for Atlantic cod were situated off the Ribble estuary, Belfast Lough, Anglesey/Colwyn Bay, Solway Firth, and the central Irish Sea (Figure 3A). Several fishers independently plotted areas in this region and off the northern coast of Wales, which increases our confidence in these data. There was broad consistency between the European whiting fishing grounds indicated by fishers and the distribution of high mean ranks of European whiting (Figure 3B), although no fishing grounds were plotted off the Ribble estuary, which had a consistent high mean rank abundance of European whiting. Fishing grounds for haddock were largely located along the Irish coast and the Solway Firth. The distribution of high haddock mean ranks was similar to the distribution of fishing grounds, although groundfish surveys indicated low abundances in the northeastern Irish Sea, where several fishers highlighted grounds of key importance (Figure 3C). No haddock fishing grounds were outlined at the low abundance stations off the English coast.

Table 1. Sediment conversions used in calculations.

British Geological Survey classification	Attributed "folk" classification
Diamicton	Mud
Gravel	Gravel/shingle
Gravelly mud	Mud
Gravelly muddy sand	Sand
Gravelly sand	Sand
Mud	Mud
Muddy sand	Mud
Muddy sandy gravel	Gravel/shingle
Rock and sediment	Hard ground
Rock or diamicton	Hard ground
Sand	Sand
Sandy gravel	Gravel/shingle
Sandy mud	Mud
Slightly gravelly mud	Mud
Slightly gravelly muddy sand	Sand
Slightly gravelly sand	Sand
Slightly gravelly sandy mud	Mud
Undifferentiated solids	Hard ground

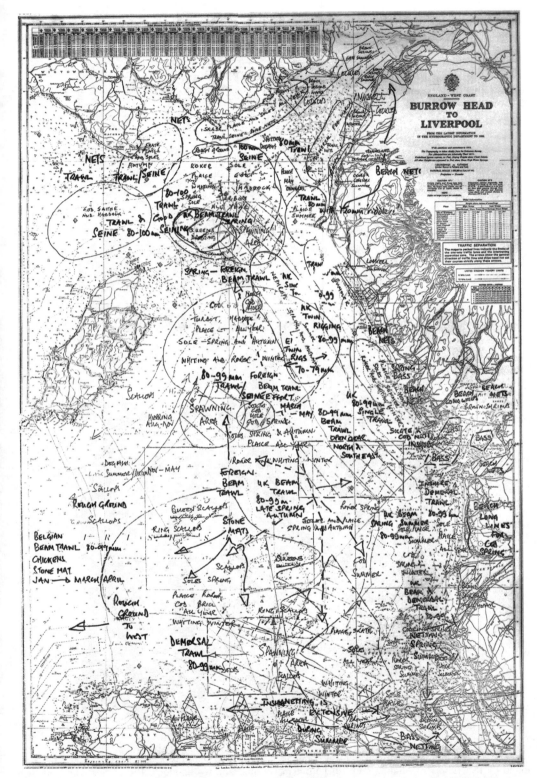

Figure 2. Chart with important fishing ground locations provided by the Fleetwood Fish Forum (approval for publication from fishers has been granted).

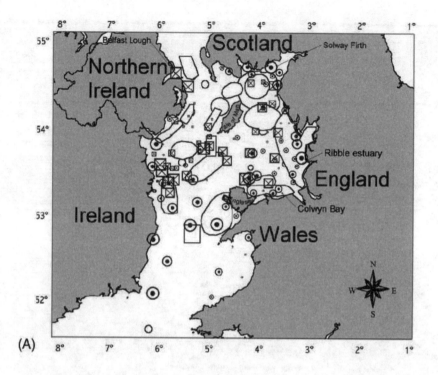

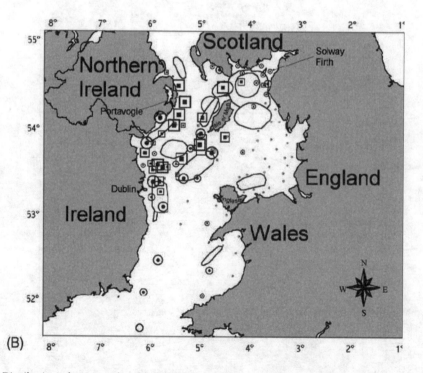

Figure 3. Distribution of mean ranks of fish abundance in the Irish Sea from 1990 to 2001 and fishing ground plotted by fishers for (A) Atlantic cod, (B) European whiting, and (C) haddock. The circles with dots in the center are CEFAS mean ranks and the squares with dots in the middle are DARD mean ranks of fish abundance.

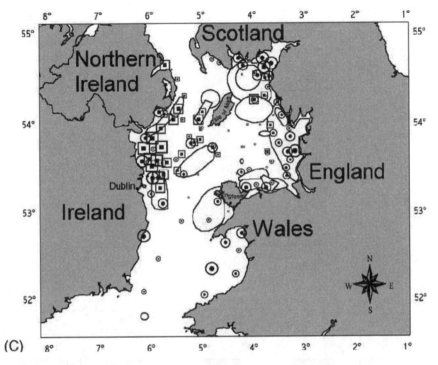

Figure 3. Continued.

Only seven of the fishers who plotted fishing grounds disclosed their identity sufficiently to facilitate a comparison between the location of their plots and records of vessel sightings by fishery protection aircraft (Table 2). Of these, only two vessels provided plots that coincided with sightings reported by aircrafts.

Questionnaires

Question 1
Atlantic cod, haddock, and European whiting were targeted by 16 of a total 39 respondents. The most important ground types stated for Atlantic cod included sand (29%) and mud (29%) (Figure 4). The most frequently stated habitat features for Atlantic cod included sand, feed (which we interpret to mean the ground that contained food for the fish), hard grounds (each 25%), wrecks and gravel (each 19%), mixed grounds, and mussel beds (each 6%). By comparison, the areas that were plotted by fishers as important grounds for Atlantic cod and overlaid with sediment maps of the British Geological Survey (BGS 2002) consisted predominantly of sand (55%) followed by mud (25%) and gravel or shingle (19%) (Figure 5). Haddock grounds contained similar proportions of sediment types. However, fishers most frequently stated hard grounds (31%) and sand (15%) as important ground types for this species. Furthermore, they named hard grounds (25%), brittle star beds (19%), feed (19%), gravel, sand, mud (13%), seaweed (we interpreted this to mean emergent growths of weed-like bryozoans), and mixed grounds (6%) as important habitat features for haddock.

Fishers responded that mud (31%) and sand (27%) were important grounds for European whiting (Figure 4). However, most of the European whiting grounds plotted

Table 2. Sightings of seven vessels associated with questionnaire respondents showing the total number of sightings of that vessel made during Department for Environment Food and Rural Affairs enforcement overflights, the number of those sightings that coincided with areas plotted on charts (number of matching sightings), the total number of grounds plotted by those fishers, and the number of those grounds in which that vessel was sighted (number of matching plots).

Fishing vessel	Total sightings	Matching sightings	Total grounds plotted by fishers	Matching plots
1	77	8	1	1
2	54	0	1	0
3	26	24	6	5
4	83	0	2	0
5	98	0	2	0
6	18	0	2	0
7	22	0	1	0

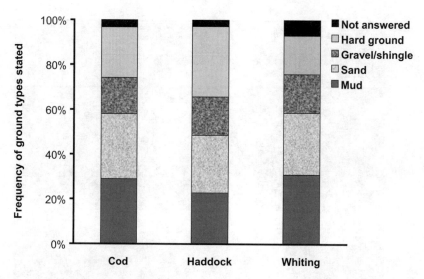

Figure 4. Frequency of important ground types as stated by fishers (*n* = 39) for Atlantic cod, haddock, and European whiting during interviews and mail questionnaires.

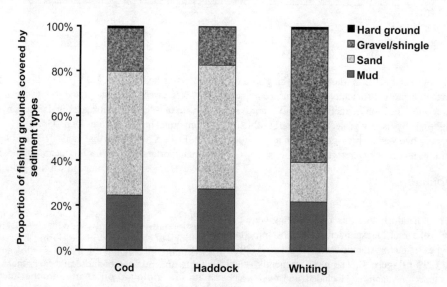

Figure 5. Proportion of fishing grounds plotted by fishers that are covered by different sediment types based on British Geological Survey data.

by fishers were characterized by gravel or shingle (60%), followed by mud (22%) and sand (18%), according to British Geological Survey maps (British Geological Survey 2002; Figure 5). The most frequently stated habitat features included hard grounds (19%), mud, sand, gravel (13%), seagrass (the respondent used the term seagrass, but we doubt that the angiosperm plant was meant given its restricted distribution in the Irish Sea; it seems more likely that he used this term for seaweed or weed-like bryozoans or hydroids), and soft corals (*Alcyonium digitatum*, 6%).

Question 2

Heavy fishing gear such as beam trawls, scallop dredges, and twin otter trawls were named as important factors that affect targeted habitats (21% of the 39 respondents). Other factors stated included fishing effort (21%), feed (15%), weather (15%), and season (13%).

Question 3

Fifty-six percent of the 39 respondents thought that fishing gear had altered their grounds. There was a high response rate to this question (96%).

Questions 4–6

A third of the respondents in the present study reported they had observed changes in catches of their target species such as a decrease in numbers (74%) and size (35%). Only 5% of the respondents stated that there was no change in catches of their target species (Table 3). Observed changes were attributed to overfishing (56%), climate change (38%), pollution (36%), changes in fishing gear (28%), and prey abundance (23%).

Discussion

Fishing Ground Locations and Distribution of Mean Ranks of Fish Abundance

While many fishers responded to questionnaires, only a proportion of these were willing to outline important fishing grounds on charts. This reluctance was related to reasons of confidentiality and due to a suspicion that this information might lead to negative management developments for fishers. For example, fishers might be concerned that the information may be used to identify areas for closure or the imposition of further restrictions to fishing (Pederson and Hall-Arber 1999).

When consulting with fishers, scientists need to modify their terminology to find common ground and to bridge social gaps. Therefore, during our survey, we used the term "ground" rather than "habitat." However, EFH does not necessarily constitute good fishing grounds (i.e., grounds that are amenable to fishing and provide economically viable catches). For example, an EFH characterized by topographically complex hard grounds also increases the risk of damage to fishing gear and could jeopardize the safety of the fishing vessel. Nevertheless, good fishing grounds should be indicative of potential EFH, as fishers are unlikely to prosecute areas that do not yield economically viable catches.

The areas of highest fish densities as identified from survey data did not always coincide with grounds plotted by fishers. This may partly reflect a local bias in the port of origin of many of the respondents that attended the fishing exhibition in Scotland. Although we targeted a fishing exhibition in Northern Ireland, the number of attendees was an order of magnitude lower (100 versus 1,000). Nevertheless, it is interesting that two Irish fishers also outlined grounds off the Solway Firth. A greater sample size, involving more fishers from Northern Ireland, would increase the balance when comparing among fishers' data and the groundfish survey data. It is possible that the spatial bias observed could be circumvented by restricting a spatial analysis of the groundfish survey data to subsets of data in the vicinity of respondents' ports.

The similarity of the fishing grounds outlined for the three different gadoids reflects, to some extent, the fact that several fishers did not distinguish among species when outlining fishing areas on the charts provided. In those cases, it was assumed that respondents fished for all their target species in the area outlined, although we recognize that it may have been a prime ground for one particular species.

The fishers' information has independently corroborated that high-density sites as plotted by the groundfish surveys are indicators of areas targeted by fishers and, therefore, are candidates for further research. The areas that were independently highlighted by several fishers in the northern Irish Sea and off Ireland and northern Wales presumably have features that consistently attract fish in sufficient numbers and quality to be of interest to fishers. Consequently, we have undertaken further surveys in these areas (Belfast Lough, Anglesey, Dublin, Ribble estuary) to investigate why they consistently attract fish.

Table 3. Responses to questions 4 (Q4), 5 (Q5), and 6 (Q6) posed in questionnaires ($n = 39$ unless stated otherwise; f = frequency of category checked; %= percentage of frequency).

Characteristic	Frequency of category checked	%
Changes over time (Q4)		
Target species	12	31
Bottom animals and plants	12	31
Habitat structure	3	8
Fish health	1	3
Bycatch	7	18
No changes	5	13
Other changes	5	13
Not answered	9	23
Changes in target species (Q5)		
No change	2	5
Increase	5	13
Decrease	29	74
Moved to other areas	5	13
Replaced by another species	2	5
Decrease in size ($n = 26$)	9	35
Other changes	1	3
Not answered	4	10
Cause of change (Q6)		
Climate	15	38
Pollution	14	36
Changes in fishing gear	11	28
Changes in prey abundance	9	23
Habitat loss	3	8
Overfishing	22	56
Other causes	4	10
Not answered	7	18

Some of the discrepancies between the fishers' charts and the groundfish survey data may also lie in the fact that there were relatively few sampling stations located between the northern Isle of Man, northwestern Scotland, and northwestern England. This is probably due to differences in the gear historically used for the CEFAS groundfish survey, a beam trawl, which is not ideally suited for use over the rough grounds around the Isle of Man. However, recent studies from the northwestern Atlantic indicate a preference of young Atlantic cod and haddock for habitats of coarse sediment interspersed with rocks (Lough et al. 1989; Gotceitas et al. 1995; Gregory and Anderson 1997; Lindholm et al. 1999); hence, these areas may have been missed or avoided during the beam trawl survey. Conversely, the groundfish survey probably includes areas that fishers normally avoid because they would catch too much "rubbish" (inert material and bycatch of nontarget species) that might clog up and damage their nets and catch during the longer commercial tows.

Although no filter was incorporated in our questionnaires to test if questions were answered truthfully (Johannes 1981; Maynou and Sardà 2001), we believe that most respondents answered the questions to the best of their knowledge. There were, however, some discrepancies between the ground types that fishers stated as important for each fish species and the sediment composition (determined from British Geological Survey data) of grounds that they plotted on charts. Large areas of the Atlantic cod and haddock fishing grounds were characterized by sand, although this sediment type was mentioned less frequently in questionnaires. Similarly, the European whiting fishing grounds were composed of a much higher proportion of gravel or shingle than was stated in questionnaires. Generally, hard grounds were named as an important habitat feature more often than would appear from the features of plotted fishing grounds, which might be explained with differences between the British Geological Survey-converted sediment classification and the categories used by fishers. For example, gravel or shingle may have been termed hard grounds by some respondents. Also, muddy sand (here termed as mud) may have been classed as sand by some fishers. Alternatively, the resolution of fishers' knowledge may exceed the sampling resolution of the British Geological Survey sediment data, which relied on interpolation between widely interspersed sample points.

Further discrepancies became evident when comparing the location of fishing grounds plotted by individual respondents with aircraft patrol sightings for the same fishing vessel. This could have several reasons. Fishers often target different species throughout the seasons, which also affects their fishing locations. When sighted by aircraft, a vessel may have been fishing for a species other than that for which fishing grounds were plotted. Also, weather conditions and fishing restrictions may have prevented respondents from fishing in their favored areas, leading to the observed discrepancies.

Maurstad (2000) highlighted that the publication of maps and other information given by fishers in a purely scientific context can put scientists into a dilemma in terms of intellectual property rights and confidentiality. Also the knowledge becomes separated from its sociological context. We decided to publish our results, however, as we feel that the quality of the charts presented here is not sufficiently accurate to pose a threat to any individual respondent's livelihood. Also, it is likely that the information volunteered is known by many fishers.

Questionnaires

Fishers named a wide range of ground types of similar importance for all three fish species. This may indicate that adult Atlantic cod, haddock, and European whiting are caught over a variety of seabed types and that they may be "habitat generalists." In a similar study in the United States, fishers indicated that they preferred to fish for whiting on fine-grained sediments, whereas other groundfish were targeted across all habitat categories (Pederson and Hall-Arber 1999).

Interestingly, three fishers stated independently that "wigs" (brittlestar beds) are an important habitat feature for haddock, especially after spawning. Although fishers suggested that haddock sought out brittle star beds to "clean themselves" after spawning, it is known that haddock feed on brittlestars as a grinding substance in their stomachs (Mattson 1992). This emphasizes the potential value of apparently obscure observations made by fishers.

A few fishers noted that weed (possibly hydroids or the widespread bryozoan *Flustra* spp.) was often found in their haddock catches, and one fisher also associated European whiting with dead men's fingers (*Alcyonium digitatum*, a soft coral). Such structures may provide fish with shelter from predators or act as foci of prey species (e.g., pandalid shrimps). These features of fish habitats are currently the subject of further investigation (Freeman et al. 2002). Similar to the findings of Pederson and Hall-Arber (1999), few fishers commented on habitat features other than ground types (see above), and such features were given in interviews rather than in mailed questionnaires. Fishers are often ignorant of species names, especially those of nontarget invertebrates, and seem unwilling to offer their own interpretation that may be proven incorrect (Mackinson 2001). It was easier to expand questions during interviews through explanations and by showing images of marine animals that fishers would recognize. In a more comprehensive survey, the provision of a standard photo card showing common marine animals could help to increase the response rate and train fishers, who are often keen to expand their knowledge of the marine environment.

More than 50% of the respondents believed that fishing gear has, in some way, altered their grounds. Many recent studies have shown that towed bottom fishing gears have altered the seabed (Jennings and Kaiser 1998). Fishers were also concerned about heavy mobile fishing gear such as scallop dredges, beam trawls, and twin otter trawls. Similarly, Collie et al. (2000) showed that scallop dredging has one of the greatest initial impacts on benthic biota. Fishers tended to attribute habitat changes to gear types that were not used by themselves. Less than a third of the respondents polled in a study in the United States believed that fishing gear had changed their grounds (Pederson and Hall-Arber 1999). This difference may be attributed to the fact that in Pederson and Hall-Arber's study, fishers were asked if the their own gear had altered the grounds; thus, more than 50% of the fishers identified mobile gear as the most important factor that affected habitats (different question). Also, fishing is more intense in Europe, and different countries tend to dominate different gear sectors (e.g., The Netherlands and Belgium operate the largest beam trawl fleets in northern Europe).

Only a few fishers commented on habitat loss through time, although many fishers stated that fishing gear smoothes seabed topography and "damages the ground." It is possible that once stated, fishers thought it unnecessary to repeat the statement in subsequent questions of the questionnaire. Also, fishers may have been unfamiliar and, therefore, uncomfortable with the term "habitat," although the meaning was explained either verbally or on enclosed information leaflets, and the word "ground" was used instead in most questions.

Although more time-consuming, questionnaire-based face-to-face interviews yielded the best data, enabled the establishment of trust between the scientist and the fisher, and allowed for elaboration of specific questions when technical terms were unclear. Our consultation with fishers has not only added to the credibility of the study and any future management decisions that may rely on its findings (Maurstad and Sundet 1994) but has also highlighted how our current knowledge can be expanded. Most importantly, it has helped us to pinpoint areas that may constitute EFH for further comprehensive habitat surveys. One drawback of using fishers' knowledge and data from groundfish surveys in order to locate possible EFH is that only trawlable areas are included. Although certain areas may be more suitable for gadoids than other trawlable areas, they may not necessarily constitute EFH. For example, a high abundance station may be located next to an EFH such as a rocky reef or wreck that was saturated with fish, such that some fish spill over into the second best habitat that is amenable to sampling with a trawl gear. Jagielo et al. (2003) found significant differences in the density of several flatfish and rockfish species in trawlable and untrawlable habitats. Additional interviews with scuba divers, sea anglers, and fishers that use fixed gears may yield information from a wider range of habitats in future research.

Further insights may be gained by an analysis of statements made in questionnaires, which are then integrated with biological data using fuzzy logic (Mackinson 2000). The integration of fishers' knowledge into science and management is a potentially invaluable tool that should not be overlooked (Pederson and Hall-Arber 1999).

Acknowledgments

The authors thank all the fishers who participated in this study. The project was funded by the Department for Environment Food and Rural Affairs (DEFRA) contract MF0805 and the European Commission contract Q5RS-2002–00787. We would like to thank Jason Hall-Spencer and an anonymous referee whose valuable suggestions helped to improve an earlier version of this manuscript.

References

Baelde, P. 2001. Fishers' description of changes in fishing gear and fishing practices in the Australian South East Trawl Fishery. Marine and Freshwater Research 52:411–417.

Benaka, L. 1999. Fish habitat: essential fish habitat and rehabilitation. American Fisheries Society, Symposium 22, Bethesda, Maryland.

British Geological Survey. 2002. Sea-bed sediments around the United Kingdom (digital data, version 1.0). Kingsley Dunham Centre, License 2003/133 British Geological Survey, Nottingham, UK.

Coleman, F., J. Travis, and A. B. Thistle, editors. 2000. Essential fish habitat and marine reserves. Bulletin of Marine Science (special issue) 66(3).

Collie, J. S., S. J. Hall, M. J. Kaiser, and I. R. Poiner. 2000. A quantitative analysis of fishing impacts on shelf-sea benthos. Journal of Animal Ecology 69:785–798.

Dayton, P., S. F. Thrush, M. T. Agardy, and R. J. Hofman. 1995. Environmental effects of marine fishing. Aquatic Conservation: Marine and Freshwater Ecosystems 5:205–232.

DEFRA (Department for Environment Food and Rural Affairs). 2000. United Kingdom sea fisheries statistics 1999 and 2000. DEFRA, The Stationery Office, London.

Ebert, D., and S. P. Filipek. 1988. Response of fish communities to habitat alternation in a small Ozark stream. Arkansas Academy of Science Proceedings 42:28–32.

Ellis, J. R., M. J. Armstrong, S. I. Rogers, and M. Service. 2002. The distribution, structure and diversity of fish assemblages in the Irish Sea. Pages 93–107 in J. D. Nunn, editor. Marine biodiversity in Ireland and adjacent waters. Ulster Museum, Belfast, Ireland.

FAO (Food and Agriculture Organization of the United

Nations). 1995. The state of the world fisheries and aquaculture. FAO, Rome.

Freeman, S. M., M. Bergmann, H. Hinz, M. J. Kaiser, and J. Bennell. 2002. Acoustic seabed classification: identifying fish and macro-epifaunal habitats. International Council for the Exploration of the Sea Council Meeting Papers, ICES-CM-2002/K:08, Copenhagen.

Freire, J., and A. García-Allut. 1999. Integration of fishers' ecological knowledge in fisheries biology and management. A proposal for the case of the artisanal coastal fisheries of Galicia (NW Spain). International Council for the Exploration of the Sea Council Meeting Papers, ICES-CM-1999/S:7, Copenhagen.

Gotceitas, V., S. Fraser, and J. A. Brown. 1995. Habitat use by juvenile Atlantic cod (*Gadus morhua*) in the presence of an actively foraging and non-foraging predator. Marine Biology 123:421–430.

Gregory, R. S., and J. T. Anderson. 1997. Substrate selection and use of protective cover by juvenile Atlantic cod *Gadus morhua* in inshore waters of Newfoundland. Marine Ecology Progress Series 146:9–20.

Jagielo, T., A. Hoffmann, J. Tagart, and M. Zimmermann. 2003. Demersal groundfish densities in trawlable and untrawlable habitats off Washington: implications for the estimation of habitat bias in trawl surveys. U.S. National Marine Fisheries Service Fishery Bulletin 101:545–565.

Jennings, S., and M. J. Kaiser. 1998. The effects of fishing on marine ecosystems. Advances in Marine Biology 34:201–352.

Jennings, S., J. K. Pinnegar, N. V. C. Polunin, and K. J. Warr. 2001. Impacts of trawling disturbance on the trophic structure of benthic invertebrate communities. Marine Ecology Progress Series 213:127–142.

Johannes, R. E., 1981. Words of the lagoon: fishing and marine lore in the Palau District of Micronesia. University of California Press, Berkeley.

Keast, A., J. Harker, and D. Turnbull. 1978. Nearshore fish habitat utilization and species associations in Lake Opinicon (Ontario, Canada). Environmental Biology of Fishes 3:173–184.

Koehn, J. D. 1993. Freshwater fish habitats: key factors and methods to determine them. Pages 77–83 *in* D. A. Hancock, editor. Sustainable fisheries through sustaining fish habitat. Australian Government Publishing Services, Canberra.

Lindholm, J. B., P. J. Auster, and L. S. Kaufman. 1999. Habitat-mediated survivorship of juvenile (0-year) Atlantic cod *Gadus morhua*. Marine Ecology Progress Series 180:247–255.

Lough, R. G., C. V. Page, D. C. Potter, P. J. Auditore, G. R. Bolz, J. D. Neilson, and R. I. Perry. 1989. Ecology and distribution of juvenile cod and haddock in relation to sediment type and bottom currents on eastern Georges Bank. Marine Ecology Progress Series 56:1–12.

Mackinson, S. 2000. An adaptive fuzzy expert system for predicting structure, dynamics and distribution of herring shoals. Ecological Modelling 126:155–178.

Mackinson, S. 2001. Integrating local and scientific knowledge: an example in fisheries science. Environmental Management 27:533–545.

Marrs, S., I. Tuck, R. Atkinson, T. Stevenson, and C. Hall. 2002. Position data loggers and logbooks as tools in fisheries research: results of a pilot study and some recommendations. Fisheries Research 58:109–117.

Mattson, S. 1992. Food and feeding habits of fish species over a soft sublittoral bottom in the northeast Atlantic. 3. Haddock (*Melanogrammus aeglefinus* (L.)) (Gadidae). Sarsia 77:33–45.

Maurstad, A. 2000. Trapped in biology. Pages 135–152 *in* B. Neis and L. Felt, editors. Finding our sea legs—linking fishery people and their knowledge with science and management. ISER Books, St. John's, Canada.

Maurstad, A., and J. Sundet. 1994. Improving the link between science and management: drawing upon local fishers' experience. International Council for the Exploration of the Sea Council Meeting Papers, ICES-CM-1994/T:20, Copenhagen.

Maynou, F., and F. Sardà. 2001. Influence of environmental factors on commercial trawl catches of *Nephrops norvegicus* (L.). ICES Journal of Marine Science 58:1318–1325.

Moore, P. G. 2003. Seals and fisheries in the Clyde Sea area (Scotland): traditional knowledge informs science. Fisheries Research 63:51–61.

Pederson, J., and M. Hall-Arber. 1999. Fish habitat: a focus on New England fishermen's perspectives. Pages 188–211 *in* L. R. Benaka, editor. Fish habitat: essential fish habitat and rehabilitation. American Fisheries Society, Symposium 22, Bethesda, Maryland.

Sardà, F., and F. Maynou. 1998. Assessing perceptions: do Catalan fishermen catch more shrimp on Fridays? Fisheries Research 36:149–157.

Symonds, D. J., and S. I. Rogers. 1995. The influence of spawning and nursery grounds on the distribution of sole *Solea solea* (L.) in the Irish Sea, Bristol Channel and adjacent areas. Journal of Experimental Marine Biology and Ecology 190:243–261.

Taylor, R., 1998. Another approach to scallop production, habitat concerns and biodiversity. Pages 111–114 *in* E. M. Dorsey and J. Pederson, editors. Effect of fishing gear on the sea floor of New England. Conservation Law Foundation, Boston.

Williams, A., and N. Bax. 2003. Integrating fishers' knowledge with survey data to understand the structure, ecology and use of a seascape off southeastern Australia. Pages 238–245 *in* N. Haggan, C. Brignall, and L. Wood, editors. Putting fishers' knowledge to work. The Fisheries Centre, University of British Columbia, Vancouver.

Delineating Juvenile Red Snapper Habitat on the Northern Gulf of Mexico Continental Shelf

William F. Patterson,[1] Charles A. Wilson, Samuel J. Bentley, and James H. Cowan

Department of Oceanography and Coastal Studies, Louisiana State University, Baton Rouge, Louisiana 70808, USA

Tyrrell Henwood

National Marine Fisheries Service, Southeast Fisheries Science Center, Pascagoula Laboratory, Pascagoula, Mississippi 39568-1207, USA

Yvonne C. Allen and Triniti A. Dufrene

Department of Oceanography and Coastal Studies, Louisiana State University, Baton Rouge, Louisiana 70808, USA

Abstract. A database of resource survey trawl samples was analyzed to determine if patterns in spatial variability of estimated density of juvenile red snapper *Lutjanus campechanus* in an approximately 15 × 10^3-km^2 area in the north–central Gulf of Mexico were consistent among years from 1991 through 2000. Areas that consistently produced high ($n = 1$), median ($n = 2$), or low ($n = 1$) estimated juvenile red snapper density during this time series then were mapped with digital side-scan sonar, and differences in acoustic reflectance of the seabed were groundtruthed with sediment analyses of boxcore samples. Spatial variability in juvenile density estimated from trawl samples ($n = 80$) in summer and fall 2001 were similar to historic patterns. Juvenile density was significantly higher in areas with shell rubble or sponge habitat, thus indicating juveniles require habitat with small-scale (cm to m) complexity. Results of this study indicate our mapping techniques were effective in delineating juvenile red snapper habitat, but future studies also should examine diet, growth, and mortality of juveniles to distinguish suitable versus essential habitats.

Introduction

One of the most pressing federal fisheries management concerns in the Gulf of Mexico (GOM) region of the United States is the overfished status of red snapper *Lutjanus campechanus*. Since the 1980s, increased knowledge of GOM red snapper life history and population dynamics, improvements in stock assessment methods, and federal legislation have resulted in changes to red snapper stock biomass rebuilding schedules and management regulations. However, two themes have remained omnipresent throughout the evolution of federal management during the past two decades. The first is that despite increased fishery regulations, estimated spawning potential ratio has remained dangerously low (<10%) for the stock (Schirripa and Legault 1999). The second persistent theme is bycatch of age-0 and age-1 red snapper by the GOM shrimp fleet has been the greatest source of mortality to red snapper (Goodyear 1995; Ehrhardt and Legault 1998; Schirripa 1998; Schirripa and Legault 1999).

Fishery biologists conjectured as to the possible role of shrimp trawl bycatch in declining red snapper abundance as early as the 1960s (Moe 1963; Bradley and Bryan 1975; Gutherz and Pellegrin 1986). Recent estimates indicated mortality of juvenile red snapper due to shrimp trawls approached 90% (prior to implementation of bycatch reduction devices [BRDs]) (Goodyear 1995). Results from simulation analyses also indicated that without significant reduction (i.e., >> 50%) of juvenile red snapper bycatch in shrimp trawls, the directed red snapper fisheries must be severely restricted or closed (Goodyear 1995; Schirripa 1998; Schirripa and Legault 1999). These results provided im-

---

[1] E-mail: wpatterson@uwf.edu; present address: Department of Biology, University of West Florida, 11000 University Parkway, Pensacola, Florida 32514, USA.

petus for the Gulf of Mexico Fishery Management Council to require BRDs in shrimp trawls to reduce bycatch of juvenile red snapper (GMFMC 1996), and BRDs became mandatory in the western GOM beginning in May 1998 (U.S. Office of the Federal Register 63:71[April 14, 1998]:18139–18147).

The decision to require GOM shrimpers to install BRDs in their nets was supported by National Marine Fisheries Service (NMFS) trawl experiments that demonstrated a BRD design, known as the EE-Fisheye, reduced juvenile red snapper bycatch by 59% (Watson et al. 1997). More recent analyses call to question whether the EE-Fisheye, or BRDs in general, will sufficiently reduce bycatch by excluding juveniles from shrimp trawls (Engaas et al. 1999; Gallaway and Cole 1999; Rogers 1999). New concerns also have been raised regarding the fate of juveniles excluded from trawls. Additional mortality may result from embolism after being brought from depth (I. Workman, NMFS Pascagoula Laboratory, personal communication) or from predation by piscivores following the net (Broadhurst 1998). In both of these cases, survival of excluded red snapper will be less than 100%, and the effectiveness of BRDs would be compromised (Crowder and Murawski 1998).

Therefore, the problem of juvenile red snapper bycatch in shrimp trawls may not have a wholly technological solution. To achieve reductions in bycatch required to increase spawning stock biomass of red snapper, it may be necessary to augment the BRD program with shrimp trawl time–area closures that would provide refuges for juvenile red snapper (Gallaway and Cole 1999; Gallaway et al. 1999). Essential to any consideration of shrimp trawl time–area closures is a fundamental understanding of juvenile red snapper habitat requirements and knowledge of where juvenile GOM red snapper essential fish habitat exists.

Juvenile red snapper (age-0 and age-1 fish) have been reported from a variety of habitats including open sand, relict shell rubble, and artificial structures with vertical relief, but results of habitat preference studies have been equivocal (Bradley and Bryan 1975; Holt and Arnold 1982; Workman and Foster 1994; Szedlmayer and Howe 1997; Gallaway et al. 1999; Szedlmayer and Conti 1999). Laboratory and small-scale in situ experiments have demonstrated juvenile red snapper display an affinity for low-relief shell rubble habitat (Szedlmayer and Howe 1997; Lee 1998; Szedlmayer and Conti 1999). Results from mesoscale (km^2) and large-scale (100 km^2) studies on the shelf, however, indicated no difference in juvenile abundance between sand-silt and shell rubble bottom types and that vertical relief was a nonsignificant factor in explaining GOM-wide variance in juvenile red snapper abundance (Workman and Foster 1994; Gallaway et al. 1999).

Our study was conducted to gain greater understanding of juvenile red snapper habitat requirements in an area on the north-central GOM continental shelf that historically has supported a range of juvenile catch rates in annual resource surveys. The objectives of our study were to (1) determine if patterns existed in the spatial and temporal variability in juvenile red snapper density estimates from NMFS trawl surveys conducted over several years; (2) map shelf areas that historically supported high, median, and low juvenile densities with digital side-scan sonar; and (3) relate geotechnical properties of the seabed to historic and contemporary juvenile red snapper density. A priori, we hypothesized the seabed in areas that historically supported high densities of juvenile red snapper would be characterized by shell rubble habitat. To test this hypothesis, we characterized the seabed of selected areas with a combination of side-scan sonar and boxcore sediment samples and then related sediment type to historic and contemporary juvenile red snapper density estimates.

Methods

Sample Area Selection

Historic juvenile red snapper catch data were obtained from the NMFS's Fall Groundfish Survey (FGS; SEAMAP Information System, National Marine Fisheries Service, Pascagoula, Mississippi), which is an annual resource survey conducted in the northern Gulf of Mexico since the early 1970s with standardized sampling gear (e.g., a single 12.8-m, four seam semiballoon shrimp trawl rigged with 2.4-m × 1-m doors, a 54.9-m bridle, and a tickler chain set 1.1 m shorter than the trawl footrope, towed at approximately 4.6 km/h). Juvenile red snapper catch data from individual trawl samples originally were sorted by time of collection (day or night; ratio of day versus night samples approximately 1:1) and imported into a geographic information system database. Trawl station locations within an area of approximately 15×10^3 km^2 off Alabama, Mississippi, and eastern Louisiana were inspected visually to determine their spatial coverage for years 1991–2000 for both day-collected and night-collected samples. We determined there was sufficient spatial replication in the annual distribution of trawl samples to permit examination of estimated juvenile red snapper density in 10' latitude by 10' longitude cells (approximately 350 km^2) as did Gallaway et al. (1999). Therefore, our original large area of the shelf was divided into 40 cells of this size. Preliminary analysis of annual differences in trawl catches within each cell revealed no consistent pattern of higher catches during day versus night sampling; thus, night and day samples were combined. After juvenile density (individuals/ha) was computed for all individual trawl samples in the combined data set, annual mean juvenile density was computed for each cell in each year of the time series.

Spatial and temporal variability in juvenile catch rates were examined to select cells that consistently produced high, median, or low juvenile densities. All cells were not sampled every year, however, and GOM red snapper year-class strength varied among years (Schirripa and Legault 1999). We attempted to account for these potential biases by standardizing annual cell-specific mean juvenile densities for a given year by scaling them to the mean density of all cells in that year. Resultant unitless values were treated as a standardized index of juvenile red snapper density. Mean index value was computed across the entire time series for cells that were sampled in at least 4 of the 10 years of the time series. These were plotted on a map of the sample region to examine visually the spatial variability in juvenile density.

Seabed Characterization

Based on results of the above analysis, the seabed of four cells that historically produced a range of juvenile red snapper density estimates was characterized by surveying an approximately 8-km^2 area at the center of each cell with digital side-scan sonar and then groundtruthing differences in acoustic reflectance with boxcore sediment samples. In adopting this strategy, we assumed the seabed of the center 8-km^2 area of each cell reflected the predominant sediment types found throughout the cell, which was based on prior studies that evidenced patterns in regional sediment characteristics were consistent over large areas (10s to 100s km^2; Parker et al. 1992; Schroeder et al. 1995; Strelcheck 2001).

Our digital side-scan system consisted of a Klein (Salem, New Hampshire) model 2260NV dual frequency (100/500 kHz) tow fish, T2100 transceiver, and a high fidelity, low loss armored single conductor coaxial tow cable. Data were acquired simultaneously on port and starboard channels at 100 and 500 kHz using the Isis sonar system (version 5.75, Triton Imaging, Inc., Watsonville, California). Data from the side-scan tow fish and transceiver were georeferenced using a C&C Technologies (Lafayette, Louisiana) Ashtec global positioning system (GPS) receiver and a SatLoc (Scottsdale, Arizona) (sub-meter accuracy) differential beacon receiver. Real-time vessel position was superimposed on a nautical chart in ArcPad (ESRI, Inc., Redlands, California) and displayed to aid navigation. Following acquisition, the 100-kHz data were postprocessed using Isis and Delphmap (version 2.00.04, Triton Imaging, Inc.), and the resulting mosaic was exported as an 8-bit unsigned (0–255) georeferenced tagged image file format file. Overlapping lines were merged in mosaic creation using maximum shinethrough to preserve the most intense acoustic return.

Boxcores (45 × 45 cm) were located using differential GPS interfaced with previously acquired digital side-scan mosaics for core placement. Twenty to 25 boxcore samples were taken in groups of 3–5 replicates at 5–7 stations in each surveyed 8-km^2 area. Station location was stratified based on acoustic reflectance intensity. For example, if a large patch of highly reflective seabed was observed, a group of cores was collected from the center of the patch. Cores also were collected near transitions between contrasting reflectance patterns. Following collection, 2 subsamples (15-cm^2 area × 4-cm depth) were taken from each core for determination of sedimentologic properties (e.g., surficial grain size, organic carbon content, and carbonate content). Grain size was determined by wet sieving samples through 63-mm mesh, analyzing the fine fraction in a Micromeritics (Norcross, Georgia) Sedigraph particle-size analyzer, analyzing the coarse fraction in a Gilson (Lewis Center, Ohio) Autosiever sonic siever, and then merging the coarse and fine data sets. Organic carbon and calcium carbonate content were calculated from loss on ignition after 4 h at 500°C and 950°C, respectively (Carver 1971).

Spatial coverage of different seabed types within each cell was estimated by relating reflectance intensity from side-scan mosaics to either sand:mud ratio or calcium carbonate content depending on predominant surficial sediments. First, 10-m^2 areas were constructed around boxcore sample locations in side-scan mosaics. Mean pixel values then were computed for each area with Imagine (Erdas, Inc., Atlanta, Georgia) image analysis software. Linear regressions were computed to relate sand:mud ratio or percent calcium carbonate content (an index of shell content) to mean pixel value (an index of acoustic reflectance) (SAS Institute, Inc. 1996). Finally, linear regressions were applied to pixel values from side-scan mosaics to estimate percent coverage of sediment with either greater than 50% sand content or greater than 30% calcium carbonate content.

Trawl Sampling

Trawl sampling was conducted in our four study cells to examine contemporary juvenile red snapper habitat utilization patterns. Survey areas within each cell were divided into 20 stations measuring 200 m (north to south) by 2 km (east to west). Five fixed stations then were randomly selected within each cell for trawl sampling on four cruises conducted between August and December 2001. Trawl sampling was conducted onboard the NMFS's R/V *Caretta*, an 18-m former commercial shrimp trawler converted for research purposes. All samples were collected between 0.5 h after sunrise and 0.5 h before sunset. Trawls were rigged following FGS protocols as described above; however, two trawls were fished simultaneously, one each on starboard and port outriggers. Prior to trawling at a given station, a Sea-Bird (Bellvue, Washington) conductivity–temperature–depth (CTD) sensor was deployed to measure salinity, temperature, dissolved oxygen, and depth. Trawls then were deployed with an approximate scope of 6:1 cable distance to water depth and were towed at approximately 4.6 km/h for 20 min across the long (east to west) axis of trawl stations. Vessel position was superimposed on a nautical chart in ArcPad to aid navigation, and distance towed was estimated from GPS coordinates of the ship's path. Area sampled was estimated by multiplying distance towed by the combined distance of trawl openings (24 m).

Catches from both trawls were combined on deck and treated as a single sample. Samples were weighed to the nearest 0.1 kg. Following subsampling for species composition analysis (not detailed here), all reef fishes (including juvenile red snapper) and sponges (phylum porifera) were selected from the sample before it was returned to the sea. Sponges were weighed to the nearest 0.1 kg and discarded. All juvenile red snapper were frozen in plastic bags and transported back to the laboratory where they were thawed, weighed to the nearest 0.1 g, and measured to the nearest mm total length (TL).

Juvenile red snapper density was computed as the number of individuals divided by ha trawled and sponge density as mass divided by ha trawled. Differences in red snapper densities among cells and sampling cruises were tested with analysis of variance (ANOVA) (SAS Institute, Inc. 1996). Prior to analysis, density data were transformed as $Y = \log(\text{density} + 1)$ to meet the assumptions of normality and homogeneity of variances. The relationship between juvenile red snapper density and sponge density in cell 11 was tested with correlation analysis (SAS Institute, Inc. 1996).

Results

The NMFS FGS sampled 513 stations during 1991–2000 in the area of the northern GOM continental shelf bounded by 89°10'W longitude to the west and 88°W longitude to the east (Figure 1). The FGS survey design was stratified by depth. Therefore, two clusters of station locations are apparent in the data, one centered around 30°N latitude and one centered around 29°20'N latitude. The number of stations sampled in a given 10'-latitude by 10'-longitude cell in a given year ranged from 0 (many cells throughout the time series) to 10 (cells 22 and 34 in 1991 and 1999, respectively). The mean number of cells (±SE) in which at least one station was sampled in a given year was 18.1 (±1.65) (range = 14 in 1991 and 1999 to 29 in 1993), but only 26 of 40 cells were sampled in at least 4 years of the time series.

Mean juvenile density index values were highest for cells 23 and 33 (Figure 2); however, the high index value for cell 33 was skewed by one sample in 1993 for

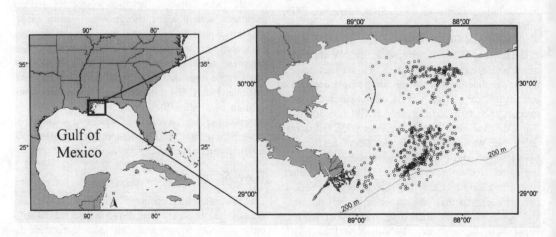

Figure 1. Map of the continental shelf in the north-central Gulf of Mexico depicting starting coordinates for trawl stations (*n* = 513) occupied by the National Marine Fisheries Service's Fall Groundfish Survey from 1991 to 2000. The 200-m isobath represents the shelf edge.

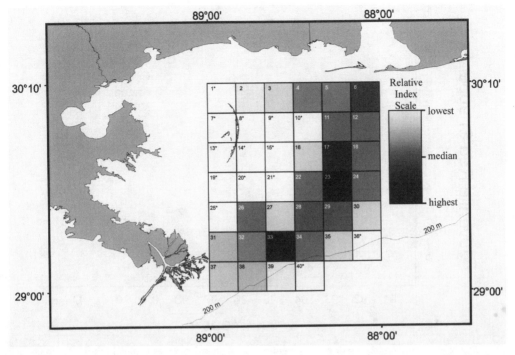

Figure 2. Map of the north-central Gulf of Mexico shelf with 10' latitude by 10' longitude cells demarcated by numbered rectangles. The relative index scale (from the figure legend) refers to annual mean juvenile red snapper among all cells for years 1991–2000 (see text for details). An asterisk following a cell's number indicates insufficient data to compute a mean index value.

which estimated juvenile density was 178.1 fish/ha. This value was 5.5-fold greater than any other trawl sample in 1993 and 2.6-fold greater than the next highest estimated density among all other samples. Therefore, cell 23 was selected as our high juvenile density cell. Cells 5 and 11 were selected randomly from a group of 10 cells whose average index values were near the median for the time series, and cell 16 was selected as our low juvenile density cell based on its low mean index value and proximity to the other three cells. Examination of juvenile density estimates for these four cells revealed cell 16 density estimates were always below the annual mean and cell 23 density estimates, while much more variable than those of cell 16, were always greater than annual mean (Figure 3). Juvenile density estimates for cells 5 and 11 also were variable, but annual mean density estimates for both were near the mean throughout the time series.

The center approximately 8-km² of cells 5 and 11 were surveyed with digital side-scan sonar on August 14–15, 2001, but due to technical difficulties, cells 16 and 23 were not surveyed until May 21–22, 2002. After side-scan sonar mosaics were created, boxcore samples from cells 5 and 11 were obtained during August 22–24, 2001, and for cells 16 and 23, during May 23–25, 2002.

Side-scan data revealed spatial heterogeneity of seabed acoustic reflectance in cells 5 and 23, but much more uniform reflectance patterns existed in cells 11 and 16 (Figure 4).

Cells 5 and 23 were characterized by regions with both high and low reflectance corresponding to high (shell rubble) and low (sand) $CaCO_3$ content, respectively (Table 1; Figure 4). Mud content of surficial sediments in both blocks was between 3% and 10% in both shell rubble and sand habitat types. Surficial sediments of shell rubble habitats contained approximately 50% $CaCO_3$ (Table 1), which consisted mostly of fragments and entire valves of the estuarine oyster *Crassostrea*, along with a range of fully marine species, including the genera *Strombus*, *Murex*, and *Oliva*. Shell rubble habitat in both blocks tended to have 1–2 m of positive relief above the surrounding seabed. In cell 5, shell rubble habitat occurred in ridges that were 100–200 m wide, kms in length, and were oriented along northwestern to southeastern axes. The orientation of shell rubble habitat in cell 23 was more irregular than in cell 5 but comprised nearly a quarter of the area surveyed.

Side-scan data from cell 11 indicated a relatively homogeneous seabed with minimal relief. All boxcore

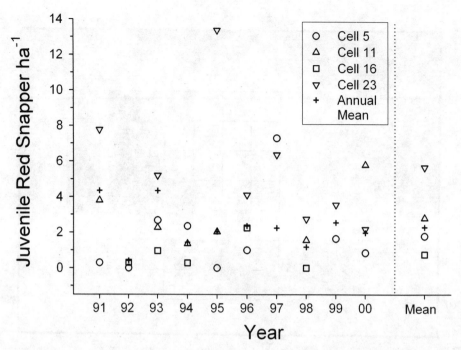

Figure 3. Annual and overall mean juvenile red snapper density estimates from the National Marine Fisheries Service's Fall Groundfish data for cells selected for seabed characterization.

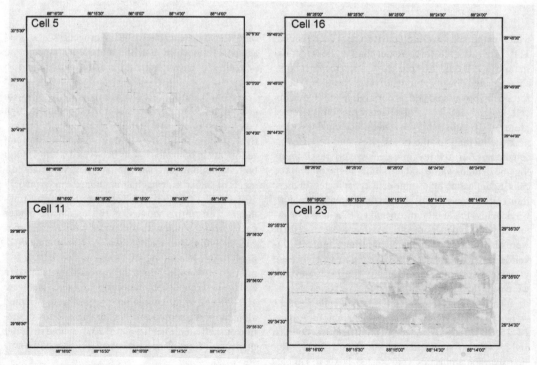

Figure 4. Acoustic reflectance mosaics of the center approximately 8-km² area of four 10' latitude by 10' longitude cells.

Table 1. Seabed geotechnical properties estimated from sediment analyses of boxcore samples taken in study cells. Percentage of survey area reported for each seabed type was extrapolated from application of linear regression equations relating sand:mud ratio (cell 16) or % CaCO3 (cells 5, 11, and 23) to reflectance (pixel) values from side-scan mosaics.

Cell	Seabed type	Mean % mud	Mean % sand	Mean % organic content	Mean % CaCO$_3$	% survey area
5	Sand	8.33	79.48	2.43	9.77	95.2
	Shell rubble (CaCO$_3$ > 30%)	7.12	38.58	3.06	51.25	4.8
11	Sand (sand:mud > 1.0)	20.23	63.63	2.62	13.52	100
16	Mud (sand:mud < 1.0)	58.22	24.51	7.75	9.52	65.7
	Sand (sand:mud > 1.0)	28.32	61.13	3.08	7.47	34.2
23	Sand	4.03	89.51	0.90	5.56	76.4
	Shell rubble (CaCO$_3$ > 30%)	3.78	45.64	3.47	47.11	23.6

samples taken in cell 11 were collected from regions with less than 30% CaCO$_3$ content and mud content of approximately 20% (Table 1). Cell 16 also was relatively low-relief and homogeneous but consisted of much muddier sediments with low CaCO$_3$ content (7–10%) (Table 1). The linear regression model computed to relate sand:mud ratio to acoustic reflectance intensity in cell 16 was

$$\text{sand:mud ratio} = 0.11 \times (\text{8-bit pixel value}) - 5.75$$
$$(F_{1;19} = 5.96, P = 0.025, R^2 = 0.36).$$

The linear regression model computed to relate carbonate content to reflectance intensity in cells 5, 11, and 23 was

$$\%\text{CaCO}_3 = 0.78 \times (\text{8-bit pixel value}) - 32.6$$
$$(F_{1;68} = 8.08; P < 0.001; R^2 = 0.66).$$

Application of these models to acoustic reflectance data allowed estimation of percent coverage of different seabed types in each cell (Table 1).

Trawl sampling cruises conducted to examine juvenile red snapper habitat utilization patterns occurred on August 19–21, September 18–20, November 6–8, and December 4–6, 2001. Mean water depths were approximately 19.5, 32.5, 36.0, and 39.5 for cells 5, 11, 16, and 23, respectively. Mean salinities measured 1 m off the bottom ranged from 33.6 to 36.3 p.s.u., and mean temperature ranged from 20.8°C to 28.7°C among all cells and cruises (Table 2). Mean dissolved oxygen generally was measured to be greater than 3.5 mg/L for all cells on all four cruises, except for cells 16 and 23 during September.

Juvenile red snapper density estimates from trawl samples were within the range of historic density estimates computed from NMFS data (Figures 3, 5). There was a significant difference in estimated juvenile density among cells (ANOVA: $F_{3;64} = 8.94$, $P < 0.001$), but neither cruise (ANOVA: $F_{3;64} = 1.69$, $P = 0.1790$) nor the interaction between cell and cruise (ANOVA: $F_{9;64} = 1.02$, $P = 0.4337$) were significant in the model. Tukey's studentized range test (a = 0.05) indicated estimated juvenile density was not significantly different among cells 5, 11, and 23, but densities in those three cells were significantly different than cell 16. In cell 11, juvenile red snapper density was significantly correlated with sponge density (Pearson's $r = 0.71$, $P < 0.001$) (Figure 6).

Total length frequency distributions indicate differences in size of juveniles captured in the different cells also existed, with two modes evident in the data (Figure 7). The first mode was between 75 and 100 mm TL in August, which shifted to between 125 and 150 mm TL by December. The second mode was between 200 and 250 mm TL in August, which shifted to greater than 250 mm TL by December. Analysis of sagittal otoliths of 20 fish sampled from each of these modes revealed otoliths of fish in the mode with shorter TL had no opaque zones (i.e., no annuli) and, thus, were judged to be age-0 juveniles (Patterson et al. 2001b). Sagittae of fish in the second mode contained one opaque and two translucent zones and, therefore, were judged to be age 1 (Patterson et al. 2001b).

In August, some age-1 fish were present in cells 5 and 11, but all fish in cell 23 appeared to be 1-year-olds. Age-0 fish were most abundant in cell 5 throughout the summer and fall but appeared to recruit to cell 11 starting in September when age-1 abundance decreased; fish sampled from cells 5 and 11 were predominantly age-0 individuals thereafter. Cell 23, on the other hand, contained mostly age-1 fish through September, but large, possibly early-spawned, age-0 fish began to recruit to this cell in November when numbers of age-1 appeared to be declining.

Table 2. Hydrographic parameters measured 1 m above the seabed during trawl sampling cruises in summer/fall 2001. Parameters are reported as the mean (SE) of 3 conductivity–temperature–depth casts made at each cell.

Cell	Date	Dissolved oxygen (mg/L)	Temperature (°C)	Salinity (psu)
5	Aug 21	No data	No data	No data
	Sep 19	3.88 (0.07)	28.7 (0.09)	33.6 (0.09)
	Nov 8	4.82 (0.12)	23.1 (0.01)	35.6 (0.01)
	Dec 4	4.85 (0.223)	20.8 (0.10)	35.1 (0.10)
11	Aug 20	No data	No data	No data
	Sep 20	4.11 (0.59)	28.4 (0.16)	35.2 (0.015)
	Nov 6	5.43 (0.14)	22.4 (0.04)	35.3 (0.01)
	Dec 5	4.99 (0.20)	21.9 (0.05)	35.9 (0.03)
16	Aug 20	3.45 (0.58)	25.0 (0.16)	36.2 (0.01)
	Sep 18	1.77 (0.03)	25.7 (0.12)	36.2 (0.03)
	Nov 7	4.83 (0.13)	23.4 (0.01)	35.7 (0.03)
	Dec 5	5.30 (0.54)	21.69 (0.17)	35.7 (0.07)
23	Aug 19	2.96 (0.19)	24.6 (0.11)	36.3 (0.02)
	Sep 18	1.84 (0.01)	25.5 (0.12)	36.1 (0.01)
	Nov 7	4.79 (0.32)	24.1 (0.01)	35.9 (0.01)
	Dec 6	5.08 (0.40)	22.7 (0.01)	36.1 (0.01)

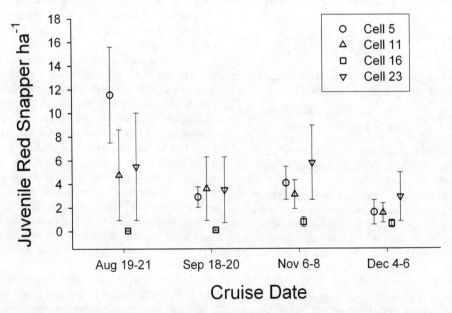

Figure 5. Mean (±SE) juvenile red snapper density (fish/ha) estimated from trawl samples taken from five stations within each of the four study cells during summer and fall 2001.

Discussion

Historic and contemporary juvenile red snapper density estimates were highest in shelf habitats with structures that provided small-scale (centimeters to meters) habitat complexity. We expected areas of the northern GOM that historically produced high numbers of juvenile red snapper in NMFS trawl samples would be characterized by low relief shell rubble habitat while areas that produced low catches were expected to have mud or sand seabed

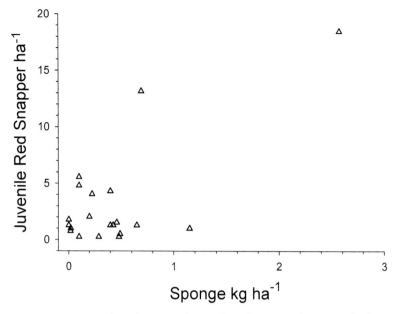

Figure 6. Scatterplot of estimated juvenile red snapper density (individuals/ha) versus estimated sponge density (kg/ha) from 20 trawl samples in cell 11 collecting during 2001 sampling cruises.

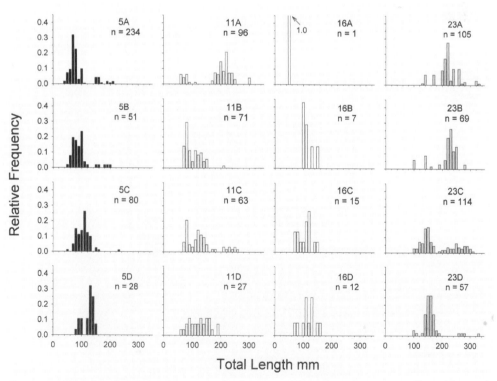

Figure 7. Total length relative frequency distributions of juvenile red snapper collected in trawl samples during 2001. Panel legends indicate cell, cruise (Cruise A, August 18–20; B, September 19–21; C, November 6–8; and D, December 4–6, 2001), and total sample size from five trawl stations.

types. Median catch areas were expected to contain a mixture of habitats. These expectations were formed based on previous findings that age-0 red snapper demonstrated a high affinity for shell rubble versus sand substrates in tank experiments (Szedlmayer and Howe 1997). However, age-0 red snapper catch per unit effort in trawl tows made on the shallow shelf (<20-m depth) off Alabama was highest from an area containing a mixture of fine sand and shell rubble sediments (Szedlmayer and Conti 1999). Our results also suggest shell rubble seabed is important habitat for juvenile red snapper, as highest juvenile densities estimated from FGS data were from cell 23, which contained the highest percentage of shell rubble habitat. However, there was no significant difference in juvenile densities among cells 5, 11, and 23 in samples collected during fall 2001. This is significant because the seabed in cell 11 was uniformly sandy with low $CaCO_3$ (i.e., shell) content. Sponge density in cell 11 was positively correlated with estimated juvenile red snapper density, thus suggesting sponges also supplied habitat complexity at a scale required by juvenile red snapper.

These results may explain equivocal results from previous juvenile red snapper habitat studies. Shelf areas containing shell rubble habitat were estimated to contain median to high juvenile densities (both historic and in 2001), supporting the conclusion of some authors that shell rubble is important for juvenile red snapper production (Szedlmayer and Howe 1997; Lee 1998; Szedlmayer and Conti 1999). However, not all studies have indicated a correlation between juvenile red snapper density and shell rubble habitat. Workman and Foster (1994) conducted video sampling during diver surveys and with a remotely operated vehicle at several sites in an approximately 350-km² area just to the west of our cells 5 and 11. The number of juvenile red snapper they observed was similar between sand-silt and shell rubble habitats, but, much like we report here, fish encountered in sand-silt habitat were associated with items providing small-scale complexity (e.g., squid eggs, worm tubes). Gallaway et al. (1999) reported density of low relief structures was not significant in explaining large-scale variability in juvenile red snapper density in northern GOM habitat suitability index models. This finding may have resulted from inadequate data on where small-scale, low-relief habitat exists. Alternatively, it may have resulted from loss of resolution due to integration of juvenile density estimates on large spatial scales (334-km² model cells) across patchily distributed low relief habitat.

Both of the latter studies highlight the importance of choosing an appropriate scale of observation for the ecological process being observed. At small spatial scales (centimeters to meters), Workman and Foster (1994) reported juvenile red snapper were associated with a variety of structures that provided small-scale structural complexity. At large spatial scales (100 km²), which likely integrated across most habitat variability, Gallaway et al. (1999) were unable to detect a significant habitat effect to explain variability in juvenile red snapper density. The meso-scale (100 m to km) on which our 2001 trawl sampling operated also likely integrated over significant patchiness in habitats and juvenile red snapper distribution. This is evidenced by the apparent uniform distribution of surficial sediments in cell 11 yet significant correlation between estimated juvenile red snapper density and estimated sponge density. Moreover, variability in juvenile density estimates was greatest for cell 5, which also had the greatest spatial heterogeneity in seabed type.

Distinct patterns in year-classes present in each block during our 2001 sampling were evident despite the fact that juvenile density was not significantly different among cells 5, 11, and 23. Age-1 fish were most abundant in cell 23, but larger age-0 fish began recruiting to this cell by late fall. It appears from these results, as has been suggested by other authors (Lee 1998; Bailey et al. 2001), that the scale of habitat complexity required by red snapper increases with fish size and age. An alternate hypothesis concerning differences in habitat utilization between the two year-classes would be that juveniles display an ontogenetic shift to deeper areas on the shelf as they grow, and hardbottom areas such those in cell 23 serve as recruitment corridors to outer shelf reefs that constitute areas of greatest natural relief in the north-central GOM (Bradley and Bryan 1975; Parker et al. 1992; Kennicutt et al. 1995; Gardner et al. 2001). Testing this hypothesis may prove difficult, however, because more than 20,000 artificial reefs (e.g., liberty ships, discarded appliances, and fabricated concrete structures) have been deployed in a 3,100-km² area off Alabama (approximately 15 km east of our study sites) designated for this purpose (Minton and Heath 2000; Patterson et al. 2001a). Therefore, traditional ontogenetic shifts in habitat utilization or recruitment pathways of reef fishes such as red snapper may have been altered by placement of artificial reef habitat (Bohnsack 1989; Cowan et al. 1999; Bailey et al. 2001; Patterson et al. 2001a).

Conversely, artificial reef permit areas off Alabama may provide a unique setting to test hypotheses concerning juvenile red snapper habitat requirements and preferences because they are avoided by shrimp trawlers out of fear of losing gear to bottom hangs in the form of artificial reefs; thus, they serve as de facto no-trawl zones. Parker et al. (1992) described shell rubble habitat analogous to that identified in our cell 23 in an approximately 20-km² area (30°02'N, 87°57'W) contained within an artificial reef area off Alabama. Strelcheck (2001) described shell ridge features similar to those in cell 5 from an approximately 25-km² area (29°59'N, 88°02'W) also within an artificial reef permit zone. Therefore, these de facto no-trawl zones may provide control habitat to test if trawls significantly affect the density and distribution of benthic invertebrate flora and fauna (e.g., soft corals, calcareous algae, sponges) that may add structural complexity not found in areas subjected to trawling.

Future studies also should address whether differences in juvenile feeding opportunity, growth, or survival exist among different seabed types, and the spatial extent of different seabed types on the shelf should be estimated. While this study provides evidence that trends in juvenile red snapper spatial distribution on the north-central GOM shelf are temporally consistent among years, and acoustic mapping techniques described herein enabled us to link high juvenile densities with habitats that provided structural complexity, further research is needed to determine what constitutes quality juvenile red snapper habitat. Studies examining diet, growth, and mortality should provide additional insight into what constitutes good versus poor juvenile red snapper habitat (i.e., what habitat conveys high versus low recruitment potential; Beck et al. 2001). However, our understanding of what habitats are essential will be incomplete without knowledge of the spatial extent of different habitats (Stunz et al. 2002). We envision an index of habitat quality may consist of the product of estimated survival in a given habitat, the density of individuals located there, and habitat dimension. Thus, habitat that exhibits only modest increases in juvenile survivorship and density will need to be expansive in order to contribute significantly to year-class strength. In contrast, habitat that dramatically increases potential survivorship and supports very high densities will be essential to juvenile production even if the total area of the habitat is small.

Acknowledgments

We thank National Oceanic and Atmospheric Administration–NMFS for providing ship time aboard the R/V *Caretta* and logistic support while at sea. We thank Butch Pellegrin and Mark Murphy for providing Fall Groundfish Survey data. We thank Drew Hopper, Jered Jackson, Melissa Woods, Kendal Falana, Roger Zirlot, James Barber, and Wendy Talyor for help in side-scan, boxcore, and trawl sampling. We thank Andy Fischer for preparing and analyzing otolith thin sections. Funding for this study was provided by Louisiana Sea Grant and Mississippi/Alabama Sea Grant, and we are especially indebted to Ron Becker for facilitating cooperation between Sea Grant programs. An earlier draft of the manuscript was greatly improved by comments of two anonymous reviewers.

References

Bailey, H. K., J. H. Cown, and R. L. Shipp. 2001. Potential interactive effects of habitat complexity and sub-adults on young of the year red snapper (*Lutjanus campechanus*) behavior. Gulf of Mexico Science 19:119–131.

Beck, M. W., K. L. Heck, Jr., K. W. Able, D. L. Childers, D. B. Eggleston, B. M. Gillanders, B. Halpern, C. G. Hays, K. Hoshino, T. J. Minello, R. J. Orth, P. F. Sheridan, and W. P. Weinstein. 2001. The identification, conservation, and management of estuarine and marine nurseries for fish and invertebrates. BioScience 51:633–641.

Bohnsack, J. A. 1989. Are high densities of fishes at artificial reefs the result of habitat limitation or behavioral preference? Bulletin of Marine Science 44:631–644.

Bradley, E., and C. E. Bryan. 1975. Life history and fishery of the red snapper (*Lutjanus campechanus*) in the northwestern Gulf of Mexico: 1970–1974. Proceedings of the Gulf and Caribbean Fisheries Institute 27:77–106.

Broadhurst, M. K. 1998. Bottlenose dolphins, *Tursiops truncatus*, removing bycatch from prawn trawl codends during fishing in New South Wales. Australian Marine Fisheries Review 60:9–14.

Carver, R. E. 1971. Procedures in sedimentary petrology. Wiley-Interscience, New York.

Cowan, J. H., Jr., W. Ingram, J. McCawley, B. Sauls, A. Strelcheck, and M. Woods. 1999. The attraction vs. production debate: does it really matter from the management perspective? A response to the commentary by Shipp, R.L. Gulf of Mexico Science 17:137–138.

Crowder, L. B., and S. A. Murawski. 1998. Fisheries bycatch: implications for management. Fisheries 23(6):8–17.

Ehrhardt, N. M., and C. M. Legault. 1998. Impact of variability in numbers and size at first recruitment on the F sub(0.1) reference point in the US red snapper fisheries of the Gulf of Mexico. ICLARM Conference Proceedings 48:337–349.

Engaas, A., D. Foster, B. D. Hathaway, J. W. Watson, and I. Workman. 1999. The behavioral response of juvenile red snapper (*Lutjanus campechanus*) to shrimp trawls that utilize water flow modifications to induce escapement. Marine Technology Society Journal 33:51–56.

Gallaway, B. J., and J. G. Cole. 1999. Reduction of juvenile red snapper bycatch in the U.S. Gulf of Mexico shrimp trawl fishery. North American Journal of Fisheries Management 19:342–355.

Gallaway, B. J., J. G. Cole, R. Meyer, and P. Roscigno. 1999. Delineation of essential habitat for juvenile red snapper in the northwestern Gulf of Mexico. Transactions of the American Fisheries Society 128:713–726.

Gardner, J. V., P. Dartnell, K. J. Sulak, B. Calder, and L.

Hellequin. 2001. Physiography and late Quarternary–Holocene processes of northeastern Gulf of Mexico outer continental shelf off Mississippi and Alabama. Gulf of Mexico Science 19:132–157.

GMFMC (Gulf of Mexico Fishery Management Council). 1996. Amendment number 9 to the reef fish fishery management plan for the shrimp fishery of the Gulf of Mexico, U.S. waters with supplemental environmental regulatory flexibility analysis and social impact assessment. GMFMC, Tampa, Florida.

Goodyear, C. P. 1995. Red snapper in U.S. waters of the Gulf of Mexico. National Marine Fisheries Service, Southeast Fisheries Science Center, Miami Laboratory, MIA-95/96–05, Miami.

Gutherz, E. J., and G. J. Pellegrin, Jr. 1986. Report on snapper–grouper mortality by shrimp trawlers in the U.S. Gulf of Mexico. National Marine Fisheries Service, Southeast Fisheries Science Center, Pascagoula, Mississippi.

Holt, S. A., and C. R. Arnold. 1982. Growth of juvenile red snapper, *Lutjanus campechanus*, in the northwestern Gulf of Mexico U.S. National Marine Fisheries Service Fishery Bulletin 80:644–648.

Kennicutt, M. C. II, W. W. Schroeder, and J. Brooks. 1995. Temporal and spatial variations in sediment characteristics on the Mississippi–Alabama continental shelf. Continental Shelf Research 15:1–18.

Lee, J. D. 1998. Diet shifts of red snapper, *Lutjanus campechanus*, with changes in habitat and fish size. Auburn University, Master's thesis, Auburn, Alabama.

Minton, V., and S. R. Heath. 1998. Alabama's artificial reef program: building oases in the desert. Gulf of Mexico Science 16:105–106.

Moe, M. A. 1963. A survey of offshore fishing in Florida. Florida Board of Conservation Technical Series 4:1–117.

Parker, S. J., A. W. Shultz, and W. W. Schroeder. 1992. Sediment characteristics and seafloor topography of a palimpsest shelf, Mississippi–Alabama continental shelf. Quaternary Coasts of the United States: Marine and Lacustrine Systems, Society for Sedimentary Geology Special Publication 48:243–251.

Patterson, W. F., J. C. Watterson, R. L. Shipp, and J. H. Cowan. 2001a. Movement of tagged red snapper in the northern Gulf of Mexico. Transactions of the American Fisheries Society 130:533–545.

Patterson, W. F. III, J. H. Cowan, Jr., C. A. Wilson, and R. L. Shipp. 2001b. Age and growth of red snapper from an artificial reef area in the northern Gulf of Mexico. U.S. National Marine Fisheries Service Fishery Bulletin 99:617–627.

Rogers, D. R. 1999. Behavior of red snapper, *Lutjanus campechanus*, in relation to trawl modifications to reduce shrimp trawler bycatch. Doctoral dissertation. Louisiana State University, Baton Rouge.

SAS Institute, Inc. 1996. Statistics, version 6.11. SAS Institute, Inc., Cary, North Carolina.

Schirripa, M. J. 1998. Status of red snapper in U.S. waters of the Gulf of Mexico: updated through 1997. National Marine Fisheries Service, Sustainable Fisheries Division (SFD), SFD-97/98–30, Miami.

Schirripa, M. J., and C. M. Legault. 1999. Status of the red snapper in U.S. waters of the Gulf of Mexico: updated through 1998. NMFS, SFD, SFD-99/00–75, Miami.

Schroeder, W. W., A. W. Schultz, and O. H. Pilkey. 1995. Late Quaternary oyster shell and sea-level history, inner shelf, north east Gulf of Mexico. Journal of Coastal Research 11:664–674.

Strelcheck, A. 2001. The influence of reef designs and nearest-neighbor dynamics on artificial reef fish assemblages. University of South Alabama, Master's thesis, Mobile.

Stunz, G. W., T. J. Minello, and P. S. Levin. 2002. A comparison of early juvenile red drum densities among various habitat types in Galveston Bay, Texas. Estuaries 25:76–85.

Szedlmayer, S. T., and J. Conti. 1999. Nursery habitats, growth rates, and seasonality of age-0 red snapper, *Lutjanus campechanus*, in the northeast Gulf of Mexico. U.S. National Marine Fisheries Service Fishery Bulletin 97:626–635.

Szedlmayer, S. T., and J. C. Howe. 1997. Substrate preference in age-0 red snapper, *Lutjanus campechanus*. Environmental Biology of Fishes 50:203–207.

Watson, J., A. Shah, S. Nichols, and D. G. Foster. 1997. Bycatch reduction device summary. Bycatch in the Southeast shrimp trawl fishery. National Marine Fisheries Service, Southeast Fisheries Science Center, SFA Task N-10.03, Miami.

Workman, I. K., and D. G. Foster. 1994. Occurrence and behavior of juvenile red snapper, *Lutjanus campechanus*, on commercial shrimp fishing grounds in the northeastern Gulf of Mexico. Marine Fisheries Review 56:9–11.

Living Substrate in Alaska: Distribution, Abundance, and Species Associations

PATRICK W. MALECHA,[1] ROBERT P. STONE,[2] AND JONATHAN HEIFETZ[3]

Auke Bay Laboratory, Alaska Fisheries Science Center, National Marine Fisheries Service, National Oceanic and Atmospheric Administration, 11305 Glacier Highway, Juneau, Alaska 99801-8626, USA

Abstract. "Living substrate" has been identified as an important marine habitat and is susceptible to impacts from fishing activities. In Alaskan waters of the North Pacific and Bering Sea, little is known about the distribution of deepwater living substrate such as sponges (phylum Porifera), sea anemones (order Actiniaria), sea whips and sea pens (order Pennatulacea), ascidians (class Ascidiacea), and bryozoans (phylum Ectoprocta). Based on 26 years of survey data (mostly from catches in bottom trawls collected between 1975 and 2000), we created living substrate distribution maps. In general, the five groups of living substrate were observed in varying densities along the continental shelf and upper continental slope. Catch per unit effort (CPUE) of sponges was greatest along the Aleutian Islands, while CPUEs of ascidians and bryozoans were greatest in the Bering Sea. Large CPUEs of sea anemones, sea pens, and sea whips were observed in both the Bering Sea and the Gulf of Alaska. Broad-scale species associations between living substrate and commercially important fishes and crabs were also identified. Flatfish (Bothidae and Pleuronectidae) were most commonly associated with ascidians and bryozoans; gadids (Gadidae; also known as cods) with sea anemones, sea pens, and sea whips; rockfish (*Sebastes* spp. and shortspine thornyhead *Sebastolobus alascanus*) and Atka mackerel *Pleurogrammus monopterygius* with sponges; crabs (*Chionoecetes* spp., *Paralithodes* spp., *Lithodes* spp., Dungeness crab *Cancer magister*, and hair crab *Erimacrus isenbeckii*) with ascidians; and other commercial fish species (sablefish *Anoplopoma fimbria*, Hexagrammidae, and Rajidae) with sea pens and sea whips. These data should provide resource managers with insight into living substrate distribution and relationships among benthic community organisms and, ultimately, with future in-depth studies, may aid in determining specific areas for habitat protection and facilitate management practices that minimize fishery impacts to living substrate.

Introduction

New regulations implementing the essential fish habitat provisions of the Magnuson-Stevens Fisheries Conservation and Management Act (NMFS 1996) have increased emphasis on protection of marine habitat. Within a Final Rule (NMFS 2002), the National Marine Fisheries Service (NMFS) identified "habitat areas of particular concern" (HAPC), which are geographic areas and types of habitat that have special importance and may require additional protection from adverse effects. Designation of HAPC locations is based on ecological importance, sensitivity, vulnerability, and rarity of the habitat. In Alaska, "living substrate" has been identified as habitat meeting the criteria for HAPC designation. The term "living substrate" applies to a vast array of structure-forming organisms that provide habitat complexity in diverse environments. Although the term "living substrate" could pertain to many organisms, we use the term to discuss our specific subject matter collectively. The subject matter of this paper, sponges (phylum Porifera), bryozoans (phylum Ectoprocta), ascidians (class Ascidiacea), anemones (order Actiniaria), and sea pens and sea whips (order Pennatulacea), are sessile, epibenthic, living substrate specifically mentioned in HAPC considerations. Other living substrate exist and are important (such as macroalgae, including eel grass *Zostera* sp. and kelp *Macrocystis* sp.; hydroids, class Hydrozoa; barnacles, class Cirripedia; bivalves, class Bivalvia; and brachiopods, phylum Brachiopoda, to name a few) but have not been included in this analysis.

Living substrate provides refuge from predation, especially for juvenile life stages, and attracts prey

---

[1] Corresponding author: pat.malecha@noaa.gov
[2] E-mail: bob.stone@noaa.gov
[3] E-mail: jon.heifetz@noaa.gov

species. A laboratory study of Alaskan species demonstrated that juvenile Pacific halibut *Hippoglossus stenolepis* and northern rock sole *Lepidopsetta polyxystra* had a behavioral affinity for sediments structured with sponges and bryozoans and that survival of age-0 Pacific halibut was positively correlated with habitat structure (Stoner and Titgen 2003). In the Bering Sea, in situ observations, revealed that adult Pacific ocean perch *Sebastes alutus* diurnally utilized a sea whip "forest," presumably as a refuge at night and to feed on swarms of euphausiids above the sea whips during the day (Brodeur 2001). In the Gulf of Alaska, juvenile rockfish *Sebastes* spp. closely associate with sponges (J. L. Freese, Auke Bay Laboratory, National Marine Fisheries Service, unpublished data), and groundfish prey species are positively associated with sea whips *Halipteris willemoesi* (Stone et al., in press). Studies in tropical waters have also shown the importance of living substrate as habitat. For example, in New Caledonia, France, fish density and biomass were correlated to the abundance and species richness of sponges and ascidians (Wantiez and Kulbicki 1995).

Associations between living substrate and crustaceans have also been documented. Off Kamchatka, Russia, young-of-the-year red king crabs *Paralithodes camtschaticus* were associated with an array of sponges, bryozoans, and hydroids (Tsalkina 1969). In the laboratory, recently molted red king crab glaucothoes preferred to settle on complex substrate (hydroids, bryozoans, algae, and plastic mesh) rather than sand (Stevens and Kittaka 1998; Stevens 2003). Associations between shrimp (Infraorder Caridea) and anemones in tropical waters have long been established (Engel 1972). In northern waters off the west coast of Sweden, associations between the anemone *Bolocera tuediae* and a lithodid crab *Lithodes maja* and shrimps have also been observed (Jonsson et al. 2001).

The vulnerability of living substrate to fishing activities is well documented. In the Gulf of Alaska, Freese et al. (1999) described trawling impacts to sponges, and Stone et al. (2005, this volume) and Freese et al. (1999) report trawling impacts to sea whips. In the Bering Sea, McConnaughey et al. (2000) identified fishing-related damage to bryozoans, anemones, and ascidians. Damage has also been documented in the South Pacific to sponge (Wassenberg et al. 2002) and bryozoans (Bradstock and Gordon 1983) and in the North Sea to ascidians (de Groot 1984).

Heifetz (2002) described the distribution, abundance, and species associations of corals (class Anthozoa) in Alaskan waters. In the present study, we provide similar information about sponges, sea anemones, sea pens and sea whips, ascidians, and bryozoans in Alaska. Like corals, these living substrate organisms provide important habitat and foraging areas for a wide range of commercially exploited fish and invertebrates such as crabs and shrimp.

Protecting habitat is an important goal of resource management and may be critical to maintaining biodiversity and sustaining fisheries. Identifying areas with high concentrations of living substrate may provide managers with needed information so that fishing practices can be modified so as to minimize impacts to those habitats. Little is known about the distribution of these abundant organisms in Alaskan waters. Kessler (1985) provides limited distribution information for some sponge, bryozoan, and ascidian species. However, information is limited as to whether they are common, uncommon, or absent north and south of the Alaska Peninsula. Even less is known regarding the distribution of sea pens, sea whips, and sea anemones in Alaskan waters. We provide distribution maps based on 26 years of fishery survey data. In addition, we identify broad-scale associations (on the scale of a single trawl pass) between living substrate and commercially important fishes and crabs. This paper provides an initial, overview description of important living substrate distribution and a possible starting point for designating HAPCs in Alaskan waters.

Methods

We used data from the NMFS Alaska Fisheries Science Center RACEBASE survey database (Stark and Clausen 1995). These surveys are mainly used to provide data for stock assessment of important groundfish species. However, early surveys were largely exploratory for commercial species and the recording of living substrate organisms was of secondary importance. Therefore, we limited our analyses to hauls that occurred from 1975 through 2000. These hauls generally covered the continental shelf and upper slope of the Gulf of Alaska, eastern Bering Sea, and the Aleutian Islands to a depth of 1,000 m (Figure 1). Survey scientists of varying expertise identified specimens to the lowest level of their particular taxonomic knowledge. However, because many of the living substrate species are not well known, identifications were often to higher taxonomic levels, such as family. Therefore, we combined observations into higher taxa to facilitate analyses. The database was queried for hauls containing at least one record of living substrate (e.g., sponge, anemone, sea pen or sea whip, ascidian, or bryozoan). For our analyses, we did not consider the sponge (genus *Suberites*), commonly inhabited by hermit crabs (family Paguridae), or pelagic ascidians of the class Thaliacea as living substrate. However, if *Suberites* or Thaliacean observations were recorded at a higher taxonomic level, they were obligatorily included in our analyses. We used the standardized RACEBASE taxonomic classifications as presented in the Species Codebook (NMFS 2001) and, for clarification, we referred to Brusca and Brusca (2003), the Integrated Taxonomic Information System (ITIS-North America 2002), and the Pennatulacea Index provided by Gary C. Williams (2003).

Our analyses include domestic and foreign survey hauls and a variety of gear types including scallop dredges and shrimp trawls. However, NMFS survey crews completed the majority of hauls with nonpelagic otter trawls. Recent survey design and methods used for the Gulf of Alaska are detailed in Britt and Martin (2000). Most NMFS trawl surveys are designed to catch fish, not sessile invertebrates, and, thus, do not capture and retain a high percentage of living substrate. Additionally, rough bottom topography precludes trawling in some areas. Therefore, many areas with high densities of living substrate are likely under sampled or not sampled at all.

For each living substrate, we computed frequency of occurrence in the database by determining the number of hauls in which a living substrate was observed. Living substrate communities were compared among the Gulf of Alaska, Bering Sea, and Aleutian Islands. For regional comparisons, we generally used the geographic designations of Alaska provided by Orth (1967). However, for the Gulf of Alaska we designated the U.S.–Canadian border as the southern boundary and the southwestern tip of the Alaska Peninsula as the western boundary. Also, hauls occurring in the Chukchi Sea and Arctic Ocean were combined with Bering Sea observations. For distribution plots, living substrate data were converted to catch per unit effort (CPUE) in kg/h. Relative abundance was calculated by scaling each CPUE observation to the largest CPUE value for that living substrate. Relative abundances were plotted with Generic Mapping Tools software (Wessel and Smith 1998).

Broad-scale associations between living substrate and commercially important species were determined using species composition information from hauls containing at least one observation of living substrate. Commercially important species were grouped into six categories: (1) flatfish (Bothidae and Pleuronectidae), (2) gadids, mostly walleye pollock *Theragra chalcogramma* and Pacific cod *Gadus macrocephalus*, (3) rockfish *Sebastes* spp. and shortspine thornyhead *Sebastolobus alascanus*, (4) Atka mackerel *Pleurogrammus monopterygius*, (5) other fish including sablefish *Anoplopoma fimbria*, greenlings (Hexagrammidae), and skates (Rajidae), and (6) crabs (*Chionoecetes* spp., *Paralithodes* spp., *Lithodes* spp., Dungeness crab *Cancer magister* and hair crab *Erimacrus isenbeckii*). For each living substrate, the CPUE for each commercial group was calculated as a percentage of total commercial CPUE. The described species associations are not fine-scale but are on the scale of a single trawl pass (rarely greater than 4.48 km; 1.5 nautical mi). Thus, fish and crabs caught in the trawl could have been distantly separated, both horizontally along the seafloor and vertically in the water column, from where the living substrate was encountered.

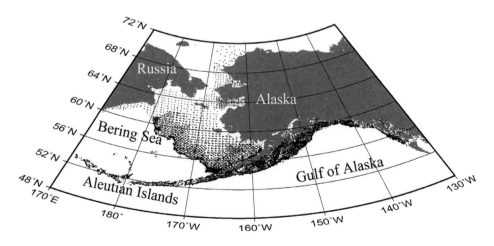

Figure 1. Map of survey hauls occurring in the Bering Sea, Gulf of Alaska, and Aleutian Islands between 1975 and 2000.

Results and Discussion

Between 1975 and 2000, there were 36,564 hauls recorded in the RACEBASE database (Figure 1). Of these hauls, 95% were made by trawls, primarily nonpelagic otter trawls but also pelagic and shrimp trawls. The number of hauls per year was greatest during the early 1980s, peaking in 1981 at 2,513 hauls. During the 1990s, the number of hauls per year was substantially lower. Despite this reduction in survey effort, the number of living substrate observations per year increased in the 1990s, peaking in 1997 (Figure 2). Survey gear and methods have not changed much over time, so this trend is not related to increased fishing power or efficiency. The increase in living substrate observations may be related to a proliferation of living substrate but could also be related to increased recording of invertebrates due to enhanced interest in benthic habitats.

Hauls were unevenly distributed among the Bering Sea (50%), Gulf of Alaska (41%), and Aleutian Islands (9%). Likewise, the number of living substrate observations were unevenly distributed among the three regions; the Bering Sea accounted for 59% of all living substrate observations, compared to 25% in the Gulf of Alaska and 16% in the Aleutian Islands (Table 1). At least one of the living substrate types (sponge, anemone, sea pen, sea whip, ascidian, or bryozoan) was recorded in 35% of all hauls. In general, the five groups of living substrate were observed along the continental shelf and upper slope in varying densities. Sponges, anemones, and ascidians were observed over five times more often than bryozoans and over eight times more often than sea whips and sea pens.

Sponges were caught in 20% of all hauls since 1975 in the RACEBASE database. An approximately equal number of sponge observations occurred in all three regions, but the percentage of living substrate observations attributed to sponges varied considerably among regions (Figure 3). In the Aleutian Islands, sponges accounted for 67% of all living substrate observations. In the Gulf of Alaska, sponges made up 43% of living substrate observations, and in the Bering Sea, 16% of living substrate observations were sponges. Sponges were by far the most numerous living substrate in Aleutian Island hauls and they were also the most numerous living substrate in Gulf of Alaska hauls. In the Bering Sea, ascidian and anemone observations were more than two times as numerous as sponge observations. Despite this fact, several large sponge catches occurred in the Bering Sea, particularly on the shelf north of the Alaska Peninsula and near St. George Island (Figure 4). Many large sponge catches occurred along the Aleutian Archipelago, including the largest sponge catch, which occurred on Petrel Bank. Sponge catches in the Gulf of Alaska were generally minor, but catches were distributed quite evenly over most of the shelf, unlike in the Bering Sea, where catches were patchier. Alaskan waters have a high diversity of sponges, many of which are difficult to identify and some are not well described. The vast majority of sponge observations in the database were recorded at the level of phylum, but the database also includes 14 genera (Table 1). Fifteen common names were also found in the database; these observations were recorded in Table 1 at the phylum level. The most common genera were *Aphrocallistes*, *Mycale*, and *Tethya*.

Anemones were recorded in 19% of all hauls since 1975 in the RACEBASE database. Although the vast majority were recorded at the order level, eight distinct

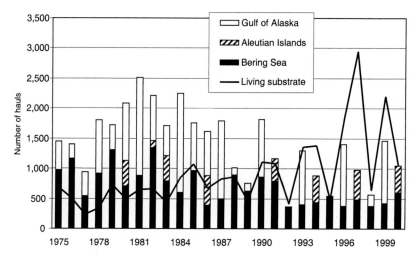

Figure 2. Time series plots of the number of survey hauls by region and living substrate observations between 1975 and 2000.

Table 1. Taxonomic names and frequency of occurrence of five living substrate groups in survey hauls between 1975 and 2000.

Taxonomic name	Frequency of occurrence			
	Aleutians	Bering Sea	Gulf of Alaska	Total
Phylum Porifera (sponges)				
Aphrocallistes vastus	97	29	253	379
Halichondria panicea	81	94	79	254
Eumastia sitiens	210		44	254
Hyalonema sp.	83		12	95
Leucandra heathi	6		1	7
Leucosolenia blanca	84	4	28	116
Mycale bellabellensis	7		12	19
Mycale loveni	159	8	174	341
Myxilla incrustans	16	1	87	104
Isodictya quatsinoensis	29		1	30
Phakettia cribrosa			13	13
Polymastia sp.	138		94	232
Polymastia pachymastia	7		3	10
Rhabdocalyptus sp.	19	15	31	65
Stylissa sp.	5		4	9
Syringella amphispicula	72			72
Tethya sp.	198	1	80	279
Other Porifera	1,349	2,101	1,633	5,083
Total Porifera	2,560	2,253	2,549	7,362
Order Actiniaria (sea anemones)				
Actinauge verrilli		22	89	111
Corallimorphus sp.			1	1
Cribrinopsis fernaldi		6	2	8
Liponema brevicornis	10	110	9	129
Metridium sp.	10	245	279	534
Metridium senile	1	100	81	182
Metridium farcimen			2	2
Paractinostola faeculenta		27	15	42
Stomphia sp.	1	10	27	38
Stomphia coccinea		5		5
Stomphia didemon			6	6
Urticina sp.		231	4	235
Urticina felina	1	69	9	79
Other Actiniaria	266	3,930	1,351	5,547
Total Actiniaria	289	4,755	1,875	6,919
Order Pennatulacea (sea pens and sea whips)				
Halipteris sp.		2	1	3
Halipteris finmarchica		8	7	15
Halipteris californica			11	11
Ptilosarcus gurneyi	1		70	71
Stylatula sp.	28	8	101	137
Stylatula gracilis		9	15	24
Virgularia sp.	1	21	18	40
Other Pennatulacea	34	176	335	545
Total Pennatulacea	64	224	558	846
Class Ascidiacea (sea squirts)				
Amaroucium sp.	72	1,021	60	1,153
Ascidia paratropa			2	2
Boltenia sp.	2	1,015	55	1,072
Boltenia ovifera	3	735	4	742
Boltenia villosa			24	24
Botrylloides sp.			1	1
Chelyosoma orientale		1		1
Halocynthia sp.	27	460	49	536
Halocynthia aurantium	81	414	63	558

Table 1. Continued.

Taxonomic name	Aleutians	Bering Sea	Gulf of Alaska	Total
Class Ascidiacea (sea squirts)				
Molgula sp.	1	28	4	33
Molgula griffithsi	23	40	11	74
Molgula retortiformis	33	6	49	88
Styela sp.		8		8
Styela rustica	54	1,323	29	1,406
Synoicum sp.	14	15	25	54
Other Ascidiacea	323	1,070	444	1,837
Total Ascidiacea	633	6,136	820	7,589
Phylum Ectoprocta (Bryozoans)				
Cellepora ventricosa		21		21
Escharopsis sp.	1	25	4	30
Eucratea loricata	31	42	23	96
Serratiflustra serrulata	52	149	13	214
Microporina articulata	1			1
Myriozoum subgracile	3		4	7
Porella compressa	15	10	1	26
Rhamphostomella costata	8	34	1	43
Other Ectoprocta	158	620	57	835
Total Ectoprocta	269	901	103	1,273
Totals				
Total taxa combined	3,815	14,269	5,905	23,989
Percentage of all observations	16%	59%	25%	100%

genera were identified, with *Metridium* the most common (Table 1). In the Gulf of Alaska and Bering Sea, anemones were the second most numerous living substrate, accounting for 32% and 33% of all living substrate observations, respectively (Figure 3). In the Aleutian Islands, anemones accounted for just 8% of living substrate observations. Distribution of anemones in all areas was quite broad with a few areas of higher concen-

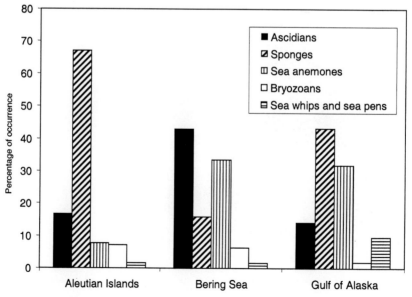

Figure 3. Percentage of occurrence of living substrate taxa recorded in survey hauls between 1975 and 2000 by geographic region.

trations on the continental shelf, especially in the Bering Sea and western Gulf of Alaska (Figure 5).

Sea pens and sea whips were the least frequently observed living substrate and occurred in a little over 2% of all hauls since 1975. On a percentage basis, they were most common in the Gulf of Alaska, where they accounted for 9% of all living substrate observations (Figure 3). In the Bering Sea and Aleutian Islands, sea pens and sea whips accounted for less than 2% of the living substrate observations. Distribution of sea pen and sea whip catches is patchy (Figure 6). The highest catches were located south of the Alaska Peninsula in Pavlof Bay and in the inside waters of southeastern Alaska. Shelikof Strait, the shelf south of Prince William Sound, and the lower shelf of the southeastern Bering Sea also had substantial concentrations of sea pens and sea whips. Most observations were recorded at the order level, as Pennatulaceans, but four genera were reported, and sea whips from the genus *Stylatula* were most common (Table 1).

Ascidians were the most numerous living substrate in the RACEBASE database and occurred in 21% of all hauls since 1975. In the Aleutian Islands, ascidians were the second most observed living substrate (Figure 3). In the Bering Sea, ascidians were the most commonly observed living substrate and were observed far more often than in either the Aleutian Islands or the Gulf of Alaska (Table 1). In the Bering Sea, ascidians accounted for 43% of all living substrate observations, compared to 17% and 14% in the Aleutian Islands and Gulf of Alaska, respectively (Figure 3). Within the Bering Sea, high-density catches were distributed over a broad area of the shelf and into Norton Sound (Figure 7). Large CPUEs were also recorded in the Arctic Ocean. A diverse mix of ascidians was reported from all three areas, with the majority of observations identified to genus or species (Table 1). In all, nine genera were observed, but four genera made up most of the observations. The most abundant ascidians were sea onions *Boltenia* sp. (a stalked ascidian), sea potatoes *Styela* sp., sea globs *Amaroucium* sp. (a compound ascidian), and sea peaches *Halocynthia* sp.

Ectoprocts or bryozoans were caught in approximately 3.5% of all hauls since 1975. Most observations occurred in the Bering Sea (Table 1), but expressed as a percentage of living substrate caught by region, bryozoans were most abundant in the Aleutian Islands, accounting for 7% of all Aleutian Island living substrate observations (Figure 3). The percentage of bryozoans in living substrate catches in the Bering Sea and Gulf of Alaska was 6% and 2%, respectively. Bryozoan catches with the largest CPUEs occurred in Bristol Bay, on the Bering Sea shelf, and in the western Aleutians near Kiska Island (Figure 8). Eight species of bryozoans were observed, although most observations were recorded at the level of phylum (Table 1).

The assemblage of commercial fish and crab species caught in association with each living substrate varied (Figure 9). Flatfish were most commonly associated with ascidians and bryozoans. In hauls with ascidians, flatfish made up 64% of the commercial species CPUE and in hauls with bryozoans, flatfish made up 55% of the commercial CPUE. Gadids were most commonly caught with sea anemones (47% of CPUE) and with sea pens and sea whips (40% of CPUE). Rockfish and Atka mackerel were most commonly caught with sponges, accounting for 28% and

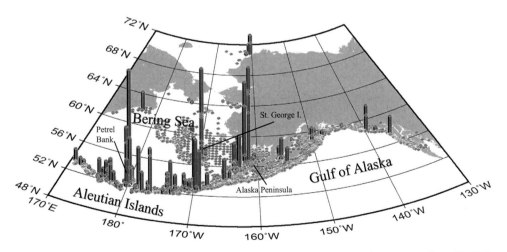

Figure 4. Relative abundance of sponges (phylum Porifera) based on catch per unit effort (CPUE) in survey hauls between 1975 and 2000. The CPUE is scaled relative to the largest value.

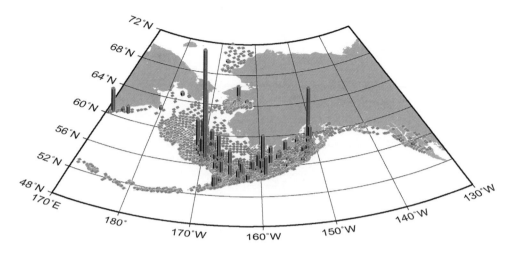

Figure 5. Relative abundance of anemones (order Actiniaria) based on catch per unit effort (CPUE) in survey hauls between 1975 and 2000. The CPUE is scaled relative to the largest value.

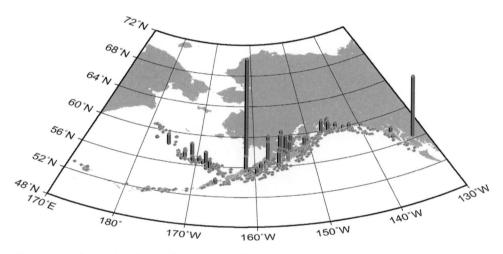

Figure 6. Relative abundance of sea whips and sea pens (order Pennatulacea) based on catch per unit effort (CPUE) in survey hauls between 1975 and 2000. The CPUE is scaled relative to the largest value.

15% of the commercial CPUE, respectively. Sponge habitat produced the most diverse crab and fish catches, as commercial CPUEs were widely distributed among several groups. In contrast, flatfish and gadids dominated the catch associated with the other living substrates, accounting for a combined total of at least 79% of the commercial CPUE. Crab and other fish species were minor percentages of the commercial catch associated with living substrate. However, crabs were most common in hauls that contained ascidians (5% of CPUE), and other fish were most common in hauls with sea pens and sea whips (6% of CPUE). As previously stated, these associations are not fine-scale. Our results do not provide definitive evidence that fish and crabs caught in the same trawl as a particular living substrate are actually utilizing the living substrate as habitat. The scale of our sampling device (one trawl pass) is far too imprecise to make such conclusions. However, it is valuable to know that a commercially important species is in the same neighborhood with a living substrate and that they may be associated on a smaller scale than a single trawl pass. Species associations with living substrate also likely vary by region. A more in-depth analysis would be needed to determine those relationships, which is outside the scope of this paper.

The living substrate distributions and species as-

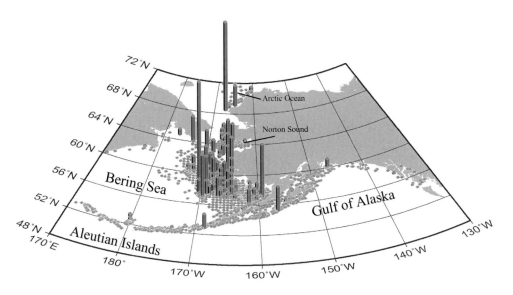

Figure 7. Relative abundance of ascidians (class Ascidiacea) based on catch per unit effort (CPUE) in survey hauls between 1975 and 2000. The CPUE is scaled relative to the largest value.

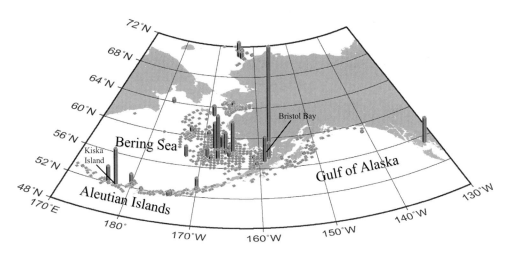

Figure 8. Relative abundance of bryozoans (phylum Ectoprocta) based on catch per unit effort (CPUE) in survey hauls between 1975 and 2000. The CPUE is scaled relative to the largest value.

sociations we describe are unrefined for several reasons. As discussed previously, the surveys that catch living substrate are not specifically designed for that purpose and may not adequately assess areas with high densities of living substrate due to rough bottom topography. Additionally, some living substrate (e.g., sponges) are more readily caught in bottom trawls than others (e.g., anemones), so their relative abundances also indicate selectivity of the survey gear. Given the fact that survey gear may not fish where living substrate may be concentrated (e.g., rocky bottom) and that the selectivity of the gear results in poor catch efficiency, survey catches of living substrate could be considered minimal. Furthermore, taxonomic identification of survey catches has been inconsistent over the years because identification of many living substrate species is difficult, and interest in benthic habitat has changed over time.

Regardless, the distribution maps and species associations we present provide broad-scale representa-

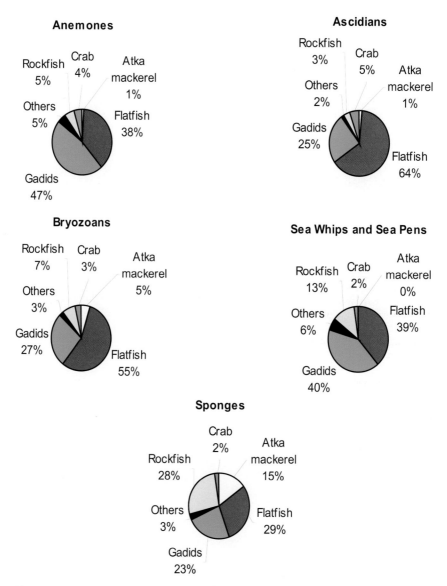

Figure 9. Associations between living substrate and commercially important fish and crab groups based on survey hauls between 1975 and 2000. For each living substrate, the percentage of catch per unit effort (CPUE) attributed to each commercial group is scaled to the overall CPUE for all commercial fish and crab groups. See text for determination of commercial groups.

tions of living substrate characteristics in Alaskan waters. These data can be used as a basis for further investigation of living substrate and its importance to benthic habitats and communities. Refinement of commercial species catch data, as it relates to the geographic distribution of specific living substrate, may facilitate identification of essential or important habitats and possible HAPC designations. The fact that taxonomic identification of living substrate is becoming routine and more precise in survey hauls bodes well for resource managers, as future decisions regarding the classification of benthic habitat may be greatly enhanced with these data.

References

Bradstock, M., and D. P. Gordon. 1983. Coral-like bryozoan growths in Tasman Bay, and their protection to

conserve commercial fish stocks. New Zealand Journal of Marine and Freshwater Research 17(2):159–163.
Britt, L. L., and M. H. Martin. 2000. Data report: 1999 Gulf of Alaska bottom trawl survey. NOAA Technical Memorandum NMFS-AFSC-121.
Brodeur, R. D. 2001. Habitat–specific distribution of Pacific ocean perch (*Sebastes alutus*) in Pribilof Canyon, Bering Sea. Continental Shelf Research 21(3):207–224.
Brusca, R. C., and G. J. Brusca. 2003. Invertebrates, 2nd edition. Sinauer Associates, Sunderland, Massachusetts.
de Groot, S. J. 1984. The impact of bottom trawling on the benthic fauna of the North Sea. Ocean Management 9(3–4):177–190.
Engel, C. P. 1972. The distribution and abundance of anemones and their commensal shrimp in Discovery Bay, Jamaica. Atoll Research Bulletin 152:1–6.
Freese, L., P. J. Auster, J. Heifetz, and B. L. Wing. 1999. Effects of trawling on seafloor habitat and associated invertebrate taxa in the Gulf of Alaska. Marine Ecology Progress Series 182:119–126.
Heifetz, J. 2002. Coral in Alaska: distribution, abundance, and species associations. Hydrobiologia 471(1–3):19–28.
ITIS-North America. 2002. ITIS: integrated taxonomic information system. Available: *www.itis.usda.gov/* (August 2003).
Jonsson, L. G., T. Lundälv, and K. Johannesson. 2001. Symbiotic associations between anthozoans and crustaceans in a temperate costal area. Marine Ecology Progress Series 209:189–195.
Kessler, D. W. 1985. Alaska's saltwater fishes and other sea life: a field guide. Alaska Northwest Publishing Company, Anchorage.
McConnaughey, R. A., K. L. Mier, and C. B. Dew. 2000. An examination of chronic trawling on soft-bottom benthos of the eastern Bering Sea. ICES Journal of Marine Science 57(5):1377–1388.
NMFS (National Marine Fisheries Service). 1996. Magnuson-Stevens fishery conservation and management act. NOAA Technical Memorandum NMFS-F/SPO-23.
NMFS (National Marine Fisheries Service). 2001. Species codebook, March 2001. Resource assessment and conservation engineering division. U.S. National Marine Fisheries Service, Alaska Fisheries Science Center, Seattle.
NMFS (National Marine Fisheries Service). 2002. Magnuson-Stevens act provisions; essential fish habitat (EFH). Final Rule. Federal Register 67(12):2343–2383.
Orth, D. J. 1967. Dictionary of Alaska place names. U.S. Geological Survey Professional Paper 567.
Stark, J. W., and D. M. Clausen. 1995. Data report: 1990 Gulf of Alaska bottom trawl survey. NOAA Technical Memorandum NMFS-AFSC-49.
Stevens, B. G. 2003. Settlement, substratum preference, and survival of red king crab *Paralithodes camtschaticus* (Tilesius, 1815) glaucothoe on natural substrate in the laboratory. Journal of Experimental Marine Biology and Ecology 283:63–78.
Stevens, B. G., and J. Kittaka. 1998. Postlarval settling behavior, substrate preference, and time to metamorphosis for red king crab *Paralithodes camtschaticus*. Marine Ecology Progress Series 167:197–206.
Stone, R. P., M. M. Masuda, and P. W. Malecha. 2005. Effects of bottom trawling on soft-sediment epibenthic communities in the Gulf of Alaska. Pages 461–475 *in* P. W. Barnes and J. P. Thomas, editors. Benthic habitats and the effects of fishing. American Fisheries Society, Symposium 41, Bethesda, Maryland.
Stoner, A. W., and R. H. Titgen. 2003. Biological structures and bottom type influence habitat choices made by Alaska flatfishes. Journal of Experimental Marine Biology and Ecology 292:43–59.
Tsalkina, A. V. 1969. Characteristics of the epifauna of the West Kamchatka shelf (from "Problems of commercial hydrobiology"). Proceedings of the All-Union Research Institute of Marine Fisheries Oceanography. (VNIRO) Trudy 65:248–257. (Fisheries Research Board of Canada Translation Series 1568, 1970).
Wantiez, L., and M. Kulbicki. 1995. Main fish populations and their relation to the benthos in a silted bay of New Caledonia, as determined by visual censuses. Cybium 19(3):223–240.
Wassenberg, T. J., G. Dews, and S. D. Cook. 2002. The impact of trawls on megabenthos (sponges) on the northwest shelf of Australia. Fisheries Research 58(2):141–151.
Wessel, P., and W. H. F. Smith. 1998. New, improved version of generic mapping tools released. EOS Transactions, AGU 79(47):579.
Williams, G. C. 2003. Index pennatulacea. Department of Invertebrate Zoology, California Academy of Sciences. Available: *www.calacademy.org/research/izg/index_intro.htm* (August 2003).

Pockmarks on the Outer Shelf in the Northern Gulf of Mexico: Gas-Release Features or Habitat Modifications by Fish?

KATHRYN M. SCANLON[1]

*U.S. Geological Survey, 384 Woods Hole Road,
Woods Hole, Massachusetts 02543, USA*

FELICIA C. COLEMAN[2] AND CHRISTOPHER C. KOENIG[3]

*Department of Biological Science, Florida State University,
Tallahassee, Florida 32306-1100, USA*

Abstract. Side-scan sonar and multibeam data and video observations from six outer continental shelf sites in the northern Gulf of Mexico reveal evidence of three distinct types of seafloor pockmarks. Two sites in the northwestern Gulf (Flower Garden Banks and the Pinnacles area off Mississippi and Alabama) have large-diameter (up to 65 m) pockmarks that are probably generated by the seepage of gas through the sediments. Three sites (Madison–Swanson, Twin Ridges, and an area north of Steamboat Lumps) have small-diameter (less than 2 m) pockmarks that occur in fine-grained sediments at water depths greater than about 100 m. These may be attributed to excavation by fish and other animals. The sixth site, Steamboat Lumps, has medium-diameter (5–25 m) pockmarks in sand with loose cobble-sized to boulder-sized rocks. Red grouper *Epinephelus morio* are associated with these pockmarks and appear to be maintaining and excavating them. Because this area may have experienced freshwater seeps in the past, the pockmarks may be the result of a combination of water seepage followed by fish excavation.

Introduction

The term "pockmark" was first applied to circular seafloor depressions by King and MacLean (1970) to describe the cone-shaped features, 15–45 m in diameter, that they observed on the Scotian Shelf in the northwestern Atlantic Ocean. Since that time, pockmarks of various sizes have been reported in seafloor and lakebed sediments around the world and in a wide variety of environments (e.g., Nelson et al. 1979; Hovland 1981; Pecore and Fader 1989; Scanlon and Knebel 1989; Colman et al. 1992; Pickrill 1993; Solheim and Elverhoi 1993; Baraza and Ercilla 1996; Paull et al. 2002; Whiticar 2002). The increased availability and usage of swath mapping technologies, such as side-scan sonar, during the past 20 years has resulted in an increased number of observations (and more accurate interpretations) of circular seabed depressions. Discussions in the geologic literature on the origins of these features have focused on resuspension of sediments by seepage of gas or water as the mechanism of formation (e.g., Hovland 1982; Harrington 1985; Kelley et al. 1994; Whiticar 2002). Possible biological mechanisms, such as whale feeding sites (Nelson and Johnson 1987) and fish burrowing (Twichell et al. 1985), have been suggested but have received little attention.

We have observed pockmarks in several locations on the outer shelf in the northern Gulf of Mexico. These pockmarks occur in a range of sizes and a variety of sediment types. Some differ in one or more aspects from the "typical" pockmarks described by Hovland and Judd (1988) in their comprehensive book on seabed pockmarks. Most importantly, we have observed red grouper *Epinephelus morio*, which are known to excavate around hard objects (Coleman and Williams 2002), associated with some of the pockmarks and blueline tilefish *Caulolatilus microps*, which are known to create burrows (Hoese and Moore 1998), associated with other pockmarks. This paper describes the occurrences of pockmarks on the outer shelf in the northern Gulf of Mexico (Figure 1) and discusses their formation.

---
[1] Corresponding author: kscanlon@usgs.gov
[2] E-mail: coleman@bio.fsu.edu
[3] E-mail: koenig@bio.fsu.edu

Methods

Side-scan sonar data were collected by the U.S. Geological Survey (USGS) at the Twin Ridges site (Figure 1) in 1997 using the EdgeTech DF 1000 system (EdgeTech, West Wareham, Massachusetts). The data were processed to remove artifacts, noise, and geometric distortions and then digitally stitched to create a continuous coverage mosaic of the approximately 150-km² study area. Details of the processing techniques can be found in Scanlon (2000). Side-scan sonar data from the Madison–Swanson and Steamboat Lumps sites (Figure 1) were collected in 2000 using the SIS 1000 system (Benthos, North Falmouth, Massachusetts). These data were processed and mosaicked in the same manner as the Twin Ridges dataset, resulting in complete coverage mosaics of approximately 300 km² at the Madison–Swanson site and 400 km² at the Steamboat Lumps site. All our side-scan sonar data resolved objects about 1 m across.

Multibeam bathymetry and backscatter data were collected at Flower Garden Banks in 1997 using a Kongsberg Simrad EM300 multibeam sonar system, at the Pinnacles area in 2000 using a Kongsberg Simrad EM1002 system, and at the Madison–Swanson and Steamboat Lumps sites (Figure 1) in 2001 using a Kongsberg Simrad EM1002 system. The parts of the multibeam surveys we used had the following spatial resolutions: Flower Garden Banks, 5 m/pixel; the Pinnacles area, 4 m/pixel; Madison–Swanson, 8 m/pixel; and Steamboat Lumps, 4 m/pixel. Details of the processing for each dataset and downloadable data are included in several reports (Gardner et al. 2002a, 2002b, 2002c), which are available online (*http://walrus.wr.usgs.gov/pacmaps/data.html*).

Sediment samples were collected using a modified Van Veen grab sampler during numerous cruises between 1997 and 2002. Grain size analyses were done in the sedimentology laboratory at the USGS in Woods Hole, Massachusetts, according to the methodology reported in Poppe and Polloni (2000). Carbonate content was determined by weight loss of 15 g bulk material after digestion with 10% hydrochloric acid.

Video observations were made during several cruises using either Deep Ocean Engineering's Mini-Phantom remotely operated vehicle (ROV) or a passive-towed camera sled (Rose Bud), each carrying a Sony Hi-8 video camera. Manned submersible dives were made using Nuytco Inc.'s DeepWorker, a one-person submarine with a Sony 3-chip digital video camera. Each camera was equipped with a parallel laser beam system for measuring objects.

Observations

We have identified pockmarks in several sites in the northern Gulf of Mexico (Figure 1) and describe each below, from west to east. Table 1 summarizes the dimensions and sediment properties for each site.

(1) Flower Garden Banks

Multibeam bathymetry data from East Flower Garden Bank revealed two areas of pockmarks, both in more than 120 m of water (Dartnell and Gardner 1999; Gardner et al. 2002a). One area is northwest of the bank, which has been uplifted by diapiric salt (Nettleton 1957; Rezak et al. 1985). The other area is east and southeast of the bank (Figure 2). Pockmarks from these two areas are up to 65 m in diameter but are only about 1 m deep. We estimate a density of about 65 pockmarks/km². The "typi-

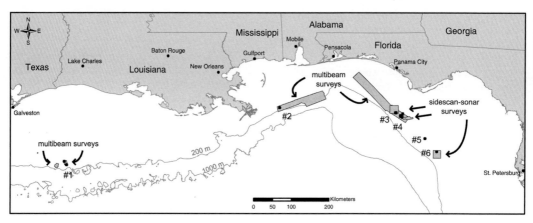

Figure 1. Map showing the locations of six occurrences of pockmarks described in the text at (1) Flower Garden Banks; (2) Mississippi and Alabama Pinnacles; (3) Madison–Swanson; (4) Twin Ridges; (5) north of Steamboat Lumps; and (6) Steamboat Lumps. Gray polygons represent areas of coverage by either multibeam or side-scan sonar data. Small dots identify the locations of close-up figures.

Table 1. Comparison of several pockmark occurrences in the Gulf of Mexico and burrows by tilefish *Lopholatilus chamaeleonticeps* near Hudson submarine canyon. Abbreviations used for data types are: sss = side-scan sonar, mb = multibeam, vid = video, and sub = submersible. Typical relief refers to the depth of the center of the pockmark relative to the surrounding seafloor.

Site number and name	Water depth (m)	Typical diameter (m)	Maximum diameter (m)	Typical relief	Density/ km²	Sediment type	Percent CaCO₃	Data types	References
(1) Flower Garden Banks	>120	35	65	1	65	Mud	25	mb	Gardner et al. (1998); Koenig (unpublished); Scanlon (unpublished)
(2) Pinnacles area	80–110	35	55	1	80			mb	Sager et al. (1992); Gardner (2002b)
(3) Madison–Swanson	>100	1	4	1		Silty to sandy clay, clayey silt	80	sss, mb, vid	J. Duggar, Project Consulting Services, Inc., Metairie, Louisiana, personal communiation; Scanlon,unpublished
(4) Twin Ridges	>100	4	8	1	600	Muddy to slightly gravelly sand	40–60	sss,mb	Scanlon (2000)
(5) North of Steamboat Lumps	97	<1	2	<1		Gravelly sand, clayey silt	97	sss,vid	Scanlon (unpublished)
(6) Steamboat Lumps	70–80	6	25	2	250	Slightly gravelly sand	~95	sss,mb, vid,sub	SSE (2001); Scanlon (unpublished)
Hudson Canyon tilefish	120–500	1.6	Up to 5	1.7	2,500	Stiff silty clay	Low	sss, sub	Twichell et al. (1985)

cal" size appears to be about 35 m in diameter, but this may be an artificially high estimate because the multibeam data cannot resolve pockmarks smaller than about 10 m. Sediment here is mud with about 25% calcium carbonate (Scanlon et al. 2003).

(2) Mississippi and Alabama Pinnacles Area
Sager et al. (1992) first noted the presence of pockmarks in side-scan records from the outer shelf off Mississippi and Alabama (Figure 1). They describe the features as being 15 m or less in diameter, 1 m or less deep, and occurring in several patches, each covering several km^2 of seafloor in water depths between 80 and 110 m. They reported the density of pockmarks to be variable, with the maximum approaching 80/km^2. Multibeam bathymetric data (Gardner et al. 2002c) collected from the same area in 2000 also show several patches of pockmarks, mostly in the same water depths, and adjacent to salt diapirs (Roberts et al. 1999; Figure 3). We found the pockmarks here to be very similar in character to those at the Flower Garden Banks site (Figure 2B). They range up to 55 m in diameter (typically 35 m), are about 1 m deep, and are of similar density. Both Sager et al. (1992) and Gardner et al. (2001, 2002c) suggest that the pockmarks resemble gas-generated pockmarks described by Hovland and Judd (1988) and, because gas is known to be present in the sediments in this region (e.g., Roberts et al. 1999), they presume they were formed by expulsion of gas from underlying sediments.

(3) Madison–Swanson
Side-scan sonar data collected from the Madison–Swanson Fishery Reserve in 2000 suggested the presence of small pockmark-like features (1–2 m in diameter) in water depths deeper than about 100 m (Figure 1) in the northwestern and southern parts of the reserve. The presumed pockmarks were near the limit of resolution of the side-scan data, making it impossible to determine pockmark density for this site. Geophysical surveys conducted at about the same time for a proposed pipeline also revealed pockmarks in the southern and southwestern sections of the reserve (J. Duggar, personal communication), adjacent to areas where our side-scan data suggested pockmarks. Multibeam bathymetry collected in 2001 (Gardner et al. 2002b) resolves only a few large pockmarks (>20 m across) in the southern part of the reserve; these data are of too low resolution to distinguish small pockmarks. Video data, collected using the Rose Bud camera sled in the fall of 2003 at several deepwater (>100 m) sites within the reserve, reveals a fine-grained sediment bottom with numerous burrows and pockmarks varying in size from a few centimeters to 2 m in diameter. The pockmarks are typically steep sided, and some of the larger ones contain a narrow burrow near the bottom of the pit (Figure 4). A blueline tilefish was observed swimming into a 1-m-diameter pockmark as the camera sled approached. The sediments in the pockmarked parts of the reserve are silty clay, clayey silt, and sandy silty clay.

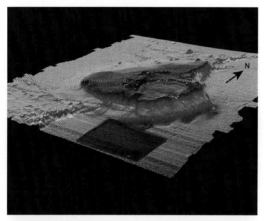

Figure 2. (Top) Oblique view (azimuth = 315°, altitude = 30°) of multibeam color-coded shaded-relief bathymetry of East Flower Garden Bank (area 1 in Figure 1), based on data from Gardner et al. (2002a). Red denotes shallow water; blue denotes deep water. Shaded rectangle identifies the area shown in the bottom panel. (Bottom) Plan view of grayscale shaded-relief (illuminated from the northwest) of an area south of the uplifted bank where pockmarks are numerous. These pockmarks are up to 65 m in diameter but only about 1 m deep (Table 1). East–west lineations are artifacts of the ship's track. High-resolution bathymetry data are courtesy of J. Gardner, U.S. Geological Survey, Menlo Park, California.

(4) Twin Ridges
Side-scan sonar data (Scanlon 2000) from the Twin Ridges area (Figure 1) show fields of small pockmarks in water depths greater than 100 m (Figure 5). The pockmarks are up to 8 m in diameter, but most are much smaller in diameter, and all are approximately 1 m deep. They are densely spaced at approximately 600/km^2. The sediment textures in

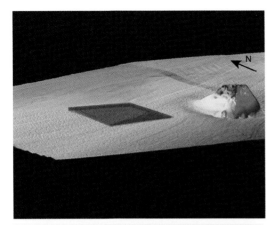

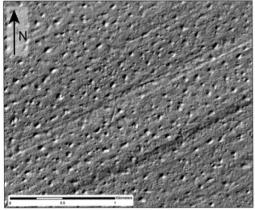

bioturbation and burrows and pockmarks of varying sizes, from a few cm to 2 m across, were documented using an ROV-mounted video camera (Figure 4). All observed pockmarks were 1 m or less in depth and typically steep sided. Some of the larger pockmarks contain a narrow burrow near the bottom of the pit (Figure 4). We could not estimate the density of pockmarks here because we did not obtain sufficient side-scan coverage. The one sediment grab sample obtained at this site was classified as gravelly muddy sand (70% sand, 15% gravel, and 15% silt plus clay), based on bulk textural analysis. However, visual inspection of the sample by one of the authors (KS) at the time the sample was collected indicated it was poorly sorted clayey silt with patches of gravelly sand. The carbonate content is 97%. These pockmarks are similar to those observed at Twin Ridges (Table 1).

(6) Steamboat Lumps

Using side-scan sonar data, we identified a small area (about 0.4 km^2) on the outer shelf in the Steamboat Lumps Fishery Reserve (Figure 1) containing about 100 pockmarks (Figure 6). The pockmarks in this area are typically about 6 m in diameter (Figure 7) but range from less than a meter to 25 m in diameter. They are 1–2 m deep and have rocks at their bottoms or embedded in the sloping sides. In some of the larger pockmarks, the rocks embedded in the sides have had the sediment surrounding them removed (Figure 8), creating

Figure 3. (Top) Oblique view (azimuth = 315°, altitude = 30°) of multibeam color-coded shaded-relief bathymetry from the Pinnacles area (area 2 in Figure 1), based on data from Gardner et al. (2002c). Red denotes shallow water; blue denotes deep water. Shaded rectangle identifies the area shown in the bottom panel. (Bottom) Plan view of grayscale shaded-relief (illuminated from the northwest) of an area west of the uplifted bank where pockmarks are numerous. These pockmarks are similar to those at Flower Garden Banks (Figure 2, Table 1). They are shallow (about 1 m) and typically about 35 m in diameter but some are as large as 55 m in diameter. Northeast–southwest lineations are artifacts of the ship's track. High-resolution bathymetry data are courtesy of J. Gardner, USGS.

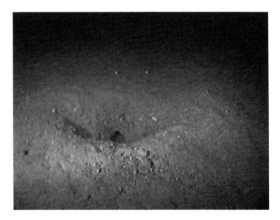

Figure 4. Image from video of a pockmark area north of Steamboat Lumps (area 5 in Figure 1), taken using a remotely operated vehicle-mounted video camera during a cruise of the National Oceanic and Atmospheric Administration ship R/V *Oregon II* in 2000. The two small bright spots in the upper middle of the image are 20 cm apart. The central depression is only about 0.5 m across, but the feature (including the mounded sediment surrounding the depression) is approximately 1 m in diameter. Note the horizontal burrow extending into the sediment near the bottom of the pockmark.

this area, based on three sediment samples, range from slightly gravelly sand to muddy sand, and the samples contain between 40% and 60% calcium carbonate. These pockmarks are similar to those observed in area 5, North of Steamboat Lumps, described below.

(5) North of Steamboat Lumps

Using side-scan sonar, we identified an area north of Steamboat Lumps (Figure 1) with small pockmarks at a water depth of approximately 100 m. Extensive

Figure 5. Side-scan mosaic of pockmarks at Twin Ridges site (area 4 in Figure 1). The pockmarks are concentrated in the southwestern half of the figure where the water depth is about 100 m. Sediment samples and lower acoustic backscatter indicate that the sediment is finer grained than it is in the northeastern half of this figure where no pockmarks occur. Side-scan sonar data are from Scanlon (2000).

a composite pockmark. Additional small groups of pockmarks occur in the same 70-m to 80-m depth range along a 15-km-long segment of the outer shelf.

Textural analyses of two sediment grab samples taken near the pockmark area in Steamboat Lumps indicate that the sediments in this area are slightly gravelly sands. The gravel-sized particles are composed of broken bits of shells and other biogenic carbonate material. Both samples contain about 95% calcium carbonate.

Submersible observations indicate that rocks exposed in the pockmarks are irregular in shape and highly rugose (Figure 9). They occur at the bottom of, or are embedded in the sides of, the bowl-shaped pockmarks, and when nudged by the submersible during one dive, some of the rocks moved, suggesting that they are not the exposed tips of kartified (partially dissolved limestone) bedrock features. Instead, they appear to be free-floating nodules within the unconsolidated sand substrate. These rocks have a similar appearance to carbonate-cemented nodules reported by others (e.g., Hovland and Judd 1988; Poppe et al. 1990; Conti et al. 2002) from environments subject to gas or freshwater seepages. A red grouper was observed in nearly every pockmark.

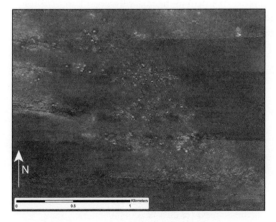

Figure 6. Side-scan mosaic of a cluster of pockmarks at Steamboat Lumps (area 6 in Figure 1). These pockmarks are up to 25 m in diameter and up to 2 m deep. Red grouper and rocks were noted in nearly all of the pockmarks that were observed during submersible dives. Side-scan data were collected by U.S. Geological Survey during a cruise of the National Oceanic and Atmospheric Administration ship R/V *Oregon II* in 2000.

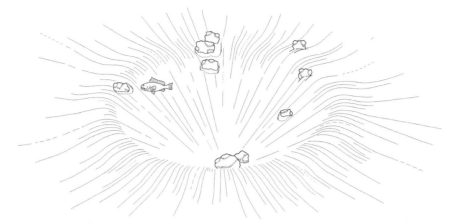

Figure 7. Sketch of a typical "excavated seep pockmark" in the Steamboat Lumps Fishery Reserve (area 6 in Figure 1), drawn from video data. Note partially excavated rocks embedded in the sides of the pockmark and loose rocks in the bottom, apparently having rolled from the sides after being excavated. Sketch courtesy of J. Zwinakis (U.S. Geological Survey).

Several high-resolution seismic-reflection profiles, collected simultaneously with the side-scan sonar data (Capone et al. 2002), crossed the pockmarked area. The profiles show a lens of sand up to about 10 m thick covering a discontinuous strong subsurface reflector. The pockmarks occur where the sand is at least several meters thick. Additional seismic profiles (Capone et al. 2002) a few km southeast of the pockmarked area show less than a meter of sand overlying the strong reflector, and side-scan data in the same area do not show pockmarks. Observations from a submersible in this area reported a flat, creviced, rocky bottom where the hard layer was exposed on the seafloor.

Discussion

The continental shelf in the northeastern Gulf of Mexico is underlain by a nearly flat, drowned, and partially dissolved carbonate (karst) platform covered by a thin veneer of Quaternary sediments (e.g., Mullins et al. 1988; Brooks and Doyle 1991; Hine 1997). In contrast, the shelf in the northwestern Gulf of Mexico has been uplifted and faulted by numerous salt diapirs and is covered in places by thick overlapping deltas of terrigenous sediment deposited during times of lower sea level (Kindinger 1989; Aharon et al. 1992; Roberts et al. 1999; Sager et al. 1999). These differences in geologic history affect significant differences in modern seafloor habitats and are

Figure 8. Photomosaic from video of pitted landscape near the rim of a large pockmark at Steamboat Lumps (area 6 in Figure 1). Numerous rocks have been partially excavated, presumably by red grouper. Further excavation may cause the pits to coalesce and the rocks to tumble into the center of the compound pockmark (video data courtesy of SSE [2001].

Figure 9. Close-up photo of rocks in a pockmark at Steamboat Lumps (area 6 in Figure 1). Note the rugosity and irregular morphology of the rock. The image is approximately 50 cm across in the mid-range (photo is courtesy of SSE[2001]).

important to discussions about the origin of the pockmarks described here. We found three distinct types of pockmarks and, thus, speculate three different mechanisms or combinations of mechanisms (gas escape, fish burrowing, and excavation of seep deposits by fish) as responsible for their formation.

Gas-Escape Pockmarks

Pockmarks in the northern and northwestern Gulf of Mexico (Pinnacles and Flower Garden Banks areas) are associated with salt diapirs and are larger in diameter than the pockmarks occurring elsewhere in the Gulf of Mexico (see Table 1). These areas are underlain by known oil and gas reservoirs, making escape of thermogenic gas from the seabed likely (Hovland and Judd 1988). Alternatively, the thick terrigenous sediments could provide a source of biogenic methane. In fact, gas bubbles have been observed escaping from sediments in the Flower Garden Banks area (Rezak et al. 1985). We do not know the sediment type at the Pinnacles area, but it is likely to be similar to the mud sampled in the pockmarked parts of the Flower Garden Banks area. These characteristics are consistent with the "typical" gas-generated pockmarks described by Hovland and Judd (1988). Although we are lacking the seismic-reflection data that would reveal gas in the sediments, the circumstantial evidence suggests that these pockmarks are probably formed by gas escaping through the sediments.

Fish Burrow Pockmarks

The small (generally 1–2 m in diameter) pockmarks observed in the northeastern Gulf of Mexico (i.e., Madison–Swanson, Twin Ridges, and the area north of Steamboat Lumps) occur in fine-grained sediments (generally silt, clay, or silty sand) and in water depths greater than about 90 m. They are very similar in size and shape to burrows from tilefish described off the East Coast of the United States (Able et al. 1982), and many of them contain a narrow burrow extending vertically or horizontally into the sediment near the bottom of the depression, as is characteristic of tilefish burrows. Larger pockmarks, like those seen in the northwestern Gulf of Mexico, are entirely absent here. Video observations reveal that these areas contain numerous bioturbation features over a continuum of sizes—pits and mounds with dimensions of a few centimeters to a few meters are common in all three areas (Madison-Swanson, Twin Ridges, and north of Steamboat Lumps).

Biological activity appears to be creating these pockmarks. Figure 4 shows a pockmark surrounded by a symmetrical mound of sediment that appears to be of the same texture as that on the nearby seabed. This suggests that an animal excavated the pockmark by carrying the sediment a short distance and depositing it at the edge. Resuspension of sediment by fluid escape, in contrast, would cause winnowing of the fine sediment and, if currents were present, asymmetrical deposits. Seismic-reflection profiles and video and direct observations produced no evidence of gas (such as changes in acoustic properties or bubbles in the water column) in the sediments in these areas.

Tilefish, a known burrower (Able et al. 1982); blueline tilefish, also a burrower (Hoese and Moore 1998); and yellowedge grouper *Epinephelus flavolimbatus*, a possible burrower (Jones et al. 1989), occur in the Gulf of Mexico in shelf-edge and deeper waters (Hoese and Moore 1998). Tilefish abundance off Georgia has been shown to be directly related to substrate composition; specifically, catch rates are higher where the silt–clay fraction in the sediments is higher (Grossman et al. 1985). Twichell et al. (1985) also showed that tilefish burrows near the Hudson Canyon off the eastern United States occurred only in clay substrate and were absent from areas with sand substrate. Tilefish apparently select habitat with fine-grained substrate, perhaps because the more cohesive sediment is better for burrowing. Based on our sediment-texture data and our side-scan sonar data, there is a similar preferential occurrence of small pockmarks in the northeastern Gulf of Mexico in the fine-grained sediments that occur on the upper slope in water depths greater than 90 m. Although we do not have direct observations of fish constructing these pockmarks, we have observed a blueline tilefish occupying one of the pockmarks. We suggest it is likely that these pockmarks, like similar ones off the eastern United States, were constructed by tilefish.

Excavated Seep Pockmarks

The pockmarks on the outer continental shelf in the Steamboat Lumps Fishery Reserve are the most interesting of the three types described here. Each pockmark contains one or more cobble-sized to boulder-sized rock "floating" in the non-cohesive sandy sediment. We speculate that the rocks may be nodules of cemented sand that formed within the unconsolidated sediment by chemical precipitation of carbonates and iron oxides from percolating water. Freshwater springs, such as those that occur on the inner shelf of western Florida (Rosenau et al. 1998), can provide an environment conducive to nodule formation (Harris 1985). The Tertiary carbonate strata, which make up the Floridan aquifer system that feeds these springs, extend offshore beneath the West Florida Shelf (Hine 1997; Miller 1997). These strata could provide a conduit for freshwater to the outer shelf, particularly during past periods of lower sea level. Submersible dives made specifically to look for freshwater springs, however, revealed no evidence of fluid (gas or water) escaping from the pockmarks. This suggests that if fluid flow is involved in the formation of these rocks, it is either intermittent or was only active in the past (perhaps at a time of lower sea level). Somewhat larger, elongate pockmarks in Eckernforde Bay, Denmark, are believed to have been created by episodic freshwater seeps (Whiticar 2002). Gas seeps are also known to be associated with the formation of carbonate cemented rocks or nodules (e.g., Conti et al. 2002; Hovland 2002), but there is no evidence of gas in the sediments in the Steamboat Lumps area.

Although the speculative freshwater spring theory suggested above may provide an explanation for the enigmatic presence of rocks within the sandy substrate, it does not fully explain the existence of the pockmarks. These pockmarks appear "active"; rock surfaces are devoid of any fine-grained sediment, and steep slopes near the angle of repose of the sandy sediment are maintained. We suggest that red grouper, observed in nearly every pockmark in the Steamboat Lumps area, actively excavate the pockmarks by removing sediment from around the rocky nodules. Unlike tilefish, which prefer to burrow in cohesive sediment, red grouper are known to inhabit areas with sandy sediment, which they remove to expose hard substrate (Coleman and Williams 2002). As sediment is removed around one nodule, additional nodules may be uncovered in the sloping sides (Figure 7) of the pockmark. With continued excavation, the nodules may become unstable and tumble downward, creating the small piles of rocks observed in the pockmark centers during submersible dives (Figure 9). It is possible that these pockmarks may have been initiated by fluid flow but now are being enlarged and maintained by the efforts of red grouper.

Sediment Volume Calculations

In the Steamboat Lumps area, a typical excavated seep pockmark, which we attribute to red grouper activities, has a depth of 2 m and a diameter of 6 m (Table 1). If we assume it is a partial sphere, we can calculate its volume using the formula

$$V_{part} = 1/6 pD(3A^2 + D^2),$$

where V_{part} is the volume of the partial sphere, D is the depth of the pockmark below the surrounding seafloor, and A is the radius of the pockmark (which is smaller than the radius of the sphere). We find that the typical pockmark in Steamboat Lumps has a volume of 32.4 m². Since the density of pockmark occurrence in Steamboat Lumps is about 250/km², the volume of sediment moved here is about 8,100 m³/km².

We made similar calculations for tilefish burrow pockmarks at Twin Ridges, using the data in Table 1 and assuming a cone shape, which more closely resembles the shape of these pockmarks than does a partial sphere. The results suggest that tilefish are responsible for moving somewhat less sediment (2,512 m³/km²) than are red grouper (8,100 m³/km²). Twichell et al. (1985) reported a remarkably similar volume of sediment (2,500 m³/km²) moved by tilefish on the continental shelf near the head of Hudson Canyon off the eastern United States. Perhaps an optimum size and spacing of burrows is consistently achieved in tilefish habitats.

In contrast, the volume of the gas-release pockmarks in the northwestern Gulf of Mexico (Flower Garden Banks and Pinnacles areas), calculated assuming a partial sphere shape, is over 30,000 m³/km². The volume of sediment we suggest was moved by gas release is, therefore, an order of magnitude greater than the volumes we calculated for the other two types of pockmarks, which we suggest are wholly or partially excavated by fish.

Possible Implications

Recognition that fish may play a role in the creation of seabed pockmarks is important for marine engineers, geologists, ecologist, fishers, and resource managers for a number of reasons. (1) The occurrence of pockmarks has been taken to imply the presence of gas in the sediment, which can make the substrate unstable (Platt 1977; Newton et al. 1980). Pockmarks that are excavated by fish rather than by escaping gas, however, may not indicate unstable substrate. This has important implications for placement of seafloor engineering projects, such as pipelines and drilling platforms. (2) Fish can be a significant "geologic agent," moving large amounts of sediment (thousands of cubic meters per square kilometers, by our calculations) and mixing sedimentary strata to depths of up to 2 m (the depth of some of the fish burrows described here). Depending on sedimentation rates, geologic materials (e.g., mineral grains, microfossils, etc.) with ages spread over tens of thousands of years could be mixed, introducing errors in dating of geologic, archaeologic, and paleoclimatic events. (3) The presence of certain types of pockmarks may suggest the presence of certain economically important fish such as red grouper or tilefish. Pockmarks with diameters of 1–2 m, such as those created by tilefish, are readily imaged with high-resolution side-scan sonar systems, and their presence can be used to delineate the extent of a particular habitat. The number of fish-excavated pockmarks per unit area could be used to estimate the populations of species known to be associated with such pits. Changes in pockmark density over time could be used to assess changes in fish populations to evaluate the health of a stock or the effectiveness of a fishery closure.

Conclusions

We describe three distinct types of pockmarks in the northern Gulf of Mexico and the mechanisms (or combinations of mechanisms) responsible for their formation: (1) gas seeps are probably responsible for the formation of the large pockmarks observed in the northern and northwestern Gulf of Mexico, (2) simple burrows made by tilefish or other organisms are the likely origin the smaller pockmarks in the fine-grained sediment on the upper slope in the northeastern Gulf of Mexico, and (3) a combination of geologic and biological processes may create the pockmarks at Steamboat Lumps, where it appears that red grouper are excavating sandy sediment around rocks that may have been deposited and cemented by gas or water seeps during periods of lower sea level. Modern seepage, if it occurs, could supplement the digging of the fish.

We suggest that excavation by fish and other organisms may be more common than previously thought and that some pockmarks in the Gulf of Mexico or elsewhere assumed to be created by gas escape may actually be excavations made by fish. Also, combinations of biological and geologic mechanisms (such as gas seepage and fish excavation) may occur, since fish and other organisms appear to be attracted to gas and water seeps (Hovland and Judd 1988). Distinguishing between modes of formation is difficult, but our data may suggest that gas-generated pockmarks attain larger diameters (tens of meters) than do fish-constructed pockmarks (one to several meters) and represent an order of magnitude greater volume of sediment removal. Also, gas-generated pockmarks may have a wider range of sizes, whereas fish-constructed pockmarks appear to be more uniform in size and density.

Acknowledgments

We thank the National Oceanic and Atmospheric Administration's Sustainable Seas Expedition for the opportunity to work offshore in the northeastern Gulf of Mexico during the summer of 2001 and National Geographic Society for permission to use photographic images from the cruise. Funding for this research was provided by the Flower Garden Banks National Marine Sanctuary (Scanlon), the Pew Marine Conservation Fellows Program (Coleman), the National Sea Grant Program (Project Number R/LR-B-51, Coleman and Koenig) and the U.S. Geological Survey (Scanlon). We thank S. Ackerman for manipulation of data and preparation of figures and D. Twichell, P. Valentine, and two anonymous reviewers for helpful reviews.

References

Able, K. W., C. B. Grimes, R. A. Cooper, and J. R. Uzmann. 1982. Burrow construction and behavior of tilefish, *Lopholatilus chamaeleonticeps*, in Hudson Submarine Canyon. Environmental Biology of Fishes 7(3):199–205.

Aharon, P., E. R. Graber, and H. H. Roberts. 1992. Dissolved carbon and d13C anomalies in the water column caused by hydrocarbon seeps on the northwestern Gulf of Mexico slope. Geo-Marine Letters 12:33–40.

Baraza, J., and G. Ercilla. 1996. Gas-charged sediments and large pockmark-like features on the Gulf of Cadiz slope (SW Spain). Marine and Petroleum Geology 13(2):253–261.

Brooks, G. R., and L. J. Doyle. 1991. Geologic development and depositional history of the Florida Middle Ground: a mid-shelf, temperate-zone reef system in the northeastern Gulf of Mexico. Pages 189–203 *in* R. H. Osborne, editor. Shoreline to abyss. Society for Sedimentary Geology, Special Publication 46, Tulsa, Oklahoma.

Capone, M. K., B. A. Swift, and K. M. Scanlon. 2002. Archive of chirp subbottom data collected during USGS cruise ORGN00005 northeastern Gulf of Mexico, 15 February–2 March 2000. U.S. Geological Survey, Open-File Report 02-45, 2 DVD-ROM set, Reston, Virginia.

Coleman, F. C., and S. L. Williams. 2002. Overexploiting marine ecosystem engineers: potential consequences for biodiversity. Trends in Ecology and Evolution 17(1):38–44.

Colman, S. M., D. S. Foster, and D. W. Harrison. 1992. Depressions and other lake-floor morphologic features in deep water, southern Lake Michigan. Journal of Great Lakes Research 18(2):267–279.

Conti, A., A. Stefanon, and G. M. Zuppi. 2002. Gas seeps and rock formation in the northern Adriatic Sea. Continental Shelf Research 22:2333–2344.

Dartnell, P., and J. V. Gardner. 1999. Sea-floor images and data from multibeam surveys in San Francisco Bay, Southern California, Hawaii, the Gulf of Mexico, and Lake Tahoe, California–Nevada. U.S. Geological Survey Digital Data Series DDS-55, CD-ROM.

Gardner, J. V., J. D. Beaudoin, J. E. Hughes Clarke, and P. Dartnell. 2002a. Multibeam mapping of selected areas of the outer continental shelf, northwestern Gulf of Mexico—data, images, and GIS. U.S. Geological Survey, Open-File Report OF02-411, Reston, Virginia.

Gardner, J. V., P. Dartnell, and K. J. Sulak. 2002b. Multibeam mapping of the West Florida Shelf, Gulf of Mexico. U.S. Geological Survey, Open-File Report OF02-005, Reston, Virginia.

Gardner, J. V., P. Dartnell, and K. J. Sulak. 2002c. Multibeam mapping of the Pinnacles region, Gulf of Mexico. U.S. Geological Survey, Open-File Report OF02-006, Reston, Virginia.

Gardner, J. V., P. Dartnell, K. J. Sulak, B. Calder, and L. Hellequin. 2001. Physiography and Late Quaternary-Holocene processes of northeastern Gulf of Mexico outer continental shelf off Mississippi and Alabama. Gulf of Mexico Science 2:132–157.

Gardner, J. V., L. A. Mayer, J. E. Hughes Clarke, and A. Kleiner. 1998. High-resolution multibeam bathymetry of East and West Flower Gardens and Stetson Banks, Gulf of Mexico. Gulf of Mexico Science 16(2):131–143.

Grossman, G. D., M. J. Harris, and J. E. Hightower. 1985. The relationship between tilefish (*Lopholatilus chamaeleonticeps*) abundance and sediment composition off Georgia. U.S. National Marine Fisheries Service Fishery Bulletin 83(3):443–447.

Harrington, P. K. 1985. Formation of pockmarks by pore-water escape. Geo-Marine Letters 5:193–197.

Harris, P. M., 1985. Carbonate cementation—a brief review. In N. Schneidermann and P. A. Harris, editors. Carbonate cements. Society for Sedimentary Geology, Tulsa, Oklahoma.

Hine, A. C. 1997. Structural and paleoceanographic evolution of the margins of the Florida Platform. In A. F. Randazzo and D. S. Jones. The geology of Florida. University Press of Florida, Gainesville.

Hoese, H. D., and R. H. Moore. 1998. Fishes of the Gulf of Mexico, 2nd edition. Texas A&M University Press, College Station.

Hovland, M. 1981. Characteristics of pockmarks in the Norwegian trench. Marine Geology 39:103–117.

Hovland, M. 1982. Pockmarks and the Recent geology of the central section of the Norwegian Trench. Marine Geology 47:283–301.

Hovland, M. 2002. On the self-sealing nature of marine seeps. Continental Shelf Research 22:2387–2394.

Hovland, M., and A. G. Judd. 1988. Seabed pockmarks and seepages: impact on geology, biology and the environment. Graham and Trotman, London.

Jones, R. S., E. J. Gutherz, W. R. Nelson, and G. C. Matlock. 1989. Burrow utilization by yellowedge grouper, *Epinephelus flavolimbatus*, in the northwestern Gulf of Mexico. Environmental Biology of Fishes 26:277–284.

Kelley, J. T., S. M. Dickson, D. F. Belknap, W. A. Barnhardt, and M. Henderson. 1994. Giant sea-bed pockmarks: evidence for gas escape from Belfast Bay, Maine. Geology 22:59–62.

Kindinger, J. L. 1989. Depositional history of the Lagniappe Delta, northern Gulf of Mexico. Geo-Marine Letters 9:59–66.

King, L. H., and B. MacLean. 1970. Pockmarks on the Scotian Shelf. Geological Society of America Bulletin 81:3141–3148.

Miller, J. A. 1997. Hydrogeology of Florida. In A. F. Randazzo and D. S. Jones. The Geology of Florida. University Press of Florida, Gainesville.

Mullins, H. T., A. F. Gardulski, E. J. Hinchey, and A. C. Hine. 1988. The modern carbonate ramp slope of central west Florida. Journal of Sedimentary Petrology 58(2):273–290.

Nelson, C. H., and K. R. Johnson. 1987. Whales and walruses as tillers of the seafloor. Scientific American 256(2):74–81.

Nelson, H., D. R. Thor, M. W. Sandstrom, and K. A. Kvenvolden. 1979. Modern biogenic gas-generated craters (sea-floor "pockmarks") on the Bering Shelf, Alaska. Geological Society of America Bulletin, Part I 90:1144–1152.

Nettleton, L. L. 1957. Gravity survey over a Gulf coast continental shelf mound. Geophysics 22:630.

Newton, R. S., R. C. Cunningham, and C. E. Schubert. 1980. Mud volcanoes and pockmarks: seafloor engineering hazards or geological curiosities? Pages 425–429 in Proceedings 12th Offshore Technology Conference. Offshore Technology Conference, Houston, Texas.

Paull, C., W. Ussler III, N. Maher, H. G. Greene, G. Rehder, T. Lorenson, and H. Lee. 2002. Pockmarks off Big Sur, California. Marine Geology 181:323–335.

Pecore, S. S., and G. B. J. Fader. 1989. Surficial geology, pockmarks, and associated neotectonic features of Passamaquoddy Bay, New Brunswick, Canada. Geological Survey of Canada, Open-file Report 2213.

Pickrill, R. A. 1993. Shallow seismic stratigraphy and pockmarks of a hydrothermally influenced lake, Lake Rotoiti, New Zealand. Sedimentology 40:813–828.

Platt, J. 1977. Significance of pockmarks for engineers. Offshore Engineer August:45.

Poppe, L. J., R. C. Circe, and A. K. Vuletich. 1990. A dolomitized shelfedge hardground in the northern Gulf of Mexico. Sedimentary Geology 66:29–44.

Poppe, L. J., and C. F. Polloni. 2000. USGS east-coast sediment analysis: procedures, database, and georeferenced displays. U.S. Geological Survey Open-File Report 00-358, CD-ROM, Reston, Virginia.

Rezak, R., T. J. Bright, and D. W. McGrail. 1985. Reefs

and banks of the northwestern Gulf of Mexico. Wiley, New York.

Roberts, H. H., R. A. McBride, and J. M. Coleman. 1999. Outer shelf and slope geology of the Gulf of Mexico: an overview. *In* H. Kumpf, K. Steidinger, and K. Sherman, editors. The Gulf of Mexico large marine ecosystem. Blackwell Scientific Publications, Malden, Massachusetts.

Rosenau, J. C., G. L. Faulkner, C. W. Hendry, Jr., and R. W. Hull. 1998. Springs of Florida. Geological Bulletin 31, revised. Florida Geological Survey online publication. Available: *www.flmnh.ufl.edu/springs_of_florida/* (June 2004).

Sager, W. W., C. S. Lee, I. R. Macdonald, and W. W. Schroeder. 1999. High-frequency near-bottom acoustic reflection signatures on hydrocarbon seeps on the northern Gulf of Mexico continental slope. Geo-Marine Letters 18:267–276.

Sager, W. W., W. W. Schroeder, J. S. Laswell, K. S. Davis, R. Rezak, and S. R. Gittings. 1992. Mississippi–Alabama outer continental shelf topographic features formed during the Late Pleistocene-Holocene Transgression. Geo-Marine Letters 12:41–48.

Scanlon, K. M. 2000. Surficial seafloor geology of a shelf-edge area off West Florida. *In* P. R. Briere, K. M. Scanlon, G. Fitzhugh, C. T. Gledhill, and C. C. Koenig. West Florida shelf: side-scan sonar and sediment data from shelf-edge habitats in the northeastern Gulf of Mexico, U.S. Geological Survey, Open-File Report 99-589. CD-ROM.

Scanlon, K. M., S. D. Ackerman, and J. E. Rozycki. 2003, Texture, carbonate content, and preliminary maps of surficial sediments of the Flower Garden Banks area, northwestern Gulf of Mexico outer shelf. U.S. Geological Survey, Open-File Report 03-02, CD-ROM.

Scanlon, K. M., and H. J. Knebel. 1989. Pockmarks in the floor of Penobscot Bay, Maine. Geo-Marine Letters 9:53–58.

Solheim, A., and A. Elverhoi. 1993. Gas-related sea floor craters in the Barents Sea. Geo-Marine Letters 13:235–243.

SSE (Sustainable Seas Expedition). 2001. Sustainable seas expedition, R/V *Gordon Gunter* cruise. National Oceanic and Atmospheric Administration, Silver Spring, and National Geographic Society, Washington, D.C.

Twichell, D. C., C. B. Grimes, R. S. Jones, and K. W. Able. 1985. The role of erosion by fish in shaping topography around Hudson Submarine Canyon. Journal of Sedimentary Petrology 55:712–719.

Whiticar, M. J. 2002. Diagenetic relationship of methanogenesis, nutrients, acoustic turbidity, pockmarks and freshwater seepages in Eckernforde Bay. Marine Geology 182:29–52.

Symposium Abstract

Symposium abstracts have been reprinted from the Abstract Volume prepared for the Symposium on Effects of Fishing Activities on Benthic Habitats: Linking Geology, Biology, Socioeconomics, and Management without any further review or editing.

Understanding the Complex Nature of Fish-Seagrass Associations

T. J. ANDERSON[1]

NOAA National Marine Fisheries Service. Santa Cruz Laboratory, Santa Cruz, California; and U.S. Geological Survey, Coastal and Marine Geology, Menlo Park, California

Seagrass beds are rarely homogenous entities. Instead, they form a mosaic that is structured at many different scales. This has important implications for fish communities. However, while seagrass beds are known to have higher abundances of fishes and greater richness of species than unvegetated habitats, few studies have identified how fish dispersions are modified by the spatial structure inherent in most habitats. In this study a multi-scaled observational (meters to 30 km) and experimental approach was used to quantify the relationship between demersal fishes and subtidal seagrass areas in Port Phillip Bay, Melbourne, Australia. While most species were correlated with seagrass, either directly (e.g. seagrass density and length) or indirectly (e.g. patchiness), seagrass alone did not explain species distributions. Instead, the association of a fish with it's 'preferred' habitat was conditional on the spatial structure of the habitat and the spatial location along the shore, and that these landscape elements operated additively, or synergistically. Additionally, a large-scale temporal dynamic both in the supply of larvae and in seagrass health and presence also operated across all scales examined. This study highlights that measuring the association between organisms and their habitat requires many levels of information, ranging from understanding individual habitat preferences at fine-scales, to understanding the spatially-explicit structure of fish and habitat at landscapes. Understanding and predicting fish assemblage structure in the face of habitat change is no simple task, and relies heavily on the integration of fine-scale empirical and landscape-level studies, but this study demonstrates it is achievable.

---

[1] E-mail: tara.anderson@noaa.gov

Geoacoustic and Geological Characterization of Juvenile Red Snapper Habitat, Northern Gulf Of Mexico Continental Shelf

S. J. BENTLEY[1]

Department of Oceanography and Coastal Sciences and Coastal Studies Institute, Louisiana State University, Baton Rouge, Louisiana

W. F. PATTERSON

Dauphin Island Sea Lab, Dauphin Island, Alabama

Y. ALLEN, W. VIENNE, AND C. WILSON

Department of Oceanography and Coastal Sciences and Coastal Fisheries Institute, Louisiana State University, Baton Rouge, Louisiana

Laboratory and small-scale in situ experiments have demonstrated juvenile red snapper display an affinity for low-relief shell-rubble habitat; however, the spatial extent and temporal variability of large-scale shell-rubble features on the Mississippi-Alabama shelf are unknown. Moreover, the seabed geology of the entire region in general is poorly known, with little significant research conducted since the 1950's. Therefore, to develop a geological understanding of quality juvenile snapper habitat in the region, we have undertaken a program of sidescan seabed mapping and seabed sampling in areas on the northern Gulf of Mexico continental shelf that historically produced high, median, and low juvenile red snapper catch rates in trawl surveys. Preliminary results of sidescan surveys and grab samples indicate highest juvenile snapper catch rates are found near irregular low-relief ridges of shell and sand, with $CaCO_3$ content to 100%. The ridges are elevated 1-2 m above the surrounding seabed and generally orient along NW-SE axes. Surrounding seabed is more typical of the Holocene transgressive sand sheet, composed of fine-medium muddy sand with shell content <10%. Most shell material found on the ridges appears to be fragments of the oyster Crassostrea (now highly encrusted by epibionts), indicating ridges are of estuarine origin, and are probably remnants of coastal shell reefs formed during the Holocene Transgression (i.e., during the past ~6000 y). Ongoing research focuses on elucidating origin of the ridges, developing a geoacoustic fingerprint for quality juvenile red snapper habitat, and examining temporal and spatial variability in juvenile snapper habitat utilization patterns.

[1] E-mail: sjb@lsu.edu

Habitat Associations of Upper Slope Rockfishes (*Sebastes* spp.) and Co-Occurring Demersal Fishes In Ascension Canyon, California

J. J. Bizarro, J. M. Field,[1] and H. G. Greene

Center For Habitat Studies, Moss Landing Marine Laboratories, Moss Landing, California

R. N. Lea

California Department. of Fish and Game, Monterey, California

J. deMarignac

Center For Habitat Studies, Moss Landing Marine Laboratories, Moss Landing, California

Due to their typical life history patterns (slow growth, late age at maturity, extreme longevity) deep-water rockfishes (Sebastes spp.) are especially susceptible to overfishing, as evidenced by recent declines in most commercially targeted stocks. To establish effective Marine Protected Areas (MPAs), the interaction between fishes and their available habitats must be determined. Our objectives were to describe habitat associations for upper slope rockfishes and co-occurring fish species within the headward part of Ascension Canyon at both large (1 to 10s of kilometers) and small (10s to 100s of meters) scales. Geologic structure and lithology were investigated using high-resolution multibeam bathymetric and backscatter data. These data were interpreted to produce habitat maps of the study area. Seafloor features and fish assemblages were then surveyed using the Delta submersible at 50-meter intervals between 200 and 350 m. Thirty-two ten minute transects were completed between two distinct, large-scale habitat types. At 200 and 250 m, stripetail (Sebastes saxicola} and greenstriped (S. elongatus) rockfishes were the dominant fish species. At 300 and 350 meters, splitnose (S. diploproa) and shortspine thornyhead (Sebastolobus alascanus) were the most abundant rockfishes. Large and small-scale habitat associations of these and several other commercially important demersal fishes were also determined.

[1] E-mail: JField@ci.pacific-grove.ca.us

Data Sets Relevant to Identification of Essential Fish Habitat (EFH) on the Gulf of Mexico Continental Shelf and for Estimation of Effects of Shrimp Trawling Gear

P. CALDWELL[1] AND P. SHERIDAN

NOAA Fisheries, Southeast Fisheries Science Center, Galveston, Texas

Our objectives were: to identify data describing habitats, shrimp trawling, and other human activities on the Gulf of Mexico continental shelf; to incorporate such data into a GIS format; and to provide preliminary experimental designs for assessment of effects of shrimp trawling on EFH. We developed 57 data layers describing habitat (benthic organism densities, sand/silt/clay, digitized sediment and biotic community maps), structures (bathymetry, State/Federal waters, safety fairways, oil and gas, artificial reefs, bottom obstructions), and fishing (patterns of shrimp fishing effort, experimental trawling sites/catches, closed waters). Best opportunities for experimental trawling in closed waters lie in southern and northwest Florida (permanent closures) and in Texas (seasonal closures). Experiments in open waters need to account for seasonal closures, ambient shrimping effort, and variations in sediments and their associated benthic communities. Cross-Gulf replication is necessary to provide a fishery-wide assessment of gear impacts. Most opportunities for replication exist at depths of 18-27 m for both sand and mud habitats. Moving to waters only as deep as 46-55 m forces experiments to become more regional and less Gulf-wide in nature. Benthic data are most dense off south Texas and Mobile Bay, less dense off Florida, and are largely absent off west Louisiana and north Texas. Non-extractive or no-take marine reserves could be used to study effects of complete cessation of trawling on habitats and fauna (estimating recovery rates of ecosystem components, conducting fishery-free gear impact studies). We present only a few options - data sets are available on CD.

---

[1] E-mail: Phil.Caldwell@noaa.gov

Habitat and Species Associations of Demersal Fish and Benthic Invertebrates in the New York Bight Apex

S. CHANG,[1] J. VITALIANO, AND F. STEIMLE

National Marine Fisheries Service, Northeast Fisheries Science Center,
James J. Howard Marine Sciences Laboratory, Highlands, New Jersey

The associations among demersal fish and benthic invertebrate species with numerous habitat variables were investigated in the data collected during the 12 Mile Dumpsite Study (12MDS) in the inner New York Bight (July 1986 to September 1989). The 12MDS study was unique because synoptic measurements were made at numerous levels of the benthic ecosystem over the 39-month study period. Also, a number of federally managed resource species spend all or part of their life cycle in the inner New York Bight and adjacent estuaries. Factor analysis and canonical correlation analysis reveal strong to moderate associations among fish species, between fish species and water and sediment quality variables, and between fish species and invertebrate prey species. Furthermore, strong to moderate associations were also found among invertebrate species and between invertebrate species and water and sediment quality variables. The approach of using multivariate statistical procedures to explore the associations between habitat variables and important resource fish species can be used to better understand the essential fish habitat relationships of these species.

---

[1] E-mail: Sukwoo.Chang@noaa.gov

Fish Landings, Discards, and Benthic Material from Otter Trawling in the Western English Channel

S. P. Cotterell[1]

Institute of Marine Studies, University of Plymouth, Drake Circus, Plymouth, United Kingdom

A fleet-stratified sampling design was employed between 1998 and 2000 to study fish discards and landing and to quantify the other incidentally caught material. The studied techniques were <12m single boat otter trawling, <12m paired demersal trawling and >12m single boat otter trawling. Trips for <12m ranged from one to three days while those for >12m boats were one to six days. These boats operated out of the four principle English ports of ICES area VIIe, western English Channel. On board the boats and prior to any sorting by the crew a sample (~40kg) of the catch was taken and all fish were identified and measured, and their fate (whether to be landed or discarded) was noted. All non-fish material was stored in the fish hold for later detailed analysis. The non-fish material was categorised as benthos, or biogenic, inorganic, or anthropogenic material. The benthos was classified, weighed and measured. Also, a system to assess its degree of damage was developed, allowing length-weight regressions to be generated for the more common invertebrate species. On average 60% (by weight) of the haul was landed, 10% was bait fish, 20% was discarded and 10% was non-fish material. Landing samples were compared to confidential catch composition figures of trip landings. British Geological Survey data was used to assess the substrate over which the trawl had passed and benthos composition was compared to historical data sets. From this study it would appear that economic overfishing would occur before irreparable benthic disturbance for these techniques.

[1] E-mail: s.cotterell@plymouth.ac.uk

Decreasing Habitat Disturbance by Improving Fish Stock Assessments: A New Method of Remote Species Identification and Quantification

D. F. Doolittle[1] and M. R. Patterson

College of William and Mary, School of Marine Science at the Virginia Institute of Marine Science, Gloucester Point, Virginia

Z.-U. Rahman

College of William and Mary, Department of Computer Science, Williamsburg, Virginia

R. Mann

College of William and Mary, School of Marine Science at the Virginia Institute of Marine Science, Gloucester Point, Virginia

A direct link exists between the quality of fisheries data and the effectiveness of fisheries management. Increasing the quality and quantity of data on which stock assessments and management decisions are based has been cited as a critical national issue (National Research Council, 2000. Improving the Collection, Management, and Use of Marine Fisheries Data. National Academy Press, Washington, D.C.). We approach the challenge of limiting deleterious habitat impacts due to fishing through the creation and demonstration of novel stock assessment and habitat visualization tools. We present here a new method of fish species identification and quantification. The technique uses a Radial Basis Function artificial neural network classifier to discriminate and enumerate selected fish species from high-resolution side scan sonar images. We demonstrate this technology onboard a Fetch! class Autonomous Underwater Vehicle (AUV) and provide examples of how such technologies could augment fisheries stock assessment as well as essential fish habitat determination. Ancillary benefits of this technology include the opportunity to simultaneously characterize surficial bottom types and document habitat utilization by species that are known to the classifier. Such side scan sonar species identification tools would significantly augment current stock assessment methods, provide new insight to habitat usage, and allow more ecologically realistic models to be constructed.

---

[1] E-mail: danield@vims.edu

Small-Scale Analysis of Subtidal Fish Guilds and Associated Habitat Characteristics along Central California

J. M. Field[1]
Moss Landing Marine Laboratories, Moss Landing, California

M. M. Yoklavich
National Marine Fisheries Service Tiburon Lab, Santa Cruz, California

G. M. Caillet
Moss Landing Marine Laboratories, Moss Landing, California

S. Bros
San Jose State University, San Jose, California

J. deMarignac
Moss Landing Marine Laboratories, Moss Landing, California

R. N. Lea
California Department of Fish and Game, Monterey, California

Recent declines in fish populations are prompting revisions and alterations to current fishery management policies. One alternative is the establishment of Marine Protected Areas (MPAs) to promote the recovery of fish stocks. However, before MPAs can be created, habitat associations of the fishes designated for protection need to be characterized to ensure that the ideal habitat can be included when MPAs are designated. Once the habitat associations of each species are known, remote-sensing technology, such as sidescan sonar, can be used to survey large-scale areas to identify potential habitat for MPAs. In the Eastern Temperate Pacific, rockfishes (*Sebates* spp.) are slow growing, have a late age-at-maturity and specific habitat affinities. These life history characteristics make them especially susceptible to fishing pressure and ideal candidates for protection through MPAs. To assess habitat associations of fishes within the Big Creek Ecological Reserve, central California, we conducted submersible dives to identify habitat at the meter scale and quantify fish populations. Multivariate statistical analysis revealed distinct habitat associations for several rockfish species. In addition, distinct seafloor features were identified as unique habitats at the meter scale.

---

[1] E-mail: JField@ci.pacific-grove.ca.us

The Sensitivity of Fish and Macro-Epifauna to Habitat Change: An Analytical Approach

S. M. Freeman[1] AND S. I. Rogers

CEFAS, Lowestoft Laboratory, Lowestoft, Suffolk, United Kingdom

Increased use of seabed resources and greater awareness of the effects of fishing on the seabed call for an urgent need to assess the extent and diversity of seabed habitats affected by such activities. Existing methods that describe and predict the distribution of benthic habitats using either substrata or depth are generally inadequate. When other factors such as tidal velocity, temperature and salinity are combined with substrata and depth, they more clearly characterize these habitats. Principal components analysis (PCA) was used to evaluate the distribution and abundance of fish and macro-epifauna using a suite of factors. Characteristic habitat types were identified and provided a mechanism for predicting their spatial extent. A new analytical approach to link species to their habitat was constructed using a combination of PCA and a generalized additive model (GAM). The method predicts the habitat preferences of an individual species based on their association with the environment. Preferences were used to describe the likelihood of a species occurring across a range of different habitats; this was called the habitat-envelope. The strength of the association between species patchiness and its habitat-envelope indicated the potential sensitivity of the species to habitat change. Generally, fish had larger habitat-envelopes and more likely to exploit a wider range of habitats than crustaceans, whereas echinoderms were more selective, and hence more sensitive to habitat change.

[1] E-mail: s.m.freeman@cefas.co.uk

Integration of Acoustic Seabed Classification and Fish Census Data for Determining Appropriate Boundaries of Marine Protected Areas

A. C. R. Gleason[1]

Rosenstiel School of Marine and Atmospheric Science, University of Miami, Miami, Florida

A.-M. Eklund

National Marine Fisheries Service, Southeast Fisheries Science Center, Miami, Florida

R.P. Reid

Rosenstiel School of Marine and Atmospheric Science, University of Miami, Miami, Florida

D.E. Harper, D.B. McClellan, AND J. Schull

National Marine Fisheries Service, Southeast Fisheries Science Center, Miami, Florida

In southern Florida, fine-scale benthic habitat maps are unavailable for water depths greater than 20 m because the depth is too great to effectively exploit traditional optical mapping methods. These deep water zones may, however, harbor diverse communities of benthic invertebrates and fish that are under-represented in most population surveys. In the Florida Keys, for example, SCUBA divers documented black grouper aggregations at 28 m depth, just seaward of a no-take zone within the Florida Keys National Marine Sanctuary, but the distribution of potential deep-water grouper spawning habitats in the Florida Keys is as yet unknown. Systematic mapping of the acoustic diversity of the sea floor (i.e. variations in response of diverse bottom types to an acoustic signal) offers a potential means for (1) identifying deep-water benthic habitats, (2) describing relationships between benthos and substrate on a regional scale, and (3) establishing effective boundaries of marine protected areas. We are currently using the sea bed classification system QTC VIEW system V to map bottom types in the vicinity of Carysfort reef, a known site of black grouper (*Mycteroperca bonaci*) aggregation. The acoustic mapping is being performed in coordination with diver-based surveys of fish populations. Mapping results guide the locations of dives, which are limited in time and scope due to water depth. Diver surveys, in turn, provide ground truth data to refine and adjust classification maps. Delimiting benthic habitats that are potential sites of grouper aggregation is critical to defining appropriate boundaries of marine reserves.

[1] E-mail:art.gleason@miami.edu

Effects of Fishing on the Mid-Atlantic Tilefish Habitat: Restructuring a Structured Habitat

V. G. Guida[1]

National Marine Fisheries Service, Northeast Fisheries Science Center, Highlands, New Jersey

P. C. Valentine

U.S. Geological Survey, Woods Hole Field Center, Woods Hole, Massachusetts

F. Almeida

National Marine Fisheries Service, Northeast Fisheries Science Center, Woods Hole, Massachusetts

The tilefish habitat on the New Jersey continental shelf and uppermost slope near Hudson Canyon is a little studied, heavily trawled region (~800 km^2) with unusual sediments and topography and substantial fisheries habitat value. During October 2001, we conducted an investigation to map the distribution of habitat types, macrofaunal associations and trawling disturbance at depths of 100-300 m. Side scan sonar was used for mapping and locating stations for video transects, using the SEABOSS drift camera vehicle, and for sediment grab sampling. Otter trawl tows were made to assess abundances and confirm the identities of organisms seen on video. Surficial sediments consisting of sand-clay mixtures, underlain by consolidated clay and producing high side scan backscatter, occurred at depths exceeding 111 m. The combination of sidescan sonar and visual observations revealed the structural complexity of the habitat. Low relief structures included hummocks, biogenic depressions, trawl marks, tilefish burrows, small burrows (1-6 cm diameter), and linear strings of cobbles and boulders. Trawl mark frequencies ranged from 100% coverage to complete absence. Benthic megafauna seen in videos at all stations (sea pens: *Virgularia* sp., sea stars: *Astropecten americanus*, cerianthid anemones, brachyuran crabs) showed no pattern with respect to trawl disturbance. Fish, e.g. spotted hake (*Urophycis regius*), were commonly seen in depressions. Areas with trawl marks had fewer depressions, small burrows, hummocks, and fewer fish than comparable ones without such marks, suggesting a negative impact on habitat value by bottom trawls. The cobble/boulder habitat supported the greatest density and diversity of fishes.

[1] E-mail: vincent.guida@noaa.gov

Development of an Electronic Logbook to Assess Shrimp-Trawl Catch, Effort, and Associated Environmental Data in Areas Fished off Florida and Texas

P. J. Rubec,[1] A. Jackson, C. Ashbaugh, and S. Versaggi

Florida Fish and Wildlife Conservation Commission, Florida Marine Research Institute, St. Petersburg, Florida

An electronic logbook (ELB) system was developed and evaluated by Florida Marine Research Institute (FMRI) biologists in collaboration with a shrimp company and an electronics firm, both based in Tampa, Florida. The software compiles shrimp-trawl catch, and effort data entered onto a computer situated on the bridge, and environmental data transferred from a conductivity-temperature-depth (CTD) data logger. A Vessel Monitoring System (VMS) was integrated with the ELB to capture geographic positions and transmit data to the shrimp-company headquarters and to FMRI. The system was evaluated on an FMRI research vessel and then on three shrimp vessels. Data obtained using the ELB system indicated that pink shrimp (Farfantepenaeus duorarum) caught off the west coast of Florida (March-June 2001) and brown shrimp (Farfantepenaeus aztecus) caught off the coast of Texas (July-September 2001) exhibited different preferences for environmental conditions. The highest mean catch rates (CPUEs) of pink shrimp occurred over sand and hard bottoms when the temperature was 20-24.9°C. Pink shrimp had no apparent preference for any particular salinity or depth range. The highest mean CPUEs of brown shrimp occurred over mud bottom, at temperatures of 25-29°C, and at salinities of 35-39.9 g/L. Brown shrimp did not exhibit a preference for any particular depth range. These results indicate that the ELB system has the potential to assist the fishery in locating areas where shrimp are abundant. Further development of the system would benefit both the shrimp fishery and fisheries management.

---

[1] E-mail: peter.rubec@fwc.state.fl.us

Linking Predator and Prey Species Dynamics in Deep-Water Reefs of the Northeastern Gulf Of Mexico

P. E. Thurman[1]

Florida Marine Research Institute, St. Petersburg, Florida

G. Dennis and K. Sulak

U.S. Geological Survey, Florida Caribbean Science Center, Gainesville, Florida

R. S. McBride

Florida Marine Research Institute, St. Petersburg, Florida

Deep-water reefs common to the Gulf of Mexico provide essential habitat for economically important species such as red snapper (Lutjanus campechanus), grouper (Epinephelus/Mycteroperca spp.), and amberjack (Seriola spp.). Recent diet studies of these predators indicate small serranids (<150mm

SL) such as roughtongue bass (Pronotogrammus martinicensis) and red barbier (Hemanthias vivanus) are an important link between zooplankton and these predators. Life history data and video population estimates are being analyzed to determine if these predators affect the size and age structure, distribution, and abundance of these prey species. Ages of 182 P. martinicensis and 80 H. vivanus were estimated by examining the ring structure on whole sagittal otoliths. Although we cannot validate the periodicity of ring formation until year-round samples have been examined, most of the fish collected during May had a ring that was either newly formed or on the margin. Individuals of both prey species grew rapidly through their first year and then growth slowed significantly. Modal ages of P. martinicensis and H. vivanus were 3 and 4 years, respectively. The oldest P. martinicensis examined was 8 years and the oldest H. vivanus was 7 years. These preliminary estimates of ages were higher than anticipated and indicate a fairly stable age-structure for these species. Previous work has suggested that changes in predator abundance can affect prey population dynamics. Therefore, monitoring the prey species could provide information that can be used to determine the status of the predators in areas of differing regulatory regimes.

---

[1] E-mail: Paul.thurman@fwc.state.fl.us

Effects of Fishing: Assessment and Recovery

Effects of Fishing on Gravel Habitats: Assessment and Recovery of Benthic Megafauna on Georges Bank

JEREMY S. COLLIE[1] AND JEROME M. HERMSEN[2]

*University of Rhode Island, Graduate School of Oceanography,
Narragansett, Rhode Island 02882, USA*

PAGE C. VALENTINE[3]

*United States Geological Survey, 384 Woods Hole Road,
Woods Hole, Massachusetts 02543, USA*

FRANK P. ALMEIDA[4]

*Northeast Fisheries Science Center, 166 Water Street,
Woods Hole, Massachusetts, 02543, USA*

Abstract. This study assessed the effects of disturbance to benthic communities and the rate of recovery in an area closed to bottom fishing. The study site was the gravel sediment habitat on the northern edge of Georges Bank, which is an important fishing ground and a nursery area for juvenile fish. On eight cruises to this area from 1994 to 2000, we collected dredge samples and photographs from sites of varying depths and with varying degrees of disturbance from otter trawling and scallop dredging. We assessed the megafaunal communities at two adjacent sites in Canadian waters, one heavily fished and the other only lightly trawled. The lightly trawled site (84-m water depth) had significantly higher numerical abundance and biomass of benthic megafauna than did the heavily fished site. There were also marked differences in community composition between the two sites: the undisturbed site was characterized by fragile species—shrimps, polychaetes, and brittle stars—that live in the complex habitat provided by colonial epifauna, which is not present at the disturbed site (80 m). We also monitored the recovery of a previously disturbed shallow area (47 m) that was closed to bottom fishing in January 1995. In the closed area (Closed Area II), we observed significant shifts in species composition and significant increases in abundance (4×), biomass (18×), production (4×), and epifaunal cover. Among the taxa that have increased are species of crabs, molluscs, polychaetes, and echinoderms. Species-dominance curves reversed following the closure, with species abundance progressively decreasing and species biomass progressively increasing, as large animals came to dominate the biomass. Results of this and prior studies have been used by the New England Fishery Management Council to designate and maintain a Habitat Area of Particular Concern for juvenile Atlantic cod *Gadus morhua*.

Introduction

The direct effects of fishing on benthic habitats are generally well understood (NRC 2002). The main effect is the physical disturbance of benthic communities by trawls and dredges. The resulting mortality includes organisms captured in fishing nets and those killed but not retained (Bergman and van Santbrink 2000). In particular, bottom fishing gear damages colonial epifaunal taxa (e.g., algae, sponges, corals, colonial tube worms, hydroids, bryozoa, among others) that provide a three-dimensional habitat for other animals (Jennings and Kaiser 1998; Hall 1999).

Second-order effects of fishing are also well documented but not well quantified. These indirect effects include nutrient resuspension, altered predator–prey dynamics, and loss of ecosystem engineers that burrow into and rework seabed materials (Coleman and Williams 2002). The mechanisms whereby bottom fishing affects fish survival are known through laboratory

---

[1] Corresponding author: jcollie@gso.uri.edu
[2] E-mail: Jerome.Hermsen@noaa.gov
[3] E-mail: pvalentine@usgs.gov
[4] E-mail: Frank.Almeida@noaa.gov

(Lindholm et al. 1999) and field studies (Tupper and Boutilier 1995), but it is difficult to extrapolate these results to the population level. Bottom fishing also reduces production of benthic infauna (Jennings et al. 2001) and megafauna (Hermsen et al. 2003), thereby reducing the energy available for fish production. Structural epifauna plays a dual role by providing a habitat for many of the small, fragile invertebrates that are important prey species (Collie et al. 1997) and also by providing juvenile fish protection from predators. Removal of this epifauna requires juvenile fish to forage for longer periods, thereby exposing them to higher levels of predation (Walters and Juanes 1993).

The response of benthic communities to fishing depends on the fishing gear, sediment type, and sensitivity of particular taxa to physical disturbance (Collie et al. 2000b). Most of the data on fishing effects come from controlled fishing experiments with before–after or control–impact (BACI) designs. These experiments benefit from a controlled experimental design but suffer from being of short duration. For reasons of practicality, the most common studies have been short-term experiments with otter trawls on sandy sediments in relatively shallow water—situations in which the size of the affected area is likely to be small (Collie et al. 2000b). More recently, trawl-impact studies are being conducted in deeper water and on more complex habitats (e.g., Freese et al. 1999). Even so, experimental fishing cannot emulate the spatial extent or intensity of commercial fishing operations. Collie et al. (2000b) found that chronic effects of bottom fishing exceeded acute effects. Only recently have there been studies to compare benthic communities over a gradient of fishing effort (e.g., Thrush et al. 1998; Bradshaw et al. 2000; Jennings et al. 2001)

Consideration of the chronic effects of bottom fishing requires understanding and measurement of recovery rates. Experimental trawl corridors often are surrounded by undisturbed populations that can quickly recolonize the disturbed areas. In contrast, on commercial fishing grounds, there may be no adjacent undisturbed areas and recovery may need to come from remnant populations or rely on longer-range transport of larvae from remote sources. Unfortunately, relatively few trawl-impact studies have measured recovery rates (Collie et al. 2000b), and those that have suffer from being of small spatial scale. Existing recovery data suggest that in some soft-sediment habitats (e.g., mud and sand), recovery can be complete in about 1 year, and it might occur more rapidly in mobile sand. However, these studies did not include complex habitats with long-lived animals that may stabilize the sedimentary substrate. Trawl-caught specimens of coldwater and deep-sea corals have been aged at tens to thousands of years old (Hall-Spencer et al. 2002). When bottom fishing damages these epifaunally rich communities, the implied recovery times are decades to millennia.

In this paper, we present results of field work conducted on gravel habitats on the northern edge of Georges Bank in the northwestern Atlantic Ocean. This gravel pavement is fished with scallop dredges and bottom trawls. The gravel is a substrate for the demersal eggs of Atlantic herring *Clupea harengus* and a nursery area for juvenile Atlantic cod *Gadus morhua* and haddock *Melanogrammus aeglefinus* (Lough et al. 1989; Valentine and Lough 1991). A portion of this habitat in U.S. waters inside Closed Area II has been designated a Habitat Area of Particular Concern for juvenile cod. In earlier papers, based on data from two cruises in 1994, we compared disturbed and undisturbed sites on the gravel pavement. Differences between sites were quantified from dredge samples (Collie et al. 1997) and bottom photographs (Collie et al. 2000a).

In January 1995, a large area (3,960 km^2) of the U.S. part of Georges Bank adjacent to the United States–Canada boundary was closed to all bottom fishing to reduce fishing mortality on demersal fish. The establishment of Closed Area II provided an excellent opportunity to measure the recovery rate of the benthic communities on gravel habitat. We have monitored the recovery with annual sampling inside and outside the closed areas and with sediment recolonization experiments. We also have compared the characteristics of benthic gravel communities over time in adjacent fished and unfished areas on the Canadian part of the bank. In this paper, we report on the time series of data collected at these sites since 1994.

Methods

Eight sampling cruises were conducted to northern Georges Bank, approximately once per year, between 1994 and 2000 (Table 1). The original sampling strategy was guided by side-scan sonar surveys conducted by the U.S. Geological Survey. Quadrants of the sea bottom, measuring approximately 5 km × 10 km, were selected for side-scan sonar surveys (Figure 1). Trawl door marks, and particularly the parallel tracks of paired scallop dredges, are visible in the sonograms and were used to classify the sites as disturbed or undisturbed. Scallop fishing effort data up to 1993 also were used to classify the degree of disturbance (Collie et al. 1997). Based on this information, we chose disturbed and undisturbed sites at 80–90-m depths on the Canadian

Table 1. Cruise dates and numbers of dredge samples collected at sampling sites 13 (80 m deep; disturbed), 20 (84 m deep; undisturbed), 17 (47 m deep; recovering, closed to mobile bottom fishing in January 1995), 17W (49 m deep; disturbed), and 18 (45 m deep; disturbed) on northern Georges Bank. See Figure 1 for the locations of these sites.

Sampling dates	Site 13	Site 20	Site 17	Site 17W	Site 18
10–13 Apr 1994	7		5		3
12–16 Nov 1994	8	6	3		6
30 Jul 1995			3		3
25–27 May 1996			6		3
18–22 Jul 1997	3	3	6	3	3
15–22 Jun 1998	3	1	3	3	3
19–23 Jun 1999	3	3	4	4	3
5 Nov 2000			3		2

side of Georges Bank and a set of sites with different disturbance histories at 40–50-m depths on the U.S. side. Even though all the study sites are located on a shallow submarine bank, we grouped the sites as "deep" and "shallow" to distinguish them in this paper.

More recent data show the fishing patterns at these sites since 1993. Data on scallop dredging and bottom trawling in the Canadian zone were derived from logbook reports and are recorded to the closest minute of longitude and latitude. These data were extracted from the Zonal Interchange Format database (J. Black, Department of Fisheries and Oceans, personal communication). Data on the locations of American fishing boats come from two different sources. Fishing locations of trawlers and scallop vessels were compiled from logbook and trip reports, with a spatial resolution of 10 min latitude and longitude (D. Stevenson, National Marine Fisheries Service, personal communication). Scallop fishing locations were also obtained from the satellite vessel monitoring program, which is manda-

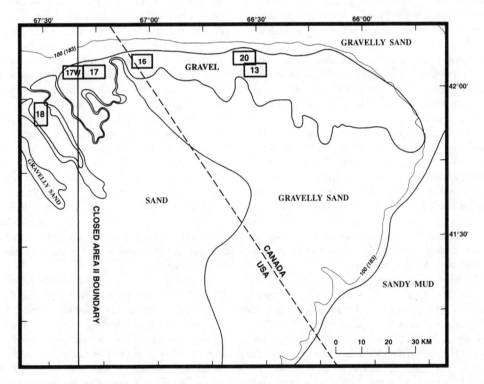

Figure 1. Location of sampling sites on northern Georges Bank. The numbered rectangles are sites on gravel habitat that were surveyed with side-scan sonar in 1994 by the U.S. Geological Survey.

tory on U.S. scallop fishing vessels. Scallop vessels are assumed to be in transit when speed is above 9.26 km/h (5 knots) and fishing when speed is less than 5 knots. Total fishing time in each 1-nm^2 cell was, therefore, estimated as the sum of vessel hours at speeds less than 5 knots (Rago and McSherry 2001).

In this study we contrast two adjacent sites, 13 (80 m) and 20 (84 m), on the Canadian side that have similar gravel sediments and similar depths (Table 1) but different fishing histories. We also compare three sites on the shallower U.S. side of the gravel pavement. Site 17 (47 m) was heavily disturbed prior to 1994 (Collie et al. 1997) but is located in the northern part of Closed Area II, which was closed to all bottom fishing in January 1995. Site 18 (45 m) is located outside the closed area. Prior to the closure, it was lightly disturbed but has attracted more fishing effort since the closure. In 1997, we initiated sampling at Site 17W (49 m), which is adjacent to the closed area and should be a good disturbed control for Site 17. All the sampling sites are located on the gravel pavement (predominantly large pebbles and small cobbles; Figure 1), though there were subtle differences in sediment composition.

At each site, samples of the benthic megafauna (animals retained on a 5-mm screen) were collected with a 1-m-wide Naturalists' dredge fitted with a quarter-inch mesh liner. We aimed for three replicate dredge samples at each site, though the number was sometimes more or less than three depending on sea condition and time constraints (Table 1). Once the site locations were established on the 1994 cruises, the same sites were resampled on subsequent cruises to establish time series.

Tow duration was kept short (30–60 s) to avoid overfilling the dredge bag with gravel and losing the sample. Once the gravel sample was brought on board, all living animals were removed and preserved in a buffered solution of formalin in seawater. The gravel was shoveled into 9-L metal pails to measure the sediment volume and discarded overboard. One 9-L subsample was sieved on a 5-mm screen to check for any animals that were missed in the initial sorting. On average, about one tenth of the total sample was sieved.

Photo transects were made at the same locations as the dredge samples and at additional locations at each site. The primary tool was a grab sampler equipped with video and still cameras (SEABOSS; Valentine et al. 2000; Blackwood and Parolski 2001). This sampler drifted with the tidal current over the sea bottom at a height of about 1 m for a duration of 15–20 min. Photo transects were also made with a remotely operated vehicle (1996, 1997, and 1999) and a submersible (1998). With each type of sampler, still photographs were taken with a downward-looking, 35-mm camera at 30–60-s intervals, depending on drift speed over the bottom. Paired lasers were used to define a linear scale in each photo.

In the laboratory, the dredge samples were sorted to species and counted. The aggregate mass of each species was measured to 1 mg after blotting excess moisture on paper towels. The 9-L sieved subsamples from each dredge tow were sorted and enumerated separately. These subsample analyses were extrapolated by the total sediment volume of the dredge tow to account for any animals missed during the initial sorting. Each final sample was standardized by the total volume of sediment collected, such that the resultant data had units of numbers/L and g/L. The combined species list was searched for taxa that could not be reliably sampled or identified. These included colonial taxa that could not be counted (presence or absence only), macrofaunal species (mainly amphipods and small polychaetes) that were too small to be retained consistently on a 5-mm sieve, and animals that could only be identified to a higher taxonomic level because of missing body parts. This filter resulted in a list of 124 selected species out of 319 taxa.

For each sample, we calculated a number of aggregate ecological indices including numerical abundance, biomass, and number of species (S). Shannon–Wiener species diversity (H') was calculated with base-2 logarithms. Pielou's evenness was calculated as $J' = H'/\log S$ (Krebs 1989). To determine the appropriate transformation for each variable, we calculated the within-replicate mean and variance at each site and year combination. The transformation that made within-replicate variance independent of the mean was selected. Biomass and abundance were log transformed; no transformation was necessary for number of species. Species diversity and evenness were exponentiated, as these indices were calculated from base-2 logarithms. The diversity index, $2^{H'}$, is the expected number of species (Krebs 1989). After appropriate transformation, analysis of variance (ANOVA) was used to test for differences among sites and cruises and before and after the area closure at the shallow sites.

Multivariate analyses were conducted with the PRIMER software package (Clarke and Warwick 1994). Bray–Curtis similarity matrices were calculated on square root-transformed species abundance data. Ordination was performed with nonmetric multidimensional scaling (MDS). The analysis of similarities (ANOSIM) routine was used to test for significant differences in species composition among sites and years. The similarity percentages (SIMPER) routine was used to determine which species accounted for most of the similarity and dissimilarity among sites. The sites were also compared with *k*-dominance curves and abundance–biomass curves.

Results

Consistent differences in the benthic megafaunal communities were found on two cruises in 1994. Numerical abundance and biomass were significantly higher at the deep sites and also significantly higher at undisturbed sites compared with disturbed sites at each depth (Collie et al. 1997). There were, however, significant interaction effects, such that the disturbance effect on abundance was greater at the deeper sites. Though the differences between sites are quite clear-cut, they depend on the interpretation of the disturbance level at each site. High-resolution data on fishing patterns are a necessary element of gear–impact research. We, therefore, determined the fishing history at each sampling site and followed the community structure over time. Because of depth × disturbance interactions in the 1994 samples, different fishing patterns, and sampling schedules, we analyzed data from the deep and shallow sites separately.

Deep Sites (80–84 m)

Since 1992, there has been scallop dredging at Site 13 but not at Site 20 (Figure 2). There has been some bottom trawling at Site 20 but always less than at Site 13, except in 1999 and 2001. These effort data are consistent with our interpretation of the side-scan sonar data in 1994, except that Site 20 has been lightly trawled, especially since 1998. Though the fishing locations are reported to the nearest minute, high intensities are evident at regular intervals, suggesting that some fishermen round to the nearest 10 min (Figure 2C). Therefore the precision of these location data are somewhat less than 1 min and they can only be used to approximate bottom fishing intensity at our sampling sites.

The biomass and abundance of benthic megafauna has remained significantly higher at Site 20 than at Site 13 (Table 2); in fact, the differences between Sites 13 and 20 appear to have increased since 1994 (Figure 3A, B). Shannon-Wiener species diversity was significantly higher at Site 13 (Table 2; Figure 3C). The reason for lower diversity at Site 20 is not that there are fewer species; there were significantly more species (S) at Site 20 (Table 2; Figure 3D). The difference between the two sites is due to the distribution of numbers of individuals among species. Evenness was significantly higher at Site 13 ($P < 0.001$), which causes species diversity to be higher there. There were also significant site × cruise interactions for diversity and evenness; these indices were significantly different between sites in 1997, 1998, and 1999, but not in 1994. At Site 20, the lower species diversity in 1997–1999 matches the high numbers during those years and is due to higher dominance in the community composition, with single species (the polychaete *Thelepus cincinnatus*) accounting for up to 50% of the individuals sampled. In contrast, at Site 13, single species accounted for, at most, 15% of the individuals. It is this high numerical dominance at Site 20 that accounts for the lower species diversity despite a greater number of species.

Multivariate analysis of the community composition data from Sites 13 and 20 included 102 species. The complete species list is not included here but can be requested from the first author. The 50 most abundant megafaunal species were listed by Collie et al. (1997). Multidimensional scaling of the similarity matrix shows a clear separation of the community composition of the two sites, as well as changes with time (Figure 4). According to the two-way crossed ANOSIM, the between-site differences were highly significant ($R = 0.995, P = 0.001$) as were differences among years ($R = 0.825, P = 0.001$). A SIMPER contrast of Sites 13 and 20 indicated an average dissimilarity of 75%. The top ten species accounting for this dissimilarity were fragile taxa such as the polychaetes *Thelepus cincinnatus* and *Potamilla neglecta*, the brittle star *Ophiopholis aculeata*, the toad crab *Hyas coarctatus* (also known as Arctic lyre crab), and six shrimp species (Table 3). The among-year differences at Site 20 were due to higher abundances of some species in 1997, 1998, and 1999, especially *T. cincinnatus*, *P. neglecta*, and *Spirontocaris spinus*. The benthic community at Site 13 was dominated by echinoderms (*Asterias vulgaris*, *Strongylocentrotus droebachiensis*) and bivalves (*Astarte* spp. and *Cyclocardia borealis*). The bivalves, in particular, have heavy shells that apparently can resist encounters with bottom fishing gear.

Shallow Sites (45–49 m)

Prior to the area closure in January 1995, Site 17 (47 m) had high levels of scallop fishing effort (Collie et al. 1997); since the closure, there has been no bottom fishing in this area except for very limited experimental fishing in 1998 (Table 4; Figure 5). Site 18 was lightly fished before the closure but has had more fishing effort directed at it since then. Site 17W is a heavily fished site just outside Closed Area II, adjacent to Site 17. The distribution of scallop fishing, and of the scallops, matches the gravel habitat, as shown in Figure 1. The effort data confirm our designation of Sites 17W and 18 as fished control sites for the recovering Site 17. Both sites have been fished with otter trawls and scallop dredges during the period our samples

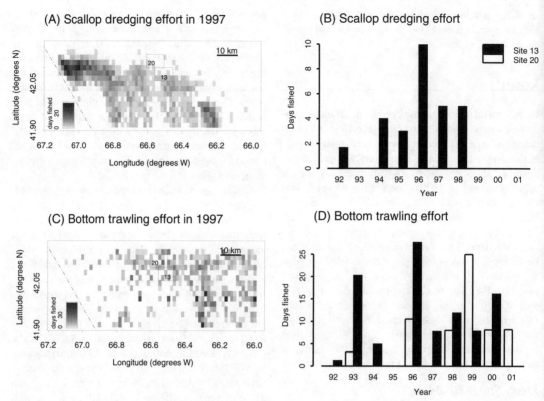

Figure 2. Spatial distribution of bottom fishing on northeastern Georges Bank. Fishing locations were reported to the nearest minute of latitude and longitude. Panel (A) shows the distribution of scallop dredging effort in 1997. The diagonal line is the Canada-USA border, and the rectangles are quadrants that have been surveyed with side-scan sonar. The numbered sites show the locations where dredge samples were taken. Panel (B) illustrates scallop dredging effort in the 1-min quadrants corresponding to dredge Sites 13 and 20; note the absence of scallop dredging at Site 20. Panel (C) shows the distribution of bottom trawling effort in 1997. Some of the highest density quadrants occur at exact 10-min intervals, which suggests that some locations have been rounded by fishermen. Panel (D) illustrates bottom trawling effort in the quadrants corresponding to dredge Sites 13 and 20. These location data were extracted from the Zonal Interchange Format database by Jerry Black, Canada Department of Fisheries and Oceans, Dartmouth, Nova Scotia.

were collected. Fishing intensity has been greater at Site 17W than at Site 18 (Table 4).

Two-way ANOVA was used to test the univariate indices from the shallow sites, with site and closure (before–after) as fixed effects. The first set of ANOVAs omitted the data from Site 17W, which was not sampled before the closure. The site × closure interaction term tests for the effect of the closure of Site 17 relative to Site 18. Prior to the closure, biomass was significantly lower at Site 17 than at Site 18 (Collie et al. 1997). After 1995, this pattern was reversed, with a significant site × closure interaction (Table 5). On average, biomass increased by a factor of 2/year (Figure 6A). By 2000, average biomass was 18 times higher inside than outside the closed area. Likewise, abundance increased significantly at Site 17 following the closure (site × closure interaction; Table 5). On average, abundance increased by a factor of 1.5/year inside the closed area (Figure 6B). In 1999, mean abundance was four times higher inside than outside; in 2000, this difference was reduced to a factor of two. The indices at Site 17W can only be compared with Sites 17 and 18 for the years 1997 through 1999 (Figure 6). According to Sidak's multiple comparison test, biomass was significantly higher at Site 17 than both Sites 17W and 18, except in 1999 when Sites 17 and 17W were not significantly different (Figure 6A). Likewise abundance was higher at Site 17 than at both Sites 17W and 18, except in 1997, when Sites 17 and 18 were not significantly different (Figure 6B).

Species diversity increased at both Sites 17 and 18, but the increase was significantly greater at Site 17 (Table 5; Figure 6C). The increase in species diversity reflects the increase in number of species per sample (Table

Table 2. Analysis of variance (ANOVA) of univariate indices at the deep Sites 13 (80 m) and 20 (84 m). See Figure 3 for graphical plots of these indices. The ANOVA table is calculated from Type-III Sums of Squares. H' is Shannon-Wiener diversity. In comparing Sites 13 and 20, the F-statistic and the probability in the site column show that Sites 13 and 20 are significantly different, with Site 20 having higher biomass, abundance, and number of species and Site 13 having higher species diversity. The data are from four cruises in the years 1994 and 1997–1999.

Source of variation	Site	Cruise	Site × Cruise	Residual
Log (Biomass)				
Mean square	9.226	0.309	0.480	0.274
F-statistic	33.647	1.218	1.750	
Probability	<0.001	0.362	0.189	
Log (Abundance)				
Mean square	38.321	0.422	0.510	0.198
F-statistic	195.974	2.130	2.577	
Probability	<0.001	0.128	0.083	
Species Diversity ($2^{H'}$)				
Mean square	340.782	9.308	54.784	10.317
F-statistic	33.031	0.902	5.310	
Probability	<0.001	0.457	0.007	
Number of species (S)				
Mean square	588.000	106.784	122.080	38.450
F-statistic	15.293	2.777	3.175	
Probability	<0.001	0.68	0.047	
Degrees of freedom	1	3	3	20

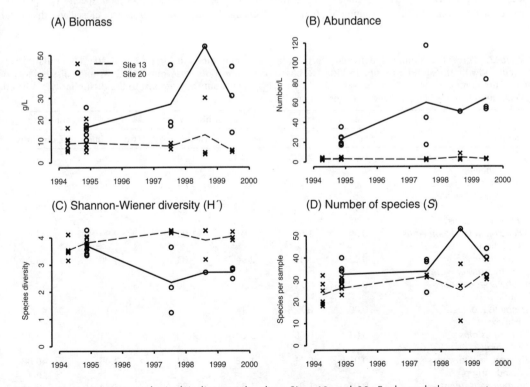

Figure 3. Univariate ecological indices at the deep Sites 13 and 20. Each symbol represents one dredge sample and the lines connect the means from each sampling cruise.

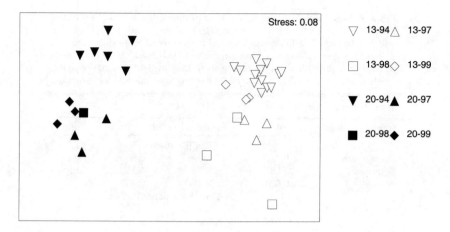

Figure 4. Nonmetric multidimensional scaling of megafaunal communities at deep Sites 13 (open symbols) and 20 (closed symbols). Each symbol represents one dredge sample identified by site and year. This ordination is based on the Bray-Curtis similarity matrix of square-root transformed data on the abundance of 102 species.

5; Figure 6D). The number of species also increased at Site 18, but the increase was greater at Site 17, as indicated by the significant site × cruise interaction term (Table 5). Evenness at Sites 17 and 18 was not significantly different, and the closure effect was insignificant. In 1999, Site 17W had significantly higher abundance and lower species diversity than did Site 18 (Figure 6B, 6D).

Multivariate analysis of the community composition data from the shallow sites included 100 species. Multidimensional scaling of the similarity matrix separated the samples by sites on the vertical axis (Figure 7); temporal shifts in species composition are also apparent along the horizontal axis of the MDS plot. Species composition changed more at Site 17 inside the closed area than it did

Table 3. Percent dissimilarity between the megafaunal communities at Sites 13 and 20. Dissimilarity was calculated with the SIMPER routine in PRIMER, based on square-root transformed data. Average dissimilarity of the entire communities (102 species) was 75%. The 10 species contributing most to the dissimilarity are listed. Mean abundance is standardized per liter of gravel.

Species	Mean abundance		Dissimilarity		Percent	
	Site 13	Site 20	Mean	MeanSD	Contribution	Cumulative
Thelepus cincinnatus (polychaete)	0.09	22.47	11.48	1.84	15.24	15.24
Brittle star *Ophiopholis aculeata*	0.04	3.23	4.91	3.45	6.52	21.75
Toad crab *Hyas coarctatus*	0.13	2.78	4.18	2.23	5.55	27.30
Dichelopandalus leptocerus (red shrimp)	0.09	1.63	3.01	1.93	3.99	31.30
Potamilla neglecta (feather duster worm)	0.02	2.32	2.94	1.03	3.90	35.20
Lebbeus groenlandicus (shrimp)	0.00	0.77	2.54	1.90	3.37	38.57
Pandalus montagui (shrimp)	0.00	0.72	2.34	1.58	3.10	41.68
Eualus pusiolus (shrimp)	0.03	0.68	2.04	1.70	2.71	44.38
Crangon septemspinosa (sand shrimp)	0.01	0.71	1.88	0.92	2.49	46.88
Spirontocaris spinus (shrimp)	0.01	1.00	1.84	0.76	2.44	49.31

Table 4. Relative fishing intensity at the shallow sites (45–49 m). Days absent from port were extracted from vessel trip reports and calculated from the date and time the vessel left and returned to port. The entire trip is assigned by the fisherman to a single location, and the data have been summed by 10-min squares (David Stevenson, National Marine Fisheries Service, personal communication). We then selected the 10-min square that contained each of our sampling sites. The scallop fishing hours come from the satellite vessel monitoring system (Rago and McSherry 2001). We summed scallop fishing hours within 3.7 km (2 nautical mi) of our sampling sites.

Gear type or year	Sampling site		
	17	17W	18
Days absent from port 1995–2000			
Bottom trawl	0	1,770	463
Scallop dredge	0	1,373	544
Scallop fishing within 3.7 km of sites			
1998	0	2,244	1,267
1999	0	234	11
2000	0	756	7

at the two sites (17W and 18) outside the closed area. Species composition at Sites 17 and 18 was most similar in 1996 when total abundance and biomass at these sites was also similar (Figure 6A,B); since 1996, species composition at these sites has diverged.

The SIMPER routine in PRIMER was used to identify the species that contributed most to the similarity and dissimilarity of species composition in samples collected at the shallow sites. Averaged across years, Site 17 had greater abundance of the brittle star *Ophiopholis aculeata*, the polychaete *Nereis zonata*, the sea star *Asterias vulgaris*, and the sea urchin *Strongylocentrotus droebachiensis* than did either of the open sites 17W and 18. Site 17W was distinguished by having greater abundance of the sand shrimp *Crangon septemspinosa* (also known as sevenspine bay shrimp) than either Sites 17 and 18. Site 18 differed by having more crabs of the species *Cancer irroratus* (known as Atlantic rock crab) than the other two sites.

Changes in epifaunal cover are apparent in the bottom photographs from Site 17 within the closed area (Figure 8). In 1994 (prior to the closure), the gravel was

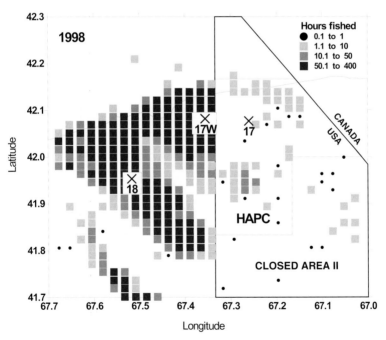

Figure 5. Spatial distribution of scallop fishing in 1998. Location data were obtained from the satellite vessel monitoring program. Total fishing time in each 1 nmi^2 cell was calculated by Rago and McSherry (2001) as the sum of vessel hours at speeds less than 5 knots. Scallop vessel activity in the closed area was for experimental fishing to assess the sea scallop population. Sites 17, 17W, and 18 are shown. HAPC = Habitat Area of Particular Concern established for juvenile cod inside the closed area.

Table 5. Analysis of variance (ANOVA) of univariate indices at the shallow Sites 17 (47 m) and 18 45 m). See Figure 6 for graphical plots of these indices. The ANOVA table is calculated from Type-III Sums of Squares. H' is Shannon-Wiener diversity. Of primary interest are the F-statistics and corresponding probabilities in the Site × Closure interaction column. Comparing the differences in biomass, abundance, species diversity, and number of species between Site 17 and Site 18, highly significant changes occurred following the closure at Site 17. The data are from eight cruises from 1994 through 2000.

Source of variation	Site	Closure	Site × Closure	Residual
		Log (Biomass)		
Mean square	0.949	20.895	16.268	0.800
F statistic	1.879	26.130	20.344	
Probability	0.280	<0.001	<0.001	
		Log (Abundance)		
Mean square	0.0405	24.902	6.917	0.347
F statistic	0.117	71.832	19.955	
Probability	0.734	<0.001	<0.001	
		Species diversity ($2^{H'}$)		
Mean square	0.916	477.689	58.713	12.839
F statistic	0.071	37.206	4.573	
Probability	0.790	<0.001	0.037	
		Number of species (S)		
Mean square	53.365	3274.569	146.487	33.511
F statistic	1.592	97.715	4.371	
Probability	0.212	<0.001	0.041	
Degrees of freedom	1	1	1	58

barren with very little epifauna. In 1996, the gravel was covered with a biogenic layer, which was being grazed by the nudibranch *Coryphella*. By 1997, we saw colonization of the gravel by sponges and hydrozoans and increased abundance of crabs and small scallops. The 1999 photo shows an increase in sponge cover, particularly colonies of *Polymastia* and *Isodictya*. Though uncommon, we also observed small colonies of *Filograna implexa*, a colonial polychaete with fragile calcareous tubes that characterized the deep, undisturbed Site 20 (Collie et al. 2000a).

Many megafaunal species increased in abundance at Site 17 following the closure in January 1995, with the differences becoming most apparent in 1997 (Figures 6, 7). Twelve species accounted for most of the dissimilarity in species composition at Site 17 and, hence, the increase in biomass and abundance following the closure (Figure 9). This species group includes three crabs (*Cancer irroratus*, *Hyas coarctatus*, and *Pagurus acadianus*), three echinoderms (*Ophiopholis aculeata*, *Strongylocentrotus droebachiensis*, and *Asterias vulgaris*), three bivalves (*Crenella glandula*, known as glandular crenella; *Astarte borealis*, known as boreal astarte; and *Placopecten magellanicus*, known as sea scallop), one gastropod (*Buccinum undatum*, known as waved whelk), one shrimp (*Dichelopandalus leptocerus*, known as bristled longbeak), and one polychaete (*Nereis zonata*). Three of the 12 species—*O. aculeata*, *H. coarctatus*, and *D. leptocerus*—were among those that distinguished the deep undisturbed Site 20 from disturbed Site 13. Hence, the presence–absence of these three fragile species can be considered indicators of disturbance and recovery. Conversely, the abundance of the crab *Cancer irroratus* and the sand shrimp *Crangon septemspinosa* (not shown in Figure 9) remained relatively constant at Site 17 following the closure; these two scavenging species appear less sensitive to bottom fishing disturbance and potentially benefit from the increased feeding opportunities associated with disturbance.

Marked shifts in species dominance occurred over time at Site 17 (Figure 10). In 1994, prior to the closure, the abundance dominance curve lay above the biomass curve because there were few large animals at this site. Following the closure, the biomass dominance curve was shifted upward, reaching its highest level in 2000. Starting in 1996, the biomass came to be dominated by *Placopecten magellanicus*, *Strongylocentrotus droebachiensis*, *Asterias vulgaris*, and *Buccinum undatum*. Shifts in the abundance dominance curve mirror the changes in species diversity and evenness (Figure 6C), with high dominance corresponding with low evenness. The abundance dominance curve

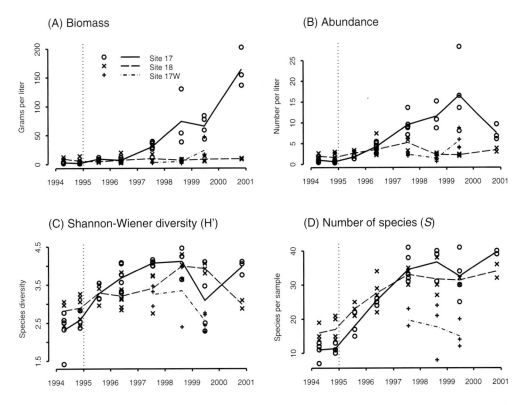

Figure 6. Univariate ecological indices at the shallow Sites 17, 17W, and 18. Each symbol represents one dredge sample and the lines connect the means from each sampling cruise. The vertical dotted lines indicate the date that Site 17 was closed to all bottom fishing. Sites 18 and 17W remained open throughout this period.

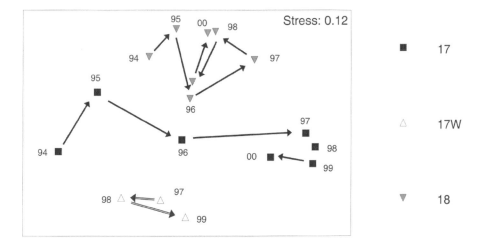

Figure 7. Nonmetric multidimensional scaling of megafaunal communities at Sites 17, 17W, and 18. To simplify the graph, each symbol represents the mean of replicate dredge samples taken at each site and cruise. This ordination is based on the Bray-Curtis similarity matrix of square-root transformed data on the abundance of 100 species.

Figure 8. Photographs of the sea floor at Site 17. The field of view in each photograph is approximately 76 cm by 51 cm (30 × 20 in). In 1994, prior to the closure, only a few burrowing anemones (ba) can be seen. In 1996, many burrowing anemones and a hermit crab (hc) can be seen in the top right corner. In 1997, sponges, bryozoans (br), a sea urchin (su), small scallop (sc), a hermit crab, and toad crab (tc) can be seen. In 1999, sponges (*Polymastia* and *Isodictya*) and a large scallop are evident. Photographs by Dann Blackwood, U.S. Geological Survey.

generally shifted down with time, except in 1999, when *Ophiopholis aculeata* was numerically dominant. The number of species found in the samples at Site 17 increased from 24 in 1994 to 54 in 2000 (Figure 10).

Discussion

Univariate Indices

We have demonstrated significant differences between sites in the abundance and biomass of benthic megafauna sampled from gravel habitats on Georges Bank. Moreover, we measured significant increases in abundance and biomass at Site 17 following its closure to bottom fishing. The most likely explanation of the differences is bottom fishing intensity. However, our study is subject to the same caveats inherent in most spatial comparisons of fishing effects in benthic communities (Hall 1999), namely that fishing effort is not experimentally controlled and that the sites may differ in characteristics other than fishing intensity. This study was limited to a small number of sites sampled systematically over time. Despite sampling constraints, the multiyear nature of the study has allowed us to document trends of change over time in the composition of benthic communities on gravel habitats.

At the deep sites (80–84 m), fishing effort was greater at Site 13 than at Site 20. Some bottom trawling effort was recorded at Site 20, so it can be classified as lightly trawled, especially after 1998. In bottom photographs, we have observed trawl cables snagged on boulders. The effort data are consistent with our personal observations of increased trawling intensity at Site 20. It is possible that, with improved navigation, fishermen are identifying trawl corridors in what was previously considered untrawlable bottom. Given the light trawling that occurred at Site 20, it appears that scallop dredging is responsible primarily for the difference in megafauna between Sites 13 and 20. This conclusion is consistent with other studies, which found that, on a per-tow basis, scallop dredges cause more disturbance than otter trawls (Collie et al. 2000b).

Why are there fewer scallops, and hence no scallop dredging, at Site 20 compared with Site 13? These two sites are adjacent and located at virtually the same depth. Therefore, they share similar bottom currents and most likely a similar supply of larval scallops. Both sites have gravel substrates, though Site 20 has slightly more cobbles and less small pebbles (Collie et al. 2000a). Site 20 is characterized by a high percent cover of hydroids and bryozoa (on which post-larval scallops are known

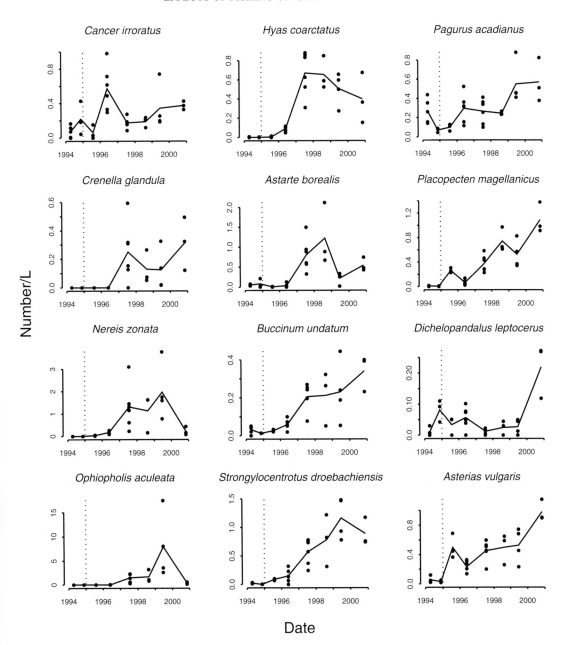

Figure 9. Abundances of 12 megafaunal species that contributed most to the changes in community composition at Site 17. Each point represents one replicate dredge sample, and solid lines connect the means at each sampling date. The dotted vertical lines indicate the date this site was closed to bottom fishing.

to attach; Thouzeau et al. 1991; Stokesbury and Himmelman 1995) and of colonies of colonial polychaete tubeworms (*Filograna implexa*). Predation rates on recently settled scallops may be greater at Site 20 due to the higher densities of invertebrate predators associated with attached epifauna (see below). As in our study, Thouzeau et al. (1991) found higher scallop densities on gravel substrates but not at sites with a high cover of *Filograna implexa*.

At the shallow sites (45–49 m), interpretation of the effort data are more clear-cut, thanks to the area closure. Since the closure, Site 17W and 18 have been subject to both scallop dredging and bottom trawling, with effort levels somewhat lower at Site 18 than 17W. Site 17W

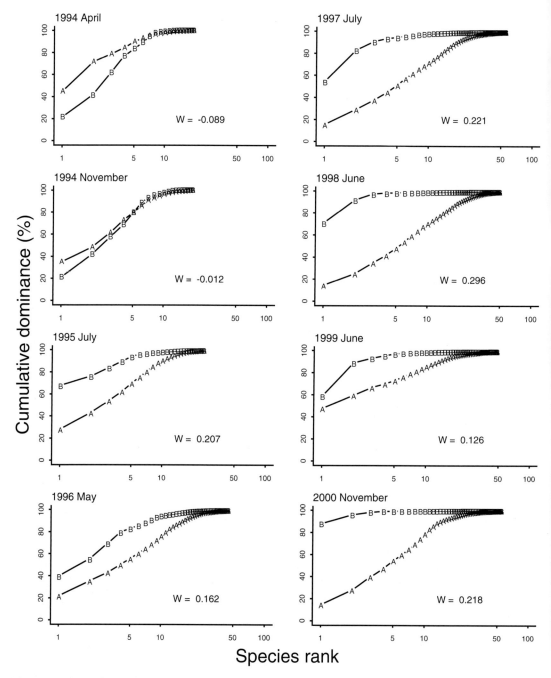

Figure 10. Cumulative dominance curves calculated from the abundance (A) and biomass (B) of benthic megafauna sampled at Site 17 on eight cruises. The W-statistic sums the differences between the biomass and abundance curves (Clarke and Warwick 1994).

was, thus, a good choice of fished control for Site 17. Site 18 remains an important control because it was sampled prior to the closure and, therefore, enables BACI comparisons with Site 17. These shallow sites are at similar depths and, thus, subject to similar current regimes. Again, there are subtle differences in the gravel substrate (Collie et al. 2000a) that could contribute to the differences in community composition.

The patterns in species diversity were inconsistent between the deep and shallow sites. At the deep sites, the difference was due primarily to greater evenness at the disturbed Site 13. The increase in diversity observed at the shallow sites was due primarily to an increase in the number of species per sample. Interestingly, the number of species per sample at Site 17 increased to about 40 following the closure, which is about the same S as at the undisturbed Site 20. It is well recognized that S depends on sample size (Krebs 1989), but we presented these results to aid in interpreting the patterns in species diversity. Species diversity is known to be an insensitive indicator of disturbance (Clarke and Warwick 1994), and we, therefore, rely more heavily on the multivariate analyses.

Multivariate Indices

Significant differences in species composition were found among sites, which can be attributed to differences in bottom fishing intensity. We also observed shifts in species composition, particularly at the deep sites (compare 1994 samples with other years at Sites 13 and 20; Figure 4) that appear unrelated to fishing intensity. These shifts could be explained by natural variability in the megafaunal populations. We did not sample at the same time each year, and though the dominant megafaunal species live longer than 1 year (Hermsen et al. 2003), they do have seasonal recruitment and mortality patterns. Seasonality could explain the lower biomass and abundance at Site 20 in November 1994 (Figure 3A, B) and the decline in abundance at Site 17 in November 2000 (Figure 6B).

Though we used the same sampling methodology over time, different groups of people picked the samples from each cruise, possibly introducing sampling variability. Likewise, we have taken pains to standardize the taxonomic lists from one cruise to the next. However, as our taxonomic skill increased with time, it is possible that we identified more species in recent cruises. We are confident that sampling and identification biases are small, considering that over 40,000 individual specimens weighing 242 kg were included in this analysis. If there were a bias in species identification, it would not affect comparisons made among sites on the same cruise.

From the differences in species composition, we can identify sensitive taxa and species that can be considered as indicator species. At the deep sites, the polychaete *Thelepus cincinnatus* accounted for the greatest dissimilarity (15%). This species builds its tubes around cobbles; physical disturbance of the cobbles will abrade and crush the tubes. One of the most apparent visual differences among the sites is that the gravel particles at Site 20 are encased and bound together by worm tubes, whereas the gravel particles at disturbed sites have a polished look. Another major visual difference is that Site 20 has a high percent cover of hydroids, bryozoans, and calcareous worm tubes. This matrix of emergent epifauna shelters fragile animals, including pandalid shrimps, *Ophiopholis aculeata*, and *Hyas coarctatus*. The horse mussel *Modiolus modiolus* (also known as northern horsemussel), is also known to be an indicator of fishing disturbance (Hall 1999; Bradshaw et al. 2002). Its large, yet relatively thin valves are vulnerable to physical damage, and its slow growth rate requires a long recovery time. Though not abundant at our sites, the mussels *M. modiolus* and *Musculus discors* (known as discordant mussel) were found at Site 20 and not Site 13.

Some of the same indicator species increased in abundance at Site 17 following the area closure, namely *Ophiopholis aculeata*, *Hyas coarctatus*, and *Dichelopandalus leptocerus*. Increases in these species coincided with increased epifaunal coverage (Figure 8) as expected from prior studies. Among the other species that increased at Site 17, some are fragile, such as the errant polychaete *Nereis zonata* and the northern red sea anemone *Urticina felina*. Other shelled animals also increased, including *Strongylocentrotus droebachiensis*, *Cancer irroratus*, and *Placopecten magellanicus*. Scavengers, such as the sea star *Asterias* do not suffer a high mortality from trawling (Ramsay et al. 2000) but may have benefited from increased feeding opportunities following the area closure. The hermit crab *Pagurus acadianus* increased in parallel with *Buccinum undatum*, one of its principal sources of housing (Figure 9). *Pagurus acadianus* was also more abundant at undisturbed Site 20 than at disturbed Site 13.

Among the taxa more resistant to bottom fishing disturbance are small, hard-shelled bivalves such as *Astarte* spp. and *Cyclocardia borealis*. Similar observations have been made in other studies, and it has been suggested that the pressure wave in front of the bottom fishing gear blows these small mollusks out of the way (Gilkinson et al. 1998). Of the six *Astarte* species found in our samples, *A. borealis* appears most sensitive to bottom fishing and *A. elliptica* the least sensitive. In the closed area (Site 17), *A. borealis* increased in abundance while *A. elliptica* fluctuated without trend. Some crustaceans (e.g., *Cancer irroratus* and *Crangon septemspinosa*), though not particularly robust, do not appear as sensitive to bottom fishing, perhaps because of a high rate of population increase coupled with their scavenging lifestyle.

The dynamics of *Placopecten magellanicus* are particularly important because it is the target of the dredge fishing that disturbed the bottom. In addition to the fishing mortality, there is substantial mortality to scallops that are crushed by the gear but not retained (Myers et al. 2000). Scallops feed on suspended matter from the water column but require a stable substrate for settlement (Stokesbury and Himmelman 1995). The chief predators of scallops are sea stars, crabs, and lobsters. Sea scallops are a dominant component of the benthic megafauna in the closed area. However, they are not solely responsible for the recovery patterns observed at Site 17. We observed numerical responses in numerous species (Figure 9), and other mollusks and echinoderms contribute to the increased biomass.

Recovery, Succession, and Dominance

We observed significant increases in abundance and biomass at Site 17 in 1997, 2.5 years after the area closure. The recovery time of gravel habitats is clearly longer than for soft-sediment communities (Collie et al. 2000a). In 2000, 5 years after the closure, we were still seeing increases in biomass and in the abundances of certain taxa. Our results so far, suggest that the recovery time of the gravel habitats is on the order of 10 years, but continued sampling is required to validate this prediction. Similar recovery rates were observed during 10 years of sampling a gravelly habitat off the Isle of Man, UK, following closure to scallop dredging (Bradshaw et al. 2000).

The pattern of succession at Site 17 can be interpreted with respect to the lifespan of the different species, as listed in Hermsen et al. (2003). We expect short-lived species to recover more quickly but to exhibit more variability in abundance ("r-selected" species). In contrast, longer-lived species should exhibit slower and steadier recovery patterns ("K-selected" species). Many species were absent or rare at Site 17 prior to the closure. Some species increased rapidly after the closure, then declined. For example, the nudibranch, *Coryphella*, was abundant at Site 17 in 1996 only. Several short-lived (on the order of up to 5 years) species first became abundant in 1997, and then their numbers leveled off (e.g., *Hyas coarctatus, Astarte borealis,* and *Crenella glandula*). Other longer-lived species (10–20 years) continued to increase throughout the time series (e.g., *Placopecten magellanicus, Buccinum undatum,* and *Asterias vulgaris*). With higher numerical densities, species abundance will be increasingly affected by competitive and predator–prey interactions.

The successional end point of the benthic megafaunal community at Site 17 is still unclear. There does not appear to be a linear succession of the community composition at Site 17 toward that at Site 20. Instead, the successional pattern seems more like the indeterminate pattern suggested by Auster and Langton (1999; Figure 5B). Site 20 is almost twice as deep as Site 17 and, therefore, has weaker bottom currents (Butman 1987); stronger bottom currents at Site 17 may reduce settlement rates of some taxa. It remains unclear to what extent the sea bottom at Site 17 will become covered with colonial epifauna. Analysis of recent bottom photos at Site 17 is not yet complete and will be the subject of future analyses.

Dominance curves from Site 17 are consistent with shifts in benthic community structure that have been observed along pollution gradients. At unpolluted sites, the biomass curve lies above the abundance curve, but with increasing pollution, the position of the curves is reversed (Clarke and Warwick 1994). Abundance–biomass (AB) curves and the difference between them (the W statistic) can, therefore, be used as indicators of pollution. Prior to the closure at Site 17, the abundance curve was above the biomass curve on both cruises in 1994. Following the closure in 1995, the biomass curve shifted above the abundance curve, and the separation between the two curves has increased since then, as reflected in the W statistic (Figure 10). Thus, the AB curves gave the earliest indicator of recovery at Site 17 following the closure. This initial response was due to an increased biomass of echinoderms (*Asterias vulgaris* and *Strongylocetrotus droebachiensis*); scallop biomass did not increase until 1996.

Links to Fish Production

The main results of this study show that there is a higher abundance and biomass of benthic megafauna at undisturbed gravel habitat sites and that the community composition is significantly different from that of disturbed sites. What are the implications of our results for the production of demersal fish? One prediction is that emergent epifauna provides juvenile fish with shelter from predators (Lindholm et al. 2001), but our study did not address this hypothesis. A second prediction is that bottom fishing reduces the abundance of prey species that are important in the diets of demersal fish. In a related analysis, Hermsen et al. (2003) estimated production from the size-frequency distributions of benthic megafauna from the same samples. They found significantly lower production of benthic megafauna at Site 13 than at Site 20. Likewise, there was an increase in production at Site 17, following the closure, to levels comparable with Site 20. The differences in production between disturbed and undisturbed sites are substantial when viewed in the context of the Georges Bank food web. Our results echo those of Jennings et al. (2001) who found reduced macrofaunal production with increasing trawling intensity in the North Sea. Taken together, these new results indicate that bottom fishing alters the flow of energy through continental shelf ecosystems.

Many of the species that were more abundant at the undisturbed sites are also important in the diets of demersal fish species. Not all benthic invertebrates in fish stomachs can be identified to species level, yet some important prey species stand out. Numerous demersal fish species, including flatfish and skates (Bowman et al. 2000), specialize on eating shrimp (*Dichelopandalus leptocerus, Pandalus montagui, Crangon septemspinosa*, and other pandalids) and crabs (*Hyas coarctatus, Cancer irroratus*, and *Pagurus* spp.). Haddock eats *Ophiopholis aculeata*, and American plaice *Hippoglossoides platessoides* also specializes in eating ophiuroids (Bowman et al. 2000). To the extent that these prey species are reduced in abundance, demersal fish must spend more time foraging, and the juveniles will be exposed to increased predation risk (Walters and Juanes 1993).

What actions should be taken to mitigate the effects of bottom fishing disturbance? The National Research Council committee that studied the effects of trawling and dredging on seabed habitats recommended that a combination of effort reduction, gear modifications, and area closures be tailored to fit specific combinations of fisheries and habitats (NRC 2002). Effort reduction—the cornerstone of fisheries management—should result in commensurate decreases in bottom fishing disturbance (Hall 1999). Sensitive habitats with long recovery times require the additional protection of area closures. There are some incentives and opportunities for "reduced-impact" fishing gears to operate within closed areas. However, bottom contact is required to catch species such as flatfish and scallops, and for these fisheries, there is limited scope to reduce bottom impacts with gear modifications.

Rotational harvest strategies may increase the yield per recruit of scallops by reducing the mortality of small scallops (Myers et al. 2000). However, the rotation times that are being considered (3–5 years) are shorter than the recovery times of gravel habitats (~10 years). The result of a rotational harvest strategy on gravel habitats could be to maintain all the areas in a chronically disturbed state. During a temporary trawl closure in the North Sea, fishing effort was displaced outside a closed area but then returned when the area was reopened (Rijnsdorp et al. 2001). The net result was a more homogeneous distribution of fishing effort and increased effort in areas that formerly were less impacted by bottom gear. From a habitat perspective, it is preferable to keep fishing effort patchy (Duplisea et al. 2002) because repeated tows of the same area cause a diminishing mortality of benthic species and large areas remain unfished. Thus, permanently closed areas of gravel habitat are preferred over temporary or rotating closures to mitigate the effects of fishing on benthic communities. However, rotating closures of other kinds of habitats (e.g., those sand and mud habitats that recover more rapidly than gravel) might be an appropriate management strategy.

These management issues are especially topical on Georges Bank as the New England Fisheries Management Council considers different closed area options as amendments to the groundfish and scallop fishery management plans (NEFMC 2003). The existing areas closed to mobile bottom fishing gear have been successful in reducing fishing mortality and have protected benthic habitat (Murawski et al. 2000). In fine-tuning the existing closed areas, the primary considerations are (1) to reduce fishing mortality on overfished stocks while allowing harvest of abundant stocks; (2) to protect vulnerable habitats; and (3) to create informative spatial comparisons that will allow the benefits and costs of the closed areas to be measured and evaluated.

Acknowledgments

We thank the many people who helped to process these samples at sea and in the laboratory. In particular, we acknowledge the contributions of Galo Escanero, Eddie Hughes, Bob Wallace, and Jim Nelligan. Allison DeLong assisted with the data analysis; Stephen Hall and Bob McConnaughey made valuable comments on an earlier draft. This research was funded with grants from the National Oceanic and Atmospheric Administration/University of Rhode Island Cooperative Marine Education and Research Program, the U.S. Geological Survey, the National Undersea Research Center, and the National Sea Grant College Program.

References

Auster, P. J., and R. W. Langton. 1999. The effects of fishing on fish habitat. Pages 150–187 in L. Benaka, editor. Fish habitat: essential fish habitat and rehabilitation. American Fisheries Society, Symposium 22, Bethesda, Maryland.

Bergman, M. J. N., and J. W. van Santbrink. 2000. Fishing mortality of populations of megafauna in sandy sediments. Pages 49–68 in M. J. Kaiser and S. J. de Groot, editors. Effects of fishing on non-target species and habitats. Blackwell Scientific Publications, Oxford, UK.

Blackwood, D., and K. Parolski. 2001. Seabed observation and sampling system. Sea Technology 42(2):39–43.

Bowman, R. E., C. E. Stillwell, W. L. Michaels, and M. D. Grosslein. 2000. Food of northwest Atlantic fishes and two common species of squid. NOAA Technical Memorandum NMFS-NE-155.

Bradshaw, C., L. O. Veale, and A. R. Brand. 2002. The role of scallop-dredge disturbance in long-term changes in Irish Sea benthic communities; a re-analysis of an historical dataset. Journal of Sea Research 47:161–184.

Bradshaw, C., L. O. Veale, A. S. Hill, and A. R. Brand. 2000. The effects of scallop dredging on gravelly seabed communities. Pages 83–104 in M. J. Kaiser and S. J. de Groot, editors. Effects of fishing on non-target species and habitats. Blackwell Scientific Publications, Oxford, UK.

Butman, B. 1987. Physical processes causing surficial sediment movement. Pages 147–162 in R. H. Backus and D. W. Bourne, editors. Georges Bank. MIT Press, Cambridge, Massachusetts.

Clark, K. R., and R. M. Warwick. 1994. Change in marine communities: an approach to statistical analysis and interpretation. Natural Environmental Research Council, Swindon, UK.

Coleman, F. C., and S. L. Williams. 2002. Overexploiting marine ecosystem engineers: potential consequences for biodiversity. Trends in Ecology and Evolution 17:40–44.

Collie J. S., G. S. Escanero, and P. C. Valentine. 1997. Effects of bottom fishing on the benthic megafauna of Georges Bank. Marine Ecology Progress Series 155:159–172.

Collie J. S., G. S. Escanero, and P. C. Valentine. 2000a. Photographic evaluation of the impacts of bottom fishing on benthic epifauna. ICES Journal of Marine Science 57:987–1001.

Collie, J. S., S. J. Hall, M. J. Kaiser, and I. R. Poiner. 2000b. A quantitative analysis of fishing impacts on shelf-sea benthos. Journal of Animal Ecology 69:785–798.

Duplisea, D. E., S. Jennings, K. J. Warr, and T. A. Dinmore. 2002. A size-based model of the impact of bottom trawling on benthic community structure. Canadian Journal of Fisheries and Aquatic Sciences 59:1785–1795.

Freese, L. P. J. Auster, J. Heifetz, and B. L. Wing. 1999. Effects of trawling on seafloor habitat and associated invertebrate taxa in the Gulf of Alaska. Marine Ecology Progress Series 182:119–126.

Gilkinson, K., M. Paulin, S. Hurley, and P. Schwinghammer. 1998. Impacts of trawl door scouring on infaunal bivalves: results of a physical trawl door model/dense sand interaction. Journal of Experimental Marine Biology and Ecology 224:291–312.

Hall, S. J. 1999. The effects of fishing in marine ecosystems and communities. Blackwell Scientific Publications, Oxford, UK.

Hall-Spencer, J., V. Allain, and J. H. Fossa. 2002. Trawling damage to northeast Atlantic ancient coral reefs. Proceedings of the Royal Society of London 269:507–511.

Hermsen, J. M., J. S. Collie, and P. C. Valentine. 2003. Mobile fishing gear reduces benthic megafaunal production on Georges Bank. Marine Ecology Progress Series 260:97–108.

Jennings, S., T. A. Dinmore, D. E. Duplisea, K. J. Warr, and J. E. Lancaster. 2001. Trawling disturbance can modify benthic production. Journal of Animal Ecology 70:459–475.

Jennings, S., and M. J. Kaiser. 1998. The effects of fishing on marine ecosystems. Advances in Marine Biology 34:201–352.

Krebs, C. J. 1989. Ecological methodology. HarperCollins, New York.

Lindholm, J. B., P. J. Auster, and L. S. Kaufman. 1999. Habitat-mediated survivorship of juvenile (0-year) Atlantic cod *Gadus morhua*. Marine Ecology Progress Series 180:247–255.

Lindholm, J. B., P. J. Auster, M. Ruth, and L. Kaufman. 2001. Modeling the effects of fishing and implications for the design of marine protected areas: juvenile fish responses to variations in seafloor habitat. Conservation Biology 15:424–437.

Lough, R. G., P. C. Valentine, D. C. Potter, P. J. Auditore, G. R. Bolz, J. D. Neilson, and R. I. Perry. 1989. Ecology and distribution of juvenile cod and haddock in relation to sediment type and bottom currents on eastern Georges Bank. Marine Ecology Progress Series 56:1–12.

Murawski, S. A., R. Brown, H.-L. Lai, P. J. Rago, and L. Hendrickson. 2000. Large-scale closed areas as a fishery-management tool in temperate marine systems: the Georges Bank experience. Bulletin of Marine Science 66:775–798.

Myers, R. A., S. D. Fuller, and D. G. Kehler. 2000. A fisheries management strategy robust to ignorance: rotational harvest in the presence of indirect fishing mortality. Canadian Journal of Fisheries and Aquatic Sciences 57:2357–2362.

NEFMC (New England Fishery Management Council). 2003. Amendment 10 public hearing document. NEFMC, Newburyport, Massachusetts.

NRC (National Research Council). 2002. Effects of trawling and dredging on seafloor habitat. National Academy Press, Washington, D.C.

Rago, P., and M. McSherry. 2001. Spatial distribution of fishing effort for sea scallops: 1998–2000. Prepared for the workshop on effects of fishing gear on fish habitat in the northeastern U.S. October 23–25, 2001, Boston, Massachusetts.

Ramsay, K., M. J. Kaiser, A. D. Rijnsdorp, J. A. Craeymeersch, and J. Ellis. 2000. Impact of trawling on populations of the invertebrate scavenger *Asterias rubens*. Pages 151–162 in M. J. Kaiser and S. J. de Groot, editors. Effects of fishing on non-target species and habitats. Blackwell Scientific Publications, Oxford, UK.

Rijnsdorp, A. D., G. J. Piet, and J. J. Poos. 2001. Effort allocation of the Dutch beam trawl fleet in response to a temporarily closed area in the North Sea. ICES, C.M. 2001/N:01, Copenhagen.

Stokesbury, K. D. E., and J. H. Himmelman. 1995. Biological and physical variables associated with aggregations of the giant scallop *Placopecten magellanicus*. Canadian Journal of Fisheries and Aquatic Sciences 52:743–753.

Thouzeau, G., G. Robert, and S. J. Smith. 1991. Spatial variability in distribution and growth of juvenile and adult sea scallops *Placopecten magellanicus* (Gmelin) on eastern Georges Bank (Northwest Atlantic). Marine Ecology Progress Series 74:205–218.

Thrush, S. F., J. E. Hewitt, V. J. Cummings, P. K. Dayton, M. Cryer, S. J. Turner, G. Funnell, R. Budd, C. J. Milburn, and M. R. Wilkinson. 1998. Disturbance of the marine benthic habitat by commercial fishing; impacts at the scale of the fishery. Ecological Applications 8:866–879.

Tupper, M., and R. G. Boutilier. 1995. Effects of habitat on settlement, growth, and postsettlement survival of Atlantic cod (*Gadus morhua*). Canadian Journal of Fisheries and Aquatic Sciences 52:1834–1841.

Valentine, P. C., D. Blackwood, and K. Parolski. 2000. Seabed observation and sampling system (SEABOSS). U.S. Geological Survey, Fact Sheet FS-142–00. Available: *http://pubs.usgs.gov/fs142-00/* (June 2004).

Valentine, P. C., and R. G. Lough. 1991. The sea floor environment and the fishery of eastern Georges Bank—the influence of geologic and oceanographic environmental factors on the abundance and distribution of fisheries resources of the northeastern United States continental shelf. U.S. Geological Survey, Open-File Report 91-439, Washington, D.C.

Walters, C. J., and F. Juanes. 1993. Recruitment limitation as a consequence of natural selection for use of restricted feeding habitats and predation risk taking by juvenile fishes. Canadian Journal of Fisheries and Aquatic Sciences 50:2058–2070.

The Effects of Area Closures on Georges Bank

JASON LINK[1] AND FRANK ALMEIDA

*National Marine Fisheries Service, Northeast Fisheries Science Center,
166 Water Street, Woods Hole, Massachusetts 02543, USA*

PAGE VALENTINE

*U.S. Geological Survey, Woods Hole Field Center, 384 Woods Hole Road,
Quisset Campus, Woods Hole, Massachusetts 02543, USA*

PETER AUSTER

*National Undersea Research Center and Department of Marine Sciences,
1084 Shennecossett Road, Groton, Connecticut 06340, USA*

ROBERT REID AND JOSEPH VITALIANO

*National Marine Fisheries Service, Northeast Fisheries Science Center,
74 Magruder Road, Highlands, New Jersey 07732, USA*

Abstract. In late 1994, substantial portions of Georges Bank were closed to commercial fishing to assist with stock rebuilding. These areas were Closed Area I (CAI), located on the western portion of the bank, and Closed Area II (CAII), on the eastern portion. After about 5 years of closure, the southern portion of CAII and the central portion of CAI, having exhibited substantial increases in biomass and density of sea scallops *Placopecten magellanicus*, were reopened to scallop fishing. Before the industry was allowed entry, we conducted surveys to monitor the recovery of benthic habitat and fauna inside both areas. Sampling sites were selected in a paired station design for an inside–outside comparison representative of major habitat types in each closed area; other stations were chosen to survey the remainder of the closed areas. At each station, we examined a suite of biotic and abiotic variables ranging from substrate type to benthos to nekton. Our results suggest few differences between the inside–outside paired stations in both closed areas for nekton and benthic species composition and species richness. Fish abundance and biomass were similar inside and outside the closed areas. However, individuals of species such as skates (*Raja* spp.), haddock *Melanogrammus aeglefinus*, and flounders (Pleuronectiformes) were generally larger inside than outside the closed areas. Additionally, habitat type was important in determining the distribution, abundance, biomass, size, and feeding ecology for some of the more benthic-oriented species studied. In CAI, the differences we observed in the suite of biotic metrics are likely a result of the high diversity of habitat types, with many of the habitat types composed of higher-relief material (e.g., cobble, gravel, etc.) in the region. The seabed in the southern portion of CAII is a relatively high-energy sand habitat of low to moderate complexity and has a relatively low vulnerability to trawling and dredging, which may explain why there were less pronounced differences in abundance or biomass across habitat types in that closed area as compared to CAI. Other parts of closed areas on the northeastern shelf may exhibit more obvious changes in the same biological metrics due to the presence of more complex habitats and increased vulnerability to bottom tending fishing gear. Those differences we observed for CAI and CAII may have implications for the population dynamics of commercially valuable benthic species, yet that question remains a major challenge.

Introduction

There has been much recent interest in the effects of fishing on habitat and nontarget species and how these effects can influence populations of economically valuable species (e.g., Jennings and Kaiser 1998; Benaka 1999; Kaiser and de Groot 2000). The general paradigm is that as habitat complexity increases, sensitivity to fishing effects increases, and changes induced by fishing activities become more extreme (Auster et

---
[1] E-mail: jlink@whsun1.wh.whoi.edu

al. 1996; Auster and Langton 1999). Fishing pressure usually supersedes ecological processes and can cause a notable decline in benthic macrofauna and habitat complexity (Jennings and Polunin 1996; Collie et al. 1997, 2000a; Thrush et al. 1998; Auster and Langton 1999; Jennings et al. 2002). If habitat effects of fishing are reduced or eliminated but insufficient suitable habitat or food remains for juvenile settlement, it may not be possible to attain sustainable levels for given fish stocks or populations. Thus, the utility of area closures (a narrower application of the marine reserve concept) merits examination in an ad hoc, in situ experimental sense.

There has also been much recent interest in marine protected areas (MPAs), area closures, or marine reserves (see Conover et al. 2000). Year-round, no-take marine reserves have been identified, particularly, as tools that can enhance both populations of exploited species and biodiversity in general (e.g., McManus 1998; Auster and Shackell 2000; Mosquera et al. 2000; NRC 2001; Roberts et al. 2001; Ward et al. 2001; Fisher and Frank 2002). However, opportunities to examine closed areas are generally limited, particularly on the large temporal and spatial scales at which fisheries operate (i.e., significant fractions of continental shelves).

The effects of fishing activities on the seafloor off New England are also receiving increased attention from fishery managers, conservationists, fishery scientists, and fishermen (e.g., Auster et al. 1996; Dorsey and Pederson 1998; Auster and Shackell 2000). Georges Bank was once the prime fishing ground on the U.S. East Coast. However, overfishing has caused the decline of the primary groundfish species (e.g., Atlantic cod *Gadus morhua*, haddock *Melanogrammus aeglefinus*, and yellowtail flounder *Limanda ferruginea*), which has led to increasingly restrictive management measures. The bank is an important nursery ground for these and other groundfish species (Smith and Morse 1985; Lough et al. 1989), and large areas have been closed to fishing to protect recovering stocks (Murawski et al. 1997; Fogarty and Murawski 1998). As groundfish populations continue to decline in areas open to fishing, there is increasing pressure to allow access to resources in the closed areas. In addition, fishery managers and other resource stakeholders are requiring information about the effects of fishing gears and the effects of closed areas on the benthic habitats of Georges Bank.

Since 1970, fishery management in New England has included seasonal closures of parts of Georges Bank to a variety of fishing gears. The boundaries of the closed areas have changed over time in response to fluctuations in spawning grounds and concerns of management. The continued deterioration of groundfish stocks (declining abundance and high exploitation rates) prompted the closure of two areas on Georges Bank and one in southern New England to fishing gears that catch groundfish (trawls, gill nets, hook and line, and scallop dredges) in December 1994 (Figure 1). The Georges Bank closed areas comprise about 10,900 km^2, roughly 25% of the total area of the bank. Closed Area II was among the most heavily fished areas on Georges Bank from 1982 to 1993. The southern part of CAII (south of 41°30'N) was reopened to commercial fishing for sea scallops *Placopecten magellanicus* during June 15–November 12, 1999, the first time in nearly 5 years. The central portion of CAI was reopened to commercial fishing for sea scallops during October 1, 1999–January 12, 2000. The closed areas on Georges Bank have been examined with respect to targeted resource species (e.g., Brown et al. 1998; Murawski et al. 2000); however, no studies have examined the entire fish and benthic communities simultaneously in these closed areas.

The objective of our study was to evaluate the effects of area closures on nekton and benthic community composition over large areas encompassing a range of habitat types. In particular, we examine a suite of abiotic and biotic metrics, before the fishery reopened in 1999, to determine if, after 4.5 years of closure, there were discernable differences inside versus immediately outside each of the closed areas.

Geologic, Oceanographic, and Benthic Attributes of Georges Bank

Georges Bank is a shallow (3–150 m) extension of the continental shelf of the northeastern United States, covering approximately 40,000 km^2 within the 100-m isobath. The bank is covered by glacial debris and is eroding; sediment is currently not transported from the continent or from adjacent shelves to the bank. The bank has strong tidal (predominantly northwestern–southeastern) currents and storm currents that transport sand from the bank into deep water. In shallow areas, tidal currents mix the water column and combine with weaker currents to form a broad, clockwise gyre (5–30 cm/s) around the bank (Butman et al. 1982; Twichell et al. 1987; Valentine and Lough 1991; Valentine et al. 1993). There is a highly energetic peak in the north with sand ridges up to 20–30 m high and an extensive gravel pavement with a relatively smooth seabed of finer-grained sediments on the deeper southern flank. Sediments are generally sand and gravel with little silt and clay (1–5% by weight) or organic carbon (<0.5% by weight) (Butman et al. 1982; Twichell et al. 1987; Valentine and Lough 1991; Valentine et al. 1993; Theroux and Wigley 1998). The macrobenthic fauna is

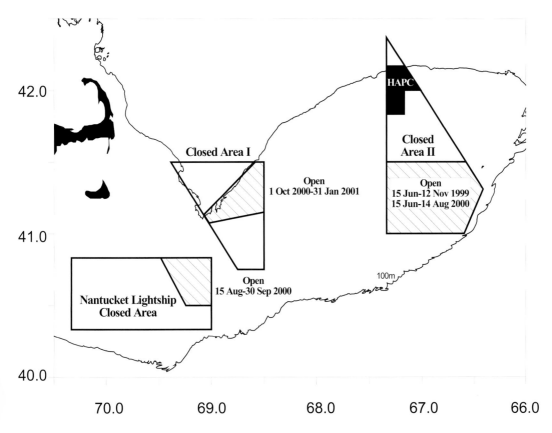

Figure 1. Map of the major Closed Areas on Georges Bank and Southern New England with areas re-opened to sea scallop fishery during 1999–2001. Latitude and Longitude are given in decimal degrees.

generally dominated numerically by crustaceans, annelids, echinoderms, and mollusks and dominated in terms of biomass by echinoderms, mollusks, crustaceans, and annelids (Theroux and Wigley 1998).

Closed Area I

The seafloor in Closed Area I covers approximately 3,960 km² with water depths ranging from 55 to 110 m. This area is affected by tidal currents that flow predominantly north–south with a mean velocity of 45–60 cm/s and by intermittent storm currents of greater strength (Butman et al. 1982; Twichell et al. 1987; Valentine and Lough 1991; Valentine et al. 1993). The sediment is highly varied, ranging from fine-grained sand in deep water, to gravel pavements and boulder piles associated with sand dunes at intermediate depths, to a mixture of coarse-grained sand and gravel at shallow depths (Butman et al. 1982; Twichell et al. 1987; Valentine and Lough 1991; Valentine et al. 1993; Theroux and Wigley 1998).

The predominant seabed features are long east–west trending sand dunes with steep faces that range up to 20 m in height (Butman et al. 1982; Twichell et al. 1987; Valentine and Lough 1991; Valentine et al. 1993; Theroux and Wigley 1998; P. Valentine, personal observation). The rippled surfaces of the dunes are maintained by tidal currents. Strong north–south flowing storm currents are responsible for forming the dunes, which are relatively stable from year to year. Large areas of the seabed are covered by pebble-gravel pavement, and some parts are characterized by mounds of boulders and cobbles of glacial origin. In areas of mixed sand and gravel, the sediment is segregated into large storm sand ripples (up to 30 cm high) that are separated by gravel troughs. Deeper parts of the area are floored with fine-grained rippled sand becoming increasingly unrippled as water depths increase.

The seafloor can be separated into three main zones. A moderate-energy zone (Zone 1) with mixed sand and gravel occupies the mid-portion of the closed area (Figures 2, 3). This zone has tidal and intermittent storm currents. It has minimal epifauna on storm sand ripples, with epifauna concentrated on gravel in

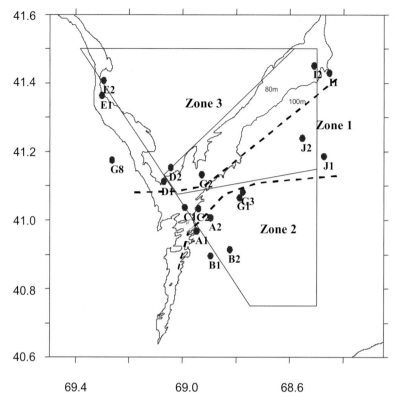

Figure 2. Map of Closed Area I on Georges Bank, with zonation and location of sampling sites occupied during the cruises to each area. Latitude and Longitude are given in decimal degrees.

ripple troughs. However, the epifauna will be covered by sand during a major storm or during trawling and dredging events. A subsequent storm will form new ripples and expose gravel in the troughs. A high-energy zone (Zone 2) with large dunes, gravel mounds, pebble gravel pavements, and rippled sand migrating over pebble gravel is located in the southern portion of the area. This zone is also affected by both tidal and intermittent storm

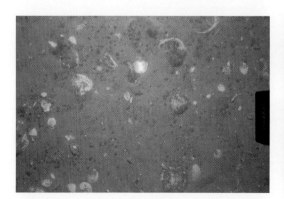

Figure 3. Sampling site J2, in Zone 1 of CAI at a depth of 65 m. The image shows gravelly sand in trough of storm sand ripple; dead mollusk shells with sparse attached epifauna, probably §hydrozoa.

Figure 4. Sampling site A2, inside Closed Area I at a depth of 68 m. The image shows starfish and shells in a gravel trough.

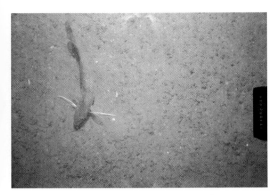

Figure 5. Sampling site I2, inside Closed Area I at a depth of 88 m. The image shows a red hake on a fine sand bottom.

currents. The zone has epifauna on hard bottoms, but epifauna are rare on rippled sand and dunes (Figure 4). Finally, Zone 3 is a low-energy zone with fine-grained sand and very infrequent weak storm currents or no currents. Epifauna are rare in this zone, but burrows are common (Figure 5).

Closed Area II

The seafloor in the southern part of Closed Area II (south of 41°30'N) covers approximately 3,880 km^2 (Figure 6). The area slopes gently southeastward with water depths that range from 35 m in the northwest to about 90 m in the southeast. It is affected by strong tidal currents that flow predominantly northwest–southeast and has mean tidal current speeds greater than 40 cm/s in the northwestern part of the study area, diminishing to 10 cm/s in the southeastern part (Butman et al. 1982; Twichell et al. 1987; Valentine and Lough 1991; Valentine et al. 1993). The area is impacted by intermittent storm currents that extend to depths of at least 90 m. The sediment is predominantly sand, with small areas of burrowed muddy outcrops in the northwestern corner (Butman et al. 1982; Twichell et al. 1987; Valentine and Lough 1991; Valentine et al. 1993; Theroux and Wigley 1998).

The seabed morphology in the northwestern part (between depths of 35–65 m) is characterized by rippled sand bedforms approximately 20–30 cm high whose crests are aligned northeast–southwest, normal to the strongest tidal flow (Butman et al. 1982; Twichell et al. 1987; Valentine and Lough 1991; Valentine et al.

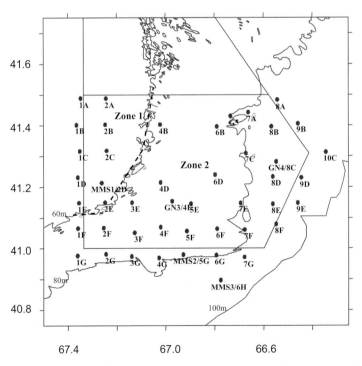

Figure 6. Map of Closed Area II on Georges Bank, with zonation and location of sampling sites occupied during the cruises to each area. Latitude and Longitude are given in decimal degrees.

1993; Theroux and Wigley 1998; P. Valentine, personal observation). Sand ripples and bedforms are generally absent below 65 m, except for those created by occasional storm currents. Movement and reworking of the sediment by bottom currents (and possibly by mobile fishing gear) winnows fine material. Sand contains less than 1% combined silt and clay throughout the 65–90-m depth interval, except at several locations in the eastern part where mud content reaches 1–3%.

This portion of the closed area can be separated into two main zones (Figure 6). First is a high-energy zone (Zone 1) in depths of 35–65 m where sand is transported back and forth on a daily basis by tidal currents and occasionally moved by storm currents (Figure 7). A low-energy zone (Zone 2) in depths of 65–90 m is affected principally by storm currents (Figure 8).

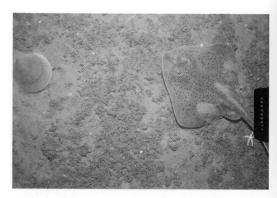

Figure 8. Sampling site 9D, outside Closed Area II at a depth of 94 m (51.4 fm). The image shows several small sea scallops ranging in size from 4.3 to 5.3 cm (1.7–2.1 in) and a few dead scallop shells.

Methods

Sampling Site Selection

The overall approach we used for both closed areas was to examine a series of paired stations inside a closed area and immediately outside the area. We examined a suite of variables to explore the hypothesis that responses inside the closure should be different (usually an expected increase) than outside the closure. Stations inside the closed area were at least 2 nm from the closed area boundary to ensure that the sites chosen were undisturbed (i.e., to minimize sampling in areas where fishing vessels may have fished illegally). In both CA I and CA II, we assumed no fishing occurred inside the buffer zone within the closed areas. The outside stations were chosen in proximity to the closed area and the related inside paired station to represent a similar depth, substrate, and overall similarity of environmental attributes for a given habitat type. We did not assume that all areas immediately outside of the closed areas were necessarily impacted continuously by bottom tending fishing gear, but we did assume that these locations were susceptible to fishing efforts over the past 4.5 years. Vessel monitoring data for the scallop and groundfish fleet has generally supported this assumption (Northeast Regional Office [NERO] of the National Marine Fisheries Service [NMFS], unpublished data). We do know that a large of amount of fishing effort is regularly concentrated proximal to the closed area boundaries (NERO, unpublished data).

Closed Area I

Sites in this area were first selected to provide a pairwise (inside versus outside the closed area) experimental design, followed by several stations within the closed area to survey and to reoccupy sites previously sampled (Figure 2). A 2 nm buffer zone (Figure 2) was established inside the closed area boundary to ensure that the sites chosen were undisturbed. A total of 18 sites were occupied during the study, 8 outside the closed area and 10 inside the area, with 7 sets of inside–outside pairs. The sites were sampled during July 6–10, 1999.

Closed Area II

In this area, as in CAI, sampling sites were first selected in order to provide a paired station experimental design to allow comparisons between sites inside and outside the closed area. Once paired stations were determined, a series of sites in grid form were selected to survey the rest of the closed area. A 3 nm buffer zone (Figure 6) was established inside the closed area boundary to ensure that the sites chosen were undisturbed.

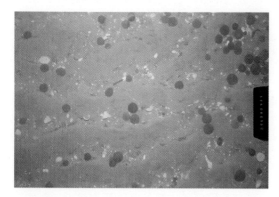

Figure 7. Sampling site 2C, inside Closed Area II at a depth of 49 m (26.8 fm). The image shows many small sand dollars about 2.8 cm (1.1 in) in diameter among sand ripples.

Grid stations were located 5 nm apart in an east–west orientation. A total of 47 sites were occupied during the study, 21 outside the closed area and 26 inside the area, with 18 sets of inside–outside pairs. These sites were sampled during June 1–11, 1999, 4 d prior to the area being reopened to sea scallop fishing.

Sampling

At each site, several habitat monitoring and sampling instruments were deployed. First, a Seabed Observation and Sampling System (SEABOSS) was deployed to quantify microhabitat distributions, microhabitat relationships of fishes, and the distribution of seabed sand ripples and dunes. Still photographs were taken at approximately 1-min intervals throughout each transect to estimate percent cover for common habitat types. The SEABOSS had two video cameras (forward and downlooking), a downlooking 35-mm camera and a modified Van Veen sediment grab sampler. Quartz halogen lights provided illumination for the video, and an electronic flash unit provided lighting for still photographs. The system was tethered and essentially "flown" over the seafloor while the support vessel was drifting. Each SEABOSS transect (1 per trawl station) collected 20 min of continuous video, 20 still photographs, and a sediment sample. In order to characterize the seabed, the top 2 cm of sediment was collected from samples obtained with the VanVeen grab.

A Smith-McIntyre spring-loaded bottom sampler was deployed to sample the benthic macrofauna. Three bottom grab samples were taken at each trawl station (see below). For each grab, samples were sieved utilizing a 0.5-mm (or 0.3-mm for selected stations) sampling screen and preserved in a 10% formalin–rose bengal solution to enhance collection and identification of invertebrates in the sample. After 24–72 h, samples were washed and transferred to a 95% EtOH–glycerin solution for storage.

In the laboratory, benthic grab samples were sorted under a dissecting microscope, and the invertebrates were removed, counted, and identified to species level where possible. Wet weight biomasses were determined for each species in a sample by blot drying the species collections on absorbent paper towels for about 3 min and weighing them to the nearest mg.

In CAII, a 15-min tow using a standardized Northeast Fisheries Science Center (NEFSC) #36 Yankee otter trawl was made at each site, at a towing speed of 6.5 km/h. In the sandy bottom of CAII, we used a trawl rigged with a rubber disk-covered chain sweep, 11 floats, 5-m ground cables, and 450-kg polyvalent trawl doors. Due to the heterogeneous bottom encountered in CAI, a 30-min haul using a trawl rigged with 41-cm-diameter rollers, and 9-m bridles was used. Catches were sorted to species, and all fish and invertebrates caught were weighed and enumerated. Length frequency data were collected on all species; biological samples, including age structures, sex and maturity stage determinations, and stomachs were collected from selected species. Stomach samples were either examined at sea (volumetrically) or individually preserved in 10% formalin for later analysis.

Analysis

A two-way analysis of variance (ANOVA) was performed to determine the effect of both inside versus outside and zonation (i.e., habitat type) of a closed area. The inside–outside comparison and comparison across habitats (i.e., zonation) were treated as fixed factors. A distinct set of analyses was conducted for each closed area. Each station was treated as a replicate for the paired inside versus outside comparison. Each paired set of stations was assumed to be independent from the other closure and nonclosure replicates. Multivariate (for multiple species simultaneously) analogues (i.e., MANOVA, not described here) suggested that there were probable differences among some species across either inside–outside or zonation. Thus, we conducted an individual ANOVA for each of the major trawl-caught fish species, and the analysis was done separately for four main biological variables. Weight (kg) per tow, numbers per tow, stomach contents (g), and fish length (cm) were used as the response variables in these ANOVAs; the first three were log transformed for purposes of normality and homoscedasticity in this analysis. We also did the same analysis for the 10 most abundant benthic species from the grabs, using both abundance and biomass (mg) per grab as the response variables. The infaunal invertebrates chosen for analysis based on abundance, frequency of occurrence, biomass, and living position (i.e., surface dwellers, burrowers, etc.). Results typically classified as marginally significant ($0.10 > P > 0.05$) were also noted.

Results

Closed Area I

Species Abundance and Biomass

We did not detect a difference in nekton and benthic species richness inside and immediately outside the closed area (Table 1). However, the species richness of benthic macroinvertebrates was different across the different habitats (i.e., zones), with the highest number of species and individuals in the low-energy sand habitat (Zone 3).

For most of the fish, there were minimal differ-

Table 1. Species richness in Closed Area I. The table shows the mean number of species collected in benthic grabs, the mean number of benthic grab individuals, and the mean number of trawl caught (fish) species. Significance values from a two-way ANOVA evaluating inside–outside and zonation are presented, if more than marginally significant, next to All Zones and In & Out as appropriate. ns = non-significant; ?* = $0.05 < P < 0.10$.

Zone	In	Out	In & Out
Mean number of benthic species			
1	64.0	72.6	68.3 ns
2	77.1	61.5	69.3 ns
3	88.7	83.7	86.2 ns
All ?*	72.4	69.8	72.3 ns
Mean number of benthic individuals			
1	1,190.0	2,040.0	1,615 ns
2	888.3	874.8	881.5 ns
3	5,355.0	3,364.0	4,359.5 ns
All ?*	1,902.3	1,838.7	1,870.5 ns
Mean number of fish species			
1	19.5	20.5	20.0 ns
2	16.3	17.3	16.8 ns
3	17.5	17.5	17.5 ns
All ns	17.6	18.3	18.0 ns

ences in biomass (mean weight per tow; Figure 9; Table 2) and abundance (mean number per tow; Figure 10) inside versus outside the closed area. Exceptions were haddock, cunner, and Atlantic cod, all of which had a greater abundance inside the closed area. Particularly, in Zones 2 and 3, haddock was an order of magnitude more abundant inside versus outside the closed area. Atlantic cod was twice as abundant inside versus outside of the closed area. As only 23% of the species studied exhibited even a marginally significant difference inside versus outside the closed area, habitat (i.e., zone) appeared to be a relatively more important factor than inside–outside the closure for affecting species abundance and biomass. For example, winter skate was more abundant in the high-energy mixed substrate habitat (Zone 1) and yellowtail flounder was more abundant in the low-energy sand habitat (Zone 3). Haddock was more abundant in Zones 2 and 3 than in the high-energy mixed substrate habitat (Zone 1).

The same pattern was true for species collected in benthic grabs (Table 3; Figure 11). Only one species exhibited a significant difference in biomass (mean weight per grab) inside versus outside the closed area (Figure 11; a higher biomass inside for *Ampharete cf. lindstroemi*). Yet habitat (i.e., zone) was an important factor for several species in terms of both abundance and biomass. In particular, the abundance and biomass of the amphipod *Unciola inermis* and the biomass of the polychaete *Aricidea catherinae* were significantly lower in Zone 1 (moderate energy, mixed substrate) compared to Zones 2 and 3 (gravel and sand substrates). Thus, habitat type apparently affected the abundance and biomass distribution of many benthic invertebrates and demersal fish species in and near CAI.

Fish Lengths

We did not detect a difference in the mean length (and length distributions) of most species inside and immediately outside the closed area (Figure 12; Table 2). The exception was haddock, which was approximately 5 cm larger inside the closed area. Length differences were apparent among habitats (i.e., zones) for four species: haddock (larger lengths in the low energy sand habitat, Zone 3), red hake (larger lengths in Zone 3), longhorn sculpin (larger lengths in Zone 2), and yellowtail flounder (shorter lengths in zone 1) (Table 2).

Stomach Contents

Of the two species examined (haddock and yellowtail flounder), there were major differences in stomach content amount between inside and immediately outside the closed area (Figure 13; Table 2). For both species, notably more food was eaten inside the closed area. Habitat (i.e., zone) also appeared to be an important factor in determining the amount of food a fish ate. Haddock and yellowtail flounder exhibited significant differences in the amount of stomach contents among zones, with both species consuming more food in the low-energy sand habitat (Zone 3).

Closed Area II

Species Abundance and Biomass

As in CAI, we did not detect a difference in nekton and benthic species richness inside and immediately outside the closed area (Table 4). The species richness of benthic macroinvertebrates was distinct across habitats (i.e., zones), with the highest number of species and individuals in the low-energy sand habitat (Zone 2).

There were few detectable differences in biomass (mean weight per tow; Figure 14; Table 5) and abundance (mean number per tow; Figure 15) inside versus outside the closed area for most of the fish. The notable exception was haddock, which showed greater biomass inside the closed area. Habitat (i.e., zone) was an important factor in determining species abundance and biomass for haddock, winter skate, and little skate, with all species more abundant in the high-energy sand habitat (Zone 1). Fourspot flounder *Paralichthys oblongus* was more abundant in the low-energy sand habitat (Zone 2). However, we did not detect differences for the remaining species across the two habitat types.

The species collected in benthic grabs followed a similar pattern, with only the polychaete *Exogone*

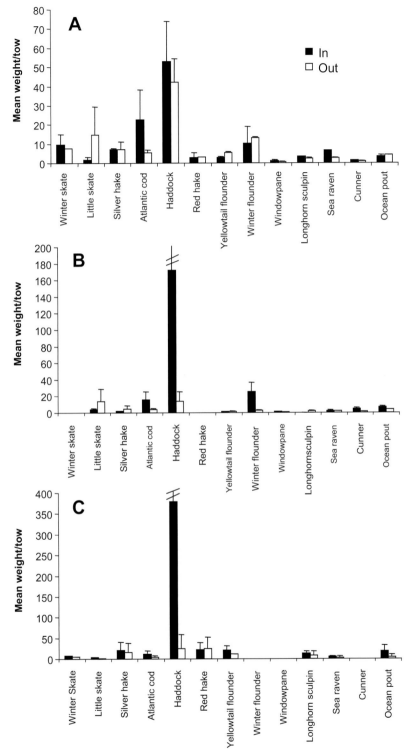

Figure 9. Biomass per tow for major macrofauna species in each zone of CAI, contrasting inside (solid bars) versus outside (open bars) (A) Zone 1, (B) Zone 2, and (C) Zone 3. Error bars represent ±1 SE. Note different scales on the axes across the zones.

Table 2. Results from ANOVA assessing the factors of Inside versus Outside (IO), Zonation (Z), and the Interaction between these factors (Int) for each of the major species in Closed Area I. The analysis was done separately for four main biological variables. The numbers (per tow), weight (kg/tow), and stomach contents (g) were log transformed for this analysis. If a factor was significantly different, an I or O for IO or the number of the zone for Z is given in parenthesis to show where the highest values for each variable were observed. Significance levels: ?* = $P < 0.10$; * = $P < 0.05$; ** = $P < 0.01$; *** = $P < 0.001$; and !** = $P < 0.0001$.

Species	Numbers			Weight			Stomach			Length		
	IO	Z	Int	IO	Z	Int	IO	Z	Int	IO	Z	Int
Winter skate *Leucoraja ocellata*					?* (1)							
Little skate *Leucoraja erinacea*												
Silver hake *Merluccius bilinearis*												
Atlantic cod				* (I)								
Haddock	?* (I)			?* (I)	?* (3)		!** (I)	!** (3)		** (I)	** (3)	
Red hake *Urophycis chuss*											*** (3)	
Yellowtail flounder		* (2)			** (3)		** (I)	!** (3)	!**		** (3)	
Winter flounder *Pseudopleuronectes americanus*												
Windowpane *Scophthalmus aquosus*												
Longhorn sculpin *Myoxocephalus octodecemspinosus*											?* (2)	
Sea raven												
Cunner *Tautogolabrus adspersus*	?* (I)			* (I)	* (2)							
Ocean pout *Zoarces americanus*												

hebes exhibiting a significant difference inside versus outside the closed area (Table 6; Figure 16; 10 times more abundant inside the closure). A few other species did show a difference across habitats (i.e., zones) in terms of both abundance and biomass (Table 6; Figure 16). In particular, the abundances and biomasses of the amphipod *Erichthonius rubricornis*, the caprellid *Aeginina longicornis* and the brittlestar *Ophiura robusta* were significantly higher in the lower-energy sand habitat (zone 2) compared to the high-energy sand habitat (zone 1).

Fish Length

We did not detect a difference in the mean length (and length distributions) for 11 of the 15 species we examined inside and immediately outside the closed area (Figure 17; Table 5). However, length differences were apparent inside versus outside for winter skate, red hake, yellowtail flounder, and winter flounder, all of which were larger inside than outside of the closed area. There were significant differences in mean length for winter skate and yellowtail flounder, both larger in the low-energy sand habitat (Zone 2), and for haddock, which was larger in the high-energy sand habitat (Zone 1) (Table 5).

Stomach Contents

There were no significant differences in stomach content amount between inside and immediately outside the closed area for most species (Figure 18; Table 5). The exceptions to this pattern were winter skate, red hake, and longhorn sculpin, all of which consumed more food inside the closed area. Habitat (i.e., zone) appeared to be an equally or even more important factor than inside–outside the closure in determining the amount of food a fish ate. Haddock and winter flounder ate more food in the high–energy sand habitat (Zone 1), whereas yellowtail flounder ate more found in the low-energy sand habitat (Zone 2).

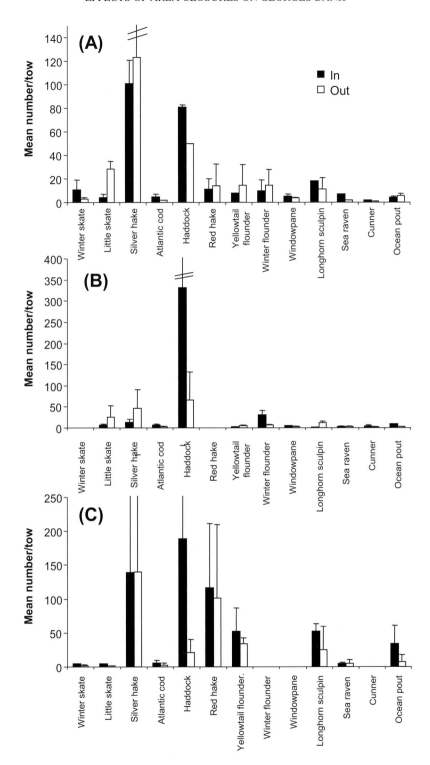

Figure 10. Abundance (number) per tow for major macrofauna species in each zone of CAI, contrasting inside (solid bars) versus outside (open bars) (A) Zone 1, (B) Zone 2, and (C) Zone 3. Error bars represent ±1 SE. Note different scales on the axes across the zones.

Table 3. Results from ANOVA assessing the factors of Inside versus Outside (IO), Zonation (Z), and the Interaction between these factors (Int) for each of the major benthic species in Closed Area I. The analysis was done separately for both abundance and biomass per grab. If a factor was significantly different, an I or O for IO or the number of the zone for Z is given in parenthesis to show where the highest values for each variable were observed.

| | Abundance | | | Biomass | | |
Species	IO	Z	Int	IO	Z	Int
Ceriantheopsis americanus						
Exogone hebes						
Goniadella gracilis						
Aricidea catherinae		?* (3)			* (3)	
Spiophanes bombyx						
Euclymene collaris						
Ampharete cf. lindstroemi		?* (3)		?* (I)		
Unciola inermis		** (3)			** (3)	
Pontogeneia inermis				.		?*
Echinarachnius parma						

Discussion

There were few detectable differences in the suite of biotic variables we studied inside versus immediately outside the closed areas on Georges Bank. Those species that exhibited significant differences inside versus outside are species that have a strong benthic affinity. Other studies (Lindholm et al. 2004) have shown that the other major macrofauna to exhibit notable differences between inside and immediately outside of the closed areas are epifaunal sponges. Fish species such as haddock, some of the flounders, some of the skates, and particularly, sessile invertebrates such as scallops (see NEFSC 2001) showed the greatest differences (usually more or larger in terms of biomass, abundance, length, and stomach contents) inside versus outside the areas. Those species for which we did not detect a significant difference were not surprising since they are highly mobile; the closed areas boundaries were not delineated to enclose the yearly ambit of the adult life phase of finfish, with the exception of the spawning period. These conclusions are generally consistent with other studies done for this area that examined a longer time series of data or more explicitly focused on particular populations (e.g., Brown et al. 1998; Murawski et al. 2000; NEFSC 2001).

Our results are in distinct contrast to much of the closed area, MPA, and marine reserves literature (e.g., Agardy 1997; Lauck et al. 1998; McManus 1998; Babcock et al. 1999; Mosquera et al. 2000; Roberts et al. 2001; Ward et al. 2001; Fisher and Frank 2002; but see Alder 1996; Allison et al. 1998; Auster and Shackell 2000; and Cote et al. 2001). Many of these studies effectively assert that any area closure is inherently valuable, principally because these studies focused on fish species with low movement rates associated with particular kinds of habitats (e.g., coral reefs, kelp forests). Few studies have focused on species that are facultative habitat users and have adult stages not necessarily tied to particular habitat features (Auster and Langton 1999). Simply closing an area irrespective of habitat considerations may not be effective, particularly when there is evidence for little, if any, response for certain types of species or habitats (Auster et al. 1996; Collie et al. 2000b; Cote et al. 2001; Kaiser et al. 2002; Kaiser 2003). Qualitative factors such as habitat type should be considered along with size and geographic location when developing new area closures. The duration of an area closure also merits consideration but should similarly not supplant the type of habitat closed.

It is possible that after 4.5 years of closure, the portions of the populations inside the closed areas on Georges Bank had "seeded" via larval and juvenile exports to the regions outside of the closed areas, making it difficult to detect any significant differences between inside and outside the closed areas (sensu Russ and Alcala 1996; Roberts 1998; McLanahan and Mangi 2000; Lindholm et al. 2001; and Strathamm et al. 2002). It is also possible that our sampling regime, always subject to logistical constraints, limited our power to detect differences. However, we generally did not see, nor have we seen across the time that the areas have been closed, a notable increase in biomass and abundance inside the closed areas for most species, as confirmed by similar patterns observed in trawl surveys across Georges Bank (NEFSC, unpublished data). Conversely, for those species for which we did detect differences, the differences are likely genuine.

Another major consideration is the amount of fishing effort outside the closed areas on Georges Bank. Cote et al. (2001) have shown that fishing intensity immediately outside a reserve, coupled with species composition, is one of the main factors determining the success of a no-take area

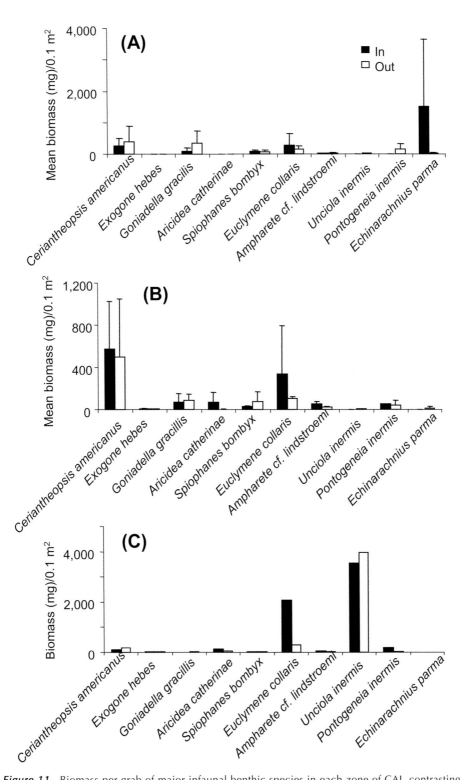

Figure 11. Biomass per grab of major infaunal benthic species in each zone of CAI, contrasting iinside (solid bars) versus outside (open bars) (A) Zone 1, (B) Zone 2, and (C) Zone 3. Error bars represent ±1 SE. Note different scales on the axes across the zones.

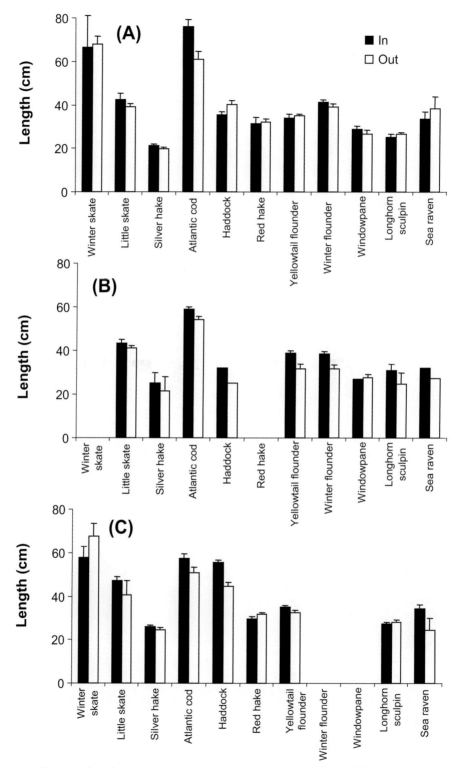

Figure 12. Mean lengths for major macrofauna species in each zone of CAI, contrasting inside (solid bars) versus outside (open bars) (A) Zone 1, (B) Zone 2, and (C) Zone 3. Error bars represent ±1 SE.

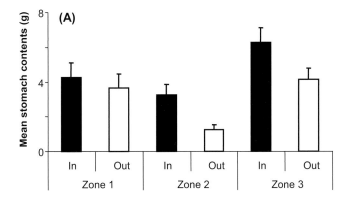

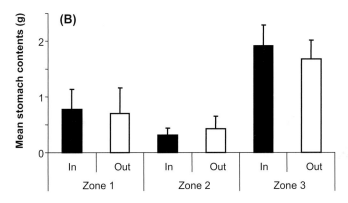

Figure 13. Mean stomach contents for (A) haddock and (B) yellowtail flounder in each zone of CAI, contrasting inside (solid bars) versus outside (open bars) the area Error bars represent ±1 SE.

closure. Auster et al. (1996) and Auster and Langton (1999) have demonstrated a high level of fishing intensity on the Georges Bank seabed. Even those species with high benthic affinities (e.g., haddock, some of the flounders, some of the skates, and particularly sessile invertebrates such as scallops) and with significant differences inside versus outside the areas have a high degree of mobility. It is probable that many of the species inside these area closures move across the boundaries into areas of much higher fishing effort. It then follows that the increased fishing mortality on these species outside a closed area might have negated many potential benefits accrued within the closed area.

Recognizing the minimal effects we generally observed on Georges Bank, it is unclear what the second and third order effects of these area closures are on nektonic macrofauna. It is not known whether the closed areas serve as a source, spill-over, reserve, or "seed bank" from which the rest of Georges Bank fish populations are re-stocked (sensu Russ and Alcala

Table 4. Species richness in Closed Area II. The table shows the mean number of species collected in benthic grabs, the mean number of benthic grab individuals, and the mean number of trawl caught (fish) species. Significance values from a two-way ANOVA evaluating inside–outside and zonation are presented, if more than marginally significant, next to All Zones and In & Out as appropriate. ns = non-significant; *** $P < 0.001$.

Zone	In	Out	In & Out
	Mean number of benthic species		
1	38.2	31.1	34.7 ns
2	58.0	59.0	58.5 ns
All ***	49.8	47.4	48.6 ns
	Mean number of benthic individuals		
1	1,119.4	767.7	943.5 ns
2	1,827.5	1,652.5	1,740.0 ns
All ***	1,532.4	1,283.8	1,408.1 ns
	Mean number of fish species		
1	17.2	18.8	18 ns
2	18.5	18.5	18.5 ns
All ns	18.1	18.6	18.4 ns

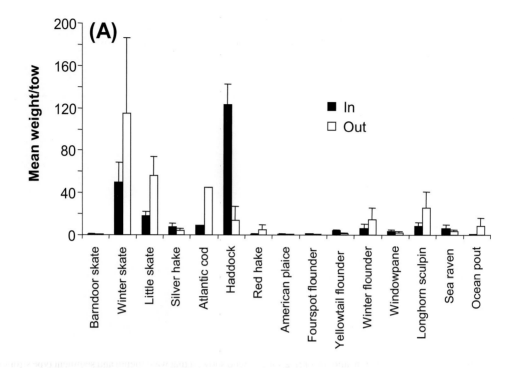
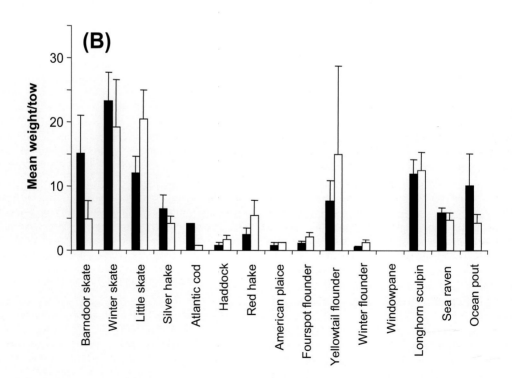

Figure 14. Biomass per tow for major macrofauna species in each zone of CAII, contrasting inside (solid bars) versus outside (open bars) (A) Zone 1 and (B) Zone 2. Error bars represent ±1 SE. Note different scales on the axes across the zones.

Table 5. Results from ANOVA assessing the factors of Inside vs. Outside (IO), Zonation (Z), and the Interaction between these factors (Int.) for each of the major species in CAII. The analysis was done separately for four main biological variables. The Numbers (per tow), Weight (kg per tow), and Stomach contents (g) were log transformed for this analysis. If a factor was significantly different, an I or O for IO or the number of the zone for Z is given in parenthesis to show where the highest values for each variable were observed.

Species	Numbers			Weight			Stomach			Length		
	IO	Z	Int	IO	Z	Int	IO	Z	Int	IO	Z	Int
Barndoor Skate												
Winter skate		* (1)			?* (1)		?* (I)					
Little skate		?* (1)			?* (1)							
Silver hake												
Atlantic cod												
Haddock		?*		?* (I)	?* (1)	*		!** (1)	**			
Red hake							** (I)					* (I)
Fourspot flounder		?* (2)										
Yellowtail flounder								* (2)		?* (I)	!**(2)	**
Winter flounder								!** (1)	**		?* (I)	
Witch flounder												
Windowpane												
Longhorn sculpin				* (I)								
Sea raven												
Ocean pout												

1996; Roberts 1998; McLanahan and Mangi 2000; Lindholm et al. 2001; Strathamm et al. 2002). Certainly this could be the case for organisms with a strong benthic affinity (e.g., haddock, yellowtail flounder, scallops), but whether this is also true for more mobile species is also unknown. Other considerations such as changes in fish growth, consumption, juvenile survival, or production due to area closures are also effectively unexplored. It is also possible that the area closures could have caused indirect or higher-order changes in predator–prey dynamics, competitor dominance, and other inter-specific interactions. Whether changes in these rates actually occurred and, if so, whether the changes resulted in population-level effects on economically valuable species is again unknown.

How has the entire benthic community been (or not been) impacted by these closures? Further analysis of the benthic infaunal community data, videos of benthic macrofauna and habitat, fish stomach (diet composition), and benthic chemistry are underway and should provide further insight. Yet of the 20 infaunal macroinvertebrate species we analyzed, only one small polychaete, *Exogone hebes,* showed significantly greater abundance and biomass inside versus outside the closed areas. This would imply a surprisingly limited effect of an area closure on the benthic infaunal community. A number of other species showed significant differences among habitat zones at each closed area. These results concur with Jennings et al. (2002), who showed that water depth and sediment type strongly influenced polychaete populations but that the intensity of trawling had little effect.

The main result from this work is that there were few differences in the nekton and benthic communities due to the closed areas. A secondary point is that habitat type (i.e., benthic substrate in terms of grain size, sediment dynamics, rugosity, and related seabed features) appears to be a relatively more important consideration than inside–outside a closed area. In fact, this was a common contributing factor for many of the, albeit few, significant results we observed. Several studies (e.g., Auster et al. 1996; Collie et al. 2000b; Jennings et al. 2002) have shown that various biotic communities and substrate types are differentially affected by fishing gears and that habitat types can be the strongest determinant of benthic community composition and recovery. In a sense, it was surprising that habitat type was relatively more important than the effects of opened and closed areas. Yet in another sense, given the dynamic nature of many of the types of habitats we examined, coupled with the relatively rapid recovery times reported for the benthic communities associated with those habitats, such a conclusion is sensible. The relative importance of habitat type when compared to the importance of area opening and closures has some intriguing implications for the future development of MPAs and marine reserves. We do not mean to imply that area closures are not important but rather: (1) that this study did not show the obvious changes between inside

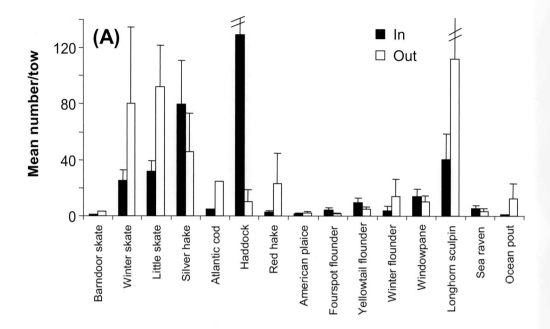

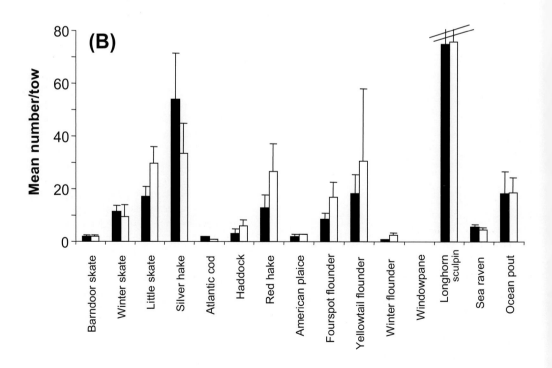

Figure 15. Abundance (number) per tow for major macrofauna species in each zone of CAII, contrasting inside (solid bars) versus outside (open bars) (A) Zone 1 and (B) Zone 2. Error bars represent ±1 SE. Note different scales on the axes across the zones.

Table 6. Results from ANOVA assessing the factors of Inside versus Outside (IO), Zonation (Z), and the Interaction between these factors (Int) for each of the major benthic species in Closed Area II. The analysis was done separately for both abundance and biomass per grab. If a factor was significantly different, an I or O for IO or the number of the zone for Z is given in parenthesis to show where the highest values for each variable were observed. See Table 2 for explanation of significance values.

	Abundance			Biomass		
	IO	Z	Int	IO	Z	Int
Exogone hebes	** (I)			** (I)		
Spiophanes bombyx						
Arctica islandica				?* (I)		
Cirolana polita						
Byblis serrata						
Corophium crassicorne						
Erichthonius rubricornis		*** (2)			*** (2)	
Unciola irrorata						
Aeginina longicornis		*** (2)			*** (2)	
Ophiura robusta		* (2)			** (2)	

and outside closed areas as found in other area closure–MPA studies and models for several possible reasons we discuss above; and (2) the type of habitat to be closed can be an important determinant influencing the response of demersal and benthic communities to an area closure.

Acknowledgments

We thank the officers and crew and of the National Oceanic and Atmospheric Administration (NOAA) FRV *Albatross IV* for their diligence and efforts to complete the two cruises. We are also indebted to the scientific and technical staff of the NMFS, U.S. Geological Survey (USGS), National Undersea Research Center (NURC) at the University of Connecticut who have participated in these cruises and processed the biological samples and data. We also thank P. Renaud for processing many of the benthic samples. We thank T. Noji, W. Gabriel, and anonymous reviewers for their comments on earlier versions of the manuscript. This work was supported by and represents a collaborative partnership of NOAA Fisheries, USGS, and NURC at University of Connecticut.

References

Agardy, T. S. 1997. Marine protected areas and ocean conservation. Academic Press, London.

Alder, J. 1996. Have tropical marine protected areas worked? An initial analysis of their success. Coastal Management 24:97–114.

Allison, G. W., J. Lubchenco, and M. H. Carr. 1998. Marine reserves are necessary but not sufficient for marine conservation. Ecological Applications 8:S79–S92.

Auster, P. J., and R. W. Langton. 1999. The effects of fishing on fish habitat. Pages 150–187 *in* L. Benaka, editor. Fish habitat: essential fish habitat and rehabilitation. American Fisheries Society, Symposium 22, Bethesda, Maryland.

Auster, P. J., R. J. Malatesta, R. W. Langton, L. Watling, P. C. Valentine, C. L. S. Donaldson, E. W. Langton, A. N. Shepard, and I. G. Babb. 1996. The impacts of mobile fishing gear on seafloor habitats in the Gulf of Maine (northwest Atlantic): implications for conservation of fish populations. Reviews in Fisheries Science 4:185–202.

Auster, P. J., and N. L. Shackell. 2000. Marine protected areas for the temperate and boreal northwest Atlantic: the potential for sustainable fisheries and conservation of biodiversity. Northeastern Naturalist 7:419–434.

Babcock, R. C., S. Kelly, N. T. Shears, J. W. Walker, and T. J. Willis. 1999. Changes in community structure in temperate marine reserves. Marine Ecology Progress Series 189:125–134.

Benaka, L., editor. 1999. Fish habitat: essential fish habitat and rehabilitation. American Fisheries Society, Symposium 22, Bethesda, Maryland.

Brown, R.W., D. Sheehan, and B. Figuerido. Response of cod and haddock populations to area closures on Georges Bank. International Council for the Exploration of the Sea, CM 1998, Copenhagen.

Butman, B., R. C. Beardsley, R. Limeburner, B. Magnell, D. Frye, J. A. Vermersh, W. R. Wright, R. Schlitz, and M. A. Noble. 1982. Recent observation of the mean circulation on Georges Bank. Journal of Physical Oceanography 12:569–591.

Collie, J. S., G. A. Escanero, and P. C. Valentine. 1997. Effects of bottom fishing on the benthic megafauna of

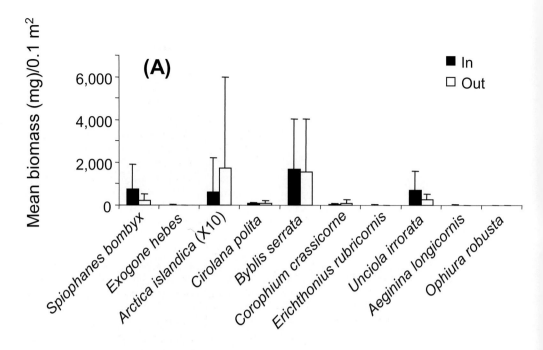
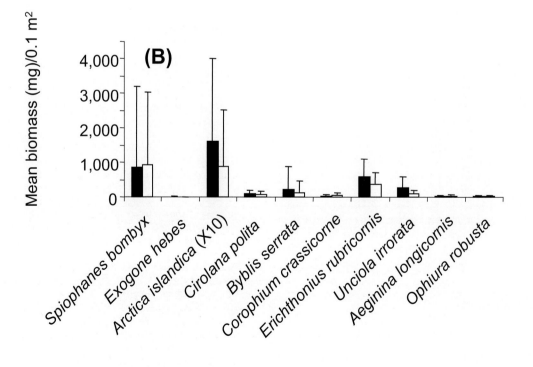

Figure 16. Biomass per grab of major infaunal benthic species in each zone of CAII, contrasting inside (solid bars) versus outside (open bars) (A) Zone 1 and (B) Zone 2. Error bars represent ±1 SE. Note different scales on the axes across the zones.

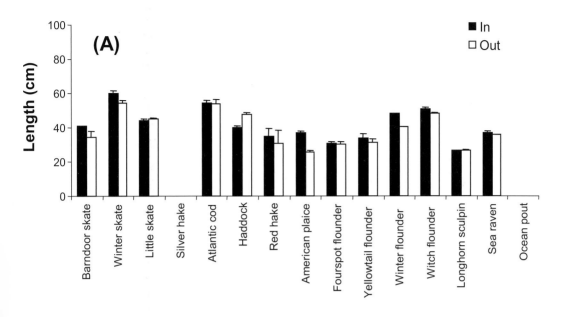
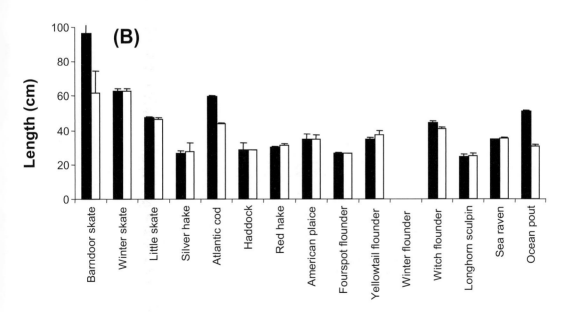

Figure 17. Mean lengths for major fish species in each zone of CAII, contrasting inside (solid bars) versus outside (open bars) (A) Zone 1 and (B) Zone 2. Error bars represent ±1 SE.

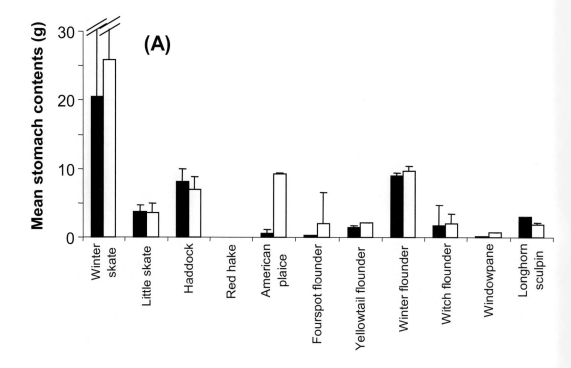
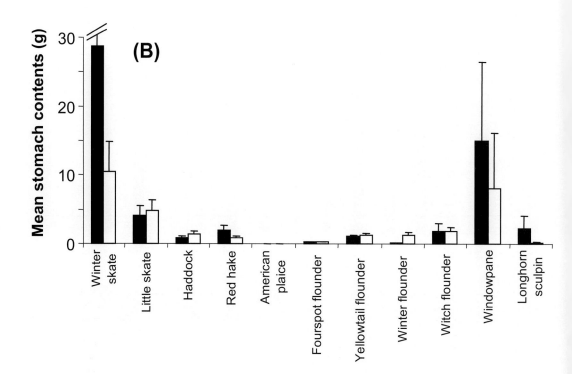

Figure 18. Mean stomach contents for major fish species in each zone of CAII, contrasting inside (solid bars) versus outside (open bars) (A) Zone 1 and (B) Zone 2. Error bars represent ±1 SE.

Georges Bank. Marine Ecology Progress Series 155:159–172.

Collie, J. S., G. A. Escanero, and P. C. Valentine. 2000a. Photographic evaluation of the impacts of bottom fishing on benthic epifauna. ICES Journal of Marine Science 57:987–1001.

Collie, J. S., S. J. Hall, M. J. Kaiser, and I. R. Poiner. 2000b. A quantitative analysis of impacts on shelf-sea benthos. Journal of Animal Ecology 69:785–798.

Conover, D. O., J. Travis, and F. C. Coleman. 2000. Essential fish habitat and marine reserves: an introduction to the second Mote symposium in fisheries ecology. Bulletin of Marine Science 66:527–534.

Cote, I. M., I. Mosqueira, and J. D. Reynolds. 2001. Effects of marine reserve characteristics on the protection of fish populations: a meta-analysis. Journal of Fish Biology 59:178–189.

Dorsey, E. M., and J. Pederson. 1998. Summary of Discussion and Recommendations from the Conference. pp. 140–143 in E. M. Dorsey and J. Pederson, editors. Effects of fishing gear on the sea floor of New England. Conservation Law Foundation, Boston Massachusetts.

Fisher, J. A. D., and K. T. Frank. 2002. Changes in finfish community structure associated with an offshore fishery closed area on the Scotian Shelf. Marine Ecology Progress Series 240:249–265.

Fogarty, M. J., and S. A. Murawski. 1998. Large-scale disturbance and the structure of marine systems: fishery impacts on Georges Bank. Ecological Applications 8(S1):S6–S22.

Jennings, S., and M. J. Kaiser. 1998. The effects of fishing on marine ecosystems. Advance in Marine Biology 34:201–352.

Jennings, S., M. D. Nicholson, T. A. Dinmore, and J. E. Lancaster. 2002. Effects of chronic trawling disturbance on the production of infaunal communities. Marine Ecology Progress Series 243:251–260.

Jennings, S., and N. V. C. Polunin. 1996. Effects of fishing effort and catch rate upon the structure and biomass of Fijian reef fisheries. Journal of Fish Biology 46:28–46.

Kaiser, M. J. 2003. Detecting the effects of fishing on seabed community diversity: importance of scale and sample size. Conservation Biology 17:512–520.

Kaiser, M. J., J. S. Collie, S. J. Hall, S. Jennings, and I. R. Poiner. 2002. Modification of marine habitats by trawling activities: prognosis and solutions. Fish and Fisheries 3:114–136.

Kaiser, M. J., and S. J. de Groot. 2000. Effects of fishing on non-target species and habitats. Blackwell Scientific Publications, Oxford, UK.

Lauck, T., C. W. Clark, M. Mangel, and G. R. Munro. 1998. Implementing the precautionary principle in fisheries management through marine reserves. Ecological Applications 8:S72–S78.

Lindholm, J. B., P. J. Auster, M. Ruth, and L. Kaufman. 2001. Juvenile fish responses to variations in seafloor habitats: modeling the effects of fishing and implications for the design of marine protected areas. Conservation Biology 15:424–437.

Lindholm, J., P. Auster, and P. Valentine. 2004. Role of a large marine protected area for conserving landscape attributes of sand habitats on Georges Bank (Northwest Atlantic). Marine Ecology Progress Series 269:61–68.

Lough, R. G., P. C. Valentine, D. C. Potter, P. J. Auditore, G. R. Bolz, J. D. Neilson, and R. I. Perry. 1989. Ecology and distribution of juvenile cod and haddock in relation to sediment type and bottom currents on eastern Georges Bank. Marine Ecology Progress Series 56:1–12.

McLanahan, T. R., and S. Mangi. 2000. Spillover of exploitable fishes from a marine park and its effect on the adjacent fishery. Ecological Applications 10:1792–1805.

McManus, J. W. 1998. Marine reserves and biodiversity: toward 20% by 2020. ICLARM Conference Proceedings 57:25–29.

Mosquera, I., I. M. Cote, S. Jennings, and J. D. Reynolds. 2000. Conservation benefits of marine reserves for fish populations. Animal Conservation 3:321–332.

Murawski, S. A., R. Brown, H.-L. Lai, P. J. Rago, and L. Hendrickson. 2000. Large-scale closed areas as a fishery-management tool in temperate marine systems: the Georges Bank experience. Bulletin of Marine Science 66:775–798.

Murawski, S. A., J.-J. Maguire, R. K. Mayo, and F. M. Serchuk. 1997. Groundfish stocks and the fishing industry. Pages 27–70 in J. Boreman, B. S. Nakashima, J. A. Wilson, and R. L. Kendall, editors. Northwest Atlantic groundfish: perspectives on a fishery collapse. American Fisheries Society, Bethesda, Maryland.

NEFSC (Northeast Fisheries Science Center). 2001. Report of the 32nd Northeast Regional Stock Assessment Workshop (32nd SAW): Stock Assessment Review Committee (SARC) consensus summary of assessments. NEFSC Center Reference Document 01-05:58–190.

NRC (National Research Council). 2001. Marine protected areas: tools for sustaining ocean ecosystems. National Academy Press, Washington, D.C.

Roberts, C. M. 1998. Sources, sinks, and the design of marine reserve networks. Fisheries 24(7):16–19.

Roberts, C. M., J. A. Bohnsack, F. Gell, J. P. Hawkins, and R. Goodridge. 2001. Effects of marine reserves on adjacent fisheries. Science 294:1920–1923.

Russ, G. R., and A. C. Alcala. 1996. Do marine reserves export adult fish biomass? Evidence from Apo Island, Central Philippines. Marine Ecology Progress Series 132:1–9.

Smith, W. G., and W. W. Morse. 1985. Retention of larval haddock *Melanogrammus aeglefinus* in the Georges Bank region, a gyre-influenced spawning area. Marine Ecology Progress Series 24:1–13.

Strathamm, R. R., T. P. Hughes, A. M. Kuris, K. C. Lindeman, S. G. Morgan, J. M. Pandolfi, and R. R. Warner. 2002. Evolution of local recruitment and its consequences for marine populations. Bulletin of Marine Science 70(Supplement):377–396.

Theroux, R. B., and R. L. Wigley. 1998. Quantitative composition and distribution of the macrobenthic invertebrate fauna of the continental shelf ecosystems of the northeastern United States. NOAA Technical Report NMFS 140.

Thrush, S. F., J. E. Hewitt, V. J. Cummings, P. K. Dayton, M. Cryer, S. J. Turner, G. A. Funnell, R. G. Budd, C. J. Milburn, and M. R. Wilkinson. 1998. Disturbance of the marine benthic habitat by commercial fishing: impacts at the scale of the fishery. Ecological Applications 8:866–879.

Twichell, D. C, B. Butman, and R. S. Lewis. 1987. Shallow structure, surficial geology, and the processes currently shaping the bank. Pages 31–37 *in* R. H. Backus and D. W. Bourne, editors. Georges Bank. Massachusetts Institute of Technology Press, Cambridge.

Valentine, P. C., and R. G. Lough. 1991. The sea floor environment and the fishery of eastern Georges Bank - the influence of geologic and oceanographic environmental factors on the abundance and distribution of fisheries resources of the northeastern United States continental shelf. U.S. Geological Survey Open-File Report 91-39, Reston, Virginia.

Valentine, P. C., E. W. Strom, R. G. Lough, and C. L. Brown. 1993. Maps showing the sedimentary environment of eastern Georges Bank. U. S. Geological Survey Miscellaneous Investigations Series, Map I-2279-B, scale 1:250,000, 1 sheet, Woods Hole, Massachusetts.

Ward, T. J., D. Heinemann, and N. Evans. 2001. The role of marine reserves as fisheries management tools: a review of concepts, evidence and international experience. Bureau of Resource Science Kingston, ACT, Australia.

Effects of Fisheries on Deepwater Gorgonian Corals in the Northeast Channel, Nova Scotia

PAL B. MORTENSEN,[1] LENE BUHL-MORTENSEN,[2] AND DONALD C. GORDON, JR[3]

Department of Fisheries and Oceans, Bedford Institute of Oceanography, Post Office Box 1006, Dartmouth, Nova Scotia B2Y 4A2, Canada

GORDON B. J. FADER[4]

Natural Resources Canada, Bedford Institute of Oceanography, Post Office Box 1006, Dartmouth, Nova Scotia B2Y 4A2, Canada

DAVID L. MCKEOWN[5] AND DEREK G. FENTON[6]

Department of Fisheries and Oceans, Bedford Institute of Oceanography, Post Office Box 1006, Dartmouth, Nova Scotia B2Y 4A2, Canada

Abstract. Video surveys in the Northeast Channel, Nova Scotia, between Georges Bank and Browns Bank in the Northwest Atlantic, were conducted to determine the distribution of deepwater corals and extent of damage from bottom fishing activities. Three gorgonian species, *Paragorgia arborea*, *Primnoa resedaeformis*, and *Acanthogorgia armata*, were observed between 190-m and 500-m depths (the maximum working depths of the video equipment). *Paragorgia arborea* and *Primnoa resedaeformis* occurred along 21 and 35 of the 52 transects, respectively, whereas *Acanthogorgia armata* was observed on only 4 transects. The colonies grow on cobbles and boulders in glacial deposits and often have both mobile and sessile associated species, including fishes. The Northeast Channel is an important fishing area targeted by otter trawling and longline and gill-net fleets. Signs of fishing impact were visible as broken live corals, tilted corals, and scattered skeletons. Lost fishing gear was often observed entangled in corals. Broken or tilted corals were observed along 29% of the transects. In total, 4% of the coral colonies observed were impacted. *Paragorgia arborea* seems to be more susceptible to breakage from encounters with fishing gear than *Primnoa resedaeformis*. This is most likely due to its larger size and less flexible skeleton. Using the results of this study, the Department of Fisheries and Oceans established a 424-km^2 coral conservation area in 2002 to protect corals from further damage from bottom fishing activity.

Introduction

Deepwater gorgonian corals are found in oceans around the world most commonly at depths of 200–1,500 m (Broch 1935; Madsen 1944; Carlgren 1945; Strømgren 1970; Genin et al. 1986; Mistri and Ceccherelli 1994) and are considered to be important components of deepwater ecosystems (Rogers 1999; Krieger and Wing 2002). It has been documented that some scleractinian corals form deepwater reefs with a highly diverse fauna of associated species (Dons 1944; Jensen and Frederiksen 1992; Mortensen et al. 1995). However, less is known about the structure and ecology of deepwater gorgonians (Heifetz 2002). In general, our knowledge about the distribution, habitat, environmental requirements, age composition, and many other biological aspects of deepwater corals, while growing rapidly, is very limited.

There is an increasing global awareness that human activities, in particular fishing and oil and gas exploration, in water deeper than 200 m may damage corals and other organisms relying on this habitat (Probert et al. 1997; Fosså et al. 2002; Reed 2002). Stable biogenic habitats are more susceptible to physical disturbance and have slower recovery rates (Collie et al. 2000; National Research Council 2002). In gen-

---
[1] Corresponding author: MortensenP@mar.dfo-mpo.gc.ca
[2] E-mail: MortensenL@mar.dfo-mpo.gc.ca
[3] E-mail: GordonD@mar.dfo-mpo.gc.ca
[4] E-mail: GFader@nrcan.gc.ca
[5] E-mail: McKeownD@mar.dfo-mpo.gc.ca
[6] E-mail: FentonD@mar.dfo-mpo.gc.ca

eral, intense bottom fishing has a strong negative effect on colonial and larger sessile benthic organisms (Sainsbury et al. 1997). Deepwater corals seem to be particularly vulnerable to encounters with fishing gear because of their arborescent growth form, and there are several documented cases that bottom trawling is extremely destructive (Koslow et al. 2001; Krieger 2001; Fosså et al. 2002). Recovery of deepwater corals from damage can be expected to be slow because of their low growth rates assumed to be on the order of 1–2 cm/year (Andrews et al. 2002; Risk et al. 2002). Studies on *Primnoa resedaeformis* indicate that this coral may reach an age of over 300 years (Risk et al. 2002), although most colonies are younger (Andrews et al. 2002).

Deepwater gorgonians occur off Atlantic Canada on the continental slope, in submarine canyons, and in channels between offshore banks (Verrill 1922; Deichman 1936; Breeze et al. 1997; MacIsaac et al. 2001). They are particularly abundant in the Northeast Channel, which crosses the outer continental shelf between Georges Bank and Browns Bank (Figure 1A). It is the deepwater connection between the North Atlantic Ocean and the Gulf of Maine and has a sill depth of about 230 m.

The morphology of the Northeast Channel (Figure 1B) can be described as a deepwater shelf break system with two major reentrants. Beyond the shelf break, deeper than 400 m, the seabed drops away steeply in a series of submarine canyons with canyon heads, internal ridges, and steep-walled, incised valleys. The seabed in the Northeast Channel is composed of thick till (unstratified glacial deposits with a mixture of gravel, sand, silt, and clay), part of the Scotian Shelf Drift Formation (Lawrence et al. 1985; Fader et al. 1988). Three ridged areas at the shelf break represent terminal moraines (till features formed at maximum extent of the glaciers; Figure 1C). The northernmost ridge is an extension of Browns Bank, the central ridge is known as Romeys Peak, and the southern ridge is called The Rips. Linear iceberg furrows (or plowmarks) and circular iceberg pits flanked by berms of boulders are widespread (Figure 1C). Sand occurs in some areas where the surface is covered with linear bedforms aligned northwest–southeast (Figure 1C), indicating transport by strong currents.

At 300–400 m, salinity and temperature range from 34.87 to 35.05 and 5.74–7.64°C, respectively (Petrie and Dean-Moore 1996). Strong semidiurnal tidal currents are generally dominant in the channel (Ramp et al. 1985; Smith and Showing 1991). Sixteen m off the bottom, the maximum current velocity at the sill is between 40 and 50 cm/s (Ramp et al. 1985). A residual inflow occurs on the northeastern side, near Browns Bank, while a residual outflow

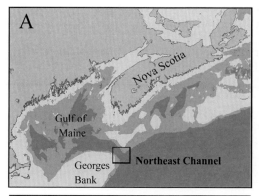

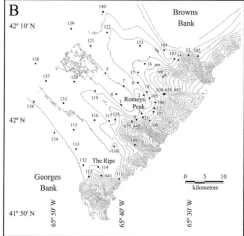

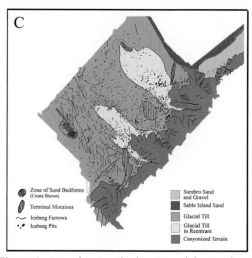

Figure 1. Map showing the location of the Northeast Channel (A). The topography of the Northeast Channel, based on multibeam bathymetry, and location of the video transects are also shown (B). Finally, the geological interpretation of the multibeam bathymetry showing the main seabed features is shown (C). Note the location of the terminal moraines as ridges at the mouth of the channel.

occurs on the southwestern side, near Georges Bank (Ramp et al. 1985).

The Northeast Channel and the two neighboring banks have supported an active fishery since the early 1800s. The early groundfish fishery was based on hook-and-line vessels out of both American and Nova Scotian ports. The fishery sought a range of groundfish species, with Atlantic cod *Gadus morhua* and haddock *Melanogrammus aeglefinus* fished on the banks and Atlantic halibut *Hippoglossus hippoglossus* often targeted in the channel itself. Currently, the groundfish resources in this area are exploited by commercial fishing vessels of a wide variety of sizes and types, ranging from small fixed gear (longline and gill-net) boats to large offshore trawlers. All these gear types have contact with the seabed, but only trawls are so-called active gear, which is moved along the seafloor. When set, longlines and gill nets are anchored at one or both ends, and there are often weights attached at intervals to keep them on the bottom. During hauling, these gears may move on the seafloor and hit obstacles such as boulders or corals. Gorgonians are commonly captured by the different bottom fishing gears. In addition to groundfish, there are several existing and emerging invertebrate fisheries, including American lobster *Homarus americanus* and Jonah crab *Cancer borealis*, within and surrounding the Northeast Channel.

In order to manage fisheries in a way that would minimize damage to deepwater corals, it is crucial not only to know their distribution and preferred habitat but also to quantify the effect of fisheries. Therefore, we conducted detailed video surveys of deepwater gorgonians in the Northeast Channel in 2000 and 2001. This paper describes the distribution of deepwater gorgonians, their habitat, the observed damage from bottom fishing gear, and the distribution of fishing effort by different gear sectors. It also describes a coral conservation area that was created by the Department of Fisheries and Oceans (DFO) in 2002 using the results from this study.

Methods

Selection of Survey Sites

The study was carried out in the outer part of the Northeast Channel. Survey sites were limited to less than 500 m because of depth restrictions with the video equipment. In total, 48 sites (Figure 2) covering a depth range of 183–498 m were selected for deployment of the gear. A detailed bathymetric map (Figure 1B) from recent mapping with multibeam echosounder by the Canadian Hydrographic Service was used as a guide. In addition, interpretation of side-scan sonar records was used to select some of the sites. Twenty-three sites were selected within areas with seabed topography and substrate most likely to support corals (i.e., at noses and ridges along the shelf edge at depths in the range of 300–500 m). Previous studies (e.g., Genin et al. 1986; Mortensen et al. 2001) have shown that rugged seafloor often supports deepwater coral communities. The remainder of the sites were selected to fill in the gaps in the study area in order to describe the geographical distribution and the upper depth limits for the corals.

Inspection of the Seabed

Video records were collected along 52 transects with *ROPOS*, a remotely operated vehicle (ROV) operated by the Canadian Scientific Submersible Facility, and Campod, a tethered video and still photo camera system (Gordon et al. 2000). Both *ROPOS* and Campod were equipped with an ultra-short baseline navigation system (ORE Trackpoint II) to provide detailed navigation data of their track along the seabed. The transects varied in length between 301 and 1,683 m with a total length of 32 km. Seven transects were video recorded with *ROPOS* (Transects 636–642, Figure 1B) in August 2001 during a cruise on the C.C.G.S. *Martha L. Black*. The transects were carried out at three locations and covered depths between 331 and 498 m. The plan for the dives was to move along 1.5-km transects on selected ridges. However, the strong tidal currents restricted the ROV to an area less than 300 m × 300 m on each dive. The Campod was deployed at 38 sites (Transects 1–141, Figure 1B) within the study area during two cruises on the C.C.G.S. *Hudson* in June 2000 and September–October 2001. Campod was deployed while the ship was drifting with the current and was kept close (1–2 m) to the seabed for at least 15 min on each transect.

Video Analysis

During subsequent analysis, video records were divided into sequences. These video sequences were mainly of 30-s duration but were made shorter when abrupt changes in the habitat occurred. Geographical positions and depth were registered at the start and end of each sequence. In total, the 52 video recorded transects were divided into 1,751 sequences for analysis. Each video sequence covered a distance of between 5 and 25 m (average = 14 m), estimated from the navigation data. Within these sequences, corals were identified and counted, and the percent cover of substrate types (i.e., sand, pebble, cobble, and boulder) was estimated following the size-classes as defined by the

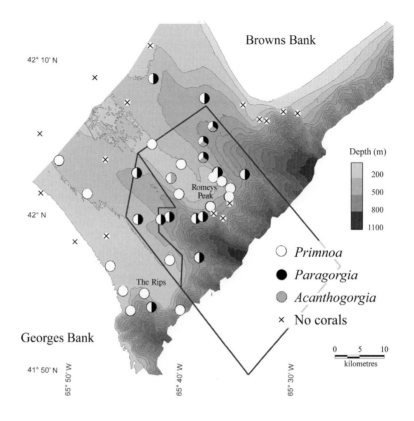

Figure 2. Map showing the general bathymetry, the distribution of corals, and the boundaries of the coral conservation area in the Northeast Channel. The line inside the conservation area divides the larger restricted bottom fisheries zone from the smaller limited bottom fisheries zone.

Wentworth scale (Wentworth 1922). Corals were classified as being: (1) intact, (2) broken, (3) tilted, or (4) dead (skeleton). Broken and tilted corals were regarded as damaged. It was not possible to determine whether dead coral skeletons were the result of natural mortality or damage from fishing gear. Colonies with different associated sessile epibionts were counted, and the percent coverage was estimated when the video cameras came close enough to provide images that could be assessed.

Analysis of Groundfish Landings

In order to understand the recent patterns and intensity of bottom fishing in the Northeast Channel area (>200 m) a review of recent fisheries activities in the area was conducted. The primary source of information was commercial landings data, including results from logbooks and dockside monitoring programs that are part of industry reporting requirements by the DFO. The data were aggregated within a grid-net of 1 latitute–longitude minute over a period of 5 years (1997–2001) to represent overall fishing effort in the region. Therefore, they do not necessarily represent the exact location of individual sets, tracks, or hauls of gear.

Results

General Distribution of Gorgonian Corals

Primnoa resedaeformis was observed along 35 of the 52 transects, while *Paragorgia arborea* was present along 21 transects (Figure 2; Table 1). *Primnoa resedaeformis* was also more abundant than *Paragorgia arborea*, indicated by its much higher average frequency of occurrence (Table 1). *Acanthogorgia armata* was observed along only four transects (6, 8, 16, and 17). The corals were more common along the shelf break and slope at the mouth of the channel than on the shelf. The most frequent occurrences of *Primnoa resedaeformis* and *Paragorgia arborea* were on the northern side of Romeys Peak and in the valley between The Rips and Romeys Peak.

The shallowest observations of *Primnoa resedaeformis* and *Paragorgia arborea* occurred at 190 and 235 m, respectively. *Acanthogorgia armata* was observed at depths between 231 and 236 m. There was a positive significant correlation ($P < 0.05$) between frequency of corals and depth. This correlation was stronger for *Paragorgia arborea* than for *Primnoa resedaeformis* ($R^2 = 0.31$ and 0.28, respectively). The highest densities were observed between 410 and 490 m. The range of both species is expected to extend well below 500 m, the operational depth limit for the video platforms used.

Description of Coral Habitat

Except for the two easternmost video transects (13 and 102), all transects are within areas characterized as till (including terminal moraines; Figure 1C) where cobble and boulder offer firm attachment sites for corals. The distribution of corals along the transects was patchy and restricted to areas with cobbles or boulders. Large colonies of *Paragorgia arborea* were observed almost exclusively on boulders, whereas smaller colonies (<25 cm) were often observed on cobbles as well. *Primnoa resedaeformis* occurred on both cobbles and boulders while *Acanthogorgia armata* was found only on cobbles. No corals were observed attached to sand or pebbles. Boulders were commonly observed along the video transects (average frequency of occurrence = 0.65; Table 1) but seldom with a seabed coverage of more than 50%. On average, for all transects, the percentage cover of cobble and boulder was 21% and 9%, respectively. However, the correlation between the frequency of corals and percentage cover of any of these two substrate types was weak, with the highest R-value for boulders ($R = 0.33$, $P < 0.05$). Forty percent of the video sequences had boulders but no corals (Table 2).

The average density for transects with corals was 0.6 colonies per 100 m^2 for *Paragorgia arborea* and 4.8 colonies per 100 m^2 for *Primnoa resedaeformis*. The highest local densities in stands were 49 and 104 colonies per 100 m^2 for *Paragorgia arborea* and *Primnoa resedaeformis*, respectively. The largest colony of *Paragorgia arborea* observed was approximately 1.70 m tall, whereas the maximum height observed for *Primnoa resedaeformis* was only 80 cm.

A parasitic colonial zoanthid anemone was the most common sessile epibiont observed on *Primnoa resedaeformis*. On average, this anemone covered approximately 60% of the surface of the 28 infected colonies, but one-third of the colonies were entirely covered. Hydroids were also frequently observed on *Primnoa resedaeformis*. Redfish *Sebastes* spp. commonly occurred in the vicinity of corals but were also observed in areas without corals. However, redfish were almost four times as common in video sequences with corals than in sequences with boulders and no corals (Table 2).

Observations of Damaged Corals

Damaged corals (broken or tilted) were observed on 29% of the transects and were not concentrated in any particular area (Figure 3; Table 1). In total, 4% of the coral colonies observed were damaged. The highest frequency of video sequences with damaged *Paragorgia arborea* (15%) was observed at Transect 117 in the valley between The Rips and Romeys Peak (Table 1). There was a positive correlation between the frequency of damaged coral colonies and the frequency of intact colonies for both *Primnoa resedaeformis* and *Paragorgia arborea*, with R-values of 0.67 and 0.70, respectively ($P < 0.001$). Lost longlines were observed loose on the seabed or entangled in corals on 37% of the transects (Table 1). Tracks on the seabed, either from longline anchors or parts of otter trawl gear, were present along three transects, while lost gill nets were observed along two transects. With one exception, longlines were only found on transects where corals were present. Photos of corals impacted by fishing gear are shown in Figure 4. In one case, a large, broken, partly-live branch of *Paragorgia arborea* was found close to the colony where it belonged. A longline colonized with hydroids and anemones was entangled in the part of the colony still standing. There was a significant positive relationship ($R = 0.61$, $P < 0.005$, $N = 46$) between the frequency of encounters with lost longlines and the frequency of broken, live *Paragorgia arborea*. The frequency of tilted corals (both *Primnoa resedaeformis* and *Paragorgia arborea*) was weakly correlated with the frequency of longline encounters. A higher percentage of *Paragorgia arborea* was damaged (broken or tilted) compared to *Primnoa resedaeformis* (7.9% versus 3.4%; Figure 5). The occurrence of coral skeletons was also much higher for *Paragorgia arborea* (20.4% versus 1.7%). The parasitic anemone was more common on damaged colonies of *Primnoa resedaeformis* than on intact ones.

Distribution of Fishing Effort

Most of the recent (1997–2001) groundfish catches by the longline, gill-net, and otter trawling fleets in the Northeast Channel are concentrated on the Georges Bank side (Figure 6). The level of activity in the Romeys Peak area and deeper portions of the channel are relatively low in comparison to adjacent areas. Within the channel, the preferred fishing grounds

Table 1. Brief description of video transects (Tr) in the Northeast Channel, including number of video sequences (Seq), frequency of occurrence of boulders, colonies of *Paragorgia arborea* and *Primnoa resedaeformis* with different statuses (intact, I; broken, B; tilted, T; and dead, D), epibionts (Epi) *Sebastes*, lost fishing gear (net and longlines) and trawl tracks. A hyphen indicates missing data.

Tr	Seq	Boulders	Paragorgia arborea					Primnoa resedaeformis					Epi	Sebastes	Net	Track	Longlines
			I	B	T	D	% damaged	I	B	T	D	% damaged					
1	83	0.64	0	0	0	0		0.33	0.01	0.01	0	1	0	-	0	0.01	0.02
2	76	0.58	0	0	0	0		0.08	0	0	0	0	0	-	0	0	0.01
3	50	0.42	0	0	0	0.02	0	0.02	0.02	0	0	33	0	-	0	0	0
4	48	0.60	0	0	0	0		0.23	0	0	0.02	2	0	0	0	0	0
5	64	0.63	0	0	0	0		0.16	0	0.06	0.08	24	0	0.03	0	0	0.03
6	51	0.55	0	0	0	0		0.18	0	0.02	0		0	-	0	0	0
7	73	0.75	0.04	0.01	0	0.01	25	0.23	0.03	0	0	3	0	0.29	0	0	0.01
8	55	0.80	0.02	0.05	0	0	0	0.27	0.04	0	0	4	0	0.27	0	0	0
9	22	1.00	0	0	0	0		0	0	0	0		0	0.27	0	0	0
10	48	0.77	0	0	0	0		0	0.02	0.05	0	100	0	0.31	0	0	0
11	32	0.13	0	0	0	0		0	0	0	0		0	0.69	0	0	0
12	15	0.20	0	0	0	0		0	0	0	0		0	0.07	0	0	0
13	38	0.82	0.34	0.03	0.08	0.14	21	0.65	0	0.24	0.08	8	0	0.58	0	0	0
14	40	0.43	0.15	0	0.03	0.05	10	0.15	0	0.05	0.03	9	0	0.38	0	0	0
15	50	0.84	0	0.02	0	0	100	0.42	0.04	0	0.02	2	0.04	0.04	0	0	0.08
16	81	0.00	0	0	0	0		0	0	0	0		0	0.03	0	0	0.00
17	39	0.87	0	0	0	0		0	0	0	0		0	0.03	0	0	0.05
18	32	0.66	0	0	0	0		0.12	0	0	0	0	0	0	0	0	0
19	26	0.81	0	0	0	0	0	0.42	0	0	0	0	0.04	0.19	0	0	0.08
20	19	0.79	0.05	0	0	0		0	0	0	0		0	0.21	0	0	0
21	40	0.50	0	0	0	0		0	0	0	0		0	0.10	0	0	0
22	23	0.13	0	0	0	0		0	0	0	0		0	0.09	0	0	0
23	97	0.69	0	0	0	0.05	0	0.14	0	0.05	0	0	0	0.08	0	0	0.02
24	29	0.31	0	0	0	0		0.21	0	0	0	0	0	0	0	0	0
25	27	0.74	0	0	0	0		0.07	0	0	0	0	0	0.41	0	0	0
26	25	0.80	0	0	0	0		0.68	0	0	0	0	0	0.04	0	0	0.04
27	35	0.83	0.03	0	0	0	0	0.26	0.03	0.03	0.09	9	0	0.20	0.03	0	0
28	19	0.47	0	0	0	0		0	0	0	0		0	0	0	0	0
29	38	0.95	0.11	0	0.05	0.05	20	0.45	0	0.03	0.05	2	0	0.11	0	0	0.03
30	34	0.85	0.28	0.03	0.15	0.03	25	0.56	0	0	0.09	0	0	0.71	0	0	0.03
31	45	0.91	0.67	0.07	0.09	0.37	13	0.39	0	0.04	0.02	3	0	0.82	0	0	0.02
32	43	0.77	0.05	0	0	0.16	0	0.40	0	0	0.12	0	0.07	0.42	0	0	0.02
33	23	0.70	0	0	0	0		0	0	0	0		0	-	-	-	-
34	37	0.70	0	0	0	0		0	0	0	0.03		0	0	0	0	0
35	35	0.46	0.03	0	0	0	0	0.03	0	0	0	0	0	0	0	0	0
36	56	0.82	0.02	0	0.02	0	50	0.07	0.05	0	0	30	0	0.04	0	0	0

Table 1. Continued.

Tr	Seq	Boulders	Paragorgia arborea						Primnoa resedaeformis					Epi	Sebastes	Net	Track	Longlines
			I	B	T	D	% damaged	I	B	T	D	% damaged						
132	15	0.67	0	0	0	0		0.13	0	0	0	0	0	0	0	0	0
133	22	0.77	0	0	0	0		0.09	0	0	0	0	0	0	0	0	0
134	23	0.74	0	0	0	0		0	0	0	0		0	0	0	0	0
135	16	0.38	0	0	0	0		0	0	0	0		0	0	0	0	0
136	31	0.23	0	0	0	0		0	0	0	0		0	0	0	0	0
137	44	0.95	0	0	0	0		0	0	0	0	0	0	0	0	0	0
138	22	0.49	0	0	0	0		0	0	0	0		0	-	-	-	-
139	24	0.48	0	0	0	0		0	0	0	0		0	-	-	-	-
140	25	0.41	0	0	0	0		0	0	0	0		0	-	-	-	-
636	30	0.83	0.50	0	0	0.07	0	0.80	0	0	0	0	0.17	0.47	0.03	0	0.03
637	69	0.80	0.77	0	0.03	0.10	2	0.90	0	0	0	0	0.03	0.57	0.01	0.01	0.07
638	55	0.98	0.24	0.02	0	0.06	6	0.74	0.04	0.06	0.04	3	0.15	0.56	0.02	0.02	0.16
639	18	0.56	0	0.11	0.06	0.17	50	0.33	0	0.06	0	8	0.44	0.44	0	0	0.22
640	25	0.88	0.13	0.04	0	0.22	40	0.61	0	0	0.09	0	0.44	0.60	0.04	0.04	0.12
641	27	0.89	0.04	0.04	0	0.04	50	0.36	0.11	0.07	0	12	0.63	0.74	0	0.15	0.07
642	32	0.78	0.63	0	0.06	0.17	5	0.97	0.25	0.34	0	7	0.28	0.50	0	0	0.03
Average		0.65	0.08	0.01	0.01	0.04	10.4	0.22	0.01	0.02	0.02	3	0.04	0.23	0.00	0.00	0.02
Sum	2,026																

Table 2. Number of video sequences with occurrence of redfish *Sebastes* spp. in relation to substratum and presence of corals and the estimated probability of occurrence in the different habitats. Frequency of occurrence (expressed as %) is given in parentheses. Estimates are based on 1,866 video sequences (each covering about 15 m) from 44 video transects.

Substratum	Habitat sequences	*Sebastes* sequences	Probability
Boulders absent	602 (32%)	58 (15%)	0.10
Corals absent	1,314 (70%)	132 (35%)	0.10
Boulders present/ corals absent	747 (40%)	90 (23%)	0.12
Boulders present	1,264 (68%)	322 (84%)	0.25
Corals present	552 (30%)	248 (65%)	0.45
Total	1,866	380	0.20

vary between specific fleet sectors, showing distinct patterns of activity. For instance, the gill-net fishery is highly localized to a small area close to Georges Bank. By comparison, the longline fishery is more widespread, representing most of the activity in the deeper portions of the channel and around Romeys Peak. Longline sets can extend for several km and are set at depths between 100 and 700 m. Current trawl fisheries generally take place in waters less than 300 m, mainly on the Georges Bank side of the channel and on the southeastern side of the Browns Bank. Invertebrate fisheries in the channel are limited to date and, in general, do not extend deeper than 300 m.

Discussion

Significance of the Observed Coral Damage

There are indications from DFO's Fisheries Observer Program that all gear types considered in this study (gill-net, trawl, and longline) may cause damage to corals within the Northeast Channel (DFO, unpublished data). However, each fleet has different coverage of observers, which may skew the results. We found few signs of damage due to trawling (i.e., co-occurrence of damaged corals and linear marks on the seabed within a transect) along the video transects, but several transects revealed longlines entangled in damaged corals. In many other places in the world, bottom trawling is known to cause severe damage to coral habitats (Pitcher et al. 1999; Koslow et al. 2001; Krieger 2001; Fosså et al. 2002). Fewer observations have been made on the effects of longlines and gill nets (e.g., Bavestrello et al. 1997; High 1998; Krieger and Wing 2002). Fishermen in Nova Scotia report that longline gear can get tangled up and catch coral when set in areas with coral "trees" (Breeze et al. 1997). This fits well with our observations of lost longlines entangled in the branches of *Paragorgia arborea* and *Primnoa resedaeformis* (Figure 4). Furthermore, a snagged longline has a free end that may move with the currents causing further entangling and damage.

Bottom trawling has a larger potential impact per

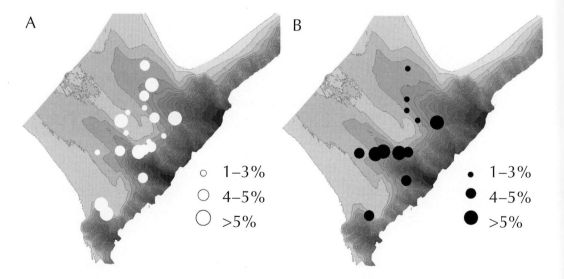

Figure 3. Distribution of damaged corals, (A) *Primnoa resedaeformis* and (B) *Paragorgia arborea*, in the Northeast Channel. The size of the circles represents the frequency of occurrence (in percent) of damaged corals.

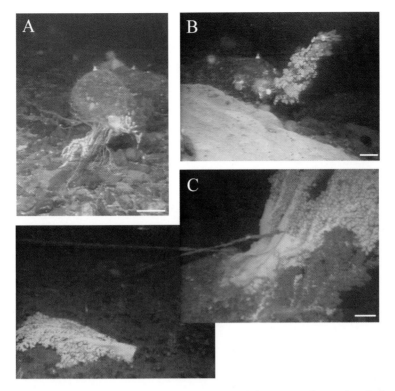

Figure 4. Video frames of corals impacted by fishing gear observed with the ROV *ROPOS*. Scale bar = 10 cm. Panel A is a tilted colony of *Primnoa resedaeformis* with entangled rope, panel B is a colony of *Primnoa resedaeformis* adjacent to a trawlmark, almost entirely covered with a parasitic anemone, and panel C is a mosaic of overlapping video frames showing a colony of *Paragorgia arborea* fouled with longline. The broken branch laying on the bottom was partly alive.

Figure 5. Pie diagrams showing the percentage composition of occurrence of intact colonies, skeletal fragments, tilted colonies, and broken, live colonies.

fishing effort compared to long-lining because the seabed area covered by a single tow of a bottom trawl is larger than a longline set, implying a higher probability of impacting corals. Shrimp trawling off Australia has been estimated to damage 10% of the sea-fans (Gorgonacea) with each pass of the trawl (Pitcher et al. 1999). A study of damage to *Primnoa* sp. in Alaskan waters shows that 30% of the corals in the trawl path were removed or broken (Krieger 2001). However, even though long-lining is much less deleterious to corals than trawling, it may still represent a serious threat in areas with high fishing intensity. Based on our observations of damage (Figure 3), combined with the distribution of fishing effort (Figure 6), we conclude that the damage to corals in the Romeys

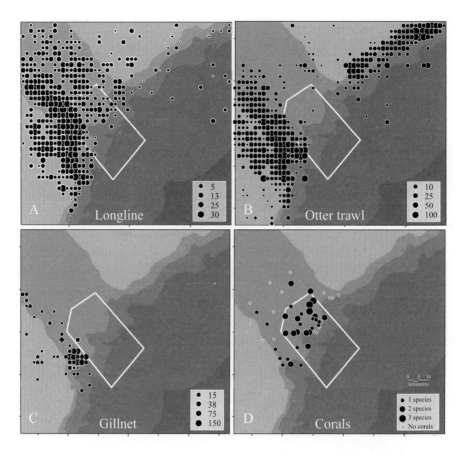

Figure 6. Distribution of groundfish catches (1997–2001) in the general region of the Northeast Channel by (A) longline, (B) otter trawl, and (C) gill net. The size of the circles represents aggregated catches given in metric tons. Also plotted on the same scale is the distribution of corals revealed from video surveys in 2000 and 2001. The boundary of the coral conservation area is indicated by the white line.

Peak area is caused mainly by longlines, since there has been little effort of other gear types during the last 5 years.

Some of the observed damage may not be caused by fishing. Natural mortality may occur due to a number of possible causes. Strong currents may cause large colonies to topple (Tunnicliffe and Syvitski 1983), especially if they are growing on cobbles or small boulders. This may apply mainly to *Paragorgia arborea* since it can reach a much larger size than *Primnoa resedaeformis*. However, in most cases the tilted colonies were also broken, indicating snagging by lines.

Historic Changes

It is difficult to judge the overall magnitude of the damage to corals since there are no detailed historical data on coral distribution and density from the area. Most likely, the fisheries have had a negative effect on coral densities since bycatch of corals has been common on longlines in the area since the late 19th century (Goode and Collins 1887; Verrill 1922; Breeze et al. 1997). The present size distribution of *Paragorgia arborea* does, however, suggest that fisheries have not affected the maximum size within the population. Verrill (1922) reports that *Paragorgia arborea* collected in the late 19th century had a maximum size of approximately 1.60cm (5 ft), which corresponds to our observed maximum height of 1.7 m. However, Verrill (1922) also reported a common size of around 1 m (3 ft), which is much larger than our observed mean size of 52 cm. This may reflect a real reduction in the number of large colonies but may also very well be due to an underrepresentation of small specimens in Verrill's study material (mainly longline bycatch). Furthermore, the statistical meaning of the term "common" is not clear, making a comparison difficult. This aspect of the corals' population structure in the Northeast Channel should be moni-

Secondary Effects of Fisheries Damage

The strong currents observed in our study area may cause erosion into the skeleton by lost longlines trapped in the corals (Figure 4). Most of the observations of *Primnoa resedaeformis* with the parasitic anemone were made in areas with a high frequency of damaged corals. Zoanthid anemones can sometimes cover entire colonies of *Primnoa resedaeformis*. Corals with tissue damage are known to be more susceptible to colonization by epibionts (Grigg 1972). The presence of epibionts can result in an increased drag, which may cause the coral to break or tilt (Weinberg and Weinberg 1979). Bavestrello et al. (1997) found that damage to colonies by fishing line, with subsequent colonization of epibionts, is the major cause of mortality for the gorgonian *Paramuriecea clavata* in the Ligurian Sea. There are several organisms living on the exposed skeletons of *Paragorgia arborea* and *Primnoa resedaeformis* (e.g., hydroids, anemones, stalked barnacles, etc.; Storm 1901). The fertility of corals may decrease as a result of tissue damage and infection by parasites, because it is size dependent (Wahle 1983) and energy is dislocated for tissue repair (Ward 1995).

Effects of Fisheries Damage on the Coral Habitat

Increased mortality among corals may weaken the benthic pelagic coupling. Deepwater corals feed on live and dead particles in the benthic boundary layer and, thus, represent an important link between the water column and seabed in areas rich in corals. As a result of the energy transferred from the water column, corals can form complex three-dimensional habitats that provide shelter and food for a variety of species. Except for some parasitic copepods, the deepwater corals do not seem to be utilized directly by many organisms (Buhl-Mortensen and Mortensen, in press). Observations of trawl-induced habitat changes on the Northwest Shelf of Australia have shown that larger (>25 cm) benthic organisms, mostly sponges, alcyonarians and gorgonians, are particularly susceptible to trawling (Sainsbury et al. 1997). Sainsbury et al. (1997) conclude that these habitat changes probably have altered the fish community, which to a large extent is habitat dependent.

The largest organisms associated with coral observed in the Northeast Channel are redfish, basketstars *Gorgonocephalus lamarkii*, and shrimp *Pandalus propinquus* (Buhl-Mortensen and Mortensen, in press). Shrimps and amphipods living on the surface and between the branches of the colonies may represent an important food source to demersal fish such as tusk *Brosme brosme*. Redfish are known to rest on the seabed, which was also confirmed by the direct observations in this study. The redfish benefit from benthic three-dimensional structures to avoid predation (Shepard et al. 1986; Pikanowski et al. 1999). In this study, we observed that redfish lean against boulders and other objects that add structure to the bottom. We also observed that when corals were present, the redfish commonly lay below, or even within, the coral colonies. This behavior protects them from both drifting away with the current and predation. Redfish are also very common on reefs of *Lophelia pertusa* off Norway (e.g., Dons 1944; Burdon-Jones and Tambs-Lyche 1960; Mortensen et al. 1995). Because redfish are not particularly active (Kelly and Barker 1961), it is reasonable that they would be more exposed to predation over a featureless bottom than a structured bottom (Pikanowski et al. 1999). Thus, a reduced abundance of gorgonian corals in the Northeast Channel may represent a threat to the local redfish population.

Timeframe of Recovery of Damaged Coral Habitats

In general, little is known about the growth rates and life history of deepwater gorgonians. In a recent study by Andrews et al. (2002), age and growth characteristics of *Primnoa resedaeformis* from Alaska were described by counting growth rings in cross-sections of the coral skeleton. These estimates were then validated using a radiometric aging technique. They found an average growth rate of 1.74 cm/year in height. The largest limb studied was approximately 112 years old and had a height of 197.5 cm.

Assuming that this growth rate is representative for other regions, it will take 46 years for *Primnoa resedaeformis* to reach a size of 80 cm, the maximum height that we observed in the Northeast Channel. Nothing is known about the recolonization potential of coral colonies in disturbed deepwater habitats.

Sainsbury et al. (1997) observed a clear decrease in abundance of both small and large benthos in trawled areas, and results indicate that the recovery time for the larger organisms may be at least 15 years and that their settlement rate is low. Similar studies are lacking from deepwater, high latitudinal, hard bottom communities, and there is no indication that recovery time is shorter in deepwater coral communities in the North Atlantic. Bavestrello et al. (1997) carried out a recolonization experiment with the gorgonian *Paramuricea clavata* in shallower water in the Mediterranean Sea. They found that its reestablishment is extremely slow even though it has very strong sexual reproduction. The high abundance of uncolonized boulders shows that it is not a lack of substratum that limits the abundance of corals in the Northeast Channel. The recruitment of the deepwater gorgonians may be low since few small, and presumably young, colonies were observed. Unlike scleractinians and a few shallow-water gorgonians, reproduction by fragmentation (Lasker 1984) is, to our knowledge, not documented for deepwater gorgonians.

Establishment of the Northeast Channel Coral Conservation Area

For several years, some sectors of the fishing industry and environmental organizations have proposed specific zones for deep-sea coral protection in the Northeast Channel. In early 2002, DFO and the fishing industry began to work more closely to address the potential impacts of fisheries on deep-sea corals, through the formation of a working group composed of major fleet sectors.

After reviewing preliminary results from the visual surveys in 2000 and 2001, the DFO proposed a "conservation area" to the working group, centered on Romeys Peak (Figure 2) because of its high abundance of gorgonian corals. A fisheries assessment was conducted, using the available databases within the DFO. Discussions with the industry working group and a public meeting led to a greater understanding of the overlap between fishing activities and areas of coral abundance. In particular, concerns over the impact of a closure on a small group of long-liners who currently fish the area were raised. Adjustments were made to the design of the conservation area, reducing the effect on some fisheries while still protecting deep-sea corals.

The resulting Coral Conservation Area, approximately 424 km^2 in size (Figure 2), was established by the DFO in June 2002 at the start of the annual fishing season (Fisheries and Oceans Canada 2002). This area will provide scientists with an opportunity to study these marine organisms in habitats not exposed to further disturbance.

The Coral Conservation Area is divided into two zones: (1) restricted bottom fisheries zone and (2) limited bottom fisheries zone. About 90% of the area, the restricted bottom zone, is closed to all bottom fishing gear (longline, gill-net, trap, and trawl), while about 10% of the area, the limited bottom fisheries zone, is open to authorized fishing, which at the present time is restricted to longline gear (Fisheries and Oceans Canada 2002).

As the first conservation measure for deep-sea corals in Canada, the area's effectiveness will be subject to ongoing review using information gathered from the fishing industry and from additional research. Based on these results, further adjustments to the design of the Coral Conservation Area are expected.

Acknowledgments

This study was funded by the DFO, the Environmental Studies Research Fund (ESRF), and the National Science and Engineering Research Council (NSERC). We thank Anna Metaxas of Dalhousie University for organizing and serving as chief scientist on the 2001 *ROPOS* cruise. The *ROPOS* team, and the crews of the C.C.G.S. *Hudson* and C.C.G.S. *Martha Black* were very helpful in collecting the video records at sea. We thank Ken Paul of the Canadian Hydrographic Service for the providing the multibeam bathymetric data of the Northeast Channel and Stan Johnson of the Oceans and Coastal Management Division for producing the general bathymetric map.

References

Andrews, A. H., E. E. Cordes, M. M. Mahoney, K. Munk, K. H. Coale, G. M. Cailliet, and J. Heifetz. 2002. Age, growth and radiometric age validation of a deep-sea, habitat-forming gorgonian (*Primnoa resedaeformis*) from the Gulf of Alaska. Hydrobiologia 471:101–110.

Bavestrello, G., C. Cerrano, D. Zanzi, and R. Cattaneo-Vietti. 1997. Damage by fishing activities to the Gorgonian coral *Paramuricea clavata* in the Ligurian Sea. Aquatic Conservation: Marine and Freshwater Ecosystems 7:253–262.

Breeze, H., D. S. Davis, M. Butler, and V. Kostylev. 1997. Distribution and status of deep sea corals off Nova Scotia. Ecology Action Centre, Marine Issues Committee, Special Publication 1, Halifax, Nova Scotia.

Broch, H. 1935. Oktokorallen des nördlichsten Pazifischen Ozeans. Det norske Vidensgaps-Akademi i Oslo I. Matematisk-Naturvidenskabelig Klasse 1935:1–53.

Buhl-Mortensen, L., and P. B. Mortensen. In press. Crustaceans associated with the deep-water gorgonian corals *Paragorgia arborea* (L., 1758) and *Primnoa resedaeformis* (Gunnerus 1763). Journal of Natural History.

Burdon-Jones, C., and H. Tambs-Lyche. 1960. Observations on the fauna of the North Brattholmen stone-coral reef near Bergen. Årbok for Universitetet i Bergen, Matematisk-Naturvitenskapelig Serie 1960(4):1–24.

Carlgren, O. 1945. Polyppdyr (Coelenterata) III. Koraldyr. Danmarks fauna 51. G.E.C. Gads Forlag, Copenhagen.

Collie, J. S., S. J. Hall, M. J. Kaiser, and I. R. Poiner. 2000. A quantitative analysis of fishing impacts on shelf-sea benthos. Journal of Animal Ecology 69:785–798.

Deichman, E. 1936. The Alcyonaria of the western part of the Atlantic Ocean. Harvard University Museum of Comparative Zoology Memoirs 53:1–317.

Dons, C. 1944. Norges korallrev. Det Kongelige Norske Videnskabers Selskabs Forhandlinger 16:37–82. (In Norwegian.)

Fader, G. B. J., E. King, R. Gillespie, and L. H. King.

1988. Surficial geology of Georges Bank, Browns Bank and Southeastern Gulf of Maine. Geological Survey of Canada Open File Report 1692.

Fisheries and Oceans Canada. 2002. Deep-sea coral research and conservation in offshore Nova Scotia. Backgrounder B-MAR-02-(5E). Fisheries and Oceans Canada. Available: www.mar.dfo-mpo.gc.ca/communications/maritimes/back02e/B-MAR-02-(5E).html (July 2002).

Fosså, J. H., P. B. Mortensen, D. M. Furevik. 2002. The deep-water coral Lophelia pertusa in Norwegian waters: distribution and fishery impacts. Hydrobiologia 471:1–12.

Genin, A., P. K. Dayton, P. F. Lonsdale, and F. N. Speiss. 1986. Corals on seamount peaks provide evidence of current acceleration over deep-sea topography. Nature (London) 322:59–61.

Goode, G. B., and J. W. Collins. 1887. The fresh halibut fishery. Pages 3–43 in G. B. Goode, editor. The fisheries and fishery industries of the United States, section V. History and methods of the fisheries. U.S. Commission of Fish and Fisheries, Washington, D.C.

Gordon, D. C., E. L. R. Kenchington, K. D. Gilkinson, D. L. McKeown, G. Steeves, M. Chin-Yee, W. P. Vass, K. Bentham, and P. R. Boudreau. 2000. Canadian imaging and sampling technology for studying marine benthic habitat and biological communities. ICES, CM 2000/T:07, Copenhagen.

Grigg, R. W. 1972. Orientation and growth form of sea fans. Limnology and Oceanography 17:2:185–192.

Heifetz, J. 2002. Coral in Alaska: distribution, abundance, and species associations. Hydrobiologia 471:19–28.

High, W. L. 1998. Observations of a scientist/diver on fishing technology and fisheries biology. Alaska Fisheries Science Center, Processed report 98-01, Seattle.

Jensen, A., and R. Frederiksen. 1992. The fauna associated with the bank-forming deepwater coral Lophelia pertusa (Scleractinaria) on the Faroe Shelf. Sarsia 77:53–69.

Kelly, G. F., and A. M. Barker. 1961. Observations on the behavior, growth, and migration of redfish at Eastport, Maine. International Commission for the Northwest Atlantic Fisheries Special Publication 3:220–233.

Koslow, J. A., K. Gowlett-Holmes, J. K. Lowry, T. O'Hara, G. C. B. Poore, and A. Williams. 2001. Seamount benthic macrofauna off southern Tasmania: community structure and impacts of trawling. Marine Ecology Progress Series 213:111–125.

Krieger, K. J. 2001. Coral (Primnoa) impacted by fishing gear in the Gulf of Alaska. Pages 106–116 in J. H. M. Willison, J. Hall, S. E. Gass, E. L. R. Kenchington, M. Butler, and P. Doherty, editors. Proceedings of the first international symposium on deep-sea corals. Ecology Action Centre, Halifax, Nova Scotia.

Krieger, K. J., and B. L. Wing. 2002. Megafauna associations with deepwater corals (Primnoa spp.) in the Gulf of Alaska. Hydrobiologia 471:83–90.

Lasker, H. R. 1984. Asexual reproduction, fragmentation, and skeletal morphology of a plexaurid gorgonian. Marine Ecology Progress Series 19:261–268.

Lawrence, P., K. W. Strong, P. Pocklington, P. L. Stewart, and G. B. Fader. 1985. A photographic atlas of the eastern Canadian continental shelf: Scotian Shelf, Grand Banks of Newfoundland. Maritime Testing Ltd., Dartmouth, Nova Scotia.

MacIsaac, K., C. Bourbonnais, E. Kenchington, D. Gordon Jr., and S. Gass. 2001. Observations on the occurrence and habitat preference of corals in Atlantic Canada. Pages 58–75 in J. H. M. Willison, J. Hall, S. E. Gass, E. L. R. Kenchington, M. Butler and P. Doherty, editors. Proceedings of the first international symposium on deep-sea corals. Ecology Action Centre, Halifax, Nova Scotia.

Madsen, F. J. 1944. Octocorallia (Stolonifera – Telestacea – Xeniidea – Alcyonacea – Gorgonacea). The Danish ingolf expedition, volume 13. Bianco Luno, Copenhagen.

Mistri, M., and V. U. Ceccherelli. 1994. Growth and secondary production of the Mediterranean gorgonian Paramuricea clavata. Marine Ecology Progress Series 103:291–296.

Mortensen, P. B., M. Hovland, T. Brattegard, and R. Farestveit. 1995. Deep water bioherms of the scleractinian coral Lophelia pertusa (L.) at 64° N on the Norwegian shelf: structure and associated megafauna. Sarsia 80:145–158.

Mortensen, P. B., M. T. Hovland, J. H. Fosså, and D. M. Furevik. 2001. Distribution, abundance and size of Lophelia pertusa coral reefs in mid-Norway in relation to seabed characteristics. Journal of the Marine Biological Association of the United Kingdom 81:581–597.

National Research Council. 2002. Effects of trawling and dredging on seafloor habitat. National Academy Press, Washington, D.C.

Petrie, B., and J. Dean-Moore. 1996. Temporal and spatial scales of temperature and salinity on the Scotian Shelf. Canadian Technical Report of Hydrography and Ocean Sciences 177.

Pikanowski, R. A., W. M. Morse, P. L. Berrien, D. L. Johnson, and D. G. McMillan. 1999. Redfish, Sebastes spp., life history and habitat characteristics. NOAA Technical Memorandum NMFS-NE-132.

Pitcher, C. R., I. R. Poiner, B. J. Hill, and C. Y. Burridge. 1999. The implications of the effects of trawling on sessile mega-zoobenthos on a tropical shelf in northeastern Australia. ICES Journal of Marine Science 57:1359–1368.

Probert, P. K., D. G. McKnight, and S. L. Grove. 1997. Benthic invertebrate bycatch from a deep-water trawl fishery, Chatham Rise, New Zealand. Aquatic Conservation: Marine and Freshwater Ecosystems 7:27–40.

Ramp, S. R., R. J. Schlitz, and W. R. Wright. 1985. The deep flow through the Northeast Channel, Gulf of Maine. Journal of Physical Oceanography 15:1790–1808.

Reed, J. K. 2002. Deep-water Oculina coral reefs of Florida: biology, impacts, and management. Hydrobiologia 471:43–55.

Rogers, A. D. 1999. The biology of *Lophelia pertusa* (Linnaeus 1758) and other deep-water reef-forming corals and impacts from human activities. International Revue of Hydrobiology 84:315–406.

Risk, M. J., J. M. Heikoop, M. G. Snow, and R. Beukens. 2002. Lifespans and growth patterns of two deep-sea corals: *Primnoa resedaeformis* and *Desmophyllum cristagalli*. Hydrobiologia 471:125–131.

Sainsbury, K. J., R. A. Campbell, R. Lindholm, and A. W. Whitelaw. 1997. Experimental management of an Australian multi-species fishery: examining the possibility of trawl-induced habitat modification. Pages 107–112 *in* E. K. Pikitch, D. D. Huppert, and M. P. Sissenwine, editors. Global trends: fisheries management. American Fisheries Society, Symposium 20, Bethesda, Maryland.

Shepard, A. N., R. B. Theroux, R. A. Cooper, and J. R. Uzmann. 1986. Ecology of Ceriantharia (Coelenterata, Anthozoa) of the northwest Atlantic from Cape Hatteras to Nova Scotia. U.S. National Marine Fisheries Service Fishery Bulletin 84:625–646.

Smith, P. C., and F. B. Schwing. 1991. Mean circulation and variability on the eastern Canadian continental shelf. Continental Shelf Research 11:977–1012.

Storm, V. 1901. Oversigt over Throndheimsfjordens fauna (med et kort). Trondhjems Biologiske Station, Meddelelser fra stationsanleggets arbeidskomite, H. Moe's Bog & Accidentstrykkeri, Trondhjem, Norway. (In Norwegian.)

Strømgren, T. 1970. Emergence of *Paramuricea placomu* (L.) and *Primnoa resedaeformis* (Gunn.) in the inner part of Trondheimsfjorden (western coast of Norway). Det Kongelige Norske Videnskabers Selskabs Skrifte 4:1–6.

Tunnicliffe, V., and J. P. M. Syvitski. 1983. Corals move boulders: an unusual mechanism of sediment transport. Limnology and Oceanography 28:564–568.

Verrill A. E. 1922. The Alcyonaria of the Canadian Arctic expedition, 1913–1918, with a revision of some other Canadian genera and species. Report of the Canadian Arctic Expedition 1913–18, volume 8 molluscs, echinoderms, coelenterates, etc., part G alcyonaria and actinaria. F. A. Acland, Ottawa.

Wahle, C. M. 1983. Regeneration of injuries among Jamaican gorgonians: the roles of colony physiology and environment. Biological Bulletin (Woods Hole 165:778–790.

Ward, S. 1995. The effect of damage on the growth reproduction and storage of lipids in the scleractinian coral *Pocillopora damicornis* (Linnaeus). Journal of Experimental Marine Biology and Ecology 187:193–206.

Weinberg, S., and F. Weinberg. 1979. The life cycle of a gorgonian: *Eunicella singularis* (Esper, 1794) Bijdragen tot de Dierkunde 48:127–140.

Wentworth, C. K. 1922. A scale of grade and class terms for clastic sediments. Journal of Geology 30:377–392.

Susceptibility of the Soft Coral *Gersemia rubiformis* to Capture by Hydraulic Clam Dredges off Eastern Canada: The Significance of Soft Coral–Shell Associations

KENT D. GILKINSON[1]

*Department of Fisheries and Oceans, Northwest Atlantic Fisheries Centre,
Post Office Box 5667, St. John's, Newfoundland A1C 5X1, Canada*

DONALD C. GORDON, JR.,[2] DAVID MCKEOWN,[3] DALE RODDICK,[4]
ELLEN L. KENCHINGTON,[5] KEVIN G. MACISAAC,[6] CYNTHIA BOURBONNAIS,[7]
AND W. PETER VASS[8]

*Department of Fisheries and Oceans, Bedford Institute of Oceanography,
Post Office Box 1006, Dartmouth, Nova Scotia B2Y 4A2, Canada*

Abstract. Soft corals (order Alcyonacea) are common on low-relief seabeds on the eastern Canadian shelf where they attach to empty shells and gravel. The effect of hydraulic clam dredging on the abundance of soft corals (specifically *Gersemia rubiformis*) was investigated on a sandy seabed at depths of 70–80 m on Banquereau, off eastern Canada. Visible soft corals were counted from video taken in two dredged and reference boxes before dredging; immediately after dredging; and 2 weeks, 1 year, and 2 years after dredging. Experimental dredging followed standard commercial practices. Dredge tracks covered approximately 53% and 68% of the surface area inside the two dredging boxes, and, although postdredging video surveys were conducted in heavily disturbed areas, no detectable effects of dredging on soft coral abundances were seen. However, the power of the analysis of variance was relatively low (0.6). Low dredge capture rates of soft corals also could partially explain the absence of an observed dredging effect on soft coral abundances. Dredge bycatch from the two dredging boxes showed that soft coral capture rates were variable and generally low (2% and 19%). In this sandy habitat, soft corals attach to empty shells at frequencies of 84% and higher. Most of the shell-attached soft corals were probably displaced from the dredge path through dredge-generated water turbulence. There are two spatial patterns of soft coral–shell associations on Banquereau. There is a low background density of soft corals (averaging 1–5 per 15 m^2) with scattered patches (15–50 m in diameter) of dense accumulations of empty shells with attached epifauna including soft corals at higher abundances. It is expected that dredging through these dense patches, which are expected to be relatively hydrodynamically stable, would result in greater bycatch and damage to soft corals and disruption of these structurally complex seabed features.

Introduction

Soft corals (order Alcyonacea) are widely distributed on the eastern Canadian continental shelf, where they occur on both hard and soft substrata attached to gravel and empty shells (MacIsaac et al. 2001). On low-relief sandy seabeds, they increase habitat structural complexity with colonies of some species growing to heights of 20 cm or more. Soft corals serve a variety of ecological functions. For example, *Gersemia antarctica* is host to the sea spider (class Pycnogonida) (Fry and Hedgpeth 1969), while newly metamorphosed basketstars *Gorgonocephalus eucnemis* live symbiotically on the polyps of *Gersemia* spp. until they reach the arm-branching stage (Mortensen 1977). Certain species of soft corals are also a food source for nudibranch mollusks (Todd 1981).

---

[1] Corresponding author: Gilkinsonk@dfo-mpo.gc.ca
[2] E-mail: Gordond@mar.dfo-mpo.gc.ca
[3] E-mail: McKeownd@mar.dfo-mpo.gc.ca
[4] E-mail: RoddickD@mar.dfo-mpo.gc.ca
[5] E-mail: Kenchingtone@mar.dfo-mpo.gc.ca
[6] E-mail: MacIsaackg@mar.dfo-mpo.gc.ca
[7] E-mail: Bourbonnaisc@mar.dfo-mpo.gc.ca
[8] E-mail: Vassp@mar.dfo-mpo.gc.ca

Soft corals occupy the same habitat as the bivalves (e.g., Arctic surfclam *Mactromeris polynyma*) that are targeted by the Canadian hydraulic clam dredge fishery. There is a considerable literature on the negative impacts of various mobile bottom fishing gears on epifauna (for recent reviews see Jennings and Kaiser 1998; Hall 1999; Kaiser and de Groot 2000). However, although hydraulic clam dredges have been used worldwide since at least the 1960s, relatively little attention has been paid to their environmental impacts, with little or no information available for deepwater, offshore habitats. Based on studies in shallow water habitats, using a variety of dredge designs, it has been shown that hydraulic dredging reduces the biomass of benthos and numbers of individuals and species with varying recovery times (Hall et al. 1990; Pranovi and Giovanardi 1994; Tuck et al. 2000). While there has been increasing concern about the impacts of bottom fishing on deepwater corals (e.g., order Gorgonacea and Scleractinia)(Krieger 2001; Willison et al. 2001; Fosså et al. 2002; Hall-Spencer et al. 2002), there has been less attention paid to shelf-dwelling, soft coral species. The soft coral *Gersemia rubiformis* is known to be susceptible to otter trawling (Prena et al. 1999).

Here, aspects of the impacts of hydraulic clam dredging on populations of soft corals are presented, based on a 3-year experiment conducted on a sandy seabed on Banquereau, eastern Canada. Immediate and longer-term (2 years after dredging) impacts on population abundances are presented, while factors associated with the susceptibility of soft corals to capture by hydraulic dredges are also discussed.

History of the Canadian Hydraulic Clam Dredge Fishery

Off the eastern coast of Canada, a hydraulic clam dredge fishery for the Arctic surfclam has been conducted on Banquereau (Scotian Shelf) and Grand Bank (off Newfoundland) since 1987. Three factory freezer vessels are used in the fishery. Working at depths of 50–80 m, the vessels use paired hydraulic dredges (see Methods for design details). The standard approach is to conduct pulse fishing in which an area of seabed is dredged until catch rates are no longer commercially viable. Industry estimates that at this point, approximately 50% of the patch of seabed has been dredged (i.e., covered with tracks). At this point, the vessel moves to a new location. Due to a combination of variable recruitment and slow growth of the Arctic surfclam, industry typically does not return to a previously dredged site to harvest commercial-size clams for a period of about 10 years. It is estimated that the total area dredged annually is on the order of several hundred km^2 (Roddick 1996). In recent years, harvest rates have ranged between 23,000 and 26,500 metric tons, with most landings coming from Banquereau, generating annual export sales of Can$50 million.

Experimental Site

The experimental site is located on Banquereau, the easternmost outer shelf bank on the Scotian Shelf, southeastern Atlantic Canada (Figure 1). Sediments are well-sorted, medium-grain sand, while the depth range is about 70–80 m. The seabed is relatively flat, while small, transient bedforms (centimeters in height) are created several times annually after the passage of severe storms. Abundant polychaete tubes, bivalve-raised burrow openings, and empty shells create small-scale structural complexity in this habitat (Gilkinson et al. 2003).

The experimental site had not been previously hydraulically dredged, while other bottom fishing activity (e.g., otter trawling, scallop dredging) occurs in the general area. Annual side-scan sonar surveys at the experimental site indicated no recent detectable activity. Industry was involved in the design of the experiment, and, as a result, the site is closed indefinitely to hydraulic clam dredging.

Methods

Experimental Design and Dredging

Three treatment boxes (100 m × 500 m) were established so that the effects of dredging and discarding could be studied separately and in combination. In addition, there were two reference boxes that were separated from the treatment boxes by at least 500 m (Figure 1). A commercial offshore clam vessel (MV *Atlantic Pursuit*) conducted the dredging. The vessel towed two hydraulic clam dredges from the stern (Figure 2). Each dredge was 4 m wide by 3.6 m long by 1 m high and weighed 12 metric tons. The dredges are essentially steel cages constructed of bars spaced 4 cm apart. In front of the dredge opening is a blade set at a cutting depth of 20 cm. A manifold directed jets of water under pressure (130 pounds per square inch or 91 metric tons/m^2 exiting the vessel) into the sediment in front of the blade, thereby liquefying the sediment. A total of 12 offset tows were made at about 4–6 km/h (2–3 knots) in alternating directions in each dredging box. Dredges were periodically retrieved, their contents dumped, and then redeployed. Bycatch and unused portions of clams were dumped continuously during dredging operations in the "dredging

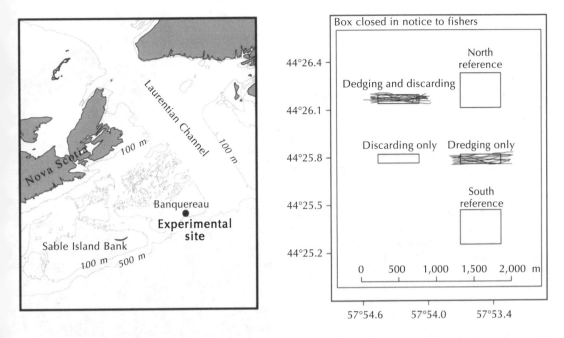

Figure 1. Location of experimental dredging site on Banquereau (left) and layout of experimental boxes (right).

and discarding" box while bycatch and unused portions of the primary target bivalve, *Mactromeris polynyma*, and several secondary target bivalve species from the "dredging only" box were stored and then dumped in the "discard only" box.

Calculation of Area of Seabed Dredged

Using enlarged side-scan sonogram images collected immediately after dredging, the total area covered by dredge tracks in each box was determined using a

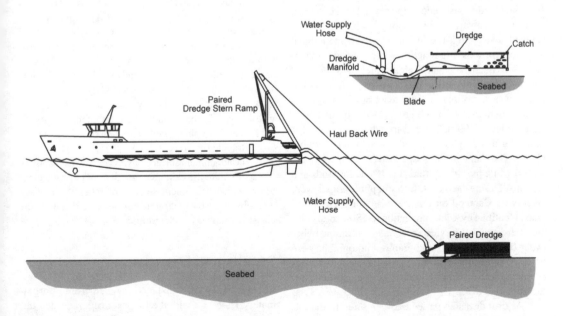

Figure 2. Details of design and deployment of a Canadian offshore hydraulic clam dredge.

planimeter and expressed as a percentage of the total area inside each box (50,000 m²).

Soft Coral Bycatch

In order to calculate dredge capture efficiencies for soft corals, dredge bycatch was quantified for each of the dredging boxes and compared with densities of soft corals based on the video surveys. In the "dredging only" box, after each of the 12 tows, dredge contents were dumped into the vessel's two hoppers. After the final tow, the total volume of each hopper was estimated by comparing the level to a diagram giving the volume at each crossbar on the side of the dredge. Known volumes of subsamples of the bycatch were taken from the conveyer belts, and all invertebrates and fish were identified and weighed. Total numbers and weights of organisms captured after 12 tows were then determined. In the case of the "dredging and discarding" box, processing of the bycatch occurred continuously, and subsamples were taken from 3 of the 12 tows.

Soft Coral Video Counts

Seven species (or species groups) of Alcyonacea have been recorded from the eastern Canadian shelf and slope regions (Breeze et al. 1997). While it was difficult to identify species from video, concurrent grab sample collections from Banquereau indicated that *Gersemia rubiformis* was the most common species at the experimental site. Visible soft corals were counted in video collected using Campod, an instrumented tripod equipped with video and still cameras (Figure 3). It was deployed by a cable from the sampling vessel, CCG *Hudson*. Continuous video of the seabed was collected using a downward-looking Sony 3-CCD (charged coupled device) high-resolution video camera mounted on the central axis. Illumination was provided by a pair of 500-W quartz halogen lights. Video was recorded continuously during drifts at 10 stations positioned randomly in each treatment and reference box at the following time periods: before dredging (1998), immediately after dredging (1998), 2 weeks after dredging (1998), 1 year after dredging (1999), and 2 years after dredging (2000). The Campod was suspended 2–2.5 m above the seabed while the ship drifted slowly due to winds and currents. During each drift, five still photographs were taken with Campod on the seabed, with each photograph identified by a ship navigation fix. Soft coral colonies were counted in video recorded continuously between the first and last photographs. Photographs were not used to count soft corals given their low densities combined with the small surface area of individual photographs (0.2 m²).

As drift distances varied between stations, the following method was used to standardize soft coral counts.

Figure 3. Profile of Campod used to conduct video counts of soft corals.

Drift distance was the distance traversed between the first and last photograph. During drifts, time and differential geographic positioning system (dGPS) position of a reference point on the vessel was recorded. For each dGPS fix, the easting (x) and northing (y) of the Campod position was determined relative to this reference point. Subsequently, polynomials were fitted to both the x and y time series with the Nth order determined by best fit. Total track length was determined by calculating the fitted x and y positions from the polynomials at 1-s intervals between the first and last photographs and then summing the distances traveled each second.

The attachment surface of soft corals was recorded. In cases where the camera angle did not permit a clear view of the attachment surface or when the surface was obscured by emergent epifauna or detritus, the classification was "indeterminate."

Results and Discussion

Average video drift distances ranged from approximately 15–60 m, with the longest drifts recorded immediately after dredging during the search for dredge tracks. Several video drifts were excluded from analyses since they fell outside the box boundaries. The majority of video drifts conducted immediately after dredging showed evidence of considerable seabed disturbance. Although there were significant differences in soft coral abundance between treatments and time periods, no significant immediate effect of dredging on soft coral abundance was detected in either dredging box (Table 1; Figure 4). However, it is noted that the power in the analysis of variance to detect a dredging effect was approximately 0.6, meaning that there was a 40% probability that a true dredging effect on soft corals (i.e., decreased abundance) was missed. Notwithstanding, there is an alternative mechanism that could explain the lack of a dredging effect.

The lack of a recorded impact on soft coral abundance could be because the water turbulence generated by the dredges displaced soft corals out of the dredge path. In the present experiment, dredge capture efficiencies for soft corals, although variable, were generally low (Table 2), despite the fact that the dredges swept a considerable portion of the seabed area inside the "dredging only" (53%) and "dredging and discarding" (68%) boxes. Capture efficiencies of 2% and 19% were derived for the "dredging only" and "dredging and discarding" boxes, respectively. It is not clear why the capture efficiency was higher in the "dredging and discarding" box compared to the "dredging only" box given the similar densities of soft corals in the two dredging boxes. Although considered unlikely, it is possible that some of the soft corals discarded from earlier tows were recaptured in subsequent tows in the "dredging and discarding" box.

Over the 3-year period of the experiment, the attachment surface of 435 soft corals could be identified. In addition, a total of 213 corals were classified as indeterminate due to a poor viewing angle. Frequencies of attachment of soft corals on empty mollusk shells (primarily disarticulated valves of bivalves) ranged from 84% to 95%. The attachment of the relatively lightweight soft corals to shell valves makes them susceptible to displacement by currents. Thus, shell-attached soft corals would also likely be displaced by the water turbulence generated by a pressure wave in front of a towed dredge, particularly when full. In addition, the high-pressure water jets directed toward the seabed in front of the dredge opening would contribute to turbulence. Combined, these two sources of turbulence were likely to blow to the side a significant proportion of the shell-attached corals in the dredge path. Shell-attached soft corals were briefly resuspended when Campod landed gently on the seabed.

The fate of discarded soft corals is not known. For those soft corals that escape capture, injury or mortality could occur through physical contact with the dredge or the water jets. Nonetheless, many of the soft corals observed immediately after dredging had expanded polyps. There were no longer-term declines in abundances of soft corals, while mean densities of soft corals increased 1 and 2 years after dredging in the "dredging and discarding" box (Figure 4).

There are two primary spatial patterns of distribution of empty shells on Banquereau. At small spatial scales (e.g., video and still photos), there is a relatively uniform pattern of dispersed shells with soft corals attached to a portion of these. One aspect of habitat structural complexity is the concept of threshold levels of complexity. Results of studies to date suggest that prey vulnerability generally decreases as environmental complexity increases (see Nelson and Bonsdorff 1990 and references therein). In the present context, a key question would be whether or not soft corals, at low densities in low-relief soft-bottom habitat, are important to either fish or invertebrates as protective cover or larval settlement surfaces or, alternatively, as predators of settling larvae. However, there is a second spatial pattern of soft coral–shell associations that is only discernible at larger spatial scales (e.g., hundreds of meters). Side-scan sonograms taken from sandy seabeds off eastern Canada exhibit patches of high acoustic backscatter which submersible surveys and direct sampling have shown to be dense concentrations of broken and whole empty mollusk shells (Amos and Nadeau 1988; Fader 1991). These dense shell patches range from 15 to 50 m in

Table 1. Analysis of variance of standardized $\log_{10}$-transformed soft coral counts (number/15 m^2). Data taken from before and immediately after dredging (time) in dredging only, dredging and discarding, and reference (treatment) boxes. An asterisk indicates statistical significance ($P = 0.05$); df = degrees of freedom; SS = sum of squares; MS = mean square; and Power = probability of detecting a true effect.

Source	df	SS	MS	F	P	Power (post hoc)
Treatment	2	0.869	0.434	3.92	0.0264*	0.61
Time	1	0.859	0.859	7.74	0.0076*	0.73
Treatment × time	2	0.163	0.082	0.74	0.4844	0.61
Error	49	5.438	0.111			

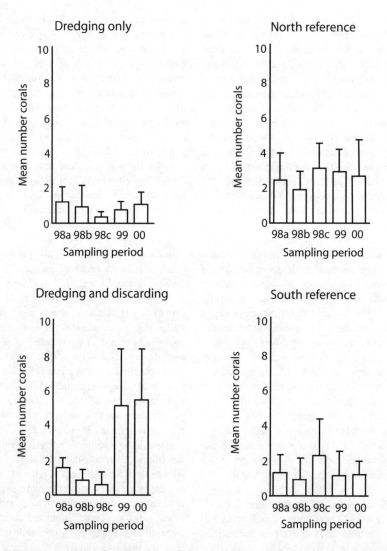

Figure 4. Mean number of soft corals per 15 m², by treatment and time period. 95% confidence intervals shown. 98a = before dredging, 1998; 98b = immediately after dredging, 1998; 98c = 2 weeks after dredging, 1998; 99 = 1 year after dredging, 1999; and 00 = 2 years after dredging, 2000.

diameter and are likely formed by hydrodynamic forces (Fader 1991). Their persistence may be attributed to hydrodynamic stability achieved by their accumulating in lower-energy seafloor depressions (Gilkinson et al. 2003) and possibly through their associated attached epifauna. A portion of one of these shell patches was observed in a video survey in the "dredging and discarding" box, and soft corals were particularly abundant attached to the shells. Combined with the soft corals, the dense packing of empty gastropod (mostly large whelks, family Buccinidae) and bivalve shells provide structural complexity on the sandy seabed. Dredging through these relatively hydrodynamically stable shell patches would likely result in greater bycatch and damage to soft corals and disruption of these structurally complex habitat features.

Acknowledgments

We thank the captains and crews of the CCG *Hudson* and the MV *Atlantic Pursuit* for their expertise and valuable role in the experiment and Eric Roe and Ian Bruce of Clearwater Fine Foods Inc., for their support and assistance. We also thank Kelly Bentham for operating Campod.

Table 2. Estimated mean densities (per m^2) of *Gersemia rubiformis* in the two experimental dredging boxes with corresponding dredge capture efficiencies. For the dredging and discarding treatment, $N = 9$, and for the dredging only treatment, $N = 10$.

Dredging treatment	Mean density (±SD)[a]	Dredging box standing crop[b]	Expected number captured[c]	Total bycatch[d]	Capture efficiency[e]
Dredging and discarding	0.13 (0.05)	6,500	4,420	845	19%
Dredging only	0.12 (0.08)	6,000	3,180	55	2%

[a] Mean predredging densities (per m^2) in the dredging boxes based on standardized counts along video drifts (*N*).
[b] Total number of soft coral colonies inside dredging box = density × box surface area (m^2), where box surface area = 50,000 m^2 (100 × 500 m).
[c] Expected number captured = proportion of box area swept by dredge × standing crop, where proportions are 0.68 for dredging and discarding box and 0.53 for dredging only box.
[d] Total bycatch = estimated total number captured in dredge after 12 passes, based on subsampling the bycatch.
[e] Capture efficiency = (total bycatch/expected capture) × 100.

Funding for this experiment was provided by the Strategic Research Fund and Clearwater Fine Foods, Inc.

References

Amos, C. L., and O. C. Nadeau. 1988. Surficial sediments of the outer banks, Scotian Shelf, Canada. Canadian Journal of Earth Sciences 25:1923–1944.

Breeze, H., D. S. Davis, M. Butler, and V. Kostylev. 1997. Distribution and status of deep-sea corals off Nova Scotia. Ecology Action Centre, Marine Issues Committee Special Publication 1, Halifax, Nova Scotia.

Fader, G. B. J. 1991. Gas-related sedimentary features from the eastern Canadian continental shelf. Continental Shelf Research 11:1123–1153.

Fosså, J. H., P. B. Mortensen, and D. M. Furevik. 2002. The deep-water coral *Lophelia pertusa* in Norwegian waters: distribution and fishery impacts. Hydrobiologia 471:1–12.

Fry, W. G., and J. W. Hedgpeth. 1969. Fauna of the Ross Sea, part 7. Pycnogonida, 1. Colossendeidae, Pycnogonidae, Endeidae, Ammotheidae. New Zealand Department of Scientific and Industrial Research Bulletin 198, New Zealand Oceanographic Institute Memoir 49.

Gilkinson, K. D., G. B. J. Fader, D. C. Gordon, Jr., R. Charron, D. McKeown, D. Roddick, E. L. R. Kenchington, K. MacIsaac, C. Bourbonnais, P. Vass, and Q. Liu. 2003. Immediate and longer-term impacts of hydraulic clam dredging on an offshore sandy seabed: effects on physical habitat and processes of recovery. Continental Shelf Research 23:1315–1336.

Hall, S. J. 1999. The effects of fishing on marine ecosystems and communities. Blackwell Scientific Publications, UK.

Hall, S. J., D. J. Basford, and M. R. Robertson. 1990. The impact of hydraulic dredging for razor clams *Ensis* sp. on an infaunal community. Netherlands Journal of Sea Research 27:119–125.

Hall-Spencer, J., V. Allain, and J. H. Fosså. 2002. Trawling damage to Northeast Atlantic ancient coral reefs. Proceedings of the Royal Society of London, Series B 269(1490):507–511.

Jennings, S., and M. J. Kaiser. 1998. The effects of fishing on marine ecosystems. Advances in Marine Biology 34:201–352.

Kaiser, M. J., and S. J. de Groot. 2000. Effects of fishing on non-target species and habitats. Biological, conservation and socio-economic issues. Fishing News Books, Oxford, UK.

Krieger, J. 2001. Coral (*Primnoa*) impacted by fishing gear in the Gulf of Alaska. Pages 106–116 *in* J. H. Willison, J. Hall, S. Gass, E. L. R. Kenchington, M. Butler, and P. Doherty, editors. Proceedings of the First International Symposium on Deep-Sea Corals. Ecology Action Centre and Nova Scotia Museum, Halifax.

MacIsaac, K., C. Bourbonnais, E. Kenchington, D. Gordon, Jr., and S. Gass. 2001. Observations on the occurrence and habitat preference of corals in Atlantic Canada. Pages 58–75 *in* J. H. Willison, J. Hall, S. Gass, E. L. R. Kenchington, M. Butler, and P. Doherty, editors. Proceedings of the First International Symposium on Deep-Sea Corals. Ecology Action Centre and Nova Scotia Museum, Halifax.

Mortensen, T. 1977. Handbook of the echinoderms of the British Isles. Dr. W. Backhuys, Uitgever, Rotterdam, The Netherlands.

Nelson, W. G., and E. Bonsdorff. 1990. Fish predation and habitat complexity: are complexity thresholds real? Journal of Experimental Marine Biology and Ecology 141:183–194.

Pranovi, F., and O. Giovanardi. 1994. The impact of hydraulic dredging for short-necked clams, *Tapes* spp., on an infaunal community in the lagoon of Venice. Scientia Marina 58:345–353.

Prena, J., P. Schwinghamer, T. W. Rowell, D. C. Gordon, Jr., K. Gilkinson, W. P. Vass, and D. L. McKeown. 1999. Experimental otter trawling on a sandy bottom ecosystem of the Grand Banks of Newfoundland: analysis of trawl bycatch and effects on epifauna. Marine Ecology Progress Series 181:107–124.

Roddick, D. L. 1996. The Arctic surfclam fishery on Banquereau Bank. Department of Fisheries and Oceans, Atlantic Fisheries Research Document 96/36, Dartmouth, Nova Scotia.

Todd, C. 1981. The ecology of nudibranch molluscs. Oceanography and Marine Biology Annual Review 19:141–234.

Tuck, I. D., N. Bailey, M. Harding, G. Sangster, T. Howell, N. Graham, and M. Breen. 2000. The impact of water jet dredging for razor clams, *Ensis* spp., in a shallow sandy subtidal environment. Journal of Sea Research 43:65–81.

Willison, J. H., J. Hall, S. Gass, E. L. R. Kenchington, M. Butler, and P. Doherty, editors. 2001. Proceedings of the First International Symposium on Deep-Sea Corals. Ecology Action Centre and Nova Scotia Museum, Halifax.

Effects of Experimental Otter Trawling on the Feeding of Demersal Fish on Western Bank, Nova Scotia

ELLEN L. KENCHINGTON,[1] DONALD C. GORDON, JR.,[2]
CYNTHIA BOURBONNAIS-BOYCE,[3] AND KEVIN G. MACISAAC[4]

*Department of Fisheries and Oceans, Bedford Institute of Oceanography,
Post Office Box 1006, Dartmouth, Nova Scotia B2Y 4A2, Canada*

KENT D. GILKINSON[5]

*Department of Fisheries and Oceans, Northwest Atlantic Fisheries Centre,
Post Office Box 5667, St. John's, Newfoundland A1C 5X1, Canada*

DAVID L. MCKEOWN[6] AND W. PETER VASS[7]

*Department of Fisheries and Oceans, Bedford Institute of Oceanography,
Post Office Box 1006, Dartmouth, Nova Scotia B2Y 4A2, Canada*

Abstract. The impact of otter trawling on demersal fish feeding was studied over 3 years (1997–1999) on Western Bank, off Nova Scotia, Canada. This site is within the 4TVW Haddock Closed Area, which has been closed to otter trawling since 1987. The seabed was gravel (mostly pebbles and cobbles), and water depth averaged 70 m. A rich and diverse benthic community of 341 taxa characterized the site. Experimental trawling (at least 12 sets along the same line per year) was conducted using an Engel 145 otter trawl with 1,250-kg otter boards and rockhopper ground gear. Twenty-two species of fish and five invertebrate species were captured. Atlantic cod *Gadus morhua*, haddock *Melanogrammus aeglefinus*, and winter flounder *Pseudopleuronectes americanus* increased significantly in the trawl catches after the first trawl pass, indicating movement into the area. Detailed analysis of the stomach contents of Atlantic cod, haddock, American plaice *Hippoglossoides platessoides*, yellowtail flounder *Limanda ferruginea*, and winter flounder showed changes in diet attributed to the trawling. Collectively, these fish preyed upon 177 taxa, with haddock having the most diverse diet and American plaice the most specialized. Two classes of effects attributed to the trawling disturbance are documented: (1) quantitative changes in the abundance of prey species consumed and (2) qualitative changes through opportunistic feeding on novel food items. All five predator species investigated demonstrated statistically significant changes in their diets. Qualitative changes in diet were seen in Atlantic cod, American plaice, and yellowtail flounder. Each of these fish species increased consumption of the horse mussel *Modiolus modiolus* (also known as northern horsemussel), with American plaice and yellowtail flounder found feeding on the mussels only after trawling commenced. Increases in consumption of the tube-dwelling polychaete *Thelepus cincinnatus* were observed across all species, with high propensity and individual selectivity within species. Our results demonstrate that repetitive otter trawling produced both quantitative and qualitative changes in the diets of demersal fish, presumably caused by changes in prey availability brought about by trawl disturbance.

---

[1] E-mail: Kenchingtone@mar.dfo-mpo.gc.ca
[2] E-mail: Gordond@mar.dfo-mpo.gc.ca
[3] E-mail: BourbonnaisC@mar.dfo-mpo.gc.ca
[4] E-mail: MacIsaacKG@mar.dfo-mpo.gc.ca
[5] E-mail: GilkinsonK@dfo-mpo.gc.ca
[6] E-mail: McKeownD@mar.dfo-mpo.gc.ca
[7] E-mail: VassP@mar.dfo-mpo.gc.ca

Introduction

Temperate, deepwater, marine gravel substrates are commonly associated with relatively high species diversity due to the relatively high habitat complexity and low frequency of natural disturbance (National Research Council 2002). Benthic organisms that live in these habitats, especially large epibenthic species, are often considered particularly vulnerable to the impacts of mobile fishing gear (e.g., Auster and Langton 1999; Collie et al. 2000; National Research Council 2002), particularly when trawled for the first time (Collie et al. 2000). Changes to the benthic habitat may also affect the associated fish communities. For some demersal fish species, the benthos has been suggested as providing both cover (e.g., Atlantic cod; Tupper and Boutilier 1995; Collie et al. 1997) and food for juveniles, each of which affects survivorship (e.g., Langton et al. 1995). Loss of structural benthos caused by trawling appears to have resulted in major, large-scale, and lasting impacts on tropical fish communities, which are difficult to reverse (Sainsbury et al. 1997). Equivalent linkages have not yet been established in temperate waters. Trawling will also leave some damaged or dead animals on the bottom. Both fish and invertebrate species (e.g., Prena et al. 1999) may scavenge upon such organisms, with the net impacts on the scavengers positive or negative depending upon whether they themselves interact with the gear.

An experiment to investigate the impacts of otter trawling on a gravel habitat on Western Bank, on the continental shelf off Nova Scotia, Canada, was carried out from 1997 to 1999. Repetitive otter trawling was conducted along an experimental line for 3 years in a row. The effects of the trawling on benthic habitat and communities will be presented elsewhere (Kenchington et al., unpublished manuscript). Here, we report first on the feeding of the dominant fish species in the catch, specifically Atlantic cod *Gadus morhua*, haddock *Melanogrammus aeglefinus*, American plaice *Hippoglossoides platessoides*, yellowtail flounder *Limanda ferruginea*, and winter flounder *Pseudopleuronectes americanus*. In addition to establishing the diets of the fish, we specifically address the question: How do those diets change with trawling disturbance?

Methods

Experimental Site and Design

The experimental site was located on Western Bank at approximately 43°45'N, 61°41'W (Figure 1). This site is within the 4TVW Haddock Closed Area, which has been closed to trawling since 1987, though it had been subjected to trawling previously (Frank et al. 2000). The area has remained open to scallop fishing, but analysis of effort data indicated that the chosen site had not been worked by the scallop fleet. Side-scan sonar surveys conducted prior to and during the experiment showed no evidence of recent (resolution approximately 2 years) seabed disturbance by fishing gear other than the experimental trawling. Therefore, we are confident that the seabed at the site had not been disturbed by mobile gear for at least 10 years when we started the experiment and that it was protected from disturbance by any commercial fishing activity for the duration of the experiment.

The surficial sediments at the site were the gravel facies of the "Sable Island Sand and Gravel" map unit, which covers a large area of the Scotian Shelf (King 1970). Side-scan sonar and video surveys indicated that the seabed had a relatively high degree of spatial homogeneity. Most sediment clasts were pebbles and cobbles, but both boulders and sand patches were also present. Water depth was relatively uniform and averaged about 70 m.

A 2 × 2 km box was laid out as the experimental frame for the benthic grab sampling. Eleven north–south lines, 200 m apart, were plotted within that box, and one was randomly selected to be the experimental (i.e., impacted, trawled) line (Figure 1, line E). Three additional lines were randomly selected as reference lines (i.e., control lines B, G, and I) for use for analyses of benthic impacts (Kenchington et al., unpublished manuscript). Grab samples from all four lines were used to estimate the abundance and biomass of benthic invertebrates used in the calculations of selectivity (see below).

The experimental trawling was carried out using an Engel 145 otter trawl with 1,250 kg polyvalent otter boards and fitted with a Scanmar net mensuration system (McCallum and Walsh 1997). As determined by Scanmar, the trawl had wing and door spreads on the order of 20 ± 2 m and 60 ± 5 m, respectively. The net was rigged with 46-cm-diameter rockhopper gear and had mesh sizes of 180 mm in the wings and belly with 130 mm in the cod end. A 30-mm-square-mesh liner was installed in the final 9 m of the 18.5-m-long cod end in order to capture organisms that may be damaged but not retained by commercial nets. This was the same gear used in the earlier Grand Banks trawling experiment (Rowell et al. 1997). The Canadian Department of Fisheries and Oceans research vessels C.C.G.S. *Needler* (1997 and 1999) and C.C.G.S. *Teleost* (1998) did the experimental trawling.

Each year, the experimental line was trawled at least 12 times in alternating directions (14 trawl sets were made in 1997). The two vessels were equipped with a differential global positioning system (dGPS). The vessel was requested to follow the same line on each set, but varying sea conditions and ship handling skills led to some deviations off the line. The trawl was shot at least 500 m outside the experimental box and not recovered until it was at least 500 m outside the other end. This ensured a relatively uniform level of seabed disturbance along the entire

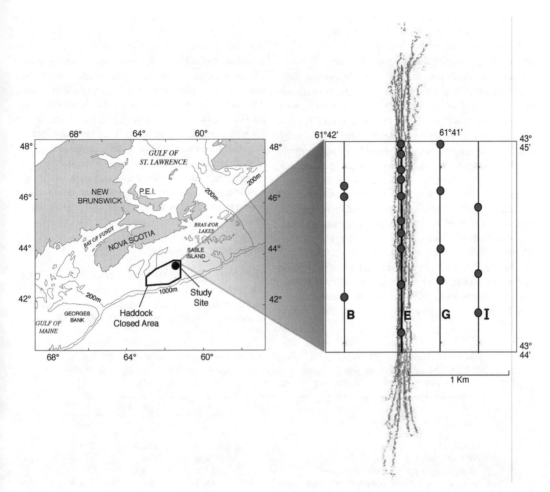

Figure 1. General location of the study site on Western Bank, Nova Scotia, with the Haddock Closed Area and the experimental layout. "E" marks the experimental (trawled) line while "B," "G," and "I" mark the three reference lines (used in the analysis of benthos). Also shown are the location of stations at which macrobenthic data were collected and the path of the otter trawl (center line of headrope) determined using Trackpoint II in 1998.

experimental line within the experimental box. Therefore, the average distance over the bottom of each set was 3.20 km (SD = 0.26). Average trawling speed was 1.8 m/s (3.5 knots), and the average time for each set was 31 min (range = 25–36 min). The vessel steamed several kilometers away from the experimental box between sets before dumping discards from the catch processing. The total time needed to carry out all sets each year ranged from 15 to 19 h This regime represents a heavy and concentrated intensity of trawling, greater than that generally applied in the past by the commercial fishery off Atlantic Canada (Kulka and Pitcher 2001).

In 1997 and 1998, the position of the otter trawl on the seabed was determined using dGPS in conjunction with a Trackpoint II acoustic tracking system (transponder mounted on the headrope). This allowed the disturbance zone to be estimated with a high degree of accuracy (within about 6 m), which is important in confirming that the fish captured in the trawl were from successive sweeps of the experimental line. The results from 1998 are shown in Figure 1.

In 1997 and 1998, fish were collected from the first seven trawl sets. In 1997, the trawl did not fish properly during the first set, but data were retained for the analyses since this set represented the pretrawl state. In 1999, fish were collected from 11 of the 12 trawl sets, with Set 11 being excluded due to the trawl not fishing properly.

Sample Processing

At the completion of each trawl set, the catch was separated by species, and the wet weight of each species was recorded. Stomach samples were collected from five focus species: Atlantic cod, haddock, American plaice, yellowtail flounder, and winter flounder (Table 1). Two stomach sampling strategies were employed for the two levels of taxonomic resolution (see below). For the coarse data set, fish stomachs were selected on a length-stratified basis within sets, while separately, a maximum of seven stomachs per focus species per set was collected for detailed analyses. These latter samples were not length-stratified. The total length of each dissected fish was recorded.

Following dissection, stomachs were individually placed in a hypersaline solution and frozen to stop digestive processes. Food material in the esophagus was pushed back into the stomach prior to dissection. Only stomachs containing a noticeable amount of ingested material were removed. Empty (constricted), everted, or ejected stomachs were recorded but not collected.

In the laboratory, entire stomachs with contents were removed from the freezer, thawed, blotted with a paper towel, and weighed. Stomachs were then opened and all contents removed and weighed together. For the length stratified samples ($N = 1,432$), stomach contents were identified coarsely to phylum or class. For the fish selected for detailed analysis ($N = 444$), prey items were identified to the lowest possible taxon, usually species, genus, or family. Where possible, identifications were made on incomplete parts or segments of animals. Sorted contents were then enumerated and weighed to the nearest 0.1 mg (wet weight). Material was fixed in 70% ethanol, labeled, and archived. The contribution of unidentified fluids and nonbiological material (e.g., gravel) to stomach contents was not measured directly nor retained. Fragments of shells and tests without either attached tissue or additional evidence that the animal was a food item were considered nonbiological and not included.

Analyses

The trawl catch weights were standardized to a 3.2-km tow distance, and only the first seven sets per year were used in statistical analyses to standardize between years. In order to determine whether the five focus fish species

Table 1. Trawl catch composition, by weight per standard trawl distance (3.2 km or 1.7 mi), summed across the first seven trawl sets in each of 1997, 1998, and 1999. Species are ordered by decreasing contribution to the catch in 1997.

Species	Weight (kg)		
	1997	1998	1999
Haddock[a]	6,547.1	2,226.4	2,073.8
Atlantic cod[a]	778.0	315.7	148.6
Atlantic herring Clupea harengus	257.6	3.8	682.1
Sea raven Hemitripterus americanus	46.3	204.9	4.9
Winter flounder[a]	41.6	9.6	212.9
Pollock Pollachius virens	9.8	0.2	3.9
Yellowtail flounder[a]	7.0	18.1	1.7
Atlantic halibut Hippoglossus hippoglossus	5.1	0	53.5
American plaice[a]	5.0	18.7	4.3
Northern shortfin squid Illex illecebrosus	1.4	0	1.9
Sea cucumber Cucumaria frondosa	0.9	6.5	0.5
Longhorn sculpin Myoxocephalus octodecemspinosus	0.8	49.1	0
Ocean pout Zoarces americanus	0.7	12.0	0
Sea anemone Tealia felina	0.5	0	0
Atlantic mackerel Scomber scombrus	0.4	0	0.7
Silver hake Merluccius bilinearis	0.3	0.2	0
Atlantic wolffish Anarhichas lupus	0.2	3.9	0.6
Seastars (family Asteriidae)	0.2	0	0.2
Bigeye sculpin Triglops nybelini	0	5.8	0
Goosefish Lophius americanus (also known as monkfish)	0	10.8	0
Redfish Sebastes sp.	0	2.6	1.2
Thorny skate Amblyraja radiata	0	4.6	2.9
Little skate Leucoraja erinacea	0	1.0	0
Atlantic spiny lumpsucker Eumicrotremus spinosus	0	0.2	0
Toad crab Hyas coarctatus	0	0.2	0
Winter skate Leucoraja ocellata	0	0	5.9
Witch flounder Glyptocephalus cynoglossus	0	0	1.9

[a] Indicates species used in analyses.

altered their feeding habits as trawling progressed, the data were divided into two time categories. Data from the first two trawl sets each year were designated Time 1, and later trawl sets were designated as Time 2. All of Set 1 and most, if not all, of Set 2 sampled fish from a seabed that had been undisturbed by trawling for at least a year. The two trawl sets of Time 1 were completed within 2 h in each year.

Comparisons between the two time periods used independent-samples t-tests. Those were preceded by the Kolmogorov–Smirnov goodness-of-fit procedure (which all data passed) and by Levene's test of homogeneity of error variances. The t-test calculations assumed either equal or unequal variances, as indicated by Levene's test, using a significance level for the latter of $\alpha = 0.05$.

This test procedure was used first to compare trawl catch weight between time periods, using trawl sets as replicates. It was also used to examine changes in mean lengths of the fish between time periods of each focus species in the catch, using individual fish as replicates. Atlantic cod was the only species of the five which showed a significant length difference between the two periods, with Time 1 capturing larger fish (mean = 50 cm) than Time 2 (mean = 41 cm) ($t = 5.633$, $P = 0.000$). As food choice changes gradually with the size of the fish in this species (Methven 1999), the mean size difference (10 cm) has potential to confound the results of other tests. Consequently, the Atlantic cod were divided into two size-classes for statistical analyses and data summaries. The chosen cut point was 45 cm, following a break in the observed size distribution. Link and Garrison (2002) used a similar length (50 cm) to distinguish medium and large Atlantic cod for their dietary studies.

The diets of Atlantic cod (two size-classes), haddock, American plaice, yellowtail flounder, and winter flounder were assessed by the percent contribution of each prey taxon to both the abundance and weight of total stomach contents within a species. Only those dietary taxa that comprised greater than 1% of the total (abundance or weight), within each focus predator species, were included in subsequent statistical analyses. In addition, the prey taxa were grouped a priori into five habitat classes (epifaunal, epifaunal and subsurface or suprasurface, infaunal, pelagic, and tube-dwelling). Prey species accumulation curves for each of the five focus species, with separate curves for the two size-classes of Atlantic cod, were constructed for the total list of prey taxa and for prey taxa constituting greater than 1% of total prey abundance and 1% of total prey weight.

The proportion of individual prey species comprising greater than 1% of total stomach-contents abundance, for each focus species, was compared with the proportion of the same species in the benthic grab samples of the area (Kenchington et al., unpublished data). Ivlev's selectivity index was calculated as (proportion of species x in stomachs – proportion of species x in grabs)/(proportion of species x in stomachs + proportion of species x in grabs). Values range from –1 to +1, with high positive values indicating selective feeding, high negative values indicating avoidance, and values around 0 being neutral (cf. Ivlev 1961). This calculation was done individually for each fish and averaged across all fish of the same focus species to get "species selectivity"; averaging across only those individuals which had eaten the particular prey taxon generated an "individual selectivity." Finally, propensity was calculated as the proportion of fish, within each focus species, consuming the prey taxon.

The question of whether or not the fish feed at certain times of the day is not directly relevant to this work. However, it is possible that certain prey taxa are differentially selected depending upon time of day, which might interact with unbalanced collection of samples between day and night in Times 1 and 2 to confound the results of later analyses. The trawl sets were therefore divided into four categories: dawn, day, dusk, and night. Very few sets were made at dawn or dusk. The independent-samples t-test procedure was used to compare the abundance and biomass of each of the prey taxa constituting greater than 1% of the total stomach contents, for each of the five focus species, between the day and night categories, using individual stomachs as replicates. A significant diel response was found in the diets of small Atlantic cod, haddock, and American plaice.

The impact of trawling on diet was assessed by t-tests of the individual prey taxa (both abundance and weight) between the two time periods. Separate t-tests were conducted on the taxa grouped by habitat category. For both, individual stomachs were the replicates. For those species where a diel effect on diet was identified (small Atlantic cod, haddock, and American plaice), the impact of trawling on diet was further assessed by t-tests confined to data from only the night sets. For haddock, day sets could also be analyzed (due to limitations on the degrees of freedom, data from day samples of small Atlantic cod and American plaice could not be tested). For these last analyses, the sample sizes within night or day categories were considerably smaller than for the entire data set, and the power to detect differences was low. These results were only used for post hoc interpretation of the responses detected in the former analyses.

With the large number of t-tests performed (~300), application of the Bonferroni procedure would require such a low level of α that the tests would have had extremely low power and could only have yielded type II errors. Conversely, using the conventional $\alpha = 0.05$ would result in multiple type I errors, approximately 15 falsely significant results being expected. As we were interested in looking at changes in individual prey items in stomachs of their predators as opposed to general questions on feeding, this number of tests could not be avoided. We, therefore, provide the probability level for each test to allow for open interpretation.

Results

With few exceptions, the research trawler stayed close to the experimental line on all sets. The average spread of all sets for a given year was on the order of 200 m, based on the center line of the trawl (Figure 1). Adding the total spread of the trawl (60 m between the doors) yields a potential sampling zone about 260 m wide. However, it is expected that benthic disturbance, as well as fish sampling, is more concentrated near the center line. Because of this lateral spread in the sampling zone, over four times the width of the trawl, each set has the potential to pass over some undisturbed bottom, but this drops with each successive set. While all fish were collected within this 260-m-wide zone, some fish in later sets may have migrated in from the surrounding area while trawling was taking place. The trawl captured 27 species: 22 fish species and 5 invertebrates (Table 1). Most of those species were solitary groundfish, although seven were schooling fish (Atlantic herring, Atlantic mackerel, Atlantic cod, haddock, pollock, silver hake, and redfish; Scott and Scott 1988). Only 13 species were common to all 3 years. The bycatch of invertebrates was small (27.95 kg over all sets), most of which was a single catch of squid caught in Set 10 in 1999 (not included in Table 1), with seastars, sea cucumbers *Cucumaria frondosa*, anenomes *Tealia felina*, and toad crabs *Hyas coarctatus* (also known as Arctic lyre crabs) also being recorded.

In total, 15,257 kg of fish was processed. Haddock dominated the catch weight followed by Atlantic cod (Table 1). Standardized fish catch weight increased significantly between Time 1 and Time 2 trawl sets across years (Figure 2). However, not all of the 22 fish species contributed to this difference. Individually, only Atlantic cod, haddock, winter flounder, and little skate increased significantly between Time 1 and Time 2. For Atlantic cod, mean catch weight per set was 16 kg in Time 1 and 62 kg in Time 2 ($t = -2.400$, $P = 0.026$). Haddock mean catch weight in Time 1 was 170 kg per set, increasing to 554 kg per set in Time 2 ($t = -2.935$, $P = 0.007$), while winter flounder catch weight averaged 6 kg per set in Time 1 and 20 kg in Time 2 ($t = -2.564$, $P = 0.018$). Little skate were rarely caught (Table 1) and were not present in Time 1 sets. These changes are suggestive of scavenging behavior, with fish moving into the trawled area after trawling commenced. In these catches, Atlantic cod was the only species that showed significant length differences between time periods, with Time 1 capturing larger fish on average than Time 2. Thus, for Atlantic cod, the scavenging behavior was also size stratified, with smaller fish moving into the area.

Fish Stomach Contents

Stomach contents were generally fresh and digestion had not proceeded far, indicating that animals were caught soon after feeding. In total 1,432 fish were processed and 11 phyla were represented in their diets (Table 2) in the coarse data set. Only 10 of these phyla were identified in the detailed data set, with no platyhelminthes being recorded. Numerically, arthropods were the dominant food source for Atlantic cod, haddock, American plaice, and yellowtail flounder, while winter flounder stomachs were dominated by annelids. Haddock were the most generalized feeders, with representatives from all 11 phyla in their stomachs. By weight, fish contributed the most to the stomach contents of the larger size-class of Atlantic cod and to those of American plaice, whereas annelids did so in winter flounder and yellowtail flounder, while haddock stomachs were primarily full of mollusks and echinoderms. Mollusks predominated in the stomachs of the smaller sized Atlantic cod. These broad classifications provide comparison with similar values widely reported in the literature (cf. Methven 1999), but they do not allow for more detailed analysis of fish feeding due to the coarseness of classification.

In the data set with the prey identified to greater taxonomic resolution, the length of Atlantic cod ranged from 26 to 78 cm, haddock from 28 to 58 cm, American plaice from 26 to 40 cm, yellowtail flounder from 24 to 36 cm, and winter flounder from 24 to 44 cm. In all, 177 prey taxa were identified from the stomachs of 444 individual fish across the five species. They included six invertebrate taxa not identified in the benthic grab data set collected from the same area (Kenchington et al., unpublished data). Four of these six were pelagic species not expected to be seen in the grab samples: two mysids (*Mysis mixta* and *Erythrops erythrophthalma*) and twp euphasiids (*Meganyctiphanes norvegica* [known as northern krill] and *Thyanoessa* sp.). The others were a benthic amphipod *Rhachotropis* sp. and the benthic cumacean *Petalosarsia declivis*. The former comprised 1.3% (by number) of the stomach contents of Atlantic cod less than 45 cm in length.

The prey species accumulation curve for each predator, using all prey items, showed Atlantic cod greater than 45 cm and American plaice to be the least saturated, while haddock, yellowtail flounder, and winter flounder were closer to the projected asymptote. This indicates that increasing sample size would have identified a greater total number of prey taxa than observed. For most of the focus species, the prey species accumulation curves for

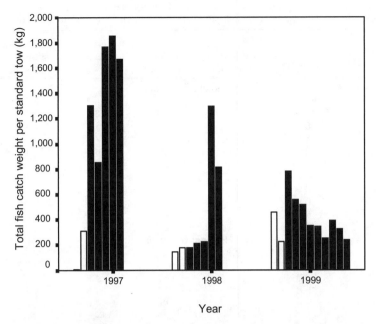

Figure 2. Total catch weight (kg) of fish, standardized to 3.2 km (1.7 nm) tow distance, for each trawl set in 1997, 1998, and 1999. Open bars indicate Time 1 sets while solid bars indicate Time 2 sets. Sets are presented chronologically.

the major food items had reached an asymptote, allowing for meaningful comparisons within and between species. The large Atlantic cod (>45 cm) and winter flounder curves were approaching the asymptote but were not saturated.

The abundances in stomachs of the major prey species, those constituting greater than 1% of the total stomach-contents abundance, are presented in Table 3 for each focus species. Of the 40 major prey taxa, 33 were arthropods or polychaetes. Echinoderms, mollusks, and fish were also included.

The size-classes of Atlantic cod differed in their diets. The smaller fish had a greater overall diversity of diet (82 versus 50 prey taxa in the larger fish) with 15 major food items consumed (Table 3). The shrimp *Eualus pusiolus* was the numerically dominant prey item. The larger Atlantic cod also consumed *E. pusiolus* in large quantities, but fish played a larger role in their diet compared to the smaller Atlantic cod. Twenty-one major prey items were identified in the larger Atlantic cod (Table 3). Haddock consumed the greatest diversity of prey, with 141 prey taxa recorded in total. Nineteen constituted greater than 1% of the total abundance. As with Atlantic cod, *E. pusiolus* was a numerically dominant item, although similar amounts of caprellids and ophiurids were consumed. American plaice had the narrowest overall diet, with only 42 species recorded, although the species accumulation curve suggested that more species would have been identified if more fish had been sampled. Ten of the 42 were taken in large quantities, with the tube-building amphipod *Ericthonius fasciatus* numerically dominant (Table 3). American plaice consumed the highest proportion of fish, by number, of any of the focus predators. Although yellowtail flounder consumed 77 prey taxa overall, only 4 of these were consumed in large quantities, with *E. fasciatus* dominating at 81% of the stomach prey abundance. Winter flounder consumed 91 prey taxa. Of these, 15 were major prey, with the dominant prey taxon being the tube-building polychaete *Thelepus cincinnatus* (Table 3). Only three prey taxa were shared by all five fish species. These were *T. cincinnatus*, *E. fasciatus*, and a composite taxon of sabellid worms that could not be identified further (Table 3).

The percent weight of major prey taxa (each constituting greater than 1% of the total stomach weight) for each of the focus fish species is listed in Table 4. In contrast to abundance, only 19 taxa met this criterion. All species accumulation curves were saturated. As for abundance, the large and small size-classes of Atlantic cod showed different patterns. The horse mussel *Modiolus modiolus* (also known as northern horsemussel), comprised the greatest weight of the stomach contents of small Atlantic cod, while fish contributed the most to the stomach weight of larger Atlantic cod (Table 4). Haddock stomach-contents weight was also dominated by

Table 2. Abundance and weight of prey in the diets of five predator species, based on data from length-stratified sampling only. N = number of fish stomachs processed; + represents percentages less than 1.

Prey category	Atlantic cod		Haddock (N = 584)	American plaice (N = 99)	Yellowtail flounder (N = 88)	Winter flounder (N = 335)
	<45 cm (N = 224)	>45 cm (N = 102)				
Percent by number						
Arthropoda	62.6	55.1	43.2	36.3	65.1	16.5
Annelida	21.0	16.2	18.4	28.1	31.6	73.0
Mollusca	6.8	8.7	4.5	3.4	+	7.2
Chordata	5.9	16.7	+	17.6	+	+
Echinodermata	2.2	1.8	31.0	14.1	1.8	+
Other Cnidaria	1.2	1.5	2.2	+	+	2.2
Brachiopoda	+	0	+	+	+	0
Bryozoa/Hydrozoa	0	0	+	0	+	0
Nemertea	0	0	+	0	0	+
Sipuncula	0	0	+	0	+	0
Platyhelminthes	0	0	+	0	0	0
Percent by weight						
Arthropoda	13.7	9.8	6.8	+	10.3	1.1
Annelida	7.1	+	16.0	16.2	68.5	59.7
Mollusca	59.8	13.8	39.7	2.7	4.2	37.2
Chordata	18.8	75.7	3.7	75.6	16.6	+
Echinodermata	+	+	32.6	4.6	+	+
Other Cnidaria	+	+	1.2	+	+	1.6
Brachiopoda	+	0	+	+	+	0
Bryozoa/Hydrozoa	+	+	+	0	+	0
Nemertea	+	0	+	0	0	+
Sipuncula	0	0	+	0	+	0
Platyhelminthes	0	0	+	0	0	0

M. modiolus, with the daisy brittlestar *Ophiopholis aculeata* making a similar contribution. Fish were the dominant taxon by weight for American plaice, while sabellid worms accounted for half of the weight in yellowtail flounder stomachs. Winter flounder stomach-contents weight was divided between three taxa: *M. modiolus*, *T. cincinnatus*, and sabellid worms. Fish prey items were common to all predators, including both large and small Atlantic cod, as was *M. modiolus*. Small, whole mussels were found in the stomachs as were the crushed shells and tissue of larger individuals. The fish in the stomachs of most species could not be identified, though sand lance (*Ammodytes* sp.), haddock, and silver hake were identified (but not individually quantified here) in Atlantic cod, in addition to unidentified fish.

In all, 341 benthic taxa have been identified in grab samples from the study site (Kenchington et al., unpublished data), 171 of which were also found in the sampled stomachs. Table 5 compares the proportions of major prey taxa found in the fish stomachs to their proportional abundance in the benthic grabs, through selectivity and propensity calculations. Only taxa that were directly and unambiguously comparable between the two data sets were analyzed. For each focus predator species, most prey taxa were not eaten in proportion to their abundance in the benthos, as species selectivities were generally negative and near −1. Mean propensity across prey items was high for yellowtail flounder (65.4%), haddock (50.3%), winter flounder (41.5%), and small Atlantic cod (38%), indicating group feeding behavior. Large Atlantic cod and American plaice had lower mean propensities across prey taxa (14.1% and 21.1%, respectively), indicating more solitary feeding patterns. Mean propensity within species was highly negatively correlated with mean individual selectivity ($R^2 = 0.917$, $P < 0.01$), indicating that individuals of those species with low mean propensity were highly selective in their diets while those which may have been group feeding were less selective as individuals. Small Atlantic cod and haddock had positive selectivity values for *M. modiolus* and *E. pusiolus*, respectively, with a large proportion of the fish sampled eating these taxa. In contrast, the selectivity values for those individuals consuming specific prey items tend to be large and positive, indicating that individual fish feeding were strongly selecting for their prey. Yellowtail flounder is the exception, with low individual selection values for two of the three major prey items. Individual selection for the dominant prey taxon, *E. fasciatus*, is relatively low despite the importance of this species in the diet (Table 3). Propensity values have a large range within fish species, despite the focus of these analyses only on the major prey

Table 3. Percentage of total stomach-contents abundance for each of Atlantic cod (2 length classes), haddock, American plaice, yellowtail flounder and winter flounder by prey taxon, showing only those taxa comprising greater than 1%. A = Arthropoda; P = Polychaeta; M = Mollusca; E = Echinodermata; and Ch = Chordata.

Prey taxa	Atlantic cod <45 cm	Atlantic cod >45cm	Haddock	American plaice	Yellowtail flounder	Winter flounder
Doll eualid *Eualus pusiolus* (A)	30.0	11.0	16.8	-	-	-
Thelepus cincinnatus (P)	17.2	8.1	3.9	6.7	1.2	23.1
Caprellidae (A)	10.4	6.6	11.4	3.7	-	3.5
Meganyctiphanes norvegica (A)	10.0	6.6	-	-	-	-
Modiolus modiolus (M)	5.3	8.8	-	-	-	1.8
Anonyx spp. (A)	5.0	3.7	-	-	-	-
Hyas coarctatus (A)	4.2	5.5	-	-	-	-
Fish species (Ch)	4.0	11.7	-	16.1	-	-
Ericthonius fasciatus (A)	3.9	1.5	2.1	26.3	80.6	12.0
Unciola spp. (A)	2.5	-	4.3	-	2.5	3.1
Ophiopholis aculeata (E)	1.9	1.8	9.3	2.3	-	-
Paradalisca cuspidata (A)	1.6	-	2.2	-	-	-
Syrrhoe crenulata (A)	1.5	-	-	-	-	-
Sabellidae (P)	1.4	1.5	3.8	16.3	5.7	15.9
Rhachotropis sp. (A)	1.3	-	-	-	-	-
Bristled longbeak *Dichelopandalus leptocerus* (A)	-	4.8	-	-	-	-
Jonah crab *Cancer borealis* (A)	-	4.4	-	-	-	-
Pagurus acadianus (A)	-	2.6	-	-	-	-
Lysianassidae (A)	-	2.2	-	-	-	-
Sevenspine bay shrimp *Crangon septemspinosa* (A)	-	1.5	-	-	-	-
Tiron spiniferum (A)	-	1.5	1.3	-	-	-
Erythrops erythropthalma (A)	-	1.5	-	-	-	-
Westwoodilla spp. (A)	-	1.0	1.3	-	-	-
Amphipoda (A)	-	1.1	1.2	-	-	-
Decapoda (A)	-	1.1	-	-	-	-
Ophiuridea (E)	-	-	12.5	6.1	-	-
Paroediceros lynceus (A)	-	-	3.5	-	-	-
Strongylocentrotus pallidus (E)	-	-	1.9	5.8	-	-
Leptocheirus pinguis (A)	-	-	1.9	-	-	3.1
Podoceridae (A)	-	-	1.4	-	-	-
Janira alta (A)	-	-	1.2	-	-	-
Phyllodoce spp.(P)	-	-	1.1	-	-	2.9
Phyllodoce maculata (P)	-	-	1.1	-	-	7.6
Cerrastoderma pinnulatum (M)	-	-	-	2.1	-	-
Maldanidae (P)	-	-	-	1.4	-	3.8
Ampharetidae (P)	-	-	-	-	-	2.7
Glycera capitata (P)	-	-	-	-	-	2.2
Notomastus latericeus (P)	-	-	-	-	-	1.5
Polychaete (P)	-	-	-	-	-	1.5
Northern white chiton *Stenosemus albus* (M)	-	-	-	-	-	1.4
Number of stomachs	70	52	126	53	51	92
Total number of prey taxa	82	50	141	42	51	91

items. While the smaller Atlantic cod size-class included a large proportion of individuals eating *E. pusiolus*, *T. cincinnatus*, and *M. modiolus*, large Atlantic cod showed low propensities, with individual fish tending to eat different prey taxa. This explains why the species accumulation curve for the major prey items of larger Atlantic cod was not saturated. Haddock showed at least two prey taxa, *E. pusiolus* and *O. aculeata*, eaten by the majority of fish, while other taxa were more irregular. Propensity was lowest across all prey taxa for American plaice, while the highest value was 96% of yellowtail feeding on *E. fasciatus*. Winter flounder were similar to haddock, with two prey items, *T. cincinnatus* and *E. fasciatus*, eaten by the majority of individuals.

Table 4. Percentage of total stomach-contents weight of Atlantic cod (two size-classes), haddock, American plaice, yellowtail flounder, and winter flounder by prey taxon, showing only those taxa comprising greater than 1%. Arthropoda (A); Polychaeta (P); Mollusca (M); Echinodermata (E); Chordata (Ch).

Species found in stomach	Atlantic cod <45 cm	Atlantic cod >45 cm	Haddock	American plaice	Yellowtail flounder	Winter flounder
Fish species (Ch)	20.9	80.0	3.6	67.5	15.3	1.8
Modiolus modiolus (M)	59.1	7.4	35.4	3.9	5.3	31.0
Hyas coarctatus (A)	8.8	2.1	-	-	-	-
Thelepus cincinnatus (P)	5.9	-	8.2	3.9	13.3	31.5
Cancer borealis (A)	-	5.0	-	-	-	-
Pagurus acadianus (A)	-	2.0	-	-	-	-
Illex illecebrosus (M)	-	1.2	-	-	-	-
Ophiopholis aculeata (E)	-	-	33.5	3.5	-	-
Sabellidae (P)	-	-	6.2	13.3	49.1	22.0
Ophiuroidea (E)	-	-	2.1	-	-	-
Eualus pusiolus (A)	-	-	1.6	-	-	-
Nephtyidae (P)	-	-	1.4	-	-	-
Strongylocentrotus pallidus (E)	-	-	-	4.1	-	-
Ericthonius fasciatus (A)	-	-	-	-	5.4	-
Lumbrineris fragilis (P)	-	-	-	-	2.4	-
Maldanidae (P)	-	-	-	-	2.3	1.6
Myxicola infundibulum (P)	-	-	-	-	-	2.8
Notomastus latericeus (P)	-	-	-	-	-	1.6
Mollusca (M)	-	-	-	-	-	1.2
Number of stomachs	70	52	126	53	51	92
Total number of prey taxa	84	51	147	43	79	93

Trawling-Induced Changes in Fish Diet

None of the major prey items of winter flounder showed a significant difference between day and night abundance or weight in the stomachs, although total abundance of stomach contents was greater at night ($t = -2.386$, $P = 0.021$). Fish were found in yellowtail flounder stomachs only at night, resulting in a weight increase of that taxon ($t = -2.129$, $P = 0.014$), but no other prey taxa showed diel differences in either abundance or weight in the stomachs of that predator. In Atlantic cod greater than 45 cm, the crabs *Pagurus acadianus* and *Cancer borealis* (known as Jonah crab) were only found in stomachs of fish caught during the day (abundance: $t = 2.683$, $P = 0.012$; $t = 2.252$, $P = 0.032$, respectively). No other prey taxa showed differences in weight or abundance at $\alpha = 0.05$. However, the proportions of stomachs of this predator group that were collected in the day and night were nearly identical for Time 1 and Time 2 (D: $N = 2$ and 2.1, respectively). Thus, the potential for diel differences to confound subsequent analyses for these three focus predators are minimal and are not considered further.

In the smaller Atlantic cod, *E. pusiolus*, *E. fasciatus*, and caprellids had lower abundances in stomachs collected from night hauls ($t = 2.768$, $P = 0.008$; $t = 3.005$, $P = 0.004$; and $t = 2.424$, $P = 0.019$, respectively). By weight, *T. cincinnatus* was the only taxon to show a diel difference, with significantly greater weight in the stomach contents by day ($t = 4.363$, $P = 0.000$). Haddock ate more *T. cincinnatus* ($t = -2.375$, $P = 0.022$), *O. aculeata* ($t = -2.139$, $P = 0.015$), *Paradalisca cuspidata* ($t = -2.443$, $P = 0.018$), and *Janira alta* ($t = -2.317$, $P = 0.025$) during the night but fewer *Strongylocentrotus pallidus* ($t = 2.937$, $P = 0.004$). Stomach-contents weights of *T. cincinnatus*, *O. aculeatea*, and *M. modiolus* were greater during the night ($t = -2.101$, $P = 0.041$; $t = -2.328$, $P = 0.023$; and $t = -2.107$, $P = 0.041$, respectively). American plaice ate fewer *T. cincinnatus* ($t = 2.396$, $P = 0.028$) and *S. pallidus* ($t = 2.207$, $P = 0.014$) at night while eating more fish, both numerically and by weight ($t = -2.650$, $P = 0.021$ and $t = -2.574$, $P = 0.014$, respectively). For these three predators and for the listed prey taxa noted here, any differences between Time 1 and Time 2 may be due to the effects of trawling but may alternatively result from confounding diel variation with the unbalanced sampling of stomachs.

Fish diets in Time 1 and Time 2 were compared in terms of both abundance and weight of each individual major prey taxon, of the prey taxa grouped into habitat classes, and of prey phyla, the latter based only on the length-stratified samples with coarse prey identification (Tables 6, 7). The tables show the results of all t-tests that produced probabilities of the null hypothesis of less than 0.05. At that level of probability, 78 tests returned "significant" results, which is more than what would be expected by chance alone (15 of approximately 300 tests).

Table 5. Selectivity and propensity for species comprising greater than 1% of total stomach-contents abundance for each of Atlantic cod (2 length classes), haddock, American plaice, yellowtail flounder and winter flounder. Arthropoda (A); Polychaeta (P); Mollusca (M); Echinodermata (E); Chordata (Ch)

Taxa found in stomachs	Species selectivity (SE)	Propensity	Individual selectivity (SE)
Atlantic cod <45 cm			
Eualus pusiolus (A)	0.350 (0.110)	0.686	0.968 (0.003)
Thelepus cincinnatus (P)	0.142 (0.089)	0.729	0.567 (0.041)
Modiolus modiolus (M)	0.411 (0.093)	0.771	0.829 (0.014)
Anonyx spp. (A)	−0.550 (0.100)	0.229	0.970 (0.007)
Hyas coarctatus (A)	−0.403 (0.110)	0.300	0.989 (0.002)
Ericthonius fasciatus (A)	−0.738 (0.049)	0.329	−0.204 (0.061)
Unciola spp. (A)	−0.873 (0.045)	0.114	0.114 (0.142)
Ophiopholis aculeata (E)	−0.784 (0.059)	0.171	0.257 (0.096)
Paradalisca cuspidata (A)	−0.691 (0.086)	0.329	0.966 (0.005)
Syrrhoe crenulata (A)	−0.715 (0.084)	0.143	0.992 (0.001)
Atlantic cod >45 cm			
Eualus pusiolus (A)	−0.582 (0.113)	0.212	0.975 (0.003)
Modiolus modiolus (M)	−0.109 (0.135)	0.462	0.932 (0.008)
Thelepus cincinnatus (P)	−0.500 (0.110)	0.288	0.734 (0.034)
Hyas coarctatus (A)	−0.846 (0.075)	0.077	0.998 (0.001)
Dichelopandalus leptocerus (A)	−0.885 (0.065)	0.058	0.999 (0.0001)
Cancer borealis (A)	−0.731 (0.100)	0.135	0.999 (0.0001)
Anonyx spp. (A)	−0.732 (0.100)	0.135	0.989 (0.003)
Pagurus acadianus (A)	−0.731 (0.100)	0.135	0.999 (0.0001)
Ophiopholis aculeata (E)	−0.887 (0.056)	0.077	0.457 (0.114)
Ericthonius fasciatus (A)	−0.912 (0.043)	0.077	0.138 (0.022)
Crangon septemspinosa (A)	−0.885 (0.070)	0.058	0.999 (0.0001)
Tiron spiniferum (A)	−0.886 (0.056)	0.058	0.969 (0.001)
Westwoodilla spp. (A)	−0.885 (0.065)	0.058	0.998 (0.004)
Haddock			
Eualus pusiolus (A)	0.432 (0.075)	0.746	0.919 (0.008)
Ophiopholis aculeata (E)	0.115 (0.052)	0.865	0.289 (0.040)
Unciola spp. (A)	−0.539 (0.051)	0.468	−0.016 (0.054)
Thelepus cincinnatus (P)	−0.410 (0.062)	0.492	0.198 (0.062)
Paroediceros lynceus (A)	0.120 (0.085)	0.579	0.931 (0.007)
Paradalisca cuspidata (A)	−0.012 (0.087)	0.508	0.946 (0.005)
Ericthonius fasciatus (A)	−0.841 (0.029)	0.310	−0.480 (0.062)
Strongylocentrotus pallidus (E)	−0.215 (0.065)	0.563	0.393 (0.036)
Leptocheirus pinguis (A)	−0.403 (0.060)	0.484	0.233 (0.048)
Tiron spiniferum (A)	−0.245 (0.077)	0.437	0.730 (0.021)
Janira alta (A)	−0.324 (0.079)	0.373	0.813 (0.016)
Westwoodilla spp. (A)	−0.399 (0.082)	0.540	0.991 (0.001)
Phyllodoce maculata (P)	−0.717 (0.056)	0.175	0.622 (0.055)
American plaice			
Ericthonius fasciatus (A)	−0.510 (0.097)	0.340	0.445 (0.053)
Thelepus cincinnatus (P)	−0.634 (0.098)	0.226	0.661 (0.076)
Strongylocentrotus pallidus (E)	−0.579 (0.108)	0.226	0.858 (0.027)
Ophiopholis aculeata (E)	−0.803 (0.072)	0.132	0.488 (0.144)
Cerrastoderma pinnulatum (M)	−0.741 (0.092)	0.132	0.964 (0.014)
Yellowtail flounder			
Ericthonius fasciatus (A)	0.433 (0.058)	0.961	0.491 (0.043)
Unciola spp. (A)	−0.394 (0.086)	0.647	−0.063 (0.090)
Thelepus cincinnatus (P)	−0.569 (0.087)	0.353	0.222 (0.078)
Winter flounder			
Thelepus cincinnatus (P)	0.300 (0.071)	0.804	0.617 (0.028)
Ericthonius fasciatus (A)	−0.252 (0.054)	0.750	−0.002 (0.040)
Phyllodoce maculata (P)	−0.237 (0.093)	0.423	0.800 (0.022)
Leptocheirus pinguis (A)	−0.485 (0.077)	0.337	0.527 (0.036)
Unciola spp. (A)	−0.677 (0.053)	0.326	−0.008 (0.063)

Table 5. Continued.

Taxa found in stomachs	Species selectivity (SE)	Propensity	Individual selectivity (SE)
Glycera capitata (P)	−0.406 (0.094)	0.304	0.951 (0.007)
Modiolus modiolus (M)	−0.263 (0.091)	0.424	0.739 (0.023)
Notomastus latericeus (P)	−0.641 (0.076)	0.196	0.834 (0.023)
Stenosemus albus (M)	−0.725 (0.064)	0.174	0.582 (0.082)

At an $\alpha = 0.01$ criterion, approximately 3 tests would be expected to yield type I errors whereas 53 tests produced lower probabilities, and at the $\alpha = 0.001$ level, no falsely significant results would be expected, yet 29 results were observed in that range. Clearly, while some of the results in Tables 6 and 7 are type I errors, most are not, though it is impossible to unambiguously separate the two groups.

For the smaller size-class of Atlantic cod, repetitive trawling resulted in an increased consumption of $E.$ $pusiolus$, $M.$ $modiolus$, $T.$ $cincinnatus$, and sabellid worms (Table 6). This change was also reflected in increased consumption of annelids, arthropods, and mollusks, though echinoderms were also consumed in greater proportion in Time 2. The infauna was not exploited in Time 1 but formed a portion of the diet in Time 2, while consumption of pelagic species increased threefold, perhaps suggesting that invertebrate scavengers as well as fish, entered the area of trawl tracks. Increased consumption of $T.$ $cincinnatus$ and annelids was reflected in increased weight of those items in the stomachs of these fish, with the weight of $T.$ $cincinnatus$ twenty times greater in the stomachs in Time 2 (Table 7; Figure 3). The diel differences in diet reported above for this size-class of Atlantic cod have the potential to confound these results. However, analyses of stomach-contents abundance of $E.$ $pusiolus$ and both the abundance and weight of $T.$ $cincinnatus$ in fish captured in night trawl sets only confirmed the differences described above, despite the low power of the tests ($t = -3.416$, $P = 0.004$; $t = -3.697$, $P = 0.002$). The response of the abundance of $E.$ $fasciatus$ and of the composite caprellid taxon between the time periods in the night trawl samples were nonsignificant, but the tests had very low power ($t = 0.557$, $P = 0.621$, adjusted power = 0.05; and $t = -0.798$, $P = 0.468$, adjusted power = 0.050, respectively). It is likely that diel feeding patterns at least contributed to the apparent differences in consumption of these two prey taxa between Time 1 and Time 2.

Larger Atlantic cod also increased consumption of $T.$ $cincinnatus$ and $M.$ $modiolus$, as well as of the hermit crab $P.$ $acadianus$. These were at the expense of fish taxa, which declined by 50%. At the phylum level, only the annelids showed a significant change between the periods, with increased consumption. Epifauna and tubicolous species were numerically more abundant in the stomachs in Time 2. The weight of $M.$ $modiolus$ in stomachs also increased in Time 2 as did the weight of annelids and mollusks (Table 7; Figure 3).

In Time 2, haddock fed on fewer individuals of $O.$ $aculeata$ and the amphipod $P.$ $cuspidata$ compared to Time 1, and this was reflected in a decline in consumption of epifaunal and shallow infaunal species. However, consumption of the polychaetes $Phyllodoce$ spp. and $T.$ $cincinnatus$ increased (Table 6). Stomach-contents weight showed an increased percentage of $M.$ $modiolus$, $T.$ $cincinnatus$ (Figure 3), and fish species in Time 2, while the total weight of the stomach contents increased as well (Table 7). Haddock were observed to have a diel feeding response to most of these prey taxa (see above). However, post hoc testing of the affected prey taxa within only day or night sets paralleled the results from the whole data set (abundance, night: $P.$ $cuspidata$, $t = 2.296$, $P = 0.04$; $Phyllodoce$, $t = -2.123$, $P = 0.041$; $T.$ $cincinnatus$, $t = -2.998$, $P = 0.006$; weight: night: $T.$ $cincinnatus$, $t = -2.998$, $P = 0.006$; day: $T.$ $cincinnatus$, $t = -2.147$, $P = 0.036$). The only exception was the failure to detect a decline in stomach-contents abundance of $O.$ $aculeata$ in Time 2, though the power of the tests was low in both day and night data sets (adjusted power 0.05 and 0.63, respectively). Thus, this one trend may not be a response to trawling.

American plaice expanded their diets in Time 2 to include new species, feeding on the polychaete $T.$ $cincinnatus$ and $O.$ $aculeata$ (Table 6). For $T.$ $cincinnatus$, this resulted in an increase in weight as well (Table 7; Figure 3). The weight of fish in American plaice stomachs dropped sharply in Time 2, reflected in a negative change in the weight of Chordata. Fish and $T.$ $cinncinnatus$ showed these same responses with the tests restricted to data from night samples, indicating that the differences between Time 1 and Time 2 were not caused by diel variability alone ($T.$ $cincinnatus$: abundance, $t = -2.348$, $P = 0.031$; weight, $t = -2.104$, $P = 0.049$; fish: weight, $t = 3.337$, $P = 0.003$).

Yellowtail flounder also fed on $T.$ $cincinnatus$ only during Time 2, resulting in significant increases in the abundance and weight of this species in their stomachs (Tables 6, 7; Figure 3). Consumption of infauna increased sixfold in Time 2, while overall stomach contents were on average 4.5 times heavier.

Table 6. Comparisons of stomach-contents abundance (by major prey taxa, phyla, and prey habitat groupings) between Time 1 and Time 2 trawl periods. Only the results of t-tests that produced probabilities ≤ 0.05 are shown. N = number of stomachs. Arthropoda (A); Polychaeta (P); Mollusca (M); Echinodermata (E)

Prey taxa	Time period	N	Mean abundance	t	P
Atlantic cod < 45 cm					
Eualus pusiolus (A)	1	4	0.3	−5.339	0.000
	2	66	4.6		
Modiolus modiolus (M)	1	4	0.3	−2.170	0.010
	2	66	0.8		
Thelepus cincinnatus (P)	1	4	0.5	−4.346	0.000
		66	2.6		
Sabellidae spp. (P)	1	4	0	−4.183	0.000
	2	66	0.2		
Pelagic	1	4	7.8	−6.446	0.000
	2	66	24.3		
Infauna	1	4	0	−6.123	0.000
	2	66	0.6		
Annelida	1	35	0.2	−5.509	0.000
	2	189	1.5		
Arthropoda	1	35	1.7	−3.171	0.002
	2	189	4.2		
Echinodermata	1	35	0.01	−2.383	0.018
	2	189	0.2		
Mollusca	1	35	0.1	−3.202	0.002
	2	189	0.5		
Atlantic cod ≥ 45 cm					
Thelepus cincinnatus (P)	1	21	0.1	−3.22	0.003
	2	31	0.7		
Modiolus modiolus (M)	1	21	0.2	−3.671	0.001
	2	31	0.7		
Pagurus acadianus (A)	1	21	0	−2.958	0.006
	2	31	0.2		
Epifauna	1	21	1.8	−2.701	0.010
	2	31	3.6		
Fish species combined	1	21	0.8	2.959	0.005
	2	31	0.4		
Tubicolous species	1	21	0.2	−3.442	0.001
	2	31	0.9		
Annelida	1	18	0.2	−3.169	0.002
	2	84	1.1		
Haddock					
Ophiopholis aculeata (E)	1	29	9.6	2.566	0.015
	2	97	4.3		
Thelepus cincinnatus (P)	1	29	0.7	−3.418	0.001
	2	97	2.8		
Phyllodoce spp. (P)	1	29	0.2	−3.729	0.000
	2	97	0.8		
Paradalisca cuspidata (A)	1	29	2.2	2.097	0.044
	2	97	1.0		
Epifauna/Subsurface	1	29	41.5	2.196	0.034
	2	97	24.9		
Annelida	1	126	2.7	−3.142	0.002
	2	458	4.7		
Chordata	1	126	0.02	−2.520	0.012
	2	458	0.1		
Echinodermata	1	126	10.2	2.623	0.010
	2	458	6.4		
Mollusca	1	126	0.8	−2.062	0.040
	2	458	1.1		

Table 6. Continued.

Prey taxa	Time period	N	Mean abundance	t	P
American plaice					
Thelepus cincinnatus (P)	1	10	0	−2.645	0.011
	2	43	0.7		
Ophiopholis aculeata (E)	1	10	0	−2.228	0.031
	2	43	0.2		
Annelida	1	33	0.3	−3.896	0.000
	2	66	1.7		
Chordata	1	33	0.3	−2.803	0.007
	2	66	1.2		
Yellowtail flounder					
Thelepus cincinnatus (P)	1	12	0	−3.840	0.000
	2	39	1.3		
Sabellidae spp. (P)	1	12	0.2	−4.416	0.000
	2	39	6.1		
Unciola spp. (A)	1	12	0.6	−3.419	0.001
	2	39	2.5		
Epifauna/Subsurface	1	12	0.8	−3.174	0.003
	2	39	2.8		
Infauna	1	12	1.3	−4.670	0.000
	2	39	8.1		
Annelida	1	19	0.9	−4.326	0.000
	2	69	6.8		
Winter flounder					
Ampharetidae spp. (P)	1	16	0.1	−3.512	0.001
	2	76	0.8		
Modiolus modiolus (M)	1	16	0.2	−2.465	0.021
	2	76	0.5		
Unciola spp. (A)	1	16	0.3	−2.650	0.010
	2	76	0.8		
Infauna	1	16	3.2	−2.240	0.030
	2	76	6.2		
Annelida	1	43	10.1	−3.042	0.003
	2	292	14.5		
Mollusca	1	43	0.4	−6.683	0.000
	2	292	1.5		

Winter flounder increased consumption of three species, including *M. modiolus*, which also increased by weight (Tables 6, 7; Figure 3). Infaunal species were twice as abundant in the stomachs in Time 2.

In all, it is particularly notable that all the focus predator species, except for winter flounder, increased consumption of *T. cincinnatus* (both numerically and by weight) between the time periods. Yellowtail flounder and American plaice did not feed on this species at all in Time 1 but did so in Time 2.

Discussion

The trawl catches taken during this experiment were consistent with expectations for this area, with haddock being the dominant species (Fisher and Frank 2002). Our results are focused on two different fish groups: the gadoids (Atlantic cod and haddock) and the flatfishes (American plaice, yellowtail flounder, and winter flounder). The gadoids are schooling fish, feeding on both benthic and pelagic species, while the flatfishes are considered solitary and are generally bottom feeders.

The benthic community on Western Bank is diverse, with a rich macrofauna of 341 taxa drawn from 12 phyla (Kenchington, unpublished data). Our study has shown that the fish included in our analyses feed on about half (~51%) of these benthic invertebrates, including deep infaunal species, indicating a strong coupling between benthic and demersal fish communities. This breadth of diet and the associated estimates of feeding preference were only obtained because of a combination of thorough taxonomic identification and the availability of comparative data sets from grab samples and fish stomachs, a combination that we recommend for future studies.

Table 7. Comparisons of stomach-contents weight (by major prey taxa, phyla, and prey habitat groupings) between Time 1 and Time 2 trawl periods. Only the results of t-tests that produced probabilities ≤ 0.05 are shown. N = number of stomachs.

Prey taxon	Time period	N	Mean weight (mg)	t	P
Atlantic cod < 45 cm					
Thelepus cincinnatus (P)	1	4	29.9	−6.323	0.000
	2	66	645.4		
Annelida	1	19	132.6	−6.598	0.000
	2	205	639.0		
Atlantic cod ≥ 45 cm					
Modiolus modiolus (M)	1	22	245.1	−3.108	0.004
	2	30	3,221.1		
Annelida	1	33	22.0	−4.973	0.000
	2	69	241.7		
Mollusca	1	33	1,014.3	−3.447	0.001
	2	69	6,411.9		
Haddock					
Modiolus modiolus (M)	1	29	147.3	−4.059	0.000
	2	97	1,887.2		
Thelepus cincinnatus (P)	1	29	45.4	−4.296	0.000
	2	97	430.5		
Fish species (Ch)	1	29	1.1	−1.974	0.050
	2	97	193.0		
Weight of stomach contents	1	29	17,257.5	−2.894	0.005
	2	97	31,782.6		
Annelida	1	126	423.3	−5.455	0.000
	2	459	1,127.3		
Echinodermata	1	126	2,489.8	2.212	0.028
	2	459	1,837.6		
Mollusca	1	126	197.6	−7.538	0.000
	2	459	3027.7		
American plaice					
Thelepus cincinnatus (P)	1	10	0	−2.609	0.013
	2	43	149.9		
Fish species (Ch)	1	10	5,366.0	2.624	0.025
	2	43	1,373.8		
Sabellidae spp. (P)	1	10	214.6	−2.342	0.023
	2	43	466.9		
Chordata	1	33	4,010.7	2.014	0.050
	2	66	2,179.9		
Yellowtail flounder					
Thelepus cincinnatus (P)	1	12	0	−4.004	0.000
	2	39	246.0		
Weight of stomach contents	1	12	2,999.2	−2.898	0.006
	2	39	13,581.2		
Annelida	1	19	213.3	−3.549	0.001
	2	69	946.6		
Winter flounder					
Notomastus latericeus (P)	1	16	13.8	−2.283	0.025
	2	76	69.9		
Maldinidae spp. (P)	1	16	20.9	−2.073	0.042
	2	76	69.1		
Modiolus modiolus (M)	1	16	419.8	−2.300	0.027
	2	76	1,352.0		
Mollusca	1	43	206.6	−8.580	0.000
	2	292	2,847.7		

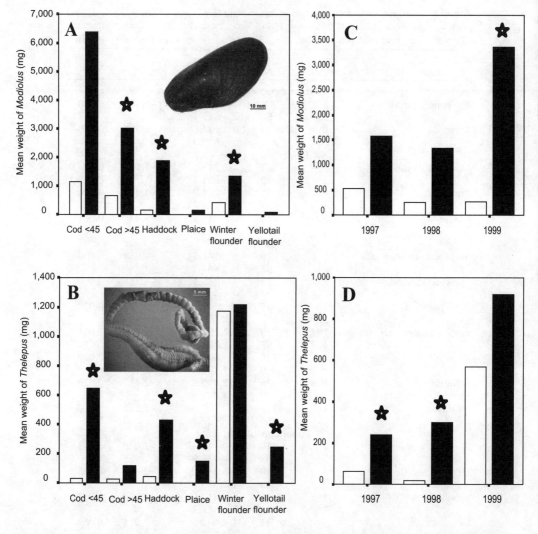

Figure 3. Mean weight (mg) in Time 1 and Time 2 of (A, C) the horse mussel *Modiolus modiolus* (also known as northern horsemussel) and (B, D) the tube-building polychaete *Thelepus cincinnatus* in (A, B) the stomachs of Atlantic cod (two size-classes), haddock, American plaice, winter flounder, and yellowtail flounder, summed across years, and (C, D) for all five predators combined in each year individually. Time 1 (open boxes) and Time 2 (shaded boxes); star indicates a significant increase (see Table 7 for probabilities associated with individual predator species; in panels C, D significance is assessed at $P < 0.01$).

Atlantic cod are visual feeders that respond to prey movement and, as adults, primarily feed on fish, crustaceans, and mollusks (e.g., Kohler and Fitzgerald 1969; Langton and Bowman 1980; Lilly and Meron 1986; Methven 1999). Typically, they feed on many taxa but with relatively few making up the majority of the diet (e.g., Klemetsen 1982). Spatial variability is high, and diets can be quite different over short distances (Methven 1999; Link and Garrison 2002). Our study is consistent with these results, with fish, *M. modiolus*, and *H. coarctatus* dominating the weight, while various amphipods and annelids rounded out the diet and were numerically abundant. However, while the percentage of fish in the diets of these Atlantic cod (approximately 60% across the two size-classes by weight) was similar to that found more generally on the Scotian Shelf (Langton and Bowman 1980), the percentage of mollusks in the diet was high (approximately 25% across the two size-classes) and was consistently so across both the taxonomically detailed and the coarser data sets. Kohler and Fitzgerald (1969) reported that mollusks increase in importance in the diet of Atlantic cod on

Western Bank during the summer months. This seems to be especially true for medium-sized Atlantic cod. Diet composition in our study suggests that the smaller Atlantic cod feed in groups while the larger Atlantic cod are highly selective in their food choices but differ among individuals in their diets.

In our experiment, the diet of Atlantic cod changed after trawling. They ate significantly more annelids and both size-classes increased their consumption of *M. modiolous*, adding significantly to its weight in their stomachs. Smaller Atlantic cod also exploited 14 infaunal species after trawling commenced that were not detected in their diet pretrawling, while the larger size-class increased consumption of epifauna and tubicolous species at the expense of fish.

Haddock are typically benthic feeders noted for the diversity of species they feed on (Wigley and Theroux 1965). Those caught on Western Bank were feeding on a variety of amphipods, annelids, echinoderms, and mollusks and had the most diverse diet among the species sampled. Haddock feed in groups (Mattson 1992), and the main prey items are typically small and slow moving. Group feeding is supported by the high propensity for major prey across individuals in our study. Haddock greatly increased the weight of their stomach contents between the two time periods. They are considered facultative consumers of fish, and a large increase (193 times) in their consumption of that taxon after trawling was noted, most likely accounting for the increased overall weight of stomach contents. Consumption of *M. modiolus* increased numerically by 13 times and was reflected in an increase of molluscan weight in the stomachs. Epifaunal and shallow subsurface prey items were significantly less frequent in the diets after trawling commenced, presumably as food choices were made in favor of fish.

American plaice are bottom feeders that have a large mouth relative to other flounders and consume a variety of prey (Langton and Watling 1990; Methven 1999). American plaice larger than 21 cm have a diet of larger organisms (Martell and McClelland 1992) such as echinoderms and fish (Gerasimova et al. 1991). Link et al. (2002) described this species as an echinoderm specialist over much of its distribution on the continental shelf of the United States. In contrast, our results show a diet that was evenly comprised of approximately 10 prey items, including fish and echinoderms as well as the polychaetes and amphipods more typically found by earlier authors in smaller American plaice. Individual diet composition was variable, and highly selected items were found in individual stomachs. The trawling disturbance had a profound effect on the diet of this species. The tube-dwelling polychaete *T. cincinnatus*, *O. aculeata*, and *M. modiolus* were new dietary items not found in the stomachs sampled from the first trawl passes, although consumption of the latter was too variable for this change in its abundance or weight to be detected as significantly different.

Yellowtail flounder diet has been well-studied on Georges Bank and in the Gulf of Maine, where amphipods and annelids are the primary constituents (Libey and Cole 1979; Langton and Bowman 1981; Langton and Watling 1990; Methven 1999; Link et al. 2002). In our study, sabellid worms comprised the largest biomass component of the stomach contents by weight, while *E. fasciatus* was numerically dominant and selected, while the yellowtail flounder diet included only seven major taxa. Collectively, the fish ate very similar prey items. Generally, fish are not a dominant feature of the diet of this species (e.g., Link et al. 2002); however, Pitt (1976) identified fish as comprising 7.5% of the total weight of the stomach contents of yellowtail flounder on Grand Bank. Our study identified twice this percentage of fish in the stomachs from Western Bank—a result seen in both the size-stratified samples and those with detailed identification of prey taxa. Trawling also affected the diet of yellowtail flounder, with a new species, the tube-dwelling sabellid worm *T. cincinnatus*, being eaten along with increased consumption of sabellid worms in general. Eighteen species living on the surface or immediately below or above it and 18 infaunal species were consumed in greater quantity once trawling commenced.

Winter flounder, like Atlantic cod, is a visual predator, feeding primarily on polychaetes, although they are considered generalist feeders (Keats 1990; Martell and McClelland 1992; Methven 1999). Mollusks are not normally a large component of their diet, but on Georges Bank, Langton and Bowman (1981) recorded 15.7% of their stomach contents as being unidentified bivalves. On Western Bank, the polychaete *T. cincinnatus* dominated both the abundance and the weight of their stomach contents, though *M. modiolus* accounted for 31% of the weight. Trawling resulted in increased consumption of 20 infaunal species, the amphipod *Unciola* sp., ampharetid polychaetes in general, and *M. modiolus*, while another polychaete, *Notomastus latericeus*, maldanid worms, and *M. modiolus* increased in terms of the weight consumed.

Detailed dietary changes associated with fishing disturbance in the Northwest Atlantic have rarely been examined through experimentation, and this study adds significantly to existing knowledge of the effects of trawling on fish feeding. Two classes of effects are documented here: (1) quantitative changes (positive or negative) to the proportion of species consumed and (2) qualitative changes through opportunistic feeding on novel food items. All five focus predator species demonstrated proportional changes in their diets associated with the trawling disturbance, while Atlantic cod, American plaice, and yellowtail flounder also exploited new diet items. Where dietary shifts were associated with the habitat of the prey, they were identified as increases in consumption of infauna and species living on or near the surface (below or above). This result is not surprising, given that trawling impacts are focused on the surface and shallow subsurface. However, only large Atlantic cod showed increased consumption of epifaunal species after trawling commenced. This result was not expected and suggests that the trawling disturbance did not alter the bioavailability of epifaunal prey species as a group for the other fish (despite their consumption). This result may be reflective of the properties of the benthic taxa comprising this group (as with the other habitat-based groupings) or could be due to the rockhopper gear reducing the expected impact on those taxa. Larger Atlantic cod also increased consumption of tubicolous species, of which five, including *T. cincinnatus*, were present in the diet.

Several of the diet changes reported here as resulting from the effects of trawling were between taxa that have different energy contents. Some shifts increased the proportion of high-energy food items, as with fish consumption in haddock, while others were toward lower-energy food, as in *M. modiolus* at the expense of fish in Atlantic cod. Changes in diet over long temporal and spatial scales have been linked to deleterious effects on fish populations (cf. Link and Garrison 2002). However, it is debatable what, if any, population-level effects a change in diet in trawl tracks would have.

Most of the invertebrate prey species that were consumed in greater amounts, or added to the diets of the fish after trawling, were abundant in the area and were epifaunal or subsurface dwellers, a result that could be expected. Trawling was seen to have an effect on the benthic community at our study location (Kenchington et al., unpublished data). A high percentage of benthic sampling stations on the experimental line showed some degree of biological damage, as determined from high-resolution photographs taken immediately after trawling. *Modiolus modiolus* and *T. cincinnatus* were among the species noted as suffering damage, and those were also prominent in the dietary changes noted here. For those two species, trawl-induced damage was clearly implicated in their increased availability to predators.

Our data show evidence of fish moving into the area to scavenge, as indicated by increasing trawl catches in areas already trawled. The catch weight of the visual predators Atlantic cod, haddock, and winter flounder increased significantly between the two periods of the trawl sets. The failure to detect significant decreases in American plaice and yellowtail flounder implies that low levels of movement into the area may also be occurring in those species.

Our study was not designed to determine the spatial limits of the changes in diets nor to quantify the behavior of the fish in responding to the trawl. Such information would be necessary to place our results into the context of ecosystem effects of trawling. However, this study does provide detailed insight into the feeding preferences of demersal fish, trophic linkages with the benthos, how those fish react to trawling disturbance both collectively and as individuals, and what benthic species are most affected.

Acknowledgments

We thank the officers and crews of the CCGS *Parizeau*, CCGS *Hudson*, CCGS *Needler*, and CCGS *Teleost* for their able assistance in carrying out this major field experiment. The Geological Survey of Canada (Atlantic) provided the Trackpoint and AGCNav systems that were critical to the success of the experiment. Technical assistance on board the CCGS *Needler* and CCGS *Teleost* was provided by the Marine Fish Division of the Department of Fisheries and Oceans, Canada, Maritimes Region. M. Moeleker, J. Ford, and D. MacDonald assisted in the processing of samples. T. Kenchington, K. Zwanenburg, and K. Frank and two anonymous reviewers provided comments on the manuscript. Funding for this project was provided by the Department of Fisheries and Oceans Environmental Sciences Strategic Research Fund.

References

Auster, P. J., and R. W. Langton. 1999. The effects of fishing on fish habitat. Pages 150–187 *in* L. R. Benaka, editor. Fish habitats: essential fish habitat and rehabilitation. American Fisheries Society, Symposium 22, Bethesda, Maryland.

Collie, J. S., G. A. Escanero, and P. C. Valentine. 1997. Effects of bottom fishing on the benthic megafauna of Georges Bank. Marine Ecology Progress Series 155:159–172.

Collie, J. S., S. J. Hall, M. J. Kaiser, and I. R. Poiner. 2000. A quantitative analysis of fishing impacts on shelf-sea benthos. Journal of Animal Ecology 69:785–798.

Fisher, J. A. D., and K. T. Frank. 2002. Changes in finfish community structure associated with the implementation of a large offshore fishery closed area on the Scotian Shelf. Marine Ecology Progress Series 240:249–265.

Frank, K. T., N. L. Shackell, and J. E. Simon. 2000. An evaluation of the Emerald/Western Bank juvenile haddock closed area. ICES Journal of Marine Science 57:1023–1034.

Gerasimova, O. V., L. K Albikovskaya, and S. P. Melnikov. 1991. Preliminary results from feeding analysis for abundant commercial fishes on the Newfoundland Bank in April–May 1991. Northwest Atlantic Fisheries Organization, NAFO SCR Document 91/125, Serial Number N2018, Dartmouth, Nova Scotia.

Ivlev, V. S. 1961. Experimental ecology of the feeding of fishes. Yale University Press, New Haven, Connecticut.

Keats, D. W. 1990. The food of winter flounder *Pseudopleuronectes americanus* in a sea urchin dominated community in eastern Newfoundland. Marine Ecology Progress Series 60:13–22.

King, L. H. 1970. Surficial geology of the Halifax-Sable Island map area. Geological Survey of Canada, Marine Sciences Paper 1, .

Klemetsen, A. 1982. Food and feeding habits of cod from the Balsfjord, northern Norway during a one-year period. Journal du Conseil International pour l'Exploration de la Mer 40:101–111.

Kohler, A. C., and D. N. Fitzgerald. 1969. Comparisons of food of cod and haddock in the Gulf of St. Lawrence and on the Nova Scotia banks. Journal of the Fisheries Research Board of Canada 26:1273–1287.

Kulka, D. W., and D. A. Pitcher. 2001. Spatial and temporal patterns in trawling activity in the Canadian Atlantic and Pacific. International Council for the Exploration of the Sea, ICES CM 2001/R:02, Copenhagen.

Langton, R., P. J. Auster, and D. C. Schneider. 1995. A spatial and temporal perspective on research and management of groundfish in the northwest Atlantic. Reviews in Fisheries Science 3:201–229.

Langton, R. W., and R. E. Bowman. 1980. Food of fifteen northwest Atlantic gadiform fishes. NOAA Technical Report NMFS SSRF-740.

Langton, R. W., and R. E. Bowman. 1981. Food of eight northwest Atlantic pleuronectiform fishes. NOAA Technical Report NMFS SSRF-749.

Langton, R. W., and L. Watling. 1990. The fish-benthos connection: a definition of prey groups in the Gulf of Maine. Pages 424–438 in M. Barnes and R. N. Gibson, editors. Trophic relationships in the marine environment. Proceedings of the 24th European Marine Biological Symposium. Aberdeen University Press, Aberdeen, UK.

Libey, G. S., and C. F. Cole. 1979. Food habits of yellowtail flounder, *Limanda ferruginea* (Storer). Journal of Fish Biology 15:371–374.

Lilly, G. R., and S. Meron. 1986. Propeller clam (*Cyrtodaria siliqua*) from stomachs of Atlantic cod (*Gadus morhua*) on the southern Grand Bank (NAFO Div. 3NO): natural prey or an instance of net feeding. International Council for the Exploration of the Sea, ICES CM 1986:36, Cophenhagen.

Link, J. S., K. Bolles, and C. G. Milliken. 2002. The feeding ecology of flatfish in the Northwest Atlantic. Journal of Northwest Atlantic Fishery Science 30:1–17.

Link, J. S., and L. P. Garrison. 2002. Trophic ecology of Atlantic cod *Gadus morhua* on the northeast US continental shelf. Marine Ecology Progress Series 227:109–123.

Martell, D. J., and G. McClelland. 1992. Prey spectra of pleuronectids (*Hippoglossoides platessoides, Pleuronectes ferrugineus, Pleuronectes americanus*) from Sable Island Bank. Canadian Technical Report of Fisheries and Aquatic Sciences 1895.

Mattson, S. 1992. Food and feeding habits of fish over a soft sublittoral bottom in the northeast Atlantic. 3. Haddock (*Melanogrammus aeglefinus* (L.)) (Gadidae). Sarsia 77:33–45.

McCallum, B. R., and S. J. Walsh. 1997. Groundfish survey trawls used at the Northwest Atlantic Fisheries Centre, 1971 to present. NAFO Scientific Council Studies 29:93–104.

Methven, D. A. 1999. Annotated bibliography of demersal fish feeding with emphasis on selected studies from the Scotian Shelf and Grand Banks of the Northwestern Atlantic. Canadian Technical Report of Fisheries and Aquatic Sciences 2267.

National Research Council. 2002. Effects of trawling and dredging on seafloor habitat. Committee of Ecosystem Effects of Fishing, Ocean Studies Board, National Academy Press, Washington, D.C.

Pitt, K. T. 1976. Food of yellowtail flounder on the Grand Bank and a comparison with American plaice. Pages 23–27 in International Commission for the Northwest Atlantic Fisheries Research Bulletin 12.

Prena, J., P. Schwinghamer, T. W. Rowell, D. C. Gordon, Jr., K. D. Gilkinson, W. P. Vass, and D. L. McKeown. 1999. Experimental otter trawling on a sandy bottom ecosystem of the Grand Banks of Newfoundland: analysis of trawl bycatch and effects on epifauna. Marine Ecology Progress Series 181:107–124.

Rowell, T. W., P. Schwinghamer, K. Gilkinson, D. C. Gordon, Jr., E. Hartgers, M. Hawryluk, D. L. McKeown, J. Prena, W. P. Vass, and P. Woo. 1997. Grand Banks otter trawling impact experiment: III. Sampling equipment, experimental design and methodology. Canadian Technical Report on Fisheries and Aquatic Sciences 2190.

Sainsbury, K. J., R. A. Campbell, R. Lindholm, and A. W. Whitelaw. 1997. Experimental management of an Australian multispecies fishery: examining the possibility of trawl induced habitat modification. Pages 107–112 in E. K. Pikitch, D. D. Huppert, and M. P. Sissenwine, editors. Global trends: fisheries management. American Fisheries Society, Symposium 20, Bethesda, Maryland.

Scott, W. B., and M. J. Scott. 1988. Atlantic fishes of Canada. Canadian Bulletin of Fisheries and Aquatic Sciences 219.

Tupper, M., and R. G. Boutilier. 1995. Effects of habitat on settlement, growth, and postsettlement survival of Atlantic cod (*Gadus morhua*). Canadian Journal of Fisheries and Aquatic Sciences 52:1834–1841.

Wigley, R. L., and R. B. Theroux. 1965. Seasonal food habits of highlands ground haddock. Transactions of the American Fisheries Society 94:243–251.

Summary of the Grand Banks Otter Trawling Experiment (1993–1995): Effects on Benthic Habitat and Macrobenthic Communities

DONALD C. GORDON, JR.[1]

*Department of Fisheries and Oceans, Maritimes Region, Bedford Institute of Oceanography,
Post Office Box 1006, Dartmouth, Nova Scotia B2Y 4A2, Canada*

KENT D. GILKINSON[2]

*Department of Fisheries and Oceans, Newfoundland Region, Northwest Atlantic Fisheries Centre,
Post Office Box 5667, St. John's, Newfoundland A1C 5X1, Canada*

ELLEN L. KENCHINGTON[3], CYNTHIA BOURBONNAIS[4], KEVIN G. MACISAAC[5],
DAVID L. MCKEOWN,[6] AND W. PETER VASS[7]

*Department of Fisheries and Oceans, Maritimes Region, Bedford Institute of Oceanography,
Post Office Box 1006, Dartmouth, Nova Scotia B2Y 4A2, Canada*

Abstract. A 3-year experiment was conducted to examine the effects of repetitive otter trawling on a sandy bottom ecosystem at a depth of 120–146 m on the Grand Banks of Newfoundland. Immediate effects were observed on habitat structure. However, these effects were relatively short-lived since the available evidence shows that the habitat recovered in about a year or less. Except perhaps for snow crabs *Chionoecetes opilio* and basket stars *Gorgonocephalus arcticus*, direct removal of epibenthic fauna by the otter trawl appeared to be insignificant because of its very low efficiency in catching benthic organisms. Immediately after trawling, the mean biomass of epibenthic organisms (as sampled with an epibenthic sled) was reduced by an average 24%. The most affected species were snow crabs, basket stars, sand dollars *Echinarachnius parma*, brittle stars *Ophiura sarsi*, sea urchins *Strongylocentrotus pallidus* and soft corals *Gersemia* spp. The immediate impacts of otter trawling on the infauna (as sampled by a large videograb) appeared to be minor and limited to a few species of polychaetes. Significant effects could not be detected on the majority of species found at the study site, including all mollusk species. All available evidence suggests that the biological community recovered from the annual trawling disturbance in a year or less, and no significant effects could be seen on benthic community structure after 3 years of otter trawling. The habitat and biological community at the experimental site are naturally dynamic and exhibited marked changes irrespective of trawling activity, and this natural variability appeared to overshadow the effects of trawling.

Introduction

Many studies have been conducted to investigate the impacts of mobile fishing gear on benthic habitat and communities (e.g., Auster and Langton 1999; Collie et al. 2000; Johnson 2002; National Research Council 2002). One of the most widely used mobile gear types is the otter trawl (Messieh et al. 1991). Most otter trawling impact experiments have been conducted in depths less than 100 m, been of limited spatial scale, and considered only a single disturbance event (Collie et al. 2000). Data on the recovery of benthic habitat and communities from trawl disturbance are particularly sparse.

Understanding the impacts of mobile gear on benthic habitat and communities is a difficult and expensive undertaking, especially in highly variable offshore marine

---

[1] Corresponding author: gordond@mar.dfo-mpo.gc.ca
[2] E-mail: gilkinsonk@dfo-mpo.gc.ca
[3] E-mail: kenchingtone@mar.dfo-mpo.gc.ca
[4] E-mail: bourbonnaisc@mar.dfo-mpo.gc.ca
[5] E-mail: macisaackg@mar.dfo-mpo.gc.ca
[6] E-mail: mckeownd@mar.dfo-mpo.gc.ca
[7] E-mail: vassp@mar.dfo-mpo.gc.ca

environments. Small-scale manipulative experiments can provide direct evidence on immediate impacts of a known disturbance event (e.g., gear type, location, or intensity) on a particular habitat. Alternatively, observational studies comparing areas with different fishing histories, or observations over time in the same area, can provide indirect evidence of longer-term impacts at the spatial scale of whole fishing grounds, though the conclusions may be equivocal as natural fluctuations may be mistakenly attributed to human impacts. With observational studies, it can be difficult to study the impacts of specific gear types since fished areas often have histories of multiple gear usage. Regardless of the experimental approach taken, it is necessary to include observations in reference areas not influenced by previous fishing disturbance.

During 1993–1995, we conducted a large-scale manipulative experiment on the effects of otter trawling on a deep sandy bottom ecosystem on the Grand Banks of Newfoundland. Our objective was to investigate the immediate impacts of a high level of repetitive trawling disturbance, the subsequent recovery, and possible medium-term impacts on benthic habitat and communities. Considerable care went into the selection of the undisturbed experimental site. A wide variety of acoustic, imaging, and sampling tools was used to examine potential effects on a broad spectrum of habitat and biological variables at different spatial scales. Results have been presented in a series of technical reports (Prena et al. 1996; McKeown and Gordon 1997; Rowell et al. 1997; Gordon et al. 2002), scientific papers (Guigné et al. 1993, Schwinghamer et al. 1996; Gilkinson et al. 1998; Schwinghamer et al. 1998; Prena et al. 1999; Kenchington et al. 2001), and a thesis (Gilkinson 1999). This paper provides an overview and synthesis of one of the most comprehensive otter trawling impact experiments conducted.

Methods

Site Selection

After careful consideration of numerous factors (Prena et al. 1996), an experimental site was selected on the Grand Banks of Newfoundland (Figure 1). Analysis of commercial fishing effort data indicated that the site had not been subjected to trawling for at least 13 years, and it was closed to all fishing activity for the duration of the experiment. Therefore, the benthic habitat and community should have been in a relatively natural state with little if any modification by human disturbance. Depth range was 120–146 m. The sandy seabed was representative of large areas of the continental shelf off Atlantic Canada. Seabed properties were relatively uniform, both spatially and temporally, potentially making it easier to detect trawling effects above natural variability. The seabed was relatively flat and featureless (except on a micro-scale) and easily sampled, and samples could be processed in a reasonable time. The benthic community was diverse with high abundance and biomass, including abundant epibenthic species. The species composition showed a relatively high level of homogeneity, making it potentially easier to detect trawling effects.

Experimental Design

Three, 13-km-long experimental corridors were established within the 20-km × 20-km experimental box, each with a different heading and parallel reference corridor 300 m to one side (Figure 1). Experimental and reference corridors were sampled with a wide variety of equipment at different times and spatial scales using the C.C.G.S. *Parizeau* (Table 1). Time constraints did not allow us to sample all corridors equally. It was our intention to sample experimental corridors both before and after trawling each year. However, this was achieved only for the side-scan sonar surveys and videograbs (Table 1). Quantitative biological sampling in reference corridors was carried out before retrawling, with the exception of epibenthic sled samples in 1994. Changes in the reference corridors were assumed to reflect natural variability.

Experimental Trawling

Each year in July, the three experimental corridors were trawled 12 times with an Engel 145 otter trawl equipped with 1,250-kg polyvalent oval doors and rockhopper footgear using the C.C.G.S. *Wilfred Templeman*. This trawl had an average door spread of 60 m and wing spread (i.e., footgear) of about 20 m. The rockhopper footgear had a diameter of 46 cm. Mesh size was 18 cm on the wings and belly of the net and 13–15 cm in the cod end. A 3-cm square mesh liner was installed in the cod end to capture organisms damaged but not retained by commercial gear. This trawling intensity was considered to be a relatively high level of disturbance, roughly equivalent to the greatest commercial trawling intensity on the Grand Banks in recent years (Kulka and Pitcher 2001).

The widths of the disturbance zones created were estimated from the navigation data to be on the order of 120–250 m depending upon year (McKeown and Gordon 1997). Cumulative trawling disturbance was greatest nearer the center lines of the trawled corridors, but the local intensity of the impact from doors, ground warps, footgear, and net was necessarily uneven within the trawled area.

Trawl Catch and RoxAnn

Trawl catch data were collected by set each year in all three corridors (Prena et al. 1999). Fish and invertebrates

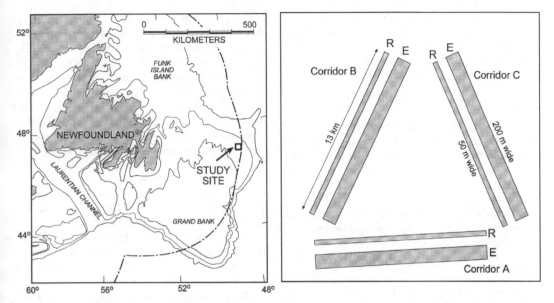

Figure 1. Location of the experimental box (20 × 20 km) on the Grand Banks of Newfoundland and the relative positions of Corridors A, B, and C (not to scale). The reference corridors (R) were 300 m to one side of the experimental (i.e., trawled) corridors (E). Ends of the corridors were about 3.5 km apart.

Table 1. Summary of the 3-year sampling program (1993–1995). EXP = experimental (trawled) corridor; REF = reference corridor; B = number of deployments of sampling gear before trawling; and A = the number of deployments of sampling gear after trawling. Area sampled is estimated for a single deployment of each gear type (i.e., one trawl set, one side-scan survey, one sled tow, one videograb station, etc.).

			Corridor A				Corridor B				Corridor C			
			EXP		REF		EXP		REF		EXP		REF	
Gear	Area sampled (m²)	Date	B	A	B	A	B	A	B	A	B	A	B	A
Otter trawl	26×10⁴ m²	Jul 93	12	0	0	0	12	0	0	0	12	0	0	0
		Jul 94	12	0	0	0	12	0	0	0	12	0	0	0
		Jul 95	12	0	0	0	12	0	0	0	12	0	0	0
RoxAnn	20×10⁴ m²	Jul 95	12	0	0	0	12	0	0	0	12	0	0	0
Sidescan	520×10⁴ m²	Jul 93	1	1	1	1	1	1	1	1	0	1	0	1
		Sep 93	1	0	0	0	1	0	0	0	1	0	0	0
		Jul 94	1	1	0	1	1	1	0	1	1	1	0	1
		Jul 95	1	1	0	0	1	1	0	0	0	0	0	0
BRUTIV	2.6×10⁴ m²	Jul 93	0	1	0	1	0	0	0	0	0	0	0	0
		Jul 95	1	2	1	0	0	0	0	0	0	0	0	0
Sled	17 m²	Jul 93	2	4	2	0	2	4	2	0	0	0	0	0
		Sep 93	3	0	4	0	1	0	1	0	0	0	0	0
		Jul 94	0	10	0	10	0	10	0	10	0	0	0	0
		Jul 95	0	10	10	0	0	10	10	0	0	0	0	0
Videograb	0.5 m²	Jul 93	5	10	5	0	5	10	5	0	0	0	0	0
		Sep 93	10	0	10	0	10	0	10	0	0	0	0	0
		Jul 94	10	10	10	0	10	10	10	0	0	0	0	0
		Jul 95	10	10	10	0	10	10	10	0	0	0	0	0
DRUMS	0.036 m²	Jul 94	10	10	10	0	10	10	10	0	0	0	0	0
		Jul 95	10	10	10	0	10	10	10	0	0	0	0	0

were separated by species and weighed. In 1995, the *Templeman* was equipped with a RoxAnn acoustic bottom classification system and collected data on all trawl sets (Schwinghamer et al. 1998). These data were subsequently processed to examine potential changes in acoustic characteristics of the seabed.

Side-Scan Sonar

Side-scan sonar surveys were run each year before and after experimental trawling (Schwinghamer et al. 1998). The results were used to provide a general qualitative description of the seabed at the experimental site, ensure that there was no recent seabed disturbance (e.g., commercial trawling, icebergs) that might influence experimental results, provide a general picture of the degree and distribution of the trawling disturbance, and provide information on the persistence of trawl marks.

BRUTIV

In 1993 and 1995, black-and-white video surveys of the seabed along Corridor A (experimental and reference) were conducted using BRUTIV (Bottom Referencing Underwater Towed Instrument Vehicle), an underwater vehicle flown a few meters above the seabed (Rowell et al. 1997; Schwinghamer et al. 1998). Field of view was about 2 m wide. The video footage was examined qualitatively for signs of trawling disturbance to habitat and epibenthic organisms.

Epibenthic Sled

Large epibenthic (i.e., surface-dwelling) organisms were collected in experimental and reference corridors with a modified Aquareve III epibenthic sled (Rowell et al. 1997; Prena et al. 1999). Each tow was about 50 m long and sampled an area of the seabed on the order of 17 m^2. The cutting blade was set at a depth of 2–3 cm. The collection box had 1-cm-diameter holes so smaller organisms were not retained. Performance was monitored by a color video camera directed backward, showing the sled opening, and tows of dubious quality (i.e., lifting off the seabed) were aborted and repeated. The location of each tow was randomly selected (McKeown and Gordon 1997).

Immediately after retrieval, the sled catch was transferred to a sorting table and washed with seawater over a 1-mm-mesh screen. Common organisms were sorted by species, counted, and weighed (Rowell et al. 1997; Prena et al. 1999). Damage was also assessed for some species. Less abundant organisms more difficult to identify were frozen and processed ashore.

Videograb

Macrobenthic communities were sampled with the Department of Fisheries and Oceans (DFO) videograb that was developed specifically for this experiment (Schwinghamer et al. 1996; Rowell et al. 1997; Gordon et al. 2000; Kenchington et al. 2001). This hydraulically actuated bucket grab, equipped with color video cameras, provides the scientific operator the ability to view the seabed, close and open the bucket remotely, and verify that the bucket closed properly prior to recovery. The area sampled was 0.5 m^2. Most previous gear-impact studies have used grabs that sampled a much smaller area (<0.2 m^2), with limited penetration on hard sand. The location of each grab was randomly selected (McKeown and Gordon 1997).

The contents of the videograb were washed with seawater over a 1-mm-mesh screen, and all retained material (i.e., organisms and gravel) was preserved in formalin. Samples were processed ashore using standard procedures (Rowell et al. 1997; Kenchington et al. 2001) including identification to the lowest possible taxon and determination of abundance and biomass by taxon. Damage was also assessed for some species. Detailed damage classification and length measurements were made on all mollusks since this group is known to leave records of encounters with fishing gear. High-resolution video imagery of the seabed, collected before the grab closed, was viewed to provide a qualitative assessment of habitat conditions.

DRUMS

In 1994 and 1995, a dynamically responding underwater matrix sonar (DRUMS) was mounted in the videograb to provide high-resolution information on small-scale structural properties of surficial sediments before and after trawling (Guigné et al. 1993, Schwinghamer et al. 1996, 1998). Acoustic images of the seabed were collected before the videograb was closed. Fractal values were calculated and compared to examine potential changes in small-scale sediment structure down to a depth of approximately 4.5 cm.

Navigation

The use of dGPS, a Trackpoint II acoustic tracking system, and the AGCNav navigation display and logging system ensured the positioning of vessels, the otter trawl, and all sampling equipment within an accuracy of 4–20 m (McKeown and Gordon 1997). It was confirmed that all samples were collected from either disturbed or reference areas as intended.

Laboratory Experiments

An experiment investigating the effects of trawl doors on mollusks buried in sediment was conducted in an ice scour research tank at Memorial University, St. John's, Newfoundland (Gilkinson et al. 1998). An instrumented test bed of sand was built and preserved specimens of six common bivalve species collected at the experimental site and placed at species-specific depths. A full-size model of a trawl door was then towed across the surface with the shoe cutting to a depth of 2 cm. Patterns of displacement and damage were recorded. Burrowing experiments with bivalves were also conducted (Gilkinson 1999) using common species collected at the experimental site with the videograb and transported live to the Northwest Atlantic Fisheries Centre. Species-specific burrowing rates and depths were determined in sediment held in flowing seawater aquaria.

Statistics

A variety of standard univariate and multivariate statistical methods were used to process the data, in particular to quantify natural variation and trawling impacts. These included analysis of variance (ANOVA) and multi-dimensional scaling (MDS) ordination. Full details are described in Schwinghamer et al. (1996, 1998), Prena et al. (1999), Gilkinson (1999), and Kenchington et al. (2001).

Summary of Results

Otter Trawl Catch

The total fish catch in the otter trawl was extremely low and averaged just 18 kg (wet weight) for a 13-km-long set over the entire experiment (Prena et al. 1999; Figure 2). Such poor catches attest to the depressed state of groundfish resources on the Grand Banks at the time of the experiment, which led to the 1992 fishing moratorium still in effect today. The dominant species caught were American plaice *Hippoglossoides platessoides* and thorny skate *Raja radiata*. Other species captured in smaller quantities were capelin *Mallotus villosus*, Arctic cod *Boreogadus saida*, and northern sand lance *Ammodytes dubius* Very few Atlantic cod *Gadus morhua* were caught. There were no statistically significant differences in fish catch with increasing trawl set or between corridors. However, the total fish catch gradually decreased each year of the experiment (Figure 2).

The invertebrate bycatch, which averaged 10 kg (wet weight) for a 13-km-long set, was dominated by snow crabs *Chionoecetes opilio*, basket stars *Gorgonocephalus arcticus*, and sea urchins *Strongylocentrotus pallidus* (Prena et al. 1999; Figure 2). Other species included soft corals *Gersemia* spp., whelks *Buccinum* spp., and hermit crabs *Pagurus* spp. Iceland scallops *Chlamys islandica* were occasionally caught. These are all large, surface-dwelling species and most have some degree of mobility. The average biomass of invertebrates captured by the trawl (per square meter of bottom sampled) was just 0.01% of that captured by the epibenthic sled, which indicates the extremely low efficiency of the otter trawl in collecting most epibenthic organisms. However, the trawl was relatively efficient in capturing snow crabs and basket stars, and their biomass decreased over the 12 trawl sets. An influx of adult male snow crabs into the trawled corridors was observed after the first six trawl sets (approximately 10–12 h). It is presumed that they were migrating into the disturbed zone to feed on damaged organisms. There were significant differences in the biomass of

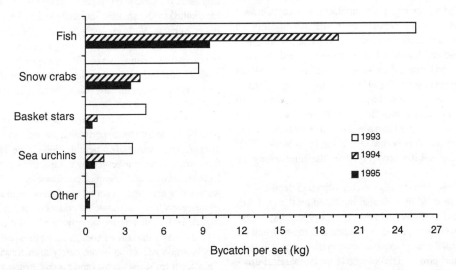

Figure 2. Otter trawl fish catch and invertebrate bycatch by year expressed as wet weight biomass per trawl set (13 km long). Average of all sets and corridors. Redrawn from Prena et al. (1999).

invertebrate bycatch among trawl set, corridor, and year (Prena et al. 1999).

Effects on Habitat

Side-scan sonar surveys covered a very large area (Table 1). At the start of the experiment, they revealed a physically uniform seabed with no evidence of any large-scale features indicating sediment transport. Some iceberg furrows were observed, but to our knowledge, never sampled. There was no evidence of fishing disturbance other than the experimental trawling. Trawl door tracks, and in some cases disturbance from footgear, were readily visible immediately after trawling and 10 weeks later. The 1993 tracks were not visible in 1994, but the 1994 tracks were faintly visible in 1995.

There was a significant increase in the RoxAnn E2 signal in the experimental corridors during trawling in 1995 (Schwinghamer et al. 1998). The E2 signal is considered to be a proxy for sediment hardness (Chivers et al. 1990; Schwinghamer et al. 1998 mistakenly attributed the E2 signal to sediment roughness). These results indicate an immediate change in seabed acoustic properties at the scale of the echo sounder footprint (about 200 m^2) due to the trawling disturbance.

Continuous video observations made with BRUTIV over large areas (Table 1) revealed that the seabed in freshly trawled corridors was lighter in color than that in reference corridors and that organisms and shell hash tended to be organized into linear features parallel to the direction of trawling (Schwinghamer et al. 1998). Door tracks and damaged organisms were also visible on occasion. High-resolution video observations made with the videograb within a very limited area indicated that untrawled seabed had a hummocky, mottled appearance with abundant organic detritus, while recently trawled seabed was generally smoother and cleaner. While some new structural features were created by the doors (i.e., furrows and berms), the visual observations showed that trawling generally reduced small-scale biological structures on the seabed surface, presumably due in part to sediment resuspension. However, there was no visible difference in seabed surface structures of trawled and reference corridors before retrawling in 1994 and 1995, suggesting habitat recovery over the intervening 12 months.

A loss in biological structure down to depths of 4.5 cm due to trawling was also indicated by the DRUMS acoustic data (Schwinghamer et al. 1996, 1998). These results were interpreted to indicate that the trawling reduced small-scale biogenic structures such as mounds, tubes, and burrows. The observed loss of biological structures may explain the increase in seabed hardness detected by RoxAnn.

The experimental otter trawling had no apparent effect on the grain size of surficial sediments (Schwinghamer et al. 1998). However, there was a suggestion of some temporal changes in the finer fractions over the 3-year experiment which are thought to be natural in origin. The trawling did not appear to affect organic carbon and nitrogen in surficial sediments in the first 2 years while concentrations were significantly lower after trawling in 1995 (Kenchington et al. 2001).

In summary, immediate physical impacts of the experimental trawling on benthic habitat were substantial. Most apparent were the furrows and berms created by the doors, but perhaps more important was the general homogenization and reduction of small scale biological structures caused by sediment resuspension and destruction by gear, which could have important implications for marine biodiversity (Thrush et al. 2001).

Effects on Organisms Sampled with the Epibenthic Sled

Most organisms collected with the epibenthic sled were large, epibenthic species, but since the blade penetrated 2–3 cm into the seabed, some shallow burrowing infauna were also collected. The total biomass of organisms collected averaged 400 g (wet weight) per m^2, and the total number of species captured was 115 (in 96 samples; Prena et al. 1999). The dominant epibenthic species at the experimental site, which comprised 95–98% of the biomass, were sand dollars *Echinarachnius parma*, brittle stars *Ophiura sarsi*, sea urchins, snow crabs, soft corals, and four mollusks (*Astarte borealis* [known as boreal astarte], *Margarites sordidus*, *Clinocardium ciliatum* [known as hairy cockle], and *Cyclocardia novangliae* [known as New England cyclocardia]). With the exception of *M. sordidus*, the mollusks were shallow, infaunal species. A tube-dwelling polychaete, *Nothria conchylega*, was also abundant in epibenthic sled samples but was not processed because of the extensive sorting time required (however, it was processed in the videograb samples.). Only sea urchins and snow crabs were commonly caught by both the epibenthic sled and the otter trawl.

Unfortunately, due to time constraints, it was not possible to sample the experimental corridors before and after trawling with the epibenthic sled (Table 1). Therefore, impacts were assessed by comparing samples collected in trawled and reference corridors. Immediately after trawling each year, the total biomass of organisms was significantly lower in trawled corridors (Prena et al. 1999; Figure 3). This difference was greatest in 1995 and averaged 24% over the entire experiment. At the species level, this biomass reduction immediately after trawling was significant for snow crabs, sand dollars, brittle stars, sea urchins, and soft corals. Actual impacts on snow crabs were probably greatest because of their rapid migration into trawled corridors (i.e., some of the snow crabs sampled after trawling were not present before). Judging by their capture rate in the otter trawl (Figure 2), there was probably a similar reduction in the biomass of basket stars, but this was not detected in the epibenthic sled data, probably because this species was poorly sampled. Using the technique of comparing regressions of the mean and standard deviation of abundance of all species (Warwick and Clarke 1993), it appears that trawling decreased the homogeneity and increased the aggregation of organisms sampled by the epibenthic sled. This presumably reflects the realignment of organisms into rows as seen in the BRUTIV video footage. Sand dollars, brittle stars, and sea urchins demonstrated significant levels of damage from trawling. In addition, the mean individual biomass of epibenthic organisms was lower in trawled corridors, suggesting size specific impacts of trawling, especially for sand dollars (Prena et al. 1999). However, no significant effects of trawling were observed on the four mollusk species commonly captured by the epibenthic sled.

The reduced biomass of epibenthic organisms in trawled corridors immediately after trawling is presumably due to several factors, including direct removal by the trawl, displacement, predation, and possibly migration of noncaptured organisms. Since it was not possible to collect epibenthic sled samples before retrawling in 1994 and 1995, there are no direct data on the recovery rates of epibenthic organisms in the trawled corridors. However, the fact that most of the affected species have some degree of mobility and that the biomass of epibenthic species in the trawled corridors was relatively constant over the 3-year experiment (Figure 3) suggests that recovery time for most affected species is on the order of a year or less.

In summary, significant immediate effects of the experimental trawling were detected on some species of large, epibenthic organisms, in particular, snow crabs, sand dollars, brittle stars, sea urchins, and soft corals. Indirect evidence suggests that recovery of these species took place within a year, most likely through migration from undisturbed areas immediately outside the narrow trawled corridors.

Effects on Organisms Sampled with the Videograb

A rich and diverse infaunal community was present at the experimental site (Kenchington et al. 2001). A total of 246 invertebrate taxa were identified in the 200 videograb samples, with an average of 68 taxa per sample. Mean biomass and abundance were on the order of 1 kg (wet weight) per m^2 and 2,100 individuals per m^2, respectively. With respect to biomass, the dominant taxa were mollusks and echinoderms. The most abundant taxonomic groups were polychaetes, mollusks, and crustaceans. Twenty-seven taxa occurred in more than 90% of the samples and accounted for 89% of the total biomass and 87% of the numerical abundance. The videograb also collected abundant epibenthic species (e.g., sand dollars and brittle stars).

Considerable temporal variability was observed in the community properties of reference corridors (Kenchington et al. 2001). These were always sampled

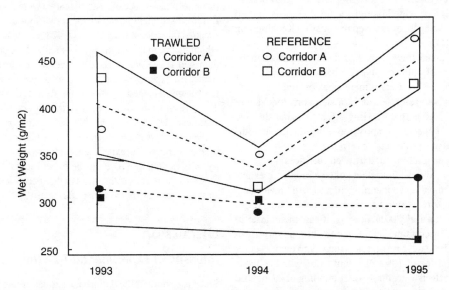

Figure 3. Total biomass of invertebrates collected by the epibenthic sled in trawled and experimental corridors (A and B only) each year of the experiment. Solid lines indicate 95% confidence intervals. Redrawn from Prena et al. (1999).

before trawling (Table 1) and, to the best of our knowledge, were not directly affected by the otter trawling. There was a clear trend of decreasing abundance and number of species in reference corridors during the 3-year experiment. Multivariate analyses (i.e., MDS) indicated temporal changes in community structure (Figure 4). These results indicate that the benthic communities at the experimental site are naturally dynamic and exhibit temporal changes irrespective of trawling disturbance. This natural variability makes it more difficult to detect the effects of otter trawling.

The immediate effects of trawling were examined by comparing community properties before and after trawling each year in the trawled corridors (Table 1; Kenchington et al. 2001). Few significant effects were observed in 1993 and 1995. However, in 1994, there was a significant drop in total community abundance immediately after trawling, with polychaetes being the most affected taxonomic group (Table 2). In the same year, there was no significant drop in the total community biomass, but sand dollars and several species of polychaetes were negatively affected. No significant immediate effects of trawling were detected on the number of species present, species diversity, or evenness. No significant damage to sand dollars or brittle stars could be detected in the videograb samples. Multivariate analyses indicated changes in community structure only in 1994. It may have been easier to detect immediate impacts in 1994 because that year the trawling disturbance was more concentrated along the center line of trawled corridors and there was a longer delay between the cessation of trawling and videograb sampling which provided more time for predators to feed on dead, damaged, or exposed organisms (Kenchington et al. 2001).

The annual trawling had no significant effect on the abundance of adult or juvenile mollusks in any year (Gilkinson 1999). Recruitment rates of bivalves were similar in both reference and trawled corridors. There was no evidence that trawling affected either the size structure or species composition of mollusks. In addition, there was no detectable difference in mollusk damage between reference and trawled corridors.

The medium-term effects of trawling were examined by comparing community properties in trawled corridors before trawling with reference corridors (Kenchington et al. 2001). Few significant medium-term effects on macrofauna abundance and biomass were observed. These were restricted primarily to eight species of polychaetes in 1994. It is not clear whether these apparent effects were due to otter trawling or the natural spatial and temporal variability in the benthic communities. However, there was a significant increase in the frequency of damaged sand dollars in trawled corridors,

especially for the larger organisms. Ten dominant species were remarkably stable during the experiment and showed no effects of either natural or trawling disturbance (Table 2). The overall effect of trawling on the benthic community was examined by MDS. After 3 years of trawling, there was considerable overlap between trawled and reference corridors (Figure 4). The largest driving force for change over the 3-year period was time (Kenchington et al. 2001).

In summary, it appears that the effects of otter trawling on the abundant and diverse macrobenthic organisms present at the experimental site, as sampled by the videograb, are relatively minor and primarily restricted to sand dollars and some species of polychaetes. A significant effect on sand dollars was also evident in the epibenthic sled data. Some of the observed drop in polychaete abundance most likely is due to predation but some could also be due to displacement outside the trawled corridors. Echinoderms (except sand dollars), mollusks, and crustaceans experienced little or no effect. The major signal observed in the data was a natural temporal trend over the 3-year experiment. The minor impacts observed appear to have shifted the benthic community in the same direction as the natural changes. The macrobenthic community appears to have recovered fully from the trawling disturbance within a year or less and no medium-term effects could be detected.

Laboratory Experiments

The results of the laboratory experiments (Gilkinson et al. 1998) indicated that small and medium-size bivalves living on or near the surface of a sandy seabed were displaced in fluidized sediment ahead of the trawl door and thereby escaped direct hits. Only 2 out of 42 recovered bivalves which had been buried in the scour path of the door were damaged. Large, near-surface bivalves, which are not common at the experimental site, were not displaced and could have been damaged by direct hits from the trawl door. These results are consistent with the observations of no significant detectable impacts of otter trawling on mollusks in the field experiment (Prena et al. 1999; Kenchington et al. 2001). Burrowing experiments indicated that the majority of bivalve taxa (70%) at the experimental site were shallow burrowers and that most of the species tested were slow to very slow burrowers (Gilkinson 1999).

Discussion

Intensity of Experimental Trawling

Since the width of the disturbance zones created was on the order of twice the distance between the trawl doors, most areas within the trawled corridors were probably swept by 3–6 sets of the 12 sets each year (i.e., 300–

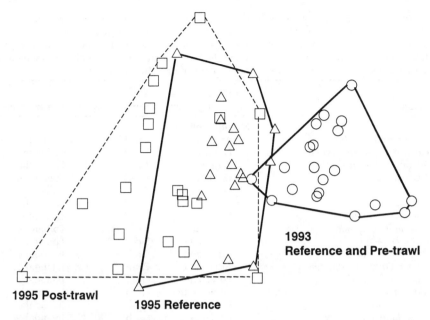

Figure 4. Two-dimensional multi-dimensional scaling configuration of the 1993 (reference and experimental pre-trawl combined) and 1995 (reference and experimental post-trawl) videograb samples based on Bray-Curtis dissimilarity from $\log_{10}$-transformed abundances of 196 species. Circles denote 1993 reference and pre-trawl collections, triangles denote 1995 reference collections, and squares denote 1995 experimental post-trawl collections. Redrawn from Kenchington et al. (2001).

600% per year). Analysis of the DFO offshore trawler observer program database from Atlantic Canada, where data was collected between 1980 and 1998 and scaled up to total effort, indicated that the detailed distribution of effort has been very patchy (Kulka and Pitcher 2001). Most trawling was concentrated in specific regions, and large areas were untrawled. The area of seabed swept by otter trawls at an intensity greater than 100% (i.e., more than once a year) was generally less than 1.5% of the total shelf area. The maximum trawling intensity calculated ranged from 141% to 644% per year (i.e., the seabed was swept by a trawl 1–6 times a year). Therefore, the trawling intensity applied in our experiment is at the upper end of the range of effort applied by the commercial fleet in Atlantic Canada in recent years. However, it should be kept in mind that this disturbance was concentrated into just a few days each year while commercial effort is generally spread out over several months or the entire year. This difference could have implications for both immediate impacts and recovery potential.

Impacted Organisms

Benthic organisms vary in their sensitivity to trawling disturbance due to factors such as life position and size. In our experiment, just 25 species out of a total of 246 taxa appeared to be negatively affected to some degree by the repetitive otter trawling (Table 2). As expected, many of these were large epibenthic species which were significantly affected each year of the experiment (Figure 3). In contrast, the effects on infaunal species were generally restricted to 1 year (1994). However, not all the immediate impacts were negative, and two infaunal species showed a significant increase in either abundance or biomass (Table 2). In addition, there were 13 common species which appeared to be unaffected by both trawling and natural variation (Table 2). These species showed little variation around their mean abundance suggesting genuine stability with time rather than a lack of detectable change. Overall, it appears that the organism most sensitive to trawling was the sand dollar, which is not surprising because it is epibenthic, abundant (~100 per m^2), attains a large size (~8 cm in diameter), and has a fragile test. In general, our results agree with the conclusions of other studies that the species most vulnerable to otter trawling are large, epibenthic, and slow-growing (e.g., Auster and Langton 1999; Collie et al. 2000) species.

Experimental Design

Our experimental design allowed three separate assessments of immediate impacts and two assessments of re-

Table 2. Common taxa at the experimental site that showed an immediate negative effect (25 taxa), a positive effect (2 taxa), and no effect (13 taxa) from repetitive otter trawling, based on analysis of samples collected by the otter trawl, epibenthic sled, and videograb. Total number of taxa was 246. While the distinction between epibenthic and infaunal is not always clear for some species, predominant life position is given.

Taxon	Phylum, class or subphylum	Sampling Gear	Life Position
Negative effect			
Chionoecetes opilio	Arthropod	Trawl, sled	Epibenthic
Gorgonocephalus arcticus	Echinoderm	Trawl	Epibenthic
Strongylocentrotus pallidus	Echinoderm	Trawl, sled	Epibenthic
Gersemia spp.	Cnidarian	Trawl, sled	Epibenthic
Buccinum spp.	Gastropod	Trawl	Epibenthic
Pagurus spp.	Crustacean	Trawl	Epibenthic
Chlamys islandica	Bivalve	Trawl	Epibenthic
Echinarachnius parma	Echinoderm	Sled, grab	Epibenthic
Ophuria sarsi	Echinoderm	Sled	Epibenthic
Protomedeia fasciata	Arthropod	Grab	Infaunal
Monoculodes intermedius	Arthropod	Grab	Infaunal
Capitella capitella	Polychaete	Grab	Infaunal
Mediomastus ambiseta	Polychaete	Grab	Infaunal
Eteone longa	Polychaete	Grab	Infaunal
Apistobranchus typicus	Polychaete	Grab	Infaunal
Chaetozone setosa	Polychaete	Grab	Infaunal
Polydora socialis	Polychaete	Grab	Infaunal
Prionospio steenstrupi	Polychaete	Grab	Infaunal
Spio filicornis	Polychaete	Grab	Infaunal
Ophelina breviata	Polychaete	Grab	Infaunal
Arcidea catherinae	Polychaete	Grab	Infaunal
Arcidea wassi	Polychaete	Grab	Infaunal
Polynoidae (juvenile)	Polychaete	Grab	Infaunal
Oligochaeta (undetermined)	Oligochaete	Grab	Infaunal
Ophelia limacina	Polychaete	Grab	Infaunal
Positive effect			
Yoldia myalis (known as oval yoldia)	Bivalve	Grab	Infaunal
Protomedeia grandimana	Arthropod	Grab	Infaunal
No effect			
Margarites sordidus	Gastropod	Sled	Epibenthic
Astarte borealis	Bivalve	Sled, grab	Infaunal
Clinocardium ciliatum	Bivalve	Sled	Infaunal
Cyclocardia novangliae	Bivalve	Sled	Infaunal
Cyrtodaria siliqua (known as northern propellerclam)	Bivalve	Grab	Infaunal
Macoma calcarea (known as chalky macoma)	Bivalve	Grab	Infaunal
Ampharete finmarchica	Polychaete	Grab	Infaunal
Lysippe labiata	Polychaete	Grab	Infaunal
Laphania boecki	Polychaete	Grab	Infaunal
Scoloplos armiger	Polychaete	Grab	Infaunal
Nephtys caeca	Polychaete	Grab	Infaunal
Group Maldanid	Polychaete	Grab	Infaunal
Priscillina armata	Arthropod	Grab	Infaunal

covery over 1 year. It also allowed assessment of possible cumulative effects of repetitive trawling over a 3-year period as well as a thorough assessment of natural variability in the benthic ecosystem. Overall, this experimental design provides a high probability of detecting any impacts of otter trawling on the benthic ecosystem at the experimental site.

In comparison to the other otter trawling experiments that have been conducted, our experiment remains unique in terms of its depth, spatial scale, repeat nature of trawling disturbance, duration, and scope of data collected. Only three other offshore, deep-water otter trawling impact experiments have been reported. Engel and Kvitek (1998) studied otter trawling effects on mixed sediment at 180 m in Monterey Bay off California by comparing two adjacent areas subjected to different levels of commercial trawling intensity. Smith et al. (2000) investigated the effects of commercial trawling on a soft-bottom habitat at 200 m in the Mediterranean Sea off Greece by comparing a heavily trawled lane with nearby untrawled reference areas. In a manipulative experiment like ours, Freese et al. (1999) examined the effects of single otter trawl sets on a hard-bottom habitat at 206–274 m in the Gulf of Alaska.

It was our intention to sample the trawled corridors with the epibenthic sled before retrawling in 1994 and 1995 as we did with the videograb. This would have provided direct information on both the immediate impacts of trawling and the recovery rate of the benthic organisms over the intervening year. However, time constraints prevented this, so we had to estimate impacts by comparing trawled and reference corridors (Table 1). This is probably the major shortcoming in our experiment, especially since the major biological effects were observed in the epibenthic sled database (Figure 3). In retrospect, we regret not collecting more epibenthic sled samples. Due to the rapid response of migrating organisms (e.g., snow crabs), it is also important to standardize the time interval between trawling and the biological sampling.

We also attempted to sample our reference corridors before trawling in case they might be affected by sedimentation. Unfortunately, we were unable to do this with the epibenthic sled in 1994 (Table 1) and, as discussed by Prena et al. (1999), this may have confounded the results due to possible emigration of mobile organisms to feed on dead or damaged organisms in the nearby trawled corridors. Ideally, it would have been valuable to have sampled the reference corridors both before and after trawling, which would have verified our assumption that the reference corridors were not affected by trawling. Again, time constraints prevented this, but it should be attempted in future experiments.

We considered whether it was necessary to identify organisms to the lowest taxonomic level, especially in the videograb samples which had 246 taxa. Analyses at higher levels of aggregation generally showed similar results (Kenchington et al. 2001), but we concluded that the effort needed to identify organisms to the lowest possible taxonomic level was justified because of the additional information it provided. For example, this experiment has provided a detailed description of the benthic community for a representative region of the Grand Banks that can be used in the future as a baseline for assessing long-term variations in community properties and biodiversity.

There is no one correct trawling regime to use in manipulative experiments because commercial fishing disturbance covers a continuum of intensity and frequency. However, the trawling disturbance regime we applied was clearly quite different from that usually expected from commercial trawling in an offshore setting. For example, the disturbance was limited to a narrow corridor (about 120–250 m), and therefore, undisturbed seabed was immediately adjacent on both sides. The rate of biological recovery observed in our experiment is probably higher than would take place under conditions of a wider disturbance zone where mobile organisms would have to travel greater distances to recolonize (either adult migration or larval settlement). In addition, it probably would have been more realistic to trawl once a month instead of 12 times within a day and a half, but this was impossible because of constraints in scheduling research vessel time. Differences in the timing of the disturbance in relation to natural biological cycles (e.g., growth, spawning, migration, etc.) could potentially affect both the immediate impacts and recovery period. And finally, commercial trawl fisheries usually operate in a given area for more than 3 years, the time span of this experiment. We may have observed some long-term changes if we had extended the length of the experiment.

This inability to completely replicate the disturbance imposed by commercial trawling constitutes the principal shortcoming of the manipulative experimental approach to studying gear impacts. However, its distinct advantages, such as knowing and controlling the exact disturbance regime and accounting for natural variability, still makes it a valuable approach to follow. In future experiments, consideration should be given to having the gear disturbance applied by a commercial fishing vessel following standard fishing practices within the constraints of an experimental approach.

Extrapolation of Results

The conclusions of this experiment are specific to the conditions of the disturbance regime employed (i.e., annual experimental trawling along narrow corridors surrounded by undisturbed seabed) and the characteristics of the benthic habitat and community at the experimental site (i.e., sandy bottom with few sessile epifauna). It cannot be assumed that the same conclusions would be reached for different disturbance regimes and different benthic ecosystems. Otter trawls are designed to capture demersal species while being towed in limited contact with the seabed, but other gear types such as scallop rakes and clam dredges are designed to collect benthic species that live on or in the seabed. Therefore, the physical disturbance they create is much greater for each square meter of the seabed contacted. As the intensity of use increases, so does the potential for lasting effects, especially if the time interval between disturbance events is less than the recovery period of the benthic ecosystem. Sandy-bottom habitats are generally subjected to considerable physical stress through waves and currents and tend to be relatively robust and self healing. Other benthic habitat types, including both mud and gravel bottoms, are more prone to physical damage. Our experiment site had very few sessile epibenthic species (the dominant one being the soft corals). Such species, which stand erect attached to the seabed, are particularly vulnerable to damage from mobile gear, and, being nonmobile and generally slow growing, will take longer to recover. These include mussels, corals, sponges, anemones, and seapens. Therefore, the conclusions of this experiment would have been quite different if it had been conducted on a different bottom or with different gear.

Possible Effects Not Detected

Despite the care that went into the design and execution of this experiment, it is possible that some trawling effects may have been missed. The lack of proof of an effect is not synonymous with proof of an absence of impacts. Factors such as incomplete sampling (Table 1) and natural variation can weaken the power of statistical analyses so that real effects are not detected (type II error). However, we feel that the chances are low that we missed any major effects of otter trawling on habitat and macrobenthic communities at the experimental site, especially since our study site had not been disturbed by trawling for at least 13 years and the sampling program was able to resolve significant levels of natural variability in the reference corridors. A possible exception could be effects on large, dispersed, epibenthic organisms such as basket stars and whelks which could have been substantially damaged by the repetitive trawling but which were not adequately sampled by either the epibenthic sled or the videograb. These are included in the list of affected species (Table 2).

Effects on large, sessile, epibenthic organisms, which are highly vulnerable to trawl disturbance, were not observed because of their scarcity at the experimental site. While some would argue that trawling before 1980 may have affected these slow-growing organisms, we feel that their absence was due primarily to the sandy habitat that favors free-living taxa. Sessile epifauna off Atlantic Canada are much more common in more stable, structurally complex gravel habitats (Kostylev et al. 2001). For example, corals have never been common on top of offshore fishing banks but are most abundant in suitable habitats below 200 m (Breeze et al. 1997). However, soft corals attached to shells were common at the experimental site and were negatively affected by trawling (Table 2). Unfortunately there are no long-term data sets on benthic communities off Atlantic Canada like those in the North Sea that have proven useful in examining temporal changes (e.g., Frid et al. 2000).

Subsamples of sediment for analysis of meiofauna and bacteria were extracted from the videograb during the experiment, but the death of our colleague Peter Schwinghamer prevented their complete analysis and interpretation. However, preliminary data from 1993 suggested an increase in bacterial numbers immediately after trawling. Because of the substantial disturbance to the seabed, there may have been potential impacts on biogeochemical fluxes between the sediment and water column which were not investigated; neither were potential impacts on the early life history stages of groundfish.

Acknowledgments

We thank our many colleagues who contributed to the development of the instrumentation, the design of this experiment, the execution of the field program, the processing of samples, and the interpretation of the results. In particular, we thank our colleague, the late Peter Schwinghamer, for his important contributions to all aspects of this experiment. We also thank the officers and crews of the C.C.G.S. *Parizeau* and the C.C.G.S. *Wilfred Templeman* for their outstanding support at sea, the Geological Survey of Canada (Atlantic) for providing the side-scan sonar support, Trackpoint, and the AGCNav navigation system, and Guigné International, Ltd. for providing DRUMS support. Funding was provided by the Department of Fisheries and Oceans, Natural Resources Canada, the Atlantic Fisheries Adjustment Program, the North Cod Science Program, and the Green Plan Sustainable Fisheries Program.

References

Auster, P. J., and R. W. Langton. 1999. The effects of fishing on fish habitat. Pages 150–187 *in* L. R. Benaka, editor. Fish habitat: essential fish habitat and rehabilitation. American Fisheries Society, Symposium 22, Bethseda, Maryland.

Breeze, H., D. S. Davis, M. Butler, and V. Kostylev. 1997. Distribution and status of deep sea corals off Nova Scotia. Ecology Action Centre, Marine Issues Committee Special Publication Number 1, Halifax, Nova Scotia.

Chivers, R. C., N. Emerson, and D. R. Burns. 1990. New acoustic processing for underway surveying (RoxAnn). The Hydrographic Journal 56:9–17.

Collie, J. S., S. J. Hall, M. J. Kaiser, and I. R. Poiner. 2000. A quantitative analysis of fishing impacts on shelf-sea benthos. Journal of Animal Ecology 69:785–798.

Engel, J., and R. Kvitek. 1998. Effects of otter trawling on a benthic community in Monterey Bay National Marine Sanctuary. Conservation Biology 12:1204–1214.

Freese, L., P. J. Auster, J. Heifetz, and B. L. Wing. 1999. Effects of trawling on seafloor habitat and associated invertebrate taxa in the Gulf of Alaska. Marine Ecology Progress Series 182:119–126.

Frid, C. L. J., K. G. Harwood, S. J. Hall, and J. A. Hall. 2000. Long-term changes in the benthic communities on North Sea fishing grounds. ICES Journal of Marine Science 57:1303–1309.

Gilkinson, K. D. 1999. Impact of otter trawling on infaunal bivalves living in sandy bottom habitats on the Grand Banks. Doctoral dissertation. Memorial University, St. John's, Newfoundland.

Gilkinson, K. D., M. Paulin, S. Hurley, and P. Schwinghamer. 1998. Impacts of trawl door scouring on infaunal bivalves: results of a physical trawl door model/dense sand interaction. Journal of Experimental Marine Biology and Ecology 244:291–312.

Gordon, D. C., Jr., K. D. Gilkinson, E. L. R. Kenchington, J. Prena, C. Bourbonnais, K. MacIsaac, D. L. McKeown, and W. P. Vass. 2002. Summary of the Grand Banks otter trawling experiment (1993–1995): effects on benthic habitat and communities. Canadian Technical Report of Fisheries and Aquatic Sciences 2416.

Gordon, D. C., Jr., E. L. R. Kenchington, K. D. Gilkinson, D. L. McKeown, G. Steeves, M. Chin-Yee, K. Benthan, and P. R. Boudreau. 2000. Canadian imaging and sampling technology for studying marine benthic habitat and biological communities. International Council for the Exploration of the Sea C.M. 2000/T:07.

Guigné, J. Y., P. Schwinghamer, Q. Liu, and V. H. Chin. 1993. High resolution and broadband processing of acoustic images of the marine benthos. Proceedings of the Institute of Acoustics 15(Part 2):237–252.

Johnson, K. A. 2002. A review of national and international literature on the effects of fishing on benthic habitats. NOAA Technical Memorandum NMFS-F/SPO-57.

Kenchington, E., J. Prena, K. D. Gilkinson, D. C. Gordon, Jr., P. Schwinghamer, T. W. Rowell, K. MacIsaac, C. Bourbonnais, D. L. McKeown, and W. P. Vass. 2001. Effects of experimental otter trawling on the macrofauna of a sandy bottom ecosystem on the Grand Banks of Newfoundland. Canadian Journal of Fisheries and Aquatic Sciences 58:1043–1057.

Kostylev, V. E., B. J. Todd, G. B. J. Fader, R. C. Courtney, G. D. M. Cameron, and R. A. Pickrill. 2001. Benthic habitat mapping on the Scotian Shelf based on multibeam bathymetry, surficial geology and sea floor photographs. Marine Ecology Progress Series 219:121–137.

Kulka, D. W., and D. A. Pitcher. 2001. Spatial and temporal patterns in trawling activity in the Canadian Atlantic and Pacific. International Council for the Exploration of the Sea C.M. 2001/R:02.

McKeown, D. L., and D. C. Gordon, Jr. 1997. Grand Banks otter trawling impact experiment: II. Navigation procedures and results. Canadian Technical Report of Fisheries and Aquatic Sciences 2159.

Messieh, S. N., T. W. Rowell, D. L. Peer, and P. J. Cranford. 1991. The effects of trawling, dredging and ocean dumping on the eastern Canadian continental shelf seabed. Continental Shelf Research 11:1237–1263.

National Research Council. 2002. Effects of trawling and dredging on seafloor habitat. National Research Council, Committee of Ecosystem Effects of Fishing, Ocean Studies Board, Washington, D.C.

Prena, J., T. W. Rowell, P. Schwinghamer, K. D. Gilkinson, and D. C. Gordon, Jr. 1996. Grand Banks otter trawling experiment: I. Site selection process, with a description of macrofaunal communities. Canadian Technical Report of Fisheries and Aquatic Sciences 2094.

Prena, J., P. Schwinghamer, T. W. Rowell, D. C. Gordon, Jr., K. D. Gilkinson, W. P. Vass, and D. L. McKeown. 1999. Experimental otter trawling on a sandy bottom ecosystem of the Grand Banks of Newfoundland: analysis of trawl bycatch and effects on epifauna. Marine Ecology Progress Series 181:107–124.

Rowell, T. W., P. Schwinghamer, M. Chin-Yee, K. D. Gilkinson, D. C. Gordon, Jr., E. Hartgers, M. Hawryluk, D. L. McKeown, J. Prena, D. P. Reimer, G. Sonnichsen, G. Steeves, W. P. Vass, R. Vine, and P. Woo. 1997. Grand Banks otter trawling impact experiment: III. Experimental design and methodology. Canadian Technical Report of Fisheries and Aquatic Sciences 2190.

Schwinghamer, P., D. C. Gordon, Jr., T. W. Rowell, J. Prena, D. L. McKeown, G. Sonnichsen, and J. Y. Guigné. 1998. Effects of experimental otter trawling on surficial sediment properties of a sandy bottom ecosystem on the Grand Banks of Newfoundland. Conservation Biology 12:1215–1222.

Schwinghamer, P., J. Y. Guigné, and W. C. Siu. 1996. Quantifying the impact of trawling on benthic habitat using high resolution acoustics and chaos theory. Canadian Journal of Fisheries and Aquatic Sciences 53:288–296.

Smith, C. J., K. N. Papadopoulou, and S. Diliberto. 2000. Impact of otter trawling on an eastern Mediterranean commercial trawl fishing ground. ICES Journal of Marine Science 57:1340–1351.

Thrush, S. F., J. E. Hewitt, G. A. Funnell, V. J. Cummings, J. Ellis, D. Schultz, D. Talley, and A. Norkko. 2001. Fishing disturbance and marine biodiversity: role of habitat structure in simple soft-sediment systems. Marine Ecology Progress Series 223:277–286.

Warwick, R. M., and K. R. Clarke. 1993. Increased variability as a symptom of stress in marine communities. Journal of Experimental Marine Biology and Ecology 172:215–226.

Effects of Chronic Bottom Trawling on the Size Structure of Soft-Bottom Benthic Invertebrates

ROBERT A. MCCONNAUGHEY,[1] STEPHEN E. SYRJALA,[2] AND C. BRAXTON DEW[3]

National Marine Fisheries Service, Alaska Fisheries Science Center,
7600 Sand Point Way Northeast, Seattle, Washington 98115, USA

Abstract. Chronic bottom trawling reduces benthic biomass, but it is generally unknown whether this represents a decrease in numbers of individuals or their mean body size. Because this distinction provides insight into the mechanisms of disturbance and recovery, we investigate the matter here. Using comprehensive historical effort data, adjacent untrawled (UT) and heavily trawled (HT) areas were identified along the boundary of a long-standing no-trawl zone in Bristol Bay, a naturally disturbed offshore area of the eastern Bering Sea. The study site was shallow (44–52 m) with a sandy substrate, ubiquitous bottom ripples, and strong tidal currents. A modified research trawl was used to collect 42 HT–UT paired samples of benthic infauna and epifauna. These data were used to compare mean sizes (kg) of 16 invertebrate taxa. On average, 15 of these taxa were smaller in the HT area, and the overall HT–UT difference in body size was statistically significant ($P = 0.0001$). However, individually, only the whelk *Neptunea* spp. ($P = 0.0001$) and Actiniaria ($P = 0.002$) were significantly smaller in the HT area after correcting for multiple tests. Mean size of red king crabs *Paralithodes camtschaticus* was 23% greater in the HT area ($P = 0.17$). Supplemental length–frequency data indicate that substantially fewer small red king crabs, rather than more large individuals, occupy the HT area ($P = 0.0001$). Finally, a large number of within-year, within-taxon comparisons of mean body size were made using 1982–2001 U.S. National Marine Fisheries Service trawl survey data collected in the same closed area. Overall, these comparisons indicate natural variability of body size in UT areas is large relative to the observed HT–UT differences due to chronic bottom trawling. Since active fishing in the HT area occurred 3 or more years before our field sampling program, our findings reflect conditions associated with an intermediate level of recovery.

Introduction

Commercial fishing with mobile gear is a rather unique form of disturbance, given its high frequency and widespread occurrence in the eastern Bering Sea (Figure 1). Recent expansions into previously unexploited (deep) and untrawlable (hard) grounds as a result of technological advances and declining catches in more traditional areas have heightened concern about possible environmental consequences. Whereas disturbance of the benthos with mobile fishing gear was once just a scientific curiosity (e.g., Caddy 1968), it is now the focus of a major international research effort with statutory mandates to consider possible adverse effects on biodiversity, seafloor habitats, and their associated fish stocks (Turner et al. 1999). Despite decades of intensive research, however, its overall impact on marine ecosystems and, in particular, fish production is largely unknown (Daan 1991; Rogers et al. 1998; Thrush et al. 1998; Auster and Langton 1999; Turner et al. 1999; NRC 2002).

Research to characterize mobile fishing gear effects is pursued using two main experimental approaches. Short-term (acute) effects are studied by comparing conditions in experimental corridors before and after a single pass or repeated passes of a trawl or dredge (e.g., Tuck et al. 1998). Occasionally, the recovery process is examined by resampling at a later date; these studies incorporate untrawled control corridors into the sampling program in order to account for natural variability during the study period (a before–after, control–impact design; Green 1979; Drabsch et al. 2001). Multiple trawled and control corridors are preferred (Lindegarth et al. 2000). This approach provides insights about the process of trawl disturbance and is the basis for most knowledge about trawling effects. Longer-term (chronic) effects are studied by comparing conditions in heavily fished and lightly fished or unfished areas and, as such, measure

---

[1] Corresponding author: bob.mcconnaughey@noaa.gov
[2] E-mail: steve.syrjala@noaa.gov
[3] E-mail: braxton.dew@noaa.gov

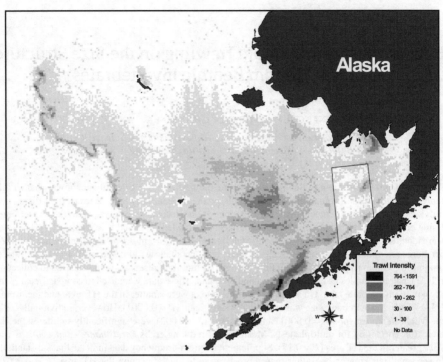

Figure 1. Bottom trawl effort in the eastern Bering Sea (1973–2001). Endpoints for 400,776 individual tows recorded by fishery observers are summarized on a 25-km² grid. The rectangular outline indicates the location of the Crab and Halibut Protection Zone 1 in Bristol Bay.

cumulative effects of fishing (e.g., McConnaughey et al. 2000). These experiments are relatively uncommon because high-quality historical trawl effort data are frequently unavailable, and their designs are often flawed because the (unfished) "control" areas have previously been fished or they are fundamentally different than the corresponding experimental units (Hall et al. 1993; NRC 2002).

Both natural and anthropogenic disturbances are relatively discrete events that disrupt ecosystem, community, or population structure and change resources, substrate availability, or the physical environment (Pickett and White 1985b; Lytle 2001). Their form varies from predictable environmental fluctuations to irregular destructive events. The magnitude of environmental effects generally increases with the intensity and spatial extent of the disturbance, while the persistence of these effects is closely related to the frequency of disturbances or, equivalently, the degree of recovery since the last event. All or parts of populations may be affected through collective effects on individuals, which are themselves influenced by characteristics such as body size, age, density of individuals, and their life histories. Community attributes such as richness, diversity, and structure may also be affected. Generally speaking, disturbances produce fragmented habitats that are structurally simple and populated by a relatively few small, short-lived, and highly productive (opportunistic) species that respond favorably to open space and newly available resources. Although subtidal communities do not respond to disturbances in a uniform way, responses of benthic systems to bottom trawling are generally consistent with disturbance theory (Pickett and White 1985b; Dayton et al. 1995). In particular, the physical effects of mobile fishing gear include winnowing, excavation, and resuspension of sediments; loss of physical features; and translocation of biogeochemical material at the sediment–water interface (Mayer et al. 1991; Fonteyne 2000). There also may be reductions of structural biota, altered relative abundances of species due to differential mortality and removals, loss of species from part of their normal range, altered predator–prey relationships, decreases and fragmentation of certain populations, and, presumably, higher-order food web and ecosystem effects (Lindeboom and de Groot 1998; NRC 2002). In general, these effects are thought to be cumulative and, thus, depend on the frequency of disturbance (NRC 2002). In some cases, effects may be delayed for a period of days or weeks (Kaiser and Spencer 1996; Mensink et al. 2000; Maguire et al. 2002) and, overall, the effects may persist for many years (Probert et al. 1997; NRC 2002).

Most biological research has targeted the specific changes in benthic invertebrate populations and community structure that occur when mobile fishing gear, particularly bottom trawls, contact the seabed. This focus on benthic invertebrates reflects their limited mobility and vulnerability to bottom-tending gear and observations that structurally complex seabeds are an important element of healthy, productive benthic systems (Kaiser et al. 1999a; Turner et al. 1999; Thrush et al. 2001) that are preferred by some (Sainsbury et al. 1997; Freese et al. 1999) but not all (Kaiser et al. 1999b; McConnaughey and Smith 2000; Else et al. 2001) groundfish. Effects are typically measured as changes in abundance (both biomass and numeric) or community structure (e.g., diversity or evenness). In some cases, mortalities according to sex and in situ density have been reported (Bergman and van Santbrink 2000). Studies that document gear effects have been (and remain) the primary research emphasis, despite a practical need for more interpretive work. This is because site-specific factors such as the resident species assemblage, substrate characteristics, water depth, gear type(s) used, and the disturbance history (natural or anthropogenic) influence the nature of the response to trawling (Collie et al. 2000).

Effects on the size structure of invertebrate populations are usually ignored. When size is addressed, it is typically limited to statements concerning gear selectivity and specific vulnerabilities (e.g., Bergman and van Santbrink 2000). In a very few cases, individual measurements of benthic invertebrates have been taken to examine possible effects along depth or established disturbance gradients (Hall et al. 1993; Collie et al. 1997), to compare epifaunal populations (Ramsay et al. 1996; Prena et al. 1999) and demersal fish diets (Kaiser and Spencer 1994) before and after experimental trawling, and to evaluate trawling effects on benthic production and infaunal size structure (Jennings et al. 2001, 2002). This general inattention to size structure is problematic given the fundamental importance of body size in biological systems (Peters 1983; Calder 1984). Without information about the size structure of populations, simple measures of biomass or numeric abundance cannot fully address the mechanism of disturbance, the functional status of the benthos after trawling, and certain key processes (e.g., production) that influence recovery dynamics.

In this paper, we compare mean body sizes of 16 benthic invertebrate taxa in a chronically trawled area of the eastern Bering Sea with those in an adjacent untrawled area. We then consider the magnitude of observed differences in the context of natural size variability in these same taxa at untrawled locations elsewhere in Bristol Bay. This study is an elaboration of earlier work that compared invertebrate biomasses in this same experimental area (McConnaughey et al. 2000).

Data and Methods

Study Area and Biological Sampling

The study area is located at the northeastern corner of a long-standing no-trawl zone in the Bristol Bay region of the eastern Bering Sea (now the Crab and Halibut Protection Zone 1, CHPZ1; Figure 2). This area is shallow (44–52 m) and is characterized by sandy substrates, ubiquitous bottom ripples, strong tidal currents (≤1 m/s [2 knots]), and a rich invertebrate assemblage (Marlow et al. 1999; Smith and McConnaughey 1999; McConnaughey et al. 2000). The history of bottom trawling there was reconstructed using foreign regulations and catch statistics, international agreements, U.S. National Marine Fisheries Service (NMFS) enforcement and surveillance reports, the NORPAC fishery observer database (1973–1996; maintained at the NMFS Alaska Fisheries Science Center, Seattle), and details of historical and more recent regulatory closures, which collectively span the 40+ year history of Bering Sea trawl fisheries. From this analysis, we identified heavily trawled (HT) and untrawled (UT) areas straddling the CHPZ1 boundary. This pattern reflects a trawl prohibition first instituted by the Japanese in 1959 (Bristol Bay Pot Sanctuary, which included the subsequently designated CHPZ1; Forrester et al. 1978) and intensive fishing for yellowfin sole *Pleuronectes aspera* along the northern boundary of the closed area.

A total of 42 HT–UT paired stations were established in 3.4-km^2 (1-nautical-mi^2) cells straddling the CHPZ1 boundary (Figure 2). Adjacent stations were paired to account for a possible east–west environmental gradient in this poorly surveyed area (D. Hill, Pacific Hydrographic Branch, National Oceanic and Atmospheric Administration Office of Coast Survey, personal communication). Each pair was sampled by one of two 38-m chartered stern trawlers (1,137-kW [1,525-hp] sister ships, the F/V *Arcturus* and F/V *Aldebaran*) using matching NMFS 83-112 survey trawls modified to improve catchability and retention of benthic fauna (McConnaughey et al. 2000). Biota were collected between 31 July and 6 August 1996 with 15-min tows at 1.5 m/s (3 knots) through the center of each preselected cell. Benthic fauna in the catch were identified to the lowest practicable taxon and were then separately weighed on a motion-compensating balance and enumerated while at sea. Acoustic net mensuration and a global positioning system were used to standardize catches according to area swept (average net width for this study was 14.7 m). For each taxon that could be both weighed and counted, mean size (kg) was calculated for each haul as the ratio of total biomass (kg) and the total number of individuals in the catch. Hermit crabs (family Paguridae) were weighed in their shells after

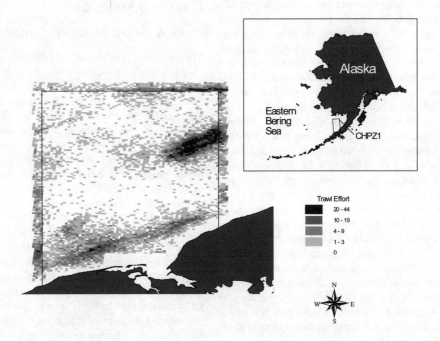

Figure 2. Bottom trawl effort for the Crab and Halibut Protection Zone 1 (CHPZ1) and vicinity. Endpoints for individual tows recorded by fishery observers are summarized on a 3.4-km² (1-nautical-mi²) grid. Effort shown within the CHPZ1 boundaries was primarily joint venture fisheries that were permitted during domestication of trawl fisheries in the early to mid-1980s. A modified research trawl was used in July–August 1996 to sample benthic invertebrates at 42 heavily trawled (HT; darker cells) and untrawled (UT; unfilled cells) station pairs located along the northeastern boundary of the CHPZ1 closed area. Given the use of endpoints only, the nonrandom orientation of high intensity effort along the CHPZ1 boundary, and an average tow length in Bristol Bay of 19.4 km (10.5 nautical mi), trawl effort in each grid cell should be considered a relative index with upward scaling of 10 or less.

removing attached biota. Carapace lengths for all specimens of red king crab *Paralithodes camtschaticus* were measured to the nearest mm using calipers.

Statistical Analysis

Heavy Trawling Effect on Body Size

Prior to analysis, catch data were lumped into larger taxonomic groups for consistency with biomass data used in previous analyses (McConnaughey et al. 2000). Not all taxa were caught in all tows. To preserve the paired-station design of the experiment, only those pairs in which a given taxon occurred in both of the samples were included in the analysis, resulting in the exclusion of some data. Observed positive correlation in the data confirmed the importance of the paired design. Statistical analysis using linear models showed no statistically significant differences between the catches of the two boats; hence, data from the two boats were pooled for analysis. In order to better align type I and type II error levels, α was set equal to 0.10 (supporting arguments by Peterman [1990], Dayton et al. [1995], Mapstone [1996], and Nester [1996]). Given the nature of the data (paired observations, skewed distribution), the Wilcoxon signed rank test was used to test whether the difference between mean body size in the HT and the UT areas differed from zero. A Bonferoni correction for multiple tests was applied (Miller 1981). In addition, a randomization test was designed to examine the overall effect, if any, of trawling on all taxa. The test was constructed to test the null hypothesis that the across-taxa average difference in HT–UT mean size was equal to zero. For each taxon, the mean size difference was calculated for the observed data and for 9,999 pseudo-random permutations of the data for each HT–UT pair. These means produced a distribution for each taxon that was used to generate a randomization test for that taxon, the results of which were consistent with the results of the Wilcoxon signed rank tests. To construct the simultaneous test of all taxa, the 10,000 means per taxon were normalized within taxa—to give equal weight to each taxon—and then summed across taxa. Thus, the 10,000 pseudo-random sums of normalized means formed the distribution for the randomization test.

To examine the individual carapace length measurements for red king crabs, simple histograms were used to look at the difference in the size distributions from the UT and HT areas. The Kolmogorov–Smirnov two-sample test was used to evaluate statistical significance.

Natural Size Variability

For comparison with experimental results, we estimated natural size variability of benthic invertebrates by examining catches at 14 of the 17 standard NMFS trawl survey stations located within the CHPZ1 closed area (Nebenzahl 2001; Figure 3). Three stations containing the northern boundary, and other stations located outside the closed area, were excluded to avoid confounding effects of commercial bottom trawling. Prior to our analysis, NMFS survey catch data were lumped into larger taxonomic groups in order to be consistent with the mean body size data from this study. For each year from 1982 to 2001, we calculated the absolute value of the pair-wise differences in mean size for all stations where a specific taxon was caught. Absolute value was used because we had no reason to expect one station would have a greater or lesser mean size than another station within the untrawled area. These differences were then collected across years to generate an empirical distribution of differences in mean size based on natural variability among stations. The location of the absolute value of each HT–UT difference within its corresponding empirical distribution was then determined.

Results

Approximately 176 ha of seabed were sampled with 84 tows of the modified bottom trawl. Overall, 12.5 metric tons of invertebrate megafauna representing 92 distinct

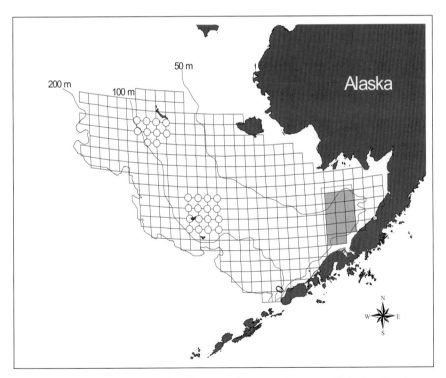

Figure 3. Sampling grid for the annual National Marine Fisheries Service (NMFS) bottom trawl survey in the eastern Bering Sea. Standard samples are collected near the center of each 37 km × 37 km (20-nautical-mi × 20-nautical-mi) survey cell and at grid cell intersections located in areas of high red king crab abundance. Each sample consists of a 30-min tow at 1.5 m/s (3 knot) using the NMFS 83-112 survey trawl. Shading indicates the location of the 14 stations whose 1982–2001 catch data were used to estimate natural variability of invertebrate body size in the untrawled Crab and Halibut Protection Zone 1.

taxa (including empty shell and egg case categories) were processed for an average catch of 71.0 kg/ha. For comparison, this is considerably more than the average catch (0.69 kg/ha) reported for 13-km sets in a 120–146-m deep, sandy area on the Grand Banks of Newfoundland (net width = 20 m; Prena et al. 1999).

Heavy Trawling Effect on Body Size

Mean individual size could be determined for 16 of the invertebrate taxa that were caught. Nine motile epifaunal taxa (five crab, two starfish, one snail, and one shrimp), four infaunal bivalves, and three sedentary epifaunal groups (anemones, a sea cucumber, and a compound colonial tunicate) were represented. Two additional taxa were excluded because of low sample size: the bivalve *Siliqua* spp. ($N = 3$ pairs) and the basket star *Gorgonocephalus caryi* ($N = 2$ pairs). Overall, 8 of 16 invertebrate taxa were caught in at least 31 pairs, while the other 8 were represented in 6–19 pairs (Table 1).

The mean sizes of the whelk *Neptunea* spp. ($P = 0.0001$) and species from the order Actiniaria ($P = 0.002$) were significantly smaller in the HT area (Table 1). The mean size of *Crangon* spp. ($P = 0.05$) was also smaller in the HT area; however, the result was not statistically significant after a Bonferoni correction for multiple tests ($\alpha' = 0.006$). Twelve additional taxa were also smaller in the HT area, albeit, not significantly smaller. Observed HT–UT differences ranged from –4% (*Asterias amurensis*, *Hyas* spp., and *Serripes* spp.) to –68% (*Tellina* spp.). Overall, the weighted-average effect for the infaunal group (–31.1%) exceeded that for the sedentary (–16.1%) and the motile (–9.1%, excluding *P. camtschaticus*) groups. The overall HT–UT difference in body size was statistically significant ($P = 0.0001$).

The mean body size of the red king crab was 23% greater in the HT area ($P = 0.17$; Table 1). This was the only exception to the overall pattern of smaller size in the HT area. An examination of carapace length (CL) frequencies indicated this mean effect was primarily due to the presence of fewer small crabs rather than more large individuals in the HT area (Figure 4; $P < 0.0001$ that there is no difference in the two CL distributions). A single pod of juveniles ($N = 588$, or >50% of all specimens caught; mean CL = 62 mm) caught in the UT area was excluded from the paired analysis because the corresponding HT sample was zero. Including those measurements in the CL distribution does not qualitatively change the conclusions.

Natural Size Variability

A total of 10,018 within-year, within-taxon comparisons of mean body size were made using 1982–2001 NMFS trawl survey data collected in the CHPZ1 closed area. Overall, these comparisons indicate natural vari-

Table 1. Mean body size (kg) of 16 invertebrate taxa sampled in the untrawled (UT) and heavily trawled (HT) areas. Percentage difference represents the HT treatment effect (HT–UT) divided by the UT value. The biomass percentage difference is after McConnaughey et al. (2000). Simultaneous consideration of all taxa based on a randomization test indicated the HT–UT differences were significantly different from zero ($P = 0.0001$). Significance level of Wilcoxon results is 0.006 after Bonferoni correction for experiment-wise error ($\alpha = 0.10$).

Taxon	Pairs	Mean body size (kg)				Biomass
		HT	UT	% difference	P	% difference
Motile						
Asterias amurensis	42	0.1007	0.1049	–4.0	0.419	10.2
Crangon spp.	34	0.0018	0.0021	–15.3	0.053	–9.2
Evasterias spp.	11	0.8584	0.9862	–13.0	0.206	38.0
Hyas spp.	31	0.0664	0.0691	–3.8	0.660	23.6
Neptunea spp.	41	0.1215	0.1496	–18.8	0.000	–29.4
Oregonia gracilis	19	0.0351	0.0373	–5.9	0.671	–19.2
Paguridae	31	0.0687	0.0737	–6.8	0.496	–32.5
Pagurus ochotensis	42	0.0642	0.0680	–5.6	0.464	–7.6
Paralithodes camtschaticus	33	1.2948	1.0536	22.9	0.170	–42.4
Sedentary						
Actiniaria	40	0.1722	0.2106	–18.2	0.002	–46.8
Aplidium spp.	7	0.2379	0.2809	–15.3	0.469	–58.4
Cucumaria spp.	13	0.4520	0.5029	–10.1	0.685	–14.9
Infauna						
Macoma spp.	8	0.0270	0.0321	–15.7	0.844	–62.7
Mactromeris spp.	6	0.0446	0.1000	–55.4	0.156	73.0
Serripes spp.	9	0.0426	0.0442	–3.7	1.000	–13.3
Tellina spp.	6	0.0262	0.0830	–68.4	0.438	–76.1

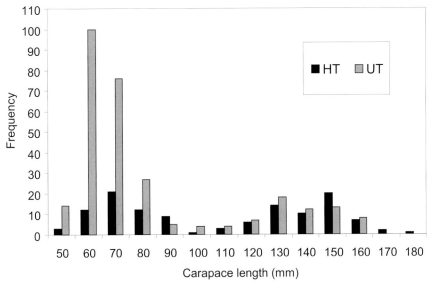

Figure 4. Histograms of red king crab carapace length frequencies from the 33 paired samples used to compare mean body sizes (kg) of crab in the untrawled (UT) and heavily trawled (HT) areas straddling the Crab and Halibut Protection Zone 1 closed area boundary.

ability of body size in untrawled areas is large relative to the observed HT–UT differences due to chronic bottom trawling (Table 2). On average (weighted by the number of comparisons), spatial differences in body size exceeded the observed trawling effect in 91% of the comparisons involving sedentary taxa, 81% of those for motile taxa, and 22% of those for infauna. In this latter case, the composite result is most applicable to the bivalve *Mactromeris* sp. ($N = 563$) because sample sizes for the other three taxa in this group ($N = 19$ total) were quite low. Overall, 78% of all spatial differences in body size exceeded the HT–UT difference observed in the 16 benthic invertebrate taxa considered.

Discussion

Body size directly affects the fitness of individuals, thereby influencing the structure and function of populations, communities, and ecosystems (Peters 1983; Calder 1984; Blanckenhorn 2000). Although easily measured, it is rarely considered in the context of mobile fishing gear effects. To our knowledge, there is only one other study set that has investigated effects of chronic trawling on body size and its possible ecological consequences (Jennings et al. 2001, 2002; Duplisea et al. 2002). In our case, the observed effects were generally consistent with theoretical expectations but were probably limited in magnitude by several factors. First, the study area has a relatively high level of natural disturbance (McConnaughey et al. 2000), and there is general consensus that sandy areas with strong tidal flow are less sensitive to mobile gear effects (Hall 1994; Thrush et al. 1998; Collie et al. 2000; Gordon et al. 2002; NRC 2002). Also, our findings probably reflect conditions associated with an intermediate stage of recovery since active fishing in the HT area declined to a very low level prior to field sampling in 1996 (only five tows during 1993–1995). Moreover, trawl intensities for Alaskan waters are low relative to other shelf areas in the United States (NRC 2002) and Europe (Rijnsdorp et al. 1998). Although the intensity level at our study area is only moderate for the eastern Bering Sea shelf as a whole, the concentrated pattern of trawling along the CHPZ1 boundary, nevertheless, suggests an intensity that is higher than would ordinarily result from more randomly oriented effort (Figure 2).

The overall HT–UT difference was statistically significant ($P = 0.0001$), reflecting the fact that 15 of the 16 invertebrate taxa we examined were smaller in the more highly disturbed HT area (two of which were individually significant at $P < 0.006$, the Bonferoni-corrected α; Table 1). The binomial likelihood of observing at least 15 of 16 taxa showing a decrease in mean size by chance alone is highly unlikely ($P = 0.0003$). Reduced body size is a common response to stress, pollution, or disturbance by a very diverse set of fauna from a variety of terrestrial and aquatic systems. These include responses to size-selective predation and organic enrichment by marine sediment assemblages (Raffaelli et al. 2000), pesticide stress on freshwater zooplankton (Hanazato 1998), land management disturbances on carabid beetles (Ribera

Table 2. The heavily trawled–untrawled (HT–UT) effect size compared to all possible within-year (1982–2001) differences in mean body size at 14 National Marine Fisheries Service (NMFS) survey stations in the Crab and Halibut Protection Zone 1 area closed to trawling. A relatively modest effect of chronic trawling on body size is indicated when natural (spatial) differences in body size frequently (i.e., high percentage) exceed the HT–UT difference, whereas a comparatively large effect is indicated by a relatively low percentage in the last column of the table. Maximum natural difference is the most extreme natural difference in mean size observed among all possible within-year pairs of 14 standard NMFS survey stations in the CHPZ1 area closed to trawling.

Taxon	HT–UT difference (kg)	Maximum natural difference (kg)	Number of comparisons	Natural difference > HT–UT difference (%)
		Motile		
Asterias amurensis	−0.0042	0.6279	1,420	97.5
Crangon spp.	−0.0003	0.0187	32	84.4
Evasterias spp.	−0.1278	3.5380	130	82.3
Hyas spp.	−0.0027	0.2005	1,388	92.7
Neptunea spp.	−0.0281	0.4561	1,324	75.2
Oregonia gracilis	−0.0022	0.0876	538	85.9
Paguridae	−0.0050	0.2556	899	91.2
Pagurus ochotensis	−0.0038	0.1802	900	91.2
Paralithodes camtschaticus	0.2412	2.3202	2,563	61.3
		Sedentary		
Actiniaria	−0.0384	0.5554	39	79.5
Aplidium spp.	−0.0430	1.9955	136	91.9
Cucumaria spp.	−0.0509	0.8843	67	94.0
		Infauna		
Macoma spp.	−0.0050	0.0230	4	75.0
Mactromeris spp.	−0.0554	0.2912	563	21.3
Serripes spp.	−0.0016	0.0594	7	42.9
Tellina spp.	−0.0567	0.0291	8	0.0

et al. 2001), and nutritional stress in Steller sea lions *Eumetopias jubatus* (Calkins et al. 1998). Overall, reduced body size is a general expectation for stressed ecosystems (Odum 1985).

When our results are considered in combination with biomass differences reported previously, it is possible to draw general conclusions about the overall status of these populations (McConnaughey et al. 2000; Table 1). In most cases, biomass was also reduced as a result of heavy trawling, suggesting a general population decline (Actiniaria, *Aplidium* spp., *Crangon* spp., *Cucumaria* spp., *Macoma* spp., *Neptunea* spp., *Oregonia gracilis*, pagurids, *Pagurus ochotensis*, *Serripes* spp., *Tellina* spp.; Table 1). In a few cases, greater overall biomass accompanied the observed body-size reduction, suggesting a proliferation of relatively small individuals in the HT area (*A. amurensis*, *Evasterias* spp., *Hyas* spp., *Mactromeris* spp.). The mean size of the red king crab was larger in the HT area. This was the only exception to the pattern of smaller individuals in the HT area. In this case, an analysis of length–frequency data indicated the difference was due to substantially fewer small crabs in the HT area than in the UT area. Since biomass in the HT area was lower than that in the UT area, the red king crab response to chronic bottom trawling appears to be fewer individuals of greater mean size.

In order to gain insight into the ecological significance of our findings, it is important to consider our results in the context of natural variability of body size. Overall, the effect of chronic trawling on body size was relatively modest when compared with natural spatial variability in a large, adjacent area closed to commercial trawling. More than three-fourths of all within-year size differences for a 20-year period exceeded the HT–UT difference we observed (Table 2). The degree to which this conclusion is applicable to other areas and times is unknown because similar studies are not available for comparison. At this point in time, we can only surmise that trawling effects on body size are additive and, thus, may augment (or offset) other processes affecting size variability in the large geographic areas exposed to mobile fishing gears (Figure 1).

Our methods did not address the mechanism by which chronic bottom trawling reduces the mean body size of benthic invertebrates, but it is likely that several factors contribute to the effect. Size-selective removals or damage by trawl gear is the simplest explanation and is consistent with the pattern of lower mortalities in smaller individuals and species of megafauna (Bergman and van Santbrink 2000). These authors conclude that smaller megafauna are primarily affected by disturbances of the sediment with possible suspension into the water column and relocation, whereas larger specimens suffer mortalities as a result of direct physical contact with the gear. Since the HT area has been disturbed repeatedly over many years, it is also reasonable to consider possible changes in habitat quantity and quality. Protective cover in the HT area is relatively sparse due to reduced epibenthic biomass that is both more patchy and less diverse than in UT areas (McConnaughey et al. 2000). This condition may be particularly limiting for larger individuals of "naked" motile epifauna (e.g., shrimp and some nonpagurid crabs) that are consumed by actively searching visually oriented, benthic-feeding groundfish, such as yellowfin sole; McConnaughey and Smith 2000). Intensive predation by starfish and crab that favor disturbed areas (McConnaughey et al. 2000; this study) may augment the effect of selective but incomplete removals of larger bivalves by trawls. For the neptunids studied here, high predation and selective removals or damage by trawling may act synergistically to induce smaller body sizes in the HT area. Trussell and Nicklin (2002) demonstrated increased shell thickness (and strength), accompanied by decreased body mass and length, in an intertidal snail *Littorina obtusata* (also known as yellow periwinkle) exposed to persistent crab predation. Inducible responses such as this require a temporally and spatially unpredictable cue, as is typical of trawling disturbances and associated scavenger influxes.

We can only speculate as to the basis of the red king crab result. Rose (1999) reported low injury rates for red king crabs passing under commercial trawl footropes, which, combined with the 3-year lapse in active trawling in the HT area, suggests that indirect or habitat-related effects may be most important in this case. Indeed, earlier work (McConnaughey et al. 2000) indicates that heavy trawling in the study area results in a more patchy and less structured habitat with lower overall diversity of sedentary taxa. For unknown reasons, these conditions may be better suited to large rather than small crabs. Moreover, if we assume that abundance and habitat suitability are positively associated, then the fact that more crabs were found in the UT area suggests that the untrawled habitat was somehow more suitable overall ($N_{UT} = 288$, $N_{HT} = 121$ for 33 sample pairs; $N_{UT} = 886$, $N_{HT} = 126$ when all 74 samples are included). Alternatively, the HT–UT difference in abundance may be entirely unrelated to the presence or absence of trawling. Historical data from NMFS trawl surveys indicate that Bristol Bay red king crabs typically are less abundant in the HT area north of the CHPZ1 boundary (58°N) and substantially more abundant to the south. If a large-scale, temporally stable, north–south, habitat-suitability gradient exists in this area and is detectable over the relatively short distances separating the HT and UT station pairs, then we might expect to catch more crabs south of the HT–UT boundary regardless of treatment effects. Consistent with the idea that suitable crab habitat exists south rather than north of 58°N is the fact that the northern boundary of the no-trawl Bristol Bay Pot Sanctuary established by the Japanese in 1959 extended northward no further than 58.17°N.

Our ability to comment on the ecological consequences of smaller invertebrates in the Bering Sea is very limited. However, some illustrative examples are possible and we feel these clearly demonstrate the complex dependencies and interdependencies that structure the benthos there. Empty shell constitutes most of the hard substrate in these soft-bottom areas. It has previously been argued that less empty gastropod shells in HT areas may have cascading consequences for hermit crabs, snail eggs, and, ultimately, for the snails themselves (McConnaughey et al. 2000). This shell not only serves as the primary substrate for snail egg attachment, but most of the sedentary megafauna in this area, including *Aplidium* spp. and the actiniarians (sea anemones) attach to shells or shell fragments. If shell size is mechanically reduced and dispersed by repeated trawling, it follows that the mean size of attached epifauna could be reduced as well. Furthermore, the occurrence of smaller living neptunid whelks (and their shells) in the HT area could exacerbate a possible shortage of pagurid shelters and may explain the observed reduction in hermit crab size.

Despite considerable research to document changes in benthic populations and communities as a result of bottom trawling, it is generally true that the mechanisms and, especially, the consequences of these disturbances are poorly understood. Even when actual vulnerabilities to mobile gear are reasonably well known, specific outcomes may still be unpredictable. For example, although whelks are reported to have very low direct mortality rates by virtue of their thick, rounded shells (Fonds 1994), they are very sensitive to chronic trawling in this area of the Bering Sea, nevertheless (McConnaughey et al. 2000; Table 1). This suggests that indirect effects on habitat are at least as important as the more direct effects of removals and injury, and perhaps more so in some cases. Clearly, improved knowledge of invertebrate biology and life histories and, in particular, their ecological roles and linkages would be beneficial since almost all research on fishing gear disturbances measures effects in terms of these species. However, in many cases, particularly in Alaskan waters, this information may not be available. As a result, unless change per se is deemed undesirable, it can be argued that we are generally unable to state whether changes due to trawling are harmful or not, and, thus, cannot provide objective technical guidance to resource managers concerned about possible adverse effects on fish and shellfish habitat. Until such time as informed management is possible, the National Research Council (NRC 2002:3) advises that:

> Fishery managers should evaluate the effects of trawling based on known responses of specific habitat types and species to disturbances by different fishing gears and levels of fishing effort, even when region-specific studies are not available. The lack of area-specific studies on the effect of trawling and dredging gear is insufficient justification to postpone management of fishing effects on seafloor habitat. The direct responses of benthic communities to trawling and dredging are consistent with ecological predictions based on disturbance theory. Predictions from common trends observed in other areas provide useful first-order approximations of fishing effects for use in habitat management. As more site-specific information becomes available on the fine-scale distribution of fishing effort and habitat distribution, those estimates should be revised.

Notwithstanding the need for precautionary management and the general consistency between mobile fishing gear effects and disturbance theory, the proposed interim strategy may be impractical. In their seminal work on patch dynamics and disturbance theory (which incidentally is included in the list of references for the National Research Council document), Pickett and White (1985a:377) argue that:

> Predictions from first principles, which treat all taxa, ecological systems, or situations as equivalent particles without an accrued history, such as are possible in physics, are not likely to be productive in the study of disturbance. Organisms, communities, populations, and ecosystems vary, have a history, and are subject to the vagaries of climatic and geological background. Thus the sorts of predictions we *can* make about disturbance are mechanistic—those that take into account the peculiarities of a particular system and situation.

If we are to advance beyond purely descriptive studies of mobile fishing gear effects, a mechanistic and scalable interpretive framework is needed. To this end, the ability to make predictions is an essential prerequisite to theory development, and to be efficient at it, we must clearly define the parameters of interest (e.g., production or diversity; Pickett and White 1985a). Although this will undoubtedly require more study and discussion, an existing body of knowledge may be applicable to the task. Allometry is the study of change in the proportions (size) of various parts of an organism as a consequence of development or growth (Peters 1983; Calder 1984). Calder refers to it as the "biology of scaling." This is a mature discipline with hundreds of empirical relationships between individual body size and a broad array of organismic properties (e.g., anatomy, physiology, and reproduction) for animal groups ranging from Protozoa to mammals. In the context of our present study, for example, allometry predicts that animal density, population production (i.e., the ratio of production to biomass, P/B), and incidence of successful defenses by prey generally increase with decreasing body size, while standard metabolic rate, ingestion rate, individual growth rate, social dominance, prey size, locomotory speed, and foraging radii/territory size generally decrease with decreasing body size. Expressions of these differences may confer competitive advantages that restructure benthic communities, and it has been argued that the main determinant of variation in species composition in epifaunal communities (specifically subtidal communities) is the degree to which residents surviving a disturbance influence invasion of the patch created (Connell and Keough 1985). For example, Ramsay et al. (1996) suggest that larger specimens of the common hermit crab *Pagurus berhardus* gain significant amounts of extra food by rapidly moving into recently-trawled areas and excluding smaller individuals that arrive later. In instances when disturbances are frequent and the supply of food is greater and nonlimiting, a broader range of crab sizes may populate the area (Ramsay et al. 1997). At a larger scale, substantial increases in abundance of North Sea flatfishes (Rijnsdorp et al. 1996) and perhaps certain Bering Sea species such as rock sole *Lepidopsetta* spp. (McConnaughey 1995) may in some way be related to sustained high levels of trawling effort and disturbance-related shifts to smaller, highly productive prey that are preferred by these small-mouthed predators. The end result of trawling disturbances and selective removals may be systems where small fish feed on small food items, as postulated by Jennings et al. (2002).

If estimates of population size are available, allometric equations can be used to predict various population rates, and if the size structure of the populations is known, these relations can be applied to describe processes at the community and ecosystem levels (Peters 1983). A preliminary effort by Daan (1991) evaluated the potential impact of fishing by comparing published P/B ratios with estimated total mortalities. Because size structure information was not available, a homogeneous assemblage of individuals was assumed. He concluded that it is unlikely that the North Sea benthic community as a whole is presently under severe pressure by the beam trawl fleet. More recently, benthic production was calculated using new allometric relationships and size data collected at locations with different levels of "relative fishing effort" (Jennings et al. 2001, 2002; Duplisea et al. 2002). These authors concluded that epifaunal production (and biomass) were not significantly related to trawling disturbance and, while the infaunal P/B ratio rose with increased disturbance, overall production decreased, reflecting the selective removal of larger individuals and increased relative production of smaller animals. This infaunal result is entirely consistent with allometric predictions, as is Daan's (1991) conclusion about minimal impacts by the beam trawl fleet. Recalling selective losses of large individuals by mobile fishing gear (Bergman and van Santbrink 2000), Peters (1983:181) states, in terms of community rates of energy or material flow:

> The importance of individual animals and size classes declines with size. Community processes are dominated by smaller species. Therefore, removal of larger species and individuals may have little effect on the rest of the community, and destruction of the smallest size classes could be disastrous.

Obviously, additional factors such as size-based reproductive value must be considered in any complete analysis of mobile fishing gear effects on benthic habitats. Unfortunately, despite the new, higher-level perspective that allometric studies provide, additional progress will be constrained by the general absence of body size data for noncommercial benthic invertebrates. It is interesting to note that this situation is not unique to the study of mobile fishing gear effects. In fact, "the significance of size is pointed out periodically, often with the criticism that zoologists fail to give this preeminent factor sufficient attention" (reviewed by Calder 1984). Fortunately, these data are readily collected and the situation is easily remedied with some additional effort.

Acknowledgments

The authors sincerely appreciate at-sea assistance from the captains and crews of the F/V *Aldebaran* and the F/V *Arcturus*, as well as the scholarly advice of two anonymous reviewers.

References

Auster, P. J., and R. W. Langton. 1999. The effects of fishing on fish habitat. Pages 150–187 *in* L. R. Benaka, editor. Fish habitat: essential fish habitat and rehabilitation. American Fisheries Society, Symposium 22, Bethesda, Maryland.

Bergman, M. J. N., and J. W. van Santbrink. 2000. Fishing mortality of populations of megafauna in sandy sediments. Pages 49–68 *in* M. J. Kaiser and S. J. de Groot, editors. The effects of fishing on non-target species and habitats: biological, conservation and socio-economic issues. Blackwell Scientific Publications, Oxford, UK.

Blanckenhorn, W. U. 2000. The evolution of body size: what keeps organisms small? Quarterly Review of Biology 75:385–407.

Caddy, J. F. 1968. Underwater observations on scallop (*Placopecten magellanicus*) behavior and drag efficiency. Journal of the Fisheries Research Board of Canada 25:2123–2141.

Calder, W. A., III. 1984. Size, function, and life history. Harvard University Press, Cambridge, Massachusetts.

Calkins, D. G., E. F. Becker, and K. W. Pitcher. 1998. Reduced body size of female Steller sea lions from a declining population in the Gulf of Alaska. Marine Mammal Science 14:232–244.

Collie, J. S., G. A. Escanero, and P. C. Valentine. 1997. Effects of bottom fishing on the benthic megafauna of Georges Bank. Marine Ecology Progress Series 155:159–172.

Collie, J. S., S. J. Hall, M. J. Kaiser, and I. R. Poiner. 2000. A quantitative analysis of fishing impacts on shelf-sea benthos. Journal of Animal Ecology 69:785–798.

Connell, J. H., and M. J. Keough. 1985. Disturbance and patch dynamics of subtidal marine animals on hard substrata. Pages 125–151 *in* S. T. A. Pickett and P. S. White, editors. The ecology of natural disturbance and patch dynamics. Academic Press, San Diego, California.

Daan, N. 1991. Theoretical approach to the evaluation of ecosystem effects of fishing in respect of North Sea benthos. International Council for the Exploration of the Sea, C.M. 1991/L27, Rome.

Dayton, P. K., S. F. Thrush, M. T. Agardy, and R. J. Hofman. 1995. Viewpoint: environmental effects of marine fishing. Aquatic Conservation: Marine and Freshwater Ecosystems 5:205–232.

Drabsch, S. L., J. E. Tanner, and S. D. Connell. 2001. Limited infaunal response to experimental trawling in previously untrawled areas. ICES Journal of Marine Science 58:1261–1271.

Duplisea, D. E., S. Jennings, K. J. Warr, and T. A. Dinmore. 2002. A size-based model of the impacts of bottom trawling on benthic community structure. Canadian Journal of Fisheries and Aquatic Sciences 59:1785–1795.

Else, P., L. Haldorson, and K. Krieger. 2001. Shortspine thornyheads (*Sebastolobus alascanus*) abundance and habitat associations in the Gulf of Alaska. U.S. National Marine Fisheries Service Fishery Bulletin 100:193–199.

Fonds, M. 1994. Mortality of fish and invertebrates in beam trawl catches and the survival chances of discards. Pages 131–146 *in* NIOZ Rapport 1994-11. Netherlands Institute for Fisheries Research, Texel.

Fonteyne, R. 2000. Physical impact of beam trawls on seabed sediments. Pages 15–36 *in* M. J. Kaiser and S. J. de Groot, editors. Effects of fishing on non-target species and habitats: biological, conservation and socioeconomic issues. Blackwell Scientific Publications, Oxford, UK.

Forrester, C. R., A. J. Beardsley, and Y. Takahashi. 1978. Groundfish, shrimp, and herring fisheries in the Bering Sea and northeast Pacific—historical catch statistics through 1970. International North Pacific Fisheries Commission Bulletin 37.

Freese, L., P. J. Auster, J. Heifetz, and B. L. Wing. 1999. Effects of trawling on seafloor habitat and associated invertebrate taxa in the Gulf of Alaska. Marine Ecology Progress Series 182:119–126.

Gordon, D., K. Gilkinson, E. Kenchington, J. Prena, C. Bourbonnais, K. MacIsaac, D. McKeown, and W. Vass. 2002. Summary of the Grand Banks otter trawling experiment (1993–1995): effects on benthic habitat and communities. Canadian Technical Report of Fisheries and Aquatic Sciences 2416.

Green, R. H. 1979. Sampling design and statistical methods for environmental biologists. Wiley, New York.

Hall, S. J. 1994. Physical disturbance and marine benthic communities: life in unconsolidated sediments. Oceanography and Marine Biology: An Annual Review 32:179–239.

Hall, S. J., M. R. Robertson, D. J. Basford, and S. D. Heaney. 1993. The possible effects of fishing disturbance in the northern North Sea: an analysis of spatial patterns in community structure around a wreck. Netherlands Journal of Sea Research 31:201–208.

Hanazato, T. 1998. Response of a zooplankton community to insecticide application in experimental ponds: a review and the implications of the effects of chemicals on the structure and functioning of freshwater communities. Environmental Pollution 101:361–373.

Jennings, S., T. Dinmore, D. Duplisea, K. Warr, and J. Lancaster. 2001. Trawling disturbance can modify benthic production processes. Journal of Animal Ecology 70:459–475.

Jennings, S., M. D. Nicholson, T. A. Dinmore, and J. E. Lancaster. 2002. Effects of chronic trawling disturbance on the production of infaunal communities. Marine Ecology Progress Series 243:251–260.

Kaiser, M. J., K. Cheney, F. E Spence, D. B. Edwards, and K. Radford. 1999a. Fishing effects in northeast Atlantic shelf seas: patterns in fishing effort, diversity and community structure. VII. The effects of trawling disturbance on the fauna associated with the tubeheads of serpulid worms. Fisheries Research 40:195–205.

Kaiser, M. J., S. I. Rogers, and J. R. Ellis. 1999b. Importance of benthic habitat complexity for demersal fish assemblages. Pages 212–223 *in* L. R. Benaka, editor. Fish habitat: essential fish habitat and rehabilitation. American Fisheries Society, Symposium 22, Bethesda, Maryland.

Kaiser, M. J., and B. E. Spencer. 1994. Fish scavenging behavior in recently trawled areas. Marine Ecology Progress Series 112:41–49.

Kaiser, M. J., and B. E. Spencer. 1996. Behavioural responses of scavengers to beam trawl disturbance. Pages 116–123 *in* S. P. R. Greenstreet and M. L. Tasker, editors. Aquatic predators and their prey. Blackwell Scientific Publications, Oxford, UK.

Lindeboom, H. J. and S. J. de Groot, editors. 1998. Impact II: the effects of different types of fisheries on the North Sea and Irish Sea benthic ecosystems. Netherlands Institute for Fisheries Research, NIOZ Rapport 1998-1, Texel.

Lindegarth, M., D. Valentinsson, M. Hansson, and M. Ulmestrand. 2000. Interpreting large-scale experiments on effects of trawling on benthic fauna: an empirical test of the potential effects of spatial confounding in experiments without replicated control and trawled areas. Journal of Experimental Marine Biology and Ecology 245:155–169.

Lytle, D. A. 2001. Disturbance regimes and life-history evolution. American Naturalist 157:525–536.

Maguire, J. A., A. Coleman, S. Jenkins, and G. M. Burnell. 2002. Effects of dredging on undersized scallops. Fisheries Research 56:155–165.

Mapstone, B. D. 1996. Scalable decision criteria for environmental impact assessment: effect size, type I and type II errors. Pages 67–80 *in* R. J. Schmitt and C. W. Osenberg, editors. Detecting ecological impacts: concepts and applications in coastal habitats. Academic Press, San Diego, California.

Marlow, M. S., A. J. Stevenson, H. Chezar, and R. A. McConnaughey. 1999. Tidally generated sea-floor lineations in Bristol Bay, Alaska. Geo-Marine Letters 19:219–226.

Mayer, L. M., D. F. Schick, R. H. Findlay, and D. L. Rice. 1991. Effects of commercial dragging on sedimentary organic matter. Marine Environmental Research 31:249–261.

Miller, R. G. 1981. Simultaneous statistical inference, 2nd edition. Springer-Verlag, New York.

McConnaughey, R. A. 1995. Changes in geographic dispersion of eastern Bering Sea flatfish associated with changes in population size. Pages 385–405 *in* B. R. Baxter, editor. Proceedings of the international symposium on North Pacific flatfish. University of Alaska Sea Grant College Program Report 95-04.

McConnaughey, R. A., K. Mier, and C. B. Dew. 2000. An examination of chronic trawling effects on soft-bottom benthos of the eastern Bering Sea. ICES Journal of Marine Science 57:1377–1388.

McConnaughey, R. A., and K. R. Smith. 2000. Associations between flatfish abundance and surficial sediments in the eastern Bering Sea. Canadian Journal of Fisheries and Aquatic Sciences 57:2410–2419.

Mensink, B. P., C. V. Fischer, G. C. Cadee, M. Fonds, C. C. ten Hallers-Tjabbes, and J. P. Boon. 2000. Shell damage and mortality in the common whelk *Buccinum undatum* caused by beam trawl fishery. Journal of Sea Research 43:53–64.

NRC (National Research Council, Committee on Ecosystem Effects of Fishing). 2002. Effects of trawling and dredging on seafloor habitat. National Academy Press, Washington, D.C.

Nebenzahl, D., compiler. 2001. 2001 bottom trawl survey of the eastern Bering Sea continental shelf. National Marine Fisheries Service, Alaska Fisheries Science Center processed report 2001-08, Seattle.

Nester, M. R. 1996. An applied statisticians creed. Applied Statistics 45:401–410.

Odum, E. P. 1985. Trends expected in stressed ecosystems. BioScience 35:419–422.

Peterman, R. M. 1990. Statistical power analysis can improve fisheries research and management. Canadian Journal of Fisheries and Aquatic Sciences 47:2–15.

Peters, R. H. 1983. The ecological implications of body size. Cambridge University Press, Cambridge, UK.

Pickett, S. T. A., and P. S. White. 1985a. Patch dynamics: a synthesis. Pages 371–384 in S. T. A. Pickett and P. S. White, editors. The ecology of natural disturbance and patch dynamics. Academic Press, San Diego, California.

Pickett, S. T. A., and P. S. White, editors. 1985b. The ecology of natural disturbance and patch dynamics. Academic Press, San Diego, California.

Prena, J., P. Schwinghamer, T. W. Rowell, D. C. Gordon, K. D. Gilkinson, W. P. Vass, and D. L. McKeown. 1999. Experimental otter trawling on a sandy bottom ecosystem of the Grand Banks of Newfoundland: analysis of trawl bycatch and effects on epifauna. Marine Ecology Progress Series 181:107–124.

Probert, P. K., D. G. McKnight, and S. L. Grove. 1997. Benthic invertebrate bycatch from a deep-water trawl fishery, Chatham Rise, New Zealand. Aquatic Conservation: Marine and Freshwater Ecosystems 7:27–40.

Raffaelli, D., S. Hall, C. Emes, and B. Manly. 2000. Constraints on body size distributions: an experimental approach using a small-scale system. Oecologia 122:389–398.

Ramsay, K., M. J. Kaiser, and R. N. Hughes. 1996. Changes in hermit crab feeding patterns in response to trawling disturbance. Marine Ecology Progress Series 144:63–72.

Ramsay, K., M. J. Kaiser, and R. N. Hughes. 1997. A field study of intraspecific competition for food in hermit crabs (*Pagurus bernhardus*). Estuarine, Coastal and Shelf Science 44:213–220.

Ribera, I., S. Dolédec, I. S. Downie, and G. N. Foster. 2001. Effect of land disturbance and stress on species traits of ground beetle assemblages. Ecology 82:1112–1129.

Rijnsdorp, A. D., A. M. Buys, F. Storbeck, and E. G. Visser. 1998. Micro-scale distribution of beam trawl effort in the southern North Sea between 1993 and 1996 in relation to the trawling frequency of the sea bed and the impact on benthic organisms. ICES Journal of Marine Science 55:403–419.

Rijnsdorp, A. D., P. I. van Leeuwen, N. Daan, and H. J. L. Heessen. 1996. Changes in abundance of demersal fish species in the North Sea between 1906–1909 and 1990–1995. ICES Journal of Marine Science 53:1054–1062.

Rogers, S. I., A. D. Rijnsdorp, U. Damm, and W. Vanhee. 1998. Demersal fish populations in the coastal waters of the UK and continental NW Europe from beam trawl survey data collected from 1990 to 1995. Journal of Sea Research 39:79–102.

Rose, C. S. 1999. Injury rates of red king crab, *Paralithodes camtschaticus*, passing under bottom-trawl footropes Marine Fisheries Review 61(2):72–76.

Sainsbury, K. J., R. A. Campbell, R. Lindholm, and A. W. Whitelaw. 1997. Experimental management of an Australian multispecies fishery: examining the possibility of trawl-induced habitat modification. Pages 107–112 in E. K. Pikitch, D. D. Huppert, and M. P. Sissenwine, editors. Global trends: fisheries management. American Fisheries Society, Symposium 20, Bethesda, Maryland.

Smith, K. R., and R. A. McConnaughey. 1999. Surficial sediments of the eastern Bering Sea continental shelf: EBSSED database documentation. NOAA Technical Memorandum NMFS-AFSC-104.

Thrush, S. F., J. E. Hewitt, V. J. Cummings, P. K. Dayton, M. Cryer, S. J. Turner, G. A. Funnell, R. G. Budd, C. J. Milburn, and M. R. Wilkinson. 1998. Disturbance of the marine benthic habitat by commercial fishing: impacts at the scale of the fishery. Ecological Applications 8:866–879.

Thrush, S. F., J. E. Hewitt, G. A. Funnell, V. J. Cummings, J. Ellis, D. Schultz, D. Talley, and A. Norkko. 2001. Fishing disturbance and marine biodiversity: the role of habitat structure in simple soft-sediment systems. Marine Ecology Progress Series 223:277–286.

Trussell, G. C., and M. O. Nicklin. 2002. Cue sensitivity, inducible defense, and trade-offs in a marine snail. Ecology 83:1635–1647.

Tuck, I. D., S. J. Hall, M. R. Robertson, E. Armstrong, and D. J. Basford. 1998. Effects of physical trawling disturbance in a previously unfished sheltered Scottish sea loch. Marine Ecology Progress Series 162:227–242.

Turner, S. J., S. F. Thrush, J. F. Hewitt, V. J. Cummings, and G. Funnell. 1999. Fishing impacts and the degradation or loss of habitat structure. Fisheries Management and Ecology 6:401–420.

Effects of Commercial Otter Trawling on Benthic Communities in the Southeastern Bering Sea

Eloise J. Brown,[1] Bruce Finney, and Sue Hills

Institute of Marine Science, University of Alaska Fairbanks, 245 O'Neill Building, Post Office Box 757220, Fairbanks, Alaska 99775, USA

Michaela Dommisse

Monash University, Department of Geography and Environmental Science, Post Office Box 11A, Clayton 3168, Australia

Abstract. The effects of commercial bottom trawling for yellowfin sole *Limanda aspera* on benthic communities were investigated in a sandy habitat exposed to high wave and tidal disturbance at 20–30 m depth in the southeastern Bering Sea. We compared an area that has been closed to commercial trawling for 10 years with an adjacent area that is now open to commercial trawling. In addition, we examined the immediate effects of experimental trawling on benthic community structure in the area closed to trawling. The fished area was characterized by reduced macrofauna density, biomass, and richness relative to the closed (unfished) area, but diversity was not different. Interannual variability of macrofauna assemblages was high in the system, yet assemblages in the two areas were distinguished using multivariate analyses and dominant taxa. After 10 years, sessile taxa (e.g., Maldanidae polychaetes) were prevalent in the closed area, and mobile scavengers (e.g., Lysianassidae amphipods) were more common in the fished area. Immediate responses of macrofauna to experimental trawling were subtle (i.e., reduced richness, absence of rare taxa, and patchy changes in assemblage biomass), but no differences were detected relative to controls for density, diversity, or total biomass. Fragile, structure-forming megafauna were rare. However, when trawled, they appeared mostly unaffected. Though we could not completely rule out other factors such as food supply from water column primary production, our results indicate that trawling altered macrofauna communities. Our findings also suggest that individual taxa respond differently to trawling but that commonly used summary measures such as total abundance do not capture these changes. Based on the functional attributes of individual taxa, hypotheses explaining different macrofaunal assemblages include bottom-up shifts caused by physical disturbance from trawling, altered benthic food webs from discards and processing waste, and top-down shifts caused by altered predator–prey interactions between *Asterias amurensis* and yellowfin sole. Stomach contents of yellowfin sole indicated that this flatfish did not target specific prey taxa; however, fish-processing waste was prominent in its diet. More ecological and bioenergetic data are required to determine how changes in benthic communities are linked specifically to the productivity of yellowfin sole.

Introduction

Mobile fishing gear has recently become the focus of international attention as scientists, fishers, and policy makers fear habitat degradation, loss of biological diversity, and the subsequent consequences to fishery resources (e.g., Dayton et al. 1995; Jennings and Kaiser 1998; Hall 1999; Norse and Watling 1999). Bottom trawling has even been compared to forest clear-cutting (Watling and Norse 1998). Yet habitat type, gear type, and gear configuration largely affect the degree to which biological communities are affected by fishing. Few experimental studies have quantified trawl effects using realistic fishing gear at the intensities imposed by commercial fleets (Collie et al. 2000).

The southeastern Bering Sea supports the largest flatfish fishery in the United States (>300,000 metric tons in 1997; NPFMC 1997; Fritz et al. 1998; NRC

---

[1] Corresponding author: browne02@student.uwa.edu.au; present address: University of Western Australia, School of Plant Biology (MO29), 35 Stirling Highway, Crawley WA 6009, Australia.

2002). Targeted primarily by commercial catcher–processors, yellowfin sole *Limanda aspera* accounted for over half the harvest at the time of this study (NPFMC 1997; Fritz et al. 1998). Despite the long history and economic importance of fishery resources in the Bering Sea, few studies on otter–trawl impacts have been conducted in this region (McConnaughey et al. 2000, 2005, this volume). However, trawl impacts on benthos in other areas (e.g., North Sea, eastern Atlantic, South Pacific) include reduced infauna density; reduced biomass and diversity; selective elimination of large, sedentary, and fragile taxa by physical removal; or damage in heavily trawled areas (e.g., Bergman and Hup 1992; Eleftheriou and Robertson 1992; Jones 1992; Collie et al. 1997; Engel and Kvitek 1998; Freese et al. 1999). In contrast, most research in shallow, high-energy regions did not find detectable changes to benthic fauna caused by trawl gear (e.g., Gibbs et al. 1980; Brylinsky et al. 1994). Populations of certain prey and commercially important species may be enhanced by some level of trawl disturbance through increased food availability from damaged organisms and the replacement of large benthic animals with smaller, opportunistic prey (e.g., Millner and Whiting 1996; Engel and Kvitek 1998).

The 1996 Magnuson–Stevens Fishery Conservation and Management Act (MSFC) requires the identification and description of essential fish habitat (EFH) for each commercially fished species within the United States and determination of any adverse impacts from fishing (U.S. Code, title 16, section 1801 et seq.; Public Law 94-265 as amended October 11, 1996). Essential fish habitat is defined as those waters and substrate necessary for spawning, breeding, feeding, or growth to maturity and, hence, encompasses an enormous range of Alaska's marine environment (DiCosimo 1999). Given the regional differences and spatial heterogeneity in benthic communities and substrate types, it is crucial that trawl impacts are evaluated specifically for the southeastern Bering Sea.

The objectives of this 2-year study were to evaluate long-term effects of commercial flatfish trawling on EFH in the southeastern Bering Sea and to investigate initial responses of the benthic community to experimental trawling. Macrofaunal and megafaunal community responses to trawling were described, and yellowfin sole prey were examined to determine potential impacts on EFH for this species. The incorporation of commercial fishing intensity and gear typical for the Bering Sea flatfish fleet ensured that the results of the experiment were relevant to local fisheries and managers. A concurrent study of physical habitat response to trawling in this study area found that surface sediments were composed primarily of fine sand with low organic content (~3 mg C/g). However, the top 3 cm of sediments in the fished zone were slightly better sorted, less variable, and contained less fine-grained sediment than those of the unfished zone (Brown et al., in press).

We focused on three hypotheses: (1) Benthic community indices including density, biomass, diversity, richness and multivariate patterns in macrofaunal assemblages do not change in response to chronic exposure to commercial trawling for flatfish; (2) Benthic community indices and multivariate patterns do not change immediately after experimental trawling; and (3) Based on (1), (2), and yellowfin sole diet composition, trawling does not affect essential fish habitat for this species.

Methods

Study Location

Since the 1950s, a single commercial groundfish fishery (yellowfin sole) has been active in the federally managed waters of the southeastern Bering Sea, within an area created in 1996 called the Nearshore Bristol Bay Closure (Figure 1a; Witherell and Pautzke 1997). The catch is primarily yellowfin sole, and bycatch rates are particularly low (NPFMC 1991). Immediately adjacent to the fishery area is the Walrus Islands State Sanctuary, where the Alaska Department of Fish and Game has prohibited all commercial fishing in state waters (out to 5 km; 3 mi) since 1989. Within approximately 19 km (12 mi) of Round Island, fishing has been prohibited since 1992 by the North Pacific Fishery Management Council (P. Koehl, Alaska Department of Fish and Game Wildlife Conservation Division, personal communication). Global positioning system (GPS)-based maps of fishing locations and gear type recorded by the National Marine Fisheries Service (NMFS) Observer Program indicate that the Round Island 12-mi (~19 km) closure has not been fished for approximately 10 years (NPFMC 1997; NRC 2002). Our study area spanned the Round Island closure and adjoining commercial fishery areas (Figure 1b) and offered a rare opportunity for comparison of two areas that were otherwise extremely similar. This study is based on two research cruises to this site with R/V *Big Valley* in 1999 and 2000.

Chronic Study Design

Long-term effects of trawling on the benthic community were evaluated based on a comparison between the Round Island closure (unfished) and commercially fished habitats. In 1999, we selected a study area where the fished and unfished areas were contiguous and had similar depths (20–30 m; mean = 26 m) and sediment

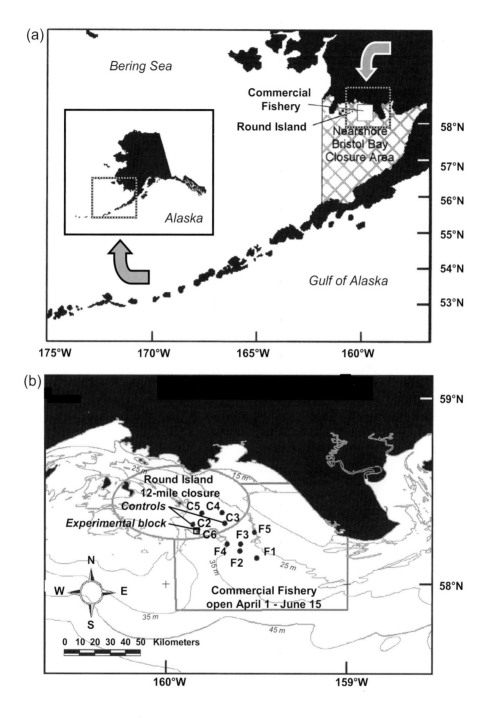

Figure 1. (a) Nearshore Bristol Bay trawl closure area showing location of Round Island (closed to fishing) and commercialy fished area, adapted from Witherell and Pautzke (1997). (b) Study area showing location of closed stations (C2–C5) within the unfished 12-mi (~19 km) Round Island closure and fished stations (F1–F5) within the commercial trawl fishery boundaries. Also shown is the 4-km² block and both controls for the experimental study.

size (fine sand) (NOAA 1988; Smith and McConnaughey 1999). Two "zones" within this area were established, as either closed or fished, using GPS-based maps of fishing effort (NPFMC 1997; Fritz et al. 1998). A 1-km-wide buffer between the zones was excluded from comparisons to avoid confounding edge effects. Nine random stations were selected: four (C2–C5) in the closed zone and five (F1–F5) in the fished zone (Figure 1b). At each station, four to six replicates for macrofauna were sampled with a 0.1-m^2 van Veen grab (Table 1). Macrofauna were sieved through a 0.5-mm screen, preserved in 10% neutral buffered formalin with 0.1% rose bengal, identified to family (or lower level for common species), and counted, excluding pelagic organisms and meiofauna (e.g., Warwick 1988; USEPA 1995; Jewett et al. 1999; Blanchard et al. 2002). Wet biomass was determined as blotted wet weights. A single qualitative video transect per station was used to characterize mobile and emergent megafauna. These transects were obtained using a monochrome high-resolution video camera on a sled towed approximately 25 cm above seafloor (Simrad Osprey model 1359, Aberdeen, Scotland). Water depth, temperature, and salinity were recorded at each station with a conductivity–temperature–depth instrument (CTD; SeaBird Electronics model SBE 19 SeaCat Profiler, Bellevue, Washington).

In 2000, sampling was repeated at sites C2–C5 and F1–F5 with some minor modifications (Table 1). An additional station was randomly selected in the unfished zone (C6), and macrofauna grab replicates were reduced from a maximum of six to two per station. Processing of macrofauna samples in 2000 followed the same protocol as in 1999 except that all animal fragments excluding heads were discarded, and whole weights were estimated for individual species to save time (e.g., Coyle 2000). Divers collected megafauna voucher specimens and duplicate sediment cores, providing additional information about surficial sediment properties at each station. A 30-min video transect in each zone was used to document the presence of megafauna and flatfish using a towed remotely operated vehicle (ROV) and wide-angle color video camera (Phantom model HD2 ROV and Sony Smart Zoom model EVI-330; Deep Ocean Engineering, San Leandro, California). The ROV was towed along the bottom, maintaining speeds of less than 0.5 m/s. The camera was mounted 40 cm above seafloor and pointed down at an angle of 8° from horizontal for the majority of the transects or 11.5° when visibility was low. Video quality was enhanced by using ambient light. Twin 15 mW lasers (Deep Ocean Engineering), spaced 10 cm apart, facilitated measurement of organism sizes and densities as well as total area covered during each strip transect (Auster et al. 1989; W. Wakefield, National Marine Fisheries Service, personal communication).

Experimental Study Design

Experimental trawling was performed by F/V *Verdal* during a 30-h period on May 26–27, 2000, and consisted of 10 tows within a 4-km^2 block inside the closed zone (centered on C6; Figures 1b, 2). This level of trawling intensity was chosen because it was realistic for this fishery (J. Gauvin, Groundfish Forum, personal communication). The commercial factory–processor was equipped with an otter trawl and footrope with east-coast style roller gear typical for the Bering Sea flatfish fleet. The catch was processed onboard for commercial sale, with factory waste discarded over-

Table 1. Field sampling design for chronic and experimental studies. Number of stations and individual samples are listed by zone for the chronic study and type for the experimental study. For total number of stations and total number of grab samples, three closed stations were included in the experiment, one as an impact before and two as controls. The number of grab samples reflects the dual roles (closed–control; closed–impact before) of those three stations. For total number of video transects, the control before also served as the closed video transect. Only impact video transects ($n = 6$) were quantitatively analyzed; closed and fished video transects were viewed for qualitative descriptions.

Year	Date	Zone	Type	Number of stations	Number of grab samples	Number of video transects
1999	Jul 20–25	Closed		4	22	4
	Jul 20–25	Fished		5	28	5
2000	May 18–Jun 1	Closed		5	10	1
	May 18–Jun 1	Fished		5	11	1
	May 18–25	Closed	Impact before	12	24	3
	May 28–Jun 1	Closed	Impact after	12	24	3
	May 18–25	Closed	Control before	2	4	1
	May 28–Jun 1	Closed	Control after	2	4	1
Totals				21	121	18

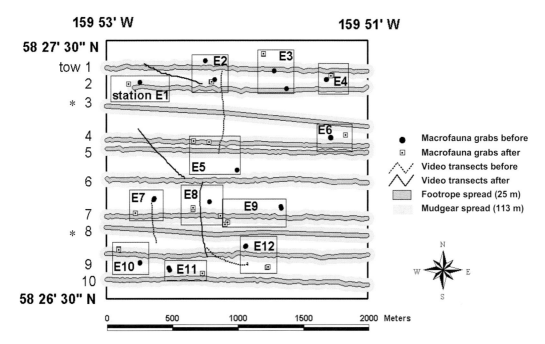

Figure 2. Location of replicate samples and trawl tows, with tow widths for footrope and mudgear from the experimental block in 2000. The center point of each impact station (E1–E12; $n = 12$) was marked with a differential global positioning system during pre-trawl sampling and revisited after experimental trawling. The areal extent of stations was determined by the radius of the boat swing when anchored and is represented by a rectangle for clarity. All tows continued ~11 km but are not shown beyond the experimental block. The position of the net was estimated during tows 3 and 8, shown by as asterisk (Brown et al., in review).

board to simulate realistic fishing practices. During the first tow, the catch was smaller than expected, so the tows were extended to an average of 11 km, ensuring the cod-end was full and sufficient for commercial sale. However, alterations to benthic habitat were not quantified beyond the 4-km² experimental block. Further details of experimental trawling, seabed characteristics, and water properties documented during the chronic and experimental studies are described in Brown et al. (in press).

Twelve impact stations (E1–E12) within the experimental block were sampled before (B) and after (A) trawling and compared to two untrawled controls (C2 and C3) sampled over the same time period (Figure 2; Table 1). Initial responses of the benthic community were determined within 1 week after experimental trawling. Macrofauna and CTD sampling at each station followed the 2000 chronic study protocol. A total of six randomly placed 30-min video transects (three before and three after trawling) were taken inside the experimental block, averaging 82 ± 3.8 cm (SE) wide and 441 ± 36 m (SE) long (Figure 2). Video documented megafaunal densities and physical alterations to the seabed. The position of each tow, grab sample, and video transect was recorded by sonar tracking (Trackpoint; ORE Offshore Systems, Falmouth, Massachusetts), with differential global positioning system (DGPS; Garmin, Olathe, Kansas). Due to equipment malfunction, the location of some pre-trawl grab samples (E5, E6, E8, and E11) and the net (tows 3 and 8; Figure 2) was estimated based on the position of R/V *Big Valley* or F/V *Verdal* during post-field analysis of Trackpoint data using ArcView geographical information system software (ESRI, Redlands, California).

Yellowfin sole (3 per tow, $n = 30$) were randomly sampled from the catch (AFSC 1999). Fish lengths and weights were recorded, and stomachs were removed (AFSC 1999), preserved in 10% buffered formalin, and eventually transferred to 50% isopropyl alcohol to prevent dissolution of $CaCO_3$. Fishes were excluded from the sample if the stomach was obviously empty or showed signs of regurgitation (stomach distended but empty), but despite initial screening, three stomachs contained no prey items. Stomach contents of the fishes ($n = 27$) were identified to lowest taxonomic category possible.

Results

General Description

A total of 86 macrofauna and 8 megafauna taxa were identified in the van Veen grab samples and video transects. The invertebrate community at the study site was typical of the southeastern Bering Sea continental shelf infauna and epifauna (e.g., Stoker 1981) and consisted of 25 families of polychaetes, 17 bivalves, 14 crustaceans, 11 gastropods, and 5 echinoderms.

Chronic Study: Macrofauna Community Indices

Macrofauna data from 1999 were tested in STATISTICA (StatSoft, Tulsa, Oklahoma) with a mixed-model nested analysis of variance (ANOVA) with three factors: station, zone, and subsample (Zar 1999). Indices per sample included density, biomass, Shannon diversity (H'), and Margalef richness (SR) and were measured per individual family (Shannon and Weaver 1949; Margalef 1958; Pielou 1969). Data were normalized per m^2 and per mean grab volume (4 L) and visually assessed for normality and equality of variances using residual plots. Natural-log or square-root transformations were performed where necessary to satisfy these ANOVA assumptions (Zar 1999). Some differences were apparent in macrofaunal community parameters between closed and fished zones in 1999 (Figure 3; Table 2). Diversity and richness were higher in the closed zone ($P < 0.05$), but density and biomass were not significantly different (Table 2). Low within-station sampling variance in 1999 (Table 2) allowed for the reduction of replicate grabs to two the following year, and the addition of station C6 in 2000 created a balanced design. Macrofauna data in 2000 were tested with a mixed-model nested analysis of covariance (ANCOVA) with station, zone, and subsample as factors (Zar 1999). Total density, biomass, and richness were significantly higher in the closed zone ($P < 0.05$), but no difference was detected in diversity (Table 3; Figure 3). The environmental parameters water depth, salinity, and temperature did not significantly alter the fit of the ANOVA models and were dropped in both years.

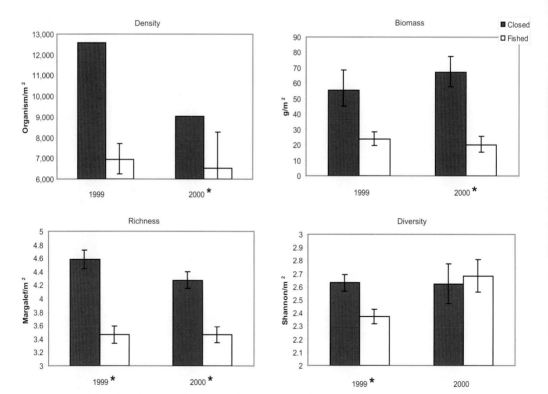

Figure 3. Ecological indices (a, density; b, biomass; c, richness; and d, diversity) for macrofauna in the chronic study, with significant results ($P < 0.05$, Tables 2, 3) marked with an asterisk. The bars indicate means from closed ($n = 4$ in 1999; $n = 5$ in 2000) and fished ($n = 5$) stations, and the vertical lines are standard errors. Density and biomass data were de-transformed.

Table 2. Summary of ANOVA between closed and fished zones from 1999, with significant results ($P < 0.05$) marked with an asterisk. Macrofaunal density and biomass data were natural-log transformed. Mean square error (MS) and degrees of freedom (df) were adjusted due to unbalanced design.

Variable	Effect	df	Adjusted df	MS	Adjusted MS	F	P-value
Density	Zone	1	6.976	6.074	2.603	2.33	0.1706
	Error (between)	7		2.539			
	Sampling (within)	41		0.163			
Biomass	Zone	1	6.944	15.272	3.045	5.02	0.0604
	Error (between)	7		2.977			
	Sampling (within)	41		0.452			
Diversity	Zone	1	6.872	1.271	0.061	20.85	0.0027*
	Error (between)	7		0.060			
	Smpling (within)	41		0.021			
Richness	Zone	1	6.974	16.497	1.539	10.72	0.0137*
	Error (between)	7		1.501			
	Sampling (within)	41		0.105			

Mean grain size was a significant factor for density ($P < 0.05$) in 2000 (Table 3). Within-station sampling variance was small compared to between-station variance, further justifying the reduction in number of replicate grabs in 2000 (Table 3).

Chronic Study: Multivariate Macrofauna Assemblages

Nonmetric multidimensional scaling (MDS) and cluster analyses described patterns in macrofauna assemblages using STATISTICA (e.g., Clarke and Ainsworth 1993; Jewett et al. 1999). Dissimilarity matrices employed the Bray–Curtis coefficient calculated for density and biomass data listed by family and station (Bray and Curtis 1957). Station groups in MDS plots were based on cluster analysis dendrograms of percent dissimilarity. Multivariate analyses did not include grain size or other environmental variables, so station groupings largely reflected biological composition. There was a clear separation (35% dissimilarity) of closed and fished zones in terms of macrofauna density, with a couple of exceptions (Figure 4a). These outliers presumably account for the MDS stress (0.130), which indicates that this model had a reasonable fit (Clarke and Ainsworth 1993). The "x" dimension in Figure 4a could be interpreted as decreasing fishing disturbance toward the right. The division by zone was less clear for biomass data (Figure 4b). Stations appeared to be grouped instead by slight variations in sediment mean grain size, which ranged from fine to very-fine sand (Brown 2003). The separation of the finer grain-sized group at 60% and the medium and coarser grain-sized groups at 50% dissimilarity show that macrofauna assemblages between groups were not very closely related in terms of biomass. Although the stress (0.143) was still acceptable, the fit for this model was not as good (Clarke and Ainsworth 1993). High interannual

Table 3. Summary of ANCOVA between closed and fished zones from 2000, with significant results ($P < 0.05$) marked with an asterisk. Macrofaunal density data were natural-log transformed, and biomass data were square-root transformed.

Variable	Effect	df	Adjusted df	MS	Adjusted MS	F	P-value
Density	Zone	1	6.921	4.202	0.365	11.51	0.0118*
	Grain size	1		13.955		38.82	0.0004*
	Error (between)	7		0.360			
	Sampling (within)	11		0.115			
Biomass	Zone	1	7.965	72.652	5.097	14.25	0.0055*
	Error (between)	8		5.017			
	Sampling (within)	11		0.746			
Diversity	Zone	1	7.957	0.016	0.097	0.17	0.6939
	Error (between)	8		0.095			
	Sampling (within)	11		0.014			
Richness	Zone	1	7.977	3.709	0.375	9.90	0.0137*
	Error (between)	8		0.368			
	Sampling (within)	11		0.029			

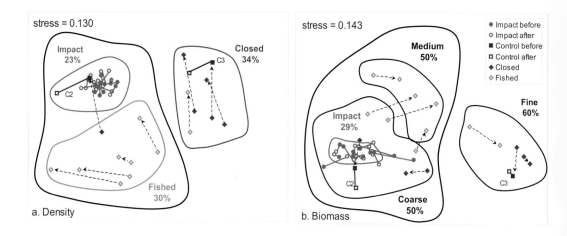

Figure 4. (a) Nonmetric multidimensional scaling of macrofauna assemblages by density, with separation of stations and percent dissimilarities from cluster analyses on natural-log transformed ($\log_e x + 1$) data from chronic and experimental studies. A dashed arrow indicates how much closed ($n = 4$) and fished ($n = 5$) stations changed from 1999 to 2000 (Table 1). Impact stations ($n = 12$) and controls ($n = 2$; C2 and C3) are connected with solid lines to indicate amount of change after experimental trawling. (b) Nonmetric multidimensional scaling of macrofauna assemblages by biomass. Group I represents finer-grained sediments (very fine sand), group III represents coarser-grained sediments (fine sand), and group II includes those stations with sediments between very fine and fine sand (Brown 2003).

variability for both density and biomass is shown by dashed arrows indicating change at each station between the 1999 and 2000 sampling periods (Figure 4).

Chronic Study: Individual Macrofauna Taxa

Common macrofauna and the total number of taxa from the closed and fished zones were different. In terms of density, tube-dwelling and carnivorous polychaetes (Spionidae and Sigalionidae) ranked first and second in the closed zone and, together, contributed 31% of the total number. Other relatively common taxa in the closed zone included fragile Nuculidae bivalves, tube-dwelling Oweniidae polychaetes, and the gastropod family Turritellidae (composed exclusively of *Tachyrhynchus erosus*). Tellinidae bivalves and tube-dwelling and burrowing amphipods (Isaeidae and Oedicerotidae) were common in the fished zone and accounted for 35% of the total number. Other top-ranking taxa in the fished zone included unidentified Bivalvia (mostly juveniles too young to identify), herbivorous Trochidae gastropods, and Lysianassidae amphipods (Brown 2003). In terms of biomass, Turritellidae ranked first in both zones but contributed 47% of the total biomass in the closed zone compared to 19% in the fished zone (Table 4). Large, deposit-feeding, and tube-dwelling polychaetes (Opheliidae, Orbiniidae, and Maldanidae) and the bivalve Astartidae were prevalent in the closed zone. Other bivalves (Tellinidae and Solenidae) and amphipods (Isaeidae and Lysianassidae) were relatively common in the fished zone. Numerous rare taxa, including the large Sternaspidae polychaete, were only sampled in the closed zone, and a few taxa were only found in the fished zone (e.g., Echinoidea and Isopoda; Table 4).

Chronic Study: Megafauna

Poor video quality in 1999 and low sample size in 2000 (Table 1) prevented statistical analysis of video data. Comparisons of megafaunal and demersal fish populations in the closed and fished zones were, therefore, limited to qualitative descriptions. The predominant species of megafauna was the seastar *Asterias amurensis*, often seen in feeding excavations measuring up to 20 cm across with exposed bivalves in the center or lying adjacent to the pits. Other common organisms included hermit crabs (Paguridae), clumps of sessile bryozoa (Diastopodidae), and hydrozoa. Rare species included other crabs (Anomura and Brachyura), anemones (*Metridium* spp.), colonial tunicates (Ascidiacea), and sponges (Porifera). Divers observed clumps of sessile organisms often attached to tubes, probably of the polychaete *Sabella* spp. (Hobson and Banse 1981). Video transects verified the presence of yellowfin sole. Megafaunal assemblages appeared similar between the closed and fished zones both years.

Table 4. Individual macrofauna taxa in chronic study ranked by percentage of total mean biomass (g/m^2). Data within zones were pooled across years for $n = 9$ closed and $n = 10$ fished stations. Functional group designations based on feeding habits include: F = suspensivore, D = sub-surface detritivore, C = carnivore, S = surface deposit feeder and O = omnivore (Feder et al 1991; Pearson 2001). Totals calculated from all families for each zone (closed/fished).

Taxa	Function	Closed			Fished		
		Rank	%	Mean ± SE	Rank	%	Mean ± SE
				Common taxa			
Turritellidae	F	1	47.3	34.49 ± 11.91	1	18.6	5.58 ± 2.74
Opheliidae	D	2	5.2	3.8 ± 1.68	9	3.5	1.06 ± 0.43
Nuculanidae	D	3	5.1	3.7 ± 1.8	3	10.4	3.13 ± 2.25
Nephtyidae	C	4	4.7	3.42 ± 0.5	2	15.1	4.54 ± 0.88
Sternaspidae	D	5	4.3	3.17 ± 1.81			
Spionidae	S	6	3.7	2.67 ± 1.16	5	4.3	1.3 ± 0.33
Orbiniidae	D	7	3.1	2.29 ± 0.55	11	3.4	1.01 ± 0.11
Nuculidae	D	8	3	2.2 ± 0.39	6	4.2	1.26 ± 0.35
Astartidae	F	9	2.6	1.89 ± 1.68	63	<1	0 ± 0
Maldanidae	D	10	2.3	1.71 ± 0.5	22	<1	0.21 ± 0.07
Tellinidae	S	12	2.2	1.59 ± 0.66	4	6.2	1.87 ± 0.65
Solenidae	F	13	2.1	1.5 ± 0.61	7	4.2	1.25 ± 0.88
Isaeidae	S	14	1.9	1.39 ± 0.34	10	3.4	1.02 ± 0.39
Lysianassidae	S	26	<1	0.2 ± 0.06	8	4.2	1.25 ± 0.31
				Rare taxa			
Cancellariidae	O	22	<1	0.3 ± 0.299			
Haustoriidae	F	40	<1	0.1 ± 0.062			
Hiatellidae	F	48	<1	0.033 ± 0.032			
Syllidae	O	51	<1	0.025 ± 0.017			
Sabellidae	F	53	<1	0.022 ± 0.014			
Ungulinidae	S	54	<1	0.017 ± 0.012			
Sabellariidae	S	60	<1	0.008 ± 0.008			
Cardiidae	S	62	<1	0.008 ± 0.007			
Myidae	F	64	<1	0.007 ± 0.004			
Lyonsiidae	F	65	<1	0.006 ± 0.006			
Enteropneusta	D	67	<1	0.004 ± 0.004			
Rissoidae	H	69	<1	0.004 ± 0.004			
Cumacea	S	76	<1	0.001 ± 0.001			
Pyramidellidae	S				59	<1	1.1 e^{-3} ± 1.1 e^{-3}
Amphictenidae	D				64	<1	4.1 e^{-4} ± 4.1 e^{-4}
Echinoidea	O				66	<1	4.7 e^{-6} ± 4.7 e^{-6}
Isopoda	S				67	<1	4.1 e^{-6} ± 4.1 e^{-6}
Totals		77	100%	72.88 ± 13.02	68	100%	30.04 ± 4.45

Experimental Study: Macrofauna Community Indices

Community indices were calculated and transformed where necessary, as in the chronic study. An asymmetrical BACI (before–after–control–impact) design evaluated immediate responses of macrofaunal communities to trawling at impact stations ($n = 12$) within the experimental block relative to controls ($n = 2$) using STATISTICA (Underwood 1994). Differences over time at each station (after minus before trawling) were determined, and a Type-II ANCOVA was performed on the paired differences with type (control or impact) as a single factor and mean grain size as a covariate. This BACI design was essentially a paired sample analysis and more powerful than an analysis of variance (Zar 1999). The macrofaunal community at one of the control sites (C2) closely resembled that of the impact stations (E1–E12). The community at the other control (C3) had higher density, biomass, diversity, and richness, probably caused by the slightly finer-grained sediment (very-fine sand; Brown et al., in press). Along with small control sample size ($n = 2$), this made error estimates high for the controls, despite transformations. Therefore, statistical comparisons between control and impact stations were unable to detect any significant changes in the mean of any macrofauna parameter after experimental trawling (Table 5; Figure 5). Grain size was significant as a covariate for biomass ($P < 0.05$; Table 5). The environmental parameters water depth, salinity, and temperature did not explain additional variability and were excluded from the analysis.

Due to the differences between the controls, nonparametric Wilcoxon matched pair tests were performed exclusively for impact stations (Conover 1999). Data before ($n = 12$) and after ($n = 12$) trawling were paired at each station to determine the mean response to trawling within the experimental block. Taxon richness was significantly reduced ($P <0.05$) after trawling (Table 6) and corresponded with a decrease in total number of taxa (Table 7), as would be expected.

Experimental Study: Multivariate Macrofauna Assemblages

Density and biomass data from the experiment were combined with chronic data sets for MDS and cluster analyses to yield a more comprehensive explanation of long-term and immediate changes to macrofauna assemblages. In the MDS plots (Figure 4), impact stations are connected with solid lines to indicate change over the two sampling periods (before and after trawling) and at controls (C2 and C3) sampled over the same time period. The tight grouping of impact stations and C2 is indicative of the limited spatial extent of the experimental block relative to the larger study area (Figures 1b, 4a) and shows that they were very closely related (23% dissimilarity) in terms of density during both time periods. Changes at impact stations after trawling were comparable to natural variability in density at both controls (Figure 4a). However, changes in macrofauna biomass were greater at some impact stations after trawling than at the controls (Figure 4b). Impact stations became more tightly grouped after trawling (29% dissimilarity), suggesting that stations became more similar in terms of macrofauna biomass, possibly through direct removal of larger taxa at some stations (Figure 4b).

Table 5. Summary of BACI ANOVA/ANCOVA between control and impact stations from 2000 experimental study, with significant results ($P < 0.05$) marked with an asterisk. Paired station data were subtracted to get the differences over time after experimental trawling and classified by type (control or impact). Density and biomass were natural-log transformed.

Variable	Effect	df	MS	F	P-value
Density	Type	1	0.097	1.71	0.2153
	Error	12	0.057		
Biomass	Type	1	0.397	1.82	0.2045
	Grain	1	1.924	8.81	0.0128*
	Error	11	0.218		
Diversity	Type	1	0.009	1.29	0.2780
	Error	12	0.007		
Richness	Type	1	0.045	0.38	0.5499
	Error	12	0.120		

Experimental Study: Individual Macrofauna Taxa

Immediately after experimental trawling, there was little change in the rankings of the most abundant taxa (Brown 2003). However, the total number of taxa decreased, and many rare taxa such as Ophiuroidea brittle stars and several bivalve families were not sampled after trawling (Table 7). Some changes were apparent in the rankings of common taxa by biomass at impact stations. For example, the common bivalves Astartidae and Nuculanidae (composed exclusively of *Yoldia* sp.) were reduced to 18th and 14th place after trawling (Table 7). Other taxa, such as the large, mobile Nephtyidae polychaetes (*Nephtys* sp.) and Isaeidae amphipods (Protomedeia sp. and *Photis* sp.), increased several places, and a few new taxa were collected. Aside from Turritellidae biomass, which increased after experimental trawling and accounted for over two-thirds of the biomass during both time periods, the mean biomass of top ranking taxa tended to decrease after experimental trawling (Table 7).

Experimental Study: Megafauna

Immediate responses of prominent megafauna to trawling were quantified exclusively from video collected inside the experimental block (Table 1). Due to small video sample size ($n = 3$ before, $n = 3$ after), inferences were drawn from means and 95% confidence intervals of megafaunal densities per m^2. Rare taxa that occurred in fewer than three transects were excluded from the data set. There was a significant reduction in density of the seastar *A. amurensis* after experimental trawling (Figure 6). Before trawling, seastars were dispersed and easily counted (Figure 7b). After trawling, individuals were frequently concentrated in large clusters, making it more difficult to estimate density (Figure 7a). Divers determined that these seastars were clearly feeding on discarded fish processing waste. Occasional dislodged hydroid clumps and rolling colonial tunicates were documented, but many clumps of sessile organisms, including hydrozoa and bryozoa, appeared unharmed after trawling. No other taxa had significant changes in density in the video data.

Experimental Study: Trawling Catch Composition

The total catch of 106 metric tons (MT) from experimental tows was extrapolated from the percent sample weight by taxa (AFSC 1999; L. Swanson, Groundfish Forum, written communication), consisting by weight of 60% yellowfin sole, 23% rock sole *Lepidopsetta bilineata*, and 12% other demersal species (Table 8). Approximately two-thirds of these fishes were retained as final product for commercial sale. The remaining 40

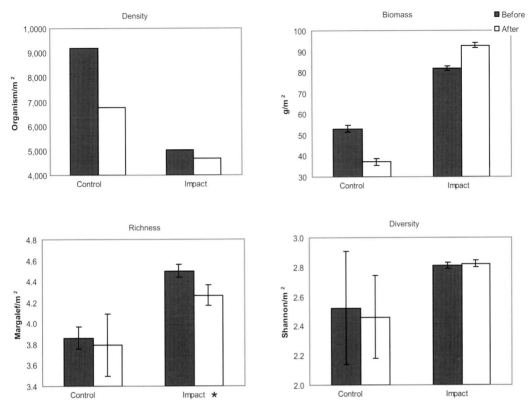

Figure 5. Ecological indices for macrofauna in the experimental study. The bars indicate means (detransformed for density and biomass) of control ($n = 2$) and impact ($n = 12$) stations before and after experimental trawling, and the vertical lines are standard errors

MT of processing waste and bycatch was discarded overboard. Bycatch included 4% demersal fishes and 1% invertebrate species, including roughly 0.5 MT of seastars plus rare invertebrate species not quantified by video (e.g., Alcyonacea and red king crab *Paralithodes camtschaticus*).

The removal and redeposition of demersal fishes and megafauna biomass was calculated based on the total area covered during experimental trawling (Table 8). This method assumed that all fishes were herded into the net between the doors (tow width = 113 m; Figure

Table 6. Wilcoxon matched pairs test on impact stations before ($n = 12$) and after ($n = 12$) experimental trawling, with significant results ($P < 0.05$) marked with an asterisk.

Variable	T	Z	P-value
Density	25	1.098	0.2721
Biomass	22	1.334	0.1823
Diversity	33	0.471	0.6379
Richness	13	2.040	0.0414*

2), that less mobile or sessile invertebrates were only caught within the mouth of the net (footrope spread = 25 m; Brown et al., in press) and that bycatch and processing waste was discarded overboard at a constant rate along the entire length of experimental tows. We estimated that 8.52 g/m² demersal fish and megafauna biomass was removed by fishing, followed by 3.02 g/m² (~35%) of bycatch and processing waste redeposited on the seafloor, assuming no loss in the water column.

Using the same method, we calculated that approximately 18% (18.7 MT) of the total catch came from inside the experimental block (Table 8) and included 88 kg of the seastar at a rate of 0.18 g/m². Since the mean density of seastars before trawling in the experimental block was 1.1 seastars/m² (Figure 6) and the average wet weight for a single seastar is 178 g (S. Jewett, University of Alaska Fairbanks, unpublished data), this indicates that experimental trawling removed less than 1% of the seastar biomass per unit area. Therefore, the reduction in seastar density observed on video was probably the result of their tightly clustered distribution after experimental trawling.

Table 7. Individual macrofauna taxa in experimental study from impact stations (n = 12) before and after trawling ranked by percentage of total mean biomass (g/m²), with functional group designations as in Table 4. Totals calculated from all families for each treatment (before–after experimental trawling).

Taxa	Function	Before Rank	%	Mean ± SE	After Rank	%	Mean ± SE
				Common taxa			
Turritellidae	F	1	66.0	57.26 ± 7.41	1	78.4	81.48 ± 11.15
Tellinidae	S	2	4.0	3.43 ± 1.16	3	2.2	2.34 ± 0.49
Nuculanidae	D	3	3.9	3.39 ± 2.32	18	0.4	0.37 ± 0.24
Solenidae	F	4	3.6	3.10 ± 1.60	5	2.1	2.21 ± 0.63
Nephtyidae	C	5	3.3	2.90 ± 0.35	2	2.7	2.80 ± 0.60
Echinarachniidae	S	6	2.9	2.51 ± 0.54	6	2.1	2.17 ± 0.39
Opheliidae	D	7	2.7	2.34 ± 0.78	4	2.2	2.30 ± 0.66
Astartidae	F	8	2.0	1.77 ± 1.24	14	<1	0.56 ± 0.14
Nuculidae	D	9	1.5	1.26 ± 0.18	9	<1	0.80 ± 0.10
Orbiniidae	D	10	1.4	1.21 ± 0.10	7	1.1	1.13 ± 0.11
Montacutidae	F	12	1.0	0.84 ± 0.06	8	1.0	1.03 ± 0.18
Isaeidae	S	14	<1	0.51 ± 0.06	10	<1	0.77 ± 0.12
				Rare taxa			
Ophiuroidea	S	45	<1	0.007 ± 0.004			
Polynoidae	C	46	<1	0.006 ± 0.006			
Thraciidae	S	50	<1	0.005 ± 0.005			
Cumacea	S	58	<1	0.002 ± 0.001			
Isopoda	S	59	<1	0.001 ± 0.001			
Hiatellidae	F	61	<1	$4.8 e^{-4} \pm 4.7 e^{-4}$			
Caprellidea	F	62	<1	$4.7 e^{-4} \pm 4.7 e^{-4}$			
Gastropteridae	C	63	<1	$4.7 e^{-4} \pm 4.7 e^{-4}$			
Kelliidae	F	65	<1	$4.3 e^{-4} \pm 4.3 e^{-4}$			
Pleustidae	S	66	<1	$3.6 e^{-4} \pm 3.6 e^{-4}$			
Dexaminidae	S	68	<1	$4.7 e^{-6} \pm 4.7 e^{-6}$			
Leptonidae	S				50	<1	0.001 ± 0.001
Lyonsiidae	F				55	<1	$5.9 e^{-4} \pm 5.9 e^{-4}$
Cancellariidae	O				56	<1	$4.7 e^{-4} \pm 4.7 e^{-4}$
Polychaeta	O				61	<1	$4.3 e^{-6} \pm 4.3 e^{-6}$
Totals:		70	100%	86.71 ± 5	63	100%	103.99 ± 12.7

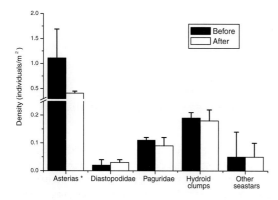

Figure 6. Mean densities prominent megafauna from 30 minute video transects before (n = 3) and after (n = 3) experimental trawling. Significant results based on 95% confidence intervals are marked with an asterisk, and a scale break from 0.25–0.3 individuals/m² facilitates comparison with less common taxa. Data were standardized per unit area based on estimates of video strip widths and lengths.

Experimental Study: Yellowfin Sole Diet Composition

A total of 49 macrofauna and pelagic prey taxa were identified from the stomachs of yellowfin sole caught during experimental trawling. Other items included fish tissue (Teleostei) and eggs, gastropod eggs, and vegetation. Unidentified, partially digested tissue was present in two-thirds of the fish and accounted for a large proportion of the prey number and biomass (Figure 8). This prey category, although an indication of recent feeding, was not included in the following summary. In order of decreasing percent frequency of occurrence of feeding fishes (n = 27; e.g., Konstantinov 1973), the dominant prey items were unidentified polychaetes, Isaeidae amphipods, Orbiniidae polychaetes, Echinarachniidae sand dollars, unidentified amphipods, and fish tissue (Teleostei). Amphipods and polychaetes were the predominant macrofauna prey in terms of numbers, and bivalves and polychaetes were prominent in terms of

Figure 7. Distribution of A. amurensis within experimental block (a) after and (b) before trawling. Densities before trawling were fairly uniform, and afterwards large clusters of over 20 seastars were feeding on processing waste. Lasers 10 cm apart were enhanced for clarity.

biomass (Figure 8). Fish tissue (Teleostei) made up a substantial portion of yellowfin sole diet in terms of percent frequency, numbers, and biomass and appeared to consist of factory trawl processing waste. Although the diet of yellowfin sole does occasionally include fish in the southwestern Bering Sea (Tokranov 1989), feeding morphology indicates that this species is not piscivorous (Lang 1992). Thus, yellowfin sole fed on many of the common macrofauna and pelagic taxa in the study site, but a significant proportion of their diet consisted of fish tissue that was most likely processing waste from experimental trawling. Nuculanidae bivalves were prominent in yellowfin sole diet and were noticeably reduced after experimental trawling (Figure 8b; Table 7).

Discussion

Did Trawling Affect the Macrofauna Community?

Clear statistical differences in benthic macrofauna communities were found, with the closed zone characterized by a richer community with higher biomass than the fished zone (Figure 3; Tables 2, 3). Although interannual variability was high in this system and patterns were not consistent across both years, distinct macrofauna assemblages between zones were reflected in station groupings and dominant taxa (Figure 4; Table 4). The following paragraphs address the question as to whether these differences were caused by chronic exposure to commercial bottom trawling.

Table 8. Total catch (metric tons; mt) from experimental trawling by F/V *Verdal* on May 26–27, 2000, calculated from 10 tows averaging ~11 km in length. Catch from within the experimental block was estimated from 10 tows of 2 km length and expressed in kg. Biomass removed per unit area was calculated from the total area covered during experimental tows per taxa (for herding fish: 10 tows of ~11-km length x 113-m width = 12.86 km^2; and for sessile and less mobile invertebrates: 10 tows of ~11-km length x 25-m width = 2.85 km^2). Bycatch is denoted with an asterisk.

Taxa	Weight (mt), entire area fished	%	Weight (kg), experimental block	Biomass removed (g/m^2)
	Flatfish			
Yellowfin sole	64.1	60	11,262	4.98
Rock sole	23.9	23	4,204	1.86
Alaska plaice *Pleronectes quadrituberculatus*	6.2		1,088	.48
Starry flounder *Platichthys stellatus*	4.4		774	.34
Pacific halibut *Hippoglossus stenolepis*	0.32	12	56	.02
	Other demersal fish			
Walleye pollock *Theragra chalcogramma*	1.3		229	0.10
Pacific cod *Gadus macrocephalus*	0.22		38	0.02
Sculpin Cottidae *	4.6		809	0.36
Sturgeon poacher *Podothecus accipenserinus**	0.09	4	16	0.01
Greenling Hexagrammidae*	0.03		4.4	0.002
Poacher Agonidae*	0.01		1.7	0.001
	Invertebrates			
*Asterias amurensis**	0.50	0.5	88	0.18
Other seastars*	0.17		30	0.06
Red king crab *Paralithodes camtschatica**	0.15		26	0.05
Barnacles *Balanus* spp.*	0.12	0.5	21	0.04
Red soft coral Alcyonacea*	0.04		6.3	0.01
Basket stars *Gorgonocephalus eucnemis**	0.002		0.42	0.001
Total catch weight	106.2	100	18,653	8.52

The randomized experimental design was appropriate to draw statistical inferences about differences between the two zones and within the experimental block (Conover 1999; Zar 1999). However, caution must be exercised when drawing general conclusions to other areas because the treatment was not spatially replicated in either chronic or experimental studies (Hurlbert 1984; Underwood 2000), and other factors such as food supply from water column primary productivity, hydrodynamic regime, larval recruitment, and behavior of adult organisms may have contributed to the observed differences (Grebmeier et al. 1988; Snelgrove and Butman 1994). Detecting human influences on natural population parameters can be hindered by temporal and spatial variability

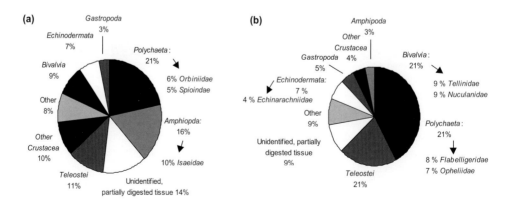

Figure 8. Yellowfin sole prey by proportional (a) abundance and (b) biomass from stomach ($n = 27$) content analysis.

(Underwood 1994). We addressed these by limiting the study to an area of similar depth and grain size and incorporating replicate subsamples and covariates into the statistical models. Statistical power to detect changes to marine ecosystems induced by fishing is greatest when sampling is combined with high-resolution spatial data on fishing effort (NRC 2002). Therefore, based on NMFS otter trawling effort (Fritz et al. 1998; NRC 2002) and surficial sediment data (Smith and McConnaughey 1999; Brown et al., in press), the study area offered an ideal opportunity for the comparison of two zones that were extremely similar except for fishing effort.

Although immediate responses to trawling were subtle, our experimental study provides further support that the differences in macrofauna community composition between zones in the chronic study were caused by long-term exposure to bottom trawling. Experimental studies are key to understanding the mechanisms that drive change in marine ecosystems and provide insight into how communities may respond to chronic bottom trawling (Watling and Norse 1998; McConnaughey et al. 2000; Kenchington et al. 2001). The results indicate that richness and total number of macrofauna taxa decreased within the experimental block and that several rare taxa were absent after trawling (Tables 6, 7). Changes in macrofauna assemblage biomass were patchy and, in some cases, greater than at the controls (Figure 4b). Yet no differences were detected relative to controls for macrofauna community indices (Figure 5), and little change was observed for the densities of the most abundant taxa immediately after experimental trawling. These patterns are consistent with those in the chronic study. Diversity did not appear to be influenced by trawling in either study. But the two parameters that were most affected by experimental trawling, overall richness (Table 6) and biomass of individual taxa (Table 7; Figure 4b), correspond with reduced overall biomass and richness observed in the fished zone (Figure 3). Although the species were not identical, several rare taxa were absent from trawled areas in both studies (Tables 4, 7). Thus, one might expect communities to adapt with chronic exposure to such a disturbance, but a reduction in density may not be manifested immediately after a single trawling event.

The differences between the closed and fished zones were clearly larger than the changes observed immediately after experimental trawling. Several explanations could account for the subtle nature of the macrofaunal response to experimental trawling. High natural variability and small control sample size could explain why no differences were detected relative to controls. Alternatively, since diversity and richness were calculated from density data, the small changes observed in the most abundant taxa immediately after experimental trawling could explain why differences were not found for these measures relative to controls. As summary measures such as total density or biomass do not reflect differential responses of individual taxa, a reduction in one taxon (e.g., Nuculanidae or Astartidae) after experimental trawling may counteract increases in other taxa (e.g., Turritellidae; Table 7). The experimental design may also have been partially responsible for masking changes. Our sampling design was unique because it quantified the effects of trawling for the entire 4-km^2 area so that some stations were outside the trawl tracks (Figure 2). Differential responses of biota to various levels of trawling disturbance (i.e., footrope, doors) may have increased variance and, thus, reduced statistical power. Changes to the physical components of the seabed (e.g., sedimentary organic content and grain size distribution) indicated that the overall effect of trawling was patchy for this bottom type because the treatment varied throughout the experimental block (Brown et al., in press). It is highly likely that the macrofauna responded to experimental trawling in a similar manner, with reductions via resuspension, displacement, or predation taking place at some stations but not at others.

Numerous recent studies in other soft-bottom habitats have associated changes in benthic community composition, density, biomass, or diversity in response to trawling (e.g., Kaiser and Spencer 1996; Engel and Kvitek 1998; Thrush et al. 1998; Tuck et al. 1998; Smith et al. 2000). By combining the results of multiple experimental trawling studies as independent replicates in a meta-analysis, Collie et al. (2000) were able to make generalized predictions about changes to and recovery of benthos after fishing in different habitat types globally. This approach directly addressed the issue of pseudoreplication that is prevalent for many studies with a single trawling treatment. These authors found that gear type, geographical region, and taxonomic class were significant factors in explaining a 55% reduction of individual taxa. A 46% decrease in total abundance and 27% decrease in number of species was observed, but these changes were not statistically significant (Collie et al. 2000). This is similar to what we found in that individual taxa respond differently to trawling but that some summary measures do not describe these changes due to counteracting responses.

Some previous studies have found no detectable trawling impact to infauna in sandy habitats. A common explanation is that organisms living in mobile sediments (i.e., sand waves) are adapted to high levels of natural disturbance from wave and tidally driven currents, and therefore, trawling in these "less sensitive" habitats causes little changes to communities (Jennings and Kaiser 1998; Collie et al. 2000; NRC 2002). However, many of these studies were conducted at depths of less than 10 m or in intertidal zones (e.g., Gibbs et al. 1980; Brylinsky et al. 1994) where wave disturbance to the seafloor increases significantly (Komar 1998). In addition, a significant component of macrofauna density (50% to 75%) is not addressed when quantified by a 1-mm screen (Olafsson et al. 1994; Brown 2003). A long-term study on sand habitat at a depth of 120–146 m concluded that any real effect of trawling on the macrofauna was minor in comparison with high natural interannual and spatial variability in the northeastern Atlantic (Kenchington et al. 2001). Another study, at 48 m in the southeastern Bering Sea, attributed reduced diversity and niche breadth of sedentary taxa to heavy trawling in a sandy habitat characterized by tidally induced ripples (McConnaughey et al. 2000). Thus, it is important to consider the sampling scale and region of individual studies before making generalizations to other habitats. We found high natural spatial and temporal variability but also documented subtle differences between macrofauna communities. Despite the study location on the inner Bering shelf in a sandy habitat exposed to high wave and tidal energy, our results indicate that commercial bottom trawling caused reduced macrofauna density, richness, and biomass, with potential consequences for ecosystem function.

Ecological Implications

In terms of overall ecosystem production or function, we can better understand how the closed and fished zone macrofauna assemblages differed by grouping taxa based on functional attributes such as feeding habits, relative motility, or reproductive requirements. Functional analysis has proven useful in determining the relative roles of different environmental variables in structuring soft-sediment benthic communities (e.g., Pearson and Rosenberg 1978; Bonsdorff and Pearson 1999; Pearson 2001). For example, distinct patterns emerge between the closed and fished zones for macrofauna grouped into functional groups based on feeding mode. The community in the fished zone was divided into relatively even proportions of four main groups in terms of biomass: carnivores, suspensivores, surface deposit feeders, and subsurface detritivores (Figure 9; Feder et al. 1991; Pearson 2001), but suspensivores contributed substantially more in the closed zone. Within this framework, hypotheses concerning how this shift may have occurred in response to fishing are possible based on feeding mode in combination with behavior, motility, or reproductive strategy.

For instance, mobile surface deposit feeders that were more common in the fished zone (e.g., Lysianassidae and Isopoda; Table 4; Figure 9) have been referred to as "bulldozers" and can cause diffusive mixing that tends to de-stabilize sediments (Thayer 1983). These small epifaunal crustaceans are also known as "sedimentary stirrers" because they escape danger by rapidly burrowing into the sediments (Myers 1977; Bromley 1996). Lysianassoid amphipods are considered widespread scavengers with the potential to consume fishing discards (Bluhm and Bechtel 2003). Their ability to avoid predation or benefit from discards may explain why trawling does not have a major impact on these highly mobile organisms. On the other hand, tube-dwelling polychaetes and bivalves prevalent in the closed zone (e.g., Maldanidae, Spionidae

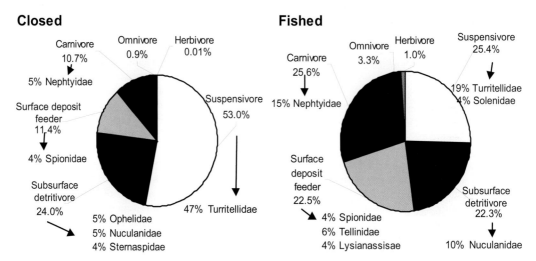

Figure 9. Functional groups according to feeding mode of closed and fished macrofauna communities. Data are from individual taxa biomass, and represent the mean of fished (*n* = 10) and closed (*n* = 9) stations, pooled across both years (1999 and 2000; Feder et al. 1991; Pearson 2001).

and Astartidae; Table 4) promote sediment stabilization through the construction of burrows. This provides resources for other small-bodied infauna below the sediment surface (Jumars et al. 1990; Levin et al. 1997) and enhances oxidation deeper within the sediments (i.e., Myers 1977; Marinelli 1992; Rysgaard et al. 1995). Sessile tube dwellers are likely to be disrupted by contact with fishing gear because they rely on their burrow to avoid predation and would become an easy target for scavengers without it. Larval development can also play an important role in an organism's ability to respond to disturbance. The dominant suspensivore in the closed zone (*T. erosus*; family Turritellidae; Figure 9) is considered discreetly motile (Feder et al. 1991; Pearson 2001) and deposits egg strings that were frequently attached to shells of *T. erosus*. Although it is technically not a brooder, this suggests that juveniles are released close to the parent and would be less likely than a highly mobile or broadcast-spawning species to recruit to a newly disturbed area in the absence of other adults. This could also explain why Neptunea gastropods and their eggs were significantly more abundant in unfished areas compared to those heavily fished further out on the Bering shelf (McConnaughey et al. 2000). Thus, differences in macrofauna communities between the closed and fished zones are consistent with expectations of macrofauna response to trawling based on an understanding of organism function.

Physical disturbance by commercial bottom trawling may have caused a bottom-up shift in the benthic ecosystem. Brown et al. (in press) found that subtle differences were present between the sediments in the closed and fished zones and that the spring commercial fishing season occurred during a period when natural sediment resuspension by waves was diminished. Because this time of year is critical for macrofaunal recruitment and overall benthic production and many macrofauna larvae demonstrate active habitat selection (Snelgrove and Butman 1994), this suggests that bottom trawling may have influenced long-term species distributions. Frequent disturbances to the seafloor associated with commercial bottom trawling in the fished zone favor opportunists (*r*-strategists) with high reproductive rates, rapid maturation, and short life-spans (e.g., Pearson and Rosenberg 1978; Olafsson et al. 1994). Sediment resuspension from trawling could also decrease food quality or smother filter feeders (Jones 1992), which could explain reduced biomass of suspensivores in the fished zone (Turritellidae, Astartidae, and Hiatellidae; Table 4). Fishing discards also have the potential to cause bottom-up shifts by altering the nature of carbon flux to the benthos, with consequences for microbial and macrofaunal communities, favoring mobile, opportunistic species over non-carnivorous, sessile organisms (e.g., Himelbloom and Stevens 1994; Dayton et al. 1995; Bluhm and Bechtel 2003). A carbon budget for the inner northeastern Bering Sea shelf indicated that annual primary production was approximately 50 g C/m^2 and annual benthic carbon demand was 27.8 g C/m^2 under Alaska coastal water (Grebmeier 1987). Our estimated benthic flux from bycatch and processing waste of 3.02 g C/m^2 is roughly 11% of the benthic demand. As several such events could occur during the commercial fishing season, processing waste could provide a supplemental source of C to benthos and transfer biomass from demersal to benthic food webs.

It is also possible that bottom trawling induced top-down shifts in the macrofauna communities by affecting predator–prey interactions with mobile megafauna and demersal fish. In a series of exclusion experiments, predatory polychaetes became more abundant in the absence of epibenthic predators (Ambrose 1984). These polychaetes tend to be more susceptible as prey because they do not form protective burrows or frequently extend large sections of their body (Ambrose 1984). Carnivorous polychaetes (Nephtyidae and Sigalionidae; Figure 9; Table 4; Brown 2003) may have been more abundant in the fished zone due to a relief of predation pressure from epibenthic predators during the commercial trawl season. For example, discards and processing waste from experimental trawling made up a substantial portion of yellowfin sole diet in this study (Figure 8). Yellowfin sole exhibits the broadest diet among the Pleuronectidae flounders in the southern Bering Sea that is ultimately determined by the distribution and relative abundance of various prey species (Tokranov 1989; Lang 1992). The consumption of processing waste implies a relief of predation for predominant yellowfin sole prey such as polychaetes and bivalves. Aggregates of *Asterias amurensis* were also seen consuming processing waste after experimental trawling. *Asterias amurensis* is a food generalist, consuming gastropods, echinoderms, crustaceans, and polychaetes and is considered a major threat to commercial bivalves in southern Australia (Feder and Jewett 1981; Ross 2002). Brewer (2003) found that in the presence of fishing discards, densities of another subtidal seastar, *Pycnopodia helianthoides*, were significantly higher, few seastars were observed excavating pits, and the bivalve population was double that of a control area. Since asteroid larvae demonstrate active habitat selection in response to chemical cues from increased prey densities (Barbeau et al. 1998; Veale et al. 2000), similar behavior by *A. amurensis* could explain the abundance of recently settled juvenile Bivalvia in the fished zone (Brown 2003) where fishing discards could provide relief from predation of *A. amurensis*. Increased recruitment of juvenile *A. amurensis* due to the presence of fishing discards could also explain why this seastar was more abundant in heavily fished areas in a separate study in the southeastern Bering Sea (McConnaughey et al. 2000). This potential of *A. amurensis* and yellowfin sole to have wide ecosystem-level effects through predation and associated bioturbation, coupled with their ability to switch feeding behavior depending on prey availability, suggests that the presence of fishing discards from bottom trawling could have a strong influence on benthic community structure in the southeastern Bering Sea.

Future Recommendations and Management Considerations for Essential Fish Habitat

The critical management question is whether the subtle changes in the macrofauna community and altered ecosystem function in the fished zone are significant in terms of EFH for yellowfin sole. Two contemporary viewpoints prevail concerning the effects of fishing on marine ecosystems. From the conservation perspective, the goal of management is to preserve biodiversity in order to maintain a healthy ecosystem (e.g., Norse and Watling 1999). But by the legal requirements of MSFC, alterations to habitat are only relevant if a federally managed, commercially harvested species is "adversely impacted" in its ability "to support a sustainable fishery" and its "contribution to a healthy ecosystem" (NOAA 2002). This study indicates that the ecosystem in this sandy habitat was altered by chronic exposure to bottom trawling. Therefore, if management objectives are to preserve biodiversity, marine protected areas appear to be a viable option for this specific region. However, if the management goal is to quantify impacts on EFH under the MSFC requirements, more ecological and bioenergetic data are required to determine how changes in benthic communities are linked to productivity of yellowfin sole and vice versa. While this study did not specifically address this question, our results demonstrate that yellowfin sole did not target any specific prey taxa that benefited from trawling, as was the case in Monterey Bay (Engel and Kvitek 1998) or the North Sea (Millner and Whiting 1996; Jennings et al. 2002). Given the different macrofauna assemblages in the closed and fished zones and the fact that fish processing waste was prominent in yellowfin sole diet, it would be useful to compare their relative energetic yield and determine which diet is more nutritious for this species. We suggest that research on impacts on EFH focus on the specific quality within a habitat that is utilized or contributed by the managed species.

Research on fishing impacts on megafauna in rocky or biogenic habitats may be less relevant for soft-bottom habitat. Although occasional sessile, structure-forming megafauna were documented by video in this study, they appeared mostly undamaged after experimental trawling, and no reduction in density was observed (e.g., Hydrozoa and Diastopodidae, Figure 6). This lack of damage may be attributed to the type of fishing gear for this fishery, which is designed to roll along the seafloor and is likely to have different effects than a beam trawl or scallop dredge (Vining and Witherell 1997). Alternatively, the sparse distribution of sessile megafauna (e.g., <0.2 hydroid clumps per m^2; Figure 6) could make it difficult to detect changes in density. As fragile megafauna are not a significant constituent of the community in this and some other soft-bottom habitats, trawling is more likely to impact features within the sediment such as sand waves, biogenic depressions, tubes, or excavations rather than high-relief epifaunal structures. Our results suggest that research evaluating the recovery process in this type of habitat should concentrate on the infauna and epifauna (including microbes, meiofauna, microfauna, and macrofauna) that inhabit the sediments and sediment surface and account for the bulk of benthic productivity (Grebmeier 1987; Schratzberger and Jennings 2002). Quantitative sampling is necessary in soft-bottom habitats because little information about infauna can be gathered from video footage of the seafloor. Most beneficial would be studies that evaluate organism function, specifically focusing on key attributes such as larval development, mobility, and ability to scavenge opportunistically (e.g., Bonsdorff and Pearson 1999; Pearson 2001). With such data, models (e.g., Ecopath, University of British Columbia Fisheries Centre, Vancouver; Okey and Pauly 1999) could be applied to evaluate system-wide trawling effects on trophic requirements, biomass flow, and functional interactions in this system.

Acknowledgments

We dedicate this research in memory of Gary Edwards and the crew of R/V *Big Valley*. We wish to thank Arny Blanchard for statistical consultation and Stephen Jewett, Brenda Konar, Howard Feder, Tom Pearson, and Peter McRoy for many insightful discussions and for providing laboratory and field equipment, along with Brenda Norcross, Chris Bubbblitz, and Dave Doudna. We are grateful to two anonymous reviewers for helpful comments on the manuscript; Max Hoberg, Ken Coyle, and the University of Alaska Fairbanks benthic lab for their taxonomic expertise; and Cathy Hegwer, Amy McKenzie, Brenda Holladay, Judy Hamilton, Boe Barnette, Meg Thornton, Sean Willison,

Buck Barbieri, and Chris Foshee for assistance with field work and laboratory analyses. Thanks to John Gauvin and the Groundfish Forum for providing a vessel and logistical support during the experimental trawling phase. We are grateful to Gary Edwards and crew of R/V *Big Valley* for their expertise and dedication during our field work. Funding for this project was provided by the North Pacific Marine Research Initiative, the Groundfish Forum, the Marine Conservation Alliance, and Alaska Sea Grant with funds from the National Oceanic and Atmospheric Administration Office of Sea Grant, Department of Commerce, under grant number NA 86RG0050 (project number RR/99-02), and the University of Alaska, with funds appropriated by the state. This research was conducted under Fish Resource Permits CF-00-029 and CF-99-029 issued by the State of Alaska Department of Fish and Game (ADF&G), Division of Commercial Fisheries, and an access permit issued by ADF&G Division of Wildlife Conservation.

References

AFSC (Alaska Fisheries Science Center). 1999. North Pacific groundfish observer manual. AFSC, North Pacific Groundfish Observer Program, Seattle.

Ambrose, W. G. J. 1984. Role of predatory infauna in structuring marine soft-bottom communities. Marine Ecology Progress Series 17:109–115.

Auster, P. J., L. L. Stewart, and H. Sprunk. 1989. Scientific imaging with ROVs: tools and techniques. Marine Technology Society Journal 23:16–20.

Barbeau, M. A., R. E. Scheibling, and B. G. Hatcher. 1998. Behavioural responses of predatory crabs and sea stars to varying density of juvenile sea scallops. Aquaculture 169:87–98.

Bergman, M. J. N., and M. Hup. 1992. Direct effects of beam-trawling on macrofauna in a sandy sediment in the southern North Sea. ICES Journal of Marine Science 49:5–11.

Blanchard, A. L., H. M. Feder, and D. G. Shaw. 2002. Long-term investigation of benthic fauna and the influence of treated ballast water disposal in Port Valdez, Alaska. Marine Pollution Bulletin 44:367–382.

Bluhm, B. A., and P. J. Bechtel. 2003. The potential fate and effects of seafood processing wastes dumped at sea: a review. Pages 121–140 *in* P. J. Bechtel, editor. Advances in seafood byproducts: 2002 conference proceedings. Alaska Sea Grant College Program, University of Alaska, Fairbanks.

Bonsdorff, E., and T. H. Pearson. 1999. Variation in the sublittoral macrozoobenthos of the Baltic Sea along environmental gradients: a functional-group approach. Australian Journal of Ecology 24:312–326.

Bray, J. R., and J. T. Curtis. 1957. An ordination of the upland forest communities of southern Wisconsin. Ecological Monographs 27.

Brewer, R. 2003. The chemosensory abilities and foraging techniques of *Pycnopodia helianthoides*. Master's thesis. University of Alaska, School of Fisheries and Ocean Sciences, Fairbanks.

Bromley, R. G. 1996. Trace fossils. Biology, taphonomy and applications. Chapman and Hall, London.

Brown, E. J. 2003. Effects of commercial otter trawling on essential fish habitat of the southeastern Bering Sea shelf. Master's thesis. University of Alaska, School of Fisheries and Ocean Sciences, Fairbanks.

Brown, E. J., B. Finney, M. Dommisse, and S. Hills. In press. Effects of commercial otter trawling on the physical environment of the southeastern Bering Sea. Continental Shelf Research.

Brylinsky, M., J. Gibson, and D. C. Gordon. 1994. Impacts of flounder trawls on the intertidal habitat and community of the Minas Basin, Bay of Fundy. Canadian Journal of Fisheries and Aquatic Sciences 51:650–661.

Clarke, K. R., and M. Ainsworth. 1993. A method of linking multivariate community structure to environmental variables. Marine Ecology Progress Series 92:205–219.

Collie, J. S., G. A. Escanero, and P. C. Valentine. 1997. Effects of bottom fishing on the benthic megafauna of Georges Bank. Marine Ecology Progress Series 155:159–172.

Collie, J. S., S. J. Hall, M. J. Kaiser, and I. R. Poiner. 2000. A quantitative analysis of fishing impacts on shelf-sea benthos. Journal of Animal Ecology 69:785–798.

Conover, W. J. 1999. Practical nonparametric statistics. Wiley, New York.

Coyle, K. O. 2000. Acoustic estimates of zooplankton biomass and distribution: application of canonical correlation to scaling of multifrequency acoustic data. Canadian Journal of Fisheries and Aquatic Sciences 57:2306–2318.

Dayton, P. K., S. F. Thrush, M. T. Agardy, and R. J. Hofman. 1995. Environmental effects of marine fishing. Aquatic Conservation: Marine and Freshwater Ecosystems 5:205–232.

DiCosimo, J. 1999. Essential fish habitat and closed areas in federal waters of the North Pacific. North Pacific Fisheries Management Council, Anchorage, Alaska

Eleftheriou, A., and M. R. Robertson. 1992. The effects of experimental scallop dredging on the fauna and physical environment of a shallow sandy community. Netherlands Journal of Sea Research 30:289–299.

Engel, J., and R. Kvitek. 1998. Effects of otter trawling in a benthic community in Monterey Bay National Marine Sanctuary. Conservation Biology 12:1204–1214.

Feder, H. M., and S. C. Jewett. 1981. Feeding interactions in the eastern Bering Sea with emphasis on the benthos. Pages 1229–1261 *in* D. W. Hood and J. A. Calder, editors. The eastern Bering Sea shelf: oceanography and resources. Office of Marine Pollution Assessment of the National Oceanic and Atmospheric Administration, University of Washington Press, Seattle.

Feder, H. M., A. S. Naidu, M. Baskaran, K. Frost, J. M. Hameedi, S. C. Jewett, W. R. Johnson, J. Raymond, and D. Schell. 1991. Bering Strait–Hope Basin: habitat utilization and ecological characterization. University of Alaska, Institute of Marine Science, Report 92-2, Fairbanks.

Freese, L, P. J. Auster, J. Heifetz, and B. L. Wing. 1999. Effects of trawling on sea floor habitat and associated invertebrate taxa in the Gulf of Alaska. Marine Ecology Progress Series 182:119–126.

Fritz, L. W., A. Greig, and R. F. Reuter. 1998. Catch-per-unit-effort, length, and depth distributions of major groundfish and bycatch species in the Bering Sea, Aleutian Islands, and Gulf of Alaska regions based on groundfish fishery observer data. National Oceanic and Atmospheric Administration, National Marine Fisheries Service, Alaska Fisheries Science Center, NMFS-AFSC-88, Seattle.

Gibbs, P. J., A. J. Collins, and L. C. Collet. 1980. Effect of otter prawn trawling on the marcobenthos of a sandy substratum in a New South Wales estuary. Australian Journal of Marine and Freshwater Research 31:509–516.

Grebmeier, J. M. 1987. The ecology of benthic carbon cycling in the northern Bering and Chukchi seas. Ph.D. Doctoral dissertation. University of Alaska, Institute of Marine Science, Fairbanks.

Grebmeier, J. M., C. P. McRoy, and H. M. Feder. 1988. Pelagic-benthic coupling on the shelf of the northern Bering and Chukchi seas. I. Food supply source and benthic biomass. Marine Ecology Progress Series 48:57–64.

Hall, S. J. 1999. The effects of fishing on marine ecosystems and communities. Blackwell Science Ltd., Oxford, UK.

Himelbloom, B. H., and B. G. Stevens. 1994. Microbial analysis of a fish waste dump site in Alaska. Bioresource Technology 47.

Hobson, K. D., and K. Banse. 1981. Sedentariate and archiannelid polychaetes of British Columbia and Washington. Department of Fisheries and Oceans, Ottawa.

Hurlbert, S. H. 1984. Pseudoreplication and the design of ecological field experiments. Ecological Monographs 54:187–211.

Jennings, S., and M. J. Kaiser. 1998. The effects of fishing on marine ecosystems. Pages 202–352 in J. H. S. Blaxter, A. S. Southward, and P. A. Tyler, editors. Advances in marine biology. Academic Press, London.

Jennings, S., M. D. Nicholson, T. A. Dinmore, and J. E. Lancaster. 2002. Effects of chronic trawling disturbance on the production of infaunal communities. Marine Ecology Progress Series 243:251–260.

Jewett, S. C., H. M. Feder, and A. Blanchard. 1999. Assessment of the benthic environment following offshore placer gold mining in the northeastern Bering Sea. Marine Environmental Research 48:91–122.

Jones, J. B. 1992. Environmental impact of trawling on the seabed: a review. New Zealand Journal of Marine and Freshwater Research 26:59–67.

Jumars, P. A., L. M. Mayer, J. W. Deming, J. A. Baross, and R. A. Wheatcroft. 1990. Deep-sea deposit feeding strategies suggested by environmental and feeding constraints. Philosophical Transactions of the Royal Society of London, Series A 331:85–101.

Kaiser, M. J., and B. E. Spencer. 1996. The effects of beam-trawl disturbance on infaunal communities in different habitats. Journal of Animal Ecology 65:348–358.

Kenchington, E. L. R., J. Prena, K. D. Gilkinson, D. C. Gordon, Jr., K. MacIsaac, C. Bourbonnais, P. J. Schwinghamer, T. W. Rowell, D. L. McKeown, and W. P. Vass. 2001. Effects of experimental otter trawling on the macrofauna of a sandy bottom ecosystem on the Grand Banks of Newfoundland. Canadian Journal of Fisheries and Aquatic Science 58:1043–1057.

Komar, P. D. 1998. Beach processes and sedimentation. Prentice-Hall, Upper Saddle River, New Jersey.

Konstantinov, K. G. 1973. The assessment of the frequency of occurrence of food components. Department of the Environment Fisheries Board of Canada, Biological Station, St. John's Newfoundland.

Lang, G. M. 1992. Food habits of three congeneric flatfishes: yellowfin sole, *Pleuronectes asper*, rock sole, *P. bilineatus*, and Alaska plaice, *P. quadrituberculatus*, in the eastern Bering Sea, 1984–1988. Master's thesis. University of Washington, School of Fisheries, Seattle.

Levin, L., N. Blair, D. DeMaster, G. Plaia, W. Fornes, C. Martin, and C. Thomas. 1997. Rapid subduction of organic matter by maldanid polychaetes on the North Carolina slope. Journal of Marine Research 55:595–611.

Margalef, D. R. 1958. Information theory in ecology. General Systems 3:36–71.

Marinelli, R. L. 1992. Effects of polychaetes on silicate dynamics and fluxes in sediments: importance of species, animal activity and polychaete effects on benthic diatoms. Journal of Marine Research 50:745–779.

McConnaughey, R. A., K. Mier, and C. B. Dew. 2000. An examination of chronic trawling effects on soft-bottom benthos of the eastern Bering Sea. ICES Journal of Marine Science 57:1377–1388.

McConnaughey, R. A., S. E. Syrjala, and C. B. Dew. 2005. Effects of chronic bottom trawling on the size structure of soft-bottom benthic invertebrates. Pages 425–437 in P. W. Barnes and J. P. Thomas, editors. Benthic habitats and the effects of fishing. American Fisheries Society, Symposium 41, Bethesda, Maryland.

Millner, R. S., and C. L. Whiting. 1996. Long-term changes in growth and population abundance of sole in the North Sea from 1940 to the present. ICES Journal of Marine Science 53:1185–1195.

Myers, A. C. 1977. Sediment processing in a marine subtidal sandy bottom community. I. Physical aspects. Journal of Marine Research 35:609–632.

NOAA (National Oceanic and Atmospheric Administration). 1988. Bathymetric map, Hagemeister Island, number 4-3, scale 1:250,000. NOAA, National Ocean Service, Washington, D.C.

NOAA (National Oceanic and Atmospheric Administration). 2002. Magnuson-Stevens Act Provisions; essential fish habitat. Federal Register 67(12):2343–2383.

Norse, E., and L. Watling. 1999. Impacts of mobile fishing gear: the biodiversity perspective. Pages 31–40 in L. Benaka, editor. Fish habitat: essential fish habitat and rehabilitation. American Fisheries Society, Symposium 22, Bethesda, Maryland.

NPFMC (North Pacific Fisheries Management Council). 1991. Environmental assessment/regulatory impact review/initial regulatory flexibility analysis for Amendment 17 to the Fishery Management Plan for the groundfish fishery of the Bering Sea and Aleutian Islands area and Amendment 22 to the Fishery Management Plan for the groundfish of the Gulf of Alaska and for a regulatory amendment to define groundfish pots. NPFMC, Anchorage, Alaska.

NPFMC (North Pacific Fisheries Management Council). 1997. Stock assessment and fishery evaluation report for the groundfish resources of the Bering Sea/Aleutian Island regions. NPFMC, Plan Team for the Groundfish Fisheries of the Bering Sea and Aleutian Islands, Anchorage, Alaska.

NRC (National Research Council). 2002. Effects of trawling and dredging on seafloor habitat. NRC, Committee on Ecosystem Effects of Fishing, National Academy Press, Washington D.C.

Okey, T. A., and D. Pauly. 1999. A mass-balances model of trophic flows in Prince William Sound: decompartmentalizing ecosystem knowledge. Pages 621–635 in Ecosystem approaches for fisheries management. Alaska Sea Grant College Program AK-SG-99-01.

Olafsson, E. B., C. H. Peterson, and W. G. J. Ambrose. 1994. Does recruitment limitation structure populations and communities of macro-invertebrates in marine soft sediments: the relative significance of pre- and post-settlement processes. Oceanography and Marine Biology: An Annual Review 32:65–109.

Pearson, T. H. 2001. Functional group ecology in soft-sediment marine benthos: the role of bioturbation. Oceanography and Marine Biology: An Annual Review 39:233–267.

Pearson, T. H., and R. Rosenberg. 1978. Macrobenthic succession in relation to organic enrichment and pollution of the marine environment. Oceanography and Marine Biology: An Annual Review 16:229–311.

Pielou, E. C. 1969. An introduction to mathematical ecology. Wiley-Interscience, Queen's University, Kingston, Ontario

Ross, D. J. 2002. Impact of introduced seastars *Asterias amurensis* on survivorship of juvenile commercial bivalves *Fulvia tenuicostata*. Marine Ecology Progress Series 241:99–112.

Rysgaard, S., P. B. Christensen, and L. P. Nielsen. 1995. Seasonal variation in nitrification and denitrification in estuarine sediment colonised by benthic microalgae and bioturbating infauna. Marine Ecology Progress Series 126:111–121.

Schratzberger, M., and S. Jennings. 2002. Impacts of chronic trawling disturbance on meiofaunal communities. Marine Biology 141:991–1000.

Shannon, C. E., and W. Weaver. 1949. The mathematical theory of communication. University of Illinois Press, Urbana.

Smith, C. J., K. N. Papadopoulou, and S. Diliberto. 2000. Impact of otter trawling on an eastern Mediterranean commercial fishing ground. ICES Journal of Marine Science 57:1340–1351.

Smith, K. R., and R. A. McConnaughey. 1999. Surficial sediments of the eastern Bering Sea continental shelf: EBSSED database documentation. National Marine Fisheries Service, NMFS-AFSC-104, Seattle.

Snelgrove, P. V. R., and C. A. Butman. 1994. Animal-sediment relationships revisited: cause versus effect. Oceanography and Marine Biology: An Annual Review 32:111–177.

Stoker, S. 1981. Benthic invertebrate macrofauna of the eastern Bering/Chukchi continental shelf. Pages 1069–1090 in D. W. Hood and J. A. Calder, editors. The eastern Bering Sea shelf: oceanography and resources. Office of Marine Pollution Assessment of the National Oceanic and Atmospheric Administration. University of Washington Press, Seattle.

Thayer, C. W. 1983. Sediment mediated biological disturbance and the evolution of marine benthos. Pages 479–625 in M. J. S. Tevesz and P. L. McCall, editors. Biotic interactions in recent and fossil benthic communities. Plenium, New York.

Thrush, S. F., J. E. Hewitt, V. J. Cummings, P. K. Dayton, M. Cryer, S. J. Turner, G. A. Funnell, R. G. Budd, C. J. Milburn, and M. R. Wilkinson. 1998. Disturbance of the marine benthic habitat by commercial fishing: impacts at the scale of the fishery. Ecological Applications 8:866–879.

Tokranov, A. M. 1989. Feeding of the yellowfin sole, *Limanda aspera*, in the southwestern part of the Bering Sea. Voprosy Ikhtiologii 29:1003–1009.

Tuck, I. D., S. J. Hall, M. R. Robertson, E. Armstrong, and D. J. Basford. 1998. Effects of physical trawling disturbance in a previously unfished sheltered Scottish sea loch. Marine Ecology Progress Series 162:227–242.

Underwood, A. J. 1994. On beyond BACI: sampling design that might reliably detect environmental disturbances. Ecological Applications 4:3–15.

Underwood, A. J. 2000. Importance of experimental design in detecting and measuring stress in marine populations. Journal of Aquatic Ecosystem Stress and Recovery 7:3–24.

USEPA (U.S. Environmental Protection Agency). 1995. Environmental monitoring and assessment program (EMAP): laboratory Methods manual - estuaries, volume 1: biological and physical analyses. USEPA, Office of Research and Development, EPA/620/R-95/008, Narragansett, Rhode Island.

Veale, L. O., A. S. Hill, and A. R. Brand. 2000. An in situ study of predator aggregations on scallop (*Pecten maximus* (L.)) dredge discards using a static time-lapse camera system. Journal of Experimental Marine Biology and Ecology 255:111–129.

Vining, I., and D. Witherell. 1997. The effects of fishing gear on benthic communities. North Pacific Fisheries Management Council, SAFE, Ecosystem Chapter, Anchorage, Alaska.

Warwick, R. M. 1988. Analysis of community attributes of the macrobenthos of Frierfjord/Langesundfjord at taxonomic levels higher than species. Marine Ecology Progress Series 46:167–170.

Watling, L., and E. A. Norse. 1998. Disturbance of the seabed by mobile fishing gear: a comparison to forest clearcutting. Conservation Biology 12:1180–1197.

Witherell, D., and C. Pautzke. 1997. A brief history of bycatch management measures for eastern Bering Sea groundfish fisheries. Marine Fisheries Review 59:15–22.

Zar, J. H. 1999. Biostatistical analysis. Prentice Hall, Upper Saddle River, New Jersey.

Effects of Bottom Trawling on Soft-Sediment Epibenthic Communities in the Gulf of Alaska

ROBERT P. STONE,[1] MICHELE M. MASUDA,[2] AND PATRICK W. MALECHA[3]

*Auke Bay Laboratory, Alaska Fisheries Science Center,
National Marine Fisheries Service, 11305 Glacier Highway, Juneau, Alaska 99801-8626, USA*

Abstract. The goal of this study was to determine if chronic bottom trawling in some of the more heavily trawled areas in the central Gulf of Alaska has altered soft-bottom marine communities. Spatial distribution and abundance of epifauna were examined at two sites that overlapped areas open to trawling and closed areas where bottom trawling had been prohibited for 11–12 years. Video strip transects of the seafloor were collected at each site from a manned submersible. Transects were bisected by the boundary demarcating open and closed areas. The positions of 155,939 megafauna were determined along 89 km of seafloor. At both sites, we detected general and site-specific differences in epifaunal abundance and species diversity between open and closed areas, which indicate the communities in the open areas had been subjected to increased disturbance. Species richness was lower in open areas. Species dominance was greater in one open area, while the other site had significantly fewer epifauna in open areas. Both sites had decreased abundance of low-mobility taxa and prey taxa in the open areas. Site-specific responses were likely due to site differences in fishing intensity, sediment composition, and near-bottom current patterns. Prey taxa were highly associated with biogenic and biotic structures; biogenic structures were significantly less abundant in open areas. Evidence exists that bottom trawling has produced changes to the seafloor and associated fauna, affecting the availability of prey for economically important groundfish. These changes should serve as a "red flag" to managers since prey taxa are a critical component of essential fish habitat.

Introduction

Diverse benthic communities on the continental shelf and upper slope of the Gulf of Alaska (GOA) support important commercial fisheries for demersal fishes (i.e., groundfish; Mueter and Norcross 2002). Combined groundfish landings from bottom trawl and longline fisheries averaged more than 202,000 metric tons per year from 1963 to 2000 (NPFMC 2000). Understanding the effects of this level of fishing effort on seafloor habitats can aid fisheries managers in developing strategies to manage fishing effects on fish habitat. The focus on fish habitat is pursuant to the essential fish habitat provisions of the Magnuson-Stevens Fishery Conservation and Management Act, as amended by the Sustainable Fisheries Act of 1996 (U.S. Department of Commerce 1996).

Previous studies worldwide have determined that bottom trawling alters seafloor habitat and directly and indirectly affects benthic communities (Jones 1992; Auster et al. 1996; Auster and Langton 1999). In addition to removing target species, bottom trawling incidentally removes, displaces, or damages nontarget species (Ball et al. 2000), changes the sedimentary properties of the seafloor (Churchill 1989), and reduces habitat complexity by physically altering biogenic structures on the seafloor (Schwinghamer et al. 1998). Such changes can lead to population level effects on species of economic importance (Lindholm et al. 1999). Ultimately, the combination of effects may result in wide-scale ecosystem change (Gislason 1994; Goñi 1998). The degree of alteration likely depends on many factors, including (1) gear type, (2) spatial and temporal intensity of trawling, (3) substrate characteristics, (4) oceanographic conditions near the seafloor, and (5) the resilience of components of benthic communities (Jones 1992; Auster and Langton 1999). These factors may be geographically specific, so generalizing the effects of trawling over broad geographical areas may not be prudent.

Gulf of Alaska bottom trawl fisheries use only otter trawls, and the gear is quite variable depending on vessel size and target species. Gear consists of five major components that either contact or potentially con-

---
[1] E-mail: bob.stone@noaa.gov
[2] E-mail: michele.masuda@noaa.gov
[3] E-mail: pat.malecha@noaa.gov

tact the seafloor: (1) the wings and bridles, (2) otter boards or doors, (3) sweeps, (4) footrope, and (5) the cod end. Door spread (i.e., total width of trawl system when fishing) may reach 110 m, but the area of the seafloor and associated epifauna contacted by the gear depends on the design of the otter boards and the configuration of protective gear (e.g., rubber disks, bobbins, chafing gear) used on the sweeps, footrope, and cod end. The morphology, behavior, and spatial distribution of epifauna are also important determinants in this interaction.

Chronic effects of fishing disturbances are difficult to distinguish from natural changes due to a lack of potential reference sites where bottom trawling has not occurred for any significant period. In April 1987, the North Pacific Fishery Management Council closed two areas near Kodiak Island, Alaska, to bottom trawling year-round (Type 1 areas). Use of scallop dredges is also prohibited in Type 1 areas. The closures are intended to rebuild severely depressed stocks of Tanner crab *Chionoecetes bairdi* (also known as southern Tanner crab) and red king crab *Paralithodes camtschaticus* by protecting juvenile habitat, areas used during molting, and migratory corridors. In addition to crab resources, the closed areas and areas immediately adjacent to them support rich stocks of groundfish including flathead sole *Hippoglossoides elassodon*, butter sole *Pleuronectes isolepis*, Dover sole *Microstomus pacificus*, rex sole *Errex zachirus*, Pacific halibut *Hippoglossus stenolepis*, arrowtooth flounder *Atheresthes stomias*, Pacific cod *Gadus macrocephalus*, walleye pollock *Theragra chalcogramma*, and several species of rockfish *Sebastes* spp. (Martin and Clausen 1995). Consequently, in areas immediately adjacent to the closed areas, bottom trawling occurs year-round, with peak activity occurring in the spring, summer, and fall for flatfish and Pacific cod and during the summer for walleye pollock.

The proximity of the closed and open areas provided a rare opportunity to investigate chronic effects of bottom trawling on a productive, deep-water (>100 m), soft-bottom marine community located on Alaska's continental shelf. Our goal was to determine if fine-scale differences in community structure exist between areas that were trawled each year and areas where bottom trawling had been prohibited for 11–12 years. Additionally, since the areas open to trawling at the study sites are among the more heavily trawled sites in the GOA (Rose and Jorgensen 2005, this volume), effects observed could be considered a "worst case scenario" for this habitat type in the GOA. In 1998 and 1999, studies were initiated to determine if changes had occurred to the infauna and epifauna community structure and the sedimentary, chemical, and biogenic properties at three sites open to bottom trawling. Previous analyses indicated that the sedimentary and chemical properties of the seafloor in areas open to trawling differed from those in the closed areas, but differences in infauna abundance and species diversity were not detected (Stone and Masuda 2003). Here, we report our findings specific to trawl-induced changes to epifaunal community structure and biogenic structures on the seafloor.

Methods

Study Area

Study sites were established along the boundaries of two area closures (Figure 1). Study sites were chosen based on two criteria: (1) the seafloor consisted of a soft-bottom substrate (i.e., sand, silt, or clay) that was relatively uniform in depth, and (2) trawling had occurred immediately adjacent to the closed area each of the preceding 5 years. The first criterion was considered necessary to reduce variation in habitat and community structure associated with depth differences between the open and closed areas at a site.

Site 1 was located in Chiniak Gully near the northeastern side of Kodiak Island (Figure 1). Commercial trawling intensity during the period 1993–1997 was estimated using the methods described in Stone and Masuda (2003) and is calculated as the maximum percentage of seafloor trawled at least once per year during that period. The estimate includes only the area of the seafloor potentially contacted by the footrope and, therefore, can be considered a conservative estimate. Maximum trawl intensity at Site 1 was estimated at 29.4% of the seafloor per year (Stone and Masuda 2003). At this site, the area open to trawling was also open to scallop dredging, and the maximum percentage of seafloor in the study area that was dredged for scallops at least once per year was estimated for the period 1993–1998. Seventeen percent was dredged in 1993, steadily declining to less than 1% in 1998 (G. Rosenkranz, Alaska Department of Fish and Game, personal communication). Strong bottom currents flow predominately from the northwest and southeast. Maximum bottom currents measured during a neap tide period in August 2001 were 0.28 m/s (R. P. Stone, unpublished data). Depth within the transect area ranged from 105 to 151 m, and the maximum depth differential along any transect was 18 m. The substrate consisted of moderately sorted, medium and fine sand (Stone and Masuda 2003).

Site 2 was located in the Two-Headed Gully southeast of Kodiak Island (Figure 1). Maximum trawl intensity was estimated at 19.4% of the seafloor per year (Stone and Masuda 2003). Moderate to light bottom currents (e.g., less than 0.28 m/s) characterize this site. Depth within the transect area ranged from 125 to 157 m, and the

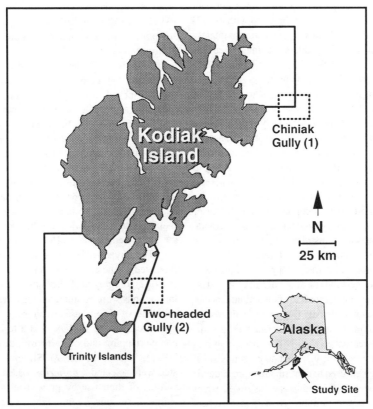

Figure 1. Map of Kodiak Island, Alaska, showing the two study sites (dashed lines) and closed areas (solid lines) where nonpelagic trawling is prohibited year-round. Site 1 is located at the Chiniak Gully. Site 2 is located at the Two-Headed Gully.

maximum depth differential along any transect was 15 m. The substrate consisted of moderately sorted, very fine sand (Stone and Masuda 2003).

Experimental Design

Two cruises aboard the Alaska Department of Fish and Game RV *Medeia* were conducted from 4 to 15 June 1998 and from 13 to 23 August 1999. The submersible *Delta* was used to record 10 video transects of the seafloor that traversed adjacent open and closed areas to bottom trawling. Ten seafloor transects were surveyed at Site 1 during both the 1998 and 1999 cruises, and 10 seafloor transects were surveyed at Site 2 in 1998. Transects were bisected by the boundary demarcating open and closed areas, parallel, 500 m apart, and 3,000 m in length. At Site 1, transects were 500 m apart each year and 250 m apart when years were combined. Transects 3 and 10 at Site 1 in 1998 were approximately 2,500 m long. Transects were purposely oriented along isobaths to minimize any biotic variation attributable to depth differences along transects. Site 1 encompassed an area of approximately 12.9 km^2, of which 14,500 m^2 and 23,500 m^2 of seafloor (0.3% of the total area) were video recorded in 1998 and 1999, respectively. Site 2 encompassed an area of approximately 10.2 km^2, of which 15,900 m^2 of seafloor (0.16% of the total area) was video recorded.

The submersible *Delta*, occupied by a pilot and scientific observer, was equipped with external halogen lights, internal and external video cameras, gyro and magnetic compasses, and sub-to-tender vessel communication. The submersible was also equipped with an acoustic transponder that allowed tracking of the submersible by the tender vessel with differential global positioning and ultra-short baseline acoustic tracking.

The submersible followed a predetermined bearing at speeds of 0.27–0.82 m/s along each transect, and its course was modified when necessary via communication with the tender vessel. Continuous contact with the seafloor maintained the external camera lens at a near constant altitude (≈80 cm). The camera was oriented with the imaging plane directed at a shallow angle of approximately five degrees from vertical. Width of the image area was approximately 0.53 m in 1998, 0.85 m for Transects 1–7 at Site 1 in 1999, and 0.63 m for Transects 8–10 at Site 1 in 1999. Image widths, recorded at the start and end of each transect and at 500-m intervals, were averaged for each transect. Images were continuously recorded on a Hi-8 videocassette recorder. The scientific observer aboard the submersible viewed the image area laterally and recorded voice observations. Data continuously displayed on the video images included real time, depth (m), and height of the camera lens above the seafloor (cm). In addition, the video camera recorded two parallel laser marks 20 cm apart, projected onto the seafloor to provide calibration for measurements of the width of the image area (i.e., transect width) and size of fauna.

In the laboratory, all epifauna (approximately more than 4 cm in any dimension) partially or fully viewed on video footage were enumerated. Epifauna abundance at Site 1 was assessed with density (number of animals per square meter), making transects of differing widths comparable. Epifauna were collected in 1999 with a 6-m shrimp trawl towed on the seafloor just outside the study sites. Trawl collections were used to confirm taxonomic identifications. Fauna were classified to species if consistent video identifications were possible (20 of 35 taxa); otherwise, epifauna were grouped at higher taxonomic levels (Table 1). Similar species could not always be distinguished from one another on video footage, and those taxa were grouped. All sea whips greater than 20 cm in height were recorded as *Halipteris willemoesi*, although some of the smaller specimens (<50 cm) were possibly *Stylatula* sp. Naticidae included both pale moonsnail *Euspira pallida* and *Crytonatica russa*. Caridea included at a minimum the following 10 species of shrimp listed in order of decreasing relative abundance: (1) arctic eualid *Eualus fabricii*, (2) Arctic argid *Argis dentata*, (3) gray shrimp *Neocrangon communis*, (4) yellowleg pandalid *Pandalus tridens*, (5) ridged crangon *Crangon dalli*, (6) barbed eualid *Eualus barbatus*, (7) Townsend eualid *Eualus townsendi*, (8) beaked eualid *Eualus avinus*, (9) Okhotsk lebbeid *Lebbeus schrencki*, and (10) Rathbun blade shrimp *Spirontocaris arcuata*. Paguridae included at least five species of hermit crabs: (1) Alaskan hermit *Pagurus ochotensis*, (2) knobbyhand hermit *P. confragosus*, (3) bluespine hermit *P. kennerlyi*, (4) armed hermit *P. armatus*, and (5) splendid hermit *Labidochirus splendescens*. Pleuronectidae (>15 cm) included Pacific halibut, flathead sole, arrowtooth flounder, butter sole, Dover sole, rex sole, and at Site 2 only, petrale sole *Eopsetta jordani*. Pleuronectidae (<15 cm) included flathead sole, arrowtooth flounder, and rex sole. Psychrolutidae included two species of fathead sculpins, tadpole sculpin *Psychrolutes paradoxus* and *Malococottus* sp.

Taxa were further assigned to mobility groups (sedentary, low mobility, and high mobility) based on their observed mobility (Table 1). Seven taxa were identified as prey items (Table 1) based on analysis of stomach contents of 10 species of groundfish commercially harvested in the study areas (Stone, unpublished data). Abundance of other key taxa was analyzed separately based on their potential importance as biotic habitat (i.e., *Protoptilum* sp. and *H. willemoesi*) or bioturbators (Paguridae). Juvenile Tanner crabs were also identified as a key taxon since their spatial distribution might provide insights into the effectiveness of the 1987 area closures.

Three types of biogenic structures (i.e., structures produced by the activity of fauna), specifically (1) burrows, (2) foraging or shelter pits, and (3) elevated feeding mounds were enumerated on 10 randomly selected segments of strip transect from each of the closed and open areas at Site 1 (1999) and Site 2. Segments were nonoverlapping and of uniform area (8.5 m^2 at Site 1 [1999], 10.6 m^2 at Site 2). Structures to be enumerated were selected *a posteriori* based on in situ observations of their use by prey species as refuge. Prey animals previously enumerated were tallied for each segment to assess their spatial relationships with biogenic structures. Biogenic structures were not enumerated at Site 1 (1998) since the relatively low numbers of prey animals observed there precluded meaningful statistical analyses.

Statistical Analysis

Epifauna Abundance and Species Diversity

Areas open and closed to bottom trawling were compared for differences in animal abundance and community structure. Total abundance of megafauna and abundances of key taxa (*Protoptilum* sp., *H. willemoesi*, Tanner crab, and Paguridae) and functional groups (S, L, H, and P; see Table 1) were compared between open and closed areas. We assessed community structure by analyzing the two components of species diversity: species richness and relative abundance of species (Magurran 1988). We used Margalef's index, a function of total numbers of species and individuals, as a measure of species richness and Simpson's index of dominance as a measure of the relative abundance of species (Clarke and Warwick 1994). Magurran (1988) notes that species richness and dominance indices are informative in environmental studies and cites several studies that demonstrated reduced species richness and

Table 1. Total number of each megafauna taxon observed on seafloor transects at Site 1 and Site 2. Taxa were assigned to the mobility groups S (sedentary), L (low mobility), and H (high mobility) based on their mobility and whether they are prey (P) for commercially harvested groundfish. The larger of the numbers by status (open or closed) is in bold.

		Site 1				Site 2	
		1998		1999		1998	
Taxon	Functional group	Closed	Open	Closed	Open	Closed	Open
Cnidaria							
Actinaria (unidentified)	S	**76**	37	**65**	37	**17**	3
Cribrinopsis fernaldi	S	212	**248**	257	**303**	**192**	31
Metridium senile	S	**400**	309	**787**	630	**69**	32
Halipteris willemoesi	S	315	**393**	**1,093**	720	143	**800**
Protoptilum sp.	S	4,935	**6,287**	14,029	**15,627**	1,852	**1,958**
Ptilosarcus gurneyi	S	**2**	0	0	0	**1**	0
Nemertinea							
Cerebratulus herculeus	L	**27**	25	**44**	38	1	1
Mollusca							
Opisthobranchia (eggs)	S	0	0	0	0	**572**	383
Tritonia diomedea	L	**1**	0	0	**1**	1	**3**
Naticidae	L	**296**	244	**427**	413	**340**	273
Naticidae (eggs)	S	112	**128**	277	**326**	68	**122**
Patinopecten caurinus	L	**133**	122	**322**	242	**30**	13
Octopus sp.	H	**1**	0	**2**	0	**4**	1
Crustacea							
Chionoecetes bairdi (juvenile)	L, P	**123**	105	275	**353**	**155**	103
Oregonia gracilis	L	**9**	8	**61**	33	**9**	3
Pandalus eous	L, P	**499**	276	**2,705**	2,087	2,510	**3,013**
Caridea (unidentified)	L, P	**4,924**	4,051	**7,055**	5,217	**3,205**	1,733
Paguridae	L	**4,948**	4,721	10,220	**10,513**	**2,153**	1,386
Echinodermata							
Asteroidea (unidentified)	L	2	2	0	0	1	1
Ctenodiscus crispatus	L	9	9	11	**13**	4	**17**
Luidia foliolata	H	**10**	8	**93**	80	0	0
Pycnopodia helianthoides	H	**97**	87	**140**	108	**27**	20
Solaster dawsoni	H	**9**	0	**3**	0	**10**	7
Gorgonocephalus eucnemis	L	71	**128**	**427**	309	24	**73**
Strongylocentrotus droebachiensis	L	10	**57**	**31**	3	**27**	4
Pisces							
Rajidae	H	**10**	4	**21**	5	**4**	2
Osteichthyes (unidentified, <20 cm)	L	**4**	3	33	**40**	**27**	2
Atheresthes stomias (juvenile)	H, P	0	0	**5,480**	4,836	0	0
Pleuronectidae (>15 cm)	H	**464**	344	650	**673**	**392**	382
Pleuronectidae (<15 cm)	H, P	**838**	700	951	**959**	**626**	524
Bathyagonus alascanus	L, P	**81**	69	349	**393**	**81**	55
Podothecus accipenserinus	L	**17**	13	19	**22**	**11**	1
Dasycottus setiger	L	**1**	0	**4**	2	**15**	12
Psychrolutidae	L	**407**	345	**323**	300	**189**	154
Lycodes sp.	L, P	**389**	384	827	**992**	**699**	573
Total (status)		**19,432**	19,107	**46,981**	45,275	**13,459**	11,685
Total (site and year)		38,539		92,256		25,144	

increased dominance in stressed communities. Damaged sea whips (i.e., not skeletons but animals with living tissue, including those dislodged from the seafloor) were grouped for all species and analyzed for differences in abundance between open and closed areas. Densities (number of animals per square meter), instead of numbers of animals, were compared at Site 1 to correct for differing transect widths.

Differences in animal abundance and species diversity between open and closed areas were tested with individual analysis of variance (ANOVA) models. The three-factor models included fixed effects status (variable indicating open or closed area), year, and transect (blocking variable) nested within year and the interaction between status and year. Models fit to Site-2 data excluded variable year and the interaction term. Dependent variables (y or $y + 1$) were Box–Cox transformed (Box and Cox 1964; Venables and Ripley 1999) if necessary with a different power parameter (λ) estimated in the range [−2,2] for each variable. Data from the two sites were analyzed separately. First-year and second-year data from Site 1 were analyzed separately if a significant interaction between status and year was detected. One-tailed t-tests were used to test for reduced species richness (Margalef's index) and increased dominance (Simpson's index) in the open area. We chose an α-level of 0.10 over the traditional α-level of 0.05 to protect against type II error (McConnaughey et al. 2000).

Biogenic Structures
First, to determine if the number of biogenic structures was significantly higher in the closed than in the open area, we fit a two-factor submodel of the ANOVA model with fixed factors (status and transect) to the number of biogenic structures. Second, to determine if prey animal abundance was positively related to the abundance of biogenic structures, we fit a two-factor analysis of covariance (ANCOVA) model to the number of prey animals with fixed factors (status and transect) and covariate (number of biogenic structures). Dependent variables (y or $y + 1$) in the two models were Box–Cox transformed (Box and Cox 1964; Venables and Ripley 1999) if necessary, and segments were treated as replicates. One-tailed t-tests of differences in status were performed. The α-level was 0.10.

Spatial Characteristics of Sea Whips
The spatial distribution of sea whips (small: *Protoptilum* sp., and large: *H. willemoesi*) on transects was treated as one-dimensional since transect length (1,500 m in closed or open area) was large compared to transect width (0.53–0.85 m). Neighbor K statistics for one-dimensional data (O'Driscoll 1998), which are based on distances between neighboring individuals, provided both tests of aggregation and spatial descriptions of individuals on transects. The test for spatial randomness of individuals on a one-dimensional transect involves computing the expected number of extra individuals within a specified distance of an arbitrary individual (O'Driscoll 1998). Tests of aggregation were computed for varying scales h in 1-m increments ($h = 1, 2, 3,..., H$, where H is transect length). Significance of tests was measured using Monte Carlo methods (O'Driscoll 1998). A significant aggregation of individuals at scale h indicates individuals have more neighbors within distance h than would be expected if individuals were randomly arranged. Following O'Driscoll (1998), we adopted an unweighted approach to edge bias, assuming no individuals occur beyond the ends of a transect. In addition to a test of spatial aggregation, neighbor K analysis provides a description of one-dimensional spatial patterns in terms of patch length and crowding (O'Driscoll 1998). Patch length is the spatial scale of clustering, and crowding is a measure of the relative number of individuals in a patch. Patch length and crowding were determined from graphs of function $L(h)$, where $L(h)$ represents the average number of extra neighbors observed within distance h of any individual than would be expected under spatial randomness. Values of patch length and crowding, which depend on inter-neighbor distances, may not be comparable among transects of differing widths. Significance of the test for spatial randomness was determined only for values of h up to $H/2$. Since bottom fishing may alter the spatial distribution of animals (Langton and Robinson 1990; Thouzeau et al. 1991; Auster et al. 1996), data from open and closed areas were analyzed separately ($H = 1,500$ m). Transects were analyzed separately, and only those transects with at least three sea whips in each 1,500-m section were tested for spatial aggregation. The α-level was 0.05.

Animal Abundance in Groves of Halipteris willemoesi
In situ observations indicated that large *H. willemoesi* (height > 80 cm) form discrete "groves" on the seafloor and that animal abundance appeared to be high there. To investigate these observations further, the locations of all *H. willemoesi* were mapped at the two study sites to delineate grove boundaries. Densities of prey taxa, large Pleuronectidae, and all taxa combined (excluding large sea whips) were compared inside and outside of groves.

Species Associations with Prey Taxa
Four species or species groups were tested for associations with prey taxa: (1) the sedentary group, (2) small sea whips (height < 20 cm, *Protoptilum* sp.), (3) *H. willemoesi*, and (4) large Pleuronectidae. The four species or species groups were chosen after data collection and prior to data analysis. We hypothesized that prey taxa would be positively associated with the sedentary group and sea whips, since they provide refuge in the form of biotic structure (i.e., structures caused by, produced by, or comprising living organisms but not those produced by the activity of living organisms) and a negative association with large Pleuronectidae based on their predator–prey relationship. Pair-wise associations be-

tween prey abundance and the abundances of four species or species groups were measured with Pearson's correlation coefficients (Snedecor and Cochran 1973). Abundance at Site 1 was assessed with density (number of animals per square meter) instead of numbers of animals. Abundance was computed by transect in open and closed areas. To satisfy the assumption of bivariate normality in the correlation test, animal densities at Site 1 were natural log transformed and animal numbers at Site 2 were left untransformed. Yearly data at Site 1 were combined. The α-level was 0.05.

Results

Epifauna Abundance and Species Diversity

We detected general and site-specific differences in epifauna abundance and species diversity between areas open and closed to bottom trawling. The relative positions of 155,939 epifauna were mapped on the seafloor at the two study sites (Table 1). Differences in epifauna abundance and species diversity between open and closed areas were generally similar for the two sites (Table 2). Total animal abundance was significantly lower (13.2%) in the open area than in the closed area at Site 2 (Figure 2) but not at Site 1 (Table 2; Figure 3). Epifauna classified as prey were 16.6% and 17.5% less abundant in the open area than in the closed area at Site 1 and Site 2, respectively (Table 2; e.g., Figure 4). Low-mobility epifauna were also significantly lower in the open area than in the closed area at both sites (10.3% and 21.7% at Site 1 and Site 2, respectively). Differences in abundances of sedentary and high-mobility animals between open and closed areas were not detected at either site (Table 2). We found no interannual differences in animal abundance at Site 1 except for fewer low-mobility animals and more high-mobility animals in the second year. This increase in high mobility animals was likely due to a strong recruitment of juvenile (young-of-year) arrowtooth flounder to the benthos, an event that apparently occurs between June and August. Abundances of two key taxa (*Protoptilum* sp. and *H. willemoesi*) were not significantly different between open and closed areas at either site (Table 2). Abundances of juvenile Tanner crabs (18–45-mm carapace width) and Paguridae were significantly lower (33.5% and 35.6%, respectively) in open than in closed areas at Site 2 but not at Site 1 (Table 2).

No difference in the abundance of damaged sea whips was detected between open and closed areas at either site (Table 2). We observed 504 damaged sea whips (1.96% of the total sea whips observed) in areas open to trawling at both sites and 439 damaged sea whips (1.97% of the total sea whips observed) in areas closed to trawling at both sites. Axial rods (i.e., skeletons) of *H. willemoesi*, however, were disproportionately more common in areas closed to trawling (262 total compared to 30 total).

In terms of species diversity, both sites exhibited significantly lower species richness (e.g., Figure 5) in the open area than in the closed area (Table 2). Simpson's index of dominance was significantly higher in the open area than in the closed area at Site 1 in 1998 only (Table 2). A significantly higher index of dominance was not found in the open area at Site 2 (Table 2). The model of Simpson's index of dominance fitted to Site-1 data had a significant interaction between the variables status and year (ANOVA, $F = 5.53 \sim F_{1,18}$, $P = 0.03$); therefore, an ANOVA model was fit to first-year and second-year data separately. Interannual differences in species diversity indices at Site 1 were not examined since transect widths differed between years, making interannual indices incomparable.

Biogenic Structures

Biogenic structures were significantly more abundant in the closed area than in the open area at Site 1 in 1999 (ANOVA, $|t| = 6.22 \sim$ Student's $t(0.05, 189)$, $P < 0.001$) and at Site 2 (ANOVA, $|t| = 10.69 \sim$ Student's $t(0.05, 189)$, $P < 0.001$) (Figure 6). Multiple R^2 for the model of number of biogenic structures was 53% at Site 1 and 44% at Site 2. Prey taxa abundance was greater in areas with greater numbers of biogenic structures at Site 1 (ANCOVA, $|t| = 2.22 \sim$ Student's $t(0.05, 188)$, $P = 0.02/2 = 0.01$) but not at Site 2 (ANCOVA, $|t| = 0.46 \sim$ Student's $t(0.05, 188)$, $P = 0.64/2 = 0.32$) (Figure 6). Multiple R^2 for this model was 58% at Site 1 and 32% at Site 2.

Spatial Characteristics of Sea Whips

No consistent patterns in spatial characteristics (patch length and crowding) of sea whips were found between open and closed areas at either site, nor between the two sites. Sea whips (*Protoptilum* sp. and *H. willemoesi*) exhibited aggregation on most transects in closed and open areas at both sites (Table 3). Patch lengths of *Protoptilum* sp. in closed and open areas of Site 1 ranged from 2 m to nearly 700 m. Crowding values of *Protoptilum* sp. in closed and open areas of Site 1 ranged from less than 1 to more than 200 sea whips. Patch lengths of *Protoptilum* sp. in closed and open areas of Site 2 ranged from 35 m to nearly 700 m. Corresponding crowding values of *Protoptilum* sp. in closed and open areas of Site 2 ranged from less than 1 to more than 60 sea whips. Median patch length of *Protoptilum* sp. in the open area was greater than in the closed area at Site 1 and vice versa for Site 2. Median crowding of *Protoptilum* sp. was greater in the closed area than in the open area at both sites.

Table 2. Summary statistics for testing differences in status (variable indicating open or closed area) for epifauna abundances and species diversity indices. Statistics include value of the F statistic or t statistic (in the case of one-tailed tests), degrees of freedom (df), P-values, and multiple R^2 (%). The percent decrease is listed in the "Open" column for those taxa that were significantly lower in the area open to trawling; an arrow indicates the direction of the index in the open area. Significance at $\alpha = 0.10$ is indicated by an asterisk.

Variable	Site 1					Site 2				
	F or $|t|$	df	P	R^2	Open	F or $|t|$	df	P	R^2	Open
Grouped taxa										
All individuals	1.32	1, 19	0.27	98		4.76	1, 9	0.06*	79	13.2
Sedentary	0.97	1, 19	0.34	98		1.70	1, 9	0.22	88	
Low mobility	3.71	1, 19	0.07*	97	10.3	31.70	1, 9	<0.001*	92	21.7
High mobility	1.71	1, 19	0.21	89		0.90	1, 9	0.37	72	
Prey	9.92	1, 19	0.005*	97	16.6	17.32	1, 9	0.002*	95	17.5
Individual taxa										
Protoptilum sp.	0.001	1, 19	0.97	97		2.30	1, 9	0.16	96	
Halipteris willemoesi	0.94	1, 19	0.35	93		0.07	1, 9	0.80	76	
Damaged sea whips	0.06	1, 19	0.81	88		1.56	1, 9	0.24	82	
Chionoecetes bairdi (juvenile)	0.04	1, 19	0.84	89		7.34	1, 9	0.02*	86	33.5
Paguridae	0.17	1, 19	0.69	89		60.81	1, 9	<0.001*	97	35.6
Species diversity										
Richness	2.37	19	0.01*	69	↓	2.83	9	0.01*	76	↓
Dominance (1998)	3.06	9	0.007*	90	↑	1.13	9	0.14	64	
Dominance (1999)	0.34	9	0.37	92						

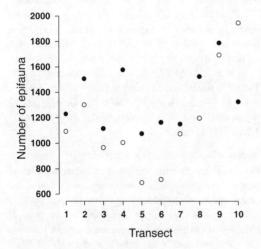

Figure 2. Abundance of megafauna per transect in the areas open (open circles) and closed (closed circles) to bottom trawling at Site 2.

Patch lengths of *H. willemoesi* in closed and open areas of Site 1 ranged from 9 m to more than 400 m. Crowding values of *H. willemoesi* in closed and open areas of Site 1 ranged from less than 1 individual to more than 35 individuals. Patch lengths of *H. willemoesi* in closed and open areas of Site 2 ranged from at least 10 m to more than 500 m. Crowding values of *H. willemoesi* in closed and open areas of Site 2 ranged from approximately 1 individual to more than 100 individuals. Median patch length and crowding of *H. willemoesi* were greater in the closed area than in the open area at Site 1 and vice versa for Site 2.

Animal Abundance in Groves of *Halipteris willemoesi*

Five groves of *H. willemoesi* were delineated at Site 1 in 1998 and 1999: four were entirely and one was partially within the closed area (Figure 7). Groves ranged from 406 m to 830 m in length. Density of *H. willemoesi* inside the groves (23 individuals per 100 m²) was nearly 70 times the density outside the groves (0.33 individuals per 100 m²). Densities of prey taxa, large Pleuronectidae (>15 cm), and all taxa combined were higher inside groves than outside groves (Table 4). Although *H. willemoesi* was fairly common at Site 2 (Table 1), no groves of large individuals (height > 80 cm) were identified there.

Species Associations with Prey Taxa

Pair-wise correlations between prey abundance and the abundances of four other species or species groups were consistently positive or negative for closed and open areas at either site but not between sites (Table 5). At Site 1, prey species abundance was positively correlated with the abundances of sedentary taxa and sea whips. Prey species abundance at Site 1 was not significantly correlated with large Pleuronectidae (>15 cm) abundance in either the closed or open area. At Site 2, no significant correlation was found between prey species abundance and abundances of sedentary taxa and sea whips. Prey species abundance was

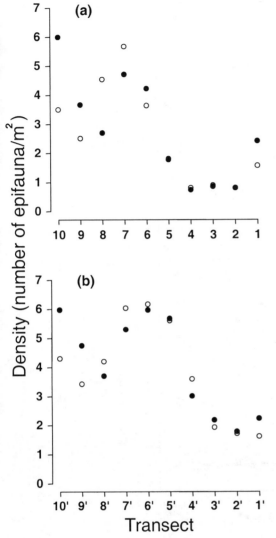

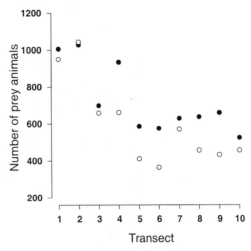

Figure 4. Numbers of prey animals per transect in the open (open circles) and closed (closed circles) areas at Site 2.

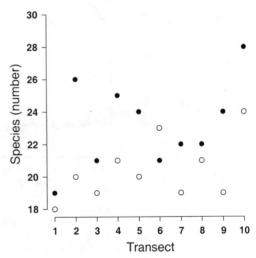

Figure 3. Densities of megafauna per transect in the open (open circles) and closed (closed circles) areas at Site 1 in (a) 1998 and (b) 1999.

Figure 5. Numbers of species per transect in the open (open circles) and closed (closed circles) areas at Site 2.

negatively correlated with large Pleuronectidae in the open area only at Site 2. Although the correlation coefficient computed between prey abundance and large flatfish abundance in the open area at Site 1 was not significant, the sign of the coefficient was also negative (Table 5).

Discussion

The use of area closures as control comparisons is a growing practice in studying the chronic effects of bottom trawling on seafloor habitat. Our in situ observations demonstrated that differences exist in the abundance and diversity of epibenthos between areas consistently bottom trawled each year and adjacent areas where bottom trawling has been prohibited for 11–12 years. These differences, which may be attributed to chronic, long-term trawling, include decreases in species richness and the abundances of low-mobility and prey species fauna at two sites. Site 1, in 1998, had higher species dominance in areas open to trawling, an indication of a stressed or disturbed community (Shaw et al. 1983). Also, at Site 2, total abundance of epifauna and the abundances of two key taxa (Tanner crab and Paguridae) were reduced in areas

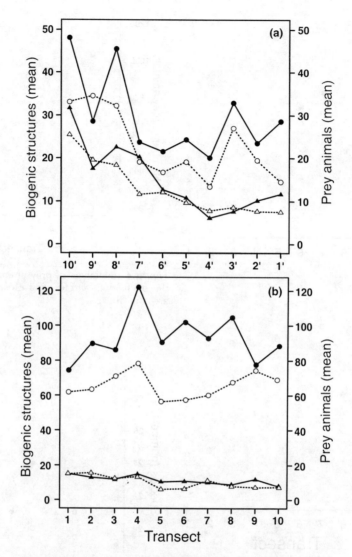

Figure 6. Mean abundance of biogenic structures (circles) in the open (dotted line) and closed (solid line) areas at (a) Site 1 (1999) and (b) Site 2. Mean abundance of prey animals (triangles) in the open (dotted line) and closed (solid line) areas at (a) Site 1 (1999) and (b) Site 2. Structures and prey were enumerated within 20 randomly selected 8.5 m^2-sections and 10.6 m^2-sections of seafloor per transect at Site 1 (1999) and Site 2, respectively.

open to trawling. Detailed examination of the distribution of biogenic structures at Site 1 indicated that the number of these structures was reduced in areas open to trawling, thereby reducing seafloor complexity in these soft-sediment communities. Our findings are in general agreement with other studies on the effects of otter trawls in soft-sediment habitats (Jennings and Kaiser 1998; Collie et al. 2000) and will have important implications in assessing the effects of ambient levels of bottom trawling on essential fish habitat in the GOA.

Our two study sites differed with respect to three factors that contribute to, and can therefore be used to predict, the magnitude of seafloor disturbance and rate of recovery. These factors are (1) fishing intensity, (2) sediment grain-size characteristics, and (3) natural disturbance regime near the seafloor (Jones 1992; Collie et al. 2000). Based on these factors, we correctly predicted that ambient levels of trawling would more adversely affect Site 2, which was characterized by finer-grained sediments in a more stable environment, than Site 1.

Table 3. Transects (% and numbers) that showed aggregation of sea whips and their associated spatial characteristics in closed and open areas. Two-year data from Site 1 were combined.

Species	Site	Area	Percent of transects with aggregation	Patch length (m) Median	Patch length (m) Range	Crowding (number) Median	Crowding (number) Range
Protoptilum sp.	1	Closed	95 (19/20)	189	2–712	20.6	0.3–204.0
		Open	89 (17/19)	212	10–697	15.8	0.4–182.4
	2	Closed	90 (9/10)	260	144–403	18.3	4.6–36.8
		Open	100 (10/10)	194	35–676	4.8	0.7–63.1
H. willemoesi	1	Closed	71 (10/14)	191	53–390	8.3	0.7–35.7
		Open	71 (10/14)	86	9–418	4.2	0.4–8.8
	2	Closed	100 (9/9)	52	14–168	1.8	1.1–9.7
		Open	100 (6/6)	234	92–523	22.6	1.6–101.8

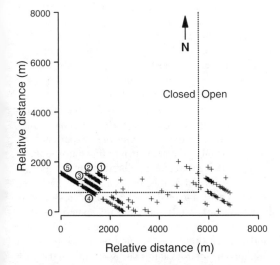

Figure 7. Locations of large (height > 80 cm) *Halipteris willemoesi* (+) on transects within the study area at Site 1 in 1998 and 1999. Five groves of large sea whips, numbered 1–5, were delineated at this site.

Table 4. Densities (number of animals/m^2) and ratio of *Halipteris willemoesi*, prey taxa, large Pleuronectidae, and all taxa combined (excluding *H. willemoesi*) inside and outside of groves of *H. willemoesi* at Site 1. Densities are from two groves delineated in 1998 and three groves delineated in 1999.

Group	Density Inside	Density Outside	Inside/ Outside
Halipteris willemoesi	0.23	0.0033	69.95
Prey taxa	2.33	1.12	2.08
Pleuronectidae (>15 cm)	0.13	0.052	2.55
All taxa	5.77	3.31	1.75

Table 5. Pearson correlation coefficients (*r*) between prey abundance and abundances of four species or species groups in closed and open areas. Two-year data from Site 1 were combined. Significance at α = 0.05 is indicated by an asterisk. Site 1: |*r*| ≥ 0.44; Site 2: |*r*| ≥ 0.63.

Group	Site 1	Site 2
	Closed	
Sedentary taxa	0.69*	–0.48
Protoptilum sp.	0.47*	–0.55
H. willemoesi	0.62*	–0.58
Large Pleuronectidae	–0.02	–0.30
	Open	
Sedentary taxa	0.65*	–0.48
Protoptilum sp.	0.52*	–0.41
H. willemoesi	0.54*	–0.50
Large Pleuronectidae	–0.35	–0.69*

We found that low-mobility taxa and prey taxa were less abundant in areas open to trawling. Many of these taxa are highly associated with seafloor structures and use these structures as refuge from predation and benthic currents. We demonstrated that prey taxa are more abundant in areas where both biogenic and biotic structures (sedentary taxa, sea whips, and *H. willemoesi* groves) were more abundant. At Site 1, abundances of prey and biogenic structures were positively associated, indicating that prey taxa may be highly dependent on these seafloor structures for refuge. Similarly, abundances of prey and sedentary taxa, including both *Protoptilum* sp. and *H. willemoesi*, were positively associated, and prey were twice as abundant inside *H. willemoesi* groves than in surrounding habitat. Since prey abundance was significantly lower in areas open to trawling at both sites and biogenic structures were less abundant in areas open to trawling (both sites), trawling may indirectly affect prey species abundance by reducing the number of biogenic structures on the seafloor. Interestingly, we detected no difference in the abundance of biotic structures between open and closed areas at either site.

Mobile fishing gear in contact with the seafloor reduces benthic complexity by leveling biogenic structures and removing the organisms that create these structures (Auster et al. 1996). Of these two mechanisms, we believe that direct removal of biogenic structures explains the reduced numbers observed in the area open to trawling at Site 1. Our in situ observations of trawl gear striations and increases in surficial total organic carbon (Stone and Masuda 2003) suggest that surface sediments are mixed by components of the trawl system. This action would tend to level sediment structures on the surface of the seafloor. Alternatively, fishing could directly remove or alter the behaviors of the fauna responsible for the structures. We found some evidence that the abundance of one abundant bioturbator (Paguridae) is affected by trawling. The foraging patterns, and hence, rate of pit digging, of Paguridae may also be altered in response to trawling disturbance (Ramsay et al. 1996).

Compared to larger fauna, prey taxa likely experience little direct mortality from bottom trawling (e.g., as bycatch or from delayed mortality due to physical damage) but rather experience mortality indirectly through the modification or removal of biogenic and biotic structures. Prey likely experience increased predation due to loss of refuge and increased exposure immediately after seafloor disturbance. We observed some evidence of the latter after disturbance by the submersible's pressure wave.

Seafloor communities at our two sites were dominated by several species of sea whips that accounted for the majority of biotic structure on the seafloor. At least two species of sea whips (*Protoptilum* sp. and *H. willemoesi*) are present within the study sites at maximum observed densities of $16/m^2$ and $6/m^2$, respectively, and provide vertical structure to this otherwise low-relief habitat. Abundances of all taxa, prey taxa, and predators (large Pleuronectidae) were higher in dense groves of *H. willemoesi* than in adjacent habitat with lower densities of *H. willemoesi*. Pacific cod and walleye pollock, although not enumerated on strip transects because they generally swim or hover just above the seafloor, also appeared to be more abundant in sea whip groves. The ecological importance of this habitat type was similarly noted by Brodeur (2001), who observed high densities of Pacific ocean perch *Sebastes alutus* within sea whip "forests" in the Bering Sea. Brodeur (2001) suggested that sea whips "may provide important structural habitat for Pacific ocean perch in an otherwise featureless environment."

Sea whips are widely distributed in the GOA and, based on bycatch records from trawl and longline fisheries, the probability of interaction with fishing gear is high (Malecha et al. 2005, this volume). Little is known about the distribution and abundance of sea whip groves, however. Furthermore, at least one species, *H. willemoesi*, is estimated to live at least 50 years (Wilson et al. 2002), so recovery time from disturbance may be substantial. We did not detect a significant difference in sea whip abundance between areas closed and open to bottom trawling, indicating that ambient levels of bottom trawling may not have affected their numbers. The spatial and size–frequency distributions (discussed below) of *H. willemoesi*, however, may have been affected by bottom trawling as evidenced by the disproportionate presence of groves in the closed area. Directed studies to determine the resistance and resilience of sea whips to bottom trawling are underway, and results of those studies will be essential to assessing the full effects of fishing on that habitat.

The density of *H. willemoesi* was not significantly different between areas open and closed to bottom trawling. When *H. willemoesi* were classified by height as medium (20–80 cm) or large (>80 cm), however, the large *H. willemoesi* accounted for a greater proportion observed in the closed area than in the open area. The large sea whips (height > 80 cm) comprised 39% (123 of 315) and 29% (313 of 1,093) of *H. willemoesi* in the closed area in 1998 and 1999, respectively, but only 6% (23 of 393) and 15% (109 of 720) of *H. willemoesi* in the open area in 1998 and 1999, respectively. This observation may be an indication that *H. willemoesi* experience reduced survival in the area open to trawling and that a shift in the size–frequency distribution of this species has occurred at this site. This effect has often been noted for targeted mobile species (reviewed in Frid and Clark 2000) but rarely for nontarget sedentary species (Bradstock and Gordon 1983). Detailed analysis of the size–frequency distribution of weathervane scallops at this site revealed that the open area had higher prerecruit abundance relative to recruit abundance than did the closed area (Masuda and Stone 2003), although we should note that *P. caurinus* is the target of a small-scale fishery in the area open to trawling at this site.

We observed little evidence of physical disturbance to the seafloor and associated fauna in the areas open to bottom trawling at Site 1 and Site 2. Obvious trawl-door furrows or striations on the substrate from ground gear were noted infrequently at Site 1 (9 of 20 transects) but more frequently at Site 2 (all transects), where fishing intensity was lower. The prominence of trawl marks on the seafloor appeared less related to trawl intensity and more related to sediment grain-size characteristics and the strength of benthic currents. Other researchers have noted the role these factors play in the rate of seafloor habitat recovery (e.g., Collie et al. 2000). Aside from damaged or dislodged sea whips (1.97% of the total observed), we observed *Metridium senile* drifting near the seafloor, often still attached to bivalve shells. We do not know if trawling activity dislodged them or if this was a previously undocumented transport mechanism for the species. Axial rods (i.e., skeletons) of large *H. willemoesi* were disproportionately more common, compared to live individuals, in areas closed to trawling. We believe that the accumulation of axial rods was simply a result of senescence. Since large *H. willemoesi* were more common in closed areas, their skeletons should tend to accumulate there. The calcified skeletons, typically lying on the seafloor, persist in seawater for at least several years (Stone, personal observations) and tend to accumulate with other debris (e.g., drift algae) within *H. willemoesi* groves. Bottom trawling may remove or bury axial rods.

Some taxa showed highly consistent patterns of abundance and distribution in areas open and closed to trawling (e.g., Paguridae and large Pleuronectidae; Table 1). Many taxa were quite small and would not have been sampled with remote techniques (e.g., trawls). Enumerating small megafauna on video allowed us to reveal important functional roles for several taxa. For example, juvenile arrowtooth flounder were very abundant in 1999 when sampling was conducted in late summer. Although not valued highly as a commercial species, the high ecological value of the species, especially at this life stage, is clear. Paguridae (hermit crabs) were highly abundant at both study sites and, surprisingly, we found no significant difference in abundance between areas open and closed to trawling. We estimate that foraging hermit crabs generated approximately one-third of the biogenic structures (pits) observed on transects at Site 1. Since hermit crabs may have small home ranges (Stachowitsch 1979) and are important bioturbators, their effect on small-scale sediment dynamics can be substantial, especially given their abundance at our two study sites.

This study provides important information that fills an existing gap in the literature on the effects of fishing on benthic habitats (Collie et al. 2000). Although more than 60 studies have been conducted on the effects of fishing on benthic habitats worldwide, few have focused on habitats deeper than 100 m (Collie et al. 2000). Only three studies have been conducted in the eastern Pacific Ocean; two of these were recently completed in Alaska. Freese et al. (1999) investigated acute effects of otter trawls on gravel and cobble habitat in deep water (>200 m) in the eastern GOA, and McConnaughey et al. (2000) examined chronic effects of otter trawls on shallow sand habitat (44–52 m depth) in the eastern Bering Sea.

There are several inherent problems associated with the use of area closures as control comparisons that potentially bias results and confound interpretations to some degree. Three potential sources of bias are specific to the Kodiak Island area closures. Firstly, we compared areas bottom trawled for approximately 30 years to areas closed to bottom trawling for only 11–12 years. Closed areas were trawled to some extent prior to 1987 so we cannot be certain that the epibenthos had stabilized to prefished conditions. Two species (*H. willemoesi* and *P. caurinus*) have life spans greater than 11–12 years, so their current abundance and spatial distribution may have been affected by fishing activity prior to 1987. Secondly, the closure areas prohibit the use of nonpelagic trawls and scallop dredges only. An unknown amount of longlining that occurs in the closure areas could have caused some disturbance to the epibenthos. Thirdly, due to record-keeping limitations of the fisheries observer program, precise locations of hauls within the open areas are unknown (i.e., designated "open" samples may have been collected in areas that were not actually trawled). Two important assumptions of this investigation are that the closed and open areas were identical at the time of the fishery closures in 1987 and that natural disturbances have equally affected those areas.

The 1987 closures in the Kodiak Island area were implemented in response to the collapse of crab stocks in the mid-1980s. For reasons unknown, crab stocks have not yet recovered, so determining the efficacy of the closures in rebuilding these stocks is not possible. We saw no juvenile or adult red king crabs within the study sites. Juvenile Tanner crabs were fairly common at the two sites, and our 2-year counts at Site 1 indicated increased abundance during that period. At Site 2, juvenile Tanner crabs were significantly more abundant in areas closed to trawling. We saw no adult Tanner crabs at the study sites. These observations indicate that at least one species intended to benefit from the habitat closures may be showing signs of recovery and possibly that the habitat closures are providing important sanctuary to that species.

In this study, we investigated the chronic effects from ambient levels of trawling at two sites within a range of soft-bottom habitat. Although we detected significant differences in epifauna abundance and species diversity between areas open and closed to bottom trawling, the magnitude of the differences do not appear to be sufficient to cause broad-scale changes to these communities. Observed differences are similar to those observed in other studies on the effects of otter trawling on soft-bottom habitat (Engel and Kvitek 1998; Prena et al. 1999; McConnaughey et al. 2000; Kenchington et al. 2001; Schwinghamer et al. 2001), but the magnitude of differences appears to be minimal compared to the effects of otter trawling in more complex habitat in the GOA (Freese et al. 1999). Evidence exists, however, that bottom trawling has produced changes to the seafloor and associated biota, affecting the availability of prey for commercially important groundfish. This should serve as a "red flag" to managers since prey taxa are a critical component of essential fish habitat.

Acknowledgments

We thank the following people who helped with this project. Ken Krieger was instrumental in initiating the work, and then he retired. Jon Heifetz, Jeff Fujioka, Jerry Pella, and Phil Rigby helped with the original study design, reviewed this manuscript, and provided many helpful suggestions. John Karinen, Ken Krieger, Jeff Regelin, and Linc Freese assisted with field operations. Dave Csepp assisted with video analysis, Bruce Wing provided taxonomic expertise, and Chris Lunsford helped with trawling intensity estimations. Craig Rose loaned us video equipment that made counting and mapping the positions of 155,939 animals possible. We are also grateful to Captain Wade Loofbourrow and the crew of the Alaska Department of Fish and Game RV *Medeia* and Delta Oceanographics for their assistance and support. The Alaska Fisheries Science Center's Auke Bay Laboratory of the National Marine Fisheries Service funded this research.

References

Auster, P. J., and R. W. Langton. 1999. The effects of fishing on fish habitat. Pages 150–187 *in* L. Benaka, editor. Fish habitat: essential fish habitat and rehabilitation. American Fisheries Society, Symposium 22, Bethesda, Maryland.

Auster, P. J., R. J. Malatesta, R. W. Langton, L. Watling, P. C. Valentine, C. L. S. Donaldson, E. W. Langton, A. N. Shepard, and I. G. Babb. 1996. The impacts of mobile fishing gear on seafloor habitats in the Gulf of Maine (northwest Atlantic): implications for conservation of fish populations. Reviews in Fisheries Science 4:185–202.

Ball, B., B. Munday, and I. Tuck. 2000. Effects of otter trawling on the benthos and environment in muddy sediments. Pages 69–82 *in* M. J. Kaiser and S. J. de Groot, editors. Effects of fishing on non-target species and habitats: biological, conservation and socio-economic issues. Blackwell Scientific Publications Ltd., Oxford, UK.

Box, G. E. P., and D. R. Cox. 1964. An analysis of transformations. Journal of the Royal Statistical Society, Series B 26:211–243.

Bradstock, M., and D. P. Gordon. 1983. Coral-like bryozoan growths in Tasman Bay, and their protection to conserve commercial fish stocks. New Zealand Journal of Marine and Freshwater Research 17:159–163.

Brodeur, R. D. 2001. Habitat-specific distribution of Pacific ocean perch (*Sebastes alutus*) in Pribilof Canyon, Bering Sea. Continental Shelf Research 21:207–224.

Churchill, J. H. 1989. The effect of commercial trawling on sediment resuspension and transport over the Middle Atlantic Bight continental shelf. Continental Shelf Research 9:841–865.

Clarke, K. R., and R. M. Warwick. 1994. Change in marine communities: an approach to statistical analysis and interpretation. Natural Environment Research Council, UK.

Collie, J. S., S. J. Hall, M. J. Kaiser, and I. R. Poiner. 2000. A quantitative analysis of fishing impacts on shelf-sea benthos. Journal of Animal Ecology 69:785–798.

Engel, J., and R. Kvitek. 1998. Effects of otter trawling on a benthic community in Monterey Bay National Marine Sanctuary. Conservation Biology 12:1204–1214.

Freese, L., P. J. Auster, J. Heifetz, and B. L. Wing. 1999. Effects of trawling on seafloor habitat and associated invertebrate taxa in the Gulf of Alaska. Marine Ecology Progress Series 182:119–126.

Frid, C. L. J., and R. A. Clark. 2000. Long-term changes in North Sea benthos: discerning the role of fisheries. Pages 198–216 *in* M. J. Kaiser and S. J. de Groot, editors. Effects of fishing on non-target species and habitats: biological, conservation and socio-economic issues. Blackwell Scientific Publications Ltd., Oxford, UK.

Gislason, H. 1994. Ecosystem effects of fishing activities in the North Sea. Marine Pollution Bulletin 29: 520–527.

Goñi, R. 1998. Ecosystem effects of marine fisheries: an overview. Ocean and Coastal Management 40:37–64.

Jennings, S., and M. J. Kaiser. 1998. The effects of fishing on marine ecosystems. Advances in Marine Biology 34:201–352.

Jones, J. B. 1992. Environmental impact of trawling on the seabed: a review. New Zealand Journal of Marine and Freshwater Research 26:59–67.

Kenchington, E. L. R., J. Prena, K. D. Gilkinson, D. C. Gordon, Jr., K. MacIssac, C. Bourbonnais, P. J. Schwinghamer, T. W. Rowell, D. L. McKeown, and W. P. Vass. 2001. Effects of experimental otter trawling on the macrofauna of a sandy bottom ecosystem on the Grand Banks of Newfoundland. Canadian Journal of Fisheries and Aquatic Sciences 58:1043–1057.

Langton, R. W., and W. E. Robinson. 1990. Faunal associations on scallop grounds in the western Gulf of Maine. Journal of Experimental Marine Biology and Ecology 144:157–171.

Lindholm, J. B., P. J. Auster, and L. S. Kaufman. 1999. Habitat-mediated survivorship of juvenile (0-year) Atlantic cod *Gadus morhua*. Marine Ecology Progress Series 180:247–255.

Magurran, A. E. 1988. Ecological diversity and its measurement. Princeton University Press, Princeton, New Jersey.

Malecha, P. W., R. P. Stone, and J. Heifetz. 2005. Living substrate in Alaska: distribution, abundance and species associations. Pages 289–299 *in* P. W. Barnes and J. P. Thomas, editors. Benthic habitats and the effects of fishing. American Fisheries Society, Symposium 41, Bethesda, Maryland.

Martin, M. H., and D. M. Clausen. 1995. Data report: 1993 Gulf of Alaska bottom trawl survey. NOAA Technical Memorandum NMFS-AFSC-59.

Masuda, M. M., and R. P. Stone. 2003. Biological and spatial characteristics of the weathervane scallop *Patinopecten caurinus* at Chiniak Gully in the central

Gulf of Alaska. Alaska Fishery Research Bulletin 10:104–118.

McConnaughey, R. A., K. L. Mier, and C. B. Dew. 2000. An examination of chronic trawling effects on soft-bottom benthos of the eastern Bering Sea. ICES Journal of Marine Science 57:1377–1388.

Mueter, F. J., and B. L. Norcross. 2002. Spatial and temporal patterns in the demersal fish community on the shelf and upper slope regions of the Gulf of Alaska. National Marine Fisheries Service Fishery Bulletin 100:559–581.

NPFMC (North Pacific Fishery Management Council). 2000. Stock assessment and fishery evaluation report for the groundfish resources of the Gulf of Alaska. NPFMC, Anchorage, Alaska.

O'Driscoll, R. L. 1998. Description of spatial pattern in seabird distributions along line transects using neighbour K statistics. Marine Ecology Progress Series 165:81–94.

Prena, J., P. Schwinghamer, T. W. Rowell, D. C. Gordon, Jr., K. D. Gilkinson, W. P. Vass, and D. L. McKeown. 1999. Experimental otter trawling on a sandy bottom ecosystem of the Grand Banks of Newfoundland: analysis of trawl bycatch and effect on epifauna. Marine Ecology Progress Series 181:107–124.

Ramsay, K., M. J. Kaiser, and R. N. Hughes. 1996. Changes in hermit crab feeding patterns in response to trawling disturbance. Marine Ecology Progress Series 144:63–72.

Rose, C. S., and E. M. Jorgensen. 2005. Spatial and temporal distributions of trawling intensity off Alaska: consideration of overlapping effort when evaluating the effects of fishing on habitat. Pages 679–690 in P. W. Barnes and J. P. Thomas, editors. Benthic habitats and the effects of fishing. American Fisheries Society, Symposium 41, Bethesda, Maryland.

Schwinghamer, P., D. C. Gordon, Jr., T. W. Rowell, J. Prena, D. L. McKeown, G. Sonnichsen, and J. Y. Guigné. 1998. Effects of experimental otter trawling on surficial sediment properties of a sandy-bottom ecosystem on the Grand Banks of Newfoundland. Conservation Biology 12:1215–1222.

Shaw, K. M., P. J. D. Lambshead, and H. M. Platt. 1983. Detection of pollution-induced disturbance in marine benthic assemblages with special reference to nematodes. Marine Ecology Progress Series 11:195–202.

Snedecor G. W., and W. G. Cochran. 1973. Statistical methods. The Iowa State University Press, Ames.

Stachowitsch, M. 1979. Movement, activity pattern, and role of a hermit crab population in a sublittoral epifaunal community. Journal of Experimental Marine Biology and Ecology 39:135–150.

Stone, R. P., and M. M. Masuda. 2003. Characteristics of benthic sediments from areas open and closed to bottom trawling in the Gulf of Alaska. NOAA Technical Memorandum NMFS-AFSC-140.

Thouzeau, G., G. Robert, and S. J. Smith. 1991. Spatial variability in distribution and growth of juvenile and adult sea scallops *Placopecten magellanicus* (Gmelin) on eastern Georges Bank (Northwest Atlantic). Marine Ecology Progress Series 74:205–218.

U.S. Department of Commerce. 1996. Magnuson-Stevens Fishery Conservation and Management Act: as amended through October 11, 1996. NOAA Technical Memorandum NMFS-F/SPO-23.

Venables, W. N., and B. D. Ripley. 1999. Modern applied statistics with S-PLUS. Springer-Verlag, New York.

Wilson, M. T., A. H. Andrews, A. L. Brown, and E. E. Cordes. 2002. Axial rod growth and age estimation of the sea pen, *Halipteris willemoesi* Kolliker. Hydrobiologia 471:133–142.

Biological Traits of the North Sea Benthos: Does Fishing Affect Benthic Ecosystem Function?

JULIE BREMNER[1] AND CHRIS L. J. FRID[2]

School of Marine Science and Technology, Dove Marine Laboratory,
University of Newcastle upon Tyne, Cullercoats, Tyne and Wear NE30 4PZ, UK

STUART I. ROGERS[3]

Centre for Environment, Fisheries and Aquaculture Science, Lowestoft Laboratory,
Pakefield Road, Lowestoft, Suffolk NR33 0HT, UK

Abstract. The impact of fishing on benthic species and habitats has been well documented, but effects on the way the ecosystem functions are less well understood. The roles performed by benthic species contribute to ecological functioning, and changes in the types of species present may have implications for the whole ecosystem. Biological traits analysis, which uses a wide range of biological characteristics, is employed to investigate the effects of fishing on the variety of roles performed by benthic taxa. Eighteen biological traits were chosen to represent aspects of the morphology, life history, feeding, and habitat use of benthic infauna from the western North Sea. Differences in relative frequency of the traits were assessed in relation to changes in fishing pressure over a 30-year period. The communities were dominated by organisms exhibiting opportunistic traits. These traits responded positively to an initial increase in fishing effort and then remained relatively stable for the remainder of the study. Traits predicted to be associated with vulnerability to fishing impacts, such as long life spans, large oocytes, and shelled body designs, decreased in proportion in response to elevated fishing effort. Those organisms with high regeneration potential and asexual reproduction also responded negatively. Traits related to feeding modes of various taxa and their interactions with the benthic habitat remained relatively stable throughout the study. We conclude that fishing has altered the biological trait composition of this benthic community in both predictable and unexpected ways over the last 3 decades. Some aspects of functioning may have been affected, while others, including those related to trophic relationships and habitat usage, have been preserved in spite of changes in taxon composition. It is not yet clear what the larger-scale implications of these trait changes may be to ecosystem functioning, fisheries management, or, indeed, the management of anthropogenic activities in the sea. What is clear is that studies of the biological trait compositions of other marine ecosystem components are now required.

Introduction

International agreements require countries to manage the marine benthos to protect biodiversity at a level that includes both species diversity and the essential functions performed by ecosystems (Secretariat of the Convention on Biological Diversity 1992; OSPAR Commission 1998). Ecosystem functioning relates to physical processes and the organisms mediating these processes. Along with species diversity and physical aspects of the habitat, it incorporates the roles played by organisms, their interactions with each other, and their interactions with their environment.

The impacts of trawl fishing on benthic habitats have been well documented, with trawling disturbing substrate surface (Watling and Norse 1998), displacing rocks (Engel and Kvitek 1998), damaging sedimentary structure (Auster et al. 1996), and moving particulate matter into the water column (Riemann and Hoffmann 1991; Pilskaln et al. 1998). The effects of trawling on biological communities are equally demonstrable, with reductions in total abundance or biomass and richness and alterations in relative species composition (Hutchings 1990; Jennings and Kaiser 1998).

How fishing affects the types of organisms inhabit-

---
[1] Corresponding author: julie.bremner@ncl.ac.uk
[2] E-mail: c.l.j.frid@ncl.ac.uk
[3] E-mail: s.i.rogers@cefas.co.uk

ing benthic systems is, however, less well understood. Structure-forming organisms, which provide a refuge from predation and perform important nursery functions for fish and invertebrate species, can be heavily impacted by the passage of trawl gear (Turner et al. 1999; Collie et al. 2000a). Direct damage also reduces large, long-lived, and fragile taxa (Hall-Spencer et al. 1999; Kaiser et al. 2000; Bradshaw et al. 2002), while indirect impacts include increases in opportunists and scavengers such as crabs and whelks that take advantage of increased dead benthos and discarded bycatch (Kaiser et al. 1998; Rumohr and Kujawski 2000).

Although changes to the types of organisms present will have implications for the functioning of benthic systems, the magnitude of the impacts is still unclear. Not all taxa respond to disturbance in the predicted manner. Different species of scavenging taxa may respond in contradictory ways to increases in fishing or may respond positively in some habitats but not others (Ramsay et al. 1996, 1998). Species defined as vulnerable to fishing may show no response or even increase in abundance in impacted systems (Frid et al. 2000).

Characteristics that make some species types vulnerable to fishing may be offset by other characteristics that impart resistance. Some upright, sessile organisms are able to withstand trawl disturbance because they also possess the ability to regenerate lost appendages or quickly recolonize postimpacted areas (Bradshaw et al. 2002). There are also indications that additional characteristics, such as flexibility (Wassenberg et al. 2002), sediment reworking activities (Coleman and Williams 2002), and sediment depth ranges (Bergman and Hup 1992), may be important in determining how such taxa respond to stress.

Biological traits analysis is a method capable of investigating the wide range of biological characteristics involved in determining how organisms respond to fishing. The approach, which originates in lotic ecosystems (Townsend and Hildrew 1994), incorporates organisms' interactions both with each other and with their environment and can be a potentially valuable tool for investigating the effects of anthropogenic disturbance at the ecosystem-functioning level (Doledec et al. 1999; Charvet et al. 2000).

Initial marine applications of biological traits analysis showed that epibenthic communities in the southern North Sea and eastern English Channel could be differentiated by the flexibility, body shape, feeding technique, method of locomotion, and degree of attachment to the substrate exhibited by resident taxa (Bremner et al. 2003; Frid et al., paper from Mini-Symposium on Defining the Role of ICES in Supporting Biodiversity Conservation, 2000). Although these differences were not investigated in relation to fishing pressure, the results of the studies indicate that the approach could also be used to elucidate the impact of trawling on the functioning of benthic ecosystems.

This paper uses the biological traits approach to investigate the potential effects of bottom trawling on an infaunal community from the western North Sea. The community has been sampled over a time period of some 30 years, and previous analysis provided evidence of fishing impacts at the species level (Frid et al. 1999). The aims of the present study were to examine whether biological trait composition of the benthic community changed over time and to investigate if these changes could be related to fluctuations in fishing pressure. Specific predictions were made about the response of each biological trait to fishing pressure, based on current ideas about the characteristics that make benthic organisms vulnerable to or tolerant of disturbance (Table 1). For example, it was predicted that traits linked to opportunistic lifestyles, such as short lifespans, early maturity, asexual reproduction, scavenging, and body regeneration, would respond positively to increased fishing, whereas long lifespans, shelled body forms, filter feeding, and large size would be impacted negatively.

Methods

Study Area

The study station is located at 55°07′N, 01°15′W, some 18.5 km (11.5 mi) off the northeastern coast of En-

Table 1. Biological trait variables and categories used to describe functional diversity in the infauna communities off the northeastern English coast and predictions about the response of each trait category to increased fishing pressure. ↑ = predicted increase; ↓ = predicted decrease; and N = no prediction or no change predicted.

Trait	Category	Prediction
Size	Small	↑
	Small–Medium	↑
	Medium	N
	Medium–Large	↓
	Large	↓
Longevity (years)	0–3	↑
	4–7	N
	8–11	↓
	12+	↓
Body design	Vermiform	↑
	Shell	↓
	Armour/Scales	↑
Reproductive mode	Asexual: budding	↑
	Asexual: fission	↑
	Sexual: planktonic development	↑

Table 1. Continued.

Trait	Category	Prediction
Reproductive mode	Sexual: direct development	↓
Number of reproductive events	Monotelic	↓
	Polytelic: annual episode	N
	Polytelic: multiple episodes	↑
Age at maturity (months)	0–12	↑
	13–24	N
	25–36	↓
	37+	↓
Egg size (mm)	0–100	↑
	101–200	N
	201–300	↓
	301+	↓
Reproductive season	Spring	N
	Summer	N
	Autumn	N
	Winter	N
Duration of reproductive season	Short episodic (0–2 months)	N
	Extended episodic (3–6 months)	↑
	Continuous	↑
Duration of planktonic development	Non-planktonic	↓
	Short (up to 1 month)	↑
	Extended (>1 month)	↑
Movement mode	Sessile	↓
	Burrow	↑
	Crawl	↑
	Glide/drift	N
	Swim	↑
Mobility	None	↓
	Low	↓
	Medium	N
	High	↑
Aggregation potential	Low	N
	Medium	N
	High	↓
Preferred substrate location (cm)	0–5	↓
	6–10	↓
	11–20	N
	21+	↑
Living habit	Tube dweller	↓
	Burrow dweller	↑
	Crevice dweller	N
	Free living	↑
Feeding mode	Filter/suspension	↓
	Deposit	↑
	Scavenger	↑
	Predator	↑
Food type	Invertebrates	N
	Carrion	↑
	Detritus	↑
	Plankton	N
	Microorganisms	N
Regeneration potential	Low	↓
	Medium	N
	High	↑

gland, UK (Figure 1). It is located in 80 m of water with predominantly silt and clay sediments. The station has been sampled annually in late winter since 1971. At least five samples of 0.1-m² area were taken on each occasion using a van Veen grab. These were passed through a 0.5-mm mesh sieve, and taxon abundance was recorded per m² (Buchanan and Warwick 1974). With the exception of 1977 and 1998, data were available from all years since 1971.

In order to overcome difficulties arising from misidentification and changes in taxonomy over the 30-year period, the taxa were analyzed at the genera level or above (Clark 2000). To simplify the analysis, the data set was reduced to the most abundant taxa by retaining only those that represented 3% or more of abundance at the station in any 2 or more years. On average, 3% of the total taxa present met this criterion in each individual year, resulting in a subset of 15 taxa (Table 2).

Fishing Pressure

The station lies within a fishing ground for Dublin Bay prawn *Nephrops norvegicus*. It is trawled principally by

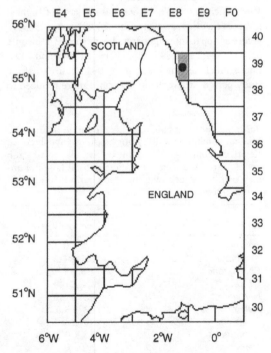

Figure 1. The location of the benthic sampling station off the northeastern coast of England, western North Sea. The shaded areas are the limits of the area in which fishing effort was measured. From International Council for the Exploration of the Sea (unpublished data).

the UK, but Danish, Dutch, and Swedish vessels have fished in the vicinity. The area is trawled for *Nephrops* spp. in the autumn, winter, and spring, although Atlantic cod *Gadus morhua*, haddock *Melanogrammus aeglefinus*, and whiting *Merlangius merlangus* (also known as European whiting) are also taken year round (Robson 1995). The fishery has landed around 1,000 metric tons locally for the past 5–6 years (MAFF 1995, 1996, 1997, 1998; DEFRA 2000). It is based primarily around otter trawling, although some beam trawling was also carried out during the 1990s.

Fishing effort data for the *Nephrops* fishery, measured as total h fished per annum (hpa) within International Council for the Exploration of the Sea rectangle 39E8, were provided by the Department for Environment, Food and Rural Affairs (Figure 2). Effort ranged from approximately 12,000–92,000 hpa (mean ± SE = 51,809.6 ± 7,435.9 hpa), equating to between 5.7 and 8.9 h/d trawled.

The fishing effort data were split into five phases, defined by changes in effort over the period 1972–2001. Effort was low during the initial years of the study, 1972 to 1981 (phase 1: mean ± SE = 37,509.3 ± 6,994.3 hpa). It then rose to moderate levels (phase 2: mean ± SE = 65,256.6 ± 13,830.1 hpa) from 1982 to 1986. The years 1987–1989 (phase 3) saw the highest levels of trawling, an average 90,259.3 hpa (±3,583.6), before a drop to more moderate levels (phase 4: mean ± SE = 63,837.2 ± 12,895.1 hpa) between 1990 and 1994. Phase 5 included the years from 1994 to the end of the study period, where effort was once more reduced to low levels (phase 5: mean ± SE = 37,563.9 ± 5,062.9 hpa).

Biological Traits Analysis

Eighteen biological traits were chosen to represent aspects of the benthic organisms' morphology, life history, and interaction with each other and their environment. Each trait was broken down into categories. For example, living habit was divided into four categories; tube-dwelling, burrow-dwelling, crevice-dwelling, and free living (Table 1). Individual taxa were then coded for their affinity to each category of the biological traits, using a fuzzy coding procedure. Fuzzy coding allows organisms to exhibit more than one category of a given trait and to exhibit these categories to varying degrees (Chevenet et al. 1994). This allows information from a variety of sources to be used and accounts for differences in trait expression within taxa (Castella and Speight 1996; Doledec et al. 1999; Charvet et al. 2000). The coding range of zero to three was adopted here, with zero being no affinity to a trait category and three being high affinity.

Fuzzy coding is particularly useful when taxa are aggregated to the genus level or above. For example, nemerteans were only identified to phylum, but fuzzy coding allowed differences among taxa within the phylum to be reflected in the trait categories. Thus, nemerteans mostly have nonplanktonic development, but may sometimes exhibit short or extended planktonic development. They can be coded 3, 1, 1 for the trait "duration of planktonic development phase."

Each taxon's trait category scores were weighted by their abundance in each year of the study, and these weighted scores were summed over all 15 taxa (Charvet

Table 2. The most abundant taxa found in the benthic infauna community off the northeastern English coast. Relative abundance is the mean abundance of each genus, expressed as a percentage of total fauna abundance and averaged over the study period. CI = confidence interval.

Taxa	Relative abundance (mean ± CI)
Abra	3 ± 0.97
Amphiura	2.5 ± 1.3
Chaetozone	4.8 ± 1
Glycera	1.6 ± 0.4
Harpinia	2.9 ± 0.6
Heteromastus	31 ± 3.9
Levinsenia	11 ± 3.8
Lumbrineris	2.3 ± 0.3
Nemertea	2.9 ± 0.6
Oligochaeta	2.2 ± 1
Ophelina	3.7 ± 1.1
Paramphinome	2.2 ± 0.6
Praxillella	9.4 ± 1.8
Prionospio	2.6 ± 0.6
Spiophanes	1.7 ± 1.1

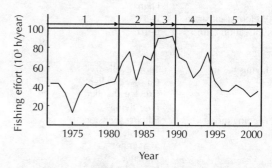

Figure 2. Fishing effort from 1972 to 2001 in International Council for the Exploration of the Sea statistical rectangle 39E8. Effort is divided into five phases: phase 1 = initial low effort (1972–1981), phase 2 = initial moderate effort (1982–1986), phase 3 = high effort (1987–1989), phase 4 = subsequent moderate effort (1990–1994), and phase 5 = subsequent low effort (1995–2001).

et al. 1998). This resulted in a year by trait table, showing the frequency of each trait category within the community for each year of the study. When no information on a particular trait was available for a taxon, zero values were entered for each trait category and the taxon did not contribute to the calculation of trait weightings.

The faunal data set was double-root transformed in order to downweight the influence of traits exhibited by very abundant taxa. It was then split into five groups according to the different phases of fishing effort. Differences in trait composition over the study period were investigated using nonmetric multidimensional scaling (MDS) on the year-by-trait table. One-way analysis of similarities (ANOSIM) was used to test for differences in biological trait composition among the fishing effort phases, and similarity of percentages (SIMPER) analysis was employed to determine the traits most responsible for any differences. The analyses were carried out using the Plymouth Routines in Marine Research package (Clarke and Warwick 1994).

Results

The benthic community at the sampling station was the variant *Brissopsis lyrifera–Amphiura chiajei* of the classic community type of *Amphiura filiformis* (Petersen and Boysen-Jensen 1911) Of the 15 most abundant taxa, 10 were polychaetes (Table 2). Nemerteans and oligochaetes were also represented, as were one genus each of mollusk (*Abra*), crustacea (*Harpinia*), and echinoderm (*Amphiura*). Most trait categories were expressed by several taxa, with the exception of long lifespans and shelled body forms, exhibited only by *Amphiura* spp. and *Abra* spp., respectively.

Based on the relative proportions of biological traits, MDS highlighted several changes in biological trait composition within the benthic community. There was evidence of a directional trend in trait composition during the early years of the study; however, the community did not follow a consistent trajectory over the whole period (Figure 3a). Differences were more marked when the years were grouped by the level of fishing effort (Figure 3b). The largest changes in trait composition occurred between the first two phases, when effort rose from low to moderate. Trait composition also altered during the mid-1980s to late 1980s, when effort became high. Changes after this period were more gradual, and even though fishing effort was similar, the community remained subtly different during the subsequent low effort phase than that encountered at the beginning of the study. Differences in trait composition were significant among all fishing phases (ANOSIM: Global $R = 0.492$, $P < 0.01$), except between phases 3 and 4 (Table 3).

The benthic community was characterized by short-lived, sessile, vermiform taxa with monotelic reproduction and extended reproductive seasons. These organisms dominated throughout the study, as did deposit feeders and animals with low regeneration potential. Proportions of these trait categories were similar within fishing phases, although, with the exception of deposit feeders, they increased between low and high effort, then reduced between phases 3 and 5 (Figures 4, 5). One or more of these trait categories were exhibited by all taxa to some degree, and *Chaetozone*, *Heteromastus*, *Levinsenia*, *Ophelina*, *Prionospio*, and *Spiophanes* strongly expressed each of the seven categories.

Organisms displaying consistent reductions in response to increased fishing effort included those exhibiting high regeneration potential, polytelic reproduction (more than one reproductive episode per year), production of large oocytes (more than 301 μm), asexual reproduction by budding, and shelled body design (Table 4). These organisms decreased in proportion between the initial low and high effort phases and increased when fishing subsequently declined during phase 5 (Figures 4b, c and 5a–c). The decrease in asexually budding organisms was the most marked of the three trait categories, and this decrease was mirrored to a lesser degree by those organisms that reproduced asexually through fission (Figure 4c).

Shelled organisms and those exhibiting high regeneration potential increased to proportions in excess of that found during the initial low effort period, whereas organisms reproducing asexually by budding failed to regain the proportions observed during the early years of the study. No single taxon exhibited each of the five trait categories negatively impacted by fishing, although oligochaetes and nemerteans both showed medium to high affinity for polytelic reproduction, large oocytes and high regeneration potential.

Other organisms displaying trends related to fishing effort were long-lived (more than 12 years) and armoured or scaled animals (Figures 4a, 5b). These trait categories decreased within the community between phases 1 and 3 (Table 5). However, they did not decrease gradually, and both categories showed a slight increase when effort initially rose at the beginning of the 1980s. As effort reduced from high to low, their proportions increased beyond those observed during the initial low effort period. Organisms that reproduced sexually with a short planktonic larval development phase, matured at 13–24 months, and produced small oocytes also increased between phases 1 and 2, although unlike the long-lived and armoured and scaled animals, they did not subsequently decrease when fishing effort became high.

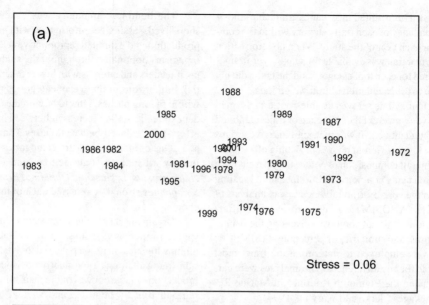

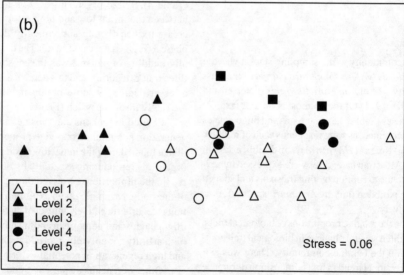

Figure 3. Multidimensional scaling of the biological trait composition of a benthic community from the northeastern English coast during the period 1972–2001. Plot (a) shows the pattern of change over the years, while plot (b) shows the relationship among the years when labeled by the phase of fishing effort (phase 1 = initial low effort, phase 2 = initial moderate effort, phase 3 = high effort, phase 4 = subsequent moderate effort, and phase 5 = subsequent low effort).

Discussion

The biological trait composition of infaunal communities off the northeastern English coast has shown a pattern of change in response to fishing pressure that has potential implications for the functioning of the ecosystem and for fisheries management. Some of these changes corresponded with current expectations about vulnerability of particular types of organisms while others did not.

Trait categories that responded according to predictions were those associated with opportunistic life history strategies. Organisms maturing relatively early, producing small oocytes, or exhibiting short pelagic development phases increased in frequency at the beginning of the 1980s, then showed less response to changing

Table 3. Comparison of biological trait composition between different fishing effort phases for northeastern English coast benthic communities by analysis of similarities (phase 1 = initial low effort, phase 2 = initial moderate effort, phase 3 = high effort, phase 4 = subsequent moderate effort, and phase 5 = subsequent low effort). n/s = not significant.

Fishing phases	R-value	P-value
Phase 1 to phase 2	0.781	<0.001
Phase 2 to phase 3	0.959	<0.05
Phase 3 to phase 4	−0.087	n/s
Phase 4 to phase 5	0.451	<0.01
Phase 1 to phase 3	0.35	<0.1
Phase 3 to phase 5	0.704	<0.05

fishing effort. Species composition of the area was relatively stable through the 1970s, where population cycles were linked to the flux of organic matter to the benthos (Buchanan and Moore 1986; Buchanan et al. 1986; Frid et al. 1996, 1999). This relationship broke down at the beginning of the 1980s when the increase in fishing led to subtle changes in species composition (Lindley et al. 1995; Frid et al. 1999). Opportunistic life history traits allowed the organisms to respond quickly to the initial increases in fishing effort and to retain dominance in the community when effort fluctuated in later years.

This dominance of opportunistic traits may have led to an alternate state of functioning within this benthic community. The increase in opportunistic traits coincided with a reduction in evenness of trait distribution during high fishing, and trait composition during the later period of low effort, though similar, did not return to that seen in the initial period of light fishing in the 1970s. The extent to which ecosystem functioning can resist fishing pressure and the forcing of alternate states is a question of much importance in marine ecology (Thrush and Dayton 2002). However, the trait composition of the community toward the end of the study appears to be situated some-

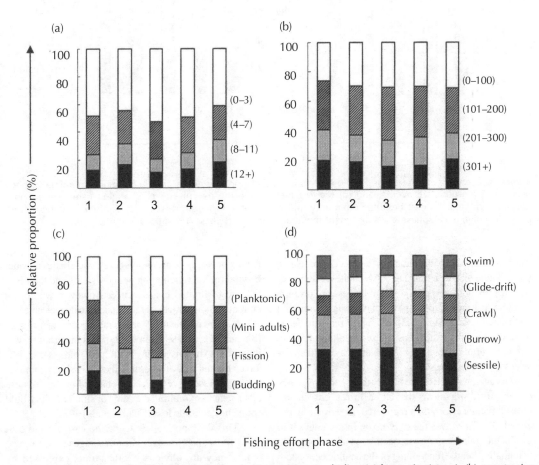

Figure 4. Comparison of selected biological trait composition, including (a) longevity (years), (b) oocyte size (μm), (c) reproductive mode, and (d) movement type, during five phases of changing fishing effort on a northeastern English coast benthic community (phase 1 = initial low effort, phase 2 = initial moderate effort, phase 3 = high effort, phase 4 = subsequent moderate effort, and phase 5 = subsequent low effort).

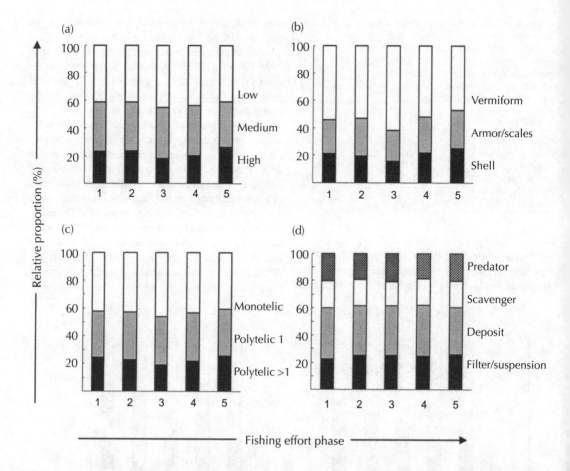

Figure 5. Comparison of selected biological trait composition, including (a) regeneration potential, (b) body design, (c) reproductive events, and (d) feeding method, during five phases of changing fishing effort on a northeastern English coast benthic community (phase 1 = initial low effort, phase 2 = initial moderate effort, phase 3 = high effort, phase 4 = subsequent moderate effort, and phase 5 = subsequent low effort).

where between that of the initial low and moderate fishing years. There may be a time lag between the reduction of fishing pressure and the response of the less opportunistic biological traits, so recovery may be happening but more slowly than the initial effects. It is too early to determine whether this system has entered an alternate state, although further monitoring of the site may yield an answer to this question.

As also predicted by life history theory, the frequency of long-lived organisms decreased in response to increased fishing effort, then recovered once trawling declined. *Amphiura* was the only taxon in the data set to exhibit this trait category, so reductions in this taxon could potentially alter the functioning of the benthic ecosystem because no other long-lived taxa were available to replace it. That ecological role and whatever ecological functions were associated with it were depressed in times of high fishing pressure. Some biological characteristics may not be vital for the functioning of ecosystems and their loss may be of limited consequence. However, we do not currently know which characteristics are dispensable and which losses may lead to considerable ecosystem change (Snelgrove et al. 1997). In light of this uncertainty, it is important to retain as many ecological roles as possible. Taxon redundancy ensures that these roles may continue to be expressed even if taxon changes occur, because other species exhibiting the same or similar traits would compensate for the loss (Chapin et al. 1997).

This does not appear to be the case for the trait "long lifespan." However, the advantages of ecological redundancy are obvious if other traits expressed by *Amphiura* spp. are considered. *Amphiura* spp. are burrowing organisms. The burrowing activities of benthic fauna facilitate nutrient recycling, aerate sediments, cre-

Table 4. A similarity of percentages analysis of the biological traits that contributed most (cumulative contribution) to differences in trait structure between fishing effort phases. Mean difference is the change in the average frequency of each trait category in the community between the fishing phases.

Trait	Category	Mean difference	Cumulative %
	Phase 1 to phase 3 (low to high effort)		
Reproductive mode	Budding	220.33	3.7
Body design	Shell	77.50	6.9
Oocyte size	301+ µm	328.67	9.6
Reproductive events	Polytelic (>1)	372.22	12.2
Maturity	0–12 months	985.44	14.8
Reproductive mode	Fission	273.28	17.3
Regeneration potential	High	289.50	19.6
Duration reproduction	Continuous	295.11	21.8
Oocyte size	201–300 µm	257.89	23.8
Movement	Crawl	178.95	25.8
	Phase 3 to phase 5 (high to low effort)		
Body design	Shell	−339.00	4.2
Longevity	12+ years	−286.67	8.1
Body design	Armour/scales	−592.00	11.8
Regeneration potential	High	−706.67	15.4
Reproductive events	Polytelic (>1)	−555.33	18.7
Longevity	8–11 years	−143.33	22.0
Oocyte size	301+ µm	−313.00	24.5
Aggregation potential	Medium	−404.33	26.7
Movement	Crawl	−364.33	29.0
Reproductive mode	Budding	−88.33	31.1
	Phase 1 to phase 2 (low to moderate effort)		
Reproductive mode	Planktonic	4,878.36	2.4
Duration reproduction	Short episodic	4,071.16	4.7
Life habit	Burrow-dweller	3,910.73	7.0
Aggregation potential	High	4,075.39	9.2
Reproductive season	Spring	3,853.43	11.5
Substrate location	11–20 cm	3,889.12	13.6
Maturity	13–24 months	4,306.73	15.7
Size	Medium-large	2,197.87	17.7
Oocyte size	0–100 µm	2,201.84	19.7
Mobility	Low	1,792.17	21.6

Table 5. Kruskal-Wallis tests of differences in proportions of selected biological trait categories between fishing effort phases.

Trait	Category	df	H	P-value
Size	Medium-large	4	17.18	<0.1
Longevity	12+ years	4	13.95	<0.01
Body design	Shell	4	9.75	<0.05
	Armour/scales	4	14.42	<0.01
Reproductive mode	Planktonic	4	17.71	<0.01
	Budding	4	11.54	<0.05
Reproductive events	Polytelic (>1)	4	13.33	<0.05
Maturity	13–24 months	4	16.63	<0.1
Oocyte size	0–100 µm	4	18.49	<0.01
	301+ µm	4	14.34	<0.01
Planktonic development	Short	4	18.35	<0.01
Mobility	Low	4	18.76	<0.01
Aggregation potential	High	4	17.00	<0.01
Living habit	Burrow dweller	4	18.59	<0.01
Regeneration potential	High	4	15.45	<0.01

ate habitat complexity, and influence hydrodynamics (Coleman and Williams 2002; Reise 2002). Reductions in abundance of *Amphiura* spp. removed their contribution to these important ecological functions. However, many taxa in the community performed some form of burrowing activities. The absence of any decrease in burrowing itself indicated that other organisms exhibiting similar traits compensated for the reduction in *Amphiura* spp.

Using a subset of the infauna community may limit the ability of the analysis to identify the responses of certain biological traits. Because large taxa tend to be rare in benthic systems (Warwick and Clarke 1996), it's likely that they are underrepresented in the analysis. Large taxa are particularly vulnerable to fishing effects, and they may have important roles in the functioning of the system. Although the individual roles played by these rare taxa may not affect functioning by themselves, their cumulative contribution may be enough to dampen the impact of fishing on traits expressed by only a single, abundant taxon. This has implications for the way the analysis is interpreted. The reduction in long-lived organisms noted in the reduced data set analyzed here is not, perhaps, as important in terms of functioning as is implied by the analysis, because the gap left by *Amphiura* spp. may be filled by the cumulative contributions of other long-lived but rare taxa.

When species composition is analyzed, the reduction of full fauna lists to a subset does not result in a loss of information (Gray et al. 1988). However, because very few species carry the same combinations of biological characteristics, reducing data sets during biological traits analysis is more likely to distort conclusions. Ideally, the full fauna list would be used, but this would require information on the biological characteristics of a large number of species. Gaps in the natural history knowledge of many benthic infauna species would prevent much trait information from being coded and compromise the ability of the analysis to reveal any patterns in trait composition. This creates a trade-off between taxa and traits; either the number of species or the range of traits included must be reduced.

As these results show, a variety of traits are involved in species' responses to fishing pressure and often not in the ways predicted. For example, Collie et al. (2000b) attributed a lack of response of *Abra* spp. to fishing to their small size. However, *Abra* spp. are also shelled organisms, and as evidenced here, this trait is negatively affected by fishing. Taxa with high powers of regeneration or asexual reproduction were predicted to respond positively, because recovery of damaged appendages allows organisms to survive the direct impact of trawls (Hill et al. 1996; Bradshaw et al. 2002) and the ability to reproduce regularly provides a survival advantage in disturbed systems (Grassle and Sanders 1973). These traits, in fact, responded negatively, possibly because regular reproduction and regeneration allow organisms to survive initial disturbance events but increase their energy requirements, making them more vulnerable to competition (Grassle and Sanders 1973; but see Hall et al. 1994).

It is clear that our knowledge of the responses of different types of organism to fishing impacts, although expanding rapidly, is far from complete. As such, retaining a wide range of traits in analyses, even at the expense of the full species list, is important at this time because it helps provide a picture of ecosystem functioning. This, ultimately, may allow us to ask which characteristics are important in determining the responses of benthic taxa to fishing.

There were no apparent changes to the relative frequencies of feeding mode and food choice traits in the present study. Increases of scavenging fish and epifauna have been found in areas subject to trawling disturbance (Hall-Spencer et al. 1999; Rumohr and Kujawski 2000), but there is little evidence of this response in opportunistic infaunal scavengers (Frid et al. 2000). Episodes of discard provision occur directly after trawling activity, and mobile fish and epifauna often arrive at the trawl site very quickly (Kaiser and Spencer 1996). This may reduce the opportunity for the smaller benthic infauna to utilize the extra food resources that become available (Frid et al. 2000). Hence, scavenging infauna may be largely unaffected by trawl discards.

If this is true, it is the combination of scavenging, location in the substrate, and mobility that determines how a species responds to fishing. Combinations of characteristics affecting species vulnerability have been encountered by other studies. Bergman and Hup (1992) found that large individuals of *Lanice* spp. and some echinoid species were less vulnerable to trawling than small ones because larger individuals built deeper tubes into which they could escape. *Ophiura* spp. live in the upper layers of the sediments, where organisms are more exposed to physical disturbances (Hall et al. 1994). Even so, this taxon can escape trawl impacts, probably because it is small enough to pass through the net (Bergman and Hup 1992).

Biological traits analysis considers traits individually, so, in this respect, it would be difficult to use the technique in its present form to determine species' sensitivities to fishing. The advantage of the approach is that, while it retains information on the species present in the community, it enables the responses of traits themselves to be examined independently of the species expressing them. This can provide an indication of the functioning of the community as a whole and how it responds to fishing.

Despite previously noted changes to species composition in the study site (Frid et al. 1999), some traits linked directly to functioning, such as burrowing and feeding, remained relatively constant over the study period. Jennings et al. (2001) also showed that trophic structure of a North Sea benthic community was resistant to fishing pressure even though species composition was altered. The implications of this are that some aspects of functioning in benthic ecosystems may well be maintained in the presence of fishing, irrespective of the identities of species performing particular roles, with reservoirs of potential replacements for those fauna that are reduced. If this is the case, it means that the functions performed by benthic communities are more robust to disturbance than the fauna performing them.

The North Sea benthos has been fished intensively for over a century (Frid et al. 2001), with rapid increases in effort occurring after the end of World War II (OSPAR Commission 2000). This system is dominated by polychaetes, an occurrence associated with trawling-impacted ecosystems (Jennings et al. 2001). Given the dominance of vermiform taxa from the onset of the study and the fishing history of the North Sea, it is likely that the system suffered some degree of impact on functioning before the 1970s. It is impossible to be sure what fishing-induced changes occurred in the system before the 1970s and how these shaped the changes occurring afterward. Biological traits analysis is a method new to marine ecology, and it would be unwise to apply the patterns emerging here to unfished systems. However, consistent patterns in the response of biological traits to human disturbance have been identified in freshwater invertebrate communities across Europe (Statzner et al. 2001). The patterns emerging from this study, particularly those that are consistent across more than one taxa, are real biological results that have relevance to the study of benthic ecosystem functioning.

Biological traits analysis has made an important contribution towards elucidating the impacts of fishing on functioning in this western North Sea benthic infauna community. The challenge now will be to determine how other infauna communities respond to fishing and whether changes in their functioning will have wider effects on other ecosystem components such as meiofauna, megafauna, and fish.

Acknowledgments

The study was funded by the School of Marine Science and Technology, University of Newcastle upon Tyne, and the Centre for Environment, Fisheries and Aquaculture Science (CEFAS) as part of JB's 3-year doctoral project. Assistance in attending the symposium was provided to JB by a conference grant from the Ecological Society of America and American Fisheries Society, to whom we are grateful. We thank Simon Thrush and an anonymous referee for valuable comments that allowed the manuscript to be substantially improved.

References

Auster, P., R. Malatesta, R. Langton, L. Watling, P. Valentine, C. Donaldson, E. Langton, A. Shepard, and I. Babb. 1996. The impacts of mobile fishing gear on seafloor habitats in the Gulf of Maine (northwest Atlantic): implications for conservation of fish populations. Reviews in Fisheries Science 4:185–202.

Bergman, M., and M. Hup. 1992. Direct effects of beam trawling on macrofauna in a sandy sediment in the southern North Sea. ICES Journal of Marine Science 49:5–11.

Bradshaw, C., L. Veale, and A. Brand. 2002. The role of scallop-dredge disturbance in long-term changes in Irish Sea benthic communities: a re-analysis of an historical dataset. Journal of Sea Research 47:161–184.

Bremner, J., S. I. Rogers, and C. Frid. 2003. Assessing functional diversity in marine benthic systems: a comparison of approaches. Marine Ecology Progress Series 254:11–25.

Buchanan, J., R. Brachi, G. Christie, and J. Moore. 1986. An analysis of a stable period in the Northumberland benthic fauna—1973–1980. Journal of the Marine Biological Association of the United Kingdom 66:659–670.

Buchanan, J., and J. Moore. 1986. A broad review of variability and persistence in the Northumberland benthic fauna—1971–1985. Journal of the Marine Biological Association of the United Kingdom 66:641–657.

Buchanan, J., and R. M. Warwick. 1974. An estimate of benthic macrofaunal production in the offshore mud of the Northumberland coast. Journal of the Marine Biological Association of the United Kingdom 54:197–222.

Castella, E., and M. Speight. 1996. Knowledge representation using fuzzy coded variables: an example based on the use of Syrphidae (Insecta, Diptera) in the assessment of riverine wetlands. Ecological Modelling 85:13–25.

Chapin, F., B. Walker, R. Hobbs, D. Hooper, J. Lawton, O. Sala, and D. Tilman. 1997. Biotic control over the functioning of ecosystems. Science 277:500–504.

Charvet, S., A. Kosmala, and B. Statzner. 1998. Biomonitoring through biological traits of benthic macroinvertebrates: perspectives for a general tool in stream management. Archiv fuer Hydrobiologie 142:415–432.

Charvet, S., B. Statzner, P. Usseglio-Polatera, and B. Dumont. 2000. Traits of benthic macroinvertebrates in semi-natural French streams: an initial application to biomonitoring in Europe. Freshwater Biology 43:277–296.

Chevenet, F., S. Doledec, and D. Chessel. 1994. A fuzzy coding approach for the analysis of long-term ecological data. Freshwater Biology 31:295–309.

Clark, R. 2000. Long-term changes in the North Sea ecosystem. Department of Marine Science and Coastal Management. University of Newcastle upon Tyne, Tyne and Wear, UK.

Clarke, K., and R. Warwick. 1994. Change in marine communities: an approach to statistical analysis and interpretation. Natural Environment Research Council, Swindon, UK.

Coleman, F., and S. Williams. 2002. Overexploiting marine ecosystem engineers: potential consequences for biodiversity. Trends in Ecology and Evolution 17:40–44.

Collie, J. S., G. Escanero, and P. Valentine. 2000a. Photographic evaluation of the impacts of bottom fishing on benthic epifauna. ICES Journal of Marine Science 57:987–1001.

Collie, J. S., S. J. Hall, M. J. Kaiser, and I. R. Poiner. 2000b. A quantitative analysis of fishing impacts on shelf-sea benthos. Journal of Animal Ecology 69:785–798.

DEFRA (Department for Environment, Food and Rural Affairs). 2000. United Kingdom sea fisheries statistics 1999 and 2000. DEFRA, London.

Doledec, S., B. Statzner, and M. Bournard. 1999. Species traits for future biomonitoring across ecoregions: patterns along a human-impacted river. Freshwater Biology 42:737–758.

Engel, J., and R. Kvitek. 1998. Effects of otter trawling on a benthic community in Monterey Bay marine sanctuary. Conservation Biology 12:1204–1214.

Frid, C., J. Buchanan, and P. Garwood. 1996. Variability and stability in benthos: twenty-two years of monitoring off Northumberland. ICES Journal of Marine Science 53:978–980.

Frid, C., R. Clark, and J. Hall. 1999. Long-term changes in the benthos on a heavily fished ground off the NE coast of England. Marine Ecology Progress Series 188:13–20.

Frid, C., R. Clark, and P. Percival. 2001. How far have the ecological effects of fishing in the North Sea ramified? Senckenbergiana Maritima 31:313–320.

Frid, C., K. Harwood, S. Hall, and J. Hall. 2000. Long-term changes in the benthic communities on North Sea fishing grounds. ICES Journal of Marine Science 57:1303–1309.

Grassle, J., and H. Sanders. 1973. Life histories and the role of disturbance. Deep-Sea Research 20:643–659.

Gray, J., M. Aschan, M. Carr, K. Clarke, R. Green, T. Pearson, R. Rosenberg, and R. M. Warwick. 1988. Analysis of community attributes of the benthic macrofauna of Frierfjord/Langesundfjord and in a mesocosm experiment. Marine Ecology Progress Series 46:151–165.

Hall, J., D. Raffaelli, and S. F. Thrush. 1994. Patchiness and disturbance in shallow water benthic assemblages. Pages 333–376 in P. Giller, A. Hildrew, and D. Raffaelli, editors. Aquatic ecology: scale, pattern and process. Blackwell Scientific Publications, Oxford, UK.

Hall-Spencer, J., C. Froglia, R. Atkinson, and P. Moore. 1999. The impact of rapido trawling for scallops, *Pecten jacobaeus* (L.), on the benthos of the Gulf of Venice. ICES Journal of Marine Science 56:111–124.

Hill, A., A. Brand, U. Wilson, L. Veale, and S. Hawkins. 1996. Estimation of by-catch composition and the numbers of by-catch animals killed annually on Manx scallop fishing grounds. Pages 111–115 in S. Greenstreet and M. Tasker, editors. Aquatic predators and their prey. Fishing News Books, Oxford, UK.

Hutchings, P. 1990. Review of the effects of trawling on macrobenthic epifaunal communities. Australian Journal of Marine and Freshwater Research 41:111–120.

Jennings, S., and M. Kaiser. 1998. The effects of fishing on marine ecosystems. Advances in Marine Biology 34.

Jennings, S., J. Pinnegar, N. Polunin, and K. Warr. 2001. Impacts of trawling disturbance on the trophic structure of benthic invertebrate communities. Marine Ecology Progress Series 213:127–142.

Kaiser, M. J., D. B. Edwards, P. Armstrong, K. Radford, N. Lough, R. Flatt, and H. Jones. 1998. Changes in megafaunal benthic communities in different habitats after trawling disturbance. ICES Journal of Marine Science 55:353–361.

Kaiser, M. J., K. Ramsay, C. Richardson, F. Spence, and A. Brand. 2000. Chronic fishing disturbance has changed shelf sea benthic community structure. Journal of Animal Ecology 69:494–503.

Kaiser, M. J., and B. E. Spencer. 1996. Behavioural responses of scavengers to beam trawl disturbance. Pages 116–123 in S. Greenstreet and M. Tasker, editors. Aquatic predators and their prey. Fishing News Books, Oxford, UK.

Lindley, J., J. Gamble, and H. Hunt. 1995. A change in the zooplankton of the central North Sea (55° to 58°N): a possible consequence of changes in the benthos. Marine Ecology Progress Series 119:299–303.

MAFF (Ministry of Agriculture, Fisheries and Food). 1995. United Kingdom sea fisheries statistics 1995. Fisheries Statistics Unit, Ministry of Agriculture, Fisheries and Food, London.

MAFF (Ministry of Agriculture, Fisheries and Food). 1996. United Kingdom sea fisheries statistics 1996. Fisheries Statistics Unit, Ministry of Agriculture, Fisheries and Food, London.

MAFF (Ministry of Agriculture, Fisheries and Food). 1997. United Kingdom sea fisheries statistics 1997. Ministry of Agriculture, Fisheries and Food, London.

MAFF (Ministry of Agriculture, Fisheries and Food). 1998. United Kingdom sea fisheries statistics 1998. Ministry of Agriculture, Fisheries and Food, London.

OSPAR Commission. 1998. Convention for the protection

of the marine environment of the north east Atlantic. OSPAR Commission, London.

OSPAR Commission. 2000. Quality status report 2000, region II—Greater North Sea. OSPAR Commission, London.

Petersen, C., and P. Boysen-Jensen. 1911. Valuation of the sea: animal life at the sea bottom, its food and quantity. Report of the Danish Biological Station 10:1–76.

Pilskaln, C., J. Churchill, and L. Mayer. 1998. Resuspension of sediment by bottom trawling in the Gulf of Maine and potential geochemical consequences. Conservation Biology 12:1223–1229.

Ramsay, K., M. Kaiser, and R. Hughes. 1998. Responses of benthic scavengers to fishing disturbance by towed gears in different habitats. Journal of Experimental Marine Biology and Ecology 224:73–89.

Ramsay, K., M. J. Kaiser, and R. Hughes. 1996. Changes in hermit crab feeding patterns in response to trawling disturbance. Marine Ecology Progress Series 144:63–72.

Reise, K. 2002. Sediment mediated species interactions in coastal waters. Journal of Sea Research 48:127–141.

Riemann, B., and E. Hoffmann. 1991. Ecological consequences of dredging and bottom trawling in the Limfjord, Denmark. Marine Ecology Progress Series 69:171–178.

Robson, C. 1995. Human activities, chapter 9. Coasts and seas of the United Kingdom. Region 5, North east England: Berwick-upon-Tweed to Filey Bay. Pages 151–158 in J. Barne, C. Robson, S. Kaznowska, J. Doody, and N. Davidson, editors. Joint Nature Conservation Committee, Peterborough, UK.

Rumohr, H., and T. Kujawski. 2000. The impact of trawl fishery on the epifauna of the southern North Sea. ICES Journal of Marine Science 57:1389–1394.

Secretariat of the Convention on Biological Diversity. 1992. The convention on biological diversity. United Nations Educational, Scientific and Cultural Organization, Paris.

Snelgrove, P., T. Blackburn, P. Hutchings, D. Alongi, J. Grassle, H. Hummel, G. King, I. Koike, P. Lambshead, N. Ramsing, and V. Solis-Weiss. 1997. The importance of marine sediment biodiversity in ecosystem processes. Ambio 26:578–583.

Statzner, B., B. Bis, S. Doledec, and P. Usseglio-Polatera. 2001. Perspectives for biomonitoring at large spatial scales: a unified measure for the functional composition of invertebrate communities in European running waters. Basic and Applied Ecology 2:73–85.

Thrush, S. F., and P. K. Dayton. 2002. Disturbance to marine benthic habitats by trawling and dredging: implications for marine biodiversity. Annual Review of Ecology and Systematics 33:449–473.

Townsend, C., and A. Hildrew. 1994. Species traits in relation to a habitat templet for river systems. Freshwater Biology 31:265–275.

Turner, S., S. F. Thrush, J. E. Hewitt, V. J. Cummings, and G. Funnell. 1999. Fishing impacts and the degradation or loss of habitat structure. Fisheries Management and Ecology 6:401–420.

Warwick, R. M., and K. Clarke. 1996. Relationships between body size, species abundance and diversity in marine benthic assemblages: facts or artefacts? Journal of Experimental Marine Biology and Ecology 202:63–71.

Wassenberg, T., G. Dews, and S. Cook. 2002. The impact of fish trawls on megabenthos (sponges) on the northwest shelf of Australia. Fisheries Research 58:141–151.

Watling, L., and E. Norse. 1998. Disturbance of the seabed by mibile fishing gear: a comparison to forest clearcutting. Conservation Biology 12:1180–1197.

The Impact of Trawling on Benthic Nutrient Dynamics in the North Sea: Implications of Laboratory Experiments

PHIL PERCIVAL[1] AND CHRIS FRID[2]

*Dove Marine Laboratory, School of Marine Science and Technology,
University of Newcastle, Cullercoats, North Shields, NE30 4PZ, UK*

ROB UPSTILL-GODDARD[3]

*Ridley Building, School of Marine Science and Technology,
University of Newcastle, Claremont Road, Newcastle upon Tyne, NE1 7RU, UK*

Abstract. The effects of trawling on North Sea sediment nutrient dynamics were quantified in microcosm experiments. North Sea sediments were collected from a fishing ground and placed into a series of nine microcosm systems. Following a stabilization period, simulated trawling events employed at moderate (trawled every other day) and heavy (trawled every day) trawl frequencies were compared to untrawled control systems. Nutrient concentrations (nitrite, nitrate, ammonium, and phosphate) were measured periodically over 98 h and resultant fluxes calculated. Trawl impact effected all the measured nutrient species. Flux rate changes following a trawl event persisted between trawl events (i.e., >48 h), and instantaneous fluxes exhibited the same basic pattern irrespective of the history of trawling. This implies that in regularly fished areas of the North Sea, nutrient dynamics may be in a permanently altered state, with enhanced fluxes of ammonium and phosphate. It is concluded that the impact of trawling and subsequent alteration to the benthic nutrient dynamic is potentially a globally significant impact on coastal nutrient dynamics and primary production.

Introduction

Offshore sediments can be highly productive due to the high levels of organic matter present (Hensen et al. 1998). The subsequent release of regenerated nutrients via molecular diffusion or biota-stimulated fluxes contributes to the productivity of the overlying waters (Hines et al. 1982). If such areas are intensively trawled, there is also the potential for considerable modification of the sediment–water column nutrient dynamic, which could have an impact on local or regional primary production (Pilskaln et al. 1998). The physical impact of trawl gear creates an immediate release of sediment pore water to surface waters (Pilskaln et al. 1998). It is well documented that nutrient concentrations within pore waters are greatly enhanced compared to adjacent overlying water (Val Klump and Martens 1981; Lavery et al. 2001). The second direct effect caused by the physical impact of trawling causes mixing of organic matter and oxygen-rich surface waters with newly exposed anoxic sediments (Mayer et al. 1991). This is the result of mobilization of oxic surface sediments through resuspension and removal by horizontal transport (Churchill 1989). These changes to the redox status of the sediment, while only temporary, could further enhance subsequent transformations of organic material and the rate of sediment–water interface exchange of inorganic nutrient fluxes.

The immediate release of nutrients from the interstitial pore water reservoir effectively makes the nitrogenous fraction more readily available for photosynthesis, given access to the photic zone (i.e., in nonstratified regions). Denitrifying bacteria associated with deeper hypoxic sediments convert remineralized forms of nitrogen into N_2 gas, which is then biologically unavailable for photosynthesis (Seitzinger 1988). However, temporarily following a trawl event, this process can no longer occur, as the pore waters are liberated into the oxygen rich water column. Studies have shown that trawl resuspension of pore water inorganic nutrients in areas of high trawl activity can contribute approximately half the nutrient requirements for photosynthesis (Pilskaln et al. 1998). How-

---

[1] Corresponding author: philip.percival@ncl.ac.uk
[2] E-mail: C.L.J.Frid@ncl.ac.uk
[3] E-mail: rob.goddard@ncl.ac.uk

ever, trawl activity in temperate regions can coincide with periods of high productivity blooms, and, as a consequence, in situ quantification and differentiation between new and regenerated nutrient sources may be difficult. Therefore, a method is needed where surface and benthic nutrient sources can be disentangled. Field measurements on high trawl impacted areas also raise important questions concerning how in situ devices can cover a sufficient temporal scale without being exposed to damage by the fishing gear. Therefore, determination of the regenerated benthic fraction of nutrient input to the water column could be better examined in laboratory microcosms where strict environmental controls and manipulations could isolate trawl-stimulated nutrient release from sediments.

In this paper, we examine the contribution by trawling to benthic nutrient dynamics in the central western North Sea. Simulated trawling events in microcosms allowed the determination of nutrient concentrations and fluxes from sediments while isolating the system from new sources of surface productivity that could prove difficult to entangle from regenerated sources in situ. The combination of control (without trawling), moderate (trawled every other day), and high trawl effort (trawled every day) regimes allowed the response of the system to varying intensities of trawling to be elucidated.

Methods

Sampling Site and Set-Up Procedure

A frequently trawled area (~100 m²) of the central, western North Sea was identified using fishing data provided by the UK Department for Environment, Food, and Rural Affairs centered on 55°13.55′N 01°27.28′W (Figure 1). Nine sediment samples (0.1 m²) were collected with a van Veen grab and transferred, with minimal disturbance to the sediment, into a series of microcosms (30 cm deep, 37.5 cm × 21 cm). Each grab was sufficiently deep to fill each microcosm to the required depth. These were then maintained under the research vessel's continuous water flow system during transportation to the laboratory. The total elapsed time from collection to laboratory establishment never exceeded 3 h. Sediments were collected from a known trawled area to limit possible alterations in nutrient release due to altered bioturbation rates from untrawled fauna.

On arrival at the laboratory, the microcosms were placed under the laboratory's regulated continuous flow of filtered seawater such that it overlaid each sediment system to approximately a 12-cm-depth water column. The seawater for the system was drawn from a coastal inlet pipe, gravimetrically settled and sand

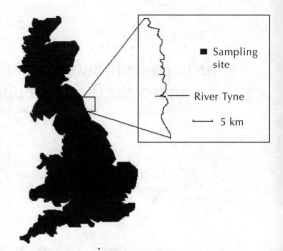

Figure 1. Location of the sampling site (box) in the central western North Sea.

filtered to remove detritus. The flow rate was maintained at a rate of 45 L/h and allowed to run to waste. Water circulation and aeration rates were controlled using a 45 degree splash plate positioned 4 cm above the water surface on to which the continuous water flow was directed. This allowed oxygen to be introduced into the system while the flow was maintained at a rate below that of visible resuspension of the sediment.

Temperature was controlled throughout the stabilization and experimental periods through use of the temperature-regulated laboratory aquarium at the ambient temperature of the bottom water at the collection site (8.5°C ± 0.5°C). The light regime was the ambient levels in the laboratory aquaria suite.

All microcosms were allowed a 10-d period for the chemical and biological processes to stabilize prior to experimention. This time period was determined to allow any redox microenvironments which may have been created during the set-up period to dissipate before excessive consumption of the sediments organic matter content could occur.

Experimental Treatments and Sampling Procedure

Three replicate microcosms were established for each of the three treatments. These were: controls (no trawl simulation), moderate trawl effort (trawled every other day), and high trawl effort (trawled every day). Trawl frequencies were selected to represent typical behavior of fishermen within International Council for the Exploration of the Sea (ICES) statistical rectangle 39E8. Popular and productive areas are often targeted for periods of up to 4 d. Within this time period, the area may be visited daily until the fisher moves to a new area (T. Catchpole, University of Newcastle Upon Tyne, personal communication).

In order to simulate a trawl event, a section of ground gear comprising a tickler chain was dragged through the sediment. The chain was dragged through the sediment at a depth of about 4 cm in a single pass at approximately 5.5 km/h. The average penetration depth of the ground gear was 4 cm and extended a maximum of 6 cm into the muddy sediment.

Prior to the start of each experiment, three 50-mL replicate samples were taken from the continuous flow water source to rule out any possible contamination (i.e., the buildup of organic matter) within the system. One h prior to experimentation, the continuous water flow system was shut off for the duration of the experiment. Following analysis, any differences in nutrient concentrations between the water source and the initial samples of the three treatments were assessed using a one-way analysis of variance (ANOVA). There were no significant differences between nutrient concentrations in the inflow and any of the treatments. Consequently, we were confident that initial nutrient concentrations were not influenced by aquarium conditions and that subsequent results would be due to the experimental treatments.

Sampling commenced 1 h prior to initial trawl simulations. A sample was then taken immediately following each trawl simulation and then hourly for the next 3 h. After 3 h, samples were collected at four hourly periods until the next trawl simulation. On each sampling occasion, 40 mL of overlying water was carefully withdrawn from the center of each microcosm's water column using clean, 50-mL plastic syringes and filtered through 0.45-μm filters into gas-tight polypropylene bottles. Samples were immediately frozen for subsequent nutrient analysis. Following analysis, nutrient fluxes were estimated. Control flux rates were calculated using the slope of the plotted concentration values and subtracting the t_0 value from the concentration at each time point per unit area per unit volume. Instantaneous fluxes were calculated from the concentration preceding a trawl event. Fluxes between trawl events were calculated using the slope of those values between trawl impacts and subtracting the concentration following a trawl event from the concentration preceding the next trawl event or end of the experiment.

Nutrient Analysis

Concentrations of nitrite (NO_2^-), nitrate (NO_3^-), ammonium (NH_4^+), and phosphate (PO_4^{3-}) were determined with an automated nutrient analyzer (San Plus; Skalar, The Netherlands) and following standard protocols (Brewer and Riley 1965; Mantoura and Woodard 1983; Kirkwood 1989).

Sediment Characterization

At the end of each experiment, three replicate core samples (7-cm diameter, 15-cm depth) were collected from each microcosm and analyzed for particle size, percentage organic matter, and porosity. Sediment particle size was determined using a standard gravity sedimentation technique (Buchanan 1984). To account for the possible presence of mineral carbon (coal), organic matter content was determined by wet digestion (Buchanan 1984). Porosity was determined by weight loss on drying (80°C).

Results

Sediment Characteristics

Sediments collected from each microcosm did not vary significantly in porosity ($W = 6$, $P = 0.081$), grain size (mean particle size = 3.5 phi for all systems), or percentage organic content ($W = 6$, $P = 0.081$) among systems.

Microcosm Nutrient Concentrations and Fluxes

All initial NO_2^- concentrations were within a range of 0.2–0.3 μmol/L and did not differ significantly between treatments or from the in-flowing seawater source (NO_2^- ANOVA, $F = 0.24$, $P = 0.867$). The observed NO_2^- concentration in the control systems increased steadily throughout the experiment 0.2 ± 0.02 to 0.6 ± 0.1 μmol/L (Figure 2A), maintaining a flux rate of 8.7 ± 0.9 μmol · m^{-2} · d^{-1} (Table 1). The first trawl event of both the moderately and heavily impacted systems gave an immediate concentration increase of 1.5 and 1.0 μmol, respectively (Figure 2B, C). The corresponding instantaneous fluxes were 166.2 ± 54.5 and 110.5 ± 7.5 μmol · m^{-2} · h^{-1} (Table 1). Conversely, the NO_2^- concentration decreased at the second trawl impact. Notably, however, this pattern was evident again for both the moderate and high trawl effort systems at 0.5 ± 0.2 to 0.3 ± 0.1 and 0.5 ± 0.1 to 0.4 ± 0.1 μmol/L, creating negative instant fluxes of -17.7 ± 6.8 and -7.3 ± 1.1 μmol · m^{-2} · h^{-1} (Table 1). Nonetheless, in the moderately trawled system, both post-disturbance flux rates displayed similar values of 22.2 ± 1.1 and 21.2 ± 1.4 μmol · m^{-2} · d^{-1}, respectively. The NO_2^- concentration in the heavily trawled system displayed a decline in the two subsequent trawl simulations (0.4 ± 0.02 to 0.2 ± 0.1 and 0.3 ± 0.04 to 0.2 ± 0.04 μmol/L; Figure 2C), giving influx values of -18.9 ± 0.5 and -12.0 ± 1.3 μmol · m^{-2} · h^{-1} (Table 1). However, while the

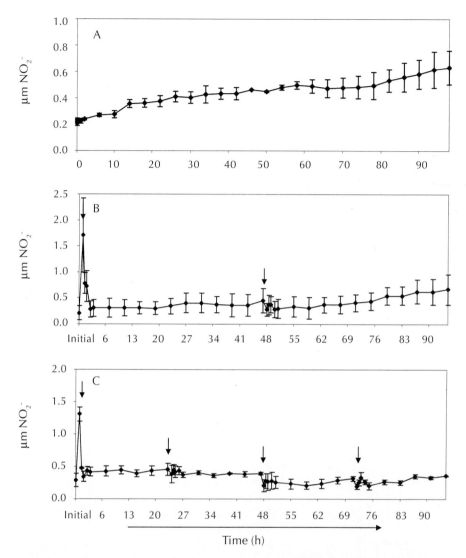

Figure 2. Concentration of nitrite (NO_2^-, μmol/L) within microcosms exposed to different trawling regimes ± SD: control (without trawling, A), moderate trawl effort (trawled every other day, B), and heavy trawl effort (trawled every day, C). Arrows denote disturbance events.

latter three trawl events stimulated a nutrient influx, the flux rates between each trawl event and to the end of the experiment all remained positive, with successive values of 2.8 ± 1.2, 1.0 ± 2.1, 8.1 ± 1.3, and 18.2 ± 1.4 μmol · m^{-2} · d^{-1} (Table 1).

Initial NO_3^- concentrations for all systems were not significantly different (range = 1.8–2.2; ANOVA, F = 2.49, $P = 0.134$). The control microcosm exhibited a net gain in concentration, increasing to 10.2 ± 1.5 μmol/L (Figure 3A) at the end of the experiment. The resultant flux was 176.4 ± 18.8 μmol · m^{-2} · d^{-1}. Within the moderately disturbed systems, the NO_3^- concentration decreased immediately following both disturbance events (4.5 ± 1.3 to 2.4 ± 1.1 and 1.6 ± 0.8 to 1.2 ± 0.6 μmol/L, respectively; Figure 3B) and, consequently, created negative instantaneous flux values of –235.0 ± 29.0 and –46.7 ± 15.7 μmol · m^{-2} · h^{-1} (Table 1). The flux rate following the first trawl event maintained a negative flux of –70.8 ± 9.3 μmol · m^{-2} · d^{-1} to the second trawl event, whereas the flux after the second trawl event remained positive to the end of the experiment 13.6 ± 6.1 μmol · m^{-2} · d^{-1}. The heavily trawled system exhibited a similar pattern of decreasing NO_3^- concentration at the first and second trawl simulations (2.2 ± 0.1 to 1.7 ± 0.6 and 0.9 ± 0.2 to 0.8 ± 0.3 μmol/L; Figure 3C). However, while the magnitude of the instantaneous flux values declined after the second

Table 1. Summary of nitrite (NO_2^-), nitrate (NO_3^-), phosphate (PO_4^{3-}), and ammonium (NH_4^+) concentrations (μmol/L) within experimental microcosms for control, moderate, and heavy trawl effort ± SD (in parentheses). Calculated instant and post-disturbance flux vales presented in $\mu mol \cdot m^{-2} \cdot h^{-1}$ and $\mu mol \cdot m^{-2} \cdot d^{-1}$, respectively. Nutr refers to the specific nutrient species. N/A denotes a non-applicable result. Min = minimum; Max = maximum.

Treatment	Nutr	Nutrient concentration					Instant flux at time of trawl event				Post-disturbance flux rate			
		Initial	Final	Min	Max	Control	1	2	3	4	1	2	3	4
Without trawling	NO_2^-	0.2 (0.02)	0.6 (0.1)	0.2 (0.02)	0.6 (0.1)	8.7 (0.9)	N/A	N/A	N/A	N/A	N/A	N/A	N/A	N/A
Moderate trawl effort	NO_2^-	0.2 (0.2)	0.7 (0.3)	0.2 (0.2)	1.7 (0.7)	N/A	166.2 (54.5)	−17.7 (6.8)	N/A	N/A	22.2 (1.1)	21.2 (1.4)	N/A	N/A
Heavy trawl effort	NO_2^-	0.3 (0.1)	0.4 (0.01)	0.2 (0.04)	1.3 (0.1)	N/A	110.5 (7.5)	−7.3 (1.1)	−18.9 (0.5)	−12.0 (1.3)	2.8 (1.2)	1.0 (2.1)	8.1 (1.3)	18.2 (1.4)
Without trawling	NO_3^-	1.8 (0.7)	10.2 (1.5)	1.8 (0.7)	10.2 (1.5)	176.4 (18.8)	N/A	N/A	N/A	N/A	N/A	N/A	N/A	N/A
Moderate trawl effort	NO_3^-	4.5 (1.3)	1.0 (0.5)	0.7 (0.8)	4.5 (1.3)	N/A	−235.0 (29.0)	−46.7 (15.7)	N/A	N/A	−70.8 (9.3)	13.6 (6.1)	N/A	N/A
Heavy trawl effort	NO_3^-	2.2 (0.1)	0.2 (0.05)	0.2 (0.06)	2.4 (0.3)	N/A	−53.0 (37.9)	−5.2 (6.6)	34.9 (10.8)	33.3 (8.6)	−80.0 (6.3)	−29.0 (0.9)	1.4 (1.0)	0.5 (9.2)
Without trawling	PO_4^{3-}	2.4 (0.3)	1.0 (0.2)	0.8 (0.2)	2.8 (0.3)	−49.2 (10.3)	N/A	N/A	N/A	N/A	N/A	N/A	N/A	N/A
Moderate trawl effort	PO_4^{3-}	2.6 (0.3)	0.6 (0.2)	0.3 (0.02)	2.6 (0.3)	N/A	−109.2 (19.2)	139.7 (33.4)	N/A	N/A	−96.7 (7.0)	−41.7 (5.4)	N/A	N/A
Heavy trawl effort	PO_4^{3-}	2.3 (0.2)	0.6 (0.1)	0.5 (0.02)	3.9 (0.4)	N/A	−65.2 (14.4)	163.3 (27.9)	200.3 (43.7)	307.7 (51.2)	−104.1 (8.0)	−147.6 (24.2)	−18.7 (6.2)	−2.3 (1.0)
Without trawling	NH_4^+	1.1 (0.3)	2.5 (0.7)	1.0 (0.1)	2.5 (0.7)	21.0 (4.5)	N/A	N/A	N/A	N/A	N/A	N/A	N/A	N/A
Moderate trawl effort	NH_4^+	1.6 (0.3)	16.3 (1.7)	1.6 (0.3)	17.3 (1.9)	N/A	438.9 (27.4)	739.7 (165.5)	N/A	N/A	192.0 (14.0)	84.6 (36.3)	N/A	N/A
Heavy trawl effort	NH_4^+	1.8 (0.3)	31.6 (1.6)	1.8 (0.3)	39.0 (2.4)	N/A	358.3 (19.1)	1,136.3 (145.3)	975.8 (81.8)	1,586.6 (140.9)	270.2 (29.8)	182.6 (25.7)	40.3 (26.3)	9.3 (11.6)

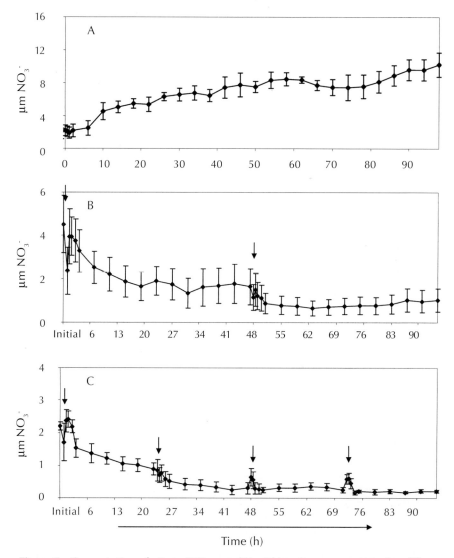

Figure 3. Concentration of nitrate (NO_3^-, μmol/L) within microcosms exposed to different trawling regimes ± SD: control (without trawling, A), moderate trawl effort (trawled every other day, B), and heavy trawl effort (trawled every day, C). Arrows denote disturbance events.

disturbance, both fluxes were again negative (-53.0 ± 37.9 and -5.2 ± 6.6 μmol · m^{-2} · h^{-1}; Table 1). Interestingly, both the third and forth trawl events elevated the NO_3^- concentration by 0.3 mmol. The instantaneous flux rates, therefore, also displayed positive fluxes of 34.9 ± 10.8 and 33.3 ± 8.6 μmol · m^{-2} · h^{-1}, respectively. Again, the magnitude of the subsequent flux rates following the later two trawls declined at 1.4 ± 1.0 and 0.5 ± 1.2 μmol · m^{-2} · d^{-1} (Table 1).

PO_4^{3-} concentrations at the start of the experiment were not significantly different from the seawater background value or between treatments (ANOVA, $F = 0.36$, $P = 0.784$). The control system displayed a constant rate of concentration decline from 2.4 ± 0.3 to 1.0 ± 0.2 μmol/L (Figure 4A) with a flux rate of -49.2 ± 10.3 μmol · m^{-2} · d^{-1}. The moderately trawled system exhibited a decrease in PO_4^{3-} concentration at the first trawl simulation of 2.6 ± 0.3 to 1.6 ± 0.1 μmol/L (Figure 4B). This resulted in a negative flux of -109.2 ± 19.2 μmol · m^{-2} · h^{-1} (Table 1). At the second disturbance, however, the PO_4^{3-} concentration immediately increased by 1.3 μmol/L. The instant flux stimulated by this second trawl event was 139.7 ± 33.4 μmol · m^{-2} · h^{-1}. The flux rates during the periods preceding and following the second

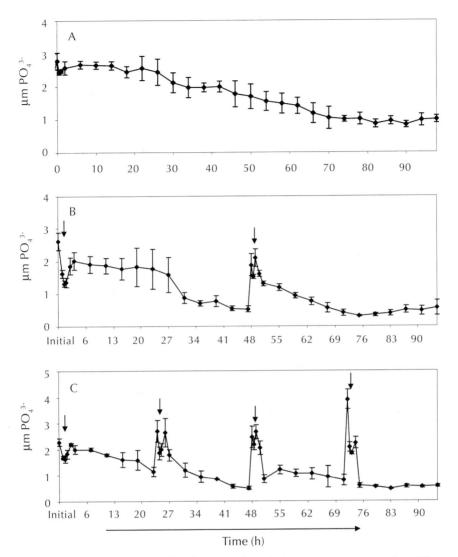

Figure 4. Concentration of phosphate (μmol/L) within microcosms exposed to different trawling regimes ± SD: control (without trawling, A), moderate trawl effort (trawled every other day, B), and heavy trawl effort (trawled every day, C). Arrows denote disturbance events.

trawl event were both negative at −96.7 ± 7.0 and −41.7 ± 5.4 μmol · m^{-2} · d^{-1}, respectively (Table 1). The heavily trawled system also exhibited a decline in concentration at the first trawl event of 0.6 μmol (Figure 4C). The following three trawl events stimulated increasingly elevated PO$_4^{3-}$ concentrations of 1.6, 2.0, and 3.1 μmol within the microcosms (Figure 4C). The corresponding instantaneous fluxes associated with each of these were −65.2 ± 14.4, 163.3 ± 27.9, 200.3 ± 43.7, and 307.7 ± 51.2 μmol · m^{-2} · h^{-1} (Table 1). From 3 h after a trawl event to the next trawl event or end of the experiment, all flux rates were negative, with values of − 104.1 ± 8.0, −147.6 ± 24.2, −18.7 ± 6.2, and −2.3 ± 1.0 μmol · m^{-2} · d^{-1}, respectively (Table 1).

Initial NH$_4^+$ concentrations within each microcosm ranged from 1.1 to 1.8 μmol/L and were not significantly different (ANOVA, $F = 1.17$, $P = 0.379$). The NH$_4^+$ concentrations within the nontrawled controls steadily increased over the duration of the experiment from 1.1 ± 0.3 to 2.5 ± 0.7 μmol/L (Figure 5A), giving a resultant flux rate of 21.0 ± 4.5 μmol · m^{-2} · d^{-1} (Table 1). Within those microcosms exposed to a moderate level of simulated trawling, the NH$_4^+$ concentration rose steeply following both trawl events from 1.6 ± 0.3 to 5.6 ± 0.5

μmol/L at time 0 and 9.9 ± 1.2 to 17.0 ± 1.0 μmol/L at time 48 (Figure 5B). The two trawl simulations produced fluxes of 438.9 ± 27.4 and 739.7 ± 165.5 μmol · m^{-2} · h^{-1}. Following a brief period of stabilization after each trawl event, the subsequent flux rates, while maintaining values above that displayed in the control system, declined at 192 ± 14.0 and 84.6 ± 36.3 μmol · m^{-2} · d^{-1}, respectively (Table 1). A similar amplitude of increase in NH_4^+ concentration was also displayed by the first two trawl events in the heavily trawled system (trawl event one: from 1.8 ± 0.3 to 5.1 ± 0.02 μmol/L; trawl event two: 7.3 ± 0.6 to 18.1 ± 0.3 μmol/L; Figure 5C). The third and forth simulated trawl impacts also increased the NH_4^+ concentration by 9.5 and 15.9 μmol, respectively, and stimulated large efflux rates of 975.8 ± 81.8 and 1,586.6 ± 140.9 μmol · m^{-2} · h^{-1}. Yet, the magnitude of each flux preceding a trawl simulation, while maintaining a positive flux, decreased, with values of 270.2 ± 29.8, 182.6 ± 25.7, 40.3 ± 26.3, and 9.3 ± 11.6 μmol · m^{-2} · d^{-1} (Table 1).

Discussion

It has often been documented that benthic nutrient cycling is controlled by the combination of bacterial decomposition of organic matter, creating steep concentration gradients, and biota-stimulated activity (Van Raaphorst et al. 1990; Widdicombe and Austen 1998), the latter acting as a catalyst to enhance sediment–water exchange through reworking of the sediment. Obviously, other factors such as storms, tidal currents, and anthropogenic factors play their part in periodically disturbing the seabed and temporarily altering local benthic dynamics. Yet, trawl activity may well represent the most intense mechanism of benthic disturbance within the North Sea. Mobile fishing gear has been quantified as the most intense factor to impact the benthic system in areas displaying a high trawl effort (Watling and Norse 1998). The results of this study clearly show that when North Sea sediments were exposed to a simulated trawl event, all the nutrient fluxes examined were impacted upon resuspension.

Following a trawl event, we have shown that an instantaneous release and temporary enhancement of nutrients within the overlying water occurs. The magnitude of this pulsed release of nutrients is also likely to be greater in situ where chain bobbins, trawl shoes, and doors penetrate deeper into the sediment, releasing interstitial pore waters from different, deeper, remineralization layers. Once released, this pulse of nutrients will be available for subsequent bacterial transformations, making the remineralized fraction available to be utilized for primary production. Blackburn (1997) predicted that the nutrient contribution to primary producers from sediment resuspension would be insignificant, as only the timing of nutrient release would be affected by sediment resuspension. Blackburn (1997) concluded that a resuspension depth of at least 2.4 cm would be required to make a significant transfer to the overlying water. Ground gear of benthic trawlers is known to penetrate to depths up to 6 cm (Duplisea et al. 2001), while trawl boards have been recorded to penetrate up to 30 cm (Krost 1990; Jones 1992). With an average penetration depth of 4 cm for the present study, we clearly demonstrated a significant increase in the quantity of some nutrients liberated into the water column. Studies have also stated that the magnitude of nutrient release is ultimately dependent on the amount and rate of organic degradation (Val Klump and Martens 1987). This is unquestionably a major factor, yet the rate of degradation can be controlled by oxygen exposure. Trawl gear penetration into the sediment extended beyond Blackburn's (1997) 2.4-cm depth required to have a significant effect, but more importantly, penetrated into what were hypoxic sediments. As a result, it is likely that nitrification and denitrification processes are stimulated, and, consequently, remineralization pathways are altered. Thus, trawl-impacted sediments can potentially contribute to the magnitude of nutrients transported to the overlying waters and not only alter the timing of release.

Once a trawl event occurred, the subsequent flux rate was altered compared to the nontrawled control sediments. Not only did trawl disturbance lead to the efflux of nutrients from the sediment, but the resuspension also acted to scavenge some nutrients from the water column. This influx was apparent for phosphate between trawl impacts and nitrate after a disturbance. Conversely, fluxes of nitrite and ammonium effluxed from the sediment. Ammonium displayed the biggest release from the sediment of all examined nutrients. This may hold particular significance to the planktonic ecosystem, as ammonium is the most biologically available nutrient utilized by phytoplankton (L'Helguen et al. 1996). We have shown phosphate influx to occur in undisturbed control sediments and between trawl events. However, an instantaneous efflux of phosphate did occur during trawl simulations. Therefore, trawl activity is effectively reversing the direction of flux across the sediment–water interface. As phosphate is also readily utilized by plankton (Karlson 1989), trawl-induced phosphate fluxes could contribute to productivity in overlying waters. Consequently, this shift could act to modify local nutrient ratios. Such shifts in local nutrient ratios can lead to altered abundance or composition of phytoplankton. These changes in nutrient dynamics inevitably hold consequences for the ecosystem and can stimulate the production of toxic phytoplankton populations (Pilskaln et al. 1998).

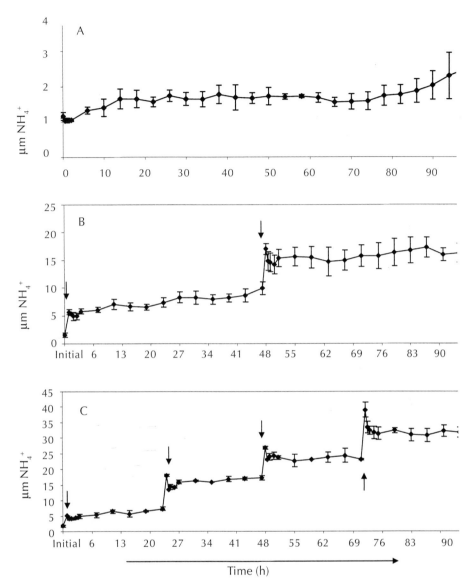

Figure 5. Concentration of ammonium (µmol/L) within microcosms exposed to different trawling regimes ± SD: control (without trawling, A), moderate trawl effort (trawled every other day, B), and heavy trawl effort (trawled every day, C). Arrows denote disturbance events.

Any flux rate change, regardless of direction across the sediment–water interface, persisted until the next trawl impact or to the end of the experiment (i.e., >48 h). Notably, the magnitude of the altered flux rate was similar between moderate and high trawl frequency systems. In other words, the change that occurred to the benthic nutrient dynamic following a trawl impact was maintained even if the time to the next impact was extended, as was the case in the moderately trawled system. However, the magnitude of successive fluxes declined. The trawl-induced instantaneous flux and change in nutrient concentration also exhibited the same basic pattern between corresponding impacts, irrespective of the trawl frequency. As successive trawl simulations occurred, the response in nutrient concentration and flux displayed a stepwise pattern.

We have demonstrated that the magnitude and timing of benthic fluxes are altered due to trawl activity. Therefore, potential impacts are likely to occur and be greatest during periods of high fishing activity. Within the North Sea, and specifically ICES rectangle 39E8, trawling increases between November and April. Consequently, trawl-induced nutrients are being loaded into the water column during winter. As a result, regenerated sources of nutrients could contribute to the system. However, loading of remineralized nutrients can potentially exert positive and negative impacts on the system. Obviously, a greater supply of nutrients can lead to a greater standing stock of plankton, thus increasing productivity (George et al. 2001). Yet, additional nutrients may also act to increase or extend the productivity bloom beyond the rate at which it can be consumed. As a result, localized eutrophication or nuisance blooms may be the result of trawl-stimulated nutrients. Isolation of new and regenerated sources of nutrients is required to determine the extent to which, if any, this phenomenon occurs. When considering the contribution of regenerated nutrients based on the findings of this investigation, we find additional daily increases of 47 and 8 $\mu mol/m^3$ of NH_4^+ and PO_4^{3-}, respectively, during periods of high trawl intensity. Stratification within the North Sea only occurs in some years and is mainly confined to central and northern regions during summer months due its shallow mean depth (74 m; Otto et al. 1990). The development of brief summer stratification would likely have little effect on trawl-induced nutrients since trawl intensity is relatively low this time. Therefore, these additional nutrients are likely to be transported directly to the euphotic zone. Therefore, it is reasonable to assume that a significant proportion of trawl-induced nutrients within the North Sea directly contribute to primary production.

The experimental treatments of this study were selected to mimic the high trawl frequency and typical behavior of the trawl fleet within the North Sea. Fishers specifically target an area of the North Sea, and within this area, effort can be as high as daily. Although different trawlers may, and often do, impact the same area, a realistic time period for one trawler to target an area is approximately 4 d before moving on to another area (Catchpole, personal communication). The results of this study, therefore, indicate that during the fishing season, North Sea sediments may well be in a permanent state of perturbation.

For strict controls to be emplaced, laboratory microcosms were determined to be the best method to isolate nutrient effects from trawl impacts. However, in order to achieve this, a time period allowing the chemical and biological processes within the sediment to recover and stabilize was needed. In situ effects, therefore, hold the potential to stimulate greater effects to the benthic nutrient dynamic. This is due to the settlement of organic debris in the form of offal and discards following a trawl event. This process was not mimicked during the experiment.

This study has increased our knowledge of how the benthic nutrient dynamic is affected by trawl activity. Certain areas, however, still need attention. Methods need to be established that could take in situ measurements over a large temporal scale without being damaged by subsequent trawl activity. Specific measurements of the organic matter returned to the sediment from trawlers and isotopic labeling of the regenerated nutrient fraction to determine the proportion utilized by plankton are two elements that are needed to accurately quantify this potentially globally significant impact to nutrient dynamics and primary production.

Acknowledgments

We thank the Master and crew of the R.V. *Bernica* for assistance with sample collection. We also thank Tom Catchpole. The study was funded by a Natural Environment Research Council studentship to P.P.

References

Blackburn, T. H. 1997. Release of nitrogen compounds following resuspension of sediment: model predictions. Journal of Marine Systems 11:343–352.

Brewer, P. G., and J. P. Riley. 1965. The automated determination of nitrate in sea water (modified to increase sensitivity, by decreasing the ammonium chloride concentration). Deep Sea Research 12:765–772.

Buchanan, J. B. 1984. Sediment analysis. Pages 41–65 *in* A. Holme and A. D. McIntyre, editors. Methods for the study of marine benthos. Blackwell Scientific Publications, London.

Churchill, J. H. 1989. The effect of commercial trawling on sediment resuspension and transport over the Middle Atlantic Bight continental shelf. Continental Shelf Research 9:841–864.

Duplisea, D. E., S. Jennings., S. J. Malcolm., R. Parker, and D. B. Sivyer. 2001. Modelling potential impacts of bottom trawl fisheries on soft sediment biogeochemistry in the North Sea. Geochemical Transactions 14:1–6.

George, T., P. George., D. Costa, and A. Theodorou. 2001. Assessing marine ecosystem response to nutrient inputs. Marine Pollution Bulletin 43(7–12):175–186.

Hensen, C., H. Landenberger., M. Zabel., and H. D. Schultz. 1998. Quantification of diffusive benthic fluxes of nitrate, phosphate, and silicate in the southern Atlantic Ocean. Global Biogeochemical Cycles 12:193–210.

Hines, M. E., H. O. William., W. Berry Lyons, and G. E. Jones. 1982. Microbial activity and bioturbation-induced oscillations in pore water chemistry of estuarine sediments in spring. Nature (London) 299:433–435.

Jones, J. B. 1992. Environmental impact of trawling on the sea-bed: a review. New Zealand Journal of Marine and Freshwater Research 26:59–67.

Karlson, B. 1989. Seasonal phosphate uptake by size-fractioned plankton in the Skagerrak. Journal of Experimental Marine Biology and Ecology 127(2):141–154.

Kirkwood, D. S. 1989. Simultaneous determination of selected nutrients in sea water. ICES, C.M. 1989/C:29, Rome.

Krost, P. 1990. The impact of otter-trawl fishery on nutrient release from the sediment and macrofauna of Kieler Bucht (Western Baltic). Berichte aus dem Institut Für Meereskunde an der Christian-Albrechts-Universität, Kiel. (In German, English summary.)

Lavery, P. S., C. E. Oldham, and M. Ghisalberti. 2001. The use of Fick's first law for predicting porewater nutrient fluxes under diffusive conditions. Hydrological Processes 15:2435–2451.

L'Helguen, S., C. Madec, and P. Le Corre. 1996. Nitrogen uptake in permanently well-mixed temperature coastal waters. Estuarine, Coastal, and Shelf Science 42(6):803–818.

Mantoura, R. F. C., and E. M. S. Woodard. 1983. Optimization of the indolpheol blue method for the automated determination of ammonia in estuarine waters. Estuarine, Coastal, and Shelf Science 17:219–224.

Mayer, L. M., D. F. Schick, R. Findlay, and D. L. Rice. 1991. Effects of commercial dragging on sedimentary organic matter. Marine Environmental Research 31:249–261.

Otto, L., J. T. F. Zimmerman, G. K. Furnes, M. Mork, R. Saetre, and G. Becker. 1990. Review of the physical oceanography of the North Sea. Netherlands Journal of Sea Research 26 (2–4): 161–238.

Pilskaln, C. H., J. H. Churchill, and L. M. Mayer. 1998. Resuspension of sediment by bottom trawling in the Gulf of Maine and potential geochemical consequences. Conservation Biology 12:1223–1229.

Seitzinger, S. P. 1988. Dentirification in freshwater and coastal marine ecosystems: ecological and geochemical significance. Limnology and Oceanography 33:702–724.

Val Klump, J., and C. S. Martens. 1981. Biogeochemical cycling in an organic rich coastal marine basin—II. Nutrient sediment–water exchange processes. Geochimica et Cosmochimica Acta 45:101–121.

Val Klump, J., and C. S. Martens. 1987. Biogeochemical cycling in an organic-rich coastal marine basin. 5. Sedimentary nitrogen and phosphorous budgets based upon kinetic models, mass balances, and stoichiometry of nutrient regeneration. Geochimica et Cosmochimica Acta 51:1161–1173.

Van Raaphorst, W., H. T. Kloosterhuis, A. Cramer., and K. J. M. Bakker. 1990. Nutrient early diagenesis in the sandy sediments of the Dogger Bank area, North Sea: pore water results. Netherlands Journal of Sea research 26:25–52.

Watling, L., and E. A. Norse. 1998. Disturbance of the seabed by mobile fishing gear: a comparison to forest clearcutting. Conservation Biology 12:1180–1197.

Widdicombe, S., and M. C. Austen. 1998. Experimental evidence for the role of *Brissopsis lyrifera* (Forbes, 1841) as a critical species in the maintenance of benthic diversity and the modification of sediment chemistry. Journal of Experimental Marine Biology and Ecology 228:241–255.

Potential Impacts of Deep-Sea Trawling on the Benthic Ecosystem along the Northern European Continental Margin: A Review

JOHN D. GAGE[1] AND J. MURRAY ROBERTS

Scottish Association for Marine Science, Dunstaffnage Marine Laboratory, Oban, Scotland PA37 1QA UK

JOHN P. HARTLEY

Hartley Anderson Ltd, Blackstone, Dudwick, Ellon, Aberdeenshire, Scotland AB41 8ER UK

JOHN D. HUMPHERY

Proudman Oceanographic Laboratory, Bidston Obervatory, Birkenhead, England CH43 7RA UK

Abstract. Little is known of the sensitivities of deep seabed communities to anthropogenic disturbance. We collate and summarize the sparse data related to effects on the seabed of deep-sea trawling and provide a new analysis of potential impacts on the benthic community in this still largely pristine environment. We concentrate particularly on the northern European continental margin. Here, the upper continental slope is now being impacted by new demersal fisheries using aggressive trawling techniques. Direct effects are caused largely by the physical impact of the trawl, where emergent sessile epifauna, such as corals and massive sponges, are particularly vulnerable to damage. Damage to coldwater coral reefs in the northeastern Atlantic is irrecoverable at ecological time scales. Large, semi-buried glacial drop stones are uprooted in heavily trawled areas, impacting their rich covering of sessile epifauna. Seabed photographs suggest that trawl scour marks on soft sediment are widespread and persist longer than those in shallow sediments because of low rates of natural sedimentation. Loss of biogenic habitat complexity in the sediment and destruction or disruption of microhabitat provided by sessile epifaunal organisms caused by dragging trawls over infaunal communities disrupts sediment habitat structure. The scale of such physical impact, including possible smothering effects caused by sediment resuspended by the trawl, is unknown. Physical damage to sediment-rooted sessile epifauna, such as glass sponges, is inferred from seabed photographs, while removal of large motile epibenthic predators may alter community structure and ecosystem processes. Indirect effects on the deep seabed sediment community probably conform to disturbance effects seen in coastal benthos. High species diversity, rich in rare species, make it vulnerable to local-scale species loss. Effects of discards arriving as food falls to the bed are unknown and may cause a shift in the balance of ecological function through expansion of natural scavenger populations. We offer a unifying hypothesis from a deep-sea perspective. Coastal seas currently reflect stages in seabed community modification that predate rigorous scientific studies. Therefore, substantial habitat alteration and change in ecosystem function caused by loss of the most vulnerable, mainly sessile epibenthic species, probably occurred a long time ago. Deep-seabed areas off Europe are now starting to experience this early, probably most visible, stage from a previously pristine state.

Introduction

Seabed Impacts from Deepwater Fishing: The Background

With the depressing pattern of serial depletion as fishermen move from one overfished stock to start the same process with another, it is arguable whether coastal fisheries have ever been completely sustainable over the long term (Pauly et al. 2002). Marine science now enters a new phase in global-scale exploration and quantification. The prevailing paradigm that pollution, and now climate change, pose the most potent risks to marine biodiversity is being qualified by the realization that historical overfishing, probably starting in the shallow seas off Europe, has already greatly

---

[1] E-mail: John.Gage@sams.ac.uk

modified marine ecosystems over very long temporal scales (Jackson et al. 2001).

The European Union's Common Fisheries Policy (CFP) has sheltered fishermen in Europe from the brutal, if natural, economic outcome of overfishing by massive programs of subsidies and grant aid. With a huge overcapacity in fishing fleets, effort has shifted from overfished stocks to new, previously under-exploited, often smaller size-classes or species, or to new fisheries in deep water beyond the continental shelf edge.

In the deep sea (defined as depths below 200 m), powerful new boats equipped with the latest navigation and fish-finding aids are moving into a pristine wilderness. Living resources in the deep sea have never before been seriously exploited except by low-impact artisanal fishermen using longlines, usually from oceanic islands. However, the true ecological cost of seabed trawling, perhaps the most destructive of man's impacts on the sea, remains largely unknown in the area where deep-sea fisheries are well-developed, on the continental margin off northern Europe (Figure 1). Here, trawlers routinely fish along the continental margin, submerged banks, and seamounts right out to the mid-Atlantic Ridge using rock hopper gear to depths in excess of 1,500 m.

One of the attractions of these new fisheries is that, because of the very limited amount of knowledge available on the basic biology of deep-sea fish necessary to inform even rudimentary stock management, little or no regulatory framework is in place to control effort. With limited funding, science has struggled to catch up to provide basic data on age structure and reproductive strategies. Available data now indicate an alarming picture of highly fragile stocks with extremely low rates of replacement (Koslow et al. 2000; Gordon 2001). These new fisheries have provided typical examples of "boom-and-bust" exploitation, with effort switching to other species when a stock becomes depleted. For the deep-sea fishery off the British Isles, significant decreases in the catch per unit effort have already been reported by the International Council for the Exploration of the Sea study group for stock assessment (Lorance and Dupouy 2001). All indicators are that in the North Atlantic, the sustainability of virtually all targeted species is in doubt (Bergstad et al. 2002; Gordon 2002). In this management climate, it is not surprising that despite the massive cost of the CFP at the time of writing, no new research has been commissioned by European Union member states to investigate the likely collateral impacts on the seabed ecosystem from trawling in the deep sea. Yet the poorly known but highly species-rich habitats on the deep seabed represent an important store of global biodiversity. Following its Green Paper (European Commission 2001), the European Commission is now starting to implement deep-cutting changes necessary for the CFP to embrace sustainability and wider ecosystem responsibility (see European Commission 2002) but makes no specific mention of deep-sea fisheries. This omission is less surprising when it is remembered that even in coastal seas, the now very substantial research effort on trawling impacts in Europe still awaits translation into effective management action (Gray 2000).

Figure 1 is a map showing the location of the northern European marine areas mentioned subsequently in this contribution.

Aims and Classification of Deep Seabed Impacts

The purpose of this paper is to organize and review available data related to trawling impacts on the deep seabed and provide informed interpretation on likely impacts on the seabed community. This information is considered in relation to data from coastal seabed areas, where relevant to our understanding of deep-water impacts. We focus particularly on the deep continental margin (the Hebridean Margin and the eastern slope of the Faroe-Shetland Channel) off Scotland that provides a substantial part of the area of the new deep-water fisheries off Europe, but we do not exclude deep water elsewhere, where relevant. However, the Scottish margin is the area we know best, with its benthic ecology relatively well known compared to some other deep-sea areas.

The sparse information available is too limited for adequate comparison with the much better understood seabed impacts from trawling in shallow seas. Available data are scattered and incomplete and have to be considered against the very high uncertainties associated with knowledge of the biology and ecological function of organisms in this still remote and inaccessible ecosystem. Because of this, information reviewed below will seem sparse and interpretation tentative. With the single exception of a comparative study of trawling impacts on species richness on the upper continental slope off New Zealand (Cryer et al. 2002), the data known to us lack the rigor possible from good experimental and statistical design. Existing information has been obtained largely as bycatch from studies with other aims. Ironically, many of these aims involve environmental studies now required of any new exploration of new deep-water oil or gas resources along the continental margin. These activities off Scotland are now subject to a stringent regulatory framework aimed to protect the deepwater environment and established fishing interests (Gage 2001).

We follow the report of the National Research Council (2002) and organize these data in relation to a categorization of the impact as *direct* or *indirect*. Direct effects include effects of physical impact of the trawl on the

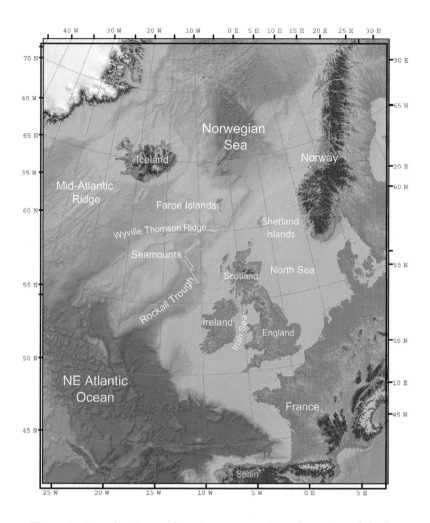

Figure 1. Map showing northern European continental margin and the location of areas mentioned in the text. The shades of blue indicate bathymetry, with the lightest shade denoting the continental shelf to 200 m depth. Depth increases rapidly along the adjacent continental margins, with the greatest depths occurring in the deep-sea basins which include the Rockall Trough. Off Scotland the Rockall Trough is punctuated by submarine banks and seamounts rising to shallower depths.

seabed and the sediment disturbance associated with it.

Indirect effects include modification to sediment properties and solute fluxes and changes in benthic community abundance and composition. The effect on the deep-sea ecosystem of addition of the large proportion of the catch that is routinely discarded by the trawler will also come under this heading. Indirect effects may occur some time after the impact. These can represent an aggregation of several separate impacts and have a cumulative expression. This breakdown seems more natural than one based on benthic habitat because the signals of impact overlap.

Direct Effects

Physical Impacts on Infaunal Communities

Soft sediments make up by far the most usual seabed trawled by both shallow and deep-sea fisheries. Unlike the usually current-swept sandy sediments of continental shelves, sediments in the deep sea characteristically are composed of a soft, muddy ooze. However, coarse, sandy sediments, typically with a high admixture of glacial erratics, occur mainly in a band just below the continental shelf edge in the Rockall Trough and Norwegian Sea. Glacial drop stones up to large boulders (Figure 2A, 2B) are found at all depths in the deep sea off northern Europe. They are most common on the upper slope, where icebergs grounded against the bottom during the Pleistocene. Areas of rock outcropping occur, for example, as steep scarps on the upper continental slope, on the flanks of seamounts, and in a few deeper areas with steep declivity experiencing very strong flow, but these areas will be too rugged to be targeted by trawlers. The "mud line" has been used to denote the depth on the continental slope where superficial sediments somewhat abruptly change to fine-grained muds with an increasingly high content of skeletal remains of tiny pelagic organisms. This seems to vary with latitude, being found at greatest depth at higher latitudes where dynamic contour-following slope currents, breaking internal waves, and storm-driven motions influence near-bed hydrodynamics below the shelf edge down to hundreds of meters (Huthnance 1986, 1995). The regional geology of the continental margin off Scotland is summarized in Stoker et al. (1993).

The physical impact on benthic habitats caused by towed gears over the seabed is perhaps the most widespread, lasting, and, consequently, best-known outcome of fishing in shallow seas worldwide. The component effects have been reviewed by Jones (1992), Jennings and Kaiser (1998), Hall (1999), Goñi (2001), and Thrush and Dayton (2002). The physical impact of the heavy rockhopper gear used in deep water will be at least as severe as that of trawls used in coastal

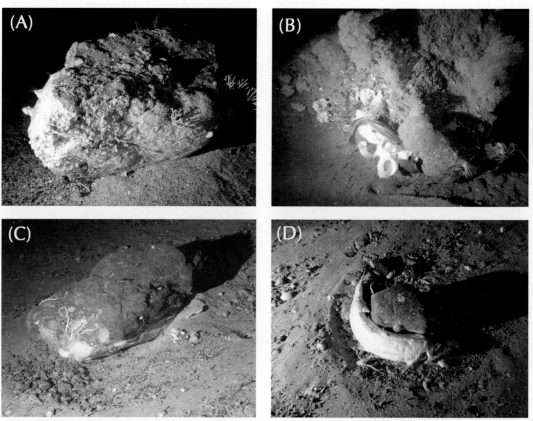

Figure 2. Seabed photographs from the eastern slope of the Faroe-Shetland Channel. (A) The glacial boulder at 425 m (60°22'51"N, 04°05'56"W) near the shelf break in the vicinity of the Schiehallion oilfield where the seabed is protected from trawling; it is heavily encrusted by sessile epifauna. Organisms visible include the colonial serpulid polychaete *Filograna implexa* and barnacle *Chirona hameri*. (B) This boulder, at 923 m, also supports an epifauna characteristic of the deep, very cold water layer in the Faroe-Shetland Channel where commercial fish species are few and, therefore, trawling very infrequent; the white sponge is *Asconema setubalense* with an Arctic rockling *Onogadus argentatus*. (C) Boulder (probably recently overturned) in a heavily trawled area at 590 m where side-scan and visual records show evidence of intense trawling. (D) A cobble in the same area as (C), with a presumably discarded fish (black arrow) being eaten by various opportunist scavengers (white arrows).

seas. The physical impact of the heavy rock hopper gear used in deep water can be expected to be at least as severe, as heavy trawls are undoubtedly responsible for the proliferation of linear marks, mostly aligned along seabed depth contours, on the continental margin. These are typically visible in side-scan records on the continental margin where trawling effort is well developed (e.g., Friedlander et al. 1999).

Such marks are seen in side-scan from the northern European continental margin. Most data have been obtained in commercial surveys for oil or gas companies. An analysis of trawl marks visible in seabed photographs taken on the Hebridean continental margin off Scotland indicates that deep-sea trawling is physically impacting the seabed to depths of more than 1,000 m (Roberts et al. 2000). The depth pattern in incidence of these marks on the seabed increases from 700 to 1,300 m. This may reflect trawl mark persistence in relation to the changing nature of the sediment and hydrodynamic processes which mold and largely control its granularity (see seabed photographs and discussion in Gage et al. 2000; Howe and Humphery 1995). In the upper slope in the Faroe Shetland Channel (Figure 1), trawls disrupt the armoring of the seabed by gravel and pebbles, which normally protect the finer sediments below from mobilization by slope and oscillatory currents.

Exact data on depth-related trawling effort off Scotland are difficult to assemble and probably of dubious accuracy. This is because trawl marks are less persistent on the coarser, less stable sediments found just below the shelf break than in the deeper, more stable muddy bottoms (Roberts et al. 2000). This is supported by data from coastal studies. Here, trawl marks on muddy sediments persisted longer than on coarser sediments (Ball et al. 2000). Acoustic studies on the Grand Banks off Canada have shown that trawling can increase small-scale hardness in topographic relief on sandy ground by reorganizing larger particles such as shells into linear patterns (Schwinghamer et al. 1998). This might also be expected on similar grounds in deep water just below the shelf edge off Scotland and Ireland where mixed, sandy sediments predominate. A smoothing of biogenic relief was also observed by Schwinghamer et al. (1998). Smoothing also was concluded to occur from study of seabed photographs from deeper levels in finer sediments in the deep water off the Hebrides (Roberts et al. 2000).

Trawl marks are sufficiently abundant on the seabed along the continental margin off Scotland that 2–12% of the seabed photographs taken on the Hebridean continental margin off Scotland showed clear marks made by trawling. These were taken using a "bed-hop" camera system in separate deployments from the ship, each covering less than 100 m^2 of seabed (Roberts et al. 2000). In an examination of 155 photographs taken in 1988 from a neighboring site between 600-m and 863-m depth on the slope a little to the south, two of six deployments undertaken showed no trawl marks. The other four showed marks in 5%, 7%, 9%, and 47% of the photographs (Lamont and Gage 1998; Roberts et al. 2000). This indicates trawling in water depths of more than 600 m on the Scottish continental slope has been producing detectable marks on the seabed for at least 10 years (Figure 3A, 3B). These may persist for 1 year, and perhaps 5 years in shallow, soft sediments (Jones 1992; Tuck et al. 1998; Palanques et al. 2001). Marks produced on the continental slope will persist for at least comparable periods.

Trawl marks appear generically in photographs on muddy sediment on the continental slope as straight-line marks less than 1 m across and with an aspect ratio of about 10:1 or more (Figure 3A). The most prominent are most likely to have been caused by the heavy trawl doors as they are pulled over the seabed. Therefore, these marks can be confidently distinguished from ancient iceberg scour marks, which are usually larger than the image size and only partially visible. Deeper gouges are sometimes seen. These may be caused when the trawler changes course, perhaps in order to follow an isobath, or when it surges in heavy seas. The gouges often have an adjacent berm and a scattering of deeper sediment or particles including clean stones or shells previously buried. In any case, the seabed immediately adjacent to the furrow created by the trawl doors is marked with a scattering of such sediment, and the normally hummocky sediment landscape that is caused by biogenic activity may be smoothed over by the trawl. Less prominent, linear marks are likely caused by the trawl foot gear, such as the large rubber or steel rollers used to prevent the trawl becoming snagged by seabed obstructions. Flattened areas of seabed marked by corrugations about 10 cm apart and orientated parallel with the deeper gouges are often visible (Figure 3B). We think these may be caused by catch protruding through meshes of the swollen net.

The presence of isolated, clean boulders and cobbles larger than 10 cm in diameter can also indicate trawling activity where they have been displaced, picked up and then subsequently dropped from nets, or disposed of from the catch by the trawler (Figure 2C). A revisit in 1999 of an area on the southern part of the Hebridean margin which had been extensively surveyed for deep-sea demersal fish stocks in the 1970s and subsequently subjected to heavy fishing effort, found evidence for a large increase in uprooted large boulders such that the nets used (not rock hopper gear) were badly damaged (Gordon 2002). It is likely that these boulders were originally semiexposed and had been dragged out by the trawl from a previously partly buried position (rates of natural sedimentation since the Pleis-

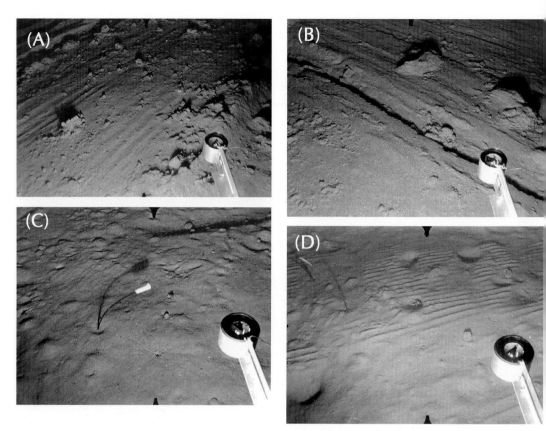

Figure 3. Oblique views of the seabed covering about 3.8 m². Photographs taken with a bed-hop camera along the Hebridean margin off Scotland, (A) and (B) at 885 m northwest of St. Kilda (58°07.33'N, 09°40.49'W) and photographed in 1988, and (C) and (D) at 1,296 (59°01.21'N, 07°54.85W) and 1,295 m (58°58.66' N, 07°57.40' W), respectively, in 1998. (A) The seabed is smoothed leaving a series of shallow marks along with a scattering of clods of sediment presumably deposited in the wake of the trawl. In both (A) and (B), numerous small brittle stars *Ophiocten gracilis* and a single sea urchin (probably *Echinus acutus* var. *norvegicus*) are visible. (B) Linear furrow about 25 cm wide and 8 cm deep made by the trawl door. (C) An intact glass sponge *Hyalonema* sp. photographed with its stalk, encrusted with epizooitic colonial zooanthids, bending over in the current. Several xenophyophores *Syringammina* sp. are visible as small globular objects, along with a single brittle star (probably *Ophiomusium lymani*). (D) The stalk of a glass sponge *Hyalonema* sp., apparently the result of the trawl net passing over it (note the corrugated seabed similar to that seen in B). The presence of the delicate xenophyophores in the track of the trawl is puzzling; it is possible that they are able to recolonize impacted areas and grow rapidly.

tocene would have only partly buried such large objects). Such alteration in seabed habitat deserves serious study. Such post-Pleistocene partial burial of boulders does not seem to occur in the more hydrodynamically energetic upper slope in the Faroe-Shetland Channel. Here, there are many single boulders and groups of boulders already exposed on the seabed surrounded by a scour moat (Figure 2A, 2C).

Physical Impacts on Epifaunal Communities

On the continental shelf, the direct physical impact resulting from passage of the trawl over the seabed is easiest to interpret for epifauna associated with hard bottom, such as that consisting of pebbles, cobbles, and boulders. Boulders are highly dispersed and, therefore, difficult to quantify as habitat. They provide a substratum for a rich variety of epifaunal organisms (Figure 2A, 2B) and must be dominating ecological features on the seabed. Studies on the outer continental shelf off Alaska showed marked displacement of boulders and damage to large sessile epifauna after a single trawl pass. Changes in density or damage to motile invertebrates were not detected (Freese et al. 1999). Uprooting a partly buried boulder will probably severely disturb these organisms because they will be unable to readjust their position. Continuing disturbance of boulders by trawls may greatly delay recolonization by sessile epifauna. In the 5-year study of Pitcher et al. (2000) in northeastern Australia, removal rates of sessile megazoobenthos ranged from 0% to 40% per trawl pass. The study showed a shift in faunal composition from sessile epibenthos to less vulnerable, usually infaunal species in trawled areas.

In deep waters at the shelf edge and upper continental slope, sessile zoobenthos that include slow-growing epifaunal animals, such as sponges and coldwater corals, have generated the most concern with regard to impacts from bottom trawling. These organisms have a keystone role in providing habitat for a large number of other organisms (Sainsbury 1988; Jensen and Frederiksen 1992; Klitgaard 1995). In the case of both sponges and coldwater corals, the vast majority of these are facultative rather than obligate associates (Jensen and Frederiksen 1992; Klitgaard 1995; Rogers 1999; Hepburn 2001). All indications are that large sessile fauna, such as massive sponges, will take many years or decades to recover (Sainsbury et al. 1997; Pitcher et al. 2000). Therefore, such habitat loss will not be quickly recoverable. The largest coldwater coral "bioherms" are likely to be thousands of years old (Mikkelsen et al. 1982). The coral structure is very slow-growing, with rates of linear extension variously estimated from 4 to 25 mm/year (Duncan 1877; Dons 1944; Mortensen and Rapp 1998; Mortensen 2001). Growth rates of deep-sea demosponges are unknown. There is anecdotal evidence from fishermen off the Faroe Islands of such huge numbers in the trawl as to endanger the ship (Klitgaard et al. 1995). Therefore, considerable habitat damage may have already occurred. Bett (2000) observed rotting sponge remains deep in the sediment of a box core sample from the bathyal Faroe Shetland Channel. Presumably, these had been pushed down by recent passage of a trawl.

Significant effort has been made to map the present-day distribution of coldwater corals, principally *Lophelia pertusa*, along the northern European continental margin (e.g., Jensen and Frederiksen 1992; Long et al. 1999; Mortensen et al. 2001; Roberts et al. 2003). Hall-Spencer et al. (2002) documented widespread trawling damage to coldwater coral reefs from commercial fishing expeditions at 840–1,300-m depths along the continental shelf edge west of Ireland and at 200-m depths off Norway. Hall-Spencer et al. observed rock hopper bycatch off western Ireland that included large pieces (up to 1 m^2) of living coral that had been torn from reefs along with a diverse array of coral-associated benthos. Further documentation of damage to coldwater coral in Norwegian waters is provided by Fosså et al. (2002). These authors conclude, from remote operated vehicle (ROV) observations and reports from fishermen, that severe damage to coral bioherms has occurred, with between 30% and 50% of reef areas being damaged.

Such reef areas have traditionally been rich fishing grounds, and studies have shown larger catches from coral than noncoral areas (Husebø et al. 2002). Anecdotal reports have claimed that trawlers use heavy gear to crush the coral and clear the area before fishing starts. These observations prompt speculation as to how much coldwater coral has already been obliterated by deep-sea trawling before being adequately mapped. Maps drawn from French surveys of the distribution of coral along the continental margin off Ireland and France made during the first three decades of the 20th century (Le Danois 1948) suggest that coldwater coral was more widespread than known today. It is possible that the current distribution of coldwater corals to the west of Scotland has been similarly degraded (Roberts et al. 2003).

Mounting public concern in Norway over damage to these ancient "bioherms" has resulted in legislation to protect several reef areas from further damage by trawlers. In the deeper waters off Scotland south of the Wyville Thomson Ridge, thickets of *Lophelia pertusa* were discovered in 1988 at about 1,000 m water depth associated with seabed mounds and named the Darwin Mounds. A revisit to the area found evidence of ongoing trawling, with one trawl mark cutting right through the coral on one mound (A. J. Wheeler et al., University College, Galway, Ireland, abstract from ICES Annual Science Conference, Oslo, NOrway, 2001). The boom-and-bust fisheries for orange roughy *Hoplostethus atlanticus* on the seamounts off Tasmania (which peak at depths in excess of about 1,000 m) provide the classic example of the probably lasting damage by trawling to a seabed spectacularly rich in suspension-feeding sessile fauna. The seamount fauna included a very slowly growing, matrix-forming colonial hard coral, *Solenomilia variabilis*, associated with the hard bottom. Trawling effectively physically removed the reef aggregate from the most heavily fished seamounts and swept away much of the other epifauna. Heavily fished seamounts had 83% less biomass than similar lightly fished or unfished sites (Koslow et al. 2001). Many of the epifaunal species appear to be undescribed and unique to the seamounts of this area (de Forges et al. 2000). A subset of seamounts, as yet unfished, now lie within a marine protected area enclosing 12 seamounts. Several seamounts are known in the northeastern Atlantic, including three off the Hebridean continental margin in the Rockall Trough. It is likely these also support luxuriant growth of fragile sessile epibenthos, including coldwater corals. Little is known of these habitats, especially the steep rugged topography on their flanks where flow is accelerated. Here, suspension-feeding organisms are likely to be well developed. It is known that at least two of the three seamounts off Scotland have already been targeted by trawlers for orange roughy (BIM 2001).

Damage to sessile epifauna associated with soft bottoms in the deep sea has been more difficult to document. Seabed photographs, from an area showing large numbers of trawl marks show the glass sponge, *Hyalonema* sp., lacking its head but leaving its stalk rooted into the sediment (Figure 3C, and compare with Figure 3D). This is suggestive of damage from the passage of a trawl (Roberts et al. 2000). Glass sponges are thought to be very slow growing, with a large specimen of the species *Rhabdocalyptus dawsoni* 1 m in length growing in a British Columbian fjord and thought to be about 220 years old (Leys and Lauzon 1998). Photographs taken in the same area show fields of the extremely fragile xenophyophore (a giant-sized rhizopod protozoan found only in the deep sea, see also Figure 3D). This suggests such organisms are particularly vulnerable to physical impact. It is unlikely any damaged specimens would remain after a trawl has passed over them. Observations from "...a bottom-moored time-lapse camera system" show these strange organisms have rapid growth compared to other deep-sea fauna (Gooday et al. 1993), perhaps even being able to dissemble and rebuild the test rapidly.

Off Maine, rock hoppers have allowed fishermen to trawl in areas previously protected by their rough topography. Such habitat supports a rich epifaunal biodiversity, with the resulting damage prompting comparisons with forest clear-cutting (Watling and Norse 1998).

Physical Impacts on Infaunal Communities

In the deep sea beyond the continental shelf edge, we can little more than speculate on impacts of soft sediments on benthic infauna. These small organisms constitute a highly diverse community on the continental slope (Grassle and Maciolek 1992). Fauna of muddy sediments create much of the structure of this habitat. Lower sedimentation rates and weaker currents will, in deeper water, lead (although not in the hydrodynamically energetic Faroe-Shetland Channel) to greater persistence of biogenic habitat structuring. This is seen worldwide as a deep-seabed landscape of biogenic pits and mounds and a wide variety of traces and imprints generated by larger seabed organisms (Heezen and Hollister 1971) down to the scale of heterogeneity created by the activities of the smallest metazoans. This small-scale habitat heterogeneity is thought to contribute, through resource and habitat partitioning, to faunal diversification in the deep sea (Jumars 1975, 1976). Disturbance of muddy benthic habitat above an ecological threshold may permanently degrade this complexity, leading to loss in biodiversity and, perhaps, functional extinction that may have severe effect on the remaining elements of the ecosystem (Thrush and Dayton 2002). However, such effects and thresholds are still obscure and remain largely untested. Seabed photographs and other data, together with our present knowledge of the biology of the burrowing organisms, suggest some resilience to the physical effects of trawling, depending on size and depth of burial. We have noticed biogenic structure occurring in trawl scours that must post-date the impact. Observations in shallow water show the larger, semiemergent organisms such as large bivalve mollusks (e.g., Rumohr and Krost 1991; Kaiser et al. 2000) and sessile epifauna (e.g., Rumohr and Kujawski 2000) are completely removed by trawling (Pitcher et al. 2000). Tank simulations of the effect of the trawl door on sand show previously buried bivalves in the scour path to be displaced to the berm and exposed, with few showing damage (Gilkinson et al. 1998). In a large-scale study of the effects of otter trawling on the sandy bed of the Grand Banks off Canada, the habitat recovered in a year from this disturbance (Gordon et al. 2002). Clearly, easy translation of these results to the deep muddy seabed is inappropriate. The Grand Banks is a high-energy environment, so such rapid recovery is not necessarily typical of even other shelf areas.

Indirect Effects

Changes in Community Characteristics

These include outcomes where there are changes in community composition, diversity, and abundance. The most immediate of these is caused by removal of large epibenthic predators, including gastropod mollusks and echinoderms such as sea stars. While such impacts on numbers of large motile species have been documented in coastal seas (e.g., Thrush et al. 1998; Smith et al. 2000), the consequences on prey populations and the food web are not well understood and are speculative in deep water. The physical disruption to habitat complexity, such as loss in burrow structures, might also modify biogeochemical fluxes between the sediment and overlying water. This might be more serious on muddy than sandy bottoms in both shallow and deep waters.

The reaction of the benthos to disturbance caused by passage of the trawl can be understood in terms of generic effects of stress on species-rich benthic communities (Gray 1989). This response consists of reduced species richness through elimination of many rare species and increased community dominance through population expansion in a few relatively tolerant species. There will also be reduction in mean body size of dominant species as a result of a demographic shift to younger animals characteristic of expanding populations.

These effects are difficult to recognize against the background of a community already in an altered state in coastal seas from trawling disturbance (Collie et al. 2000). Comparisons between heavily and more lightly trawled areas show that chronic fishing disturbance causes significant and widespread changes in soft-bottom community structure and production (Kaiser et al. 2000; Jennings et al. 2001a, 2001b). Because of the key role that disturbance is thought to play in controlling local-scale biodiversity (Connell 1977; Huston 1979), trawling disturbance arguably might have a positive effect on local diversity. This is because disturbance reduces competition, allowing more species sharing similar resources to coexist. Thrush and Dayton (2002) point out that such direct competition has been difficult to demonstrate as an important process in soft-sediment communities and may, from theoretical considerations, cause decline in diversity among multi-tropic level systems. Perhaps the most important effect is a regional-scale homogenization of habitat by removal of surface features and the generally epifaunal species that contribute to biodiversity.

Some studies have clearly shown alteration in community composition of infauna and reduction in species richness by removal of rare species. It is not clear whether these changes are a result of repeated physical disturbance or a consequence of changes in composition and ecological function of larger species, including fish. The same studies show overall loss of species, particularly the rare ones, and dramatic decrease in biomass. But overall benthic productivity may be unaffected, even if population turnover becomes more rapid (Jennings et al. 2001a). Consequently, there may be some overall biodiversity loss, but trophic structure may be more resilient to chronic disturbance by seabed trawling (Jennings et al. 2001b) even if other aspects of ecosystem functioning are affected (Jennings and Kaiser 1998; Hall 1999).

The data available from continental shelf seas suggest benthic communities of more exposed, sandy seabed have greater resilience and recover faster than those of finer-grained bottoms (Collie et al. 2000). The major study of seabed impacts of otter trawling on a sandy bottom on the Grand Banks, Newfoundland, found little evidence of long-term change, with the greatest effects seen immediately after trawling (Kenchington et al. 2001; Gordon et al. 2002). However, data from muddy bottoms, which will have the greatest similarity to the continental slope below the mud line, are relatively few (Collie et al. 2000). The effect of trawling on a previously unfished Scottish sea loch showed the classic response over 18 months of increased total abundances with decreasing species diversity and evenness (Tuck et al. 1998). Changes in community composition allowed identification of the most resilient and sensitive species. In another study of a previously protected inshore muddy-bottom habitat over 1 year in the Gullmarsfjord (west coast of Sweden), no significant differences were found in community composition, although there was a marked increase in variability in benthic composition after trawling (Lindegarth et al. 2000).

Rich trawling grounds in coastal seas, such as the southern North Sea and Irish Sea, are often revisited and the same seabed trawled more than once a year. Here fishery-mediated changes in fish assemblages by, for example, removal of top predators or increase in scavenging species will impact the benthos (Hall 1999).

Comparable data from the deep sea are nonexistent. However, from the very limited evidence of seabed photographs of old trawl marks (recognizable as such from the smoothed edges of the marks) it is possible to see the unmistakable evidence in the form of mounds of faecal ejecta of the activity of bioturbators in the sediment below (see Figure 4a in Roberts et al. 2000). This suggests these animals, provided they had not migrated to or colonized the area after the impact, had been little affected, although the rate of recovery is unknown.

The deep-sea communities of the continental slope are characterized by high species diversity and evenness (Grassle and Maciolek 1992). There are few data from analysis of experimentally disturbed areas in the deep sea to see whether this species-rich community will respond in a similar way to coastal seas. Experiments using trays of defaunated sediment placed on the deep seabed suggest recovery by larval settlement will be slower than would be expected in shallow water (Grassle 1977). Enrichment of the defaunated sediment may elicit a faster response but often with heavy dominance by presumed opportunist species (Grassle 1977; Desbruyères et al. 1980; Levin and Smith 1984; Smith 1985a, 1986; Grassle and Morse-Porteous 1987).

Thrush and Dayton (2002) suggest that effects of trawling on deep-sea biodiversity will be exacerbated compared to that of shallow water because of life history adaptations of slow growth rates, extreme longevity, delayed reproduction, and low mortality. Actually, the mass of evidence suggests that such traits are by no means the general rule and that a range of life history adaptations broadly comparable to those in shallow-water biota occur (Gage 1991; Gage and Tyler 1991). Another argument is that deep-sea communities will be particularly vulnerable to anthropogenic disturbance, such as that of seabed trawling, because they rarely encounter natural disturbance (Dayton et al. 2002). This overlooks the widespread occurrence of natural disturbance in the deep ocean caused, for example, by benthic storms or seasonal mass deposition of phytodetritus (Gage and Tyler 1991).

One detailed study of disturbance that may be close to the physical impact of a deep-sea trawl supports prediction of a broadly similar response to that described above from shallow coastal communities. The *DISCOL* (*DIS*turbance and Re*COL*onisation) study in a manganese nodule province in the abyssal Southeast Pacific (Thiel et al. 2001) experimentally lowed the seabed with recovery monitored by revisits over several years. Large megafauna that were directly impacted were eliminated, but after 7 years, even sessile species were present again in the impacted area (Bluhm 2001). Smaller size-classes were less seriously impacted, with abundances recovering rapidly but species richness being depressed and community composition and spatial heterogeneity still showing effects after 7 years (Borowski 2001).

Other Indirect Outcomes of the Physical Impact of Trawling

These would include the secondary impact caused by resuspension of fine sediment and its subsequent redeposition. A study in the northwestern Mediterranean showed water turbidity in the water column increased by a factor of up to 3 for 4–5 d after trawling (Palanques et al. 2001).

Virtually nothing is known of the sensitivities of deep-sea organisms to increased sedimentation. In the *DISCOL* study, animal densities declined, even in the seabed areas not directly impacted by the lowing but almost certainly experiencing the deposition from the plume of disturbed sediment (Bluhm 2001). Some organisms, such as massive sponges and coldwater corals, arguably may be sensitive to smothering by increased sedimentation (e.g., Rogers 1999).

Analogies with effects of periodic benthic storms on the deep seabed community might be explored. On the continental rise off Nova Scotia, these occur with sufficient frequency (Hollister et al. 1984) as to be experienced several times within the life span of even the smaller metazoan size-classes. The deep seabed elsewhere may periodically also experience intense episodes of deposition of resuspended sediments caused by benthic storms driven by eddy kinetic energy transmitted from the surface. A changed seabed community composition, with reduced species richness and uncharacteristically high dominance by a small number of very abundant species as described by Thistle et al. (1985) is typical of disturbed sediment. Relating this to the seabed impact of deep-sea trawling is complicated by the difficulty of separating any response to periodic sediment erosion and redeposition from that of the strong currents.

Studies on the Grand Banks have shown that trawling reduces the abundance of flocculated organic matter, perhaps as a result of the disturbance breaking it up into smaller particles and dispersing it over a wider area (Schwinghamer et al. 1998). A similar effect may occur in response to passage of a trawl on the continental margin off Scotland, where the seabed below about 600 m is supposed, where current energy allows, to experience a blanketing deposition of phytodetritus in late spring to summer (Bett 2000). Another, more speculative, consequence of sediment disturbance by trawling is release of interstitial solutes. This will enhance the normal biogeochemical recycling of organic material back into the water column. It has been suggested that this greatly increases water column productivity in heavily fished coastal areas (Pilskaln et al. 1998).

Effects of Discards on Deep-Sea Benthic Community

All fisheries are associated with a discarded bycatch that is usually thrown straight back into the sea on the fishing grounds. Only in coastal seas will a proportion of these organisms survive. Kaiser and Spencer (1996) found more than 60% of certain species (e.g., sea stars, hermit crabs [family Paguridae], and mollusks) survived return to the sea, although many are damaged, and those unable to repair damage may die (Bergmann et al. 2001). Little is known about the effects of discards on the seabed community, and these will be difficult to separate from effects of the physical impact of the trawl.

In the deep sea fisheries, only a small proportion of species from the catch tend to be selected for the market, resulting in a large proportion of the catch being discarded. Probably nearly all these organisms will reenter the sea as dead biomass. Those not taken by seabirds at the surface or intercepted by scavengers or predators in the water column will drop to the seabed as highly localized, and relatively massive, food falls (Connolly and Kelly 1996). The amount of discarded biomass, let alone the proportion arriving at the seabed, is unknown. Deep-water species tend to be more fragile than taxonomically equivalent species in shallow seas. This, in combination with decompression effects on biochemical processes at the cellular level, probably results in virtually all invertebrate fauna captured from depths greater than about 1,000 m being dead on deck after recovery from depth.

In marked contrast to shallow seas where specialist scavengers are rare, a specialist scavenging community is well developed in the deep sea (reviewed by Britton and Morton 1994 and Gage 2003), and scavengers become increasingly important with increasing depth (Stockton and DeLaca 1982). Smith (1985b) suggested scavengers contribute 11% of the energy requirement of a bathyal benthic community. On the continental margin off northern Europe, it is difficult to see how the natural input of large packages of organic material as dead animals to the deep-sea bed could be quantitatively enhanced by deep-sea trawler discards. This is because the process amounts to some permanent removal (retained catch) no more than an increase in rate of natural recycling (return of discards as dead biomass).

It is possible that such an increased rate of natural recycling may increase the size and importance of scavenger populations already in place, but there appear to be no data to support this. Deep-sea scavengers are typically attracted to baited traps set on the bottom and include organisms, such as fish and lysianassid amphipods, with well-developed motility and sensory ability to locate highly scattered food falls. Among fish, hagfish (class Myxini) are thought to contribute significantly to removal of carrion on the continental slope (Martini 1998). The northern cutthroat eel *Synaphobranchus kaupi*, not itself a species targeted by trawling, is one of few species where there is solid evidence for opportunist scavenging (Gordon and Mauchline 1996; Marques 1998). Many of the fish species targeted by deep-sea fisheries, such as grenadiers of the genus *Corypaenoides* (family Macrouridae) are also thought to be opportunist scavengers. The importance of this is debatable (see Merrett and Haedrich 1997). The impact of discards on less motile invertebrate that are probably opportunist scavengers, such as lithodid crabs and certain species of sea urchins (class Echinoidea) and brittle stars (class Ophiuroidea), is less certain.

Conclusions

Much of the information and interpretation above is speculative or uncertain because of the limited scope of studies or size of area investigated. However, we take this opportunity to offer the basis for a unifying, and hopefully testable, hypothesis.

Much of what we, as deep-sea biologists, have observed in the study of effects of fishing is a response to a sharply increasing scale and intensity of impact that lacks the starting point of a pristine ecosystem. Because of this, it seems possible that vulnerable and sensitive organisms may have already become depleted, or lost altogether, before detection of change in coastal inshore communities was possible. This is in line with conclusions on alterations of coastal ecosystems caused by historical overfishing (Jackson et al. 2001). The relatively huge area of the deep-sea continental margins and the recentness of industrial-scale trawling provide the opportunity to more clearly understand earlier stages of change without the need to disentangle these from other anthropogenic impacts. This presupposes that the range of lifestyle and life history adaptation in deep-sea benthos is not fundamentally different from that of benthos in pristine shelf communities, a proposition now supported by the mass of evidence. We also believe that, at least off northern Europe, this is now only historically true. This is because the seabed already shows an altered composition, with shift away from large-bodied, long-lived, slow-growing (and often sessile) epifaunal components. These organisms probably provide additional habitat complexity by providing microhabitat for smaller organisms as well as a role in ecosystem function by, for example, providing nursery areas for fish (e.g., see Bradstock and Gordon 1983). As an example of one of the most vulnerable of such organisms occurring in shallow seas off Europe, the large, extremely fragile, erect colonial bryozoan *Pentapora fascialis* may have been long eliminated by trawling, except in sheltered rocky pockets. The largest specimen recorded measured about 2 m across, and colonies are known to shelter a broad range of smaller associated fauna (Hayward and Ryland 1999). Such organisms provide an analog to the now endangered role of coldwater corals and sponges in deeper water.

Acknowledgments

We thank Dr. John Gordon, Dunstaffnage Marine Laboratory, for helpful advice regarding current deep-water fishing activity along the European continental margin, BP Exploration Ltd for permission to use ROV photographs of the seabed, and Steve Gontarek for providing the map used in Figure 1.

References

Ball, B., B. Munday, and I. Tuck. 2000. Effects of otter trawling on the benthos and environment in muddy sediments. Pages 69–82 in M. J. Kaiser and S. J. Groot, editors. Effects of fishing on non-target species and habitats. Blackwell Scientific Publications, Oxford, UK.

Bergmann, M., D. J. Beare, and P. G. Moore. 2001. Damage sustained by epibenthic invertebrates discarded in the *Nephrops* fishery of the Clyde Sea area, Scotland. Journal of Sea Research 45:105–118.

Bergstad, O. A., J. D. M. Gordon, and P. Large. 2002. Is time running out for deep-sea fish? Available: www.ices.dk/marineworld/deepseafish.asp (March 2005).

Bett, B. J. 2000. Signs and symptoms of deepwater trawling on the Atlantic margin. Pages 107–118 in Man-made objects on the seafloor. Society for Underwater Technology, London.

BIM (Bord Iascaigh Mhara). 2001. Deepwater programme—2001 Executive Report. An Bord Iascaigh Mhara, Co., Dublin.

Bluhm, H. 2001. Re-establishment of an abyssal megabenthic community after experimental physical disturbance of the seafloor. Deep-Sea Research Part II 48:3841–3868.

Borowski, C. 2001. Physically disturbed deep-sea macrofauna in the Peru Basin, southeast Pacific, revisited 7 years after the experimental impact. Deep-Sea Research Part II 48:3809–3839.

Bradstock, M., and D. P. Gordon. 1983. Coral-like bryozoan growths in Tasman Bay, and their protection to conserve commercial fish stocks. New Zealand Journal of Marine and Freshwater Research 17:159–163.

Britton, J. C., and B. Morton. 1994. Marine carrion and scavengers. Oceanography and Marine Biology: An Annual Review 32:369–434.

Collie, J. S., S. J. Hall, M. J. Kaiser, and I. R. Poiner. 2000. A quantitative analysis of fishing impacts on shelf-sea benthos. Journal of Animal Ecology 69:785–798.

Committee on Ecosystem Effects of Fishing. 2002. Effects of trawling and dredging on seafloor habitat. National Academy Press, Washington, D.C.

Connell, J. H. 1977. Diversity in tropical rain forests and coral reefs. Science 199:1302–1309.

Connolly, P. L., and C. J. Kelly. 1996. Catch and discards from experimental trawl and longline fishing in the deep water of the Rockall Trough. Journal of Fish Biology 49 (Supplement A):132–144.

Cryer, M., B. Hartill, and S. O'Shea. 2002. Modification of marine benthos by trawling: towards a generalization for the deep ocean? Ecological Applications 12(6):1824–1839.

Dayton, P. K., S. Thrush, and F. C. Coleman. 2002. Ecological effects of fishing in marine ecosytems of the United States. Pew Oceans Commission, Arlington, Virginia.

Dayton, P. K., S. F. Thrush, M. T. Agardy, and R. J. Hofman. 1995. Environmental effects of marine fishing. Aquatic Conservation: Marine and Freshwater Ecosystems 5:205–232.

de Forges, B. R., J. A. Koslow, and G. C. B. Poore. 2000. Diversity and endemism of the benthic seamount fauna in the southwest Pacific. Nature (London) 405:944–947.

Desbruyères, D., J. Y. Bevas, and A. Khripounoff. 1980. Un cas de colonisation rapide d'une sédiment profound. Oceanologica Acta 3:285–291.

Dons, C. 1944. Norges korallrev. Det Kongelige Norske Videnskabers Selskabs Forhandlinger 16:37–82.

Duncan, P. M., 1877. On the rapidity of growth and variability of some *Madreporaria* on an Atlantic Cable with remarks upon the rate of Foraminiferal Deposits. Annals and Magazine of Natural History 20:361–365.

European Commission. 2001. The future of the common fisheries policy, volumes I and II. Office for Official Publications of the European Communities, Green Paper, Luxembourg.

European Commission. 2002. Council regulation (EC) number 2371/2002 of 20 December 2002 on the conservation and sustainable exploitation of fisheries resources under the Common Fisheries policy. Official Journal of the European Communities L 358/59. Available: http://europa.eu.int./comm/fisheries/news_corner/press/inf02_61_en.htm (March 2005).

Fosså, J. H., P. B. Mortensen, and D. M. Furevik. 2002. The deep-water coral *Lophelia pertusa* in Norwegian waters: distribution and fishery impacts. Hydrobiologia 471(Special Issue):1–12.

Freese, L., P. J. Auster, J. Heifetz, and B. L. Wing. 1999. Effects of trawling on seafloor habitat and associated invertebrate taxa in the Gulf of Alaska. Marine Ecology Progress Series 182:119–126.

Friedlander, A. M., G. W. Boehlert, M. E. Field, J. E. Mason, J. V. Gardner, and P. Dartnell. 1999. Sidescan sonar mapping of benthic, trawl marks on the shelf and slope off Eureka, California. U.S. National Marine Fisheries Service Fishery Bulletin 97:786–801.

Gage, J. D. 1991. Biological rates in the deep sea: a perspective from studies on processes in the benthic boundary layer. Reviews in Aquatic Science 5(1):49–100.

Gage, J. D. 2001. Deep-sea benthic community and environmental impact assessment at the Atlantic Frontier. Continental Shelf Research 21:957–986.

Gage, J. D. 2003. Food inputs, utilization, carbon flow and energetics. Pages 601–668 *in* P. A. Tyler, editor. Ecosystems of the deep oceans, chapter 11. Elsevier, Amsterdam.

Gage, J. D., P. A. Lamont, K. Kroeger, G. L. J. Paterson, and V. J. L. Gonzalez. 2000. Patterns in deep-sea macrobenthos at the continental margin: standing crop, diversity and faunal change on the continental slope off Scotland. Hydrobiologia 440:261–271.

Gage, J. D., and P. A. Tyler. 1991. Deep-sea biology: a natural history of organisms at the deep-sea floor. Cambridge University Press, Cambridge, UK.

Gilkinson, K. D., M. Paulin, S. Hurley, and P. Schwinghamer. 1998. Impacts of trawl door scouring on infaunal bivalves: results of a physical trawl door model /dense sand interaction. Journal of Experimental Marine Biology and Ecology 224:291–312.

Goñi, R. 2001. Fisheries effects on ecosystems. Pages 117–133 *in* C. F. C. Sheppard, editor. The seas at the millennium: an environmental evaluation, chapter 115, volume 3, global issues and processes. Pergamon, Oxford, UK.

Gooday, A. J., B. J. Bett, and D. N. Pratt. 1993. Direct observation of episodic growth in an abyssal xenophyophore (Protista). Deep-Sea Research Part I 40:2131–2143.

Gordon, D. C., K. D. Gilkinson, E. L. R. Kenchington, J. Prena, C. Bourbonnais, K. MacIsaac, D. L. McKeown, and P. W. Vass. 2002. Summary of the Grand Banks otter trawling experiment (1993–1995): effects on benthic habitat and communities. Canadian Technical Report of Fisheries and Aquatic Sciences 2416.

Gordon, J. D. M. 2001. Deep-water fisheries at the Atlantic Frontier. Continental Shelf Research 21:987–1003.

Gordon J. D. M. 2002. Deep water demersal fisheries. Joint Nature Conservation Committee (JNCC) Fisheries Reports. Available: http://www.jncc.gov.uk/marine/page-2525 (March 2005).

Gordon, J. D. M., and J. Mauchline. 1996. The distribution and diet of the dominant benthopelagic, slope-dwelling eel, *Synathobranchus kaupi*, of the Rockall

Trough. Journal of the Marine Biological Association of the United Kingdom 76:493–503.

Grassle, J. F. 1977. Slow recolonization of deep-sea sediment. Nature (London) 265:618–619.

Grassle, J. F., and N. J. Maciolek. 1992. Deep-sea species richness and local diversity estimates from quantitative bottom samples. American Naturalist 139:313–341.

Grassle, J. F., and L. S. Morse-Porteous. 1987. Macrofauinal colonization of disturbed deep-sea environments and the structure of deep-sea benthic communities. Deep-Sea Research 20:643–659.

Gray, J. S. 1989. Effects of environmental stress on species rich assemblages. Biological Journal of the Linnean Society 37:19–32.

Gray J. S. 2000. Effects of trawling on the coastal environment: the need for management action. Marine Pollution Bulletin 40:93.

Hall, S. J. 1999. The effects of fishing on marine ecosystems and communities. Blackwell Scientific Publications, Oxford, UK.

Hall-Spencer, J., V. Allain, and J. H. Fosså. 2002. Trawling damage to northeast Atlantic anient coral reefs. Proceedings of the Royal Society, London, Series B 269:507–511.

Hayward, P. J., and J. S. Ryland. 1999. Cheilostomatous Bryozoa, part 2 Hippothooidae—Celleporoidea: notes for the identification of British species. In R. S. K. Barnes and J. H. Crothers, editors. Synopses of the British fauna (new series) 14. Field Studies Council, Shrewsbury UK, on behalf of the Linnean Society of London and the Estuarine and Coastal Sciences Association.

Heezen, B. C., and C. D. Hollister. 1971. The face of the deep. Oxford University Press, New York.

Hepburn, L. 2001. Assessment of the macrofaunal diversity associated with the cold-water coral, Lophelia pertusa at the Darwin Mounds, Rockall Trough. Master's thesis. University of Aberdeen, Aberdeen, UK.

Hollister, C. D., A. R. M. Nowell, and P. A. Jumars. 1984. The dynamic abyss. Scientific American 250:42–53.

Howe, J. A., and J. D. Humphery. 1995. Photographic evidence for slope-current activity, Hebrides Slope, NE Atlantic Ocean. Scottish Journal of Geology 30:107–115.

Husebø, Å., L. Nøtestad, J. H. Fosså, D. M. Furevik, and S. B. Jørgensen. 2002. Distribution and abundance of fish in deep-sea coral habitats. Hydrobiologia 471(Special Issue):91–99.

Huston, M. 1979. A general hypothesis of species diversity. American Naturalist 113:81–101.

Huthnance, J. M. 1986. Rockall slope current and shelf edge processes. Proceedings of the Royal Society of Edinburgh 88B:83–101.

Huthnance, J. M. 1995. Circulation, exchange and water masses at the ocean margin: the role of physical processes at the shelf edge. Progress in Oceanography 35:353–431.

Jackson, J. B. C., M. X. Kirby, W. H. Berger, K. A. Bjorndal, L. W. Botsford, B. J. Bourque, R. H. Bradbury, R. Cooke, J. Erlandson, J. A. Estes, T. P. Hughes, S. Kidwell, C. B. Lange, H. S. Lenihan, J. M. Pandolfi, C. H. Peterson, R. S. Steneck, M. J. Tegner, and R. R. Warner. 2001. Historical overfishing and the recent collapse of coastal ecosystems. Science 293:625–638.

Jennings, S., T. A. Dinmore, D. E. Duplisea, K. J. Warr, and J. E. Lancaster. 2001a. Trawling disturbance can modify benthic production processes. Journal of Animal Ecology 70:459–475.

Jennings, S., and M. J. Kaiser. 1998. The effects of fishing on marine ecosystems. Advances in Marine Biology 34:201–352.

Jennings, S,. J. K. Pinnegar, N. V. C. Polunin, and K. J. Warr. 2001b. Impacts of trawling disturbance on the trophic structure of benthic invertebrate communities. Marine Ecology Progress Series 213:127–142.

Jensen, A., and R. Frederiksen. 1992. The fauna associated with the bank-forming deep-water coral Lophelia pertusa (Scleractinia) on the Faroe shelf. Sarsia 77:53–69.

Jones, J. B. 1992. Environmental impact of trawling on the seabed; a review. New Zealand Journal of Marine and Freshwater Research 26:59–67.

Jumars, P. A. 1975. Environmental grain and polychaete polychaete species diversity in a bathyal benthic community. Marine Biology 30:253–266.

Jumars, P. A. 1976. Deep-sea species diversity: does it have a characteristic scale? Journal of Marine Research 34:217–246.

Kaiser, M. J., K. Ramsay, C. A. Richardson, F. E. Spence, and A. R. Brand. 2000. Chronic fishing disturbance has changed shelf sea benthic community structure. Journal of Animal Ecology 69:494–503.

Kaiser, M. J., and B. E. Spencer. 1996. The effects of beam-trawl disturbance on infaunal communities in different habitats. Journal of Animal Ecology 65:348–358.

Kenchington, E., J. Prena, K. D. Gilkinson, D. C. Gordon, P. Schwinhamer, T. W. Rowell, K. McIsaac, C. Bourbonnais, D. L. McKeown, and P. W. Vass. 2001. Effects of experimental otter trawling on the macrofauna of a sandy bottom ecosystem on the Grand Banks of Newfoundland. Canadian Journal of Fisheries and Aquatic Sciences 58:1043–1057.

Klitgaard, A. 1995. The fauna associated with outer shelf and upper slope sponges (Porifera, Demospongiae) at the Faroe Islands, northeastern Atlantic. Sarsia 80:1–22.

Klitgaard, A., O. S. Tendal, and H. Westerberg. 1995. Mass occurrences of large sponges (Porifera) in Faroe Island (NE Atlantic) shelf and slope areas: characteristics, distribution and possible causes. Pages 129–142 in L. E. Hawkins, S. Hutchinson, A. C. Jensen, and J. A. Williams, editors. Responses of marine organisms to their environments. University of Southampton, Southhampton, UK.

Koslow, J. A., G. W. Boehlert, J. D. M. Gordon, R. L. Haedrich, P. Lorance, and N. Parin. 2000. Continental slope and deep-sea fisheries: implications for a fragile ecosystem. ICES Journal of Marine Science 57:548–557.

Koslow, J. A., K. Gowlett-Holmes, J. K. Lowry, T. O'Hara, G. C. B. Poore, and A. Williams. 2001. Seamount benthic macrofauna off southern Tasmania: community struicture and impacts of trawling. Marine Ecology Progress Series 213:111–125.

Lamont, P. A., and J. D. Gage. 1998. Dense brittle star population on the Scottish continental slope. Pages 377–382 *in* R. Mooi and M. Telford, editors. Echinoderms: San Francisco. Proceedings of the Ninth International Echinoderm Conference, Balkema, Rotterdam, The Netherlands.

Le Danois, E. 1948. Les profondeurs de la mer. Payot, Paris.

Levin, L. A., and C. R. Smith. 1984. Response of background fauna to disturbance and enrichment in the deep sea: a sediment tray experiment. Deep-Sea Research 36A:1897–1915.

Leys, S. P., and N. R. J. Lauzon. 1998. Hexactinellid sponge ecology: growth rates and seasonality in deep water sponges. Journal of Experimental Marine Biology and Ecology 230:111–129.

Lindegarth, M., D. Valentinsson, and M. Hanson. 2000. Effects of trawling disturbances on temporal and spatial structure of benthic soft-sediment assemblages in Gullmarsfjorden, Sweden. ICES Journal of Marine Science 57:1369–1376.

Long, D., J. M. Roberts, and E. J. Gillespie. 1999. Occurrences of *Lophelia pertusa* on the Atlantic margin. British Geological Survey Technical Report WB/99/24, Keyworth, Nottingham, UK.

Lorance, P., and H. Dupouy. 2001. CPUE abundance indices of the main target species of the French deep-water fishery in ICES sub-areas V-VII. Fisheries Research 51:137–149.

Marques, A. 1998. A note on the diet of *Synaphobranchus kaupi* (Pisces: Synathobranchidae) from the Porcupine Seabight, north-east Atlantic. Journal of the Marine Biological Association of the United Kingdom 78:1385–1388.

Martini, F. H. 1998. The ecology of hagfishes. Pages 57–77 *in* J. M. Jorgensen, J. P. Lornholt, R. E. Weber, and H. Malte, editors. The biology of hagfishes. Chapman and Hall, New York.

Merrett, N. R., and R. L. Haedrich. 1997. Deep-sea demersal fish and fisheries. Chapman and Hall, London.

Mikkelsen, N., J. S. Erlenkeuser, J. S. Killingley, and W. H. Berger. 1982. Norwegian corals: radio-carbon and stable isotopes in *Lophelia pertusa*. Boreas 11:163–171.

Mortensen, P. B. 2001. Aquarium observations on the deep-water coral *Lophelia pertusa* (L., 1758) (Scleractinia) and selected associated invertebrates. Ophelia 54:83–104.

Mortensen, P. B., M. T. Hovland, and J. H. Fosså. 2001. Distribution, abundance and size of *Lophelia pertusa* reefs in mid-Norway in relation to seabed characteristics. Journal of the Marine Biological Association of the United Kingdom 81:581–597.

Mortensen, P. B., and H. T. Rapp. 1998. Oxygen and carbon isotope ratios related to growth line patterns in skeletons of *Lophelia pertsa* (L.) (Anthozoa, Scleractinia): implications for determining linear extension rates. Sarsia 83:433–446.

National Research Council. 2002. Effects of trawling and dredging on seafloor habitat. National Academy Press, Washington, D.C.

Palanques, A., J. Guillen, and P. Puig. 2001. Impact of bottom trawling on water turbidity and muddy sediment of an unfished continental shelf. Limnology and Oceanography 46:1100–1110.

Pauly, D., V. Christensen, S. Guenette, T. J. Pitcher, U. R. Sumailia, and C. J. Walters. 2002. Towards sustainability in world fisheries. Nature (London) 418:689–695.

Pilskaln, C. H., J. H. Churchill, and L. M. Meyer. 1998. Resuspension of sediment by bottom trawling in the Gulf of Maine and potential geochemical consequences. Conservation Biology 12:1223–1229.

Pitcher, C. R., I. R. Poiner, B. J. Hill, and C. Y. Burridge. 2000. Implications of the effects of trawling on sessile megazoobenthos on a tropical shelf in northeastern Australia. ICES Journal of Marine Science 57:1359–1368.

Roberts, J. M., S. M. Harvey, P. A. Lamont, J. D. Gage, and J. D. Humphery. 2000. Seabed photography, environmental assessment and evidence for deep-water trawling on the continental margin west of the Hebrides. Hydrobiologia 441:173–183.

Roberts J. M., D. Long, J. B. Wilson, P. B., Mortensen, and J. D. Gage. 2003. The cold-water coral *Lophelia pertusa* (Scleractinia) and enigmatic seabed mounds along the north-east Atlantic margin: are they related? Marine Pollution Bulletin 46:7–20.

Rogers, A. D. 1999. The biology of *Lophelia pertusa* (Linnaeus 1758) and other deep-water reef-forming corals and impacts from human activities. International Review of Hydrobiology 84:315–406.

Rumohr, H., and P. Krost. 1991. Experimental evidence of damage to benthos by bottom trawling with special reference to *Arctica islandica*. Meeresforschung 33:340–345.

Rumohr, H., and T. Kujawski. 2000. The impact of trawl fishery on the epifauna of the southern North Sea. ICES Journal of Marine Science 57:1389–1394.

Sainsbury, K. J. 1988. The ecological basis of multispecies fisheries and management of a demersal fishery in tropical Australia. Pages 349–382 *in* J. A. Gulland, editor. Fish population dynamics. John Wiley, London.

Sainsbury, K. J., R. A. Campbell, R. Lindholm, and W. Whitelaw. 1997. Experimental management of an Australian multispecies fishery: examining the possibility of trawl-induced habitat modification. Pages 107–112 *in* E. K. Pikitch, D. D. Huppert, and M. P. Sissenwine, editors. Global trends: fisheries management. American Fisheries Society, Symposium 20, Bethesda, Maryland.

Schwinghamer, P., D. C. Gordon, T. W. Rowell, J. Prena, D. L. McKeown, G. Sonnichsen, and J. Y. Guigne. 1998. Effects of experimental otter trawling on surficial sediments properties of a sandy-bottom ecosystem on the Grand Banks of Newfoundland. Conservation Biology 12:1215–1222.

Smith, C. J., K. N. Papadopoulou, and S. Diliberto. 2000.

Impact of otter trawling on an eastern Mediterranean commercial trawl fishing ground. ICES Journal of Marine Science 57:1340–1351.

Smith, C. R. 1985a. Colonization studies in the deep sea: are results biased by experimental design? Pages 183–190 in P. E. Gibbs, editor. Proceedings of the nineteenth European Marine Biology Symposium. Cambridge University Press, Cambridge, UK.

Smith, C. R. 1985b. Food for the deep sea: utilization, dispersal and flux of nekton falls at the Santa Catalina Basin floor. Deep-Sea Research 32:417–442.

Smith, C. R. 1986. Nekton falls, low-intensity disturbance and community structure of infaunal benthos in the deep sea. Journal of Marine Research 44:567–600.

Stockton, W. L., and DeLaca, T. E. 1982. Food falls in the deep sea: occurrence, quality and significance. Deep-Sea Research 29:157–169.

Stoker, M. S., K. Hitchen, and C. C. Graham. 1993. United Kingdom Offshore regional report: the geology of the Hebrides and West Shetland shelves and adjacent deep-water areas. Her Majesty's Stationary Office, for the British Geological Survey, London.

Thiel, H., G. Schriever, A. Ahnert, H. Bluhm, C. Borowski, and K. Vopel. 2001. The large-scale environmental impact experiment DISCOL—reflection and foresight. Deep-Sea Research Part II 48:3869–3882.

Thistle, D., J. Y. Yingst, and K. Fauchald. 1985. A deep-sea benthic community exposed to strong bottom currents on the Scotian Rise (Western Atlantic). Marine Geology 66:91–112.

Thrush, S. F., and P. K. Dayton. 2002. Disturbance to marine benthic habitats by trawling and dredging: implications for marine biodiversity. Annual Review of Ecology and Systematics 33:449–473.

Thrush, S. F., J. E. Hewitt, V. J. Cummings, P. K. Dayton, S. J. Turner, G. Funnell, R. Budd, C. Milburn, and M. R. Wilkinson. 1998. Disturbance of the marine benthic habitat by commercial fishing: impacts at the scale of the fishery. Ecological Applications 8:866–879.

Tuck, I. D., S. J. Hall, M. R. Robertson, E. Armstrong, and D. J. Basford. 1998. Effects of physical trawling disturbance in a previously unfished sheltered Scottish sea loch. Marine Ecology Progress Series 162:227–242.

Watling, L., and E. A. Norse.1998. Disturbance of the seabed by mobile fishing gear: a comparison to forest clearcutting. Conservation Biology 12:1180–1197.

Immediate Effects of Experimental Otter Trawling on a Sub-Arctic Benthic Assemblage inside Bear Island Fishery Protection Zone in the Barents Sea

TINA KUTTI,[1] TORE HØISÆTER,[2] AND HANS TORE RAPP[3]

Department of Biology, University of Bergen, Post Office Box 7800, 5020 Bergen, Norway

ODD-BØRRE HUMBORSTAD,[4] SVEIN LØKKEBORG,[5] AND LEIF NØTTESTAD[6]

Institute of Marine Research, Post Office Box 1870 Nordnes, 5817 Bergen, Norway

Abstract. Marine epibenthic fauna in a sub-arctic ecosystem have been studied at 100-m depths within the Fishery Protection Zone around Bear Island, Barents Sea. The immediate effects of intensive experimental otter trawling are described. A BACI (Before–After and Control–Impact) design was used to quantify the trawling impact. Replicate samples were collected using a *Sneli* epibenthic sled equipped with a video camera and an integrated trawl instrumentation positioning system. The benthic assemblage was characterized by a small-scale patchy distribution of the fauna. We found a numerical domination by ophiuroids, polychaetes, bivalves, cirripedes and echinoids, while echinoids and cirripedes dominated the biomass. Trawling seemed to affect the benthic assemblage mainly through resuspension of surface sediment and through relocation of shallow burrowing infaunal species to the surface of the seafloor, and 25 h after trawling, we found a significant increase in the abundance and biomass of a majority of the infaunal bivalves. The number of species sampled remained constant after trawling, as did the diversity based on numerical abundance. However, diversity based on biomass data was significantly higher after trawling. Although multivariate analyses indicated that the composition in the samples taken after trawling were somewhat different from the majority of those taken before trawling, no evidence of dramatic changes in the composition of the fauna due to trawling was found. Almost no dead or broken benthic animals were found in the epibenthic samples.

Introduction

The Barents Sea is a very productive, sub-arctic ecosystem that covers an area of 1.4 million km^2 (Sakshaug et al. 1994) and has total annual yields of demersal fish of about 1.3 million tons (Nakken 1998). The most important commercially exploited demersal fish species are Atlantic cod *Gadus morhua* and haddock *Melanogrammus aeglefinus*, with otter trawls accounting for 85% of the catch (Nakken 1998). Earlier studies have shown the Barents Sea ecosystems to be vulnerable to anthropogenic disturbances (Klungsøyr et al. 1994), but so far no studies have investigated the impact of trawling on the benthic ecosystems. As part of a 2-year project investigating the short-term and longer-term effects of trawling on benthic fauna in the Barents Sea, immediate effects of otter trawling were investigated in 2000. Based on the results from recent studies, we expected that the experimental trawling could cause immediate changes in the faunal composition through (1) resuspension of surface sediment and displacement of specimens by the trawl (Kaiser and Spencer 1996; Bergman and Santbrink 2000; Hanson et al. 2000), (2) bycatch and removal of specimens by the trawl (Collie et al. 1997; Prena et al. 1999; Ball et al. 2000), (3) immigration of scavenging species into the recently trawled area (Kaiser and Spencer 1994; Ramsay et al. 1996, 1998), and (4) physical damage and mortality inflicted on the organisms by the trawl gear (Freese et al. 1999; Prena et al. 1999). This report focuses on the immediate effects of intensive experimental otter trawling.

---

[1] E-mail: tina.kutti@bio.uib.no
[2] E-mail: tore.hoisater@bio.uib.no
[3] E-mail: hans.rapp@bio.uib.no
[4] E-mail: oddb@imr.no
[5] E-mail: svein.lokkeborg@imr.no
[6] E-mail: leif.noettestad@imr.no

Methods

Study Site

The investigation of short term effects consisted of experimental trawling and sampling of bottom fauna before and after the trawling. The study was conducted between 10 and 24 May 2000, 9 nautical miles west of Bear Island, at 74°30'N, 18°12'E (Figure 1). The site has been protected from commercial fishing activity since 1978 and is adjacent to important trawling grounds in the Barents Sea. The bottom substrate at the experimental site (85–101 m) is dominated by shell debris mixed to varying degrees with finer sediment, and at several locations in the experimental site, aggregations of boulders (0.1–1 m in diameter) are present (Humborstad et al. 2004).

Experimental Trawling

The experimental trawling was carried out by a commercial trawler equipped with otter boards weighing 2,300 kg each, rock hopper gear (21 in, 19 m), and a door

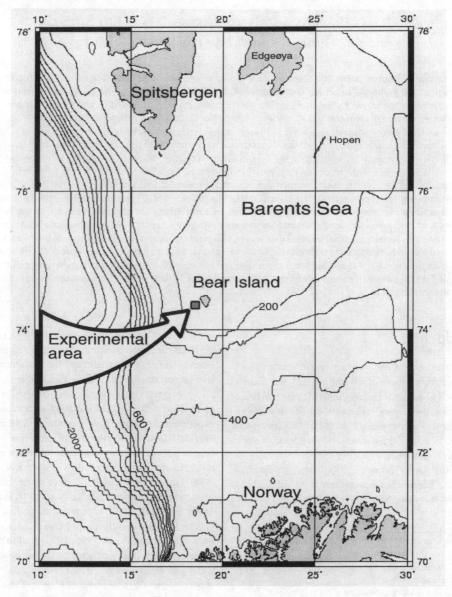

Figure 1. Map of the Barents Sea, showing the locality of the experimental site. Isobaths are shown every 200 m.

spread of 140 m. The position of the trawl was monitored with a SIMRAD integrated trawl instrumentation (ITI; Engås et al. 2000). To simulate intensive trawl disturbance, one 200-m-wide transect was trawled 10 times along the same center line. This intensity was based on the average trawl effort west of Bear Island estimated to twice annually (data set from the Norwegian Directorate of Fisheries, see Salthaug and Godø 2001). With an uneven distribution of trawling, a higher effort would be expected in some areas (Rjinsdorp et al. 1998; Kaiser et al. 2002). Towing speed was approximately 3.6 knots, and the experimental trawling was completed in 20 h.

Epibenthic Sled Sampling

Benthic organisms were collected using a *Sneli* epibenthic sled (Sneli 1998), with a sample width of 70 cm. The sled had been equipped with an ITI positioning sensor and a video camera, which made it possible to supervise sampling so that tows of dubious quality could be aborted. Video and ITI data were also used to estimate the actual area sampled in each haul, and abundance and biomass of the sled catch were converted to individuals and g/m^2. The duration of each tow was 2 min, and the speed during tows was 1 knot. For the purpose of this study, all animals larger than 5 mm were sorted from the samples and preserved in 4% buffered formalin in seawater. In the laboratory, the organisms were transferred to ethanol, identified to lowest possible taxonomic level, counted, and weighed (ethanol wet weight).

Experimental Design

A 2-year study was implemented to test the hypothesis that experimental trawling may cause changes in the benthic faunal assemblage (Underwood 1996). This entailed sampling of the fauna before experimental trawling and immediately after trawling as well as follow-up sampling at six monthly intervals for 2 years. We report here on the changes immediately after trawling.

Within the experimental site, five parallel transects, each roughly 2.25 km in length, were aligned east–west (Figure 2). One transect was treated with intensive (Ti) and one with moderate experimental trawling (Tm), while three served as control transects (C_1–C_3). To ensure that the whole length of the transect was sampled, each transect was divided into 10 250-m blocks. The samples for this study were taken from transects Ti and C_2. Two random samples were collected in each block in the intensively trawled transect, one before (Ti_B) and one immediately (15 h) after (Ti_A). Samples were also collected in the control transect (C_2) both before and after trawling.

As no proper pilot sampling could be fitted into the program and no prior quantitative knowledge of the benthic assemblage in the area existed, the patchiness of the assemblage was assumed to be similar to the level found on the Grand Banks of Newfoundland (Prena et al. 1999; in which a comparable sampling technique had been used). Given this assumption, prospective power analyses showed that a sample size of $N = 10$ should have 80% power to detect a 20% reduction in total biomass.

Data Analysis

Diversity was assessed by number of species per sample and expected number of species per 125 individuals, as estimated by Hurlbert's modification of Sanders's rarefaction index (Hurlbert 1971). Further, Shannon's diversity index (H´) ($\log_e$) (Shannon and Weaver 1949), Simpson's diversity index (D´) (Simpson 1949), and Pielou's measure of evenness (J) (Pielou 1977) were estimated for both numerical abundance and biomass data. This was done to document different effects of trawling on abundance and biomass and to enable inclusion of colonial species (i.e., bryozoans, sponges, and hydroids) in the diversity measures. Changes in total abundance and biomass from before to after trawling were examined using one-way analysis of variance (ANOVA). Changes in abundance and biomass of all species individually and changes in diversity were investigated using a Mann–Whitney U-test. In all analyses α was set to 0.10 to reduce the risk of committing type II statistical errors (Peterman 1990; Thrush et al. 1995; McConnaughey et al. 2000). The analyses of variance and U-tests were performed in Statistica (version 6.0, StatSoft). Abundance–biomass comparisons were done with ABC-curves (Warwick 1986). Similarities between stations were calculated using the Bray–Curtis similarity coefficient (Bray and Curtis 1957). Dendrograms (hierarchical cluster analysis, group average linkage) and MDS ordinations for all stations were constructed based on the similarity matrices. The matrices were then tested for differences between groups of samples using ANOSIM (Clarke and Warwick 2001). For the multivariate analyses, data were double-square-root transformed (Field et al. 1982). The ABC-curves (Clarke and Warwick 1994) and multivariate analyses were calculated using PRIMER (version 5; Clarke and Gorley 2001). Retrospective power analyses were performed using PASS (version 6.0, NCSS Statistical Software) on the nonsignificant results of the ANOVA. This was done to determine whether the power of the test was sufficient to detect the minimum effect size (20% reduction in total biomass).

Results

Faunal Composition

A total of 163 species (and 16 higher taxa) were identified. The assemblage was numerically dominated by ophi-

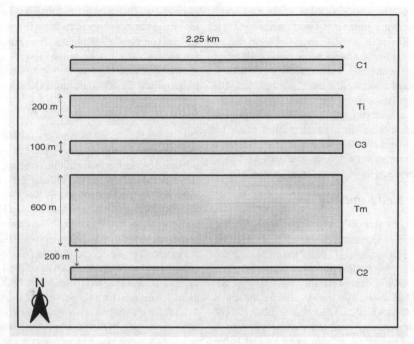

Figure 2. Relative position, transect division, and orientation of the trawled transects (Ti and Tm) and control transects (C_1–C_3) used in the experiment. Transect widths are given on the left.

uroids (24%), polychaetes (15%), bivalves (13%), cirripedes 12%, and echinoids (11%). *Ophiura sarsii* (Ophiuroidea) was the most abundant species, comprising 23% of all individuals found, followed by *Balanus balanus* (Cirripedia), *Nothria conchylega* (Polychaeta), *Nuculana minuta* (Bivalvia), *Strongylocentrotus droebachiensis* (Echinoidea), and *Pagurus pubescens* (Crustacea) (Figure 3A). Three species, *Strongylocentrotus droebachiensis*, *S. pallidus*, and *Balanus balanus*, dominated the biomass of the assemblage (Figure 3B). The benthic assemblage was characterized by a small-scale patchy distribution of the fauna (e.g., the estimated density of *Strongylocentrotus* spp. varied from 0.2 to 8.0 individuals/m² in the different samples).

Community Analyses

The cluster analysis of the Ti_B and Ti_A abundance data showed that at a cutoff level of roughly 72% similarity, three clusters are present (Figure 4). One cluster consists of three "before" stations, a second cluster of two "before" stations, and a third cluster of five stations sampled after trawling. The remaining stations were connected to the central dendrogram singly or in small, loosely interconnected subgroups with no apparent system. A two-dimensional MDS ordination plot (stress value = 0.19) using the same similarity matrix confirmed the impression from the dendrogram. In spite of this modest degree of clustering, an ANOSIM test of the "before" versus "after" stations had a sample statistic of 0.164, which had a significance level of 1.6%. Further analysis of the similarity matrix showed that the similarities between Ti_A hauls were significantly higher than those between Ti_B hauls. The similarity using biomass data resembled the pattern of the abundance data, although the grouping of "before" and "after" stations was less pronounced. A separate multivariate analysis including data from control transect C_2 confirmed that these stations were much more heterogeneous than either the "before" or "after" stations from the trawl transect. Since the stations on the C_2 transect were sampled only once on this cruise, we decided to concentrate on the direct comparison between Ti_B and Ti_A for assessing the immediate effects of trawling on the benthic fauna.

The mean number of species per sample and the mean of the expected number of species per 125 individuals were similar for the "before" and "after" samples (Table 2; Figure 3). The three diversity measures based on abundance data (Table 2) also were largely unaffected by the trawling, and no effects of the trawling could be inferred from the ABC-curves (Figure 5). However, while the biomass curve for the "before" samples was dominated by three heavy species (*Strongylocentrotus droebachiensis*, *S. pallidus*,

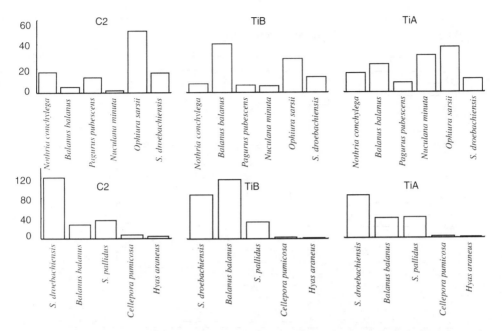

Figure 3. (A) Abundance of the six most common species in each transect sampled (C_2, Ti_B, and Ti_A). Values are given as individuals/10 m². (B) Biomass of the five most dominating species from all three transects. Values are given as g WW/7 m². S = *Strongylocentrotus*.

and *Balanus balanus*), the corresponding curve for the "after" samples indicated a much more even distribution of relative biomasses. While no significant effects of trawling on the diversity of the samples were found using numerical abundance, a clear effect of trawling was seen using biomass data. The stations sampled after trawling had a significantly higher diversity (as measured by all three diversity indices, Table 2) than did the same transects sampled before trawling (U-test, $P < 0.002$).

We did not detect a difference in total abundance or biomass of benthic species when comparing samples from before and after trawling. Retrospective power analyses that were carried out using data of total biomass in the Ti_A, Ti_B and C_2 samples revealed that the power of the experiment to detect a minimum effect size of a 20% reduction in biomass was 0.23 ($\alpha = 0.1$). A sample size of $N = 56$ would have been required to document a 20% reduction with a power of 0.8 (Table 1).

Abundance and Biomass of Individual Species

Changes in abundance and biomass have been investigated for all species, but only the large and consistent changes are described here. Many infaunal bivalves increased in number and biomass after trawling (Table 3). The common bivalves *Nuculana minuta*, *Astarte crebricostata*, *A. elliptica*, *Ciliatocardium ciliatum*, *Serripes groenlandicus* and *Thracia* sp. were all found with a significantly higher abundance and biomass in the benthic assemblage after trawling (Mann–Whitney U-test, $P < 0.1$). The epifaunal bivalve *Hiatella* sp. was the only common bivalve that we found with reduced numbers and significantly lower biomass after trawling (U-test, $P < 0.1$). It lives in association with the cirripede *Balanus balanus*, which also was found with an insignificantly lower biomass after. In general, epibenthic sled samples showed that few species reduced in abundance after trawling. However, a significant reduction (U-test, $P < 0.01$) was observed in the abundance of the shrimp *Spirontocaris spinus*, whose mean density decreased from 20 to 5.5 shrimps/30 m² after trawling.

Many of the larger invertebrates found in high numbers in our samples are potential scavengers, such as a hermit crab *Pagurus pubescens*, the brittle star *Ophiura sarsii* and the two species of the sea urchin genus *Strongylocentrotus*. Only *Ophiura sarsii* increased in abundance (Figure 3) and in biomass after trawling (U-test, $P < 0.1$), indicating an increased mean weight per individual. The other potential scavengers showed no apparent response to the trawling activity.

Physical Damage to Benthic Organisms

Very few benthic animals with physical injuries were found in the sled catches. The few injuries found were

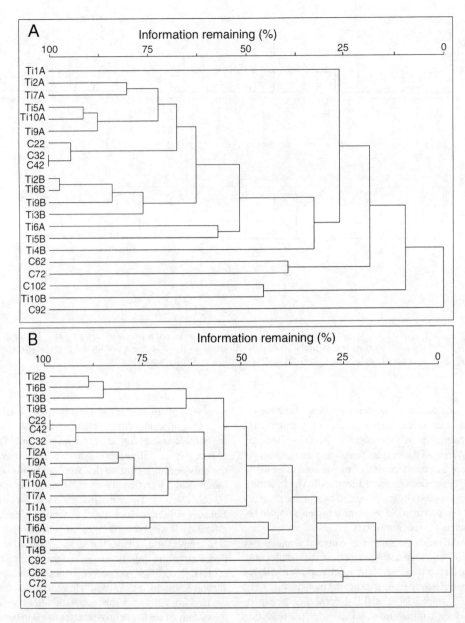

Figure 4. Dendrogram showing grouping of the "before" (B) and "after" (A) stations, based on the results of cluster analyses (hierarchical, group average sorting) of faunal abundance. The dashed line shows approximately 72% similarity. Classifications are based on Bray-Curtis similarity measure, with double square-root-transformed data.

confined to the sea urchins *Strongylocentrotus* spp., and the bivalves *Thracia* spp., *Ciliatocardium ciliatum* and *Serripes groenlandicus*. The damaged individuals were found in only three of the sled hauls from the trawled transect (and one from the untrawled). The percentage of damaged individuals of *Strongylocentrotus* spp. was never higher than 4% and much lower for the bivalves.

Discussion

Methods and Experimental Area

The Fishery Protection Zone around Bear Island was considered suitable for the conduction of an experimental study of trawl impact, as the area is both undisturbed yet suitable for fishing. The seabed, however, turned out to be rather

Table 1. Abundance-diversity and biomass-diversity of the 14 stations sampled before and after trawling. Diversity is given as Shannon's diversity index (H'), Simpson's diversity index (D'), and Pielou's evenness measure (J). Stations are arranged within each transect in falling order with the stations of highest diversity index (H'), calculated on abundance data, first.

Stations	Abundance			Biomass		
	H'	D'	J	H'	D'	J
Ti5$_B$	3.183	0.929	0.781	1.436	0.611	0.340
Ti4$_B$	2.981	0.902	0.744	1.518	0.655	0.369
Ti6$_B$	2.717	0.885	0.669	1.230	0.562	0.288
Ti9$_B$	2.620	0.868	0.681	1.474	0.697	0.377
Ti3$_B$	2.300	0.804	0.591	1.160	0.594	0.286
Ti2$_B$	2.272	0.786	0.581	1.074	0.446	0.267
Ti10$_B$	2.246	0.735	0.574	1.124	0.463	0.274
Ti9$_A$	2.938	0.901	0.751	2.019	0.769	0.502
Ti5$_A$	2.676	0.876	0.659	1.712	0.721	0.409
Ti2$_A$	2.604	0.862	0.636	1.823	0.724	0.434
Ti10$_A$	2.540	0.859	0.667	1.815	0.761	0.453
Ti7$_A$	2.537	0.853	0.615	1.594	0.690	0.373
Ti1$_A$	2.515	0.779	0.603	1.842	0.687	0.431
Ti6$_A$	2.484	0.841	0.656	1.658	0.719	0.414
Average (Ti$_B$)	2.617	0.844	0.660	1.288	0.575	0.314
Average (Ti$_A$)	2.613	0.853	0.655	1.780	0.724	0.431

heterogeneous, the fauna was a mixture of hard-bottom and soft-bottom species, and this complicated the interpretation of the results. By taking samples over a larger area using an epibenthic sled, the effects of the patchy distribution of fauna was reduced, but due to the limited number of samples taken, the detection of minor changes in faunal composition was not possible. The retrospective power analysis (Table 1) indicated that the number of sled hauls per transect was insufficient. However, this power analysis was based only on total biomass, and conclusions based on individual groups of species were still possible, as we reported for the bivalves.

Biomass was assessed as an additional measure of quantity in the diversity estimations because (1) larger animals are presumed to be more vulnerable to physical disturbance than are small ones, (2) biomass is the only means of incorporating colonial animals in a diversity index, (3) construction of ABC-curves will reveal the

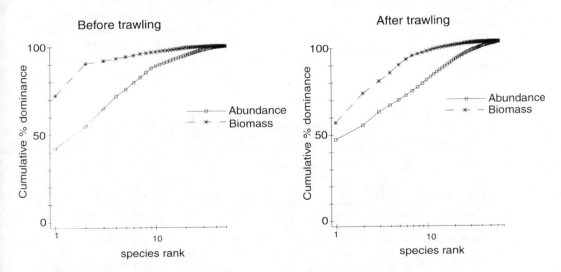

Figure 5. ABC-curves from the intensively trawled transect that summarizes the results from seven stations sampled before (top) and after (bottom) trawling. Data are per m^2 of seabed sampled.

Table 2. Retrospective calculations of the power of the experiment to detect the pre-specified effect size of 20% reduction in biomass using observed variance. Minimum detectable effect size using observed sample size and variance is also shown along with the required sample size to detect a biomass reduction of 20% using the observed variance.

	Before trawling		After trawling				Min detectable
N	Mean	SD	Mean	SD	α	power	difference
7	35.75	17.19	28.60	12.79	0.05	0.143	20%
7	35.75	17.19	28.60	12.79	0.10	0.229	20%
7	35.75	17.19	15.61	12.79	0.10	0.803	56%
71	35.75	17.19	28.60	12.79	0.05	0.803	20%
56	35.75	17.19	28.60	12.79	0.10	0.803	20%

more drastic alterations in community composition, and (4) some changes in diversity caused by the experimental treatment might be easier to detect. Pielou (1975) stated that any measure of quantity can be used in Shannon's and Simpson's indices. In our case, biomass added interesting information and revealed changes where indices based on abundance did not.

Changes in Community Composition

In general, the changes in community composition that resulted from trawling were moderate and partly obscured by the mosaic nature of the bottom substrate. However, comparing the results from the abundance–diversity and biomass–diversity calculations, it is clear that some changes in the community composition did occur following trawling. While there were no changes from before to after the trawling for the abundance diversities (Table 2), there was a marked (and highly significant) increase in the biomass diversities after trawling. That this was mainly due to the influence of the three dominating species (in biomass) was seen when the measures were recalculated without these (especially *B. balanus* but also *Strongylocentrotus* spp.). When the indices were calculated without the three species, all three measures approached the values based on abundance. Although the reduction in the biomass of cirripedes after trawling was not statistically significant, we believe it was evident. The inferred reduction could have been caused by trawl bycatch, a change in the catchability of the fauna through burial by resuspended sediment, or a result of overturning of boulders by the trawl.

The most pronounced immediate effect of trawling was the large increase in abundance of all the dominating infaunal bivalves following trawling (Table 3). Trawl gear causes resuspension of surface sediment as well as relocation of shallow burrowing organisms living near the sediment seabed surface. This would allow a larger fraction of the population to come within reach of the *Sneli* sled immediately after trawling (Gilkinson et al. 1998). Similar observations (i.e., increased density of burrowing bivalves after trawling) have been made in sandy bottom habitats on the west coast of Sweden and the North Sea (Kaiser and Spencer 1996; Bergman and van Santbrink 2000; Hansson et al. 2000).

Contrary to what we expected, immigration of scavenging species into the experimental area following trawling was not observed, although many potential scavengers were found in high numbers in the benthic samples. The increase in individual size of *Ophiura sarsii* after

Table 3. Summarization of the mean abundance (individuals/m) and biomass (g WW/m^2) of 7 dominating bivalves found in the treated transect before and after trawling. Also indicated is the percent increase from before to after trawling.

	Abundance			Biomass		
	Mean			Mean		
Species	Ti_B	Ti_A	%	Ti_B	Ti_A	%
Nuculana minuta	1.82	4.41	140	3.63	11.20	210
Astarte elliptica	0.16	0.33	110	0.63	1.35	120
Astarte crebricostata	0.03	0.20	690	0.21	3.58	1,620
Serripes groenlandicus	0.08	0.20	160	0.62	4.31	600
Ciliatocardium ciliatum	0.01	0.03	310	0.07	6.19	8,470
Thracia sp.	0.02	0.09	390	0.01	0.07	500
Hiatella sp.	0.81	0.19	-80	6.60	2.22	-70

trawling could, however, be explained by an immigration of larger specimens into the trawled area. A possible explanation for the low response may be the low number of damaged organisms found in the area after trawling. Also, our study was initiated 15 h after trawling, which could be too soon to detect an increase in the abundance of slower moving, benthic scavenging invertebrates. In a study conducted by Ramsay et al. (1998), the density of hermit crabs (*Pagurus* spp.) did not increase until 2 and 3 d after trawling, while samples only 20 h after trawling showed no increase.

Based on the results of earlier investigations (Freese et al. 1999; Prena et al. 1999), we expected that a substantial proportion of the echinoids and mollusks (that were found in large number in the sled samples) would be damaged or broken as a result of the trawling. In the experimental area, side-scan images showed that the trawl doors had caused deep scour furrows in the sediment (Humborstad et al. 2004). These scour furrows were also visible in several of the video recordings from the sled sampling. The lack of damage to displaced infaunal bivalves was, therefore, rather surprising. One explanation could be that the small size of the bivalves allowed them to be displaced rather than damaged by the trawl doors (Gilkinson et al. 1998). The lack of damage on the epifauna sea urchin (*Strongylocentrotus* spp.) was also surprising. Using the same type of foot gear, 10% of damaged individuals of *S. palllidus* were found in experimentally trawled corridors in sandy substrate by Prena et al. (1999). The lack of damage in our experiment might be explained by the type of ground gear we used, perhaps in combination with the type of bottom substrate. The rock-hopper gear has an off-center rope that prevents it from rolling along the bottom (Engås and Godø 1989), and when the gear is dragged along the bottom, it pushes a heap of sediment in front and to the side of the trawl like a plow, and physical damage on organisms could be minimized. However, many other factors can influence the impact of trawl disturbance on fauna, including the frequency of the disturbance, the time period between trawling and sampling (delayed mortality), the type of trawl gear used, and the nature of the habitat examined. The large, erect epifauna in the experimental area (i.e., hydroids, bryozoans and poriferans) was patchily distributed; therefore, detection of minor changes in abundance was impossible.

Conclusions

"Intensive" trawling, as performed in this experiment, caused a substantial increase in the abundance and biomass of a number of infaunal bivalves as measured by an epibenthic sled. This indicates a displacement of the animals toward the seabed surface. A marked increase in biomass diversity after trawling was caused by reduction in the biomass of the cirripede *Balanus balanus* and, to a lesser degree, the echinoid *Strongylocentrotus droebach-iensis*, presumably caused by the trawling (bycatch and displacement). We detected very little physical injuries or mortality on any of the animals caught 20 h after trawling. Neither did we detect any migration of scavenging species into the trawled area during this time interval.

Acknowledgments

This study is a part of the project "Environmental Effects of Fishing," funded by the Research Council of Norway. The authors thank all who assisted us with data collection and data processing. The manuscript has benefited from remarks by two anonymous referees.

References

Ball, B., B. Munday, and I. Tuck. 2000. Effects of otter trawling on the benthos and environment in muddy sediments. Pages 69–82 *in* M. J. Kaiser and S. J. de Groot, editors. The effects of fishing on non-target species and habitats: biological, conservation and socio-economic issues. Blackwell Scientific Publications, Oxford, UK.

Bergman, M. J. N., and J. W. van Santbrink. 2000. Fishing mortality of populations of megafauna in sandy sediments. Pages 49–68 *in* M. J. Kaiser and S. J. de Groot, editors. The effects of fishing on non-target species and habitats: biological, conservation and socio-economic issues. Blackwell Scientific Publications, Oxford, UK.

Bray, J. R., and J. T. Curtis. 1957. An ordination of the upland forest communities of southern Wisconsin. Ecological Monographs 27:325–349.

Clarke, K. R., and R. N. Gorley. 2001. PRIMER, version 5: user manual/tutorial. PRIMER-E, Plymouth, UK.

Clarke, K. R., and R. M. Warwick. 1994. Relearning the ABC-taxonomic changes and abundance biomass relationships in disturbed benthic communities. Marine Biology 118:739–744.

Clarke, K. R., and R. M. Warwick. 2001. Change in marine communities: an approach to statistical analysis and interpretation, 2nd edition. PRIMER-E, Plymouth, UK.

Collie, J. S., G. A. Escanero, and P. C. Valentine. 1997. Effects of bottom fishing on the benthic megafauna of Georges Bank. Marine Ecology Progress Series 155:159–172.

Engås, A., and O. R. Godø. 1989. Escape of fish under the fishing line of a Norwegian sampling trawl and its influence on survey results. Journal du Conceil 45:269–276.

Engås, A., O. R. Godø, and T. Jørgensen. 2000. A comparison between vessel and trawl tracks as observed by the ITI trawl instrumentation. Fisheries Research 45:297–301.

Field, J. G., K. R. Clarke, and R. M. Warwick. 1982. A practical strategy for analysing multispecies distribution patterns. Marine Ecology Progress Series 8:37–52.

Freese, L., P. J. Auster, J. Heifetz, and B. L. Wing. 1999. Effects of trawling on seafloor habitat and associated invertebrate taxa in the Gulf of Alaska. Marine Ecology Progress Series 182:119–126.

Gilkinson, K., M. Paulin, S. Hurley, and P. Schwinghamer. 1998. Impact of trawl door scouring on infaunal bivalves: results of a physical trawl door model/dense sand interaction. Journal of Experimental Marine Biology and Ecology 224:291–312.

Hansson, M., M. Lindegarth, D. Valentinsson, and M. Ulmestrand. 2000. Effects of shrimp-trawling on abundance of benthic macrofauna in Gullmarsfjorden, Sweden. Marine Ecology Progress Series 198:191–201.

Humborstad, O. B., L. Nøttestad, S. Løkkeborg, and H. T. Rapp. 2004. RoxAnn bottom classification system, sidescan sonar and video-sledge: spatial resolution and their use in assessing trawling impacts. ICES Journal of Marine Science 61:53–63.

Hurlbert, S. H. 1971. The nonconcept of species diversity: a critique and alternative parameters. Ecology 52:577–586.

Kaiser, M. J., J. S. Collie, S. J. Hall, S. Jennings, and I. R. Poiner. 2002. Modification of marine habitats by trawling activities: prognosis and solutions. Fish and Fisheries 3:114–136.

Kaiser, M. J., and B. E. Spencer. 1994. Fish scavenging behaviour in recently trawled areas. Marine Ecology Progress Series 112:41–49.

Kaiser, M. J., and B. E. Spencer. 1996. The effects of beam-trawl disturbance on infaunal communities in different habitats. Journal of Animal Ecology 65:348–358.

Klungsøyr, R., R. Sætre, L. Føyn, and H. Loeng. 1995. Man's impact on the Barents Sea. Arctic 48:279–296.

McConnaughey, R. A., K. L. Mier, and C. B. Dew. 2000. An examination of chronic trawling effects on soft-bottom benthos of the Bering Sea. ICES Journal of Marine Science 57:1377–1388.

Nakken, O. 1998. Past, present and future exploitation and management of marine resources in the Barents Sea and adjacent areas. Fisheries Research 37:23–35.

Peterman, R. M. 1990. Statistical power analysis can improve fisheries research and management. Canadian Journal of Fisheries and Aquatic Sciences 47:2–15.

Pielou, E. C. 1975. Ecological diversity. Wiley, New York.

Pielou, E. C. 1977. Mathematical ecology. Wiley, New York.

Prena, J., P. Schwinghamer, T. W. Rowell, D. C. Gordon, Jr., K. D. Gilkinson, P. W. Vass, and D. L. McKeown. 1999. Experimental otter trawling on a sandy bottom ecosystem of the Grand Banks of Newfoundland: analysis of trawl bycatch and effects on epifauna. Marine Ecology Progress Series 181:107–124.

Ramsay, K., M. J. Kaiser, and R. N. Hughes. 1996. Changes in hermit crab feeding patterns in response to trawling disturbance. Marine Ecology Progress Series 144:63–72.

Ramsay, K., M. J. Kaiser, and R. N. Hughes. 1998. Responses of benthic scavengers to fishing disturbance by towed gears in different habitats. Journal of Experimental Marine Biology and Ecology 224:73–89.

Rijnsdorp, A. D., A. D. Buys, F. Storbeck, and E. G. Visser. 1998. Micro-scale distribution of beam trawl effort in the southern North Sea between 1993 and 1996 in relation to the trawling frequency of the sea bed and the impact on benthic organisms. ICES Journal of Marine Science 55:403–419.

Sakshaug, E., A. Bjørge, B. Gulliksen, H. Loeng, and F. Mehlum. 1994. Structure, biomass distribution, and energetics of the pelagic ecosystem in the Barents Sea: a synopsis. Polar Biology 14:405–411.

Salthaug, A., and O. R. Godø. 2001. Standardisation of commercial CPUE. Fisheries Research 49:271–281.

Shannon, C. E., and W. Weaver. 1949. The mathematical theory of communication. University of Illinois, Urbana.

Simpson, E. H. 1949. Measurement of diversity. Nature (London) 163:688.

Sneli, J. A. 1998. A simple benthic sledge for shallow and deep-sea sampling. Sarsia 83:69–72.

Thrush, S. H., J. E. Hewitt, V. J. Cummings, and P. K. Dayton. 1995. The impact of habitat disturbance by scallop dredging on marine benthic communities: what can be predicted from the results of experiments? Marine Ecology Progress Series 129:141–150.

Underwood, A. J. 1996. Detection, interpretation, prediction and management of environmental disturbances: some roles for experimental marine biology. Journal of Experimental Marine Biology and Ecology 200:1–27.

Warwick, R. M. 1986. A new method for detecting pollution effects on marine macrobenthic communities. Marine Biology 92:557–562.

Preliminary Results on the Effect of Otter Trawling on Hyperbenthic Communities in Heraklion Bay, Cretan Sea, Eastern Mediterranean

PANAYOTA T. KOULOURI[1] AND COSTAS G. DOUNAS[2]

Institute of Marine Biology of Crete, Post Office Box 2214, GR-71003, Heraklion, Crete, Greece

ANASTASIOS ELEFTHERIOU[3]

Department of Biology, University of Crete, Post Office Box 2208, GR- 71409, Vassilika Vouton, Heraklion, Crete, Greece

Abstract.— Although the benthopelagic fish species are a focus of commercial exploitation, relatively little attention has been paid to the small-sized invertebrates (0.5–20 mm) living on or very close to the seabed, thus inhabiting the same biotope, known also as hyperbenthos. Recently, interest in this faunal group has increased, as many demersal fish and epibenthic crustaceans have been found to feed on hyperbenthic animals for at least part of their lives. Otter trawls, the most common gear used for demersal fishing, result in significant disturbance of the sediment–water interface. Animals that are disturbed by the passage of a trawl may become more available to predators and scavengers. We have been unable to find any reports of studies of the impacts of towed fishing gears, including otter trawls, on hyperbenthos. We studied these effects on the Mediterranean continental shelf in Heraklion Bay (Cretan Sea) using a novel apparatus to simulate the contact of otter trawl groundrope with the seabed. A modified three-level hyperbenthic sledge was used for collecting disturbed (groundrope present) and undisturbed (without groundrope) macrofaunal samples at a towing speed typical of the local commercial vessels. Observations were made before and during the trawling season in an area being actively fished. The preliminary results reported here indicate that trawling causes significant changes in the structure of hyperbenthic communities.

Introduction

Otter trawls, the most common gear used for demersal fishing, have been reported to disturb the seabed sediment to depths of up to a few centimeters below the sediment–water interface (Lindeboom and de Groot 1998). Many studies have shown that benthic infauna and epifauna can be exposed, damaged, or killed by the passage of a trawl and that this may lead to increased opportunistic feeding by invertebrate and fish predators (Jennings and Kaiser 1998; Hall 1999; Kaiser and de Groot 2000). However, no reports on the effects of demersal fishing gears, particularly otter trawls, on hyperbenthic fauna were found.

The term "hyperbenthos" is defined as the 0.5–20-mm bottom-dependent animals which perform, with varying amplitude, intensity, and regularity, seasonal or daily vertical migrations above the seabed (Brunel et al. 1978). Hyperbenthos can be distinguished as the permanent hyperbenthos (amphipods, cumaceans, decapods, isopods, mysids, pycnogonids, and tanaids) and near-bottom zooplankton (subdivided into mesozooplankton: copepods, crustacean larvae, chaetognaths, and polychaete larvae; and macrozooplankton: ctenophores and postlarval fishes) (Mees and Jones 1997). Thus, the hyperbenthos comprises a broad assemblage of various organisms related by their distribution in space and not by phylogeny or functional attributes. There are considerable practical difficulties in quantitatively sampling these often highly mobile and small animals. They may be caught occasionally by conventional sampling gears (e.g., grabs, and corers; Eleftheriou and Holme 1984) but are not retained in the net of traditional fisheries assessment gears (Hall 1999). Furthermore, observations of hyperbenthic animals in the stomach contents of predatory fish gives only qualitative information on the presence of these groups of organisms in the benthic environment (Labropoulou and

[1] E-mail: yol72@imbc.gr
[2] E-mail: kdounas@imbc.gr
[3] E-mail: telef@imbc.gr

Eleftheriou 1997). This lack of quantitative information contrasts with the importance of hyperbenthos to fisheries, as the hyperbenthic communities act as a food resource for many commercially exploited demersal fish and epibenthic crustaceans (Carrassón et al. 1997; Martin and Christiansen 1997; Cartes and Maynou 1998).

The aim of this study was to investigate the effects of otter trawling on the hyperbenthic community of the continental shelf of Heraklion Bay (Cretan Sea, Greece, Eastern Mediterranean). The study utilized a new sampling device (Koulouri et al. 2003) that simulates the disturbance of the seabed (and hyperbenthic organisms) caused by the passage of an otter trawl groundrope over the sediment surface.

Methods

Heraklion Bay fishing ground, which occupies an area of 110 km², is located on the northern coast of Crete (Cretan Sea, Greece, Eastern Mediterranean) between 35°20'N and 35°28'N and 25°'02'E and 25°20'E (Figure 1). The experimental site was selected within the fishing ground along the 50-m isobath, which coincides with a traditional fishing lane where most of the local commercial trawling is concentrated. Only one sedimentary facies, mud, is distinguished at the study site (Chronis et al. 2000). Polychaetes are the dominant macrobenthic infaunal group, in terms of species numbers, abundance, and biomass, followed by mollusks and crustaceans (Karakassis and Eleftheriou 1997). The dominant benthopelagic fish species of commercial value are: *Mullus barbatus* Linnaeus 1758, *M. surmuletus* Linnaeus 1758, *Serranus hepatus* (Linnaeus 1758), and *S. cabrilla* (Linnaeus 1758) (Kallianiotis et al. 2000). The commercial trawling season runs from the beginning of October until the end of May. Our field experiments were carried out during daylight hours on two sampling occasions, before (27 September 2001) and a week after (7 October 2001) the beginning of commercial trawling. On both occasions, the experimental site was surveyed with an underwater towed video sledge. Recent trawling activity was evident by fresh marks on the seabed, indicated by hard edges, uncovered lighter-gray sediments, and flat areas with no sedimentary features. Before the beginning of commercial trawling, older marks, indicated by softer edges with numerous bioturbation features such as burrows and mounds, were visible. It should be noted that since the beginning of the trawling season (1 October 2001), two trawlers were observed operating within the experimental site.

The sampling gear constructed and used for the experiment was a modified version of Towed Trawl Simulator Sledge (TTSS2, Dounas et al. 2002; Koulouri et al. 2003) fitted with an otter trawl groundrope (as used by local trawlers) in contact with the seabed (Figure 2). Three nets (0.5-mm mesh size) with three doors attached to a metal frame were added in order to permit sampling of macrobenthic fauna at three levels (5–30 cm, 31–56 m, and 57–82 m) above the sediment surface. An electro-

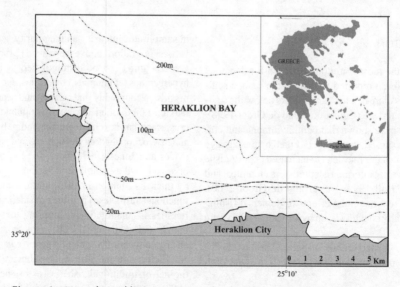

Figure 1. Map of Heraklion Bay (Crete Island, Greece, Eastern Mediterranean) showing the location of the station occupied during this study (bold dashed line shows 1,852-m (1-nautical-mi) trawling limit according to the Greek Fisheries Legislation (i.e., all bottom trawling must occur offshore of this line).

mechanical system allowed the opening (horizontal position) and closing (vertical position) of the three doors simultaneously, and an odometer was used to measure the distance traveled by the gear while in contact with the bottom. The otter trawl groundrope (1.8-m length, 6.5-cm diameter, 2-kg/m weight in water) was supported by two lightweight metal arms attached to the sides of the sledge and was positioned at a distance of 1.5 m in front of the sledge. A video camera attached to the sledge visually covered the performance of the underwater apparatus and showed that resuspension ahead of the sampling nets was caused solely by the groundrope and that the sediment cloud was no higher than the upper sampling net. The sledge was towed by the RV *PHILIA* at a normal trawling speed of ~1 m/second (2 knots). The ratio of length of wire paid out to depth was 2:1.

Profiles of standard hydrographic parameters (temperature, photosynthetically active radiation, chlorophyll a, salinity, density, and dissolved oxygen) were measured in the water column at the 50-m isobath on both sampling occasions using a conductivity/temperature/depth profiler (CTD; SBE-25, Sea-Bird Electronics, Inc., Washington).

Experimental sledge tows during daylight hours were made before and after the beginning of the trawling season. Bearing in mind the homogeneity of the bottom habitat, samples were collected on three disturbed (groundrope present) and three undisturbed (without groundrope) replicate tows along the 50-m isobath on 27 September 2001, prior to the trawling season. No attempt was made to sample exactly on the same positions, and therefore, the sampling can be considered to have been randomly distributed along the trawl track. The experimental site was sampled again (four disturbed and four undisturbed rep-

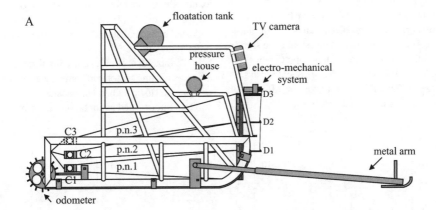

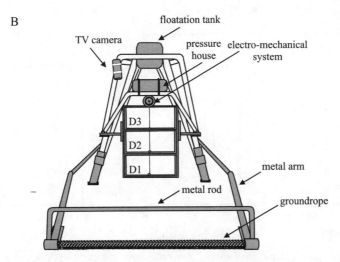

Figure 2. (A) Lateral schematic view of the TTSS2. p.n. = plankton net; C = collectors; D = doors. (B) Front schematic view of the TTSS2 (from Koulouri et al. 2003).

licate tows) on 7 October 2001, after the trawling season had commenced. Material collected was fixed with 10% formalin on board immediately after collection and sorted under a dissecting microscope, and the organisms were identified to the major taxonomic groups and counted for each replicate tow.

Densities of each taxon of the macrofauna were standardized separately to the number of individuals/m^2 of seabed for the three nets in each tow. Densities of the three nets were summed for each experimental sledge tow. Averaged densities of animals from the replicates of disturbed and undisturbed tows before and after the trawling season were calculated. The statistical significance of differences in the densities of the major taxonomic groups was assessed using the Mann–Whitney U-test.

In order to investigate the similarity between different tows, cluster analysis was performed using the Bray-Curtis similarity index (Bray and Curtis 1957) and the group average linkage method (Clarke and Warwick 1994). Data for the total number of individuals/m^2 were transformed to the square root prior to analysis. An analysis of similarity test (ANOSIM) was performed to investigate the significance of any differences found (Clarke and Green 1988; Clarke and Warwick 1993). The PRIMER statistical software package (Plymouth Marine Laboratory, Plymouth, UK) was used for the above data analyses.

Results

Data of the hydrographic parameters measured before and after the beginning of the trawling season are similar, indicating relatively stable conditions (Figure 3). No storm event took place during the short period between the two sampling occasions.

A total of 10,065 individuals were identified to 27 major taxa from all the experimental sledge tows. The averaged densities of the major taxa collected from the undisturbed and disturbed sampling tows in both sampling occasions are shown in Tables 1 and 2, respectively. The TTSS2 with groundrope present collected animals from a wide range of taxonomic groups (27), while the TTSS2 without groundrope collected a much smaller number of taxa (13). Gastropoda, Polychaeta, Mysidacea, Amphipoda, and Cumacea were the most abundant groups in the disturbed samples. Comparison of the averaged densities of the major taxa shows that most of the taxonomic groups collected from the disturbed (Table 2) tows were greater than those collected in the undisturbed (Table 1) tows at least by one order of magnitude.

The abundances of several animal groups were lower after the trawling season had commenced, while some others appeared to be higher. The densities of seven disturbed and four undisturbed major taxonomic groups varied sig-

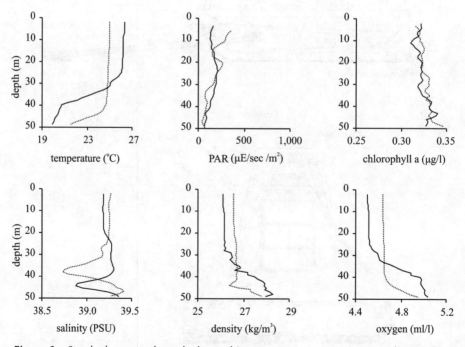

Figure 3. Standard water column hydrographic measurements (temperature, photosynthetically active radiation [PAR], chlorophyll a, salinity, density, oxygen) before (27 September 2001; black line) and after (7 October 2001; dashed line) the beginning of the commercial trawling season in Crete.

Table 1. Averaged densities (individuals/m^2 ± SD) of the major taxa collected from the undisturbed experimental sledge tows before (27 September 2001) and after (7 October 2001) the beginning of the commercial trawling season in Crete (n = number of tows). Statistical significance of differences in densities between the two sampling occasions was determined with Mann-Whitney U-test. n.s. = not significant.

Taxa	Before (n = 3)	After (n = 4)	P
Cnidaria	0.10 ± 0.05	0.03 ± 0.02	n.s.
Crustacea (larvae)	0.52 ± 0.30	0.23 ± 0.05	n.s.
Copepoda	0.11 ± 0.05	0.20 ± 0.05	<0.05
Decapoda	<0.01	<0.01	n.s.
Mysidacea	<0.01	0.02 ± 0.02	n.s.
Cumacea	<0.01	0.03 ± 0.02	<0.05
Isopoda	<0.01	<0.01	n.s.
Amphipoda	0.01 ± 0.01	0.03 ± 0.02	n.s.
Chaetognatha	0.08 ± 0.03	0.07 ± 0.02	n.s.
Echinodermata	0.03 ± 0.02	<0.01	<0.05
Thaliacea	0.04 ± 0.02	<0.01	<0.05
Fish larvae	0.02 ± 0.02	<0.01	n.s.
Eggs	0.01 ± 0.01	0.03 ± 0.02	n.s.

Table 2. Averaged densities (individuals/m^2 ± SD) of the major taxa collected from the disturbed experimental sledge tows before (27 September 2001) and after (7 October 2001) the beginning of the commercial trawling season in Crete (n = number of tows). Statistical significance of differences in densities between the two sampling occasions was determined with Mann-Whitney U-test. n.s. = not significant.

Taxa	Before (n = 3)	After (n = 4)	P
Porifera	0.01 ± 0.02	0.02 ± 0.04	n.s.
Cnidaria	0.08 ± 0.07	0.41 ± 0.17	<0.05
Ctenophora		<0.01	n.s.
Nemertini	0.02 ± 0.02		n.s.
Sipuncula	0.10 ± 0.12	0.01 ± 0.02	n.s.
Gastropoda	6.25 ± 1.37	1.12 ± 0.71	<0.05
Scaphopoda		0.09 ± 0.12	n.s.
Bivalvia	1.05 ± 0.19	0.60 ± 0.21	<0.05
Polychaeta	3.43 ± 0.38	1.43 ± 0.57	<0.05
Crustacea (larvae)	2.91 ± 0.65	3.65 ±1.59	n.s.
Copepoda	0.36 ± 0.12	3.17 ± 0.59	<0.05
Ostracoda	0.33 ± 0.33	0.10 ± 0.07	n.s.
Decapoda	1.53 ± 0.40	1.82 ± 1.24	n.s.
Mysidacea	10.55 ± 1.53	6.82 ± 3.15	n.s.
Cumacea	4.33 ± 1.66	7.38 ± 1.64	<0.05
Tanaidacea	0.01 ± 0.02	0.02 ± 0.03	n.s.
Isopoda	0.21 ± 0.12	0.21 ± 0.10	n.s.
Amphipoda	5.77 ± 1.10	5.67 ± 1.67	n.s.
Pycnogonida	0.08 ± 0.07	0.06 ± 0.06	n.s.
Chaetognatha	0.76 ± 0.27	1.37 ± 0.29	<0.05
Echinodermata	1.22 ± 0.50	0.44 ± 0.28	n.s.
Appendicularia	0.07 ± 0.09	0.02 ± 0.02	n.s.
Ascidiacea	<0.01		n.s.
Thaliacea	0.01 ± 0.02	0.03 ± 0.03	n.s.
Fish larvae	0.10 ± 0.06	0.09 ± 0.07	n.s.
Pisces	0.03 ± 0.03	0.05 ± 0.03	n.s.
Eggs	0.28 ± 0.12	0.13 ± 0.06	n.s.

nificantly (Tables 1, 2, $P < 0.05$) between the two sampling occasions. In particular, the average densities of Gastropoda, Bivalvia, and Polychaeta (Figure 4a, 4b, 4c) collected from the disturbed tows decreased significantly after the beginning of the trawling period. On the other hand, the densities of Decapoda, Mysidacea, Amphipoda, and crustacean larvae appeared not to be affected by commercial trawling (Figures 4d, 5a, 5c, 5d). Abundances of Cumacea and Copepoda were significantly greater after the beginning of the trawling season in both disturbed and undisturbed sampling tows (Figures 5b, 5e), while numbers of Chaetognatha increased only in the disturbed tows (Figure 5f).

The similarity dendrogram based on the density matrices of the major taxonomic groups (Figure 6) showed a clear separation of the samples from disturbed and undisturbed tows comprising two major groups. Furthermore, within each group of tows, the two sampling occasions were also separated in two distinct subgroups, indicating that the benthic community has been altered, probably as a result of commercial trawling activity in the area. The results of the ANOSIM test revealed that the above four groups of tows were significantly different.

Discussion

It has been suggested that the impact of otter trawling is largely restricted to the disturbance caused by the doors (Hall 1999). However, the effect of the groundrope in a trawling rig should not be underestimated, as it accounts usually for more than 90% of the area of contact of the trawl gear with the seabed (Lindeboom and de Groot 1998). The groundrope used in these experiments penetrated the muddy sediment of Heraklion Bay less than 1 mm (Dounas et al. 2002). Observations made on the Heraklion Bay fishing grounds using the new TTSS2 sampling gear to simulate sediment disturbance by otter trawl groundrope showed significant perturbations of small-sized benthos living on or very close to the sediment–water interface.

Hyperbenthic sampling (TTSS2 without groundrope) in the study area revealed low activity of animals above the seabed. Most hyperbenthic species seem to be very closely associated with the sediment during the day, undertaking vertical migrations during darkness (Sainte-Marie and Brunel 1983, 1985; Kaartvedt 1986). However, sediment disturbance experiments using the

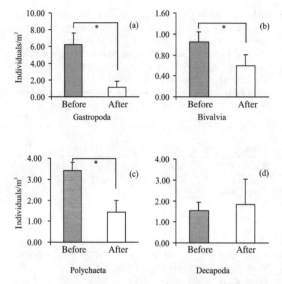

Figure 4. Averaged densities (individuals/m² ± SD) of four major epibenthic taxa collected from the "disturbed" tows before and after the beginning of the commercial trawling season. An asterisk indicates that statistically significant differences were determined by the Mann–Whitney U-test ($P < 0.05$).

TTSS2 and the otter trawl groundrope disturbed not only a large number of epibenthic animals but also rich hyperbenthic and zooplanktonic fauna living on or a few centimeters above the bottom during daylight.

Changes observed in the abundance of certain taxonomic groups a week after the start of the trawling period in the area cannot be attributed to any natural disturbance event, as calm weather conditions prevailed throughout the sampling period. This is consistent with the similarity between the hydrographic conditions measured on the two sampling occasions.

A large number of benthic animals may have been displaced, damaged, or killed by the passage of trawl gear, generating a food source for predators and scavengers (Britton and Morton 1994; Kaiser and Ramsay 1997; Groenewold and Fonds 2000). Hence, it was not unexpected to observe significant decreases in the abundance of the major epibenthic groups such as gastropods, bivalves, and polychetes shortly after the beginning of the trawling season. On the contrary, the densities of some hyperbenthic and zooplanktonic groups, such as cumaceans, copepods, and chaetognaths, increased markedly. These increases probably were due to immigration of animals from the neighboring untrawled

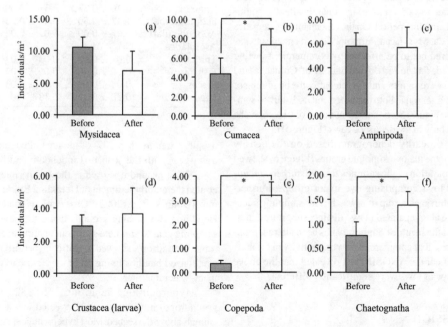

Figure 5. Averaged densities (individuals/m² ± SD) of six major hyperbenthic (a, b, c) and zooplanktonic (d, e, f) taxonomic groups collected from the "disturbed" tows before and after the beginning of the commercial trawling season. An asterisk indicates that statistically significant differences were determined by the Mann–Whitney U-test ($P < 0.05$).

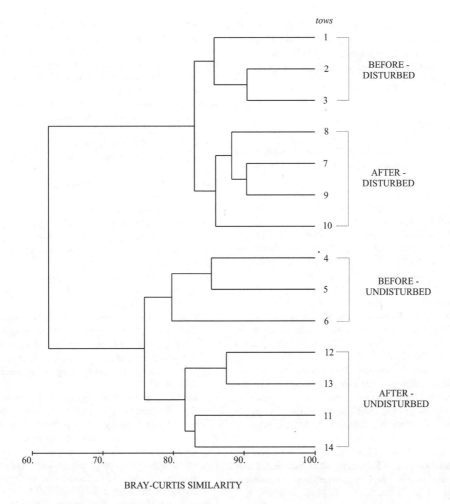

Figure 6. Similarity dendrogram based on density matrices of the major taxa collected from the "undisturbed" and "disturbed" tows before and after the beginning of the commercial trawling season. Statistical significance of differences among the four groups of tows was determined with an ANOSIM test ($R = 0.861$; $P < 0.001$).

areas into the fishing ground where there is increased availability of food and to the removal of large predators by the fishing gear (Kaiser and Spencer 1994; Ramsay et al. 1997; Prena et al. 1999). Other groups, such as decapods, mysids, amphipods, and crustacean larvae, appeared to be unaffected, although they also may be attracted by the increased availability of food resource caused by trawling, but they were probably also more vulnerable to predatory pressure from demersal benthivorous fish (Labropoulou and Eleftheriou 1997).

Current knowledge does not allow the prediction of the long-term effects resulting from groundrope disturbance on the epibenthic, the hyperbenthic, or even the zooplanktonic community. The development of novel direct sampling techniques (such as TTSS2) may assist in overcoming problems related to the inaccessibility of the hyperbenthos, which cannot be performed by standard sampling equipment. Ongoing research in the area using TTSS2 combined with conventional benthic and pelagic sampling and stomach content analyses of benthopelagic fish is expected to give further information on the effects of otter trawling on small invertebrates and on the resulting ecosystem responses.

Acknowledgments

This work was carried out in the framework of the project "Development of a new method for the quantitative measurement of the effects of otter trawling on benthic nutrient fluxes and sediment biogeochemistry," which was financed by the European Commission (Directorate General XIV, Studies for the Support of Common Fisheries Policy). The authors acknowledge the technical support provided by the captain and the crew of R/V *PHILIA*. We are also grateful for comments on the manuscript made by C. Arvanitidis.

References

Bray, J. R., and J. T. Curtis. 1957. An ordination of the upland forest communities of southern Wisconsin. Ecological Monographs 27:220–249.

Britton, J. C., and B. Morton. 1994. Marine carrion and scavengers. Oceanography and Marine Biology: An Annual Review 32:369–434.

Brunel, P., M. Besner, D. Messier, L. Poirier, D. Granger, and M. Weinstein. 1978. Le traîneau suprabenthique Macer-GIROQ: appareil amélioré pour l' échantillonage quantitatif étagé de la petite faune nageuse au voisinage du fond. Internationale Revue der Gesampten Hydrobiologie 63:815–829.

Carrassón, M., J. Matallanas, and M. Casadevall. 1997. Feeding strategies of deep-water morids on the western Mediterranean slope. Deep-Sea Research I 44:1685–1699.

Cartes, J. E., and F. Maynou. 1998. Food consumption by bathyal decapod crustacean assemblages in the western Mediterranean: predatory impact of megafauna and the food consumption-food supply balance in a deep-water food web. Marine Ecology Progress Series 171:233–246.

Chronis, G., V. Lykousis, C. Anagnostou, A. Karageorgis, S. Stavrakakis, and S. Poulos. 2000. Suspended particulate matter and nepheloid layers over the southern margin of the Cretan Sea (N. E. Mediterranean): seasonal distribution and dynamics Progress in Oceanography 46:143–162.

Clarke, K. R., and R. H. Green. 1988. Statistical design and analysis for a "biological effects" study. Marine Ecology Progress Series 46:213–226.

Clarke, K. R., and R. M. Warwick. 1993. Similarity-based testing for community pattern: the 2-way layout with no replication. Marine Biology 118:167–176.

Clarke, K. R., and R. M. Warwick. 1994. Change in marine communities: an approach to statistical analysis and interpretation. Natural Environment Research Council, Plymouth Marine Laboratory, Plymouth, UK.

Dounas, C., I. Davies, P. Hayes, C. Arvanitidis, and P. Koulouri. 2002. Development of a new method for the quantitative measurement of the effects of otter trawling on benthic nutrient fluxes and sediment biogeochemistry. European Commission, Study Project 99/036, Final Report, Hellenic Centre for Marine Research, Crete, Greece.

Eleftheriou, A., and N. A. Holme. 1984. Macrofauna techniques. Pages 140–216 in N. A. Holme and A. D. McIntyre, editors. Methods for the study of marine benthos. Blackwell Scientific Publications, Oxford, UK.

Groenewold, S., and M. Fonds. 2000. Effects on benthic scavengers of discards and damaged benthos produced by the beam-trawl fishery in the southern North Sea. ICES Journal of Marine Science 57:1395–1406.

Hall, S. J. 1999. The effects of fishing on marine ecosystems and communities. Blackwell Scientific Publications, Oxford, UK.

Jennings, S., and M. J. Kaiser. 1998. The effects of fishing on marine ecosystems. Advances in Marine Biology 34:201–352.

Kaartvedt, S. 1986. Diel activity patterns in deep-living cumaceans and amphipods. Marine Ecology Progress Series 30:243–249.

Kaiser, M. J., and S. J. de Groot. 2000. The effects of fishing on non-target species and habitats. Blackwell Scientific Publications, Oxford, UK.

Kaiser, M. J., and K. Ramsay. 1997. Opportunistic feeding by dabs within areas of trawl disturbance: possible implications for increased survival. Marine Ecology Progress Series 152:307–310.

Kaiser, M. J., and B. E. Spencer. 1994. Fish scavenging behaviour in recently trawled areas. Marine Ecology Progress Series 112:41–49.

Kallianiotis, A., K. Sophronidis, P. Vidoris, and A. Tselepides. 2000. Demersal fish and megafaunal assemblages on the Cretan continental shelf and slope (NE Mediterranean): seasonal variation in species density, biomass and diversity. Progress in Oceanography 46:429–455.

Karakassis, I., and A. Eleftheriou. 1997. The continental shelf of Crete: structure of macrobenthic communities. Marine Ecology Progress Series 160:185–196.

Koulouri, P., C. Dounas, and A. Eleftheriou. 2003. A new apparatus for the direct measurement of otter trawling effects on the epibenthic and hyperbenthic macrofauna. Journal of the Marine Biological Association of the United Kingdom 83:1363–1368.

Labropoulou, M., and A. Eleftheriou. 1997. The foraging ecology of two pairs of congeneric demersal fish species: importance of morphological characteristics in prey selection. Journal of Fish Biology 50:324–340.

Lindeboom, H. J., and S. J. de Groot. 1998. Impact II. The effects of different types of fisheries on the North Sea and Irish Sea benthic ecosystems. Netherland Institute for Sea Research, NIOZ-Rapport 1998–1/RIVO-DLO Report C003/98, Texel, The Netherlands.

Martin, B., and B. Christiansen. 1997. Diets and standing stocks of benthopelagic fishes at two bathymetrically different midoceanic localities in the northeast Atlantic. Deep-Sea Research I 44:541–558.

Mees, J., and M. B. Jones. 1997. The hyperbenthos. Oceanography and Marine Biology: An Annual Review 35:221–255.

Prena, J., P. Schwinghamer, T. W. Rowell, D. C. Gordon, Jr., K. D. Gilkinson, W. P. Vass, and D. L. McKeown. 1999. Experimental otter trawling on a sandy bottom ecosystem of the Grand Banks of Newfoundland: analysis of trawl bycatch and effects on epifauna. Marine Ecology Progress Series 181:107–124.

Ramsay, K., M. J. Kaiser, P. G. Moore, and R. N. Hughes.

1997. Consumption of fisheries discards by benthic scavengers: utilization of energy subsidies in different marine habitats. Journal of Animal Ecology 66:884–896.

Sainte-Marie, B., and P. Brunel. 1983. Differences in life history and success between suprabenthic shelf populations of *Arrhis phyllonyx* (Amphipoda, Gammaridea) in tow ecosystems of the Gulf of St. Lawrence. Journal of Crustacean Biology 3:45–69.

Sainte-Marie, B., and P. Brunel. 1985. Suprabenthic gradients of swimming activity by cold-water gammaridean amphipod Crustacea over a muddy shelf in the Gulf of Saint Lawrence. Marine Ecology Progress Series 23:57–69.

The Effect of Different Types of Otter Trawl Ground Rope on Benthic Nutrient Releases and Sediment Biogeochemistry

Costas G. Dounas[1]

Institute of Marine Biology of Crete, Post Office Box 2214, GR-71003, Heraklion Crete, Greece

Ian M. Davies and Peter J. Hayes

Fisheries Research Services Marine Laboratory, Post Office Box 101, 375 Victoria Road, Torry, Aberdeen, AB11 9DB, UK

Christos D. Arvanitidis and Panayota T. Koulouri

Institute of Marine Biology of Crete, Post Office Box 2214, GR-71003, Heraklion Crete, Greece

Abstract. The effects of seabed disturbance by otter trawling on the rate of nutrient regeneration from the sediment to the overlying water column and on the biogeochemical sediment zonation have never been studied directly. These effects may have important implications for nutrient supply to the pelagic ecosystem, on primary production over the continental shelf where most trawling activity is concentrated, and, consequently, on fish production and fisheries management. The labile fractions of sedimentary organic matter, the main source of energy to benthic environments, are usually concentrated on or near the sediment surface. Consequently, the most important impacts of trawling on sediment biogeochemistry arise from the artificial disturbance of the sediment surface layer. The trawl ground rope contributes most to this process and accounts for more than 90% of the total contact area of the trawl with the seabed in some cases. Therefore, a range of ground ropes was selected for trawling simulation experiments. The study site was the continental shelf of Heraklion Bay (Eastern Mediterranean, Cretan Sea). Nutrient releases induced by ground ropes were measured via a sampling schedule that included six deployments of a towed trawl simulator sledge. Five different ground ropes, ranging from 6.5 to 9.5 cm in diameter and 2–6 kg/m weight (in water), were used in order to estimate potential quantitative differences in their effects on nutrient releases arising from the contact of the ground rope with the muddy seabed. Average net releases of particulate organic carbon and total solids disturbed by the two heavier ground ropes were significantly greater than those from the other three, lighter ground ropes. Releases of phosphate, inorganic nitrogen compounds, chlorophyll a, and chloroplastic pigment equivalent showed no significant differences between ground ropes. The results indicate that almost all the biologically active compounds at the sediment surface are resuspended by a single passage of the simulating gear. These observations also imply that the upper, extremely thin layer of sediments contains a considerable reservoir of dissolved and particulate nutrients in concentrations that are much higher than in the immediately underlying surface sediment layers.

Introduction

Continental shelf environments, where otter trawling activity is concentrated, receive about half their nutrients for primary production through recycling of nutrients from the sediment (Nixon 1981). This input is derived from the remineralization of organic material within sediments, followed by upward physical transport driven by bioturbation and molecular diffusion as well as by recycling of nutrients within the water column. The labile fractions of sedimentary organic matter, responsible for most of the sedimentary metabolism, are usually concentrated near the sediment–water interface (Fanning et al. 1982; Mayer et al. 1991; Anderson et al. 1994). Trawling is expected to affect both the timing (in a seasonal sense) and the rate of nutrient fluxes to the water column, introducing a much more episodic pattern to the

---

[1] E-mail: kdounas@imbc.gr

releases (Mayer et al. 1991; Pilskaln et al. 1998). The trawl gear ground rope contributes most to these processes, as it may account, in some cases (e.g., shrimp and trawls for *Nephrops* sp.), for more than 90% of the total area of contact with the sediment surface (Lindeboom and de Groot 1998).

Quantitative studies of sediment disturbance by fishing gear are very limited (Churchill 1989; Black and Parry 1994, 1999). The impacts of towed fishing gears and, particularly, of otter trawling on sediment biogeochemistry have not been studied directly. Research conducted between 1999 and 2002 in Heraklion Bay, Crete, showed that the use of an experimental benthic sledge adapted to incorporate one or more lengths of otter trawl ground rope can effectively simulate the passage of trawl ground ropes across the seafloor (Dounas et al. 2002). The apparatus enabled direct measurements of the amount of sediment and concentrations of nutrients raised into suspension per unit length of trawl track or seabed surface area. This paper reports the first results from a series of field experiments conducted on the Cretan continental shelf with the new sampling technique. The objective of this part of the project was the quantitative estimation of the relative impact of otter trawl ground ropes of different sizes and weights on the sediment biogeochemistry.

Methods

Field experiments were conducted on 16–17 September 2001 in water 50 m deep in Heraklion Bay (35° 24'10N–25°06'00E), where local commercial trawling activity is concentrated (Figure 1). The study area is supporting a major trawling fishery for several commercial fishes, such as *Mullus barbatus* Linnaeus 1758 and *Mullus surmuletus* Linnaeus 1758 (Kallianiotis et al. 2000). Surficial bottom sediments in the area are classified as mud, as the fine-grained fraction (<63 mm) makes up more than 90% of the sediment (Folk 1968). The remaining coarse-grained constituents are mainly shell fragments and debris of bivalves, gastropods, echinoderms, serpulid polychaetes, foraminifera, ostracods, and sponge needles (Chronis et al. 2000). During underwater towed-video operations, trawling activity was evidenced by several scrape marks on the seabed made by the boards and wires of the trawls and by general flattening of the microtopography, probably caused by trawl nets and ground ropes.

The towed benthic sledge described by Shand and Priestley (1999) was used to simulate ground rope disturbance and was equipped with underwater video, electrical release systems, and water sampling bottles attached to its frame (Dounas et al. 2002). The sledge was modified to incorporate one or more lengths of otter trawl ground ropes between the forward parts of the runners of the sledge in contact with the sediment surface. Water samples were collected in six 2-L bottles mounted horizontally in pairs at 2.5 cm, 17.5 cm, and 30 cm (adjustable) above the seabed, 1.5 m behind the ground rope. The bottles were in front of any disturbance caused by the sledge and in the middle of the sediment cloud caused by the ground rope. The sledge was used for the direct quantitative measurement of the

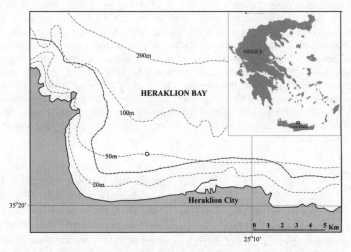

Figure 1. Map of Heraklion Bay (Crete Island, Greece, Eastern Mediterranean) showing the location of the station occupied during this study (bold dashed line shows 1,852-m (1-nautical-mi) trawling limit according to the Greek Fisheries Legislation; i.e., all trawling must occur offshore of this line).

amount of particulate material and nutrient concentrations raised into suspension by the passage of the ground rope. Monitoring by video camera attached to the sledge showed that resuspension ahead of the sampling bottles was caused solely by the ground rope, and the sediment cloud was no higher than the top of the 30-cm sampling bottle. RV *Philia* towed the sledge at normal trawling speed (~1 m/s; ~2.2 knots), and the bottles were fired remotely and simultaneously from the surface by acoustic or electro-mechanical systems. A complete technical description of the towed trawl simulator sledge (TTSS) is in preparation. It should be noted that sampling has been randomly carried out in the study area. The possibility of overlapping between successive sledge tows, especially at the moment of bottles firing, is considered highly unlikely. Five different ground ropes were used in order to estimate the effect of the type of ground rope on the amounts of sediment particles resuspended and nutrients released by the passage of the ground rope across the seabed. The sampling schedule included six deployments of TTSS for each type of ground rope.

The technical characteristics of the various combinations of ropes and chains used in each ground rope setting are given in Table 1. Type A ground rope included a light rope weighed with a chain, a setting most commonly used by the local fishermen in their otter trawling gears. The ropes used in types B and C had a much larger diameter than those in type A, while the chain type and weight was kept the same. Types D and E used the same types of rope as in settings B and C, respectively, but were much more heavily weighted with twice and three times the weight of chain that was used in ground ropes A–C. Chains were tied along the ground ropes at short intervals.

Upon recovery of the TTSS, the water samples were filtered through Whatman GF/F glass fiber filters (Whatman International Ltd., Maidstone, UK) in order to estimate chlorophyll a, phaeopigments, particulate organic carbon (POC), and particulate organic nitrogen (PON), as well as solids concentrations by routine oceanographic methods. Water sub-samples (200 mL) were collected from the filtrate for nutrient analysis (PO_4, SiO_2, NH_4, NO_2, NO_3), frozen immediately, and stored at $-20°C$ until analysis in the shore laboratory. Chlorophyll a and chloroplastic pigment equivalent (CPE) were determined according to the fluorometric method of Yentsch and Menzel (1963), using a Turner 112 fluorometer (Sequoia-Turner Corporation, Mountain View, California). The POC and PON concentrations were analyzed using a Perking Elmer CHN 2400 analyzer (PerkinElmer, Inc., Shelton, Connecticut), and nitrate (NO_3), nitrite (NO_2), ammonium (NH_4), phosphate (PO_4), and silicate (SiO_2) concentrations were determined using a Beckmann DU65 spectrophotometer (Beckman Coulter, Inc., Fullerton, California) and standard manual oceanographic colorimetric methods (Strickland and Parsons 1972).

Reference concentrations were measured in water samples collected by using TTSS but omitting the ground rope. These data collections were interspersed between deployments of TTSS with ground ropes. Therefore, the variability of control conditions were taken into account. By knowing the towing speed (1 m/s; 2.2 knots) and subtracting the reference concentrations, the average net releases of inorganic nutrients and particles/m^2 of sediment disturbance for each type of ground rope were estimated. Mann-Whitney's U-test was used to assess statistically the significance of differences between total net releases of the various nutrient compounds derived from the application of the five different ground ropes.

Results and Discussion

Figures 2 and 3 show the average net releases into the water column (per m^2 of sediment disturbed) of inorganic phosphate, total inorganic nitrogenous compounds, chlorophyll a, CPE, POC, and total solids (with C/N ratio) caused by the various types of ground ropes tested. Average net releases of POC and total solids disturbed by the two heavier ground ropes (types D and E) were significantly greater than those arising from the other three lighter ground ropes ($P < 0.05$). Releases of phosphate, inorganic nitrogen compounds, chlorophyll a, and CPE showed no significant differences ($P > 0.05$) between ground ropes. This indicates that the amounts of inorganic nutrient and chlorophyllous particulate substances resuspended by trawling are independent of the type (weight, diameter) of ground rope, at least within the range of ground rope settings used in these experiments. Reference water samples (Table 2) and particles resuspended by ground rope types A, B, and C gave similar C/N ratios ranging from 8 to 10. The two heavier ground ropes seemed to resuspend a slightly lower proportion of labile organic particulate material, with average C/N ratio values of

Table 1. Technical characteristics of the five different types of groundropes used in field experiments.

Type	Diameter (cm)	Weight (kg/m)	Weight in water (kg/m)
A	6.5	3.2	2
B	8.0	4.0	2
C	9.5	4.9	2
D	8.0	6.3	4
E	9.5	9.5	6

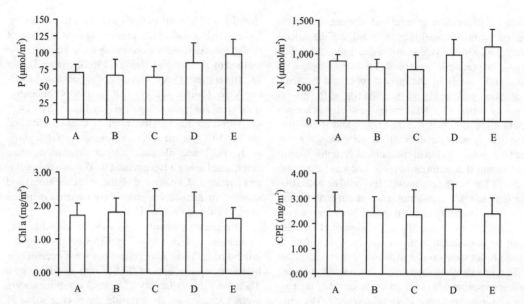

Figure 2. Net releases (±SD; $n = 6$) of inorganic phosphate (P) and total nitrogen (N), chlorophyll a (Chl-a), and chlorophyllus pigment equivalent (CPE) per m^2 of sediment disturbed by five types of ground ropes (Table 1).

about 12 (Figure 3), possibly through dilution of organic matter by a greater resuspension of inorganic particles.

By applying a surface sediment porosity value of 63% and a dry density for silt of 2.6 g/cm^3 (Dounas et al. 2000), the average penetration depth into the sediment of the different types of ground ropes used was estimated. The measured average rate of 356.7 g/m^2 of the solid material resuspended by the type A ground rope is equivalent to the resuspension of a depth of only 0.37-mm seabed sediment. Similar calculations for the heaviest ground ropes used (types D and E) indicate 0.45-mm and 0.63-mm penetration depth, respectively. For the larger diameter ground ropes (types B and C), the estimated mean depth penetration was rather less than for type A, ranging from 0.26 to 0.28 mm. This may reflect the combination of similar force exerted on the seabed by all three types of rope and the increased resistance to penetration acting on the larger diameter ropes.

Traditionally, studies aiming to estimate the direct mortality of benthic invertebrates caused by the passage of trawl gear consider the distance between the net wings plus the distance over which the bridles scrape over the seabed as being disturbed. In a typical otter trawling set-up, the width of the gear's bottom contact (ground rope plus bridles) varies between 20 and 40 m (Lindeboom and de Groot 1998; Pilskaln 1998). While the passage of an otter trawl net is generally considered to have a minor physical and visual impact on the soft sedimentary seabed (e.g., flattening of the normally uneven sediment surface), the main physical effect appears to be the tracks left in the sediment by the trawl doors. Nevertheless, the impact of the doors alone represents only a small fraction of the total sediment surface affected by the trawling rig, as it is usually less than 10% of the width of the ground rope path (Lindeboom and de Groot 1998). The labile fractions of sedimentary organic matter, responsible for most of the sedimentary metabolism, are usually concentrated near the sediment–water interface (Mayer et al. 1991). Consequently, studies of the impact of trawling on sediment biogeochemistry should not disregard the inorganic nutrient releases and the resuspension of other particulate and biologically active substances arising from the contact of the ground rope with the seabed.

Results from this study indicate that inorganic nutrient releases from sediment disturbance seem independent of the ground rope type in terms of weight and diameter, at least in the range of ground rope settings applied. However, although not statistically significant, there may be a tendency for increasing inorganic nutrient release rates as the ground rope penetrates deeper in the sediment. This tendency is not observed in the case of chlorophyll a and phaeopigments release rates. These observations may imply that an upper extremely thin layer of shelf sediments contains a considerable reservoir of dissolved and particulate nutrients in concentrations that are much higher than in the underlying surface sediment layers or the overlying water.

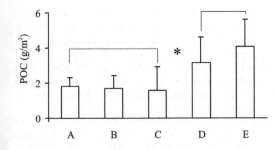
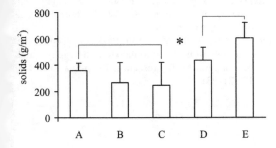
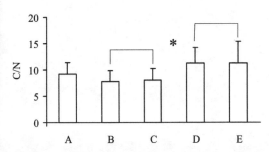

Figure 3. Net releases (± SD; $n = 6$) of particulate organic carbon (POC), suspended solids, and C/N ratio of organic particles resuspended per m^2 of sediment disturbed by five types of ground ropes (Table 1). Statistically significant ($P < 0.05$) differences, denoted by an asterisk, between lighter (types A, B, and C) and heavier (types D and E) ground ropes were determined with the Mann–Whitney U-test.

Acknowledgments

This work was carried out in the framework of the project, "Development of a new method for the quantitative measurement of the effects of otter trawling on benthic nutrient fluxes and sediment biogeochemistry," which was financed by the European Commission (Directorate General XIV, Studies for the Support of Common Fisheries Policy). We acknowledge the technical assistance of C. Christodoulou, S. Zivanovic, and H. Dafnomili and the practical support of the captain M. Kokos and the crew of R/V *Philia*.

References

Anderson, R. F., G. T. Rowe, P. F. Kemp, S. Trumbore, and P. E. Biscay. 1994. Carbon budget for the mid-slope depocenter of the Middle Atlantic Bight. Deep-Sea Research II 41:669–703.

Black, K. P., and G. D. Parry. 1994. Sediment transport rates and sediment disturbance due to scallop dredging in Port Phillip Bay. Memoirs of the Queensland Museum 36:327–341.

Black, K. P., and G. D. Parry. 1999. Entrainment, dispersal, and settlement of scallop dredge sediment plumes: field measurements and numerical modelling. Canadian Journal of Fisheries and Aquatic Sciences 56:2271–2281.

Chronis, G., V. Lykousis, C. Anagnostou, A. Karageorgis, S. Stavrakakis, and S. Poulos. 2000. Sedimentological processes in the southern margin of the Cretan Sea (NE Mediterranean). Progress in Oceanography 46:143–162.

Churchill, J. H. 1989. The effect of commercial trawling on sediment resuspension and transport over the Middle Atlantic Bight continental shelf. Continental Shelf Research 9:841–864.

Dounas, C. G., I. M. Davies, P. J. Hayes, C. D. Arvanitidis, and P. T. Koulouri. 2002. Development of a new method for the quantitative measurement of the effect of otter trawling on benthic nutrient fluxes and sediment biogeochemistry. European Commission, Study Project Number 99/036, Final Report, Hellenic Centre for Marine Research, Heraklion Bay, Crete, Greece.

Fanning, K. A., K. L. Carder, and P. R. Betzer. 1982. Sediment resuspension by coastal waters: a potential mecha-

Table 2. Reference concentrations of total inorganic nitrogen (N) and phosphate (P), chlorophyll a (Chl. a), chloroplastic pigment equivalent (CPE), particulate organic carbon (POC) and C/N ratio from water samples collected by TTSS at different heights above seabed ($n = 6$; ± SD).

Height above seabed (cm)	N (µM)	P (µM)	Chl.a (µg/L)	CPE (µg/L)	POC (µg/L)	C/N
2.5	1.9 ± 0.1	0.06 ± 0.05	0.05 ± 0.02	0.11 ± 0.04	391 ± 91	9.9 ± 0.5
17.5	1.8 ± 0.1	0.05 ± 0.01	0.05 ± 0.01	0.11 ± 0.04	258 ± 61	9.7 ± 1.4
30.0	2.1 ± 0.3	0.04 ± 0.02	0.04 ± 0.02	0.10 ± 0.04	217 ± 37	9.7 ± 1.1

nism for nutrient re-cycling on the ocean's margins. Deep-Sea Research 29:953–965.

Folk, R. L. 1968. Petrology of sedimentary rocks. Austin, Hemphills.

Kallianiotis, A., K. Sophronidis, P. Vidoris, and A. Tselepides. 2000. Demersal fish and megafaunal assemblages on the Cretan continental shelf and slope (NE Mediterranean): seasonal variation in species density, biomass and diversity. Progress in Oceanography 46:429–455.

Lindeboom, H. J., and S. J. de Groot. 1998. Impact-II. The effects of different types of fisheries on the North Sea and Irish Sea benthic ecosystems. Netherlands Institute for Sea Research, NIOZ-Rapport 1998–1/RIVO-DLO Report CO 03/98, Texel.

Mayer, L. M., D. F. Schick, R. H. Findlay, and D. L. Rice. 1991. Effects of commercial dragging on sedimentary organic matter. Marine Environmental Research 31:249–261.

Nixon, S. W. 1981. Remineralisation and nutrient cycling in coastal marine ecosystems. Pages 111–138 *in* B. Neilson and L. E. Cronin, editors. Estuaries and nutrients. Humana Press, Clifton, New Jersey.

Pilskaln, C. H., J. H. Churchill, and L. M. Mayer. 1998. Resuspension of sediment by bottom trawling in the Gulf of Maine and potential geochemical consequences. Conservation Biology 12:1223–1229.

Shand, C. W., and R. Priestley. 1999. A towed sledge for benthic surveys. Fisheries Research Services, Scottish Fisheries Information Pamphlet 22, Aberdeen, UK.

Strickland, J. D. H., and T. R. Parsons. 1972. A practical handbook of seawater analysis. Fisheries Research Board of Canada Bulletin 167.

Yentsch, C. S., and D. W. Menzel. 1963. A method for the determination of phytoplankton chlorophyll and phaeophytin by fluorescence. Deep-Sea Research 10:221–231.

Trawl Fishing Disturbance and Medium-Term Macroinfaunal Recolonization Dynamics: A Functional Approach to the Comparison between Sand and Mud Habitats in the Adriatic Sea (Northern Mediterranean Sea)

FABIO PRANOVI[1] AND SASA RAICEVICH[2]

*Dipartimento di Scienze Ambientali, Università Ca' Foscari,
Campo della Celestia 2737/B, 30122 Venice, Italy*

SIMONE LIBRALATO[3]

*Instituto Nazionale di Oceanografia e di Geofisica Sperimentale,
Borgo Grotta Gigante 42/c, 34010 Sgonico (TS), Italy*

FILIPPO DA PONTE[4]

*Dipartimento di Scienze Ambientali, Università Ca' Foscari,
Campo della Celestia 2737/B, 30122 Venice, Italy*

OTELLO GIOVANARDI[5]

*Istituto Centrale per la Ricerca Scientifica Applicata al Mare (ICRAM),
Località Brondolo, 30015 Chioggia, Italy*

Abstract.—Demersal fishing gear affects seabed habitats both directly and indirectly and modifies the processes and dynamics of benthic communities. At present, scientific attention is focused on using the functional approach to better understand the constraints that drive the recolonization of benthic fauna subjected to fishing disturbance. The Northern Adriatic Sea has an extensive trawlable area that is intensively exploited by a variety of trawling gear. Among these, the rapido, a beam trawl used to catch flatfishes on mud substrate and pectinids on sand, appears to have the greatest impact on the benthic habitat. We used various functional indicators (diversity indices, community structure, trophic groups, production analysis, and exergy) to compare the medium-term (9-months) macroinfaunal recolonization processes in sand and mud habitats treated with a single experimental rapido haul. Recolonization was found to be a community-wide process. The early successional stages in the sand habitat were dominated by scavenging organisms that peaked in abundance 7 d after the treatment. In the mud habitat, the trend in scavenger activity was less distinct, although a peak in abundance was recorded 1 month after the treatment. The functional analyses revealed that complete recovery required at least 9 months in both habitats. Finally, data collected on the fishing ground near the sand experimental area were analyzed in order to investigate the chronic disturbance caused by commercial trawling. The fishing ground samples showed a higher spatial heterogeneity than the sand experimental samples. Total abundance, total biomass, and production values in the fishing ground were comparable with the lowest recorded values in the sand experimental area, and the exergy differences suggested that the fishing ground's benthic community remained in an early successional stage.

---

[1] Corresponding author: fpranovi@unive.it
[2] E-mail: sasaraic@unive.it
[3] E-mail: slibralato@ogs.trieste.it
[4] E-mail: pippodp@hotmail.com
[5] E-mail: o.giovanardi@icram.org

Introduction

The study of the macroinfaunal recolonization dynamics in soft-sediment habitats has a relatively long and well-developed history in the field of marine benthic ecology (Pearson and Rosenberg 1978; Thistle 1981; Hall et al. 1996). Considerable attention has been paid to the examination of the recovery process following both natural and human-induced disturbances, and a variety of conceptual models have been developed that assess the role of disturbance in the population and community dynamics of benthic habitats (Grassle and Grassle 1974; Pearson and Rosenberg 1978; Dittmann et al. 1999). Moreover, recolonization studies have resulted in the development of a schema in which an early successional stage dominated by opportunistic and r-strategy species is followed by further stages with mainly k-selected species (Grassle and Grassle 1974; McCall 1977; Pearson and Rosenberg 1978; Rhoads et al. 1978). This classic successional model was elaborated further by correlating the colonization modes of invertebrate larvae, postlarvae, or adults (Santos and Simon 1980; Smith and Brumsickle 1989; Whitlach et al. 1998) with the intensity (Probert 1984; Thrush et al. 1996) and timing of the disturbance (Zajac and Whitlach 1982a; Bonsdorff and Österman 1985). However, this schema was not valid for all soft-sediment communities, because in some cases the recolonization process was more dependent on location and time than on conformance with any distinct successional stages (Zajac and Whitlach 1982b). This is especially the case when the benthic community is composed of opportunistic, stress-tolerant species, with its natural state resembling an early successional stage of the above-mentioned schema (Zajac and Whitlach 1982b; Gamenick et al. 1996).

As has already been highlighted by Zajac et al. (1998), the spatial scale is an important factor in assessing recolonization dynamics and stage of succession among soft-sediment infaunal communities. Studies demonstrating the changes induced in benthic community structure by fishing activity such as trawling frequently rely on experiments covering relatively small areas. The small scale of these studies makes it more difficult to apply the results on a real fishery scale. Moreover, due to their short-term and acute nature, these results do not reflect the chronic (long-term) disturbance caused by fishing activity on fishing grounds (Collie et al. 1997; Thrush et al. 1998; Kaiser et al. 2000; Jennings et al. 2001). Nevertheless, the experimental approach resulting from these small studies is the only one available for carrying out functional studies of recolonization dynamics. The above approach, combined with large-scale temporally and spatially collected data from the fishing grounds, permits interesting insights regarding the coexistence of and interactions between different successional stages of benthic communities, especially in areas with a high level of fishing pressure such as the Adriatic Sea.

The Northern Adriatic Sea, particularly in the western portion of the basin, is an extensive area of trawlable seabed characterized by the presence of mud substrate inshore and sand offshore. These areas are exploited by otter trawls, hydraulic dredges, and rapido trawls owned by the Italian commercial fleet (Ardizzone 1994).

In this paper, we compare the recovery patterns of benthic communities on mud and sand substrates after a single experimental rapido-trawl disturbance. To better understand the processes involved in recolonization and to evaluate the possibility of distinguishing different successional stages, we carried out the analysis by applying and comparing various functional indicators, including trophic guilds, production, diversity indices, and exergy storage, which is a measure of the energy and information embedded in ecosystem components (Jørgensen 2000). In addition, samples collected from a fishing ground located near the sand experimental area allowed us to obtain information about the chronic effects of trawl-fishing activity and to compare these with the acute effects at our trawl site.

Methods

Trawling experiments were carried out at two locations in the Northern Adriatic Sea: an area of sand substrate offshore (depth 23 m; 11 nautical miles (NM, 1 NM ≈ 1.8 km) east of the Venice Lagoon) and an area of mud substrate inshore (depth 11 m; 1 NM from the Sile River mouth) (Figure 1). The sand trawling experiment began in late November of 1999, and the mud trawling experiment began in late November of 2000.

Studies of recolonization processes in the benthic community after an experimental manipulation of the bottom by fishing activity are often hindered by the lack of undisturbed areas (Jennings and Kaiser 1998). Selection of suitable experimental sites in an area subjected to high fishing pressure represented an important objective of our research. To avoid any interference due to non-experimental fishing activity, we situated our mud study area near a site of longline mussel culture (Da Ponte 2001) and our sand study area near a shipwreck (Pranovi et al. 2000). However, the presence of such submerged structures can, in some cases, modify the behavior of the local benthic communities (Jennings and Kaiser 1998).

Preliminary side-scan sonar surveys were carried out in both study areas to confirm that trawl tracks were absent at the start of the experiment. Moreover, the side-

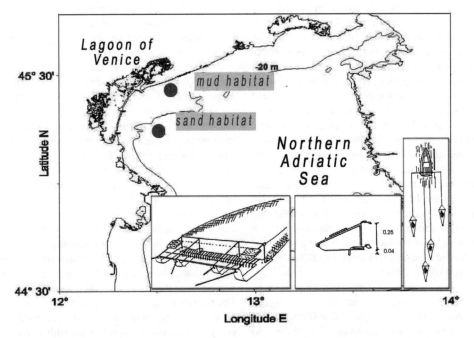

Figure 1. Location of the experimental areas in the Northern Adriatic Sea, and a description of the rapido trawling gear.

scan sonar made it possible to assess the magnitude of the fishing pressure exerted in the fishing grounds close to the experimental areas. The side-scan sonar surveys were repeated on each sampling day to confirm the absence of exploitation by local fishermen during the entire experiment.

A before–after, control–impact (BACI) sampling scheme was used. This scheme consisted of sampling the benthic community before and after the experimental disturbance. Samples were collected both inside (impact) and outside (control) the disturbed area. The disturbance consisted of a single experimental haul of a rapido trawl, a demersal gear used in the Northern Adriatic Sea to catch flatfishes (*Solea* spp. and European flounder *Platichthys flesus*) in the inshore mud habitat and scallops (*Aequipecten opercularis* and *Pecten jacobaeus*) in the offshore sand habitat (for details, see Hall-Spencer et al. 1999 and Pranovi et al. 2000). Sampling was accomplished in accordance with the temporal scheme shown in Table 1. Furthermore, for the sand habitat, six samples were collected in the surrounding fishing ground prior to the start of the experiment. This additional sampling station was located about 0.5 NM from the sand experimental area (both sites had the same depth and bottom particle size). The main ecological difference between the two sites was the presence or absence of rapido commercial fishing activity.

Scuba divers visually identified collection sites within the trawl track and collected samples using a water-lift sampler (mesh size, 1 mm, 0.3-m^2 sampling area, 25-cm depth). The samples were preserved in a freezer at –20°C. Upon thawing, the samples were treated with rose bengal dye, and the organisms were then separated and classified to the genus or species level when possible. All of the organisms were counted, and the total wet weight (±0.01 g) was recorded for each taxon.

We calculated diversity indices and biodiversity indices for each sample with PRIMER 5.0 software. To

Table 1. Temporal pattern (time = number of days after trawling) and number of replicates collected in the mud and sand habitats, in both the treated and control (in parentheses) areas.

Time (days)	Number of samples	
	Mud habitat	Sand habitat
Before	6	6
0	1	4
1	6	
7		5 (2)
9	6 (4)	
29	6 (4)	
30		5 (2)
90	5 (4)	6 (2)
270	6 (4)	6 (2)
900		6 (4)
Fishing ground		6

calculate diversity, the classical indices by Shannon and Weaver (1963) and Margalef (1957) were used. The Shannon index provides an evaluation of diversity by considering the proportion of individuals belonging to the same species, but without taking into account the absolute number of species. The Margalef index combines the total number of species and the total number of individuals. Biodiversity indices evaluate the diversity of the species assemblage by considering the taxonomic differences among the species. We chose the following biodiversity indices: (1) taxonomic diversity (the average taxonomic distance between any two organisms randomly chosen from the sample), (2) taxonomic distinctness (D^*) (the average path length (the number of taxonomic passages needed to reach a common level) between any two randomly chosen individuals of different species), and (3) the variation in taxonomic distinctness (Λ) (i.e., the variance of D^*) (Warwick and Clarke 1995; Clarke and Warwick 2001).

To analyze the trophic structure of the benthic community during each stage of recovery, we assigned sampled individuals to five trophic guilds (filter feeders, detritus feeders, herbivores, carnivores, and omnivores) based on the morphology of the feeding apparatus, feeding mode and nature, and origin of the food, in accordance with the trophic classifications and observations made by Fauchauld and Jumars (1979) and Desrosiers et al. (1986, 2000). Moreover, based on the trophic preferences, organisms belonging to carnivorous and omnivorous groups (which can act as scavengers) were extracted and then regrouped according to the different taxa (Gastropoda, Ophiuroidea, Polychaeta Errantia, Macrura Natantia, Macrura Reptantia, and Anomura).

Because of the important role community production plays in regulating marine ecosystem function, the evaluation of trawling effects on benthic community production is an important subject, although rarely studied (Duplisea et al. 2002; Jennings et al. 2002). According to Odum (1969), the production by biological communities can be used for evaluating growth and development of the ecosystem. In order to investigate the temporal pattern of benthic production in relation to the various recovery stages, experimental data were used to estimate production by applying the equation given by Brey (1990):

$$\log(P_s) = a + b_1 \log(B_s) + b_2 \log(\overline{W}_s),$$

where production (P_s) for a given species s is expressed as a function of the species' biomass (B_s) and of the mean individual weight (W_s), which is estimated as the ratio of the biomass and abundance. The parameters a, b_1, and b_2 for the main benthic taxonomic groups are given by Brey (1990). Even though this method does not provide an accurate estimation of the production of each species, it allows for an accurate evaluation of the total production of a given benthic community (Brey 1990).

In order to synthesize the biomass and composition changes in the benthic community into one index, while also accounting for each species' contribution to ecosystem complexity, we applied the exergy index. In fact, the degree of complexity of an ecosystem can be measured by an examination of its exergy (Müller 1997). Exergy, a concept derived from thermodynamics, is the amount of work that can be obtained from a system when it comes into contact equilibrium with a set of energy reservoirs (Fath and Cabezas 2004). Exergy reflects the quality of the energy, so it can be considered to be a measurement of the distance from thermodynamic equilibrium: self-organizing ecosystems tend to store and increase their exergy and, therefore, this concept measures the growth and the development of an ecosystem towards maturity (Jørgensen and Nielsen 1998). The reference state for the calculation of exergy is the ecosystem at thermodynamic equilibrium, which is the ecosystem without living forms and with all energy and compounds forming an "inorganic, primitive soup." For a chemical system, exergy can be calculated as

$$\mathrm{Ex} = RT \sum_{i=0}^{N} \left[C_i \log_e \left(\frac{C_i}{C_i^{eq}} \right) \right],$$

where C_i and C_i^{eq} are the concentrations of chemical i in the system and at equilibrium, R is the gas constant, and T is temperature in Kelvin (Jørgensen and Nielsen 1998). In a biological system, exergy represents a measurement of the energetic and information content of the living organisms (Marques et al. 1997). The exergy can be estimated by means of

$$\mathrm{Ex} = RT \sum_{i=0}^{N} (C_i \, \beta_i),$$

where C_i is the biomass concentration of species i in the system, and b_i is the weighting coefficient expressing the information contained within the species, which is estimated through its number of genes (Jørgensen et al. 1995). The genes-based method can be used to estimate the complexity and organization of organisms: that is, the information content embedded in the biomass (Jørgensen et al. 1995; Fonseca et al. 2000; Debeljak 2002). Although the genes-based method is not widely accepted since there are concerns and debate about the coefficients that should be used (Fonseca et al. 2000; Debeljak 2002), exergy is often applied as an ecosystem thermodynamic orientor in structurally dynamic models (Müller and Leupelt 1998; Jørgensen 2000; Fath et al.

2001) and in some applications concerned with empirical data (e.g., Marques et al. 1997). Moreover, although the complementarity of exergy with other, better-known ecosystem properties (such as those proposed by Odum [1969], Ulanowicz [1986], and Schneider and Kay [1994]) has been demonstrated (Fath et al. 2001), the use of exergy for studying successional processes and ecosystem development seems to provide better results (Marques et al. 1997; Jørgensen et al. 2002).

We calculated exergy over time after experimental fishing, and used the weighting coefficients given by Fonseca et al. (2000). The exergy concept was applied in order to evaluate the stress suffered by the benthic community and the recolonization process after experimental trawling.

The disturbed community could be viewed as existing in a sort of chemical and biological contact equilibrium together with the undisturbed community, which acts as a reservoir for both chemicals and species. Accordingly, we computed the difference between the exergy of the system (exergy of the community inside the trawl track, Ex_t^I) and the reservoir (exergy of the community outside the trawl track, Ex_t^E):

$$\Delta Ex_t = Ex_t^I - Ex_t^E$$

Such a difference (ΔEx_t) should show a decrease in absolute value over time if the impacted community is recovering. The above equation makes it possible to evaluate changes in the exergy amount in the benthic community with regard to the external factor (i.e., the fishing activities) plus the different recovery stages (Jørgensen and Nielsen 1998).

Statistical Analyses

The recolonization data from the mud and sand habitats were analyzed separately by use of two-way analyses of variance (analysis of variance, ANOVA) (Smith 2002). Although the two factors in the ANOVA were the before–after (BA) cases and the control–impact (CI) cases, different types of ANOVA tests can be applied (Underwood 1994; Smith 2002). Indeed, on the basis of differences in the BACI scheme (asymmetrical or not), in the type of disturbance (pulse or press), and in the selection of samples (randomly chosen or not), the ANOVA results can differ (for details, see Underwood 1994 and Smith 2002). In the present study, we used BA and CI as fixed factors, and the time factor (T) was nested in BA (T[BA]). The replicate samples were randomly chosen and treated as subsamples (Underwood 1994). The adopted experimental protocol (a single experimental haul in a previously undisturbed site) led to a "pulse effect" (i.e., an effect detectable in part of the "after" samples). Therefore, in accordance with Underwood (1994), the ANOVA analysis was used to test the interactions between CI and T(BA) (i.e., the term "T[BA] × CI" in the ANOVA table). Since the exergy index was computed as the difference between control and impact area over time, exergy was the only case in which the ANOVA analysis was used to test the interaction between BA and time (i.e., the term "T[BA]"; in the ANOVA table).

Because the experimental scheme adopted was very conservative (the impact area was treated by a single haul), the statistical differences in the ANOVA analyses were considered to be significant at P-values less than 0.1, to emphasize the possible existence of effects due to the fishing activity (Thrush et al. 1995). Moreover, to detect when the pulse effect was recognizable, the differences between the treated and control samples were tested at each sampling time with a t-test at three different levels of significance ($P < 0.1$, $P < 0.05$, and $P < 0.01$).

The differences in community structure among the treatments were assessed by use of a Bray–Curtis similarity matrix, performed on square-root transformed data, followed by a multidimensional scaling (MDS) ordination (Kruskal and Wish 1978).

The variability among the samples in each treatment group was analyzed via the multivariate dispersion index (MDI) and the relative dispersion values (Warwick and Clarke 1993). All of the analyses concerning the community structure were conducted with PRIMER 5.0 (Plymouth Marine Laboratory, UK).

Results

Mud Habitat

A total of 135 taxa were collected during all sampling activity in the mud habitat; 75 different taxa were collected prior to the experimental rapido haul. In terms of abundance, the macrobenthic community was dominated by *Lumbriconereis impatiens* (Annelida, Polychaeta) (>20%), *Corbula gibba* (Mollusca, Bivalvia) (>14%), and Gammaridae (Arthropoda, Crustacea) (about 5%) prior to the haul. The same three taxa were also dominant (*Corbula gibba* [14%], Gammaridae [13%], and *Lumbriconereis impatiens* [9%]) immediately after the experimental haul.

In terms of total biomass, the mud macrobenthic community was dominated by *Trachythyone elongata* (Echinodermata, Holoturoidea) (19%), *Philine aperta* (Mollusca, Gastropoda) (15%), and *Parthenope massena* (Arthropoda, Crustacea) (14%) before the trawl disturbance. Immediately after the experimental haul, the dominant species in terms of biomass were *Trachythyone elongata* (20%), *Ethusa mascarone* (Arthropoda, Crustacea) (20%), and *Philine aperta* (16%).

The application of the MDS did not seem to produce a clear pattern (Figure 2), as highlighted by the high stress (measure of the goodness of the obtained ordination) values (0.20). The lack of a definite pattern was due to the total number of replicates and the heterogeneity within the mud habitat. Nevertheless, 9 months (270 d) after the experimental haul, the community seemed to recover, as demonstrated by the convergence of control and treatment samples in the MDI results (Figure 3A).

Most of the considered diversity and biodiversity indices showed a small, nonsignificant increase in the treated area immediately after the experimental haul. Indices peaked 1 month later and then decreased (see below). However, the ANOVA analyses showed statistically significant differences (T[BA] × CI) in total abundance and in detritivores, carnivores, Macrura Natantia, and Anomura

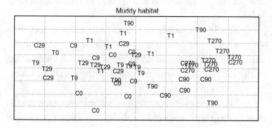

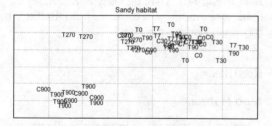

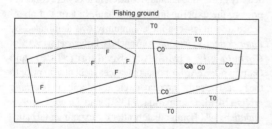

Figure 2. Multidimensional scaling ordination of the macroinfaunal abundance data collected in the mud habitat (top), sand habitat (middle), and sand fishing ground (bottom) in the Northern Adriatic Sea. Sample points are coded in terms of location (C, control; T, treatment; F, fishing ground) and the number of days after experimental trawling (0, 1, 7, 9, 29, 30, 270, or 900 d). Months and year of sampling are reported in Table 1.

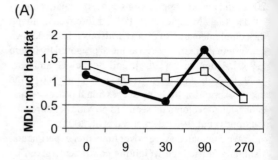

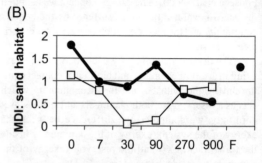

Figure 3. Multivariate dispersion index (square = control area; circle = treated area) calculated for macroinfaunal communities in the (A) mud habitat and (B) sand habitat (F = fishing ground). Months and year of sampling are reported in Table 1.

abundances ($P < 0.05$), as reported in Table 2. Furthermore, the ANOVA analyses showed statistically significant differences (T[BA]) for exergy (Table 2).

The t-test applied to each sampling time showed significant differences between the control and treated areas in terms of total biomass 9 d after trawling ($P < 0.1$) and in total abundance 29 d after trawling ($P < 0.01$), while no other significant differences were detected (Figure 4A–C). Corresponding to the increase in total abundance 29 d after trawling, the Shannon index showed a sharp decrease in both the control and treated areas (Figure 4E).

Taxonomic diversity (D) showed quite stable values over time, with the exception of 29 d after trawling, when a decrease was observed in both control and treated areas (Figure 5A). No clear trend was shown either by taxonomic distinctness (D*; Figure 5C) or the variation in taxonomic distinctness (Λ; Figure 5E), but the latter index evidenced a significantly higher value ($P < 0.1$) in the treated site than in the control site 29 d after trawling.

Regarding the trophic guilds (Figure 6A–E), the community was dominated by detritus feeders, carnivores, and omnivores, whereas the filter feeders and herbivores were less important in terms of abundance. A significant decrease in abundance immediately after the haul (before trawling versus 1 d posthaul) was observed only in the case of

Table 2. Results of the two-way analysis of variance of trophic guild abundances for the sand habitat (SS = sum of squares; MS = mean square). Only the significant analyses are reported, and values with level of significance $P < 0.1$ indicated (*).

Source of variation	df	SS	MS	F	P	SS	MS	F	P
		Total abundance				**Detritivorous feeders abundance**			
Before vs. After = BA	1	694.1	694.1	0.040		8,301.7	8,301.7	0.643	
Control vs. Impact = CI	1	235,030.6	235,030.6	13.386		160,304.3	160,304.3	12.421	
BAxCI	1	14,367.0	14,367.0	0.818		6,421.3	6,421.3	0.498	
Time(Before or After) = T(BA)	3	853,192.7	284,397.6	16.198	0.0000*	786,342.6	262,114.2	20.309	0.0000*
T(BA)xCI	2	327,916.1	163,958.1	9.338	0.0006*	191,931.1	95,965.6	7.436	0.0022*
Error	32	561,852.1	17,557.9			412,993.8	12,906.1		
Total	41	10,692,722.6				5,535,675.6			
		Predators abundance				**Macrura Natantia abundance**			
Before vs. After = BA	1	71.1	71.1	0.228		323.1	323.1	3.305	
Control vs. Impact = CI	1	24.8	24.8	0.079		8.4	8.4	0.086	
BAxCI	1	139.7	139.7	0.447		0.0	0.0	0.0	
Time(Before or After) = T(BA)	3	2834.3	944.8	3.025	0.0438*	488.7	162.9	1.666	0.1938
T(BA)xCI	2	2,431.6	1,215.8	3.892	0.0307*	764.6	382.3	3.911	0.0302*
Error	32	9,995.2	312.4			3,128.2	97.8		
Total	41	80,967.2				4,960.9			
		Anomura abundance				**Exergy storage**			
Before vs. After = BA	1	0.4	0.4	0.633		5,137,353	5,137,353.2	1.169	
Control vs. Impact = CI	1	0.7	0.7	1.105		1,659,702	1,659,702.2	0.378	
BAxCI	1	0.4	0.4	0.599		7,790,471	7,790,470.9	1.773	
Time(Before or After) = T(BA)	3	4.3	1.4	2.298	0.0962*	39,584,794	13,194,931.3	3.003	0.0448*
T(BA)xCI	2	4.3	2.2	3.447	0.0441*	9,954,333	4,977,166.4	1.133	0.3347
Error	32	20.0	0.6			140,587,600	4,393,362.5		
Total	41	33.9				204,714,253			

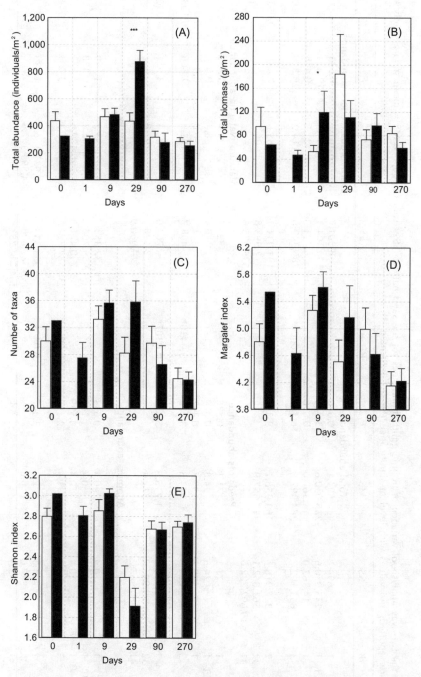

Figure 4. Temporal patterns of the (A) total abundance, (B) total biomass, (C) number of taxa, (D) Margalef index, and (E) Shannon index calculated for the macroinfaunal community of the mud habitat (mean + SE). Open bars represent data from the control area; shaded bars represent data from the area treated with a single rapido haul. The number of days after trawling (0, 1, 9, 29, 90, 270 d) is reported on the x-axis. The differences between the treated and control samples were tested by use of t-tests at three different levels of significance: $P < 0.1$ (*), $P < 0.05$ (**), and $P < 0.01$ (***).

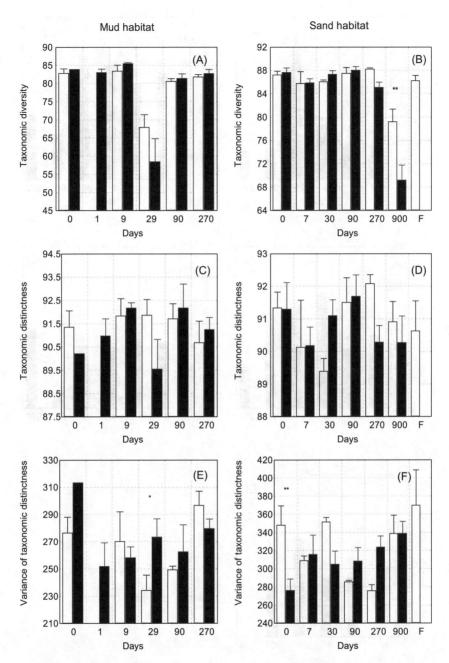

Figure 5. Biodiversity indices (mean + SE) calculated for the benthic macroinfaunal communities of mud and sand habitats in the Northern Adriatic sea: (A) taxonomic diversity of the mud community, (B) taxonomic diversity of the sand community, (C) taxonomic distinctness of the mud community, (D) taxonomic distinctness of the sand community, (E) variance in taxonomic distinctness of the mud community, and (F) variance in taxonomic distinctness of the sand community. Open bars represent data from the control areas; shaded bars represent data from the areas treated with a single rapido haul.

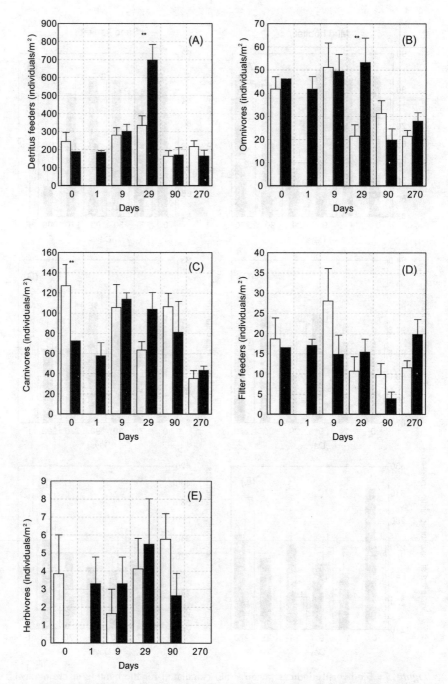

Figure 6. Temporal patterns in abundance (mean + SE) of trophic guilds in the mud habitat: (A) detritus feeders, (B) omnivores, (C) carnivores, (D) filter feeders, and (E) herbivores. Open bars represent data from the control area; shaded bars represent data from the area treated with a single rapido haul.

the carnivores ($P < 0.05$), which then showed a recovery. The abundance of detritus feeders and omnivores 29 d after trawling was significantly higher ($P < 0.05$) in the treated area than in the control area, while no differences were detectable for either the filter feeders or the herbivores.

The total abundance of scavengers (Figure 7A) showed two peaks, 9 and 29 d after trawling, when the more evident difference between the control and treated areas occurred. The scavengers were dominated by Gastropoda (Figure 7B) and Macrura Natantia (Figure 7E), both of which peaked 29 d after trawling. Moreover, some remarkable differences resulted among the different taxonomic groups: 90 d after the experimental haul, gastropod abundance was significantly higher in the treated area than in the control area ($P < 0.05$), whereas abundances of Macrura Reptantia and Macrura Natantia were significantly lower in the treated area than in the control area ($P < 0.1$). No significant differences were found either immediately after the experimental haul or at the

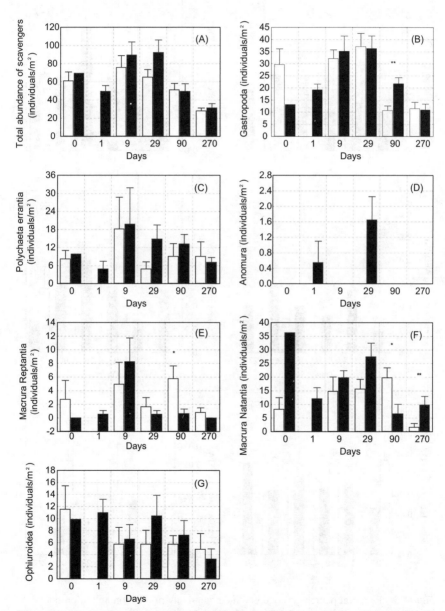

Figure 7. Temporal patterns in abundance (mean + SE) of potential scavenger taxa in the mud habitat: (A) all scavenger taxa, (B) Gastropoda, (C) Polychaeta Errantia, (D) Anomura, (E) Macrura Reptantia, (F) Macrura Natantia, and (G) Ophiuroidea. Open bars represent data from the control area; shaded bars represent data from the area treated with a single rapido haul.

end of the experiment (270 d) for any group except Macrura Natantia, which showed a significantly higher abundance in the treated area after 270 d ($P < 0.05$).

Total production (Figure 8A) in both the treated and control sites showed a similar trend, with maximum production occurring 29 d after the experimental disturbance. This trend was mainly due to the production of detritus feeders and carnivores (Figure 8B, D). The other groups showed no clear patterns. Moreover, 29 d after the experimental haul, the omnivores (Figure 8C)

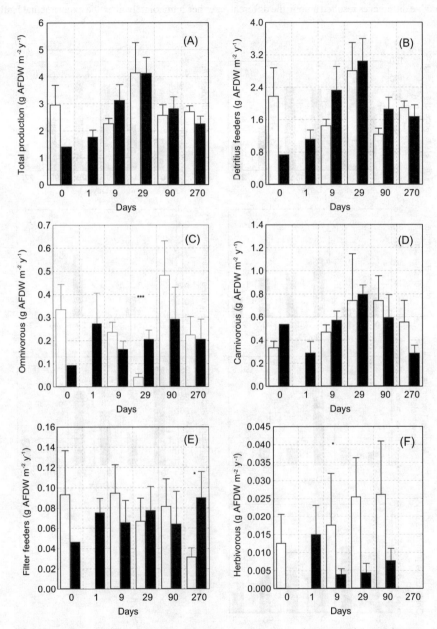

Figure 8. Temporal patterns in total benthic community production and in production by trophic guilds for the mud habitat: (A) total production, (B) detritus feeders, (C) omnivores, (D) carnivores, (E) filter feeders, and (F) herbivores. Units are in grams of ash-free dry weight (AFDW) per square meter per year.[5] Open bars represent data from the control area; shaded bars represent data from the area treated with a single rapido haul.

showed a significantly higher production ($P < 0.01$) in the treated area than in the control area, similar to the production differences exhibited by filter feeders 270 d after the haul ($P < 0.1$). Nine days after trawling, herbivores showed significantly higher densities in the control sites than in treated sites ($P < 0.01$).

Exergy (Figure 9) showed some oscillations, with positive and negative values of decreasing amplitudes. Immediately after trawling, a negative difference between the control and treatment areas was observed, which subsequently diminished at 1 d posthaul (but was still negative) and became positive only at 9 d posthaul. However, the lowest negative value was observed 29 d after trawling. At 90 and 270 d after trawling, the difference between the control and treated areas became smaller in absolute value, although still fluctuating between positive and negative values.

Sand Habitat

A total of 207 taxa were collected during all sampling activity in the sand habitat. Before experimental trawling occurred, the community comprised a total of 95 taxa. The prehaul macrobenthic community showed no dominant species in terms of abundance; some of the main taxa collected were Amphipoda (Arthropoda, Crustacea) (6%), Syllidae (Annelida, Polychaeta) (5%), *Hiatella artica* (Mollusca, Bivalvia) (5%), and *Anapagurus brevicarpus* (Arthropoda, Crustacea) (5%). The same pattern was recognizable immediately after the experimental haul. The four most common taxa accounted for only 20% of the total abundance (*Hiatella artica*, Amphipoda, *Galathea intermedia*, and *Paguristes oculatus* [the latter two taxa are both Arthropoda, Crustacea]).

Prior to the experimental disturbance, the total biomass of the sand community was dominated by *Callista chione* (Mollusca, Bivalvia), which accounted for 66%. *Levicardium oblongum* (Mollusca, Bivalvia), *Hexaplex trunculus* (Mollusca, Gastropoda), and *Sipunculus nudus* (Sipunculida) accounted for another 20%. Immediately after the experimental haul, three of these species (*Callista chione* [48%], *Laevicardium oblongum* [31%], and *Sipunculus nudus* [5%]) together made up more than 80% of the total biomass.

The results of the MDS application (stress = 0.19) were similar to those for the mud habitat (Figure 2), even though the replicates belonging to the control and treated areas 900 d after trawling were clustered together and were different from all the other samples. The MDI results for the sand habitat were similar to those of the mud habitat as well (Figure 3B).

The ANOVA analyses showed statistically significant differences for total abundance and for the abundances of carnivores, omnivores, Macrura Natantia, and scavengers, as reported in Table 3. Furthermore, the ANOVA analyses showed statistically significant differences (BA × CI) for exergy (Table 3).

The *t*-test applied to each sampling time showed significant differences between the control and treated areas at different stages in the recolonization process. Total abundance (Figure 10A) peaked twice (7 and 270 d after trawling) and was significantly lower in the treated area than in the control area 90 d after trawling ($P < 0.05$). Total biomass increased in the control area during the first month and decreased thereafter (Figure 10B). Significant differences in total biomass between treated and control areas were evident 30 and 90 d after the experimental haul ($P < 0.05$ and $P < 0.1$, respectively). In general, the other indices decreased at the end of the experiment (900 d), with a significant difference between the control and treated areas (number of taxa and the Shannon index, $P < 0.05$; Margalef index, $P < 0.01$) (Figure 10C–E). Moreover, the number of taxa and the Margalef index showed significantly lower values in the treated area than in the control area 30 d after trawling ($P < 0.1$ and $P < 0.05$, respectively).

The taxonomic diversity in the treated area showed a trend similar to that seen in the control area, although the treated area had a lower value 900 d after trawling ($P < 0.05$) (Figure 5B). Moreover, the variation in taxonomic distinctness (Λ) showed a significant reduction ($P < 0.05$) immediately after the haul (Figure 5F).

The benthic community abundance was dominated by detritivores, omnivores, and carnivores (Figure 11A–C). The control area showed a relative maximum abundance in most of the groups at 90 d posthaul. The treated area showed a slight (nonsignificant) decrease in abundance in all of the groups immediately after trawling, with the exception of the filter feeders (Figure 11D).

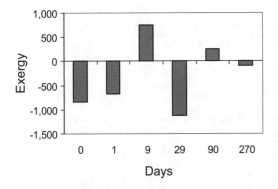

Figure 9. Temporal pattern in exergy for the mud habitat, calculated as the difference between the exergy values of the benthic communities sampled inside and outside the rapido-trawl track.

Table 3. Results of the two-way analysis of variance of trophic guild abundances for the mud habitat (SS = sum of squares; MS = mean square). Only the significant analyses are reported, and values with level of significance $P < 0.1$ indicated (*).

Source of variation	df	SS	MS	F	P
		Total abundance			
Before vs. After = BA	1	140.1	140.1		
Control vs. Impact = CI	1	839.8	839.8		
BAxCI	1	56,671.1	56,671.1		
Time(Before or After) = T(BA)	3	98,798.3	32,932.8	1.725	0.1874
T(BA)xCI	2	173,337.7	86,668.9	4.541	0.0208*
Error	25	477,177.0	19,087.1		
Total	34	7,640,349.6			
		Predators abundance			
Before vs. After = BA	1	941.8	941.8		
Control vs. Impact = CI	1	1,365.0	1,365.0		
BAxCI	1	11,619.2	11,619.2		
Time(Before or After) = T(BA)	3	2,729.2	909.7	0.773	0.5200
T(BA)xCI	2	7,270.4	3,635.2	3.089	0.0633*
Error	25	29,421.5	1,176.9		
Total	34	316,560.0			
		Total scavengers abundance			
Before vs. After = BA	1	1,124.4	1,124.4		
Control vs. Impact = CI	1	63.2	63.2		
BAxCI	1	229.5	229.5		
Time(Before or After) = T(BA)	3	24,377.7	8,125.9	3.084	0.0456*
T(BA)xCI	2	27,110.0	13,555.0	5.144	0.0135*
Error	25	65,872.9	2,634.9		
Total	34	644,432.8			
		Mixed feeders abundance			
Before vs. After = BA	1	13.7	13.7		
Control vs. Impact = CI	1	603.8	603.8		
BAxCI	1	3,038.8	3038.8		
Time(Before or After) = T(BA)	3	12,578.7	4192.9	3.375	0.0341*
T(BA)xCI	2	14,852.1	7426.1	5.977	0.0076*
Error	25	31,061.4	1242.5		
Total	34	359,103.4			
		Macrura Natantia abundance			
Before vs. After = BA	1	338.8	338.8		
Control vs. Impact = CI	1	712.4	712.4		
BAxCI	1	1091.4	1,091.4		
Time(Before or After) = T(BA)	3	2671.6	890.5	2.083	0.1279
T(BA)xCI	2	4,094.7	2047.3	4.789	0.0173*
Error	25	10,687.3	427.5		
Total	34	56,528.2			
		Exergy storage			
Before vs. After = BA	1	1,394,554,250	1,394,554,250.1	2.586	0.1204
Control vs. Impact = CI	1	2,050,937,621	2,050,937,621.1	3.803	0.0624*
BAxCI	1	2,060,597,894	2,060,597,894.1	3.821	0.0619*
Time(Before or After) = T(BA)	3	260,853,350	86,951,116.6	0.161	0.9214
T(BA)xCI	2	581,642,328	290,821,164.2	0.539	0.5898
Error	25	13,480,821,631	539,232,865.2		
Total	34	70,741,820,106			

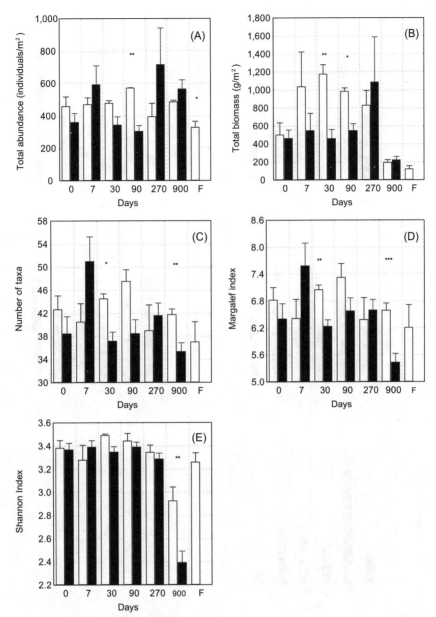

Figure 10. Temporal patterns of the (A) total abundance, (B) biomass, (C) number of taxa, (D) Margalef index, and (E) Shannon index calculated for the macroinfaunal community of the sand habitat (mean + SE). Open bars represent data from the control area; shaded bars represent data from the area treated with a single rapido haul. The number of days after trawling (0, 7, 30, 90, 270, 900 d) is reported on the x-axis. F indicates data for samples collected in the fishing ground adjacent to the study site.

Seven days after the experimental haul, the abundance of both omnivores and carnivores was significantly higher ($P < 0.1$) in the treated area than in the control area. Abundances showed a greater decrease in the treated area than in the control area: the presence of omnivores was significantly ($P < 0.1$) lower after 30 d.

At 90 d, fewer detritivores ($P < 0.01$) and carnivores ($P < 0.1$) were present in treated samples than in the control samples.

The largest taxonomic groups of potential scavengers in the sand habitat were Anomura and Macrura Natantia. The total abundance of scavengers (Figure

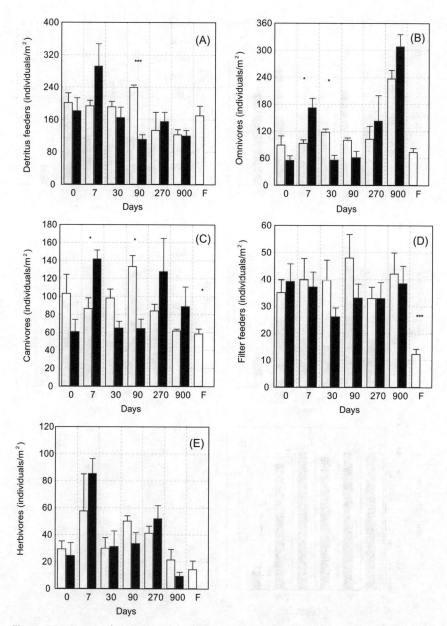

Figure 11. Temporal patterns in abundance (mean + SE) of trophic guilds in the sand habitat: (A) detritus feeders, (B) omnivores, (C) carnivores, (D) filter feeders, and (E) herbivores. Open bars represent data from the control area; shaded bars represent data from the area treated with a single rapido haul.

12A) 7 d after trawling was significantly higher in the treated area than in the control area ($P < 0.1$). This difference was mainly due to the higher abundance of Macrura Natantia in the treated area after 7 d ($P < 0.05$) (Figure 12F). Gastropods were characterized by a significantly lower abundance in the treated area 30 d after the experimental haul, and a higher presence in the treated area than in the control area 900 d after trawling (Figure 12B) ($P < 0.05$ and $P < 0.01$, respectively).

There was a distinct difference in the trends for total production (Figure 13A) between the control area (which reached a maximum 30 d after trawling) and the treated area (which had more stable values). Production values were significantly lower in the treated area versus the control area 30

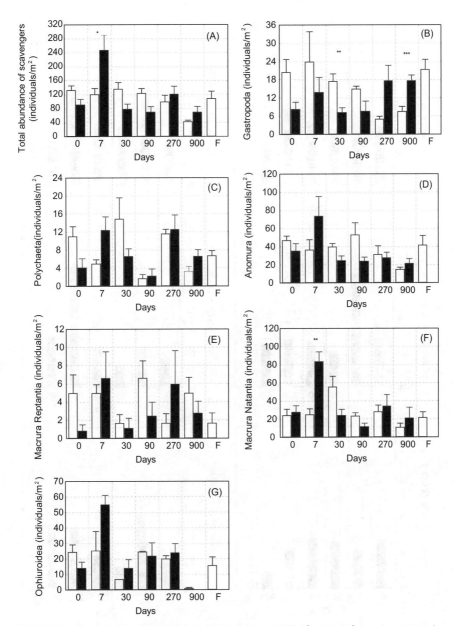

Figure 12. Temporal patterns in abundance (mean + SE) of potential scavenger taxa in the sand habitat: (A) all scavenger taxa, (B) Gastropoda, (C) Polychaeta Errantia, (D) Anomura, (E) Macrura Reptantia, (F) Macrura Natantia, and (G) Ophiuroidea. Open bars represent data from the control area; shaded bars represent data from the area treated with a single rapido haul.

and 90 d after trawling ($P < 0.1$). However, the different taxonomic groups exhibited various types of delay in response to trawling (Figure 13B–F). Production of herbivores was higher in the treated area than the control area immediately after the haul ($P < 0.01$); production of omnivores was higher in the treated area at 7 d posthaul ($P < 0.01$). The production of carnivores was significantly higher in the treated than the control area at the end of the experiment (270 d [$P < 0.1$] and 900 d [$P < 0.05$]). The filter feeders seemed to be the dominant group, based on their contribution to total production. For the filter feeders, a significantly lower value was observed for the treated area than for the control area at 7 and 30 d after the experimental haul ($P < 0.05$).

Exergy was negative throughout almost the entire

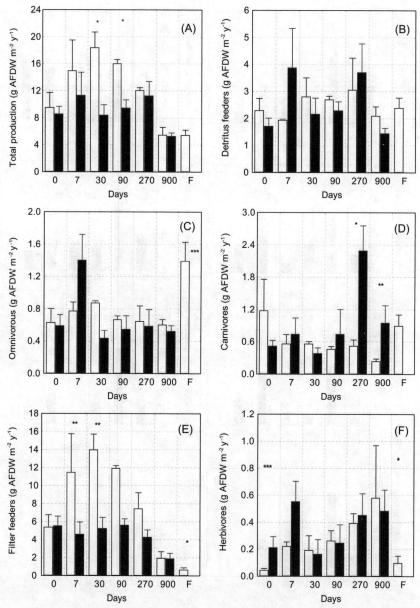

Figure 13. Temporal patterns in total benthic community production and in production by trophic guilds for the sand habitat: (A) total production, (B) detritus feeders, (C) omnivores, (D) carnivores, (E) filter feeders, and (F) herbivores. Units are in grams of ash-free dry weight (AFDW) per square meter per year. Open bars represent data from the control area; shaded bars represent data from the area treated with a single rapido haul.

experiment, with the maximum difference (i.e., the greatest absolute difference between the exergy storage in the control and treated areas) (Figure 14A) observed 30 d after trawling. This was similar to the results obtained for the mud habitat. However, exergy differences showed a more distinct trend for the sand habitat. In fact, this difference decreased asymptotically over time, being both positive and close to zero 900 d after trawling.

Fishing Ground (Sand Substrate)

In terms of abundance, the main taxa of the fishing ground community (composed of a total of 90 taxa)

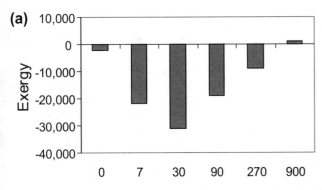
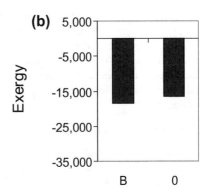

Figure 14. (A) Temporal pattern in exergy in the sand habitat, calculated as the difference between the exergy values of the benthic communities sampled inside and outside the rapido-trawl track. (B) The chronic effects of trawling on the fishing ground benthic community were evaluated by calculating the difference between the exergy of the the fishing ground and the exergy of the experimental area before trawling (B) and just after trawling (0).

were *Anapagurus brevicarpus* (10%), *Lucinella divaricata* (Mollusca, Bivalvia) (8%), Gammaridae (Arthropoda, Crustacea) (6%), and *Aspidosiphon müelleri* (Sipunculida) (6%). In terms of biomass, *Sipunculus nudus* was the dominant species (24%). Cnidaria (8%), *Laevicardium oblongum* (7%), and *Labidoplax digitata* (Echinodermata, Holoturoidea) (7%) were the other important taxa.

The results of the MDS ordination (stress = 0.16) suggested that the macrobenthic community of the fishing ground was quite different from that of the experimental area, both before and immediately after trawling, and showed a medium-high level of dispersion (Figure 2).

The comparison between the samples collected in the fishing ground and those collected in the experimental area before and immediately after the haul allowed for a comparison between the acute and chronic effects of trawling. The values for the fished area were comparable to those in the experimental area with regard to the number of taxa, diversity, and taxonomic indices (Figures 5, 10), with the exception of total abundance ($P < 0.1$) (Figure 10A).

For trophic guild abundance, comparison of samples taken before the experimental haul with those from the fishing area highlighted the fact that the abundance of filter and carnivorous feeders was significantly lower ($P < 0.01$ and $P < 0.1$, respectively) in the fished area (Figure 11C, D). No significant differences were detected concerning the other trophic guilds. Based on the results of the *t*-test, there were no remarkable differences in scavenger abundance between the samples taken in the fished area and those taken before the experimental haul. Total estimated production was lower in the fished area than in the control area (Figure 13A), with significant values in terms of filter-feeder production (Figure 13A, E).

Exergy was lower in the fishing ground (Figure 14B) than in the undisturbed sand experimental area. Thus, the observed difference was negative, with values similar to those recorded in the experimental area 7 and 90 d after trawling.

Discussion

Based on the data collected, we present a brief description of the macrobenthic community of the mud and sand habitats. The community of the mud habitat was characterized by a lower number of taxa than was found in the sand habitat. In the mud habitat, most of the dominant species in terms of abundance and biomass belonged to the detritus feeder group (such as *Corbula gibba*, *Trachythyone elongata*, and Gammaridae). In the sand habitat, the community was dominated (particularly in terms of biomass) by filter feeder species, such as the bivalves *Callista chione* and *Levicardium oblongum*.

The macrobenthic community of the fishing ground had a species richness similar to that recorded in the control of the sand experimental area, but the species composition and relative abundances were different. Moreover, for total abundance and total biomass, some of the dominant species (such as *Lucinella divaricata*, *Aspidosiphon müelleri*, *Sipunculus nudus*, and *Labidoplax digitata*) in the fishing ground belonged to the detritivores group rather than the filter feeder group, which was prominent in the sand control area.

Whereas differences in community structure between the mud and sand habitats could be attributed to edaphic and environmental features (depth, bottom grain size, hydrodynamic characteristics, natural disturbance frequencies, etc.), the sand and fishing ground habitats were characterized by similar environmental features, and therefore the community differences could be attributed to the chronic disturbances caused by commercial fishing activity.

Recolonization Dynamics

The data collected made it possible to compare the recolonization of two different habitats—an offshore sand habitat and an inshore mud habitat—that were each treated with a single rapido haul. For long-term experiments, season could be an important factor in determining the recovery process (Zajac and Whitlach 1982a, 1982b), but since the experimental hauls were done at the beginning of the winter season, the first 3 months involved the same seasonal conditions. Thus, the changes recorded within the first 3 months were ascribed as consequences of the trawling disturbance.

The results indicated that both the immediate effects and the medium-term responses were community-wide.

The magnitude of the immediate response (i.e., the change in the abundance or biomass [Figures 4, 10] and also in the sample dispersion [Figure 3]) of the organisms to fishing disturbance varied significantly in relation to the studied habitat and among different taxa, as highlighted also by Collie et al. (2000). As previously described by other authors (Boesch and Rosenberg 1981; Thiel and Schriever 1990; Kaiser and Spencer 1996; Auster and Langton 1999), both the recolonization and the long-term community composition depend on numerous factors, such as stability (local level of disturbance) of treated areas, changes in local oceanographic regimes, and the tolerance of organisms to physical changes. In fact, during the recovery process recorded in the study areas, it was possible to recognize significant differences in a shared successional pattern between the sand and mud habitats.

In general, after the initial disturbance, the early stages of recolonization in both habitats were dominated by scavengers. The greatest differences between treated and control areas were evident 1 month after the disturbance (see Figures 9, 14), and the recovery process seemed to require at least 3 months before a community structure similar to that of the control area could be observed.

After more detailed analysis, significant differences in recovery time between the two habitats were detectable. For the sand habitat, the presence of scavengers was more limited over time (the peak was recorded only 1 week after the haul) than in the mud habitat, where the scavengers were abundant for 1 month following the haul and were then exceeded by an abundance of detritus feeders.

The exergy analysis highlighted the fact that two different recovery dynamics were acting in the two habitats. For the sand habitat, the recovery process evolved in the manner of a slow, wide wave, requiring 9 months for the entire process to be completed. This was confirmed by the ANOVA results, which showed a press effect of the disturbance (i.e., an effect that persists during the entire time series) 90 d after the treatment (Underwood 1994). When an ANOVA was performed on the extended data (all samples collected 900 d after the treatment), the results showed a pulse effect, meaning that recovery had occurred. However, for the mud habitat, the recovery process seemed to take place more quickly, and was similar to a series of rapid, narrow waves, as confirmed by the ANOVA results (based on both time series to 90 and 270 d after the treatment, a pulse effect was recorded). According to Kaiser et al. (2002), the latter result could be due to the fact that the mud community, given its location in the inshore zone (1 NM from the coastline), was more exposed to natural disturbances and was therefore better adapted to a recovery process than the sand community, which was located on a deeper bottom and was further away from the coastline (about 11 NM).

The exergy analysis made it possible to identify the presence of overshooting phenomena (after periods of 1 week and 3 months, respectively) that involved the entire benthic community. Overshoots of total numerical density (i.e., an abundance increase that reached values higher than those of the reference state) inside the plots occurred frequently and were a well-described phenomenon during the recovery process (Beukema et al. 1999). Substantial overshoots in the initial phases of faunal recovery in azoic sediments have already been observed by several authors (e.g., for genus *Capitella*; Grassle and Grassle 1974; McCall 1977; Jaramillo et al. 1987; Ragnarsson 1995; Beukema et al. 1999). The examples from the literature all refer to the presence of small worm species that is generally considered to be an opportunistic species and that responds to the favorable situations created by a disturbance. The overcompensation that takes place at a community level as part of the recovery process has been less thoroughly described and is probably less understood.

Community development (i.e., succession, in the broad sense of the word) can be defined as a process of species replacement. The probability of the replacement of one species by another determines the rate and direction of the succession. The analysis of recolonization after disturbances makes it possible to assess the processes that regulate and structure communities, and to detect the successional stages and their underlying causes.

The data collected seem to suggest that, in both study areas, the recolonization process was due mainly to migration from adjacent, undisturbed sources rather than to juvenile or larval settlement, which does not occur during the winter, as confirmed by the absence of these stages in analyzed samples. In addition, juveniles and larva were absent in our biomass data. A more likely hypothesis is that the two phenomena have roles of similar importance at differing stages of the recovery process. It can therefore be concluded that, at least in the early stages, the recolonization process depended to a great extent on the natural stock of colonizers already prevailing in the area. As has already been highlighted by Zajac and Whitlach (1991), the community composition and the season determine which mechanism will dominate during the recovery process among disturbed communities.

Connel and Slatyer (1977) postulated three contrasting models in the benthic successional process: facilitation, tolerance, and inhibition, all of which are characterized by interspecific interactions. Lu and Wu (2000) reported that the replacement of the early dominant species was commonly observed during the successional stages. The same pattern also was observed in both mud and sand habitats during this study. Our observations fit well into the facilitation model, in that the early colonizers modify the substratum, making it less favorable for their own survival but more favorable for the establishment of their successors. For example, scavengers consume the carrion and induce the transformation of organic matter, thus facilitating the establishment of the detritus feeders (see mud habitat succession in Results section).

As discussed previously, the early recolonization stages were dominated by scavengers, or more specifically, by species that could potentially act as scavengers. Is it possible to define these species as opportunists? Thistle (1981) remarked that it is difficult to decide which mechanism drives the occurrence of temporary overcompensation in disturbed patches: (1) rapid and temporary responses by adapted species (i.e., the real opportunists, those with rapid reproduction and dispersal process) to the presence of temporary extra resources in disturbed patches (such as a better food supply or the presence of decaying organisms), or (2) the presence of species that are not specialized in the above manner, but are merely able to survive better in the absence of adult congeners and other competitors for resources such as space and food. The latter species should not be called opportunists, though they temporarily show abundance patterns that resemble those of the real opportunists. This fact would seem to be consistent with the results of the present study. As has already been described by Pranovi et al. (1998), the species that act as scavengers are presumably non-opportunistic species that exhibit an opportunistic type of behavior under certain conditions. In addition, another possible confirmation of this behavior could be the fact that the same taxa may or may not act as scavengers when their community structure and limiting conditions change. This is the case with regard to the ophiuroid group, which shows scavenging behavior only when in a sand habitat. Food may have been limited for this group at the sand site, but when carrion became available, the ophiuroids moved in large numbers to exploit this valuable additional resource. It goes without saying that, at sites where food is always plentiful (e.g., due to the presence of rainfall from the suspended cultured mussels), the taxa would have little motivation to migrate into treated areas (Ramsay et al. 1998).

As has been observed by Kenny and Rees (1996), disturbance events (either natural or anthropogenic) affect the long-term stability of the sediment and maintain an early successional stage in the benthic communities. Environments with high levels of disturbance, such as fishing grounds where the fishing effort is patchily distributed (Jennings et al. 2002), result in a mosaic of successional faunal patches for which the rate of conversion forward by succession exists in relative equilibrium with the conversion backward by disturbance (Wittaker and Levin 1977; Thistle 1981). Subsequently, as suggested by Hall (1994), "a sufficiently large proportion of the habitat will be at the same point on a successional trajectory." All this could explain the wide distribution of scavengers in areas subjected to high fishing pressure (such as the Northern Adriatic Sea). As has been proposed by Britton and Morton (1994), the scavenger taxa experience a tradeoff between the impact suffered due to fishing activity and the advantage produced by the increased food supply (carrion). An example of the above phenomenon can be seen by comparing Macrura Reptantia (crabs) and Macrura Natantia (shrimps) in the Adriatic Sea. Both groups can act as scavengers, but the latter seems to be more affected by trawling activity (Raicevich 2000). This results in a differential distribution of the species, with crabs being more abundant than shrimps on the fishing grounds.

Comparison Among Different Types of Indicators

Total abundance and total biomass can give an initial picture of the recolonization pattern and can provide a comparative view of what happens in both the treated and the control areas, but these indicators are less informative in describing recolonization processes. However, in accordance with Kaiser et al. (2002), the insensitivity of diversity indices (number of species, Shannon index, and Margalef index) to subtle community changes was evident. The same results were observed for taxonomic indices.

The multivariate approach facilitates the description of changes in community structure, first by comparing species composition, and second by analyzing the heterogeneity rate and the dissimilarity percentages at each sampling period for the different temporal situations. However, if more detailed information is required for inferences about the recovery processes, a functional indicator must be applied, such as the abundances of trophic guilds (Kaiser et al. 2002). Guilds make it possible to (1) identify which of the groups is responsible, in either a positive or a negative manner, for an observed trend in the recovery and (2) identify the relevant information about which factors caused the trend. For example, we observed increases in potential scavengers 1 week or 1 month after the experimental haul (Figures 7, 12) that were caused by the presence of carrion and food supply. Another example is the peak in detritus feeders that was observed in the mud habitat 1 month after the treatment (Figure 6), and which was probably linked to a facilitation process. The scavengers attracted by the food supply in the treated area initiate a food transformation process, which facilitates the subsequent arrival of species that exploit other niches (e.g., food of an inferior quality and complexity). The above procedure also allows analysis of the responses of the different communities to the same level of stress, as well as the responses of the same taxa or trophic guilds under different environmental conditions.

Obviously, the critical point in functional analysis is the allocation of the species to different trophic guilds. Collection of detailed ecological data concerning each individual species in a particular habitat is required—a time-consuming process—and this information is often lacking. However, our results show that the use of data from the literature (sometimes from habitats different than those analyzed in our study; see Methods) can yield an approximate but useful analysis of the trophic guilds involved in the recolonization process.

Another important aspect that must be considered is benthic productivity. This process can be directly or indirectly affected by trawling activity (Jennings et al. 2002; Kaiser et al. 2002). For this problem, our results indicate that, in the sand habitat, no significant variation in the total benthic production was observed in the treated area during the first 3 months after the haul. However, in the sand control area, and in both the treated and the control areas for the mud habitat, peaks in productivity were recorded after 1 month (Figure 8). At present, there is no clear explanation for this phenomenon; it is possible that the control areas obtained some benefits from the recovery process in the treated areas. The period of time involved in the process (1 month) would suggest that scavengers were present.

Finally, exergy is an interesting index for the analysis of recovery dynamics in different habitats and/or comparison of areas subjected to different disturbances. In fact, exergy distinguished the different stages of recovery, clearly identifying the overcompensation events and discriminating between the recolonization dynamics observed in the two habitats.

The functional modifications in the benthic community that were observed in the experimental area (acute stress) can be used to examine data collected in the fishing ground to better understand the effects of chronic stress. Indeed, changes induced by continuous fishing activity are impossible to analyze by means of the experimental approach and in the absence of historical data describing a pristine reference condition.

Chronic Disturbance Effects

The diversity, total biomass, and production values in the fishing ground were comparable with the lowest values recorded in the sand experimental area (Figures 10, 13). In some cases, the fishing ground indices were lower than those seen during the acute phases of the disturbance.

Evaluation of the changes induced in the benthic community by chronic fishing disturbance is often hampered by the poor resolution of the data and by the lack of an accurate time series (Kaiser et al. 2000). In the present study, analysis of data from the fishing ground surrounding the experimental area and comparison with the experimental data allowed us to make inferences concerning the chronic effects of rapido trawling on sand habitats and the implications of the experimental data for ecosystem processes. The comparison by means of a multivariate approach showed that the macrobenthic community structure was not similar between the sand experimental area and the fishing ground; a higher spatial heterogeneity was recorded in the fishing ground (Figure 2).

Data from the fishing ground showed total abundance and total biomass values comparable with the lowest recorded in the experimental area (Figure 10). The multivariate analysis and data are in accordance with theoretical predictions (Lambshead et al. 1983) and experimental data (Kaiser et al. 2000), which indicate that the relatively large-bodied fauna are removed by repeated bottom fishing. The elimination of large-bodied species appeared to be confirmed by the low abundance of filter feeders (which are mainly composed of bivalves, such as *Callista chione*) recorded in the fishing ground (Figure 11). Scavengers were well represented on the fishing ground, making up about 35% of the total abundance (about 28% in the control samples). This corroborates the findings of Collie et al. (1997) that scavenging organisms tend to dominate benthic communities in heavily exploited areas.

As has already been highlighted by other authors (Jennings et al. 2001, 2002), the trawling impact induces a reduction in total community production, which is largely due to a reduction in the biomass of large-bodied species (D. E. Duplisea, Maurice Lamontaigne Institute, unpublished). Total production was low on the fishing ground, and was comparable with the lowest value recorded in the experimental area. Only the omnivores (one of the main potential scavenger groups) showed a significantly higher value in the fishing ground than those recorded in the experimental area; omnivore production was comparable with the peak recorded in the treated area 1 week after trawling.

Finally, exergy suggested that the benthic community in the fishing ground remained in an early successional stage, with a value comparable to those recorded in the experimental area after 1 week to 1 month.

Conclusions

Disturbance has been identified as potentially the most important mechanism structuring localized benthic communities (Menge et al. 1999). Disturbance removes individuals from the community, thereby creating opportunities for the colonization of new species, which eventually brings about a change in the community structure. Thus, disturbance rewinds the successional clock, driving the community away from its equilibrium (Sousa 1984; Reice 1994). The community development trajectory following a disturbance is a result of the dynamics between the length of time that has elapsed since the previous disturbance, the strength of the disturbance, the historical composition of the individual species in the area, and the resource supply rates (Grime 1979; Tilman 1988).

Disturbance caused by fishing activity results in much more than a partial defaunation of the area. In fact, it induces changes in the community structure that are probably mediated by modifications in the food supply (e.g., carrion) and the bottom features (such as sediment texture, geochemical cycles, etc.). In particular, the presence of carrion and/or the changes in food availability (e.g., prey vulnerability) seem to play an important role in producing changes that drive the early successional stages of the recovery process.

The functional approach applied in this study made it possible to highlight the fact that the recovery process requires about 9 months in mud and sand habitats of the Northern Adriatic Sea. This fact has important implications for a heavily fished basin like the Northern Adriatic Sea, where the rapido fishing grounds are trawled 8–10 times per year (Pranovi et al. 2000). As a result, according with Jennings et al. (2002), a patchy distribution of fishing effort is urgently required to allow a mosaic of successional stages to develop and interact with each other. The presence of early successional stages in the fishing grounds of the Northern Adriatic Sea indicates that the macrobenthic community in the area is not in a stable state. Once the disturbance ceases, the community can complete the recovery process toward equilibrium.

This study allowed us to verify the applicability of a set of indicators and to assess their performance (i.e., the capacity to detect the effects of fishing disturbance). The community indicators, such as production and exergy, proved to be more sensitive than the diversity indices. Diversity indicators are probably not specifically adapted to the analysis of fishing impact, but in association with a functional approach like trophic guild analysis, they can provide a better description of community changes and can thus help to detect the main causes of disturbance.

Acknowledgments

We thank two anonymous referees for the constructive critiques that improved this article.

References

Ardizzone, G. D. 1994. An attempt of a global approach for regulating the fishing effort in Italy. Biologia Marina Mediterranea 1:109–113.

Auster, P. J., and R. W. Langton. 1999. The effect of fishing on fish habitat. Pages 150–187 in L. Benaka, editor. Fish habitat: essential fish habitat and rehabilitation. American Fisheries Society, Symposium 22, Bethesda, Maryland.

Beukema, J. J., E. C. Flach, R. Dekker, and M. Starink. 1999. A long-term study of the recovery of the macrozoobenthos on large defaunated plots on a tidal flat in the Wadden Sea. Journal of Sea Research 42:235–254.

Boesh, D. F., and R. Rosenberg. 1981. Response to stress in marine benthic communities. Pages 179–200 in G. W. Berret and R. Rosenberg, editors. Stress effects on natural ecosystems. Wiley, Chichester, UK.

Bonsdorff, E., and C. S. Österman. 1985. The establishment, succession and dynamics of a zoobenthic community—an experimental study. Pages 287–297 in P.-E. Gibbs, editor. Proceedings of the 19th International Conference ICEE/EMBS. Cambridge University Press, Cambridge, UK.

Brey, T. 1990. Estimating productivity of macrobenthic invertebrates from biomass and mean individual weight. Meeresforschung 32:329–343.

Britton, J. C., and B. Morton. 1994. Marine carrion and scavengers. Oceanography and Marine Biology: An Annual Review 32:369–434.

Clarke, K. R., and R. M. Warwick. 2001. A further

biodiversity index applicable to species lists: variation in taxonomic distinctness. Marine Ecology Progress Series 216:265–278.

Collie, J. S., G. A. Escanero, and P. C. Valentine. 1997. Effects of bottom fishing on the benthic megafauna of Georges Bank. Marine Ecology Progress Series 155:159–172.

Collie, J. S., S. J. Hall, M. J. Kaiser, and I. R. Poiner. 2000. A quantitative analysis of fishing impacts on shelf-sea benthos. Journal of Animal Ecology 69:785–798.

Connel, J. H., and R. O. Slatyer. 1977. Mechanism of succession in natural communities and their role in community stability and organization. American Naturalist 3:1119–1144.

Da Ponte, F. 2001. La pesca a strascico come fattore di disturbo ecologico in Laguna di Venezia ed in Alto Adriatico. Master's thesis. University of Venice, Venice.

Debeljak M. 2002. Applicability of genome size in exergy calculation. Ecological Modelling 152:103–107.

Desrosiers, G., D. Bellan-Santini, and J. C. Brêthes. 1986. Organisation trophique de quatre peuplements de substrats rocheux selon un gradient de polution industrielle (Golfe de Fos, France). Marine Biology 91:107–120.

Desrosiers, G., C. Savenkoff, M. Olivier, G. Stora, K. Juniper, A. Caron, J.-P. Gagnè, L. Legendre, S. Mulsow, J. Grant, S. Roy, A. Grehan, P. Scaps, N. Silverberg, B. Klein, J.-E. Tremblay, and J.-C. Therriault. 2000. Trophic structure of macrobenthos in the Gulf of St. Lawrence and on the Scotian Shelf. Deep-Sea Research 47:663–697.

Dittmann, S., C. Gunther, and U. Scleier. 1999. Recolonization of tidal flats after disturbance. Pages 175–192 in S. Dittmann, editor. The Wadden Sea ecosystem: stability, problems, and mechanisms. Springer, Berlin.

Duplisea, D. E., S. Jennings, S. Malcolm, R. Parker, and D. B. Sivyer. 2001. Modelling potential impacts of bottom trawls on soft-sediment biogeochemistry in the North Sea. Geochemical Transactions 14:1–6.

Fath, B. D., and H. CAbezas. 2004. Exergy and fisher information as ecological indices. Ecological Modelling 174:25–35.

Fath, B. D., B. C. Patten, and J. S. Choi. 2001. Complementarity of ecological goal functions. Journal of Theoretical Biology 208:493–506.

Fauchauld, K., and P. A. Jumars. 1979. The diet of worms: a study of polychaete feeding guilds. Oceanography and Marine Biology: An Annual Review 17:193–284.

Fonseca, J. C., J. C. Marques, A. A. Paiva, A. M. Freitas, V. M. C. Madeira, and S. E. Jørgensen. 2000. Nuclear DNA in the determination of weighing factors to estimate exergy from organism biomass. Ecological Modelling 126:179–189.

Gamenick, I., A. Jahn, K. Vopel, and O. Giere. 1996. Hypoxia and sulphide as structuring factors in a macrozoobenthic community on the Baltic Sea shore: colonisation studies and tolerance experiments. Marine Ecology Progress Series 144:73–85.

Grassle, J. F., and J. P. Grassle. 1974. Opportunistic life histories and genetic system in marine benthic polychaetes. Journal of Marine Research 32:253–284.

Grime, J. B. 1979. Plant strategies and vegetation processes. Wiley and Sons, Chichester, UK.

Hall, J. J., M. R. Robertson, and S. E. Thrush. 1996. Patchiness and disturbance in shallow-water assemblages. Pages 333–376 in P. S. Giller, A. G. Hildrew, and D. Raffaelli, editors. Aquatic ecology. Blackwell Scientific Publications, Oxford, UK.

Hall, S. J. 1994. Physical disturbance and marine benthic communities: life in unconsolidated sediment. Oceanography and Marine Biology: An Annual Review 32:179–239.

Hall-Spencer, J. M., C. Froglia, R. J. A. Atkinson, and P. G. Moore. 1999. The impact of rapido trawling for scallops, *Pecten jacobaeus* (L.), on the benthos of the Gulf of Venice. ICES Journal of Marine Science 56:111–124.

Jaramillo, E., R. A. Croker, and E. B. Hatfield. 1987. Longterm structure, disturbance, and recolonization of macroinfauna in a New Hampshire sand beach. Canadian Journal of Zoology 65:3024–3031.

Jennings, S., T. A. Dinmore, D. E. Duplisea, K. J. Warr, and J. E. Lancaster. 2001. Trawling disturbance can modify benthic process. Journal of Animal Ecology 70:459–475.

Jennings, S., and M. J. Kaiser. 1998. The effects of fishing on marine ecosystems. Advances in Marine Biology 34:201–351.

Jennings, S., M. D. Nicholson, T. A. Dinmore, and J. E. Lancaster. 2002. Effects of chronic trawling disturbance on the production of infaunal communities. Marine Ecology Progress Series 243:251–260.

Jørgensen, S. E. 2000. Application of exergy and specific exergy as ecological indicators of coastal areas. Aquatic Ecosystem Health and Management 3:419–430.

Jørgensen, S. E., and S. N. Nielsen. 1998. Thermodynamic orientors: exergy as goal function in ecological modelling and as an ecological indicator for the description of ecosystem development. Pages 64–86 in F. Müller and M. Leupelt, editors. Eco targets, goal function and orientors. Springer-Verlag, Berlin.

Jørgensen, S. E., S. N. Nielsen, and H. Mejer. 1995. Emergy, environ, exergy and ecological modelling. Ecological Modelling 77:99–109.

Jørgensen, S. E., P. Verdonschot, and S. Lek. 2002. Explanation of the observed structure of functional feeding groups of aquatic macro-invertebrates by an ecological model and the maximum exergy principle. Ecological Modelling 158:223–231.

Kaiser, M. J., J. S. Collie, S. J. Hall, Jennings, S., and I. R. Poiner. 2002. Modification of marine habitats by trawling activities: prognosis and solutions. Fish and Fisheries 3:114–136.

Kaiser, M. J., K. Ramsay, C. A. Richardson, F. E. Spencer, and A. R. Brand. 2000. Chronic fishing disturbance has changed shelf benthic community structure. Journal of Animal Ecology 69:494–503.

Kaiser, M. J., and B. E. Spencer. 1996. The effects of beam-trawl disturbance on infauna communities in different habitats. Journal of Animal Ecology 65:348–358.

Kenny, A. J., and H. L. Rees. 1996. The effects of marine gravel extraction on the macrobenthos: results 2 year post-dredging. Marine Pollution Bulletin 32(8/9):615–622.

Kruskal, J. B., and M. Wish. 1978. Multidimensional scaling. Sage Publications, Beverley Hills, California.

Lambshead, P., H. Platt, and K. Shaw. 1983. The detection of differences among assemblages of marine benthic species based on an assessment of dominance and diversity. Journal of Natural History 17:859–874.

Lu, L., and R. S. S. Wu. 2000. An experimental study on recolonization and succession of marine macrobenthos in defaunated sediment. Marine Biology 136:291–302.

Margalef, R. 1957. La teorìa de la informaciòn en ecologia. Memorias de la Real Academia Ciencias y Artes de Barcelona 32:373–449.

Marques, J. C., M. A. Pardal, S. N. Nielsen, and S. E. Jørgensen. 1997. Analysis of the properties of exergy and biodiversity along an estuarine gradient of eutrophication. Ecological Modelling 102:155–167.

McCall, P. L. 1977. Community patterns and adaptive strategies of the infaunal benthos of Long Island Sound. Journal of Marine Research 35:221–266.

Menge, B. A., B. A. Daley, and J. Lubchenco. 1999. Top-down and bottom-up regulation of New Zealand rocky intertidal communities. Ecological Monographs 69:297–330.

Müller, F. 1997. State-of-the-art in ecosystem theory. Ecological Modelling 100:135–161.

Müller, F., and M. Leupelt. 1998. Eco targets, goal function and orientors. Springer-Verlag, Berlin.

Odum, E. P. 1969. The strategy of ecosystem development. Science 164:262–270.

Pearson, T. H., and R. Rosenberg. 1978. Macrobenthic succession in relation to organic enrichment and pollution of the marine environment. Oceanography and Marine Biology: An Annual Review 16:229–311.

Pranovi, F., O. Giovanardi, and G. Franceschini. 1998. Recolonization dynamics in areas disturbed by the bottom fishing gears. Hydrobiologia 375/376:125–135.

Pranovi, F., S. Raicevich, G. Franceschini, M. G. Farrace, and O. Giovanardi. 2000. "Rapido" trawling in the Northern Adriatic Sea: effects on benthic communities in an experimental area. ICES Journal of Marine Science 57:517–524.

Probert, P. K. 1984. Disturbance, sediment stability, and trophic structure of soft-bottom communities. Journal of Marine Research 42:893–921.

Ragnarsson, S. A. 1995. Recolonization of intertidal sediments: the effect of patch size. Pages 269–276 in A. Elefteriou, A. D. Ansell, and C. J. Smith, editors. Biology and ecology of shallow coastal waters. Olsen and Olsen, Fredensborg, Denmark.

Raicevich, S. 2000. La pesca con il rapido in Nord Adriatico: effetti sulle comunità bentoniche ed implicazioni gestionali. Master's thesis. University of Venice, Venice.

Ramsay, K., M. J. Kaiser, and R. N. Hughes. 1998. Responses of benthic scavengers to fishing disturbance by towed gears in different habitats. Journal of Experimental Marine Biology and Ecology 224:73–89.

Reice, S. R. 1994. Nonequilibrium determinants of biological community structure. American Scientist 82:424–435.

Rhoads, D. C., P. L. McCall, and J. Y. Yingst. 1978. Disturbance and production on the estuarine seafloor. American Journal of Science 66:577–586.

Santos, S. L., and J. L. Simon. 1980. Marine soft-bottom community establishment following annual defaunation: larval or adult recruitment? Marine Ecology Progress Series 2:235–241.

Schneider, E. D., and J. J. Kay. 1994. Life as a manifestation of the second law of thermodynamics. Mathematical and Computer Modelling 19:25–48.

Shannon, C. E., and W. Weaver. 1963. The mathematical theory of communication. University of Illinois, Urbana.

Smith, C. R., and S. Brumsickle. 1989. The effects of oatch size and substrate isolation on colonization modes and rates in an intertidal sediment. Limnology and Oceanography 34:1263–1277.

Smith, E. P., 2002. BACI design, Pages 141–148 in A. H. El-Shaarawi and W. Piegorsch, editors. Encyclopedia of environmetrics, volume 1. Wiley, Chichester, UK.

Sousa, W. P. 1984. The role of disturbance in natural communities. Annual Review of Ecology and Systematics 15:353–391.

Thiel, H., and G. Schriever. 1990. Deep-sea mining environmental impact and the DISCOL project. Ambio 19:245–250.

Thistle, D. 1981. Natural physical disturbances and communities of marine soft bottoms. Marine Ecology Progress Series 6:223–228.

Thrush, S. F., J. E. Hewitt, V. J. Cummings, and P. K. Dayton. 1995. The impact of habitat disturbance by scallop dredging on marine benthic communities: what can be predicted from the results experiments? Marine Ecology Progress Series 129:141–150.

Thrush, S. F., J. E. Hewitt, V. J. Cummings, P. K. Dayton, M. Cryer, S. J. Turner, G. A. Funnell, R. G. Budd, C. J. Milburn, and M. R. Wilkinson. 1998. Disturbance of the marine benthic habitat by commercial fishing: impacts at the scale of fishery. Ecological Applications 8:866–879.

Thrush, S. F., R. B. Whitlatch, R. D. Pridmore, J. E. Hewitt, V. J. Cummings, and M. R. Wilkinson. 1996. Scale-dependent recolonization: the role of sediment stability in a dynamic sandflat habitat. Ecology 77:2472–2487.

Tilman, D. 1988. Plant strategies and the dynamics and structure of plant communities. Princeton University Press, Princeton, New Jersey.

Ulanowicz, R. E., 1986. Growth and development: ecosystems phenomenology. Springer-Verlag, New York.

Underwood, A. J. 1994., On beyond BACI: sampling designs that might reliably detect environmental disturbances. Ecological Applications 4:3–15.

Warwick, R. M., and K. R. Clarke. 1993. Increased variability as a symptom of stress in marine communities. Journal of Experimental Marine Biology and Ecology 172:215–226.

Warwick, R. M., and K. R. Clarke. 1995. New "biodiversity" measures reveal a decrease in taxonomic distinctness with increasing stress. Journal of Experimental Marine Biology and Ecology 129:301–305.

Whitlach, R. B., A. M. Lohrer, S. F. Thrush, R. D. Pridmore, J. E. Hewitt, V. J. Cummings, and R. N. Zajac. 1998. Scale-dependent benthic recolonization dynamics: life stage-based dispersal and demographic consequence. Hydrobiologia 375/376:217–226.

Wittaker, R. H., S. A. Levin. 1977. The role of mosaic phenomena in natural communities. Theoretical Population Biology 12:117–139.

Zajac, R. N., and R. B. Whitlach. 1982a. Responses of estuarine infauna to disturbance. I. Spatial and temporal variations of initial recolonization. Marine Ecology Progress Series 10:1–14.

Zajac, R. N., and R. B. Whitlfach. 1982b. Responses of estuarine infauna to disturbance. II. Spatial and temporal variations of initial recolonization. Marine Ecology Progress Series 10:15–27.

Zajac, R. N., and R. B. Whitlach. 1991. Demographic aspects of marine, soft sediment patch dynamics. American Zoologist 31:808–820.

Zajac, R. N., R. B. Whitlach, S. F. Thrush. 1998. Recolonization and succession in soft-sediment infaunal communities: the spatial scale of controlling factors. Hydrobiologia 375/376:227–240.

Short-Term Effects of the Cessation of Shrimp Trawling on Texas Benthic Habitats

PETER SHERIDAN[1] AND JENNIFER DOERR

*National Oceanic and Atmospheric Administration (NOAA) Fisheries,
Southeast Fisheries Science Center, 4700 Avenue U, Galveston, Texas 77551, USA*

Abstract. We compared sediments and benthos of two adjacent zones of the middle Texas coast, one of which was closed to shrimp trawling for 7 months. We hypothesized that the no-trawling zone would experience accumulation of fine surficial sediments, leading to increased proportions of silt, clay, and organic matter and decreased proportions of rubble and sand. We also hypothesized that cessation of trawling would affect benthic community structure, directly or indirectly leading to altered types and densities of dominant taxa between zones. During June 2001, divers collected benthos and sediment cores from 32 sites in each zone, using random stratified sampling based on previously mapped sediments. Benthic organisms were identified to family or higher taxonomic levels, counted, and weighed. Our study indicated that the predicted accumulation of fine materials over the 7-month closure did not occur, as we found no sedimentary differences that could not be referenced to sampling design. This lack of change was likely due to the short closure period and to the shallow, sand-dominated nature of the study zones, wherein winter storms, summer tropical cyclones, and seasonally reversing coastal currents more likely influence long-term sediment structure than does presence or absence of shrimp trawling. Densities and biomasses of most abundant taxa and major taxonomic groups were similar between zones, although there were significant differences between zones for nemerteans, amphipods, mactrid clams, and spionid polychaetes. Our data indicate that ambient shrimp trawling effort during winter and spring off the middle Texas coast had little impact on small benthic organisms. A better way to determine whether community and ecosystem structure (and function) might be different from what we now see is to make comparisons between areas closed to all extractive uses for an extended period of time (years to decades) and areas open to all maritime users.

Introduction

Fishing can affect the structural components of habitat and the biotic communities dependent on those habitats. Auster and Langton (1999) summarized literature that indicated mobile fishing gear, such as shrimp trawls, can reduce habitat complexity by removing, damaging, or killing epifauna and infauna, by mixing or smoothing surface sediments, and by removing biogenic structures such as tubes, reefs, burrows, and pits. Postimpact recovery of habitat and community structure and function was difficult to measure and predict since timing, severity, and frequency of impacts all interact with habitat type and recovery processes.

There are very few areas of the northern Gulf of Mexico continental shelf that are closed to the penaeid shrimp fishery, which operates bottom trawls from shoreline to 90-m depths or more (Nance 1993). Opportunities to examine trawl impacts to benthic habitats and communities without fishing gear disturbance are, thus, limited in space or time. In 2000, the Texas Parks and Wildlife Department enacted a seasonal closure of nearshore waters to all shrimp trawling from Corpus Christi Fish Pass south to the U.S.–Mexican border. This closure provided a window of opportunity to study short-term effects of the cessation of shrimp trawling on sediments and benthos. The Southern Shrimp Zone (SSZ) encompasses waters from the shoreline out 9.3 km for the period 1 December to the opening of the season for brown shrimp *Farfantepenaeus aztecus* (formerly known as *Penaeus aztecus*), usually during 1–15 July, each year. A complementary Northern Shrimp Zone (NSZ) extends from Corpus Christi Fish Pass north to the Texas–Louisiana border and permits daytime shrimp trawling during that time period.

---

[1] Corresponding author: pete.sheridan@noaa.gov; present address: NOAA Fisheries, Southeast Fisheries Science Center, 3500 Delwood Beach Road, Panama City Florida 32408, USA.

Our project describes a comparison of benthic community and habitat characteristics between these two adjacent areas after cessation of shrimp trawling in the SSZ for 7 months.

Trawling results in resuspension of fine sediments and may result in current-driven, cross-shelf transport of fines to offshore locations (Churchill 1989). We hypothesized that the no-trawling SSZ would experience accumulation of fine surficial sediments, leading to increased proportions of silt, clay, and organic matter and decreased proportions of rubble and sand. Trawling can also impact benthic community composition, either directly, by disturbing, injuring, or killing organisms, or indirectly, by altering their habitat characteristics (Thrush et al. 1998; Auster and Langton 1999). We hypothesized that cessation of trawling would lead to altered types and densities of dominant benthic taxa between zones.

Methods

The dividing line between SSZ and NSZ runs perpendicular to the coast at Corpus Christi Fish Pass (27.675°N, 97.000°W). We designated two blocks on either side of the dividing line as our research areas: a block 18.5 km long and 9.3 km wide at the northern end of the SSZ, and an adjoining block of the same dimensions at the southern end of the NSZ (Figure 1). Because of their immediate proximity, most environmental or human-induced disturbances such as storms, currents, and fishing were believed to have similar impacts equally in each block. The NSZ block was immediately adjacent to a tidal inlet and may have experienced tide-induced differences in water quality (e.g., temperature, salinity, and turbidity). There are no indications that sediments in the area have higher than expected concentrations of trace elements or contaminants other than in the immediate vicinity of oil and gas production platforms (White et al. 1983; Montagna and Harper 1996).

The SSZ was closed to all shrimp trawling from 1 December 2000–1 July 2001. We collected sediment and benthos samples from each block during 18–22 June 2001. We used a geographic information system (GIS) database to develop a stratified sampling procedure that was based on a 1976 survey of Texas coastal sediments conducted at sites 1.6 km apart (White et al. 1983; Sheridan and Caldwell 2002). Five sediment texture classes (Folk 1980) were noted within two broad categories: sand, silty sand, or sand–silt–clay (hereafter referred to as sand-dominated), and sandy silt or clayey silt (hereafter referred to as fines-dominated). Sampling effort was allocated among the five textural classes in proportion to the area covered within each block, with at least two sites per class. Each block was divided into 1-km × 1-km grids, and grids were chosen randomly for each sediment class. The center of each grid became a sampling site, except for the clayey silt class, which only appeared in one grid per block and, thus, contained two sites separated by about 1 km in each (Figure 1). Latitude and longitude of each site were preselected using the GIS database. Some sites were moved approximately 500 m from their preselected locations due to the presence of oil or gas production platforms and their potential to cause benthic community changes within that range (Montagna and Harper 1996). We expected to sample 28 sand-dominated sites and 4 fines-dominated sites per block.

Sites were located via vessel-mounted global positioning systems. Divers collected sediment and benthos cores from 32 sites at 5–20-m depths in each block. Single sediment samples were collected at each site with a 5-cm diameter corer to a depth of 10 cm. The top and bottom of the sediment corer were capped prior to diver ascent. The top 5 cm of sediments were stored on ice or refrigerated for 5 d, then frozen until analysis. Sediment processing included rubble–sand–silt–clay ratios (Folk 1980) and organic content (Dean 1974). Benthos at each site was sampled with four 10-cm-diameter cores taken to depths of 10 cm, which were later pooled to form a single sample. To prevent animal loss during diver ascent, the top of each corer was covered with 0.3-mm mesh, and the bottom of each corer was capped on removal from the substrate. Samples were rinsed through a 0.5-mm-mesh sieve, stored on ice to relax animals, then preserved in 10% formalin containing rose bengal (Birkett and McIntyre 1971). Intact organisms or anterior ends were counted. Funding constraints limited identifications to family level for polychaetes and mollusks and to order level for other groups. We felt this would not compromise the value of our study, since marked community changes are still detectable with aggregated taxonomic data (Somerfield and Clarke 1995). The 10-cm-diameter corer is not effective in sampling large, deep-dwelling, structure-building, or other long-lived taxa; however, this gear is effective for quantifying sediment characteristics and densities of small, shallow-dwelling, short-lived organisms (Holme 1971; Blomqvist 1991). These organisms are also likely to be disturbed by trawl passage, and disruption of benthic communities might disrupt ecosystem energy flow (Jennings et al. 2002). Fishes, decapods, squillids, and chaetognaths were excluded entirely, and polychaete fragments not attributable to any family were excluded from biomass analyses. Animal groups were then blotted dry

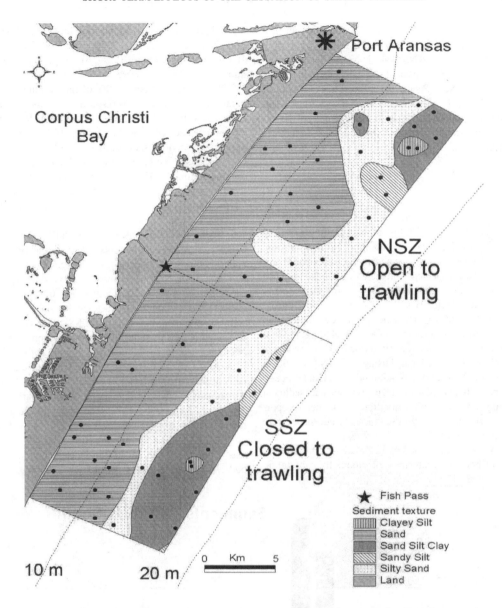

Figure 1. Benthic sampling sites relative to sediment characteristics and depth contours at the junction of the Northern Shrimp Zone (NSZ) and Southern Shrimp Zone (SSZ) off Texas.

(mollusks were not removed from their shells) and weighed to the nearest 0.1 mg.

Zone-related sediment and benthos differences were examined by one-way analysis of variance (ANOVA) with balanced cell sizes ($N = 32$ samples per block). We compared sediment components, densities of numerical and biomass dominants (those with ≥2% of either total), major taxonomic groups (e.g., polychaetes or mollusks), and number of taxa. All data were found to be heteroscedastic using Levene's test (see review by Day and Quinn 1989) and were transformed prior to ANOVA using arcsine for sediment proportions or $\log_{10}(x + 1)$ for benthic density and biomass. Tables and figures present nontransformed data for density or biomass dominants only (a complete taxonomic list is available from the senior author). All analyses were conducted using Statistica (StatSoft, Inc. 1997).

We examined shrimp trawling effort statistics collected by National Oceanic and Atmospheric Administration (NOAA) Fisheries for the area to determine the level of effort prior to the first closure and determine whether shrimp trawling was occurring in NSZ during SSZ closure. Effort is tabulated as 24-h d fished within statistical subareas and 9-m depth strata (Patella 1975). The dividing line between SSZ and NSZ falls in NOAA Fisheries Statistical Subarea 20, with about 40% of the shoreline falling into the open NSZ (Patella 1975; National Ocean Survey Chart 11300). The seaward extent of each zone (9.3 km) encompasses all of the 1–9-m stratum and most of the 10–18-m depth stratum used by NOAA Fisheries to tabulate trawling effort.

Results

Among sediment characteristics, mean percent sand was significantly ($P = 0.037$) higher in the Southern Shrimp Zone (Figure 2), in part because more Northern Shrimp Zone sites were actually located in fines-dominated habitats than were expected from the GIS sediment map. We were expecting to sample four fines-dominated sites per block, but we actually sampled seven fines-dominated sites in the NSZ. Discrepancies from the sediment map were likely due to a variety of factors including: (1) less-accurate position data from 1976 when original sediment sampling sites were located using LORAN C (White et al. 1983); (2) use of different sampling gear and processing methods; and (3) actual changes in surficial sediment composition. No significant differences in mean proportions of rubble, silt, clay, or sediment organic matter were detected between SSZ and NSZ (Figure 2).

We recorded a total of 12,572 individuals and 83 taxa of benthic organisms, coincidentally subtotaling 6,286 individuals and 73 taxa each in NSZ and SSZ. Total benthic biomass was 136.1 g, with NSZ yielding 104.9 g and SSZ yielding 31.2 g. Among major groups, numbers and biomass were dominated by polychaetes (58.0% of total number, 17.9% of total biomass) and bivalves (11.7% and 55.4%, respectively). Among polychaetes, families Spionidae, Magelonidae, and Paraonidae each composed more than 10% of the total number of individuals, but each was less than 1% of the total biomass. The bivalve family Mactridae was the dominant biomass (47.6% of the total) and ranked sixth in numerical abundance (5.1% of the total individuals). Clypeasteroida was a distant second in total biomass (7.8% of the total).

With few exceptions, densities and biomasses of major groups and dominant taxa were similar between zones (Tables 1, 2). Polychaetes were significantly more numerous in the SSZ, while bivalve biomass was significantly higher in the NSZ (ANOVA, $P < 0.05$). Among dominant taxa, spionid polychaetes and nemerteans were significantly more numerous in the SSZ, mactrid clams and amphipods were significantly more numerous in the NSZ, and mactrid clam biomass was significantly higher in the NSZ (ANOVA, $P < 0.05$). Spionid polychaetes and nemerteans were found at all 32 stations within each block. Amphipods and mactrid clams were patchily distributed. Twelve of 32 sites in the NSZ had no amphipods, and 2 of the remaining 20 sites yielded 196 of 303 individuals (64.7%). In contrast, all but one site in the SSZ had at least 1 amphipod and no site had more than 25 amphipods. As for mactrid clams, 53% of 646 individuals were

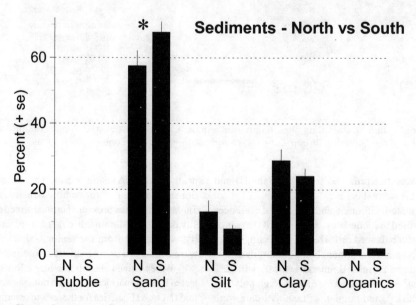

Figure 2. Mean sediment characteristics of Northen (N) and Southern (S) Shrimp Zones. Asterisk indicates significant difference (ANOVA, $P < 0.05$).

Table 1. Mean density (number/314.4 cm^2) and standard error (SE) of major taxonomic groups and dominant taxa in the Northern Shrimp Zone (NSZ) and Southern Shrimp Zone (SSZ) off Texas during June 2001. $N = 32$ sites per zone. Asterisk indicates ANOVA; $P < 0.05$.

Organisms	NSZ Density	SE	SSZ Density	SE
Major taxonomic groups				
Polychaeta (P)	98.7	9.7	129.2*	9.0
Crustacea (C)	16.9	4.9	15.4	2.0
Gastropoda (G)	7.1	1.3	5.6	1.2
Bivalvia (B)	30.5	11.4	15.5	1.8
Miscellaneous (M)	40.7	7.1	30.0	3.7
Total taxa	22.9	1.2	23.8	0.9
Dominant taxa				
Spionidae (P)	14.9	2.3	30.0*	3.5
Paraonidae (P)	16.8	3.1	24.0	3.2
Capitellidae (P)	20.7	2.7	15.2	1.5
Lumbrineridae (P)	7.7	1.1	10.5	1.3
Nephtyidae (P)	5.4	2.0	4.4	2.1
Cossuridae (P)	2.9	0.6	4.8	1.0
Mactridae (B)	20.2*	11.3	0.0	0.0
Tellinidae (B)	6.1	1.5	9.7	1.8
Nassariidae (G)	4.8	1.0	3.4	1.0
Amphipoda (C)	9.5*	4.3	8.3	1.2
Nemertinea (M)	6.4*	0.8	9.1	0.8
Phoronida (M)	4.8	1.2	3.6	0.9

Table 2. Mean biomass (blotted wet, mg/314.4 cm^2) and standard error (SE) of major taxonomic groups and dominant taxa in the Northern Shrimp Zone (NSZ) and Southern Shrimp Zone (SSZ) off Texas during June 2001. $N = 32$ sites per zone. Asterisk indicates ANOVA; $P < 0.05$.

Organisms	NSZ Density	SE	SSZ Density	SE
Major taxonomic groups				
Polychaeta (P)	105.4	11.2	124.4	26.8
Crustacea (C)	2.3	1.0	2.1	0.4
Gastropoda (G)	37.2	8.4	25.1	6.5
Bivalvia (B)	255.9*	70.1	29.5	13.4
Miscellaneous (M)	152.6	35.1	68.7	29.9
Dominant taxa				
Spionidae (P)	5.4	1.3	10.1	2.4
Terebellidae (P)	12.3	5.6	2.7	2.0
Glyceridae (P)	12.9	4.3	7.5	2.8
Lumbrineridae (P)	20.0	3.8	31.1	8.6
Onuphidae (P)	20.1	5.8	24.0	6.2
Paraonidae (P)	2.9	0.9	30.3	26.3
Mactridae (B)	205.6*	69.8	0.0	0.0
Tellinidae (B)	32.6	13.5	15.7	13.1
Lucinidae (B)	5.5	3.5	8.2	4.0
Veneridae (B)	24.2	20.4	4.3	2.1
Nassariidae (G)	26.0	5.7	18.6	6.3
Ophiurida (M)	54.0	23.5	6.8	3.3
Actiniaria (M)	15.3	11.2	28.9	28.8
Clypeasteroida (M)	61.1	29.9	10.4	10.4
Nemertinea (M)	11.3	2.5	14.8	3.9

found at a single NSZ station, 83% were found at three stations, and 98% were found at seven stations. Only one mactrid clam was collected in the SSZ. We do not know whether these patchy distributions were due to experimental design or biological interactions.

Shrimp trawling was recorded in the vicinity of our sampling sites. Trawling effort in Statistical Subarea 20 at 1–18-m depths during December 2000–June 2001 was 358 24-h d (NOAA Fisheries, unpublished data) and should have been restricted to the NSZ (there were no reported violations of the SSZ in Statistical Subarea 20; R. Riekers, Texas Parks and Wildlife Department, personal communication). By way of comparison, preclosure effort in Statistical Subarea 20 during the December–June periods of 1990–1999 and at 1–18-m depths averaged 1,224 24-h d for NSZ and SSZ combined, with a range of 261–5623 24-h d. Even though trawling was recorded within the NSZ, it is unknown whether trawl tows directly impacted any given benthic sampling site since NOAA Fisheries only records statistical subarea and depth strata fished.

Discussion

Bottom trawling resuspends sediments, particularly silts and clays, that may or may not be transported away from a trawling region, depending upon such factors as compaction, current speed and direction, or storm frequency (Churchill 1989; Lindeboom and de Groot 1998; Pilskaln et al. 1998). Thus, it is difficult to predict whether or not trawling will change surface sediment texture at any given locale. Previous research has reported both outcomes. Lindeboom and de Groot (1998) reported that sand-dominated substrates were less likely to be physically affected by passage of bottom-tending beam and otter trawls than were fines-dominated substrates, particularly if sand habitats were located in areas of current-induced or storm-induced turbulence. These authors did note, however, that fine surficial sediments were removed from hard-packed sands by trawling at times. In contrast, Van Dolah et al. (1991) found no consistent differences in sediment texture between sand-dominated substrates that were either open or closed to commercial shrimp trawling. Engel and Kvitek (1998) found no significant differences in sediment characteristics between a lightly trawled and a heavily trawled fishing ground. Schwinghamer et al. (1998) could not induce significant changes in sediment grain size by experimental trawling over sand bottoms. Our study indicates that the predicted accumulation of fine materials in the SSZ over the 7-month closure did not occur, as we found no sedimentary differences that could not be referenced to sampling design. This lack of change was likely due to the short closure period and the shallow, sandy nature of the study blocks, wherein winter storms, summer tropical cyclones, and seasonally reversing coastal currents more likely influence long-term sediment structure than does presence or absence of shrimp trawling.

Most research conducted elsewhere using cores or grabs in sand-dominated habitats indicates little or no impact to small benthic organisms. Van Dolah et al. (1991) reported no consistent differences in shallow, sand-bottom benthic communities between trawled and nontrawled areas before and after a shrimp trawling season. Engel and Kvitek (1998) recorded few significant differences in infaunal taxa in annual surveys of a lightly trawled area versus a heavily trawled area over a 3-year period. Brylinski et al. (1994) found some short-term effects of trawling on nematode densities but no other short-term or long-term benthic community impacts. Jennings et al. (2002) could not link changes in infaunal production or size structure to changes in beam trawl effort. Thrush et al. (1998) found that trawling pressure resulted in significant decreases in densities of certain infaunal indicators captured by coring (densities of echinoderms, diversity, number of species, and polychaete to mollusk ratio) but not in others (densities of deposit feeders, long-lived surface dwellers, scavengers, small opportunists, and total individuals). Even fewer significant changes were recorded using grab or suction dredge sampling (Thrush et al. 1998). Our data are in agreement with these studies and indicate that ambient shrimp trawling effort off the middle Texas coast has little impact on small benthic organisms. This is a shallow, sand-dominated ecosystem that is frequently disturbed by strong cold fronts in winter and by tropical cyclones in summer. Thus, benthic communities have likely adapted in response to physical disruption (Posey et al. 1996).

Shrimp trawling effort can be intense, especially in spring and summer when juvenile penaeid shrimps move from estuaries to the shallow continental shelf (Nance 1993). Shrimp trawling effort according to NOAA Fisheries is recorded as 24-h d fished in each of four Statistical Subareas (numbered 18–21) along the Texas coast. Statistical Subarea 20 encompasses our study zones and traditionally receives the highest trawling effort in depths less than 20 m (annual mean effort during 1990–1999 = 40 h/km^2; Sheridan and Caldwell 2002). The expected frequency of shrimp trawl passage over any particular area of bottom can be estimated as follows. Texas offshore vessels typically tow four 18.3-m headrope nets at an average speed of 4.6 km/h (NOAA Fisheries, unpublished data). Nets actually spread to about 70% of headrope length at this speed (Watson et al. 1984), so the area swept by a vessel approximates 0.25 km^2/h or 6 km^2/24-h d. A total of 358 24-h d were expended in NSZ during the closure (NOAA Fisheries, unpublished data); thus, a total of 2,148 km^2 were impacted by trawling if effort was expended uniformly throughout the NSZ. Our NSZ study block covered 172 km^2, encompassing about half of the available NSZ in Statistical Subarea 20, and thus, experienced half of the total area trawled (1,074 km^2). The expected frequency of trawl passage over any given sample site is, thus, the total area trawled (1,074 km^2 per 7 months) divided by the total area in the study block (172 km^2), which equals 6.2 per 7-month study period or approximately once per month given uniform effort. At this level of activity, ambient trawling effort during winter and spring months has little or no apparent effect on sediments or benthos in the shallow waters off central Texas.

It is possible that we might have had a better chance of detecting trawl-induced community change if we had employed finer-scale identification of benthic organisms. For example, Vanderklift et al. (1996) noted that identification to species level permitted greater ability to separate a stress-related "signal" from the background, habitat-related "noise." However, it is not always necessary to identify organisms to species in order to detect community change (Ferraro and Cole 1990, 1992; Somerfield and Clarke 1995), and it certainly is not always cost-effective if significant changes are evident at family or order levels. Although we found few significant zone-related differences in density or biomass, particularly among the most abundant polychaete and mollusk families, it remains possible that changes could have occurred at the species level.

Our sampling gear addressed the smaller, short-lived, shallow-dwelling benthos. Large, deep-dwelling, structure-building, or other long-lived taxa may have been otherwise affected by the 7-month closure. It has been noted by others, however, that shallow sand communities have evolved in naturally stressful environments (Posey et al. 1996), and it is communities in deeper waters and muddier substrates that are more likely to change when stress such as trawling is removed (Churchill 1989; Auster and Langton 1999). Sampling of the larger infauna and epifauna would best be conducted with other methods such as large-area grabs, video transects, or sonar, as employed elsewhere (e.g., Engel and Kvitek 1998; Thrush et al. 1998).

It is also possible that removing trawlers from the SSZ led to increased densities of fishes and decapods that affected benthic communities through increased bioturbation or predation. We know of no contemporary, directed assessments of the density of mobile macrofauna in the research blocks off Texas. However, we detected few differences between research blocks, so even if macrofaunal densities changed there was little basis for arguing such impacts. Engel and Kvitek (1998) found no detectable differences in catch per unit effort of fishes in a study of lightly trawled versus heavily trawled fishing grounds and, coincidentally, found few differences in benthic organisms.

We conclude that we were unable to detect significant sediment or benthic community changes resulting from the cessation of shrimp trawling for 7 months off the middle Texas coast. We concentrated our efforts on the smaller organisms whose disruption might have altered community energy flow to higher trophic levels, and it is possible that other (larger) groups of organisms may have been affected differently. The question of whether this temporal refuge has any lon-term impact on ecosystem structure (or function) is moot, however, since the closed area is reopened annually for the 5 months of heaviest shrimp trawling effort (Nance 1993). Any faunal "recovery" to the unstressed state is, thus, eliminated. A better way to determine whether community and ecosystem structure (and function) might be different from what we now see is to make comparisons between areas closed to all extractive uses for an extended period of time (years to decades) and areas open to all maritime users.

Acknowledgments

Funds for this project were provided by NOAA Fisheries, Southeast Fisheries Science Center, under the Essential Fish Habitat program. Enthusiastic assistance was provided by S. Nañez-James, S. Hillen, P. Caldwell, G. Gitschlag, B. Rhame, and G. Myers (all NOAA Fisheries, Galveston, Texas), G. Sims (National Aeronautics and Space Administration), and Captains J. Kenworthey (M/V Tropic Star) and D. Lehman (M/V Li'l Star). J. Carlson, T. Minello, and two anonymous reviewers provided constructive comments for this manuscript.

References

Auster, P. J., and R. L. Langton. 1999. The effects of fishing on fish habitat. Pages 150–187 in L. Benaka, editor. Fish habitat: essential fish habitat and rehabilitation. American Fisheries Society, Symposium 22, Bethesda, Maryland.

Birkett, L., and A. D. McIntyre. 1971. Treatment and sorting of samples. Pages 156–168 in N. A. Holme and A. D. McIntyre, editors. Methods for the study of marine benthos. Blackwell Scientific Publications, Oxford, UK.

Blomqvist, S. 1991. Quantitative sampling of soft-bottom sediments: problems and solutions. Marine Ecology Progress Series 72:295–304.

Brylinski, M., J. Gibson, and D. C. Gordon, Jr. 1994. Impacts of flounder trawls on the intertidal habitat and community of the Minas Basin, Bay of Fundy. Canadian Journal of Fisheries and Aquatic Sciences 51:650–661.

Churchill, J. H. 1989. The effect of commercial trawling on sediment resuspension and transport over the Middle Atlantic Bight continental shelf. Continental Shelf Research 9:841–864.

Day, R. W., and G. P. Quinn. 1989. Comparisons of treatments after an analysis of variance in ecology. Ecological Monographs 59:433–463.

Dean, W. E. 1974. Determination of carbonate and organic matter in calcareous sediments and sedimentary rocks by loss on ignition: comparison with other methods. Journal of Sedimentary Petrology 44:242–248.

Engel, J., and R. Kvitek. 1998. Effects of otter trawling on a benthic community in Monterey Bay National Marine Sanctuary. Conservation Biology 12:1204–1214.

Ferraro, S. P., and F. A. Cole. 1990. Taxonomic level and sample size sufficient for assessing pollution impacts on the Southern California Bight macrobenthos. Marine Ecology Progress Series 67:251–262.

Ferraro, S. P., and F. A. Cole. 1992. Taxonomic level sufficient for assessing a moderate impact on macrobenthic communities in Puget Sound, Washington, USA. Canadian Journal of Fisheries and Aquatic Sciences 49:1184–1188.

Folk, R. L. 1980. Petrology of sedimentary rocks. Hemphill Publishing Company, Austin, Texas.

Holme, N. A. 1971. Macrofauna sampling. Pages 80–130 in N. A. Holme and A. D. McIntyre, editors. Methods for the study of marine benthos. Blackwell Scientific Publications, Oxford, UK.

Jennings, S., M. D. Nicholson, T. A. Dinmore, and J. E. Lancaster. 2002. Effects of chronic trawling disturbance on the production of infaunal communities. Marine Ecology Progress Series 243:251–260.

Lindeboom, H. J., and S. J. de Groot. 1998. IMPACT-II. The effects of different types of fisheries on the North Sea and Irish Sea benthic ecosystems. Netherlands Institute for Sea Research, NIOZ Report 1998-1, and Netherlands Institute for Fisheries Research, RIVO-DLO Report C003/98, Texel and Ijmuiden, The Netherlands.

Montagna, P. A., and D. E. Harper, Jr. 1996. Benthic infaunal long-term response to offshore production platforms in the Gulf of Mexico. Canadian Journal of Fisheries and Aquatic Sciences 53:2567–2588.

Nance, J. M. 1993. Biological review of the 1992 Texas closure. NOAA Technical Memorandum NMFS-SEFSC-325.

Patella, F. J. 1975. Water surface area within statistical subareas used in reporting Gulf Coast shrimp data. Marine Fisheries Review 37(12):22–24.

Pilskaln, C. H., J. H. Churchill, and L. M. Mayer. 1998. Resuspension of sediment by bottom trawling in the Gulf of Maine and potential geochemical consequences. Conservation Biology 12:1223–1229.

Posey, M., W. Lindberg, T. Alphin, and F. Vose. 1996. Influence of storm disturbance on an offshore benthic community. Bulletin of Marine Science 59:523–529.

Schwinghamer, P., D. C. Gordon, Jr., T. W. Rowell, J. Prena, D. L. McKeown, G. Sonnichsen, and J. Y.

Guigne. 1998. Effects of experimental otter trawling on surficial sediment properties of a sandy-bottom ecosystem on the Grand Banks of Newfoundland. Conservation Biology 12:1215–1222.

Sheridan, P., and P. Caldwell. 2002. Compilation of data sets relevant to the identification of essential fish habitat on the Gulf of Mexico continental shelf and for the estimation of the effects of shrimp trawling gear on habitat. NOAA Technical Memorandum NMFS-SEFSC-483.

Somerfield, P. J., and K. R. Clarke. 1995. Taxonomic levels, in marine communities, revisited. Marine Ecology Progress Series 127:113–119.

StatSoft, Inc. 1997. Statistica, release 5.1. Tulsa, Oklahoma.

Thrush, S. F., J. E. Hewitt, V. J. Cummings, P. K. Dayton, M. Cryer, S. J. Turner, G. A. Funnell, R. G. Budd, C. J. Milburn, and M. R. Wilkinson. 1998. Disturbance of the marine benthic habitat by commercial fishing: impacts at the scale of the fishery. Ecological Applications 8:866–879.

Vanderklift, M. A., T. J. Ward, and C. A. Jacoby. 1996. Effect of reducing taxonomic resolution on ordinations to detect pollution-induced gradients in macrobenthic infaunal assemblages. Marine Ecology Progress Series 136:137–145.

Van Dolah, R. F., P. H. Wendt, and M. V. Levisen. 1991. A study of the effects of shrimp trawling on benthic communities in two South Carolina sounds. Fisheries Research 12:139–156.

Watson, J. W., I. K. Workman, C. W. Taylor, and A. F. Serra. 1984. Configurations and relative efficiencies of shrimp trawls employed in southeastern United States waters. NOAA Technical Report NMFS 3.

White, W. A., T. R. Calnan, R. A. Morton, R. S. Kimble, T. G. Littleton, J. H. McGowen, H. S. Nance, and K. E. Schmedes. 1983. Submerged lands of Texas, Corpus Christi area: sediments, geochemistry, benthic macroinvertebrates, and associated wetlands. University of Texas, Bureau of Economic Geology, Austin.

Effect of Caribbean Spiny Lobster Traps on Seagrass Beds of the Florida Keys National Marine Sanctuary: Damage Assessment and Evaluation of Recovery

AMY V. UHRIN[1] AND MARK S. FONSECA[2]

*National Oceanic and Atmospheric Administration, National Ocean Service,
Center for Coastal Fisheries and Habitat Research, 101 Pivers Island Road,
Beaufort, North Carolina 28516, USA*

GREGORY P. DIDOMENICO[3]

Monroe County Commercial Fishermen, Inc., Post Office Box 501404, Marathon, Florida 33050, USA

Abstract. In the Florida Keys, traps for spiny lobsters (also known as Caribbean spiny lobster) *Panulirus argus* are often deployed in seagrass beds. Given that several hundred thousand traps may be deployed in one fishing season, the possibility exists for significant impacts to seagrass resources. The question was whether standard fishing practices observed in the fishery actually resulted in injuries to seagrass. This study was designed to measure the degree of injury to seagrass as a function of trap deployment duration (soak time) and habitat type (seagrass species) and the recovery of seagrass following trap removal. Aspects of the deployment and retrieval process were not examined. Sampling grids composed of 30 3-m × 3-m squares were arbitrarily established within each of three monospecific seagrass beds (two of *Thalassia testudinum* and one of *Syringodium filiforme*) near Marathon, Florida. Five squares within each grid remained trap-free (controls) while the remaining squares each received a single trap. Five traps from each grid were randomly removed at each of five soak times (ranging from 1 to 24 weeks). Immediately before deployment and following trap removal, seagrass short shoot densities were recorded and compared among controls and treatments. Both seagrass species exhibited significantly decreased shoot densities after 6-week and 24-week soak times. *Thalassia testudinum* densities within the 6-week and 24-week treatments had returned to control densities 4 months after trap removal, while densities of *S. filiforme* remained significantly decreased at the end of 24 weeks. We conclude that traps must be recovered within a 6-week period, beyond which injury to seagrass beds is predicted, with long lasting effects to beds of *S. filiforme*. Within the limits of these testing parameters, it appears that standard fishing practices (typically < 5-week soak time) should not result in a significant injury to seagrass beds in the Florida Keys.

Introduction

The 1996 reauthorization of the Magnuson–Stevens Fishery Conservation and Management Act (U.S. Code, section 16, line 1801 et seq.) called for revisions to regional fishery management plans (FMPs) that included amendments for the recognition and preservation of essential fish habitat (EFH). Revised FMPs were to include sections on minimizing the impact of both fishing and nonfishing human activities on EFH. If fisheries are to survive, it is critical that EFH is safeguarded against activities that have the potential to physically damage or disrupt their normal ecological function.

In 1998, an estimated 150,000 commercial fishing vessels and permits were active within the U.S. Exclusive Economic Zone (NOAA 1998). Mobile fishing gears have the capacity to cover large expanses of the seafloor with a concomitant potential for widespread habitat injury. As such, the effect of these gear types on the benthos has been the focus of historical observation and recent studies concerned with commercial fishing gear impacts (de Groot 1984; Fonseca et al. 1984; Peterson et al. 1987; van Dolah et al. 1987; Guillén et al. 1994; Watling and Norse 1998; Meyer et al. 1999; Kaiser et al. 2000; Thrush

---

[1] Corresponding author: Amy.Uhrin@noaa.gov
[2] E-mail: Mark.Fonseca@noaa.gov
[3] E-mail: gregdi@voicenet.com; present address: Garden State Seafood Association, 212 West State Street, Trenton, New Jersey 08608, USA.

et al. 2001). Until quite recently, the effect of passive gears, such as traps, has received scant attention, as these gears are often assumed to have little impact on the benthos (Eno et al. 2001; Matthews 2003; A. Quandt, University of the Virgin Islands, unpublished report; R. S. Appeldoorn, M. Nemeth, J. Vasslides, and M. Shärer, unpublished report to the Caribbean Fishery Management Council, Hato Rey, Puerto Rico). However, passive gears may pose a threat of cumulative habitat injury when they become lost, such as through gear malfunction, carelessness, or storm events. Annual estimates of commercial trap losses range from 2,500–630,000 within various fisheries (king crab *Paralithodes camtschatica*, Smolowitz 1978a; Dungeness crab *Cancer magister*, Smolowitz 1978a; American lobster *Homarus americanus*, Smolowitz 1978a, 1978b; O'Hara 1992; Chopin et al. 1996; Carr and Harris 1997; Caribbean spiny lobster, Florida Fish and Wildlife Conservation Commission, Florida Wildlife Research Institute, unpublished data). Even so, quantification of the type and extent of damage inflicted by this lost gear has rarely been quantitatively addressed.

In this study, we focus on gear impacts associated with the fishery for spiny lobsters *Panulirus argus* (also known as Caribbean spiny lobsters) in the Florida Keys. Approximately 90% of Florida's commercial lobster harvest comes from Monroe County in the Florida Keys (Hunt 1994; Milon et al. 1998), making this area a potential target for habitat injuries resulting from trap use. For the 2002 season, certificates for 414,856 spiny lobster traps were issued in Monroe County. According to estimates calculated from surveys of commercial lobster fishermen during the 1999 season, monthly trap loss ranged from 2.6% to 5.6% during nonstorm months (Florida Fish and Wildlife Conservation Commission, Florida Wildlife Research Institute, unpublished data). Normal trap loss increases when major storms occur, as with Hurricane Irene in October 1999 (25.2% trap loss for October; Florida Fish and Wildlife Conservation Commission, Florida Wildlife Research Institute, unpublished data). Given that lobster traps have a benthic footprint of approximately 0.6 m^2, normal trap losses (3.8% monthly average loss, 1999 nonstorm months) over an 8-month lobster fishing season create the potential for upwards of 62,000 m^2 of injured habitat, excluding secondary injuries such as those that may result from scouring by trot lines (formed by linking traps together with a common line) or recovery of the traps (which sometimes results in traps being dragged across the bottom). When storm events are factored in, this potential increases.

A large portion of spiny lobster traps are fished in seagrass beds (Ault et al. 1997; Matthews 2003). These beds function as habitat, nurseries, feeding grounds, settlement sites, and refuge areas for a large number of other ecologically and commercially important marine organisms (Zieman 1982; Zieman and Zieman 1989). Trap deployment periods (soak times), on average, increase from 7 to 25 d as the season progresses (Matthews 2001). The possibility exists for set (i.e., stationary) traps to have a negative impact on seagrass, exclusive of secondary impacts, if left in place long enough, but that time frame is unknown. To our knowledge, there are no studies quantifying the time frame for trap-induced injury to seagrasses. If there is some critical period of time beyond which a set trap negatively impacts the seagrass, knowledge of this time frame could be used by commercial fishermen to minimize habitat damage in the course of fishing and by managers to guide the imposition of fishing regulations concerning seagrass habitat damage. Moreover, quantification of the time frame for trap-induced injury to seagrasses would help guide emergency response actions to limit significant habitat injury in the event of a storm.

This study establishes the threshold soak time beyond which seagrasses exhibit significant levels of sustained injury and performs preliminary calculations of the recovery trajectory for said injuries. To establish these values, the following null hypotheses were examined: (1) there are no differences in seagrass short shoot density among controls (no trap) and treatments of a given soak time; (2) there are no differences in the response of seagrass species to various soak times; and (3) provided that injury was sustained, there are no differences in the recovery trajectories exhibited as a function of seagrass species.

Methods

Study Area and Sampling Design

This study was conducted between the months of August 2001 and June 2002 in the inshore waters of the Gulf of Mexico and Atlantic Ocean near Marathon, Monroe County, Florida (Figure 1). The inner insular shelf area consists of a number of inshore and offshore coral reefs, scattered mangrove islands, and seagrass beds. The dominant seagrass species is *Thalassia testudinum*, but beds may be interspersed with *Halodule wrightii* and *Syringodium filiforme*.

Three seagrass beds were chosen based upon species composition, bed size, and water depth. Because we wished to test for differences in response to trap soak time among seagrass species, monospecific beds were chosen. Beds also had to be large enough to fully encompass the sampling grid without intrusion of other seagrass species or other habitat types. Lastly, we required beds of water depth 3 m or deeper in order to

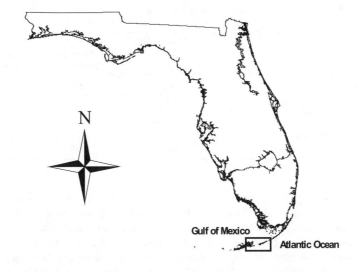

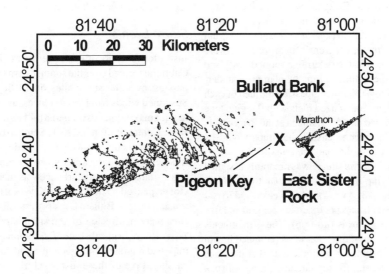

Figure 1. Map of the waters surrounding Marathon, Florida Keys. The study sites are indicated by an X.

mimic actively fished depths. Two sites consisted of monospecific *T. testudinum* (East Sister Rock and Pigeon Key) and the third consisted of *S. filiforme* (Bullard Bank). Although we desired to examine effects to a third seagrass species, *H. wrightii*, we were unable to locate monospecific beds of adequate size and depth. The two sites of *T. testudinum* were further characterized as having either firm (East Sister Rock) or soft (Pigeon Key) substrate based upon sediment compressive strength (penetrometer test; Blum 1997) and shear strength (vane shear test; Blum 1997) measurements taken during the site selection process (Table 1). We were unable to find adequate habitat of *S. filiforme* with firm substrate; therefore, only one site of *S. filiforme* was established, in the soft (Table 1), silt substrate typically associated with this species in this region. At each site, a 15-m × 18-m grid, subdivided into 30 3-m × 3-m squares, was arbitrarily established and its location recorded using a differential global positioning system (Trimble GPS Pathfinder Pro XR, ESRI, Redlands, California). Grid corners were designated by 0.6-m rebar stakes hammered into the substrate. The perimeter of the grid was marked by stretching a polypropylene rope between the rebar stakes with square subdivisions designated with lead core lines clipped to the perimeter lines. Prior to trap deployment, the seagrass within each 3-m × 3-m square was thoroughly surveyed. A 0.61-m × 0.91-

Table 1. Mean (SE) sediment compressive strength and sediment shear strength for the three study sites prior to trap deployment. $N = 30$

Site (species)	Depth (m)	Compressive strength (kg/cm^2)	Shear strength (kg/cm^2)	Sediment type
East Sister Rock (*Thalassia*)	4.3	0.19 (0.04)	0.03 (0.004)	Firm
Pigeon Key (*Thalassia*)	3.4	0.02 (0.01)	0.006 (0.004)	Soft
Bullard Bank (*Syringodium*)	3.1	0.0002 (0.001)	0.0024 (0.0026)	Soft

m PVC quadrat (representing a trap footprint) was centered within each square of the grid and examined for cover using the Braun–Blanquet technique (Fourqurean et al. 2001). The quadrat was divided into 54 10-cm × 10-cm squares, and presence–absence of seagrass shoots was recorded from these 54 squares to yield a more precise spatial description of seagrass within the footprint of the trap. Three random, replicate short shoot counts, sediment compressive strength (kg/cm^2) measurements, and sediment shear strength (kg/cm^2) measurements were made within the quadrat to determine shoot density and the firmness of the sediment. Finally, plan view, underwater digital photographs encompassing the area of each quadrat were taken. Galvanized nails with flagging tape attached were pushed into the sediment at the four corners of the quadrat in order to ensure that traps would be placed in the exact area that had been surveyed.

Each 3-m × 3-m grid square was randomly assigned one of the following five treatments (soak times): (1) 1 week, (2) 2 weeks, (3) 4 weeks, (4) 6 weeks, and (5) 24 weeks, with five replicates per treatment per grid and five squares acting as controls (no trap). These treatments represent the amount of time that traps remained in situ before removal, beginning from the date that the traps were deployed. We specifically targeted a range of soak times that would encompass those exhibited by the industry (7–25 d; Matthews 2001). With assistance from members of Monroe County Commercial Fishermen, Inc. (MCCF), 25 spiny lobster traps were deployed at each site. The tops of the traps were removed prior to deployment to prevent the traps from being actively fished. Traps were centered within grid squares by divers using lift bags. Traps were placed in the exact location where pre-deployment quadrat survey data had been collected. At each specified soak time, the outside corners of the traps were marked with rebar stakes by divers. The traps were then removed from their respective squares and placed outside the grid using lift bags. Care was taken that traps were lifted without disturbing areas beyond the footprint. The area falling within the footprint of the trap was evaluated using the previously described techniques. Rebar stakes were established in a similar fashion for the control squares to mimic trap footprints. These were also evaluated at each specified soak time. Upon completion of the surveys, traps were retrieved from the site by members of MCCF. In trap footprints where a significant decline in seagrass shoot density was observed (via preliminary Student's t-tests), treatment and control footprints were monitored for recovery at 4 and 8 months following trap removal.

Storm Events

Just after the start of our experiment, Tropical Storm Gabrielle formed over the southeastern Gulf of Mexico, passing over the study sites on 14 September 2001. Sustained winds from the southwest averaged 58.3 km/h, with maximum gusts up to 81.4 km/h (National Data Buoy Center Station SMKF1, Somberro Key, Florida). The storm occurred during the scheduled 2-week soak time retrieval. Underwater video documentation of the storm-induced trap disturbance was made 5 d after the storm passed. The Pigeon Key (soft substrate, *T. testudinum*) and Bullard Bank (soft substrate, *S. filiforme*) sites were unaffected by the storm's passing. At East Sister Rock (firm substrate, *T. testudinum*), a number of traps had migrated throughout the grid due to the storm's passing. Five of these displaced traps were randomly chosen, and the associated footprint was surveyed, as were the original control footprints, using the previously described techniques. Squares that escaped damage were resurveyed, traps were redistributed throughout these squares (new time zero) by divers, and sampling events were rescheduled based upon the remaining soak times.

A second storm, Hurricane Michelle, impacted the three sites on 4 November 2001. Sustained winds from the northeast averaged 63.4 km/h, with maximum gusts up to 91.1 km/h (National Data Buoy Center Station SMKF1). Grid lines at East Sister Rock were again ripped up, and all of the traps that had been previously redistributed throughout the grid were lost, rendering the site unusable. Further sampling at East Sister Rock was discontinued. Pigeon Key and Bullard Bank were again unaffected (no trap loss or movement).

Data Analysis

Student's *t*-tests were used to contrast control and treatment short shoot densities at each soak time and recovery period. Normality was tested using the Shapiro-Wilk test (SAS Institute, Inc. 1999). The folded form of the F statistic was used to test for homogeneity of variance (SAS Institute, Inc. 1999). Where variables did not meet the above assumptions (i.e., *T. testudinum*, 4 weeks; *S. filiforme*, 24-week treatment after 4 months recovery), nonparametric tests were employed (PROC NPAR1WAY; SAS Institute, Inc. 1999). All statistical analyses were performed using SAS version 8.0 (SAS Institute, Inc. 1999). Recovery trajectories were plotted by species.

Results

Sediment compressive strength and sediment shear strength were significantly lower at the Pigeon Key site when compared to East Sister Rock (Table 1, $P < 0.0001$ for both variables), confirming our initial qualitative judgment that these sites differed in terms of sediment strength. Unfortunately, due to Hurricane Michelle, the East Sister Rock site was so badly damaged that it was not viable for the experiment. Thus, sediment strength was eliminated as a potential factor affecting level of injury and recovery for *T. testudinum*. Sediment measurements at Bullard Bank indicated that sediment at this site was indeed soft bottom and that compressive strength and shear strength at this site were significantly lower than at Pigeon Key (Table 1, $P < 0.0001$ and $P = 0.0005$, respectively).

After 1 week of trap deployment, *S. filiforme* exhibited a significant, yet unsustained decline in short shoot densities versus controls ($P = 0.004$; Figure 2). For both species, the first sign of sustained seagrass injury occurred at 4 weeks, when an approximately 20% decline in seagrass short shoot densities was observed. By 6 weeks, shoot densities of both species exhibited significant declines compared to control values (*T. testudinum*, $P = 0.005$; *S. filiforme*, $P = 0.0003$; Figure 2). Significantly decreased densities were observed at 24 weeks as well (*T. testudinum*, $P = 0.028$; *S. filiforme*, $P = 0.0002$; Figure 2).

Visual observations of *T. testudinum* in the footprints after traps were removed at 6 and 24 weeks revealed a number of necrotic or missing blades. However, the underlying root-rhizome system appeared to sustain

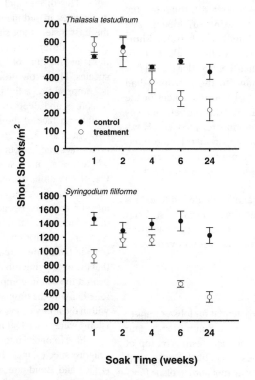

Figure 2. Seagrass short shoots/m² (means; error bars = SEs; $N = 5$) as a function of lobster trap soak time.

no physical damage; a casual observation of the subsurface rhizomes indicated that they remained intact. On the other hand, some treatment plots of *S. filiforme* were observed to have severed rhizomes with individual pieces and sections of rhizome exposed and resting on the surface of the sediment.

Four months after traps had been removed, *T. testudinum* densities in the 6-week and 24-week treatment footprints had recovered to control values ($P = 0.193$ and 0.176 respectively; Figures 3, 4). However, 8 months after trap removal, densities of *S. filiforme* within the 6-week treatments had not recovered to control values ($P = 0.001$; Figure 3). Similarly, the 24-week *S. filiforme* treatments continued to exhibit significantly decreased shoot densities 4 months after traps had been removed ($P = 0.009$; Figure 4). Shoot presence–absence surveys revealed that for injured footprints of *T. testudinum*, regrowth appeared to occur from injury centers and edges, while for *S. filiforme*, regrowth predominantly commenced from the edges surrounding the trap injury footprint.

Storm Events

Damage to the East Sister Rock site by Tropical Storm Gabrielle was moderate; grid lines were ripped up, and a number of traps had migrated throughout the grid, uprooting seagrass, shearing off blades, and leaving behind trails of denuded substrate. Short shoot densities in the footprints of the displaced traps were significantly lower than control densities ($P < 0.001$). No apparent damage to the rhizomes had occurred. No quantification of trap movement or the level of damage incurred via trap movement was made. We were unable to monitor for recovery. However, because there was no apparent rhizome damage, we can speculate that recovery rates under these circumstances may be similar to those observed for our (nonstorm) treatments.

The passing of Hurricane Michelle led to extensive movements of the remaining traps at East Sister Rock such that the traps were completely removed from the vicinity of the study site and never recovered.

Discussion

This study demonstrates the potential for lobster traps to negatively impact seagrass beds in South Florida. Our observations show that lobster traps resting on top of seagrass for extended periods cause blades to become broken or abraded, which may disrupt normal blade function. Blades are also crushed into the underlying anoxic sediments, likely suffocating the plants and leading to eventual senescence of above-ground biomass. In addition, shading by traps may reduce light availability, thereby inhibiting photosynthetic function. Seasonal changes in biomass for *S. filiforme* (Short et al. 1993) have been linked to availability of incident radiation, and light reduction has been shown to cause decreases in leaf biomass in *T. testudinum* (Lee and Dunton 1997). Lee and Dunton (1997) observed that above-ground biomass for *T. testudinum* declined more rapidly than did below-ground biomass as a result of shading. However, once leaf material was depleted, root biomass declined rapidly, as transport of oxygen from blades to roots became compromised.

Although we observed a significant decline in short shoot densities of *S. filiforme* after 1 week, this injury was not sustained, and these treatments were able to recover quickly from this disturbance. After a 4-week soak time, there was a substantial decline (~20%) in short shoot densities for both seagrass species, although not statistically significant. Both species sustained significantly decreased shoot densities when lobster traps were left in situ for 6 weeks or longer.

Trap-induced injuries sustained by seagrasses varied between the two species, as evidenced by the difference in degree of short shoot loss and recovery rates. This differential response may be due in part to differences in sediment strength, but with the loss of the East Sister Rock site, we could not test this. However, the differences may also result from differences in the architecture of the root-rhizome systems of each species. The below-ground component of *T. testudinum* is comprised of a tightly interwoven matrix of robust rhizomes and deep roots that can extend meters below the surface of the sediment. In contrast, roots and rhizomes of *S. filiforme* are quite shallow and oftentimes can be found resting on top of the sediment where they are more susceptible to physical disturbance. Our observations of damaged and broken rhizomes of *S. filiforme* confirm this and suggest injury to the rhizome apical meristems. Because seagrass growth is meristem dependent, when rhizome apical meristems are damaged, recovery of injuries will require vegetative growth from outside the injured area, as we observed. Observations that the underlying rhizome system for *T. testudinum* appeared to sustain comparatively little physical damage together with the observed regrowth of this species from within the injury center would indicate that rhizome meristems were spared injury.

Physiological integration is known to occur in some seagrass species, including *T. testudinum* and *S. filiforme* (Libes and Boudouresque 1987; Tomasko and Dawes 1989; van Tussenbroek 1996; Terrados et al. 1997; A. Schwarzschild, Department of Environmental Sciences, University of Virginia, unpublished data). Thus, the quick

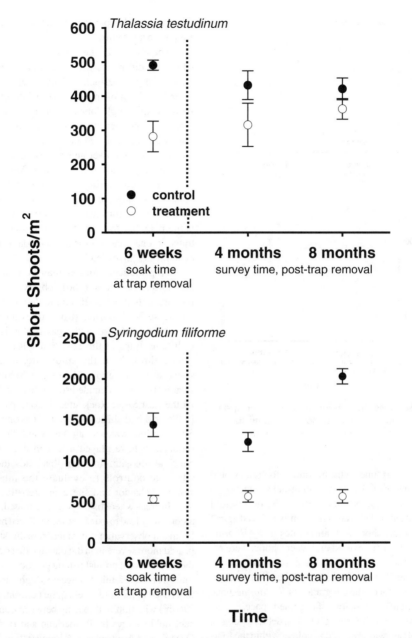

Figure 3. Seagrass short shoots/m² (means; error bars = SEs; N = 5) for the 6-week soak time and its subsequent recovery at 4 and 8 months post trap removal.

recovery observed in our site of *T. testudinum* may also have been assisted by the translocation of nutrients from nearby healthy shoots to damaged ones via the intact root-rhizome system. In shading experiments, Tomasko and Dawes (1989) found that blade production rates and organic constituent levels of shaded shoots were equal to unshaded controls located along the same rhizome. However, when the shaded shoots were physically separated (rhizome severed) from unshaded neighbors, blade production and organic levels were significantly decreased (Tomasko and Dawes 1989). In the current study, it appears that clonal integration was maintained for *T. testudinum*. In contrast, the observed rhizome damage in treatments of *S. filiforme* and the regrowth commencing

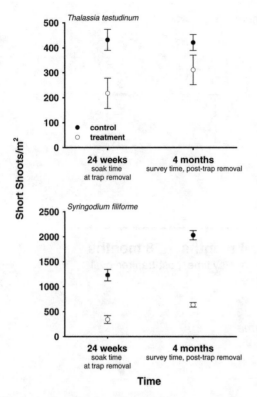

Figure 4. Seagrass short shoots/m² (means; error bars = SEs; N = 5) for the 24-week soak time and its subsequent recovery at 4 months post trap removal.

from the edges surrounding the trap injury footprint would suggest disruption of clonal integration and a reduction in the capacity for these plants to recover. Schwarzschild (unpublished data) observed significantly reduced apical growth and short shoot formation rates in *S. filiforme* when rhizomes were severed and when shoots were removed from the rhizome; in addition, fragments of rhizomes exhibited low survival. Both of these latter observations suggest that clonal integration of *S. filiforme*, once damaged, would result in loss of occupied space.

At the site for *S. filiforme*, biological debris (sponges, drift macroalgae) was observed to collect within the footprint of the trap after removal. Shading effects from this debris may also have contributed to the slowed recovery of *S. filiforme* observed in the present study. J. Holmquist (University of California, Santa Barbara, unpublished data) found that experimental gaps created in canopies of *T. testudinum* acted as sinks for algae and that algal accumulation in these gaps resulted in slowed seagrass recovery. Although high abundances of drift macroalgae were observed on numerous sampling occasions, fouling the buoy lines and rebar stakes, algae did not accumulate in the trap footprints at our site for *T. testudinum*, possibly because

the canopy at that particular site was not high enough to elicit such a response.

There may also be cumulative effects of traps for which we did not test. For example, fishing practices vary among fishermen and may depend to a certain extent on the movement of local lobster populations. Areas with a high lobster density may be fished by multiple fishermen, thereby increasing the number of traps that are deployed in a small area. Also, traps in one area are often set and retrieved several times before moving on to other locations. Moreover, damage that may occur during the setting and retrieval process, something for which we did not test, may increase the risk for localized injuries. As a result, there is the potential for trap-induced injuries to be concentrated in one area and for repeated damage, issues that bear further investigation.

Trap-induced injuries may be more severe during storms. Eno et al. (2001) observed crustacean and mollusk pots in Great Britain to "bounce up and down on the seabed" during strong tides and large swell. We observed that after the passing of Tropical Storm Gabrielle, traps at the East Sister Rock site had migrated throughout the study site, often becoming stacked end-to-end and leaving behind trap-wide swaths (0.6 m) of denuded substrate up to 2 m in length. At the East Sister Rock site, all 20 traps that remained in the grid had shifted due to the passing of the storm. Site location, water depth, and wind direction and duration may have played a role in the level of disturbance at this site; however, this study did not contain adequate controls to evaluate the interplay of the storm's behavior and site characteristics.

In areas where fishing is concentrated, the passing of a storm may lead to extensive trap-induced habitat damage. From our observations, damage from this storm event appeared more severe than did damage that occurred during the simulation of normal fishing practices (4–6-week soak times). However, this damage was highly localized, as the traps used in the study were quite concentrated (25 traps/270 m²) with minimal spacing between traps (2.5 m). Recent field surveys by P. Sheridan and R. Hill (National Oceanic and Atmospheric Administration, National Marine Fisheries Service, unpublished data) found, on average, 25 traps per 100,000 m² of seagrass, with a mean distance of 84 m between traps. Quantitative assessments of the degree of movement of traps during storm events and the potential damage to seagrass from storm-induced movement of traps warrants further study. However, from the data we have presented here, we posit that *S. filiforme* may suffer more from these types of shoot-shearing effects than would *T. testudinum*.

We conclude that traps must be recovered within a 6-week period, beyond which significant injuries to beds of

T. testudinum and *S. filiforme* within trap footprints are predicted. However, removal of traps within 4 weeks is preferred, as a substantial (~ 20%) decline in shoot densities had already begun in that amount of time. Therefore, within the limits of our testing parameters, it appears that standard fishing practices with this passive gear (typically < 5-week soak time; Matthews 2001) will reduce shoot densities of the aforementioned species but should not result in statistically significant or long-term injuries to these seagrass beds in the Florida Keys. However, in the event of an approaching storm, it appears that removal of traps, to the extent practicable, would reduce the amount of injury to seagrass. If pre-storm removal is not possible, search and recovery efforts for lost traps should be completed within 6 weeks or significant seagrass habitat injuries can occur.

Acknowledgments

This study was funded by the National Oceanic and Atmospheric Administration (NOAA), National Marine Fisheries Service. Essential Fish Habitat support to M. Fonseca from the Southeast Fisheries Science Center and Regional Office as well as additional direct support from the National Center for Coastal Ocean Service. We appreciate the field assistance of C. Addison, C. Bonn, K. Hammerstrom, B. Irwin, A. Iarocci, K. Kirsch, S. Meehan, and A. Poray. We benefited from discussions with K. Hammerstrom, K. Kirsch, T. Matthews, S. Meehan, A. Schwarzschild, and S. Slade. The manuscript was improved by comments from K. Hammerstrom, P. Marraro, G. Thayer, and two anonymous reviewers. The views contained in this manuscript do not necessarily reflect the view of NOAA.

References

Ault, J., J. Serafy, D. DiResta, and J. Dandelski. 1997. Impacts of commercial fishing on key habitats within Biscayne National Park. Rosenstiel School of Marine and Atmospheric Science, Annual Report for Cooperative Agreement Number CA-5250–6-9018. University of Miami, Miami.

Blum, P. 1997. Physical properties handbook: a guide to the shipboard measurement of physical properties of deep-sea cores. Ocean Drilling Program, Technical Note 26, College Station, Texas. Available: *www-odp.tamu.edu/publications/tnotes/tn26* (April 2004).

Carr, H. A., and J. Harris. 1997. Ghost fishing gear: have fishing practices during the past few years reduced the impact? Pages 141–151 *in* J. M. Coe and D. B. Rogers, editors. Marine debris: sources, impacts, and solutions. Springer-Verlag, New York.

Chopin, F., Y. Inoue, Y. Matsushita, and T. Arimoto. 1996. Sources of accounted and unaccounted fishing mortality. Pages 41–47 *in* Solving bycatch: considerations for today and tomorrow: proceedings of the solving bycatch workshop. Alaska Sea Grant College Program, Report 96-03, Fairbanks.

de Groot, S. J. 1984. The impact of bottom trawling on benthic fauna of the North Sea. Ocean Management 9:177–190.

Eno, N. C., D. S. MacDonald, J. A. M. Kinnear, S. C. Amos, C. J. Chapman, R. A. Clark, F. St. P. D. Bunker, and C. Munro. 2001. Effects of crustacean traps on benthic fauna. ICES Journal of Marine Science 58:11–20.

Fonseca, M. S., G. W. Thayer, and A. J. Chester. 1984. Impact of scallop harvesting on eelgrass (*Zostera marina*) meadows: implications for management. North American Journal of Fisheries Management 4:286–293.

Fourqurean, J. W., A. Willsie, C. D. Rose, and L. M. Rutten. 2001. Spatial and temporal pattern in seagrass community composition and productivity in south Florida. Marine Biology 138:341–354.

Guillén, J. E., A. A. Ramos, L. Martínez, and J. L. Sánchez Lizaso. 1994. Antitrawling reefs and the protection of *Posidonia oceanica* (L.) Delile meadows in the western Mediterranean Sea: demand and aims. Bulletin of Marine Science 55:645–650.

Hunt, J. H. 1994. Status of the fishery for *Panulirus argus* in Florida. Pages 158–168 *in* B. Phillips, J. Cobb, and J. Kittaka, editors. Spiny lobster management. Fishing News Books, London.

Kaiser, M. J., K. Ramsay, C. A. Richardson, F. E. Spence, and A. R. Brand. 2000. Chronic fishing disturbance has changed shelf sea benthic community structure. Journal of Animal Ecology 69:494–503.

Lee, K. S., and K. H. Dunton. 1997. Effects of in situ light reduction on the maintenance, growth and partitioning of carbon resources in *Thalassia testudinum* Banks ex König. Journal of Experimental Marine Biology and Ecology 210:53–73.

Libes, M., and C. F. Boudouresque. 1987. Uptake and long-distance transport of carbon in the marine phanerogam *Posidonia oceanica*. Marine Ecology Progress Series 38:177–186.

Matthews, T. 2001. Trap-induced mortality of the spiny lobster, *Panulirus argus*, in Florida, USA. Marine and Freshwater Research 52:1509–1516.

Matthews, T. 2003. Distribution of trap fishing and effects on habitats in coral reef ecosystems. Final Report for NOAA/NMFS contract number NFFN7400–2-00021. Florida Fish and Wildlife Conservation Commission, Florida Marine Research Institute, Marathon.

Meyer, D. L., M. S. Fonseca, P. L. Murphey, R. H. McMichael, Jr., M. M. Byerly, M. W. LaCroix, P. E. Whitfield, and G. W. Thayer. 1999. Effects of live-bait shrimp trawling on seagrass beds and fish bycatch in Tampa Bay, Florida. U.S. National Marine Fisheries Service Fishery Bulletin 97:193–199.

Milon, J. W., S. L. Larkin, D. J. Lee, K. J. Quigley, and C. M. Adams. 1998. The performance of Florida's spiny lobster trap certificate program. Florida Sea Grant College Program, Florida Sea Grant Report 116, University of Florida, Gainesville.

NOAA (National Oceanic and Atmospheric Administration). 1998. Ecological effects of fishing by S. K. Brown, P. J. Auster, L. Lauck, and M. Coyne. NOAA, NOAA's State of the Coast Report, Silver Spring, Maryland. Available: *http://state-of-coast.noaa.gov/bulletins/html/ief_03/ief.html* (April 2004).

O'Hara, K. J. 1992. Marine debris: taking out the trash. Pages 81–90 *in* R. H. Stroud, editor. Stemming the tide of coastal fish habitat loss: proceedings of a symposium on conservation of coastal fish habitat, Baltimore, Maryland, March 7–9, 1991. National Coalition for Marine Conservation, Inc., Savannah, Georgia.

Peterson, C. H., H. C. Summerson, and S. R. Fegley. 1987. Ecological consequences of mechanical harvesting of clams. U.S. National Marine Fisheries Service Fishery Bulletin 85:281–298.

SAS Institute, Inc. 1999. SAS, version 8.0. SAS Institute, Inc., Cary, North Carolina.

Short, F. T., J. Montgomery, C. F. Zimmermann, and C. A. Short. 1993. Production and nutrient dynamics of a *Syringodium filiforme* Kütz seagrass bed in Indian River Lagoon, Florida. Estuaries 16:323–334.

Smolowitz, R. J. 1978a. Trap design and ghost fishing: an overview. Marine Fisheries Review 40(5–6):2–8.

Smolowitz, R. J. 1978b. Trap design and ghost fishing: discussion. Marine Fisheries Review 40(5–6):59–67.

Terrados, J., C. M. Duarte, and W. J. Kenworthy. 1997. Is the apical growth of *Cymodocea nodosa* dependent on clonal integration? Marine Ecology Progress Series 158:103–110.

Thrush, S. F., J. E. Hewitt, G. A. Funnell, V. J. Cummings, J. Ellis, D. Schultz, D. Talley, and A. Norkko. 2001. Fishing disturbance and marine biodiversity: role of habitat structure in simple soft-sediment systems. Marine Ecology Progress Series 221:255–264.

Tomasko, D. A., and C. J. Dawes. 1989. Evidence for physiological integration between shaded and unshaded short shoots of *Thalassia testudinum*. Marine Ecology Progress Series 54:299–305.

van Dolah, R. F., P. H. Wendt, and N. Nicholson. 1987. Effects of a research trawl on a hard- bottom assemblage of sponges and corals. Fisheries Research 5:39–54.

van Tussenbroek, B. I. 1996. Integrated patterns of turtle grass, *Thalassia testudinum* Banks ex König. Aquatic Botany 55:139–144.

Watling, L., and E. A. Norse. 1998. Disturbance of the seabed by mobile fishing gear: a comparison to forest clearcutting. Conservation Biology 12:1180–1197.

Zieman, J. C. 1982. The ecology of seagrasses of south Florida: a community profile. U.S. Fish and Wildlife Service, USFWS/OBS-82/25, Washington, D.C.

Zieman, J. C., and R. T. Zieman. 1989. The ecology of the seagrass meadows of the west coast of Florida: a community profile. U.S. Fish and Wildlife Service Biological Report 85(7.25).

Symposium Abstract

Symposium abstracts have been reprinted from the Abstract Volume prepared for the Symposium on Effects of Fishing Activities on Benthic Habitats: Linking Geology, Biology, Socioeconomics, and Management without any further review or editing.

The Effectiveness of Marine Protected Areas on Fish and Benthic Fauna: The Georges Bank Closed Area II Example

F. ALMEIDA[1]

National Marine Fisheries Service, Northeast Fisheries Science Center, Woods Hole, Massachusetts

P. VALENTINE

U.S. Geological Survey, Woods Hole Field Center, Woods Hole, Massachusetts

R. REID AND L. ARLEN

National Marine Fisheries Service, Northeast Fisheries Science Center, Highlands, New Jersey

P. AUSTER

NOAA, National Underwater Research Program, University of Connecticut, Groton, Connecticut

J. CROSS AND V. GUIDA

National Marine Fisheries Service, Northeast Fisheries Science Center, Highlands, New Jersey

J. LINDHOLM

NOAA, National Underwater Research Program, University of Connecticut, Groton, Connecticut

J. LINK

National Marine Fisheries Service, Northeast Fisheries Science Center, Woods Hole, Massachusetts

D. PACKER, J. VITALIANO, AND A. PAULSON

National Marine Fisheries Service, Northeast Fisheries Science Center, Highlands, New Jersey

In late 1994, a substantial portion of eastern Georges Bank was closed to commercial fishing (Closed Area II) to assist with stock rebuilding. After about five years of closure, the southern portion of CAII (south of 41°30'), exhibited a substantial increase in biomass and density of sea scallops, Placopecten magellanicus, and was reopened to the scallop fishery. Before the industry was allowed entry into this area, we conducted a survey to monitor the recovery of benthic habitat and fauna inside CAII. Sampling sites were selected in a paired station design for an inside/outside comparison; a grid design was used to monitor the remainder of the inside area. At each station, we conducted video transects, collected still photos, CTD casts, and sediment samples for physical and chemical analysis. A Smith-McIntyre bottom sampler was then used to sample the benthic community, followed by an otter trawl. Trawl catches were sorted to species and all fish and invertebrates were weighed, enumerated, and measured. Stomach contents, maturity observations, and age structures were collected for selected species at each station. Our results suggest limited differences between the inside/outside paired stations for species composition, community diversity, species richness, and trophic ecology. Fish abundance and biomass was also similar inside and outside the area; however, most individuals of a species were larger inside than outside. The lack of other major differences is likely a result of the fact that the seabed in the southern portion of CAII is a relatively high-energy sand habitat of low to moderate complexity and has a relatively low vulnerability to trawling and dredging. Other parts of closed areas on the northeast shelf may exhibit stronger gradients for the same metrics due to the presence of higher complexity gravel habitats and increased vulnerability to bottom tending fishing gear. The subtle differences in the size structure of fish species we observed in CAII may have significant implications for the population dynamics of commercially valuable species.

---

[1] E-mail: frank.almeida@noaa.gov

The Impact of Scallop Dredging on the American Lobster *Homarus americanus* in the Baie des Chaleurs, Canada

P. ARCHAMBAULT[1] AND L. GENDRON

Department of Fisheries and Oceans Canada, Institut Maurice-Lamontagne, Mont-Joli, Québec, Canada

Lobster fishers in eastern Canada often complain that scallop dredging is responsible for local declines in lobster landings of the American lobster through destruction of lobster habitat. In Baie des Chaleurs, although scallop dredging is restricted to depths over 18 m to 27 m, depending on the season, it nevertheless occurs in areas where lobster is known to be present at certain times of the year. The aim of this project was to determine to what extent scallop fishery spatially overlaps lobster grounds and to examine the impact of scallop dredging on lobster habitat, more specifically in terms of loss of bottom complexity. Such a loss could affect the survival of lobster at different stages of its life, especially cryptic juvenile stages. The study was concentrated in two localities, for which fisheries managers have received site-specific request to assess the link between these two fisheries. Seasonal adult lobster distribution was examined from lobster fishing activity and from off-season experimental fishing. Location of dredging activities was obtained from scallop fishers logbooks. Habitat, in overlapping areas, was characterized using an acoustic device. Furthermore, abundance of juvenile and adult lobsters was evaluated along transects running across the overlapping area. Additionally, experimental dredging was performed at one depth and a 'Before-After-Control-Impact design' was used to identify the immediate impact of the scallop dredge on habitat complexity and benthic community. Results will be discussed in relation to the possible mechanisms explaining how lobster landings could be affected by scallop dredging activity.

---

[1] E-mail: archambaultp@dfo-mpo.gc.ca

Effect of Shrimp Trawling on Snow Crab Resource in the Northwest Atlantic

G. BROTHERS[1]

Fisheries and Oceans Canada, Fisheries Management Branch, Northwest Atlantic Fisheries Centre, St. John's, Newfoundland, Canada

J. J. FOSTER

Aquaprojects Inc., St. John's, Newfoundland, Canada

The decline of the Northern Atlantic Cod stock and favorable environmental factors have led to an increase in the Northern Pink Shrimp (Pandalus borealis) Total Allowable Catch from 37,000 MT in 1996 to 112,000 in 2002. As well, an additional 365 new, <20 meter vessels have been added to the existing fleet of 13, >50 meter vessels involved in harvesting the resource. Shrimp and Snow Crab (Chionoecetes opeilio) are known to cohabit the same area, and as such, many crab fishers have expressed concern that shrimp trawling may be having a negative impact on the crab resource. In 2001, a two-phase study was begun to determine the interaction between shrimp trawling and the crab resource. Phase one of the study was conducted in a small area (0.5 x 4 miles) cohabited by crab and shrimp. The experimental design called for three fishing trips to be undertaken, the first directing for snow crab, the second directing for shrimp, and the third directing for crab. Crabs sampled were examined to determine 'new' and 'old' leg losses and then released 10 miles from the study area. Phase two of the study which was undertaken in 2002, consisted of three, five-day shrimp trawling trips carried out in an area 5 x 10 miles where shrimp and crab cohabit. The shrimp trawl had three retainer bags attached underneath the trawl and behind the footrope to capture the crab that passed over and under the trawl footrope. 12,000 crab captured in the retainer bags were examined for 'new' and 'old' leg losses and then released 10-miles from the study area. Analysis of 'old' and 'new' leg losses were compared before and after trawling (phase I) and after trawling and at various times of the year (phase II), and phase one and two data were also compared. Results presented (with confidence limits) that cover both phases indicate a low percentage of recent leg loss, suggesting that shrimp trawling did not adversely impact crab encountered during the two-phase study.

---

[1] E-mail: brothersg@dfo-mpo.gc.ca

The Impact of Oyster Dredging on Blue Cod in New Zealand

G. D. CARBINES[1]

National Institute of Water and Atmospheric Research, Dunedin, New Zealand

Little is known about the potential impact of dredging on the growth and abundance of demersal fishes. Observations of blue cod (Parapercis colias) and oyster *Ostrea chilensis* fishing patterns indicate that dredging by the oyster fishery reduced localized catches and changed fishing patterns of blue cod fishers in Foveaux Strait, southern New Zealand. Towed underwater videos were then used to confirm the impact of dredging on habitat complexity and numbers of blue cod. An analysis of the diet and growth of blue cod from undisturbed biogenic reefs and reefs modified by oyster dredging further showed that diet complexity and growth of juvenile blue cod are reduced by dredging for oysters. However, stabilizing dredged habitat with fresh processed oyster shells shows promising signs of regeneration of blue cod populations in only a few years.

---

[1] E-mail: Carbines@storm.cri.nz

Impacts to Coral Reef Benthos from Lobster Trap Gear in the Florida Keys National Marine Sanctuary

M. CHIAPPONE[1]

Center for Marine Science and NOAA's National Undersea Research Center, University of North Carolina at Wilmington, Key Largo, Florida

D. W. SWANSON

Division of Marine Biology and Fisheries, Rosenstiel School of Marine and Atmospheric Science, University of Miami, Miami, Florida

S. L. MILLER

Center for Marine Science and NOAA's National Undersea Research Center, University of North Carolina at Wilmington, Key Largo, Florida

Growth in the Florida Keys fisheries for spiny lobster *Panulirus argus* and stone crab (Menippe mercenaria) has resulted in increased numbers of traps and environmental impacts. During 1998 alone, the stone crab and spiny lobster fisheries were estimated to utilize a total of 750,000 traps and 540,000, respectively. Impacts from gear are exacerbated when traps are lost due to severe storms. This study evaluated the distribution, density, and impacts to coral reef sessile invertebrates from lobster trap gear at 117 sites in the Florida Keys National Marine Sanctuary during 2000 and 2001. Sites were stratified according to benthic habitat type and fishing protection and encompassed 13 of the Sanctuary's 23 no-fishing zones. Diver surveys using transects were performed to document the type, length, and number of biota impacted by lost gear. Surveys yielded 86 incidences of gear totaling nearly 380 m, consisting mostly of buoy lines and wood slats. Densities of gear among the three habitat types ranged from 0.11 to 0.86 incidences/100 m^2, with four to eight times greater gear density in patch reefs compared to other habitats. The distribution of lobster trap gear did not differ significantly between protected and fished sites. Lobster trap gear, especially buoy lines, caused partial mortality or complete mortality to 152 sessile invertebrates. Relative to hook-and-line gear effects, lobster trap gear impacted sessile invertebrates varied less among the organisms considered. Gorgonians (39%) and scleractinian corals (24%) were the most commonly affected, followed by sponges (17%), colonial zoanthids (13%), and milleporid hydrocorals (7%).

---

[1] E-mail: chiapponem@uncwil.edu

Spatial Distribution and Benthic Impacts from Hook-and-Line Fishing Gear in the Florida Keys National Marine Sanctuary

M. CHIAPPONE[1]

Center for Marine Science and NOAA's National Undersea Research Center, University of North Carolina at Wilmington, Key Largo, Florida

D. W. SWANSON

Division of Marine Biology and Fisheries, Rosenstiel School of Marine and Atmospheric Science, University of Miami, Miami, Florida

S. L. MILLER

Center for Marine Science and NOAA's National Undersea Research Center, University of North Carolina at Wilmington, Key Largo, Florida

The spatial distribution and impacts to coral reef benthos from hook-and-line fishing gear were assessed at 117 sites spanning 2000 km in the Florida Keys National Marine Sanctuary during the summers of 2000 and 2001. Sites were stratified random with respect to habitat type and fishing protection. Surveys encompassed patch reef, spur and groove, and hard-bottom habitat types from 3 m to 12 m depth within and adjacent to 13 of the Sanctuary's 23 no-fishing zones. Diver surveys using transects were performed to document the type, length, and number of biota impacted by hook-and-line gear. From surveys of 34,000 m^2 of benthic habitat, 361 incidences of gear totaling nearly 465 m were documented, yielding a domain-wide density of 1.06 incidences/100 m^2. Gear densities ranged from 0.82 to 1.35 incidences/100 m^2 among the habitat types. In patch reef and spur and groove habitats, no significant differences were detected in the distribution of gear between protected and fished sites, while protected areas in the hard-bottom habitat yielded more gear than expected. Hook-and-line gear caused partial mortality or complete mortality to 434 sessile invertebrates. Organisms with upright morphologies such as gorgonians (47%), sponges (18%), and milleporid hydrocorals (18%) were the most frequently affected. Organism density, gear density, and gear length are some of the factors influencing gear impacts. For the habitats surveyed, hook-and-line gear is spatially pervasive in the Florida Keys, indicates a pattern of non-compliance with no-fishing regulations, and represents a low-level stressor to sessile reef invertebrates.

[1] E-mail: chiapponem@uncwil.edu

Effects of Fishing on the Benthic Habitat and Fauna of Seamounts on the Chatham Rise, New Zealand

M. R. CLARK,[1] A. A ROWDEN, AND S. O'SHEA

National Institute of Water & Atmospheric Research, Wellington, New Zealand

Major deepwater trawl fisheries occur for orange roughy on seamounts in New Zealand waters. These seamounts are often small, and trawling can be concentrated in a very localised area. Seamount habitat is thought to be productive, but also fragile, and there is growing concern from fisheries managers, environmental groups, and the fishing industry about effects of fishing on biodiversity and ecosystem productivity. This has prompted research to examine the nature and extent of deepwater trawling impact on seamount habitat in New Zealand. Results are presented from a recent survey where video and still imagery were applied to classify benthic habitat, and a new robust epibenthic sled used to sample the deepwater fauna. The study took place on the Chatham Rise where a group of 8 seamounts in close proximity allowed for a spatially unconfounded comparison of replicated fished and unfished seamounts. Commercial fisheries data were analyzed to determine the amount of trawling on each. Similarities within, and differences between, fished and unfished seamounts were identified for distribution of trawl gear modification of habitat; extent of live coral; macroinvertebrate assemblage composition, taxonomic distinctness and size spectra . This study provided information to help plan management strategies and develop effective management practices to allow both conservation and exploitation of seamounts, although more research is required. In May 2001, 19 seamounts throughout the New Zealand region, including several features on the Chatham Rise, were closed to bottom trawling as a precautionary measure.

[1] E-mail: m.clark@niwa.cri.nz

Effects of Smooth Bottom Trawl Gear on Soft Bottom Habitat

C. L. COGSWELL
CR Environmental, Inc., Falmouth, Massachusetts

B. HECKER
Hecker Environmental, Woods Hole, Massachusetts

A. MICHAEL
Allan D. Michael and Associates, Magnolia, Massachusetts

F. MIRARCHI
Boat Kathleen A. Mirarchi, Inc., Scituate, Massachusetts

J. RYTHER, JR.
CR Environmental, Inc., Falmouth, Massachusetts

D. STEVENSON
National Marine Fisheries Service, Gloucester, Massachusetts

R. VALENTE
SAIC, Newport, Rhode Island

C. WRIGHT
CR Environmental, Inc., Falmouth, Massachusetts

In January 2001, Boat Kathleen A. Mirarchi, Inc. and CR Environmental, Inc. were awarded a NOAA Cooperative Research Project to study smooth bottom fishing gear-induced habitat impacts to soft bottom habitats in Massachusetts Bay off Scituate, MA. To date, most of the research on effects of otter trawling on the seafloor has focused on long-term cumulative changes to gravel bottom or rocky substrate and few studies have looked at trawl effects on soft bottom habitat types. Using local fishermen's knowledge, two 'soft' bottom fishing areas, a lightly trawled area (Little Tow) and a more heavily trawled area (Mud Hole) were selected for the study. Replicate reference and experimental trawl corridors were established in 120 to 140 ft of water. For the July 2001 'immediate impact study' the bottom and water column were characterized before and after repetitive passes with a smooth bottom otter trawl. Areas were surveyed for benthic organisms, sediment surface characteristics, water column parameters, epifauna and infauna, and fish and fish prey. An expansion of the project was funded by NOAA for 2002 to investigate 'chronic trawling impacts' by continuing to trawl the experimental trawl corridors twice a week from early August into November. Surveys along reference and experimental trawl lanes measured parameters similar to those investigated in 2001 and were conducted in July 2002 (pre-chronic trawling), in September 2002, and again in November 2002 to study the cumulative effects of chronic trawling. Cooperative Research funded by NOAA/NMFS.

Effects of 135 Years of Oyster (*Ostrea Chilensis*) Fishing on the Benthic Habitat, Associated Macrofaunal Assemblages, and Sediments of Foveaux Strait, Southern New Zealand

H. J. Cranfield[1] and K. P. Michael

National Institute of Water and Atmospheric Research Ltd, Kilbirnie, Wellington, New Zealand

G. Carbines

National Institute of Water and Atmospheric Research Ltd, Dunedin, New Zealand

D. P. Gordon, B. Manighetti, A. Dunn, and A. A. Rowden

National Institute of Water and Atmospheric Research Ltd, Kilbirnie, Wellington, New Zealand

Management of the oyster fishery, and understanding of the impact of this longstanding fishery on the benthic environment, has been facilitated through periodic surveys. Fishers' and institutional fishing records and the results of biological, acoustic, and sediment surveys have been analysed to show how historical changes to benthic habitat relate to fishing. The seafloor once consisted of bioherms, hundreds of metres wide and many kilometres long, aligned with the tide, separated by similarly wide swaths of relict pebble-gravel sediment. The macrofauna of bioherms was dominated by bryozoa, (over 200 species), and bivalve molluscs, (over 60 species). Oysters were localised on this habitat alone which was also important for blue cod, *Parapercis colias*. Much biohermal epifauna was removed as bycatch of the oyster fishery and oysters were subsequently depleted locally more rapidly. Bioherm habitat was important in the formation of biogenic sediments and the recruitment, growth, and survival of both oysters and blue cod. The expansion of relict pebble gravel seafloor with the erosion of biohermal sediments relates directly to areal expansion of fishing as oyster beds were serially depleted. Mytilid bivalves and styelid tunicates are identified as early colonisers of regenerating bioherms, and helical circulation patterns in the tidal flow are implicated in the formation of these linear structures within which fine sediments again begin to accumulate. Regeneration of habitat and rebuilding of oyster and blue cod populations in the absence of oyster dredging suggest that MPAs and rotational fishing could be effective in conserving both habitat and fisheries.

---

[1] E-mail: j.cranfield@niwa.cri.nz

The Theoretical and Methodological Basis of Estimations of the Human-Made Influences (Fishing and Constructing) on the Benthic Habitats

V. A. Emelyanov[1]

First Deputy Chief of the Natural Academy of Sciences of Ukraine, and Institute of Geologycal Sciences of the National Academy of Sciences of Ukraine, Kyiv, Ukraine

In the last few decades, human-made influences have considerably increased on the upper part of lithosphere within the World Ocean's bounds as the benthic habitat. Some scientific directions have attempted to solve numerous benthic habitat problems related to increasing fishing and construction activities. But it is impossible to do this effectively within the limits of traditional scientific directions, resulting in the need to incorporate more ecological sciences with these traditional scientific approaches to investigations and estimations of the growing problems and the search for solutions. In particular, more and more explorations have concentrated their attention on the many-sided investigations of the compound characteristics of benthic habitats as a marine geo-ecological system. But many terminological, conceptual, methodological and other general questions have arisen from these studies. Without answers to these questions, it is difficult to solve many specific problems related to human-made influences on benthic habitats, as well as the creation and steady development of marine and coastal areas. In this paper, some new terms are presented, with their treatments, basic conceptions and approaches, that are more applied in the modern benthic habitat's investigations and become an important component of the theory and methodology of a new scientific directionÑmarine geo-ecology, of studying and solving many benthic habitat problems related to fishing and construction activities.

[1] E-mail: evasea2002@yahoo.com

Impacts of Mobile Fishing Gear on Sponges and Gorgonian Corals in the Gulf of Alaska

J. L. FREESE[1]

NOAA National Marine Fisheries Service, Auke Bay Laboratory, Juneau, Alaska

Research carried out in deep water on the continental shelf in the eastern Gulf of Alaska (GOA) has shown that gorgonian corals and erect sponges provide significant components of the complex habitat in that area. These organisms are susceptible to impacts by mobile fishing gear, and are slow to recover from damage once disturbed. This poster presents an overview of these studies, and also presents results of surveys from a submersible vehicle aimed at identifying and characterizing sites that may be deemed Habitat Areas of Particular Concern.

[1] E-mail: linc.freese@noaa.gov

Effects of Fishing on Organic Carbon Content of Sand Habitats on Georges Bank

V. G. GUIDA[1] AND A. PAULSON

National Marine Fisheries Service, Northeast Fisheries Science Center, Highlands, New Jersey

P. C. VALENTINE

U.S. Geological Survey, Woods Hole Field Center, Woods Hole, Massachusetts

L. ARLEN

National Marine Fisheries Service, Northeast Fisheries Science Center, Highlands, New Jersey

A 4.5 year closure to fishing of an area on Georges Bank provided an opportunity to compare physical and chemical characteristics of sand habitats from areas that had not been subjected to fishing with adjacent areas that had been fished. Sediment cores (6-15 cm deep) taken by Van Veen grab sampler in June 1999 were sectioned into 1 cm depth segments and analyzed for Total Organic Carbon (TOC, particulate plus interstitial), and for grain size. Grain size was the most important factor influencing TOC, which correlated positively with mud content. Where similar grain size distributions occurred at nearby stations inside the area closed to fishing and outside, TOC values were significantly higher in the upper sediment layers of inside (unfished) stations. Comparing TOC between inside-outside station pairs with similar grain sizes revealed two distinct patterns, suggesting two distinct mechanisms for TOC depletion. In the first, TOC of the upper 2 cm of the fished station was depleted compared to the unfished station. This probably reflects advection of depositional organic matter upon resuspension by fishing. In the second pattern, the sediment column from the fished station was depleted in TOC relative to unfished sediments to a depth of 5 cm or more. This pattern may reflect an overall increase in remineralization resulting from vertical redistribution of labile organic substrates and oxidants from the surface by fishing turbation. Which mechanism predominates may depend upon bottom hydrology, the rate and composition of organic matter deposition, and the texture and dynamics of the sediments.

[1] E-mail: vincent.guida@noaa.gov

A Before-After-Control-Impact Study of the Sea Scallop Fishing Grounds of Georges Bank

K. D. E. STOKESBURY[1] AND B. HARRIS

School of Marine Science and Technology, University of Massachusetts Dartmouth, New Bedford, Massachusetts

A Before-After-Control-Impact (BACI) study is the optimal environmental impact experimental design. The null hypothesis is "an impact resulted in no biological damage". Fisheries management often relies on time series of data but unless there is a control all before-after comparisons must assume homogeneity over time, an assumption that has been found invalid time and again. We surveyed the historic scallop fishing grounds of Georges Bank that have been closed to mobile gear since 1994. We employed a BACI design with a 1-year set of baseline observations, two experimental areas that were exposed to intense fishing pressure, two control areas with no fishing, and one control with constant fishing. Within each experimental area we conducted a high-resolution video survey using a multistage design with stations separated by 0.85 nautical miles. The video survey was based on sea scallop densities to obtain a 5% to 15% level of precision for the normal and negative binomial distributions, respectively. Mounted on the pyramid were two video cameras and several lights. Four quadrat images (2.8 m^2) of the sea floor including counts and sizes of scallops, other macroinvertebrates and benthic fishes and sediment types, were relayed in real time to the surface. These images were video taped and the exact position (latitude and longitude from differential GPS) depth, and time. During all surveys the same stations were sampled. Changes in species composition, density and distribution macroinvertebrates and groundfish, and in sediment structure will be compared.

[1] E-mail: kstokesbury@UMassD.Edu

Impacts of Scallop Dredging on Marine Bottom Complexity and Juvenile Fish Habitat

F. Hartog[1] and P. Archambault

Institut Maurice Lamontagne, Ministère des Pêches et des Océans, Mont-Joli, Québec, Canada

L. Fortier

Département de biologie, Université Laval, Ste-Foy, Québec, Canada

Dredging for scallops is known to reduce habitat complexity by homogenizing the sediments structure and by the removal of epibenthic organisms. Large bivalves such as scallops and their shells provide secondary substrate and physical structure adding to the complexity of the bottom. A complex habitat may enhance survival and growth of juvenile fishes by providing refuges from predation, abundance of prey and shelters from water flow. The Magdalen Islands shelf, in the Gulf of Saint-Lawrence, supports a fishery for Giant scallops (*Placopecten magelanicus*) and is believed to be a nursery area for juvenile Atlantic cod (*Gadus morhua*). Four scallop beds are still fished while three have been closed to fishing for 4, 10 and 12 years. During the summer 2002, three locations closed to dredging will be compared to three dredged locations in order to detect fishing impacts on epifauna and fish habitat. Bottom complexity and epifauna diversity and species abundances will be assessed from photographic sampling. Demersal fishes associated with the bottom will be sampled with fine mesh experimental nets. A complexity index will integrate sediment features, biogenic structures and patchiness values. Hypotheses are that unfished locations will be more complex and that juvenile fish and emergent benthic species will be more diverse and abundant at these locations. Epifauna diversity, abundances and assemblages will also be compared from fish and unfished sites. Differences in epifauna and fish assemblages will be examined.

---

[1] E-mail: hartogf@dfo-mpo.gc.ca

Community and Life History Divergence of Colonial Hydroids (Cnidaria, Hydrozoa) from Heavily Trawled Scallop Grounds in the Bay of Fundy, Eastern Canada

L. M. Henry[1]

Department of Biology, Dalhousie University, Halifax, Nova Scotia, Canada

Adverse effects of mobile bottom-fishing gear on communities of colonial invertebrates (sponges, anthozoans, hydrozoans and bryozoans) are rarely examined, and no studies have determined if sub-lethal damage caused by this gear impairs the life histories (e.g., sexual reproduction, growth competitive ability) of these animals. Colonial hydroids were used as a model group to investigate these issues at the community, population and colony-levels from heavily fished scallop grounds in the Bay of Fundy, eastern Canada. An MDS ordination of hydroid communities collected from the shells of 109 live scallops (*Placopecten magellanicus*) and 136 small boulders revealed a moderately strong divergence between these two assemblages: epilithic communities were comprised of more runner and vine-shaped annual species, while epizoic taxa typically had more arborescent morphologies with perennial lifespans. RAPD-PCR genetic techniques of 414 colonies revealed that epilithic populations of the upright macrobenthic hydroid *Sertularia cupressina* were dominated by fewer genotypes than those on live scallops. Epilithic colonies were damaged more often, less abundant, less often fertile and comprised of fewer, smaller and less fecund modular units than those on epizoic substrates. Field and lab experiments are planned to test the hypothesis that higher incidences of damage to colonies on small boulders versus live scallops explain community and life history divergence between epilithic and epizoic assemblages. The implications of divergent communities and impaired sexual reproduction will be discussed to emphasize the importance of considering 'less obvious' effects of bottom-fishing on marine benthos.

---

[1] E-mail: lhenry@is2.dal.ca

Analyzing the Effects of Trap Fishing in Coral Reef Habitats: Methods and Preliminary Results

R. L. HILL[1] AND P. F. SHERIDAN
NOAA National Marine Fisheries Service, Southeast Fisheries Science Center, Galveston, Texas

R. S. APPELDOORN
Department of Marine Sciences, University of Puerto Rico-Mayagüez, Lajas, Puerto Rico

T. R. MATTHEWS
FWC Florida Marine Research Institute, Marathon, Florida

K. R. UWATE
Division of Fish and Wildlife, U.S. Virgin Islands Department of Planning and Natural Resources, St. Thomas, Virgin Islands

Trap fishing is common near coral reefs in Florida and the U.S. Caribbean but little is known about the effects of these stationary gears on targeted habitats. This cooperative study between NOAA Fisheries, local resource agencies, academic researchers, and the fishing industry is investigating the effects of traps on coral reef and reef-associated habitats in the Florida Keys (lobster and stone crab traps) and in Puerto Rico and the U.S. Virgin Islands (fish and lobster traps). The initial stages of the project are underway; they include: 1) mapping the distribution of traps, 2) quantifying trap densities by habitat, and 3) quantifying damage to corals and other structural organisms. Preliminary findings from the Caribbean suggest that a relatively small percentage (<20%) of the traps set in shallow water (< 30 m) actually contact hard corals, gorgonians, or sponges. In these limited findings, damage occurred mainly to hard corals and was patchy, at a scale less than the total trap foot print. Continued research will assess whether these preliminary findings are representative of coast-wide trap fisheries and will provide more precise data on trap fishing intensity by habitat type, seasonal movement of traps among habitats, and potential for gear impacts to associated habitat components such as seagrasses, macroalgae, and sponges. A better understanding of how trap fishing affects essential fish habitats like coral reefs is integral to the development of sustainable fisheries and improved resource management.

---

[1] E-mail: ron.hill@noaa.gov

A Comparison of Habitat Structure in Fished and Unfished, Mobile and Immobile Sand Habitats on Georges Bank (Northwest Atlantic)

J. B. LINDHOLM[1]

National Undersea Research Center at the University of Connecticut, Groton, Connecticut; and NOAA's Stellwagen Bank National Marine Sanctuary, Scituate, Massachusetts

P. J. AUSTER

National Undersea Research Center at the University of Connecticut, Groton, Connecticut

P. VALENTINE

U.S. Geological Survey, Woods Hole, Massachusetts

Fishing has been described as the dominant anthropogenic impact to marine ecosystems worldwide. One subset of impacts is caused by fishing with mobile bottom-contact gear (e.g., scallop dredges, bottom trawls) on seafloor habitat and associated taxa. Mobile fishing gear reduces seafloor habitat complexity through the removal of emergent fauna that provide structure (e.g., erect sponges), the removal of structure-building megafauna that produce pits and burrows (e.g., crabs, fish), and the smoothing of bedforms (e.g., sand waves). In this study we compared the relative abundance of microhabitat features (the scale at which individual fish associate with seafloor habitat) inside and outside of a large closed area on Georges Bank (closed in December 1994 and sampled in June 1999). A total of 32 stations were selected in a paired sampling design inside and outside of the closed area in sand habitats. Video and still photographic transects were conducted at each station using the Seabed Observation and Sampling System. Seven common (i.e., flat sand, rippled sand, sand with emergent fauna, bare gravel, gravel with emergent fauna, shell, shell fragment) and two 'rare' (sponges, biogenic depressions) microhabitat types were compared separately. Analyses were conducted for 'mobile sand' habitats (< 60 meters water depth) and for 'immobile sand' habitats (> 60 meters). Results showed no significant differences in the relative abundance of the common microhabitat types between fished and unfished areas in mobile or immobile sand habitats. However, in immobile sand habitats sponges and biogenic depressions were numerically more abundant inside the closed area.

---

[1] E-mail: james.lindholm@noaa.gov

Changes in the Benthic Invertebrate Assemblage following the Establishment of a Protected Area, the "Plaice Box"

G. J. PIET,[1] J. CRAEYMEERSCH, AND A. D. RIJNSDORP

Netherlands Institute for Fisheries Research (RIVO), Department of Biology and Ecology, IJmuiden, The Netherlands

The ecosystem in the south-eastern North Sea is affected considerably by various sources of both natural and anthropogenic origin. The effect of a reduced beamtrawling effort on the benthic invertebrate assemblage could be studied from changes in the assemblage following the establishment of a protected area, the 'plaice box'. This area was established in 1989 and closed for all vessels with an engine power over 300 Hp, which constitutes the main part of the beamtrawling fleet. At first it was only effective part of the year but since 1995 the box was closed during the whole year. In order to be able to distinguish between this effect and that of potentially confounding influences from natural origin relevant environmental variables were incorporated in the analyses. The observed effects of the closure of an area on the benthic assemblage are discussed in the context of potential management measures and how to predict and assess their effectiveness and possible side effects.

[1] E-mail: g.j.piet@rivo.wag-ur.nl

The Effectiveness of Marine Protected Areas on Fish and Benthic Fauna: The Georges Bank Closed Area I Example

R. REID[1]

National Marine Fisheries Service, Northeast Fisheries Science Center, Highlands, New Jersey

F. ALMEIDA

National Marine Fisheries Service, Northeast Fisheries Science Center, Woods Hole, Massachusetts

P. VALENTINE

U.S. Geological Survey, Woods Hole Field Center, Woods Hole, Massachusetts

L. ARLEN, J. CROSS, AND V. GUIDA

National Marine Fisheries Service, Northeast Fisheries Science Center, Highlands, New Jersey

J. LINK

National Marine Fisheries Service, Northeast Fisheries Science Center, Woods Hole, Massachusetts

D. MCMILLAN

National Marine Fisheries Service, Northeast Fisheries Science Center, Highlands, New Jersey

S. MURASKI

National Marine Fisheries Service, Northeast Fisheries Science Center, Woods Hole, Massachusetts

D. PACKER, J. VITALIANO, AND A. PAULSON

National Marine Fisheries Service, Northeast Fisheries Science Center, Highlands, New Jersey

In late 1994, a substantial portion of western Georges Bank was closed to commercial fishing (Closed Area I) to assist with stock rebuilding. After about five years of closure, CAI, exhibited a notable increase in biomass and density of sea scallops, Placopecten magellanicus, and was reopened to the scallop fishery. Before the industry was allowed entry into this area, we conducted a survey to monitor the recovery of benthic habitat and fauna inside CAI. Sampling sites were selected in a paired station design for an inside/outside comparison; other stations were chosen to monitor the remainder of the inside. At each station, we conducted video transects, collected still photos, CTD casts, and sediment samples for physical and chemical analysis. A Smith-McIntyre bottom sampler was then used to sample the benthic community, followed by an otter trawl. Trawl catches were sorted to species and all fish and invertebrates were weighed, enumerated, and measured. Stomach contents, maturity observations, and age structures were collected for selected species at each station. Our results suggest notable differences between paired stations for a suite of biotic and abiotic metrics ranging from grain size to fish biomass. The reason for major differences is likely a result of the high relief, cobble habitat type in the region. The differences we observed for CAI may have notable implications for the population dynamics of commercially valuable species.

---

[1] E-mail: robert.reid@noaa.gov

Physical and Biological Effects of Shrimp Trawling on Soft Sediment Habitats in the Gulf of Maine

A. W. Simpson[1] and L. Watling

University of Maine, Darling Marine Center, Walpole, Maine

Mobile gear fisheries are a pervasive source of disturbance in marine habitats that can directly alter both the physical and biological structure of the benthic environment. In the Gulf of Maine, muddy bottoms are intensively trawled for northern shrimp during a seasonal winter fishery. We collected sediment samples from trawled and untrawled areas every 80 to 120 days over an 18-month period. Detailed bulk density measurements from sediment x-radiographs reveal that shrimp trawling may alter the sedimentary 'landscape'. Our findings suggest that in areas where biogenic disturbance is high due to the activities of large burrowing megafauna such as fish and crustaceans, discerning impacts of shrimp trawling on the structure of infaunal communities is difficult; however, trawling appears to reduce the overall density of large burrows.

---

[1] E-mail: anne.simpson@umit.maine.edu

Ecological Consequences of Lost Habitat Structure for Commercially Significant Flatfishes: Habitat Choice and Vulnerability to Predators

A. W. Stoner[1] and C. L. Ryer

Alaska Fisheries Science Center, NOAA National Marine Fisheries Service, Hatfield Marine Science Center, Newport, Oregon

R. A. McConnaughey

Alaska Fisheries Science Center, NOAA National Marine Fisheries Service, Seattle, Washington

Numerous field studies, both descriptive and experimental, have shown that fishing gear can have a negative impact on the structural complexity of benthic environment. Impacts in high-relief habitats such as coral reefs, hard-bottom, seagrasses, and cobble are well documented. Soft-bottom habitat can also contain physical structure created by different bedforms, sessile invertebrates such as sponges, anemones, soft corals, and bryozoans, and the empty shells of molluscs. Recent laboratory experiments with Alaska flatfishes show that age-0 and age-1 fish have a strong behavioral affinity for sediments structured with sand waves, sponges, bryozoans, and bivalve shells. Responses were stronger in juvenile Pacific halibut than rock sole. The presence of structured habitat also affected the survivorship of age-0 fishes in the presence of a piscivorous predator, but habitat-mediated predator-prey interactions varied with prey species. Comparisons of trawled and untrawled locations in the Gulf of Alaska and the Bering Sea reveal that densities and biomass of sponges, anemones, bryozoans, gastropod shells, soft corals, and other biota providing structure for small fishes decrease with fishing activity. It follows that loss of structured habitat in low-relief shelf environment can have both direct and indirect impacts on the function of habitat for demersal fishes, particularly during their first year of life. We need a better understanding of how structural complexity in soft-bottom environment influences abundance and recruitment of fishes and invertebrates, and better characterization of habitat features is probably required.

---

[1] E-mail: al.stoner@noaa.gov

Bottom Trawling Effects on Cerianthid Burrowing Anemone Aggregations and Acadian Redfish Habitats in Mud to Muddy Gravel Seabeds of the Stellwagen Bank National Marine Sanctuary Region, Gulf of Maine (Northwest Atlantic)

P. C. Valentine[1]

U.S. Geological Survey, Woods Hole, Massachusetts

J. B. Lindholm

NOAA's Stellwagen Bank National Marine Sanctuary, Scituate, MA; and National Undersea Research Center at the University of Connecticut, Groton, Connecticut

P. J. Auster

National Undersea Research Center at the University of Connecticut, Groton, Connecticut

Burrowing cerianthid anemones (Cerianthus spp.) occur in mud, gravelly mud, and muddy gravel in the Gulf of Maine. The non-retractable anemone tubes commonly extend 15 cm above the seabed, form dense aggregations (up to 10's m^{-2}), and provide habitat for Acadian redfish (Sebastes fasciatus). Video sampling shows that anemones are common in untrawlable areas such as the mud bases and the muddy tops of gravel banks but are less common than expected in heavily fished mud basins. Video observations were conducted in August 2001 in two settings: 'gravel window' areas on mud basin floors where cobbles and boulders on the tops of gravel mounds are almost covered by mud; and the mud floor and mud to muddy gravel walls of a long narrow basin. Video imagery shows trawling occurs predominantly in mud around the gravel windows and on the mud floor and lower walls of the narrow basin. Video transects across these features ranged up to more than a kilometer in length. Quantification of 20-meter seabed segments shows that trawling is least intense and anemones and redfish are most common on the gravel windows and on the muddy gravel middle and upper basin walls. These observations suggest that either: (1) trawling on mud has modified the distribution of cerianthids by direct removal; (2) trawling has modified the mud seabed so cerianthid recruitment has declined significantly; (3) untrawled mud in open basins is not conducive to cerianthid recruitment. Variations in the distribution of cerianthids may have important implications for the successful recruitment of redfish.

[1] E-mail: pvalentine@usgs.gov

Why Fishing Gear Impact Studies Don't Tell Us What We Need to Know

L. WATLING[1] AND C. SKINDER
Darling Marine Center, University of Maine, Walpole, Maine

The late 1990s saw several comprehensive reviews of the impact of mobile fishing gear on benthic communities published in the scientific literature. In particular, the review of Auster and Langton offered several tables detailing the results of individual studies. We have updated this review and examined the studies for their predictive value. That is, we ask, can the studies that have been done be used in very different geographic areas, or in unexamined habitats, to assess potential impacts of mobile fishing gear? We suggest that most of the studies conducted to date are very good at telling us what has happened, but will give limited or inaccurate information about what will happen, or perhaps has happened in an unstudied area. The lack of predictive capability of most studies results from the fact that they have relied on an examination of spatial patterns rather than understanding the underlying processes which result in the benthic community structure observed. In some studies it has been concluded that fishing gear will have no measurable impact in some habitats. In this paper we take a first principles approach and argue that were certain variables measured, such as sediment food quality, and were the studies done at the appropriate scale, impacts that were missed would have been seen. Changing the way trawling studies are conducted will offer greater potential for predictive capability.

[1] E-mail: watling@maine.edu

Reduction of Species Diversity in a Cobble Habitat Subject to Long-Term Fishing Activity

L. WATLING[1] AND A. PUGH

Darling Marine Center, University of Maine, Walpole, Maine

Studies dealing with the impact of mobile fishing gear have, for the most part, been conducted in areas with low fishing disturbance and are conducted using the BACI design. There have only been a few studies where bottoms that have been fisihed routinely are compared with neighboring unfished areas. We examined an area in 100 m water in the Gulf of Maine that had been fished with otter trawls for white hake and compared the epifaunal community with that of an adjacent area that was too rough for fishing even with rock-hopper gear. In both areas, boulders with relatively flat surfaces were chosen for scraping and suction-sampling with an ROV. There were far more species at the untrawled site as compared with the trawled site. All other measures of diversity, including species-area curves, showed a reduction in diversity at the trawled site of about 50%. Cluster analysis showed almost no similarity between the two sites. For the most part, the presence of large, tree-like hydroids and bryozoans, present at the unfished site and absent at the fished site, provided additional habitat for many of the smaller invertebrates sampled.

[1] E-mail: watling@maine.edu

Ecological Footprints of Scotian Shelf Groundfish Fisheries

K. C. T. ZWANENBURG,[1] M. SHOWELL, AND S. WILSON

Marine Fish Division, Bedford Institute of Oceanography, Dartmouth, Nova Scotia, Canada

We examine by-catch (non-directed catch) of commercial and non-commercial demersal fish species for a number of Scotian Shelf fisheries over the past two decades to determine their relative impacts. Although by-catch of commercial species is readily available (landings), by-catch of non-commercial species exists for only a small subset of fisheries. Landings data give conservative estimates of fishery impacts because they record only commercial species, while observing catches is costly but estimates non-commercial by-catch. We compare impacts of fisheries as derived from landings and on-board observer data. The impacts are cast as ecological footprints with the number of by-catch species defining breadth and the rates of by-catch defining depth of the footprints. From landings data, the proportions (by weight) caught as directed catch ranges from less than 1% (narrow footprints) to as high as 100% (broad footprints) while by-catch rates in these same fisheries range from near 0% (shallow footprints) to near 100% (deep footprints). True breadth and depth (in species space) of ecological footprints can; however, only be derived from observed catches. These are available for only a small subset of fisheries. The Atlantic halibut (Hippoglossus hippoglossus) fishery catches 40 additional species with 13 at a rate of 1% or more of the total halibut caught. These analyses provide a classificatory framework useful for allocating additional investigative efforts to particular fisheries with broad or deep footprints. Long-term impacts of by-catches can also be estimated by hindcasting the potential cumulative impacts of fisheries based on current by-catch profiles.

[1] E-mail: zwanenburgk@mar.dfo-mpo.gc.ca

Comparison of Effects of Fishing with Effects of Natural Events and Non-Fishing Anthropogenic Impacts on Benthic Habitat

Comparison of Effects of Fishing with Effects of Natural Events and Non-Fishing Anthropogenic Impacts on Benthic Habitats

HAN J. LINDEBOOM[1]

*Alterra-Texel and Royal NIOZ,
Post Office Box 167, 1790 AD Den Burg, The Netherlands*

Abstract. The effects of demersal fisheries include destruction of habitats and shifts in species composition or age structure of populations and numbers of benthic invertebrates and fishes. Natural events and other anthropogenic impacts may lead to similar effects. A striking phenomenon found in the marine realm is the possibility of regime shifts. Periods with high numbers or biomasses suddenly alternate with periods with lower numbers. Both in the North Sea and in the North Pacific Ocean, sudden, large-scale changes or regime shifts were observed in 1977–1978 and 1988–1989, hinting at possible large-scale causes of these phenomena. In this article, both observations and possible causes are described. Human-induced pollution may also lead to declines in populations. Tri-butyl-tin (TBT) leading to imposex in whelks *Buccinum undatum* is an example. However, despite high TBT loadings in the Dutch coastal zone, fisheries are still seen as a major cause of the whelk's local extinction. It is argued that the trends that we observe in the marine ecosystem are the result of a very complex interplay between natural and human-induced causes, the final result being an integrated summation of the effects of manageable and nonmanageable factors. Comparing the direct effects of fisheries, sand extraction, and oil or gas extraction in the Dutch part of the North Sea, it can be concluded that, presently, the impact of fisheries on the benthic fauna is a thousand times higher than that of sand extraction and a hundred thousand times higher than that of the oil and gas exploration. Since eutrophication and pollution cannot be expressed per unit area, a direct comparison with effect of fisheries is impossible. However, in general, pollution effects are local, whereas fisheries effects cover the whole North Sea. In conclusion, both natural events and fisheries are the major driving forces of changes observed in the marine ecosystem in the temperate zone.

Introduction

Fishing with bottom-towed gear leads to destruction of habitat types, such as shellfish banks and beds of sea grass *Zostera marina*, coral reefs, and grounds of maerl *Calcareous rhodophytes*, and to shifts in species composition and age structure of benthic populations and communities (Kaiser and de Groot 2000). Natural events and other anthropogenic impacts can lead to similar effects. To establish the long-term impact of fisheries and to be able to predict the possible outcome of fisheries management options, it is necessary to distinguish between the effects of fisheries and these other effects. In this article, the variability of the marine ecosystem is described and possible causes and implications for fisheries are discussed. At the Royal Netherlands Institute of Sea Research, the direct effects both of fisheries and of oil and gas exploration on the benthic fauna on the North Sea bottom were studied and the collected data used to compare these effects.

(Natural) Variability

Marine ecosystems are not in steady state but exhibit continuous changes in production, biomass, and species composition. What is the cause of these variations: natural processes or impacts of human activities? Our awareness and scientific understanding of this variability has increased over the last few decades. For example, long-term data sets on phytoplankton, zooplankton, macrofauna, fish, and birds have been collected in the Wadden Sea and North Sea. Until recently, these data sets were mainly applied to demonstrate the effects of human use and misuse on the ecosystem

---

[1] E-mail: han.lindeboom@wur.nl

(Cadée and Hegeman 1993; Hickel et al. 1993; Beukema et al. 1996). However, when the various data sets are combined, it is striking that certain changes are very sudden and not gradual as one would expect from a gradually increasing human impact. The algal biomass in the western Wadden Sea doubled between 1976 and 1978, followed by the macrobenthos in 1980 (Lindeboom 2002a). The breeding success of Eider ducks *Somateria mollissima* suddenly increased by several orders of magnitude (Swennen 1991). For the North Sea, sudden changes in the phytoplankton and zooplankton species composition and shifts in macrofauna assemblages and benthic respiration rates were reported for the late 1970s (Austen et al. 1991). Other rapid changes in the North Sea ecosystem were observed in the late 1980s when the breeding success of Eider ducks once again collapsed, coinciding with a decrease in standing stocks of mussels *Mytilus edulis* (known as blue mussels) and cockles *Cerastoderma edule* (Swennen 1991). The amount of plaice *Pleuronectes platessa* in the Dutch coastal zone started to decline, while data from the Dutch Institute for Fisheries Research indicate increased numbers of scaldfish *Arnoglossus laterra* and lesser weever *Trachinus vipera* (Lindeboom 2002a). Catches of horse mackerel *Caranx trachurus* near Norway went up threefold in 1988–1989 (Reid et al. 2001). Both in 1977–1978 and 1988–1989, striking changes in biomass and species composition were observed all around the North Sea (Lindeboom 2002a), and the same happened at other places in the northern hemisphere. The stocks of Atlantic cod *Gadus morhua* in the eastern Atlantic collapsed in the late 1980s. Hare and Mantua (2000), who analyzed 100 environmental time series, described major regime shifts in the North Pacific in 1977 and 1989.

All these data sets, both from the North Atlantic and North Pacific, indicate coinciding, sudden large-scale changes or regime shifts, hinting at a possible large-scale cause of the phenomena observed.

But there is more. Striking phenomena found in many data sets indicate cyclic changes. Three-year maxima have been observed in the abundance of *Noctiluca scintillans*, while a 6-year cycle in the amount of shrimps *Crangon crangon* landed around the North Sea (Lindeboom 2002a) and the recruitment success of shellfish were found (Beukema et al. 1996). Gray and Christie (1983) analyzed plankton data from the North Atlantic and found evidence of 3–4, 6–7, and 10–11 year cycles, whereas benthic data suggested 6–7, 10–11, and 25–30 year cycles. The El Niño-Southern Oscillation (ENSO) cycle is well-known and results in the nearly complete failure of fisheries in South American waters and many other ecological deviations worldwide every 4–7 years.

Despite an increasing number of examples for many areas around the world, cyclic behavior in coastal seas is not undisputed. Is it really the result of complex physical–biological interactions or just a statistical feature of data sets? In freshwater systems and tree rings analyses, interannual and decadal cyclicity has been well documented. Growing evidence for this behavior in sediments (see review, Pike and Kemp 1997) and corals (subclass Hexacorallia; Lough and Barnes 1997) and shellfish growth (Witbaard 1996) have been reported for marine systems. The number of papers suggesting links with solar activity is increasing. (Haigh 1996; Pap et al. 2002; Rind 2002; Solanki 2002). However, very long data sets also indicate alternations of periods with clear cyclic behavior with periods with no patterns at all. Whether predictable cyclicity really exists remains a major scientific question.

Bergman and Lindeboom (1999) concluded that many marine data series longer than 20 years indicate interannual and decadal variability, among which the following phenomena may be observed: sudden rapid changes, gradual changes (e.g., in the direction of trends), changes in monthly or seasonal variability, and changes in dominance of species and cyclic variation. What causes these phenomena?

Links with changes of short-term or large-scale weather patterns, including wind, winter or summer temperatures, or rainfall have been suggested. A shift in storm frequencies or wind directions might cause changes in sediment water exchange or mixing (Lindeboom 2002a, 2002b). For the North Sea, several data sets indicate shifts in sea surface temperature, in mean wind direction, and in mean sea level at about the same time that the sudden changes were observed.

The North Atlantic Oscillation, a periodic change in atmospheric pressure between Iceland and Portugal, determines the strength of the prevailing westerlies in the North Atlantic. This, in turn, affects the ocean surface currents and, hence, the movement of water toward northwestern Europe, in particular into the North Sea (Rogers 1984). In a continuous plankton recorder, data effects of these changing westerlies were identified (Fromentin and Planque 1996; Beare and McKenzie 1999).

In temperate regions, the occurrence of cold winters strongly influences the species composition of intertidal benthic communities (Beukema et al. 1996). Therefore, it is not unlikely that climatic features could trigger the observed phenomena. Indeed, for The Netherlands, significant sudden changes in mean annual temperature and annual number of foggy days were recorded in 1989. However, meteorologists still dispute whether this is real or just a statistical phenomenon in the data sets.

For the regime shifts in the North Pacific, Hare and Mantua (2000) have concluded that the 1977 shift was evident in both the climatic and biological data, indicating that ecosystem responses to climate variability were a major driving force. However, the 1989 shift was less pervasive, mainly found in the biological data, and not a simple return to pre-1977 conditions. They concluded that the large marine ecosystems of the North Pacific appear to filter climate variability strongly and respond non-linearly to environmental forcing.

But it is not this simple. We are dealing with a very complex system, and many researchers have come to different conclusions about the causes of change in marine ecosystems.

Philippart et al.(2000) analyzed the phytoplankton data sets from the western Wadden Sea and came to the conclusion that species shifts observed around 1977–1978 and 1989 related to a different nutrient status in the research area. Before 1978, there was a mesotrophic, phosphorus (P)-limited growth; between 1978 and 1987, there was a eutrophic, nitrogen (N)-limited growth; and since 1988, there is a eutrophic, P-limited growth, all leading to different algal dominance. Philippart (2000) relates the nutrient shifts to humanly influenced nutrient inputs via the rivers. But although human-induced eutrophication certainly plays an important role, the coinciding phytoplankton composition shift observed in 1977–1978 by Hickel et al. (1993) near Helgoland, an area where the human nutrient inputs shifted much later, hints at other driving forces.

Human-induced pollution is another factor influencing the status of the marine realm. After the discovery of the effects of dichloro-diphenyl-trichloro-ethane (DDT) and poly-chlorine-biphenyls (PCBs), local and global bans came into force, and with success. For example, the PCB-induced breeding failure of common seals *Phoca vitulina* in the western Wadden Sea in the 1960s and 1970s has stopped, and this seal population is now among the highest reproducing in the world (Reijnders and Brasseur 2003). A more recent example is tributyltin (TBT) used in antifouling paints on ships. It was shown that TBT could cause imposex (female snails getting male sex organs) in sea snails, and a worldwide ban is going to be imposed by the International Maritime Organization soon. But was TBT the major cause of the decline of the whelks *Buccinum undatum* (known as waved whelks) in the Dutch coastal zone? Mensing (1999) and ten Hallers-Tjabbes et al. (2003) proved the direct relationship between TBT and imposex in female whelks, demonstrated the disappearance of whelks from the Dutch coastal zone, and found high frequencies of imposex further offshore. They demonstrated that imposex occurs more frequently in areas with permanently mixed water columns compared to areas with summer stratification, the hypothesis being that during stratification the TBT in surface water does not reach the benthic fauna. But when comparing the lethal effects of TBT with those of the beam trawl fisheries in the study area, Mensing (1999) could find few whelks without visible damage caused by the tickler chains of the trawls, and he concluded that despite the TBT problem, the very high fishing intensity was still the major cause of the whelks decline and local extinction in the Dutch coastal zone.

Fisheries is another human action seen as a major driving force of marine change. In 1883, Olsen produced the Piscatorial Atlas depicting a huge oyster bed in the central North Sea. In the accompanying text, he wrote,

> The oyster is so great a delicacy and so well known that any comment is needless. Hundreds of oyster culture establishments are multiplying the supply around the [British] Isles, and yet not equal to the task of supplying the demand: but Mother Sea has yet in store a bed or beds of 200 mi [370 km] in length, and varying even to 70 mi [130 km] in width, situated between Heligoland and the Dogger Bank, or Bothy Gut. Oyster vessels are now being added to the Great Grimsby fishing fleet.

In 1936, the last oysters *Ostrea edulis* (known as edible oysters) were caught commercially, and since the 1970s, no live oysters have been found in the area: 20,000 km² of oyster beds were gone. However, despite clear signs of overfishing, it is questionable whether fisheries were the only cause of the decline. Most likely changes in hydrography have played an important role, while diseases like *Bonamia* may also have influenced the oyster population. Unfortunately, there are no data for the central North Sea to further examine this.

When comparing the effects of natural storm events with the effect of fisheries, unpublished data indicate that the energy put into the sediment at 20-m depth by a 10-min storm with a wind force of 9 Beaufort just about equals the energy input of one passing 12-m beam trawl. Dutch government officials quoted these data to stress the relative harmlessness of fisheries. However, the ecosystem has had ample time to adapt to storm effects, whereas no organism can adapt to decapitation or fractionation by the passing of the heavy iron chains of the beam trawl.

Of course, fisheries have a major impact on the marine ecosystem by removing commercial species, killing unwanted bycatch, and damaging habitats. In areas with high fishing intensities, such as the North

Sea, the damaging impact of fisheries on stocks, age distribution, and habitat characteristics is undisputed (Lindeboom 2000). As for the whelks, the disappearance of large, slowly reproducing fish, such as rays *Raja* spp., skates *Raja batis*, and the greater weever *Trachinus draco*, from the southeastern North Sea can be mainly attributed to fisheries. But whether decreasing trends should always be linked to fisheries, and whether the removal of fisheries pressure will always reverse the trends, is very unlikely.

From the above, it can be concluded that both natural processes and human actions can lead to large variations and changes in marine ecosystems.

However, it is often very difficult, if not impossible, to establish the real cause–effect relationships in the complex coastal ecosystem. Most likely, different factors can have interfering or enhancing effects, and the (local) human disturbances are another complication in the analyses. The picture we observe is the result of a very complex interplay between many natural and man-induced driving forces; it is an ever-changing puzzle with many different configurations (Figure 1).

After the collapse of the North Sea fish stocks at the end of the 19th century, the International Council for the Exploration of the Sea was established. However, a hundred years of intensive scientific research has not led to a much better understanding of the causes of this collapse. I would argue that underestimation of the ecosystem complexity, the hunt for a direct or linear cause–effect explanation, and the lack of adequate analysis techniques for nonlinear responses to environmental changes are major contributors to this failure.

Long-term data series can help to answer questions on the variability of coastal ecosystems, and the continuous collection of data, often hampered by limited funding, should be strongly supported. These long-term data sets, in combination with the results of experimental laboratory and field studies, are necessary to answer the question of whether we are looking at explainable phenomena with clear, though often complex, cause–effect relationships or at more or less chaotic and unpredictable behavior of the marine ecosystem.

These complex relationships and uncertainties about possible causes should not become an excuse for nonmanagement of human actions. In the marine system, there are nonmanageable features such as climate, currents, or food availability and manageable features such as fisheries and sand or gas extraction. These can all be in a system-influencing state leading to the phenomena that we observe. This is illustrated in Table 1.

The argument made is that uncertainty about the natural changes should never be a reason not to act in case the effects of manageable actions are likely to enlarge undesired developments. For this approach, the precautionary principle, as introduced in 1992 in the United Nations Convention on Environment and Development (United Nations 1992), provides suitable advice: "Where there are threats of serious or irreversible damage to the environment lack of full scientific certainty should not be used as a reason for postponing cost effective measures to prevent degradation."

Comparing the Impacts of Different Uses on the Benthic Ecosystem

The Dutch part of the North Sea is a heavily used area. Fisheries, oil and gas exploration, and sand extraction use the marine resources, while shipping, the military, tourists, and the cable and pipeline industry use available space. The increasing demand for food, building materials, energy, and living space puts an accelerating pressure on this vulnerable environment, and knowledge about the effects of the different uses is indispensable for future management.

Over the last decade, the direct effects of the different uses in the Dutch part of the North Sea have been well studied, and enough data are now available to assess and compare these effects. In this article, this is done by calculating the relative benthic fauna damage index (RBDI). This RBDI is the observed direct effect multiplied by the surface area influenced.

The effect is defined as the percentage of animals killed, or measurably hampered in their normal development, in the area which is directly influenced by the user. The surface area is the area (in km^2) where users have operated. In an original Dutch version of the RBDI, I included the restoration time, being the interval between the time the effect was caused and the time that the killed animals have been replaced by new offspring. However, this restoration time is the most difficult to estimate. New animals will only reappear after the next successful breeding season. Therefore, it is likely that the restoration time for sand extraction, which kills all animals, is longer than the restoration time for fisheries, which kills 21% of the benthos per trawl passage. In the original Dutch paper, I used 1 year restoration time for fisheries and gas extraction and 3 years for sand and gravel extraction. However, since these are disputable figures, I excluded them in this paper. Of course, immigration from other areas is always possible, but this is not seen as restoration. Apart from that, if a 20–100-year-old shellfish is killed, it will take 20–100 years before the original

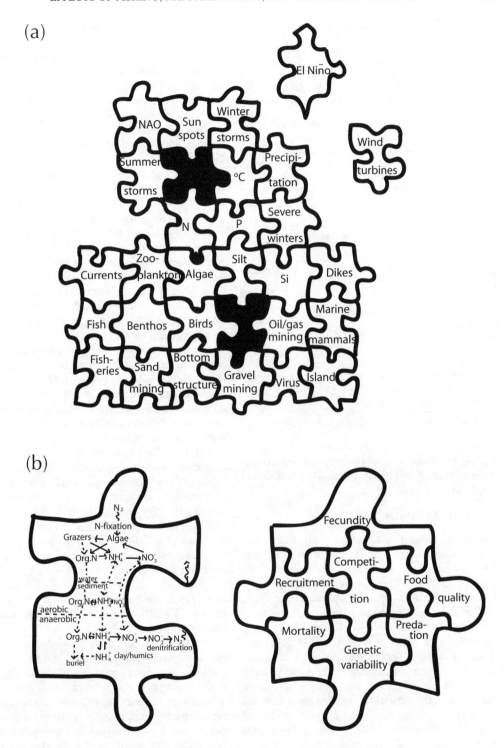

Figure 1. (a) The variable ecosystem that we observe is the result of the interplay of many natural and human-induced processes. To understand the possible cause and effect relationship, we have to unravel a continuously changing puzzle with no finite boundaries and many configurations and with new, missing, or nonfitting pieces. (b) Each piece in itself contains a very complex set of processes and interactions. Here, the pieces for nitrogen (N) and benthos from (a) are elaborated.

Table 1. Four example scenarios for the effects of non-manageable and manageable features that drive the marine ecosystem. In scenario (a), all arrows point down, leading to a rapid decline of the parameter involved. In scenario (b), the human influence ceased leading to a much slower decline. The difference between scenarios (a) and (c) could be a regime shift caused by changing, non-manageable driving forces. Depending on the human influence, this could result in an equilibrium (scenario c) or an increase (scenario d). In reality, there are many more driving forces and complex interrelationships and many more scenario possibilities (see also Figure 1).

Feature	Scenario			
	a	b	c	d
Non-manageable				
Climate	↓	↓	↑	↑
Ocean currents	↓	↓	↑	↑
Food	↓	↓	↑	↑
Predation	↓	↓	↑	↓
Diseases	↓	↓	↑	↑
Manageable				
Fisheries	↓	—	↓	—
Sand extraction	↓	—	↓	—
Gas exploitation	↓	—	↓	↑
Eutrophication	↓	—	↓	↑
Habitat destruction	↓	—	—	—
Total effect	↓	↓	—	↑

situation has been restored. If future research indicates that the restoration time of these human impacts differs significantly, this parameter could be included in the RBDI again.

The Direct Impacts

Both the 12-m and the 4-m beam trawl fisheries cause a direct mean mortality of 21% of the benthic fauna over the width of the fish track (Bergman and Lindeboom 1999). Thus, the effect factor is 0.21. When the same area is trawled a second time, again, 21% of the remaining fauna is killed, leading to an effect factor of 0.16. In 1994, the Dutch part of the North Sea was fished on average at least 1.36 times (Lindeboom and de Groot 1998). Using an effect factor of 0.18 (assuming that half of the trawling takes place in an already trawled area, leading to a lower average impact), an annual fishing intensity of 1.36, and a surface area of the Dutch part of the North Sea of 50,000 km^2, the minimal RBDI is $0.18 \times 1.36 \times 50,000 = 12,240$.

The estimate for the fishing intensity is questionable. Does the 1.36 times include the entire fleet, including all Eurocutters, and are re-flagged or foreign vessels included? A calculation starting with the number of Dutch vessels leads to the following. At a speed of 10 km/h, 16 fishing h/d, and 45 fishing weeks of 3.5 d each, one vessel with two 12-m beam trawls fishes 600 km^2/year. For 250 vessels which spend approximately 60% of their fishing time in the Dutch part of the North Sea, this leads to a fished surface area of 90,000 km^2. If we include the fleet of 200 vessels with 4-m beam trawls which spend about 80% of their time on the Dutch shelf, annually, 130,000 km^2 is fished (2.3 times the entire surface area). With an effect factor of 0.16 (the same area is fished more than one time), the maximum RBDI is 20,800. However, foreign and re-flagged ships are not included in these calculations. The effects of pollution or "ghost nets" lost by the fisheries, which cannot be expressed in surface areas, are also not included in the RBDI.

Now, with only dumping of cuttings with attached water based muds allowed, gas and oil exploration has measurable but not significant effects up to a distance of 25 m from the platform. In this area, less than 10% of the benthic animals disappeared. Thus, the effect factor is 0.1 or smaller. At a distance of 100 m, no effects could be detected (Daan and Mulder 1993). So, an area of 2,000–32,000 m^2 is influenced per drilling. In the last decade, the number of drillings on the Dutch shelf was between 25 and 45 per year. At 25 drillings, with an effect factor of 0.1 over a distance of 25 m, the minimum RBDI is 0.005. At 45 drillings per year, with an effect factor of 0.1 over 100 m from the platform, the maximum RBDI is 0.14.

Before 1993, dumping of drill cuttings with attached oil based mud (OBM) was allowed. The effects of this type of dumping could be established up to 2 km from the platform, and certain species appeared to be very sensitive (Daan and Mulder 1996). Not all species were killed, and the effect factor was estimated at 0.5 (R. Daan, Netherlands Institute for Sea Research, personal communication). Between 1981 and 1992, 226 OBM drillings were executed at 113 different sites (annual mean = 11.3 sites). The maximum area influenced per site was 12.5 km^2. This leads to an RBDI for the oil and gas exploration before 1993 of $0.5 \times 11.3 \times 12.5 = 70$.

In the Dutch part of the North Sea shelf, during the last 5 years, sand for building and beach nourishment was extracted over an area of 12 km^2 (22.5 million m^3 to a depth of 2 m) per year. During extraction, all animals were killed and the effect factor was 1. Adjacent to the extraction area, the settling of reworked sediment and silt may also have had an effect, with an effect factor comparable with the effects of water-based muds (0.1). However, since the surface area of this effect is unknown, it was excluded in the RBDI calculations, which amounted to $1 \times 12 = 12$.

In several large development plans, an increase

in the annual sand extraction is foreseen. For concrete and cement, 16–40 million m³ is needed annually, increasing the RBDI between 8 and 20.

For the extension of the Rotterdam harbor (Maasvlakte 2), about 800 million m³ is needed, and for an artificial island (airport in the sea), about 1,165 million m³. If this sand is extracted in 7 years over a depth of 2 m, the annual RBDI for sand extraction would increase between 57 and 83, respectively. If all these plans are executed at the same time, the RBDI would increase between 148 and 160.

Since there is no gravel extraction in the Dutch part of the North Sea at present, the RBDI is 0.

Shipping has effects in three different ways: dredging of harbor entrances in the North Sea, direct disturbance of the seafloor, and pollution. Large dredging operations to increase the depth of the Euro and IJ channels, The Netherlands, were executed over the last years, but since most of the material was used for building sand or beach nourishment, the direct effects of these activities are already included in the sand extraction estimate. The seafloor disturbance by ships screws will not be too large, leading to an estimated RBDI much smaller than 1. The effects of pollution cannot be assessed in km² and are not included in the RBDI of much smaller than 1.

At present, there are no known effects of military activities on macrofauna, so the RBDI is estimated at much smaller than 1. This is likely to be an underestimation since there are several ballistic ranges in the North Sea where ammunition debris ends up on the seafloor. Although there are no indications that this leads to detectable effects, more research and comparison with "clean" areas is needed.

New pipelines and cables are still being added to the existing system. The area influenced annually is approximately 1 km² or less. All animals are killed or buried, and the effect factor is 1, leading to an RBDI of smaller than 1 to 1.

The effects of pollution and eutrophication cannot be expressed in surface areas. However, with the exception of TBT, there is no clear evidence that this has led to significant measurable effects on benthic organisms in the Dutch part of the North Sea. For TBT, used as anti-fouling on ships, and introduced into the marine environment by shipping, fisheries, and oil and gas exploration, new legislation has been prepared, and a worldwide ban is expected in the very near future.

Direct effects of other pollutants on benthic animals have not been positively established for the Dutch sector of the North Sea. However, there is evidence that eutrophication has led to an increase in the biomass of benthic organisms in the German coastal zone (Hickel et al. 1993) and the Dutch Wadden Sea. For the Dutch sector of the North Sea, it remains important to stay alert for the possible effects of pollutants (e.g., endocrine disruptors) and to take measures if effects are suspected. A major problem could be small but chronic effects which, at present, are not obvious because of the large fisheries impact. Another problem could be accidental spilling or ship collisions. Since pollution, eutrophication, and accidents cannot be estimated per surface area, this has not been included in the RBDI analyses. However, since all these factors could be the result of all human uses mentioned before, this does not influence the comparison values of the relative RBDIs.

Table 2 indicates that of all human uses in the Dutch part of the North Sea, fisheries has by far the largest direct effects on the benthic fauna. Expressed in percentage of the minimal and maximal relative benthos damage index, fisheries causes more than 99.9% of the total direct effects. Sand extraction accounts for less than 0.1% and gas exploration for less than 0.01%. In the future it is expected that, due to fleet reduction, the effects of fisheries will decrease while the implementation of plans for port additions or an island for an airport at sea will considerably increase the RBDI for sand extraction.

In the medium term, if large-scale changes in the North Sea benthos are observed and if direct effects of human actions are suspected to be a major driver, fisheries is very likely the major cause. Possibly, pollution and reclamation of the major estuaries before 1980 were other major driving forces, but these cannot be estimated in detail or per unit area. The Netherlands Institute for Sea Research data series on rare species caught by fishermen between 1948 and 1988 indicate large declines, first in slow-growing fishes (such as sharks *Elasmobranchi*, rays, and skates) followed by declines in epifauna and infauna (such as lobsters *Homarus pagurus*, crabs *Cancer pagurus*, shellfish [class Bivalvia], sea snails [class Gastropoda], and anemones [order Actiniaria]. Philippart (1998) found a close relationship between the

Table 2. Relative benthos damaging index (RBDI) for the Dutch part of the North Sea.

Activity	RBDI
Fisheries	12,240–20,800
Gas and oil (after 1993)	0.005–0.14
Gas and oil (before 1993)	≅ 70
Sand extraction	12
Sand extraction after 2005	160–172
Gravel mining	0
Shipping	<<1
Military activities	<<1
Pipelines and cables	<1–1

pattern of animal disappearance and the fishing activities first with otter trawls and after 1960 with beam trawls. Although natural processes may also have contributed to the observed declines, fisheries remains the most likely causal factor.

Conclusions

Long-term data series indicate that the variability of the marine ecosystem is very large and characterized by sudden changes, regime shifts, reversing trends, and cyclic behavior. The causes for these changes are a complex interplay between natural and man-induced factors, whereby climate variability could play a major triggering role.

The effects of this natural variability, especially when numbers of organisms decrease, can be similar to the effects of fisheries, pollution, eutrophication, and sand or gas extraction.

In the Dutch part of the North Sea, the direct effects of demersal fisheries are 1,000 times larger than the direct effects of sand extraction and at least 100,000 times larger than the effects of gas and oil extraction.

References

Austen, M. C., J. B. Buchanan, H. G. Hunt, A. B. Josefson, and M. A. Kendall. 1991. Comparison of long-term trends in benthic and pelagic communities of the North Sea. Journal of the Marine Biological Association of the United Kingdom 71(1):179–190.

Beare, D. J., and E. McKenzie. 1999. Temporal patterns in the surface abundance of *Calanus finmarchicus* and *C. helgolandicus* in the northern North Sea (1958–1996) inferred from continuous plankton recorder data. Marine Ecology Progress Series 190:241–251.

Bergman, M. J. N., and H. J. Lindeboom. 1999. Natural variability and the effects of fisheries in the North Sea: towards an integrated fisheries and ecosystem management? Pages 173–184 in J. S. Gray, W. Ambrose, Jr., and A. Szaniawska, editors. Biochemical cycling and sediment ecology. Kluwer, Dordrecht, The Netherlands.

Beukema, J. J., K. Essink, and H. Michaelis. 1996. The geographic scale of synchronized fluctuation patterns in zoobenthos populations as a key to underlaying factors: climatic or man-induced. ICES Journal of Marine Science 53:964–971.

Cadée, G. C., and J. Hegeman. 1993. Persisting high levels of primary production at declining phosphate concentrations in the Dutch coastal area (Marsdiep). Netherlands Journal of Sea Research 31:147–152.

Daan, R., and M. Mulder. 1993. A study on possible short-term environmental effects of WBM cutting discharges in the Frisian Front area (North Sea). Netherland Institute for Sea Research, Rapport 1993-5, Texel.

Daan, R., and M. Mulder. 1996. On the short-term and long-term impact of drilling activities in the Dutch sector of the North Sea. ICES Journal of Marine Science 53:1036–1044.

Fromentin, J.-M., and B. Planque. 1996. *Calanus* and environment in the eastern North Atlantic. II. Influence of the North Atlantic oscillation on *C. finmarchicus* and *C. helgolandicus*. Marine Ecology Progress Series 134:111–118.

Hare, S. R., and N. J. Mantua. 2000. Emperical evidence for North Pacific regime shifts in 1977 and 1989. Progress in Oceanography 47:103–145.

Hickel, W., P. Mangelsdorf, and J. Berg. 1993. The human impact in the German Bight: eutrophication during three decades (1962–1991). Helgolaender Meeresuntersuchungen 47:243–263.

Gray, J. S., and Christie, H. 1983. Predicting long-term changes in marine benthic communities. Marine Ecology Progress Series 13:87–94.

Haigh, J. D. 1996. The impact of solar variability on climate. Science 272:981–984.

Kaiser, M. J., and S. J. de Groot, editors. 2000. Effects of fishing on non-target species and habitats. Biological, conservation and socio-economic issues. Blackwell Scientific Publications, Oxford, UK.

Lindeboom, H. J. 2000. The need for closed areas as conservation tools. Pages 290–302 in M. J. Kaiser and S. J. de Groot, editors. Effects of fishing on non-target species and habitats. Blackwell Scientific Publications, Oxford, UK.

Lindeboom, H. J. 2002a. Changes in coastal zone ecosystems. Pages 447–455 in G. Wefer, H. Berger, K.-E. Behre, and E. Jansen, editors. Climate development and history of the North Atlantic realm. Springer, Berlin.

Lindeboom, H. J. 2002b. The coastal zone: an ecosystem under pressure. Pages 51–86 in J. G. Field, G. Hempel, and C. P. Summerhayes, editors. Science 2020, an IOC/SCOPE/SCOR outlook on future marine sciences. Island Press, Washington, D.C.

Lindeboom, H. J., and S. J. de Groot, editors. 1998. The effect of different types of fisheries on the North Sea and Irish Sea benthic ecosystem. Netherlands Institute for Sea Research, Rapport 1998-1, RIVO-DLO report C003/98, Texel.

Lough, J. M., and D. J. Barnes. 1997. Several centuries of variation of skeletal extension, density and calcification in massive Porites colonies from the Great Barrier Reef: a proxy for seawater temperature and a background of variability against which to identify unnatural change. Journal of Experimental Marine Biology and Ecology 211:29–67.

Mensing, B. P. 1999. Imposex in the common whelk *Buccinum undatum*. Doctoral dissertation. University of Wageningen, Wageningen, The Netherlands.

Olsen, O. T. 1883. The piscatorial atlas of the North Sea, English and St. George's Channels. Taylor and Francis, London.

Pap, J., C. Frohlich, J. Kuhn, S. Sofia, and R. Ulrich. 2002. A discussion of recent evidence for solar irradiance variability and climate. International Solar Cycle Study (ISCS) 29:1417–1426.

Philippart, C. J. M. 1998. Long-term impact of bottom fisheries on several by-catch species of demersal fish and benthic invertebrates in the south-eastern North Sea. ICES Journal of Marine Science 55:342–352.

Philippart, C. J. M., G. C. Cadée, W. van Raaphorst, and R. Riegman. 2000. Long-term phytoplankton-nutrient interactions in a shallow coastal sea: algal community structure, nutrient budgets, and denitrification potential. Limnology and Oceanography 45:131–144.

Pike J., and A. E. S. Kemp. 1997. Early Holocene decadal-scale ocean variability recorded on Gulf of California laminated sediments. Paleoceanography 12:227–238.

Reid, P. C., M. D. Borges, and E. Svendsen. 2001. A regime shift in the North Sea circa 1988 linked to changes in the North Sea horse mackerel fishery. Fisheries Research 50(1–2):163–171.

Reijnders, P. J. H., and S. M. J. M. Brasseur. 2003. Veränderungen in Vorkommen und Status der Bestände von Seehunden und Kegelrobben in der Nordsee—Mit Anmerkungen zum Robbensterben, 2002. Pages 330–339 in L. Lozán, E. Rachor, K. Reise, J. Sündermann, and H. von Westernhagen, editors.Warnsignale aus der Nordsee: Neue Folge. Vom Wattenmeer bis zur offenen See. Wissenschaftliche Auswertungen, Hamburg.

Rind, D. 2002. Climatology—the sun's role in climate variations. Science 296(5568):673–677.

Rogers, J. C. 1984. The association between the North Atlantic Oscillation and the Southern Oscillation in the northern hemisphere. Monthly Weather Review 112:1999–2015.

Solanki, S. K. 2002. Solar variability and climate change: is there a link? Astronomy and Geophysics 43(5):9–13.

Swennen, C. 1991. Fledgling production of Eiders (*Somateria mollissima*) in The Netherlands. Journal of Ornithology 132:427–437.

ten Hallers-Tjabbes, C. C., J.-W. Wegener, A. G. M. van Hattum, J. F. Kemp, E. ten Hallers, T. J. Reitsema, and J. P. Boon. 2003. Imposex and organotin concentrations in *Buccinum undatum* and *Neptunea antiqua* from the North Sea: relationship to shipping density and hydrographical conditions. Marine Environmental Research 55(3):203–233.

United Nations. 1992. Rio declaration on environment and development. Annex 1 in Report of the United Nations conference on environment and development. United Nations, Report A, conf 151/26, vol 1, Rio de Janeiro.

Witbaard, R. 1996. Growth variations in *Arctica islandica* L. (Mollusca): a reflection of hydrography related food supply. ICES Journal of Marine Science 53:981–987.

Extrapolating Extinctions and Extirpations: Searching for a Pre-Fishing State of the Benthos

LEONIE A. ROBINSON[1] AND CHRIS L. J. FRID

School of Marine Science and Technology, University of Newcastle upon Tyne,
Dove Marine Laboratory, Cullercoats, North Shields NE30 4PZ, UK

Abstract. The long history of mechanized fishing and exploitation of marine ecosystems began before ecological studies were undertaken. Scientists are constrained when trying to establish baseline or "pre-fishing" levels of ecological condition and quality. In this study, the potential for using paleoecological records to reconstruct a pre-fishing benthic assemblage of the entire North Sea is examined. Comparison of the historic record with contemporary records reveals that seven species from three phyla are absent in modern North Sea communities. The ability to establish whether these disappearances are a result of fisheries disturbance based on the relative vulnerability of the extirpated species is also explored.

Introduction

The North Sea benthos have been subjected to a multitude of anthropogenic impacts in recent times. International agreements now require the preservation and maintenance of biodiversity and an ecosystem approach to management of the marine environment (e.g. United Nations Convention on Biodiversity, European Commission (EC) Species and Habitats Directive; see Skjoldal et al. 1999 for review). A key tenet of the North Sea ecosystem approach is the requirement to show a causal relationship between a manageable activity (such as fishing) and the impact it is suggested to have in the marine ecosystem. To be able to provide this level of scientific advice, baseline data, representing a pristine or unimpacted state of the ecosystem, are highly desirable. In addition, such data would allow for an appreciation of the scale of natural variability in a particular ecosystem prior to impact. Without this information, it is difficult to quantify fully the relative magnitude of anthropogenic disturbances (Underwood 1992, 1994; Dayton et al. 1998; Pearce 1999; Strain and Macdonald 2002).

In considering localized activities such as aggregate dredging, sewage sludge dumping, and colliery waste disposal, the area of impact is often relatively contained and small when compared to the scales at which major ecosystem drivers operate (Mayer et al. 1991; North Sea Task Force 1992; Johnson and Frid 1995; Barnes and Frid 1999). Therefore, it is comparatively easy to find unimpacted areas that are close in location and, thus, likely to be exposed to similar background levels of external drivers, so forming appropriate "controls." This allows us to establish the deviation from the control or "pristine" state and for society to make informed decisions about acceptable levels of change and appropriate implementation of management.

However, when we consider the ecosystem-level impacts related to commercialized fishing, both a long, sustained history and the large spatial scale over which fishing has operated complicate our search for the "pre-fishing" state. In the North Sea, fishing has been mechanized for over 100 years, and it is now widely accepted that there are very few, if any, significant areas that are unfished yet also comparable to the fished habitat (Lindeboom and de Groot 1998; Tuck et al. 1998; Hall 1999; Hoffman and Dolmer 2000). Thus, it is extremely difficult to find appropriate control areas to evaluate any long-term effects, especially as it is believed that sustained fishing impacts have brought about permanently altered ecosystem states in some areas (MacDonald et al. 1996; Collie 1998; Watling and Norse 1998; Jackson et al. 2001; Thrush and Dayton 2002).

Fishing with towed bottom gears is a potential cause of long-term change in the community structure of North Sea benthos (Dayton et al. 1995; Jennings and Kaiser 1998; Kaiser and de Groot 2000). Impacts are thought to result from both direct and indirect effects of the removal of target and incidental catch species, the additional mortality sustained by the bycatch, the discard-

---

[1] Corresponding author: L.Robinson@marlab.ac.uk; present address: FRS Marine Laboratory, Post Office Box 101, 375 Victoria Road, Aberdeen AB11 9DB, UK.

ing process, and seabed habitat disturbance due to trawling and dredging (Kaiser and Spencer 1994, 1995; Thrush et al. 1995; Frid et al. 1999; Hall 1999; Kaiser and de Groot 2000). At present, however, our understanding of the effects of fishing on benthic systems can be largely traced to a limited number of short-term, spatially restricted studies. Primarily, these studies have been either experimental manipulations of fishing disturbance in the field or "same time" comparisons of sites with known differences in chronic fishing pressure (for reviews see Collie et al. 2000; Kaiser and de Groot 2000; Thrush and Dayton 2002). Although these studies provide interesting and desired insights into the short-term effects, they do not further our understanding of long-term changes or our subsequent appreciation of the pristine or unimpacted state.

There have been long-term changes in benthic communities of the North Sea over the last century, but the impacts of towed fishing gear are only one of several likely causes. A number of anthropogenic activities plus changes in primary production, organic enrichment, and climatic and hydrographic fluctuations are also likely to have had effects on the benthos (Reise 1982; Kröncke 1990, 1995; Clark and Frid 2001; Kröncke and Bergfeld 2001). There is little doubt that bottom gears impact communities and habitats in the short-term, but further evidence is required to determine whether long-term changes have actually occurred and, if they have, whether impacts can be clearly and exclusively matched to fishing-induced disturbance.

In the North Sea a number of studies have compared contemporary benthic samples with historic data (Reise 1982; Riesen and Reise 1982; Reise and Schubert 1987; Kröncke 1990; Frid et al. 2000; Rumohr and Kujawski 2000). In some cases, historic data originate after the increase in intensive trawling activity that followed the widespread application of powered vessels in the 1920s (Lindeboom and de Groot 1998). Thus, they do not represent a comparison of contemporary data with that of data from a pre-fishing reference state. Where historic data are available from the late 19th and early 20th centuries, disappearances or extirpations of some species were found, but there is no clear explanation of why the specific combination of species would be vulnerable to fishing activities. As such, it is hard to attribute these changes to increased trawl activity.

Changes in the abundance or density of "vulnerable" taxa may act as indicators of benthic disturbances, including the effects of bottom trawling (MacDonald et al. 1996; Rogers et al. 1999; Collie et al. 2000; ICES 2000, 2004). However, the question must be asked: Have any such vulnerable benthic species become extirpated from heavily fished areas over the period in which fisheries disturbance has been occurring? Habitat destruction by trawling may already have caused unnoticed extirpations (Safina 1998), but without historic baseline data, it is impossible to establish if any disappearances have occurred and, thus, to then consider whether fishing impacts are the causal factor.

In this study, records of the life traces and biofacies of four benthic phyla from the heavily fished southern North Sea (Rijnsdorp et al. 1998; Frid et al. 2000) were used to reconstruct the assemblage in recent history. This data is compared with contemporary species records for the area and any disappearances noted. Those species found to be extirpated in the contemporary data were assessed for vulnerability using criteria that had been derived a priori. The potential for using the ecological and morphological characteristics of extirpated species to distinguish the causal mechanism of their extirpation is explored.

Methods

Reconstructing the Pre-Fishing State

In order to reconstruct a pristine state of the North Sea benthos that would represent the system prior to mechanized fishing but also post major environmental change, it would be necessary to find records from communities following the last major glaciation and prior to the mechanization of the fishing industry in the 1920s. The end of the late glacial period is dated to 9,000–10,000 years ago (Norton 1978) and the North Sea reached its current depth and areal configuration around 7,000–8,000 years ago (Norton 1978; Davis 1995; Graham et al. 2003).

A preliminary search of the historical records of North Sea benthos revealed a lack of comprehensive data for fossilized benthic macrofauna from this area. An actuo-paleontological guide to the benthic communities of the southern North Sea was, however, found (Schäfer 1972). Within this encyclopedic book, Schäfer describes how the ecology of recent, extant animated communities can be used to help understand the significance of fossilized remains. In doing so, he also gives examples of the life traces (e.g., burrows and fossilized hard parts) and biofacies remains of marine species that had been studied over the late 19th and early 20th centuries in the shallow sediments of the southern North Sea. From these descriptions, it was, thus, possible to derive a database of preserved benthic invertebrate species found in historic southern North Sea assemblages between 1890 and 1960.

When compiling the database, it was necessary to make a number of assumptions. First, the spatial resolution of the distribution of the historic assemblage is delimited on a regional scale, being restricted to the shallow southern North Sea, south of the 50-m-depth contour (see Figure 1). However, some of the species described are only thought to exist at depths greater than 100 m (e.g., *Calocarides coronatus*) (Ngoc-Ho 2003), and so it must be assumed that a number of the species studied were from a wider area. For this reason, contemporary records were also examined for extirpations at the scale of the whole North Sea. It is not possible to distinguish the habitat depth and spatial limits of individual species from the records, and so we have assumed that all species in the database could have existed in any suitable substrate type in both inshore and offshore areas.

Second, the age of the biofacies and life traces cannot be refined further than that they are recorded from shallow excavations of Holocene sediments. Geological and archaeological investigations attribute the upper 10 m

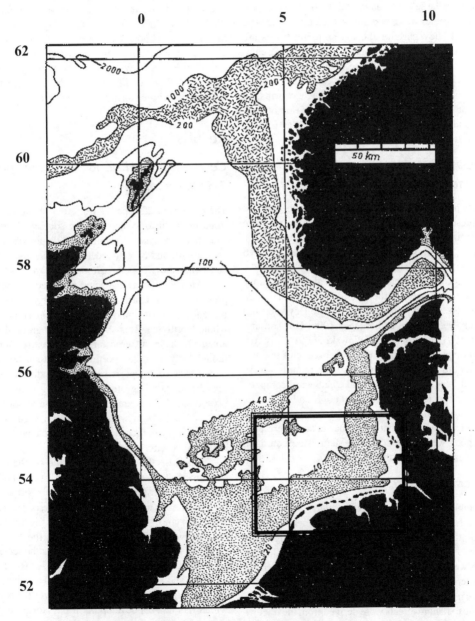

Figure 1. The North Sea, with the location of Schäfer's study area (between 53–55°North, 4–9°East) indicated by a double-lined box (adapted from Schäfer 1972). Longitude and latitude are given on the *x* and *y*-axes, respectively. Shaded areas emphasize depth contours.

of the sampled sediment to post-glacial Holocene deposits laid down in the last 5,000 years (DTI 2002). Some of the life traces could, however, be as recent as those deposited in the last 100 years over the late 19th and early 20th centuries (Schäfer 1972; G. Hertweck, Senckenberg Institute, Germany, personal communication). We, therefore, assume that the record is post-glacial but that it potentially covers a period over which fishing was developing in the North Sea.

Due to the selective nature of the fossilization process, the species described are restricted to organisms with hard body parts or those whose life traces are represented by hard structures. This left representatives of four phyla, three whose body parts lend themselves to good preservation (Mollusca, Echinodermata, and Crustacea), and annelid polychaete worms, in which at least certain major groups, by reason of their tubes and burrows or hard structures such as jaws or hooks, are well represented.

Comparison of Historic and Contemporary Communities

Species present in the historic record were checked for their presence in contemporary records, and names were standardized to those given in Howson and Picton (1997), reducing the likelihood of missing a species due to changes in nomenclature. The species represented in the historic record were then compared with contemporary records from the southern and whole North Sea (Christiansen 1969; Künitzer et al. 1992; Haywood and Ryland 1995; Kühne and Rachor 1996; Howson and Picton 1997; Rumohr and Kujawski 2000). Only forward comparisons—the presence–absence of the historic record species in contemporary records—were made. It is not justifiable to make backward comparisons, as a species present now but not found in paleontological records may be unrepresented due solely to poor preservation of that particular taxa. Where possible, notes were also made on the status (rare, common, etc.) of each species in the assemblage.

Vulnerability Criteria

Vulnerability criteria were derived a priori from the literature for (1) benthic taxa vulnerable to extinction and (2) benthic taxa vulnerable specifically to a fishing effect (Table 1). Criteria for vulnerability generic to extinction events were derived from studies that have analyzed the characteristics of benthic fauna known to have gone extinct on a global scale (see references in Table 1). Criteria specific to fishing effects were derived from the reviews of fishing effects on benthic invertebrates, which highlight characteristics of taxa that have been shown to be vulnerable in the short-term and which may, therefore, increase vulnerability to eventual extinction (see references in Table 1).

On comparing the historic species list with contemporary records, taxa found to be extirpated were then checked against the vulnerability criteria using information derived from the literature and scored for each criteria present (Squires 1965; Christiansen 1969; Schäfer 1972; Künitzer et al. 1992; Haywood and Ryland 1995; Kühne and Rachor 1996; Howson and Picton 1997; Ingle 1997; Rumohr and Kujawski 2000; Ngoc-Ho 2003) (a detailed breakdown of the historic–contemporary comparison of taxa at the species level is available at the European Fisheries Ecosystem Plan Project's Web site, *www.efep.org/appendix/Robinson_Frid_2003.pdf*). Only taxa extirpated at the scale of the whole North Sea were examined.

Results

Comparison of Historic and Contemporary Taxa Inventories

The remains of 220 species from four phyla were identified in the historic record of the southern North Sea. This list comprised 70 mollusks, 70 polychaetes, 43 crustaceans, and 37 echinoderms (species list available at *www.efep.org/appendix/Robinson_Frid_2003.pdf*).

All 70 polychaete species, recorded as a result of preserved jaws, chaetae, tubes, or burrows, were also present in the contemporary fauna of the southern and whole North Sea. This shows that this group does not appear to have suffered large-scale extirpations (although only a limited portion of the total polychaete fauna leave fossilized remains). Secondly, it suggests that the methodology for dealing with issues such as taxonomic changes are robust; if not, a major shift in the species present would have been seen. Thus, it was possible to proceed to consider the changes in the other phyla with more confidence.

Based on the available data, the southern North Sea appears to have lost around 10% of the molluscan and crustacean species present in the historic record (Table 2). As it was not clear from the source material whether some of the historic records of life traces could have originated from a wider area in the North Sea, contemporary records for distributions at the scale of the whole North Sea were also examined. A loss of 4% of mollusks and 2% of crustaceans was still apparent (Table 2). This suggests actual extirpations over the period in which fishing was developing in the North Sea in the order of 2–5% in these groups.

The pattern for echinoderms is more complicated. The data suggest extirpation of 43% of the historic record

Table 1. Vulnerability criteria derived a priori from the literature for (1) taxa vulnerable to extinction and (2) taxa vulnerable specifically to a fishing effect.

Criteria	References
Criteria for general vulnerability to extinction	
1 Specialized habitat	Carlton et al. 1991; Carlton 1993; Roberts and Hawkins 1999; Jackson 2001
2 Vulnerable/sensitive habitat	Carlton et al. 1991; Carlton 1993; Roberts and Hawkins 1999; Jackson 2001
3 Specialized feeding habit	Carlton et al. 1991; Carlton 1993; Clark 1994
4 Low reproductive success and dispersal	Carlton et al. 1991; Roberts and Hawkins 1999; Jackson 2001
5 Low mobility as an adult	Carlton et al. 1991; Roberts and Hawkins 1999
6 Geographic range small	Carlton 1993; Roberts and Hawkins 1999; Jablonski 2001; Jackson 2001; Roy et al. 2001
7 Nearshore distribution	Carlton et al. 1991; Carlton 1993
Criteria for vulnerability specific to a fishing effect	
1 Sessile, erect epifauna	Bergman and Hup 1992; MacDonald et al. 1996; Bergman and van Santbrink 2000; Collie et al. 2000
2 Sessile, filter/suspension feeder	Bergman and Hup 1992; Bergman and van Santbrink 2000; Collie et al. 2000
3 Fragile body structure (morphology)	Bergman and Hup 1992; MacDonald et al. 1996; Bergman and van Santbrink 2000; Collie et al. 2000; Ramsay et al. 2001
4 Near-surface infauna	Bergman and Hup 1992; MacDonald et al. 1996; Bergman and van Santbrink 2000; Thrush and Dayton 2002
5 Exploited by fishery	Reise 1982; Roberts and Hawkins 1999; Jackson 2001; Steneck et al. 2001
6 Unable to regenerate from fragments	MacDonald et al. 1996; Ramsay et al. 2001

from the southern North Sea but only an 8% loss from the whole North Sea (Table 2). Eight of the taxa extirpated from the southern North Sea were holothurians, and there appears to be no obvious reason why this group should lose 80% of the historic species from this area but not the greater North Sea. It is more likely that the original descriptions of the life traces were from a wider area in the North Sea. Twelve out of the 16 echinoderms extirpated from the southern North Sea were also found to have a rare status in the historic record, potentially justifying the greater losses in this region (see www.efep.org/appendix/Robinson_Frid_2003.pdf).

Considering the losses of mollusks, crustaceans, and echinoderms from the greater North Sea, these extirpations will include species adjustments caused by a multiplicity of factors, including changes in hydrography, productivity, climate, species invasions or replacements, and any anthropogenic drivers. In order to examine the possible contribution of fishing, the extent to which the extirpated species have characteristics predisposing them to fishing impacts was examined. As it must be assumed that some of the records may have been collected across a wider area of the North Sea (see Methods), any species found to have disappeared only from the southern North Sea were not considered further.

Vulnerability to Fishing Impacts and Extirpations

A priori consideration of the literature led to the identification of six factors that would pre-dispose benthic taxa to extirpation by the impacts of bottom trawling. A further seven factors would make taxa more vulnerable to general extinction events (Table 1).

Table 2. A comparison of historic and contemporary species inventories. Numbers represent total species recorded for each phylum.

		Southern North Sea		Whole North Sea	
Species	Historic record	Contemporary	% disappearance	Contemporary	% disappearance
Crustacea	43	39	9%	42	2%
Echinodermata	37	21	43%	34	8%
Mollusca	70	63	10%	67	4%
Polychaeta	70	70	0%	70	0%

Vulnerability scoring showed that for the general extinction factors, extirpated species scored between one (*Pasiphea* sp., *Cardium fasciata*, and *Lacuna divaricata*) and four (*Parastichopus tremulus*, *Leptosynapta bergensis*, and *Echinocucumis hispida*) out of a possible seven criteria (Table 3). None of the characteristics associated with vulnerability to general extinction were shared by all of the extirpated species. One was met by five of the seven taxa—a "nearshore distribution." A further two criteria, "low reproductive success and dispersal" and "low mobility as an adult" were met by four out of the seven species. Low mobility is a characteristic that would also impart vulnerability to fishing with bottom gears.

For the criteria on specific vulnerability to fishing effects, the extirpated taxa scored between one (*Pasiphea* sp.) and four (*Cardium fasciata*) out of a possible six criteria. All of the extirpated taxa met criterion 7, they were "unable to regenerate from fragments," whilst six out of seven had a "fragile morphology," and four shared a "near surface infaunal" habit (Table 3).

Summary

A total of 220 taxa from the Crustacea, Echinodermata, Mollusca, and Annelid polychaetes were recorded in the historic record of the southern North Sea based on a database derived from the paleontological studies of Schäfer and colleagues (Schäfer 1972). When compared with records for contemporary distribution, a total of seven taxa, or 3.2% of the reconstructed assemblage, were found to have become extirpated from the whole North Sea. The polychaetes appear robust with 100% of historic taxa present in the modern North Sea, but the Crustacea, Echinodermata, and Mollusca have all lost between 4% and 10% of their taxa.

Characteristics of the seven species found to have been extirpated from the North Sea were examined to consider whether fishing vulnerability prevailed over a general extinction risk of taxa. Overall, there was no difference between the total number of criteria held collectively by species for characteristics that impart vulnerability specific to a fishing effect and that for characteristics that impart vulnerability to a general extinction event (Table 3).

Discussion

There are limited or no benthic assemblages that resemble their pre-fishing state (Dayton et al. 1995; Jackson 2001). The location of a benchmark of the "normal" state of these systems is, however, highly desirable if we are to make informed judgments on the effects of

Table 3. Species extirpated from the whole North Sea scored against a priori derived vulnerability criteria for (1) general extinction and (2) sensitivity specific to a fishing effect.

Criteria	Crustacea	Echinodermata			Mollusca		
	Pasiphea sp.	*Parastichopus tremulus*	*Leptosyna patabergensis*	*Echincucumis hispida*	*Cardium fasciata*	*Bittium reticulata*	*Lacuna divaricata*
General extinction							
Specialized habitat						√	
Vulnerable habitat						√	
Specialized feeding habit							
Low reproductive success and dispersal	√	√	√	√			
Low mobility as an adult		√	√	√	√		
Geographic range small		√	√	√			
Nearshore distribution		√	√	√		√	√
Total	1	4	4	4	1	3	1
Specific fishing effect							
Sessile, erect epifauna							
Sessile filter/suspension feeder					√		
Fragile morphology		√	√	√	√	√	√
Near-surface infauna		√	√	√	√		
Exploited by fishery							
Unable to regenerate from fragments	√	√	√	√	√	√	√
Total	1	3	3	3	4	2	2

fishing (Dayton et al. 1998; Jackson et al. 2001). Fossil records provide the only real mechanism for reconstructing the pre-fishing benthos, but an extensive search of the North Sea records provided little evidence of comprehensive fossilized benthic macrofaunal data. Records were, however, found from the paleontological work of Schäfer (1972), providing a qualitative database of preserved macrofauna from the southern North Sea covering a period over which fishing has developed in this area (19th to early 20th centuries). Evidence has been presented to show that for those taxa that preserve well, it is possible to reconstruct an assemblage comparable with contemporary taxonomic records (www.efep.org/appendix/Robinson_Frid_2003.pdf).

Good documentation of the presence of taxa from three phyla, the Echinodermata, Mollusca, and Crustacea, was found. Particular taxonomic groups from the Annelid polychaetes were also represented. Regrettably, however, many soft-bodied taxa such as the sponges (Porifera), soft corals, sea pens, and sea fans *Cnidarian anthozoa* do not preserve well. If, as theory predicts, taxa with such fragile morphologies are particularly vulnerable to, and thus likely to have been extirpated during the period of, mechanized fishing, the potential for using the fossil record to hindcast a pre-fished state in future work is much reduced.

Although the levels of extirpation found in this study (up to 10% of a phyla) do not compare with the mass extinctions of the geologic past (each removing approximately 50% of abundant marine genera; Jablonski 2001), it is important to consider the implications of such losses. Overall, phyla-level changes are consistent with a fisheries impact effect; more robust or mobile taxa with higher population turnover rates suffer lower rates of extirpation (Crustacea and polychaetes) while more vulnerable taxa with lower mobility and population turnover rates suffer higher rates of extirpation (mollusks and echinoderms). In addition to this, however, the availability of actual species-level data allowed for an examination of whether a direct causal link to fishing could be shown based on the ecological and morphological characteristics of the extirpated species.

The change in abundance of vulnerable taxa has been suggested as an indicator of the direct effects of bottom trawling on the benthos (MacDonald et al. 1996; Safina 1998; ICES 2000). Vulnerability is defined based on a mixture of ecological, morphological, and life history characteristics such as living habit, morphological fragility, and reproductive strategy. Although sensitive species such as sessile, erect epifauna or large, fragile-shelled infauna are more vulnerable in the short-term to the direct impacts of trawls, the long-term implications of such effects are less clear. Following many years of fishing disturbance, benthic communities will develop into completely different associations (Dayton et al. 1998), but many marine invertebrates are considered to be relatively robust to total extinction from a system due to life history characteristics such as large population size, broad geographic range, and good dispersal of larvae (Carlton et al. 1991; Tegner et al. 1996).

In this study, a novel approach using vulnerability criteria based on ecological and morphological characteristics held by the individual extirpated species did not provide convincing evidence that fishing was the only causal driver involved in their extirpation from the North Sea. There may be a number of explanations for this, including the possibility that the specific species found in this study have been extirpated from the study area due to changes in other drivers such as climate or nutrient enrichment (Clark and Frid 2001; Kröncke and Bergfeld 2001). Alternatively, the approach used here did not select for the appropriate characteristics of animals that would make them specifically vulnerable to a physical fishing impact. For example, some of the short-term experimental studies have suggested potential size selectivity effects on the proportion of populations killed by trawls (Bergman and van Santbrink 2000; Duplisea et al. 2002).

The application of an objective-based ecosystem approach to environmental management throws up a number of challenges to science. Among these are the assessment of appropriate pre-fishing conditions and the development of metrics that can establish causality between an activity and its disturbance. The fossil record holds limited promise in providing a pre-fishing state of the benthos as a whole, partly due to the underrepresentation of many of the soft-bodied taxa in the preserved historic record. However, there are a number of available data sets from the late 19th and early 20th centuries, a period over which fishing was developing into a fully commercialized industry in the North Sea (Schäfer 1972; Reise 1982; Riesen and Reise 1982; Reise and Schubert 1987; Kröncke 1990; Frid et al. 2000; Rumohr and Kujawski 2000). It seems timely that a meta-analysis of all species extirpations be undertaken to more thoroughly evaluate whether there is potential to establish causality between fishing and changes in benthic communities based on the vulnerability associated with the ecology, morphology, and life history characteristics of those species.

Acknowledgments

The authors would like to thank Julie Bremner, Odette Paramor, and Catherine Scott for valuable contributions to the thinking behind this paper. The comments of two anonymous referees and the editors were also gratefully received and incorporated in the final version. This work was supported financially by the University of Newcastle upon Tyne, UK. Presentation of the work was supported by a conference grant to LAR from the Ecological Society of America and American Fisheries Society.

References

Barnes, N., and C. L. J. Frid. 1999. Restoring shores impacted by colliery spoil dumping. Aquatic Conservation: Marine and Freshwater Ecosystems 9:75–82.

Bergman, M. J. N., and M. Hup. 1992. Direct effects of beam trawling on macrofauna in a sandy sediment in the southern North Sea. ICES Journal of Marine Science 49:5–11.

Bergman, M. J. N., and J. W. van Santbrink. 2000. Fishing mortality of populations of megafauna in sandy sediments. Pages 49–68 in M. J. Kaiser and S. J. de Groot, editors. The effects of fishing on non-target species and habitats. Biological, conservation and socio-economic issues. Blackwell Science, Oxford, UK.

Carlton, J. T. 1993. Neoextincions of marine invertebrates. American Zoologist 33(6):499–509.

Carlton, J. T., G. J. Vermeij, D. R. Lindberg, D. A. Carlton, and E. C. Dubley. 1991. The first historical extinction of a marine invertebrate in an oceanic basin: the demise of the eelgrass limpet *Lottia alveus*. Biological Bulletin 180:72–80.

Christiansen, M. E. 1969. Crustacea—Decapoda, Brachyura. Marine invertebrates of Scandinavia (2). Uiniversitetsforlaget, Oslo, Norway.

Clark, K. B. 1994. Ascoglossan (=Sacoglossa) molluscs in the Florida Keys: rare marine invertebrates at special rish. Bulletin of Marine Science 54:900–916.

Clark, R. A., and C. L. J. Frid. 2001. Long-term changes in the North Sea ecosystem. Environmental Reviews 9:131–187.

Collie, J. S. 1998. Studies in New England of fishing gear impacts on the sea floor. Pages 53–62 in E. M. Dorsey and J. Pederson, editors. Effects of fishing gear on the sea floor of New England. Conservation Law Foundation, Boston.

Collie, J. S., S. J. Hall, M. J. Kaiser, and I. R. Poiner. 2000. A quantitative analysis of fishing impacts on shelf-sea benthos. Journal of Animal Ecology 69:785–798.

Davis, S. J. M. 1995. The archaeology of animals, 2nd edition. Routledge, London.

Dayton, P. K., M. J. Tegner, P. B. Edwards, and K. L. Riser. 1998. Sliding baselines, ghosts, and reduced expectations in kelp forest communities. Ecological Applications 8(2):309–322.

Dayton, P. K., S. F. Thrush, M. T. Agardy, and R. J. Hofman. 1995. Viewpoint. Environmental effects of marine fishing. Aquatic Conservation: Marine and Freshwater Ecosystems 5:205–232.

DTI Strategic Environmental Assessment of the North Sea. 2002. SEA 2 and 3, technical report. CEFAS, Lowestoft, UK.

Duplisea, D. E., S. Jennings, K. J. Warr, and T. Dinmore. 2002. A size-based model of the impacts of bottom trawling on benthic community structure. Canadian Journal of Fisheries and Aquatic Sciences 59:1785–1795.

Frid, C. L. J., S. Hansson, S. A. Ragnarsson, A. Rijnsdorp, and S. A. Steingrimsson. 1999. Changing levels of predation on benthos as a result of exploitation of fish populations. Ambio 28:578–582.

Frid, C. L. J., K. G. Harwood, S. J. Hall, and J. A. Hall. 2000. Long-term changes in the benthic communities in North Sea fishing grounds. ICES Journal of Marine Science 57(4):1303–1309.

Graham, M. H., P. K. Dayton, and J. M. Erlandson. 2003. Ice ages and ecological transitions on temperate coasts. Trends in Ecology and Evolution 18(1):33–40.

Hall, S. J. 1999. The effects of fishing on marine ecosystems and communities. Blackwell Science, Oxford, UK.

Haywood, P. J., and J. S. Ryland, editors. 1995. Handbook of the marine fauna of northwest Europe. Oxford University Press, Oxford, UK.

Hoffman, E., and P. Dolmer. 2000. Effect of closed areas on distribution of fish and epifauna. ICES Journal of Marine Science 57:1310–1314.

Howson, C. M., and B. E. Picton, editors. 1997. The species directory of the marine flora and fauna of the British Isles and surrounding seas. Ulster Museum and the Marine Conservation Society, Belfast and Ross-on-Wye, UK.

ICES (International Council for the Exploration of the Seas). 2000. Report of the working group on ecosystem effects of fishing activities. ICES, CM 2000/ACE: 2, Copenhagen.

ICES (International Council for the Exploration of the Seas). 2004. Report of the Working Group on Ecosystem Effects of Fishing Activities. ICES Advisory Committee on Ecosystems CM 2004/ACE:03, Ref D, E, G, Copenhagen.

Ingle, R. 1997. Crayfishes, lobsters and crabs of Europe. Chapman and Hall, London.

Jablonski, D. 2001. Lessons from the past: evolutionary impacts of mass extinctions. Pages 5393–5398 in Colloquium of the National Academy of Sciences, The Future of Evolution, Irvine, California.

Jackson, J. B. C. 2001. What was natural in the coastal oceans? Pages 5411–5418 in Colloquium of the National Academy of Sciences, The Future of Evolution, Irvine, California.

Jackson, J. B. C., M. X. Kirby, W. H. Berger, K. A. Bjorndal, L. W. Botsford, B. J. Bourque, R. H. Bradbury, R. Cooke, J. Erlandson, J. A. Estes, T. P. Hughes, S. Kidwell, C. B. Lange, H. S. Lenihan, J. M. Pandolfi, C. H. Peterson, O. R. Bjørnstad, and B. T. Grenfell. 2001. Historical overfishing and the recent collapse of coastal ecosystems. Science 293:638–643.

Jennings, S., and M. J. Kaiser. 1998. The effects of fishing on marine ecosystems. Advances in Marine Biology 34:203–314.

Johnson, L. J. and C. L. J. Frid. 1995. The recovery of benthic communities along the County Durham coast after cessation of colliery spoil dumping. Marine Pollution Bulletin 30:215–220.

Kaiser, M. J., and S. J. de Groot, editors. 2000. Effects of fishing on non-target species and habitats. Blackwell Scientific, Oxford, UK.

Kaiser, M. J., and B. E. Spencer. 1994. Fish scavenging behaviour in recently trawled areas. Marine Ecology Progress Series 112:41–49.

Kaiser, M. J., and B. E. Spencer. 1995. Survival of by-catch from a beam trawl. Marine Ecology Progress Series 126:31–38.

Kröncke, I. 1990. Macrofauna standing stock of the Dogger Bank. A comparison: II. 1951–1952 versus 1985–1987. Are changes in the community of the northeastern part of the Dogger Bank due to environmental changes? Netherlands Journal of Sea Research 25(1/2):189–198.

Kröncke, I. 1995. Long-term changes in North Sea benthos. Senckenburgiana Maritima 26(1/2):73–80.

Kröncke, I., and C. Bergfeld. 2001. Review of the current knowledge on North Sea benthos. Synthesis and new conception of North Sea research (SYCON), number 12. Zentrum für Meeres- und Klimaforschung der Universität Hamburg, Hamburg University, Hamburg, Germany.

Kühne, S., and E. Rachor. 1996. The macrofauna of a stony sand area in the German Bight (North Sea). Helgoländer Meeresuntersuchungen 50:433–452.

Künitzer, A., D. Basford, J. A. Craeymeersch, J.-M. Dewarumez, J. Dorjes, G. C. A. Duineveld, A. Eleftheriou, C. Heip, P. Herman, P. Kingston, U. Niermann, E. Rachor, H. Rumohr, P. A. W. J. de Wilde. 1992. The benthic infauna of the North Sea: species distribution and assemblages. ICES Journal of Marine Science 49(2):127–143.

Lindeboom, H. J., and S. J. de Groot. 1998. Impact II. The effects of different types of fisheries on the North Sea and Irish Sea benthic ecosystems. NIOZ Rapport 1998-1, Netherlands Institute for Sea Research, Texel, The Netherlands.

MacDonald, D. S., M. Little, N. C. Eno, and K. Hiscock. 1996. Disturbance of benthic species by fishing activities: a sensitivity index. Aquatic Conservation: Marine and Freshwater Ecosystems 6(4):257–268.

Mayer, L. M., D. F. Schink, R. H. Findlay, and D. L. Rice. 1991. Effects of commercial dragging on sedimentary organic matter. Marine Environmental Research 31(4):249–261.

Ngoc-Ho, N. 2003. European and Mediterranean Thalassinidea (Crustacea, Decapoda). Zoosystema 25:439–555.

North Sea Task Force. 1992. North Sea quality status report for 1992. North Sea Task Force, Copenhagen.

Norton, P. E. P. 1978. The history and future of the North Sea benthos. Proceedings of the Royal Society of Edinburgh 76B:193–200.

Pearce, J. 1999. Historic reconstruction of ecological effects. Marine Pollution Bulletin 38(4):233–234.

Ramsay, K., M. Bergmann, L. O. Veale, C. A. Richardson, M. J. Kaiser, S. J. Vize, and S. W. Feist. 2001. Damage, autotomy and arm regeneration in starfish caught by towed demersal fishing gears. Marine Biology 138:527–536.

Reise, K. 1982. Long-term changes in the macrobenthic invertebrate fauna of the Wadden Sea: are Polychaetes about to take over? Netherlands Journal of Sea Research 16:29–36.

Reise, K., and A. Schubert. 1987. Macrobenthic turnover in the subtidal Wadden Sea: the Norderaue revisited after 60 years. Helgoländer Meeresunters 41:69–82.

Riesen, W., and K. Reise. 1982. Macrobenthos of the subtidal Wadden Sea: revisited after 55 years. Helgoländer Meeresunters 35:409–423.

Rijnsdorp, A. D., A. M. Buys, F. Storbeck, and E. G. Visser. 1998. Micro-scale distribution of beam trawl effort in the southern North Sea between 1993 and 1996 in relation to the trawling frequency of the sea bed and the impact on benthic organisms. ICES Journal of Marine Science 55:403–419.

Roberts, C. M., and J. P. Hawkins. 1999. Extinction risk in the sea. Trends in Ecology and Evolution 14:241–246.

Rogers, S. I., D. Maxwell, A. D. Rijnsdorp, U. Damm, and W. Vanhee. 1999. Fishing effects in northeast Atlantic shelf seas: patterns in fishing effort, diversity and community structure. IV. Can comparisons of species diversity be used to assess human impacts on demersal fish faunas? Fisheries Research 40:135–152.

Roy, K., D. Jablonski, and J. W. Valentine. 2001. Climate change, species range limits and body size in marine bivalves. Ecology Letters 4:366–370.

Rumohr, H., and T. Kujawski. 2000. The impact of trawl fishery on the epifauna of the southern North Sea. ICES Journal of Marine Science 57:1389–1394.

Safina, C. 1998. Renewing the world's fisheries: why the next century must be one of care and renewal of ocean fisheries. People and the Planet 7(2):10–13.

Schäfer, W. 1972. Ecology and palaeoecology of marine environments, 2nd edition. Oliver and Boyd, Edinburgh, UK.

Skjoldal, H. R., S. vanGool, H. Offringa, C. vanDam, J. Water, E. Degre, J. Bastinck, J. Pawlak, H. Lassen, M. Svelle, H.-G. Nilsen, and H. Lorentzen. 1999. Workshop on ecological quality objectives (EcoQOs) for the North Sea. TemaNord, The Hague, The Netherlands.

Squires, H. J. 1965. Decapod crustaceans of Newfoundland, Labrador and the Canadian Eastern Arctic. Fisheries Research Board of Canada, Manuscript Report Series (Biological) Number 10, Ottawa.

Steneck, R. S., M. J. Tegner, and R. R. Warner. 2001. Historical overfishing and the recent collapse of coastal ecosystems. Science 293:629–638.

Strain, P. M., and R. W. Macdonald. 2002. Design and implementation of a programme to monitor ocean health. Ocean and Coastal Management 45(6–7):325–355.

Tegner, M. J., L. V. Basch, and P. K. Dayton. 1996. Near extinction of an exploited marine invertebrate. Trends in Ecology and Evolution 11(7):278–280.

Thrush, S. F., and P. K. Dayton. 2002. Disturbance to marine benthic habitats by trawling and dredging: implications for marine biodiversity. Annual Review of Ecological Systems 33:449–473.

Thrush, S. F., J. E. Hewitt, V. J. Cummings, and P. K. Dayton. 1995. The impact of habitats disturbance by scallop dredging on marine benthic communities: what can be predicted from the results of experiments? Marine Ecology Progress Series 129(1–3):141–150.

Tuck, I. D., S. J. Hall, M. R. Robertson, E. Armstrong, and D. J. Basford. 1998. Effects of physical trawling disturbance in a previously unfished sheltered Scottish Sea loch. Marine Ecology Progress Series 162:227–242.

Underwood, A. J. 1992. Beyond BACI: the detection of environmental impacts on populations in the real, but variable, world. Journal of Experimental Marine Biology and Ecology 161:145–178.

Underwood, A. J. 1994. On beyond BACI: sampling designs that might reliably detect environmental disturbances. Ecological Applications 4:3–15.

Watling, L., and E. A. Norse. 1998. Disturbance of the seabed by mobile fishing gear: a comparison to forest clearcutting. Conservation Biology 12(6):1180–1197.

Symposium Abstract

Symposium abstracts have been reprinted from the Abstract Volume prepared for the Symposium on Effects of Fishing Activities on Benthic Habitats: Linking Geology, Biology, Socioeconomics, and Management without any further review or editing.

Using Side-Scan Sonar to Assess the Impact and Persistence of Natural and Anthropogenic Disturbance to Low-Relief Oyster Habitats in Coastal Louisiana

Y. ALLEN,[1] C. WILSON, H. ROBERTS, AND J. SUPAN

Oyster Geophysics Program, Department of Oceanography and Coastal Sciences, SC&E, Louisiana State University, Baton Rouge, Louisiana

Traditional methods used to assess oyster reef distribution and condition are only able to provide subjective point information, which is often poorly georeferenced. Maps of oyster habitat in shallow waters are therefore typically extremely generalized, giving few details about the true distribution, character and dynamics of reefs. Sidescan sonar offers a significant advantage for oyster reef assessment in the turbid waters of coastal Louisiana. We used sidescan sonar in ultra-shallow (<2m) waters to completely image over 19,000 ha in Louisiana estuaries in advance of an impending freshwater diversion project. We also conducted four years of intense annual surveys in a more restricted area (320 ha) with a diversity of reef types and culture intensity to examine natural and anthropogenic impacts on oyster reef extent and character. Our intensive surveys identified older stable reefs that had not been actively worked. Shell abundance and structure on these reefs were high, but oyster meat productivity was low. Areas of intense oyster culture were characterized by low relief reefs that frequently showed distinct evidence of scarring from dredging and other anthropogenic sources. Smaller scars caused by oyster dredging typically healed through the within time period of our study while larger anthropogenic scarring did not diminish over the four years. We also deployed the sonar towfish over an area immediately before and after both seeding and harvesting to establish a quantitative relationship with sonar reflectance. These relationships can be further used to predict the impact of harvesting and seeding on the extent oyster habitat.

---

[1] E-mail: allenyc@lsu.edu

Survey of Fishing Gear and Fiber Optics Cable Impacts to Benthic Habitats in the Olympic Coast National Marine Sanctuary

M. S. BRANCATO[1] AND C. E. BOWLBY

Olympic Coast National Marine Sanctuary, Port Angeles, Washington

In September 2000 the Olympic Coast National Marine Sanctuary (Sanctuary) initiated a long-term monitoring program designed to assess impacts to the seafloor and the benthic communities from different intensities of commercial bottom trawling and the placement of two fiber optics cables on the seafloor in the Sanctuary. Survey sites were selected based on side scan and bathymetry data and bottom trawling records from Washington Department of Fish and Wildlife, Oregon Department of Fish and Wildlife, National Marine Fisheries Service and vessel traffic tracking information collected by the Sanctuary. We conducted our first two years of monitoring using the Delta submersible equipped with underwater cameras, box core and a benthic suction device (slurp gun). In addition, a shipboard bottom grab was used to collect bottom samples. Four distinct habitat types were monitored along low and high intensities of bottom trawling both along the buried cable route and parallel to the route. The underwater surveys were conducted at depths of 120 to 330 meters along silt/clay, sand, gravel/cobble or boulder with mixed sediments. Physical, chemical, and biological parameters were monitored.

---

[1] E-mail: mary.sue.brancato@noaa.gov

Shrimp and Crab Trawling Impacts on Estuarine Soft-Bottom Organisms

L. B. Cahoon[1]
Department of Biological Sciences, UNC Wilmington, Wilmington, North Carolina

M. H. Posey
Center for Marine Science, UNC Wilmington, Wilmington, North Carolina

W. H. Daniels
02 Tate Road, Belhaven, North Carolina

T. D. Alphin
Center for Marine Science, UNC Wilmington, Wilmington, North Carolina

This project addressed some possible impacts of trawling for crabs and shrimp in North Carolina estuaries on populations of organisms associated with soft-bottom habitats. The organisms of interest included benthic microalgae, demersal zooplankton, and macrobenthic infauna, encompassing the lower trophic levels in the benthic food chain and the essential trophic coupling that supports estuarine fishery production. The approaches used in this project included sampling before and after experimental trawling at several estuarine locations, sampling in areas actively trawled and areas closed to trawling, and sampling during several seasons over two years to address seasonal and inter-annual effects. Sampling began in February, 1999, and ended in November, 2000 at six locations in the Pamlico River Estuary. Experimental trawling had no significant effect on the biomass of benthic microalgae, no consistent effect on the abundance of demersal zooplankton, and only a slight but non-significant effect on the abundances of benthic macrofaunal animals. Benthic microalgae were significantly more abundant in untrawled locations than in trawled locations, with strong seasonal variation as well. Abundances of demersal zooplankton were not significantly or consistently different between untrawled and trawled locations. There were higher abundances of benthic macrofauna in trawled locations than at untrawled locations, but only at certain times of the year. Species dominance was fairly consistent between trawled and untrawled areas, with only a few exceptions. While inter-annual variation and substrate did have an effect, seasonal variation was far stronger and seemed to have an overriding effect. We conclude that direct, negative impacts of trawling activity on these soft-bottom organisms are small relative to other sources of population variability. The soft-bottom communities we studied experience considerable natural disturbance in these broad, shallow estuarine ecosystems. Although trawling per se does not seem to have a consistent effect on estuarine soft-bottom benthos, there are interesting differences between trawled and untrawled habitats that merit further investigation.

[1] E-mail: Cahoon@uncwil.edu

Did Bottom Trawling in Bristol Bay's Red King Crab Brood-Stock Refuge Contribute to the Collapse of Alaska's Most Valuable Fishery?[1]

C. Braxton Dew[2] and Robert A. McConnaughey[3]

National Marine Fisheries Service, Alaska Fisheries Science Center, P.O. Box 15700, Building 4, Seattle, WA 98115-6349.

The 1976 U.S. Magnuson–Stevens Fishery Conservation and Management Act effectively eliminated the no-trawl zone known as the Bristol Bay Pot Sanctuary, located in the southeastern Bering Sea, Alaska. Implemented by the Japanese in 1959, the boundaries of the Pot Sanctuary closely matched the well-defined distribution of the mature-female brood stock population of red king crab *Paralithodes camtschaticus*, affording a measure of protection to the reproductive potential of the stock. In 1980, the point at which the commercial harvest of Bristol Bay legal-male red king crab reached an all-time high after a decade-long increase, domestic bottom trawling in the broodstock sanctuary began in earnest with the advent of a U.S.–Soviet joint-venture fishery for yellowfin sole *Limanda aspera*. In the first year of trawling in the Pot Sanctuary, the Bering Sea–Aleutian Islands (BSAI) red king crab bycatch increased by 371% over the 1977–1979 average; in 1981, the BSAI bycatch increased another 235% over that of 1980, most of which were mature females. As the number of unmonitored domestic trawls in the broodstock area increased rapidly after 1979 and anecdotal reports of "red bags" (trawl cod-ends plugged with red king crab) began to circulate, the proportion of males in the mature population (0.25 in 1981 and 0.16 in 1982) jumped to 0.54 in 1985 and 0.65 in 1986. It is unlikely that normal demographics caused this sudden reversal in sex ratio. Our hypothesis is that sequential, sex-specific sources of fishing mortality were at work. Initially, there were 10 years (1970–1980) of increasing, male-only exploitation in the directed pot fishery, followed by a drastic reduction in the male harvest after 1980 (to zero in 1983). Then, beginning around 1980, there was an increase in bottom trawling among the highly aggregated, sexually mature female brood stock near the western end of the Alaska Peninsula, an area documented by previous investigators to be the most productive spawning, incubation, and hatching ground for Bristol Bay red king crab. There has been considerable discussion about possible natural causes (e. g., meteoro-logical regime shifts, increased groundfish predation, epizootic diseases) of the abrupt collapse of the Bristol Bay red king crab population in the early 1980s. Our discussion focuses on the association between record harvests of male crabs in the directed fishery, the onset of large-scale commercial trawling within the population's primary reproductive refuge, and the population's collapse.

[1] To be published in Ecological Applications, 2005.
[2] E-mail: braxton.dew@noaa.gov; new address: 3233 Bayview Drive, Kodiak, AK 99615, USA.
[3] E-mail: bob.mcconnaughey@noaa.gov

Comparative Evaluation of Natural and Trawling Sediment Disturbance via Short-Lived Radionuclides, in situ Monitors and Remote Sensing Techniques in the Pamlico River Estuary, North Carolina

J. E. Frank[1] and D. R. Corbett

Department of Geology, East Carolina University, Greenville, North Carolina

T. West and L. Clough

Department of Biology, East Carolina University, Greenville, North Carolina

W. Calfee

Coastal Resource Management Program, East Carolina University, Greenville, North Carolina

Seabed disturbance by bottom trawling has emerged as a major concern related to the conservation of essential fish habitat and water quality. Bottom sediments directly affect water quality by releasing nutrients when freshly deposited organic matter is remineralized. Resuspension and subsequent transport of bottom sediments disturbed by natural physical mixing (e.g. wind) of overlying waters or anthropogenic interactions (i.e. trawling) results in the advective release of dissolved constituents (NH4, NO3-NO2, PO4) from interstitial waters into overlying surface waters. Our study attempts to delineate natural resuspension and transport of surface sediments from trawling disturbances in South Creek, a shallow tributary of the Pamilco River, North Carolina. Our study site encompasses two similar areas, both containing a trawled and untrawled region (~100,000 m^2 per region). Within each region, concentrations of total suspended solids, dissolved nutrients and surface sediment inventories of ^{234}Th and ^{7}Be were quantified several days before and after a controlled trawling event. In addition, meteorological information (wind speed, direction, etc.) was collected in close proximity to the study site. Our first set of experiments, July and October 2001, suggest that trawling plays a minor role in sediment resuspension relative to natural wind events. Work to be conducted during summer 2002 will incorporate satellite imagery (AVHRR SeaWiFS) and *in situ* monitoring devices (current velocity, CTD, turbidity) to further constrain the importance of natural vs. trawling induced resuspension. We hope our techniques will provide the basis for operational monitoring, and provide 'real-time' information to resource managers.

---

[1] E-mail: jef0926@mail.ecu.edu

Fishing and Environmental Disturbance Indicators in a Shrimp Fishing Ground at the Mexican Central Pacific

E. Godinez-Dominguez[1]

Centro de Ecologia Costera, Universidad de Guadalajara, Jalisco, México; and Departamento de Biologia Animal, Biologia Vegetal y Ecologia, Universidad de A Coruña, A Coruña, España

J. Freire

Departamento de Biologia Animal, Biologia Vegetal y Ecologia, Universidad de A Coruña, A Coruña, España

G. González-Sansón

Centro de Investigaciones Marinas, Universidad de la Habana, La Habana, Cuba

This paper examines the concurrent effects induced by trawl shrimp fisheries, natural seasonal dynamics and interannual processes as ENSO events on a soft bottom macroinvertebrate community. Short-term effects were evaluated during an initial period of two years when five trawl cruises were carried out in successive closed and open fishing seasons coinciding with the main hydroclimatic periods. In each cruise seven sites along 100 km of coastline were selected and four depths were sampled (20, 40, 60 and 80 m). A series of community structural descriptors used frequently to determine the ecological effects of fishing disturbances were employed: ABC curves, W-statistic, normalized species size distribution as biomass spectra, spatial segregation index, Shannon diversity index, species richness and biomass. Inter-annual effects were analysed with data from semi-monthly cruises in 2 sites and the same four depths from 1995 to 1998. Theoretical predictions of the effects of fishing in the behaviour of the statistical indices used were tested. Results show a strong evidence that fishing has produced a state of chronic disturbance in the macroinvertebrate community. Short-term fishing effects could be masked by natural seasonal and interannual environmental changes. Results of short-term effects are not in agreement with the fishing disturbance theories. The trends found could evidence interannual effects associated to El Ni-o and La Ni-a events. The complexity of the sources of variability in a exploited community forces managers to adopt a more widely adaptive approach which should be focused on understanding the community structural process through temporal and spatial gradients, and to use several structural indices to evaluate critically their performance as indicators of fishing disturbance.

---

[1] E-mail: egodinez@mail2.udc.es

Benthic Perturbations from Walrus Foraging: Are They Similar to Trawling?

C. V. Jay[1]

U.S. Geological Survey, Alaska Science Center, Anchorage, Alaska

L. C. Huff

NOAA-University of New Hampshire, Joint Hydrographic Center, Chase Ocean Engineering Lab, Durham, New Hampshire

R. A. McConnaughey

National Marine Fisheries Service, Alaska Fisheries Science Center, Seattle, Washington

The Pacific walrus roots through soft sediment and feeds on a wide variety of benthic organisms, and in the process, they resuspend sediment and disturb much of the fauna in their path. Some of the consequences of walrus foraging may be similar to those produced from bottom trawling. In a preliminary study, we are using side-scan sonar images to identify walrus foraging tracks (furrows) in soft sediments to measure the distribution of foraging patches and the level of foraging effort of walruses in an area within Bristol Bay, Alaska. The approximate area affected and level of mechanical disturbance to the sea floor from walrus foraging are contrasted to those that might be expected to be produced from bottom trawling.

---

[1] E-mail: chad_jay@usgs.gov

Controversy about Trawling and Santa Maria Key's Causeway Effects on Seagrass

A. Quirós Espinosa,[1] M. E. Perdomo López, and R. Arias Barreto

Centro de Estudios y Servicios Ambientales; Ministerio de Ciencia, Tecnologia y Medio Ambiente, Santa Clara, Villa Clara, Cuba

In the winter of 1989, building of a causeway to Santa Maria key was started, 48 kilometers from the coast, in the central region of Cuba. Before the initial actions, the hydrological influence zones of this road were determined, and some ecological aspects were included in the executive project. At the same time, a monitoring program was started, based on an initial assessment of the phytobenthos (biomass, diversity and functional groups). The monitoring took place in the springs of 1990, 1994 and 2002. Different sampling stations were located out of the causeway hydrological influences. In 1995, fishing activities began, using big trawls. The fishing zones are located in areas not influenced by the causeway, permitting comparison between the effects of fishing and the causeway on the phytobentos, mainly on seagrass. Between 1990 and 1994 no significant changes were observed in the control stations of the causeway monitoring. A different situation was observed in 2002: the seagrass density decreased significantly and the algae diversity increased considerably as consequence of a lower spatial competition by Thalassia testudinum. A graphic model of the phenomenon was presented, including the effect on fishes, other fauna, vegetation, transparency, photosynthesis and sediment retention, that shows a case of positive feedback.

[1] E-mail: cisam@civc.inf.cu

Scaling of Natural and Anthropogenic Disturbance on the New York Bight Shelf: Implications for Tilefish Communities of the Shallow Continental Slope

M. C. Sullivan[1] and R. K. Cowen

Rosenstiel School of Marine and Atmospheric Science, Department of Marine Biology and Fisheries, University of Miami, Miami, Florida

K. W. Able

Marine Field Station, Institute of Marine and Coastal Sciences, Rutgers University, Tuckerton, New Jersey

M. P. Fahay

National Marine Fisheries Service, Northeast Fisheries Science Center, Sandy Hook Laboratory, Highlands, New Jersey

Trawling is a pervasive feature of continental shelf environments worldwide. However, commonalities within individual systems and among habitat types and species are rarely clear cut. Our recent work in the New York Bight apex examined the potential impact of mobile fishing gear within a storm-dominated shelf system. Over medium-grain sand sites, scallop dredging was shown to have minimal short-term effects (hours-day) on the abundance of a common, juvenile flatfish (Limanda ferruginea) and its benthic prey (gammarid amphipods, cumaceans). Longer-term impact signatures (months-year), however, were completely obscured by intense physical forcing during the fall (hurricanes) and winter (northeasters) months. Clearly, the intersection between habitat type and prevailing physical regime plays a critical role in defining the susceptibility of marine ecosystems to anthropogenic stress. Thus, an area of immediate concern is the shallow continental slope where natural disturbance is minimal and chronic trawling activity disproportionately high. The tilefish, Lopholatilus chamaeleonticeps, a long-lived, benthic excavator, plays a fundamental role in creating heterogeneity in these low variability habitats and has been compared to other 'ecosystem engineers' which alter bottom types favorably for members of lower trophic levels. Tilefish and their commensal associates (i.e. Helicolenus dactylopterus, Anthias spp.) appear particularly vulnerable to significant direct (as bycatch) and indirect (habitat alteration) impacts by mobile bottom fishing gear. Ongoing work is investigating the consequences of chronic trawling, as well as individual trawling events, on tilefish communities in these remote underwater habitats.

[1] E-mail: msullivan@rsmas.miami.edu

Impacts of Trawling and Wind Disturbance on Water Column Processes in the Pamlico River Estuary, North Carolina

T. L. WEST[1]

Department of Biology, East Carolina University, Greenville, North Carolina

D. R. CORBETT

Department of Geology, East Carolina University, Greenville, North Carolina

L. M. CLOUGH

Department of Biology, East Carolina University, Greenville, North Carolina

M. W. CALFEE

Coastal Resource Management Program, East Carolina University, Greenville, North Carolina

J. E. FRANK

Department of Geology, East Carolina University, Greenville, North Carolina

Comparatively little attention has been given to the indirect impacts of bottom trawling in soft bottom ecosystems. Our study assesses the relative effects of bottom trawling and wind disturbance on water column nutrient (NH_4, NO_3-NO_2, PO_4) loading and primary productivity (chlorophyll a concentration, bacterial abundance) in concert with measurements of sediment resuspension and transport (Frank, et al.). We are working in South Creek, a subtributary of the Pamlico River Estuary, NC (USA) that has been closed to trawling for 15 years, and is characterized by large surface area, shallow depth, and wind-driven tides. Trawling experiments in July and October 2001(repeated in 2002) were carried out during two different wind regimes: (a) days of consistent wind direction interrupted by an abrupt 180° shift in direction (July); and (b) continuously shifting wind speed and direction (October). Two replicate areas, each containing a trawled and an untrawled site ~100,000 m^2, were sampled for 4 days prior to, and immediately after a trawling event. We were unable to detect a significant difference in any water column parameter (oxygen, nutrients or chlorophyll a) that could be attributed exclusively to trawling. Instead, changes appear to be driven by a variable wind regime. Interestingly, measures of photosynthetic active radiation indicate that 60-80% of incident light is absorbed within the upper 0.5m of the water column. These findings imply that South Creek and physically similar regions of the Pamlico River Estuary are light-limited systems in which wind-forced mixing events are primarily responsible for sediment resuspension from the benthos.

---

[1] E-mail: west@mail.ecu.edu

Extrapolation of Local and Chronic Effects of Fishing and Non-Fishing Events to Significant Regions and Time Scales

Spatial and Temporal Scales of Disturbance to the Seafloor: A Generalized Framework for Active Habitat Management

SIMON F. THRUSH,[1] CAROLYN J. LUNDQUIST, AND JUDI E. HEWITT

*National Institute of Water and Atmospheric Research,
Post Office Box 11-115, Hamilton, New Zealand*

Abstract. The direct effects of marine habitat disturbance by commercial fishing have been well documented. However, the potential ramifications to the ecological function of seafloor communities and ecosystems have yet to be considered. Ecological research has demonstrated that natural disturbance processes play an important functional role in seafloor ecosystems by affecting spatial heterogeneity. When the space and time scales of human disturbance are greater than those the natural ecosystems are adapted to, changes in community structure and function are inevitable. Changes restricting the size, density, and distribution of organisms can lead to functional extinction and threaten resilience at all levels of biological diversity. This is particularly true in soft-sediment ecosystems, where the organisms create much of the heterogeneity within the habitat and also play crucial roles in many processes. Thus, there is a need to develop a scientific framework for the management of seafloor habitats, focusing on sustaining fisheries and maintaining biodiversity. Simple heuristic models can indicate disturbance regimes that, through their frequency, extent, or intensity, could result in catastrophic changes across the seafloor landscape . Our model implies that when disturbance is infrequent relative to recovery time and only a small proportion of the landscape is affected, the system is stable, but when the disturbance frequency is shorter than the recovery time and/or a large proportion of the system is disturbed, the system may flip into an altered state. Once features have been lost, it may not be a simple matter of reducing the disturbance regime to ensure their recovery. Even such a simple model emphasizes the need to understand the scales of mobility and the processes affecting recovery. We need to carefully and explicitly consider the implications of alterations of these ecosystems; they may not only reflect loss of conservation and natural heritage values but also loss of opportunity and the ecosystem services provided by diverse and heterogeneous seafloor ecologies.

Introduction

Over the last 20 years, fishing industries have begun to be subjected to the same degree of environmental scrutiny as many other industries. Studies of the direct effects of disturbance to seafloor habitats and communities have indicated the potential for profound effects (see reviews by Dayton et al. 1995; Jennings and Kaiser 1998; Watling and Norse 1998; Auster and Langton 1999; Hall 1999; Kaiser and de Groot 2000). Despite the often significant challenges in conducting these impact studies, some consistent responses to habitat disturbance have emerged across broad spatial scales (Thrush et al. 1998; Kaiser et al. 2000; McConnaughey et al. 2000; Cryer et al. 2002). Common changes in seafloor communities that occur across a variety of habitat types include reduced habitat structure, lower diversity, and the loss of large and long-lived sedentary species. This knowledge has clearly informed the debate over fisheries management and marine conservation and also raises the profile for both scientists and resource managers to understand interactions between broad-scale habitat disturbance and the functioning of benthic ecosystems.

In this paper, we develop a simple framework, based on basic ecological principles, that can help identify conditions that may lead to qualitatively different seafloor landscapes and thus help communication among scientists, managers, and stakeholders (see Done and Reichelt 1998). We use a modified version of a simple ratio-based model that was initially developed by Turner et al. (1993) to investigate the impact of disturbance at differing spatial and temporal scales on

---

[1] Corresponding author: s.thrush@niwa.cri.nz

forested landscapes. The model simplifies many of the complexities of disturbance–recovery dynamics and the potential for recovery processes to change over space or time scales in a nonlinear fashion. Nevertheless, consideration of disturbance within a space and time ratio framework provides an indication of disturbance regimes that, through their frequency, extent, or intensity, could result in catastrophic changes across the seafloor landscape.

The Value of Seafloor Habitats and Communities

Any decisions that involve opening new areas for exploitation or changing industry practice inevitably involve balancing values. There are usually important social, ethical and economic considerations, but it is also important to consider the possible threats to ecological function. By definition these functions sustain marine ecosystems and facilitate the provision of ecosystem goods and services. If these natural functions fail, many of the social and economic considerations become academic.

Soft-sediment seafloor habitats account for about 70% of the surface of the planet; they are diverse and heterogeneous across many scales of sampling. Even deep-sea basins that were once considered constant and uniform exhibit high levels of both local biogenic complexity (e.g., Jumars and Eckman 1983; Levin and Gooday 1992) and regional diversity (Levin et al. 2001). Throughout the marine environment, organisms that inhabit sediments often create much of the habitat structure, ranging from the micro-scale changes around individual animal burrows to the formation of extensive biogenic reefs. This biologically created heterogeneity and three-dimensional structure both above and below the sediment–water interface influences a number of important ecological processes.

These ecological processes include purely biological functions, such as providing refugia and juvenile habitat for many exploited populations, providing settlement sites for larvae and juveniles, and serving as a source of food for fish and other predators (e.g., Sainsbury 1988; Tupper and Boutilier 1995; Lindholm et al. 1999). Other important ecological processes occur as a result of interactions between hydrodynamics and seafloor organisms. These interactions influence sediment stability and the transport of food, larvae, sediments, and chemicals (e.g., Eckman and Nowell 1984; Frechette et al. 1989; Wildish and Khistmanson 1997). Typically, animals increase the particle exchange between water and sediment by a factor of 2–10 (Graf 1999). The density and spatial arrangement of organisms are important in determining the magnitude of the effect (e.g., Green et al. 1998; Nikora et al. 2002).

Below the surface of the sediment, animals continue to play important ecological roles by influencing biogeochemical processes such as the transformation and exchange process of organic matter and nutrients (Herman et al. 1999). For example, about one-third to half of the nutrients that fuel primary production in the sea above continental shelf environments may be derived from the sediment (Pilskaln et al. 1998). Continental shelf sediments, while occupying only 7% of the marine environment, are responsible for 52% of the global organic matter mineralization (Middleburg et al. 1997). Similar to organism–hydrodynamics interactions, larger organisms appear to be particularly important in influencing these biogeochemical processes (Thayer 1983; Sandnes et al. 2000).

These results highlight the fundamental ecological value of seafloor communities in terms of biodiversity and ecosystem process. The current threat to benthic communities from habitat disturbance by fishing is that a decrease not only in the density of individual organisms but also in the size of individuals and their spatial extent may mean we risk losing important ecological processes; that is, fishing disturbance may result in organisms becoming functionally extinct (Dayton et al. 1998).

The Disturbance Ecology of Seafloor Communities

Basic ecological research on marine benthic communities can provide some useful insights into disturbance. Many types of processes disturb seafloor ecosystems, ranging in size from the small bite marks in the sediment created by feeding fish to processes that impact larger areas, such as storms or volcanic eruptions. Spatial heterogeneity in community structure is related to the spatial extent and/or the frequency of disturbance events; in order for disturbance to create patchiness, it must be small relative to the colonization potential of the benthic community but not so small as to enable the adjacent assemblage to quickly infill the disturbed patch. The key to predicting the effect lies in the scale of disturbance in space and time and the scales of ecological response to that disturbance (recovery or succession). Empirical studies of small-scale disturbance processes cannot simply be directly scaled up to indicate the effects of broader-scale disturbances because the relationship between heterogeneity and disturbance is nonlinear (Kolasa and Rollo 1991).

Over much of the continental shelf, where most fishing is conducted, natural disturbance events covering a large area are rare. The types of large natural disturbance phenomena that potentially impact a particular area of the continental shelf are highly dependent on location. Storms rarely stir the seabed below about 60 m, although there are exceptions. Green et al. (1995), for example, reported on the impact at 25-m depth in the North Sea of a force 9 gale associated with long-period waves generated by hurricane-force winds in the Norwegian Sea. The storm resulted in a significant wave height exceeding 3 m for more than 40 h, with a maximum recorded wave height of 11.82 m. Despite the severity of the storm, the authors calculated that maximally only the top 1 cm of sediment was eroded. The distribution of iceberg scour is limited to the polar continental shelves, and hyperpycnal sediment flows across the continental shelf are also restricted to areas of extreme terrigenous sediment loading (Foster and Carter 1997; Wheatcroft 2000). Toxic algal blooms and anoxic/hypoxic events are other broad-scale disturbance phenomena that are also restricted to specific locations. The frequency of disturbance events is also important; typically in nature, the larger the event, the less frequently it will occur (Bak 1994). When the space and time scales of human disturbance are greater than those that the natural ecosystems are adapted to, then changes in community structure and function are inevitable.

Despite the problems with direct extrapolation, it is insightful to consider what we have learned from disturbance–recovery experiments in a variety of benthic habitats. Understanding recovery processes allows us to consider the potential for long-term chronic degradative changes as the cumulative effects of individual disturbance events spread across the seafloor landscape. Rates of recovery may be expected to be dependent on the spatial scale of disturbance because of the increased distance colonists need to travel to reach the disturbed area (this is particularly important for species with limited mobility) or because of interactions between the spatial scale of disturbance and habitat stability (Connell and Keough 1985; Hall et al. 1994; Thrush et al. 1996; Zajac et al. 1998). Species composition is also important, with communities dominated by large, slow-growing organisms with poor dispersal ability exhibiting slow recovery (e.g., Beukema et al. 1999; Thrush and Whitlatch 2001). Recovery may also be habitat dependent due to differences in the available pool of colonists, productivity, depth, or temperature (e.g., Pearson and Rosenberg 1978; Sousa 2001). The ecological intensity of response will also be determined by the resident species; even low-intensity disturbance can have significant effects on sensitive species (Jenkins et al. 2001).

While we have some understanding of the factors influencing recovery rates, it is important to note that there is still an appreciable amount of variability in the relative importance of the various factors that influence the recovery processes (Thrush and Whitlatch 2001; Whitlatch et al. 2001). Fishing is a large-scale disturbance and has both immediate and chronic effects. Disturbances that lead to biotic impoverishment tend to be chronic, while those that lead to enhanced diversity tend to be transient and localized (Caswell and Cohen 1991). The important point is that we should be able to gain some understanding of the effects of any disturbance to the seafloor by understanding the space and time scales over which the event occurs and how they interact with the rate of recovery of the seafloor community. This emphasizes the value of quantifying the extent, spatial distribution, and frequency with which different types of fishing gear are dropped on to or dragged over the seabed.

Scales of Fishing Disturbance

Unfortunately, data on the frequency and area of the seafloor swept by fishing gear are usually limited. The data available are often based on broad-scale fisheries management units not necessarily related to the spatial variation in seafloor habitats or biodiversity (National Academy of Sciences 2002). The spatial distribution of fishing effort on the seafloor is patchy, reflecting the relative availability of the target species. This patchiness in the distribution of fishing disturbance is a typical characteristic of ecological data, and the interpretation of such data is dependent on the scale of observation. Patchiness inevitably requires an increased sampling effort, either to generate precise estimates of the mean (by averaging over the variability) or to sample at fine resolution and characterize the spatial structure. High-resolution characterization of fishing disturbance tends to reveal fine-scale patchiness in habitat disturbance; this patchiness changes the spatial extent, frequency, and magnitude of disturbance (e.g., Rijnsdorp et al. 1998; Pitcher et al. 2000a). Different fisheries have different fishing methods and may use different types of gear that also influence the potential for and nature of disturbance to the seafloor.

Obviously, even defining the level of disturbance is a complex issue; nevertheless, it is useful to consider a few different examples to gauge the magnitude of possible effects. National Academy of Sciences (2002) reported that, on average, 56% of the New England region was swept by trawl gear more than once a year, and scallop dredges also covered a smaller but not insubstantial area. These data were based on recorded landings from 0.5° latitude × 0.5° longitude statistical blocks. Friedlander et al. (1999) recorded trawl tracks in side-scan sonar images of the seafloor to show that a typical fishery in Northern California trawled across the same section of seafloor an average of 1.5 times per year, with selected areas trawled as often as three times per year. For a fishery for scampi *Metanephrops challengeri* on the continental slope off New Zealand (200–600-m water depth), Cryer et al. (2002) used data derived from the GPS coordinates for the start and end of each trawl to calculate that an average 2,100 km^2 was swept each year by trawlers. These statistics suggested that within the study area, trawlers swept about 20% of the upper continental slope each year, although about 80% of all scampi trawls were made in an area of about 1,200 km^2. In some areas, the extent and frequency of disturbance can be even more extreme. Pitcher et al. (2000b) identifies one harbor in Hong Kong where every square meter of the seafloor is trawled three times a day. In general, it is apparent that the extent, patchiness, and frequency with which fishing gear is dragged across the seafloor are variable. To understand the ecological consequences of this disturbance, we need to consider the space and time scales of disturbance relative to the time scales of recovery of the affected benthic communities.

Managing Habitat Disturbance: The Model

Without a doubt, we need to manage habitat disturbance to the seafloor. Zoning the oceans is seen as an important tool to allow fisheries to occur embedded in functioning ecosystems (e.g., Pauly et al. 2002). Understanding the implications of different disturbance regimes on the seafloor should provide insight into how these regimes may be balanced in terms of spatial extent or frequency to maintain ecologically valuable seafloor communities. Simple heuristic models can indicate disturbance regimes that, through their frequency, extent, or intensity, could result in catastrophic changes across the seafloor landscape.

We created a simple patch model as a conceptual framework for examining the effects of fishing disturbance on benthic communities. This model is a simplistic representation of benthic community successional dynamics from a pioneering community to a mature community, created to demonstrate the effect of varying size and frequency of trawling disturbance on the proportion of mature structured habitat remaining within a landscape.

The model simplistically plays out the effects of repeated disturbance events on a seafloor habitat that is structured by certain archetypal communities. These archetypes are defined by structural attributes; they are cartoon communities defined by their slowest-growing representatives that will, finally, dominate community structure. The temporal dimension is considered by the frequency of disturbance, while the spatial dimension is considered by the ratio of size of the disturbed area to size of the habitat. The spatial dimension of the model is scale invariant, allowing for comparisons among spatial scales, while the temporal dimension is presented as a frequency (events per year) for ease of interpretation. The temporal dimension can be considered at various time scales either by changing the duration of the time step in the model or by calculating the scale-invariant ratio of the frequency of disturbance to recovery time.

The assumptions on which the model is based have important implications for the interpretation of its results. The model presents a simple representation of how the space and time scales of disturbance interact with recovery of seafloor communities. The aim is not to predict the consequences of a specific disturbance regime on a specific habitat but to provide a simple heuristic tool to facilitate greater consideration of disturbance regimes and seafloor communities in fisheries management. Models of this type make very broad generalizations about how ecosystems work. They highlight this generality and often help to identify gaps in our understanding. Key assumptions of the model include: (1) disturbance homogenizes habitat within the entire model cell and resets the cell to the pioneer stage; (2) equal colonization potential by all successional stages of all cells within the landscape; (3) equal likelihood of disturbance of all cells in the landscape regardless of disturbance history; and (4) fishing disturbance is the only disturbance operating at the scale of the model. These assumptions can be relaxed in order to give a more realistic interpretation of community dynamics; however, the simplest version is a conservative approach for assessing habitat and community response to fishing disturbance.

The model consists of a 100 × 100 cell grid, with each of 10,000 cells containing a habitat of successional stages. Successional stages last for discrete periods of time (represented in years) that reflect the average times of succession from a pioneer community immediately following a disturbance through intermediate stages to a mature final successional stage. The model commences with all cells containing a habitat of the final successional stage of a particular archetypal community; different model versions simulate different archetypal communities. Disturbances are created during each time step, after which affected cells are converted into successional stage 1, representing a pioneer community of immediate post-disturbance state. The successional stage of each non-disturbed cell is increased by 1 after each time step in the model. Disturbances are created at random locations within the model landscape to represent fishing disturbance. The spatial dimension of disturbance is calculated as the number of disturbed cells relative to the number of cells in the entire landscape. The temporal dimension of disturbance is defined as the frequency of disturbance events per time step (year).

The model simulates three different community archetypes: a fast recovering community (dominated by short-living species that characterize an infaunal community not dominated by large, deep-burrowing fauna), an intermediate community (reflecting an infaunal community dominated by larger infauna or disperse epifauna; we also use it to reflect a bivalve-dominated community that would provide hard substrate for colonization to create a biogenic reef community), and a relatively slow recovering community (such as a fully developed biogenic reef dominated by bryozoans) (Table 1). The three archetype communities thus have different numbers of successional stages reflecting increased habitat complexity. These archetype communities have recovery times of 2, 6, and 15 years, respectively. It is important to note that these archetype communities are largely based on our knowledge of shallow water coastal communities.

The model simulates a range of space and time scales of disturbance (frequency = 0–12.5 disturbances/year; spatial extent = 0–100% of the landscape disturbed). In each simulation, a total of 1,100 combinations of different space and time scales of disturbance are run for 100 time steps (years). Mean and variance of the number of model cells in each successional stage are calculated at each time step.

Model Results

Contour plots illustrate how disturbance over different space and times scales affect the persistence of the three archetypal communities (Figure 1). Contours of each community represent the maintenance of either 50% (Figure 1a) or 90% (Figure 1b) of the mature community. These plots illustrate that there is only a small proportion of the domain where disturbance can be both spatially extensive and frequent and a reasonable proportion of the mature community can be retained. Over most of the domain, high-intensity disturbance can only occur over a small spatial extent, or conversely, extensive disturbance can only occur infrequently. These constraints become more severe for the slow-to-mature archetypal communities. Note the model simply portrays the proportion of the landscape in the mature successional stage, it does not indicate that large, infrequent disturbances or small, frequent disturbances can be considered to be the same.

Another way to consider the model results is to consider the relative proportions of the landscape occupied by the different successional stages of the three archetypal communities under specific spatial and temporal disturbance regimes (Figure 2). We present average proportions of different successional stages for combinations of disturbance frequency (0.25/year, 1/year, and 4/year) and extent (4%, 25%, and 81% of the landscape) for the three archetypal communities. Figure 2 shows the ever increasing dominance of early successional stages with increasing spatial extent and frequency of disturbance. Mature stage assemblages are only eliminated at both the highest extent and frequency of disturbance (81% and 4/year) for the archetypal short-lived community that takes 2 years to reach maturity. For the intermediate community archetype that takes 6 years to reach maturity, less than 25% of the

Table 1. Community dynamics of three archetype benthic communities.

Community Type	Successional stage	Age of cell	Examples
Short-lived infauna	Pioneer	1	Small polychaetes and amphipods
	Mature infaunal community	2+	Small mollusks, ophiuriods, and large polychaetes
Intermediate community[a]	Pioneer	1	Small polychaetes and amphipods
	Colonization by mature stage dominants	2–5	Large suspension feeding bivalves, juvenile burrowing crustaceans, juvenile sponges, and large crustaceans
	Mature stage	6+	Adult bivalves and encrusting fauna (sponges, hydroids, tunicates), adult large burrowing species and other species affiliated with large burrows
Biogenic reef	Pioneer	1	Small polychaetes and amphipods
	Hard substrate colonization	2–5	Large suspension feeding bivalves, juvenile burrowing crustaceans, juvenile sponges, large crustaceans
	Colonization by mature stage dominants	6–14	Erect bryozoans and sponges colonizing hard substrate
	Mature biogenic reef	15+	Mature individuals forming complex biogenic reef

[a] Deep bioturbators or organisms creating habitat structure.

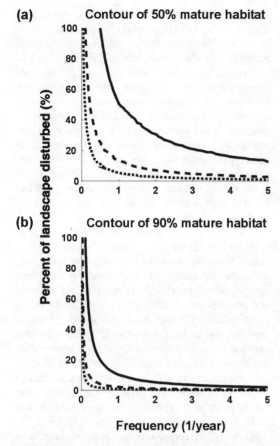

Figure 1. The effect of changing the space and time scales of disturbance on the proportion of the modeled landscape in the mature successional stage for a short-lived infaunal community reaching maturity in 2 years (solid line); a community reaching maturity in 6 years (dashed line); and a community reaching maturity in 15 years (dotted line). Area under each contour represents a community of at least (a) 50% mature stage in the model landscape and (b) 90% mature stage.

landscape is occupied by the mature successional stage when 25% of the habitat is disturbed one time or more a year or when 81% of the landscape is disturbed at least once each 4 years. For the slowest recovering community where mature successional stages occur after 15 years, mature stages would be eliminated under these disturbance regimes.

Discussion

The model provides a very general assessment of how seafloor landscapes dominated by different community types will change as a result of changes in the disturbance regime. Here, we argue that these changes are largely wrought by habitat disturbance associated with fishing, but it would be possible to add other types of seafloor disturbance to the model. That is, we could subject the model landscape to either multiple anthropogenic disturbances or include natural disturbance phenomena with space and time scales comparable to fishing.

The model illustrates how increases in the disturbance regime result in loss of late-successional-stage species, especially those that take on the order of 10–15 years to recover. The community archetypes represented in the model reflect species that can add significant structural complexity to the seafloor and influence ecosystem processes. The model provides no support for a recent meta-analysis of fishing impact studies that concluded that intertidal cockle communities are the most heavily impacted (Collie et al. 2000), as we might expect comparatively high rates of recovery in these intertidal communities (e.g., Hall and Harding 1997). Output from the model indicated that as the age to maturity of the late successional stage increases, the extent and frequency of disturbance increasingly restricts the spatial extent of the mature stage. This highlights the extreme sensitivity to disturbance of deep-water and polar habitats with very slow rates of recovery (e.g., communities dominated by coldwater corals). The biology of these species is largely unknown, but they appear to have exceptionally slow growth and low reproductive rates, with individual colonies being hundreds to thousands of years old (e.g., Druffel et al. 1995). Large reefs about 40 m tall have been reported; they are marine equivalents of the oldest terrestrial forests (Squires 1965; Roberts 1997). Even where deep-sea communities are dominated by small, burrowing infauna, recovery from large-scale disturbance may be of the order of hundreds to thousands of years (Young et al. 2001).

The model illustrates how changes in communities may not only be related to the spatial extent and frequency of fishing but also to temporal history, supporting the conclusions of Collie et al. (1997) and Bradshaw et al. (2002). Where communities are dominated by species that live for 10 or more years and require about 10 years to re-establish following disturbance, the long-term physical effect of bottom fishing is to homogenize the substratum, restrict species diversity, and create communities dominated by disturbance-tolerant species (Veale et al. 2000). Site history is an important problem when interpreting studies of the impacts of fishing gear. Some European waters have a long history of fishing by bottom trawling (de Groot 1984; Frid et al. 2000; Kaiser and de Groot 2000; Pranovi et al. 2000). There is also a long history of transformation of marine coastal ecosystems in the western Atlantic (Steele and Schumacher 2000; Jackson 2001) and eastern Pacific (Dayton and Tegner 1984;

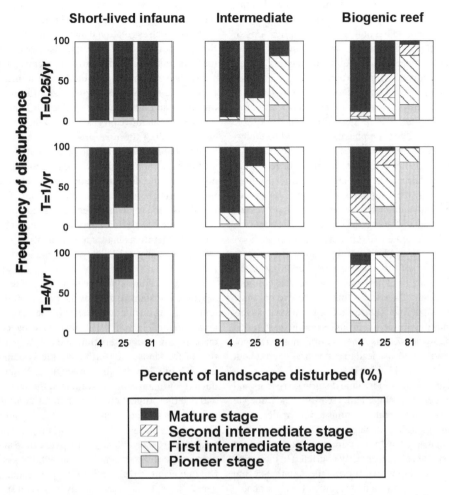

Figure 2. Proportion of model landscape occupied by different successional stages of the three archetypal communities under different frequencies and extents of disturbance.

Dayton et al. 1998). Thus, in some areas, the identification of effects is hampered by the potential historical loss of late-stage, large, and long-lived species. Information generated from comparatively pristine systems is important to understanding effects in potentially degraded ones. Assessments of environmental effects centered only on the study of disturbed areas have been described by Loehle (1991) as analogous to early psychologists defining human behavior largely from studies of the mentally ill.

Encapsulating generalities about disturbance and recovery processes in the model inevitably highlights the need for more sophisticated approaches and better empirical data. The model increases some aspects of our understanding of disturbance recovery processes and provides a picture of independent disturbance events at the landscape scale. It does not take into account any scale dependency in recovery or the potential to increase the variability in response with increasing scale. The model also assumes a constant source of colonists that are capable of settlement and growth within all model cells; there are no intrinsic processes such as facilitation or inhibition (e.g., inhibition of colonization due to disturbance-related changes in sediment stability, particle size, or settlement sites or the facilitation of colonization through provision of settlement sites). In reality, disturbance effects are also likely to be influenced by the landscape attributes of the disturbance and the spatial structure of populations and communities. These complex ecological processes are likely to make our model conservative, in other words, the model represents a minimal ecological response to different disturbance regimes. For communities that take longer to achieve maturity, these effects are likely to be exacerbated. Except in the most heavily fished areas, fishing will not directly disturb all of the seafloor. We need to gain an understanding of the implications of this habitat fragmentation, the scope of such refugia to provide colonists, and how potential recovery is limited by the frequency of disturbance.

It is generally acknowledged that our understanding of the ecology of seafloor populations, communities, and ecosystems is limited. But even the simple model described here emphasizes the need to understand the scales of disturbance and the processes affecting recovery. It is critically important for ecologists to define and describe key species in seafloor communities, including the roles that such species play in ecological, hydrodynamic, and biogeochemical processes, and, in particular, the role of organism size, density, and spatial arrangement. Even more fundamental, however, is basic natural history information encompassing life history and the physiological ecology of the species that may form the basis for archetypal communities, especially the species that are slow to colonize and grow that define mature successional stages. Better information on the disturbance regime associated with fisheries as well as other natural and anthropogenic factors would also lead to better predictions about the consequences of habitat disturbance.

The model takes individual disturbance events and translates these to changes in the seafloor landscape via community recovery rates. It aims to inform decisions and encourage fruitful debate, and it is particularly relevant to managing habitats within a risk-assessment framework. It is not necessarily restricted to one community type or region. By altering the data that define the disturbance regime and the nature of seafloor habitats and communities, the model could be easily applied across or within different regions. Deciding how much of the mature successional stage to maintain in the landscape may be influenced by knowledge of the role of the community in affecting biodiversity or ecosystem processes and the potential risk imposed by loss of communities with slow recovery rates. Lack of knowledge or concern about conservative simplifications in the model should result in precautionary management decisions such as maintaining a higher proportion of the landscape in mature community. In particular, the model indicates that to maintain any proportion of the mature stage where greater than 1–2 years is needed to reach maturity, compromises will need to be reached over the spatial extent and/or temporal frequency of disturbance. This may seem an obvious statement, but models provide for a more informed discussion on the relative consequences of different disturbance regimes.

There are a number of habitat management options aimed at reducing the threat to marine biodiversity, ranging from the creation of marine protected areas, to the spatial and temporal management of areas of the seafloor (such as rotational harvesting), to gear limitation or modification. These options have common elements of reducing habitat disturbance and actively managing the seafloor. Simple models built on basic ecological principles can contribute to refining scientific questions and management strategies. This should be viewed as an iterative process; as we gain greater knowledge regarding the relationships among disturbance regime, habitat structure, and biodiversity for different seafloor communities, we will be able to improve models. At present, such information is not yet available for many marine soft-sediment communities. These relationships need to be assessed across sampling scales, and we need to define operational measures to treat habitat structure as landscape elements characterized by variations in patch density, size, and spatial arrangement.

There is some evidence that a loss of resilience in ecological systems increases the potential for the system to shift to an alternative state (Scheffer et al. 2001). Our ratio model implies that when disturbance is infrequent relative to recovery time and only a small proportion of the landscape is affected, the system is stable and exhibits low variance over time. When the frequency of disturbance is similar to the recovery time and a large proportion of the habitat is affected, the system can still be stable but exhibits higher variance (Turner et al. 1993). However, when the disturbance frequency is shorter than the recovery time and a large proportion of the system is disturbed, the system may flip into an altered state. For example, frequent disturbance may extirpate functionally important species that provide nursery habitats for exploited populations, resulting in the loss of these valuable ecosystem services. Once these features have been lost, it may not be a simple matter of reducing the disturbance regime to ensure their recovery. We need to carefully and explicitly consider the implications of alterations of these ecosystems; they may not only reflect loss of conservation and natural heritage values but also loss of opportunity and ecosystem services provided by diverse and heterogeneous seafloor ecologies.

Acknowledgments

This research was supported by C01X0212. S.F.T. thanks the organizers of the symposium for the opportunity to present and discuss this work. Earlier drafts of the manuscript were improved by comments from Paul Dayton, Drew Lohrer, and one anonymous reviewer.

References

Auster, P. J., and R. W. Langton. 1999. The effects of fishing on fish habitat. Pages 150–187 *in* L. Benaka, editor. Fish habitat: essential fish habitat and rehabilitation. American Fisheries Society, Symposium 22, Bethesda, Maryland.

Bak, P. 1994. Self-organised criticality: a holistic view of nature. Pages 477–496 *in* G. Cowan, D. Pines, and D. Meltzer, editors. Complexity: metaphors, models and reality. Addison-Wesley, Reading, Massachusetts.

Beukema, J. J., E. C. Flach, R. Dekker, and M. Starink. 1999. A long-term study of the recovery of the macrozoobenthos on large defaunated plots on a tidal flat in the Wadden Sea. Journal of Sea Research 42:235–254.

Bradshaw, C., L. O. Veale, and A. R. Brand. 2002. The role of scallop-dredge disturbance in long-term changes in Irish Sea benthic communities: a re-analysis of an historical dataset. Journal of Sea Research 47:161–184.

Caswell, H., and J. E. Cohen. 1991. Communities in patchy environment: a model of disturbance, competition and heterogeneity. Pages 92–122 *in* J. Kolasa and S. T. A. Pickett, editors. Ecological heterogeneity. Springer-Verlag, New York.

Collie, J. S., G. A. Escanero, and P. C. Valentine. 1997. Effects of bottom fishing on the benthic megafauna of Georges Bank. Marine Ecology Progress Series 155:159–172.

Collie, J. S., S. J. Hall, M. J. Kaiser, and I. R. Poiner. 2000. A quantitative analysis of fishing impacts on shelf-sea benthos. Journal of Animal Ecology 69:785–798.

Connell, J. H., and M. J. Keough. 1985. Disturbance and patch dynamics of subtidal marine animals on hard substrata. Pages 125–151 *in* S. T. A. Pickett and P. S. White, editors. The ecology of natural disturbance and patch dynamics. Academic Press, Orlando, Florida.

Cryer, M., B. Hartill, and S. O'Shea. 2002. Modification of marine benthos by trawling: towards a generalization for the deep ocean? Ecological Applications 12:1824–1839.

Dayton, P. K., and M. J. Tegner. 1984. The importance of scale in community ecology: a kelp forest example with terrestrial analogs. Pages 457–483 *in* P. W. Price, C. N. Slobodchikoff, and W. S. Gaud, editors. A new ecology: novel approaches to interactive systems. Wiley, New York.

Dayton, P. K., M. J. Tegner, P. B. Edwards, and K. L. Riser. 1998. Sliding baselines, ghosts, and reduced expectations in kelp forest communities. Ecological Applications 8:309–322.

Dayton, P. K., S. F. Thrush, T. M. Agardy, and R. J. Hofman. 1995. Environmental effects of fishing. Aquatic Conservation: Marine and Freshwater Ecosystems 5:205–232.

de Groot, S. J. 1984. The impact of bottom trawling on benthic fauna of the North Sea. Ocean Management 9:177–190.

Done, T. J., and R. E. Reichelt. 1998. Integrated coastal zone and fisheries ecosystem management: generic goals and performance indices. Ecological Applications 8:110–118.

Druffel, E. R. M., S. Griffin, A. Witter, E. Nelson, J. Southon, M. Kasgarian, and J. Vogel. 1995. *Gerardia*: bristlecone pine of the deep-sea? Geochimica et Cosmochimica Acta 59:5031–5036.

Eckman, J. E., and A. R. Nowell. 1984. Boundary skin friction and sediment transport about an animal-tube mimic. Sedimentology 31:851–862.

Foster, G., and L. Carter. 1997. Mud sedimentation on the continental shelf at an accretionary margin—Poverty Bay, New Zealand. New Zealand Journal of Geology and Geophysics 40:157–173.

Frechette, M., C. A. Butman, and W. R. Geyer. 1989. The importance of boundary-layer flows in supplying phytoplankton to the benthic suspension feeder, *Mytilus edulis* L. Limnology and Oceanography 34:19–36.

Frid, C. L. J., K. G. Harwood, S. J. Hall, and J. A. Hall. 2000. Long-term changes in the benthic communities on North Sea fishing grounds. ICES Journal of Marine Science 57:1303–1309.

Friedlander, A. M., G. W. Boehlert, M. E. Field, J. E. Mason, J. V. Gardner, and P. Dartnell. 1999. Side-scan sonar mapping of benthic trawl marks on the shelf and slope off Eureka, California. U.S. National Marine Fisheries Service Fishery Bulletin 97:786–801.

Graf, G. 1999. Do benthic animals control the particle exchange between bioturbated sediments and benthic turbidity zones? Pages 153–159 *in* J. S. Gray, W. Ambrose, Jr., and A. Szaniawska, editors. Biogeochemical cycling and sediment ecology. Kluwer Academic, Dordrecht, The Netherlands.

Green, M. O., J. E. Hewitt, and S. F. Thrush. 1998. Seabed drag coefficients over natural beds of horse mussels (*Atrina zelanadica*). Journal of Marine Research 56:613–637.

Green, M. O., C. E. Vincent, I. N. McCave, R. R. Dickson, J. M. Rees, and N. D. Pearson. 1995. Storm sediment transport: observations from the British North Sea shelf. Continental Shelf Research 15:889–912.

Hall, S. J. 1999. The effects of fisheries on ecosystems and communities. Blackwell Scientific Publications, Oxford, UK.

Hall, S. J., and M. J. C. Harding. 1997. Physical disturbance and marine benthic communities: the effects of mechanical harvesting of cockles on non-target benthic infauna. Journal of Applied Ecology 34:497–517.

Hall, S. J., D. Raffaelli, and S. F. Thrush. 1994. Patchiness and disturbance in shallow water benthic assemblages. Pages 333–375 *in* A. G. Hildrew, P. S. Giller, and D. Raffaelli, editors. Aquatic ecology: scale, pattern and processes. Blackwell Scientific Publications, Oxford, UK.

Herman, P. M. J., J. J. Middelburg, J. Van de Koppel, and C. H. R. Heip. 1999. Ecology of estuarine macrobenthos. Advances in Ecological Research 29:195–231.

Jackson, J. B. C. 2001. What was natural in the coastal oceans? Proceedings of the National Academy of Sciences of the United States of America 98:5411–5418.

Jenkins, S. R., B. D. Beukers-Stewart, and A. R. Brand. 2001. Impact of scallop dredging on benthic megafauna: a comparison of damage levels in captured and non-captured organisms. Marine Ecology Progress Series 215:297–301.

Jennings, S., and M. J. Kaiser. 1998. The effects of fishing on marine ecosystems. Advances in Marine Biology 34:203–314.

Jumars, P. A., and J. E. Eckman. 1983. Spatial structure within deep-sea benthic communities. Pages 399–451 in G. T. Rowe, editor. The sea. Wiley, New York.

Kaiser, M. J., and S. J. de Groot, editors. 2000. The effects of fishing on non-target species and habitats. Blackwell Scientific Publications, Oxford, UK.

Kaiser, M. J., K. Ramsay, C. A. Richardson, F. E. Spence, and A. R. Brand. 2000. Chronic fishing disturbance has changed shelf sea benthic community structure. Journal of Animal Ecology 69:494–503.

Kolasa, J., and C. D. Rollo. 1991. Introduction: the heterogeneity of heterogeneity: a glossary. Pages 1–23 in J. Kolasa and S. T. A. Pickett, editors. Ecological heterogeneity. Springer-Verlag, Berlin.

Levin, L. A., R. J. Etter, M. A. Rex, A. J. Cooday, C. R. Smith, J. Pineda, C. T. Stuart, R. R. Hessler, and D. Pawson. 2001. Environmental influences on regional deep-sea species diversity. Annual Review of Ecology and Systematics 32:51–93.

Levin, L. A., and A. J. Gooday. 1992. Possible roles for xenophyophores in deep-sea carbon cycling. Pages 93–104 in G. T. Rowe and V. Pariente, editors. Deep-sea food chains and the global carbon cycle. Kluwer Academic, Dordrecht, The Netherlands.

Lindholm, J. B., P. J. Auster, and L. S. Kaufman. 1999. Habitat-mediated survivorship of juvenile (0-year) Atlantic cod *Gadus morhua*. Marine Ecology Progress Series 180:247–255.

Loehle, C. 1991. Managing and monitoring ecosystems in the face of heterogeneity. Pages 144–159 in J. Kolasa and S. T. A. Pickett, editors. Ecological heterogeneity. Springer-Verlag, Ecological Studies 86, Berlin.

McConnaughey, R. A., K. L. Mier, and C. B. Dew. 2000. An examination of chronic trawling effects on soft-bottom benthos of the eastern Bering Sea. ICES Journal of Marine Science 57:1377–1388.

Middleburg, J. J., K. Soetaert, and P. M. J. Herman. 1997. Empirical relationships for use in global diagenetic models. Deep-Sea Research 44:327–344.

National Academy of Sciences. 2002. The effects of trawling and dredging on seafloor habitat. National Academy of Sciences, Washington, D.C.

Nikora, V., M. O. Green, S. F. Thrush, T. M. Hume, and D. Goring. 2002. Structure of the internal boundary layer over a patch of horse mussels (*Atrina zealandica*) in an estuary. Journal of Marine Research 60:121–150.

Pauly, D., V. Christensen, S. Guenette, T. J. Pitcher, U. R. Sumaila, C. J. Walters, R. Watson, and D. Zeller. 2002. Towards sustainability in world fisheries. Nature (London) 418:689–695.

Pearson, T. H., and R. Rosenberg. 1978. Macrobenthic succession in relation to organic enrichment and pollution of the marine environment. Oceanography and Marine Biology: Annual Review 16:229–311.

Pilskaln, C. H., J. H. Churchill, and L. M. Mayer. 1998. Resuspension of sediments by bottom trawling in the Gulf of Maine and potential geochemical consequences. Conservation Biology 12:1223–1224.

Pitcher, C. R., I. R. Poiner, B. J. Hill, and C. Y. Burridge. 2000a. Implications of the effects of trawling on sessile megazoobenthos on a tropical shelf in northeastern Australia. ICES Journal of Marine Science 57:1359–1368.

Pitcher, T. J., R. Watson, N. Haggan, S. Guenette, R. Kennish, U. R. Sumaila, K. Wilson, and A. Leung. 2000b. Marine reserves and the restoration of fisheries and marine ecosystems in the South China Sea. Bulletin of Marine Science 66:543–566.

Pranovi, F., S. Raicevich, G. Franceschini, M. G. Farrace, and O. Giovanardi. 2000. Rapid trawling in the northern Adriatic Sea: effects on benthic communities in an experimental area. ICES Journal of Marine Science 57:517–524.

Rijnsdorp, A. D., A. M. Buys, F. Storbeck, and E. G. Visser. 1998. Micro-scale distribution of beam trawl effort in the southern North Sea between 1993 and 1996 in relation to the trawling frequency of the sea bed and the impact on benthic organisms. ICES Journal of Marine Science 55:403–419.

Roberts, M. 1997. Coral in deep water. New Scientist 155:40–43.

Sainsbury, K. J. 1988. The ecological basis of multispecies fisheries and management of a demersal fishery in tropical Australia. Pages 349–382 in J. A. Gulland, editor. Fish population dynamics the implications for management. Wiley, New York.

Sandnes, J., T. Forbes, R. Hansen, B. Sandnes, and B. Rygg. 2000. Bioturbation and irrigation in natural sediments, described by animal-community parameters. Marine Ecology Progress Series 197:169–179.

Scheffer, M., S. Carpenter, J. A. Foley, C. Folke, and B. Walker. 2001. Catastrophic shifts in ecosystems. Nature (London) 413:591–596.

Sousa, W. P. 2001. Natural disturbance and the dynamics of marine benthic communities. Pages 85–130 in M. D. Bertness, S. D. Gaines, and M. E. Hay, editors. Marine Community Ecology, Sinauer Associates, Sunderland, Massachusetts.

Squires, D. F. 1965. Deep-water coral structure on the Cambell Plateau, New Zealand. Deep-Sea Research 12:785–788.

Steele, J. H., and M. Schumacher. 2000. Ecosystem structure before fishing. Fisheries Research 44:201–205.

Thayer, C. W. 1983. Sediment-mediated biological disturbance and the evolution of marine benthos. Pages 479–625 in M. J. S. Tevesz, editor. Biotic interactions in recent and fossil benthic communities. Plenum, New York.

Thrush, S. F., J. E. Hewitt, V. J. Cummings, P. K. Dayton, M. Cryer, S. J. Turner, G. Funnell, R. Budd, C. Milburn, and M. R. Wilkinson. 1998. Disturbance of the marine benthic habitat by commercial fishing: impacts at the scale of the fishery. Ecological Applications 8:866–879.

Thrush, S. F., and R. B. Whitlatch. 2001. Recovery dynamics in benthic communities: balancing detail with simplification. *In* K. Reise, editor. Ecological comparisons of sedimentary shores. Springer-Verlag, Berlin.

Thrush, S. F., R. B. Whitlatch, R. D. Pridmore, J. E. Hewitt, V. J. Cummings, and M. Maskery. 1996. Scale-dependent recolonization: the role of sediment stability in a dynamic sandflat habitat. Ecology 77:2472–2487.

Tupper, M., and R. G. Boutilier. 1995. Effects of habitat on settlement, growth and postsettlement survival of Atlantic cod (*Gadus morhua*). Canadian Journal of Fisheries and Aquatic Sciences 52:1834–1841.

Turner, M. G., W. H. Romme, R. H. Gardner, R. V. O'Neill, and T. K. Kratz. 1993. A revised concept of landscape equilibrium: disturbance and stability on scaled landscapes. Landscape Ecology 8:213–227.

Veale, L. O., A. S. Hill, S. J. Hawkins, and A. R. Brand. 2000. Effects of long-term physical disturbance by commercial scallop fishing on subtidal epifaunal assemblages and habitats. Marine Biology 137:325–337.

Watling, L., and E. A. Norse. 1998. Disturbance of the seabed by mobile fishing gear: a comparison to forest clearcutting. Conservation Biology 12:1180–1197.

Wheatcroft, R. A. 2000. Oceanic flood sedimentation: a new perspective. Continental Shelf Research 20:2059–2066.

Whitlatch, R. B., A. M. Lohrer, and S. F. Thrush. 2001. Scale-dependent recovery of the benthos: effects of larval and post-larval stages. Pages 181–199 *in* J. Y. Aller, S. A. Woodin, and R. C. Aller, editors. Organism–sediment interactions. University of South Carolina Press, Columbia.

Wildish, D., and D. Khistmanson. 1997. Benthic suspension feeders and flow. Cambridge University Press, Cambridge, UK.

Young, D. K., M. D. Richardson, and K. B. Briggs. 2001. Turbidites and benthic faunal succession in the deep sea: an ecological paradox? Pages 101–117 *in* J. Y. Aller, S. A. Woodin, and R. C. Aller, editors. Organism–sediment interactions. University of South Carolina Press, Columbia.

Zajac, R. N., R. B. Whitlatch, and S. F. Thrush. 1998. Recolonisation and succession in soft-sediment infaunal communities: the spatial scale of controlling factors. Hydrobiologia 376:227–240.

Muddy Thinking: Ecosystem-Based Management of Marine Benthos

CHRIS L. J. FRID,[1] LEONIE A. ROBINSON, AND JULIE BREMNER

*Dove Marine Laboratory, School of Marine Science and Technology,
University of Newcastle upon Tyne, Cullercoats, North Shields, NE30 4PZ, UK*

Abstract. Globally, a number of states and parties are developing ecosystem-based approaches to environmental management. For the North Sea, the Convention for the Protection of the Marine Environment of the North-East Atlantic and North Sea Conference initiatives have identified a number of metrics for possible use in managing the benthos. We argue that the development of this framework needs to recognize that it is impacting activities which can be managed and that science should concentrate on the development of both robust decision support (performance) metrics and environmental state (descriptive) metrics in order to inform this management. A number of case studies are used to illustrate the strengths and failings of some of the proposed metrics. Performance metrics should be linked closely to the impacting activity. Thus, changes in the metric can immediately trigger a management response. We consider various proposed metrics and comment on their utility in the context of managing fisheries effects. Descriptive metrics, such as diversity indices, are useful for identifying patterns in community structure and assessing the potential consequences of impacts. However, they do not directly link changes to particular activities, making it difficult to assign causality and, therefore, apply management. To date, much of the focus of these considerations has been on taxonomic-based measures. We go on to consider the metrics that can also be developed to assess the functioning of the ecosystem. We conclude that, at present, there are few robust metrics to describe either the environmental state or the extent of fishing impacts for benthic systems. There is, therefore, an urgent need for scientific advances in this area.

Introduction

The Convention on Biological Diversity (CBD), signed at the 1992 United Nations Conference on Environment and Development in Rio, provides the principle framework for international efforts to protect natural resources. The CBD definition of biological diversity recognizes two components of biological diversity as needing protection: the biological richness of the system and the preservation of the ecological complexes of which they are part.

In the Northeast Atlantic area, the ecosystem approach to management of the marine environment is being advanced by Convention for the Protection of the Marine Environment of the North-East Atlantic and the North Sea Ministerial Conference through the development of an ecological quality objective (EcoQO) approach. The Scheveningen workshop (Skjoldal et al. 1999) defined ecological quality (EcoQ) as "an overall expression of the structure and function of aquatic systems" and the EcoQO as "the desired level of EcoQ relative to the reference level." In turn, the reference level was defined as "the level of EcoQ where anthropogenic influence on the ecological system is minimal."

From these definitions, it follows that the starting point for the development of ecosystem approaches to environmental management is to define the "overall . . . structure and function" desired for the ecosystem being considered. From this, one can derive levels for various measures of the system. The definition of the appropriate measures will vary among systems and depend on the priority given to each issue. The visualization of the desired ecosystem is a societal decision. Having developed such a view of the desired ecosystem structure and function, formal metrics for the EcoQOs can be developed.

According to the International Council for the Exploration of the Sea (ICES) criteria (ICES 2001), metrics for EcoQOs should be:

(1) relatively easy to understand by nonscientists and those who will decide on their use;

---

[1] Corresponding author: c.l.j.frid@ncl.ac.uk

(2) sensitive to a manageable human activity;

(3) relatively tightly linked in time to that activity;

(4) easily and accurately measured, with a low error rate;

(5) responsive primarily to a human activity, with low responsiveness to other causes of change;

(6) measurable over a large proportion of the area to which the EcoQ metric is to apply; and

(7) based on an existing body or time-series of data to allow a realistic setting of objectives.

There would, therefore, appear to be no need to routinely develop EcoQOs for every component of every ecosystem. Rather, EcoQO development should proceed in an integrated manner such that EcoQOs are in place for the key aspects identified by the stakeholders and that the EcoQOs are mutually consistent. Nor is there any need for EcoQOs to be set at the reference level, where anthropogenic influences are minimal. In fact, this would imply no use of environmental services such as waste treatment or food production.

It appears that, in some cases, scientists are already advancing candidates for EcoQO metrics (Skjoldal et al. 1999). In many cases, these appear to be based on parameters routinely measured (or favored) by the proponent rather than the preferred societal view of the approach. In many cases, the measures proposed are complex indices, the behaviors of which are poorly linked to human actions and are not readily amenable to management measures. Similarly, many of the measures being advanced cover the same aspects of the ecosystem (e.g., a large number of diversity and evenness measures). In addition, it is difficult to see how many of these could be made operational. How does one manage the diversity or species composition of an ecosystem? Of course one cannot. It is impacting activities which can be managed, and in order to assist in that management process, the task is for science to develop the tools needed by the managers. These tools are metrics that give a good, direct measure of the extent to which an activity is impacting the system and requires a directed management response (i.e., the measures fully meet the ICES [2001] criteria). Such metrics form a class we can refer to as performance or decision support metrics, since they directly inform us of the performance of the management regime. However, they are likely to be costly to measure at high frequency or large spatial scales, and if each measure is to comply with ICES criteria 2, 3, and 5, the number of metrics is also likely to be high. As a minimum, the number required will be at least as great as the number of human activities to be managed.

A more pragmatic approach, therefore, would be to develop a suite of performance metrics which are monitored at relatively infrequent intervals over the whole area and in areas of perceived problems. A second group of metrics, based on more integrated aspects of environmental state, such as diversity or total biomass, could then be used to give an overview of the system status, and changes in these could be the trigger for additional measurements of the performance metrics. By their very nature, this second class of integrated metrics are poorly linked to specific manageable activities, and, thus, they are referred to as descriptive or environmental state metrics.

This two-tier approach is analogous to human health monitoring. Health professionals routinely measure pulse, respiration, temperature, and blood pressure. These are simple and cheap to monitor, and changes in these parameters do not directly trigger intervention but usually prompt additional diagnostic tests. The latter equate to our performance metrics and the former to the descriptive metrics.

In the remainder of this paper, we examine the utility of this approach with respect to the management of fishing impacts on benthic systems. So what aspects of the benthos are appropriate to form the basis of these metrics? Two aspects certainly requiring consideration are the composition of the benthic fauna and its ecological functioning. In the former, appropriate measures would seem to be total abundance (biomass), species richness, diversity, and the distribution of individuals among taxa. The latter would include the balance among different trophic groups, the presence of habitat forming species, and the mix of life history strategies present. We formally compare the performance of a range of potential descriptive and performance metrics using two data sets that cover the infaunal benthos and the epibenthos, respectively. The former is a 30-year time series at a single benthic station which has been subject to varying levels of fishing effort in that period; the latter series covers 40 stations sampled in 1 year that vary in the intensity of fishing impacts.

Methods

Data Sets

Infauna

A 30-year time series of infaunal abundance data was available from a fixed station, situated approximately 18.5 km (11.5 mi) off the northeastern coast of England (55°07′N, 01°15′W) and in 80 m of water (Figure 1; station P in Frid et al. 1999). The station was initially sampled in January 1971, and the data set analyzed here covers the period 1972–2001. Samples from

1977 and 1998 were missing due to weather and operational constraints. In March of each year, five van Veen grabs were taken at the site and sieved onboard over a 0.5-mm mesh (Frid et al. 1996). Samples were sorted, identified (to species level where possible), and enumerated. Data used are total genera abundance (number individuals/m^2) for the whole sampled community (a total of 176 genera). Analysis at the genus level avoided any problems due to errors of misidentification at the species level or changes in taxonomy leading to problems with homonyms.

The station lies within a fishing ground for *Nephrops norvegicus*, where otter trawling is the predominant technique (Figure 1; Frid et al. 1999). Fishing effort data for the area (ICES statistical rectangle 39E8) were obtained from the Centre for Environment, Fisheries, and Aquaculture Science (CEFAS), Lowestoft, UK, and swept area calculated as an index of trawling pressure on the benthos. Mean annual swept area (km^2/year), was calculated by multiplying total annual effort hours trawling by the area swept per hour (average width of trawl [km] × distance per hour [km/h]). Five phases corresponding to the level of fishing effort and stage in development of the fishery were

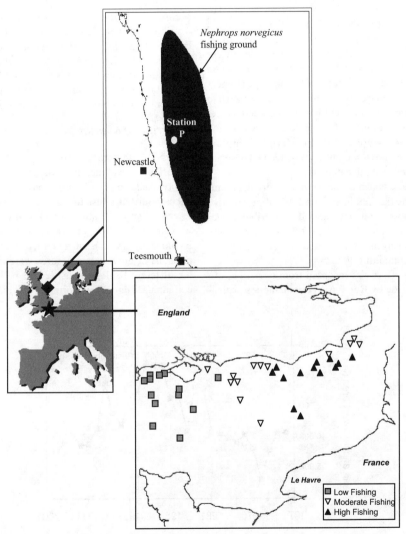

Figure 1. Location of the infaunal and epifaunal sampling areas in the North Sea and English Channel. The inset shows the location of Station P in relation to the fishing ground in northeastern England for *Nephrops norvegicus* (black shaded area). The bottom map shows the distribution of fishing effort at the 40 epibenthic trawl stations in the English Channel.

identified. These phases were: low effort 1972–1981 (effort between 3,000 and 11,000 km²/year); medium effort 1982–1986 (effort between 11,000 and 19,000 km²/year); high effort 1987–1989 (effort over 20,000 km²/year); medium effort, posthigh 1990–1995 (effort between 11,000 and 19,000 km²/year); and low effort, posthigh 1996–2001 (effort between 7,000 and 11,000 km²/year) (see Figure 2). Behavior of the metrics was examined with respect to these five phases of fisheries activity.

Epifauna

Data on epifauna biomass from 40 subtidal stations in the eastern English Channel (ICES division 7d) were gathered by the crew of the RV *Corystes* during August 1998 as part of the CEFAS flatfish beam trawl survey (Figure 1). Trawls of 30-min duration, covering around 15,000 m² (Ellis and Rogers 2000) were carried out using a 4-m beam trawl with a 40-mm stretched cod end, chain mat, and flip-up rope (Rogers et al. 1998). The macro-invertebrates were identified to species where possible and biomass recorded as wet weight (kg) caught per h. The whole catch was sorted when samples were small. For larger samples, subsamples of known weight were sorted and biomass raised to that of the full catch weight.

Only those taxa most dominant at individual sampling stations (ranked in the top 90% of cumulative biomass at any station) or those widely distributed over all stations (present at more than 50% of stations) were retained for analysis. This resulted in a reduced epifaunal data set containing 40 taxa.

Fishing effort was recorded from overflight observations made by Royal Air Force fisheries protection patrols. Effort was calculated as an index of the activity of vessels in an approximately 1.3-km² box centered on each sample station (Rogers et al. 2001). The index of fishing effort ranged from 1.5 to 192.5. In order to test the ability of the metrics to indicate fishing impacts, the data were separated into groups representing low, moderate, and high fishing intensity. The data were ranked and the top and bottom thirds separated from the middle values, thus forming the low intensity (0–16), moderate intensity (16.1–56), and high intensity (56.1–193) levels. The epifauna data were grouped according to fishing effort and metrics calculated for each.

Given the size of the survey area, differences in sediment and hydrography could potentially affect the distribution of benthic species. Therefore, this possible confounding effect must be considered in interpreting the analyses. However, previous analysis of the epibenthos of these stations has found links between local community structure and both the physical environment and fishing (Kaiser et al. 1999).

Descriptive Metrics

Univariate Measures

For all univariate measures, data were tested for normality (Anderson-Darling test) and homogeneity of variances (Bartlett's test for normally distributed data, Levenes test in all other cases), and transformations were applied where appropriate. Transformations used included $\log_{10}(x + 1)$, square root, or fourth root. Comparisons of univariate measures were made between the different fishing intensities for both infauna and epifauna. Differences in normally distributed data were

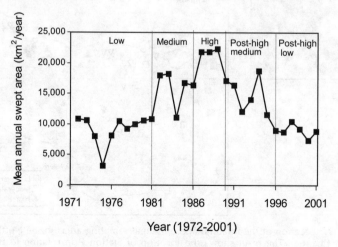

Figure 2. Fishing effort data (mean annual swept area; km²/year) for the area of the infaunal sampling station on the Northumberland coast (ICES statistical rectangle 39E8).

examined using one-way analysis of variance (ANOVA), while Kruskal–Wallis tests were employed for non-normal data (MINITAB 1998). The univariate measures tested were: total abundance or biomass, species richness (S), and Shannon diversity (H'). Based on ecological theory, we predicted a priori that all of these measures would be lowest during the highest periods of fishing effort.

Graphical Distribution: K Dominance

To examine the differences in dominance of taxa for each fishing level, K-dominance plots were produced using PRIMER (2001; Plymouth Routines in Multivariate Ecological Research; Clarke and Warwick 1994). For each fishing level, the rank of a taxon was based on the average of all samples in that level. We predicted a priori that communities would be most dominated by a few taxa at the highest level of fishing effort.

Multivariate Measures: Multi-Dimensional Scaling Ordination and Analysis of Similarity

Nonmetric multidimensional scaling (MDS) plots were produced using PRIMER (2001), to illustrate variations in community structure for the different fishing levels. Ordinations were based on Bray-Curtis similarities of samples. To reduce the effect of dominant taxa, the data were square root transformed. An analysis of similarity (ANOSIM; Clarke and Warwick 1994) test was performed to examine the significance of differences in genera composition among the different phases of fishing. A priori predictions were that interannual variation would be greatest during periods of high effort fishing and that communities in different phases would be clearly and significantly distinct in their composition.

Performance Metrics

Indicator Taxa

Indicator taxa are taxonomic-based performance metrics. They are based on the theory that certain characteristics can impart vulnerability or resistance to fishing in benthic organisms. For example, scavenging starfish and crabs are known to respond positively to fishing in benthic communities (Kaiser et al. 1998; Rumohr and Kujawski 2000; Bradshaw et al. 2002), while large bivalves and slow moving or sessile, upright species are thought to be vulnerable to the passage of the trawl gear (Hall-Spencer et al. 1999; Bradshaw et al. 2002; Wassenberg et al. 2002). Criteria for identifying indicator organisms were formulated a priori based on the results of these studies. Positive indicators were identified as those taxa that were mobile predators or scavengers, whereas negative indicators were immobile filter feeders or soft-bodied or shelled taxa. Selections of indicator taxa were identified in each of the data sets (Table 1).

Functioning of Ecosystem

Benthic community composition by trophic groups and by functional groups are two metrics proposed to indicate functioning in marine ecosystems (Pearson and Rosenberg 1987; Bonsdorff and Pearson 1999).

Trophic Groups

Based on information retrieved from literature sources, the taxa in each data set were classified as deposit feeders, filter and suspension feeders, opportunists and scavengers, or predators. The fauna were then grouped by their feeding mode and biomass (epifauna) or abundance (infauna) and summed for each trophic group. The expectation was that opportunists and scavengers or predators will benefit from fishing activities while filter and suspension feeders will be negatively impacted. There is no consensus on the response of the deposit feeders, as individual taxa may show opposing trends.

Functional Groups

Functional groups were determined by further defining feeding modes into relative mobility groups (after Fauchald and Jumars 1979). For the epifauna, each of the trophic groups was scored according to its mobility; those identified as immobile, low mobility, moderate mobility, or high mobility were separated. Trophic groups of infauna were re-classified as either mobile or

Table 1. Taxa identified (from literature sources) as potential positive and negative indicators of fishing impacts in the Northumberland infaunal series and the English Channel epibenthic data set.

Predicted direction of change as a result of fishing effects	Infauna	Epifauna
Increase	Drilonereis	Asterias rubens
	Eteone	Cancer pagurus
	Glycera	
	Glycinde	
	Goniada	
	Gyptis	
	Nephtys	
	Nereis	
	Phyllodoce	
Decrease	Abra	Porifera
	Ampharete	Pentapora foliacea
	Brissopsis	Pecten maximus
	Cerianthus	
	Echinocardium	
	Phaxas	
	Sabella	
	Virgularia	

immobile. Our expectation was that mobile groups would benefit from fishing activities, while immobile groups would be vulnerable to fishing mortality.

For each of the performance metrics, differences in abundance (infauna) and biomass (epifauna) among the fishing levels were assessed using one-way ANOVA or Kruskal–Wallis tests. Non-normal data were transformed prior to analysis.

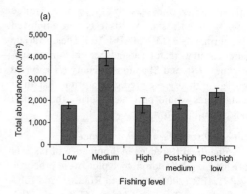

Results

Descriptive Metrics

Infauna

Total infaunal abundance increased significantly between the 1970s and mid-1980s reaching a peak in a period of medium effort trawling (1982–1986). At this time, average abundance was 3,951 individuals/m². Between 1987–1990, however, abundance dropped back down to around the levels found in the 1970s. This drop in total abundance did coincide with the 3 years of highest fishing effort, but although abundance varied significantly (one-way ANOVA, total abundance: $F = 12.79$, $P < 0.001$), the overall pattern was not consistent with our a priori prediction (Figure 3a). Richness of taxa (S) was significantly different among the five phases of fishery development, and this was the only measure that showed a pattern consistent with our a priori prediction (one-way ANOVA, genera richness: $F = 5.81$, $P = 0.002$; Figure 3b). Diversity of genera was also lowest during the period of high fishing effort (1987–1990), but the overall differences over the 30-year period were not significant.

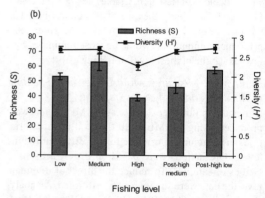

Figure 3. Variation in mean (±SE) (a) total abundance and (b) richness (S) and Shannon diversity (H') of infauna off Northumberland during five phases of fishery development. no. = number.

The large, deposit-feeding polychaete worm, *Heteromastus*, was the most dominant taxon throughout the whole 30-year period. However, during the years of highest effort trawling, *Heteromastus* made up, on average, 45% of the total abundance of the infaunal community. Abundance was distributed most evenly among taxa in the 1970s (a period of low effort trawling), when *Heteromastus* accounted for less than 25% of the total abundance of the community. The main difference in cumulative dominance of the infaunal community was during the period of highest effort trawling. All other periods showed similar patterns (Figure 4a).

Significant differences in square root-transformed genera composition for the 5 phases of fishing effort were found (ANOSIM, global $R = 0.65$, $P < 0.001$) (Figure 4b, c). It is possible to see a temporal succession in composition of taxa across the 30-year period, but the patterns are complex (Figure 4c). During the first year of high effort trawling (1987), composition of taxa changed significantly from the previous years, and a period of high interannual variation then followed, consistent with patterns seen in highly disturbed communities (Warwick and Clarke 1993). The two periods of low effort fishing (before and after the high-effort phase in the 1970s and mid- to late 1990s) had significantly distinct taxonomic composition (ANOSIM, R statistic = 0.80, $P < 0.001$). This could be due to a number of factors other than fishing but may imply that the reduction in fishing effort following a high level of exploitation does not, certainly in the short term, lead to a return to the earlier community composition (Figure 4b).

Epibenthos

Intensively fished stations were dominated by a few high-biomass taxa, the most dominant taxon comprising 80% of the total biomass, on average (Figure 5a). Biomass was distributed more evenly among taxa in the low fishing group, with the most dominant taxon making up 60% of total biomass. Moderately fished stations had the most evenly distributed biomass; the most dominant taxon accounted for less then 40% of the total biomass. *Ophiothrix fragilis* was the most domi-

(a)

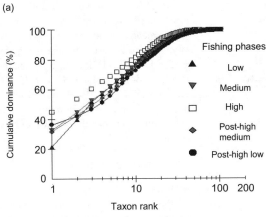

(b)

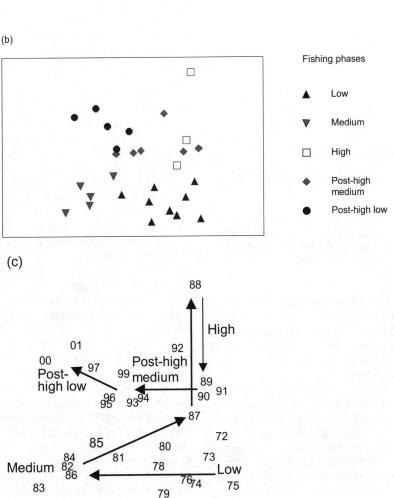

(c)

Figure 4. Changes in (a) cumulative dominance of infaunal taxa and (b) and (c) community composition, as represented by nonmetric multi-dimensional scaling ordination, of infauna off Northumberland (stress = 0.16). Panel (b) shows the phases in fishing effort, while panel (c) illustrates the temporal succession of community composition over the 30-year period. Numbers represent the years sampled (1972–2001).

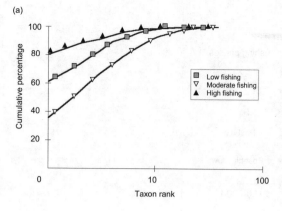

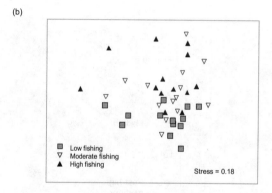

Figure 5. Changes in (a) cumulative dominance of epifaunal taxa and (b) community composition, as represented by nonmetric multidimensional scaling ordination, of epifauna at English Channel stations experiencing three levels of fishing effort.

nant species at each fishing level, although the second most dominant varied with fishing effort, being *Alcyonidium digitatum* (12% of total biomass) at low fishing, *Asterias rubens* (19% of total biomass) at moderate levels, and *Aequipecten operclaris* (8% of total biomass) at high fishing.

Lightly fished stations grouped apart from those fished heavily on the MDS plot, although the moderately fished stations did not (Figure 5b). Differences in taxon composition existed among the three groups (ANOSIM, global $R = 0.093$, $P = 0.1$), and stations subject to low effort differed significantly from the most intensively fished stations (ANOSIM, $R = 0.231$, $P < 0.001$).

Total biomass was lowest in the moderately fished stations and increased between the low and high effort groups (Figure 6a), although these differences were not significant.

Stations subject to low effort had the lowest taxon richness of the three groups (mean ± SE = 10.62 ± 0.87 taxa/h). Richness was significantly higher in the high-effort stations (one-way ANOVA, $F = 3.73$, $P = 0.033$), with a mean of 14 taxa/h (±1.14 SE). Shannon diversity was also lowest in the least-fished stations (mean ± SE = 1.01 ± 0.11), increasing in the most intensively fished stations (mean ± SE = 1.25 ± 0.15; one-way ANOVA, $F = 3.01$, $P = 0.061$). However, taxon richness and diversity were both higher in the moderately fished stations than either the low or high effort groups (Figure 6b).

Performance Metrics

Infauna

Of the eight genera proposed as negative indicators (potentially vulnerable to trawling), three showed significant differences among fishing periods. These were the bivalves *Abra* and *Phaxas* and the polychaete worm *Ampharete* (one-way ANOVA, *Abra*: $F = 3.79$, $P = 0.016$; Kruskal-Wallis, *Phaxas*: $H = 11.08$, $P = 0.026$; *Ampharete*: $H = 12.48$, $P = 0.014$). Of these, none responded consistently to fishing in the direction we would predict for a vulnerable genus (Figure 7a).

Of the nine genera proposed as positive indicators (responding favorably to trawling disturbance), seven showed significant differences among the fishing phases (one-way ANOVA, *Glycera*: $F = 4.18$, $P = 0.01$; *Nephtys*: $F = 5.58$, $P = 0.003$; Kruskal-Wallis, *Eteone*: $H = 21.37$, $P < 0.001$; *Glycinde*: $H = 14.50$, $P = 0.006$; *Goniada*: $H = 11.27$, $P = 0.024$; *Gyptis*: $H = 10.38$, $P = 0.034$; *Nereis*: $H = 12.79$, $P = 0.012$), but none responded in the direction we would expect for a positive indicator. Examples of some of the abundance patterns of positive indicators are shown in Figure 7b.

All four trophic groups were represented in the infaunal samples, and each had significant differences in abundance among fishing phases (one-way ANOVA, deposit feeders: $F = 12.40$, $P = 0.001$; filter and suspension feeders: $F = 9.34$, $P < 0.001$; scavengers: $F = 6.52$, $P = 0.001$; predators: $F = 4.04$, $P = 0.012$). Filter and suspension feeders were present in their lowest numbers in the period of highest fishing effort but did not show a consistent pattern with development of the fishery (Figure 8a). Deposit feeders were most represented in terms of both species number (45% of total species number over the 30-year period) and abundance (Figure 8b), and, as predicted, there was no clear response to changes in fishing effort. Scavengers were present in their highest numbers in the medium-effort phase as the fishery expanded, but they did not show a consistent positive response to increasing fishing effort (Figure 8c). Predators increased in abundance from the lowest fished period to the medium-effort phase and stayed at similar levels in the period of highest fishing effort (Figure 8c).

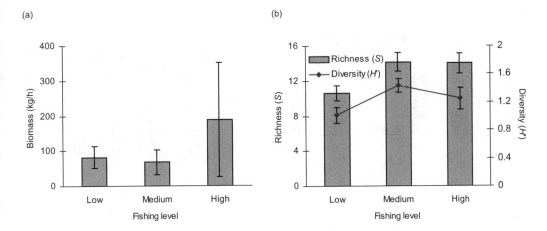

Figure 6. Variation in mean (±SE) (a) total biomass and (b) richness (*S*) and Shannon diversity (*H'*) of epifauna at English Channel stations experiencing three levels of fishing effort.

Six out of the eight possible functional groups were present in the infaunal system. With the exception of the sessile filter and suspension feeders, significant differences in abundance among fishing levels occurred in all functional groups. Sessile filter and suspension feeders were extremely rare in the infaunal data set. Although differences in abundance among fishing levels were significant for immobile deposit feeders, there was no clear negative pattern related to fishing (Kruskal-Wallis, $H = 12.85$, $P = 0.012$) (Figure 9a). Both mobile filter and suspension feeders and mobile deposit feeders responded significantly as expected, with no clear pattern related to fishing (one-way ANOVA, mobile filter and suspension feeders, $F = 11.83$, $P < 0.001$; mobile deposit feeders: $F = 4.33$, $P = 0.009$; Figure 9b). There were no immobile scavengers and predators in the infaunal data set, and variation in abundance for mobile scavengers and predators was the same as for total scavengers and predators, as described above (Figures 8c, 9c).

Epibenthos

Sponges exhibited significant differences in biomass that were related to the level of fishing effort (Kruskal-Wallis, Hadj = 7.75, df = 2, $P < 0.05$). Biomass was highest in the low-effort stations and reduced considerably in the moderately and highly fished stations (Figure 10a). There was low biomass of *Pentapora foliacea* in both the low and moderately fished stations (low fishing, 0.02 ± 0.01 kg/h; moderate fishing, 0.43 ± 0.42 kg/h), and although none at all were present in the high effort stations, differences among the groups were nonsignificant. Differences in biomass of *Pecten maximus* were also nonsignificant.

Biomass of *Asterias rubens* was highest in the moderately fished stations (Figure 10b). These and the

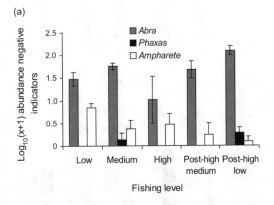

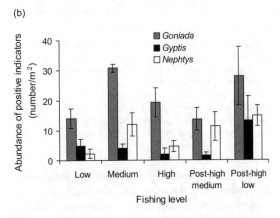

Figure 7. Variation in mean abundance (±SE) of (a) negative infaunal indicators and (b) positive infaunal indicators off Northumberland during five phases of fishery development.

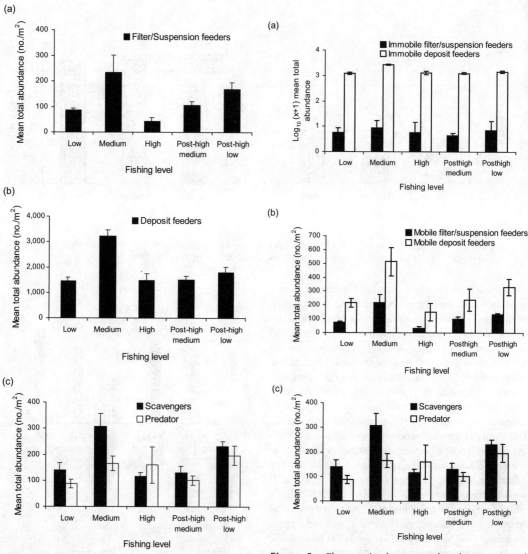

Figure 8. Changes in the mean abundance (±SE) of infaunal organisms by trophic group off Northumberland during five phases of fishery development. Panel (a) shows groups predicted to be vulnerable to fishing (filter and suspension feeders), panel (b) shows those for which no prediction was possible (deposit feeders), and panel (c) shows groups predicted to benefit from fishing (scavengers and predators). no. = number.

Figure 9. Changes in the mean abundance (±SE) of infauna by functional groups off Northumberland during five phases of fishery development. Panel (a) shows groups predicted to be vulnerable to fishing (immobile filter and suspension feeders and deposit feeders), (b) those for which no prediction was possible (mobile filter and suspension feeders and deposit feeders), and (c) those predicted to benefit from fishing (mobile scavengers and predators).

highly fished stations exhibited higher biomass than the least fished ones; however, these differences were not significant. The biomass of *Cancer pagurus* decreased nonsignificantly among each of the fishing effort levels.

Biomass of all trophic groups was higher in the stations fished heavily than those fished lightly. No deposit feeders were present in the low-effort stations, and both deposit feeders and scavengers increased

gradually among the three levels. Predator biomass was highest in the moderately fished stations, whereas filter and suspension feeder biomass was lowest. However, there were no significant differences in biomass among the groups in any case (Figure 11a, b).

Eight out of the possible 16 functional groups were present in the epifauna data set (Figure 12). Signifi-

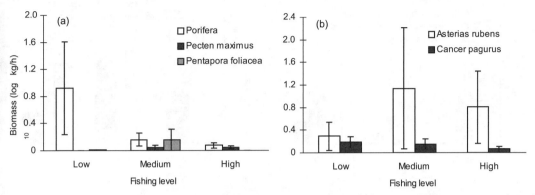

Figure 10. Variation in mean biomass (±SE) of (a) negative epifaunal indicators and (b) positive epifaunal indicators at English Channel stations experiencing three levels of fishing effort.

cant differences in biomass of immobile filter and suspension feeders (one-way ANOVA, $F = 3.09, P = 0.057$), moderately mobile scavengers (Kruskal-Wallis Hadj = 9.30, $P = 0.010$), and highly mobile scavengers (one-way ANOVA, $F = 2.79, P = 0.075$) existed among the fishing effort levels. Immobile filter and suspension feeders decreased among the three levels from a mean of 27.36 kg/h (±6.03 SE) in the least-fished stations to a mean of 7.39 kg/h (±2.09SE) in the most intensively fished stations (Figure 12a). The highest biomass of both moderately and highly mobile scavengers was found in the moderately fished stations (Figure 12b). In the least fished stations, moderately mobile scavengers exhibited very low biomass (0.09 kg/h ± 0.06SE). This increased between low and high effort levels. Highly mobile scavengers, in contrast, decreased in biomass between low and high effort levels.

Discussion

In this paper, we have examined the behavior of various metrics of ecosystem status in response to spatial and temporal changes in fishing activities. We have considered both infaunal and epifaunal benthos and sought to identify robust measures which can be applied across ecosystem components. Many of the metrics of the state of the ecosystem, proposed for use as management tools (within the EcoQO approach being developed in Europe to implement an ecosystems-based approach to manage-

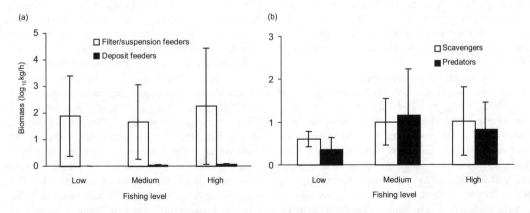

Figure 11. Changes in the mean biomass (±SE) of epifauna by trophic group at English Channel stations experiencing three levels of fishing effort, including (a) groups predicted to be vulnerable to fishing (filter and suspension feeders) and those for which no prediction was possible (deposit feeders) and (b) groups predicted to benefit from fishing (scavengers and predators).

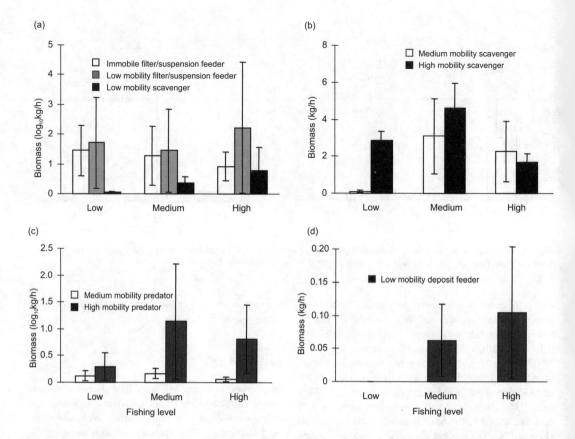

Figure 12. Changes in the mean biomass (±SE) of epifauna by functional group at English Channel stations experiencing three levels of fishing effort, including (a) groups predicted to be vulnerable to fishing (sessile and low mobility filter and suspension feeders and low mobility scavengers); those predicted to benefit from fishing, including (b) medium and high mobility scavengers and (c) medium and high mobility predators; and (d) those for which no prediction was possible (low mobility deposit feeders).

ment of the North Sea), fail the ICES (2001) criteria. These failures are primarily due to the lack of linkage between response being measured and a manageable human activity. Therefore, we have advocated a two-stage approach.

Performance metrics are metrics that are linked closely to the impacting activity, such that changes in the value of the metric can immediately trigger a management response. Metrics in this category are often expensive to monitor at large spatial and temporal scales. We have shown that, as far as the impacts of fishing are concerned, some of these metrics do show a response (Table 2). However, there is no obvious candidate group of metrics that perform well for both benthic infauna and epifauna. Of the species identified from the literature as being good indicators of fishing impacts, none of the infauna and less than 33% (depending on the category of indicator) of the epifaunal species actually responded significantly in the manner predicted. This is an inherent problem of the indicator-species approach. Species that respond positively in one region may not respond at all (or show the opposite pattern in another area; e.g., see Bustos-Baez and Frid 2003). In this study, we restricted ourselves to taxa widely recorded as showing a response, but in terms of the utility of the approach, it is a factor that must be considered. In trying to assess functional changes in the system in response to fishing, the trophic group approach was less successful than the indicator approach for the epifauna but slightly more successful with the infauna. Functional groups, including life history information, were the most promising approach for both infauna and epifauna. However, the response followed the prediction in less than 50% of the functional groups identified. There is, therefore, a major challenge for science: the need to identify robust performance metrics for fishing disturbance (i.e., aspects of the system that change in a predictable manner in response to fishing and not other causes).

Table 2. Summary of the performance of various potential Decision Support (Performance) and Environmental State (Descriptive) metrics when applied to the effects of fishing on infaunal and epibenthic assemblages. For performance to be judged satisfactory (S) there must have been a statistically significant change in the direction predicted by accepted theory. Non-significant changes and those which, although significant, did not conform to theory, were judged unsatisfactory (U). A question mark indicates a significant difference broadly in line with theoretical expectations but in a complex pattern. For performance metrics, the proportion of groups meeting the test is also given as a percent and as raw scores. MDS = multidimensional scaling.

Metric	Performance with infauna	Performance with epibenthos
Performance Metrics		
Indicators negative	U	? 33% (1/3)
Indicators positive	U	U
Trophic groups	? 25% (1/4)	U
Functional groups	? 33% (2/6)	? 29% (2/7)
Descriptive Metrics		
Total abundance/ biomass	U	U
Species richness	S	U
Species diversity	U	U
K-dominance	?	?
MDS	?	?

Descriptive metrics, in contrast, are often routinely measured at present so that time series and cost-effective monitoring programs are available. They are not, however, tightly linked to specific manageable activities, so a change in the metrics cannot trigger a direct management response. For example, diversity may increase or decrease in the face of a number of different human impacts. Descriptive metrics are useful for identifying patterns in community structure and assessing the potential consequences of impacts. We have shown the utility of many of these metrics as indicators of ecosystem health (Table 2). In general, it is the more complex, graphical, and multivariate, approaches that most clearly show change. Interpreting these changes within a theoretical framework is difficult or impossible. The superiority of these approaches in detecting pollution effects has long been recognized (Warwick and Clarke 1994). Their usefulness, however, lies in their cost effectiveness and the availability of time series and historic data rather than their ability to directly drive management measures. Changes in these metrics should be seen as triggering focused programs of investigation to establish the cause of the change, occasioning the implementation of monitoring schemes based on performance metrics. One of the issues that such programs need to address is the actual ecological, as opposed to statistical, significance of the measured effect. These techniques can highlight very subtle changes in the composition of the assemblage, and these changes may be of no ecological concern. Development of monitoring protocols must, therefore, have due regard for the sensitivity of these techniques.

In this study, we have used two series of benthic data, one covering epibenthos at 40 stations varying in their levels of fishing activity and one of infauna at a single station known to have been subjected to varying levels of fishing over the last 30 years. It is, therefore, reassuring to see the same patterns emerging. In both cases, some simple univariate measures, multivariate, and graphical representations of the assemblage structure yielded useful information about the systems' exploitation status. All these techniques also respond well to changes driven by pollution (Clarke and Warwick 1994), dredging (Kenny and Rees 1994, 1996), and natural disturbances (Hall et al. 1994).

The ecological literature is filled with any number of metrics proposing to provide information on the health of the ecosystem. There is little doubt that the investigations that have developed and used these approaches have furthered our understanding of ecosystem dynamics and processes. These are not, however, the criteria we need to be concerned with when it comes to engaging in the management of activities to protect ecosystem integrity. Regulatory authorities need to be sure they are monitoring the environment in a cost-effective manner, measuring the appropriate things at the right places and times; it is not their role to provide data for the science community. Management measures will always cost those being managed; otherwise they would already be doing the required actions for economic reasons. It is essential, therefore, that managers have a high degree of confidence that the management regime imposed will bring about the desired outcome. The science community needs, as a matter of urgency, to develop robust metrics of ecosystem health and of the impact of specific activities, such as bottom trawling, for use as management tools. Indicator taxa and, in particular, measures of functional status, hold some promise.

Acknowledgments

The thinking presented here has been developed thanks to stimulating exchanges with, among others, Silvana Bustos-Baez, Simon Greenstreet, Odette Paramor, Jake Rice, Stuart Rogers, Catherine Scott, and Mark Tasker. This work was supported financially by the European Commission (study grant number Q5RS-2001-011685), University of Newcastle upon Tyne, UK, and a studentship to JB from CEFAS, UK. Presentation of this work was supported by conference grants to JB and LAR from the Ecological Society of America and American Fisheries Society. Fishing effort data for both regions were provided by CEFAS, Lowestoft, whose assistance is gratefully acknowledged. The manuscript was considerably improved thanks to comments provided by the two reviewers; we thank them.

References

Bonsdorff, E., and T. Pearson. 1999. Variation in the sublittoral macrozoobenthos of the Baltic Sea along environmental gradients: a functional group approach. Australian Journal of Ecology 24:312–326.

Bradshaw, C., L. O. Veale, and A. R. Brand. 2002. The role of scallop-dredge disturbance in long-term changes in Irish Sea benthic communities: a re-analysis of an historical dataset. Journal of Sea Research 47:161–184.

Bustos-Baez, S., and C. Frid. 2003. Using indicator species to assess the state of macrobenthic communities. Hydrobiologia 496:299–309.

Clarke, K. R., and R. M. Warwick. 1994. Change in marine communities: an approach to statistical analysis and interpretation. Natural Environment Research Council, Swindon, UK.

Ellis, J., and S. Rogers. 2000. The distribution, relative abundance and diversity of echinoderms in the eastern English Channel, Bristol Channel and Irish Sea. Journal of the Marine Biological Association of the United Kingdom 80:127–138.

Fauchald, K., and P. Jumars. 1979. The diet of worms: a study of polychaete feeding guilds. Oceanography and Marine Biology: an Annual Review 17:193–284.

Frid, C. L. J., J. B. Buchanan, and P. R. Garwood. 1996. Variability and stability in benthos: twenty-two years of monitoring off Northumberland. ICES Journal of Marine Science 53:978–980.

Frid, C. L. J., R. A. Clark, and J. A. Hall. 1999. Long-term changes in the benthos on a heavily fished ground off the NE coast of England. Marine Ecology Progress Series 188:13–20.

Hall, S. J., D. G. Raffaelli, and S. F. Thrush. 1994. Patchiness and disturbance in shallow water benthic assemblages. Pages 333–375 in D. Raffaelli, editor. Aquatic ecology—scale, pattern and process. Blackwell Scientific Publications, Oxford, UK.

Hall-Spencer, J., C. Froglia, R. J. A. Atkinson, and P. G. Moore. 1999. The impact of Rapido trawling for scallops, *Pecten jacobaeus* (L.), on the benthos of the Gulf of Venice. ICES Journal of Marine Science 56:111–124.

ICES (International Council for the Exploration of the Sea). 2001. Report of the ICES Advisory Committee on Ecosystems. ICES, Copenhagen.

Kaiser, M. J., D. B. Edwards, P. J. Armstrong, K. Radford, N. E. L. Lough, R. P. Flatt, and H. D. Jones. 1998. Changes in megafaunal benthic communities in different habitats after trawling disturbance. ICES Journal of Marine Science 55:353–361.

Kaiser, M. J., S. I. Rogers, and J. R. Ellis. 1999. Importance of benthic habitat complexity for demersal fish assemblages. Pages 212–223 in L. R. Benaka, editor. Fish habitat: essential fish habitat and rehabilitation. American Fisheries Society, Symposium 22, Bethesda, Maryland.

Kenny, A. J., and H. L. Rees. 1994. The effects of marine gravel extraction on the macrobenthos—early post-dredging recolonization. Marine Pollution Bulletin 28(7):442–447.

Kenny, A. J., and H. L. Rees. 1996. The effects of marine gravel extraction on the macrobenthos—results 2 years post-dredging. Marine Pollution Bulletin 32(8–9):615–622.

MINITAB. 1998. MINITAB, version 12.22. Minitab, Inc., State College, Pennsylvania.

Pearson, T., and R. Rosenberg. 1987. Feast and famine: structuring factors in marine benthic communities. Pages 373–398 in J. Gee and P. Giller, editors. Organization of communities past and present. Blackwell Scientific Publications, Oxford, UK.

PRIMER. 2001. PRIMER, version 5.2.2. PRIMER-E Ltd., Plymouth, UK.

Rogers, S., A. D. Rijnsdorp, U. Damm, and W. Vanhee. 1998. Demersal fish populations in the coastal waters of the UK and continental NW Europe from beam trawl survey data collected from 1990 to 1995. Journal of Sea Research 39:79–102.

Rogers, S. I., J. R. Ellis, and J. Dann. 2001. The association between arm damage of the common starfish, *Asterias rubens*, and fishing intensity determined from aerial observation. Sarsia 86:107–112.

Rumohr, H., and T. Kujawski. 2000. The impact of trawl fishery on the epifauna of the southern North Sea. ICES Journal of Marine Science 57:1389–1394.

Skjoldal, H. R., S. vanGool, H. Offringa, C. vanDam, J. Water, E. Degré, J. Bastinck, J. Pawlak, H. Lassen, M. Svelle, H.-G. Nilsen, and H. Lorentzen. . 1999. Workshop on ecological quality objectives (EcoQOs) for the North Sea. TemaNord, The Hague, The Netherlands.

Warwick, R. M., and K. R. Clarke. 1991. A comparison of some methods for analysing changes in benthic community structure. Journal of the Marine Biological Association of the United Kingdom 71(1):225–244.

Warwick, R. M., and K. R. Clarke. 1994. Relearning the ABC—taxonomic changes and abundance biomass relationships in disturbed benthic communities. Marine Biology 118(4):739–744.

Wassenberg, T., G. Dews, and S. D. Cook. 2002. The impact of fish trawls on megabenthos (sponges) on the north-west shelf of Australia. Fisheries Research 58:141–151.

Linking Fine-Scale Groundfish Distributions with Large-Scale Seafloor Maps: Issues and Challenges of Combining Biological and Geological Data

TARA J. ANDERSON[1]

National Marine Fisheries Service, U.S. Geological Survey and University of California Santa Cruz, Southwest Fisheries Science Center, Santa Cruz Laboratory, 110 Shaffer Road, Santa Cruz, California 95060, USA

MARY M. YOKLAVICH[2]

National Oceanic and Atmospheric Administration, National Marine Fisheries Service, Southwest Fisheries Science Center, Santa Cruz Laboratory, 110 Shaffer Road, Santa Cruz, California 95060, USA

STEPHEN L. EITTREIM[3]

U.S. Geological Survey, 345 Middlefield Road, Menlo Park, California 94025, USA

Abstract.— Groundfishes are an important fishery resource on the West Coast of the United States, but their population sizes have undergone dramatic declines in recent years. A number of area-based management and assessment strategies have been suggested to help rebuild and monitor these populations. Most groundfish species have strong affinities with specific substratum types, resulting in spatially patchy distributions. Hence, incorporating information on the types and amounts of seafloor substrata present (i.e., habitat availability) into sample design and biomass assessment of groundfish populations could increase the precision and accuracy of fish density and, consequently, population abundance estimates. The success of using habitat availability as a proxy for fish abundance, however, is contingent on the ability to identify those measurable habitat characteristics (e.g., substratum type, depth, relief, etc.) that fish respond to, precisely estimating fish densities within those habitats, and accurately characterizing and delineating these same characteristics across large areas (i.e., seafloor substratum maps). Characterizing seafloor substratum over a large area is not an exact process, but rather, it commonly uses remotely collected information (e.g., acoustic data, sediment samples, and local geology) to infer the seafloor characteristics. As a consequence, combining estimates of fine-scale fish density per unit area of habitat and the amount of each habitat type to generate a population abundance estimate will reflect the combination of the uncertainty and error in both estimates. If sampling uncertainty or error is large for either estimate (error and uncertainty around the largest mean will be the most critical), then the final population abundance estimate might be of little use to managers. We examine a case study in which an in situ groundfish survey, conducted in an area where a detailed seafloor substratum map was available, suggested that maps—even with suboptimal resolution—could be used to increase precision in estimates of fish density. In considering the issues and challenges encountered in linking geological and biological data, it is vital to determine the level of resolution required in the seafloor substratum map, which will depend on the degree of habitat specificity to which the organism responds. Further considerations include whether the mapping technology and methodology can achieve this level of resolution and, finally, whether this sampling approach is cost effective.

---

[1] Corresponding author: tara.anderson@noaa.gov
[2] E-mail: mary.yoklavich@noaa.gov
[3] E-mail: seittreim@usgs.gov

Introduction

Groundfishes, in particular rockfishes *Sebastes* spp., on the West Coast of the United States have supported a commercial fishery since the early 1900s and a substantial recreational fishery since the 1950s (Love et al. 2002). However, fishing pressure along with adverse environmental conditions over the last 3 decades have reduced stocks of many rockfish species (e.g., bocaccio *Sebastes paucispinis*, cowcod *S. levis*, Pacific ocean perch *S. alutus*, widow rockfish *S. entomelas*, canary rockfish *S. pinniger*, darkblotched rockfish *S. crameri*, and yelloweye rockfish *S. ruberrimus*) to below acceptable fishery levels (Pacific Fishery Management Council 2002a). Stock assessment is the cornerstone of all fisheries management approaches, where recruitment and adult numbers are estimated using a range of fishery-dependent and fishery-independent methods and used to model the population dynamics of targeted species. Recent assessments of populations of bocaccio, darkblotched rockfish, canary rockfish, and yelloweye rockfish indicated that these stocks were at less than 25% of their unexploited size and, as a consequence, were declared overfished. The bycatch of these species in other fisheries alone would account for the allowable take in 2002 and 2003 (Pacific Fishery Management Council 2002a). These findings recently led to the closure of all targeted groundfish fisheries on the shelf by the Pacific Fishery Management Council for California, Oregon, and Washington (Pacific Fishery Management Council 2002b).

Groundfish species are not randomly distributed over the seafloor but rather are associated with specific substratum types such as soft sediments, cobbles, boulders, and bedrock (Stein et al. 1992; Yoklavich et al. 2000, 2002; Love et al. 2002). These abiotic substratum types, in part, define groundfish "habitat" (i.e., "the locality in which a plant or animal naturally grows or lives," Oxford English Dictionary 1989). We recognize that biotic components of habitat (e.g., plants and invertebrates that provide structural refuge for fishes) and a range of environmental factors ("environment" is "the conditions under which a thing lives," Oxford English Dictionary 1989), such as geographic range, water depth, and food availability, also are important in defining where an organism lives, but in this paper, we focus our discussions on substratum type (sediment facies) as an important component of, and covariate with, groundfish habitat. Terminologies, such as habitat, substratum, and environment, are used commonly in biology, geology, oceanography, and chemistry, to name a few disciplines, but are frequently used with different intent within context and discipline. In writing this paper for a multidisciplinary audience, we have become more aware of the subtle differences in terminology among disciplines, which in itself highlights an additional challenge to linking interdisciplinary research.

Area-based management, in which abiotic habitats (e.g., substratum types) that can support large numbers of groundfishes are considered as a proxy for the populations themselves, is now receiving more attention. The National Oceanic and Atmospheric Administration (NOAA), National Marine Fisheries Service (NMFS), and fishery management councils are mandated to identify, describe, and protect essential fish habitats under the Magnuson-Stevens Fishery Conservation and Management Act. At the state level, California's Marine Life Management Act and Marine Life Protection Act explicitly recognize that habitats, not just species populations, must be managed (Weber and Heneman 2000). Where the relationship between a species and its habitat can be reliably determined (e.g., mean and SE) and the distribution of those habitats can be measured (e.g., by seafloor mapping ± SE), then management decisions can be made about the magnitude of threats to a species and what management actions can be taken. Recent and proposed designations of marine protected areas (MPAs) are being used to protect critical habitats as a means either to complement catch-based fisheries management (Rowley 1994; Hastings and Botsford 1999; Murray et al. 1999; Mangel 2000a, 2000b; Weber and Heneman 2000) or to conserve marine biodiversity and marine resources that are used by a wide range of stakeholders (Bohnsack 1998; Weber and Heneman 2000). This move toward managing areas rather than single species implies that seafloor mapping and inventory of characteristics such as substratum type must be explicitly linked to the distribution of exploited species.

Mobile organisms such as groundfishes can be specific in their choice of habitats. Their local distribution can be modified by the type, structure, and patchiness of the habitat. In addition, organism–habitat relationships may occur at a number of different spatial scales ranging from m, through tens to hundreds of m, to the scale of hundreds of km (Wiens 1989). Understanding and predicting fish distribution and abundance in space, therefore, requires an implicit understanding of associations with habitat over these scales. Including information on fish–habitat associations in area-based population assessments also is likely to be fundamental to making more accurate and precise density estimates. Similarly, for other spatially explicit management approaches, such as implementation of MPAs, fish–habitat information is important to determine what proportion of the population is receiving protection and to monitor the effectiveness of the MPA (Yoklavich et al. 2002). The successful inclusion of habitat information into area-based fishery management approaches and other spatially explicit approaches such as MPAs will be dependent on obtaining sound estimates of fish densities at the scale of sampling and on the ability to scale up those densities to the spatial extent of the biological population or management area. This presents a considerable challenge.

The simplest approach to estimating population abundance is to define the area occupied by the species and sample fish densities in an unbiased way over that area. Fish densities could then be multiplied by the total area occupied to yield an estimate of population abundance. Here, the sample universe (i.e., the total study area that is to be extrapolated to) should be defined (e.g., fish living in a specified area or in a specified depth range) and then subdivided into a set of possible samples from which an unbiased subset is selected (see Underwood 1998). There are many ways of achieving unbiased sampling, such as systematic, random stratified in space, or complete random sampling, and each of these aims to distribute samples over the area of inference so that all areas within the sample universe have an equal probability of being sampled (see Legendre and Legendre 1998).

Bias is anathema to this sampling model, but it is important to realize that bias may not occur only in the sample allocation procedure but also in the methodological application itself. Many methods have been used to sample groundfish densities or catchability. Visual surveys have been used to estimate fish densities in a range of habitats and provide measures of fine-scale relationships between a species and habitat type. The success of visual surveys is contingent on their ability to sample fish in proportion to their availability and across the range of habitats that they occupy. Visual surveys undertaken by scuba, for example, are limited to nearshore, shallow habitats (<30 m) and might not be able to sample the deeper portion of the population. Deepwater visual fish–habitat surveys require more technically elaborate sampling equipment, such as manned submersibles (Pearcy et al. 1989; Krieger 1993; Yoklavich et al. 2000, 2002) and remotely operated vehicles (ROVs; O'Connell and Carlile 1994; Adams et al. 1995), and might incur biases due to the presence of illuminated vehicles and differences in fish detectability with different equipment.

Trawl surveys (both fishery dependent and independent) also are widely used to estimate fish densities. However, trawl effectiveness is strongly dependent on substratum type. Trawl gear can get tangled, damaged, and lost in complex, rocky habitats and, consequently, a potentially important subset of groundfish habitat cannot be adequately sampled with this type of gear (Jagielo et al. 2003). Fishery hook-and-line surveys, on the other hand, provide an alternate means of sampling abundance in high-relief habitats. However, hook-and-line surveys do not sample densities directly; they actually sample catchability. Converting catchability to density is itself subject to error as fish are "enticed" from an unknown distance, and catchability may vary with time and local fish density (see Ralston et al. 1986). Clearly, as with any type of survey tool, there are many methodological biases that need to be considered in addition to potential biases in sample allocation.

If unbiased sampling can be achieved by careful design and methodological consideration, then calculating the population mean (within a given area) and its associated error is straightforward. For example, if the total study area is 1,000 m^2, with an estimated mean density of 5 fish per 10 m^2 and SE of 2 fish per 10 m^2, then the predicted extrapolated mean abundance is simply a constant multiplied by the density (5), where the constant is the multiplicative extrapolation factor k (1,000/10 = 100). The standard error of this estimate (SE_{est}) is $k \times SE$, or 100 × 2. In other words, the predicted extrapolated abundance of fish in a 1,000 m^2 area is 500 ± 200 individuals.

$$SE_{est} = (\text{total area/sample area}) \times SE$$
$$= 1,000/10 \times 2$$
$$= 200.$$

Two advantages of this sampling approach are that no information is required about habitat and that the fish density estimates will be unbiased. The primary disadvantage of this approach is that where fish distributions are patchy among habitat types, many samples will be required to improve precision of the density estimate and, hence, the accuracy of the total abundance estimate (Creese and Kingsford 1998).

A more efficient and precise way of estimating fish densities might be to explicitly incorporate information about the habitat (e.g., substrata used by fishes) into a multistaged sample procedure (Cochran 1977). Stratification by habitat allows for sampling effort to be differentially allocated based on variability in fish densities with habitats and does not require that all habitats be sampled in proportion to their availability. However, in order to extrapolate fish densities to the sample universe, the relative amounts of each habitat must be known. In a stratified sampling design, the habitat-area estimates are themselves estimated with uncertainty and error and, consequently, the SEs must be calculated differently. Here, for example, estimates of both fish density and habitat-area are known for three habitat types (a, b, and c). Habitat a covers 280 m^2 ± 60 m^2 and contains 6 ± 2 fish per 10 m^2. Habitat

b covers 400 m² ± 40 m² and contains 2 ± 1 fish per 10 m². Habitat c covers 320 m² ± 20 m² and contains 9 ± 4 fish per 10 m².

The standard error can be estimated by first calculating the three multiplicative k factors and by rescaling the SEs by the sample unit size.

k = estimated area/sample unit area;
SE_k = SE/sample unit size:

k_a = 280/10 = 28; SE_{k_a} = 60/10 = 6;
k_b = 400/10 = 40; SE_{k_b} = 40/10 = 4;
k_c = 320/10 = 32; SE_{k_c} = 20/10 = 2.

The combined error is then calculated using the formula for error of a compound quantity:

$$SE = \sqrt{mean_1^2 \times SE_1^2 + mean_2^2 \times SE_2^2}.$$

Thus, the error for the Habitat a abundance estimate, 6 × 28 = 168, is:

$$SE = \sqrt{k_a^2 \times SE_{k_a}^2 + 6^2 \times 2^2}$$
$$= \sqrt{28^2 \times 6^2 + 6^2 \times 2^2}$$
$$= 168.$$

Similarly, the abundance estimates for the other habitats are 80 ± 160 fish in Habitat b and 288 ± 73 fish in Habitat c. These abundance estimates can be combined to yield a total abundance of 168 + 80 + 288 = 536 individuals, with an SE of 243:

$$SE_{total} = \sqrt{SE_a^2 + SE_b^2 + SE_c^2}$$
$$= \sqrt{168^2 + 160^2 + 73^2}$$
$$= 243.$$

Three points are apparent from these calculations. First, any multiplication of variables measured with error will itself have a large error. Second, it is more important to reduce error on the variable with the largest mean. Third, to successfully combine fish density estimates with habitat availability, the fusion of scientific methods from biology and geology will be required at scales both relevant to the organism and appropriate to the management measures. If the aim is to extrapolate fish densities to the larger sample universe, the amount of each substrata present within this sample universe must be estimated as precisely as possible. Defining the sample universe, however, will depend on the spatiotemporal range of the organism. Where a larger "regional" sample universe is critical, determining the relationship between groundfishes and habitat over fishery management scales, such as Washington–Oregon–California–Mexico, would require habitats to be sampled over the species latitude and depth ranges and should be measured in conjunction with changes in broader range chemical and physical oceanography. Regardless of the spatial scale, the initial estimate of fish density will be greatly improved by incorporating a habitat-stratified sampling design. However, the accuracy of the final estimate extrapolated to the sample universe will now depend largely on the level of error and uncertainty in the habitat-area estimate. How much uncertainty and error is present in the habitat-area estimates and how are these estimates derived? To understand this, one first needs to know how seafloor substratum maps are produced.

Multibeam and side-scan sonar methods are the primary tools used to map and infer the surficial geology of large areas (one to hundreds of km) of the seafloor. Multibeam and side-scan sonar operate in different ways to characterize the seafloor (Miller et al. 1997). Multibeam sonar is generally mounted on the vessel and measures seafloor depths over a wide swath in addition to acoustic backscatter (reflectivity) if suitably equipped. Navigation is precise because a differential geographic positioning system is used in combination with corrections for vessel movement including heave, pitch, and roll. In contrast, side-scan sonar systems are usually towed at a depth behind a vessel and record acoustic backscatter (reflectivity) of the bottom. Side-scan sonar systems have relatively large navigational uncertainties in their spatial position due to "layback" error and cross-track ocean currents. Side-scan sonar does not collect bathymetric data.

An advantage of multibeam systems is that they provide good spatially referenced depth measurements, but the spatial resolution of the measurements is dependent on the beam resolution and the water depth. Each beam will average the depth signal over a greater area at increased depth, and at increased angles from nadir. Although side-scan sonar systems are hampered by poor absolute positioning of each data point, represented by a pixel, higher image resolution (more pixels per m²) can be achieved by flying the system close to the bottom.

Both methods can provide clues to the geological composition of the seafloor. Acoustic backscatter, or the amount of acoustic energy that is reflected back to the receiver, can be used to infer shapes and textures of the seafloor (Urick 1983; Blondel and Murton 1997). For example, hard materials, such as rock, reflect more sound than do soft materials, such as mud; and rough surfaces, which reflect energy at an angle away from the incident wave (Blondel and Murton 1997), generate more backscatter than do smooth surfaces. Relief can also be measured directly using multibeam bathymetry or inferred using side-scan shadow length and sensor height off bottom (e.g., Yoklavich et al. 1997).

Translating acoustic data into seafloor maps is not a trivial procedure. The physics of sound wave propagation in water, in combination with reflective and absorptive properties of the surficial and buried substratum, are complex and present an initial technical challenge. Subjective visual interpretations of black-and-white acoustic images have been commonly employed to identify seafloor features (e.g., Eittreim et al. 2002; Yoklavich et al. 2002). The subjectivity of this approach has led to increased efforts, using models and expert systems, to objectively define seafloor characteristics by interpreting the acoustic signal using a set of rules and conditions (e.g., Mitchell and Hughes Clarke 1994; Dartnell 2000).

Given the potential uncertainties in acoustic interpretation, groundtruthing using direct visual observations, seafloor samples, seismic profiles, or photography (Burrough 1986; Gardner et al. 1991; Tlusty et al. 2000; Bax and Williams 2001) is an important component of seafloor mapping. Groundtruthing can be, and is commonly, used to "train" the interpretation of the backscatter signatures. It can also be, but less commonly, used to "verify" the interpretation of the existing seafloor map. It is preferable that samples that are to be used to train interpretations should always be collected in conjunction with acoustic surveys. In contrast, samples used to verify the existing interpreted maps can be collected at any time, as they provide an independent estimate of map accuracy. However, seafloor groundtruthing samples are costly to collect and, hence, many groundtruthing efforts are opportunistic and might not be optimal at measuring map accuracy.

The next level of complexity lies in interpreting the acoustic and geological information as groundfish habitat. Fishes respond to their habitat at a range of spatial scales, usually in species-specific, idiosyncratic ways. For example, a fish species might perceive boulders and bedrock as contrasting habitats and use them in different ways (O'Connell and Carlile 1993). Successfully linking this species to one or the other habitat will be contingent on the ability of the seafloor substratum map to reliably resolve these two habitats. A disjunction in scales of resolution between the seafloor substratum maps and the habitats perceived and used by the fishes could be considerable. For example, assume species a was associated with boulder habitats at a density of 5 ± 2 fish per 10 m^2 but was not found over bedrock. If a study area contained 500 m^2 of hard substratum, of which 100 m^2 was composed of boulders, the estimated fish abundance would be contingent on the ability of the mapping process to distinguish boulders from bedrock. If the true 100-m^2 area of boulder area could be reliably classified, then the estimated abundance of the population would be 50 ± 20 fish. However, if boulders could not be reliably distinguished from bedrock, then it would have to be assumed that either the entire 500-m^2 area was composed of bedrock and, hence, it would be surmised that there were no fish present or, conversely, that the area was entirely composed of boulders and, hence, the abundance would be estimated at 250 ± 100 fish. Incorporating this magnitude of uncertainty and error into population abundance estimates is unlikely to be valuable to resource management.

There is a clear need, therefore, to reconcile the resolution of seafloor substratum maps with the resolution at which fishes perceive and respond to their habitat if we are to reduce this uncertainty. We demonstrate some of the issues and challenges related to linking fine-scale biological information of important fishery species with a seafloor substratum map using a real example in southern Monterey Bay, California.

Case Study

This study is not meant to represent either a blueprint or an example of combining "optimal" technologies from both the fish sampling and geological mapping perspectives. Instead, it represents a real situation where in situ fish–habitat data (*Delta* submersible survey) have been sampled over an area of seafloor that also was geologically mapped (multibeam survey) in independent studies. We use this opportunity to illustrate several issues and challenges related to linking biological and geological data. Our study area covers a 12-km × 10-km area in the vicinity of Italian and Portuguese ledges in southern Monterey Bay off Point Piños in central California (Figures 1, 2). This general region has been subject to an important groundfish fishery that operated commercially since the mid- to late 1800s, which initially targeted a generic "red rockfish" species later identified as including a variety of rockfish species such as bocaccio, widow rockfish, yellowtail rockfish *S. flavidus*, vermilion rockfish *S. miniatus*, and canary rockfish (Phillips 1939). A sizeable recreational fishery also has existed since the 1970s.

A fine-scale in situ groundfish–habitat survey was conducted using the *Delta* submersible in October of 1993. Thirty-three georeferenced, visually censused strip transects (2 m wide × 15 min duration) were allocated to different substrata based on available bathymetry, seafloor sediments (Galliher 1932), and side-scan surveys (H. G. Greene, Moss Landing Marine Laboratories, and Yoklavich, unpublished data). The initial groundfish–habitat survey was intended to determine the relationship between demersal rockfishes and fine-

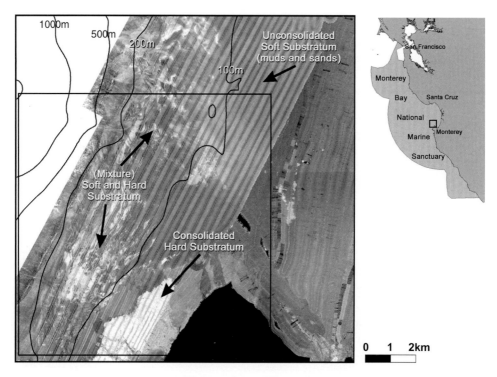

Figure 1. Raw acoustic multibeam backscatter image (left) of Italian and Portuguese ledges, Monterey Bay National Marine Sanctuary (MBNMS), central California. Three broad-scale substrata types were reliably distinguished from the acoustic image: consolidated hard substratum (white); unconsolidated soft sediments (sands and muds; gray); and mixed areas (areas with a mixture of both hard and soft substratum). The dark areas are areas where there was no mapping. Diagonal lines = trackline noise. The scale of the acoustic image is in the bottom right of the image. Depths greater than 600 m were not adequately sampled by the Simrad EM1000 and are consequently masked; depth contours are depicted for 100 m, 200 m, 500 m and 1,000 m. Inset (right) depicts the central California coastline and the location of MBNMS. The square box indicates the location of the biological study.

scale substrata (i.e., potential habitats). Transects were conducted from the starboard side of the *Delta*. Fish within each transect were identified and counted in situ by the observer and were also recorded with an external video camera mounted on the starboard side of the submersible. Postprocessing of audio (observer's counts and identification) and videotape were used to categorize and demarcate fine-scale habitat types within each transect based on the primary (>50% cover) and secondary (>20% cover) habitat protocol of Yoklavich et al. (2000) and Stein et al. (1992). Groundfish–habitat relationships were evaluated at both a broad scale (transects within strata) and a fine scale (patches within transects). From this we ascertained the habitat components that each species responded to and at which scales these responses were important and measurable.

A year later, in 1995, the 12-km × 10-km study area was acoustically surveyed as part of a larger geological survey of the Monterey Bay National Marine Sanctuary aimed at characterizing the seafloor geology and substrata (Eittreim et al. 2002). Multibeam echo sounding was conducted using a hull-mounted Kongsberg Simrad EM1000 multibeam bathymetric system that recorded bathymetry (5-m resolution) and multibeam acoustic backscatter (2.5-m resolution). Geological seafloor characterizations were interpreted based on the acoustic backscatter signature and auxiliary seafloor samples such as seismic reflection profiles, sediment samples, and drop-camera photography (Anima et al. 2002; Eittreim et al. 2002). Six seafloor geological types, such as the Purisima and Monterey formations, were characterized and delineated (Eittreim et al. 2002).

The provision of this seafloor map presented the opportunity to use the in situ visual strip transects collected from the *Delta* submersible to determine how accurately the seafloor map depicted groundfish habitats. Three substratum categories (hard, hard mixed

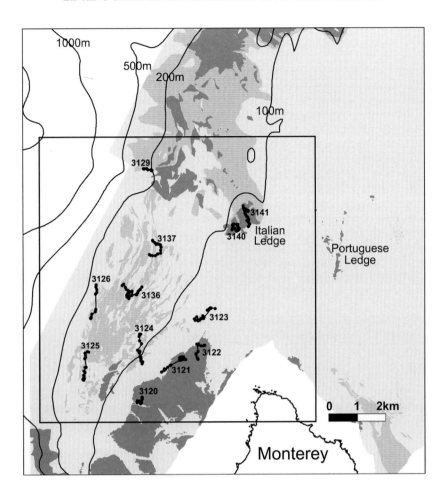

Figure 2. Seafloor map of the three broad-scale substratum types in the vicinity of Italian and Portuguese ledges, Monterey Bay National Marine Sanctuary, central California, with track lines of 12 georeferenced *Delta* submersible dives (dive numbers 3120–3141). Fish abundances and finer-scale habitats were estimated from 33 strip transects. Hard substratum are depicted as dark gray areas, mixed substratum (i.e., areas of hard mixed with soft) are depicted as medium gray areas with hatching, and soft substratum (i.e., areas of contiguous soft sediments) are depicted by light gray areas. The square box indicates the 10-km × 12-km sample universe of the biological study.

with soft, and soft substratum) could be reliably depicted from the seafloor map (Figure 2): 69.6% of the study site (120 km² seafloor) was comprised of contiguous soft substratum and was described as such by visual observations; 19.9% was comprised of hard substratum mixed with soft substratum observed as low-relief patches of cobbles within a matrix of mud; and 10.5% was comprised of hard substratum observed as complex rock outcrops, boulder, and sand patches (Figures 2, 3).

Groundtruthing data, collected from the biological and multibeam mapping surveys, verified the three interpreted substratum categories (including several strata interfaces) and their geographic locations. However, these data did not coincide with all of the backscatter interpretations and verified only a small number of the substrata boundaries seen in the backscatter image. This is important because verifying the position of habitat edges is a critical step to obtaining an accurate estimate of the area comprising each substratum. As we have little independent data to verify the exact location of edges, we have little means to determine the potential error around the estimates of the three substratum categories. Therefore, we use mean estimates of the area occupied by the three substratum categories, but we cannot provide an independent estimate of error around these means. Consequently, the accuracy (cf. precision) of the calculated estimates is unknown.

While fish survey transects were sampled from the three preidentified substratum categories (i.e., hard, mixed, soft), video analysis within transects revealed finer scales of seafloor heterogeneity than those identified from the seafloor maps (Figure 3). Importantly, many fish species responded in different ways to this finer-scale complexity. For the purpose of this case study, we will present three examples of groundfish–habitat associations to illustrate both the range in fish–habitat associations and the corresponding level of resolution in the seafloor map required to gain useful population estimates. Sanddabs *Citharichthys* spp. were highest in transects placed in soft and mixed substrata. At finer scales (within-transects), sanddabs were found in soft-sediment patches, both within contiguous soft-substratum transects and, to a lesser degree, in the soft-sediment matrix of mixed substratum transects (Figure 4a). Halfbanded rockfish *S. semicinctus* were found primarily in transects placed in the mixed substratum. However, at finer scales, unlike sanddabs, halfbanded rockfish were aggregated over discrete habitat patches containing cobbles and boulders surrounded by soft-sediments (Figure 4b). The three broad-scale substrata (hard, mixed, and soft) adequately described the distribution and abundance of both sanddabs and halfbanded rockfish. These three substrata could also be reliably distinguished and quantified from the multibeam seafloor substratum map. Consequently, incorporating these three substrata into the biological sampling design and estimation procedure should yield more precise estimates of both sanddab and halfbanded rockfish population sizes than sampling or estimation without regard to these substrata.

In contrast, densities of squarespot rockfish *S. hopkinsi* were highest in transects placed in hard substratum. At finer scales, squarespot rockfish were strongly aggregated in rock-boulder patches (i.e., patches containing >50% bedrock and >20% boulders) and boulder-sand patches (>50% boulders, >20% sand; Figure 4c). Unlike cobbles and boulders in the previous examples, different types of hard substratum within a single outcrop could not be resolved from the seafloor substratum map. Consequently, where differences in habitat heterogeneity within a rock outcrop are large, we would expect this to be reflected in high between-transect variation in fish densities. An optimal solution would be to introduce another level of stratification, in which rock-boulder and boulder-sand could be distinguished from other habitats within rock outcrops. However, the lack of acoustic contrast between different types of hard substratum means that this would not be achieved with the existing seafloor map. Although stratification using the three broad-scale habitat substrata would still generate a more precise density estimate than a simple random sample design, the benefits for squarespot rockfish would be less than for those species whose habitat associations corresponded to acoustically distinct seafloor characteristics.

Given that species-specific habitat preferences are the norm, what are the consequences of ignoring within-transect habitat heterogeneity? The answer will depend, in part, on the spatial complexity of the seafloor and how homogeneous the transects themselves are. If transects within a broad habitat differ in their within-transect habitat structure, this will be manifested in a highly variable (imprecise) estimate of species density. Importantly, the size of this effect also will be strongly dependent on the behavior of each individual species. Sanddabs, although associated with contiguous soft sediments and soft sediments within mixed habitats, are usually solitary or found in small groups. Differences in abundance between transects, due to differences in sediment availability, are, therefore, unlikely to be large. In contrast, many rockfishes such as the halfbanded rockfish and squarespot rockfish can form

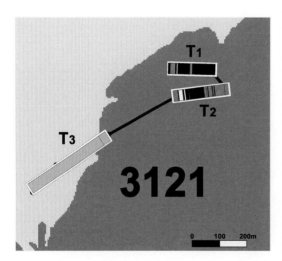

Figure 3. Substrata types in the area around dive 3121. The background habitat was derived from the acoustic map while the three rectangles represent the three strip transects (depicted as T1, T2, and T3) conducted during dive 3121. Transects within the acoustically inferred hard substratum are heterogeneous at finer scales of resolution, determined by visual observation during the *Delta* manned submersible dives. The hard substrata (T1 and T2) were, at the fine scale, composed of areas of rock (dark fill), boulders (hashed), cobbles (white), and soft sediment (gray). In contrast, transects run within soft substrata (T3) were relatively homogeneous areas of soft sediment (gray).

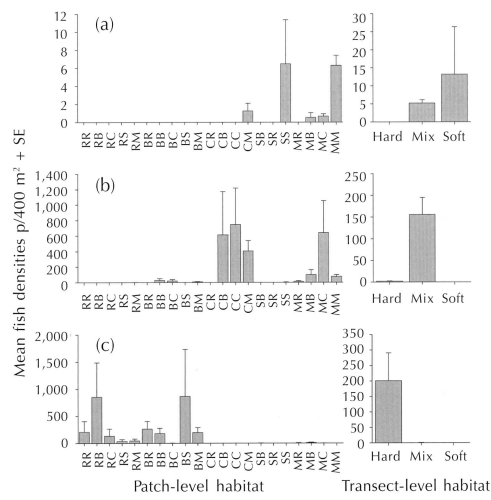

Figure 4. Fish densities for patch (within-transect habitat types) and transect habitats (coarse-resolution strata) for three groundfish species (sanddabs, a; halfbanded rockfish, b; and squarespot rockfish, c). The left series of graphs depict densities of fishes in patch habitats defined by a primary and secondary habitat type, where R = rock, B = boulders, C = cobbles, S = sand, and M = mud habitats. For example, patches with a primary habitat of rock (>50% of the patch) and a secondary habitat (>20% of the patch) of boulders would represent a rock–boulder patch and would be depicted as RB. The right series of graphs depicts densities of fishes within the three broad-scale substratum.

large aggregations around very specific habitat features (e.g., small patches of cobbles). Consequently, small errors in estimating the areal coverage of these features will be magnified by this group behavior and, therefore, translated into large errors in the overall abundance of these species.

Issues and Challenges

Even though the biological and geological approaches used in this case study were not designed explicitly to link fine-scale fish distributions with seafloor substratum maps to estimate total population abundance, the data were of sufficient resolution to potentially be useful. Abundance estimates of species that were associated with the three substrata, readily distinguished in the seafloor map, are likely to be greatly improved by incorporation of these 3 substrata into the sampling design and, ultimately, the statistical estimation of the mean density. In contrast, improvement of abundance estimates was less for those species that occupied specific types of hard-rock habitat. This is due largely to the inability of the equipment used in the mapping survey to detect finer-scale habitat boundaries within the broader category of hard substratum even though some groundfish species were associated with habitats at finer scales of resolution. It must be recognized, of course, that the acoustic data used in developing the geological classification of Monterey Bay was not designed to resolve the fine scales important to groundfish species, which emphasizes that detailed substratum maps may not provide the necessary information or resolution to achieve all objectives.

As some fishes responded to habitat features at spatial scales less than those resolved by commonly used acoustic systems, we were challenged to find solutions. A first approach would be to generate higher-resolution maps. The development of remote mapping technologies to characterize finer-resolution seafloor maps is an area of active research (e.g., high-resolution multibeam, improved laser line scan, LIDAR), but today's solutions are expensive and cover limited areas. An alternative approach is to use the existing suboptimal seafloor substratum maps and conduct formal groundtruthing surveys to differentiate and delineate the finer-scale seafloor characteristics. A grid of nonoverlapping transects using visual observations either over the extent (i.e., the length and width) of the area or at selected areas of interest, could enable internal boundaries and interfaces of finer-scale habitats to be delineated without incurring the cost of remapping the entire region. This method, however, relies on precise navigation of the groundtruth survey system, such as submersible, towed video camera, or ROV.

These options must each be considered in the context of the question to be answered. The original intent of our case study was to assess whether information from seafloor substratum maps could be used to obtain more precise estimates of groundfish abundance within a particular area. To resolve this question, it was necessary to distinguish between the types of soft sediment, hard mixed with soft sediment, rocky outcrops, and boulders within bedrock habitats. This level of resolution was dictated by the fishes' associations, identified from in situ fish–habitat observations. With the exception of the boulder–bedrock distinction, the existing map fulfilled these requirements. From the biological standpoint, while it was then necessary to estimate the mean groundfish density and its associated SE within each of these three broad-scale substratum types (i.e., hard, mixed, soft), it is important to note that the biological sampling (e.g., of each organism) does not require precise georeferencing; as long as the sampling unit (in this case, the transect) was conducted within the broad-scale substratum category, then the true position of the sample is irrelevant for generating a classical estimator of the mean. Precise sample positioning is only required if using a geostatistical estimation process such as kriging (Burrough 1986).

A profitable approach, from a practical standpoint, may be to identify the minimum geological and biological data requirements required to answer the question. In this particular example, it was sufficient to know the relative proportions of the three broad-scale substratum types. Even the initial side-scan sonar information, despite its limited positioning ability, could provide this information. The densities of fishes that occupied soft-sediment and mixed substrata were generally less variable than those in hard substratum. Consequently, less biological sample effort would be required in soft and mixed habitats to achieve a given level of precision. The inability of the existing map to distinguish between bedrock and boulder could be countered in two ways: (1) higher-resolution mapping or grid-coverage groundtruthing in selected areas of interest (e.g., complex rock outcrops) could enable more sample strata to be defined; or (2) more biological samples (i.e., submersible transects) could be allocated within the hard sample stratum to reduce the SE of the mean estimate. The cost effectiveness of submersible time versus mapping operations should then be evaluated.

Summary

Area-based population assessment holds considerable promise if fish respond to coarse-resolution habitats such as hard, mixed, and soft substrata. Empirical models of groundfish-habitat associations, in conjunction with seafloor substratum maps, could improve area-based management of these resources. Successful application of such models and maps may be more difficult when the arrangement of substrata is complex. For example, cobble habitat may lie within and, hence, be indistinguishable from bedrock habitat. If groundfish respond to the complex structure of the rock matrix itself, then this is of less importance because the broad-scale "hard" classification will be an adequate sample stratum. However, if fish respond to specific habitats such as rocks, boulders, or cobbles within this stratum, then delineating these internal boundaries would be a useful enterprise but might present a major challenge. Alternatively, it might be more cost efficient to allocate more samples to complex strata to reduce the variability of the fish-density estimate. The success with which fine-scale biology can be linked with large-scale seafloor maps will, to some degree, be contingent on the complexity of the fish-habitat association and on our ability to discern this fine-scale complexity across large-scale management areas. However, even in a less-than-perfect world, routine incorporation of geological and biological information is likely to provide more precise answers to fisheries management questions than either approach in isolation.

Acknowledgments

We would like to thank all those who helped in the biological and geological surveys, especially G. Cailliet, R. Lea, M. Love, G. Moreno, R. Parrish, P. Reilly, L. Snook, R. Anima, B. Edwards, W. Lee, A. Stevenson, and F. Wong; the Monterey Bay National Marine Sanctuary for logistical support during the mapping survey; and R. Slater, D. Slater, and C. Ijames of *Delta* Oceanographics. Many thanks go to M. H. Carr, G. Cochrane, S. Ralston, S. Sogard, and especially C. Syms, two anonymous reviewers, and the editors of these proceedings for their comments on various aspects of this manuscript. This multidisciplinary research was partially supported by NOAA's National Undersea Research Program, West Coast and Polar Undersea Research Center, University of Alaska, Fairbanks (Grants UAF-92-0063 and UAF-93-0036), and a postdoctoral fellowship to Tara Anderson funded jointly by NMFS, U.S. Geological Survey, University of California, Santa Cruz, and the National Marine Protected Areas Center Science Institute.

References

Adams, P. B., J. L. Butler, C. H. Baxter, T. E. Laidig, K. A. Dahlin, and W. W. Wakefield. 1995. Population estimates of Pacific coast groundfishes from video transects and swept-area trawls. U.S. National Marine Fisheries Service Fishery Bulletin 93:446–455.

Anima, R. J., S. L. Eittreim, B. D. Edwards, and A. J. Stevenson. 2002. Nearshore morphology and late Quaternary geologic framework of the northern Monterey Bay Marine Sanctuary, California. Marine Geology 181:35–54.

Bax, N. J., and A. Williams. 2001. Seafloor habitat on the south-eastern Australian continental shelf: context, vulnerability and monitoring. Marine Freshwater Research 52:491–512.

Blondel, P., and B. J. Murton. 1997. Handbook of seafloor sonar imagery. Wiley, West Sussex, UK.

Bohnsack, J. A. 1998. Application of marine reserves to reef fisheries management. Australian Journal of Ecology 23:298–304.

Burrough, P. A. 1986. Principles of geographical information systems for land resources assessment. Clarendon Press, Monograph on Soils and Resources Survey 12, Oxford, UK.

Cochran, W. G. 1977. Sampling techniques. Wiley, Wiley Series in Probability and Mathematical Statistics, New York.

Creese, R. G., and M. J. Kingsford. 1998. Organisms of reef and soft substrata intertidal environments. Pages 167–193 *in* M. Kingsford and C. Battershill, editors. Studying temperate marine environments: a handbook for ecologists. Canterbury University Press, Christchurch, New Zealand.

Dartnell, P. 2000. Applying remote sensing techniques to map seafloor geology/habitat relationships. Master's thesis. San Francisco State University, San Francisco.

Eittreim, S. L., R. J. Anima, and A. J. Stevenson. 2002. Seafloor geology of the Monterey Bay area continental shelf. Marine Geology 181:3–34.

Galliher, E. W. 1932. Sediments of Monterey Bay, California. Report of the State Mineralogist 28(1):42–79.

Gardner, J. V., M. E. Field, H. Lee, and B. E. Edwards. 1991. Ground-truthing 6.5-kHz side scan sonographs: what are we really imaging? Journal of Geophysical Research 96:5955–5974.

Hastings, A., and L. W. Botsford. 1999. Equivalence in yield from marine reserves and traditional fisheries management. Science 284:1537–1538.

Jagielo, T., A. Hoffman, J. Tagart, and M. Zimmermann. 2003. Demersal groundfish densities in trawlable and untrawlable habitats off Washington: implications for estimation of the trawl survey habitat bias. U.S. National Marine Fisheries Service Fishery Bulletin 101:545–565.

Krieger, K. J. 1993. Distribution and abundance of rockfish determined from a submersible and by bottom trawling. U.S. National Marine Fisheries Service Fishery Bulletin 91:87–96.

Legendre, P., and L. Legendre. 1998. Numerical ecology, 2nd English edition. Elsevier BV, Amsterdam.

Love M. S., M. Yoklavich, and L. Thorsteinson. 2002. The rockfishes of the northeast Pacific. University of California Press, Berkeley.

Mangel, M. 2000a. On the fraction of habitat allocated to marine reserves. Ecology Letters 3:15–22.

Mangel, M. 2000b. Trade-offs between fish habitat and fishing mortality and the role of reserves. Bulletin of Marine Science 66:663–674.

Miller, J. E., J. Hughes Clarke, and J. Patterson. 1997. How effectively have you covered your bottom? Hydrographic Journal 83:3–10.

Mitchell, N. C., and J. E. Hughes Clarke. 1994. Classification of seafloor geology using multibeam sonar data from the Scotian Shelf. Marine Geology 121:143–160.

Murray, S., R. Ambrose, J. Bohnsack, L. Botsford, M. Carr, G. Davis, P. Dayton, D. Gotshall, D. Gunderson, M. Hixon, J. Lubchenco, M. Mangel, A. MacCall, D. McArdle, J. Ogden, J. Roughgarden, R. Starr, M. Tegner, and M. Yoklavich. 1999. No-take reserve networks: sustaining fishery populations and marine ecosystems. Fisheries 24(11):11–25.

O'Connell, V. M., and D. W. Carlile. 1993. Habitat-specific density of adult yelloweye rockfish *Sebastes ruberrimus* in the eastern Gulf of Alaska. U.S. National Marine Fisheries Service Fishery Bulletin 91:304–309.

O'Connell, V. M., and D. W. Carlile. 1994. Comparison of a remotely operated vehicle and a submersible for estimating abundance of demersal shelf rockfishes in the eastern Gulf of Alaska. North American Journal of Fisheries Management 14:196–201.

Oxford English Dictionary. 1989. Oxford English Dictionary, 2nd edition. Oxford University Press, Oxford, UK.

Pacific Fishery Management Council. 2002a. Status of the Pacific coast groundfish fishery through 2002 and recommended acceptable biological catches for 2003: stock assessment and fishery evaluation. Pacific Fishery Management Council, Portland, Oregon.

Pacific Fishery Management Council. 2002b. Inseason adjustments adopted for 2003 groundfish fishery. Pacific Council News 26(2):1, 8. Available: www.pcouncil.org/newsletters/2002/summer02.pdf

Pearcy, W. G., D. L. Stein, M. A. Hixon, E. K. Pikitch, W. H. Barss, and R. M. Starr. 1989. Submersible observations of deep-reef fishes of Heceta Bank, Oregon. U.S. National Marine Fisheries Service Fishery Bulletin 87:955–965.

Phillips, J. B. 1939. The rockfish of the Monterey wholesale fish markets. California Fish and Game 25:214–225.

Ralston, S., R. M. Gooding, and G. M. Ludwig. 1986. An ecological survey and comparison of bottom fish resource assessments (submersible versus handline fishing) at Johnson Atoll. U.S. National Marine Fisheries Service Fishery Bulletin 84:140–155.

Rowley, R. J. 1994. Case studies and reviews: marine reserves in fisheries management. Aquatic Conservation: Marine and Freshwater Ecosystems 4:233–254.

Stein, D. L., B. N. Tissot, M. A. Hixon, and W. Barss. 1992. Fish-habitat associations on a deep reef at the edge of the Oregon continental shelf. U.S. National Marine Fisheries Service Fishery Bulletin 90:540–551.

Tlusty, M. F., J. E. Hughes Clarke, J. Shaw, V. A. Pepper, and M. R. Anderson. 2000. Groundtruthing multibeam bathymetric surveys of finfish aquaculture sites in the bay d'Espoir Estuarine Fjord, Newfoundland. Marine Technology Society Journal 34:59–67.

Underwood, A. J. 1998. Experiments in ecology: their logical design and interpretation using analysis of variance. Cambridge University Press, UK.

Urick, R. J. 1983. Principles of underwater sound, 3rd edition. McGraw-Hill, New York.

Weber, M. L., and B. Heneman. 2000. Guide to California's Marine Life Management Act. Common Knowledge Press, Bolinas, California.

Wiens, J. A. 1989. Spatial scaling in ecology. Functional Ecology 3:385–397.

Yoklavich, M., G. Cailliet, R. N. Lea, H. G. Greene, R. Starr, J. deMarignac, and J. Field. 2002. Deepwater habitat and fish resources associated with the Big Creek Ecological Reserve. CALCOFI Reports 43:120–140.

Yoklavich, M., R. Starr, J. Steger, H. G. Greene, F. Schwing, and C. Malzone. 1997. Mapping benthic habitats and ocean currents in the vicinity of central California's Big Creek Ecological Reserve. NOAA Technical Memorandum NMFS-SWFSC 245.

Yoklavich, M. M., H. G. Greene, G. M. Caillet, D. E. Sullivan, R. N. Lea, and M. S. Love. 2000. Habitat associations of deep-water rockfishes in a submarine canyon: an example of a natural refuge. U.S. National Marine Fisheries Service Fishery Bulletin 98:625–641.

Spatial and Temporal Distributions of Bottom Trawling off Alaska: Consideration of Overlapping Effort when Evaluating the Effects of Fishing on Habitat

CRAIG S. ROSE[1] AND ELAINA M. JORGENSEN

National Oceanic and Atmospheric Administration, National Marine Fisheries Service, Alaska Fisheries Science Center, Seattle, Washington, 98115, USA

Abstract. The spatial and temporal distribution of fishing effort is a critical component in assessing the effects of bottom fishing gear on benthic habitats. Fishing effort influences habitat unevenly in space and time, creating areas swept once or several times interspersed with areas unaffected by fishing gear. When initial sweeps by fishing gear damage or remove habitat features, subsequent overlapping sweeps may produce less additional damage or removal effects than comparable fishing would produce on unaffected habitat. Therefore, assuming fishing effort is uniform, with minimal overlap, may produce errors in estimating the area and effects of fishing on habitat. The distribution of bottom trawl effort off Alaska from 1997 to 2001 was used to examine changes in the estimated distribution and intensity of bottom trawl sweeps due to (1) the size of blocks used to accumulate data and (2) the assumptions used to model the distribution of fishing within blocks. The use of larger (20 × 20 km and 60 × 60 km) analytical blocks underestimated the amount of effort that overlapped previous tows by averaging effort from high-intensity locations over larger areas. Even with the smallest block size (5 × 5 km), use of statistical distributions that did not allow variation within blocks of fishing intensity (uniform) or the probability of being swept (Poisson) underestimated overlap relative to a distribution which modeled the contagious nature of fishing effort (negative binomial). Comparison of these models with the observed distribution of the 2001 fishery for Atka mackerel *Pleurogrammus monopterygius* indicated even more overlap in the fishery than even a highly contagious ($N = 2$) negative binomial distribution could emulate. Fishing distributions with reduced overlap overestimated habitat reductions when applied to a simple model of fishing effects and also overestimated the total area subject to being swept by gear and the average frequency of sweeping.

Introduction

Fishing gear can affect marine benthic habitats in ways that may limit the productivity of marine fish species. A growing body of literature on this process provides increasing detail on how habitat features may be affected, and an understanding of ways that changes in these features affect fish populations is emerging more slowly (Dayton et al. 1995; Dorsey and Pederson 1998; Collie et al. 2000; Johnson 2002; National Research Council 2002). A need to synthesize this information for policy use has motivated efforts to assess the significance of such effects using both qualitative and quantitative tools (MacDonald et al. 1996; Bergman and van Santbrink 2000; Pitcher et al. 2000; National Research Council 2002). Scientists carrying out these investigations face the difficult task of extrapolating results of experiments on small spatial scales to processes occurring over much larger scales.

Experiments on small spatial scales have measured changes in benthic features that were swept by fishing gear. The number of gear sweeps causing observed effects was an important aspect of these experiments. Where single gear tracks were studied (e.g., Van Dolah et al 1987; Freese et al. 1999), the number of sweeps was obviously one. Studies using multiple passes have estimated the effective number of sweeps either by striving to sweep the study site fully during each pass (e.g., Moran and Stephenson 2000) or with a measure of fishing intensity, the total effort (sum of area swept by the gear during all passes) divided by the area of the study site (e.g., Prena et al. 1999).

One approach to expanding the results of such studies to larger areas is to estimate the distribution of sweep rates, the proportions of seafloor swept by fishing

---

Corresponding author: craig.rose@noaa.gov

gear a given number (n) of times (for each $n = 1, 2, 3, 4,...$) during a time period. Applying and summing the estimated effects on benthic features to the areas exposed to each sweep rate could provide a first approximation of the total effects over that period (recognizing that a number of processes, especially recovery, would also need to be addressed). Fishing intensity may provide a reasonably precise measure of sweep rate for controlled fishing within an experimental site with dimensions not much greater than those of the gear itself. However, unadjusted fishing intensity may not be adequate to estimate the distribution of sweep rates over larger scales with uncontrolled fishing effort.

The quality and resolution of available data on fishing effort vary widely. Worldwide, effort data are compiled and reported in blocks with areas varying across more than two orders of magnitude (Bergman and van Santbrink 2000; Pitcher et al. 2000; National Research Council 2002). For example, McConnaughey et al. (2000) accumulated effort into blocks of 3.4 km^2, while Bergman and van Santbrink (2000) used blocks of approximately 2,000 km^2. Such differences in block size may influence both perceptions and analyses of habitat effects due to fishing, particularly where the largest units of effort, trawl tows, have a scale of kilometers in length by tens to hundreds of meters in width.

When fishing effort distributions have been analyzed at smaller blocks, including plotting of actual fishing tracks (i.e., Mirarchi 1998; Pitcher et al. 2000), non-uniform patterns of varying fishing intensity appear at each successively finer scale of resolution. However, the implicit, or often explicit, assumption in the use of effort statistics has been that fishing intensity is uniform at all scales smaller than the analytical block. Thus, all locations within a block are treated as receiving equal effort levels. A notable exception was Pitcher et al. (2000), who used the variation of fishing intensity between blocks to estimate the distribution of effort within their analytical blocks. Their comparison of this distribution with one generated assuming random and uniform tow distributions indicated that the effort had an aggregated (contagious) distribution and certainly nothing resembling a uniform distribution. Kulka and Pitcher (2001) avoided the problems with gridded data by using a geographic information system (GIS) technique to represent effort on a continuous scale.

Patterns of fishing are likely to be patchy at any scale larger than the width of the gear, forming a patchwork of isolated and overlapping locations with different sweep rates. When fishing effort data are aggregated into blocks, failure to account for this variability of sweep rates could well produce significant differences in how the effects of fishing gear on seafloor habitats are perceived and modeled.

Bottom trawlers have fished the waters off Alaska for almost 50 years (Forrester et al. 1978), a short history relative to many of the world's major fishing areas. This was primarily foreign effort until the 1980s when the fisheries were taken over by domestic vessels. An extensive observer program, commenced in the late 1970s, has provided detailed data on fishing distribution. Patterns of fishing locations have been affected by the transition to domestic vessels, changes in target species and their relative values, and management actions, including area closures and gear regulations.

The goals of this paper are to (1) describe the spatial and temporal variation of recent (1997–2001) fishing effort in the waters off Alaska, and (2) compare the effects of different block sizes on the perceptions and analyses of fishing distributions, particularly as these relate to the effects of fishing gear on benthic habitat features. These choices include aggregating effort at three different scales (5 × 5 km, 20 × 20 km, and 60 × 60 km) and different assumptions about within-block effort distributions.

Methods

Bottom trawl records from the North Pacific Fisheries Observer Database, collected between 1997 and 2001, were compiled into a GIS. This includes effort from a wide range of vessel types (<18 to >90 m in length, catcher–processor and catcher only), targeting a variety of demersal and semi-demersal species in widely varying habitats. All of these bottom trawlers used otter trawls. Available observations included end location, date, time, duration, and vessel length. Towing speed was only consistently available for 1999. The effort for each tow was assigned to a grid of 5-km × 5-km blocks, based on which block contained the tow's end location. Because tows were commonly longer than 5 km, it was inevitable that some effort was misallocated to adjacent grid blocks. Finite tow lengths limited the scale of such error to neighboring blocks, and reciprocal exchanges of effort between adjacent blocks would further limit the effect of such errors on this analysis, particularly where adjacent blocks had similar effort intensities. While misallocated effort not cancelled by such reciprocal exchanges may have altered the intensities of individual blocks, it was not considered likely to significantly affect general fishing distributions or any of the Alaska-wide summations. Lack of more detailed gear information precluded separate treatment of fishing targeted at different species, so all bottom trawl effort was assumed equivalent.

The swept area of each tow was calculated by multiplying its duration by an average speed and trawl width by vessel class (Table 1). Average speed was computed from the 1999 observer records, the only year when towing speed was recorded consistently. Average speeds were 1.7 m/s (3.3 knots) for vessels under 38 m and 1.85 m/s (3.6 knots) for vessels over 38 m. Average trawl widths were derived from questionnaire responses on headrope and sweep lengths collected in 2001 (Rose, unpublished data). Trawl net width was calculated as 55% of headrope length, and sweep area was calculated assuming a 15-degree angle of attack (sweep spread = 2 × 0.259 × sweep length). These conversions were derived from conversations with regional trawl manufacturers and trawl captains. The resulting average widths were 166 m for vessels over 38 m ($n = 11$) and 90 m for those under 38 m ($n = 21$). The proportions swept by each gear component were estimated to be 79% for the sweeps, 19% for the net, and 2% for the doors for vessels under 38 m; and 85% for the sweeps, 14% for the net, and 1% for the doors for vessels over 38 m. Due to prevailing rough seafloor in the Aleutian Islands, much shorter sweeps are used there to reduce the chance of hang-ups. A trawl width of 50 m was used for vessels of all sizes in that area based on information from trawler captains.

Haul records were only collected when observers were on board. While vessels over 38 m were required to have observers aboard at all times, smaller vessels were only required to have observers at least 30% of the time. Thirty randomly selected annual vessel logbooks were examined to estimate the percentage of tows made with an observer aboard. As the resulting average was 32%, swept areas for vessels under 38 m were divided by 0.32. This adjustment fails to account for differences in fishing locations between observed and unobserved trips. Vessels smaller than 18 m (60 ft) in length, which make up a very small proportion of the Alaska trawl fleet and an even smaller proportion (<1%) of the catch, are exempted from observer coverage and are not represented in this analysis.

We divided the total swept area (km^2) during tows each year in each 5-km × 5-km block by the block area (25 km^2) to derive a fishing intensity index. If fishing effort were uniformly distributed across a block (assuming no or minimal overlapping tow coverage), this index would be the number of times that each location in the block was swept by bottom trawl gear in a single year. The inverse of this value represents the average time between trawl sweeps (in years) for locations in each block. Annual fishing intensity indices were calculated for each 5-km × 5-km block and year, as well as a 5-year average. To allow examination of the effects of accumulating effort statistics into larger blocks, intensity values were then recombined and averaged into 20-km × 20-km and 60-km × 60-km grids. We calculated the percentages of shelf area (relative to a total shelf area [0–500 m] of 1,168,618 km^2) and total effort represented in blocks with varying ranges of fishing intensity.

Three statistical distributions, uniform, Poisson, and negative binomial, were used to model the distribution of effort within blocks. Unless fishing distribution within blocks is considered, the distribution implied is uniform within the block. Under this distribution, it is assumed that no trawl tracks overlap until the entire area of the block has been swept once. An intensity value of 0.5 using this distribution would imply that half of the block's area is swept once each year, while the other half is not swept. An intensity of 5 would imply that every location within the block is swept five times annually (Figure 1A). Such a scenario is unlikely to occur during actual groundfish trawling operations.

Another distribution might be represented by the random distribution of effort within a block, with all locations having equal probabilities of being swept. Something like this could occur where independent fishers do not have sufficient knowledge of where within a block the fish may be concentrated, or the patchiness scale of the fish is much larger than the block dimensions. This process can be modeled using the Poisson distribution (Johnson and Kotz 1969). Given an average number of sweeps by location, which corresponds to the fishing intensity statistic,

Table 1. Trawl component widths used in estimating area coverages (m). Data for headrope length and sweep length from Rose (Alaska Fisheries Science Center, unpublished data). Trawl net width estimate calculated using the equation: estimate = 0.55 × headrope length. Sweep width estimate calculated using the equation: estimate = 2 × sin(15°) × sweep length. Large trawlers are vessels greater than 38 m in length.

Trawler type	Headrope length	Sweep length	Trawl net width	Sweep width	Door width	Total width
Large trawlers	44	269	24 (14%)	139 (85%)	3 (1%)	166
Small trawlers	32	134	18 (19%)	69 (79%)	3 (2%)	90
Aleutian trawlers	44	44	24 (48%)	23 (46%)	3 (6%)	50

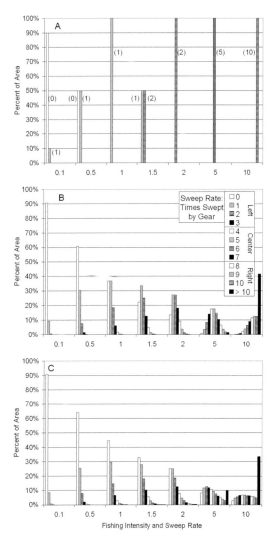

Figure 1. Distribution of sweep rates (number of times locations are swept by fishing gear in a unit time period) as a function of fishing intensity (sum of swept area in a block/block area) under assumed (A) uniform, (B) Poisson, and (C) negative binomial ($N = 2$) distributions of fishing effort within blocks.

distribution may be useful. Groundfish are not evenly distributed at spatial scales used for accumulating effort information (5 × 5 km or larger), and fishers target locations considered more likely to have high fish densities. The information used to make this judgment may come from experience, sonar readings, the behavior of other vessels, catch rates of previous tows, or any of many other ways that fishers select haul locations. By the same token, rough seabeds, wrecks, or regulatory closures may give some locations a lower probability of being fished. This patchy variation within a block of the probability of being swept by a trawl is termed a contagious distribution of fishing effort, which can be modeled statistically by the negative binomial distribution (Figure 1C). It can be represented with two parameters, the average frequency of sweeps (fishing intensity) and N, which ranges from 1 to infinity. As N approaches infinity, the negative binomial approaches the Poisson distribution (equal probabilities). Lower values of N model more patchy distributions. For this exercise, $N = 2$ was used to generate distributions with relatively strong contagion, just short of the most contagious available ($N = 1$). The negative binomial is a discrete function, allowing only positive integer values of N (Johnson and Kotz 1969). A continuous approximation, allowing values below 1, would have been desirable for this exercise, but none was identified.

To compare the statistical distributions to an example of fishing distribution from a real fishery, the distribution of sweep rates were calculated from vessel tracking information during the 2001 fishery for Atka mackerel *Pleurogrammus monopterygius* in the Aleutian Islands. These data were compiled in an analysis of fishing overlap by The Groundfish Forum, a fishing industry organization, and were made available for the current analysis. Regulations required vessel monitoring systems to be installed on all vessels fishing for Atka mackerel in the Aleutians. These systems produced position fixes for each vessel at time-varying intervals, averaging 20 min. All of the vessel tracks were compared with vessel logs and speed data to identify periods of towing. Tow tracks were defined as the lines connecting all fixes from the first fix before the start of the tow to the first fix after the end of the tow. Consequently, estimates of total swept area were biased slightly upward. Tow tracks were entered into a GIS and expanded 25 m to either side to create 50-m-wide polygons for each haul. The total area of all polygons within each block of a 5-km × 5-km grid were calculated and divided by 25 to provide fishing intensity measures comparable to those described earlier. Within each of these 5-km × 5-km grids, a 100-m × 100-m lattice of points was established to provide a systematic sample of 10,000 locations. This number was considered more than adequate

the Poisson distribution generates the proportions of the block that were swept n times with $n = 0, 1, 2, 3, 4, 5, 6,...$, assuming the probability of each unit of effort sweeping any location is equal across all locations within a block (Figure 1B). While the fishing process under such conditions would not technically be random, the Poisson distribution was considered useful in representing non-coordinated fishing with consistent sweep probability across each block.

For most groundfish fisheries, a negative binomial

to estimate the distribution of sweep rates, while not being computationally excessive. These points were compared to all of the tow polygons to determine how many times each of the points was swept. The distribution of sweep rates ($n = 0, 1, 2, 3, 4, 5,...$) across points estimated the distribution of seafloor sweep rates by the Atka mackerel fishery. This distribution was compared with distributions generated by applying the three statistical distributions to the estimated fishing intensity values for all a 5-km × 5-km blocks where fishing effort was detected.

To compare the Alaska-wide effects of using different block sizes for effort accumulations and different assumptions about effort distributions within blocks, the Poisson and negative binomial distributions were applied to all of the fished blocks from each of the three grids of different-sized blocks. The distributions of sweep rates for each block were summed to give a distribution of sweep rates across the Alaska shelf. These were then divided by the total area of the shelf to yield the percentage of area swept n times in the 5-year period. The percent of fishing effort associated with each sweep rate was calculated by multiplying the rate-specific area percentages by the rate and dividing by the total fishing effort.

To examine the effect of differences in estimated sweep rate distributions on modeling of habitat effects, the shelf-wide distribution of sweep rates generated above was applied to a simple model of habitat effect. Habitat effect models (i.e. Bergman and van Santbrink 2000; Pitcher et al. 2000) commonly acknowledge that the first sweep of a particular location causes the greatest mortality or removal effect and that subsequent sweeps cause progressively smaller effects. The usual structure for this assumption is that the same proportion of the remaining features are removed with each pass so that

$$R = \sum P_n E^n,$$

where R = the proportion of the habitat feature remaining in the area, n = the sweep rate (number of sweeps in 5 years), P_n = the proportion of the total area swept n times, and E = the sensitivity of the habitat feature to the fishing gear (the proportion remaining after being swept once). The equation was evaluated for each block and summed across all blocks. (This summation makes the implicit assumption that the quality and sensitivity of habitat is equivalent for all blocks and that no recovery of features occurs, both substantial simplifications.) This equation was applied to the sweep rate distributions resulting from the six combinations of block size and Poisson or negative binomial distributions.

Results

The distribution of average annual fishing intensity values (1997–2001) for bottom trawlers is displayed in Figure 2. The highly patchy distribution of fishing intensity shows considerable structure that cannot be sufficiently represented by either Alaska-wide (0.086 sweeps/year) or regional (eastern Bering Sea 0.10 sweeps/year, Aleutian Islands 0.08 sweeps/year, Gulf of Alaska 0.05 sweeps/year) average values.

Different regions of the Alaskan shelf have very different bathymetric (Figure 3) and habitat characteristics, and this is reflected in the distributions of fishing activity. The eastern Bering Sea has a very productive, broad sedimentary shelf with gradual habitat changes. It received the most intensive and widespread fishing. Two large closures, covering the eastern section of Bristol Bay and the waters surrounding the Pribilof Islands, create gaps in the effort distribution. Areas of highest intensities include the north side of Unimak Island, where Pacific cod *Gadus macrocephalus* and walleye pollock *Theragra chalcogramma* are seasonally abundant close to the ports of Dutch Harbor and Akutan and the central shelf, where fisheries for yellowfin sole *Limanda aspera*, rock sole *Lepidopsetta bilineata*, and flathead sole *Hippoglossoides elassodon* dominate.

The Gulf of Alaska has a wider variety of habitats, with a seafloor made up of a complex combination of banks, gullies, and slope. Fishing intensities were highest in gullies with depths of 100–200 m and along the outer edge of the shelf. Because a much larger number of vessels deliver their catches to Kodiak, Alaska, intensity is more concentrated in the vicinity of Kodiak Island. The eastern Gulf (east of 140°W) was closed to bottom trawling in 1997.

The Aleutian Islands region consists of a narrow shelf broken by passes with strong tidal currents. The seafloor has a high proportion of hard, rough substrates, limiting trawling on a local scale. Closures within 18.5–37 km (10–20 nautical mi) of rookeries of Stellar sea lions *Eumetopias jubatus* and haulouts were implemented during the late 1990s. While similar closures were established in the Bering Sea and Gulf of Alaska, a relatively high proportion of the available seafloor was closed in the Aleutian Island region because the narrow shelf puts nearly all of the region within 37 km (20 nautical mi) of land and, hence, vulnerable to closure.

Figure 4 shows annual fishing intensity distributions for an area at the intersection of the Aleutian Island, eastern Bering Sea, and Gulf of Alaska regions. While a general pattern of areas containing higher and lower intensities remains at a very broad scale, interannual variation was apparent at all smaller scales. Because of

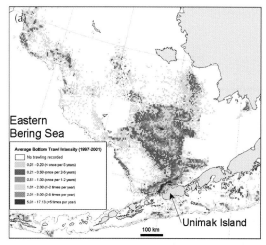
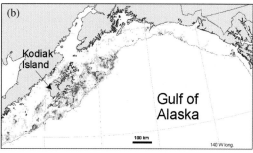
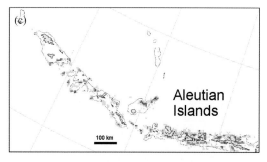

Figure 2. Geographic distribution of the average (1997–2001) bottom trawl fishing intensity (sum of swept area in each block/block area) in Alaska waters.

these annual shifts, averaging across years may have obscured structure that could be important in evaluating fishing effects. Each individual year shows higher maximum intensities and more blocks with no fishing compared to the 5-year average intensities.

Effects of Block Size

Comparison of fishing intensity distributions generated from aggregating effort into different-sized blocks (Figure 5, top) indicated that lower resolution (larger block size) expanded the total area perceived as vulnerable, while underestimating the area subjected to high fishing intensities. Proportions of unfished blocks (76% for 5-km, 43% for 20-km, and 1% for 60-km blocks) and blocks with intensities above two decreased with larger block size, while the proportion of low-intensity blocks increased. Fishing effort concentrated into relatively few high-intensity, 5-km × 5-km blocks was averaged across the whole area of the 20-km and 60-km blocks that contained them. Calculating the distribution in terms of proportion of effort (Figure 5, bottom), showed that, at larger block sizes, effort associated with high-intensity locations was deemphasized and that effort was shifted to lower-intensity categories.

Alternative Statistical Distributions to Model Sweep Rates

In comparing the three distributions (Figure 1), the most notable differences as one moves from uniform to Poisson to negative binomial are: (1) a higher proportion of area not swept, (2) an increase in higher sweep rates, and (3) a more skewed distribution, increasing emphasis on higher sweep rates. All of these trends are magnified at higher fishing intensities. Thus, as one moves toward more patchy distributions, there is a net increase in area not swept or swept few times, with that effort being redistributed to locations which have already been swept several times. This shift is stronger at higher intensities.

Comparison with Distributions Measured During a Fishery

The effort pattern for the fishery for Atka mackerel shows a great deal of heterogeneity both among and within blocks (Figure 6). Especially within the high-intensity blocks, fishing was concentrated at some locations while a substantial portion of the block went untouched. Both the distributions of seafloor area subjected to different sweep rates (Figure 7A) and the proportions of effort associated with each sweep rate (Figure 7B) indicated that the uniform distribution did not generate results comparable with the observed distributions. Both the Poisson and negative binomial simulations compared more closely with the measured pattern of sweep rates, but neither achieved the high percent of area not swept, the low proportion of area swept once, or the high proportion of area swept more than 10 times generated from the actual position data. The negative binomial distribution with a near maximum level of contagion ($N = 2$) was more similar to the completely non-contagious Poisson than to the fishery results (area swept once: 5.5% Poisson, 5.0% negative binomial, 3.3% fishery). Reducing N to 1

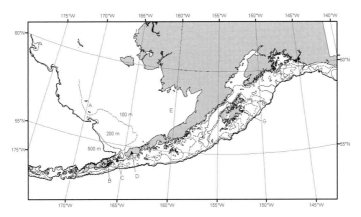

Figure 3. Bathymetry of Alaska continental shelf with locations of features mentioned in text: (A) Pribilof Islands, (B) Dutch Harbor, (C) Akutan, (D) Unimak Island, (E) Bristol Bay, (F) Kodiak Island, and (G) Kodiak.

only brought the proportion of the area swept once down to 4.6%. This indicates that a statistical distribution capable of greater levels of contagion will be needed to model fisheries like this. Combinations of Poisson distributions or a continuous approximation of the negative binomial may have such flexibility.

Alaska-Wide, 5-Year Sweep Rates

When used to estimate the distribution of sweep rates for the whole Alaska shelf over the 5-year period, differences in block size produced effects on sweep-rate estimates consistent with the results above (Figure 8). Results with the smallest block size and within-block patchiness (negative binomial model) indicated that 10.6 % of the Alaska shelf was swept at least once in 5 years, 6.2 % was swept more than once, and only 2.4 % of shelf area was swept an average of more than once per year. However, use of 20-km × 20-km and 60-km × 60-km blocks increased the proportion swept once to 13.3% and 16.4 %, respectively, with 6.7% and 7.2 %, respectively, subjected to more than one sweep. Larger block sizes also attributed more of the non-swept proportions to locations within fished blocks, as opposed to locations in blocks that were subjected to no effort.

Accommodating variable probabilities of being swept within blocks (Poisson versus negative binomial distributions) had effects with similar direction to small block sizes but much smaller magnitudes. While the 60-km × 60-km blocks estimated 55% more area swept than the comparable 5-km × 5-km value and the 20-km × 20-km estimate was 25% higher than the 5-km × 5-km value, the Poisson model for the 5-km × 5-km resolution yielded an estimate only 10% higher than that with the negative binomial for the same resolution. As above, these differences resulted from the redistribution and averaging of effort from discrete locations with high fishing intensities across larger areas with lower sweep rates.

Effects on a Model of Habitat Reduction

A habitat effect model was applied to the Alaska-wide fishing intensity distributions generated above to assess their effects on estimates of mortalities or removals of habitat features. The apparent shift of effort from smaller areas with more repeated sweeps to larger areas with less overlap increased estimated reductions of habitat features (Figure 9). The results were more sensitive to larger block size than to assumptions about within-block effort distribution, consistent with the smaller shifts of effort indicated in the previous section. While the magnitude of habitat effects was a function of the proportion remaining after each sweep of the gear (sensitivity), the proportional differences due to block size and within-block models were relatively consistent. When high proportions remained after a single sweep, there was little difference noted in the overall reductions, as it took many sweeps to alter the habitat from its original state. However, when little habitat remained after a single sweep, substantial differences were estimated. In that case, overlap was a very important factor, as successive sweeps had less additional effect and any increase in the area swept even once resulted in a nearly proportional increase in habitat reductions.

Discussion

Analysis of the distribution of bottom trawl tows off of Alaska indicated that accumulating such data by regions

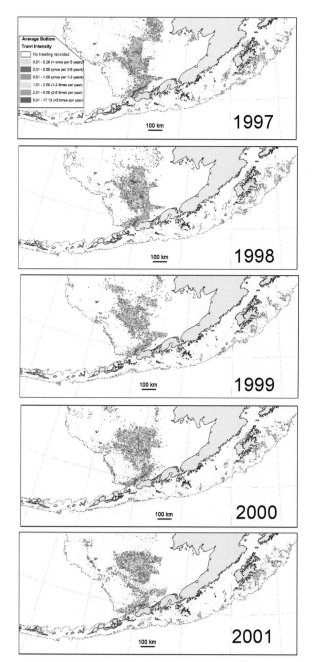

Figure 4. Annual variability in the geographic distribution of bottom trawl fishing intensity (sum of swept area in each block/block area) in Alaska waters.

or large analytical blocks or failing to account for variation of sweep rates within blocks may yield a biased perception of the extent, proportion, and frequency of fishing gear sweeping seafloor habitats. The annual average of the total area swept by bottom trawl hauls in Alaska waters from 1997–2001 was sufficient to sweep 8.6 % of the shelf area (0–500 m) if no overlap of those trawl tracks occurred. However, considerable overlap did occur, as indicated both by reporting blocks with more trawled area than the area of the block and by

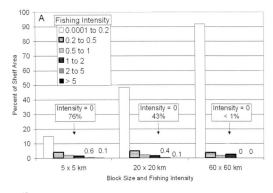

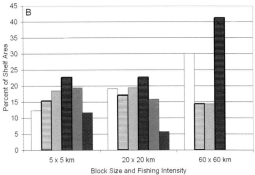

Figure 5. Effect of block size on the percent of shelf area (A) and fishing effort (B; total area swept by bottom trawls) in blocks with different fishing intensities (sum of swept area in block/block area).

application of the most plausible of three models of towing distribution within blocks. Accounting for this overlap indicated smaller swept areas or less frequent sweeping than those implied by the unadjusted regional statistic. It was estimated with the smallest block size that only 2.4 % of the shelf area averaged sweep rates above once per year, while 89.4 % was not swept at all during the whole 5-year period.

Analyzing data by blocks accounted for some of the overlap, especially when effort data was accumulated into relatively small blocks. The perceived distribution of effort from the same data set differed substantially when accumulated by 60-km × 60-km, 20-km × 20-km, and 5-km × 5-km blocks. Larger blocks resulted in less of the overlap being detected and, hence, overestimated the total area of seafloor affected and underestimated the proportion of areas subjected to high fishing intensity.

Only a few studies applying fishing effort data to assess the potential for habitat effects have explicitly accounted for any variation of sweep rates at scales smaller than analytical blocks. Pitcher et al. (2000) used variation between blocks to model the distribution within blocks, while Kulka and Pitcher (2001) used a GIS technique to create a surface describing the probability of effort levels on a continuous (non-gridded) scale. Most studies treat blocks as units with consistent effort, although many acknowledge the occurrence and potential importance of heterogeneity on smaller scales (i.e., Bergman and van Santbrink 2000; National Research Council 2002). The resulting assumption of uniform distribution of effort within blocks is highly unlikely to hold under most fishing situations. Two statistical distributions were applied here to examine the consequences of more plausible fishing behaviors. Both of these used random, independent applications of effort, with one (Poisson) having a constant probability across the block of being swept and the other (negative binomial) allowing the probability of being swept to vary (contagious distribution). As fish distributions vary across smaller scales than most analytical blocks and other factors influence the choice of fishing locations on a local scale, the latter distribution was considered more representative of common fishing behavior.

Fishing patterns observed during the 2001 Atka mackerel fishery indicated higher levels of overlap than those estimated by any of three statistical approaches for modeling within-block distributions. Increased overlap was indicated by a pattern of higher proportions of area not swept by trawls and area of high fishing intensity (i.e., swept more than 10 times per year) combined with lower proportions subjected to single or other low sweep rates. More overlap also increased the area not swept at all during fishing. The observed distribution was most similar to that modeled with the negative binomial, with increasing differences from those generated using the Poisson and uniform. This fishery had highly aggregated fishing effort due to restricted fishing areas and highly localized fish distributions and was considered likely to represent an extreme among fisheries. Similar exercises with a range of fisheries and contagious statistical distributions (i.e., combinations of Poisson and negative binomial distributions) will be needed to find the most appropriate models and parameters. Fortunately, increased vessel position monitoring should make appropriate data available from a wider range of fisheries.

Changes in analytical block size resulted in greater differences in estimated fishing distribution than differences between applications of the Poisson and negative binomial models. While assuming constant sweep probability within blocks (Poisson model) resulted in a moderate underestimation of overlap frequency, the use of large analytical blocks caused greater errors under both models. When the resulting sweep rates were applied to a simple model of effects on habitat features, habitat reduction estimates increased with larger block size and under the constant probability model. Absolute differences in these reduction estimates were much greater when high values were used for the parameter representing the

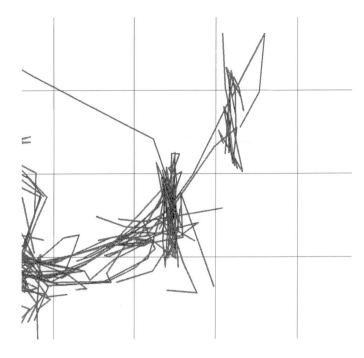

Figure 6. Distribution of area swept by trawls during the 2001 Atka mackerel fishery at a site in the Aleutian Islands. Grid spacing is 5 km.

reduction of habitat due to a single gear sweep. However, proportional changes in reduction estimates were not very sensitive to changes in the effect of a single sweep.

Overlaps in fishing distributions are important in analyzing the habitat effects of fishing because subsequent sweeps are assumed to produce sequentially smaller absolute effects than the initial sweep. For fixed habitat features, this is a reasonable assumption, as less of the feature is available for removal or damage during subsequent sweeps. This effect was detected by Moran and Stephenson (2000), who monitored living substrates during a sequence of fishing efforts. Overlap considerations may be less important for effects that are distributed away from the gear's sweep, such as resettling of suspended sediments, or for effects on highly mobile animals, which could redistribute themselves between sweeps. Long-term effects will include an interaction of periodic gear effects with habitat recovery.

A major motivation for this analysis was to provide scientists using fishing effort data to assess habitat effects of fishing gear, especially those whose data is only available on relatively coarse scales, with an example of the effects of this scale on such analyses. Even with relatively small analytical blocks, effort concentrations on a finer scale are likely to generate a greater amount of overlap than they are able to detect. Analyses that do not fully account for sweep overlaps are vulnerable to overestimating the total area of seafloor affected and the average sweep rate while underestimating sweep rates at locations experiencing the highest fishing intensity. Such errors are likely to result in overestimates of resulting habitat effects. Improved analytical techniques are needed to better account for the resulting errors in assessments of habitat effects.

Acknowledgments

The authors would like to thank the Groundfish Forum for making the Atka mackerel fishery data available and Sergio Henricy of SBH Company for his work in analyzing that data. The work of countless observers in accumulating this very useful data set always deserves recognition. Insightful comments from three anonymous reviewers greatly improved this paper.

References

Bergman, M. J. N., and J. W. van Santbrink. 2000. Mortality in megafaunal benthic populations caused by trawl fisheries on the Dutch continental shelf in the North

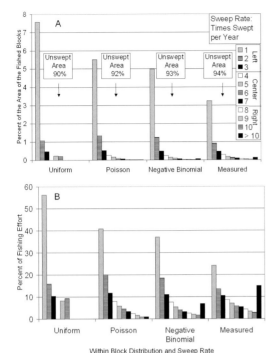

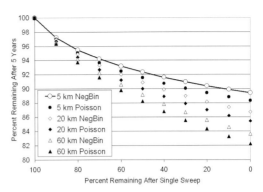

Figure 9. Percent of hypothetical habitat features remaining after 5 years, considering the effects of block size, assumed statistical distribution of fishing effort within blocks, and sensitivity of the feature to a single sweep by a bottom trawl. Estimates were based on the fishing intensity distribution for bottom trawling on the continental shelf off Alaska (1997–2001) and disregarded habitat recovery.

Figure 7. Percent of seafloor area (A) and fishing effort (B) at locations with different sweep rates (number of times a location was swept by a trawl) during the 2001 Atka mackerel fishery. Estimations from statistical modeling based on the fishing intensity (sum of swept area in each block/block area) in each 5-km × 5-km block were compared to measured sweep rates.

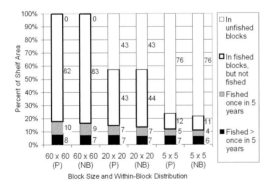

Figure 8. Comparison of the effects of block size and assumed statistical distribution of fishing effort within blocks on estimating percentages of the continental shelf off Alaska swept by bottom trawling. P = determined using Poisson distribution, NB = determined using negative bionomial distribution.

Sea in 1994. ICES Journal of Marine Science 57:1321–1331.

Collie, J. S., S. J. Hall, M. J. Kaiser, and I. R. Poiner. 2000. A quantitative analysis of fishing impacts on shelf-sea benthos. Journal of Animal Ecology 69:785–798.

Dayton, P. K., S. F. Thrush, M.T. Agardy, and R. J. Hofman. 1995. Environmental effects of marine fishing. Aquatic Conservation: Marine and Freshwater Ecosystems 5:205–232.

Dorsey, E. M., and J. Pederson, editors. 1998. Effects of fishing gear on the sea floor of New England. MIT Sea Grant Publication 98-4, Conservation Law Foundation, Boston.

Forrester, C. R., A. J. Beardsley, and Y. Takahashi. 1978. Groundfish, shrimp and herring fisheries in the Bering Sea and northeast Pacific—historical catch statistics through 1970. International North Pacific Fisheries Commission Bulletin 37.

Freese, L., P. J. Auster, J. Heifetz, and B.L. Wing. 1999. Effects of trawling on seafloor habitat and associated invertebrate taxa in the Gulf of Alaska. Marine Ecology Progress Series 182:119–126.

Johnson, N. L., and S. Kotz 1969. Discrete distributions. Wiley, New York.

Johnson, K. 2002. A review of national and international literature on the effects of fishing on benthic habitats. NOAA Technical Memorandum NMFS-F/SPO-57.

Kulka, D. W., and D. A. Pitcher 2001. Spatial and temporal patterns in trawling activity in the Canadian Atlantic and Pacific. ICES, CM 2001/R:02, Copenhagen.

MacDonald, D. S., M. Little, N. C. Eno, and K. Hiscock. 1996. Disturbance of benthic species by fishing activities: a sensitivity index. Aquatic Conservation Marine and Freshwater Ecosystems 6:257–268.

McConnaughey, R. A., K. L. Mier, and C. B. Dew 2000. An examination of chronic trawling effects on soft-bottom benthos of the eastern Bering Sea. ICES Journal of Marine Science 57:1377–1388.

Mirarchi, F. 1998. Bottom trawling on soft substrates. Pages 80–84 *in* E. M. Dorsey and J. Pederson, editors. Effects of fishing gear on the sea floor of New England. MIT Sea Grant Publication 98-4, Conservation Law Foundation, Boston.

Moran, M. J., and P. C. Stephenson. 2000. Effects of otter trawling on macrobenthos and management of demersal scalefish fisheries on the continental shelf of northwestern Australia. ICES Journal of Marine Science 47:510–516.

National Research Council (NRC). 2002. Effects of trawling and dredging on seafloor habitat. National Academy Press, Washington D.C.

Pitcher, C. R., I. R. Poiner, B. J. Hill, and C. Y. Burridge 2000. Implications of the effects of trawling on sessile megazoobenthos on a tropical shelf in northeastern Australia. ICES Journal of Marine Science 57:1359–1368.

Prena, P., P. Schwinghamer, T. W. Rowell, D. C. Gordon, K. D. Gilkinson, W. P. Vass, and D. L. McKeown. 1999. Experimental otter trawling on a sandy bottom ecosystem of the Grand Banks of Newfoundland: analysis of trawl bycatch and effects on epifauna. Marine Ecology Progress Series 181:107–124.

Van Dolah, R. F., P. H. Wendt, and N. Nicholson 1987. Effects of a research trawl on a hard-bottom assemblage of sponges and corals. Fisheries Research 5:39–54.

Hydraulic Clam Dredge Effects on Benthic Habitat off the Northeastern United States

DAVID H. WALLACE[1]

Wallace and Associates, 1142 Hudson Road, Cambridge, Maryland 21613-3234, USA

THOMAS B. HOFF[2]

*Mid-Atlantic Fishery Management Council, Room 2115 Federal Building,
300 South New Street, Dover, Delaware 19904-6790, USA*

Abstract. Hydraulic clam dredges have been used for fishing off the East Coast of the United States since World War II. Fisheries for Atlantic surfclam *Spisula solidissima* and ocean quahog *Arctica islandica* occur in waters less than 80 m deep, where storms and currents tend to dominate the disturbance regime. These fisheries cover only a relatively small area (less than 400 km^2) of the shelf that is sandy. Therefore, natural events (storms, currents) have more effect on the Atlantic surfclam and ocean quahog essential fish habitat (EFH) than regulated hydraulic clam dredges. A workshop on the effects of fishing gear on marine habitats concluded that hydraulic clam dredges were not a major concern relative to otter trawls and scallop dredges. The Mid-Atlantic Fishery Management Council concurred with the conclusions of that workshop. The council concluded that any EFH impacts would be temporary and minimal. In addition, the council evaluated seven alternatives to minimize any potential impacts of hydraulic clam dredging and concluded that the "no action alternative" should be maintained and that no management measures were necessary (MAFMC 2003).

Introduction

The Sustainable Fisheries Act of 1996 significantly altered the requirement of fishery management plans (FMPs) to address habitat issues. The 2002 final rule (NOAA 2002) establishes guidelines to assist the regional Fishery Management Councils in the description and identification of essential fish habitat (EFH), the identification of adverse effects to EFH, and the identification of actions required to conserve and enhance EFH. The council's FMP Amendment 12 (MAFMC 1998) was partially approved in 1998. The section that dealt with fishing gear impacts to EFH was disapproved by the National Marine Fisheries Service (NMFS) because it contained insufficient consideration of the FMP's compliance with the requirements for minimizing the adverse effects caused by fishing of Atlantic surfclam *Spisula solidissima* and ocean quahog *Arctica islandica*.

A workshop on the effects of fishing gear on the marine habitat of the northeastern United States was hosted by NMFS and the New England and Mid-Atlantic councils (NMFS 2002). The participants in the workshop included scientists from the National Oceanic and Atmospheric Administration, states, academia, and environmental organizations, as well as international gear experts and industry representatives.

Presentations and discussions focused on: (1) descriptions of each fishery and gear type; (2) the effects of the fishery on the environment; (3) the strength of the evidence of those effects; and (4) potential management implications.

A description of the evolution and current use of the hydraulic clam dredge for the Atlantic surfclam and ocean quahog fisheries was discussed at the workshop (Wallace 2002) and is summarized here.

Hydraulic Clam Dredge Gear Description and Operation

Hydraulic clam dredges have been used in the Atlantic surfclam fishery for more than 5 decades and in the ocean quahog fishery since its inception in the early 1970s (MAFMC 1977). These dredges are designed to: (1) capture 80–95% of the target species; (2) produce a low bycatch of other species; and (3) retain few undersized clams.

---

[1] E-mail: dhwallace@aol.com
[2] E-mail: THoff@MAFMC.org

The Atlantic surfclam and ocean quahog fisheries' hydraulic dredge vessels are equipped with a 3-m-wide dredge and pumps that inject seawater into the seafloor to liquefy the sediment. A trailing horizontal blade is used to remove the clams, which then pass back into the product holding area of the dredge. The dredge weight and clams are carried on wide runners, which distribute pressure on the ocean floor. The industry tows dredges only over sandy bottom. The dredge is towed at about 3.7 km/h in order not to exceed the rate at which the sand bottom is liquefied.

Harvesting Practices

In 1990, an individual transferable quota (ITQ) system in the clam fishery was implemented. The federally managed clam fishery was transformed from a time allocation system with no consolidation of effort to an ITQ system that allowed consolidation to reduce the overcapitalization in the fishery. With the ITQ system in place, the obsolete vessels were voluntarily removed from the fishery, and there no longer was the constant race to fish.

Atlantic Surfclam Habitat

Atlantic surfclams live mostly in sandy bottom, which is routinely disturbed and reworked by storms and, in some locations, strong bottom currents (Auster and Langton 1999). Ocean quahogs live at slightly greater depths, mostly in fine sand substrates, which are less affected by natural physical disturbances. Atlantic surfclams and ocean quahogs are not found in commercial quantities in gravel or mud habitats or in depths greater than 80 m.

Data compiled by NMFS from the logbooks of all Atlantic surfclam and ocean quahog fishing vessels (logbooks are a requirement for all fishing in federal waters) show that the amount of bottom area directly impacted by the hydraulic clam dredge fleet in 2000 was less than 400 km^2 (Table 1). An additional 50 km^2 were dredged in state waters of New Jersey and New York for a total of less than 450 km^2.

Hydraulic Clam Dredges: Impacts and Recovery

Participants at the NMFS 2001 Fishing Gear Effects Workshop concluded that the direct physical effects of hydraulic clam dredging are two types (NMFS 2002). First, a trench about 20 cm deep is left behind the dredge, and windrows of sediment and organisms are formed on either side of the trench. Second, sediments are re-suspended during dredging.

The direct biological impacts of hydraulic dredges vary depending on whether organisms are hard bodied like clams or soft bodied like amphipods or polychaetes. However, the biological effects of hydraulic dredging of a benthic habitat are not fully known, and the workshop noted that more study is needed (NMFS 2002), particularly for soft-bodied organisms that are moved by the dredge or pass through the dredge and are deposited back on the seafloor. Often, after an area is dredged, scavengers move in rapidly and eat broken clams and soft-bodied organisms that are redistributed to the sediment water interface. The workshop (NMFS 2002) concluded that the temporal scale of effects varied depending on the background energy of the environment, with recovery of physical structure ranging from days in high energy environments to months in low energy environments.

Alternatives for Managing Adverse Effects from Fishing

According to the final EFH rule (NOAA 2002), fishery management options to minimize any adverse effects to EFH from fishing may include but are not limited to: (1) fishing equipment restrictions, (2) time–area closures, and (3) harvest limits assuming there is evidence that a fishing activity adversely effects EFH in a manner that is more than minimal and not temporary in nature.

With regard to hydraulic dredging, the gear effects workshop report (NMFS 2002) stated:

> The effectiveness of the Individual Transferable Quota (ITQ) management program since 1990 and the opinion that the two resources are underfished led the panel to conclude that reductions in effort are probably not practicable. Nor is it likely that gear substitutions or modifications are practical since the current gear is highly efficient at harvesting clams. Therefore, spatial area management seems to be the only practicable approach to minimizing gear impacts, if necessary.

The council, using input developed by the workshop (NMFS 2002), noted that there was sufficient information to conclude that hydraulic clam dredges could have a negative effect on EFH if the gear was fished improperly or in the wrong sediment type (MAFMC 2003). The council also evaluated seven alternatives that focused mostly on spatial area management (MAFMC

Table 1. The estimated area in federal waters towed by hydraulic clam dredges in 2000 (from National Marine Fisheries Service vessel logbook data). Hours at sea per year data from clam vessel logbook data, excludes Maine ocean quahog fishery

Area	Quahogs	Surfclams
Hours at sea per year	28,440	19,907
Setting and hauling gear (25%)	7,110	4,977
Hours fished per year	21,330	14,930
Average speed/tow (km/h)	3.7	3.7
Total distance towed (m)	7.901×10^7	5.530×10^7
Average dredge width (m)	2.8	2.8
Area towed per year (m^2)	2.216×10^8	1.548×10^8
Area towed per year (km^2)	221.58	155.04

2003). The seven alternatives were fully evaluated in terms of biological, economic, and social impacts as well as the effects to EFH and protected species. Finally, a practicability analysis of the alternatives was included. After consideration of the results and recommendations of the workshop (NMFS 2002), the council concluded that any adverse effect of the hydraulic clam dredge on EFH would be minimal and temporary in nature, stating a preference for the no action alternative (MAFMC 2003). There were no public comments or support for any other alternatives.

Discussion and Implications

Atlantic surfclams and ocean quahogs are generally found in shallow water on high-energy sandy bottom. Natural factors (i.e., storm events and bottom currents) in this shallow ocean environment often have significant effects on bottom sediment (Auster and Langton 1999). The federal fishery covers less than 400 km^2/year (Table 1).

The Atlantic surfclam industry learned years ago that hydraulic clam dredging in rocky areas would result in damaged gear or loss of the dredge. When the clam fleet was fishing under a time allocation system during the 1970s and 1980s (6 h of fishing every other week), some captains would attempt to fish in rocky areas. If only damage occurred, repair could take days to weeks. If the gear was lost, replacement could take weeks to months. The cumulative effects of fishing in non-sandy areas would be lost fishing time, significant additional expense, and, at best, a relatively low catch. The low financial reward for catching Atlantic surfclams in rocky areas has not been worth the expense. With the implementation of the ITQ management system in 1990, quota owners now have a long-term stake in the health of the Atlantic surfclam stocks.

The council, in Amendment 13 (MAFMC 2003), concluded that there may be potential adverse effects on EFH from the hydraulic clam dredge but concurred with the 2001 workshop panel (NMFS 2002) that as the clam fishery is currently prosecuted, any adverse impacts are temporary and minimal in nature. The 2001 workshop (NMFS 2002) concluded that hydraulic clam dredges were not a major concern relative to otter trawls and scallop dredges.

References

Auster, P., and R. Langton. 1999. The effects of fishing on fish habitat. Pages 150–187 *in* L. Benaka, editor. Fish habitat: essential fish habitat and rehabilitation. American Fisheries Society, Symposium 22, Bethesda, Maryland.

MAFMC (MidAtlantic Fishery Management Council). 1977. Atlantic surfclam and ocean quahog fishery management plan. MAFMC, Dover, Delaware.

MAFMC (MidAtlantic Fishery Management Council). 1998. Amendment 12 to the Atlantic surfclam and ocean quahog fishery management plan. MAFMC, Dover, Delaware.

MAFMC (MidAtlantic Fishery Management Council). 2003. Amendment 13 to the Atlantic surfclam and ocean quahog fishery management plan. MAFMC, Dover, Delaware.

NMFS (National Marine Fisheries Service). 2002. Workshop on the effects of fishing gear on marine habitats of the northeastern United States. Northeast Fisheries Science Center, Reference document 02-01, Boston.

NOAA (National Oceanic and Atmospheric Administration). 2002. Magnuson–Stevens Fishery Management Conservation Act provision: essential fish habitat (EFH). Federal Register 67(12):2343–2383.

Wallace, D. H. 2002. Presentation to the workshop on the effects of fishing gear on the marine habitat of the northeast U.S. National Marine Fisheries Service, Boston.

Symposium Abstract

Symposium abstracts have been reprinted from the Abstract Volume prepared for the Symposium on Effects of Fishing Activities on Benthic Habitats: Linking Geology, Biology, Socioeconomics, and Management without any further review or editing.

The Spatial Extent and Nature of Mobile Bottom Fishing Methods within the New Zealand EEZ, 1989–90 to 1998–99

S. J. Baird,[1] N. W. Bagley, B. A. Wood, A. Dunn, and M. P. Beentjes

National Institute of Water and Atmospheric Research, Kilbirnie, Wellington, New Zealand

Temporal-spatial representation of fishing effort distribution for the main mobile bottom fishing methods used in New Zealand waters was investigated using 10 years of commercial effort data, from 1989-90 to 1998-99. Tow position data were used to map the changes in fishing patterns for fisheries using otter trawls on the bottom by collating the number of fishing operations and the area swept into 22 km2 blocks. The intensity of effort varied between fishing years: many 22 km2 blocks were trawled more than 10 times, representing a swept area of more than 10 km2. In most fishing years a median of 2 tows were made in each block (with the third quartile at 4-6 tows) and the maximum number of tows in a block was 370. Swept area values were scaled to vessel power and graphic representations of these data indicated areas trawled by heavier ground gear. Transects of selected areas for each fishing year showed large differences in the monthly spread of effort. Analyses of data for other otter trawl effort (predominantly inshore) and shellfish dredge effort are based on larger fishery areas because fine-scale position data were not collected. At this scale, spatial and temporal relationships between fisheries with different gear types were evident. Ground gear components used in the main otter trawl and dredge fisheries are described. The requirements for consistent data collection and the application of this work to a wider understanding of the impact of fishing in New Zealand waters are discussed.

---

[1] E-mail: s.baird@niwa.cri.nz

Deepwater Trawl Fisheries Modify Benthic Community Structure in Similar Ways to Fisheries in Coastal Systems

M. Cryer,[1] B. Hartill, and S. O'Shea

National Institute of Water and Atmospheric Research, Auckland, New Zealand

Off north-eastern New Zealand, the Bay of Plenty continental slope supports bottom trawl fisheries for gemfish *Rexea solandri*, hoki *Macruronus novaezelandiae*, tarakihi *Nemadactylus macropterus*, and, most recently scampi (a burrowing, deep-water lobster, *Metanephrops challengeri*). Excellent information has been collected since 1988 on the distribution of trawling effort in these fisheries, including the start and finish location of each trawl tow with a precision of 1 minute of latitude and longitude. Using a GIS, we linked these data to information on the invertebrate bycatch of 66 research trawls, and explored the extent to which the composition of our bycatch (as one index of benthic community structure) could be explained by the frequency of trawling at a given site. Using multivariate ordination techniques, we explained up to 65% of variation in the distribution of species among samples, more than half of which was attributable to our indices of trawling (mainly for scampi and gemfish). Qualitatively, the inferred effects of deep-water trawling were similar to those of coastal fisheries; increasing fishing activity was associated with reductions in species richness, diversity, and the abundance of large or fragile taxa. The gross quality of information on fishing effort has hitherto been a major constraint on our understanding of the effects of fishing. This study is one example of the way good quality information at the right (fine) scale can further that understanding, but comprehensive information on the distribution of fishing effort may also allow extrapolation of experimental studies to the wider scale of fisheries management.

[1] E-mail: m.cryer@niwa.cri.nz

Detecting the Effects of Fishing on Seabed Community Diversity: Importance of Scale and Sample Size

M. J. Kaiser[1]

School of Ocean Sciences, University of Wales-Bangor, Anglesey, United Kingdom

I investigated the importance of the extent of area sampled to the observed outcome of comparisons of the diversity of seabed assemblages in different areas of the seabed that experience either low or high levels of fishing disturbance. Using a finite data set, within each disturbance regime, samples of the benthic communities were pooled at random. Thus while individual sample size increased with each additional level of pooled data, the number of samples decreased accordingly. Detecting the effects of disturbance on species diversity was strongly scale dependent. Despite increased replication at smaller scales, disturbance effects were more apparent when larger, but less numerous, samples were collected. The detection of disturbance effects was also affected by the choice of sampling device. Disturbance effects were apparent when using pooled anchor-dredge samples, but were not apparent for pooled beam-trawl samples. A more detailed examination of the beam-trawl data emphasised that a whole community approach to the investigation of changes in diversity can miss responses in particular components (e.g. decapod crustacea) of the community. The latter may be more adversely affected by disturbance in comparison with the majority of the taxa found within the benthic assemblage. Further, the diversity of some groups (e.g. echinoderms) actually increased with disturbance. Experimental designs and sampling regimes that focus on diversity at only one scale may miss important disturbance effects that occur at larger or smaller scales.

[1] E-mail: m.j.kaiser@bangor.ac.uk

Spatial and Temporal Patterns in Trawling Activity in the Canadian Atlantic and Pacific

D. W. Kulka[1]

Department of Fisheries and Oceans, St. John's, Newfoundland, Canada

D. A. Pitcher

Spatial Metrics Atlantic, Dartmouth, Nova Scotia, Canada

GIS was used to spatially analyze trawling in Canadian Atlantic and Pacific waters as part of a program to assess the effect of trawling on benthic habitats. Data for 1980-2000 (Atlantic) and 1994-2000 (Pacific) in the form of geo-referenced fishing set locations were used to spatially describe effort location. The results are a series of maps depicting the spatial distribution of trawling intensity. Temporal changes and patterns in trawling intensity are described. In the Atlantic, trawl grounds are patchy and complex covering between 8 and 38% of the shelf in any year although actual trawled bottom area is much smaller. Spatial patterns of trawling changed dramatically over the time sequence analyzed but locations of high intensity trawling were similar from one year to the next. The spatial patterns were most stable during the 1980's while the greatest changes occurred during the early 1990's. There were numerous persistent areas of trawling spread mainly along the shelf edge and between the banks. A substantial portion (shallow and shoreward) of the shelf was consistently unfished. In the Pacific, the trawl locations were more consistent but the observed timeframe was much shorter (1994-2000). Trawl grounds comprised a string of partially joined patches along the shelf edge off Vancouver Island, three patches within the southern Queen Charlotte Sound, south and east of Queen Charlotte Island at deeper locations and on the shelf edge north and west of Queen Charlotte Island. The results, a first step in quantifying trawl effects, provide precise information on extent and intensity of bottom disturbance due to trawling.

---

[1] E-mail: Kulkad@dfo-mpo.gc.ca

A GIS Routine for Assessing Designs that Sample an Area of Fish or Lobster Traps

G. A. Matthews,[1] R. L. Hill, and P. F. Sheridan

National Marine Fisheries Service, Galveston, Texas

Traps used in tropical fish and lobster fisheries may harm shallow reef habitats that have been identified as essential fish habitat for a number of federally managed species. We are evaluating the effects of these traps on benthic habitats in the Florida Keys, Puerto Rico, and the U.S. Virgin Islands by examining spatial distribution of traps and quantifying damage to structural benthos. ArcView is being used to organize and visualize the data and to choose some sampling parameters. An automated routine tests how well different transect widths represent the actual population density of traps among the area's habitats. Custom scripts register the trap population density, the trap sample density by benthic habitat, and their cumulative differences. Trap locations (latitude and longitude) and benthic habitat polygons form the bases over which the user fits a rectangle to encompass the known trap population area. The user can then establish sub-rectangles of varying widths as 'test-transects' for sampling. A test-transect is stepped across the trap area in as many increments as desired up to a limit set by a required minimum 1-m offset distance. The number of traps in each transect is obtained, along with the density distribution among the benthic habitats. Using the cumulative differences in percentage distributions for habitats between test-transects and the trap population area, an optimal minimum transect width can be established to acceptably represent the distribution of traps among habitats. This will help focus underwater research on specific habitats (corals) where damage is most likely.

---

[1] E-mail: geoffrey.matthews@noaa.gov

Essential Fish Habitat (EFH) in Alaska: Issues in Consistency and Efficiency When Using Geographical Information Systems (GIS) to Evaluate Effects to EFH

R.F. REUTER[1]

National Marine Fisheries Service, Alaska Fisheries Science Center, Seattle, Washington

C.C. COON

North Pacific Fisheries Management Council, Anchorage, Alaska

J.V. OLSON AND M. EAGLETON

National Marine Fisheries Service, Alaska Regional Office, Habitat Conservation Division, Anchorage, Alaska

The NMFS Alaska Region is presently drafting an Environmental Impact Statement (EIS) to evaluate a reasonable range of alternatives to develop the mandatory Essential Fish Habitat (EFH) provisions. The process of updating habitat definitions and creating maps that describe EFH have led to basic questions about the standardization of map criteria. Although national EFH GIS standards are suggested, there are regional differences, which will require specific standards for map criteria, such as projections, map features, and categories. Currently within the Alaska Region, research and GIS data are summarized by management areas (e.g., Gulf of Alaska, Bering Sea and Aleutian Islands), which share common features such as depth, but do not necessarily overlap between areas. For an Alaska Region project, such as the EFH EIS, an overall standard to make map production efficient is needed. Therefore, interpretation of the data sets and maps will be consistent. The standardization will assist NMFS scientists delineate EFH habitat types, depth strata, and species distributions. Additionally, the overlays will provide background layers to assist fishery resource managers in identifying possible effects to EFH. Thus, the completed GIS EFH product will allow NMFS to visually present how management decisions may appear in a geographic reference. This poster presents an example of resolving these issues.

---

[1] E-mail: rebecca.reuter@noaa.gov

Spatial Distribution of Fishing Activity for Principal Commercial Fishing Gears Used in the Northeast Region of the United States, 1995–2000

D. K. STEVENSON[1]

U.S. National Marine Fisheries Service, Northeast Regional Office, Gloucester, Massachusetts

Numbers of fishing trips, days-at-sea, and fishing days made by federal commercial fishing vessel permit holders were compiled from logbook data and assigned to 10 minute 'squares' of latitude and longitude to show spatial distribution patterns for 18 individual gear types and 3 major gear categories in the northeast United States (North Carolina - Maine) during 1995-2000. Principal gear types included in the analysis were otter trawls (fish), lobster pots, handlines, otter trawls (shrimp), hydraulic clam dredges, scallop dredges, quahog dredges, sink gill nets, and bottom longlines. GIS plots of days-at-sea (otter trawls and scallop dredges) and fishing time (clam dredges) accounted for geographical variations in trip duration and thus provided unbiased distributions of fishing effort for mobile bottom gear types. The distributions of scallop dredge days-at-sea in 1998 and 1999 closely resembled plots of fishing activity that were derived on a much finer time and spatial scale from vessel tracking system signal data (McSherry and Rago 2001). Overlays of ten minute squares that accounted for 90% of all days-at-sea or fishing time on bottom sediment data for the northeast U.S. continental shelf showed that most scallop dredging takes place on sand and gravel bottom, clam dredging in sand, and otter trawling on a variety of bottom types.

[1] E-mail: David.Stevenson@noaa.gov

Structure and Use of a Continental Slope Seascape: Insights for the Fishing Industry and Marine Resource Managers

A. WILLIAMS,[1] B. BARKER, R. J. KLOSER, N. J. BAX, AND A. J. BUTLER

CSIRO Marine Research, Hobart, Tasmania, Australia

Benthic habitats of the upper continental slope seabed (~300-700 m depth) off SE Australia are being surveyed for the first time in response to the needs of regional, ecosystem-based, marine management plans being developed under Australia's Oceans Policy, and increased commercial fishery reliance on fishes that inhabit the slope seascape. We developed substratum maps of the Big Horseshoe Canyon - one of the region's prime fishing grounds - using multi-beam acoustic backscatter data, and target-sampled with video cameras and a range of physical samplers. In upper-slope depths, a patchy mosaic of habitats is formed of sloping terraces of muddy substrata and rubble patches that support a sparse benthic epifauna, together with low-relief rocky ridges formed by outcropping claystones and limestones that support communities of erect epifauna dominated by sponges. Many sedentary adult individuals of two key commercial species (pink ling and ocean perch) shelter in a range of microhabitats provided only by the rocky habitats. Video shows that bottom trawls 'hook-up' on rocky substratum, turning and moving loose pieces - an observation acknowledged by commercial fishermen who also report that boulders and 'slabs' are removed and redistributed. This is evidence of a fishing impact that is, at least in part, irreversible. The question then is, how much impact will adversely affect long-term fishery productivity and conservation values? We discuss this question with respect to managers needs for both the detailed understanding and fine-scale mapping of habitats provided by scientific survey, and the fishing industry's knowledge of broad-scale habitat distributions that enables extrapolation to a regional fishery scale.

[1] E-mail: alan.williams@csiro.au

Social and Economic Issues and Effects

Perspectives on an Ethic toward the Sea

STEPHEN R. KELLERT[1]

Yale University, School of Forestry and Environmental Studies,
New Haven, Connecticut 06511, USA

Abstract.—This paper argues that beyond good science, management technology, and regulatory policies, long-term sustainable fisheries conservation depends on developing an ethic toward the marine environment based on a greatly expanded understanding of human self-interest. This ethic will require the recognition of how human physical and mental well-being relies on a diverse array of values and benefits derived from the marine environment. Conversely, this ethic will necessitate the realization that the degradation of the marine environment inevitably reduces human physical, material, emotional, intellectual, and even moral and spiritual welfare. This ethic of enlightened self-interest derives from far more than a narrow material and economic calculus, also including a greatly expanded notion of utility that emphasizes human dependence on the sea to enhance human creativity, problem solving, intellectual ability, affective capacity, moral understanding, and more. This perspective treats human dependence on the marine environment as rooted in human biology and evolution; as humans evolved, so did their dependence on the marine realm as a source of various adaptive benefits in the struggle to survive as individuals and societies. This human genetic affinity for nature, including the sea, is referred to as biophilia. The notion of biophilia is linked to nine biologically based values that, when adaptively expressed, support an ethic of stewardship for the marine environment. These nine values and their relation to an ethic toward the sea are nonetheless weak genetic tendencies, requiring adequate learning, experience, and cultural support to become functionally and adaptively expressed. An illustration is provided of radically altered values and ethical perspectives toward large cetaceans during the 20th century to indicate how profoundly, effectively, and rapidly an ethic toward aspects of the marine environment can develop under particular historical and cultural circumstances.

Introduction

The typical assumption is that better science and more effective management technologies, matched by more forceful law and policy, will foster sustainable fisheries conservation. I am doubtful these strategies alone can ever produce the desired outcome of a more ecologically healthy and sustainable fisheries. My skepticism stems from the impression that despite extraordinary investments in scientific research and management technology and the formulation and passage of new policies and laws during the past half-century, fisheries conservation has not substantially improved, and perhaps even worsened. For example, Carl Safina (1995:5) suggests that, "numerous [commercial] fish populations [are at] extremely low levels . . . and the fishing industry increasingly extracts fish faster than these populations can reproduce." A Food and Agriculture Organization of the United Nations (1995) report similarly concludes, "no country can be viewed as generally successful in fisheries management." And more recently, a Scripps study (Pew Oceans 2002) reports there is "overwhelming evidence that the unintended consequences of fishing on marine ecosystems are 'severe, dramatic, and in some cases irreversible,' and that many current fishing activities are harming the very ecosystems on which fishing depends and this [situation] is worsening."

Many believe these judgments and evaluations reflect the influence of factors beyond scientific and managerial control such as expanding human populations, material consumption, economic and commercial globalization, misguided and short-term politics, etc. These are all undoubtedly relevant, but I would contend that sustainable fisheries conservation will never be achieved while lacking fundamental shifts in human values and ethics, particularly the evolution of a marine ethic derived from a vastly expanded understanding of human self-interest viewed over the long run. Lest this seem hopelessly idealistic, I would suggest that both the dismal record of contemporary fisheries management and the evidence of extraordinary change that can occur when

---

[1] E-mail: stephen.kellert@yale.edu

human values radically change recommend at least consideration of this often-neglected strategy.

As suggested, a new ethic for protecting and sustaining the marine environment can emerge from a greatly expanded understanding of human individual and collective self-interest (Kellert 2003). Later in the paper, the case of marine mammals, particularly large whales, will be offered to illustrate how radically altered policies can result from changing values and perceived interests relating to the sea. This illustration will suggest that an ethic toward the marine environment depends on better recognition of how much human physical and mental welfare relies on a deep and diverse array of inherent affinities for the sea, particularly its living biota. This ethic of enlightened self-interest stems from far more than a narrow material and economic calculus, reaching much farther and deeper into the human dependence on the sea as a fundamental basis for achieving a satisfying and fulfilled humanity. An ethic toward the marine environment that builds from material and economic calculations is clearly essential, but this focus is too restricted, carrying with it the seeds of its own defeat by implying a moral justification for degrading the marine environment when superior material and economic rewards are achieved or projected. The marine ethic advanced here expands our notions of self-interest beyond the narrow confines of material and economic benefit, advocating a utility rooted in the human dependence on the sea for bodily comfort, physical health, intellectual capacity, emotional affiliation, spiritual relation, and more. This perspective contends that our dependence on the marine environment remains rooted in human biology and explains how, during the long course of our species evolution, the sea conferred adaptive benefits essential in the struggle to persist as individuals and as societies (Kellert and Wilson 1993; Kellert 1997). This ethical viewpoint echoes the views of E. O. Wilson (1993:37) when he suggests,

> In the end, decisions concerning preservation and use of biodiversity will turn on our values and ways of moral reasoning. A sound ethic . . . will obviously take into account the immediate practical uses of species, but it must reach further and incorporate the very meaning of human existence . . . A robust, richly textured, anthropocentric ethic can instead be made based on the hereditary needs of our own species, for the diversity of life based on esthetic, emotional and spiritual grounds.

A range of biologically based values of nature that form the basis for this expanded ethic of stewardship for the marine environment will be described. Before turning to this description, I want to begin with a "big fish" story that illustrates how an ethic toward the marine environment can become powerfully revealed, profoundly affect human behavior, and develop far more rapidly than generally assumed based on greatly expanded understanding of human interest and affinity for the marine world.

This example considers marine mammals, not fish, of course, but largely perceived and treated as fish for centuries. Norris (1978:320) aptly describes the story of their exploitation as "the greatest fishery the world has ever known." While historic, it is not exactly ancient history; as recently as the 1960–1961 fishing season, some 15% of the world's "fish catch" consisted of whale meat. Although few people in 1960 actually viewed whales as fish, from the perspective of how they were managed and culturally perceived, they were often treated more like fish than mammals.

The historic scale of the whale fishery was so vast and unrelenting that Norris (1978:321) concluded, "no other group of large animals has had so many of its members driven to the brink of extinction." Three distinct periods marked the exploitation of the large whales: coastal whaling prior to the 17th century, the sperm whale fishery during the 18th and 19th centuries, and the 20th century innovations, such as the explosive harpoon, pneumatic lance, fast catcher boats, and factory ships, that allowed any whale to be exploited no matter how fast or remotely located. The height of this third wave occurred during the 1930s when as many as 70,000 large whales were killed in some years.

A number of factors contributed to the decline of the large whales, most factors of a commercial and technological character (Kellert 1996). Widespread biological ignorance also prevailed, especially erroneous assumptions regarding the inexhaustibility of the resource, deficient understanding of whale ecology and behavior, and a tendency to manage every species in much the same way. Other factors included the difficulty of regulating whales in the open ocean and the practice of treating animals as the property of everyone and no one and, thus, fair game for all. Many nations also possessed independent whaling operations utilizing highly efficient harvesting technologies, which offered few incentives for limiting the exploitation of whales or maintaining effective management controls.

Despite all these influences, prevailing values were the underlying factor in the demise of the large whales, particularly an inordinate emphasis on what will later be described as utilitarian, dominionistic, and negativistic values and, conversely, the near absence of more sympathetic perspectives such as esthetic, scientific, naturalistic, humanistic, and moralistic values of whales. I am not suggesting the illegitimacy of the former values but rather their lack of moderation and balanced consideration with other equally legitimate and beneficial basic perspectives of whales and the marine environment. The story of whaling illustrates a dysfunctional and exaggerated emphasis on a narrow utilitarian view of marine resources. This focus was exacerbated by large capital investments in whaling ships and technology, the tendency to reinvest surplus profits in other areas of economic return, and excessive greed. A strong dominionistic view of whales also treated their conquest as an affirmation of the human capacity to master the largest creature the world had ever known in the most inhospitable of all environments, the open ocean. A dysfunctional negativistic viewpoint largely regarded whales as monstrous and lacking the capacities for thought and feeling. These prevailing sentiments encouraged a lack of empathy and compassion for whales and helped to economically justify and morally legitimize their excessive exploitation and even biological annihilation.

Yet, a profound transformation in values and ethics toward whales occurred during the latter part of the 20th century, particularly in several Western European and North American nations. Indeed, the plight of the whale became a powerful metaphor and cause for the modern environmental movement. An extraordinary emergence of scientific, esthetic, naturalistic, humanistic, and moralistic perspectives on whales occurred, accompanied by a corresponding decline in a view of these creatures from the vantage point of their material value, conquest, and fear. This profound shift in what might be called valuational chemistry formed the basis for the development of a new ethic of concern and stewardship for these and other marine mammals.

A number of factors fueled this revolutionary shift in values, behaviors, and ethics (Lavigne et al. 2000). A dramatic decline in populations of most species and the near extinction of some created a sense of impending catastrophe. The whale emerged as a powerful symbol of the excessively destructive practices of modern society toward the natural environment. Many were horrified at the prospect of our species extinguishing the largest creature of all time, and many lamented the presumed suffering experienced by harpooned whales. These moralistic and humanistic sentiments encouraged the rapid growth of a new breed of aggressive and highly influential nongovernmental environmental organizations.

A pronounced change in scientific appreciation of whales also resulted from a surge of study and technical understanding of the marine environment following the development of new technologies during and after World War II. Startling information regarding marine mammal biology and behavior greatly affected public attitudes of whales, particularly the extraordinary intelligence of many cetaceans, their complex social lives, advanced communications abilities (e.g., the "singing" of humpback whales), and presumed strong social ties (e.g., the "kindness" of killer whales), etc. These scientific discoveries supported an impression among many that a basic commonality connected people and whales. In addition, the naturalistic appreciation of whales was stimulated by the development of new recreational opportunities for viewing these creatures in their natural habitats. Whale watching, although only initiated in the 1950s, by the end of the century had become a global industry yielding more than US$1 billion annually and involving more than nine million participants in some 90 countries. Growing interest and concern for whales was additionally stirred by their frequent appearance in popular publications, films, and even musical recordings, and despite nearly three decades of extensive media coverage, whales remain widely portrayed in television, film, print, and sound media.

This extraordinary change in appreciation of whales led to significant changes in governmental law and policy. The most innovative change was the passage of the 1972 U.S. Marine Mammal Protection Act (MMPA) which, among other things, declared marine mammals to be an "esthetic and recreational as well as economic" resource. A number of features of the MMPA illustrate the change in values toward whales and other marine mammals, including:

- a shift from maximum sustainable yield to "optimum population" standards for managing marine mammals, including consideration of their contribution to the "productivity . . . and health of [marine] ecosystems" and the nonconsumptive uses of these mammals;

- an emphasis on the management of individual population stocks rather than entire species;

- a stress on ecosystem management, including not allowing marine mammals "to diminish beyond the point at which they cease to be a significant functioning element of the ecosystems of which they are a part;"

- the placing of a moratorium on the "taking" of whales, with certain exceptions; and

- the humane treatment of marine mammals, including preventing their harassment in the wild, maintaining captive conditions marked by decent treatment, and an emphasis on education and science as much as entertainment.

Despite continuing ambivalence in perceptions of whales among elements of the resource-using public and other nations such as Norway and Japan, a profound value and ethical transformation occurred toward these sea creatures in a remarkably short time period. For many people and nations, whales now possess moral standing that demands they be treated with kindness, compassion, and ethical regard. They may not yet have achieved the moral standing and associated rights of cats and dogs—creatures awarded individual names, the right to live among us and, most of all, not be eaten—but whales come fairly close for some elements of the public.

As noted, the emergent ethical standing of whales reflects the rise of highly appreciative esthetic, naturalistic, scientific, humanistic, symbolic, and moralistic values and a moderating decline in more materialistic, exploitative, domineering, and fearful perspectives. What is the moral of this tale? Simply that an analogous shift can occur as profoundly and perhaps as rapidly in human values and ethical relations toward other elements of the marine environment. This will require, however, a vastly transformed view of people's self-interest rooted in an expanded recognition of how human physical, emotional, intellectual, and even spiritual well-being is affected by our relations to the sea. As will be described, the diverse values noted in perceptions of whales reflect the richness of the human reliance on nature as a source of adaptive fitness and security. When all these values occur in balanced and functional relation, they offer the basis for an environmental ethic rooted in a greatly expanded realization of self-interest that encompasses a conventional understanding of material and commodity advantage but also the adaptive significance of nature and the sea as a source of beauty, affection, intellect, spiritual inspiration, and more. Together, these values comprise a web of relational dependency that rationalizes and supports an ethic of care for the marine environment. The example of the whale reveals that an expanded appreciation of nature can foster radical changes in ethical and moral relationships and, in turn, trigger pronounced alterations in regulatory and legal policies, often in a far shorter time than presumed possible.

This greatly expanded realization of self-interest emerges from a deeper understanding of the biological basis for the human affinity for nature, a perspective that is called biophilia (Wilson 1984; Kellert and Wilson 1993; Kellert 1997). In my own work, biophilia is tied to the various ways people are genetically inclined to attach value and derive benefit from the natural world. The functional expression of these values reveals how the human experience of nature serves as an anvil on which our fitness can be forged and, conversely, how we inevitably diminish our physical and mental well-being by substantially degrading the natural environment. Yet, these inherent inclinations to value nature are weak genetic tendencies, highly contingent on adequate experience, learning, and sociocultural support, to become fully and functionally manifest. Our inherent tendencies to value nature and the sea are greatly influenced by human choice, creativity, and free will. Reflecting the influence of learning and culture, each value is highly variable among individuals and groups, but as expressions of human genetics, their adaptive expression is bound by biology.

All values of biophilia described below have been noted in the discussion of whales. What follows, therefore, are brief descriptions of each value as a biologically based rationalization of human self-interest for developing an enhanced ethical position toward the marine environment.

Utilitarian

This value emphasizes the contribution of the marine environment to human material and physical security in the form of foods, medicines, building materials, and other commodities. The marine world has long functioned as an indispensable source of human sustenance and security. Marine organisms provide the largest source of food protein for countries like Japan, Indonesia, and Ghana, and worldwide, the sea yields some 95 million tons of food annually, worth more than $100 billion (Norse 1993; Pimentel et al. 1997). Employing this narrow economic calculus, Pimentel and others reported in 1997 that the natural environment generated more than $300 billion (roughly 5%) to the American economy and nearly $3 trillion (11%) to the global economy. This utilitarian dependence will also increase greatly in the future as only an estimated 15% of the world's species, and a smaller proportion of the marine environment, have been identified and assessed for their potential utility. As a consequence, rapid developments in systematics, biochemical prospecting, and genetic engineering will likely generate a revolution in new product development derived from exploiting an infinite array of marine biological solutions to survival fashioned over millions of years of evolutionary trial and error.

Even in the absence of necessity, people exercise and indulge their utilitarian dependence on the sea. They do this because activities like casting for finfish, digging for shellfish, or diving for corals offer a wealth of opportunity for achieving physical and mental well-being. Beyond the obvious material gains, people extract with skill a portion of their sustenance from the land and the sea and, in so doing, harvest physical and mental fitness and affirm their ability to sustain themselves with cunning and craft.

Scientific

Understanding, studying, and learning from nature, including the marine environment, builds intellectual and cognitive capacity. Humans possess an undeniable need to know and understand their world with authority—a tendency, independent of culture and history, where intellectual prowess is developed through study and observation of natural processes and diversity. This value exists among all people, even so-called primitive peoples who often possess an extraordinary knowledge of their terrestrial and aquatic environments, much of this understanding acquired independent of any obvious and immediate material utility (Diamond 1993; Nelson 1993). What the natural world offers all peoples, both primitive and modern, is a nearly limitless stage for developing critical thinking skills, problem solving abilities, and analytical capacities (Bloom et al. 1956). Observing, examining, and understanding the sea provides an array of challenging opportunities for acquiring knowledge, developing understanding, and honing evaluative aptitudes. These cognitive capacities can be advanced in other learning contexts, like modern computers, but natural diversity offers an accessible, familiar, and always stimulating context for pursuing intellectual competence.

The sea provides us with an extraordinary array of opportunities for study, exploration, and the exercise of curiosity and imagination. Its intellectual fascination is revealed in the endless capacity of children to explore and be absorbed by the beach and seashore. The value of studying and observing the sea has become especially apparent today as we become aware of how little we know about the most basic features and functions of the marine realm. Even the largest creatures, the cetaceans, have had 15% of their known species discovered in the 20th century alone (Wilson 1992). The vastness of life in the ocean represents an intellectual treasure awaiting empirical and systematic inquiry, whether by scientists, snorkelers, divers, boaters, or adventure travelers.

Esthetic

The natural world is one of humanity's greatest sources of esthetic attraction and inspiration. Few experiences in our life exert as powerful and persistent an effect on us as the beauty and physical attraction of nature. Even the most hardened person, if suddenly exposed to a beautiful ocean sunset or a profusion of coral life, is usually unable to resist or repress feelings of esthetic wonder and fascination. This irresistible urge, like all genetically encoded responses, developed because it fostered a range of adaptive benefits linked to the consistency and intensity of its response. The recognition of beauty in nature and the sea engenders an enhanced awareness and understanding of harmony, balance, and grace (Hersey 1999). People discern unity and order in certain marine features, and these esthetic responses inspire and instruct. Natural beauty functions as a design model whereby, recognizing the prototype, we can potentially capture analogous excellence and refinement in our own lives. The ideal offers a template and a means where we can discover clues to a more rewarding existence through mimicry and ingenuity. Perceiving beauty and harmony in a waving fan coral, an angelfish, a sinuous dolphin, or a mighty whale offers a glimpse of perfection in a world where chaos, frailty, and shortcoming are far more normative.

The human esthetic preference for nature also increases the likelihood of obtaining safety, sustenance, and security (Appleton 1975; Heerwagen and Orians 1993). People esthetically favor landscapes with water, that enhance sight and mobility, that offer the chance to discern danger or locate shelter, and that have bright flowering colors—all features which have been instrumental in human survival over evolutionary time. And it is remarkable how many of these features prominently occur in the marine realm, especially along the shore and coastal areas rich in the life-sustaining combinations of light and nutrients. Finally, the esthetic attraction to the sea fosters curiosity, imagination, and creativity. We are drawn to the beauty of nature because it represents the most information-rich environment we will ever encounter (Wilson 1984). A tide pool and a coral reef initially attract us by their immense esthetic appeal but, with cultivation and effort, we are inevitably drawn into ever-increasing possibilities for exploration, discovery, and creativity.

Humanistic

Nature is a powerful stimulus and source of emotional bonding and attachment (Searles 1959; Katcher and Wilkins 1993). This often occurs through companionship with other creatures but also by strongly identifying with particular plants, seascapes, and landscapes. These natural elements offer the chance for experiencing intimacy, connection, and a sense of kinship. With rare exceptions, people crave affiliation and companionship. By strongly identifying with elements of the natural world, they can achieve a feeling of relationship, a way to express and seemingly receive affection. Bonding and affiliation constitute critical pathways for humans to develop the capacities for cooperation and sociability, evolutionarily important for a social species like our own. People covet responsibility for others and, in turn, gratefully receive their affection and allegiance. Caring and being cared for by another creature and, more generally, exercising stewardship for the natural environment, provides a means to express loyalty and devotion. These benefits accrue under normal circumstances but often become particularly noticeable in times of crisis and disorder. A nurturing relation to the natural world can be both physically and mentally restorative. When distressed, people seek the therapeutic and healing powers of flowers, gardens, the ocean, and the seashore. These feelings of emotional connection are often evident in people's affection for particular creatures like whales, dolphins, manatees, and seals but also can become manifest toward more biologically remote creatures and habitats.

Dominionistic

People often hone their physical and mental fitness through subduing and mastering the natural environment (Ewart 1989; Shepard 1996). By outwitting and outcompeting nature, they can develop and refine their capacities for resolving challenge and adversity. The marine world has always offered a highly coveted context for developing these more competitive traits and abilities. People may no longer rely on besting prey or eluding menacing marine predators for their survival, but the strengths and prowess obtained from challenging the land and the sea continue to offer critical pathways for fostering physical and mental competence. Moreover, a sense of self-confidence and self-esteem can frequently derive from successfully testing oneself in the marine world. By demonstrating the ability to function under difficult circumstances, people emerge more certain of themselves. By coping with adversity, they affirm the ability to confront risk, master uncertainty, and deal with the unknown. These benefits can explain why so many people, especially the young, in the absence of necessity and often at great personal risk, deliberately pursue challenge and risk in the marine environment and often speak so highly of the experience (Kaplan and Talbot 1983; Kellert and Derr 1998; Kellert 2002). The challenge and competition afforded by the sea remains one of its main attractions. Sailing, kayaking, scuba diving, deep-sea fishing, and more continue to feed the human desire for growth and development through facing risk, mastering challenge, and resolving adversity.

Moralistic

The natural world serves as a source of deep moral and spiritual inspiration (Berry 1988; Kellert and Farnham 2002). Life's extraordinary variability is reflected in approximately 1.7 million scientifically classified species, an estimated 10–100 million living species, and the disappearance of 99% of all species that ever existed. Despite this incredible variability, an astonishing commonality also appears to underlie much of life on earth—most living creatures share analogous molecular and genetic structures, similar circulatory and reproductive features, parallel body parts, and more. A remarkable web of relationships connects a crab on the ocean floor, a tuna in the deep sea, a wandering albatross, or a person reclining in a modern cruise ship. This commonality offers us a glimpse of unity, underlying order, and a cornerstone for spiritual and religious belief. We give shape and definition to our existence by discerning universal patterns in creation. We achieve spiritual comfort and moral cohesion through a shared conviction of underlying meaning and order in the natural world. By nurturing the belief that at the core of our existence exists a fundamental logic and relation, we are able to transcend our individual separateness and aloneness. This moral and spiritual relation often inspired by the sea is reflected in the words of John Steinbeck, commenting on life in the marine realm (1941:93):

> It seems apparent that species are only commas in a sentence, that each species is at once the point and the base of a pyramid, that all life is related . . . And it is a strange thing that most of the feeling we call religious, most of the mystical out crying which is one of the most prized and used and desired reactions of our species, is really the understanding and the attempt to say that man is related to the whole thing, related inextricably to all reality, known and unknowable… That all things are one thing and that one thing is all things—a plankton, a shimmering phosphorescence on the sea and the spinning planets and an expanding universe, all bound together by the elastic string of time.

When people divine spiritual and moral connection with creation, they usually lessen their tendency to harm and destroy its constituent parts. The inclination to protect the sea derives as much from moral and spiritual inspiration as from any economic materialism or regulatory mandate.

Naturalistic

People mine physical, intellectual, and emotional ore from exploring and immersing themselves in the rich tapestry of shapes and forms of the marine environment. In the process, they achieve physical fitness, mental acuity, and an expanded capacity for curiosity, imagination, and discovery. Advantages accrue in encounters with the ocean that enhance one's ability to react quickly, resolve new and challenging situations, and consume with efficiency. Any marine creature or habitat can represent an opportunity for intensive exploration, revelation, and discovery, but especially for examination of the most conspicuous organisms encountered through such activities as snorkeling, scuba diving, fishing, and more. A timeless and boundless feeling emerges from deep absorption and penetrating experience of the lives and habitats of ocean organisms. Moreover, the greater the immersion, typically the greater the sense of feeling intensely alive, aware, and attune. Thus, the otherwise dull and colorless ocean becomes vivid and textured, the amorphous aquatic vegetation emerges loaded with definition and relief, the stillness of the sea becomes replaced by a cacophony of sound and sensation, and the sameness of the water is transformed into an expressive and diverse medium. A sharpened awareness and appreciation results, often liberating, exhilarating, and generating a clearer sense of priorities and resolve. In addition to the gains in physical health this active involvement in the marine environment can produce, important emotional and intellectual gains also result from an enhanced inclination to exercise one's curiosity, imagination, and exploratory drive.

Negativistic

The ocean world is a source of some of our deepest fears and anxieties. Sharks, sting rays, jellyfish, great waves, fierce storms, a wild ocean sea, etc. frequently provoke great fright, and often with little provocation. Fear and aversion of the marine world can provoke irrational responses, but more typically this inclination is a functional aspect of our inherent behavior. Human well-being depends on skills and emotional tendencies originating in a healthy distancing from potentially injurious elements in nature (Ulrich 1993). Lacking this awareness, we often behave naively, ignoring our inevitable vulnerability before uncertain and powerful natural forces, constructing structures where they do not belong along barrier beaches or tidal wetlands, venturing into risky situations such as boating in a stormy sea, or swimming in shark-infested waters.

Moreover, our anxieties and fears of the marine world do not always produce contemptuous or destructive responses. Deference to and respect for the ocean arises as much from recognizing its capacity to defeat and destroy us as much as from feelings of affection and kinship. Awe combines reverence with wonder and fear. The ocean stripped of its power—a shark or killer whale swimming aimlessly in an empty, featureless tank—usually becomes little more than a focus of superficial amusement and condescension. Marine species and habitats utterly subdued provoke little deep or lasting admiration, humility, or respect.

Symbolic

Nature and the sea offer an invaluable source for human imagination, communication, and thought (Shepard 1978; Lawrence 1993). The imagery of nature represents raw material for expediting the exchange of information and understanding. We accomplish this through metaphor, analogy, and abstraction and by using the media of language, story, myth, fantasy, and dream. As symbol, nature continues to be instrumental in language acquisition, confronting basic issues of identity and selfhood, and ordinary communication. Symbolizing nature has been a device used by all peoples to address, in a disguised and tolerable way, many of life's conflicts and dilemmas. This occurs in children's stories, fairy tales, legends, myths, cultural totems, and taboos. In our culture, we think of Jonah and the whale, the aquatic creatures of Aesop's fables, the more recent classics of Jules Verne and Herman Melville (1941), and much more. Through this literature, like Ishmael in *Moby Dick,* we see in our relation to the sea "The precise situation of every mortal that breathes; [how] he, one way or other, has this Siamese connection with a plurality of other mortals." Symbolizing nature assists in everyday communication and discourse from the imagery of the street to the metaphors of the marketplace to great oratory and debate. We refer to a "whale of a good time," "clamming up," "pool sharks," "feeling crabby," and the virtue of telling "big fish" stories. The marine environment offers a substrate for symbolic creation not unlike the way the ocean's genetic variability provides a biochemical template for laboratory discovery. Each exploitation of the marine realm helps us to mold and fabricate solutions to life's many and never-ending challenges.

Conclusion

All the biophilic values—the inherent inclination to affiliate with the natural world—reflect varying strands of human reliance on nature and the marine environment for physical and mental reward. When adaptively expressed, each contributes a vital thread to a strong ethical garment of care and stewardship for the land and the sea. This ethic relies less on charity or kindness and more on the self-interested realization of how human health and vitality over the long run emerges from the quality of our experience of the terrestrial and marine worlds. The human relationship to the sea helps shape our identity and offers an unrivaled potential for nourishing the human body, mind, and spirit.

We have ineffectually relied on the promise of only better science, technology, and law to achieve the elusive goal of ecologically sustainable fisheries. This strategy has not worked and will inevitably fail without the addition of an informed, appreciative, and, most of all, ethically motivated resource user and citizenry. The current scale of global fisheries exploitation and benthic habitat destruction requires a basic change in our values and ethics toward the marine environment. Lest this vision seem impossible and unrealistic, the recent history of the whale suggests it may be not only feasible but practical and occur far more rapidly than assumed. A fundamental shift in basic perceptions of the sea constitutes a highly relevant strategy for advancing significant and lasting change in fisheries conservation. But this shift must originate in a greatly expanded understanding of human self-interest and biological dependence on the sea, one that recognizes in our diverse connections to the marine realm a moral posture that confers a host of physical, mental, and spiritual rewards. In this way, we can come to see in our relation to the sea an essential part of what it means to be fully and satisfyingly human. As Henry Beston remarked (1971:vi), inspired by his experience of the sea:

> Nature is a part of our humanity, and without some awareness and experience of that divine mystery man ceases to be man. When the Pleiades and the wind in the grass are no longer a part of the human spirit, a part of very flesh and bone, man becomes, as it were, a kind of cosmic outlaw, having neither the completeness and integrity of the animal nor the birthright of a true humanity.

References

Appleton, J. 1975. The experience of nature. Wiley, London.

Berry, T. 1988. The dream of the earth. Sierra Club Books, San Francisco.

Beston, H. 1971. The outermost house: a year of life on the great beach of Cape Cod. Ballantine Books, New York.

Bloom, B., M. Engelhart, E. Furst, W. Hill, and D. Krathwohl. 1956. Taxonomy of educational objectives, handbook I: the classification of educational goals—cognitive domain. Longman, New York.

Diamond, J. 1993. New Guineans and their natural world. In Kellert and Wilson (1993).

Ewart, A. 1989. Outdoor adventure pursuits: foundations, models, and theories. Publishing Horizons, Scottsdale, Arizona.

Food and Agriculture Organization of the United Nations (FAO). 1995. The state of world fisheries and aquaculture. FAO, Rome.

Heerwagen, J., and G. Orians. 1993. Humans, habitats, and aesthetics. Pages 138–172 in Kellert and Wilson (1993).

Hersey, G. 1999. The monumental impulse: architecture's biological roots. MIT Press, Cambridge, Massachusetts.

Kaplan, S., and J. Talbot. 1983. Psychological benefits of a wilderness experience. In I. Altman and J. Wohlwill, editors. Behavior and the natural environment. Plenum, New York.

Katcher, A., and G. Wilkins. 1993. Dialogue with animals: its nature and culture. Pages173–200 in Kellert and Wilson (1993).

Kellert, S. 1996. The value of life: biological diversity and human society. Island Press, Washington, D.C.

Kellert, S. 1997. Kinship to mastery: biophilia in human evolution and development. Island Press, Washington, D.C.

Kellert, S. 2002. Experiencing nature: affective, cognitive, and evaluative development in children. In P. Kahn, Jr., and S. Kellert, editors. Children and nature: psychological, sociocultural, and evolutionary investigations. MIT Press, Cambridge, Massachusetts.

Kellert, S. 2003. Human values, ethics, and the marine environment. In, D. Dallmeyer, editor. Values at sea: ethics for the marine environment. University of Georgia Press, Athens.

Kellert, S., and V. Derr. 1998. National study of outdoor wilderness experience. Yale University School of Forestry and Environmental Studies, National Outdoor Leadership School, and Student Conservation Association, New Have, Connecticut.

Kellert, S., and T. Farnham, editors. 2002. The good in nature and humanity: connecting science, religion, and spirituality with the natural world. Island Press, Washington, D.C.

Kellert, S. R. and E. O. Wilson, editors. 1993. The biophilia hypothesis. Island Press, Washington, D.C.

Lavigne, D., V. Scheffer, and S. Kellert. 2000. The changing place of marine mammals in American thought. In, J. Twiss and R. Reeves, editors. Marine mammals. Smithsonian Press, Washington, D.C.

Lawrence, E. 1993. The sacred bee, the filthy pig, and the bat out of hell: animal symbolism and cognitive biophilia.Pages 301–344 in Kellert and Wilson (1993).

Melville, H. 1941. Moby Dick; or the white whale. Dodd, Mead & Co., New York.

Nelson, R. 1993. Searching for the lost arrow: physical and spiritual ecology in the hunter's world. Pages 201–228 in Kellert and Wilson (1993).

Norris, K. 1978. Marine mammals and man. Pages 315–325 in H. Brokaw, editor. Wildlife and America. Council on Environmental Quality, Washington, D.C.

Norse, E. 1993. Global marine biological diversity. Island Press, Washington, D.C.

Pew Oceans. 2002. The ecological effects of fishing in marine ecosystems in the United States. Pew Charitable Trust, Philadelphia.

Pimentel, D., C. Wilson, C. McCullum, R. Huang, P. Dwen, J. Flack, Q. Tran, T. Saltman, and B. Cliff. 1997. "Environmental and economic benefits of biodiversity. BioScience 47:747–757.

Safina, C. 1995. The world's imperiled fish. Scientific American 273:46–53.

Searles, H. 1959. The nonhuman environment. International Universities Press, New York.

Shepard, P. 1978. Thinking animals: animals and the development of human intelligence. Viking Press, New York.

Shepard, P. 1996. The others: how animals made us human. Island Press, Washington, DC.

Steinbeck, J. 1941. Log from the Sea of Cortez. J. J. Appel, Mamaroneck, New York.

Ulrich. R. 1993. Biophilia, biophobia, and natural landscapes. Pages 73–137 *in* Kellert and Wilson (1993).

Wilson, E. O. 1984. Biophilia: the human bond with other species. Harvard University Press, Cambridge, Massachusetts.

Wilson, E. O. 1992. The diversity of life. Harvard University Press, Cambridge, Massachusetts.

Wilson, E. O. 1993. Biophilia and the conservation ethic. Pages 31–41 *in* Kellert and Wilson (1993).

Getting to the Bottom of It: Bringing Social Science into Benthic Habitat Management

BONNIE J. MCCAY[1]

*Department of Human Ecology, Cook College, Rutgers, the State University,
55 Dudley Road, New Brunswick, New Jersey 08901, USA*

Abstract. Social science has a small but growing place within marine science and policy. Legal and political imperatives for social impact assessment have increased attention to the need for social research, and the high frequency of conflict over and resistance to habitat and fisheries management measures, including marine protected areas, suggests that there is much to be done. The methodology of social impact assessment is changing to incorporate notions such as vulnerability and tools such as mapping. In the larger sense, social science helps identify and formulate "sea changes" in the knowledge base and institutional framework for marine science and conservation. These are major developments in the understanding of the human dimensions of fisheries and marine environments that have begun to influence both research and policy. The "sea changes" identified are increased attention to more cooperative, participatory, and community-oriented research and management regimes on the one hand and increased reliance on exclusive property rights and markets on the other.

Introduction

In this paper, I discuss the underrepresentation of the social sciences in the discussion of benthic habitat management and the rise of serious federal government investment in remedying the situation.

In the second part of the paper, a brief introduction is provided on approaches to social impact assessment that incorporate concepts of vulnerability as well as geographic mapping tools, which are particularly important to management decisions intended to protect particular places or habitats. The primary role of social science in marine resource management has been "impact assessment." This role is essential, given legal mandates and some, if not all, goals of management. Within the context of the Magnuson Act and federal waters fisheries, those legal mandates—especially the Sustainable Fisheries Act of 1996—bring to the forefront both the need for major changes in management to protect essential fish habitats and the requirement that the needs of fishing communities be taken into account when making management decisions. The question becomes how decisions made to "minimize to the extent practicable" the adverse effects of fishing on marine environments can be better informed by socioeconomic analyses.

In the last part of this paper, I focus on important developments in the understanding of the human dimensions of fisheries and marine environments that have begun to influence both research and policy. The major "sea changes" contrasted are (a) increased attention to more cooperative, participatory, and community-oriented research and management regimes; and (b) increased reliance on exclusive property rights and markets. Both have implications for how benthic habitats are managed.

Social Science at the Symposium

A good way to introduce the range and substance of the social sciences is to focus on particular social scientists and their work. The program for the Symposium on Effects of Fishing Activities on Benthic Habitats (Malakoff 2002) included five abstracts that are clearly identified as socioeconomic and that serve this purpose. Although they cannot represent the entire universe of socioeconomic studies of fisheries and marine ecosystems, which is very diverse, they indicate some of the significant directions to be taken and questions to be asked.

Spatial Analysis, Economics, and Fishers' Behavior

James Wilen of University of California–Davis and James N. Sanchirico of Resources for the Future have taken

---

[1] E-mail: mccay@aesop.rutgers.edu

the lead in bringing spatial analyses into economics, focusing on marine reserves (Sanchirico and Wilen 1999, 2001). At this symposium, J. E. Wilen (University of California, Davis, abstract from Symposium on Effects of Fishing Activities on Benthic Habitats, 2002) addressed the question of what leads to particular spatial distributions of fishing mortality. He showed the importance of modeling the behavior of fishers responding to economic incentives (price changes) in answering the question, which is central to the task of both identifying the sources and scale of problems created by fishing activities and predicting the consequences of marine protected area management.

Social and Environmental Ethics

Ethics are central to human culture and behavior. Although they can only be part of the values, motivations, and intentions that cause human actions—competing, for instance, with assessments of costs and benefits—they deserve to be taken into account far more than they have to date. Stephen Kellert speculated on the evolution of a new environmental ethos where self-interest might be broadened in recognition of dependence on marine and other natural systems (see Kellert 1997, 2005). One might add the increased recognition of social as well as environmental ethics in marine conservation; social ethics may include notions of procedural justice, including the right of those affected by policy matters to participate meaningfully in the construction of such policy. The issue of greater public participation in marine conservation and fisheries management policy is addressed below, as one of the "sea changes" in marine policy.

Cultural Systems of Knowledge, Identity, and Power

Kellert's remarks concerned ostensibly universal or widespread changes in human ethics. Particular cultures and sub-cultures develop their own ethical systems that are embedded in particular heritages and experiences and are framed in relation to their identity. This is very true of the people and communities involved in commercial, recreational, and subsistence fisheries. Michael Chiarappa, an environmental historian, addressed the question of how and why commercial and native fishers of Lake Michigan persist in fishing, against many odds, by exploring the roles of traditional ecological knowledge and the formulation of vernacular environmental ethics, based, in part, upon an extensive oral history project (Chiarappa and Szylvian 2003; M. J. Chiarappa, Western Michigan University, abstract from Symposium on Effects of Fishing

Activities on Benthic Habitats, 2002). The traditional and experience-based environmental knowledge of commercial and recreational fishers is particularly important for marine resource conservation but, as yet, imperfectly incorporated into management, as discussed below as yet another important "sea change" in marine policy that is directly relevant to the social sciences.

Social and Economic Impacts of Regulatory Change

Social scientists are very active in the tasks of "impact assessment," which are legislatively mandated for many fisheries management regimes. In the United States, the Magnuson–Stevens Fishery Conservation and Management Act, the Regulatory Flexibility Act, and the National Environmental Policy Act (NEPA) mandate impact assessment. However, the infrastructure for social and economic impact assessment for fisheries is primitive because fishery data sets have been created mainly for biological stock assessment purposes, and other data, such as census data, are often at scales that are not commensurate with the scales of fisheries and marine ecosystems, a topic to which I will turn below. Astrid Scholz and colleagues reported on their efforts to use existing fisheries and socioeconomic data to create a spatially explicit analytic framework for assessing the socioeconomic effects of different management scenarios on the Pacific coast (Scholz et al. 2005). Their work shows the value of geographic information systems in accomplishing this goal (Ecotrust 2003).

Legal Frameworks for Environmental Management

As noted above, legislation mandates some kinds of research. Legislation—and the court decisions that interpret the legislation—also frames and channels scientific activity. Alison Rieser (University of Maine School of Law, abstract from Symposium on Effects of Fishing Activities on Benthic Habitats, 2002) explored changing interpretations of key phrases in the essential fish habitat provision of the 1996 Sustainable Fisheries Act: "best available science" and "adverse impacts." How they are interpreted and acted upon is very important because the essential fish habitat provision, together with bycatch provisions of the 1996 act, brought the rudiments of ecological management into federal fisheries management in the United States. Rieser has also contributed to analyses of the contentious role of property rights in relation to environmental conservation (Rieser 1991, 1997), another of the matters to which I will refer below in the discussion of "sea changes."

Underrepresentation of Social Science

The six [including this one] papers in the symposium program for the "socioeconomics" domain indicate extremely important arenas for social science work in relation to benthic habitat management and fisheries: how people utilize space and respond to changes in access rules; the role of ethics and experience-based or traditional knowledge; legal mandates, "mandated science," and policy. However, it is also telling that whereas they represented about 20% of the invited abstracts, they were only about 4% of the published submitted abstracts in the symposium program. For one reason or another, social scientists do not seem to see themselves as meaningfully involved in the topic, even though the subtitle of the symposium is "Linking Geology, Biology, Socioeconomics, and Management." This speaks to a major, persistent issue: the underrepresentation of social science in marine fisheries and conservation despite the fact that the primary control when dealing with wild and unpredictable natural systems is the behavior of people. (Note that the term "social science" has meanings that differ depending on user and context. Properly, it incorporates economics as well as sociology, political science, policy science, parts of anthropology and psychology, geography, and law, but in practice, a distinction is often made between economics and "other social sciences." I use it in the more inclusive sense.)

Too Little, Too Late? The Marine Protected Areas Case

Creating a marine protected area (MPA) is one of the favored policy responses to evidence of damage to marine benthic environments from fishing activities. I follow Agardy et al. (2003) in accepting "marine protected area" as a broad category that includes everything from no-take reserves to temporary fishing closures and zoning.

Major reports on MPAs by the National Research Council (2001) and the International Union for Conservation of Nature and Natural Resources (Salm et al. 2000) underscore the importance of social science research, but Patrick Christie and 16 co-authors, most of whom have done this kind of work in southeast Asia, the Caribbean, and the United States, observed that when the impacts of MPAs are studied, they are mostly biological impacts. Moreover, the social impacts, when treated at all, are typically reported by biologists not trained in social science. They point out that when social scientists are called upon to examine the consequences of establishing MPAs (or related questions), it is typically too late in the design process to make a difference (Christie et al. 2003.).

Redressing the Imbalance

The underrepresentation of the social sciences in marine affairs is well known. Steps are being taken to redress it. The Science Advisory Board of the National Oceanic and Atmospheric Administration (NOAA) commissioned a study of the place, role, and importance of social science within NOAA's line agencies (NOAA 2003). Two significant steps at the national level are the National Marine Fisheries Service (NMFS) program to increase its social science staff and the Social Science initiative of the National Marine Protected Areas Center (NMPAC) at Santa Cruz, California.

The NMFS is implementing a plan to greatly enhance social science expertise through hirings at headquarters and in the regional science centers, through pre-doctoral and postdoctoral fellowships, and through targeted research programs, including special initiatives to help the agency and the regional fishery management councils meet their requirements Agency social scientists at NMFS headquarters have played a major role in interagency efforts to define social impact assessment standards and, more recently, to develop draft manuals for social science research on fishing communities.

The NMFS has much greater representation of social sciences than do the other NOAA line agencies (NOAA 2003a). National Ocean Service includes the marine sanctuary and marine estuary programs; it also hosts the NMPAC, which sponsored a workshop on social science and marine protected areas in April 2002 and is completing a national agenda for social science research on MPAs (National Marine Protected Areas Center 2003) that, not surprisingly, includes a call for more social science and more social scientists.

There are other ways that social scientists are incorporated into the fisheries and oceanography establishment, including the appointment of social scientists to important advisory committees, such as the Science Advisory Board of NOAA and the Ocean Studies Board of the National Research Council, as well as regional committees such as those for the New England Fishery Management Council and the Atlantic States Marine Fisheries Commission.

Impact Assessment: Vulnerability, Space, and Place

The primary role of social science in marine resource management has been "impact assessment." Within the context of the Magnuson–Stevens Act and federal waters fisheries, legal mandates—especially the Sustainable Fisheries Act of 1996—bring to the forefront the need for major changes in management not only to protect essential fish habitats and reduce bycatch but also to consider fishing communities when making management decisions.

Interventions in federal waters also must undergo NEPA analyses of environmental impacts. Human communities are essential components of the environment under consideration in a NEPA analysis, under the rubric "human environment." Moreover, the Magnuson–Stevens Act also calls for consideration of "fishing communities" in National Standard 8 as of 1996. It defines "fishing community" as one that is substantially dependent on or substantially engaged in the harvest and processing of fishery resources to meet its social and economic needs; vessel owners, operators, crew members, and processors based in such a community are included (section 3[16]). These are place-based communities, that is, specific, contiguous geographic locations where fishermen or those associated with the fishing industry live and work. Social scientists also recognize occupational communities, which may comprise networks of interaction and shared identity that cut across place-based communities (J. Olson, paper presented at the American Anthropological Association Annual Meeting, 2000). Although the families of fishermen who work out of a port town live in many different towns in the area, they may be thought to constitute such a community (B. Oles, paper presented at the American Anthropological Association Annual Meeting, 2002). Communities of interest and virtual communities are also identified by social scientists (NRC 1999a, 1999b). In these communities, people share common interests and activities that may or may not be associated with particular places. In this sense, members of organizations such as the Recreational Fishing Alliance, East Coast long-liner vessel owners, and owners of ocean quahog individual transferable quotas (ITQs) may be thought to constitute communities. Finally, epistemic communities (Haas 1992) from around specific management institutions and issues, involving agency personnel, management body staff and appointees, and representatives of various stakeholder and public groups who participate regularly.

Social and economic impact analysis (SIA) is a well-defined field of applied social research. In our own recent SIA work (discussed below), we have tried to make it more manageable and more directly relevant to the task of identifying likely impacts of management alternatives on people and their businesses and communities (McCay et al. 2002).

In 2002, the Mid-Atlantic Fishery Management Council (hereafter, the council) sought to create an SIA for an amendment to the fishery management plan for a complex of species, including squid *Loligo pealeii* and *Illex illecebrosus*, Atlantic mackerel *Scomber scombrus*, and butterfish *Peprilus triacanthus* (McCay et al. 2002). Among the alternatives, the council planned to consider the closing of several essential fish habitat (EFH) areas to use of otter trawls at the head of the Hudson Canyon and in deep parts of the outer continental shelf between the 91-m and 259-m isobaths, previously identified as the Habitat Area of Particular Concern (HAPC) for tilefish *Lopholatilus chamaeleonticeps*. As it turned out, neither of these closures was adopted by the council.

The SIA followed the standard practice of describing the various place-based communities involved (typically, well-known fishing ports) in terms of socio-demographic data gleaned from the U.S. census and other sources and from visits to the ports and interviews with key informants (e.g., McCay and Cieri 2000). However, the study was guided by an explanatory framework, developed in the social sciences and used in an earlier fisheries SIA (Wilson and McCay 1998) that emphasizes the vulnerability of individuals, business enterprises, and communities to changes in management rules or related circumstances. In other words, a particular regulatory change can have drastically different effects on different individuals, groups, businesses, and communities because of their different degrees of vulnerability to change. For example, one fish-processing business may be highly diversified, another narrowly specialized. One fisheries group (often known in terms of gear-types or species-focus) may be particularly vulnerable to the regulatory change because of a recent history of other regulatory changes that have reduced their alternatives and capacity to respond. One community may have few jobs other than fishing, whereas another may offer reasonable alternatives. In some communities, fishing is protected by local zoning and other support against economic pressures to convert waterfront and moorage space to other uses, but in others, the forces of coastal gentrification may be so great and unchecked that even a small change in the regulations affecting fishing may lead waterfront owners to sell out, resulting in irreversible loss of working waterfront. For the study, each place-based fishing community and each management alternative, including the EFH alternatives, was studied for effect on (a) expected volume of fish landings; (b) the flexibility of businesses (i.e., their ability to respond to changes in fish availability or markets);

(c) safety at sea; (d) available alternatives; and (e) community support. This approach allowed us to factor into the analyses important trends such as suburbanization and gentrification of coastal communities as well as the strategies of fishery businesses.

An important tool in the research was the ability to use National Marine Fisheries Service data on reported fish landings (the Vessel Trip Report, or VTR data) to map spatial fishing patterns for different communities—both place-based and fishery-oriented (i.e., gear-type). There are problems in using these data, but at an aggregate level, they can be very useful in delineating uses of fishing space and their linkages to human communities (St. Martin 2001, 2004). A system was developed that allowed us to identify the importance of proposed EFH restricted areas to the bottom trawl fishers of one or another community on the basis of their recent history of fish landings in those areas. In other words, one could say that closing the tilefish HAPC to otter trawls would disproportionately hurt Montauk, New York, fishing enterprises more than those of Pt. Pleasant, New Jersey—an analysis that could be extremely important to the politics and social impacts of resource management.

Moreover, when discrepancies were found in the VTR data for the most recent year and interviews with key informants in the communities, the study was able to do analyses of a number of years of VTR data. This analysis tended to corroborate input from informants and underscored the importance of long time series and historical approaches to the topic, both of which are difficult to substantiate. For example, the pattern of squid fishing in recent years is affected not only by changes in the availability and abundance of squid but also by closures of large areas to small-mesh gear. As a result, the squid fishing enterprises of the region were vulnerable to proposed EFH alternatives because of this preexisting limitation, which is part of the cumulative effects of environmental and socio-political change.

These gear restricted areas were not part of our SIA mandate but were very much part of the human dimensions system that influenced how people would respond to other closures. There were other important contextual factors that would be lost in a traditional SIA. For example, a lawsuit filed by an environmental group was underway concerning the tilefish HAPC. Moreover, other lawsuits had underscored the fact that fishery management councils had to meet the procedural requirements of NEPA—mainly environmental impact assessments—as well as the existing requirements of the 1996 Sustainable Fisheries Act that strengthened essential fish habitat and bycatch provisions and introduced a "fishing community" national standard, hence the council's consideration of multiple alternatives and its decision to employ social scientists beyond its own staff members to help sort out the complicated context for people in the affected fishing community.

Sea Changes in Marine Policy

Sea changes, or subtle changes in marine fisheries policy, have important implications for how we respond to information about the effects and consequences of fishing activities on benthic and other habitats. Most widely accepted is the shift toward more ecological management, with stronger emphases on biodiversity and habitat protection, a shift identified as part of postmodernism in natural resource management (McCay 2000c). A postmodernist approach gives less weight to utilitarian values than to land-ethic values. But it also appreciates diversity and contingency of objectives and goals and emphasizes the fundamentally cultural nature of human enterprises such as conservation. Accordingly, ecosystem management is not antithetical to but tightly interwoven with social concerns.

Two major institutional sea changes can be discerned (McCay 2000a, 2001). One involves the increased application of property rights to marine environments and their products, including the use of individual transferable quotas and other tradable or marketable allocations of resource rights (McCay 1995; National Research Council 1999). It also includes the acquisition of private property to protect marine environments. The other is about community, including a shift toward greater involvement of resource users and other interested stakeholders in various aspects of marine conservation and fisheries management, ranging from research to management deliberations to implementation and enforcement. The institutional frameworks for marine resource and habitat management are changing accordingly and are reflected, albeit unevenly, in policy and practice in the United States, Canada, Europe, Australia, New Zealand, and other parts of the world.

Creating Commodities Out of Fishing Rights

A fundamental feature of marine ecosystems is that, by and large, they are not subject to private property rights; they are treated as common property systems with greater or lesser restrictions on what happens within them. The absence of exclusive, marketable property rights in fish or bottom lands or fish habitat may be thought to reduce incentives on the part of individuals to protect them or enhance their value, hence the need for governments to be involved in managing them. The federal National Marine

Sanctuary program and National Estuarine Research Reserves are examples of efforts in this direction, examples that are plagued with many problems, including those of effective public participation.

An alternative is to create or utilize private property rights. The most direct way is to buy private property in marine systems where they exist. Shellfish beds are often held as fee simple private property or, more often, as leaseholds or grants from the state. The Nature Conservancy has long followed a strategy of purchasing private lands in order to conserve and protect species and habitats. In October 2002, it made its first move in the marine sector, acquiring 9,160 ha of an oystering company's bottom land in Great South Bay, Long Island, New York (Heilprin 2002). The acquisition will be used to establish marine reserves in the waters of Long Island and is the first representation of a larger strategy to use shellfish leaseholds for the purpose of restoring the ecology of submerged lands (The Nature Conservancy 2002).

Although people often claim and defend rights to valued fishing grounds, and formal and informal examples of sea tenure are well known, especially in small-scale fisheries (Cordell 1989), privatizing underwater lands does not make sense for most marine fisheries given the mobility and unpredictable migratory paths of finfish and crustaceans. Moreover, in the United States, outright privatization of underwater lands is curtailed by very strong claims of public rights to fishing, backed by the legal doctrine of public trust, which pertains to navigable and tidal waters and submerged lands (McCay 1998). Even the Long Island case mentioned earlier is subject to claims from clammers not to be excluded from the area purchased.

Beginning in the early 1980s, quasi-privatization of fishing rights through ITQs emerged as an important tool of marine fisheries management in New Zealand, Canada, Iceland, Australia, and—beginning in 1990—the United States. Owners of fishing boats become owners of shares of a quota that they may trade, buy, and sell. The overall quota is established by a government body, but decisions about the use of ITQs are made by those who hold them. The system reduces pernicious outcomes of traditional fisheries management such as derby fishing, and it appears to create incentives to reduce fishing capacity without having to have government buy-back programs (National Research Council 1999). In specific cases, the use of ITQs in fisheries management may affect the extent to which fishing activities disturb benthic habitats because ITQ fisheries can be more selective about where and when fishing takes place (cf. Dewees and Ueber 1990). However, in the Sustainable Fisheries Act of 1996, the U.S. Congress imposed a moratorium on new ITQ systems, leaving in place only that for surfclams *Spisula solidissima* and ocean quahogs *Arctica islandica* in the mid-Atlantic–southern New England region, for wreckfish *Polyprion americanus* in the south Atlantic, and for Pacific halibut *Hippoglossus stenolepis* and sablefish (also known as black cod) *Anoplopoma fimbria* in the north Pacific. The moratorium was lifted in 2003, but the ITQ issue remains very controversial, largely because of concern for the industry reorganization and concentration of ownership that take place (McCay 2000b; Apostle et al. 2002; Shotton 2002).

Creating commodities out of the right to fish—or exercise other powers—in marine environments is a kind of decentralization. Management regimes with individual fishing quotas (IFQs), one example of privatized property rights at sea, delegate many, but not all, decisions to the individuals or firms that hold exclusive fishing rights. Individual fishing quotas and other commoditized systems (including tradable trap permits, tradable days at sea, and of course, leaseholds or fee simple property in fishing grounds) may or may not involve cooperation, co-management, and community, the other sea change trajectory to be discussed below.

Decentralization, Co-Management, and the Movement toward Community

Cooperation, co-management, and community comprise a bundle of concepts and approaches that challenges the established institutional framework of top-down, command-and-control, agency-led management of marine fisheries and ecosystems. These terms are used in arguments for more decentralized, place-based, and social-minded approaches to management, where some power and responsibility is delegated to people who are closer to the scene.

Co-management has long been the practice in state and federal agencies where the government's major role has been that of partner with an industry group or other user group (Weber 2002). It became a term of art in the 1970s and 1980s (Pinkerton 1989). A series of court cases that required genuine sharing of resources between tribal groups and state citizens in the Pacific Northwest led to increased tribal power and responsibility for salmonids and other natural resource management, in cooperation (co-management) with state and federal agencies (Lee 1993; Singleton 1999). The idea has been extended to many institutions that involve cooperation among agencies, scientists, resource users, and others (McCay and Jentoft 1996).

The term "co-management" is often watered down

as synonymous with any resource-user participation in management planning, program implementations, and research. In other words, it can mean the same as public participation. Instead, it is useful to think of co-management as one form of public participation. Co-management is where there is recognized and balanced sharing of power and responsibility.

One can think of a graduated scale of relationships between a governing agency and the public. When developing protected areas or fisheries restrictions, there may be virtually no relationship, except that an agency makes a decision and notifies the resource users or other interested members of the public. At the other extreme, decisions are made exclusively by members of the public or resource user group. Most cases in the developed and democratic world are in between. They tend to be ones where public or user-group involvement is advisory or consultative. The government agency retains exclusive power to make regulations (within parameters that are defined politically and legally) but is required or finds it helpful to consult with those involved in order to obtain better information or to increase the acceptability and legitimacy of the decisions it makes. Greater political legitimacy can pay off in terms of improved compliance with the regulations. Public participation in the deliberative processes involving marine reserves or sanctuaries in the United States is usually advisory, as it is for most fishery management issues. The critical political and social question concerns whether those involved feel that they are being heard and have some effect on decisions. Agency personnel may ask for advice but have no intention of using it, or they may ask with the intention of revising their decisions accordingly, if appropriate. People involved are often very sensitive to this difference, upon which can rise or fall the success of a program of public participation (Hance et al. 1988; see Pomeroy and Hunter, in press, for an MPA case study).

Chaos and Experience-Based Knowledge

Another sea change to which social scientists have contributed is increased recognition of the dynamic, difficult-to-predict, and seemingly chaotic nature of many marine processes. The major social science contributions have been to emphasize the importance of cognitive models or world views toward an understanding of how people relate to the natural world and to each other, and to draw out institutional and management implications of chaos-like natural processes (Wilson 2002).

An economist, James Wilson, played a significant and controversial role in highlighting the likelihood of chaotic processes in marine fisheries. His work challenges the authority of stock assessment models based on deterministic processes and has been used to call for parametric tools in fisheries management including area closures rather than precise restrictions such as quotas (Wilson and Kleban 1992; Acheson and Wilson 1996). The anthropologist M. Estellie Smith demonstrated the importance of chaos-like ideas in the thinking of New England commercial fishermen, which explained their resistance to the messages of government scientists about the status of groundfish stocks (Smith 1990).

More recently, anthropologists have developed techniques for elucidating cultural cognitive models, differences that help explain conflicts generated by coastal and marine environmental change (Kempton et al. 1995; Paolisso and Maloney 1999; Paolisso et al. 2000; Blount 2002). For example, a key question is whether there are differences in how fishermen, environmentalists, and scientists think about the role of fishing activities in the natural world. Are those differences due to different interests, to different ways people experience nature, or to more profound cultural forces linked to social class, education, ethnicity, etc.? Finally, how do such differences play out in the policy process, and how are they mitigated?

Cooperation in Research

Whether or not chaotic forces are at play, these critiques have highlighted the existence of important differences in how environmental systems work and have underscored the need for greater cooperation among professional scientists, managers, resource users, and others.

One impediment to implementing an ecosystem approach to marine fisheries is limited spatial and temporal knowledge about marine ecosystems and fish populations. Numerous studies have documented the value of applying experience-based knowledge (EBK) (e.g., Johannes 1984; Pálsson 1995; Conkling and Ames 1997; Hall-Arber and Pederson 1999). A few studies have explored the utilization of fishermen's knowledge or user participation in fisheries science (e.g., Dobbs 2000; Harms and Sylvia 2001; Harte 2001; Neis and Felt 2001; Maurstad 2002) and have shown that EBK of fishermen has contributed significantly to the knowledge about fish stocks needed for management decisions (Weeks 1995; Conkling and Ames 1997; Stephenson et al. 1999). Pederson and Hall-Arber (1999) found that fishers in New England have and can communicate useful knowledge of the seafloor, habitat structure, and fish distribution. Neis et al. (1996) found that EBK data from fishers contributes to management by (1) providing additional indices for use in stock assessments and scientific debates; (2) providing data on responses by fishers to management measures and on the status of poorly understood species; (3) suggesting novel hypotheses; and (4) enhancing long-term legitimacy of the management regime.

Agencies are required to use the "best available scientific information," which may include other stakeholders who claim the right and wherewithal to have a say in what that information is and whether it is to be trusted and relied upon and who may wish to cooperate in research. An important sea change in marine policy in the United States is increased support for industry–government–academic cooperation in fisheries research. The revival of cooperative research, using fishing vessels as research platforms and going beyond, to involve fishermen directly in data gathering, priority setting, and other tasks, brings many different perspectives to the table. One is the frequently heard complaint of fishermen about the scientists' insistence on randomly sampled sites rather than sites selected in part because they are known to be where the fish are. Science is one of the major arenas where differences in how people think and what they know about the natural world plays out and where social learning may take place, facilitating rational and legitimized exchanges of information that can increase the success of management.

Community Conservation: A Counter Narrative

For many decades, a stronger focus on local communities and their engagement with natural resources, endangered species, and fragile, valued environments has been evident throughout the world. The terms "community-based forestry," "community-based natural resource management," and "community-based fisheries management" are now in the lexicon of development and conservation specialists. Fisheries and marine conservation come to this a bit late, still dominated by emphases on government and—more recently—markets as sources of control and governance (McCay and Jentoft 1998).

As noted above, at one end of the public participation scale are situations where the governing agency has little power or chooses not to exercise its power. Decisions are made directly or indirectly by industries or other groups—such as recreationists, developers, and environmentalists. This end of the spectrum may be viewed negatively as agency capture or positively as community-based management. Agency capture is what is meant by the phrase, "foxes in the henhouse." Community-based management is a more positive perspective on the management capacity of resource users, local residents, and others who are close to the resource or ecosystem of interest.

A community-based management system is one where local communities or groups of resource users assume many of the responsibilities for designing, implementing, monitoring, and enforcing management rules and, in return, reserve some exclusive rights to the resources for themselves. It can be opposed to fortress conservation, or attempts to separate humans from other species, reserving parks, reserves, and other places for wild nature (Adams and Hulme 2001). This method has also been called the fences and fines approach (Wells et al. 1992) and coercive conservation (Peluso 1993). It has dominated conservation thinking in terrestrial contexts, including "...the U.S. idea of a national park as a pristine or wilderness area, and the British notion of a nature reserve that is managed intensively" (Adams and Hulme 2001). It is a powerful, often dominant narrative about conservation. In marine policy, it appears in the ongoing movement for no-take reserves in areas where fishing activities are believed to harm benthic habitats.

A counter narrative (cf. Roe 1991) is that of community conservation (Adams and Hulme 2001). It focuses on alternatives that give local people access to places, resources, policy processes, and revenues associated with protected areas management. This narrative arose in the 1980s and 1990s in various national and international forums, mainly as concerns about the effects of terrestrial protected areas on the people who live around them or have other claims on the land (Adams and Hulme 2001). A second dimension of this counter narrative addresses human needs and development (Wells et al. 1992), and this emerged as a powerful concern in fisheries, initially in India, through the International Collective in Support of Fish Workers (www.icsf.org) and the more recent emergence of a World Forum of Fisher Peoples (www.wffp.org).

In addition, the community conservation narrative implicitly acknowledges the trump card of local residents: that without their participation and agreement, it is very costly to enforce regulations. For terrestrial wildlife, this is proving particularly true in developing countries where budgets for wildlife and parks enforcement are small, where subsistence needs are great, and where corruption is rampant. Similarly, for marine wildlife, regulation enforcement is extremely difficult, and many of the biodiversity hot spots are areas where dependence on marine resources is high because of poverty and underdevelopment. If people are asked to make changes and sacrifices in the name of conservation biodiversity, they should get some tangible, direct benefits, and ecological benefits are often indirect and long term. Community conservation projects often seek to combine both kinds of benefits, linking conservation with income-producing activities such as eco-tourism.

For marine fishers and protected area proponents, both narratives are evident. One emphasizes the need to sharply curtail or end their activities in order to restore and protect valued species and environments. The other claims the importance of keeping people in the decision-making system to enlist local expertise and knowledge and to assure minimal cheating and increased compliance. A great deal of social science research is involved in helping make this happen. However, privileging the community conservation approach has its pitfalls too. These include overly enthusiastic appraisals of a few well-known cases, ineffectiveness in delivering conservation goals, the potential that some groups will be marginalized if they are not clearly part of place-based communities, and whether it is cost-effective (for sub-Saharan African examples, see Adams and Hulme 2001; for community forestry examples, see J. McCarthy, Department of Geography, Pennsylvania State University, unpublished administrative report to the Ford Foundation, State College, Pennsylvania).

Increased public participation has other unintended consequences. A more participatory approach to marine conservation can open the door to other processes. As Adams and Hulme (2001) say, for wildlife and natural resource issues in Africa, increased community participation can open the door to

> a debate about and affirmation of rights to resources, or the awakening of political consciousness. Fortress conservation is an inherently authoritarian approach in which both means and end were clear, fixed, and expedient. By contrast, community conservation is underpinned by much more democratic ideas which conceptually undermine the foundations of the fortress. (Adams and Hulme 2001, page 21)

Therefore, it is not surprising that there is resistance to community conservation on the part of governing agencies and scientists. On the one hand, assertions of claims to rights and political action can lead to healthy human communities that are able to develop sustainable relationships with the natural resources on which they depend (Pinkerton and Weinstein 1995). On the other hand, community conservation is a difficult and time-consuming process with uncertain outcomes, and the benefits are often difficult to see when resources are rapidly declining.

Conclusion

Social impact analysis is an important developing aspect of fisheries and benthic habitat protection. Social impact analyses will be subjected to greater critical attention, especially as agencies are subjected to greater public and legal review (National Academy of Public Administration 2002; National Research Council 2002). Unfortunately, social research is too often truncated and inadequate to the task of understanding the most important dimensions of protected area or space-based fisheries management. For instance, a recent report on MPAs worldwide (Roberts et al. 2001) emphasized the positive biological consequences of MPAs and reported growing social acceptance. However, it gave little consideration to the conflicts and instances of local resistance surrounding many of the MPAs that were surveyed (Trist 1999; cf. Sandersen and Koester 2000; Christie et al. 2003).

Social scientists are brought into the fishing and benthic habitat discussion late or not at all, and the definition of appropriate social science research seems to be narrowed to pro forma descriptions of local communities and resource uses and other matters such as the question of stakeholders' views on MPAs (Suman et al. 1999). As important as such studies of attitudes and views are, little attention has been given to the conditions behind such views and how different views, interests, resources, and institutional frameworks can lead to different outcomes (see National Center for Marine Protected Areas 2003).

Factors such as conflict and resistance are extremely important to the success and failure of regimes for marine conservation. Seldom are social scientists asked to engage in studies required to understand reasons behind the situations of conflict or local support for MPAs. However, progress is being made, including studies of the adoption of and resistance to MPAs in the Philippines (White et al. 2002), studies of stakeholder involvement in planning in California (Pomeroy and Hunter, in press) and New Zealand (Wolfenden et al. 1994), and analysis of fisherman-created and co-managed MPAs in the Gulf of California (R.Cudney-Bueno, A. Sáenz, J. Torre, and L. Bourillón, paper presented at the American Anthropological Association Annual Meeting, 2002). The research topic is enriched by its position within more general inquiries into conditions for and against collective action and conflict resolution concerning common property (McCay and Acheson 1987; Ostrom 1990), the conditions that foster compliance and noncompliance (Kuperan and Sutinen 1998; Sutinen and Kuperan 1999), and the sources and consequences of inequality in the distribution of wealth, risks, and the costs and benefits of different ways to manage natural resources (e.g., Johnston 1994, 1997).

Community orientation, as one of the sea changes

influencing marine policy, derives, in part, from recognition that many of the world's nature reserves could not be protected because of the scarcity of resources for enforcement and the desperate need to utilize local natural resources for existence. Some kind of engagement of local communities in the work of conservation was critical. In addition, critical appraisals of so-called tragedies of the commons recognize the presence and value of community institutions for managing common resources (McCay and Acheson 1987; McKean 1992). When trying to create effective institutions for managing fisheries and protecting marine ecosystems, people are engaged in social action, itself deserving of research attention. One might call that action comedies of the commons. Comedy, in Greek drama, is "The drama of humans as social rather than private beings, a drama of social actions having a frankly corrective purpose" (McCay and Acheson 1987, citing M. E. Smith, paper presented at the Annual Meeting of the Society for Applied Anthropology, 1984). Sometimes such community action works and sometimes it does not, but this perspective underscores the importance of going beyond individuals and markets to explain and shape the outcomes of efforts to protect the environment.

Lastly, the diversity of social science approaches, needs better recognition. Too often, social sciences are relegated to the narrow, although important, domain of impact assessment. Yet social scientists also engage in studies of conflict and institutional change as well as studies of cooperation and co-management and studies into the production of knowledge and the diversity of world views. These tools should be central to society's attempts to improve how we use and manage our marine environments.

Acknowledgments

I appreciate the invitation from the organizing committee of the Symposium on the Effects of Fishing Activities on the Benthic Habitat, and particularly thank James Thomas for encouraging me to further develop my talk into this paper. The insights of two anonymous reviewers influenced how I revised the paper. Colleagues in the Rutgers Fisheries Project, including Teresa Johnson, Kevin St. Martin, Bryan Oles, and Douglas Wilson, contributed greatly to the ideas in the paper. However, no one besides myself is responsible for them. Support for research that influenced the paper came from a New Jersey Agricultural Experiment Station Hatch Act project; the Mid-Atlantic Fishery Management Council; the National Oceanic and Atmospheric Administration, through a Cooperative Marine Education and Research cooperative agreement between the Northeast Fisheries Science Center and Rutgers University; and National Science Foundation.

References

Acheson, J. M., and J. Wilson. 1996. Order out of chaos: the case for parametric fisheries management. American Anthropologist 98(3):579–594.

Adams, W., and D. Hulme. 2001. Conservation and community: changing narratives, policies and practices in African conservation. Pages 9–23 in D. Hulme and M. Murphree, editors. African wildlife and livelihoods: the promise and performance of community conservation. Heinemann, Portsmouth, New Hampshire.

Agardy, T., P. Bridgewater, M. P. Crosby, J. Day, P. K. Dayton, R. Kenchington, D. Laffolley, P. McConney, P. A. Murray, J. E. Parks, and L. Peau. 2003. Dangerous targets? Unresolved issues and ideological clashes around marine protected areas. Aquatic Conservation: Marine and Freshwater Ecosystems 13(4):281–372.

Apostle, R., B. J. McCay, and K. Mikalsen. 2002. Enclosing the commons: individual transferable auotas in a Nova Scotia fishery. Institute of Social and Economic Research, Memorial University of Newfoundland, St. John's.

Blount, B. 2002. Keywords, cultural models, and representation of knowledge: a case study from the Georgia Coast (USA). Coastal Anthropology Resources Laboratory, Department of Anthropology, University of Georgia, Occasional Publication Number 3, Athens.

Chiarappa, M. J., and K. M. Szylvian. 2003. Fish for all: an oral history of multiple claims and divided sentiment on Lake Michigan. Michigan State University Press, East Lansing.

Christie, P., B. J. McCay, M. L. Miller, C. Lowe, A. T. White, R. Stoffle, D. L. Fluharty, L. T. McManus, R. Chuenpagdee, C. Pomeroy, D. O. Suman, B. G. Blount, D. Huppert, R.-L. V. Eisma, E. Oracion, K. Lowry, and R. B. Pollnac. 2003. Toward developing a complete understanding: a social science research agenda for marine protected areas. Fisheries 28(12):22–26.

Conkling, P., and E. P. Ames. 1997. Cod and haddock spawning grounds in the Gulf of Maine: from Grand Manan Channel to Ipswitch Bay. Island Institute, Rockland, Maine.

Cordell, J., editor. 1989. A sea of small boats. Cultural Survival, Inc., Boston.

Dewees, C., and E. Ueber, editors. 1990. Effects of different fishery management schemes on bycatch, joint catch, and discards. California Sea Grant College, University of California, Report Number T-CSGCP-019, Summary of a National Workshop, La Jolla.

Dobbs, D. 2000. The great gulf: fishermen, scientists, and the struggle to revive the world's greatest fishery. Island Press, Washington, D.C.

Ecotrust. 2003. Groundfish fleet restructuring information and analysis project. Report, April 2003. Available: www.ecotrust.org/gfr/ or www.inforain.org/gfr/ (July 2003).

Haas, P. M. 1992. Epistemic communities and interna-

tional policy coordination: introduction. International Organization 46(1):1–35.
Hall-Arber, M., and J. Pederson. 1999. Habitat observed from the decks of fishing vessels. Fisheries 24(6):6–13.
Hance, B. J., C. Chess, and P. M. Sandman. 1988. Improving dialogue with communities: a risk communication manual for government. New Jersey Department of Environmental Protection, Division of Science and Research, Trenton.
Harms, J., and G. Sylvia. 2001. The economics of cooperative fishery research: A survey of U.S. west coast groundfish industry and scientists. Paper presented at IIFET 2000: Microbehavior and Macroresults. Oregon State University. Corvallis, Oregon.
Harte, M. 2001. Opportunities and barriers for industry-led fisheries research. Marine Policy 25(2):159–167.
Heilprin, J. 2002. Nature Conservancy goes underwater in land acquisition to help protect oyster habitat. Available: *http://www.enn.com/news/wire-stories/2002/10/10222002/ap_48767.asp* (July 2004).
Johannes, R. E. 1984. Marine conservation in relation to traditional life-styles of tropical artisanal fishermen. The Environmentalist 4:30–35.
Johnston, B. R., editor. 1994. Who pays the price? The sociocultural context of environmental crisis. Island Press, Washington, D.C.
Johnston, B. R., editor. 1997. Life and death matters: human rights and the environment at the end of the millennium. Altamira Press, Walnut Creek, California.
Kellert, S. 1997. Kinship to mastery: biophilia in human evolution and development. Island Press, Washington, D.C.
Kellert, S. R. 2005. Perspectives on an ethic toward the sea. Pages 703–711 in P. W. Barnes and J. P. Thomas, editors. Benthic habitats and the effects of fishing. American Fisheries Society, Symposium 41, Bethesda, Maryland.
Kempton, W., J. Boster, and J. Hartley. 1995. Environmental values in American culture. MIT Press, Cambridge, Massachusetts.
Kuperan, K., and J. G. Sutinen. 1998. Blue water crime: legitimacy, deterrence and compliance in fisheries. Law and Society Review 32(2):309–338.
Lee, K. 1993. Compass and gyroscope: integrating science and politics for the environment. Island Press, Washington, D.C.
Malakoff, D. 2002. Trawling's a drag for marine life, say studies. Science 298:2123.
Maurstad, A. 2002. Fishing in murky waters—ethics and politics of research on fisher knowledge. Marine Policy 26(3):159–166.
McCay, B. J. 1995. Social and ecological implications of ITQs: an overview. Ocean and Coastal Management 28(1–3):3–22.
McCay, B. J. 1998. Oyster wars and the public trust: property, law and ecology in New Jersey history. University of Arizona Press, Tucson.
McCay, B. J. 2000a. Sea changes in fisheries policy: contributions from anthropology. Pages 201–217 in E. P. Durrenberger and T. D. King, editors. State and community in fisheries management: power, policy, and practice. Bergin and Garvey, Westport, Connecticut.
McCay, B. J. 2000b. Resistance to changes in property rights or, why not ITQs? Pages 39–44 in R. Shotton, editor. Use of property rights in fisheries management. Proceedings of the FishRights99 Conference, Mini-course lectures and core conference presentations. FAO Fisheries Technical Paper 404/1, Rome.
McCay, B. J. 2000c. Post-modernism and the management of natural and common resources. Common property resource digest 54:1–6.
McCay, B. J. 2001. Environmental anthropology at sea. Pages 254–272 in C. Crumley, editor. New directions in anthropology and environment. Altamira Press, Walnut Creek, California.
McCay, B. J., and J. M. Acheson, editors. 1987. The question of the commons: the culture and ecology of communal resources. University of Arizona Press, Tucson.
McCay, B. J., and M. Cieri. 2000. Fishing ports of the mid-Atlantic. Report to the Mid-Atlantic Fisheries Management Council, Dover, Delaware.
McCay, B. J., and S. Jentoft. 1996. From the bottom up: participatory issues in fisheries management. Society and Natural Resources 9(3):237–250.
McCay, B. J., and S, Jentoft. 1998. Market or community failure? Critical perspectives on common property research. Human Organization 57(1):21–29.
McCay, B. J., B. Oles, B. Stoffle, E. Bochenek, K. St. Martin, G. Graziosi, T. Johnson, and J. Lamarque. 2002. Social impact assessment, amendment 9, squid, atlantic mackerel, and butterfish FMP. Report to the Mid-Atlantic Fishery Management Council, Dover, Delaware.
McKean, M. A. 1992. Success on the commons: a comparative examination of institutions for common property resource management. Journal of Theoretical Politics 4(3):247–281.
National Academy of Public Administration. 2002. Courts, Congress, and constituencies: managing fisheries by default. National Academy of Public Administration, Washington, D.C:
National Marine Protected Areas Center. 2003. Social science research Strategy for protected areas. National Marine Protected Areas Center, MPA Science Institute, Santa Cruz, California. Available: *http://www.mpa.gov/mpabusiness/mpacenter/social_sci_strategy.pdf* (July 2004)
National Research Council. 1999. Sharing the fish: toward a national policy on individual fishing quotas. National Academy Press, Washington, D.C.
National Research Council. 2001. Marine protected areas: tools for sustaining ocean ecosystems. National Academy Press, Washington D.C.
National Research Council. 2002. Science and its role in the National Marine Fisheries Service. National Academy Press, Washington, D.C.
Neis, B., and L. Felt, editors. 2001. Finding our sea legs: linking fishery people and their knowledge with sci-

ence and management. ISER Books, St. John's, Newfoundland.

Neis, B., L. Felt, D. C. Schneider, R. Haedrich, J. Hutchings, and J. Fischer. 1996. Northern cod stock assessments: what can be learned from interviewing resource users. Canadian Journal of Fisheries and Aquatic Sciences 56:1944–1963.

NOAA 2003a. Social science research within NOAA: review and recommendations. Final Report to the NOAA Science Advisory Board by the Social Science Review Panel, Washington, D.C.

Ostrom, E. M. 1990. Governing the commons: the evolution of institutions for collective action. Cambridge University Press, Cambridge, UK.

Pálsson G. 1995. Learning by fishing: practical science and scientific practice. Pages 85–97 in S. Hanna, and M. Munasinghe, editors. Property rights in a social and ecological context: case studies and design applications. The Beijer Institute, Stockholm.

Paolisso, M., and R. S. Maloney. 1999. Recognizing farmer environmentalism: nutrient runoff and toxic dinoflagellate blooms in the Chesapeake Bay region. Human Organization 59(2):209–221.

Paolisso, M., R. S. Maloney, and E. Chambers. 2000. Cultural models of environment and pollution. Anthropology News 41(2):48–49.

Pederson, J., and M. Hall-Arber. 1999. Fish habitat: a focus on New England fishermen's perspectives. Pages 188–211 in L. Benaka, editor. American Fisheries Society, Symposium 22, Bethesda, Maryland.

Peluso, N. O. 1993. Coercing conservation? The politics of state resource control. Global Environmental Change June:199–217.

Pinkerton, E., editor. 1989. Cooperative management of local fisheries: new directions for improved management and community development. University of British Columbia Press, Vancouver.

Pinkerton, E., and M. Weinstein. 1995. Fisheries that work: sustainability through community-based management. The David Suzuki Foundation, Vancouver.

Pomeroy, C., and M. Hunter. In press. The Channel Islands marine reserve process: the role of the social sciences. In O. T. Magoon, H. Converse, B. Baird, and M. Miller-Henson, editors. American Society of Civil Engineers, California and the World Ocean '02, Reston, Virginia.

Rieser, A. 1991. Ecological preservation as a public property right: an emerging doctrine in search of a theory. Harvard Environmental Law Review 15(2):393–433.

Rieser, A. 1997. Property rights and ecosystem management in U.S fisheries: contracting for the commons? Ecology Law Quarterly 24(4):813–832.

Roberts, C. M., J. A. Bohnsack, F. Gell, J.P. Hawkins, and R. Goodridge. 2001. Effects of marine reserves on adjacent fisheries. Science 294:1920–1923.

Roe, E. 1991. Development narratives, or making the best of blueprint development. World Development 19:287–300.

Salm, R.V., J. Clark, and E. Siirila. 2000. Marine and coastal protected areas: a guide for planners and managers. International Union for Conservation of Nature and Natural Resources, Washington D.C.

Sanchirico, J. N., and J. Wilen. 1999. Bioeconomics of spatial exploitation in a patchy environment. Journal of Environmental Economics and Management 37:129–150.

Sanchirico, J. N., and J. Wilen. 2001. A bioeconomic model of marine reserve creation. Journal of Environmental Economics and Management 42(3):257–276.

Sandersen, H. T., and S. Koester. 2000. Co-management of tropical coastal zones: the case of the Soufriere Marine Management Area, St. Lucia, WI. Coastal Management 28:87–97.

Scholz, A. J., M. Mertens, D. Sohm, C. Steinback, and M. Bellman. 2005. Place matters: spatial tools for assessing the socioeconomic implications of marine resource management measures on the Pacific Coast of the United States. Pages 727–744 in P. W. Barnes and J. P. Thomas, editors. Benthic habitats and the effects of fishing. American Fisheries Society, Symposium 41, Bethesda, Maryland.

Shotton, R., editor. 2002. Case studies on the effects of transferable fishing rights on fleet capacity and concentration of quota ownership. FAO Fisheries Technical Paper 412, Rome.

Singleton, S. 1999. Constructing cooperation: the evolution of institutions of comanagement. University of Michigan Press, Ann Arbor.

Smith, M. E. 1990. Chaos in fisheries management. Maritime Anthropological Studies (MAST) 3(2):1–13.

Stephenson, R. L., D. G. Aldous, and D. E. Lane. 1999. An in-season approach to management under uncertainty: the case of the SW Nova Scotia herring fishery. ICES Journal of Marine Science 56:1005–1013.

St. Martin, K. 2001. Making space for community resource management in fisheries. The Annals of the Association of American Geographers 91(2):122-142..

St. Martin, K. 2004. Geographic information systems in marine fisheries science and decision making. Pages 237–258 in Geographic information systems in fisheries. W. L. Fisher and F. J. Rahel, editors. American Fisheries Society, Bethesda, Maryland.

Suman, D., M. Shivlani, and W. Milon. 1999. Perceptions and attitudes regarding marine reserves: comparison of stakeholder groups in the Florida Keys National Marine Sanctuary. Ocean and Coastal Management 42(12):1019–1040.

Sutinen, J. G., and K. Kuperan. 1999. A socioeconomic theory of regulatory compliance in fisheries. International Journal of Social Economics 26(1/2/3):174–193.

The Nature Conservancy. 2002. Leasing and restoration of submerged lands: strategies for community-based, watershed-scale conservation. The Nature Conservancy, Washington, D.C. Available: *http://nature.org/files/lease_sub_lands.pdf* (August 2002).

Trist, C. 1999. Recreating ocean space: recreational consumption and representation of the Caribbean marine environment. Professional Geographer 51:376–387.

Weber, M. L. 2002. From abundance to scarcity: a history

of U.S. marine fisheries policy. Island Press, Washington, D.C.

Weeks, P. 1995. Fisher scientists: the reconstruction of scientific discourse. Human Organization 54(4):429–436.

Wells, M., I. K. Brandon, and L. Hannah. 1992. People and parks: linking protected areas with local communities. World Bank., Washington, D.C.

White, A. T., A. Salamanca, and C. A. Courtney. 2002. Experience with marine protected area planning and management in the Philippines. Coastal Management 30:1–26.

Wilson, D., and B. J. McCay. 1998. Social and cultural impact assessment of the highly migratory species management plan and the amendment to the Atlantic billfish management plan. Prepared for the Highly Migratory Species Office, National Marine Fisheries Service, National Oceanic and Atmospheric Administration, The Ecopolicy Center, New Brunswick, New Jersey.

Wilson, J. A. 2002. Scientific uncertainty, complex systems, and the design of common pool resources. Pages 323–359 in E. Ostrom, T. Dietz, N. Dolsak, P. C. Stern, S. Stonich, and E. U. Weber, editors. The drama of the commons. National Academy Press, Washington, D.C.

Wilson, J. A., and P. Kleban. 1992. Practical implications of chaos in fisheries: ecologically adapted management. MAST/ Maritime Anthropological Studies 1(1):66–78.

Wolfenden, J., F. Cram, and B. Kirkwood. 1994. Marine reserves in New Zealand: a survey of community reactions. Ocean and Coastal Management 25:31–51.

Place Matters: Spatial Tools for Assessing the Socioeconomic Implications of Marine Resource Management Measures on the Pacific Coast of the United States

ASTRID J. SCHOLZ[1], MIKE MERTENS,[2] DEBRA SOHM,[3] AND CHARLES STEINBACK[4]

*Ecotrust, 721 Northwest Ninth Avenue,
Portland, Oregon 97209, USA*

MARLENE BELLMAN[5]

*Department of Fisheries and Wildlife, Oregon State University,
104 Nash Hall, Corvallis, Oregon 97331, USA*

Abstract. Fishery management measures, such as the reduction of excess fishing capacity, and conservation measures, such as networks of marine protected areas, have considerable socioeconomic impacts. Users of marine resources—commercial and recreational fishermen, boaters, divers, and others—experience direct and indirect costs and benefits from such measures, notably foregone earnings and changing economic opportunities. In this paper, we present results and tools from a 2-year project to build an integrated, spatially explicit analytical framework for assessing management options such as fleet reductions and area closures on the West Coast of the United States, using the groundfish fleet as an example. After developing an extensive relational database comprising fishery-dependent, ecological, and socioeconomic data, we built a regional geographic information system (GIS) and assessed the relative impacts of five management scenarios. The results are spatially explicit and specific to particular communities, gear groups, fishing fleets, and ecological habitats, thus allowing decision makers to consider a range of issues that present themselves in management situations. The GIS makes for an intuitive interface that allows for participatory and consensus-oriented approaches to fishery management.

Introduction

The challenges facing West Coast fisheries are emblematic of the difficulties associated with shifting marine resource management to an ecosystem perspective (i.e., the explicit consideration of habitat and bathymetric associations of the fish species in question, as well as their assemblages). Legacies of past decisions create considerable structural challenges, notably in the form of overcapacity, while relatively new fishery management and conservation tools, such as area closures, require a reinterpretation of existing data sets, if not the collection of new data altogether. These challenges are compounded by the "catching up" required of socioeconomic considerations in fishery management, which historically has been dominated by stock assessments and biological models. Yet the larger the challenges facing fishery managers, the more important socioeconomic concerns become. A recent National Academy of Sciences study on marine protected areas went so far as to say that successful implementation of ecosystem-based management measures is impeded by the lack of attention to, and careful consideration of, the socioeconomic implications of proposed measures (NRC 2001). In this paper, we introduce an approach for assessing socioeconomic impacts of both structural changes and routine management measures on the West Coast, using the example of the groundfish fishery.

Commercial fisheries off the coasts of Washington, Oregon, and California are undergoing dramatic changes. Whether pelagic species, groundfish, or salmon *Oncorhynchus* spp., landings and revenues have been

---

[1] Corresponding author: ajscholz@ecotrust.org
[2] E-mail: mike@ecotrust.org
[3] E-mail: dsohm@ecotrust.org
[4] E-mail: charles@ecotrust.org
[5] E-mail: mbellman@orst.edu

declining for most fisheries over the last 10 years. In this paper, we focus on the groundfish fishery, which comprises more than 80 species of soles (e.g., Dover sole *Microstomus pacificus* and petrale sole *Eopsetta jordani*), flounders (e.g., arrowtooth flounder *Atheresthes stomias*), rockfish *Sebastes* spp., sablefish *Anoplopoma fimbria*, and Pacific hake *Merluccius productus* (also known as Pacific whiting) managed by the Pacific Fishery Management Council (PFMC). Commercial fisheries for groundfish date back to the 19th century and have been prosecuted with gear types including hook and line, pots, traps, and trawl nets. These gear types have evolved considerably over time and are characterized by one main distinction between mobile and fixed gear types: typically, trawl gear is used with relatively high volume, low value target species such as Pacific hake, while hook-and-line caught rockfish command some of the highest market values. After a period of expansion following the passage of the Magnuson-Stevens Fisheries Conservation and Management Act of 1976 and the subsequent "Americanization" of the fishery, landings of groundfish have declined over the past 20 years. Some species have declined to levels that trigger stringent rebuilding plans and bycatch avoidance measures, affecting the rest of the groundfish management complex. These measures, in turn, threaten the economic viability of the fishery, which was declared a federal disaster in 2000.

In response to the worsening crisis, the PFMC adopted a strategic plan, "Transition to Sustainability," that lays out a set of management priorities to safeguard the future of the fishery (PFMC 2000a). The top priorities set forth in the plan are the reduction of the groundfish fleet by at least 50% in each fishery sector and the viable balancing of ecological and economic considerations for the fishery. Implicit in this plan is a consensus among fishermen, scientists, and managers that the expansion of the fleet in the 1970s has resulted in a fleet that is too large to profitably harvest the allocations deemed biologically sustainable. In the 3 years since the adoption of the strategic plan, however, the council agenda has been dominated by the increasingly more urgent day-to-day business of responding to stock assessments and the harvest restrictions they necessitate.

This mismatch between short-term management needs and strategic considerations for restructuring the fishery was further exacerbated by the area closures first implemented with the 2002 in-season closures of large sections of the continental shelf. Essentially, this is an acknowledgment that traditional trip limit management is not effective at reducing bycatch of some of the most threatened species of rockfish such as canary rockfish *Sebastes pinniger* and darkblotched rockfish *Sebastes crameri*. Instead, the fishery is increasingly managed on the basis of time and area closures designed to minimize the chance of encountering populations of threatened species. In a considerable departure from single-species-driven management, these measures may well be the first wave of ecosystem-based management principles and may constitute early steps towards ecosystem-based fishery management plans (Field et al. 2001).

While spatiotemporal management measures, such as shelf closures, appear better aligned with ecosystem mandates for marine resource management (Ecosystems Principles Advisory Panel 1999; NRC 2001), they pose considerable challenges to the management process. Rather than having the leisure to design ecosystem management principles for fisheries from first principles, federal and state agencies have to act now, within the current paradigm of fishery management. This is further complicated by the absence of technical and technological capacity in many management regions to address spatial and temporal aspects of fishery management. In particular, fishery-dependent and independent data that have been collected on the West Coast for over 20 years are not typically mined or interpreted in spatially explicit ways. This is part of a larger issue identified, for example, in a recent National Research Council study. In reference to legally mandated essential fish habitat assessments and novel management tools such as marine protected areas, the study notes that "NMFS [National Marine Fisheries Service; now National Oceanic and Atmospheric Administration Fisheries] and its partner agencies should integrate existing data […] to provide geographic databases for major fishing grounds" (NRC 2002:3).

In addition to the lack of spatial integration of existing data, there is the added problem of lacking or insufficient economic and other "soft" data about coastal communities and the effects of fishery management measures on them. With the groundfish crisis unfolding, the socioeconomic data deficits are well known and documented (PFMC 2000b) but have hardly been addressed in the interim. The problem at hand, therefore, is fourfold: declining stocks, excess capacity, lack of integrated databases for assessing spatiotemporal management measures, and lack of socioeconomic data for assessing effects on coastal communities. In this paper, we report on a model we developed to address this nexus of issues, with the aim of providing decision-support tools for managers and communities on the West Coast.

In the following section, we outline a spatial framework for linking ecological, fisheries, and socioeconomic information that utilizes existing, readily available data. We then discuss the application of this approach to the overcapacity issue in the West Coast groundfish fishery and present results from a static comparison of different management scenarios.

The OCEAN Framework

In order to meet the fourfold challenge facing ecosystem-based fisheries management—managing declining stocks in the presence of overcapitalized fleet with data that are not sufficiently spatially explicit for ecosystem-based management or do not exist at all—we designed a framework for jointly addressing these issues. The OCEAN (Ocean Communities Economic/Ecological/Equity ANalysis) framework comprises a set of linked, spatially explicit models. The three Es—economics, ecology, and equity—represent Ecotrust's mission to promote economic development that is both ecologically sustainable and equitable (Scholz 2003). It is rooted in the growing literature of marine geographic information system (GIS) models that are being developed to address a host of oceanographic, coastal, and fisheries issues and problems (Kruse et al. 2001; Breman 2002; Valavanis 2002; Green and King 2003). The OCEAN framework is essentially a meta-analytical tool for combining a range of data, using a relational database architecture and GIS as the "common currency."

The centerpiece of our analysis is the modeling of data that are already available in spatially explicit formats, as well as the spatial interpretation of other, not yet spatially explicit, information. The challenge is to organize data from diverse sources, in diverse formats, and of varying quality and to integrate them into a single framework. We began work on OCEAN by reviewing existing sources of data and compiling them into one relational database. Where necessary, we built new geographic models to spatially interpret data, especially those pertaining to the distribution of fishing effort. Combining bathymetry and habitat information with fishing effort and species distributions then formed the basis for analyzing which vessels fish where, with what gear, and targeting what species. To this, in turn, we added an economic model for assessing the relative socioeconomic impacts of different management scenarios. We present here the first, static, "version 1.0" of OCEAN.

Data Sources

Fishery-dependent data on landings, revenues, and vessels are collected by the three states in our study area (Washington, Oregon, and California) and stored in the Pacific Fisheries Information Network (PacFIN) of the Pacific States Marine Fisheries Commission (Sampson and Crone 1997). For the project on groundfish fleet restructuring presented here, we obtained 14 years of data summarized to individual vessels by port, gear, species, and year for all vessels fishing for groundfish. The availability and quality of data for different fishing sectors varied considerably. The trawl fishery is best documented, with at-sea logbooks augmenting the information on catch and landings that is reported port side in the landing tickets. Trawl logbooks are spatially explicit, with trawl set points recorded for individual trawls (typically referenced by 10-min blocks). Trawl duration is also recorded, thus providing a measure of effort. The trawl logbook data, however, have two major limitations. First, although skippers record trawl endpoints, these are not transcribed into the PacFIN database. Since there is as yet no comprehensive vessel monitoring system in place on the West Coast, estimating the precise extent of trawl activity remains rather difficult. We tried to ameliorate this with a model based on the trawl set points and trawl duration. Essentially, we constrain vectors of possible trawl directions by using habitat and bathymetric considerations for each recorded tow. The result is a density map of probable tow tracks. This model is not discussed further here, but forms—together with a more detailed discussion of the effort model built on the landing receipts—the basis for a forthcoming paper. Secondly, although the same vessel identifiers are used in both data collection efforts, there remain considerable gaps between the logbook and landing receipts record sets (Fox and Starr 1996; Sampson and Crone 1997). For our analysis, we used a record set provided by PacFIN in which the records were already matched up, thus subsuming any associated uncertainty.

No such logbooks exist for the fixed gear sectors of the fishery, making landing receipts the only source of information. These are less spatially explicit and typically contain no measure of fishing intensity or effort. With the exception of California, where all landings are recorded in 10-min blocks, the spatial unit of PacFIN landing receipts are statistical areas defined by the now defunct International North Pacific Fisheries Commission (INPFC). There are only 12 INPFC areas for the entire West Coast from Cape Flattery in Washington to the Mexico border, each covering thousands of square miles. We developed an iterative algorithm (described below), drawing on all the data assembled in OCEAN to make the landing receipts more spatially explicit. This, in turn, is a prerequisite for considering the socioeconomic implications of management measures, such as the in-season shelf closure, which affect vessels that used to fish in the now closed areas.

Bathymetry and other data on oceanographic characteristics were obtained from the National Oceanic and Atmospheric Administration (NOAA), the U.S. Geological Survey (USGS), and state agencies such as the California Department of Fish and Game. One key component for ecosystem management is habitat and the consideration of the impact of fishing activities on different parts of the seafloor. The continental shelf in our study area has been the subject of considerable habitat mapping efforts, such as the USGS habitat GIS for the Monterey Bay National Marine Sanctuary (Wong and Eittreim 2001) and the ongoing effort in support of NOAA Fisheries' Essential Fish Habitat Environmental Impact Statement (*www.nwr.noaa.gov/1sustfsh/ groundfish/eis_efh/efh/*). Using known habitat associations for various fish species, as well as the depth constraints on particular types of fishing gear, habitat data can be used to relate fishing effort to particular areas.

The scientific surveys conducted by NOAA Fisheries over the past 25 years are a major source of fishery-independent data. We obtained all available years of shelf and slope surveys from the NOAA Alaska Fisheries Science Center. The NOAA research vessels using trawl gear record the total number, size and age distribution, and weight of fish sampled along fixed transects (Lauth 2000; Weinberg et al. 2002). Because of the consistency of the sampling protocol the trawl surveys generate a comprehensive picture of species abundance, at least along the trawl transects. One obvious limitation of the NOAA surveys is that they are conducted in summer months, using trawl gear. There are few remedies for this situation. In other areas, notably waters off Alaska, it would be possible to use data from surveys conducted by the International Pacific Halibut Commission using fixed gear. This would generate more representative species distribution maps. In the surveys, species and abundance (number of fish per species) are recorded for each trawl start point. We extracted individual records of species targeted in the commercial fishery, normalized these by total effort, and generated species-specific density maps. Following an approach developed by NOAA's Biogeography Group (NOAA's National Centers for Coastal Ocean Science and National Marine Sanctuaries Program 2002), we summarized these to 9-km × 9-km analysis units and derived single, cumulative species distribution maps for each target species.

The final component of OCEAN concerns the linkage between fishing activity and coastal communities. The obvious points of contact are the landing ports, where vessels sell their catch to fish buyers and processors. Together with other marine services and businesses, processing is a major contributor to income generated in coastal communities. This aspect of socioeconomic impacts is already captured in a regional input–output model used by the PFMC to assess the economic impacts of fishery policy (Jensen 1996, 1998). The Fisheries Economic Assessment model (FEAM) belongs to a class of regional input–output models that treat the economic activity in a region as a set of interconnected sectors. This is a class of modeling pioneered by Wassily Leontief to describe the U.S. economy (see Hewings 1985 for an overview). Each dollar generated in one sector has a "multiplier effect" because it generates economic activity in other sectors. For example, fish are landed, and the vessel is paid a price per pound for its catch. Out of this exvessel revenue, crew shares, maintenance, and moorage costs and other expenses are paid, which in turn generate personal income and revenues for the port district and other marine-related businesses. The FEAM estimates these effects for the two primary sectors affected by fishing activity (i.e., harvesters [fishermen and their families] and processors). We summarized these model outputs in a set of spreadsheets which we integrated into OCEAN. This allowed us to consider the income impacts of changes in landings in a port resulting from particular management scenarios.

A key limitation of the FEAM analysis is that it is static in nature and only provides an incomplete snapshot in time. It is premised on the landings and revenues generated by the fishing fleet but is silent on alternative sources of revenues in coastal communities, such as tourism. Unlike other regional input–output models, FEAM is not designed to assess employment effects. Furthermore, there are a host of considerations over and beyond economic impacts that are of importance to coastal communities and managers but are not yet routinely assessed. For example, the lifestyle aspects of fishing communities are important (Hanna et al. 2000), as are concerns about the social and cultural resilience of ports and towns in response to the structural changes in the fishery (Langdon-Pollock 2002). By way of addressing these concerns, and to lay the groundwork for more in-depth analysis of coastal communities in future applications of OCEAN, we incorporated census statistics as well as qualitative information (Scholz 2003).

Methods

Our analytical approach centers on the spatial association of data. This kind of analysis has been used in other marine applications of GIS, for example, to assess the location of fishing effort close to shore (Caddy and Carocci 1999) or to detect trends in global fishery statistics (Watson and Pauly 2001). The OCEAN approach operates at an intermediate, regional scale, with explicit consideration of the socioeconomic impacts in coastal communities. Conceptually, OCEAN is a multilayered information system comprising geographic and other data in a "smart map" environment. The system can be queried from within any one data layer (e.g., to find particular vessels or gear groups fishing in a habitat of interest or to generate the exvessel revenues associated with a particular species). Information can be manipulated on map-based user interfaces, and results are summarized in map formats.

The bulk of our analytical work consists of assembling information for the groundfish fishery and developing models for integrating and interpreting data that, more often than not, were recorded at widely disparate spatial scales, if at all. A central part of OCEAN is a fishing effort sub-model that we describe in some detail here. In the next section, we present results from an analysis of coast-wide impacts of various capacity reduction scenarios.

Fishing Effort Sub-Model

A central analytical challenge facing managers on the West Coast is to determine the spatial extent of trawl and fixed gear fisheries in order to gauge where in the ocean fishing effort is concentrated. This, in turn, is key for assessing the socioeconomic implications of reducing certain sectors of the fleet and the effects of area closures and for determining the likely habitat interactions of particular gear types. In the absence of a comprehensive observer program (data from the first year of West Coast observer program are expected in early 2003) or vessel monitoring systems, there is considerable uncertainty about where vessels using gear types other than trawl gear are fishing.

The OCEAN effort sub-model essentially consists of a sequence of steps, programmed in ArcINFO, which successively constrain each landing record and subsequently apportion catch and revenue to equal area analysis units (9-km × 9-km blocks) based on probability of fishing activity in an area. In contrast to multivariate analysis used in terrestrial applications, which generally predicts what happens in a particular location (e.g., Hargrove and Hoffman 2000), we try to predict the location for known entities. Given the data volume and processing constraints, we are evaluating a statistical approach for future iterations of this project, especially if they involve fine-scale (e.g., daily) data. The following steps characterize this process; Figure 1 shows a flow chart of the model:

(1) Each PacFIN record contains information on the gear used, species caught, landing port, vessel information, and 1 of 12 statistical management areas where the catch originated;

(2) Impose a maximum range from the landing port that a vessel is likely to have fished, given its length and gear type used. This is currently derived from expert witness testimonies, pending more formal studies of fishing behavior on the West Coast;

(3) Impose depth restrictions on fishing gear used and target species. There are limits to the depth from which West Coast trawlers can haul their nets, or in what depth various fixed gear types are used; similarly, different species of fish have known ranges of bathymetric associations;

(4) Compare this to the species distribution densities derived from the fishery-independent surveys. Some areas are associated with higher frequencies of the target species in question, making it more likely that a fishing vessel would have gone there for its catch;

(5) Within that maximum range, weight the species density clusters inversely by distance from port. This is a "friction of distance" idea; because travel is costly, vessels tend to fish closer to port even if they are slightly less likely to encounter the target species;

(6) Impose habitat restrictions on fishing gear used. Trawlers do not operate in high-relief areas, but these same areas tend to be frequented differentially by vessels using hook and line gear;

(7) Apportion pounds caught and associated revenue to fishing blocks within the maximum range based on probabilities derived from distance from port, targeted species densities, habitat restrictions, and previous activity;

(8) Repeat for all records, and map the resulting distribution of fishing activity. In principle, this can be normalized by number of records associated with an area or, in the case of trawlers, number and duration of tows made there, to provide a measure of effort.

The result of these computations is a GIS data set consisting of the distribution of pounds per species and associated revenues caught with different gear types in different areas on the continental shelf (see Figure 2). It is important to note that this constitutes a spatial reinterpretation of historic data. While it is conceivable to turn this into a predictive model, a major confounder lies in the absence of behavioral models of the fishery. In other words, there are few, if any, known rules that describe fleet behavior, and most economic models that attempt this are based on simplistic assumptions about rational actors and individual profit maximizing considerations. Also, the model in its current form is deterministic, effectively attributing catch and revenues to particular locations. We are currently working on a probabilistic version of the model.

Sensitivity of the OCEAN Analysis

There are two kinds of uncertainty associated with the OCEAN analysis. First, as a meta-model, it is only as good as the various inputs. In the section on data sources and throughout the text, we discuss some of the known problems with the data used in our analysis. These include possible transcription errors and omissions in the PacFIN data,

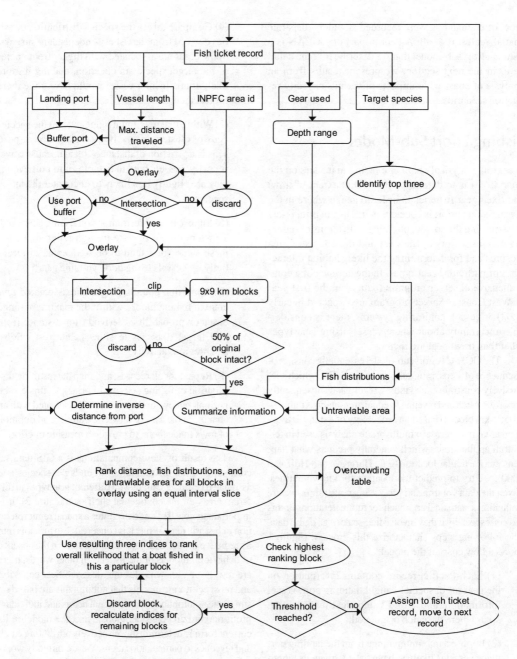

Figure 1. Effort model flow chart.

instrument and interpretation errors in the bathymetry data, temporal mismatches between current industry infrastructure and that assumed in the FEAM model, and the manifold issues around census data. Since testing or improving any or all of these data sources was beyond the scope of our project, we took these inputs as a given. The second kind of uncertainty pertains to the model we constructed using these data inputs, notably the fishing effort model.

Short of observing whether vessels actually fish where our model predicts they fish, there are few ways to test the accuracy of our model. Therefore, we tested the relative importance of various parameters and assumptions on the outcome of the model. The remainder of this section details a sensitivity analysis of the spatial interpretation of fishing effort.

Due to processing constraints, we performed a sensi-

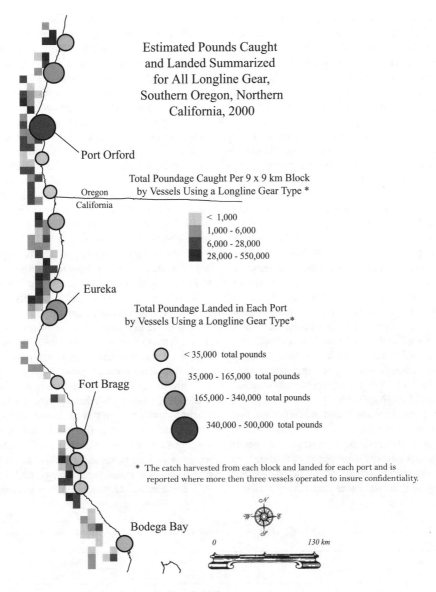

Figure 2. Sample output of effort model.

tivity analysis on a subset of the entire data set, using line item (i.e., per trip) landing receipts for the port of Eureka, California, in 2000. This port is interesting for two reasons: (1) the fleet is fairly well stratified in terms of gear types used and vessel sizes represented, and (2) it is located in California, where landing receipts for all gear types are referenced by 10-min management blocks, thus providing some measure of accuracy against which to compare our results. There are known problems with the California logbooks because of the widespread strategic reporting that took place in earlier years. Also, landing receipts are filled out by fish buyers and processors. Since these may or may not verify the catch location with the skipper, there is some inherent uncertainty as to their accuracy that we could not control for.

As apparent from Figure 1, there are two distinct sets of variables that are used to predict fishing activity: data inputs used to predict the area where fishing is likely to occur (constraining inputs) and parameters used to predict how catch and effort is spatially distributed within that area (weighting parameters).

We analyzed the constraining variables with respect to their effects on the model results by removing data inputs one at a time, re-running the model, and comparing the difference of total probable fishing area after each iteration. To do so, we adapted a technique described in Crosetto and Crosetto (2002). We quantified the differences between these iterations in terms of the total number of blocks per record. In other words, in the control scenario (with all variables used), each record is associated with a fishing area measured in a number of spatial units (blocks). Removing a particular data input (e.g., distance from port) results in a change in the area associated with that record and, consequently, with the number of blocks. We then calculated the mean number of blocks per record and determined the variance, standard deviation, percent change from the control scenario (i.e., using all variables), and the coefficient of variation (median centered) for the entire fleet as well as for individual gear types (Table 1).

As expected, of the variables tested, the maximum distance from port had the greatest constraining effect. This varies for different fishing gears. For example, results for trawl gear are less affected by distance from port, probably due to their greater range associated with the larger vessel sizes. Furthermore, in some cases, such as hook and line, the model appears sensitive to management area. This is most likely a result of the small sample size. However, no parameter by itself influenced the results of the model substantially. For this reason, any error originating from the data inputs used for constraining probable fishing areas is not propagated through the model.

Both the gear-depth association and the distance from port are assumed variables, derived from expert testimonials. Because of the sensitivity to these two parameters, the model could be greatly improved by calibrating these parameters with empirical data, especially as observer coverage for the study area becomes more readily available.

We tested the sensitivity of the model to the second set of variables (i.e., those used to weight the distribution of catch and effort within the predicted fishing areas) by adjusting parameters incrementally to identify ranges of greatest effect. We analyzed the results from this analysis by summing the total pounds apportioned to each block based on the probability of effort occurring in any given block. The sum of the squared variation and the coefficient of variation were then calculated for each iteration. Figure 3 shows the variation of the effects of adjusting individual parameters (based on the coefficient of variation). Individual parameters were adjusted from no-effect to four times the effect (relatively), while other parameters were held constant (in this case at a relative weight of 100). Results are based on variation from the zero-based effect (relative weight of 1).

While Figure 2 does not explain the importance of any given parameter in relation to the others, it does describe the magnitude of variation resulting from adjustment of the given case (i.e., how much the results change in response to changes in one parameter). The effort model is most sensitive to weights given to the overcrowding potential. A minor change in the weight of the parameter results in a dramatic difference in the spatial distribution of catch. Additionally, variance resulting from adjusting the overcrowding parameter begins to decrease after weights exceed three times that of other parameters. This is due to a maximization of the effects of overcrowding resulting from equal distribution of catch. That is, the effects of overcrowding become so influential on the model that any activity results in a shift in behavior for all areas associated with a record, and these effects begin to cancel each other out.

The model is also quite sensitive to changes in the

Table 1. Effects of selectivity removing various data inputs. INPFC = International North Pacific Fisheries Commission.

Variable removed from analysis	Min	Max	Mean number of blocks	SD	Percent difference	Coefficient of variation
Control (no parameters removed)	3	223	107.74	52.98	-	-
Trawl gears	9	223	130.37	34.63	-	-
Hook and line	2	43	18.97	11.54	-	-
Longline	4	36	26.48	7.67	-	-
INPFC management area	9	375	149.36	80.10	138.63%	64.60%
Trawl gears	42	375	181.44	58.56	139.17%	17.10%
Hook and line	9	67	26.44	15.66	139.35%	254.80%
Longline	12	44	33.77	5.09	127.50%	152.80%
Distance from port	43	1,107	188.61	90.77	126.28%	229.50%
Trawl gears	43	929	226.03	59.80	124.57%	107.60%
Hook and line	43	117	55.76	28.07	210.89%	409.60%
Longline	43	117	50.53	22.39	149.66%	464.90%
Gear/depth association	4	269	150.64	50.09	79.87%	98.30%
Trawl gears	34	269	168.31	37.59	74.46%	60.30%
Hook and line	4	269	67.20	69.81	120.52%	45.70%
Longline	14	159	89.00	30.09	176.11%	17.90%

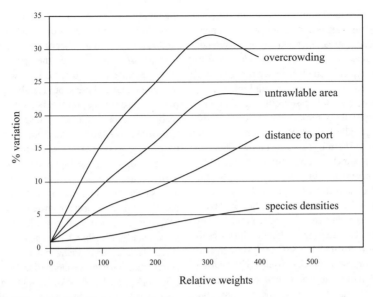

Figure 3. Coefficient of variation of weighting parameters.

relative weight given to untrawlable areas. It appears, however, that the variation of effects that this parameter has on the results tends to flatten out after it has been weighted to three times that of the other parameters. This is due, in part, to the limited number of analysis units containing untrawlable areas. The magnitude of importance of these areas quickly becomes limited by other parameters (such as overcrowding).

The sensitivity analysis helps in determining the parameters that influence the results of the model the most and, therefore, warrant the most attention in terms of obtaining empirical data for the purpose of calibration. From our analysis, it is apparent that different assumptions made about overcrowding of specific fishing grounds and the distance a boat will travel from port will result in considerably different distributions of catch and effort. Extra care, therefore, should be taken in empirically verifying these assumptions.

Application to West Coast Groundfish Fleet Restructuring

We now turn to the analysis of a suite of coast-wide management scenarios related to the restructuring of the West Coast groundfish fleet. The starting point for the analysis in Ecotrust's Groundfish Fleet Restructuring (GFR) project (2001–2003) was the strategic plan that the PFMC adopted in 2000, the first priority of which is the reduction of fishing capacity by at least 50% in each sector (PFMC 2000a). Fishing capacity is notoriously difficult to measure, since it is a combination of number and size of vessels, their technical efficiency (i.e., fish-hold capacity, horse power, volume, and efficiency of nets and other gear), and the time commitment of fishermen (Smith and Hanna 1990; Federal Fisheries Investment Task Force 1999; Gréboval 1999). Of these factors, only the number and size (length) of vessels is well documented on the West Coast. While technical efficiency is generally assumed to have increased over time, estimating the physical capacity of the fleet is hampered by the absence of comprehensive data about specific vessel characteristics outlined above. What we do know about the fleet is almost entirely limited to fish tickets and logbooks, which record mainly how much has been landed. This, however, is increasingly determined by market and regulatory factors and is not an accurate reflection of the true capacity of the fleet—only of what it is allowed to land. In other words, landings are not so much an indicator of capacity as of what the fleet is allowed to catch, whereas capacity is a measure of what it could potentially catch. The National Oceanic and Atmospheric Administration is working on quantitative measures of the capacity of U.S. fishing fleets (Offices of Science and Technology and Sustainable Fisheries 2001), but a comprehensive set of capacity measures is not yet available.

Capacity Estimates

On the West Coast, the PFMC's Scientific and Statistical Committee (SSC) Economics Subcommittee (SSC 2000) has estimated the rate of the capacity utilization (i.e., the ratio of catch to capacity) in the groundfish fishery. Using landings from earlier, less-constrained periods together with current fleet sizes, the underlying capacity of the fleet can be inferred. The assumption is that vessels were fishing at or close to their capacity in earlier, relatively unconstrained time periods. In the case of the West Coast groundfish fishery, a distinction is typically made between pre-1994 and post-1994 fishing seasons when the present limited entry regime was implemented and vessels either qualified for limited entry (LE) permits or remained in the open access (OA) fishery. Using the 1995–1998 participation in the fishery by fleet sectors and permit status, in other words, open access, limited entry trawl, limited entry nontrawl/sablefish, and limited entry nontrawl/nonsablefish, vessels were assigned LE or OA status in earlier years. This assumes considerable temporal homogeneity of the fleet and basically codes vessels in earlier years by the fleet sector they participated in later. Considering the landing histories of each fleet sector prior to 1994, the SSC analysis then estimated the number of vessels needed in each of the earlier years to catch 2000 harvest targets defined for each sector. The resulting capacity utilization rates are an expression, therefore, of how many vessels fishing at relatively unconstrained capacities would have been needed to catch the allocations for each fleet sector in 2000. The SSC analysis found capacity utilization rates ranging from 6% in the open access to around 40% in the limited entry trawl fleet sectors (SSC 2000). Thus, there is considerable excess capacity in the West Coast groundfish fleet, as is the case for many commercial fisheries.

We applied the same logic to our data set and derived the number of vessels in each of the four fleet sectors in 2000. Table 2 summarizes our and the SSC's findings as well as the inferred number of vessels needed in each sector from the SSC study. In the open access and limited entry trawl fleets, we derived somewhat higher capacity utilization rates. These should not be read as an improvement of the capacity problem but rather a reflection of the incongruities between the two data sets. Also, we decided to use our 2000 figure for the number of distinct vessels in the open access (614) fleet rather than the 1995–1998 average (980, which was higher than the SSC's average of 910) because we believe that this reflects trends in the open access fleet better. Our capacity utilization rates for the limited entry nontrawl fleet are lower than those derived in the SSC study. We believe this is a function of our data set: since PacFIN distributes catches of rockfish across vessels and areas, our annualized data do not allow for sufficient distinction between targeted and bycatch harvest strategies for sablefish and rockfish caught in the LE fixed gear fleet. We also found a high degree of overlap between vessels in the LE nontrawl sector: most of these fish for both sablefish and rockfish, and only 16 vessels target sablefish exclusively. The SSC inferred a need for 40 nontrawl LE vessels total, of which 15 are needed to harvest the sablefish target. In our reduction scenarios, therefore, we apply the reduction logic to the entire LE nontrawl sector, and treat the 16 sablefish-only vessels separately.

We used these capacity utilization estimates to identify a subset of vessels for removal from the fleet based

Table 2. Summary of Groundfish Fleet Restructuring (GFR) and Scientific and Statistical Committee (SSC) capacity calculations. The numbers in bold are used in the subsequent scenario analysis; table contains double counts of vessels that occur in more than one size category in information received from PacFIN. This is one of many factors that makes accurate counts of vessels challenging. OA = open access; LE = limited entry.

	GFR			SSC		
Fleet sector	Number of vessels	Number of distinct vessels	Capacity utilization estimates	Number of vessels	Inferred number of vessels needed	Capacity utilization estimates
OA (2000)	1,524					
OA with groundfish landings > 0.25 MT (2000)	713	**614**	7% (low); 14% (high)		50 (low); 100 (high)	
OA with groundfish landings > 0.25 MT (1996-1998 average)	983			910	50 (low); 100 (high)	5.5% (low); 11% (high)
LE trawl (2000)	452	**244**	**45.5%**	274	**107**	39%
LE non-trawl (2000)						
Non-sablefish	279	**177**	**14.3%**	232	**40**	17.2%
Sablefish	228	176	6.6%	164	**15**	9.1%
Sablefish exclusive		**16**	**100% (assumed)**			

on a number of different criteria that simulate some prominent policy considerations:

(a) Reducing all excess capacity. Given the estimates of how many vessels are needed to harvest the 2000 harvest targets, we consider what the fishery would look like if only the highest producing vessels up to the level needed were harvesting;

(b) Reducing capacity by 50% in each sector. This is the PFMC priority articulated in the strategic plan. Since the council has not yet identified reduction mechanisms for all sectors, we randomly selected half of the vessels in each. So far, the council has only identified a reduction mechanism for the trawl sector, a part government, part industry financed buyback of qualifying trawl vessels, which are to be identified in a reverse auction. Reducing capacity by 50% in each sector while preserving fleet diversity takes into consideration the strategic plan goal to preserve fleet diversity, which we interpret here as preserving the proportions of different vessel lengths present in each fleet sector in the base year, 2000;

(c) Reducing capacity in each sector while preserving economic viability. This reduces capacity not to or by a given percentage but, rather, preserves all those vessels that currently achieve a minimum level of exvessel revenues from groundfish landings in each sector (i.e., irrespective of where they are operating with regard to capacity estimates). We set these revenue levels, somewhat arbitrarily, as follows: LE trawl, more than US$50,000; LE nontrawl/nonsable, more than $10,000; LE nontrawl/sable, more than $20,000; and OA, more than $5,000. In other words, for a vessel to remain in, for example, the LE trawl sector, it has to have at least $50,000 groundfish exvessel revenues in the base year, 2000.

We then compared the "before" and "after" effects of removing vessels according to these criteria, using the 2000 fishery as a baseline. The immediate effect is to diminish revenues and landings associated with vessels exiting the fleet. Since total landings are a function of allowable harvest limits, and since these would not necessarily change in a capacity reduction exercise, a redistribution of landings and revenues (and with it the associated income and other community impacts) along the coast takes place. In the medium to long term, vessels remaining in the fleet would harvest the difference resulting from the reduction to the harvest targets. Thus, the immediate impacts are not the ultimate outcome of a capacity reduction. They do, however, indicate the order of magnitude of the wealth transferred between vessels exiting and vessels remaining in the fleet.

Vessels remaining in the fleet are better off since they compete with fewer boats for the total allowable catch. Their trip limits increase, and they bring more landings and revenues into their port—presumably up to the total of coast-wide landings and revenues from groundfish in 2000. The difference between the "before" and "after" levels of landings and revenues, therefore, accrues to vessels remaining in the fleet. Indeed, to the extent that the remainder of the fleet is successfully managed to achieve stock rebuilding objectives, total allowable catches may even increase. Those ports associated with vessels removed from the fleet, however, experience a corresponding decline in landings and revenues. For them, the "after" effect of our various scenarios may be permanent and can be interpreted as a cost associated with a fleet reduction measure.

Summary of Capacity Reduction Scenarios

The results of the four numeric reduction scenarios are summarized in Table 3 below. We focus on two central aspects of fleet reductions: (1) the landings, exvessel revenues, and associated income that are redistributed from the vessels exiting the fishery to the remainder of the fleet and (2) the effects on overall size and composition of the fleet. There are many more issues that can be explored with OCEAN than can be discussed in this paper. In particular, we only compare the coast-wide results of the scenarios. It is possible to examine the effects down to the level of individual ports (Scholz 2003), which is of considerable interest to our community partners.

From Table 3, it is apparent that even at the coast-wide scale, different reduction schemes have substantially different effects on the fleet. There are three notable results. First, the effects of removing all excess capacity and removing 50% of capacity are remarkably similar. Recall that scenario 1 removes the entire excess capacity, and only the numbers of vessels per fleet sector needed to harvest the 2000 targets remain in the fleet. The overall effect is comparable to that of a 50% capacity reduction, both in terms of the initial reductions, landings, and revenues and the amount of income effectively redistributed from the exiting vessels to those remaining. Indeed, since scenario 1 was selected for the highest producers, this occurs at a smaller initial decline in exvessel revenues than in the random selection process of scenario 2. To the extent that income impacts can be thought of as the "cost" of capacity reduction, a coast-wide reduction of fleet capacity can thus be achieved for between $70 million and $75 million. Coast-wide, the effect on fleet composition is most pronounced when removing all excess capacity to the level of only 50 distinct open access vessels. Not surprisingly, the share of the smallest vessel class, VS 1, drops.

Secondly, selecting for a diverse remaining fleet im-

Table 3. Summary of fleet reduction scenarios. OA = open access; LE = limited entry.

Value	Base (2000)	Scenario 1 (50 OA remaining)	Initial value of fleet remaining after capacity reduction Scenario 1 (100 OA remaining)	Scenario 2 (random)	Scenario 3 (diversity)	Scenario 4 (viability)
Coast-wide landings (pounds)	272,390,187	123,131,582	123,772,655	132,480,150	153,934,597	181,145,380
Change from base		−55%	−55%	−51%	−43%	−33%
Coast-wide revenues (US$)	62,141,810	36,127,210	37,274,029	30,509,611	43,664,904	47,744,959
Change from base		−42%	−40%	−51%	−30%	−23%
Coast-wide income impacts (US$)	138,961,151	63,832,802	65,180,144	68,244,427	90,997,105	101,667,573
Income redistributed (change from base; US$)	0	75,128,349	73,781,007	70,716,723	47,964,046	37,293,578
Income change	0%	−54%	−53%	−51%	−35%	−27%
Implied multiplier (US$/pound)	0.51	0.52	0.53	0.52	0.59	0.56
Number of vessels[a]	2,427	986	1,011	1,212	1,464	1,499
LE trawl	642	344	344	311	378	553
LE non-trawl, non sablefish	422	143	143	210	255	152
LE sablefish exclusive	25	25	25	11	17	21
Open access	1,339	174	499	680	814	773
Fleet diversity (%) of total fleet						
VS1	39%	26%	37%	40%	39%	38%
VS2	37%	29%	27%	37%	36%	32%
VS3	18%	35%	28%	17%	17%	23%
VS4	5%	9%	7%	3%	6%	6%
VS5	0%	0%	0%	0%	0%	0%
VS6	1%	1%	1%	3%	2%	1%

Note: Revenue, landings and income estimates are based on the unique vessels identified in the capacity calculation. Each vessel, however, has multiple instances as a function of making landings in multiple ports and using multiple gears over the course of a year. The numbers reported in this table report these per port "vessel-gear instances." So in the base year, there were, for example, 642 gear-port combinations of the 244 vessels in the LE trawl sector.

[a]

poses significantly lower costs (in terms of the income redistributed) on the coast. In other words, reducing fleet capacity by 50% in each vessel size-class in each sector achieves the same reduction of vessels but at the smaller redistributive "price" of around $48 million. Also, consider the meaning of the multiplier: each pound landed has an income "footprint." The fleet remaining after scenario 3 has a larger income footprint than the other scenarios. In other words, each pound caught generates more income than the same pound caught in a differently configured fleet. The total amount of income redistributed from the vessels exiting to those remaining is around $50 million. Since the number and sizes of vessels remaining in the fleet have a different geographic distribution than in scenarios 1 and 2, the effects of this scenario are also distributed differently. In comparison to the random reduction of 50% in each fleet sector, the overall fleet composition remains the same but with more vessel instances (1,464 versus 1,212) and, thus, with more associated income and jobs.

Finally, scenario 4 suggests that economic viability may be a useful consideration in designing capacity reduction measures. Recall that this scenario is based on some explicit and not entirely realistic assumptions about levels of exvessel revenues derived from groundfish needed to "make a living." Since these economic constraints can be translated into vessels to select for removal from the fleet, there are clear effects on the size, composition, and distribution of the remaining fleet. It would be interesting to examine the economic viability criterion in conjunction with numeric reduction targets. Interestingly, the particular set of economic viability criteria we chose had the effect of increasing the share of vessels in the 18–24-m (60–80-ft) range (VS 3). This illustrates the fleet composition effects of fleet reductions, which can be explicitly considered in the GFR framework.

The local implications of these scenarios differ along the coast. Figure 4 shows the amount of income generated by fishing in 2000, aggregated by port group from south to north. To the right of the base column for each port group are the income impacts of each scenario (i.e., the amount of harvester and processor income generated by the vessels remaining after the reduction). As is evident from the graph, some scenarios (notably the economic viability one) result in some ports maintaining income levels at pre-reduction levels (e.g., the Monterey Bay area or ports in the Eureka area). Also, the income effects suggest that economic viability concerns may be more important in some ports than in others. For example, Eureka; Coos Bay, Oregon; and Newport, California, areas fare better in terms of income associated from landings by the remainder of the fleet under the economic viability scenario than the fleet diversity one, whereas there is little difference for Astoria, Oregon, or the Northern Puget Sound, Washington, area. Recall, however, that the economic viability criteria were set rather low, thus retaining more vessels in the fleet than under more realistic constraints.

Figure 4 is properly interpreted as the level of income generated in each port group immediately after each reduction scenario is implemented, by the vessels that were making landings in those ports before. The dynamic response of the fishery to each scenario cannot be inferred. In particular, it is not clear how the landings formerly accruing to the vessels exiting the fishery would be allocated among the remainder of the fleet. This offsetting effect on port-reduction income levels could be approximated by making some assumptions about the remainder of the fleet. For example, one could reasonably assume that vessels remaining in the fleet would harvest the now "surplus" allocation according to the same proportions as they did before. Alternatively, one could impose some new allocation rules such as gear requirement on the remainder of the fleet. The fleet composition effects summarized for the entire coast in Table 3 are considerably more pronounced at the local level, where some scenarios eliminate entire vessel size and gear classes in some ports.

It is important to note that our analysis assumes that the total possible harvest remains unchanged (i.e., that there is no net reduction in the harvest allocations in conjunction with a fleet reduction). Specifically, for our 2000 baseline this means that the remainder of the fleet is catching the same total poundage as the fleet prior to the reductions. In light of the increasingly more stringent measures necessitated by rebuilding plans and other considerations, there may be a concomitant reduction of the overall harvest. In that case, there would be income impacts in addition to the redistribution effect between exiting and remaining vessels we consider here.

Effects of Area Closures

Another application of OCEAN lies in assessing the ecological, economic, and community effects of area closures. These have recently come to more prominence in the groundfish fishery with large closures off the West Coast to protect vulnerable species, notably the Cowcod Closure Areas in effect off Southern California since 2000 and the in-season shelf closures put into effect in the federal fishery in June and September 2002. Similar time–area closures continue in the 2003 fishery and, given the slow rebuilding rates of many rockfishes and other species of concern, are likely to shape fishery management for many years to come.

Implicit in area-based management measures is a displacement effect on fishing vessels. Depending on the size and depth covered by closure area, some vessels may be induced to exit the fishery. For example, the 2002 in-season shelf closures affected depth between 100 and 250 fath-

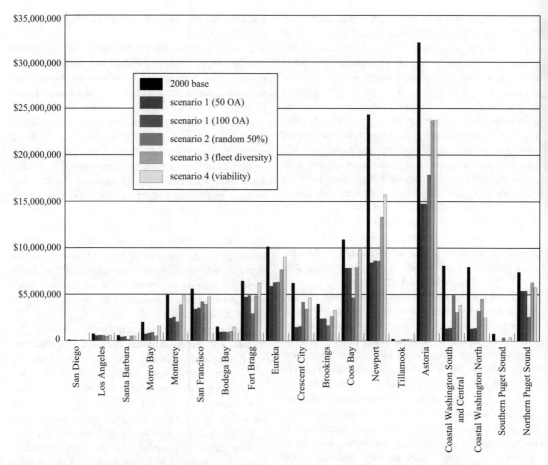

Figure 4. Summary of local impacts of scenarios along the coast.

oms; fishing farther offshore, to the west of the closure area is only feasible for a subset of the fishing fleet, vessels with sufficiently large engines and deepwater gear. Not all of the vessels that used to fish in the closure area will successfully relocate closer inshore. As an initial estimate of the potential displacement effect, in economic terms, we investigated the 2002 in-season closures in our framework. Figure 5 shows the extent of the shelf closure.

Using the 2000 fleet and effort distributions as a baseline, we identified the number and types (by gear, size, and species targeted) of vessels that fished in the closure area. Assuming that the same vessels would have fished there in 2002, we then computed the coast-wide income impacts associated with the landings initially lost due to the closure. Again, since this is a static analysis, we did not consider the adaptive effects, and, consequently, the estimates constitute the upper bound of the wealth effect. For the total coast, the income impacts generated by landings outside the closure area amount to around $115 million, and, thus, the closure potentially results in lost income on the order of $22 million if vessels were permanently displaced. More interestingly, the effects of the shelf closure vary along the coast, since fishing in the shelf areas is of varying importance for different ports, gear groups, and fishing vessels. The effect of the closure, therefore, varies accordingly, as shown in Figure 6.

While ports farther north generally have higher total groundfish landings, the relative effect of the shelf closure area varies considerably. For example, while Eureka and Crescent City, California, appear relatively unaffected, Newport and Astoria experience somewhat of an impact, and a more pronounced effect appears in Coos Bay and northern coastal Washington. Even more revealing is the consideration of the percentage of landings and revenues derived from the shelf closure area, as shown in Figure 7. Although total income derived from groundfish is small in ports like Santa Barbara, they do account for over 50% of landings and revenues derived from groundfish there. In other words, what fishing there is for groundfish is

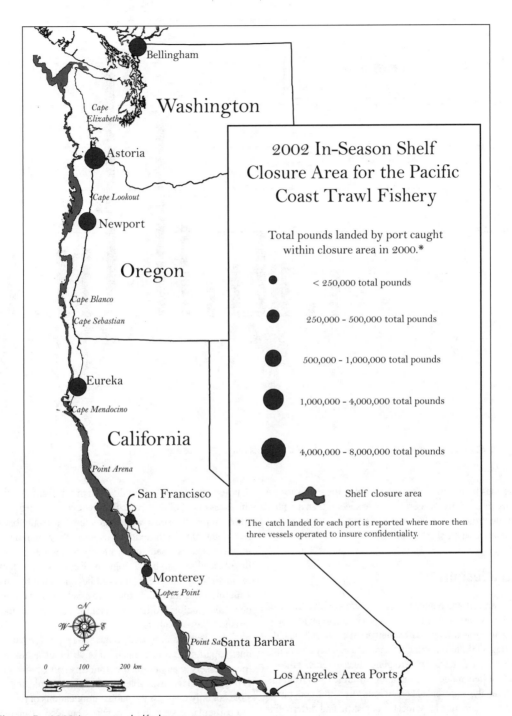

Figure 5. 2002 in-season shelf closure area,

highly dependent on the closed areas, which comprise much of the area around the historically productive Channel Islands, California.

Another important aspect of the geographical differ-ence of reliance on the shelf closure areas emerges from the difference between landings and revenues. For example, less than 5% of groundfish landings in Newport come from the shelf, but these account for almost 20% in exvessel

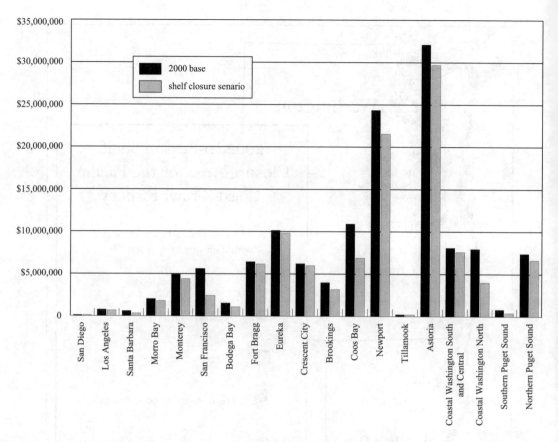

Figure 6. Income impacts before and after 2002 in-season shelf closure.

revenues in this port. This suggests that the shelf closure areas yield relatively more valuable species than do other fishing grounds. Again, there is a geographic differential in reliance on the shelf closure area.

Conclusions

The analysis and results presented here are illustrative of the kind of assessments that can be conducted with a spatially integrated analytical framework such as OCEAN. By linking a GIS to real-time fishery data and economic impact models, we were able to generate estimates of the effects of fleet reductions and area closures. Even from the static analysis presented here, it is apparent that geography matters: scenarios have location-specific and differential effects in different parts of the coast, on differently composed fleets (by size and gear types), and by the relative reliance on particular species or fishing grounds.

The approach taken here makes use of currently available data and harnesses them in a framework that can be accessed by decision makers and communities directly.

A potentially important application of this framework is the assessment of particular fleet reduction strategies—for example, the coast-wide estimates of income that is redistributed from the exiting vessels to the remainder of the fleet is an estimate for an amount that would have to be financed in a buyback program. Similarly, the effects of a fishing quota program could be simulated by constructing rules for which kinds of vessels (based on revenue and landing profiles, ownership, and other characteristics) buy out others.

The first actual capacity reduction measure, a part industry, part government financed buyback of qualifying trawl vessels, was designed before this sort of spatially explicit model was available. In principle, it would be possible not only to assess the habitat implications of removing particular vessels in the buyback but also to analyze whether the trawl buyback accomplishes the essential fish habitat objectives being developed in an as yet unrelated policy process. We are cautiously optimistic that as spatially explicit models such as OCEAN become more commonplace in fishery management, decision makers and stakeholders will seize upon them for investigating syner-

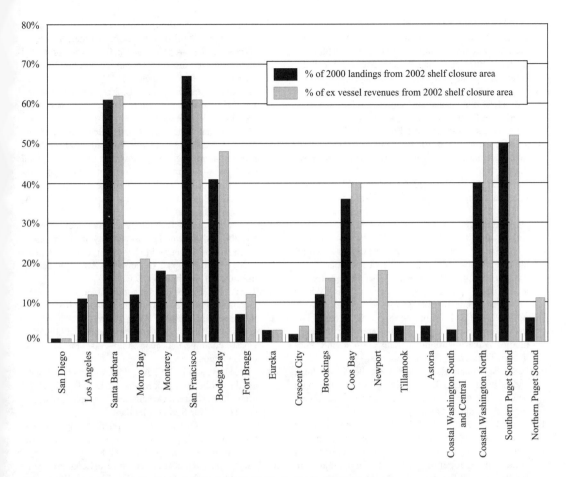

Figure 7. Landings and revenues from inside 2002 shelf closure area.

gistic effects (e.g., between habitat protection and increasing the economic feasibility of the fleet).

While there are many conceivable extensions of this approach to predictive modeling, an immediate benefit of OCEAN is that it makes visible the existing data and, thus, helps identify gaps and problems with current information sources. In particular, it remains to be seen if the spatial interpretation of historical landing receipts can be validated using the forthcoming observer data on the West Coast or a future vessel monitoring system. The kind of close spatial scrutiny of fishery management measures may also indicate the reprocessing of other data, notably the commercial trawl logbooks.

Acknowledgments

We would like to thank our project partners at the Pacific Marine Conservation Council, and colleagues on the West Coast and elsewhere who reviewed this project in various earlier forms. We are indebted to the over 100 fishermen, scientists, managers, and industry observers who kindly provided background and other information on the fishery and who made numerous valuable contributions to the research design. We thank Marc Mangel for his suggestions and three anonymous reviewers for their thoughtful and constructive criticism. This project was funded by the David and Lucile Packard Foundation, NOAA Fisheries' Northwest Regional office, NOAA's National Marine Protected Area Science Center, the Marisla Foundation, and Oregon Sea Grant.

References

Breman, J., editor. 2002. Marine geography: GIS for the oceans and seas. ESRI Press, Redlands, California.

Caddy, J. F., and F. Carocci. 1999. The spatial allocation of fishing intensity by port-based inshore fleets: a GIS application. ICES Journal of Marine Science 56:388–403.

Crosetto, M., and F. Crosetto. 2002. Optimised resource allocation for GIS-based model implementation. European Commission, Joint Research Centre, Institute for Systems, Informatics and Safety, Methodologies for Information Analysis Unit, Ispra, Italy.

Ecosystems Principles Advisory Panel. 1999. Ecosystem-based fishery management: a report to Congress. U.S. Department of Commmerce, Washington, D.C.

Federal Fisheries Investment Task Force. 1999. Federal fisheries investment task force: report to Congress. National Oceanic and Atmospheric Administration, Washington, D.C.

Field, J. C., R. C. Francis, and A. Strom. 2001. Toward a fisheries ecosystem plan for the Northern California Current. CalCOFI (California Cooperative Oceanic Fisheries Investigations) Reports 42:74–87.

Fox, D. S., and R. M. Starr. 1996. Comparison of commercial fishery and research catch data. Canadian Journal of Fisheries and Aquatic Sciences 53:2681–2694.

Gréboval, D., editor. 1999. Managing fishing capacity: selected papers on underlying concepts and issues. FAO Fisheries Technical Paper 386.

Green, D. R., and S. D. King, editors. 2003. Coastal and marine geo-information systems: applying the technology to the environment. Kluwer Academic Publishers, Dordrecht, The Netherlands.

Hanna, S., H. Blough, R. Allen, S. Iudicello, G. Matlock, and B. J. McCay. 2000. Fishing grounds: defining a new era for American fisheries management. Island Press, Washington, D.C.

Hargrove, W. W., and F. M. Hoffman. 2000. An analytical assessment tool for predicting changes in a species distribution map following changes in environmental conditions. Cooperative Institute for Research in Environemental Sciences, Banff, Alberta, Canada.

Hewings, G. 1985. Regional input–output analysis. Sage Publications, Beverly Hills, California.

Jensen, W. 1996. Pacific Fishery Management Council West Coast fisheries economic assessment model. William Jensen Consulting, Vancouver, Washington.

Jensen, W. 1998. Notes on using the "FEAM" economic impact model: the practitioner's approach. William Jensen Consulting, Vancouver, Washington.

Kruse, G. H., N. Bez, A. Booth, M. W. Dorn, S. Hills, R. N. Lipcius, D. Pelletier, C. Roy, S. J. Smith, and D. Witherell, editors. 2001. Spatial processes and management of marine populations. Alaska Sea Grant, Fairbanks.

Langdon-Pollock, J. 2002. West Coast marine fisheries community descriptions. Pacific States Marine Fisheries Commission, Seattle.

Lauth, R. R. 2000. The 1999 Pacific West Coast upper continental slope trawl survey of groundfish resources off Washington, Oregon, and California: estimates of distribution, abundance, and length composition. Alaska Fisheries Science Center, Seattle.

NRC (National Research Council). 2001. Marine protected areas: tools for sustaining ocean ecosystems. National Academy Press, Washington, D.C.

NRC (National Research Council). 2002. Effects of trawling and dredging on seafloor habitat. National Academy Press, Washington, D.C.

NOAA's National Centers for Coastal Ocean Science and National Marine Sanctuaries Program. 2002. Interim product: a biogeographic assessment off North/Central California to support the joint management plan review for Cordell Bank, Gulf of the Farrallones and Monterey Bay National Marine Sanctuaries. NOAA, Silver Spring, Maryland.

Offices of Science and Technology and Sustainable Fisheries. 2001. Identifying harvest capacity and over-capacity in federally managed fisheries: a preliminary qualitative report. National Oceanic and Atmospheric Administration, Washington, D.C.

PFMC (Pacific Fishery Management Council). 2000a. Groundfish fishery strategic plan "Transition to Sustainability." PFMC, Portland, Oregon.

PFMC (Pacific Fishery Management Council). 2000b. West Coast economic data plan 2000–2002. PFMC, Portland, Oregon.

Sampson, D. B., and P. R. Crone. 1997. Commercial fisheries data collection procedures for U.S. Pacific Coast groundfish. National Oceanic and Atmospheric Administration, Silver Spring, Maryland.

Scholz, A. 2003. Groundfish fleet restructuring information and analysis project, final report. Ecotrust. Available: www.ecotrust.org (July 2003).

SSC (Scientific and Statistical Committee, Economics Subcommittee). 2000. Overcapitalization in the West Coast groundfish fishery—background, issues, and solutions. Pacific Fishery Management Council, Portland, Oregon.

Smith, C. L., and S. S. Hanna. 1990. Measuring fleet capacity and capacity utilization. Canadian Journal of Fisheries and Aquatic Sciences 47(11):2085–2091.

Valavanis, V. D., editor. 2002. Geographic information systems in oceanography and fisheries. Taylor and Francis, New York.

Watson, R., and D. Pauly. 2001. Systematic distortions in world fisheries catch trends. Nature 414:534–536.

Weinberg, K. L., M. E. Wilkins, F. R. Shaw, and M. Zimmerman. 2002. The 2001 Pacific West Coast bottom trawl survey of groundfish resources: estimates of distribution, abundance, and length and age composition. Alaska Fisheries Science Center, Seattle.

Wong, F. L., and S. E. Eittreim. 2001. Continental shelf GIS for the Monterey Bay National Marine Sanctuary. U.S. Geological Survey, Washington, D.C.

When Do Marine Protected Areas Pay? An Analysis of Stylized Fisheries

HAROLD F. UPTON[1] AND JON G. SUTINEN

*Environmental and Natural Resource Economics Department, Kingston Coastal Institute,
1 Greenhouse Road, University of Rhode Island, Kingston, Rhode Island 02881, USA*

Abstract. In this paper, we analyze the bioeconomic consequences of using marine protected areas (MPAs) to manage fisheries. Marine protected area proponents assert that the potential benefits of MPAs include larger population biomass and associated fishery harvests. Most analytical models have considered only direct changes in harvest patterns resulting from decreases in fishing mortality following establishment of an MPA. Yet recent studies show that mobile fishing gear can decrease fish habitat complexity and alter benthic species composition. Protection of the services provided by the environment in which fish live, grow, and reproduce may be the most important fishery benefit associated with MPAs. Our main objective is to examine the conditions under which MPAs are likely to increase net benefits. Our analysis is based on a stylized bioeconomic model in which fishing has an impact on fish habitat. We examine two distinct sub-categories: (1) fishing effort that impacts the habitat of its target species, and (2) fishing effort that impacts the habitat of species targeted by other fisheries. The model incorporates the movements of the fish population between open and closed areas and the response of the fishing effort level in the area remaining open. Equilibrium stock, catch, and profit levels are compared for open access and the level of effort at which net revenue is maximized. Habitat effects are then incorporated and compared to cases where only stock effects occur. Sensitivity analysis was used to examine a wide range of fishing costs, MPA sizes, damage functions, and migration rates. Our model shows that consideration of habitat may be the main reason to use MPAs as a fishery management tool. The curvature of the habitat damage function is also shown to be an important factor when considering the potential use of MPAs.

Background

Marine protected areas (MPAs; MPA is defined in this paper as a permanent, no-take area for the fishery in question) involve the restriction of specific activities in specific geographic areas over a period time, usually indefinitely. Marine protected areas are also referred to as reserves, sanctuaries, refuges, or closed areas (we use the terms "closed area" and "MPA" interchangeably in this paper), often depending on the purpose for their establishment. The term "MPA" is commonly used in reference to the prohibition of fishing activities, although restriction of a wide range of activities such as mining, oil drilling, and recreation are also common. (We only consider the restriction of fishing activities, although non-fishing impacts on marine fish habitat may be considered with similar methods.) Closed areas are among the oldest and most commonly employed fisheries management strategies, although usually on a temporary or seasonal basis rather than a permanent basis.

The popularity of MPAs may originate from the concept's simplicity and proponents' claims that MPA establishment will satisfy multiple objectives. In addition to general non-consumptive benefits such as the conservation of biodiversity and ecosystem integrity, MPA proponents claim that benefits include greater fish population biomass, greater harvest, decreased risk of stock collapse, and less burdensome enforcement and information needs. However, many of these claims have not been substantiated in actual practice.

As a fishery management tool, MPAs are thought to sustain or increase harvest by supplementing surrounding fisheries through the export of early life stages such as eggs, larvae, juveniles, and adult fish. By decreasing fishing mortality within the reserve, the population segment within the MPA will likely increase. Reasoning follows that as fish move from relatively high population densities within the reserve to relatively low densities outside the reserve, they become available to the fishery. Recruitment outside the reserve may be increased through the export of planktonic egg and larval stages with currents.

---

[1] E-mail: harry.upton@dfw.state.or.us

Marine protected areas may also protect fish habitat from potentially destructive fishing activities. Fishing has a variety of effects on marine habitats and ecosystems depending on the type of gear, the level of fishing effort, and the spatial extent of fishing (NRC 2002). Trawls and dredges can reduce habitat complexity by removing or damaging the biological and physical structures of the seafloor (NRC 2002). It is likely that the loss of benthic habitat complexity and diversity have a detrimental effect on the survival of early life stages of commercially important species such as juvenile Atlantic cod *Gadus morhua* (Lindholm et al. 2001).

The potential of MPAs has received greater attention during the last decade as problems associated with overfishing and environmental degradation have increased in many parts of the world. In 2000, Executive Order 13158 was issued to direct federal agencies to strengthen and expand the national system of marine protected areas. Both the World Conservation Union and the parties to the UN Convention on Biological Diversity have called for a closure of 20% of the world's nearshore habitat.

However, the evidence with respect to the biological and economic benefits of reserve creation is not conclusive. Although closed areas are used in many fisheries, management measures including time–area closures in over 50 fisheries in 11 countries produced little clear evidence of improved resource conservation (OECD 1997). It was difficult to interpret the extent of positive or negative results because of the concurrent use of several management measures and closures that varied in extent and duration. Many studies that promote MPAs are qualitative in nature and only consider biological elements of the fishery.

Literature Review

A growing number of studies have attempted to simulate probable outcomes that follow MPA establishment. Polacheck (1990) used an age-structured biological model to investigate outcomes associated with a two-area spatial model with movement between areas. His work showed that a closed area led to reduction of fishing mortality, substantial increases in spawning biomass, and stock recovery (Polacheck 1990). He stressed that yield per recruit is a complex function of the size of the reserve, fishing mortality rates, and movement rates. Net increases in yield per recruit were small in most cases (Polacheck 1990). He assumed that effort before and after the establishment of the reserve was fixed, and economic elements of the fishery were not considered.

Holland and Brazee (1996) modeled the biological system with a modified Leslie population matrix that incorporated impacts of a reserve on recruitment and harvest over time. Unlike Polacheck, the stock recruitment function was modeled, but the economic system was limited to equating the level of harvest to the value of benefits. The study assumed that fishing effort is fixed and a stock that is homogenous except for differences between reserve and non-reserve stocks. They applied their simulation to Gulf of Mexico red snapper *Lutjanus campechanus* and found that marine reserves are likely to sustain or increase yields for moderate to heavily fished fisheries but probably will not improve yields for lightly fished fisheries.

Conrad (1999), Pezzey et al. (2000), Hannesson (1998), and Sanchirico and Wilen (2001) developed deterministic bioeconomic models with fishing effort that responds to MPA creation. Conrad also considered stochasticity while both Conrad and Hannesson also added seasonality. All these efforts used a regulated, open-access specification that achieved equilibrium when economic rents were dissipated. Hannesson, Conrad, and Pezzey et al. also used an optimal (maximum economic yield) specification to make comparisons to open access. Brown and Roughgarden (1997) considered an optimal case, but under the specific circumstances associated with a common larval pool and an adult metapopulation associated with limited space during the adult phase. All these efforts used similar lumped parameter logistic growth functions to model the fish population. Conrad, Hannesson, and Pezzey et al. used homogenous stock as did Holland and Brazee (1996), but Sanchirio and Wilen took a metapopulation approach that considered the possibility of heterogeneous fishing grounds. That enabled them to consider various ecological structures such as density-dependent and source-sink patch configurations. Their results showed that MPAs can be used to increase both stock and harvest when the stock has been overfished.

Conrad (1999) compared a deterministic model of perfect management and an MPA alternative and showed that MPAs do not increase benefits. Only when he included stochasticity in the model did the lower stock variability associated with MPAs make them a more attractive management tool. Hannesson (1998) obtained similar results with a deterministic model of marine reserves. His results suggest that marine reserves would have to be very large to achieve harvest and stock effects that are equivalent to an optimally controlled fishery (Hannesson 1998). However, he stressed that in the absence of additional controls, fishing effort will increase outside the reserve and dissipate rents. If the MPA size is closely suited to the stock and economic characteristics of the fishery, then the resulting equilibrium may exhibit similar stock levels and catches as the optimally managed cases. However, net social benefits will be lower in the regulated open-access case due to the overuse of fishing inputs. Pezzey et al. (2000) used a similar bioeconomic model, but they made ad hoc modifications to the stock growth function to more closely model mobile larvae and sessile adults that are characteristic of coral reef fisheries. Results generally agreed with previous studies that showed catch can be increased when the stock equilibrium level has been fished below maximum sustainable yield. Holland and Brazee (1996) also found improvements in catch were possible in moderate to heavily fished fisheries. Hannesson also showed that increases in stock and harvest were possible for fisheries originally fished to low levels.

Previous studies have attempted to evaluate changes in harvest patterns resulting from MPA establishment. Yet, the greatest benefits associated with MPAs may be related to conservation of habitat and ecological integrity. Habitat of sufficient quality and quantity is a necessary environmental input to the natural processes associated with reproduction and growth. Integration of environmental elements with bioeconomic models has been attempted since the 1970s. Isard (1972) outlined the potential use of input-output models that included the principal elements of economic and ecological systems. Lynne et al. (1981) integrated wetlands as a factor related to the growth of blue crab *Callinectes sapidus*. Lynn used the extent of wetlands in a region as an argument in the blue crab growth function. Kahn (1987) examined shifts of the biological growth function by assuming the population carrying capacity changes with environmental quality. In an associated empirical study, Kahn and Kemp (1985) attempted to relate changes in submerged aquatic vegetation in the Chesapeake Bay (Maryland) to changes in population productivity of striped bass *Morone saxatilis*. Swallow (1990) studied the relationship between renewable natural resource production and a nonrenewable element of the environment. In an empirical example, he considered the relationship between shrimp, a renewable resource, and wetlands, a nonrenewable resource that is developed over time (Swallow 1994). Barbier and Strand (1998) used a similar approach by modeling the relationship between tropical shrimp fisheries and mangrove systems. They assumed that the shrimp population carrying capacity is directly related to the spatial coverage of mangroves in near shore areas.

MacCall (1990) examined the relationship between the environment and fish populations. He stated that modeling environmental changes with a logistic growth function will likely involve changes to both the intrinsic growth rate and carrying capacity parameters. Generally, these studies recognize that biological parameters are often assumed to be fixed, when they actually vary in response to environmental fluctuations. Changes in survivability of specific life stages due to changes in the quality and quantity of habitat can be equated conceptually with changes in logistic growth parameters. These changes can then be related to their impact on populations and the economic systems that utilize them.

The Model

Although ecosystem functions such as the quality and quantity of fish habitat are gaining more attention, a fully integrated bioeconomic treatment of MPA creation that considers habitat as well as the associated fish population has remained elusive. Habitat is composed of a web of interacting physical and biological components. Associated human management and economic and social systems are complex as well.

Our study examines situations where the fishing effort that is used to target marine species also damages marine fish habitat. We also consider situations where effort from a second fishery degrades marine fish habitat of the species targeted by the first fishery. Therefore, habitat and the associated fish population are modeled as renewable resources that are impacted by fishing effort. Habitat has both renewable and nonrenewable qualities depending on timeframe and feature. One might argue that all habitats are renewable, as even filled wetlands can be restored. Natural regeneration is also likely for coral reefs, mangrove systems, sea grasses, and benthic organisms, although recovery time may vary from months to centuries. The model includes density-dependent movement of the population among areas and fishing effort that depends on profitability, the management system, and resource availability. This is accomplished by combining the MPA and habitat-related elements of the studies cited in the previous section. While attempting to include the essential elements of interest, we have presented the simplest model possible in order to retain analytical tractability and clarity.

We begin by using an MPA model of a targeted biological population and economic system as developed in several of the preceding examples (Hannesson 1998; Conrad 1999; Pezzey et al. 2000; Sanchirico and Wilen 2001). A logistic growth specification is used to model population dynamics, with K (carrying capacity, measured in biomass) set at unity; X as the stock size (measured in biomass), and r as the intrinsic growth rate (measured in the rate of growth in biomass). Harvest is identified as AEX, with A, the catchability coefficient (the fraction of stock that is caught by a defined unit of fishing effort), normalized at unity (Hannesson 1998), and where E represents effort (measured in numbers of vessels or fishing days) and X represents stock size.

$$dX/dt = rX(1 - X) - EX \quad (1)$$

Changes in population size (biomass) depend on the difference between population growth as represented by logistic growth and harvest in equation (1). The level of harvest links the economic and biological elements of the model.

Two specifications are used to model the economic system. Open access is defined as an equilibrium level of stock and harvest at which resource rent is equal to zero (Gordon 1954). (Resource rents are the returns to fishermen in excess of the returns that are necessary to keep them in the fishery. Resource rent is the surplus remaining after the full opportunity costs of production have been met. It is equal to the difference between returns in the fishery and the return necessary to keep fishermen employed in the fishery.) Therefore, we assume that entry of fishing effort occurs until resource rent is dissipated or, conversely, exit occurs until negative marginal profits of the fleet reach zero. At this point, fishing effort will neither enter nor exit the fishery because fishermen can gain equal returns from their best opportunity outside the fishery. Profits or net benefits are defined as the difference between the product of price (P) and harvest, and total cost (the product of a cost constant, C) and the level of fishing effort (E) squared. Therefore, unit costs are modeled to increase with increasing effort resulting in an upward sloping marginal cost curve:

$$dE/dt = PEX - CE^2 \quad (2)$$

Open-access equilibrium is defined as the level of effort at which average revenue is equal to marginal costs. The economic maximum yield is defined as the level of fishing effort that maximizes profits.

These two specifications attempt to define two institutional extremes or bounds within which fishery effort levels will occur. Each provides benchmark outcomes associated with levels of catch, stock, and habitat effects. By setting the price, catchability coefficient, and carrying capacity at one, results cannot be considered in absolute, but rather in relative terms. Simplifying assumptions include a constant output price and homogenous fishing ground.

The MPA is incorporated by closing a fraction of the total area to fishing while allowing the target species to move among areas. The probability of movement from one area to the other depends on the movement rate that is characteristic of the species, the size of the closed area, and the number of individuals in the area or population density (Hannesson 1998). Movement could also be modeled in a number of other ways, depending on underlying ecological assumptions related to the organism's life history and seasonal migration. For example, a single species population can be modeled as a source with net growth and sink in when reproduction fails to keep pace with deaths (Pulliam 1988). The habitat is assumed spatially heterogeneous with net movement of individuals from the source to the sink environment.

The relative density of the individuals in adjacent areas determines net movement between areas. Density-dependent habitat selection refers to the influence of population density on habitat suitability (MacCall 1990). Realized habitat suitability is dependent on the density of individuals and habitat characteristics. Fretwell (1972) was one of the first to propose the idea of an "ideal free distribution" where individuals will move to new areas until no individual can move to a better habitat. In other words, as the density of individuals increase in a habitat, habitat suitability decreases due to density-dependent effects (Fretwell 1972). The "ideal free distribution" occurs when there is equal realized suitability among all habitats (MacCall 1990). MacCall pointed out that the "ideal free distribution" is conceptually similar to Ricardian Rents, where farmers utilize land at the extensive margin. A density-dependent system was also used by Hannesson (1998), Sanchirico and Wilen (2001), and Conrad (1999) to model movement.

The change in density for areas outside the MPA is the product of the size of the MPA (m), the movement of organisms (n), and the difference in densities of organisms inside and outside the MPA ($X_i - X_o$). Equilibrium population density inside the reserve is dependent on movement as well as harvesting outside the MPA. Equation (1) is modified to include movement to the open area as shown in equation (3). An additional equation (4) is added to the bioeconomic system to represent population density inside the MPA and movement to the open area:

$$dX_o/dt = rX_o(1 - X_o) - EX_o + mn(X_i - X_o) \quad (3)$$
$$dX_i/dt = rX_i(1 - X_i) - (1 - m)(n)(X_i - X_o) \quad (4)$$
$$E/dt = PEX_o - CE^2 \quad (5)$$

Equation (4) can be solved for X_i in terms of X_o, resulting in equation (6).

$$X_i = -0.5[(1 - m)n/r - 1] + \{0.25[(1 - m)n/r - 1]^2 + (1 - m)nX_o/r\}^{0.5} \quad (6)$$

Equation (6) is then be substituted into equation (3). Then effort (E) and stock (X_o) can be solved for both open-access and the economic maximum using equations (3) and (5). These relationships were programmed in MATLAB to solve for the open-access and economic maximum solutions. The open-access equilibrium requires the following stock and economic equalities:

$$rX_o(1 - X_o) + mn(X_i - X_o) = EX_o, \text{ and}$$
$$PX_o = 2CE.$$

The optimal solution is determined by maximizing equation (5), subject to the stock constraint:

maximize $PEX - CE^2$, (from 5)
subject to $rX_o(1 - X_o) + mn(X_i - X_o) = EX_o$. (from 3)

Many of the most important habitat features such as structure and ecological relationships are biological in nature. The impact of fishing on habitat depends on the gear type, its use, and the vulnerability of the habitat it contacts. Fish habitat can be envisioned as a living natural resource such as mangroves, eelgrass *Zostera marina*, benthic communities, or coral reefs. These elements are renewable in the sense that they grow and may recover from negative impacts.

We propose the addition of a second renewable resource component to the model that represents habitat quality. Its two main elements include biological growth or recovery and damage from fishing effort. The biological component of benthic marine habitat may be composed of a large number of species with different life histories and strategies. However, it is likely that their population growth can be explained by logistic properties with slow growth during initial colonization, subsequent exponential growth as unused resources such as space are utilized, and slowing growth as high densities are reached and crowding occurs. Therefore, we assume that habitat can be represented by a logistic growth function, although we recognize the complexity of habitat both with respect to biological diversity and ecological relationships among components.

Habitat damage is related to the level of the damaging activity. Habitat damage or vulnerability of a habitat type might be calibrated according to a damage coefficient that is related to habitat and gear types. The damage coefficient is analogous to the catchability coefficient that can be interpreted as the vulnerability of a fish species or group to a specific gear type. Total damage is an increasing function of fishing effort, but the marginal damage function may either increase or decrease with increasing effort. We use the following logistic functional form for habitat and general form of the damage function to represent changes in habitat quality and quantity:

$$dH/dt = r_h H(1 - H) - D(q,E,H),\quad(7)$$

where H represents habitat population density level, r_h is the generalized intrinsic habitat growth rate or regeneration rate, and the carrying capacity of habitat is set at unity. The damage function, $D(q,E,H)$, consists of q (characteristic of the gear type), E (fishing effort), and H (level of habitat density).

Physical damage from repeated trawls in the same area will likely kill diminishing numbers of organisms and produce smaller levels of damage. The resulting total damage function increases at a decreasing rate, and the marginal damage function decreases at a decreasing rate with increasing effort (Figure 1). Low levels of effort are assumed to have a greater marginal impact because many areas are still being swept by gear for the first or second time. Once an area has been swept, subsequent effort produces a smaller marginal impact because the most vulnerable habitat features have already been removed or damaged. We use an exponential function of the following form that is similar to the form proposed by the National Research Council trawl study to represent mortality of habitat components (NRC 2002):

$$dD/dt = H(1-e^{-qE}),\quad(8)$$

where D is habitat damage, q is the damage coefficient that was introduced in equation (7), and E is fishing effort. Habitat damage is bounded by 0 and 1, with higher values associated with habitats that are highly vulnerable to a specific fishing gear type.

Another candidate damage function increases at an increasing rate within a range of fishing effort. Damage functions of this type are commonly used as examples in environmental economics where the marginal impact of a unit of pollution becomes greater as the total level of pollution increases. Eventually, the marginal impacts decrease as the resource such as water becomes so polluted that further damage is not possible. Marginal damages may increase at threshold levels of an activity or associated pollutant concentration. A threshold level of habitat disturbance could occur if additional levels of effort cause greater damage because the initial passes of

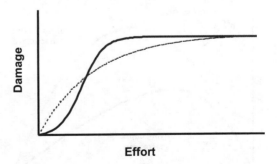

Figure 1. Total habitat damage as a function of fishing effort. The dotted line represents decreasing marginal damages (from equation 8). The solid curve (from equation 9) represents marginal damages that initially increase then decrease as all vulnerable habitat is destroyed.

gear weaken habitat resilience without complete destruction. Subsequent passes cause greater levels of mortality from which the habitat attributes that support fishery production cannot readily recover. Further damage from additional units of fishing effort will cause less marginal damage because habitat cannot be further degraded by the gear in question. A logistic function of the following functional form is a candidate damage function with these characteristics (Figure 1):

$$dD/dt = 1/(1 + e^{-aE}). \quad (9)$$

The effort identified in the damage function may simultaneously remove individuals targeted by the fishery and impact population growth parameters related to habitat quality (Figures 2, 3). Another general case involves habitat degradation that is caused by activities that are exogenous to the fishery, such as a shore-related activity or a fishery targeting a different species (Figures 4, 5). If effort impacts the habitat of the same species that it targets, the resulting equilibrium is determined by the relationship between effort and both habitat damage and fishing mortality. If effort impacts habitat of a different species, bioeconomic equilibrium is determined by fishing mortality of the target species, and degradation of the second species' habitat is incidental. The same is also true of other economic sectors that might impact marine fish habitat, such as power generation, dredging, or aquaculture.

It is likely there is a functional relationship between habitat quality and both the stock intrinsic growth rate and carrying capacity. MacCall (1990) concluded that habitat suitability and marginal changes in reproductive value are manifested as changes in the intrinsic rate of increase of the population. We as-

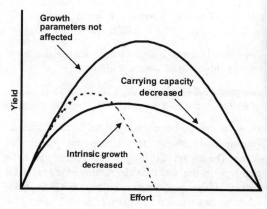

Figure 3. Sustainable yield as a function of effort (from equations 1, 7 and 8) for the single-fishery case. Fishing effort impacts both stock and habitat related parameters with a decreasing marginal damage function.

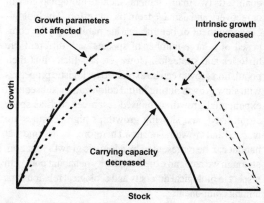

Figure 4. Population growth curves (from equation 1) resulting from exogenous impacts on growth function parameters r and K.

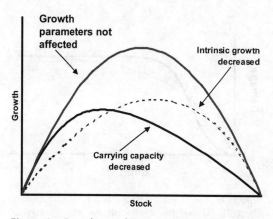

Figure 2. Growth as a function of stock (from equations 1, 7 and 8) for the single-fishery case. Fishing effort impacts both stock and habitat-related parameters with a decreasing marginal damage function.

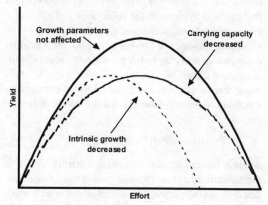

Figure 5. Sustainable yield (derived from equation 1) as a function of effort with shifts in yield resulting from exogenous impacts on grow parameters r and K.

sume that habitat is directly proportional to these population growth parameters. Equation (7) is solved for habitat (H) in terms of effort, and the expression is inserted into the population growth function such as equations (1) and (3). We assumed a linear relationship between growth parameters and habitat, although other functional relationships are likely given specific circumstances:

$$dX/dt = r_x(H)X(1 - X/K(H)). \qquad (10)$$

MacCall (1990) proposes use of the constant slope model that shifts both the density-independent r and the density-dependent parameter K with changes in habitat quality. He suggests that use of a constant r and variable K for a variable habitat logistic model has been misleading due to semantics. At low densities, it is likely that different habitats exhibit different intrinsic growth rates due to different habitat quality or suitability. At higher population densities, population growth is slowed because of decreasing habitat suitability that is related to declining availability of environmental resources such as space, prey, and other biological and physical requirements. Several cited studies have shifted only carrying capacity (Kahn and Kemp 1985; Sainsbury 1988; Barbier and Strand 1998). We modified each parameter individually and both parameters with changes in habitat quality and quantity and found that our basic results were similar in each case, especially within intermediate stock levels. General stylized results were similar when each or both parameters were varied and reported as such. We recognize that actual parameters will be affected to varying degrees and that the relative shifts associated with each parameter, such as the shifts in Figures 2 and 3, will reflect specific elements of a given problem.

Results

We used this stylized model to explore likely outcomes related to the use of MPAs as a fishery management tool when fishing effort affects habitat. The model compares the use of MPAs under open access and at the economic maximum (maximum economic yield; MEY) over a wide range of costs, migration rates, habitat damage functions, and MPA sizes. In our model, low-cost fisheries under open access result in a system equilibrium that is fished beyond the biological maximum to low stock levels, but high-cost, open-access, and MEY fisheries are fished below or near the biological maximum at relatively high stock levels. High-cost fisheries have relatively high stock levels because the associated bioeconomic equilibrium supports lower effort levels. Profit, harvest, stock, and effort levels are employed as benchmarks to make comparisons among different scenarios related to management institution and habitat-damage function type.

First, the single-fishery case using MPAs without effort-related habitat damage is discussed. As reported in previous studies, MPAs are found to increase total stock size for both open access and MEY fisheries because they provide protection within the closed area (Figure 6). However, as the migration rate is increased above intermediate levels, stock and harvest benefits decrease and approach the levels that are found without closures. Migration is driven by the difference between population densities in the open and closed areas. As the migration rate increases, the population responds more quickly to the density gradient between open and closed areas. When the population density in the closed area approaches that of the open area, the fishery approaches the open access and MEY levels of stock and harvest without closures (Figures 7, 8).

Notable gains to harvest and profit were obtained in open-access cases when costs were low, migration rates were set at intermediate levels, and a relatively high proportion of the area was closed to fishing (Figure 9a). High-cost, open-access fisheries provided higher stock levels with closures but lower harvest and profit levels when compared to the no-closure alternative. Marine protected areas were also found to decrease harvest and profits for fisheries managed at MEY (Figures 9b, 10).

In open-access cases where stock is driven to low levels, MPAs allow for stock recovery within the closed area. At equilibrium, migration into the open area becomes greater than the surplus production that was harvested when the area was open. However, if initial stock levels are relatively high, such as high-cost, open-access fisheries or fisheries managed at their economic optimum, the resulting migration to the open area is less than that produced when the area was open. Migration is lower in this case because the density

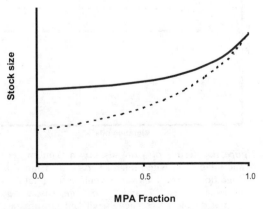

Figure 6. Total stock size as a function of MPA fraction for open-access (dotted line) and maximum economic yield cases (from equations 3 and 5).

gradient between open and closed areas is smaller, resulting in lower density-dependent migration to the open area (Figure 9).

When fishing effort is modeled to degrade habitat and decrease associated population parameters, the resulting stock, harvest, and profit levels depend on the magnitude and curvature of the damage function. We modeled damage functions to decrease habitat-related population parameters by a range of 0% to 50% of undis-

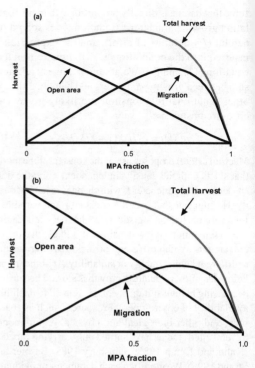

Figure 9. Contributions of migration and the open area to harvest as a function of MPA fraction with no habitat effects (from equations 3 and 5) for the open-access, one-fishery case (a) and maximum economic yield, one-fishery case (b). In both cases, the sum of the harvest from the open area and harvest resulting from migration from the closed area are equal to the top curve, total harvest.

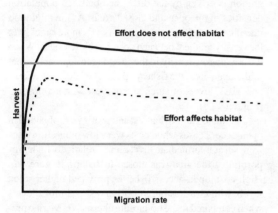

Figure 7. Results of a one-fishery, open-access case, with plots of harvest as a function of migration rate with closure of half of the total area (from equations 3, 5, 7, and 9). For comparison, gray lines that denote the harvest level without closures are provided. The lower gray line characterizes the case in which effort affects habitat, and for the upper gray line, effort does not affect habitat.

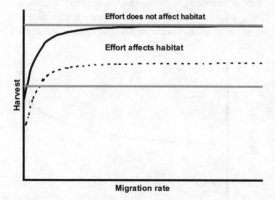

Figure 8. Results of a one-fishery, maximum economic yield case, with plots of harvest as a function of migration rate with closure of half of the total area (from equations 3, 5, 7, and 8). For comparison, gray lines that denote harvest levels without closures are provided. The lower gray line characterizes the case in which effort affects habitat, and for the upper gray line, effort does not affect habitat.

turbed levels as a function of the level of fishing effort. The shifts in the growth and sustainable yield relationships are shown in Figures 2 and 3. In addition to the downward shifts of these curves, the level at which maximum sustainable yield occurs and the shape of the production function are altered (M. J. Fogarty, presentation to the 6th National Stock Assessment Workshop, 2000). As expected, habitat impacts resulted in lower stock, effort, catch, and profit levels when compared to cases without habitat degradation. When MPAs were established, stock size generally increased as in the previous example (Figure 6).

Yet, several notable differences from previous studies also became apparent, especially in the case of decreasing marginal damages. For low-cost, open-access fisheries with low associated stock sizes, closures provided a greater relative increase to harvest and profit levels (Figure 11). The disproportionate increases are the result of the combined influence of stock and habitat effects. However, it should be noted that when habitat

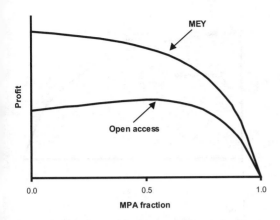

Figure 10. Profit as a function of MPA fraction for the one fishery case (from equations 3 and 5). Maximum Economic Yield and open access are considered with no habitat impacts.

degradation is a factor, closures never provide the benchmark harvest level associated with the no-degradation case. For relatively high-cost, open-access fisheries and fisheries managed at MEY, both with relatively high initial stock sizes, it was shown that MPAs can increase harvest and profit levels (Figure 11). This result is in stark contrast to the previous case that does not consider habitat impacts.

Also in contrast to previous results is the case where migration rate is increased to high levels. The MPA protects habitat within the closed area and increases the stock growth function parameters. Although stock improvements are dissipated at high migration rates, habitat-related increases in production are maintained (Figures 7, 8).

Changes in habitat are related to an increase in population growth parameters that provide an upward shift in the stock growth curve in the closed area. Stock effects are associated with movement along the stock growth and sustainable yield curves that also increase in the case of overfishing. Therefore, the habitat effect is positive in the closed area. However, increased harvest from the area remaining open will also support more fishing effort. This may further decrease habitat-related population growth parameters in the open area. The net gain or loss associated with habitat effects in both the closed and open areas depends on the strength and curvature of the relationship between habitat and fishing effort.

Net habitat-related benefits depend on whether the habitat gains associated with the closure are greater than the habitat losses in the area that remains open. We believe that the habitat damage function is characterized by decreasing marginal damages with increasing fishing effort (NRC 2002). In this case, most damage occurs at

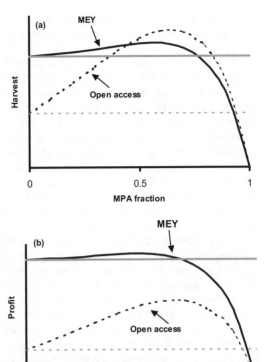

Figure 11. Harvest (a) and profit (b) as a function of MPA fraction with decreasing marginal habitat damages (from equations 3, 5, 7, and 8) for the one-fishery case comparing maximum economic yield and open access. For comparison, the dotted gray horizontal line represents open-access harvest (a) or profit (b) with no closures, and the solid gray horizontal line represents maximum economic yield harvest (a) or profit (b) with no closures.

low effort levels with decreasing damage as marginal units of effort are added. Vulnerable habitat has already been removed or damaged by the first units of effort that swept the area. The closure provides a large increase in habitat quality to undisturbed levels, but effort increases in the area remaining open create little additional damage. Therefore, the habitat benefits in the closed area are significant and more than offset the additional damages of greater effort in the open area. The increases in the MEY case shown in Figure 11 are a habitat effect of both sufficient magnitude and correct curvature to offset stock effects.

If damages are small at existing levels of effort and increase marginally with fishing effort, the habitat gains associated with the closure are likely to be smaller than the habitat losses in the area that remains open. In Figure 1, the solid logistic curve below the inflection point is the corresponding segment of the damage function. In this case, most damage will occur at relatively high levels of fishing effort with increasing marginal damage as units of effort are added. The closure provides a relatively small increase in habitat quality to undisturbed levels, but increasing effort in the area remaining open will create greater additional damage. Therefore, the habitat benefits in the closed area are relatively small and may not offset additional damages of greater effort in the open area. In this case, MPAs provide lower increases in net benefits under open access and negative net benefits for the MEY solution (Figure 12). Since habitat benefits are small or negative, they do not offset the decrease in net benefits of the MEY case shown in the first example without habitat effects.

The increasing marginal damage function will eventually begin to level off and exhibit decreasing marginal damages when most vulnerable habitat has already been damaged. In Figure 1, this segment corresponds to the solid logistic curve above the inflection point of the total damage function. At this level of fishing effort, gains in the closed area will be high relative to losses in the open area where vulnerable habitat has already been lost. However, if effort can be reduced to levels where the marginal damage function is increasing, an efficient solution can be found with the management of fishing effort.

Outcomes associated with MPA creation are determined by a combination of stock and habitat effects (Table 1). If the net habitat effect is positive, then stock and habitat effects complement each other when the stock has been overfished, on the right side of the sustainable yield curve. In these cases, MPAs will yield net benefits. However, on the left side of the harvest maximum, habitat and stock effects are working against each other, with decreases from the stock effect and increases associated with habitat improvement. The strength of the habitat effect relative to the stock effect will determine whether the MPA can improve net benefits at relatively high stock levels such as the MEY case.

If the net habitat effect is negative, then the habitat effect is working against the stock effect on the right side of the sustainable yield curve and complementing the negative stock effect to the left of the harvest maximum. Net benefits from closed areas depend on the relative strengths of the habitat and stock effects in the case where the stock has been overfished. However, when initial stock levels are high, such as the MEY case, MPA establishment will decrease net benefits because both stock and habitat effects are negative (Table 1). The relative magnitudes of these effects will vary in individual circumstances.

There are many other plausible scenarios in addition to the one fishery case outlined above. Another general case might involve habitat degradation that is

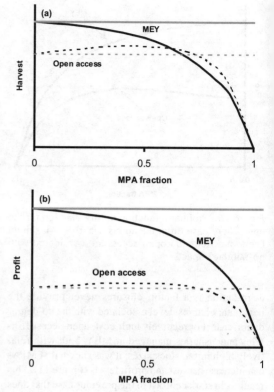

Figure 12. Harvest (a) and profit (b) as a function of MPA fraction with increasing marginal damages (from equations 3, 5, 7, and 9) for the one-fishery case comparing maximum economic yield and open access. For comparison, the dotted gray line represents open-access harvest (a) or profit (b) with no closures, and the solid gray line represents maximum economic yield harvest (a) or profit (b) with no closures.

caused by activities that are exogenous to the fishery, such as a shore-related activity or a fishery targeting a different species. Figures 4 and 5 show the result of a constant change in the level of growth parameters. A likely case involves two fisheries, one that degrades habitat of another fishery but does not affect its own habitat. One might imagine cases such as two fisheries, both of which affect habitat and target the same or different species, and two fisheries that target the same species, one of which affects habitat. In addition, MPAs might selectively restrict fisheries depending on habitat effects or restrict all fisheries. Many of the same elements of the first model apply, except that the habitat damage associated with effort affects the growth function of the second fishery. Therefore, the two fisheries are linked by habitat equation (11) that determines the impact of the first fishery on habitat and equation (12) that incorporates the impact of changing habitat quality on the growth function of the second fishery:

Table 1. Sign of stock and habitat effects that determine harvest and associated benefits after MPA establishment. An asterisk indicates that the net result depends on the relative magnitude of stock and habitat effects although at intermediate migration levels our model provided these results.

Relationship between effort and habitat damage	Low-stock (overfished), open-access example	High stock levels, MEY example
No habitat-related damage	Positive (stock effect positive)	Negative (stock effect negative)
Habitat related damage with decreasing marginal damages	Positive (stock effect positive) (habitat effect positive)	Positive*(stock effect negative) (habitat effect positive)
Habitat related damage with increasing marginal damages	Positive* (stock effect positive) (habitat negative)	Negative (stock effect negative) (habitat effect negative)

$$dH_2/dt = r_h H_2(1 - H_2) - D(q_1, E_1, H_2); \quad (11)$$
$$dX_2/dt = r_2(H_2)X_2(1 - X_2/K(H_2)) - E_2 X_2, \quad (12)$$

where H_2 is the habitat of the second fishery, q_1 is the damage coefficient of the first fishery, and E_1 is effort in the first fishery. Equation (12) is the growth function of the second fishery with growth parameters that are dependent on habitat quality that are determined in equation (11). The level of habitat damage is dependent on the level of effort in the first fishery. The second fishery does not influence the effort level in the first fishery, but habitat of the second fishery is influenced by effort of the first fishery.

We consider a scenario where the first fishery impacts habitat of the second fishery and both fisheries are closed within the MPA. Establishment of the MPA will result in three primary impacts. The first fishery will exhibit a stock effect or movement along its growth curve. Net benefits of this effect will depend on the initial stock and harvest levels of the fishery. As shown earlier, with no habitat effects, net benefits depend on whether the stock was harvested to the left or the right of the harvest maximum before the closure.

The second fishery will exhibit changes in both stock and habitat with MPA establishment. The stock effect will depend on the initial stock and harvest levels of the second fishery before the closure. Habitat of the second fishery will improve in the closed area, but the magnitude of the effect will depend on the level of damage before the closure and the curvature of the damage function that relates effort of the first fishery and habitat of the second fishery. The habitat effect will cause an upward shift in the growth curve of the second fishery within the reserve that is associated with the absence of habitat degradation caused by the first fishery. However, fishing effort of the first fishery will increase outside the reserve, resulting in a negative habitat effect. The net effect, as in the single fishery case, will depend on the shape and magnitude of the habitat damage function. Decreasing marginal damages will favor the closure, while increasing marginal damages will favor effort management.

Net benefits from both fisheries will depend on the tradeoff between losses to the first fishery and gains to the second fishery associated with habitat improvement. The result depends on management of each fishery with respect to stock level and the habitat damage function. Given the assumption that we start with two identical fisheries, Table 2 provides the net benefits for each fishery and both fisheries.

Discussion

We recognize that our results are dependent on the biological and economic parameters used, especially the magnitude and curvature of the damage functions. Yet several generalizations became apparent as we compared different scenarios. Our study and previous studies indicate that MPAs can be used to achieve objectives associated with increasing stock size and harvest when fisheries are fished beyond their maximum harvest levels. By adding the habitat dimension, we found that MPAs may further increase harvests when overfishing has not occurred, depending on the magnitude and curvature of the habitat and effort relationship. Damage functions with high initial damage and decreasing marginal damages favor the use of MPAs. Yet in these cases, the additional net benefits of MPAs in fisheries managed at MEY were small and required closure of a large fraction of the total area. On the other hand, even when MPAs decrease net benefits, they do not fall off sharply with closures of moderate sizes between 10% and 30% of total area.

However, in all cases, rationalization of fishing effort that is associated with the MEY alternative has the greatest impact on net benefits when compared to open access.

The performance of MPAs depends on management objectives and evaluation criteria. Economists often consider economic profits or social welfare to make comparisons among policy alternatives. We recognize that there are other MPA-related objectives. Our model

Table 2. Two fishery cases examining the changes in net benefits associated with establishment of an MPA when habitat damages are exogenous to the affected fishery. Fishery I affects the habitat of Fishery II, and both fisheries are closed within the MPA. Growth and economic parameters of both fisheries are identical with intermediate migration rates. Maximum economic yield (MEY) and open-access cases are considered for each case. Net benefits are defined as the difference of revenue and cost. An asterisk indicates that sign depends on the relative magnitude of the fisheries.

Fishery I	Fishery II	Marginal damage	Net benefits Fishery I	Net benefits Fishery II	Total benefits
MEY	Open access	Decreasing	-	+	+*
MEY	MEY		-	+	+*
Open access	MEY		+	+	+
Open access	Open access		+	+	+
MEY	Open access	Increasing	-	+	+*
MEY	MEY		-	-	-
Open access	MEY		+	-	+*
Open access	Open access		+	+	+

considers the most basic elements of a bioeconomic MPA system that is used as a fisheries management tool. Pezzey et al. (2000) pointed out that there are additional consumer and producer benefits associated with higher harvests that are made possible by MPAs. Even under open-access scenarios, higher harvest levels provide greater consumer and producer benefits. We have attempted to capture producer benefits under open access by assuming a simple, upward-sloping industry marginal cost curve. Although net benefits exist under open access and increase under most MPA scenarios, improved stock levels and harvest are associated with greater investment in fishing effort that dissipates resource rent at the bioeconomic equilibrium. Although greater levels of catch may be obtained with MPAs, one also needs to consider the opportunity cost of resources used as inputs, such as capital and labor. It is likely that these inputs could be used more productively in another sector. When effort is associated with environmental degradation, less is generally better. Since less fishing effort is always employed at the economic maximum, moving in that direction appears to be logical. The bonus associated with this course of action is the resource rent that would be gained by fishery participants. Although beyond the scope of this paper, institutional arrangements that provide incentives to conserve habitat as well as stocks are needed, rather than use of traditional measures that create inefficiencies or "high-cost fisheries."

Marine protected areas may also constrain fishing operations and require fishing in less productive areas or impose greater costs on the fishing industry. Although our model considers the potential shifts in effort among areas, it does not consider the potential loss of revenues or costs associated with these shifts that would arise due to heterogeneity of fishing grounds. For example, MPAs may displace vessels from relatively close to more distant fishing grounds. Vessel costs such as fuel and other inputs are likely to increase with travel to more distant areas. These considerations are not included in this model and are likely to make closed areas less attractive.

Marine protected areas may also be useful in cases where an activity unrelated to the primary fishery degrades fish habitat, such as another fishery, coastal development, dredge disposal, or sand and gravel mining. In these cases, similar downward shifts of the growth function will occur due to lower related population parameters. Similar models of habitat and fisheries have been used by Barbier and Strand (1998) and Kahn and Kemp (1985) to estimate the economic losses associated with habitat losses or degradation. In these cases, the habitat effects of a closure will always be positive since fishing effort of the primary fishery is no longer related to habitat quality. However, closures cannot be based solely on the basis of benefits to the fishery. Economic efficiency now depends on the tradeoffs between the fishery and the activity in question. This is the basic environmental economics case where marginal benefits of the activity in question must be balanced against marginal costs associated with damages to the fishery. One might imagine many different scenarios where closures are specific to fishery, gear, vulnerable habitat, activity other than fishing, a specific fraction of an area, etc., but the underlying principles are similar to these basic fishery examples. The magnitude and curvature of the marginal damage functions relative to marginal benefits will determine whether closure or a reduction in the level of the activity is warranted.

As shown above, the vulnerability of habitat and its linkage to resource productivity are important elements of the problem. The shape and magnitude of the habitat damage function determine, in part, whether MPA creation will increase harvest. If we can determine basic characteristics of these interactions, such as curvature, we can infer the direction or sign of the habitat effect and the resulting impact of the closure on harvest. Another interesting consideration involves threshold levels where habitat impacts may be especially large at specific effort, pollution, or degradation levels. These areas of the damage function will likely determine the potential utility of MPA establishment. These results also hinge on economic elements of the fishery, such as the relative level of fishery costs to the price of outputs, and ecological factors, such as migration rates.

Extensions to this work need to recognize that fish movement is more complex than the density-dependent framework used in this paper. Fish also have different habitat requirements according to life stage and season. Fish habitat and distribution and the related distribution of fishing effort are patchy in nature. Further work needs to superimpose patchiness and the spatial elements of life history attributes.

The concept of habitat may be inadequate when considering these factors and related elements of the biological and physical environment. Fishery resource productivity may also be closely linked to diversity, ecological integrity, and predator-prey relationships. Ultimately, the context of what constitutes habitat may need to be broadened to the community or ecosystem level to include these factors.

References

Barbier, B. E., and I. Strand. 1998. Valuing mangrove-fishery linkages. A case study of Campeche, Mexico. Environmental and Resource Economics 12:151–166.

Brown, G., and J. Roughgarden. 1997. A metapopulation model with private property and a common pool. Ecological Economics 22:65–71.

Conrad, J. M.1999. The bioeconomics of marine sanctuaries. Journal of Bioeconomics 1:205–217.

Fretwell, S. 1972. Populations in a seasonal environment. Princeton University Press, New Jersey.

Gordon, H. S. 1954. Economic theory of a common-property resource: the fishery. Journal of Political Economy 62:124–142.

Hannesson, R. 1998. Marine reserves: what would they accomplish? Marine Resource Economics13:159–170.

Holland, D. S., and R. J. Brazee. 1996. Marine reserves for fisheries management. Marine Resource Economics 11:157–171.

Isard, W. 1972. Ecologic-economic analysis for regional development; some initial explorations with particular reference to recreational resource use and environmental planning. Free Press, New York.

Kahn, R. J. 1987. Measuring the economic damages associated with terrestrial pollution of marine ecosystems. Maine Resource Economics 4:193–209.

Kahn, R. J., and W. M. Kemp. 1985. Economic losses associated with the degradation of an ecosystem: the case of submerged aquatic vegetation in Chesapeake Bay. Journal of Environmental Economics and Management 12: 246–263.

Lindholm, J., B., P. J. Auster, M. Ruth, and L. Kaufman. 2001. Modeling the effects of fishing and implications for the design of marine protected areas: Juvenile fish responses to variations in seafloor habitat. Conservation Biology 15(2):424–437.

Lynne, G. D., P. Conroy, and F. J. Prochaska. 1981. Economic valuation of marsh areas for marine production processes. Journal of Environmental Economics and Management 8:175–186.

MacCall, D. A. 1990. Dynamic geography of marine fish populations. University of Washington Press, Seattle.

NRC (National Research Council). 2002. Effects of trawling and dredging on seafloor habitat. National Academy Press, Washington, D.C.

OECD (Organization for Economic Co-operation and Development). 1997. Towards sustainable fisheries: economic aspects of the management of living marine resources. Organization for Economic Co-operation and Development, Paris.

Pezzey, J. C. V., C. M. Roberts, and B. T. Urdal. 2000. A simple bioeconomic model of a marine reserve. Ecological Economics 33:77–91.

Polacheck. T. 1990. Year around closed areas as a management tool. Natural Resource Modeling 4:327-54.

Pulliam, H. R. 1988. Sources, sinks, and population regulation. American Naturalist 132:652–661.

Sainsbury. K. J. 1988. The ecological basis of multispecies fisheries, and management of a demersal fishery in tropical Australia. Pages 349–375 in J. S. Gulland, editor. Fish population dynamics implications for Management. Wiley, New York.

Sanchirico, J. N., and J. E. Wilen. 2001. A bioeconomic model of marine reserve creation. Journal of Environmental Economics and Management 47:257–276.

Swallow, S. K. 1990. Depletion of the environmental basis for renewable resources: the economics of interdependent renewable and nonrenewable resources. Journal of Environmental Economics and Management 19:281–296.

Swallow, S. K. 1994. Renewable and nonrenewable resource theory applied to coastal agriculture, forest, wetland, and fishery linkages. Marine Resource Economics 9:291–310.

Symposium Abstract

Symposium abstracts have been reprinted from the Abstract Volume prepared for the Symposium on Effects of Fishing Activities on Benthic Habitats: Linking Geology, Biology, Socioeconomics, and Management without any further review or editing.

Development of a West Coast Cooperative Research Program, Working Together toward Better Information

J. BLOESER[1]

Pacific Marine Conservation Council, Arcata, California

The Pacific Marine Conservation Council (PMCC) is presently working with federal and state agencies, scientists and fishermen to develop a West Coast Groundfish Cooperative Research Program. This program will provide a clearinghouse for ongoing cooperative research projects. It will also house the primary source of information on research priorities, funding and contacts for interested scientists and fishermen. The need for systematically combining the expertise of fishermen with the scientific rigor of researchers has clearly emerged in the evolution of our Rockfish Rebuilding Campaign, launched in 2001. Cooperative research programs provide a unique opportunity for those interested in fisheries to collectively resolve complex issues. Through collaboration, federal agencies and fisheries managers benefit from the experience, equipment, and insights of fishermen, while fishermen participate in designing and conducting the research to gather data for superior management of the resource. PMCC strongly believes that sustainable fisheries depend upon implementing standardized methods of collecting, analyzing and applying the 'experiential' data of fishermen to fisheries science. Cooperative research also has the potential to improve communication and trust while elevating the level of scientific understanding.

---

[1] E-mail: Jennifer@pmcc.org

Promoting Environmental Awareness and Developing Conservation Harvesting Technology for the Fishing Industry

G. BROTHERS[1]

Fisheries and Oceans Canada, Fisheries Management Branch, Northwest Atlantic Fisheries Centre, St. John's, Newfoundland, Canada

The Fisheries Diversification Program, a Canada / Newfoundland Cooperation Agreement, has four components. One of which deals with Environmental Awareness and Conservation Technology. During the past two years several projects have been carried out jointly with the fishing industry. They include; Awareness of Gillnet Environmental Impact, Cod by-catch in American Plaice Gillnets, Impact of Scallop Fishing on Lobster Habitat workshop, By-Catch of Juvenile Groundfish and Pelagic in Shrimp Trawls, American Plaice By-Catch on Cod Longlines, Size Selectivity in Yellowtail Bottom Trawls, effect of Chaffing Gear on Codend Selectivity, and Crab by-catches in Scottish Seines. The methodology used and results obtained in each of these projects will be depicted in a poster session. Results obtained during some of the projects have produced changes in the way commercial fishing is carried out and managed.

---

[1] E-mail: Brothersg@dfo-mpo.gc.ca

Occupational Endurance and Contested Resources: Managing the Cultural and Economic Tensions of Lake Michigan's Commercial Fishery

M. J. CHIARAPPA[1]

Western Michigan University, Great Lakes Center for Maritime Studies, Kalamazoo, Michigan

The establishment of commercial fishing in the Great Lakes in the 19th century introduced unprecedented economic considerations and ecological effects to the basin's benthic habitats. By the late 19th/early 20th century, the alarming effects of overfishing this region's benthic speciesÑmost of which were the principal target species (lake whitefish, lake trout, perch, walleye, chub) of the Great Lakes commercial fisheryÑwas increasingly apparent to fisheries scientists and policy makers. The Great Lakes commercial fishery managed to weather these stock fluctuations until the end of World War II when the combined ecological effect of overfishing, sea lamprey predation, and a disproportionate alewife population dangerously reduced benthic dwelling species. With the exception of lake trout, benthic species have recovered over the past 50 years. But commercial fishers, particularly Lake Michigan's, have not necessarily enjoyed the benefits of this recovery. As Great Lakes states revised their fisheries policies to enhance sport fisheries, commercial fishers faced stringent restrictions in their harvesting technologies, in their fishing areas, and in the implementation of quotas. Today, in spite of highly contested policy debates, Lake Michigan's benthic fisheries are comprised of two constituencies that visibly utilize benthic habitats to maintain distinct cultural prerogatives and economic goals. Euroamerican commercial fishers draw on their cultural life, traditional ecological knowledge (TEK), and a legacy of economic adaptation to maximize their limited share of the benthic realm. Native American fishers, motivated by the maintenance of treaty fishing rights, are using the benthic realm for economic empowerment and the recovery of tradition-bound cultural/ecological awareness. Looking at TEK and the formulation of vernacular environmental ethics, this paper will consider the benefits and problems that have accrued to each group's attempt to utilize Lake Michigan's benthic habitat to achieve historical and contemporary goals.

---

[1] E-mail: michael.chiarappa@wmich.edu

Building a Database for Benthic Fisheries Using Tourist Income

G. C. LANE[1]

Clearwater Marine Aquarium, Clearwater Florida

Tourist dollars and programming can be utilized for real science. The Marine Life Adventures Program at the Clearwater Marine Aquarium has been developing programs and curriculums that both teach hands-on science and help contribute to the building of a database of juvenile fisheries in the Clearwater Harbor (St Joseph's Sound) area of western Florida. To date, they have cataloged nearly 150 species in a program that samples benthic habitats twice a day almost every day of the year. At present, there are nearly 100 sample sites that cover the entire range of benthic habitats found in the Harbor. These sites range from a depth of 1.2 to 5.0 meters. The most amazing thing is that all this is cheerfully paid for by tourist dollars and produces a product that is scientifically sound and has been utilized by Universities, Planners, and local governments. We believe that it is the first program of it's kind in the world, and can be a model for cash-strapped and budget hungry institutions to help offset some of the costs of obtaining this type of data.

[1] E-mail: mlatrips@cmaquarium.org

Impacts of Marine Reserves: How Fishermen Behavior Matters

J. E. WILEN[1]

Department of Agricultural and Resource Economics, University of California, Davis, California

By a wide margin, most of what we think we know about the impacts of marine reserves on fisheries has been derived by analytical and simulation modeling rather than with hard empirical evidence. Most of that analytical modeling, in turn, has been done by biologists, focusing on aspects of the system with which they are most familiar, namely biological mechanisms. The most important findings derived from biological modeling of marine reserves are that dispersal mechanisms are critical to the kinds and magnitudes of impacts of closed areas. These findings have largely been derived from models that make simplifying assumptions about fishing mortality. This paper argues that the spatial distribution of fishing mortality is as important as biological dispersal mechanisms to the ultimate impact of reserves. Moreover, the spatial distribution of fishing effort is determined by economically motivated decisions not typically considered by biological modelers. We predict, using data from the Northern California red sea urchin fishery, how the distribution of effort is likely to change in response to closed areas, and how that behavioral response is important to the ultimate impacts of closed areas. We argue that failure to account for the economic determinants of fishing effort bias conclusions about reserves, and we discuss the nature of those biases.

[1] E-mail: wilen@primal.ucdavis.edu

Determinations of "To the Extent Practicable" Phrase in U.S. Law and Other Legal Issues Concerning Fishing Effects

Symposium Abstract

Symposium abstracts have been reprinted from the Abstract Volume prepared for the Symposium on Effects of Fishing Activities on Benthic Habitats: Linking Geology, Biology, Socioeconomics, and Management without any further review or editing.

The Legal Requirement to Address Fishing Effects on Essential Fish Habitat: Thresholds, Qualifiers, and the Burden of Proof

A. Rieser[1]

University of Maine School of Law, Portland, Maine

The 1996 Sustainable Fisheries Act (SFA) requires fisheries managers to describe and identify the essential fish habitat of all managed fish stocks in the U.S. EEZ. The law also requires managers to 'minimize to the extent practicable adverse effects on such habitat caused by fishing.' In addition to the practicability language, the SFA's fishing effects mandate is qualified by the Act's overall requirement that conservation and management measures 'shall be based upon the best scientific information available.' The National Marine Fisheries Service's (NMFS) 1997 EFH guidelines, requiring councils to mitigate when there was evidence of 'identifiable adverse effects,' gave little additional understanding of the threshold for 'adverse effects' that would trigger the need for mitigation measures. In the first generation of EFH amendments, the practicability standard did not come into play because most councils concluded there was inadequate scientific information regarding the seafloors and fishing gear impacts in their region to warrant the development of mitigation measures, whether practicable or not. Conservation groups challenged the Secretary's approval of five of the regional councils' EFH amendments, claiming the councils had not taken into account the growing scientific consensus that bottom trawling and dredging can have significant ecological effects. Government lawyers defending the Secretary's action convinced the court that the amendments were adequate because the agency interpreted the SFA's EFH and 'best available science' standards to require site-specific scientific information that particular fishing practices or gears are having identifiable impacts on particular habitats within the council's region. In deferring to this interpretation, the court thus established that the 1996 provisions require more detailed scientific evidence to cross the threshold for 'adverse effects' than was then available for each region. More important, however, was the court's conclusion that the decision-making process for approving the EFH amendments was deficient under the National Environmental Policy Act (NEPA) because NMFS failed to take a 'hard look' at the environmental consequences of fishing practices and gears. NEPA required the agency to analyze a broader range of feasible alternatives for protecting EFH from fishing activities than the status quo alternatives recommended by the councils. Under a court-approved settlement agreement, NMFS and the councils must prepare new environmental impact statements that will improve their EFH identifications and consider a range of alternative approaches to the fishing effects question. This second generation of EFH actions is likely to differ significantly from the first for several reasons. First, NMFS has directed the councils to consider all scientific information currently available regarding fishing effects, a body of literature that has increased considerably since 1997-98 when the first EFH amendments were prepared and approved. Moreover, NEPA does not have a 'best available science' requirement, a standard NMFS used implicitly to justify its limited efforts to require protective EFH amendments. The councils will also need to consider the National Academy of Sciences' 2002 report, 'The Effects of Trawling and Dredging on Seafloor Habitat,' which concludes that seafloor habitat should and can be effectively protected from gear impacts in the absence of site-specific information. The report describes a comparative risk analysis process that NMFS and the councils can use in the face of scientific uncertainty. This risk analysis process is a form of structured decision-making the National Academy of Sciences' panel on science and the Endangered Species Act recommended agencies use when conservation decisions must be made with incomplete information and where conflicting social values are at play. There is also a strong possibility that Congress will soon amend the EFH and fishing effects mandate as well as the 'best available science' requirement. The 107th Congress is considering competing bills that would define in more detail how science-based fishery management decisions are to be made in the face of uncertain scientific information. Thus, the next round of EFH amendments and bycatch provisions are likely to be reviewed for approval under very different thresholds, qualifiers, and burdens of proof.

---

[1] E-mail: rieser@maine.edu

Minimzing the Adverse Effects of Fishing on Benthic Habitats: Alternate Fishing Techniques and Policies

Impacts of Fishing Activities on Benthic Habitat and Carrying Capacity: Approaches to Assessing and Managing Risk

MICHAEL J. FOGARTY[1]

National Oceanic and Atmospheric Administration, National Marine Fisheries Service, Northeast Fisheries Science Center, Woods Hole, Massachusetts 02543, USA

Abstract. The need to consider the direct and indirect effects of fishing activities on the productivity of marine populations, communities, and ecosystems is now widely appreciated. Fishery management strategies have traditionally centered on controlling the direct effects of harvesting on the productivity of exploited stocks. Recognition of the need to consider the broader ecosystem effects of fishing has focused attention on management strategies also designed to preserve vulnerable habitats, conserve biodiversity, and protect ecosystem goods and services. Here, it is shown that if fishing activity simultaneously results in removal of target species and in habitat degradation affecting carrying capacity, fundamental changes in the shape of the production function result, and the corresponding biological reference points are altered relative to the case where fishing only results in removal of biomass. Further, the time scales of recovery (or indeed whether recovery is possible) from an over-exploited state may depend substantially on the recovery of habitat and benthic productivity. If these effects remain unrecognized, the risk to the population(s) and the ecosystem as a result of fishing are exacerbated. Managing risk in this context entails evaluating alternative hypotheses concerning the habitat effects of fishing and tailoring management approaches to the specific risks engendered by different fishing practices. Tactical management strategies available to address these risks include (1) overall effort controls, (2) changes in gear design or fishing practices, and (3) use of various forms of spatial management strategies including marine protected areas and regions zoned for the use of particular gear types only. A full evaluation of risk will involve assessment of additional stressors in relation to fishing-related impacts on marine ecosystems.

Introduction

Marine conservation and resource management addresses a broad spectrum of objectives, including preservation of biodiversity, maintenance of ecosystem structure and function, habitat protection, and sustainable resource use. The development of management techniques to meet these diverse requirements presents difficult challenges. Research efforts to understand the direct and indirect effects of fishing on marine ecosystems have intensified with the goal of developing ecosystem-based approaches to fishery management. The importance of habitat-related considerations in these efforts is reflected both in the rapidly expanding scientific literature on this topic (Jones 1992; Dayton et al. 1995; Jennings and Kaiser 1998; Thrush et al. 1998; Auster and Langton 1999; Hall 1999; Langton and Auster 1999; Kaiser and de Groot 2000; Kaiser et al. 2003) and legislative mandates such as the essential fish habitat provisions of the Sustainable Fisheries Act of 1996 in the United States (Benaka 1999; Fluharty 2000; Rosenberg et al. 2000).

The effects of fishing on marine ecosystems have elicited concerns dating back to at least the 14th century (Anonymous 1921; Sahrhage and Lundbeck 1992). In Britain, acts passed in Parliament in 1350 and 1371 called for strict observance of "ancient" statutes directed to the preservation of the early life stages of fishes (Anonymous 1921). In 1376, a petition was placed before the Commons to ban the Wondyrchoun, a type of trawl/dredge, because "the great and long iron of the Wondyrchoun presses so hard on the ground when fishing that it destroys the . . . plants growing on the bottom under the water, and also the spat of oysters, mussels, and other fish, by which the large fish are accustomed to live and be nourished" (Anonymous 1921). In 1499, the use of trawls was prohibited in Flanders in response to observations that "the trawl scraped and ripped up everything it passed

---

[1] E-mail: mfogarty@whsun1.wh.whoi.edu

over in such a way that it rooted up and swept away the seaweeds which served to shelter the fish; it robbed the beds of their spawn or fry."

These statements reflect a remarkably intuitive understanding of ecological factors potentially affecting fish production including both trophic and habitat-related considerations. Further evidence of longstanding concern over fishing practices deemed to be destructive to habitat includes bans imposed on the use of trawls in estuarine waters in the Netherlands in 1583 and in Britain in 1631 where the use of "traules [sic] [was] forbidden as well as of other nets which shall not have the meshes of the size fixed by law and orders," and the establishment of trawling as a capital offense in France in 1584 (Anonymous 1921).

The scientific study of fishing impacts on habitat and benthic communities, in contrast, is comparatively recent and rigorous comparisons between fished and nonfished areas with respect to habitat impacts were not initiated until the mid-20th century (e.g., Graham 1955). The impact of different types of fishing gears on marine ecosystems remains an important source of concern that has intensified with the development of new harvesting technologies and expansion of areas subject to intense harvesting activity.

Specification of quantitative objectives for fishery management has traditionally centered on the direct effects of removal of biomass from the population(s). Although broader community-level impacts of harvesting mediated through trophic interactions have also been incorporated in management-oriented models (e.g., Anderson and Ursin 1977; May et al. 1979; Mercer 1982; May 1984; Daan and Sissenwine 1991), factors related to habitat and productivity have received much less attention in the development of management models. Sainsbury (1988, 1991) explicitly incorporated the effects of habitat availability in production models for fish assemblages on the northwestern continental shelf of Australia. MacCall (1990) considered mechanisms of habitat selection and the implications for the development of spatially explicit harvesting strategies. Walters and Juanes (1993) developed models of predation risk in relation to habitat complexity and availability. Saila et al. (1993) constructed models of the effects of destructive fishing practices on productivity and biodiversity of coral reef systems. Hayes et al. (1996) explored the linkages between habitat, recruitment processes, and the implications for management. Habitat-related considerations now also have been directly incorporated in models of marine protected areas (e.g., Mangel 2000; Lindholm et al. 2001; Rodwell et al. 2003).

Recognition of the critical role of habitat is reflected in the specification of the essential fish habitat (EFH) requirement of the Sustainable Fisheries Act (Public Law 104-297, 104th Congress, 11 October 1996). Identification and protection of EFH is required under the act. Essential fish habitat is defined as

> those waters and substrates necessary to fish for spawning, breeding, feeding or growth to maturity. For the purposes of interpreting the definition of essential fish habitat, "Waters" include aquatic areas and the associated physical, chemical, and biological properties that are used by fish and may include aquatic areas historically used by fish where appropriate; "substrate" includes sediment, hard bottom, structures underlying the waters, and associated biological communities; "necessary" means the habitat required to support a sustainable fishery and the managed species; contribution to a healthy ecosystem and "spawning, breeding, feeding or growth to maturity" covers a species full life cycle. (SFA)

This broad definition provides a clear recognition of the importance of habitat in fishery production. The act further sets qualitative guidelines for protective action. The critical step in identifying essential fish habitat is keyed to the nature of the available information. Four levels of information have been recognized for this purpose, in increasing level of resolution and relevance to management (Table 1). The first two levels reflect the recognition that habitat requirements shape the distribution patterns of marine organisms and that inferences on habitat requirements can be drawn from observed patterns of abundance in space and time. The second two levels reflect the recognition that a full understanding of the importance of habitat to marine populations requires information on how habitat affects the components of production (recruitment, growth, and mortality). Traditional forms of fishery management advice are linked explicitly to the concept of biological production. Accordingly, a full specification of habitat effects on fisheries will require information at these upper information tiers to permit the development of management strategies with explicit consideration of habitat structure and function.

Scientific uncertainty concerning the effects of harvesting at the population, community, and ecosystem levels translates to risk which must be considered in the development of management strategies for marine systems (e.g., Fogarty et al. 1992, 1996). If the production characteristics of the stock depend on the relationship between fishing, habitat, and carrying capacity, and these factors are ignored, the risk to the population is potentially high. Here, I explore the implications of habitat dependence on fishery productivity in relation to the establishment of biological reference points and assessment of risk to the population. I focus specifically on the case where fishing activity results directly in habitat loss affecting the production of target species. Impacts on

Table 1. Specification of information levels for essential fish habitat under the Sustainable Fisheries Act of 1996 in the United States.

Information level	Specification
Level 1	Distribution data are available for some or all of the geographical range of the species. Either systematic presence–absence data or opportunistic observations of the location of various life stages may be used to infer habitat use.
Level 2	Habitat-related densities of the species are available. Geographical information on the density or relative abundance of a species at each life stage may be used to assess habitat value compared with the overall species distribution.
Level 3	Growth, reproduction, or survival rates within habitats are available. The success of the species in a given habitat, based on growth, reproduction, and survival rates, is used as a proxy for productivity.
Level 4	Production rates by habitat are available. Direct assessments of production rates as a function of habitat type. Location quality and quantity are used to determine the habitat essential for a sustainable fishery and for the species contribution to a healthy ecosystem

carrying capacity for a target resource species are used as a vehicle to examine the management implications of habitat loss and degradation. I argue that it is critical that habitat-related considerations be framed in terms of production and biological reference points if this information is to be fully integrated into management. It is recognized that full specification of an ecosystem-based approach to fishery management will entail a much broader set of considerations concerning fishing effects on marine systems, including impacts on nontarget species in general and the potential for changes in trophic structure in response to species-selective harvesting patterns. The narrower focus adopted here is intended to illustrate just one aspect of the requirements for ecosystem-based management.

Carrying Capacity and Biological Reference Points

Effective integration of information on the importance of habitat in a fishery management context entails the development of conceptual and analytical models to explore the effects of fishing on productivity of a resource species. Harvesting activity can affect habitat-related productivity in several distinct ways, ranging from reduction in structural complexity to effects on benthic production and prey availability to benthivores (e.g., Jennings and Kaiser 1998; Hall 1999; Kaiser et al. 2003). The former includes consideration of loss of shelter-providing structures with attendant effects on survivorship at one or more life history stages while the latter concerns patterns of energy flow and availability to higher trophic levels. If fishing activity itself affects the carrying capacity of the system, the development of biological reference points such as the level of maximum sustainable yield (MSY) and the corresponding level of fishing effort resulting in MSY (f_{MSY}) are directly affected. Specification of these reference points (or suitable proxies) to guide fishery management are required under the provisions of the Magnuson-Stevens Fishery Management and Conservation Act in the United States and form the basis for management in many other national and international regulatory frameworks.

In the following, a simple heuristic model will be used to illustrate key issues in the specification of biological reference points for management when fishing activity directly affects habitat and carrying capacity. The logistic production model provides the point of departure:

$$B_{t+1} = \left[1 + r\left(1 - \frac{B_t}{K}\right)\right]B_t - qf_tB_t,$$

where B_t is the biomass at time t, r is the intrinsic rate of increase of the population, K is the carrying capacity or equilibrium virgin population biomass, q is the catchability coefficient, and f_t is the standardized fishing effort at time t. The following analysis builds on that of Sainsbury (1988, 1991) to explore the equilibrium relationship between yield and fishing effort and related reference points when harvesting directly affects the carrying capacity of the target species. The direct and indirect effects of harvesting on habitat and community structure may also affect the intrinsic rate of increase of the population with direct implications for the shape of the production function and biological reference points

(MacCall 1990; Fogarty 2000) but will not be further explored here.

Consider the case where carrying capacity declines as a function of fishing effort according to

$$K_t = K_o e^{-\alpha f_t},$$

where K_o is the carrying capacity in the absence of fishing and α is a coefficient reflecting the impact of fishing effort on carrying capacity. A unit of fishing effort results in both removal of biomass as catch and reduction in carrying capacity. This relationship implies that degradation of the carrying capacity is initially rapid and then decreases. This is consistent with a mechanism where repeated passes of fishing gear over habitat have a progressively lesser effect on carrying capacity (Figure 1a). In some instances, fishers may deliberately attempt to reduce structural complexity of benthic systems in the initial passes of the gear.

Substitution of this expression into the logistic production model above results in a fundamental change in the shape of the production function and in the resulting biological reference points. The equilibrium relationship between yield (Y) and fishing effort is given by

$$Y = qK_o e^{-\alpha f_t} \left[\frac{r - qf}{r} \right] f.$$

If fishing activity results in a degradation of the carrying capacity as specified above, not only is the level of sustainable yield reduced, but the level of fishing effort resulting in maximum sustainable yield is lower relative to the case where carrying capacity is not impacted (Figure 1b). This stands in marked contrast to the case where the carrying capacity changes due to factors other than fishing (Figure 1c), for example, as a shift in climatic conditions or degradation of habitat due to anthropogenic impacts other than fishing itself. In this case, the level of sustainable yield decreases under a less favorable environmental regime or degraded habitat conditions, but the f_{MSY} level is unchanged.

Alternative specifications for the effect of fishing on habitat degradation may be necessary to accommodate patterns such as threshold effects or other trajectories of habitat change within increasing fishing effort. For example, if the loss in carrying capacity is initially low but accelerates before again declining, a model such as

$$K_t = \frac{K_o}{1 + e^{-b + af_t}}$$

(where b and a are coefficients) may be appropriate (Figure 2a). Substitution of this logistic model into the production function again results in a shift in the position of the limit

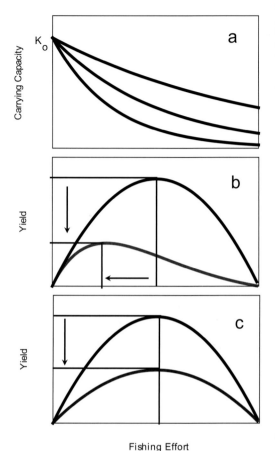

Figure 1. (a) Loss in carrying capacity with increasing fishing effort for three levels of harvesting impact on fishing activity; (b) comparison of production models for standard logistic harvesting model (upper curve) and production model (lower curve) with exponential decay in carrying capacity with fishing activity; and (c) comparison of production function for logistic harvesting model under two levels of carrying capacity with changes unrelated to fishing activity. Arrows indicate direction of change in maximum sustainable yield (MSY) and level of fishing effort resulting in MSY (f_{MSY}).

reference point relative to the case where fishing does not alter carrying capacity. The equilibrium yield is given by

$$Y = \frac{qK_o}{1 + e^{-b + af}} \left[\frac{r - qf}{r} \right] f$$

(Figure 2b). This production curve differs from that incorporating the exponential model in exhibiting a more rapid decline in yield at moderate-high levels of fishing effort (c.f. Figure 1b).

The exposition above shows that explicit consider-

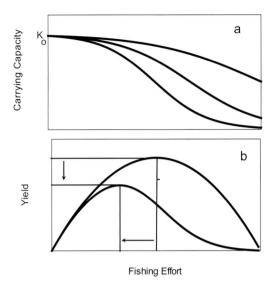

Figure 2. (a) Change in carrying capacity with increasing fishing effort with logistic decay function and (b) comparison of production models for standard logistic harvesting model (upper curve) and production model (lower curve) with logistic decay in carrying capacity with fishing activity. Arrows indicate direction of change in maximum sustainable yield (MSY) and evel of fishing effort resulting in MSY (f_{MSY}) from standard logistic model to habitat model. K_o = carrying capacity in the absence of fishing.

ation of habitat effects of fishing results in changes in key biological reference points. Inferences concerning the recovery of over-exploited populations are also directly affected. Several critical factors will determine the time scale of recovery (or, indeed, whether recovery is possible). First, if the damage to habitat and carrying capacity is caused through loss of biogenic structures or impacts on benthic communities which provide important prey resources for the target species, the time scale of recovery will be a function of the generation time(s) of the these organisms and the dynamics of the target species. If the generation time of the organisms providing biogenic structure is greater than that of the resource species, the dynamics of the former will dominate the recovery process. For some shelter-providing species such as corals, the recovery time may be very long.

When geological features are disturbed, recovery may not be possible if structures are permanently destroyed (e.g., disruption of boulder fields or cobble beds). In other instances, the time scales of geological processes will dominate. For example, if bedforms such as sand waves are modified by mobile fishing gear, their rate of formation will play a critical role in recovery times of resource species depending on these structures.

Structures of this type can be important in the overall expenditure of energy of fishes with potential consequences for growth and reproductive output. These considerations are very different than those under traditional models of exploited species where inferences concerning recovery times are based on the dynamics of the resource species alone and where recovery is implicitly assumed to always be possible at rates governed by the internal dynamics of the target species (unless alternate stable states exist for the target species [e.g., Collie and Spencer 1994]).

In cases where recovery is possible but is dominated by the dynamics of the habitat system (e.g., the generation time of structure-forming organisms or geological processes controlling bedforms), it will be necessary to modify the production models described above to reflect time delays in recovery from fishing effects on habitat. A full analytical treatment of this problem will entail modeling the population dynamics of the shelter-forming organisms (see Sainsbury 1988, 1991) or the geological processes as appropriate. An example of a simple stochastic simulation to reflect time delays in rebuilding carrying capacity is provided in Figure 3. In cases where the damage to the habitat is permanent or species with extremely long generation times are affected, discernible recovery may not be evident on the time scales (annual to decadal) typically considered in fishery management.

Uncertainty and Risk

If fishing directly affects carrying capacity but this effect is unrecognized, the resulting model mis-specification will contribute to risk to the population because of the possibility of inadvertently setting fishing mortality rates too high. For consistency with the arguments developed above, risk is here defined as the probability that carrying capacity and production will be degraded by fishing below some specified acceptable level. If it is assumed that recovery from over-exploitation is possible as in traditional harvesting models but, in fact, long-term or permanent damage is done to the habitat, the risks are substantially exacerbated because the consequences of errors in specification of management limits are much greater.

In the absence of direct information on habitat effects of fishing, it may be difficult to distinguish among alternative hypotheses concerning fishing effects on carrying capacity given the uncertainties that result from the intrinsic variability in fish populations and measurement errors in fishery and population parameters. An illustration of these issues is provided in Figure 4, where two alternative models are examined for a hypothetical

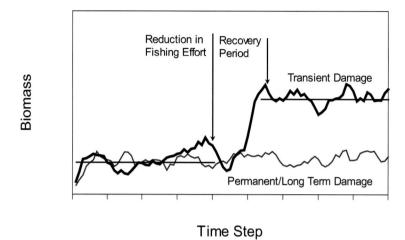

Figure 3. Biomass trajectories in stochastic simulation incorporating habitat effect of fishing when there is a delay of 10 time units due to the generation time of benthic organisms providing shelter (thick solid line) and the recovery period for this case and for the case where long-term or permanent damage is done to structural features of the habitat and recovery either does not occur or occurs on very long time scales (thin solid line). The effect of a reduction in fishing effort mid-way through the sequence is illustrated.

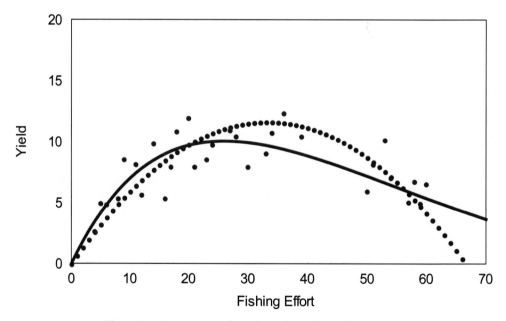

Figure 4. Comparison of two hypotheses (models) of the relationship between equilibrium yield and fishing effort fit to a simulated series generated by adding random variation to a model incorporating harvesting effects on carrying capacity (solid line). The dotted line is the fit using a logistic harvesting model with no consideration of fishing effects on habitat.

series of yield and fishing effort data. The models represent situations where (1) fishing affects carrying capacity according to the exponential decay model and (2) fishing has no impact on K. The corresponding estimates of MSY, f_{MSY}, and F_{MSY} for these models are provided in Table 2. If the potential for habitat effects of fishing are unrecognized and carrying capacity is actually affected by fishing effort, the potential for setting the limiting level of fishing effort (f_{MSY}) too high is great. In this hypothetical example, f_{MSY} would be set 31% too high and F_{MSY} would be set too high by nearly 100%. The MSY level would be set 17% too high. In the context of risk assessment, the choice of which model formulation to accept will be guided by diagnostic characteristics of the model fits and decisions concerning the degree of risk aversion to be adopted. Clearly, the most risk-averse approach would be to assume some form of habitat impact of fishing activity. The use of a generalized production function, such as the Pella and Tomlinson (1969; Quinn and Deriso 1999) model, could implicitly capture the changes in production characteristics due to changes in fishing effects on carrying capacity to reduce the potential for mis-specification of biological reference points. However, the problem of dealing with the recovery dynamics and whether fishing effects are reversible would remain.

Risk Evaluation

A quantitative evaluation of risk in the context of the models described above would entail specification of the form of the impact of fishing on carrying capacity (exponential, logistic, etc.) and estimation of the appropriate parameters relating fishing effort to K. The simple heuristic models above consider a single fishery and habitat. In practice, it will be necessary to consider multiple gear types operating over a number of habitat types. The potential for habitat damage by fishing gear is a function of the habitat characteristics (structural complexity, natural disturbance regimes, associated organisms, etc.), the type of fishing gear employed, and the levels of fishing effort exerted. The development of a habitat classification scheme that can accommodate an assessment of vulnerability to disturbance by fishing gear is a necessary prerequisite for further analysis. Auster (1998) provided qualitative assessment of the relative impact of mobile fishing gears (trawls and dredges) with consideration of the level of fishing activity and the vulnerability of the habitats on flat sand and mud, sand waves, biogenic structures, shell aggregates, pebble-cobble, pebble-cobble with sponge cover, partially buried or dispersed boulders, and piled boulders habitats. The National Research Council (NRC 2002) synthesis examined effects of fishing on selected habitat types including submerged aquatic vegetation, sand–hard–biogenic (including coral and oyster reefs), muddy–sand, and gravel. Gear types considered included scallop dredges, oyster dredges, otter trawls, beam trawls, roller-rigged otter trawls, and roller frame shrimp trawls (see Table 3.1 of NRC [2002] for specific impacts in different habitats). Stephan et al. (2000) assessed the potential impacts of different gear types on submerged aquatic vegetation, a critical habitat at one or more life history stages for many exploited marine species.

The development of information matrices of this type is an essential step in evaluating the potential risk to habitat under different types of fishing activity including damage to the physical structure of the environment and impacts on biotic components of the system. Further development of risk assessment requires an evaluation of the differential impacts of fishing gears in different environments that can be represented in quantitative form or at least rank order of importance. An assessment of this type was made for the northeastern continental shelf of the United States for sand, mud, and gravel habitats for dominant gear types in this region including hydraulic and scallop dredges, otter trawls, sink gill nets, bottom longlines, pots, and traps (Anonymous 2002a). This evaluation further allowed consideration of natural disturbance

Table 2. Biological reference points for two hypotheses concerning production characteristics in relation to fishing activity: Fishing (1) does impact carrying capacity of the target species and (2) has no impact on carrying capacity. For this hypothetical case, the production model for hypothesis 1 has the parameters $r = 1.0$, $K = 100$ units, and $q = 0.01$. For hypothesis 2, the parameters are $r = 1.0$, $K = 70$ units, and $q = 0.015$. MSY = maximum sustainable yield; f_{MSY} = level of fishing effort resulting in MSY; and F_{MSY} = fishing mortality rate resulting in MSY.

Reference point	Habitat hypothesis 1	Habitat hypothesis 2	Percent difference
MSY	10.0	11.7	17.0
f_{MSY}	26	34	30.8
F_{MSY}	0.26	0.51	96.2

regimes (high and low energy environments; see Table 3). In the following, I will use this assessment for the northeastern region to illustrate the main points in risk evaluation and management. Further information on habitat effects of mobile fishing gear in this region is provided by Auster et al. (1996).

Hydraulic dredges used to harvest bivalves are employed principally on sand substrates, and the impact on biological and physical structures is considered to be potentially high (Table 3). These dredges utilize high pressure water jets to fluidize sediments and facilitate capture of infaunal bivalves. Scallop dredges deployed in sand habitats were judged to have high impact on both physical and biological structures in low energy environments. However, impact on biological structure in high energy environments was thought to be low because of adaptations of organisms in these environments to high rates of natural disturbance. Such an evaluation must, of course, be made on a case-by-case basis in relation to benthic community composition (M. Cryer, National Institute of Water and Atmosphere, personal communication).

Otter trawl impacts were considered to be potentially high on gravel substrates and associated communities. Moderate effects in sand were projected, except for high energy environments, where the effect on physical structures was scored as low. In mud habitats, trawls were considered to have high potential impact on physical structure in high energy environments and moderate effects on biological and physical features in low energy systems.

Static fishing gears, including sink gill nets, bottom longlines, pots, and traps, were judged to have relatively low impact in all habitats because the ecological footprint of these gear types is relatively small relative to that of mobile fishing gear. Habitat damage can occur with these gear types, particularly during setting and retrieval, and other impacts such as entanglement of nontarget species such as marine mammals must be carefully considered.

Evaluations of this type can provide the basis for devising tactical management strategies using tools appropriate to the assessment of risk to different populations and habitats. The simple visual presentation of an increasing gradation of risk attributable to different gear types in various habitats (Table 3) can be a useful device in conveying potential fishing impacts and options to managers.

Table 3. Assessment of impact of different gear types in mud, sand, and gravel substrates in the Northeastern United States based on expert opinion and evaluation of existing information in the literature (Anonymous 2002a). The potential impacts for biological and physical structures are scored separately on a scale of 1 (lowest impact) to 3 (highest impact). Effects in sand and gravel habitats are further partitioned in low (L) and high (H) energy environments for scallop dredge and otter trawl impacts. Cells with no number code were not scored in the exercise.

Habitat	Gear type				
	Hydraulic dredge	Scallop dredge	Otter trawl	Gill net or longline	Pot or trap
Mud					
Biological structure				1	1
Low energy			2		
High energy					
Physical structure				1	1
Low energy			2		
High energy			3		
Sand					
Biological structure	3			1	1
Low energy		3	2		
High energy		1	2		
Physical structure	3			1	1
Low energy		3	2		
High energy		3	1		
Gravel					
Biological structure				1	1
Low energy		3	3		
High energy			3		
Physical structure				1	1
Low energy		3	3		
High energy			3		

Risk Management

Following the assessment of risks of application of different gear types in different environments, evaluation of different management options designed to reduce risk is necessary. The management tools available to minimize impacts of fishing activities on habitat include reductions in overall fishing effort, modifications to fishing gears or fishing practices to reduce seabed impacts, forms of spatial management including zonal application of fishing gears in which the use of certain gears is restricted in some habitats, and the designation of marine reserves in which all extractive activities would be prohibited.

Control of Fishing Effort

Reduction in overall levels of fishing effort has the potential to reduce habitat damage while also addressing other issues such as economic efficiency and optimizing biological production of exploited resources. Effort reductions can be affected by the imposition of measures to reduce overcapitalization through programs designed to reduce participation in the fishery, imposition of constraints on fishing time through fishing seasons or limitations on days at sea, and other measures (NRC 2002). Clearly, the value of reduction in fishing effort for the purpose of habitat protection will depend on whether recovery from fishing-induced damage is possible. Differential allocation of permissible fishing effort (e.g., total allowable trawl hours) in habitats of different vulnerabilities is under consideration as a management option (T. Noji, National Marine Fisheries Service, personal communication).

In the context of the production models considered earlier, the effect of effort reductions on habitat will be strongly dependent on the nature of the effort–carrying capacity relationship. For example, if carrying capacity declines according to the exponential model described above, the proportional change in habitat increases at a decreasing rate at high levels of fishing effort (Figure 5). Under these conditions, and depending on the magnitude of the impact parameter (a), large-scale decreases in fishing effort may be necessary to effect habitat protection. If further spatial controls are not implemented, impacts on vulnerable habitats may be exacerbated. Effort reductions can potentially provide beneficial effects by reducing expansion into previously unexploited or lightly exploited regions (NRC 2002). These regions may be serving as de facto refuge areas. In the case where some form of spatial management is employed (see below), accompanying effort reductions will be valuable to prevent displacement of effort from a closed area to remaining open fishing zones with adverse consequences for the system as a whole.

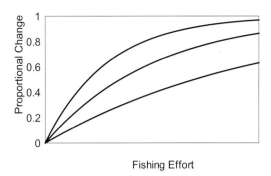

Figure 5. Proportional change in carrying capacity as a function of changes in fishing effort for three values of the habitat-impact parameter α of the exponential model.

Gear Design, Modification, and Use

Conservation engineering has traditionally focused on approaches to improve the selectivity of fishing gear to minimize the retention of small fish of the target species and of nontarget organisms (e.g., Van Marlen 2000; Fonteyne and Polet 2002). Attention to reducing impacts of fishing gear on structural features of the habitat itself is comparatively recent. For species in which catches are not dependent on the herding effect of the disturbance caused by otter doors and ground cables, it is possible to use soft trawl doors in which fabric kites are employed to replace the heavy traditional otter doors and, thereby, reduce seabed disturbance (Anonymous 2002b). The use of soft trawl doors eliminates the need for ground cables and bridles which also result in habitat disturbance in conventional trawls. Fabric devices incorporated into the body of the net can also be used to generate lift to reduce contact of the net itself on the seafloor. Sweepless raised footrope trawls have also been designed to reduce bottom contact and catch of nontarget species (Pol 2000; M. V. Pol, H. O. Milliken, H. Y. McBride, and H. A. Carr, Massachusetts Division of Marine Fisheries, unpublished). Attempts to replace ground cables and chains on otter trawls with other devices designed to stir target species from benthic habitats using electric fields and high pressure water jets have received some attention (Van Marlen 2000) but have not yet proved practicable. These techniques and others (e.g., use of sound) deserve further evaluation (Valdemarsen and Suuronen 2003).

Alteration in fishing practices such as changes in towing speed and duration can also be employed to reduce habitat impacts (NRC 2002). Where feasible, the use of pelagic trawls rather than bottom-tending otter trawls can be effective. In Alaska's fishery for walleye pollock *Theragra chalcogramma*, the largest single-species fishery in the United States, allocation strategies favoring pelagic trawl gear were used to reduce bycatch of king crabs (*Paralithodes* spp.) and tanner crabs (*Chionoecetes* spp.) and in response to concerns over habitat damage by bottom trawls (North Pacific Fishery Management Council 1999).

Finally, the use of precise habitat mapping information derived from side-scan sonar and other sources, coupled with controls on fishing effort and catches, has proven effective in minimizing impacts on bottom communities in the scallop fishery on Brown's Bank off Nova Scotia (Manson and Todd 2000) by allowing finer-scale targeting of effort to high-density locations. Acoustic assessment and identification of habitat types can be used in conjunction with existing high precision satellite navigation systems to allow placement of effort near, but not directly on, sensitive and highly productive habitats (Valdemarsen and Suuronen 2003).

Zonal Application of Fishing Gear

The assessment of risk associated with different fishing gears in different habitats can be used to consider strategies of zoned use. Examples of zoned gear usage to minimize conflict between different gear and fleet sectors have also demonstrated beneficial secondary impacts on habitat (e.g., Kaiser et al. 1999). Prohibition of certain gears in vulnerable habitats (e.g., nursery habitats and coral habitats) as a protective measure is a common feature of many management systems. Considerations derived from the risk matrix described earlier (Table 3) will be used to illustrate some key points.

The high risk associated with the use of otter trawls in gravel habitats (Table 3) indicates the importance of restricting use of trawls in this habitat. Vulnerable gravel habitats on Georges Bank are currently designated as Habitat Areas of Particular Concern in recognition of the importance of these areas for survival and growth of juvenile Atlantic cod *Gadus morhua* and haddock *Melanogrammus aeglefinus* (Lindholm 1999). The level of risk associated with the use of otter trawls in sand and mud habitats is considered to be less (Table 3), and other management options for otter trawl fisheries could be considered in these habitats. It is interesting to note that as early as the turn of the 20th century, Alexander et al. (1915) had proposed that otter trawling be restricted to areas south of 41 degrees off the northeastern United States, which are dominated by sand and mud habitats.

High potential impacts of scallop gear in sand and gravel habitats and for hydraulic dredges in sand also suggest the utility of spatial restrictions. Because the impacts are judged to be high within the dominant habitats exploited using these gears, the use of marine reserves might be the preferred management option. If it can be shown that recovery rates are relatively rapid in unconsolidated sand habitats, it might be possible to devise an effective strategy employing a rotational closure system where fishing effort is sequentially deployed in restricted areas followed by closures on time scales determined by the recovery dynamics of the physical and biological components of the system.

The risk of habitat damage using static fishing gears was judged to be relatively low (Table 3), and spatial restrictions may not be necessary. It is possible that use of these gear types as a replacement for mobile, bottom-tending fishing gears under certain circumstances would be desirable. For example, the use of fish traps could be evaluated in areas where trawling might be restricted. As noted above, any such switch must also be carefully evaluated with respect to potential impacts on other components of the ecosystem such as protected species.

Marine Reserves

The use of marine reserves has generated considerable recent interest as a tool for marine conservation and resource management (e.g., Fogarty 1999; Murray et al. 1999; Fogarty et al. 2000). The use of closed areas as a fishery management tool has an extensive history, and the current focus on marine reserves can be viewed as an extension of these longstanding approaches to meet broader conservation objectives. The use of no-take marine reserves has been advocated as a hedge against uncertainty in our understanding of ecosystem structure and function (Lauck et al. 1998) and in our ability to control harvest rates (Mangel 1998). The use of marine reserves holds the potential to reduce fishing mortality rates, protect habitat, and preserve ecosystem integrity. While there is substantial evidence of increases in biomass, mean size, and biodiversity within reserves, the effects on adjacent areas through spillover effects are less well documented, although information is accruing. The importance of reserves as source areas for adjacent sites open to harvesting and other activities can be critical for the resilience of these systems. However, transfer rates between open and closed areas through larval dispersal and other mechanisms can be difficult to determine and are subject to considerable uncertainty.

In the context of the production models employed above, it is necessary to expand the model structure to include transfer rates between areas or stock components to fully evaluate the potential utility of a reserve. To illustrate, consider the simple case where the population comprises two components, a segment that is closed to fishing and that exports biomass to a recipient population segment that is open to harvesting. The biomass in the reserve area at time $t + 1$ ($B_{1,t+1}$) is then

$$B_{1,t+1} = \left[1 + r_1\left(1 - \frac{B_{1,t}}{K_1}\right)\right]B_{1,t} - \delta B_{1,t}$$

where δ is the exchange rate between the reserve and the open area, and all other terms are defined as before, after subscripting for the spatial component. The biomass in the area subject to exploitation is given by

$$B_{2,t+1} = \left[1 + r_2\left(1 - \frac{B_{2,t}}{K_2}\right)\right]B_{2,t} + \delta B_{2,t} - qf_t B_{2,t}$$

where, again, all terms are subscripted for area and defined as above. In the following, I will consider the case where the carrying capacity is directly affected by fishing according to the exponential decay model

$$K_{2,t} = K_{2,o} e^{-\alpha f_t}$$

where, again, all terms are defined as before, after specifying subscripts to designate area.

If both population components have identical intrinsic rates of increase and carrying capacity in the absence of harvesting, and the transfer rate from the closed area to the open area is $\delta = 0.2$, the maximum obtainable yield from the management system employing the reserve is comparable to that for a management system in which both population segments are subject to harvest but exhibits substantially enhanced resilience to exploitation with increasing levels of fishing effort (Figure 6). Other choices for model structures and exchange rates could result in high yield levels in general with the use of reserves when habitat is affected by fishing activities (see Rodwell et al. 2003). The relatively high sustained yields at high fishing effort levels in the present analysis provide a clear hedge against uncertainty and risk if the relationship between fishing effort and carrying capacity can be specified and if the exchange rate between the closed area and the open area can be reliably estimated. An understanding of dispersal dynamics is critical in this regard. If a "donor" area is inadvertently open to exploitation while a recipient area is instead closed, the overall resilience of the system at high levels of fishing effort will be compromised.

Comparative Risk Assessment

Carrying capacity can also be affected by non-fishing-related causes, including changes in the physical or chemical environments (e.g., regime shifts in temperature and other factors), and by other anthropogenic disturbances (e.g., siltation of coastal areas through runoff, pollution effects, mineral extraction, etc.). These different stressors hold different implications for the potential for recovery from disturbance, the time scale of recovery (if recovery is possible), and the management options possible to control risk (Table 4).

For impacts on ecological structure, a key consideration for whether recovery will be possible and the time scale for recovery will depend on whether alternate stable states exist. For example, if harvesting results in a change in the relative abundance of a predator–prey complex, the nature of the functional feeding response of the predator and the numerical response of the prey will determine the recovery prospects. Situations in which an unstable lower equilibrium point exists may result in a system in which a catastrophic collapse could occur and recovery would not be possible without a re-colonization event. In other circumstances, a stable lower equilibrium point could result in the case where a lower productivity level would be maintained until a stochastic event resulted in a shift to a higher production domain (e.g., Collie and Spencer 1994). Similar considerations hold for stability characteristics resulting from changes in benthic community structure and production or changes in species and communities producing biogenic structure. For changes in carrying capacity as a result of other anthropogenic impacts (e.g., pollution events), recovery processes will be determined by factors such as whether the impact is chronic or acute and the nature of abatement or remedial strategies. In some cases, the recovery strategy might entail replacement of lost habitat (e.g., replanting of seagrass meadows, transplanting of coral, rebuilding of reef structural backbone) or other ecologi-

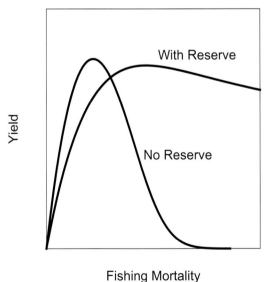

Figure 6. Comparison of predicted yields as a function of fishing mortality for a management system employing a marine reserve protecting a source region and a nonspatially structured harvesting system (i.e., no reserve).

Table 4. Evaluation of alternative hypotheses for factors affecting carrying capacity and the recovery characteristics under these mechanisms and the management options under each hypothesis. Mechanisms considered include the effect of a regime shift in the physical environment, changes in trophic structure as a result of species-selective harvesting patterns or differential mortality, impacts of fishing activity on benthic production, geological structures (structural habitat loss), and biogenic structures and other sources of anthropogenic impact (non-fishing habitat loss).

Value	Physical regime shift	Trophic structure change	Carrying capacity hypothesis				Non-fishing habitat loss
			Reduced benthic production	Physical structural habitat loss	Biogenic structural habitat loss		
Recovery characteristic	Recovery possible	Dependent on nature of biological interactions; alternate stable states possible	Recovery possible; alternate stable states possible	Recovery may be possible if geological structures permanently destroyed	Recovery possible; alternate stable states possible		Recovery possible
Time scale of recovery	Time scale of physical forcing	Dependent on generation times of interacting species; time to strong recruitment	Dependent on generation times of benthic organisms	Dependent on time scale of geological processes or habitat engineering	Dependent on generation times of emergent epibenthic organisms		Dependent on rate of abatement or habitat engineering
Management actions	Adjust biological reference points if intrinsic rate of increase affected	Adaptive manipulation of abundance of predators and/or competitors	Reduce fishing effort; modify gear to minimize bottom contact; restrict mobile gear in vulnerable substrates; marine reserves	Reduce fishing effort; modify gear to minimize bottom contact; restrict mobile gear in vulnerable substrates; marine reserves	Reduce fishing effort; modify gear to minimize bottom contact; restrict mobile gear in vulnerable substrates; marine reserves		Eliminate/minimize source of habitat loss; habitat replenishment program

cal engineering initiatives. For a coral reef example, see Koenig et al. (2005, this volume).

The application of comparative risk assessment (e.g., NRC 2002) to evaluate the impact of multiple stressors in exploited systems and to differentiate among distinct effects of fishing on exploited communities will be required to develop ecosystem-based management strategies. For managing risk associated with different effects of fishing on exploited systems, some combination of one or more of the tactical management elements (effort control, conservation engineering, and spatial management) will be required with the appropriate mix of management options determined by the nature of the impact of different gear types in different habitat.

Changes resulting from regime shifts or other low-frequency alterations in the physical environment may require different management responses depending on the impact on the system dynamics. For example, if the intrinsic rate of increase in the target species is affected, it may be necessary to change the limiting levels of fishing effort and the appropriate targets. In contrast, if carrying capacity alone is altered, the limiting levels of fishing effort would remain unchanged (although the corresponding MSY level would be changed). Similar considerations hold for changes due to anthropogenic alteration of the environment derived from causes other than fishing.

Discussion

Fishing activity can have direct effects on carrying capacity through changes in habitat quality and availability. If fishing simultaneously results in removal of the target species and degradation of carrying capacity, a fundamental shift in the production dynamics and corresponding reference points for management results. If an impact of this type occurs but remains unrecognized, the target population may be inadvertently overexploited. Further, the inferences for recovery trajectories and time scales of recovery may be fundamentally different if habitat is adversely affected relative to the case where fishing only results in removal of target organisms. If we are to fully understand the importance of habitat protection in exploited systems and how best to develop appropriate management strategies, it is critical that we focus on the relationship between habitat and production. Scientific advice incorporating the habitat perspective can be most effectively integrated into existing management structures if it is cast in terms of production and targets and limits to exploitation.

With respect to the production models illustrated above, the options available for management essentially involve controlling fishing effort (f_t), the habitat impact parameters (e.g., a in the exponential model), or the catchability coefficient (q) for different species in the complex. As noted earlier, control of fishing effort is desirable from a number of ecological and economic points of view. Modifications of gear design, fishing strategies, and behaviors, or changes in the allocation of gear types in different habitats, all have the potential to affect the habitat-impact parameters in the submodels relating fishing effort to carrying capacity. Changes in the catchability coefficient (affected through changes in gear design or fishing practices) can be critical in controlling impacts on trophic structure and the indirect effects that accompany changes in predator–prey and competitive communities. The use of spatial management strategies affects the area over which the habitat impact parameters apply.

Sainsbury (1988, 1991) has demonstrated that incorporating measures of habitat in production models is feasible using information derived from monitoring programs for both target species and benthic organisms. Information on fish abundance and habitat availability (two size-classes of biogenic structures) derived from areas open and closed to fishing over a 6-year period on the northeastern continental shelf of Australia was sufficient to test hypotheses concerning the role of habitat in fish production against alternatives specifying biological interactions among the fish community and standard single-species production models. The habitat-impact model was found to be most consistent with the observed population trajectories (see also Sainsbury et al. 1997). The use of fishery closures in general to experimentally determine the role of habitat characteristics in the production dynamics of exploited species provides an important avenue for parameterizing production models with explicit consideration of habitat effects of fishing. Clearly, the use of an adaptive experimental approach will be beneficial in fully utilizing information derived from fishery closures.

Examples of cases where fishing exerts direct effects on carrying capacity exist for a diverse set of systems. For example, harvesting of oysters results in both removal of the target species and disruption and degradation of oyster reefs (e.g., Breitburg et al. 2000). Chemical cues from oyster reef structures mediate settlement of oyster larvae, and the degradation of the reefs results in a loss in settlement habitat and, ultimately, the overall carrying capacity. Oyster reefs also serve as essential habitat for fish species and other organisms (Breitburg et al. 2000). The large-scale reduction in oyster abundance in areas such as the Chesapeake Bay undoubtedly reflect the combined effects of overexploitation, habitat destruction, and outbreaks of pathogens. Remedial action to rebuild habitat includes the reconstruction of oyster reefs using fossil shell and contemporary shell stock.

Impacts of harvesting on carrying capacity need not be limited to disruption of benthic systems, and the effects can be both subtle and complex. For example, the removal of Pacific salmon by high seas fisheries reduces the return of salmon to the natal streams where the fish spawn and die. The salmon carcasses provide an essential source of nutrients in the stream and river ecosystems and have an important role in the overall productivity and carrying capacity of these systems. Thus, the loss of biomass through harvesting can have broader and longer-term consequences for the stream ecosystems supporting salmon populations.

In the context of risk assessment for marine systems, we require an evaluation of factors such as the problem of measurement error in estimation of abundance and vital population rates, the effects of intrinsic variability in marine ecosystems, and the implications of our incomplete understanding of the factors controlling populations, communities, and ecosystems. A key issue in the latter is the resulting uncertainty in specifying appropriate model structures. It is demonstrated here that model mis-specification can be an important component of risk. Measurement error and high levels of natural variability contribute to the difficulty in distinguishing among alternative hypotheses and models of system structure and dynamics. Risk management involves the evaluation of reference points under different model formulations and the selection of the most effective management tools to minimize risk. The considerable effort now being devoted to assess impacts of fishing on benthic systems provides an important basis for devising management strategies. This information will be most readily accommodated in fishery management if it is expressed in a form consistent with other types of management advice with respect to targets and limits to exploitation.

Acknowledgments

I am grateful to Jim Thomas for the invitation to attend the symposium and to Martin Cryer, Thomas Noji, and an anonymous referee for many helpful comments on this paper.

References

Alexander, A. B., H. F Moore, and W. C Kendall. 1915. Otter trawl fishery. Report of the United States Fisheries Commission, 1914, Appendix VI, Washington, D.C.

Anderson, K. P., and E. Ursin. 1977. A multispecies extension to the Beverton and Holt theory of fishing, which accounts for phosphorus circulation and primary production. Meddelelser fra Danmarks Fiskeri- og Havundersogelser 7:319–435.

Anonymous. 1921. The history of trawling. Fish Trades Gazette March 19: 22–69.

Anonymous. 2002a. Workshop on the effects of fishing gear on marine habitats off the Northeastern United States October 23–25, 2001 Boston, MA. Northeast Fisheries Science Center, Reference document 02-01, Woods Hole, Massachusetts.

Anonymous. 2002b. NMFS National Standing Working Group on Fishing Technology. U.S. Department of Commerce, Silver Spring, Maryland.

Auster, P., and R. Langton. 1999. The effects of fishing on fish habitat. Pages 150–187 in L. Benaka, editor. Fish habitat: essential fish habitat and rehabilitation. American Fisheries Society, Symposium 22, Bethesda, Maryland.

Auster, P. J. 1998. A conceptual model of the impacts of fishing gear on the integrity of fish habitats. Conservation Biology 12:1198–1203.

Auster, P. J., R. J. Malatesta, R. W. Langton, L. Watling, P. C. Valentine, C. L. S. Donaldson, E. W. Langton, A. N. Shepard, and I. G. Babb. 1996. The impacts of mobile fishing gear on seafloor habitats in the Gulf of Maine (Northwest Atlantic): implications for conservation of fish populations. Reviews in Fisheries Science 4(2):185–202.

Benaka, L. R. 1999. Fish habitat: essential fish habitat and rehabilitation. American Fisheries Society, Symposium 22, Bethesda, Maryland.

Breitburg, D. L., L. D. Coen, M. W. Luckenbach, R. Mann, M. Psoey, and J. A. Wesson. 2000. Oyster reef restoration: convergence of harvest and conservation strategies. Journal of Shellfish Research 19:371–377.

Collie, J. S., and P. D. Spencer. 1994. Modeling predator–prey dynamics in a fluctuating environment. Canadian Journal of Fisheries and Aquatic Sciences 51(12):2665–2672

Daan, N., and M. P. Sissenwine. 1991. Multispecies models relevant to management of living resources. ICES Marine Science Symposium 193.

Dayton, P. K., S. F. Thrush, M. T. Agardy, and R. J. Hofman. 1995. Environmental effects of marine fishing. Aquatic Conservation: Marine and Freshwater Ecosystems 5:205–232.

Fluharty, D. 2000. Habitat protection, ecological issues, and implementation of the Sustainable Fisheries Act. Ecological Applications 10:325–337.

Fogarty, M. J. 1999. Essential habitat, marine reserves, and fishery management. Trends in Ecology and Evolution 14:133–134.

Fogarty, M. J. 2000. Habitat loss, carrying capacity and fishery management. Pages 28–29 in National Marine Fisheries Service. Proceedings of the 6th NMFS National Stock Assessment Workshop (NSAW). National Marine Fisheries Service, Silver Spring, Maryland.

Fogarty, M. J., J. A. Bohnsack, and P. Dayton. 2000. Marine reserves and fishery management. Pages 283–300 in C. Sheppard, editor. Seas at the millennium: an environmental evaluation. Elsevier, Amsterdam.

Fogarty, M. J., R. K. Mayo, L. O'Brien, F. M. Serchuck, and A. A. Rosenberg. 1996. Assessing uncertainty and risk in exploited marine populations. Reliability Engineering and System Safety 54:183–195.

Fogarty, M. J., A. A. Rosenberg, and M. P. Sissenwine. 1992. Fisheries risk assessment: sources of uncertainty. Environmental Science and Technology 26:440–447.

Fonteyne, R., and H. Polet. 2002. Reducing the benthos by-catch in flatfish bean trawling by means of technical modifications. Fisheries Research (Amsterdam) 55:219–230.

Graham, M. 1955. Effect of trawls on animals of the sea bed. Papers in Marine Biology and Oceanography 3(Supplement):1–6.

Hall, S. J. 1999. The effects of fishing on marine ecosystems and communities. Blackwell Scientific Publications Ltd., Oxford, UK.

Hayes, D. B., C. P. Ferreri, and W. W. Taylor. 1996. Linking fish habitat to their population dynamics. Canadian Journal of Fisheries and Aquatic Sciences 53(Supplement 1):383–390.

Jennings, S., and M. J. Kaiser. 1998. The effects of fishing on marine ecosystems. Advances in Marine Biology 34:201–352.

Jones, J. B. 1992. Environmental impact of trawling on the seabed: a review. New Zealand Journal of Marine and Freshwater Research 26:59–67.

Kaiser, M. J., J. S. Collie, S. J. Hall, S. Jennings, and I. A. Poiner. 2003. Impacts of fishing gear on marine benthic habitats. Pages 197–218 in M. Sinclair and G. Valdemarsson, editors. Responsible fisheries in the marine ecosystem. FAO and CABI Publishing, Rome.

Kaiser, M. J., and S. J. de Groot. 2000. Effects of fishing on non-target species and habitats. Blackwell Scientific Publications, Oxford, UK.

Kaiser, M. J., F. E. Spence, and P. J. B. Hart. 1999. Fishing gear restrictions and conservation of benthic habitat complexity. Conservation Biology 14:1512–1525.

Koenig, C. C., A. N. Shepard, J. K. Reed, F. C. Coleman, S. D. Brooke, J. Brusher, and K. M. Scanlon. 2005. Habitat and fish populations in the deep-sea *Oculina* coral ecosystem of the western Atlantic. Pages 795–805 in P. W. Barnes and J. P. Thomas, editors. Benthic habitats and the effects of fishing. American Fisheries Society, Symposium 41, Bethesda, Maryland.

Langton, R. W., and P. J. Auster. 1999. Marine fishery and habitat interactions: to what extent are fisheries and habitat interdependent? Fisheries 24(6):14–21.

Lauck, T., C. W. Clark, M. Magel, and G. R. Munro. 1998. Implementing the precautionary principle in fisheries management through marine reserves. Ecological Applications 8(Supplement):S72–S78.

Lindholm, J. B., P. J. Auster, M. Ruth, and L. Kaufman. 2001. Modeling the effects of fishing and implications for the design of marine protected areas: juvenile fish responses to variations in seafloor habitat. Conservation Biology 15:424–437.

Lindholm, J. B. X. 1999. Habitat-mediated survivorship of juvenile Atlantic cod (*Gadus morhua*): fish population responses to fishing induced alteration of the seafloor in the Northwest Atlantic and implications for the design of marine reserves. Doctoral dissertation. Boston University, Boston.

MacCall, A. 1990. Dynamic geography of marine fish populations. Sea Grant Press, Seattle.

Mangel, M. 1998. No-take areas for sustainability of harvested species and a conservation invariant for marine reserves. Ecology Letters 1:87–90.

Mangel, M. 2000. Trade-offs between fish habitat and fishing mortality and the role of reserves. Bulletin of Marine Science 66:663–674.

Manson, G., and B. J. Todd. 2000. Revolution in Nova Scotia scallop fishery: seabed maps turn hunting into harvesting. Fishing News International 39:20–22.

May, R. M., editor. 1984. Exploitation of marine communities. Springer-Verlag, Berlin.

May, R. M., J. R. Beddington, C. W. Clark, S. J. Holt, and R. M. Laws. 1979. Management of multispecies fisheries. Science 205:267–277.

Mercer, M., editor. 1982. Multispecies approaches to fisheries management advice. Canadian Special Publication of Fisheries and Aquatic Sciences 59.

Murray, S. N., R. F. Ambrose, J. A. Bohnsack, L. W. Botsford, M. H. Carr, G. E. Davis, P. K. Dayton, D. Gotshall, D. R. Gunderson, M. A. Hixon, J. Lubchenco, M. Mangel, A. MacCall, D. A. McArdle, J. C. Ogden, J. Roughgarden, R. M. Starr, M. J. Tegner, and M. M. Yoklavich. 1999. No-take reserves networks: sustaining fishery populations and marine ecosystems. Fisheries 24(11):11–25.

North Pacific Fishery Management Council. 1999. Environmental assessment/regulatory impact review/initial regulatory flexibility analysis for amendment 57 of the FMP for the groundfish fishery of the Bering Sea and Aleutian Islands Area to prohibit the use of nonpelagic trawl gear in directed pollock fisheries. North Pacific Fishery Management Council, Anchorage, Alaska.

NRC (National Research Council). 2002. Effects of trawling and dredging on seafloor habitat. National Academy Press, Washington, D.C.

Pella, J. J., and P. K. Tomlinson. 1969. A generalized stock production model. Inter-American Tropical Tuna Commission Bulletin 13:419–496.

Pol, M. 2000. Effect of the removal of the chain sweep from a whiting raised footrope trawl. Commonweath of Massachusetts, Division of Marine Fisheries Conservation Engineering Report.

Quinn, T. J., II, and R. B. Deriso. 1999. Quantitative fish dynamics. Oxford University Press, New York.

Rodwell, L. D., E. B. Barbier, C. M. Roberts, and T. R. McClanahan. 2003. The importance of habitat quality for marine reserve-fishery linkages. Canadian Journal of Fisheries and Aquatic Sciences 60:171–181.

Rosenberg, A., T. E. Bigford, S. Leathery, R. L. Hill, and K. Bickers. 2000. Ecosystem approaches to fishery management through essential fish habitat. Bulletin of Marine Science 66:535–542.

Sahrhage, D., and J. Lundbeck 1992. A history of fishing. Springer-Verlag, Berlin.

Saila, S. B., V. Lj. Kocik, and J. W. McManus. 1993. Modelling the effects of destructive fishing practices on tropical coral reefs. Marine Ecology Progress Series 94:51–60.

Sainsbury, K. 1988. The ecological basis if multispecies fisheries, and management of a demersal fishery in tropical Australia. Pages 349–382 *in* J. A. Gulland, editor. Fish population dynamics, 2nd edition. Wiley, New York.

Sainsbury, K. J. 1991. Application of an experimental approach to management if a tropical multispecies fishery with highly uncertain dynamics. ICES Marine Science Symposium 193:301–320.

Sainsbury, K. J., R. A. Campbell, R. Lindholm, and A. W. Whitelaw. 1997. Experimental management of an Australian multispecies fishery: examining the possibility of trawl- induced habitat modification. Pages 107–112 in E. K. Pikitch, D. D. Huppert, and M. P. Sissenwine, editors. Global trends: fisheries management. American Fisheries Society, Symposium 20, Bethesda, Maryland.

Stephan, C. D., R. L. Peuser, and M. S. Fonseca. 2000. Evaluating fishing gear impacts to submerged aquatic vegetation. Atlantic States Marine Fisheries Commission, Habitat Management Series 5, Washington, D.C.

Thrush, S. F., J. E. Hewitt, V. J. Cummings, P. K. Dayton, M. Cryer, S. J. Turner, G. A. Funnell, R. G. Budd, C. J. Milburn, and M. R. Wilkenson. 1998. Disturbance of the marine benthic habitat by commercial fishing: impacts at the scale of the fishery. Ecological Applications 8:866–879.

Valdemarsen, J. W., and P. Suuronen. 2003. Modifying fishing gear to achieve ecosystem objectives. Pages 197–218 *in* M. Sinclair and G. Valdemarsson, editors. Responsible fisheries in the marine ecosystem. FAO and CABI Publishing, Rome.

Van Marlen, B. 2000. Technical modifications to reduce the by-catches and impacts of bottom-fishing gears. Pages 253–268 *in* M. J. Kaiser and S. J. de Groot, editors. Effects of fishing on non-target species and habitats. Blackwell Scientific Publications, Oxford, UK.

Walters, C. J., and F. Juanes. 1993. Recruitment limitation as a consequence of natural selection for use of restricted habitats and predation risk taking by juvenile fishes. Canadian Journal of Fisheries and Aquatic Sciences 50:2058–2070.

An Alternative Paradigm for the Conservation of Fish Habitat Based on Vulnerability, Risk, and Availability Applied to the Continental Shelf of the Northwestern Atlantic

JOSEPH DEALTERIS[1]

Department of Fisheries, University of Rhode Island, Kingston, Rhode Island 02881, USA

Abstract. A conservation paradigm is proposed based on the premise that the priority for habitat protection is directly related to habitat vulnerability and risk and inversely related to habitat availability. This paradigm is applied to three regions of continental shelf in the northwestern Atlantic: southern New England, Georges Bank, and the mid-Atlantic. Habitat is interpreted to include the physical substrate and the fauna and flora associated with that substrate. The vulnerability of a habitat to the effects of fishing is a measure of the potential reduction in ecological services that may result from the disturbance of the habitat due to the passage of fishing gear over the seabed. Sand substrates are considered less vulnerable to the effects of fishing than are gravel or mud substrates due to the lack of emergent fauna and infauna in the sand substrate. Risk of fishing disturbance, based on 11 years of fishing effort data for mobile fishing gears within each area, is assessed by estimating the area impacted by each gear type annually relative to the area of a statistical unit. The relative availability of each habitat type within each region was estimated from the sediment distribution data. In the southern New England region, the less vulnerable sand habitat is dominant, and the more vulnerable and spatially limited mud habitat is found only in a small depression on the outer shelf. Mean fishing disturbance (risk) is nearly equal in the two habitats. Therefore, some protection is suggested for the mud habitat due to its more limited availability and vulnerability. On Georges Bank, areas of vulnerable and spatially limited gravel habitat are at risk due to intense fishing. Therefore, these habitats should have the highest priority for protection. The less vulnerable sand habitat dominant on Georges Bank is less impacted by fishing gear and, therefore, has a lower priority for protection. In the mid-Atlantic region, the less vulnerable sand habitat is again dominant, and the more vulnerable and spatially limited mud habitat is on the slope. Mean fishing disturbance (risk) is very high in both habitats as compared to the other two areas. Therefore, some protection of the more vulnerable mud habitat in this region is suggested as well. The proposed paradigm, although easily implemented with relatively limited data, does not address the heterogeneity in the habitat due to the large areas represented by limited physical data nor the effects of multiple encounters between the fishing gear and habitat. The methodology does provide a rational approach to prioritizing large-scale habitat protection that is consistent with our ability to enforce habitat closures.

Introduction

Habitat protection has emerged as an important fisheries management issue in the last decade. The essential fish habitat (EFH) amendments to the Sustainable Fisheries Act designates EFH as those waters and substrate necessary for fish to spawn, feed, and grow. Current efforts to implement these amendments have resulted in vast areas of United States continental shelf seabed being identified as EFH for the over 700 species managed by the National Marine Fisheries Service (NMFS) and eight regional fishery management councils under the authority of the Secretary of Commerce (www.nmfs.noaa.gov/habitat/habitatprotection/). The legislation specifies that adverse impacts to EFH be minimized to the extent practicable. Habitat areas of particular concern have been identified to highlight specific habitat areas with extremely important ecological functions or those that are especially vulnerable to degradation. Implementation of the EFH amendments has been difficult, as the regulations have already been challenged in the courts by both environmentalists and fishermen (L. A. McGee, New England Fishery Manage-

---

[1] E-mail: jdealteris@uri.edu

ment Council, personal communication). Musick (1999) proposed that a more pragmatic approach to conserve habitat would be based on habitat vulnerability, utilization, and availability. I have interpreted these terms as follows. The vulnerability of a habitat to the effects of fishing is a measure of the potential reduction of ecological services that may result from the disturbance of the habitat due to the passage of the fishing gear over the seabed. Utilization is the risk of a habitat being affected by fishing and is related to the frequency that the area has been fished in the recent past, is currently fished, or may be fished in the future given new regulations or restrictions. Availability of the habitat refers to relative availability of a substrate with particular physical characteristics within a given region.

The effects of trawling and dredging on the seabed have been investigated for more than 40 years (Graham 1955; Caddy 1973; Peterson et al. 1987; Thrush et al. 1995) and, more recently, have been summarized in several review papers (Auster and Langton 1999; Hall 1999; DeAlteris et al. 2000). Most of the primary research papers have reported on the ecological consequences of bottom trawling and dredging based on direct measurements of the impacts on the benthos, although some more recent investigations have considered the effects of fishing on sediment resuspension and transport (Churchill 1989; Schwinghamer et al. 1998; DeAlteris et al. 1999). The seabed habitat for fish is composed of two components: the physical substrate, generally composed of unconsolidated sediments but also may include rock, and the benthic fauna and flora that are associated with the substrate and provide ecological services to fish, specifically food and shelter.

The surficial sediments of the Northwest Atlantic continental shelf range from stable bedrock outcrops in the Gulf of Maine to coarse sand and gravel lag deposits on Georges Bank to dynamic sands off the mid-Atlantic to a depositional "mud patch" south of Nantucket Shoals (Milliman et al. 1972; Poppe et al. 1989; Poppe and Pollini 2000). The physical stability of unconsolidated seabed sediments is a function of grain size and contemporary hydrodynamic environment. The "mud patch" is the result of the deposition of fine silt-clay sediments winnowed from Georges Bank forming a stable, cohesive mud environment. Stable, coarse sand and gravel sediments remain on portions of Georges Bank due to this winnowing of glacial deposits. In the mid-Atlantic, the dynamic sand sediments originated as fluvial deposits at lower stands of sea-level and are actively being reworked by bottom currents forming sand waves and wave ripples that reflect contemporary sediment transport processes driven by the combined effects of uniform unidirectional currents and oscillatory wave-generated currents. Early studies of the continental shelf surficial sediment distribution described large areas of relatively homogenous sediment characteristics (Milliman et al. 1972). However, more recent investigations of the region using higher-resolution sampling and acoustic imaging techniques have identified considerable heterogeneity in sediment characteristics within the same area (Poppe et al. 1989).

Bottom contact, mobile gear fisheries of the northeastern and mid-Atlantic coastal regions of the United States (Northwest Atlantic continental shelf) are dominated by bottom trawls but also include scallop dredges and hydraulic clam dredges (Clark 1998). Habitat vulnerability to fishing gear impacts has been directly related to structural complexity of the habitat in terms of physical rugosity of the substrate and the magnitude and abundance of vertically developed organisms on the surface of or within the substrate. The fauna and flora of the benthic environment have evolved through behavioral and morphological adaptations to the unique physical conditions of that environment (Krebs 1994). Stable, relatively immobile substrates allow for the development of larger, sessile epifaunal species in coarse sediments and larger, tube-building infaunal species in fine sediments. These habitats have high fish utilization because they provide shelter for juvenile fish and food for other fish. Trawls and dredges cause substantial impact on stable gravel and mud substrate environments with vertically developed benthic fauna and flora by scraping sessile epifauna from its hard substrate, rolling over cobble and boulders, and disturbing larger infaunal organisms in soft mud substrates (Currie and Parry 1996; Collie et al. 1997; Kaiser et al. 1998). The time required for these habitats to recover to their original condition is long, and sometimes the damage is permanent in the case of boulder fields that are leveled (Hall 1999). In contrast, the benthos of relatively dynamic sediment substrates (sands) are primarily smaller infaunal species and some larger epifaunal species but are generally rapid colonizers and are adapted to a mobile sediment seabed (Hall 1999). On dynamic substrates, impacts of mobile, bottom contact fishing gear are minimal, temporarily smoothing bedforms, resuspending sediments, disturbing small benthic infaunal organisms, and injuring some of the larger epifaunal species (Gibbs et al. 1980; Bergman and Hup 1992; Eleftheriou and Robertson 1992). The recovery time of these habitats is relatively rapid, ranging from days to weeks after the cessation of fishing (Hall 1999). In summary, when structurally complex habitats on relatively stable substrates are disturbed by fishing gear, the ecological services provided by that habitat are reduced; whereas when a less stable substrate, one that is naturally disturbed and has a more spatially heterogeneous and impoverished fauna, is disturbed by fishing gear, the impacts are difficult to detect (Kaiser and Spencer 1996).

Thus, the relative stability of the substrate can be used as a proxy for the vulnerability of the habitat, that is sandy substrates, being the least stable, are the least vulnerable, and mud and gravel substrates, being the most stable, are the most vulnerable.

Most previous studies have investigated the ecological consequences of the fishing gear on a localized basis, but habitat protection must be implemented over relatively large areas to be ecologically effective and enforceable (Halpern 2003). The purpose of this study was to evaluate a conservation paradigm based on the premise that the priority for habitat protection is directly related to habitat vulnerability and risk and inversely related to habitat availability. Resource managers require information that will allow for the rational identification and prioritization of habitat protection based on existing data. The proposed conservation paradigm is applied to the three regions of the continental shelf in the northwestern Atlantic, and the results of these analyses are used to make recommendations for the protection of particular habitats within each area.

Methods

Study Areas

Three regions of the Northwest Atlantic characteristic of the New England and mid-Atlantic regions of the United States (Figure 1) were investigated. Each study region, approximately 175 km in length and 50 km in width, was partitioned into 10' (10-minutes of latitude or longitude) squares based on latitude and longitude, and each square (180 km^2) was represented by a single data point located in its geographic center. A National Ocean Survey hydrographic database was used for water depth, and data available within each 10' square were averaged and contoured in Surfer (Golden Software 1994).

Habitat Vulnerability and Availability

Habitat vulnerability was based on sediment mean grain size. Bottom sediment mean grain size data (mm) consisted of data points unevenly distributed over the study areas (Poppe and Pollini 2000). Data available for each study region within each 10' square were averaged and then contoured using Surfer (Golden Software 1994). The density of sediment data within each 10' square was variable within and between study regions. Portions of Georges Bank had the highest density of data, while data were sparse in the mid-Atlantic region. Averaging of multiple data values within a 10' square resulted in the masking of sediment heterogeneity within a 10' square. Habitat availability was estimated within each study region based on the relative abundance of each sediment type within each area: very fine to medium sand sediments (0.06–0.50 mm), mud sediments (<0.06 mm), and coarse sand and gravel sediments (>0.50 mm).

Habitat Risk

The risk of disturbance or the utilization of habitat by mobile fishing gear was based on the conversion of fishing effort to area swept. While the analysis allows

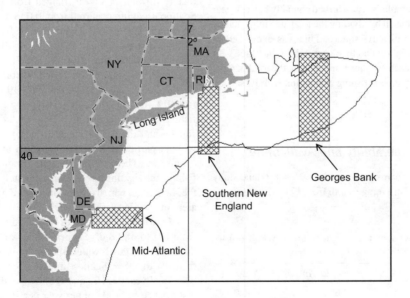

Figure 1. Map the northeast region of the United States showing three study regions: southern New England, Georges Bank, and mid-Atlantic.

for the estimation of the total area impacted by the fishing gear within a single statistical square, it does not account for the differences in the ecological impacts between a single encounter or multiple encounters between the habitat and the gear. For all three study regions, effort data were obtained from the NMFS Northeast Fisheries Science Center database. It was collected by port agents (dockside samplers) who interviewed vessel captains landing catches at principal fishing ports. While these data underestimate the total fishing effort and are subject to erroneous reports of fishermen, until the recent development of vessel tracking systems and log books, this was the only data available. The data included days fished by gear type within each 10' statistical square for southern New England data from 1983 to 1993; for Georges Bank data from 1982 to 1993; and for the mid-Atlantic data from 1982 to 1991. The data were sorted in Excel (Microsoft 2000) by area and gear type: all bottom trawls including fish, shrimp, and pair trawls; bay and sea scallop dredges; and surf clam–ocean quahog dredges. The average annual days fished by each gear type was determined per 10' square. The area impacted annually by each gear type in each 10' square was estimated based on an assumed 18-h fishing day and fishery-specific towing speeds and gear widths (Table 1). These values were estimated from personal observation and are believed to be representative for the New England–mid-Atlantic fleet. There is considerable variability in the swept width of trawls depending on the vessel size; however, the widths of the scallop and clam dredges are reasonably standardized.

The number of times an area equivalent to a 10' square was cumulatively disturbed annually by each gear type was estimated by dividing the total area impacted by the total area of a 10' square. This was expressed as percent disturbance and then summed for the three gear types within each 10' square to estimate the total percent disturbance by mobile fishing gear within each 10' square.

Results

Habitat Vulnerability and Availability

The 37 data squares in the southern New England study region ranged in average depth from 17 to 421 m (Figure 2a), and sediments ranged from medium sands (0.50–0.25 mm) to silt (0.06–0.02 mm) (Figure 3a). There was a large area of fine sediments in the mid-shelf region referred to as the "mud patch." The 78 data squares on Georges Bank ranged from 22 to 1,732 m in depth (Figure 2b), and sediments ranged from pebble gravel (4.00 mm) to very fine silt (<0.01 mm) (Figure 3b). The 29 data squares in the mid-Atlantic study region ranged in depth from 13 to 1,343 m (Figure 2c), and sediments ranged from coarse sand (0.50 mm) to medium silt (<0.03 mm) (Figure 3c). The relative distribution of sediment types within each study region was determined based on the frequency of sediment types represented in each 10' square. In the southern New England study region (Figure 4a), 73% of the sediment types were medium and very fine grain-sized sand (0.50–0.06 mm), and the remaining 27% were mud (<0.06 mm). In the Georges Bank study region (Figure 4b), 83% of the sediment types were medium and very fine grain-sized sand (0.50–0.06 mm), 4% were coarse sands and gravels (>0.50 mm), and the remaining 13% were mud (<0.06 mm). In the mid-Atlantic study region (Figure 4c), 83% of the sediment types were medium and very fine grain-sized sand (0.50–0.06 mm), and the remaining 17% were mud (<0.06 mm).

Habitat Risk

In the southern New England study region, trawling was the dominant fishing activity. Trawling effort occurred from 0.1 to 359.7 d/year per 10' square. The seabed disturbance by trawls within a particular square, that is an area equivalent to a 10' square that was completely trawled in 1 year, ranged from 0% to 469%. Scallop dredging occurred from 0.0 to 61.6 d/year per 10' square. The seabed disturbance by scallop dredges within a particular square ranged from 0% to 32%. Hydraulic clam dredging effort occurred from 0.0 to 28.0 d/year per 10' square. The seabed disturbance by hydraulic dredges area within a particular 10' square ranged from 0% to 2%. When all gear types were combined, fishing effort ranged from 0.1 to 408.0 d/year per 10' square, and the seabed disturbance by all gear types within a particular 10' square ranged from 0% to 481% (Figure 5a). There were three concentrations of intense fishing activity, inshore, at approximately 60 m, and at the shelf edge, which corresponded to the three intense areas of trawling activity.

Trawling in the Georges Bank study region ranged from 0.0 to 489.8 d/year per 10' square. The seabed disturbance by trawls within a particular 10' square ranged from 0% to 638%. Scallop dredging occurred from 0.0 to 225.6 d/year per 10' square, corresponding to seabed disturbance of 0–118% per year per 10' square. Clam dredging occurred from 0.0 to 9.0 d/year per 10' square, corresponding to seabed disturbance of 0–1% per year

Table 1. Tow speed (km/h) and mouth width (m) for trawl, scallop dredge, and clam dredge.

Gear	Speed (km/h)	Width (m)
Trawl	4.63	40.0
Scallop dredge	7.41	10.0
Clam dredge	2.78	3.7

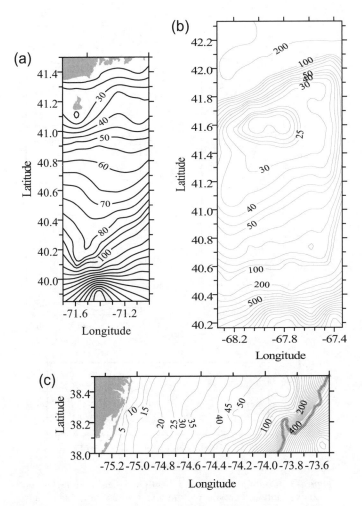

Figure 2. Contoured bathymetry (m) for: (a) southern New England, (b) Georges Bank, and (c) mid-Atlantic study regions.

per 10' square. When all gear types were combined, fishing effort ranged from 0.0 to 639.4 d per 10' square, corresponding to seabed disturbance of 0–712% per year per 10' square (Figures 5b). There were two areas that were fished most intensely, and these corresponded to the two areas of intense trawling activity.

Trawling effort in the mid-Atlantic region ranged from 0.0 to 2170.9 d/year per 10' square. Seabed disturbance by trawls within a particular 10' square ranged from 0% to 2,828%. Scallop dredging effort ranged from 0.0 to 5803.5 d/year per 10' square, corresponding to seabed disturbance of 0–3,025% per year per 10' square. Clam dredging occurred from 0.0 to 470.4 d/year, corresponding to seabed disturbance of 0–30% per year per 10' square. When all fishing gear types are combined, fishing effort per 10' square per year ranged from 0.0 to 7,138.8 d, corresponding to seabed disturbance of 0–4,438% per year per 10' square (Figure 5c). There were three areas of intense fishing activity, which corresponded to areas of intense trawling activity.

Synthesis of Vulnerability, Risk, and Availability

In the southern New England study region (Figure 6a), the less vulnerable sand habitat is most available (73%) and the more vulnerable and less available mud habitat (27%) is found only in a small depression on the outer shelf. Mean fishing disturbance (risk) for each of the two habitat types is nearly equal at 50%. In the Georges Bank study region (Figure 6b), the more vulnerable gravel habitat is least available (4%). Additionally, it is at risk due to intense fishing activity (mean fishing disturbance of 110%). The more vulnerable mud habitat is also spa-

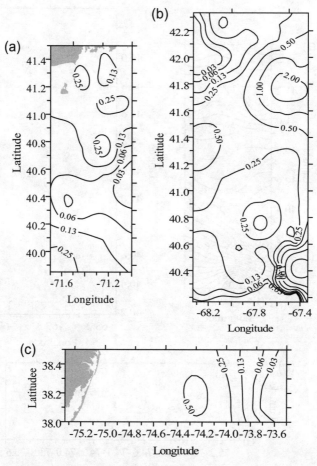

Figure 3. Contoured mean sediment grain size (mm) (>0.5 mm is coarse sand and gravel, <0.5 mm and >0.06 mm is sand, and <0.06 mm is mud): (a) southern New England, (b) Georges Bank and (c) mid-Atlantic study regions.

tially limited (12%), but it has a substantially lower level of fishing disturbance (mean fishing disturbance of 50%). The sand habitat is less vulnerable, most available (84%), and less disturbed by fishing (mean fishing disturbance of 90%). In the mid-Atlantic study region (Figure 6c), the less vulnerable sand habitat is most available (83%). The more vulnerable mud habitat on the slope is spatially limited (17%). Fishing disturbance (risk) is higher in both the sand and mud habitats as compared to the other two study regions (mean fishing disturbance of 610% in the sand habitat and 340% in the mud habitat).

Discussion

A conservation paradigm is proposed based on the premise that the priority for habitat conservation is inversely related to habitat availability and directly related to habitat vulnerability to and risk of fishing disturbance. Fish habitat on the seabed off the continental shelf of the northeastern United States was evaluated along three strip transects for vulnerability to and risk of fishing disturbance and the relative availability. One transect crossed Georges Bank, another extended south across southern New England shelf, and the last extended east across the mid-Atlantic shelf. The vulnerability of a habitat to the effects of fishing was interpreted to be a measure of the reduction in ecological services provided by a habitat that may result from the disturbance of the substrate due to the passage of the fishing gear over the seabed. Sand substrates were considered to be less vulnerable habitats to the effects of mobile bottom contact fishing gear, while gravel and mud substrates were considered to be more vulnerable habitats to the effects of mobile bottom contact fishing

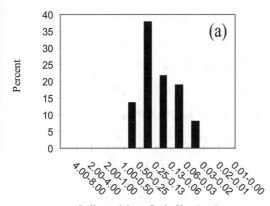

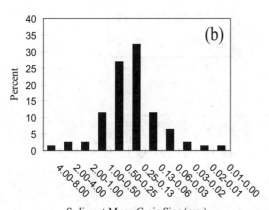

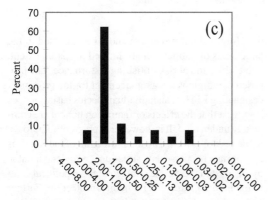

Sediment Mean Grain Size (mm)

gear. The risk of fishing disturbance was estimated based on the frequency of disturbance by mobile fishing gear. The relative availability of each habitat type within each study region was estimated from the sediment distribution data.

Data for each strip transect was divided into 10' squares based on latitude and longitude. Fishing effort data for scallop dredges, hydraulic clam dredges, and bottom trawls within each square was averaged over an 11-year period from the early 1980s to early 1990s, and the area impacted by each gear type was estimated. Seabed disturbance due to fishing was not evenly distributed but reflected the intentional targeting of areas due to fishery resource abundance and accessibility to the area by the fishing gear. Some squares were intensely fished, and large numbers of other squares were minimally fished.

Coarse sand and gravel sediments are unique to the Georges Bank study region but only represent a small portion of the overall area. The sand sediments are nearly ubiquitous throughout the southern New England, Georges Bank, and mid-Atlantic study regions, with relatively small areas of mud habitat. The southern New England study area has the largest percentage of mud habitat, exceeding the other areas by almost twofold.

In the southern New England study region, the less vulnerable sand habitat is dominant, and the more vulnerable and spatially limited mud habitat is found only in a small depression on the outer shelf. Mean fishing disturbance (risk) is nearly equal in the two habitats. Therefore, some protection is suggested for the mud habitat due to its more limited availability and vulnerability. On Georges Bank, areas of vulnerable and spatially limited gravel habitat are at risk due to intense fishing. Therefore, these areas should have the highest priority for protection. Sand habitat is dominant on Georges Bank, is less vulnerable, and is less impacted by fishing gear and, therefore, has a lower priority for protection. In the mid-Atlantic study region, the less vulnerable sand habitat is again dominant, and the more vulnerable, but limited, mud habitat is on the slope. Mean fishing disturbance (risk) is very high in both habitats as compared to the other two study regions. Therefore, some protection of the more vulnerable mud habitat is suggested in this region.

Implementation of the proposed paradigm only requires data on fishing effort and sediment type at a reso-

Figure 4. Relative availability of habitat type as determined by sediment mean grain size (mm; >0.05 mm is coarse sand and gravel, <0.05 mm and >0.06 mm is sand, and <0.06 mm is mud) in (a) southern New England, (b) Georges Bank, and (c) mid-Atlantic study regions.

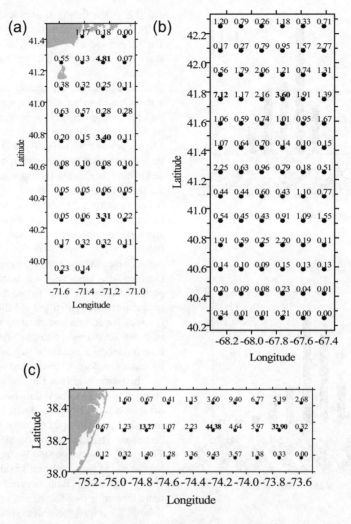

Figure 5. Percent disturbance (X100) of the seabed of an area equivalent to 10' square annually by all fishing gear types per 10' square in: (a) southern New England, (b) Georges Bank, and (c) mid-Atlantic study regions. Points in bold indicate areas where fishing is most intense.

lution consistent with the scale of potential management action. This analysis was conducted with data binned into 10' squares because fishing effort data were available at this level of resolution. Sediment distribution data were available at much higher resolution in some limited areas but not in others. By averaging high-resolution data into a single data point representative of a 10' square, heterogeneity in the sediment distribution is masked, and small but unique habitats may not be identified or protected. However, our ability to protect vulnerable habitats of limited geographic extent can be compromised by our ability to enforce closures of these areas. In addition, the ecological benefits of protected areas increase directly with the size of the area, and therefore, habitat protection should be directed to the large areas of habitat that are limited in spatial extent, vulnerable, and at risk of fishing disturbance. The proposed paradigm is not a replacement for the process of designating EFH on an individual species basis nor does it suggest that the effects of fishing on habitat are completely understood. It is, however, a pragmatic approach to the task of prioritizing the protection of habitat. It is appropriate in sparse data situations, is understandable by all constituencies, and only requires the political commitment to set standards for habitat protection. Further applications of this methodology could potentially include more sediment or habitat types represented in smaller geographic units if the data were available.

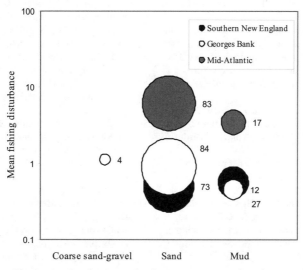

Figure 6. Comparison of risk (mean percent disturbance of the seabed), vulnerability (sediment type: coarse sand and gravel, sand, and mud), and availability (% coverage) for: southern New England, Georges Bank, and mid-Atlantic study regions.

References

Auster, P., and R. Langton. 1999. Indirect effects of fishing. Pages 150–187 *in* L. Benaka, editor. Fish habitat: essential fish habitat and rehabilitation. American Fisheries Society, Symposium 22, Bethesda, Maryland.

Bergman, M. J. N., and M. Hup. 1992. Direct effects of beam trawling on macrofauna in a sandy sediment in the southern North Sea. ICES Journal of Marine Science 49:5–11.

Caddy, J. F. 1973. Underwater observations in tracks of dredges and trawls and some effects of dredging on a scallop ground. Journal of the Fisheries Research Board of Canada 30:173–180.

Churchill, J. H. 1989. The effect of commercial trawling on sediment resuspension and transport over the Middle Atlantic Bight continental shelf. Continental Shelf Research 9:841–864.

Clark, S. H., editor. 1998. Status of fishery resources off the northeastern United States for 1998. NOAA Technical Memorandum NMFS-NE-115.

Collie, J. S., G. A. Escanero, and P. C. Valentine. 1997. Effects of bottom fishing on the benthic mega-fauna of Georges Bank. Marine Ecology Progress Series 155:159–172.

Currie, D. R., and G. D. Parry. 1996. Effects of scallop dredging on a soft sediment community: a large-scale experimental study. Marine Ecology Progress Series 134:131–150.

DeAlteris, J. T., L. G. Skrobe, and K. Castro. 2000. Effects of mobile bottom fishing gear on biodiversity and habitat in offshore New England waters. Northeastern Naturalist 7(4):379–394.

DeAlteris, J. T., L. Skrobe, and C. Lipsky. 1999. The significance of seabed disturbance by mobile fishing gear relative to natural processes: a case study in Narragansett Bay, Rhode Island. Pages 224–237 *in* L. Benaka, editor. Fish habitat: essential fish habitat and rehabilitation. American Fisheries Society, Symposium 22, Bethesda, Maryland.

Eleftheriou, A., and M. R. Robertson. 1992. The effects of experimental scallop dredging on the fauna and physical environment of a shallow sandy community. Netherlands Journal of Sea Research 30:289–299.

Gibbs, P. J., A. J. Collins, and L. C. Collett. 1980. Effect of otter prawn trawling on the macrobenthos of a sandy substratum in a new South Whales estuary. Australian Journal of Marine and Freshwater Research 31:509–516.

Golden Software. 1994. Surfer. Golden Software, Inc., Golden, Colorado.

Graham, M. 1955. Effect of trawling on the animals of the seabed. Papers in marine biology and oceanography. Deep Sea Research Supplement 3:1–6.

Hall, S. J. 1999. The effects of fishing on marine ecosystems and communities. Blackwell Scientific Publications Ltd., Oxford, UK.

Halpern, B. S. 2003. The impact of marine reserves: do reserves work and does size matter? Ecological Applications 13(1):S117–S137.

Kaiser, M., D. Edward, P. Armstrong, K. Radford, N. Lough, R. Flatt, and H. Jones. 1998. Change in megafaunal benthic communities in different habitats

after trawling disturbance. ICES Journal of Marine Science 55:353–361.
Kaiser, M., and B. Spencer. 1996. The effects of beam trawl disturbance on infaunal communities in different habitats. Journal of Animal Ecology 65:348–358.
Krebs, C. 1994. Ecology. Harper-Collins, New York.
Microsoft. 2000. Excel. Microsoft Corporation. Redmond, Washington.
Milliman, J. D., O. H. Pilkey, D. A. Ross. 1972. Sediments of the continental margin off the eastern United States. Geological Society of America Bulletin 84:1315–1334.
Musick, J. 1999. Part two, essential fish habitat identification, Pages 41–42 *in* L. Benaka, editor. Fish habitat: essential fish habitat and rehabilitation. American Fisheries Society, Symposium 22, Bethesda, Maryland.
Peterson, C. H., H. C. Summerson, and S. R. Fegley. 1987. Ecological consequences of mechanical harvesting of clams. U.S. National Marine Fisheries Service Fishery Bulletin 85(2):281–298.

Poppe, L. J., and C. F. Pollini. 2000. USGS east coast sediment analysis: procedures, database, and georeferenced displays. U.S. Geological Survey, Open-File Report 00-358, Woods Hole, Massachusetts.
Poppe, L. J., J. S. Schlee, B. Butman, and C. M. Lane. 1989. Map showing the distribution of surficial sediments, Gulf of Maine and Georges Bank. U.S. Geological Survey, Miscellaneous Investigation Serial Map I-1986-A.
Schwinghamer, P., D. Gordon, T. Rowell, J. Prenna, D. McKeown, G. Sonnichsen, and J. Guigne. 1998. Effects of experimental otter trawling on surficial sediment properties of a sandy-bottom ecosystem on the Grand Banks of Newfoundland. Conservation Biology 12(6):1215–1222.
Thrush, S. F., J. E. Hewitt, V. J. Cummings, and P. K. Dayton. 1995. The impact of habitat disturbance by scallop dredging on marine benthic communities: what can be predicted from the results of experiments? Marine Ecology Progress Series 129:141–150.

Habitat and Fish Populations in the Deep-Sea *Oculina* Coral Ecosystem of the Western Atlantic

CHRISTOPHER C. KOENIG[1]

Department of Biological Science, Florida State University, Tallahassee, Florida 32306-1100, USA

ANDREW N. SHEPARD[2]

*NOAA Undersea Research Program, University of North Carolina at Wilmington,
5600 Marvin Moss Lane, Wilmington, North Carolina 28409, USA*

JOHN K. REED[3]

Harbor Branch Oceanographic Institution, 5600 U.S. 1 North, Fort Pierce, Florida 34946, USA

FELICIA C. COLEMAN[4]

Department of Biological Science, Florida State University, Tallahassee, Florida 32306-1100, USA

SANDRA D. BROOKE[5]

Oregon Institute of Marine Biology, Post Office Box 5389, Charleston, Oregon 97420, USA

JOHN BRUSHER[6]

National Marine Fisheries Service, 3500 Delwood Beach Road, Panama City, Florida 32408, USA

KATHRYN M. SCANLON[7]

U.S. Geological Survey, 384 Woods Hole Road, Woods Hole, Massachusetts 02543, USA

Abstract.— The growth form of the scleractinian ivory tree coral *Oculina varicosa* (also known as fused ivory tree coral) that occurs on the shelf edge off Florida's eastern coast is unique for this species. Here, the branching coral colonies coalesce into thickets supporting high vertebrate and invertebrate biodiversity and high densities of economically important reef fish. In 1984, the South Atlantic Fishery Management Council took the first step to protect the area from trawling and other disruptive bottom activities. Despite these protective measures, however, there is evidence that trawling has damaged previously intact coral habitat. In this paper, we describe results from mapping studies conducted in 2001 and improvements to reef fish populations that have occurred in the last few years. We find that less than 10% of the area contains intact *Oculina* coral thickets, which we continue to attribute primarily to trawling. In addition, we find increased grouper density and male abundance inside the protected area, suggesting population recovery, and the appearance of juvenile speckled hind *Epinephelus drummondhayi* (family Serranidae), suggesting nursery function for this and possibly other commercially important species.

---

[1] E-mail: koenig@bio.fsu.edu
[2] E-mail: sheparda@uncwil.edu
[3] E-mail: jreed@hboi.edu
[4] E-mail: coleman@bio.fsu.edu
[5] E-mail: sbrooke@oimb.uoregon.edu
[6] E-mail: john.brusher@noaa.gov
[7] E-mail: kscanlon@usgs.gov

Introduction

Deep-sea coral species are subject to both increased interest and increased pressure (Malakoff 2003). The ivory tree coral *Oculina varicosa* (also known as fused ivory tree coral) is a case in point. This species, common in small (<30 cm), isolated, shallow water (2–30 m) colonies from the West Indies to North Carolina and Bermuda, occurs off Florida's eastern coast in deep (60–120 m), species-unique reefs as 2-m high azooxanthellate thickets on the slopes and crests of pinnacles (Reed et al. 1982; Reed 2002). These reefs, the *Oculina* Banks, extend 67 km along the outer shelf (Avent et al. 1977; Virden et al. 1996; Figure 1). Healthy *Oculina* reefs support a diverse invertebrate assemblage (Reed et al. 1982; Reed 2002), dense populations of fishes (G. Gilmore, Dynamac Corporation, Kennedy Space Center, unpublished data), and important spawning sites for many economically important reef fish species (Gilmore and Jones 1992; Koenig et al. 2000).

Interest in these unique *Oculina* thickets set in motion a series of protective measures by the South Atlantic Fishery Management Council (SAFMC), starting in 1984 with the designation of the *Oculina* Habitat Area of Particular Concern (OHAPC; 316 km^2) to prohibit the use of bottom gear. In 1994, the area became the Experimental *Oculina* Research Reserve (EORR), extending the prohibition to bottom fishing for 10 years to explore the use of marine protected areas (MPAs). In 2000, the OHAPC was expanded to 1,029 km^2. More recently (2003), the EORR closure was extended indefinitely.

Our 1995 observations in the EORR confirmed previous findings by Reed (1980) of extensive coral rubble (Koenig et al. 2000). We also found trawl damage to coral habitat known to be intact 20 years earlier and severely reduced reef fish populations. Jeff's Reef, a small (4-ha) area in the southern EORR (Figure 1), appeared to be the only intact area, although the biomass and number of economically important fish were much lower than they had been in the 1970s. The objectives of this study were to estimate the relative proportion of intact and rubble *Oculina* habitat on high relief sites in the OHAPC based on knowledge that intact coral habitat occurred predominantly on high relief and to evaluate reef fish use of both natural and artificial structure in the EORR.

Methods

The data presented here were derived from the National Oceanic and Atmospheric Administration's 2001 Islands in the Stream expedition. In 8 d (30 August–6 September), we completed 13 remotely operated vehicle (ROV) dives (Phantom S4, National Undersea Research Center, Wilmington, North Carolina) and 16 submersible dives (Clelia, Harbor Branch Oceanographic Institution, Ft. Pierce, Florida) in the EORR and other portions of the OHAPC (Figure 1), producing more than 70 h of underwater videography. The ROV made line transects over large areas of the seafloor to determine the relative abundances of coral habitat on the ridges and pinnacles. The submersible made belt transects to quantify habitat type and fish density within habitat types.

Habitat Condition

We used a base map derived from side-scan sonar images (Scanlon et al. 1999) and the ship's echo sounder to set sampling sites. Collection site coordinates and transect lengths were determined with differential global positioning system navigation (Magnavox MX 200 global positioning system [GPS], accurate to within ± 5 m) and ArcView software. Plots of the submersible tracks and specific sample sites were made with the Integrated Mission Profiler (Florida Atlantic University, Boca Raton) linked to the ship's GPS.

Remotely Operating Vehicle Sampling

The ROV was tethered near the bottom to a 100-kg down weight by two parallel lines: (1) the first 20 m of the ROV umbilical, extending beyond the down weight to allow limited ROV movement at approximately 1.85 km/h (1.0 knot) northerly (i.e., with the Florida Current); and (2) a 19-m polypropylene tension-relief line to tow the ROV. The remaining ROV umbilical extended vertically to the vessel and was attached along its length. Transect positions were recorded while the ROV was under way to allow location of changes in geomorphology, habitat, and depth.

Manned Submersible Sampling

Habitat types identified by the ROV were further characterized by submersible with an underwater video (Insite-Tritech high sensitivity [0.0003 lux], high-resolution monochrome 1.23-cm charge coupled device). Video imagery on statistically random belt transects was recorded with laser-equipped downward-looking and forward-looking cameras. The downward-looking camera's two parallel lasers (25 cm apart) provided the scale for standardizing quadrat size and measuring coral colonies. The forward-looking camera's three inline lasers (10-cm intervals) provided the scale (two adjacent parallel beams) and distance (the third beam converged on the other two at 5 and 10 m), allowing determination of transect width at a selected distance from the camera. Laser dots were visible at approximately 5 m in the lower half of the camera's field of view.

Quadrats were derived from 16 to 20 randomly selected video frames from each downward-looking transect. Each quadrat was standardized relative to the laser metric and overlain with 100 randomly distributed dots to determine percent cover of each habitat type. The mean percent cover was calculated for each transect (averaging randomly selected frames) and compared within a given site for each habitat type using analysis of variance (ANOVA; arcsine transformation). A Shapiro–Wilk test for normality and Duncan's Multiple Range test were used to iden-

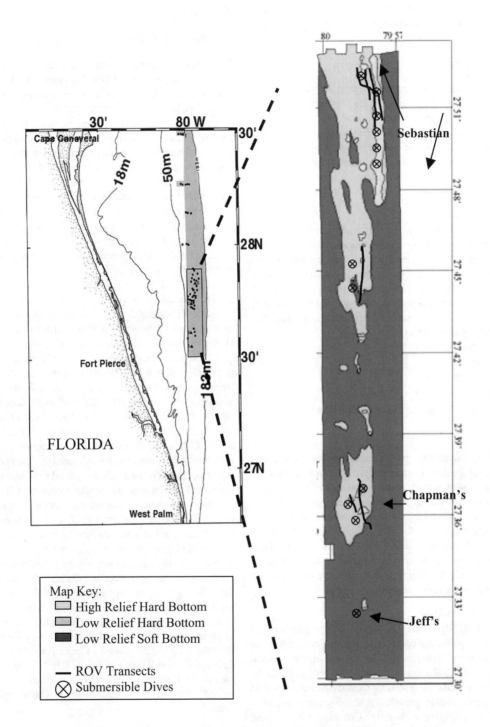

Figure 1. Map of *Oculina* Banks Habitat Area of Particular Concern (OHAPC, shaded area in left locator map), includes the Experimental *Oculina* Research Reserve (EORR, inset box on left, expanded on right). Dots in OHAPC are historic dive sites visited in the 1970s and 1980s. Expanded EORR shows hard and soft bottom, high and low relief (Scanlon et al. 1999), and location of 2001 remotely operated vehicle (ROV) transects and submersible dives.

tify homogeneous subsets among transect means. Randomly selected coral colonies within each transect were measured.

Fish Densities

Fish densities (numbers per ha) were determined during submersible belt transects in each habitat type. Transects were run without lights to avoid affecting fish behavior. Belt transect quantification of fish populations provides a statistical basis for spatial and temporal comparisons, measuring relative rather than absolute abundance and requiring that interannual comparisons account for temporal activity patterns.

Natural Habitat

Estimating belt transect area from submersible videos required determining the effective distance (D), the camera's horizontal angle of view ($A = 92°$), and the length (L) of the transect. The effective distance is the distance from the camera within which fish are counted and identified with high certainty rather than the limits of visibility (typically < 5 m, which was used as the standard distance; fish occurring beyond 5 m were excluded).

The field of view width (W) at distance D was calculated by:

$$W = 2 [\tan (0.5A)] (D).$$

The area of the transect (TA) was calculated by:

$$TA = (L \times W) - 0.5(W \times D).$$

Estimating the transect area allowed calculating the average density and standard error of observed fish species within each habitat type. Species tending to follow or circle the submersible (e.g., greater amberjack and almaco jack) were *not* counted each time they appeared on the video. Rather, their total abundance was determined by observers in the submersible. Density differences among habitats were determined for numerically dominant species, economically important groupers, and pelagic species using ANOVA.

Habitat Modules

Reef balls (Reef Ball Foundation, www.reefball.org) ($N = 105$)—perforated concrete domes 1 m across and 0.7 m high—were deployed in 2000 to simulate the size and aspect of *Oculina* coral colonies and serve as larval recruitment surfaces, centers for *Oculina* thicket restoration through transplant growth, and structure replacement for reef fish (Figure 2; Koenig, Coleman, Brooke, and Brusher, unpublished data). They were distributed among nine areas (each 500 m^2) in clusters of 5, 10, or 20, with three replicates of each cluster size in a randomized block design to determine the most efficient density for attracting fish.

Results

Habitat Condition

Seven ROV line transects were made over high relief pinnacles and ridges in the EORR (Chapman's Reef: $N = 3$; Sebastian Reef: $N = 4$). Transect lengths ranged from 424 to 2,867 m, covering 7,645 m of high-relief seafloor (Figure 1). Three coral cover levels were identified: (1) dense—relatively undisturbed, large live and dead coral thickets with multi-scale structural complexity; (2) sparse—small colonies widely distributed in expanses of consolidated (rubble with identifiable coral branching) and unconsolidated rubble (fine coral debris), providing little structural complexity; and (3) no coral cover—sand, rock, and unconsolidated rubble, providing essentially no structural complexity. The relative proportion of each coral cover type was estimated as the proportional distance traversed by the ROV over that habitat type.

Of the total high relief pinnacle ridges transected, 464 m (6%) contained dense coral cover, 302 m (4%) contained sparse cover, and 6,877 m (90%) contained no cover. The only additional dense thickets identified during this study (Jeff's Reef having been located in 1995) were approximately 4 ha on the western bank of Chapman's Reef (Chapman's Reef West), one of three banks in Chapman's Reef (Figure 1). The only sparse habitats occurred on the south face of Chapman's Reef East and on the slope bases of Jeff's Reef and Chapman's Reef West. Three additional random transects covering 2,041 m of high relief just north of the EORR within the OHAPC revealed only unconsolidated rubble. Sparsely distributed, small (5–20-cm diameter) colonies of *Oculina* were associated with some of the rubble and with large boulders on low relief rocky bottom.

Sixteen belt transects were made in the EORR with the submersible ($N = 8$ at Jeff's Reef, $N = 5$ at Chapman's Reef West, $N = 3$ at Sebastian Reef; Figure 1), revealing four habitat types: intact live coral, intact dead coral, coral rubble, and bare rock and sand (Figure 3). Intact live coral only occurred on Jeff's Reef and Chapman's Reef West. Sebastian Reef was mostly coral rubble. Within each reef, the mean live coral coverage varied considerably among transects (ANOVA, $P < 0.01$). For Jeff's Reef, mean live coral coverage ranged from 9% to 21% and for Chapman's Reef West, 7% to 22% (Table 1). Coral colony diameter on Chapman's Reef West ranged from 8 to 143 cm, with a mean of 47.4 cm (SE = 4.75 cm, $N = 43$). Coral colony size on Jeff's Reef was not measured due to a laser malfunction.

Fish Populations

Natural Habitat

Population densities for the dominant fish species correlated highly with habitat type (Figure 4). Only one

Figure 2. Habitat modules deployed on Sebastian Reef within the Experimental *Oculina Research Reserve* off the east coast of Florida in 2000. A wooden cross attached to the top of each reef ball with jute line (both substances being biodegradable) provided sufficient drag to make the reef balls land upright on the bottom.

economically important species was observed on coral rubble (Table 2). Highly cryptic juvenile speckled hind *Epinephelus drummondhayi* associated with intact habitat at average densities of 3–5 per ha. Male gag *Mycteroperca microlepis* occurred on Jeff's Reef.

Habitat Modules

Surveys of reef ball clusters occurred thirteen months after reef ball deployment. Surveys were easy to do because each cluster covered a small area (12.6-m radius). The mean species richness and abundance of economically important fish were greater for reef ball densities of 10 per cluster than for 5 but did not increase further at densities of 20 per cluster (Figure 5; Table 3). Male gag and scamp *Mycteroperca phenax* occurred near reef ball clusters.

Submersible observations around habitat modules and coral-transplant modules revealed module pieces missing and littering the bottom, suggesting impact by strong

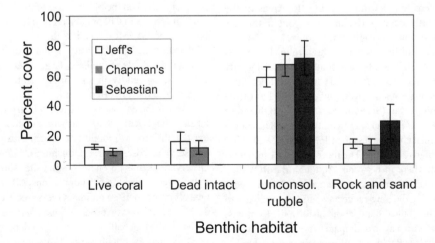

Figure 3. Percent cover of habitat types in intact coral habitat (Jeff's and Chapman's West Reef) and in unconsolidated rubble habitat (Sebastian Reef) within the Experimental *Oculina* Research Reserve off the eastern coast of Florida. Bars = standard error.

Table 1. Mean live coral cover and standing dead coral cover (in parentheses) determined from belt transects made with submersible on Jeff's Reef and Chapman's Reef in the *Oculina* Experimental Research Reserve. Thick horizontal lines indicate homogenous groupings of live coral (based on Duncan's multiple range test).

Reef	Transect number							
	1	2	3	4	5	6	7	8
Jeff's Reef	8.9 (8.4)	9.2 (3.2)	11.1 (17.7)	12.9 (8.4)	13.6 (7.4)	13.6 (10.4)	20.0 (46.2)	21.3 (49.8)
Chapman's Reef	7.0 (12.7)	7.7 (0)	9.2 (16.0)	11.0 (24.8)	22.3 (727.1)			

mechanical means. Apparent trawl tracks in the rubble were noted near the damage.

Discussion

Habitat Characterization

During this study, we specifically targeted high-relief sites in the OHAPC known in the 1970s to have either intact coral thickets (e.g., Jeff's Reef and Chapman's Reef) or extensive coral rubble (e.g., Sebastian Reef; Reed 1980; Koenig et al. 2000). We used direct observation rather than acoustic methods because the latter does not distinguish among live coral, dead intact coral, or unconsolidated rubble. Rubble is a major component on high-relief features. The concern is that so few high relief sites had intact thickets. Indeed, about 90% of the habitat surveyed was unconsolidated rubble; less than 10% contained intact coral colonies. No additional coral thickets were found within the EORR. Areas of the OHAPC north of the EORR known to contain thickets 20 years ago contained only coral rubble.

Ten percent intact coral on high relief features is likely a high estimate. If one assumes the EORR's only intact habitat is on Jeff's Reef and Chapman's Reef West, a lower estimate results. Roughly 3% (947 ha) of the EORR is high relief (Scanlon et al. 1999) and therefore suitable for *Oculina* thicket growth. With only 8 ha known to contain intact habitat, less than 1% of the intact habitat occurs on high relief sites. The more accurate estimate is likely somewhere in between. Although the ROV line transects targeted areas that once supported *Oculina* thickets, transposing old long-range navigation coordinates to GPS introduces uncertainty about historic site locations, and there was no way to anticipate which features would contain rubble and which would contain intact colonies.

In intact habitat, live coral coverage was less than half that of dead standing coral coverage, and both types of coverage were highly variable among transects. Observations of small coral colonies within coral rubble (primarily on high relief sites, occasionally on low relief sites) and extensive coral colonies on 60-year-old shipwrecks just outside of the EORR (M. Barnette, National Marine Fisheries Service, personal communication) suggest that coral colonization and growth occur but are insignificant. The presence of small dead standing colonies in low relief sites suggests that these may be marginal sites for survival.

Fish Populations

In the past 30 years, the size, age, and proportion of male gag and scamp have declined throughout the southeastern United States (Coleman et al. 1996; McGovern et al. 1998; Koenig et al. 2000). The results of this study suggest that protecting aggregation sites and resident populations within MPAs can help reestablish historical fish populations. Indeed, gag and scamp, including males, occur on coral thickets within the EORR but not on sites outside of the EORR. They also suggest some nursery function, based on the observation of juvenile speckled hind on Jeff's and Chapman's reefs. This is significant because the SAFMC considers this species threatened (Coleman et al. 2000). Density estimates of small fish or young individuals of typically larger species are probably low, especially in structurally complex habitats where these fish are often cryptic.

Unlike the typical artificial reef, which provides habitat and attracts reef fish to areas where neither previously existed, the reef ball modules replace destroyed habitat and serve as bases for reestablishing *Oculina* thickets. Observations thus far on restoration sites show promise only for reestablishing fish populations. All grouper species observed in 1980 on intact reefs (see Koenig et al. 2000), except warsaw grouper *Epinephelus nigritus*, associated with reef balls 1 year after their deployment. The reef balls may eventually support spawning, based on the presence of both gag and scamp males (typical of spawning aggregation sites, per Coleman et al. 1996), with scamp exhibiting presumed courtship behavior (described in Gilmore and Jones 1992).

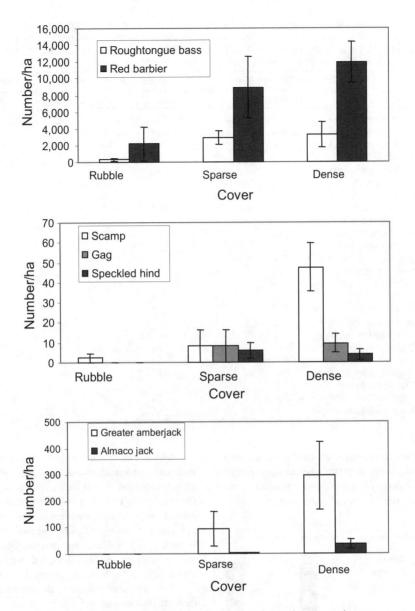

Figure 4. Mean population densities of (A) dominant basses (Anthiinae; roughtongue bass *Holanthias martinicensis* and red barbier *Hemanthias vivanus*), (B) dominant groupers (Epinephelinae; scamp, gag, and speckled hind), and (C) pelagic species (greater amberjack *Seriola dumerili* and almaco jack *S. rivoliana*) in three levels of coral habitat condition. Bars = standard errors. Scamp density in intact habitat was significantly greater ($P = 0.05$) than in other habitats.

Possible Causes of Habitat Decline

Natural, wholly unmanageable events that damage coral include extreme temperatures (Fitt et al. 2001), excessive nutrient input (Szmant 2002), strong currents (Lugo et al. 2000), and disease (Porter et al. 2001). *Oculina* is relatively tolerant of changes in temperature and nutrient and sediment input that occur during episodic deep-sea upwelling events (Reed 1983), although this tolerance may not persist in the face of global warming or increased nutrient loads associated with ocean dumping. Although no studies of

Table 2. Comparison of mean densities of species observed in intact habitat (Jeff's Reef and Chapman's Reef) and unconsolidated coral rubble (Sebastian Reef; presumably coral destroyed by trawling) within the Experimental *Oculina* Research Reserve. An asterisk indicates an economically important species.

Species	Jeff's Reef Number/ha	SE	Chapman's Reef Number/ha	SE	Sebastian Reef Number/ha	SE
Red barbier	7,301	2,757	18,082	3,429	277	277
Roughtongue bass	1,211	410	6,424	3,560	141	89
Greater amberjack*	298	182	284	187		
Yellowtail reeffish *Chromis enchrysurus*	60	30	277	111	53	43
Almaco jack*	55	29				
Scamp*	48	11	47	27		
Blue angelfish *Holacanthus bermudensis*	39	13	204	72	6	6
Bank butterflyfish *Chaetodon aya*	27	10	110	35	19	12
Gag*	16	7				
Reef butterflyfish *Chaetodon sedentarius*	13	6	102	50	17	11
Specked hind*	3	3	5	5		
Tattler *Serranus phoebe*	3	3	27	12	44	12
Spotfin butterflyfish *Chaetodon ocellatus*	2	2				
Porgy *Calamus* spp.*			10	10		
Wrasse bass *Liopropoma eukrines*			17	13		
Soapfish *Rypticus* spp.			17	13		
Wrasse Labridae					42	26
Purple reeffish *Chromis scotti*			36	18		
Snapper *Lutjanus* spp.*					6	6

disease have been conducted in the *Oculina* banks, virulent pathogens would be expected to cause extensive damage to ahermatypic reefs like *Oculina* rather than selective elimination of some reefs but not adjacent reefs.

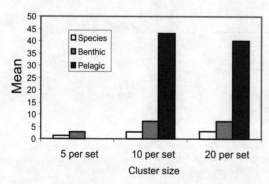

Figure 5. Mean number of species and individuals (benthic and pelagic) of economically important reef fish associated with three reef ball densities, 5 per 500 m^2, 10 per 500 m^2, and 20 per 500 m^2 set over unconsolidated coral rubble in Sebastian Reef within the Experimental *Oculina* Research Reserve off the east coast of Florida There were 3 replicates of each set.

Most of the evidence for *Oculina* habitat destruction points to human-induced impacts. While it is possible that World War II exchanges between U.S. and German vessels west of the OHAPC (Cremer 1986) caused some damage, these encounters ended about 60 years ago, allowing sufficient time for habitat recovery. Indeed, U.S. freighters sunk near the OHAPC by German U-boats in the 1940s support dense *Oculina* thickets on their decks (Barnette, personal communication).

Trawling continues to be the greatest manageable threat to the *Oculina* reefs (Koenig et al. 2000). Bottom trawling and dredging worldwide result in severe coral damage (Jones 1992; Rogers 1999; Fosså et al. 2000; Koslow et al. 2000; Richer de Forges et al. 2000), requiring long recovery times (Dayton et al. 2002; Johnson 2002). These have occurred off Florida's eastern coast for years, involving both foreign and domestic fleets. Foreign trawling stopped in the late 1970s with development of the U.S. Exclusive Economic Zone.

The extent to which domestic trawling persists in the *Oculina* banks is unknown. However, circumstantial evidence suggests that it does to some degree. Trawlable high relief bottom features where *Oculina* normally occurs show little evidence of coral recolonization, while

Table 3. Comparison of reef fish found on Sebastian Reef in the Oculina Experimental Research Reserve, associated with three different densities of reef balls deployed in coral restoration experiments. Asterisk indicated economically important species.

Species	5 reef balls/cluster		10 reef balls/cluster		20 reef balls/cluster	
	Number/1,500 m²	%	Number/1,500 m²	%	Number/1,500 m²	%
Greater amberjack*			109	37.72		
Roughtongue bass	7	41.18	120	41.52	53	21.9
Red barbier			1	0.35	25	10.33
Almaco jack*			20	6.92	20	8.26
Scamp*	3	17.65	15	5.19	14	5.79
Wrasse			1	0.35	10	4.13
Blue angelfish			3	1.04	5	2.07
Reef butterflyfish			4	1.38	3	1.24
Red snapper* *Lutjanus campechanus*			6	2.08	2	0.83
Snowy grouper* *Epinephelus niveatus*	2	11.76			2	0.83
Speckled hind*					3	1.24
Tattler	1	5.88			2	0.83
Red porgy* *Pagrus pagrus*	2	11.76			2	0.83
Sharpnose puffer *Canthigaster rostrata*					1	0.41
Queen angelfish *Holacanthus ciliaris*					1	0.41
Bank butterfly	1	5.88	2	0.69		
Short bigeye *Pristigenys alta*			2	0.69		
Twospot cardinalfish *Apogon pseudomaculatus*			2	0.69		
Spinycheek soldierfish *Corniger spinosus*			2	0.69		
Sharpnose puffer *Canthigaster rostrata*			1	0.35		
Bank sea bass *Centropristis ocyurus*	1	5.88				

untrawlable wrecks in the same area support dense thickets. The incidence of trawling is sufficiently high that the SAFMC requires local trawlers to use vessel monitoring systems. The council did not alter the penalties for trawling, however, which currently are relatively light (i.e., confiscated catch and moderate fines) and viewed by violators as a business expense (anonymous commercial fisherman, personal communication). This differs significantly in the Florida Keys National Marine Sanctuary, where, based on the National Marine Sanctuaries Act (U.S. Code, Title 16, chapter 32, section 1431 et seq., as amended in Public Law 106-513, November 2002), those guilty of destroying coral habitat—for whatever reason—are subject to fines substantial enough to cover the costs of habitat restoration or mitigation.

While surveillance and enforcement are important to management of MPAs, compliance indicates that extractive users perceive MPA boundaries as fair and equitable. This typically results from knowledge of the natural resources that occur within reserve boundaries and the ecological and economic benefits derived from their protection. Education clearly provides the most efficient, cost-effective, and powerful stimulus to habitat protection.

Acknowledgments

We thank J. McDonough (National Oceanic and Atmospheric Administration [NOAA] National Ocean Service [NOS]), T. Potts (National Undersea Research Center [NURC], University of North Carolina, Wilmington [UNCW]), and S. Orlando (NOS) for organizing the NOAA Islands in the Stream cruise. L. Horn (NURC/UNCW) served as expert pilot for the ROV. M. Barnette (NOAA National Marine Fisheries Service [NMFS], St. Petersburg, Florida) provided information on *Oculina* growth on shipwrecks. G. Gilmore (Dynamac Corporation, Kennedy Space Center) provided significant historical perspective. E. Proulx (NMFS, retired) participated in many discussions on enforcement issues. We thank the NMFS Panama City Laboratory, for support, particularly H. Kumpf (Acting Director), L. Barger, C. Palmer, and A. David. NOS and NMFS Southeast Fisheries Science Center supplied funding for the "Islands in the Stream" OHAPC study, especially for the use of Harbor Branch Institution's ship and submersible. We are grateful to two anonymous reviewers whose comments improved the manuscript.

References

Avent, R. M., M. E. King, and R. H. Gore. 1977. Topographic and faunal studies of shelf-edge prominences off the central eastern Florida coast. Internationale Revue gesamten Hydrobiologie 62:185–208.

Coleman, F. C., C. C. Koenig, and L. A. Collins. 1996. Reproductive styles of shallow-water grouper (Pisces: Serranidae) in the eastern Gulf of Mexico and the consequences of fishing spawning aggregations. Environmental Biology of Fishes 47:129–141.

Coleman, F. C., C. C. Koenig, G. R. Huntsman, J. A. Musick, A. M. Eklund, J. C. McGovern, R. W. Chapman, G. R. Sedberry, and C. B. Grimes. 2000. Long-lived reef fishes: the grouper–snapper complex. Fisheries 25:14-21.

Cremer, P. 1986. U-boat commander: the battle of the Atlantic through a periscope. Berkley Books, New York.

Dayton, P. K., S. Thrush, and F. C. Coleman. 2002. The ecological effects of fishing in marine ecosystems of the United States. The Pew Oceans Commission, Arlington, Virginia.

Fitt, W. K., B. E. Brown, M. E. Warner, and R. P. Dunne. 2001. Coral bleaching: interpretation of thermal tolerance limits and thermal thresholds in tropical corals. Coral Reefs 20:51–65.

Fosså, J. H., P. B. Mortensen, and D. M. Furevik. 2000. The deep water coral *Lophelia pertusa* in Norwegian waters; distribution and fishery impacts. First International Symposium on Deep Sea Corals 25.

Gilmore, R. G., and R. S. Jones. 1992. Color variation and associated behavior in the epihepheline groupers, *Mycteroperca microlepis* (Goode and Bean) and *M. phenax* Jordan and Swain. Bulletin of Marine Science 51:83–103.

Johnson, K. A. 2002. A review of national and international literature on the effects of fishing on benthic habitats. NOAA Technical Memorandum NMFS-F/SPO-57.

Jones, A. D. 1992. Environmental impact of trawling on the seabed a review. New Zealand Journal of Marine and Freshwater Research 26:59–67

Koenig, C. C., F. C. Coleman, C. B. Grimes, G. R. Fizhugh, K. M. Scanlon, C. T. Gledhill, and M. Grace. 2000. Protection of fish spawning habitat for the conservation of warm temperate reef fish fisheries of shelf-edge reefs of Florida. Bulletin of Marine Science 66:593–616.

Koslow, J. A., G. W. Boehlert, J. D. M. Gordon, R. L. Haedrich, P. Lorance, and N. Parin. 2000. Continental slope and deep-sea fisheries implications for a fragile ecosystem. ICES Journal of Marine Science 57:548–557.

Lugo, A. E., C. S. Rogers, and S. W. Nixon. 2000. Hurricanes, coral reefs and rainforests: resistance, ruin and recovery in the Caribbean. Ambio 29:106–114.

Malakoff, D. 2003. Cool corals become hot topic. Science 299:195.

McGovern, J. C., D. M. Wyanski, O. Pashuk, C. S. I. Manooch, and G. R. Sedberry. 1998. Changes in the sex ratio and size at maturity of gag *Mycteroperca microlepis*, from the Atlantic coast of the southeastern United States during 1976–1995. U.S. National Marine Fisheries Service Fishery Bulletin 96:797–807.

Porter, J. W., P. Dustan, W. C. Jaap, K. L. Patterson, V. Kosmynin, O. W. Meier, M. E. Patterson, and M. Parsons. 2001. Patterns of spread of coral disease in the Florida Keys. Hydrobiologia 460:1–24.

Reed, J. K. 1980. Distribution and structure of deep-water *Oculina varicosa* coral reefs off central eastern Florida. Bulletin of Marine Science 30:667–677.

Reed, J. K. 1983. Nearshore and shelf-edge *Oculina* coral reefs: the effects of upwelling on coral growth and on the associated faunal communities. Pages 119–124 *in* M. E. Reaka, editor. The ecology of deep and shallow coral reefs. National Oceanic and Atmospheric Administration, Symposia Series for Undersea Research, volume 1, Rockville, Maryland.

Reed, J. K. 2002. Deep-water *Oculina* coral reefs of Florida: biology, impacts, and management. Hydrobiologia 471:43–55.

Reed, J. K., R. H. Gore, L. E. Scotto, and K. A. Wilson. 1982. Community composition, structure, areal and trophic relationships of decapods associated with shallow- and deep-water *Oculina varicosa* coral reefs. Bulletin of Marine Science 32:761–786.

Richer de Forges, B., J. A. Koslow, and G. C. Poore. 2000. Diversity and endemism of the benthic seamount fauna in the southwest Pacific. Nature 405:944–947.

Rogers, A. D. 1999. The biology of *Lophelia pertusa* (Linnaeus 1758) and other deep-water reef forming corals and impacts from human activities. International Review of Hydrobiology 84:315-406.

Scanlon, K. M., P. R. Briere, and C. C. Koenig. 1999. *Oculina* bank: sidescan sonar and sediment data from a deep-water coral reef habitat off east-central Florida. U.S. Geological Survey, Open File Report 99-10, Woods Hole, Massachusetts.

Szmant, A. M. 2002. Nutrient enrichment on coral reefs: is it a major cause of coral reef decline? Estuaries 25:743–766.

Virden, W. T., T. L. Berggren, T. A. Niichel, and T. L. Holcombe. 1996. Bathymetry of the shelf-edge banks, Florida east coast. 1. National Oceanic and Atmospheric Administration, National Geophysical Data Center, National Marine Fisheries Service, Beaufort, North Carolina.

The Impact of Demersal Trawling on Northeast Atlantic Deepwater Coral Habitats: The Case of the Darwin Mounds, United Kingdom

ANDY J. WHEELER[1]

*Department of Geology and Environmental Research Institute,
Donovan's Road, University College Cork, Cork, Ireland*

BRIAN J. BETT,[2] DAVE S. M. BILLETT,[3] DOUG G. MASSON,[4] AND DANIEL MAYOR[5]

Southampton Oceanography Centre, European Way, Southampton SO14 3ZH, UK

Abstract. Deepwater corals form reefs and carbonate mounds that are important biological habitats along the European continental margin. Recent mapping of these features has highlighted significant habitat impact resulting from demersal trawling. With the current expansion of European deepwater fisheries, the potential for further coral habitat damage will increase. Seabed observations (100kHz side-scan sonar, still, and video imagery) are presented here that document trawling impacts on the Darwin Mounds, a field of small, coral-topped mounds at c.1,000 m water depth in the northern Rockall Trough. Comparisons between trawled and nontrawled mounds are startling. Trawl marks are clearly visible on side-scan sonar records, with visual imagery showing higher abundance of dead coral and coral rubble at trawled sites compared to untrawled sites. Some of the seabed in the Darwin Mound areas has been intensely trawled, with local areas at a scale resembling the distance between trawl doors being 100% trawled. Some areas show evidence for multiple trawling events. Coral habitat destruction can occur on a scale that impacts the coral growths on entire coral mounds. The conflict between deepwater fisheries and habitat protection in the European Atlantic Margin is discussed.

Introduction

Recent years have seen significant scientific and public attention focused on the occurrence of deepwater coral ecosystems (e.g., Edwards 2000; Irish Skipper 2001; Montgomery 2001; Siggins 2001; Urquhart 2001; Clarke 2002; Dybas 2002). These communities represent important biological habitats of high biodiversity (Jensen and Frederiksen 1992; Rogers 1999) in water depths between c. 50 and 1,100 m on the European continental margin (see Zibrowius 1980; Rogers 1999; ICES 2003 and references therein) and elsewhere (see Cairns 1979; Reed 1980 for examples of regional studies). This paper concentrates exclusively on European examples. The presence of the framework-building corals *Lophelia pertusa* and *Madrepora oculata* enables the development of carbonate mounds and reefs varying in height from a few meters (e.g., Masson et al. 2003; Wheeler et al. 2005b) to several hundred meters (e.g., Henriet et al. 1998; De Mol et al. 2002; Kenyon et al. 2003). The deepwater coral ecosystems may have a role as fisheries nurseries and refuges (Rogers 1999), indicators of hydrocarbon seepage (Hovland 1990; Hovland et al. 1994, 1998; Hovland and Thomsen 1997; Henriet et al. 1998) and reservoirs of biodiversity (Jensen and Frederiksen 1992; Rogers 1999).

Recent studies have detailed the destruction of deepwater coral habitats resulting from the activity of demersal trawling (Fosså et al. 2002; Hall-Spencer et al. 2002). The detrimental effects of trawling on benthic communities is well documented (for reviews see: Auster et al. 1996; Jennings and Kaiser 1998; Hall 1999; Collie et al. 2000), with studies of impacts on coral communities showing damage to coral and sponge species and a decrease in the abundance of invertebrates and fish (e.g., Bradstock and Gordon 1983; Van Dolah et al. 1987; Probert et al. 1997; Koslow et al. 2001; Fosså et al. 2002; Hall-Spencer et al. 2002). Destruction of Euro-

---

[1] E-mail: a.wheeler@ucc.ie
[2] E-mail: bjb@soc.soton.ac.uk
[3] E-mail: dsmb@soc.soton.ac.uk
[4] E-mail: dgm@soc.soton.ac.uk
[5] E-mail: dxm@soc.soton.ac.uk

pean deepwater coral reefs was first documented in detail at the Storegga shelf break off Norway, where *Lophelia* reefs occur at relatively shallow water depths (300–400 m) (Fosså et al. 2002). The decline in inshore and shallow-sea fish stocks has resulted in increasing fishing pressure on the deep waters of the European Atlantic margin. Large, ocean-going, demersal trawlers are now operating in areas where deepwater coral ecosystems are likely to be encountered. Low relief coral mounds and reefs, such as those described in the present contribution, may be at particular risk from the heavy trawl gear operated by such vessels. There is already some indication that coral systems on large carbonate mounds may also be at risk (see Wheeler et al. 2005b, for example).

Here, we present information from geo-acoustic and visual mapping of the seafloor that appears to indicate the direct destruction of coral habitat by deep-sea demersal trawling activity. These observations were made in the Darwin Mounds area, named after the research vessel RRS *Charles Darwin* (Masson and Jacobs 1998; Bett 1999), a field of some hundreds of small coral-topped mounds in the northern Rockall Trough (Bett 2001; Masson et al. 2003). The location of the Darwin Mounds and the particular area covered by the observations presented here are illustrated in Figure 1.

Survey Techniques

The Darwin Mounds site was first detected using Southampton Oceanography Centre's (SOC) TOBI deep-tow side-scan sonar (30kHz) system in the summer of 1998 (Masson and Jacobs 1998), with initial photographic surveys using the SOC WASP vehicle carried out shortly thereafter (Bett 1999). Further TOBI mapping and photography was carried out in 1999 (RRS *Charles Darwin* cruise 119), with some additional photography undertaken in 2000 (RRS *Charles Darwin* cruise 123). The bulk of the observations reported here were made during RRS *Discovery* cruise 248 (Bett et al. 2001) in the summer of 2000.

Seabed mapping was carried out using a Geoacoustic dual frequency (100 and 410 kHz) high-resolution side-scan sonar. The towfish was flown 50 m above the seabed at 100 kHz and 10 m off the seabed at 410 kHz. Initial towfish navigation was calculated by layback from the ship's position (differential global positioning system). Side-scan sonar data were processed using SOC's PRISM software (Le Bas and Hühnerbach 1999). During this process, towfish navigation was refined to produce an optimum side-scan sonar mosaic to a 50-m navigational accuracy as confirmed by comparison with features observed on other seabed survey data sets. Ground truthing of the sonar imagery was undertaken using SOC's SHRIMP (Seabed High Resolution Imaging Platform) vehicle (*www.soc.soton.ac.uk/OED/index.php?page=sh*). The video footage obtained from SHRIMP deployments was split into 30-s windows and benthic organisms identified and quantified. Various other seabed features (e.g., trawl marks) were also recorded by time of occurrence (and, hence, position). Coral cover (percent live, dead, and coral rubble) was estimated every 15 s (approximately the time it takes for one video screen to pass the field of view) and averaged for each 30-s period. Live coral refers to coral frameworks where polyps or a colored fleshy covering (usually pinkish-orange) to the coral exoskeleton were observed. Dead coral refers to coral exoskeletons where no polyps or colored fleshy covering were observed and corals appear white to gray. Coral rubble refers to broken coral fragments that may be alive, although usually dead, and have been formed by either natural degradation processes or mechanical damage by fishing bottom gear. The SHRIMP navigation was based on layback from the ship's position (i.e., knowledge of water depth, length of cable deployed, and assumption of the vehicle following the ship's track). Comparison with the operation of a similar vehicle (SOC WASP system; Huggett 1987) tracked using an ultrashort baseline acoustic navigation system suggests that SHRIMP is likely to (90% of the time) be located within 60 m of the ship's track when operated at 1,000 m (see Bett 1999).

Study Location

The Darwin Mounds are relatively small, discrete, coral-colonized features that occur between 900 m and 1,060 m water depth. They are characteristically ovoid in shape, measuring up to 75 m across, and have a maximum topographic elevation of some 5 m. Mound height tends to decrease from north to south within the area. The most southerly mounds appear to have limited coral growth. The corals occur on the rim of features that may have both positive and negative relief (Masson et al. 2003). Further to the south, there is a large area of pockmarks having similar dimensions to the mounds. These observations have led Masson et al. (2003) to suggest that the Darwin Mounds are fluid escape features, with both mounds and pockmark sharing a common origin. The mounds form on a contourite sand drift, where fluid escape produces small "sand volcanoes"; the pockmarks form in softer sediments where this sand layer is absent. Mound height may be a function of both (1) the degree of sand emplacement by fluid escape and (2) the subsequent entrapment of sediments by colonizing fauna (coral and associated organisms). Bottom water temperature in the Darwin Mounds area is around 8°C with a salinity of 35.0 ppt (Bett

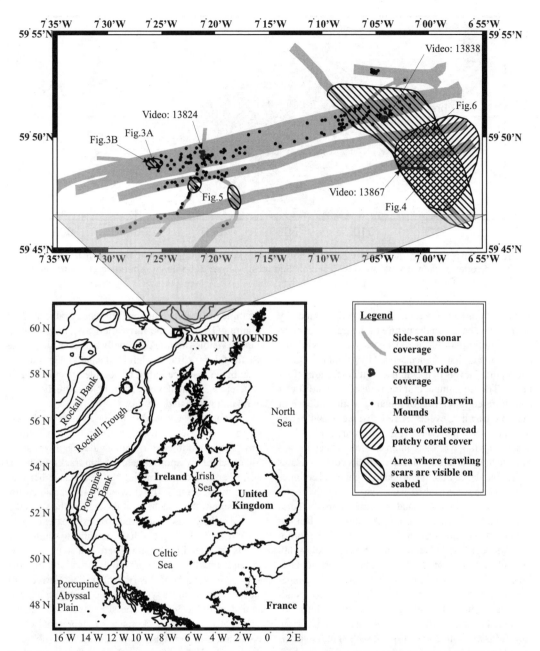

Figure 1. Location map showing the Darwin Mounds area in the northern Rockall Trough (Northeastern Atlantic), the areas surveyed by high-resolution side-scan sonar, the distribution of "patchy" coral cover, and regions where trawling impacts have been detected. The locations of video transects and figures showing examples of side-scan sonar imagery are also shown.

1999; Bett et al. 2001). Maximum bottom current speeds recorded during the RRS *Discovery* cruise were 35 cm/s (Masson et al. 2003). Figure 1 illustrates the distribution of individual Darwin Mounds imaged during the present side-scan sonar survey.

Trawling Impacts on the Darwin Mounds

Abundant evidence of demersal trawling across the Darwin Mounds and on the intervening seabed is apparent.

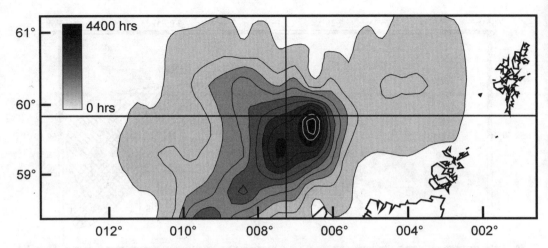

Figure 2. Trawling intensity in the vicinity of the Darwin Mounds (hours fished by French trawlers landing catch in Scotland, data from Fisheries Research Services, Marine Laboratory, Aberdeen, UK; adapted from Gubbay et al. 2002).

The trawl marks are presumed to have been made by otter trawls, based on the nature of the trawl mark (including outer parallel furrows cut by the trawl doors and shallower disturbance caused by groundline gear) and reported levels of trawling activity in the area (Figure 2). Trawling impact (at the time of the study) was concentrated in the east of the area, with evidence of multiple trawling events. A few discrete trawl marks were also seen in the west of the study area (Figure 1). Where trawling was more intense, mound "health" appears to have suffered. "Healthy" in this context relates to the abundance of undisturbed, upright coral colonies which, when video truthed, were live; "unhealthy" coral mounds show a proportional increase in broken coral rubble and dead coral. Figure 3A shows two Darwin Mounds: the example on the left of the figure has probably not been trawled and, although a relatively small example, is typical of "healthy" mounds. The mound is irregular in shape, with a double ridge internal arrangement of high backscatter areas (dark tones) corresponding to individual coral colonies. The mound's long axis is aligned with the direction of residual bottom current flow (Masson et al. 2003). Typical seabed photographs from "healthy" mounds are presented in Figure 4A–F. Running diagonally across the image is a lineation identified as a furrow cut into the seabed by an otter trawl door. Fainter lineations attributed to the net and its groundline gear can also be seen. When compared with the nontrawled mound to the left of the image (see also Figure 5), the trawled mound clearly shows reduced overall backscatter and fewer small intense backscatter "spots" which we interpret as individual coral colonies. Figure 4G shows a typical seabed photograph from a trawled seabed area. Figure 3B shows another example of a trawled mound (reference is also made to Figure 4G). Again, the furrow left by a trawl door is clearly visible running diagonally across the image, as are the fainter striations left by the net and groundline gear. In this instance, subtle backscatter variations probably represent patches of coral rubble where the former coral mound existed.

Some of the seabed in the Darwin Mound areas has been intensely trawled, with up to 28 individual trawl marks recorded during one video deployment (c. 3-h observation, approximately 5-km track). Local areas, at a scale corresponding to the distance between trawl doors, being 100% trawled. Side-scan sonar imagery shows that mechanical damage to the seabed is caused by both trawl doors and, to a lesser extent, by the net and groundline gear with the potential to smash erect corals that stand in its path. There is also evidence of multiple trawling events in various directions. Figure 6 shows an example of this type of seabed viewed with side-scan sonar and a typical seabed photograph is presented in Figure 4G. On the side-scan sonar image, small patches of high backscatter may represent isolated coral colonies, dense accumulation of coral rubble, or dropstones. Video ground truthing of the side-scan sonar coverage reveals numerous long, straight furrows (c.30 cm wide and 10–20 cm deep) and associated parallel lineations interpreted as trawl door scars and marks left by groundline gear and nets. Dimensions of trawl marks viewed on the side-scan sonar are often considerably larger (up to several meters across). This may be because the side-scan sonar shows the gross area of seabed disturbance that includes the trawl mark and disturbed and possible redistributed sediment adjacent to the mark. In some cases, the side-scan sonar seems to be

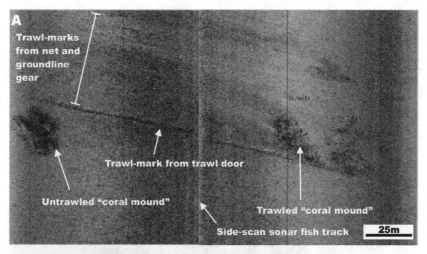
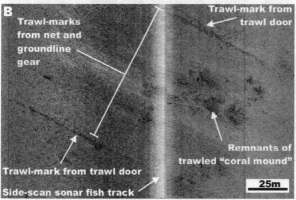

Figure 3. (A) Side-scan sonograph showing a "healthy" nontrawled Darwin Mound (center left), a trawl mark (diagonally across the image), and a trawled mound (center right) with reduced backscatter suggesting a decrease in the abundance of coral colonies (the dark spots); (B) Side-scan sonograph showing a trawl mark (diagonally across the image) and a fainter backscatter impression of a former mound. Backscatter probably identifies areas of coral rubble. Dark tones represent high backscatter typical of the presence of coral colonies. Acoustic shadows appear white. Faint vertical lines are processing artifacts for navigational and scaling purposes.

imaging trawl marks that have been infilled by sediment that is acoustically in contrast to the surrounding undisturbed seabed (e.g., Figure 7). Despite the intensity of trawling in some areas, "healthy" coral mounds still exist. Figure 5 shows such a mound with evidence of a trawl that passed close by, representing a "near miss."

Seabed areas associated with a "stippled" side-scan sonar acoustic facies are also common in this area and appear to correlate with patchy coral cover by small colonies, areas of coral rubble, and iceberg dropstones (Figure 1). This form of coral colonization may occur in coarser substrata (i.e., presence of cobbles and boulders at the seabed) where coral colonization is not restricted to the sandy sediments of the mounds. Evidence of intensive trawling in the "stippled" side-scan sonar acoustic facies is also present, with coral rubble contributing to this backscatter pattern (Figure 7).

A detailed comparison of the biological communities of trawled and nontrawled areas was not possible as a result of navigational uncertainties at small scales. Instead, we have characterized the seabed into five facies on the basis of biological characteristics based on video observations (see Figure 8): (1) "sediment facies," the general background environment of the Darwin Mounds area; (2) "Xenophyophore facies," areas with elevated densities of the giant protozoan xenophyophore

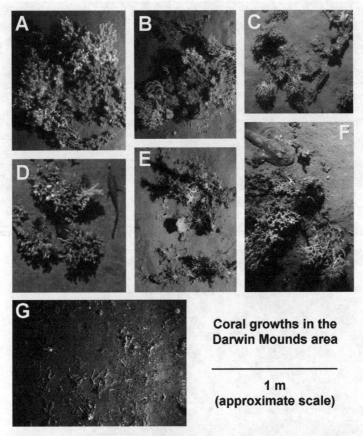

Figure 4. Seabed photographs from the Darwin Mounds area. Upper six images show typical erect coral growth forms; lower image shows scattered, smashed coral fragments assumed to result from the passage of a deepwater trawl.

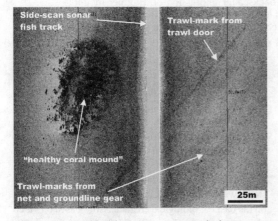

Figure 5. A "healthy" (i.e., nontrawled) Darwin Mound with evidence of a trawl mark that passed close by. Dark tones represent high backscatter typical of the presence of coral colonies. Acoustic shadows appear white.

Syringammina fragilissima, often located adjacent to mounds (Bett 2001); (3) "coral rubble facies," areas with a high percentage of broken coral fragments; (4) "dead coral facies," areas with a high percentage of dead coral; and (5) "live coral facies," areas with a high percentage of living coral. Examples of the "live coral facies" and "coral rubble facies" are shown in Figure 4. These video stills also illustrate the difference between nontrawled areas, where live corals provide significant seabed relief and potential refugia for fish species, and trawled areas in which dead coral and coral rubble provide only low relief.

The relative abundance of these five seabed facies, as recorded in three video transects, is illustrated in Figure 8. Note that the relative abundance of general seabed facies (sediment [A]; Xenophyophores [B]; coral [C–E]) is variable between the three areas (camera stations 13824, 13838, and 13867). The apparently high abundance of coral in the "trawled patchy" area

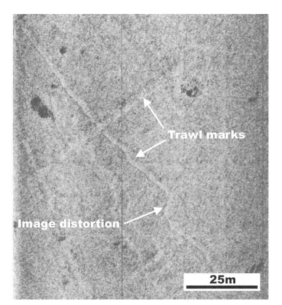

Figure 6. Sonograph from the Darwin Mounds area showing evidence of multiple trawling events in various directions. High backscatter areas may represent isolated coral colonies or accumulations of coral rubble. This area has been 100% trawled. Dark tones represent high backscatter typical of the presence of coral colonies. Acoustic shadows appear white.

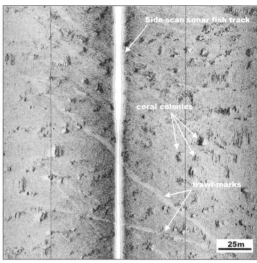

Figure 7. An example of the "stippled" side-scan sonar acoustic facies typified by a widespread patchy coral cover resulting from the presence of small (5 m across) coral colonies standing proud of the seafloor but not forming discrete Darwin Mounds. Coral rubble and iceberg dropstones may also contribute to this backscatter pattern. Dark tones represent high backscatter typical of the presence of coral colonies. Acoustic shadows appear white. Lighter-toned lineations crossing the image represent sediment-filled furrows caused by multiple trawling events.

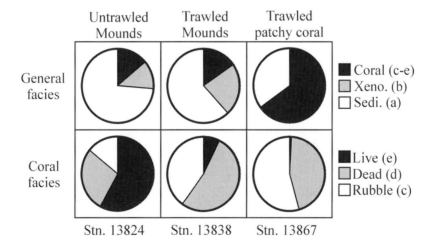

Figure 8. The relative abundance of seabed facies (see text for details) recorded in three SHRIMP video transects (RRS *Discovery* cruise 248) in the Darwin Mounds area. Each summary represents some 2–4 h of video survey at each of the camera stations (stn.). See Figure 1 for locations of the trawled and nontrawled areas studied (Xeno. = xenophyophore; Sedi. = sediment).

relates to the frequent occurrence of isolated coral colonies (live, dead, or rubble) rather than the aggregated coral growth observed on mounds. Similarly, the absence of the Xenophyophore facies is to be expected, as high densities of these protists are particularly associated with areas immediately surrounding mounds. When only coral facies are considered, there are very major differences between the three areas studied: living coral communities predominate in the nontrawled mound areas (station 13824) whereas dead coral or coral rubble is overwhelmingly dominant in the other two areas (stations 13838 and 13867). Video footage from the trawled mounds area shows distinct areas of disturbed coral rubble some 60 m across, suggesting that trawler impact can occur at the scale of entire mounds.

The Current Status of Habitat Conservation Measures

Damage to deepwater coral habitats by fishing activity in European waters is not restricted to the Darwin Mounds (Bett 2000; Roberts et al. 2000; Fosså et al. 2002; Hall-Spencer et al. 2002; Wheeler et al. 2005a). Indeed, it is also worth noting that such destruction is not a new phenomenon and can certainly be traced to the early decades of the 20th century in the Biscay and Porcupine areas to the southwest of Ireland (Teichert 1958). Trawling impact on deepwater coral communities from Norwegian waters, associated with fisheries for redfish (*Sebastes* spp.), is well documented (Fosså et al. 2002). Up to 50% of Norwegian coral habitat was impacted before a general ban on bottom trawling in known coral reef areas was implemented in 1999 under the Norwegian Sea Fisheries Act. Subsequently, two areas (the Sula Ridge and the Iver Ridge) were closed to bottom trawling in 1999 and 2000, respectively, and a further two reefs, the Tisler and Røst reefs, were closed in June 2003. One additional coral reef, located in the Trondheims fjord, is protected according to the Environmental Protection Act. The first marine reserve designated to protect deepwater coral, in particular *Oculina*, was established off Florida in 1984 (Reed 2002).

Despite fundamental differences, primarily in accessibility and the nature of impacts, some lessons may be learned from longer established shallow-water coral reef system management practice. Like deepwater coral reefs, fishing is one of the major human-induced factors impacting the ecology and diversity of shallow-water coral reef systems (e.g., Ginsburg 1993; Polunin and Roberts 1993; Birkeland 1997; McClanahan et al. 1999). Crosby et al. (2002) point out that effective shallow-water coral reef management strategies include representation from the science and management communities along with other stakeholders. The active involvement of the fishing community in the management process is fundamental to successful protection and can be achieved when fishermen understand that the conservation measures may increase fishing yields in surrounding areas and have a positive effect on the sustainability of the fisheries. Furthermore, Christie et al. (2002) point out that as coral reefs are a component of a broader ecosystem, there is a need to include individual marine protected areas, especially if they are small scale, within broader management frameworks that lead to overall reduction in fishing effort.

Framework-building corals, e.g. *Lophelia pertusa* and *Madrepora oculata*, within the exclusive economic zones of European nations may be protected under Annex I of the Habitats Directive (Natura code 1170). *Lophelia pertusa* is also listed under the Convention on International Trade in Endangered Species (CITES) Appendix I (Council Regulation [EC] number 338) and *Lophelia* spp. under CITES Appendix II (EC number 397) (CITES appendices can be found at http//:www.cites.org/eng/append/index.shtml). However, all Scleractinia are listed here and, as there is no direct evidence that *Lophelia* is specifically endangered, this is slightly misleading. As a result of the data presented here and in other initiatives (e.g., ICES 2001, 2002, 2003), the United Kingdom government has indicated to the European Commission that it will be proposing the Darwin Mounds site as a Special Area of Conservation (SAC) under the European Union (EU) Habitats Directive. This immediately posed difficulties as it conflicted with the existing EU Common Fisheries Policy (CFP). However, following revision of the CFP an emergency ban on bottom trawling in the Darwin Mounds area was implemented in August 2003, with the European Parliament finally voting for a permanent ban in February 2004. However, legal issues regarding the designation of habitat protection areas are complex (Long and Grehan 2002). Under the EU Common Fisheries Policy, the United Kingdom government does not have jurisdiction to exclude fishing activity from areas outside the 12 nautical mile limit and has also drawn attention to the need for the commission to exercise its sole competency in fisheries management in EU waters in regulating fishing in the area of the Mounds (ICES 2003). Nevertheless, enforcing exclusion of fishing activity from such remote areas may have practical limitations, especially with respect to policing by state vessels. One option may be monitoring fisheries activity near the Darwin Mounds using the satellite-based VMS (vessel monitoring systems) (Marrs and Hall-Spencer 2002). Gubbay et al. (2002) further discuss the options for the management of offshore Special Areas of Conservation, including the Darwin Mounds.

Acknowledgments

The authors would like to acknowledge the help of all aboard RRS *Discovery* cruise 248; RRS *Charles Darwin* cruises 112, 119, and 123; and RV *Colonel Templar* cruise 01/98 as well as Seatronics Ltd., Tim Le Bas, Maxim Kozachenko, and funding provided by EU 5th Framework projects ("Atlantic Ecosystem Study [ACES]," contract number EVK3-CT-1999-00008) and "Environmental Controls on Carbonate Mound Formation [ECOMOUND]," contract number EVK3-1999-00061) and the Irish Marine Institute. RRS *Discovery* cruise 248 was funded by Southampton Oceanography Centre; other cruises referred to were funded by the Atlantic Frontier Environmental Network (AFEN) and the United Kingdom Department of Trade and Industry. Anthony Grehan and Vikram Unnithan provided critical comments to this manuscript that are greatly appreciated.

References

Auster, P. J., R. J. Malatesta, R. W. Langton, L. Watling, P. C. Valentine, C. L. S. Donaldson, E. W. Langton, A. N. Shepard, and I. G. Babb. 1996. The impacts of mobile fishing gear on seafloor habitats in the Gulf of Maine (northwest Atlantic): implications for conservation of fish populations. Reviews in Fisheries Science 4:185–202.

Bett, B. J. 1999. RRS *Charles Darwin* cruise 112C leg 2, 19 May–24 Jun 1998. Atlantic Margin Environmental Survey: seabed survey of deepwater areas (17th round Tranches) to the north and west of Scotland. Southampton Oceanography Centre, Cruise Report 25, Southampton, UK.

Bett, B. J. 2000. Signs and symptoms of deep-water trawling on the Atlantic Margin. Pages 107–118 *in* Manmade objects on the seafloor. Society for Underwater Technology, London.

Bett, B. J. 2001. UK Atlantic Margin Environmental Survey: introduction and overview of bathyal benthic ecology. Continental Shelf Research 21:917–956.

Bett, B. J., D. S. M. Billett, D. G. Masson, and P. A. Tyler. 2001. RRS *Discovery* cruise 248, 07 Jul–10 Aug 2000. A multidisciplinary study of the environment and ecology of deep-water coral ecosystems and associated seabed facies and features (The Darwin Mounds, Porcupine Bank and Porcupine Seabight). Southampton Oceanography Centre, Cruise Report 36, Southampton, UK.

Birkeland, C., 1997. Life and death of coral reefs. Chapman Hall, New York.

Bradstock, M., and D. P. Gordon. 1983. Coral-like bryozoan growths in Tasman Bay, and their protection to conserve commercial fish stocks. New Zealand Journal of Marine and Freshwater Research 17:159–163.

Cairns, S. D. 1979. The deep-water scleractinia of the Caribbean Sea and adjacent waters. Studies on the Fauna of Curacao and Other Caribbean Islands 67:1–341.

Christie, P., A. White, and E. Deguit. 2002. Starting point or solution? Community-based marine protected areas in the Philippines. Journal of Environmental Management 66(4):441–454.

Clarke, T., 2002. Fishing scars Atlantic reefs: trawlers threaten cold-water corals thousands of years old. Nature Science Update Feb 2002.

Collie, J. S., S. J. Hall, M. J. Kaiser, and I. R. Poiner. 2000. A quantitative analysis of fishing impacts on shelf-sea benthos. Journal of Animal Ecology 69:785–798.

Crosby, M. P., G. Brighouse, and M. Pichon. 2002. Priorities and strategies for addressing natural and anthropogenic threats to coral reefs in Pacific Island nations. Ocean and Coastal Management 45:121–137.

De Mol, B., P. Van Rensbergen, S. Pillen, K. Van Herreweghe, D. Van Rooij, A. McDonnell, V. Huvenne, M. Ivanov, R. Sweenan, and J. P. Henriet. 2002. Large deep-water coral banks in the Porcupine Basin, southwest of Ireland. Marine Geology 188:193–231.

Dybas, C. L., 2002. Survival of fish, deep-sea corals may be linked. The Washington Post November 11, 2002:A9.

Edwards, R., 2000. Smashing up the seabed: rare coral mounds are being wrecked by deep-sea fishing. New Scientist 167:15.

Fosså, J. H., P. B. Mortensen, and D. M. Furevik. 2002. The deep-water coral *Lophelia pertusa* in Norwegian waters: distribution and fishery impacts. Hydrobiologia 471:1–12.

Ginsburg, N. R., 1993. Global aspects of coral reefs: health, hazards and history. Rosenstiel School of Marine and Atmospheric Science, University of Miami, Miami.

Gubbay, S., M. C. Baker, and B. J. Bett. 2002. The Darwin Mounds and Dogger Bank. Case studies of the management of two potential "Special Areas of Conservation" in the offshore environment. World Wide Fund for Nature, UK.

Hall, S. J., 1999. The effects of fishing on marine ecosystems and communities. Blackwell Scientific Publications, Oxford, UK.

Hall-Spencer, J., V. Allain, and J. H. Fössa. 2002. Trawling damage to Northeast Atlantic ancient coral reefs. Proceedings of the Royal Academy of Sciences B 269:507–511.

Henriet, J.-P., B. De Mol, S. Pillen, M. Vanneste, D. Van Rooij, W. Versteeg, P. F. Croker, P. M. Shannon, V. Unnithan, S. Bouriak, P. Chachkine, and the Porcupine-Belgica 97 shipboard party. 1998. Gas hydrate crystals may help build reefs. Nature 391:648–649.

Hovland, M. 1990. Do carbonate reefs form due to fluid seepage? Terra Nova 2:8–18.

Hovland, M., P. F. Croker, and M. Martin. 1994. Fault-associated seabed mounds (carbonate knolls?) off western Ireland and north-west Australia. Marine and Petroleum Geology 11:232–246.

Hovland, M., P. B. Mortensen, T. Brattegard, P. Strass, and K. Rokoengen. 1998. Ahermatypic coral banks off mid-Norway: evidence for a link with seepage of light hydrocarbons. Palaios 13:189–200.

Hovland, M., and E. Thomsen. 1997. Cold-water corals—are they hydrocarbon seep related? Marine Geology 137:159–164.

Huggett, Q. J. 1987. Mapping of hemipelagic versus turbiditic muds by feeding traces observed in deep-sea photographs. *In* P. P. E. Weaver and J. Thomson, editors. Geology and geochemistry of abyssal plains. Blackwell Scientific Publications, Oxford, UK.

ICES (International Council for the Exploration of the Sea). 2001. Initial report of the study group on cold water corals in relation to fishing. ICES, Study Group Report, Copenhagen.

ICES (International Council for the Exploration of the Sea). 2002. Report on the study group on the mapping of cold water Coral. ICES, CM 2002/ACE:05, Copenhagen.

ICES (International Council for the Exploration of the Sea). 2003. Report of the study group on mapping the occurrence of cold water corals. ICES, Study Group Report, Copenhagen.

Irish Skipper. 2001. Beauty beneath the waves. Irish Skipper 6 December 2001.

Jennings, S., and M. J. Kaiser. 1998. Impact of fishing on marine ecosystems. Advances in Marine Biology 34:201–352.

Jensen, A., and R. Frederiksen. 1992. The fauna associated with the bank-forming deepwater coral *Lophelia pertusa* (Scleractinia) on the Faroe Shelf. Sarsia 77:53–69.

Kenyon, N. H., A. M. Akhmetzhanov, A. J. Wheeler, T. C. E. van Weering, H. de Haas, and M. K. Ivanov. 2003. Giant carbonate mud mounds in the southern Rockall Trough. Marine Geology 195:5–30.

Koslow, J. A., K. Gowlett-Holmes, J. K., Lowry, T. O'Hara, G. C. B. Poore, and A. Williams. 2001. Seamount benthic macrofauna off southern Tasmania: community structure and impacts of trawling. Marine Ecology Progress Series 213:111–125.

Le Bas, T. and V. Hühnerbach. 1999. P.R.I.S.M. Processing of remotely-sensed imagery for seafloor mapping operators manual, version 3.1. Southampton Oceanography Centre, Southampton, UK.

Long, R., and A. Grehan. 2002. Marine habitat protection in sea areas under the jurisdiction of a coastal member state of the European Union: the case of deep-water coral conservation in Ireland. International Journal of Marine and Coastal Law 17:235–261.

Marrs, S., and J. M. Hall-Spencer. 2002. UK coral reefs. Ecologist 32(4):36–37.

Masson, D. G., B. J. Bett, D. S. M. Billett, C. L. Jacobs, A. J. Wheeler, and R. B. Wynn. 2003. The origin of deep-water, coral-topped mounds in the northern Rockall Trough, Northeast Atlantic. Marine Geology 192:215–237.

Masson, D. G., and C. L. Jacobs. 1998. RV *Colonel Templer* cruises 01 and 02/98, 22 Apr–18 May, 20 May–18 Jun 1998. TOBI surveys of the continental slope north and west of Scotland. Southampton Oceanography Centre Cruise Report, AFEN UKCS 17th Round Atlantic Margins Environmental Survey Data CD-ROM, Southampton, UK.

McClanahan, T. R., N. A.Muthiga, A. T. Kamukuru, H.Machano, and R. W. Kiambo. 1999. The effects of marine parks and fishing on coral reefs of northern Tanzania. Biological Conservation 89:161–182.

Montgomery, D. 2001. Coral reefs threatened by fishing. The Scotsman 16 March 2001.

Polunin, N. V. C., and C. M. Roberts. 1993. Greater biomass and value of target coral-reef fishes in two small Caribbean marine reserves. Marine Ecology Progress Series 100:167–176.

Probert, P. K., D. G. McKnight, and S. L. Grove. 1997. Benthic invertebrate bycatch from a deep-water trawl fishery, Chatham Rise, New Zealand. Aquatic Conservation: Marine and Freshwater Ecosystems 7:27–40.

Reed, J. C. 1980. Distribution and structure of deep-water *Oculina varicosa* coral reefs off central and eastern Florida. Bulletin of Marine Science 30:667–677.

Reed, J. K. 2002. Deep-water *Oculina* coral reefs of Florida: biology, impacts, and management. Hydrobiologia 471:43–55.

Roberts, J. M., S. M. Harvey, P. A. Lamont, J. D. Gage, and J. D. Humphery. 2000. Seabed photography, environmental assessment and evidence for deep-water trawling on the continental margin west of the Hebrides. Hydrobiologia 441:173–183.

Rogers, A. D. 1999. The biology of *Lophelia pertusa* (Linnaeus 1758) and other deep-water reef-forming corals and impacts from human activity. International Review of Hydrobiology 84:315–410.

Siggins, L., 2001. "Victor" reveals underwater wonder world of coral gardens on Ireland's deep seabed. The Irish Times August 16:3.

Teichert, C., 1958. Cold- and deep-water coral banks. The Bulletin of the American Association of Petroleum Geololgists 42:1064–1082.

Urquhart, F., 2001. Scientist set to uncover new secrets of coral world. The Scotsman 17 March 2001.

Van Dolah, R. F., P. H. Wendt, and N. Nicholson. 1987. Effects of a research trawl on a hard-bottom assemblage of sponges and coral. Fisheries Research 5:39–54.

Wheeler, A. J., T. Beck, J. Thiede, M. Klages, A. Grehan, F. X. Monteys, and Polarstern ARK XIX/3a Shipboard Party. 2005a. Deep-water coral mounds on the Porcupine Bank, Irish margin: preliminary results from Polarstern ARK-XIX/3a ROV cruise. Pages 393–402 *in* A. Freiwald and J. M. Roberts, editors. Cold-water corals and ecosystems. Springer-Verlag, Berlin.

Wheeler, A. J., M. Kozachenko, A. Beyer, A. Foubert, V. A. I. Huvenne, M. Klages, D. G. Masson, K. Olu-Le Roy, and J. Thiede. 2005b. Sedimentary processes and carbonate mounds in the Belgica Mounds province, Porcupine Seabight, NE Atlantic. Pages 571–603 *in* A. Freiwald and J. M. Roberts, editors. Cold-water corals and ecosystems. Springer-Verlag, Berlin.

Zibrowius, H., 1980. Les scléractiniaires de la Méditerranee et de l'Atlantique nord-oriental. Monaco, Mémoires de l'Ínstit Oceanographique 11.

Fishing Impacts on Irish Deepwater Coral Reefs: Making a Case for Coral Conservation

ANTHONY J. GREHAN[1] AND VIKRAM UNNITHAN[2]

Department of Earth and Ocean Sciences, National University of Ireland, Galway, Ireland

KARINE OLU-LE ROY[3]

DRO/Département Environnement Profond, Institut Français de Recherche pour l'Exploitation de la Mer, Centre de Brest, BP 70-29280, Plouzane, France

JAN OPDERBECKE[4]

IFREMER Sub-sea Robotics, Navigation and Vision Service, Institut Français de Recherche pour l'Exploitation de la Mer, Zone portuaire de Bregaillon, BP 330, 83507 La Seyne sur Mer, France

Abstract.—Deepwater coral reefs, formed principally by the azooanthelate scleractinian *Lophelia pertusa*, occur off the western coast of Ireland in water depths of 500 to 1,200 m. They are found in association with provinces (clusters) of giant carbonate mounds which rise 10 to 300 m above the seafloor. These reefs are home to a rich associated invertebrate and fish fauna. Pressures on the coral habitat are increasing with trawling in Norway, estimated to have damaged 30% to 50% of known reefs and significantly impacted coral locations west of Scotland. Concerns over potential further damage to corals prompted a consortium of Irish and European scientists to begin a detailed European Union-funded environmental baseline study (the Atlantic Coral Ecosystem Study) of the coral ecosystem where it occurs along the Atlantic Margin. In summer 2001, a French-Irish-European Union research mission CARACOLE (Carbonate Mound and Cold Coral Research) visited five deepwater coral locations in the Irish Porcupine Seabight and Rockall Trough. High-resolution video and close-up digital stills taken with the French VICTOR remotely operated vehicle revealed the extensive and spectacular nature of the coral reef formations. Evidence of fishing activity was confined to imaging of static gears (gill and tangle nets) used to fish for monkfish or anglerfish *Lophius* spp. and hake *Merluccius merluccius* lost on the side of mounds. No evidence of trawl-related damage to the corals was obtained even though coral bycatch has been reported from several deepwater trawl surveys in Irish water over the years. However, recent expansion of a deepwater trawl fishery, principally for orange roughy *Hoplostethus atlanticus*, highlights the urgent need for more focused investigations of fishing impacts and the rapid implementation of conservation measures to protect the corals. The designation of coral Special Areas of Conservation under the European Union Habitats Directive and their subsequent management will provide a rigorous test of the commitment of the European institutions to improve environmental integration implicit in the recent reform of the European Common Fisheries Policy.

Introduction

Irish Deepwater Coral Reefs

While deepwater corals have been known to occur off the Irish coast since the last century (Le Danois 1948), their extent (Wilson 1979; Zibrowius 1980; Frederiksen et al.1992; Rogers 1999; Roberts et al. 2003) and their potential importance as a key structural element in the European deepwater biotope (Jensen and Frederiksen 1992; Rogers 1999; Freiwald 2002; Freiwald et al. 2002) has only recently become apparent. Advances in side-scan and multibeam mapping technology combined with improved in situ exploration capabilities (principally remotely operated vehicles [ROVs] and other imaging platforms) have revealed a hitherto unexpected realm of coral colonies, reefs, and giant bioherms (Mortensen et al.1995; Henriet et al.1998; Hovland et al.1998; Freiwald 2002; Freiwald et al. 2002). Deepwater corals have high intrinsic value as an out-

---

[1] E-mail: anthony.grehan@nuigalway.ie
[2] E-mail: vikram.unnithan@ucd.ie; present address: Department of Geology, University College Dublin, Belfield, Dublin 4, Ireland.
[3] E-mail: kolu@ifremer.fr
[4] E-mail: Jan.Opderbecke@ifremer.fr

standingly biodiverse example of Irish and European natural marine heritage (Rogers 1999). In addition, deepwater corals provide important goods and services in providing spawning grounds and refugia for juvenile fish of commercially important fish species (Rogers 1999), as a major source and sink of carbonate, as a potential paleo-climate indicator important for the study of global climate change (Ecomound 2002), and as a marine equivalent of rainforests, in having potential as a source of novel biocompounds for use in medicine and by the pharmaceutical and biotechnology industries (Grehan 2000; Witherell and Coon 2001).

Concerns about the potential environmental damage to corals during activities such as oil and gas exploration (Rogers 1999) and documented evidence of large-scale damage to reefs off Norway due to destructive fishing practices (Fossä et al. 2000) have brought sharply into public focus the urgent need to develop a sustainable management strategy to protect the coral ecosystem (Grehan et al. 2002). Fears that this unique ecosystem could be irreparably damaged in the short term prompted a consortium of European scientists to begin a detailed European Union (EU)-funded environmental baseline study, the Atlantic Coral Ecosystem Study (ACES), of the coral ecosystem along the Atlantic Margin (Freiwald 2000; Grehan et al. 2001).

Irish Deepwater Fisheries

Exploitation of Irish shelf and slope fish stocks in the depth range where corals are typically found has been dominated by the French deepwater fleet since the 1980s, principally exploiting orange roughy *Hoplostethus atlanticus*, black scabbard *Aphanopus carbo*, blue ling *Molva dypterygia* and grenadier *Coryphaenoides rupestris*, particularly off the northwestern coast (Gordon 2001). Further south, there is a mixed fishery primarily for hake *Merluccius merluccius*, monkfish or anglerfish *Lophius spp.* and megrim *Lepidorhombus whiffiagonis*, exploited by vessels from Spain, France, and Ireland, as well as a number of flag of convenience vessels ("flagships") (Marine Institute Stock Book 2001). These vessels use trawls, gill nets, and tangle nets as well as long-lining. In recent years, the larger vessels in the Irish whitefish fleet have moved away from traditional white fish species (Atlantic cod *Gadus morhua* and hake) toward the exploitation of non-quota deepwater species and have been particularly successful in targeting orange roughy with catches increasing from 3 metric tonnes in 2000 to over 2,200 metric tonnes in 2001 (Marine Institute Stock Book 2002). Long-line fisheries for the Portuguese dogfish *Centroscymnus coelolepis*, deepwater cod *Mora moro*, and blue ling have also recently been developed (BIM 2001). There is a small but growing Spanish and Irish pot fishery principally for the deepwater red crab *Chaceon affinis*. Four vessels exploited this species as a seasonal targeted fishery during 2002 off the western coast of Ireland (M. Robertson, Trinity College, Dublin, personal communication).

Deepwater Fishing Effort

The typical effort involved in fishing deepwater species in terms of quantities of gear deployed in the case of static gears and in the robust design of gear used in deepwater bottom trawling is substantial and generally is restricted to vessels of 24 m or above (BIM 2001). Typical deployments for each of the commonly used deepwater fishing techniques off the west coast of Ireland are as follows:

- long-liners, predominantly targeting hake, typically deploy between 100 and 120 lines, each equipped with 85 hooks, spaced 3 m apart. Sets with between 8,000 and 9,600 hooks average some 28 to 35 km in length (European Commission 1994; D. Rihan, Bord Iascaigh Mhara, personal communication);

- gill nets, again used for hake, are typically 50 m long × 12 m high and are shot in strings of 700 nets. A typical shoot would fish an area of some 35 km (European Commission 1994; Rihan, personal communication);

- tangle nets, used principally to catch monkfish, are deployed in strings of up to 500 nets (50 m long × 5 m high) over 24 km. The vessels usually work around 3–4 strings, totaling 75 to 100 km of gear. These nets may be left in place for 1 to 2 weeks at a time (European Commission 1994; Rihan, personal communication);

- baited pots used to catch the deepwater red crab are set 50 m apart in 100-pot strings, weighted with large anchors at each end. Each vessel fishes between 300 and 600 pots per fishing day, working two sets of gear on alternate days (Robertson, personal communication);

- demersal trawl fisheries for deepwater species deploy trawls fitted with heavy rock-hopper gear, kept open by otter boards typically weighing in excess of 1,000 kg each. The trawls are towed at 5 to 8 km/h with the otter boards set at some 60–70 m apart. Gear is worked for approximately 4 h/haul, sweeping 20 to 30 km of seabed. There are regularly four to five hauls per day, so an average trip of 10 d can cover up to 100 km^2 of seafloor (Hall-Spencer et al. 2002).

A more robust type of bottom trawling gear has been developed for fishing orange roughy. The technique involves the sophisticated use of bottom sonar to map target pinnacles and sea mounts prior to fishing. The trawls are fitted with high-resolution headline transducers so that the actual position of the net with respect to the bottom can be constantly monitored. These vessels are also fitted with high-resolution color sounders, and many use systems such as the SIMRAD CM 60 Chart mapping system to generate topographic seabed maps in real time. Orange roughy tend to aggregate near the summit of topographical highs. Fishing an acoustically identified stock involves shooting the trawl to pass close to the summit of the peak, then allowing the net to sink quickly to drive the orange roughy onto the seafloor before towing off into deep water. This is a high-risk technique, as under shooting the trawl during the initial pass over the summit will result in snagging of the net on the side of the pinnacle or seamount. Only certain tracks on each pinnacle or seamount are suitable for successful operation of this technique (BIM 2001; Andrae 2002).

Deepwater Fishing Impacts

While it is likely that all the fishing techniques mentioned above will have some impact if deployed in areas of coral, bottom trawling is undoubtedly the practice with the most potential to cause collateral damage. Allegations of deliberate destruction of corals off the western coast of Ireland, similar to the Norwegian practice described by Fosså et al. (2000), were made at an ACES scientist–stakeholder workshop in 2000 (Grehan et al. 2002).

It was with a sense of urgency that the French–Irish–EU ROV research mission, CARACOLE (Carbonate Mound and Cold Coral Research), was undertaken in 2001. Five deepwater coral locations were mapped in detail, primarily to describe the biology and geology of the carbonate mounds and deepwater coral assemblages but also to determine the level of anthropogenic impacts, particularly from fishing activity (Olu-Le Roy 2001). The latter aspect is reported here.

Study Area

Irish deepwater coral reefs have been found associated with raised topographical features called carbonate mounds (Hovland et al. 1994; Henriet et al. 1998; Kenyon et al. 1998; De Mol et al. 2002) located to the west of Ireland (Figure 1). There are a number of mound clusters fringing the upper continental slope of the Rockall Trough and Porcupine Seabight (Croker and O'Loughlin 1998). In the Porcupine Seabight, two major mound provinces have been identified (De Mol et al. 2002): the Hovland-Magellan mound province on the northern slope of Porcupine Seabight and the Belgica mound province on the eastern slope of Porcupine Seabight (Figure 1). In the Rockall Trough, two major mound clusters on the southeastern and southwestern margin have been studied: the Pelagia mound province on the southeastern Rockall Trough and the Logachev mound province on the southwestern Rockall Trough (Haas et al. 2000).

These mounds occur in water depths of 500–1,200 m and vary from small structures of a few meters to over 300 m in height (Kenyon et al. 1998; De Mol et al. 2002). Densest living coral cover occurs on the summits of mounds where current flow is generally highest. Current speeds in excess of 40 cm/s have been recorded close to mounds by moored and lander deployed current meters (White 2001). Temperature and salinity at these locations is typically in the range of 6°C to 11°C and 35.2‰ to 35.6‰, respectively (White 2001).

Study Sites

Five coral and mound locations were studied in detail during the CARACOLE cruise (Figure 1): Thérèse Mound in the Belgica Mound Province; Propellor and Perserverence mounds in the Hovland/Magellan Mound Province; the R1 Mound complex in the Pelagia Mound Province; and the R2 Mound Complex in the Logachev Mound Province. An additional upslope, Site C, in the Connemara oil and gas field was investigated for evidence of suspected fluid migration (Figure 1).

Methods

During summer (July 30 to August 15, 2001), an institut français de recherche pour l'exploitation de la mer (IFREMER) remotely operated vehicle (ROV) was deployed from the research vessel *l'Atalante* at each of the study locations to carry out detailed geo-referenced video and multi-beam mapping.

The VICTOR 6000 ROV is a modular, remote-controlled system rated for scientific work down to 6,000 m. It is capable of continuous 24-h survey and comes equipped with manipulators, video, and accurate undersea navigational capabilities. The later uses an inertial navigation system consisting of a fiberoptic gyro compass (an IXSea (France) Octans) and a doppler log (an RD Instruments (USA) Workhorse 600) to supplement position fixing using a Posidonia (IXSea, France) ultra-short baseline position system linked to the differential Global Positioning System of the research vessel *l'Atalante*.

A total of 162 h of video was recorded during 10 dives (mean = 16 h/dive; duration = 8–32 h) that mapped over 100 km of seafloor at the study locations. Detailed

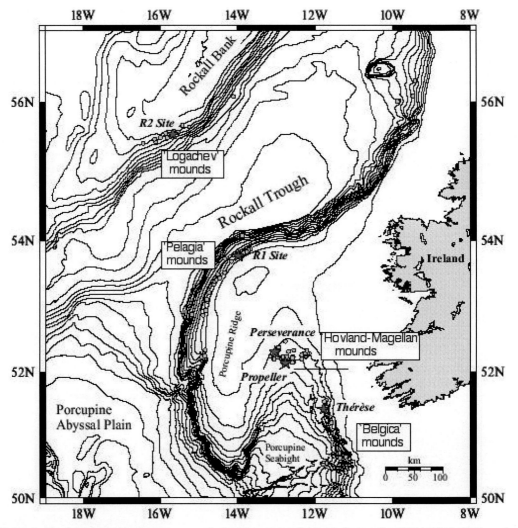

Figure 1. The location of carbonate mound sites investigated (stars) during the CARACOLE 2001 cruise. Filled circles highlight carbonate mound locations identified on the basis of seismic data (Croker and O'Loughlin 1998). Bathymetric contour interval = 200 m.

micro-bathymetry using an ROV-mounted multi-beam (Reson (USA) Seabat 8101/240 kHz) was carried out at the Thérèse Mound and R2 sites. Typically, video surveys were flown at an altitude of 2 to 3 m over bottom with a vehicle speed of 1.5 km/h, while multi-beam surveys were flown at an altitude of 15 to 20 m above bottom with a vehicle speed of 3 km/h. Dive transects were planned to maximize reconnaissance on and around mound targets.

High-resolution (3-chip color) video was routinely recorded on S-VHS tapes in PAL format and occasionally (during spectacular passages) on broadcast-quality Betacam tapes. Digital still photographs taken with a Hytec VSPN3000 megapixel zoom camera were stored as JPEGs. Dive commentary and data were logged using Adelie and Bio-Ocean software packages developed by IFREMER for dive video management and habitat characterization and analysis. Post-cruise dive reconstruction was supported by Adelie software that permitted synchronization of video playback with ArcView (ESRI, USA) geographic information system display of dive navigation data so that the ROV position could be visualized on the dive track during tape review. Real-time video mosaicing was performed using IFREMER Matisse proprietary software.

Time, position, and type of anthropogenic impact (including lost fishing gear) was documented and included in dive logs entered into the Biocean database.

Results

The Occurrence of Deepwater Corals

The main framework forming corals observed during the ROV surveys at all sites were *Lophelia pertusa* and *Madrepora oculata* (Figure 2A). Occasionally, the solitary coral *Desmopyllum cristagalli* was seen to form pseudo-colonies (Figure 2B). The predominant *Lophelia pertusa* commonly occurs as small (c. 20–30 cm high), single, bush-like colonies attached to drop stones and other hard substrates (Figure 2C) on and adjacent to mounds. When environmental conditions are suitable, particularly on the summits and upper flanks of mounds where current flow is highest (White 2001), individual colonies coalesce and form coral thickets (up to a meter in height). Dense thickets form extensive reefs (75% to 100% coral cover, Figure 2D) extending over several hundred m^2 at some of the sites (i.e., Thérèse and R2). These reefs are composed of both living and dead coral, with the living portion confined to the outer part of the framework structure. At some of the mounds (R1, Propellor, and Perserverance), living corals thickets and reefs were restricted to small areas (tens of meters) while more plentiful dead coral was observed either in dead stands or fragmented and forming coral rubble.

Both living and dead coral supported a variety of epifauna and mobile megafauna whose composition varied depending on location. Site-to-site differences were often readily apparent (e.g., Thérèse Mound assemblages were characterized by a high abundance of the glass sponge *Aphrocallistes bocagei*; Figure 3A), while the R2 site corals supported high densities of epibiont comatulid criniods (Figure 3B). Over 35 fish taxa were recorded during the dives (P. Lorance, JFREMER, France, personal communication), with *Neocyttus helgae*, *Nezumia aequalis*, *Phycis blennoides*, *Lepidion eques*, *Mora moro*, and northern cutthroat eel *Synaphobranchus kaupii* among the most common species.

Anthropogenic Impacts

Areas of Coral Habitat
No evidence of coral damage attributable to trawling activity was found in areas of coral at any of the five mound

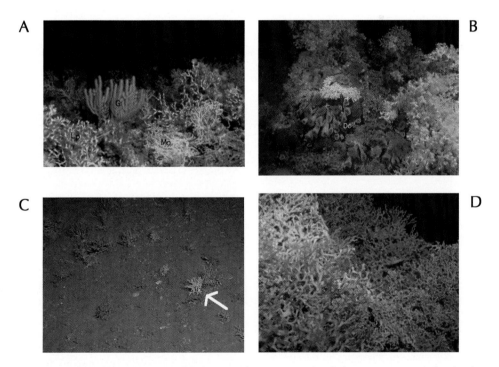

Figure 2. Examples of deepwater coral habitat at Therese Mound, off the western coast of Ireland (images copyright of IFREMER 2001). (A) The two principal framework-building species: *Lophelia pertusa* (Lp) and *Madrepora oculata* (Mo). Also visible is a soft coral gorgonian (G) species. Field of view is approximately 1 m. (B) Psuedo-colonies of the solitary coral *Desmophyllum cristagalli* (Dc). Field of view is approximately 1.5 m. (C) Small individual colonies of *Lophelia pertusa* (arrow) are common along the base and adjacent to mounds. Field of view is approximately 2 m. (D) Dense coral thickets form reefs on the upper slope and summits of some mounds. Field of view is approximately 1 m.

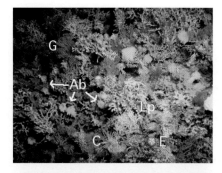

Figure 3. Examples of inter-site variability in the deepwater coral habitat (images copyright of IFREMER 2001). (Top) Deepwater coral assemblages at Thérèse Mound are dominated by abundant *Aphrocallistes bocegi* (Ab), *Lophelia pertusa* (Lp), and gorgonians (G). Echinoderms are also plentiful: echinoid (E) and pencil urchin *Cidaris cidaris* (C). Field of view is approximately 2 m. (Bottom) High densities of epibiont comatulid crinoids (C) are often a feature of assemblages of the R2 deepwater coral *Lophelia pertusa* (Lp) and *Madrepora oculata* (Mo). Field of view is approximately 1 m.

locations. Physical damage to corals due to their removal by scientific dredge sampling was observed on Thérèse Mound (Table 1). Dredge scars resulting from samples (St. 54907#2 and #3; Billett 2000) taken 2 years previously (May 1999) were still plainly visible both on video (Figure 4A) and on the multi-beam-generated micro-bathymetric map of Thérèse Mound (Figure 4B). The presence of fishing activity in coral areas was apparent through several sightings of a lost gill net (Table 1; Figure 5A, 5B) draped along the side of Thérèse Mound and another lost net on Propellor Mound (Table 1). A tangle net (Figure 5C) was imaged a number of times at the R1 site (Table 1) in the vicinity of living coral habitat and was probably responsible for the loss of a CTD located and recovered using VICTOR.

At Perserverance Mound, specimens of the decapod *Paromola cuvieri* were seen to carry man-made objects: one carried two small, shiny (aluminum?) sheets (Table 1) while another carried what appears to be a ball of mono-filament deployed in characteristic fashion as a defence above its carapace (Table 1; Figure 6A). Remnants of nets or long-lines were also seen at Perserverance and R1 (Table 1).

Non-coral Habitat Areas

Trawl scars were visible in the vicinity of Perserverance and R1 mounds (Table 1) and were common in the mud at the shallower Connemara site (Table 1), indicating that this area is actively trawled. The remains of an old trawl net, possibly discarded with bycatch as suggested by the presence of a plastic cup (Table 1; Figure 6B), was observed during the Connemara dive transect.

Discussion

Management Issues

It is widely accepted that enlightened fisheries management should be predicated on an integrated fisheries–ecosystem management approach that ensures both the long-term sustainability and viability of fish stocks and the maintenance of habitat integrity (Pope and Symes 2000; Covey and Laffoley 2002; Long and Grehan 2002; Mellett 2002; Sinclair et al. 2002). There is a now a pressing need to adopt such an approach in the management of deep-sea fisheries. More fishermen are examining the feasibility of exploiting deepwater fish species as traditional inshore fish stocks become depleted and as improved electronic charts and advances in sonar equipment become available. An intense debate has ensued as to whether deepwater fish species can ever be sustainably exploited or will be gradually fished out through serial depletion of local sub-populations (Koslow et al. 2000; Clarke 2001; Gordon 2001; Pauly et al. 2002; Roberts 2002). The potential impact on the deep-sea ecosystem is of equal concern due to the large bycatch of non-commercial fish and invertebrates (Roberts 2002) and the potential for habitat destruction caused by the use of robust deep-sea trawls fitted with rock-hopping gear (Probert et al. 1997; Fosså et al. 2000; Koslow et al. 2001; Hall-Spencer et al. 2002; ICES 2002; Roberts 2002).

The importance of areas of topographically complex habitat in the deep sea, such as deep, coldwater coral reefs, gorgonian gardens and sponge belts, in the maintenance of healthy fish stocks, has only recently begun to receive attention (Baker et al. 2001; Witherell and Coon 2001; Fosså et al. 2002; Thrush and Dayton 2002). Coldwater coral habitats formed by the framework-constructing species *Lophelia pertusa* are areas of high biodiversity (Le Danois 1948; Jensen and Frederiksen 1992; Mortensen et al. 1995; Rogers 1999). Suspension feeding invertebrates such as sponges (Porifera), anemones (Actinia), Bryozoa,

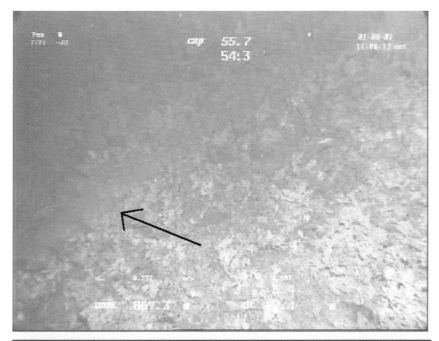
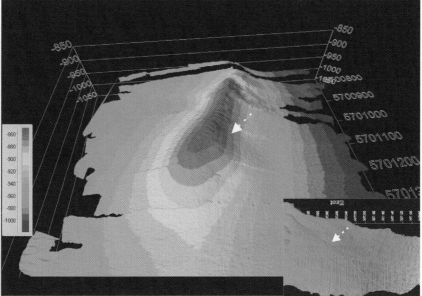

Figure 4. Evidence of past dredge sampling of corals on Thérèse Mound (images copyright of IFREMER 2001). (A) Video grab image from the forward-looking camera showing a dredge scar through coral at the Thérèse Mound (Table 1). The arrow indicates the location of the dredge scar. Field of view is approximately 3 m. (B) 3-D image of the Thérèse Mound rendered using swath (micro) bathymetry data acquired during the CARACOLE cruise. The arrow shows the location of the dredge scars. The inset image (right bottom) is a detailed view of the dredge scars on the western flank of the Thérèse Mound.

polychaetes (Annelida), bivalves (Molluska), gorgonians, and antipatharians (Cnidaria) use the dead coral framework as a hard substrate for attachment while many more invertebrates, including errrant polychaetes, various decapods such as squat lobsters (Galatheidae), squat lobsters, starfish, and urchins, occupy the spaces created by the inter-

Table 1. Locations of anthropogenic impact and activity observed during CARACOLE 2001 at the study sites in the Porcupine Seabight (Thérèse Mound, T; Propellor Mound, P1; Perserverance Mound, P2), in the Rockall Trough (R1 and R2), and in the Connemara field (C).

Dive number and site	Latitude (°N)	Longitude (°W)	Depth (m)	Notes
123-1				
T	51°25.5050′	011°46.1420′	944	Rope
T	51°25.6700′	011°45.9840′	939	Double rope
T	51°25.5560′	011°46.1080′	932	Double rope
T	51°25.7820′	011°46.3450′	954	Rope
T	51°25.7780′	011°46.5610′	957	Rope
T	51°25.8060′	011°46.6300′	956	Rope
T	51°25.6190′	011°46.4800′	900	Net
124-2				
T	51°25.7043′	011°46.3607′	867	Dredge mark (Figure 4)
T	51°26.0040′	011°46.3950′	976	Gill net
125-3				
T	51°25.7130′	011°46.3071′	898.9	Gill net draped over coral; fish (Figure 5A, B)
126-4				
P1	52°08.5740′	012°45.4890′	755	Net
127-5				
P2	52°17.8920′	013°02.8140′	645	Trawl mark
P2	52°17.9160′	013°02.8230′	646	Pile of old rope
P2	52°18.0600′	013°02.8060′	630	*Paramola* with aluminum sheet
P2	52°19.0380′	013°03.2260′	625	Trawl mark
P2	52°19.0590′	013°03.0630′	626	Trawl mark
P2	52°18.7380′	013°01.7190′	628	*Paramola* with ball of monofilament (Figure 6A).
128-6				
C	53°04.4380′	012°37.0220′	376	Trawl mark
C	53°04.4380′	012°36.7750′	380	Trawl mark
C	53°04.4170′	012°36.7420′	379	Trawl mark
C	53°04.2940′	012°36.4640′	379	Trawl mark
C	53°04.2120′	012°36.34200′	376	Trawl mark
C	53°04.1660′	012°36.2260′	373	Trawl mark
C	53°04.0210′	012°36.0590′	373	Trawl mark
C	53°04.4830′	012°32.6530′	366	Discards and piece of old net (Figure 6B)
C	53°04.3670′	012°32.5990′	366	Trawl mark
C	53°04.3420′	012°32.6010′	366	Trawl mark
C	53°04.1970′	012°32.5970′	366	Trawl mark
129-7				
R1	53°46.5910′	013°58.3170′	829	Trawl mark
R1	53°46.7460′	013°56.1050′	788	Double rope with crabs
R1	53°46.7580′	013°55.8140′	779	Old rope with *Cirripedes*
R1	53°46.6330′	013°55.4340′	754	Snagged ball of monofilament on rock
130-8				
R1	53°46.6220′	013°55.6450′	776	CTD snagged on old ropes
R1	53°46.6970′	013°55.9170′	787	Two crabs caught in tangle net (Figure 5C)
131-9				
R2	55°29.6248′	015°47.4834′	800	No impacts noted
132-10				
R2	55°29.6248′	015°47.4834′	800	No impacts noted

twining coral branches. It seems likely that coldwater coral habitat benefits fish stocks through increased food web complexity and the provision of refugia offering protection to spawning fish and nursery areas for juveniles. The nature of the functional relationship between coral assemblages and fish species has yet to be deciphered, although

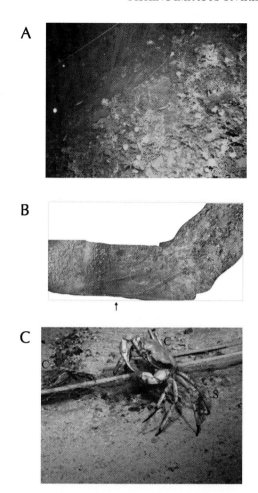

Figure 5. Evidence of fishing activity in and adjacent to areas of coral habitat (images copyright of IFREMER 2001). (A) A video grab image from the forward-looking camera showing a lost gill net on the western side of the Thérèse Mound (Table 1). Note the abundant colonies of *Lophelia pertusa* and large hexactinellid sponge *Aphrocallistes bocagei* and small gadoid-like fish. Upslope is toward the right of the viewing direction. The field of view is approximately 3.5 m. (B) A video mosaic created with Matisse software from downward-looking video camera footage. In this image, the deeper section is toward the left, while the crest of the mound is further toward the right (east). This image highlights the spatial extent of the gill net shown in (A). The arrow shows the viewing direction of (A). The field of view is approximately 10 m. (C) A digital still image of a lost tangle net at the R1 site (Table 1), showing a pair of deepwater red crabs *Chaceon affinis* (C) in the process of becoming entangled as a scorpion fish (S; family Scopaenidae) looks on. The field of view is approximately 0.5 m.

one study shows higher abundance of redfish *Sebastes* spp. over *Lophelia* coldwater coral reefs than in adjacent non-reef areas (Fosså et al. 2002).

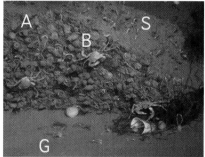

Figure 6. Evidence of fishing discards (images copyright of IFREMER 2001). (Top) High-resolution digital still image of the decapod *Paromola cuvieri* holding a ball of monofilament line (see arrow) in defensive pose in front of a barnacle encrusted drop-stone at Perserverance Mound (Table 1). The dropstone is ~30 cm in height and the field of view is approximately 0.75 m. (Bottom) High-resolution digital still image of a section of a discarded trawl net at the Connemara site (Table 1) with bycatch. Numerous anemones (A), bivalves, serpulid tubes (S), galatheids (G) and *Bathynectes* sp. (B) decapod crabs are visible in addition to a plastic cup. The plastic cup is approximately 0. in height.

Coldwater coral habitats are formed over long periods, with some Norwegian reefs estimated to be more than 8,500 years old (Fosså et al. 2002). *Lophelia pertusa* in the Northeast Atlantic grows slowly with a linear extension rate of only 0.4 to 2.5 cm/year (Mortensen and Rapp 1998; Freiwald 2002; Freiwald et al. 2002). Restoration of damaged coral habitat is likely to be a very slow process if it proves possible at all. Unfortunately, considerable documentation of trawl damage to reefs now exists. In Norway, between 30% and 50% of known reefs have been impacted, and anecdotal reports claim that trawlers often use the gear, wires, chains, and trawl doors to deliberately crush corals and clear an area before fishing begins (Fosså et al. 2002). Along the eastern Florida shelf of the United States, deepwater *Oculina* coral reefs have been impacted by industrial and recreational fisheries (Reed 2002) while gorgonian coral has been impacted in the

Gulf of Alaska (Krieger 2001). In New Zealand and Australia, Probert et al. (1997) and Koslow et al. (2001) have described heavy coral and benthic bycatch during fishing for orange roughy over seamounts.

Current Status of Fishing Impacts on Irish Corals

It was with some relief that the coral areas surveyed during the CARACOLE cruise, reported above, were found to be relatively undamaged. No evidence of recent impacts was found except for dredge scars and limited breakage beneath the gill net found on Thérèse Mound. Whether the patchy distribution of corals at some of the other mound sites (e.g., R1) is natural or due to the direct (removal) or indirect (increased sedimentation) impact of historical fishing is not possible to determine in the absence of a pre-fishing environmental baseline. The presence of robust coral populations at some of the sites surveyed and the absence of apparent recent damage should not lull us into a false sense of security. There are several reports of coral bycatch in Irish waters: Hall-Spencer et al. (2002) point to accidental bycatch of coral during French exploratory deepwater trawling in Irish waters, and coral bycatch has been reported during Irish deepwater trawl surveys (BIM 2001) and anecdotally by deepwater fishing skippers (Mellett 2002). Coral bycatch upon retrieval of gill nets and long lines has also been observed during Irish Naval Service fisheries inspections (M. Mellett, Irish Naval Service, personal communication). This suggests that time may be running out to protect pristine examples of Irish deepwater coral habitat.

Future Threats to Irish Coral

The recent, rapid expansion of the Irish orange roughy trawl fishery in 2001 is of major concern and reinforces the need for the urgent introduction of coral conservation measures. This fishery utilizing the high-risk fishing technique described above is potentially very destructive to coral habitats. Orange roughy fishing in the southern hemisphere, particularly in Australia and New Zealand, has had a major impact on seamount coral ecosystems (Probert et al. 1997; Koslow et al. 2000, 2001). Almost 90% of corals have been removed in some areas (Koslow et al. 2001), which has prompted both Australia and New Zealand to create a network of seamount marine protected areas (Koslow 2001). The recent introduction of total allowable catch quotas for several deepwater fish species in European Community waters (European Community 2002a) will help reduce the potential damage caused by unregulated fishing. Nevertheless, the very nature of the orange roughy fishery, which puts a premium on the identification and exploitation of virgin stocks, suggests that the capacity for collateral habitat damage during exploratory fishing will remain high. In mitigation, many of the mound sites surveyed are steep-sided with slope angles in excess of 20 degrees. Successful trawling requires slope angles less than 20 degrees (Andrae 2002), which means that the mounds themselves may confer a certain degree of natural protection from trawling impacts. That said, given the fragile and ancient nature of Irish coral reefs (Hall-Spencer et al. 2002), even a relatively short period exposed to this type of fishing impact would be catastrophic for the long-term viability of the coral habitat in its present physical configuration. The future effects of such habitat damage to local fish stocks are simply not known at this time. There is a clear need for a better understanding of the role played by benthic communities and ecosystem-level processes in the resilience of fish stocks (Thrush and Dayton 2002). The discovery of lost gill and tangle nets and the discarded net at the Connemara site suggest that any effect habitat damage might have on fish populations may be compounded by the impact of ghost fishing.

Conservation Initiatives

While coldwater coral marine protected areas have been declared by several countries (United States, Reed 2002; Norway, Fosså et al. 2002; Canada, DFO 2002; and Australia and New Zealand, Koslow 2001), no offshore coral sites have been protected in European Community waters to date, although there are proposals to designate coral conservation areas in the United Kingdom (Gubbay and Bett 2002) and Ireland (Long and Grehan 2002). In the European Community (EC), the primary mechanism for nature conservation is the application of the EC Habitats Directive (European Economic Community 1992, 1997), which ensures the protection of terrestrial and aquatic species and habitats. This is less than comprehensive for the deep sea but does allow for protection of biogenic reefs, which includes coldwater corals. The Habitats Directive supports the establishment of a European network (known as Natura 2000) of protected areas called Special Areas of Conservation (SACs). Designation of SACs by themselves, however, will not be sufficient to protect deepwater coral reefs (Long and Grehan 2002).

Irish coldwater corals are found between the Irish 12-mile Territorial Sea and 200-mile Exclusive Fisheries Zone (EFZ) in waters where Ireland has previously ceded legal competence in relation to the utilization and management of its fishery resources to the EU. The latter discharges the obligation to conserve, manage, and develop marine resources in the Irish EFZ through the Common Fisheries Policy (CFP) (Long and Curren 2000). For this reason, deepwater coral may only be protected from the activities of fishing through the medium of Community Law (Long and Grehan 2002). Specifically, protection will require the adoption of a fisheries technical conservation regulation by the European Council of Fisheries Ministers. This regulation will be binding on all member states and will be enforced by the regulatory agencies in Ireland against all vessels that operate in the Irish EFZ (Long and Grehan 2001). The principal shortcoming for implementation of this requirement is that proposed regulations can only be adopted following a majority vote by the Council of Fisheries Ministers representing each member state. This leaves the process open to considerable delay. Notwithstanding problems with rapid implementation, Long and Grehan (2002) have proposed a three-strand approach for the protection of deepwater corals under Irish jurisdiction: (1) designation of coral Special Areas of Conservation under the EU Habitat's Directive; (2) support of fishery-related management measures through community binding technical conservation measures under the CFP; and (3) adequate provision of resources to ensure compliance with conservation regulations.

The CFP, created in 1983, governs the conservation, management, and exploitation of living aquatic resources (fisheries) and aquaculture throughout the European Community. The failure of the policy over the past 20 years to prevent overfishing and adequately integrate environmental considerations has been widely criticized (Pope and Symes 2000; Symes and Pope 2000; European Commission 2001). Recently, in the second of two major policy reforms (the first in 1992), the EU Council of Fisheries Ministers met in Brussels between December 16 and 20, 2002, and agreed upon new measures (European Community 2002b) that go a long way in addressing the shortcomings identified in earlier policies. The objectives of the CFP have been reviewed to ensure the sustainability of exploited living resources through the application of sound scientific advice and the precautionary principal in fisheries management. The CFP firmly integrates community policy on sustainable development taking account of environmental, economic, and social aspects in a balanced manner. If a serious threat to the conservation of resources or to the marine ecosystem is posed by fishing, emergency measures can be taken by the European Commission for a period of up to 6 months. In a move to improve control and enforcement, satellite vessel monitoring (VMS) of all vessels over 18 m (rather than over 24 m) is mandatory since January 1, 2004. Regional Advisory Councils are to be established by the council to improve participation by fishermen and other stakeholders in CFP decision making. The council has already ratified support for a Community Action Plan to integrate environmental protection requirements into the CFP (European Commission 2002). The action plan aims to progressively implement an ecosystem-based management approach in the CFP and regards conservation of marine ecosystems as central to an environmentally integrated fisheries policy. It encourages member states to "fulfill, within the shortest deadlines, their obligations concerning the nature protection Directives, especially those regarding the designation and management of marine areas forming part of the Natura 2000 framework" (European Commission 2002).

Conclusions

Some coldwater corals form unique, physically complex and biodiverse habitats that are long-lived and fragile. It is likely that restoration of impacted (including coral) marine habitat will become a costly priority in the future (Pitcher 2001; Pauly et al. 2002). It follows that known outstanding and pristine examples of coldwater coral habitat should be protected both from an environmental and economic perspective as soon as possible. Based on our findings during the CARACOLE, we now know there are areas of relatively undamaged coral habitat in Irish waters. However, the full extent of coral resources must be mapped, and targeted studies are required to assess the level of trawl impact over the entire distribution of the coral habitat (see also ICES 2002). The designation of coral Special Areas of Conservation (SAC) must proceed immediately. Implementation of SACs and subsequent management will provide a rigorous test of the integration of environmental considerations in the reformed EC Common Fisheries Policy. It will also present a unique opportunity within the European Community to demonstrate how commercial fishing can be continued within protected areas provided it is made compatible with conservation requirements.

Acknowledgments

We would like to thank the captain and crews of the *l'Atalante* and VICTOR ROV (IFREMER) for their assistance during the CARACOLE cruise. We acknowledge the support of the (Irish) Marine Institute in facilitating Irish participation. AJG is funded by the EU 5FP Atlantic Coral Ecosystem Study (contract number EVK3-CT-1999-0000), while VU is funded by the EU 5FP Geomound (contract number EVK3-CT-1999-00016). Ronan Long (Law Department, National University of Ireland, Galway) and Dominic Rihan (Bord Iascaigh Mhara) made helpful comments on parts of the manuscript, and Martin Robinson, Zoology Department, Trinity College, Dublin, kindly identified the decapods visible in the images.

References

Andrae, D. 2002. Seamount fisheries—myths and realities. Fishing Boat World September 2002:27–28.

Baker, C. M., B. J. Bett, D. S. M. Billett, and A. D. Rogers. 2001. An environmental perspective. Pages 1–68 in World Wildlife Fund/The World Conservation Union (IUCN)/World Commission on Protected Areas, editors. The status of natural resources on the high-seas. WWF/IUCN, Gland, Switzerland.

Billett, D. G. 2002. Temporal and spatial variability of benthic communities on the Porcupine Abyssal Plain and in the Porcupine Seabight. Southampton Oceanography Centre, RRS Challenger Cruise 142, 19 April to 19 May 1999, Cruise Report 30, Southhamptom, UK.

BIM (Bord Iascaigh Mhara). 2001. BIM deepwater programme 2001. Bord Iascaigh Mhara, executive report, Dublin.

Clarke, M. 2001 Are deepwater fisheries sustainable? The example of orange roughy (*Hoplostethus atlanticus*) in New Zealand. Fisheries Research 51:123–135.

Covey, R., and D. d'A. Laffoley. 2002. Maritime state of nature report for England: getting onto an even keel. English Nature, Peterborough, England.

Croker, P. F., and O'Loughlin. 1998. A catalogue of Irish offshore carbonate mud mounds. In Proceedings of Carbonate Mud Mounds and Cold Water Reefs Conference, 7 to 11 February 1998, University of Gent, Gent, Belgium.

De Mol, B., P. van Rensbergen, S. Pillen, K. van Herreweghe, D. van Rooij, A. McDonnell, V. Huvenne, M. Ivanov, R. Swennen, and J. P. Henriet. 2002. Large deepwater coral banks in the Porcupine basin, southwest of Ireland. Marine Geology 188:193–231.

DFO (Department of Fisheries and Oceans). 2002. Minister Thibault announces 2002 Georges Bank groundfish management plan. DFO, Press release, Ottawa.

European Commission. 1994. Fishing with passive gear in the community: the need for management, its desirability and feasibility. European Commission COM (94) 235:1–47, Brussels.

European Commission. 2001. Communication from the commission to the council and the European parliament—element of strategy for the integration of environmental protection requirements into the Common Fisheries Policy. European Commission COM 2001/0143, Brussels.

European Commission. 2002. Communication from the commission setting out a community action plan to integrate environmental protection requirements into the Common Fisheries Policy. European Commission COM/2002/0186, Brussels.

European Community. 1997. Council directive 97/62/EEC on the conservation of natural habitats and of wild fauna and flora—amendment. 08/11/1997. Official Journal of the European Union, Brussels.

European Community. 2002a. Fixing for 2003 and 2004 the fishing opportunities of deepsea fish stocks. Council Regulation 2340/2002. Official Journal of the European Union, Brussels.

European Community. 2002b. On the conservation and sustainable exploitation of fisheries resources under the Common Fisheries Policy. Council Regulation 2371/2002. Official Journal of the European Union, Brussels.

European Economic Community. 1992. Council directive 92/43/EEC on the conservation of natural habitats and of wild fauna and flora. 22/07/1992. Official Journal of the European Union, Brussels.

Fosså, J. H., P. B. Mortensen, and D. M. Furevik. 2000. *Lophelia*-korallrev langs norskekystem, Forekomst og tilstand. Fisken og Havet 2, Havforsknings Instituttet 1-94.

Fosså, J. H., P. B. Mortensen, and D. M. Furevik. 2002. The deepwater coral *Lophelia pertusa* in Norwegian waters: distribution and fishery impacts. Hydrobiologia 471: 1–12.

Frederiksen, R., A. Jensen, and H. Westerberg. 1992. The distribution of the scleractinian coral *Lophelia pertusa* around the Faroe Islands and the relation to internal tidal mixing. Sarsia 77:157–171.

Freiwald, A. 2002. Reef-forming cold-water corals. Pages 365–385 in G. Wefer, D. Billett, D. Hebbeln, B. B. Jørgensen, M. Schlüter, and T. van Weering, editors. Ocean margin systems. Springer-Verlag, Berlin.

Freiwald, A., V. Hühnerbach, B. Lindberg, J. P. Wilson, and J. Campbell. 2002. The Sula reef complex, Norwegian Shelf. Facies 47:179-200.

Gordon, J. D. M. 2001. Deepwater fisheries at the Atlantic frontier. Continental Shelf Research 21:987-1003.

Grehan, A. J. 2000. Irish deepsea biodiversity. Page 12 in M. Costello, editor. A framework for an action plan on marine biodiversity in Ireland. Marine Resource Series 14, Dublin.

Grehan, A. J., A. Freiwald, and the ACES Consortium. 2001. The Atlantic Coral Ecosystem Study (ACES): forging a new partnership between scientists and principal stakeholders. Pages 95–105 in J. H. M. Willison, J. Hall, S. E. Gass, E. L. R. Kenchington, M. Butler, and P. Doherty, editors. Proceedings of the First International Symposium on Deep Sea Corals, Halifax, Canada. Ecological Action Centre, Nova Scotia Museum of Natural History, Halifax.

Grehan, A. J., R. Long, and M. Mellett. 2002. Deepwater coral conservation—a primer for the development of an Irish integrated ocean management dtrategy. Pages 386–393 in J. Feehan and F. Convery, editors. Achievement and challenge: Rio + 10 in Ireland. Environmental Institute, University College, Dublin.

Gubby, S., and B. J. Bett. 2002. The Darwin Mounds and the Dogger Bank—case studies of the management of two potential "Special Areas of Conservation" in the offshore environment. World Wildlife Fund North-East Atlantic Programme Report, Bremen, Germany.

Haas, H., A. Grehan, and M. White. 2000. Cold water corals in the Porcupine Bight and along the Porcupine and Rockall Bank margins. Cruise Report R.V. *Pelagia*

Cruise M2000 (64PE165). Netherlands Institute for Sea Research, Texel.

Hall-Spencer, J., V. Allain, and J. H. Fosså. 2002. Trawling damage to Northeast Atlantic ancient coral reefs. Proceedings of the Royal Society London 269:507–511.

Henriet, J. P., B. De Mol, S. Pillen, M. Vanneste, D. van Rooij, W. Versteeg, P. F. Crocker, P. M. Shannon, V. Unnithan, S. Bouriak, and P. Chachkine. 1998. Gas hydrate crystals may help build reefs. Nature 391:64–649.

Hovland, M., P. M. Croker, and M. Martin. 1994. Fault-associated seabed mounds (carbonate knolls?) off western Ireland and north-west Australia. Marine Petroleum Geology 11:232–246.

Hovland, M., P. B. Mortensen, T. Brattegard, P. Strass, and K. Rokoengen. 1998. Ahermatypic coral banks off mid-Norway: evidence for a link with seepage of light hydrocarbons. Palaios 13:189–200.

ICES (International Council for the Exploration of the Sea). 2002. Report of the Advisory Committee on Ecosystems 2002 to EC DG-Fish on identification of areas where cold-water corals may be affected by fishing. ICES, CM 2002/ACE:5, Copenhagen.

Jensen, A., and R. Frederiksen. 1992. The fauna associated with the bank-forming deepwater coral *Lophelia pertusa* (Scleractinia) on the Faroe shelf. Sarsia 77:53–69.

Kenyon, N. H., M. K. Ivanov, and A. M. Akmetzhanov. 1998. Cold water carbonate mounds and sediment transport on the northeast Atlantic margin. Preliminary results of the geological and geophysical investigations during the TTR-7 cruise of R/V *Professor Logachev* in co-operation with the CORSAIRES and ENAM2 programmes July–August, 1997. UNESCO, Intergovernmental Oceanographic Commission Technical Series 52, Paris.

Koslow, J. A. 2001. Fish stocks and benthos of seamounts. Pages 43–54 *in* H. Thiel and A. J. Koslow, editors. Managing risks to biodiversity and the environment on the High Sea, including tools such as marine protected areas—scientific requirements and legal aspects. Proceedings of the Expert Workshop held at the International Academy for Nature Conservation, Isle of Vilm, Germany, 27 February to 4 March 2001, Bundesamt für Naturschutz, Bonn, Germany.

Koslow, J. A., G. W. Boehlert, J. D. M. Gordon, R. L. Haedrich, P. Lorance, and N. Parin. 2000. Continental slope and deepsea fisheries: implications for a fragile ecosystem. ICES Journal of Marine Science 57:548–557.

Koslow, J. A., K. Gowlett-Holmes, J. K. Lowry, T. O'Hara, G. C. B. Poore, and A. Williams. 2001. Seamount benthic macrofauna off southern Tasmania: community structure and impacts of trawling. Marine Ecology Progress Series 213:111–125.

Krieger, K. J. 2001. Coral (*Primnoa*) impacted by fishing gear in the Gulf of Alaska. Pages 106–116 *in* J. H. M. Willison, J. Hall, S. E. Gass, E. L. R. Kenchington, M. Butler, and P. Doherty, editors. Proceedings of the First International Symposium on Deep Sea Corals, Halifax, Canada, 30 July–3 August, 2000. Ecological Action Centre–Nova Scotia Museum of Natural History, Halifax.

Le Danois, E. 1948. Les profundeurs de la mer. Payot, Paris.

Long, R., and P. Curren. 2000. Enforcing the common fisheries policy. Blackwell Science, Oxford, UK.

Long, R., and A. Grehan. 2002. Marine habitat protection in sea areas under the jurisdiction of a coastal member state of the European Union: the case of deepwater coral conservation in Ireland. The International Journal of Marine and Coastal Law 17:235–261.

Marine Institute Stock Book. 2001. The stock book. Report to the Minister for the Marine and Natural Resources. Annual Review of Fish Stocks in 2001 with Management Advice for 2002. Marine Institute, Galway, Ireland.

Marine Institute Stock Book. 2002. Report to the Minister for the Marine and Natural Resources. Annual Review of Fish Stocks in 2002 with Management Advice for 2003. Marine Institute, Galway, Ireland.

Mellett, M. 2002. Integrated ocean management provides a framework for the adoption of an ecosystem approach and the application of precaution. Master's thesis. National University of Ireland, Cork.

Mortensen, P. B., M. Hovland, T. Brattegard, and R. Farestveit. 1995. Deep water bioherms of the scleractinian coral *Lophelia pertusa* (L.) at 64°N on the Norwegian shelf: structure and associated megafauna. Sarsia 80:145–158.

Mortensen, P. B., and H. T. Rapp. 1998. Oxygen and carbon isotope ratios related to growth line patterns in skeletons of *Lophelia pertusa* (L) (Anthozoa, Scleractinia): implications for determination of linear extension rates. Sarsia 83:433–446.

Olu-Le Roy, K. 2001. Carbonate mound and cold coral research. N/O Atalante and ROV VICTOR, CARACOLE Cruise Report, 30 July to 15 August, 2001. Institute francais de recherche pour l'exploitation de la mer (IFREMER), France.

Pauly, D., V. Christensen, S. Guenette, T. J. Pitcher, U. Rashid, C. J. Walters, R. Watson, and D. Zeller. 2002. Towards sustainability in world fisheries. Nature 418:689–695.

Pitcher, T. J. 2001. Fisheries managed to rebuild ecosystems? Reconstructing the past to salvage the future. Ecological Applications 11:601–617.

Pope, J. G., and D. Symes. 2000. An ecosystem-based approach to the common fisheries policy: defining the goals. English Nature, Peterborough, England.

Probert, P. K., D. G. McKnight, and S. L. Grove. 1997. Benthic invertebrate bycatch from a deepwater trawl fishery, Chatham Rise, New Zealand. Aquatic Conservation of Marine and Freshwater Ecosystems 7:27–40.

Reed, J. K. 2002. Deepwater *Oculina* coral reefs of Florida: biology, impacts and management. Hydrobiologia 471:43–55.

Roberts, C. M. 2002. Deep impact: the rising toll of fishing in the deep sea. Trends in Ecology and Evolution 17(5):242–245.

Roberts, J. M., D. Long, J. B. Wilson, P. B. Mortensen, and J. D. Gage. 2003. The cold-water coral *Lophelia pertusa* (Scleractinia) and enigmatic seabed mounds along the north-east Atlantic margin: are they related? Marine Pollution Bulletin 46:7–20.

Rogers, A. D. 1999. The biology of *Lophelia pertusa* (Linnaeus 1758) and other deepwater reef-forming corals and impacts from human activities. International Revue of Hydrobiology 84:315–406.

Sinclair, M., R. Arnason, J. Csirke, Z. Karnicki, J. Sigurjonnsson, H. Rune Skjoldal, and G. Valdimarrsson. 2002. Responsible fisheries in the marine ecosystem. Fisheries Research 58:255–265.

Symes, D., and J. G. Pope. 2000. An ecosystem-based approach to the common fisheries policy: achieving the objectives. English Nature, Peterborough, England.

Thrush, S. F., and P. K. Dayton. 2002. Disturbance to marine benthic habitats by trawling and dredging: implications for marine biodiversity. Annual Review of Ecology and Systematics 33:449–473.

White, M. 2001. Hydrography and physical dynamics at the NE Atlantic margin that influence the deep water cold coral reef ecosystem. National University of Ireland, Department of Oceanography, EU ACES-ECOMOUND Internal Report, Galway.

Wilson, J. B. 1979. The distribution of the coral *Lophelia pertusa* (L.)[*L. prolifera* (Pallas)] in the north-east Atlantic. Journal of the Marine Biological Association of the United Kingdom 59:149–164.

Witherell, D., and C. Coon. 2002. Protecting gorgonian corals off Alaska from fishing impacts. Pages 117–125 *in* J. H. M. Willison, J. Hall, S. E. Gass, E. L. R. Kenchington, M. Butler, and P. Doherty, editors. Proceedings of the First International Symposium on Deep Sea Corals, Halifax, Canada, 30 July- 3 August, 2000. Ecological Action Centre/Nova Scotia Museum of Natural History, Halifax, Nova Scotia.

Zibrowius, H. 1980. Les Scléractiniaires de la Méditerranée et de l'Atlantique nord-oriental. Memoirs de l'Institut Oceanographique Foundation, Prince Albert I Monaco 11:1–247.

Symposium Abstract

Symposium abstracts have been reprinted from the Program and Abstract Volume for the Symposium on Effects of Fishing Activities on Benthic Habitats: Linking Geology, Biology, Socioeconomics, and Management and were not edited.

Biological and Socio-Economic Implications of a Limited-Access Fishery Management System

R. E. BLYTH[1] AND M.J. KAISER

School of Ocean Sciences, University of Wales-Bangor, Anglesey, United Kingdom

G. EDWARDS-JONES

School of Agricultural and Forest Sciences, University of Wales-Bangor, Bangor, Gwynedd, United Kingdom

P. J. B. HART

Department of Biology, University of Leicester, Leicester, United Kingdom

Marine reserves are considered to be effective conservation tools in tropical waters, but to date few studies have determined the economic and biological implications of limited access fishery management systems in temperate zones. The Inshore Potting Agreement (IPA), a fishery management system operated off the south coast of the United Kingdom, was conceived to reduce conflict between fishers that operate towed bottom-fishing gears and fishers that operate static gears. This system has operated on a voluntary basis since 1978, and covers an area of 480km^2. In this study, an interview survey of fishers, associated industry members and interested parties determined the economic implications of the IPA. Long-term recreational angling records from within and outside the area of the IPA were analysed to determine possible biological benefits for large-bodied fishes. The results suggest that the long-term maintenance of the IPA is likely to have greater economic and social benefits for local communities than if the area was open to all fishing activities.

---

[1] E-mail: osp818@bangor.ac.uk

Results of a Workshop on the Effects of Fishing Gear on Benthic Habitats off the Northeastern United States

L. A. CHIARELLA,[1] D. K. STEVENSON, AND C. D. STEPHAN

National Marine Fisheries Service, Northeast Regional Office, Gloucester, Masschusetts

R. N. REID AND J. E. MCCARTHY

National Marine Fisheries Service, Northeast Fisheries Science Center, Highlands, New Jersey

M. W. PENTONY

New England Fishery Management Council, Newburyport, Masschusetts

T. B. HOFF
Mid-Atlantic Fishery Management Council, Dover, Delaware

C. D. SELBERG
Atlantic States Marine Fisheries Commission, Washington, DC

K. A. JOHNSON
National Marine Fisheries Service, Office of Habitat Conservation, Silver Spring, Maryland

A panel of experts in the fields of benthic ecology, fishery ecology, geology, fishing gear technology and operations were convened in October 2001 to assist the Northeast Region's fishery management councils in evaluating the effects of fishing gear on local benthic habitats and identifying potential management measures. The panel expressed greatest overall concern about impacts from otter trawls and scallop dredges to structure forming organisms. Gravel habitat was considered to be most at risk from gear impacts, followed by sand and mud habitats. In some circumstances the extent of impact in each habitat varied based on the environment's energy level (high vs. low energy). In general, bottom tending mobile gear was of greater concern than fixed gear. Clam dredges were rated as having the least effect of the mobile gears because of the limited geographic area and the rapid recovery rates of the high energy sand environment in which they are fished. Scallop dredges were rated as having large effects in the gravel and sand habitats in which they are fished. Panelists had the greatest difficulty reaching consensus on the impacts of otter trawls due to their widespread use over a large variety of habitat types as well as the numerous gear configurations employed. The three primary management measures proposed to reduce fishing gear impacts included effort reductions, spatial closures, and gear modifications.

---

[1] E-mail: Lou.Chiarella@noaa.gov

Changes in the Epibenthos Assemblages of the North Sea following the Establishment of a Protected Area, the "Plaice Box"

J. A. CRAEYMEERSCH[1] AND G. J. PIET
Netherlands Institute for Fisheries Research

In 1989 a protected area in the south-eastern North Sea was established: the 'plaice box'. Data of the by-catch of annual beam trawl surveys carried out since 1985 will be used to determine the effect of the changes in fishing effort. A first analysis showed significant changes in the species composition after the 'closure' of the box. Changes, however, also occurred in the reference area (although in other species), suggesting that in addition to changes in impact by bottom fishing gear, other (climatic?) variables may have been involved. We will present the results of further analysis using multivariate techniques. Changes in species composition in the box area and in a reference area will be related to changes in fishing effort, environmental variables and climate.

---

[1] E-mail: johan@rivo.wag-ur.nl

Fishing for Shellfish in an Internationally Important Nature Reserve: Do Current Policies Achieve Their Objectives?

B. J. Ens[1]
Alterra, Texel, The Netherlands

A. C. Smaal
RIVO-CSO, Yerseke, The Netherlands

J. De Vlas
RIKZ, Haren, The Netherlands

The Dutch Wadden Sea is a nature reserve of international significance. Fishing for shellfish is allowed as long these activities do not cause significant harm to the natural values of the area. In 1993 this objective was implemented in a new shellfishing policy via two management policies: closed areas and food reservation. Thus, 26% of the intertidal mudflats are permanently closed for fishery to restore important habitats, particularly intertidal mussel beds and seagrass beds. To prevent food shortages for shellfish eating birds, mainly oystercatchers and eider ducks, caused by shellfish fishery, fishing for shellfish is not allowed when shellfish stocks are below a threshold value. In 2003, this new shellfishing policy must be evaluated. To this end, a major research program was initiated. It includes testing the hypothesis that mechanised fishing for cockles has long-term negative effects on the recruitment of cockles and other bivalves mediated by a loss of fine sediments. It also includes detailed investigations whether declining numbers of oystercatchers and recent high mortality among eider ducks can be related to food shortages and if so, whether these food shortages are linked to the current shellfishing practices. While the program relies heavily on massive long-term monitoring of shellfish stocks, shellfish fishery (including continuous registration of all fishing activities), benthic habitats and bird numbers in combination with mathematical modelling, some field experiments are also conducted. Most notable is an experimental test of the hypothesis put forward by the fishermen that fishing on mussel seedbeds helps to stabilise these beds. Preliminary results of the project will be discussed.

---

[1] E-mail: b.j.ens@alterra.wag-ur.nl

Developing a Fisheries Ecosystem Plan for the North Sea

C. L. J. Frid[1] and C. L. Scott
University of Newcastle, Newcastle upon Tyne, England, UK

M. F. Borges
Instituto Portugês de Investigação das Pescas e do Mar, Lisboa, Portugal

N. Daan
Netherlands Institutes for Fisheries Research, IJmuiden, The Netherland

T. S. Gray and J. Hatchard
University of Newcastle, Newcastle upon Tyne, England, UK

L. Hill
Instituto Português de Investigação das Pescas e do Mar, Lisboa, Portugal

O. A. L. Paramor
University of Newcastle, Newcastle upon Tyne, England, UK

G. J. Piet
Netherlands Institutes for Fisheries Research, IJmuiden, The Netherlands

S. A. Ragnarsson
Marine Research Institute, Reykjavik, Iceland

W. Silvert
Instituto Português de Investigação das Pescas e do Mar, Lisboa, Portugal

L. Taylor
Marine Research Institute, Reykjavik, Iceland

Considerable effort is being directed in many countries towards achieving sustainable exploitation of fisheries resources, protection of the ecosystem, safeguarding biological diversity and the promotion of sustainable fishing industries. Fisheries Ecosystem Plans (FEPs) are seen as one way of delivering these simultaneous management objectives while matching to the biological realities of the underpinning resources. In January 2002 the European Union commissioned an innovative project to marry together socio-economic theory with ecological understanding of the marine systems under study. The first phase in the process has been the development and initiation of links with the stakeholders to obtain their opinions on their preferred management regimes. The second phase is the characterisation of the biological and physical-chemical environment of the North Sea, which supports the fishery, leading to the development of a conceptual model of the North Sea food web. This is intended to supply a FEP for the North Sea which is acceptable to the stakeholders balanced with achieving protection of the ecosystem. In this poster we present the initial results that indicate that combinations of models can evaluate the ecological consequences of alternative management strategies and the interaction of different sets of fishes with varying life history characteristics. Preliminary data from the stakeholders has been both aggregated (to obtain an overall preference ranking of all stakeholders) and disaggregated (to identify the preference rankings of particular countries and different sectors of stakeholders). The stakeholders appear to appreciate the effort to factor in their preferences into the construction of a FEP and are open to further discussion in this ongoing project.

---

[1] E-mail: c.l.j.frid@ncl.ac.uk

The Ocean Habitat Protection Act: Overdue Protection for Structurally Complex Seafloor Habitats

H. Gillelan[1]
With Marine Conservation Biology Institute, Arlington, Virginia, and Oceana, Washington, DC

Until the mid-1980s, bottom trawls were used only on relatively flat ocean bottoms where the net would not snag on vertical structures such as corals, boulders, shipwrecks, and rock pinnacles. Large, heavy roller and rockhopper gear now enables bottom trawls to access areas of the oceans that were previously safe havens for marine life and fragile habitats. There is increasing scientific consensus that trawling in structurally complex habitats is one of the most destructive types of bottom fishing because of the long-term damage it causes to the diversity, species abundance, and ecological processes dependent on these habitats. Commercially and recreationally important fish, such as rockfish, haddock, Atlantic cod, snappers/groupers, and American lobster, and other types of marine life depend on structurally complex habitat during different stages of their lives. The young of species can show far greater survival rates where the seabed is complex. Where bottom structure has been damaged by bottom trawlers, diversity and the health of fish populations are negatively impacted. The Ocean Habitat Protection Act, by limiting roller and rockhopper gear to an 8-inch diameter, would protect these essential, structurally complex habitats by removing the gear that allows access to the habitats. Many states and several federal Fishery Management Councils have passed regulations that have begun to address this threat by restricting use of this gear in designated areas or fisheries. However, the patchwork of existing regulations often applies only to certain fisheries and leaves unprotected large areas of sensitive deep-sea corals, sponge beds, and other aggregations of geologic and biogenic structures. The restriction proposed by this bill is one shown to have been effective at reducing trawling in these habitats on the West Coast. Implementation of this gear restriction would maintain biodiversity and healthy seafloor habitats, and would assist many depleted species in recovery to sustainable levels.

[1] E-mail: hannah@mcbi.org

Reducing Seabed Contact of Bottom Trawls

P. He[1]

University of New Hampshire, Durham, New Hampshire

Typical bottom trawls leave tracks when they are towed over the seabed due to trawl doors and bobbins, or other roller gears in contact with the seabed. Reducing contacting points of doors or bobbins can reduce tracks left by trawls and impact of trawling on benthic system. Footgear or sweep of a bottom trawl may consist large steel bobbins in order to roll over rough grounds to protect netting from damaging. We examined if the number of bobbins on an offshore shrimp trawl was necessary for maintaining trawl geometry and stability, and preventing the gear from damaging when fishing for shrimps off Newfoundland and Labrador. Through flume tank tests, we were able to balance a commercial shrimp trawl when the number of bobbins was reduced from the original 31 to nine. Reducing the number of bobbins on the footgear also reduced drag of the trawl by 12%, resulting in savings on fuel. Preliminary sea trials indicated that the trawl with less number of bobbins on its footgear may result in the footgear intermittently being lifted off bottom, but this may not necessarily result in reduction in catch of shrimps. The trawl rigged with less bobbins on its footgear was more likely to incur damage when fishing under rough sea and seabed conditions. We are continuing the project in Newfoundland and in New England with the concept of semi-pelagic shrimp trawls with either trawl doors off the seabed while leaving the trawl on the bottom, or with the bottom-contacting trawl doors and a off-bottom "sweepless" trawl.

[1] E-mail: Pingguo.He@unh.edu

Rapid Build-Up of Fish Biomass, but Still Declining Coral Reefs: Why a Marine Fishery Reserve Designation Is Not Enough for the Protection of Reef Epibenthic Communities

E. A. Hernández-Delgado[1] and A. M. Sabat

University of Puerto Rico, Department of Biology, Coral Reef Research Group, San Juan, Puerto Rico

Reef fish communities in Culebra Island (27 km off northeastern Puerto Rico) have declined significantly in recent years. In 1999 the government of Puerto Rico established the Luis Pe-a Channel Marine Fishery Reserve (LPCMFR) with the objective of restoring local fisheries. Random stationary visual censuses and permanent line intercept transects have been used since 1996 to document the long-term changes in the coral reef fish and epibenthic communities assemblages before and after designation. A preliminary comparison of data from 1999 and 2002 shows a 38% increase in mean fish species richness/census. A dramatic increase in the abundance (2,539%) and in the biomass 26,618% of the yellowtail snapper, *Ocyurus chrysurus*, was observed. Also, a significant increase in the abundance (414%) and in the biomass (868%) of the schoolmaster, *Lutjanus apodus*, was documented. In spite of that, a significant decline of epibenthic communities was observed between 1997 and 2001, including coral species richness (31%), colony abundance (24%), and % coral cover (39%). Also, a 175% mean increase in macroalgal cover was documented. This decline was attributed to a combination of long-term indirect cascade effects of spearfishing, low densities of *Diadema antillarum*, and to coral disease outbreaks. These results suggest that although a MFR can be an excellent management tool to restore depleted fish stocks, the recovery fish communities alone is not enough to prevent further coral reefs decline associated to acute severe mortality caused by disease. Active restoration, in combination with MFRs, is recommended to recover coral reef epibenthic communities.

---

[1] E-mail: coral_giac@yahoo.com

Using Ideal Free Distribution Theory to Identify Potential Marine Protected Areas

H. Hinz[1], and M. J. Kaiser, M. Bergmann

School of Ocean Science, University of Wales-Bangor, Anglesey, United Kingdom

S. I. Rogers

The Centre for Environment, Fisheries and Aquaculture Science, Lowestoft Laboratory, Lowestoft, Suffolk, United Kingdom

There is increasing interest in the use of marine protected areas (MPAs) as tools to achieve the goal of ecosystem management. MPAs could counteract these effects by limiting the impact on areas important for feeding, shelter, spawning and migration. Thus MPAs should include features that enable fish to reach maturity and thus contribute to the spawning stock. But how do we find the most suitable habitats for fish species in large marine areas? Ecological theory (ideal free distribution) suggests that fish will be most abundant in areas that have the most favourable habitat characteristics for that species. Fish stock assessment data could be used to assess where fish species consistently tend to aggregate. In this study we examined stock assessment data collected for three flatfish species (plaice, sole and lemon sole) from 134 stations in the English Channel over nine years. Juvenile fish were excluded from the analysis. The fish abundance data for each year was ranked and the mean rank calculated. The coefficient of variation of the mean rank score had the least variability at stations with the highest mean rank scores and also at those with the lowest rank scores. Stations that had a mid-range mean rank had the greatest inter-annual variation. It would appear that stations with the highest mean rank abundance consistently attract fish. Such areas may be the prime focus for potential MPAs

---

[1] E-mail: h.hinz@bangor.ac.uk

The Path towards Ecologically Sustainable Fisheries: A Case Study in the Great Barrier Reef World Heritage Area

D. HUBER[1]

Great Barrier Reef Marine Park Authority, Fisheries Issues Group, Townsville, Queensland, Australia

Australia's Great Barrier Reef is the largest complex of reefs and islands in the world and it supports the most diverse ecosystem known to man. As a result of its unique status the Great Barrier Reef was included on the World Heritage List in 1981. The Great Barrier Reef Marine Park Authority (GBRMPA) is responsible for managing this vast and complex ecosystem in accordance with World Heritage values. The Authority must manage and conserve this unique system for future generations, whilst providing reasonable multi-use access to tourism, traditional hunting by indigenous people, recreational, charter and commercial fishing. Achievement of these goals requires a multi-jurisdictional, multi-sectoral and multi-agency approach. Trawling, line fishing and netting are the major forms of commercial fishing activity within the Marine Park. There are also smaller trap and harvest fisheries. All commercial fishing activity is managed by the Australian State of Queensland on a day-to-day basis. However, in line with its broader mandate to ensure the maintenance of World Heritage values, the Authority has a co-management role for commercial fisheries in the Marine Park. To achieve this the GBRMPA uses Zoning Plans, Management Plans, permits, regulations and education. Members of the Authority are also represented on various State based Management Advisory Committees. Of all commercial fishing in the World Heritage Area, trawling for prawns and scallops has the biggest impact on the benthic environment. Some 550 vessels have access to nearly 172,000 km^2 of Marine Park. The fishery has undergone a major and highly controversial restructure in the past two years. Amongst the achievements was a significant reduction and capping of fishing effort, the closure of an additional 96,000 km^2 of Marine Park, the mandatory use of turtle excluder devices and bycatch reduction devices throughout the park, restrictions on the targeting of species and the use of a Vessel Monitoring System whilst trawling. The trawl fishery of the Great Barrier Reef World Heritage Area has still a long way to go to achieve ecological sustainability but concessions won over the past two years present a major step along this path.

---

[1] E-mail: dorthea@gbrmpa.gov.au

Monitoring Changes in the Fully Protected Zones of the Florida Keys National Marine Sanctuary

B. D. KELLER[1]
Florida Keys National Marine Sanctuary, Marathon, Florida

The Florida Keys National Marine Sanctuary is a 9,850 km2 marine protected area managed by the U.S. National Oceanic and Atmospheric Administration and the State of Florida. A comprehensive management plan was implemented in 1997 to protect and conserve marine resources of the Florida Keys. One aspect of the management plan is the creation of a network of 24 fully protected zones (marine reserves). An ongoing monitoring program is designed to determine effects of 'no-take' protection on heavily exploited fishes and invertebrates, benthic communities, and human activities. Data on the abundance and size of fish, spiny lobster, and queen conch; algal cover, diversity and recruitment; and zone usage are collected from fully protected zones and adjacent reference sites. Preliminary results indicate increases within fully protected zones in the number and size of heavily exploited species such as spiny lobster and certain reef fishes. Slower growing benthic species such as corals and sponges have not shown significant changes with fully protected zones, possibly because the zoning plan was implemented less than five years ago.

---

[1] E-mail: brian.keller@noaa.gov

Identification and Evaluation of Indicators for Environmental Performance of European Marine Fisheries

L.-H. LARSEN[1]
Akvaplan-niva, The Polar Environmental Centre, Tromso, Norway

A. ZENETOS AND N. STREFTARIS
National Centre for Marine Research, Athens, Greece

Fishing activities are known to have significant effects not only on target species, but also on the wider marine environment. This is a result of the incidental catch of non-target species, and physical disturbance of benthic habitats and communities from bottom fishing gear. Indirect effects of forced ecosystem changes are also known from intensive fishing areas. The Central Fisheries Policy (CFP) of the European Union, implemented some twenty years ago, is now facing a major challenge because its contribution to sustainable, environmental performance within European marine fisheries is not well defined or explained. A scoping study has been undertaken for the identification of environmental performance indicators for the fisheries in the greater European (EU + EFTA) waters. Of the 23 indicators, two are directly relevant to fishing and benthic habitat concerns. These are 'Loss of fishing gear' and 'Physical damage to habitats and species'. The report compiles results on habitat and species alterations from more than 100 individual national and European wide research projects on the interactions between fisheries and the environment. These studies collectively resulted in more than 1000 individual scientific publications. The methodology of the scoping study is presented, including criteria for selection of indicators. The data compilation on one indicator 'physical damage to habitats and species' and other indicators relevant to this symposium are described and illustrated. Examples of habitat alterations are given from European (North Sea benthic trawling) and Arctic (Barents Sea shrimp trawling) fisheries and implications for management practice and the CFP are presented.

---

[1] E-mail: LHL@akvaplan.niva.no

Decision Framework for Describing and Identifying EFH, Mitigating Fishing Impacts, and Designating HAPC in Federal Fishery Management Plans

G. B. PARKES,[1] H. B. LOVETT, AND R. J. TRUMBLE

MRAG Americas, Inc, Tampa, Florida

Environmental impact statements (EISs) require development of a range of reasonable alternatives for the proposed action and a comparative analysis of the environmental and economic impacts of the alternatives. This analysis is a type of risk assessment. In the case of designating essential fish habitat (EFH), three separate ranges of alternatives are required: those to designate EFH for each species and life stage managed by the region in question; methods to reduce or mitigate adverse fishing impacts; and alternatives for the designation of habitats of particular concern (HAPC). A decision framework was designed to (a) facilitate the appropriate identification of alternatives for each of the three suites of actions; (b) incorporate the required criteria (per the EFH Final Rule); and (c) to frame the comparative assessment of the alternatives. The decision framework thus incorporates the following factors. For EFH, four levels of information: (1) distribution data for some or all portions of the geographic range of the species; (2) habitat-related densities of the species where available; (3) growth, reproduction, or survival rates within habitats where available; and (4) production rates by habitat. For methods to reduce or mitigate adverse fishing impacts: (1) does the fishing activity have an adverse impact; (2) is the adverse impact minimal; and (3) is the adverse impact temporary? For designation of HAPC: (1) the ecological importance provided by the habitat; (2) the sensitivity of the habitat to human-induced environmental degradation; (3) the extent that development activities are or will be stressing the habitat; and (4) the rarity of the habitat type.

---

[1] E-mail: graemeparkes@compuserve.com

The Characteristics and Function of Commercial Fishing Gears: How These Relate to Their Effects on Seafloor Habitats and the Pursuit of Ways to Minimize Effects

C. S. ROSE[1]

NOAA National Marine Fisheries Service, Alaska Fisheries Science Center, Seattle Washington

Of the factors that influence the effects of fishing gear on benthic ecosystems, the characteristics of the fishing gear may be as significant as the characteristics of the physical habitat or the community of organisms affected. While the effects of different fishing gear components on benthic ecosystems may have general similarities, variations in particular gear characteristics, such as force of bottom contact, and component dimensions, could cause profound differences in the severity of such effects. Few studies have addressed such distinctions and much of the relevant information has been collected incidental to studies with other goals, such as during research of fishing gear selectivity and related fish behavior. This has limited the usefulness of gear selection and modification in reducing seafloor effects. This paper will categorize and describe the identified effects of fishing, noting which characteristics of fishing gear components affect the severity of each effect. We will also describe the main components of the major classes of fishing gears and relate them to these significant characteristics. This background will be used to motivate and describe several concepts of gear modifications to reduce seabed effects.

---

[1] E-mail: craig.rose@noaa.gov

Long-Term, Large-Scale Biological Surveys: A Necessary Component of Fishery and Ecosystem Management

C. Syms[1]

University of California, Santa Cruz, Department of Ecology and Evolutionary Biology and Long Marine Laboratory, Santa Cruz, California

Abundance, size, and distribution information is essential for successful fisheries and ecosystem management. Population censuses are rarely possible, so sampling programs must be employed to infer poopulation parameters. Recent advances in marine habitat mapping techniques have generated an interest in shifting from traditional sampling and inference methods toward incorporating habitat maps and fish-habitat associations to increase precision and, hopefully, accuracy of sample estimates. The success of this approach is strongly conditioional on knowledge of uncertainties in the system. Fish populations and habitat associations are variable in time and space, habitat boundaries are either imprecisely known, or precisely known only over a limited extent. In California, long-term, large-scale monitoring of fish populations and habitat variables using annual SCUBA surveys indicates that the primary recourse in the face of these uncertainties is to reduce reliance on habitat classification, by retaining coarse scale habitat resolution, and increase reliance on design-based inference. This need highlights the disparity between sampling scales of fish, habitat, and the need of resource manager.

---

[1] E-mail: syms@biology.ucsc.edu

Approach to Evaluating Fishing Effects on EFH off Alaska

D. Witherell[1] and C. Coon

North Pacific Fishery Management Council, Anchorage, Alaska

The North Pacific Fishery Management Council is in the process of evaluating fisheries for adverse effects on essential fish habitat, and over the coming year, will be considering additional measures to minimize these effects to the extent practicable. The steps used in the evaluation include 1) description of the fisheries, 2) distribution and intensity of the fisheries, 3) description of the habitat effected, 4) summary of scientific studies applicable to each fishery, and finally, 5) evaluation of each fishery to see if effects on habitat are more than minimal and more than temporary in nature. For each of the principal fisheries managed under federal fishery management plans, we first described the gear used, fishing methods, and habitats potentially impacted (including living substrates and prey species). We then mapped the distribution and intensity of fishing effort in each fishery, and compared this with available information on benthic habitats to determine the extent of effects. Using this information, along with the literature on fishing gear impacts, the fisheries will be evaluated against criteria for determining adverse effects. For those fisheries having adverse effects, management measures will be proposed to minimize effects to the extent practicable. Management measures will also be proposed to mitigate cumulative effects of fisheries, and may be proposed to reduce effects of fisheries that have not been determined to have significant adverse effects on fish habitat. The fishery evaluation and identification of potential management measures will be completed by October 2002. The poster reviews the fishery evaluation process and development of alternatives to minimize adverse effects of fisheries on essential fish habitat off Alaska.

---

[1] E-mail: David.Witherell@noaa.gov

What Next? What Have We Learned? What More Do We Need to Know? What Should We Act on Right Now?

Moderated Panel and Open Discussion

The following is a transcript of the last session of the symposium; "What next? (after the symposium) What have we learned? What more do we need to know? What should we act on right now?" The goal of this session was to summarize, to obtain broad input on what needed to be done to address individual aspects of "Effects of Fishing on Benthic Habitat" issues. The panel members, known for their expertise, represented the areas of fishery management, the fishing industry, environmental conservation, and marine and benthic habitat science. Following initial panel discussions of the questions above, the audience was invited to participate with comments and in raising additional issues of concern. The speakers were allowed to edit their remarks for clarity and correctness. The resulting document was reviewed by the moderator. The resulting report that follows reflects both the tone and the serious content of the discussion.

DATE: November 14, 2002
TIME: 2:00–4:12 pm
PLACE: Doubletree Westshore Hotel
          4500 West Cypress Street
          Tampa, Florida
REPORTED BY: Rebecca S. Witt, RMR, CRR
                  Notary Public, State of Florida
                  at Large

PANEL MODERATOR
        Mike Sissenwine, NOAA/NMFS Science Director, Northeast Fisheries Science Center

PANEL MEMBERS
        Clarence Pautzke, North Pacific Research Board, formerly NOAA/NMFS/Deputy Assistant
            Administrator for Fisheries and North Pacific Fishery Management Council
        Nils Stolpe, Garden State Seafood Association
        John Gauvin, Groundfish Forum, Inc.
        Elliott Norse, Marine Conservation Biology Institute
        Simon Thrush, National Institute of Water and Atmospheric Research of New Zealand

PROCEEDINGS

DR. SISSENWINE: Let's get started now. This is the final session of the Symposium on Effects of Fishing Activities on Benthic Habitat: Linking Geology, Biology, Socioeconomics and Management. It is a long title, but I think we all know what we've been talking about for the last few days and have a real sense of the significance and importance of it. In fact, the title does relay the complexity of the topic.

I want to make a comment about the purpose of this symposium. I might state the purpose in the context of possibly a test of a hypothesis. And the hypothesis is that with more information and better communications, we ought to be able to converge towards consensus solutions to some very difficult and complex issues. That's the reason to get together in an interdisciplinary group like this, to present the latest information that's available and to try to communicate with each other, and that's what we're going to continue to do this afternoon.

Realistically, we have to acknowledge that the alternative hypothesis, which is that even with perfect information and perfect communication, we're going to all disagree about a lot of things because we have different values. And probably the truth lies somewhere in between these two hypotheses. Let's at least continue for the moment on the track of the positive sense of how we can benefit from information and from improved communication.

I think we have to do that to the maximum extent possible, so that we can actually allow that communication and quality information to really identify those issues that are ones of value and which are in a different category in terms methodology for finding ultimate solutions. In the end, there will probably be a need for old fashion politics, or litigation, or whatever, but let's hope that science can inform these processes.

In fact, I think we have received a tremendous amount of information and we have done a lot of communicating. We have heard about 50 papers presented, without concurrent sessions, so we all had the opportunity to hear all the papers. We've seen over a hundred posters, very high quality posters, that are every bit as important as the papers. The sessions have defined issues in terms of habitat and fisheries and the status of management systems and the livelihoods of people. They've also set the stage with respect to natural changes and advances in the technology we need to measure changes. We have also had sessions to help understand how chronic and how acute events affect benthic habitats; to link fisheries and benthic habitats; to compare effects of fisheries to natural factors and to nonfishing anthropogenic factors; to consider and connect time and space scales; to consider social and economic issues, including issues of values and of ethics; and, of course, also to consider how the long arm of the law plays into the issues we're talking about and the brilliant tale we heard about just a little while ago. And finally, we also had sessions that consider some of the options for mitigating effects of fishing on benthic habitats.

This afternoon, we have a panel to address some of the key questions posed right at the beginning of the symposium. The questions are what have we learned?, what more do we need to know?, what do we know enough about to act on right now?

I can't resist making a comment about the first question. I think it's been an excellent meeting and therefore I think that anyone who has not learned something from the symposium is either all-knowing or a very slow learner! I doubt there are very many in the first category, and so I suspect that most people will acknowledge they've learned quite a bit from this symposium.

Anyway, I think it's probably time for me to stop editorializing and turn to our panel and let them address the questions and make whatever other comments they want to make.

The panel was selected to be a cross section of the perspectives on the issue of fisheries and benthic habitat. Some are advocates for the critters, if I can characterize fish and worms that way; some are advocates for the fishing industry, food, recreation dollars; and some are scientists who are probably advocates for data and research, which cost dollars. We'll let them speak for themselves. And to the extent they want to characterize where they're coming from, that's fair enough. They may or may not represent the extremist views on some of these issues, but I do think they all represent very effectively and very articulately the perspectives for which they've been chosen, and in that context, I think we've got an excellent panel that should make this discussion a very valuable one and a very lively one.

So let me just turn to them. We're not going to go through lengthy introductions again because for the most part, they've already been introduced through various papers.

I'm just going to take people in the order they're sitting at the table, which is random as far as I'm concerned. I don't know whether somebody else had grand designs, but to my knowledge, it's random.

The first person to have his chance to respond to questions is Clarence Pautzke from the North Pacific Research Board and formerly the North Pacific Fishery Management Council.

DR. PAUTZKE: I have learned that there is a vibrant research community out there helping us to understand what is happening on the ocean floor. We need research methods and technologies that will help us understand the effects of fishing on bottom habitat. There is a wide range of instrumentation to do so, but it is very expensive. As I showed you earlier, a Delta submersible can cost about $10,000 per day. And the scales that need to be studied are vast—from macrofauna to meiofauna between the grains of sand. So the scales range from the study of ocean currents to water flow through the sediments, which depends on the natural porosity of those sediments. And these are scales that we're not used to looking at. They are scales that a benthic biologist might be interested in, but at the policy level we're not used to looking at that much detail. Policy makers will need to focus on some intermediate level (i.e. localized areas that are sensitive and need protection). We need to move away from the problem of essential fish habitat being defined as everywhere.

We also learned that fishing is very patchy and that patchiness may help mitigate against longterm fishing impacts on the sea floor. And this argues against concentrating the fleet into a specific area, which can easily happen if large areas are closed for reasons such as bycatch or to reduce interference with protected species. When contemplating closures, we need to assess the cumulative impacts of all the closures before we move onto the next one. Closed areas may be used to achieve a wide spectrum of policy goals such as protecting biodiversity, decreasing bycatch, saving target fish stocks, protecting long-lived corals and sponges, maintaining a pristine area, or maintaining the health of the ecosystem, however that's defined. Mediating these disparate policy goals, as Drs. Steele and Wilen noted, requires effective interaction of all the stakeholders. Everyone becomes involved, not only user groups in the local areas, but often the policy debate ranges all the way up through Congress and sometimes they take action to override what managers are doing. So new regulations have to be carefully crafted, if new closures with high economic costs are being contemplated.

The last point I would like to emphasize is that the council system is probably the most immediately accessible pathway toward achieving additional fish protections. In considering various approaches to protecting marine habitat, through national parks, sanctuaries, and so on, I think probably the most immediately accessible one is the council system.

What more do we need to know? We need to continue the advance of technologies. This of course goes beyond the various types of instrumentation shown at this symposium. We need to move beyond bathymetrics and make sure we are obtaining biological observations. Submersible observations are great and they tell us about the ocean bottom and the associated flora and fauna, but we need to move our information base from presence/absence data on Level 1 toward Level 4 information that is envisioned in the Secretary of Commerce guidelines on habitat. That will be expensive and take a lot of time to get there, to be able to associate production levels with a certain type of habitat and predict that an incremental increase of 20% protection will provide a commensurate increase in fish production.

We also, as was noted by Dr. Wilen, need to define the benefits and costs in our policy analysis. What is the goal of the proposed closure? Is it to save the resource from bungling fishery managers? Is it to protect a sensitive area such as a coral garden? Or is it to protect habitat for some other target species? We need to be very clear about our purposes and carefully assess the costs and benefits of such closures. And then you have to start getting buyin to the proposed closure. That's what the whole National Environmental Policy Act (NEPA) process is about. The NEPA process does not necessarily require you to protect areas, but it does require analyses of the policy trade-offs in closing or not closing a particular area to fishing activities. And this requires a good analytical process. I thought a good example of that was the analysis that was done by the Ecotrust representative. Several other talks have hit on that, too. I think we need to stay away from the sorts of emotional pictures that were shown by some of the speakers and stay with a truly analytical, structured approach.

What should we act on right now? I think we need a national plan for research. And I am not referring to the more general national plan that the National Oceanic and Atmospheric Administration (NOAA) puts out, but rather a plan for concentrated research in a particular area such as the Aleutians. All areas need specific plans, with dollar amounts attached so that long-term planning can occur. We need to determine the level of resolution and how much it is going to cost, and then try to garner the funds necessary to accomplish the plan over time towards more information on sea floor habitat rather than just one experiment here, one experiment there, and so on and so forth. We need an integrated analysis of what's down there. We need to be aware of the impacts of fishing activities on the sea floor, and whether those disturbances are actually constraining fish production. We need to know if fish habitat issues are constraining fish production or is it constrained by some other factors? For example, off Alaska in the 1970s, the foreign fleets fished down Pacific Ocean perch stocks, taking hundreds of thousands of tons, all the way from the Bering Sea down to Washington and Oregon. Since then, we have maintained very low catch levels, essentially only bycatch levels of two to three thousand tons in other target groundfish fisheries. Despite that conservative approach, nothing reinvigorated the stock for years until all of a sudden the environmental factors came together and the stock shot up. Now it is very robust compared to where it was left when the foreign fishery was closed. So we have to know if habitat is really the constraining factor before we close a large portion of an area to fishing to achieve some goal of sustainable fisheries.

And the last thing I'm going to say here is just a piece of advice for the budding conservationists out there, those fresh faces that we need to come tell resource managers what needs to be done. I fully understand that those on the environmental side of the spectrum need to forcefully represent that viewpoint, but when you come up before me or the regional fishery management councils, and I've sat in over 130 council meetings over the years and listened to extensive public testimony, show me the pictures of broken coral, show me the pictures of the bycatch, show me the pictures of the trawl tracks. That's meaningful. But don't show me inflammatory pictures of dead wolves and babies and pups, or combines out in the wheat field that have nothing to do with the fisheries policy under consideration. Doing so just marginalizes your whole argument and may make it a whole lot less compelling. Thank you very much.

DR. SISSENWINE: Thank you, Clarence. Our next panelist is Nils Stolpe of the Garden State Seafood Association. Nils?

MR. STOLPE: Thanks, Mike. My name is Nils Stolpe. I work for the Garden State Seafood Association, a trade group of commercial fishermen primarily in New Jersey. And I do odd jobs for other fishing groups on the East Coast, mostly in the northeast.

First off, I have to basically agree what Clarence said. We've learned that dredge and trawl fishing can cause changes in some bottom communities that are directly impacted by the gear and that we can identify and measure some of these changes. We've also learned that some of these changes can provide some really dramatic photo opportunities, but that's a little bit off topic, so I won't get into that.

As far as what we need to know, most importantly we need to know what, if any, are the effects of these changes on the fisheries. This seems to contradict Professor Rieser's statements, but quoting from the National Research Council's report on trawling and dredging effects, "Existing fish habitat legislation is intended to protect the ecological function of habitat to support the fish populations."

We need to know what the impacts of these effects are on the fish stocks. Of course, we also need to know what the effects of these changes are on the bottom communities that are being trawled. That's common sense. We also need to know how these trawling and dredging effects compare to other perturbations of the sea floor, whether they're natural or anthropogenic.

We talked quite a bit in the last three days about George's Bank. I'd like to add some material in a paper by Ellie Dorsey, whose contributions to marine conservation were recognized by Dr. Norse on Tuesday. In a Masachusetts Institute of Technology Sea Grant/Conservation Law Foundation publication, "The Effects of Fishing Gear on the Sea Floor off New England," in her discussion on shallow areas on George's Banks, Ellie wrote that those areas consisting of ridges and dunes composed of medium to coarse sand migrated at variable rates up to 60 meters in three months. Compared to that level of natural perturbation, I can't imagine any level of fishing with any kind of gear that's going to have any noticeable impact. It's just not going to show up relative to the natural "background" perturbations.

When you consider the spectrum of high-energy environments, in the mid-Atlantic, which I think Dr. D'Alteris said was 83% sandy bottom, storms must be considered. The entire area, including the shallows that are commercially fished most intensively, is usually pounded by at least one nor'easter, and more commonly by two or three or four, every year. Again, what's the impact of trawling in these areas compared to chronic exposure to the naturally occurring high energy perturbations due to storms?

We need to know the proportions of particular areas that are actually impacted by trawls and dredges. While it's useful to know how much fishing takes place over a certain time in a 10-mile or a 10-minute or a 10 whatever size block—and you've seen a lot of that in the last three days—where and when that fishing takes place is what actually determines the impacts on the bottom.

Is the fishing spread evenly over the whole area, either spatially or temporally? Usually it's not because fishermen tend to fish where the fish are and they tend to fish when the fish are there.

As has been pointed out by several previous speakers, when left to their own devices, fishermen tend to fish the same areas on a day-by-day or season-by-season and sometimes even on a generation-by-generation basis.

I'm sure that you've all seen statements similar to "trawling and dredging cover an area the size of Antarctica, Australia or various other small continents, seven or fourteen or however many times every year." The assumption seems to be that fishermen go out and randomly start dragging their nets and trawls around, doing their best to cover the entire ocean bottom. That's not the case. Fishermen tend to fish in the same places. Often they try to follow the same tracks, and with modern navigation devices, they can do this quite precisely. So there isn't the widespread impact of trawling and dredging that we all have read about. Of course, this means that the amount of trawling or dredging in particular areas can be much more intense.

I was going to show a slide of a printout generated from a vessel tracking system on a scallop boat illustrating a day's worth of scallop fishing. If I recollect, the fishing was confined to a 10-mile square and in that square the boat covered maybe 10% of the area. But if you analyzed that boats fishing at a low enough resolution, it would appear as if the entire bottom of that 10-mile block was directly impacted by dredging.

As far as what we should do right now, the obvious answer, and Clarence covered that, is we need to more research. We need to connect the obvious and the not so obvious changes caused by the passage of trawls and dredges across the bottom to their actual effects on fisheries and on ecosystems.

Looking at a picture of a smashed hunk of coral or a crushed lobster or something similar shows what's happening on the micro scale, but what does that mean when you project it to a meaningful level? We don't know, but we certainly should. And we should know sooner rather than later.

One of the not so obvious things we should be doing is at least starting work on developing a public policy that recognizes that we can't continue to harvest the hundred million or so tons of seafood from the world's oceans each year that we've been harvesting for years without having some effect on various ocean ecosystems. Some members of the conservation community seem intent on trying to equate sustainable fishing with no impact fishing. That's not the case. I don't know of any method of harvesting fish, other than doing it gently and one at a time, that is not going to have some effect on the habitat, on the environment. We're getting to the point where we should be considering a public dialogue concerning the trade-offs that we're willing to accept in terms of habitat change for what level of seafood production we expect.

I was surprised to learn that global fish production is a hundred million tons or so a year. That's higher than pork, chicken, or beef production. And we certainly have no problem with accepting environmental perturbations in beef farms or chicken farms.

So this is something that we should be considering. Do we want a pristine ocean and low levels of inefficient fishing or do we want an ocean where areas will be subjected to the necessary impacts of higher levels of production?

Recently, we've heard quite a bit about the precautionary approach as applied to fisheries. In a number of fisheries in a number of areas we're at the point now where our fishing communities are more at risk than the marine resources they depend on. While it might not be that obvious, there's a threshold level of business activity that has to be maintained to keep a commercial fishing port. And considering the cutbacks in harvest and in efficiency that have been put in place over the past 10 or so years, and considering those that are being threatened, a number of U.S. ports might very well be on the verge of falling below that threshold. One of my concerns—particularly in the northeast, the southeast, and the mid-Atlantic, though I don't know how it is on the West Coast—is that once a fishing or fishing-dependent company located on the waterfront goes out of business, it's never going to come back. Once the fishing industry infrastructure goes away, I doubt that we're ever going to get it back.

So, for those of you that know that it does have some value, I'd like you to think about applying a precautionary approach to the fishing industry. Thank you.

DR. SISSENWINE: Our next panelist is John Gauvin from the Groundfish Forum.

MR. GAUVIN: Thanks, Mike. I'd like to first agree with Clarence. I think he did a great job of summarizing the direction for moving to evaluate the science and make it accessible to everyone.

In my time involving fisheries, I have devoted a lot of effort to making science more accessible to fish managers. I used to work for the South Atlantic Fishery Management Council and I had some success at making them understand fairly difficult and complex concepts, making them evaluate the scientific tradeoffs as applied to economic concepts for decision making. And so I put a great deal of value on making science work in the process. And I think we need to do that with this issue of fishing impacts on benthic communities. I think we need to establish some terms of reference for the scientific concepts. I think these terms have to be clear and understandable and accessible to fishery managers and useful. And they have to be useful in the management context that managers face, which is to somehow make them measurable against other concepts they have to put in their management tool bag to understand how to make the right decisions.

And so I'm looking in this scientific information for how we're going to get there, what we're going to use, and I was heartened to have some papers touch on what I don't understand fully yet, but it's sort of a measure of benthic productivity. And I think Simon Jennings and others attempted to do that and that concept is one to guide you. There were papers that discussed just exactly what you decide on, what invertebrates, what indicators, whatever, and the difficulties of selecting them. But, nonetheless, I think that's what we need to have, some measure that will help us relate effects of fishing to the health and productivity of fishery management plan (FMP) species. If we are unable to go in that direction, then I think we need to suggest that managers prepare fishery management plans for benthic invertebrates because that way I think they'll explicitly set goals that are tangible. That's what an FMP makes you do. It allows you to explore the range of tradeoffs involved with your management decisions.

So I think we need to clarify our terms. I used somewhat loosely the term "loaded terms." I guess what I meant by that in my talk earlier was that two or three different scientists even won't necessarily get the same concept for when they first hear the term or start to talk about the term.

I think with these, when we get our terms of reference and our scientific metrics together, we have to make sure that they're measurable against very tangible concepts that fishery managers are going to face. I can tell a fishery manager as a staffer or as a consultant or as somebody writing a letter to the council what the impacts of closing the pollock or flatfish fishery would be in terms of revenue per year, in terms of employment. I may not be able to tell the exact impacts on different communities with the current set of data we have, but that's a pretty tangible piece of information.

I think we need to get our information about the benefits of protecting benthos in somewhat of a comparable metric. Even if it's not if it's quantitative, it has to be scaleable somehow. And so we need to get there and we need to get there in a hurry because at least in this country in the current situation, as you heard from the paper by Allison Rieser today, we're in the throes of developing an environmental impact statement (EIS) for how to manage the essential fish habitat (EFH) mandate. And decisions have got to be made that are practical and should reflect science in the best way it can. So I think that we're going to have a hard time doing this.

I think we also have to start out with reasonable steps. The now famous wheat field and combine image, my thought on that is that if that's the standard of what we're looking for then when one of the speakers said, is this really what we want, if we can't show fishery managers the importance of why we don't want that, you're likely to get the answer, yes, we do want it. I mean, fishery managers have been, I think, very troubled by the fishing industry's inability to, and the fishery management's bad record on keeping sustainable fisheries. If you showed them the image of what looks to be a productive area, even if it's highly modified, it's a better outcome than something that's not productive.

So in balancing the concepts of protecting benthos, we just have to make it understandable to managers and the process.

I want to give a couple of examples of processes that have gone well and then ones that haven't gone well from my experience. With the Bycatch Reduction Mandate in Alaska, I think we've done a great job. I think clear goals have been set. Clarence was pretty involved in setting some of those goals for our council (North Pacific Fishery Management Council), helping them move forward. And basically, the industry knew, you're catching too much bycatch or, your rates are too high. If we had sort of a, you know, something you could wrap your arms around and say, okay, well, we have this much discard in a year and we'll reduce it by this amount and that will meet the goal. I don't know. Some people translate the goals into meals or whatever you want. I mean, we have some metric. And I think the industry has and is doing a really good job working on that. We've had some success.

In a process that occurred actually down here while I was involved with fishery management, fishery managers were faced with allocation decisions between allocation to recreational and commercial fishermen. And I see, as an example, where science failed to get information in the process in a manner that policymakers understood. We had discussions. We had council members walking around saying what exactly is "consumer surplus." If we have a surplus of consumers, why do we need any more of them? People are saying, "I don't know about you, but that shadow price thing looks a little illegal."

Then we started talking about nonmarket valuation with hedonic pricing. Whoa, this stuff just didn't get into the system. Basically, we ended up with a process that, in my view, it's just my view, it ended up being determined by boat numbers and dollars and not necessarily something that resulted in the best outcome for the nation.

So we want to avoid that. We want to make this process work. We want to protect fisheries and meet the EFH mandates.

So where do we go from here? One of the things I learned, the things that I threw on the table that were working out in Alaska may not really make sense. This idea of rotating closures really needs to be looked at.

I also, I guess you know, I've been focusing on coral protection and I'm more impressed with the ecological function of tube worms at this point given what they do in their environment.

So this scientific information is percolating around, and I think we need to come up with some metrics to make it make sense. I think we need a stakeholder's process that allows the industry and the public to present ideas on how to identify areas given the holes in the data. Gear modification ideas are going to come best from industry and from guys like Craig Rose. That process, I think, will be better if we take allocation out of this whole issue. You're not going to lose the fish or fishing rights. Maybe if there's a gear modification or change, you're the one that gets to do it.

Mainly, I think we need to set clear and reasonable goals. I don't think a noimpact standard is where we want to go. It's not realistic. And I don't think the process will end up with a useful product if we try to say you have to harvest fish with no impact on the environment. I don't think any other industry could meet that standard. Thank you.

DR. SISSENWINE: Our next panelist is Elliott Norse.

DR. NORSE: Thanks, Mike.

I've heard three presentations, all of which directly or indirectly allude to some images. Perhaps we touched a nerve there. And that, of course, is a bad thing because I think what we need to do is ask ourselves whether we want to keep all decisions about the ocean within the priesthood. We need ask whether we want to confine it. Should we let the public know about what we're doing if we think they wouldn't like it? Should we look for examples from other countries, other disciplines, other practices, or should we rule out anything but our own perspectives because they're not relevant?

You know, one of the funniest things I saw here was something that wasn't said. It was an accident, I'm sure. Alison showed this, but John Steele didn't say it. The National Research Council's report said, "The absence of sitespecific or geographicspecific information should *not* be used as an excuse not to reach a broad understanding, make generalizations and act." Think of the extreme one could take this to. If one knows what the Gulf coast of the United States is like, one could say, well, that study may have been true on the coast of Alabama, but this is Mississippi, and things are really different here. And that's what we heard here when we were told that we don't have studies from Alaska, we only have studies from other places and they're not relevant.

I beg to differ. The diversity of life is enormous. The diversity of geography is substantial. The complexity in ocean systems is real. Scientists have seen similar ecosystems in different places functions somewhat differently. I don't dispute that, but many of the sea's processes are broadly applicable, even universal. And the patterns that are a result of our studies from those processes are pretty robust.

There are things that make a whole lot of sense. If you physically impact organisms on the sea floor, they will show a response. Those responses are things that many people don't like but some other people don't care about, but in any event, the seafloor organisms respond. And these responses are true in many different kinds of ecosystems in many places in the world, and that is a matter of growing concern. And if I may allude to Steve Kellert's talk here, whether your ethics are utilitarian or dominionistic on one hand, whether they are spiritual, moral or scientistic on the other hand, the repeatedly demonstrable fact that the seabed is impacted is more and more difficult to dispute.

One of the most important things that's happened here is that hundreds of people—not tens the way it would have been six years ago—but hundreds of people from a diverse group of disciplines have come from around the world and talked about bentic impacts of fishing as never before. And patterns are emerging, and it's becoming clearer that certain processes are key.

Therefore, we should call this meeting *Requiem for the Red Herring*, because I think some of the red herrings that I called attention to are less successful in misleading people than they used to be, and that gladdens my heart. I think one that has died is the idea that what fishing does has no effect and that people are really, we're just harvesters. I think the idea that we can think of marine environment or a fish population in isolation from its environment and its environment in isolation from people—and not only coastal communities, but all people—is on its way out. Undoubtedly, there will always be a few people who are in denial, and these few diehards will die hard because that's what diehards do, but I think it's over, whether or not everyone realizes it.

Now, I have a couple of observations here. The first is that I think geology is trumped by biology, I think biology is trumped by economics, and economics is trumped by sociology, and that to me is one of the most profound findings at this conference. We can know what the physical structure of the sea floor is and yet not be able to predict with a hundred percent certainty what the biological communities are going to be. There are a variety of reasons for this. Some of them are currents, some of them are history, including human history. So the geology is a beginning, but it is not an end.

The biology is very, very important and yet we see that similar looking organisms in different can places have nonidentical, though perhaps similar responses.

Economics drives an awful lot of what we've been talking about, to the point where a lot of people think that the oceans are a subset of fisheries. And I tend to think, unreasonably, perhaps, in the eyes of fishing interests, that fisheries are just *one* thing we do in the oceans. There are other values for the oceans. But the thing that really impressed me most of all is that human values, human interests, human processes are what drives what happens to the oceans. And if what we want to do is manage oceans, all we really can do only is manage ourselves. The oceans are going to do what they do. We are going to affect them the way we affect them. It's not all in our hands, but we can take responsibility for the piece that is in our hands.

And I must tell you, until this time, I have despaired because I haven't seen us taking that responsibility. Now I'm seeing growing signs of it here, portents of change.

Now, what should we do with our new awareness? One answer I hear is we need to do more research, we need to have more committees, and we need to have more reports. You know, I'm a scientist. I do research, I sit on committees, and I love reports. My office is absolutely filled with them, some of which I actually wrote myself. But those are to inform an ongoing process, to stimulate it and inform it and refine it. They are not a *substitute* for it. More of the same gets you more of the same, and I don't think we want that. I don't think we need that. I think we need to move forward. And what I'm seeing here, whether everybody likes it or not, is we're going to move forward.

DR. SISSENWINE: Thank you, Elliott. Our final panel member is Simon Thrush. Simon is from the National Institute of Water and Atmospheric Research in New Zealand.

DR. THRUSH: I have been asked to represent the scientist, so let me start by commenting on science and uncertainty. We scientists are trained to focus on what we don't know. We're trained to think about uncertainty. As part of the scientific process we need to be critical of the way studies are designed and results interpreted. But the aim of this process, no matter how arduous, is to come up with robust information. So it may seem like we are slow learners.

This conference has provided a good opportunity to step back from this process and look at what is emerging about the environmental effects of fishing on benthic habitats. We know that there are effects on ecological systems as a result of fishing. We know that there's going to be some variation in effects from place to place and from time to time. We know that the first cut is going to be the deepest and that this has some particularly important implications for new fisheries and the detection of effects in fisheries that have been operating for some time.

We know that there needs to be a better balance in the involvement of different groups in managing fisheries and sustaining the ecosystems that support them. This includes sciences that are not usually considered central to fisheries management. For example, oceanography clearly contributes to our understanding of broad-scale variations in productivity and recruitment, but it also provides an important context for assessing seafloor disturbance regimes. Mapping is also incredibly important. We need to know what's out there and how it is changing, but we also need to develop ways of classifying this information and putting it into a meaningful ecological context.

We have heard from other conference speakers that social sciences are very important in affecting change. We were told there is "too little too late," but I think the same can be said about ecology. Ecology provides us with the tools to understand ecosystem function and how these processes are modified by fishing. This is very important because it enables us to relate the patterns that we can map to the processes that make the system function.

I would like to generalize a point raised at the beginning of the meeting by Jake Rice; there are some huge global inequities when it comes to fishing and the ability to manage resources. I think it's important, therefore, not to think solely of highly complex, technical and expensive fixes to these problems because that won't be appropriate in many parts of the world. This is especially important because a lot of fishing activity is being exported out of the developed world and is being pursued in other parts of the planet that don't have the resources to manage as well as might be possible in countries like the U.S., Canada, New Zealand, or Australia.

So what should we act on now? Any kind of action is only going to succeed if people want to move forward, and I get the sense from this meeting that people really do want to move forward. One great step forward for fisheries management would be to get precise records of where all commercial fishing gear is deployed and lifted from the sea bed. The technology to do this is available and in the scheme of things it's not very expensive. This would enable us to get over arguments about scaling in terms of effort and disturbance, how much of the seabed is impacted and how much of it isn't. This should provide important information about human disturbance to a large proportion of the planet. As others at the conference have said, we need criteria so that management decisions can be made and those decisions can be assessed. In this regard we need good measures of disturbance associated with natural phenomena, fishing and other anthropogenic impacts.

We know that these ecosystems are complex and some questions difficult to answer. But we do know that we can qualitatively assess the risk of environmental degradation due to fishing, just as we can for other kinds of impacts on the environment. Fishing is no different. Such assessment procedures need to be holistic and applied in an adaptive kind of framework. I think this an important step in moving away from using uncertainty and lack of perfect knowledge as an excuse for inaction.

I think we also need some new research initiatives to improve the knowledge base. Over the last ten years or so, we've learned a tremendous amount about the environmental effects of fishing. But we need to keep moving forward. This means supporting research that isn't going to fulfill the needs of management today, but it should work towards fulfilling the needs of society in the next decades. These research programs should be large-scale, representative, and include measures of the ecosystem function. There are some issues about design that we will need to consider very carefully—designs should be based on the knowledge of the seabed landscape. Research programmes should be broad scale, but important natural history information and ecological process must be nested within these studies.

Thank you.

DR. SISSENWINE: Thank you, Simon, and thanks to all the panelists for a lot of food for thought. None of them are all knowing, and all of them are better than slow learners because everyone brought forward very important things that they learned and benefitted from this conference, as I'm sure we all have, and all of them had some suggestions as to the ways forward.

I think it's now time to give a lot of patient people, a few hundred of them sitting here, a chance. If any of you want to come forward and take your shot at responding to the questions that we put to the panelists, as well as perhaps asking followup questions to the panelists. The ground rule is that you've got to come forward, give your name and your affiliation, and speak into that microphone.

DR. MILLER: My name is Margaret Miller. I'm a benthic ecologist with the Southeast Fisheries Science Center (National Marine Fisheries Service [NMFS]). I wanted to express a concern about some things that I think we know that haven't been discussed at this conference; specifically, the indirect effects of fishing. I think we know that trophic cascades exist and we know that they can be altered by human harvesting activities. I think we know with reasonable certainty that there are mass balance effects in ecosystems that fishing, the removal of biomass, has on ecosystems. It seems to me, although a couple of those invited speakers yesterday touched on this issue, that this has really been neglected and we haven't really examined what we know about indirect effects of fishing at this conference, and that's been a great disappointment to me. I don't know why this subject has been neglected, but one hypothesis would be that those sorts of effects don't have a technological fix. The only fix to those sorts of problems is a reduction of harvest I don't know that that's the case. But I wanted to bring it up because I think that indirect effects need to be acknowledged and critically examined in terms of what we know and what we don't know and how different regimes of harvest impact ecosystems in that way.

DR. SISSENWINE: Thank you. I'll ask if any of the panelists want to respond to the comment. Have we neglected the impact of trophic interactions, basically, in terms of their changes on ecosystems, and if so why? Go ahead, Elliott.

DR. NORSE: I think it's a broader question, if I can interpret correctly what you said. Not only trophic interactions, but trophic cascadesS. Not just who eats whom, but what the consequences of who eats whom. That's one of the things that we haven't really talked about. We started to talk about succession. What a fascinating idea, in the meaning of fisheries biologists, talking about succession. That's ecology. I'm thrilled. What I haven't heard yet is conversation about metapopulation dynamics and island biogeography. When I heard the disagreement between Simon and Michel Kaiser, I asked myself what's the difference between the North Sea, which has been very heavily impacted for a long time, and ecosystems in some places in New Zealand and some places elsewhere in the world that we haven't quite impacted yet? I mean, is it impacted or unimpacted places that we should be studying and taking our lessons from? And what occurs to me is that both impacted and unimpacted ecosystems have important things to tell us because impacted ones may have been deprived of sources of recolonization, and what we may be seeing is a function of a very different metapopulation dynamic than is true in another place. And I want to sow this little seed so that the next time we come together we have eight papers on metapopulation dynamics.

One more thing. We've seen papers on biogeochemistry. Yet another crucial point. We're talking about carbon content. We're talking about digestibility in food found in carbon. We're talking about nitrogen flux. And that's good, but iron and silicon and nitrogen and phosphorus are important elements, too.

So I think the genie is out of the bottle. We are seeing broader thinking, and I think this conference is going to stimulate still broader thinking, and that's a good thing.

DR. SISSENWINE: Let's give Bill a chance to get in on this.

DR. SILVERT: I'm Bill Silvert from Instituto de Investigação das Pescas e do Mar in Portugal. I'd like to congratulate the organizers on a very high level symposium and all the speakers are looking to the future. I'd like to bring things to a lower level, lower the level of discussion and look backwards.

When I went on my first benthic cruise over a quarter of a century ago, we studied things that you couldn't see in their natural environment. We brought up drags full of mud and sorted them on deck. Nowadays, of course, we've progressed. We have underwater video, we have remotely operated vehicles and all kinds of great things. And the end result has been that practically all the discussion that I've heard at this symposium about the animals and plants that live on the bottom has been on epibenthic macrofauna. We've lost sight of the stuff that's down there in the mud. I've heard very few discussions of what's in the mud, very few discussions of meiofauna, benthic bacteria, and of a very important group of organisms that don't poke their heads up very much, namely bioturbating polychaetes and other bugs and worms that live in the bottom. These are really important. This is for many organisms the basis of the benthic food chain. Of course, there is detritus coming down from the surface. But I think it's remarkable that at the same time that people can study the water column by looking at smaller and smaller and smaller organisms, I'm talking microbial loop and so on, those of us in benthic ecology are talking and looking at quite literally larger and larger organisms.

I think it's very important to pay attention to what goes on under the water–sediment interface. There's been almost no mention of something that Elliott brought up, which is geochemistry. There were a couple of mentions of benthic oxygen and the depth of the anoxic layer, but the general nutrient fluxes in the sediments and so on are extremely important as are the organisms that live there.

I've seen experiments where a single polychaete was removed from samples, and in no time at all a rich, flourishing bottom turned into a desert, a wasteland. We really need to look at these things and give some concern to what's going on in the bottom. We have to look at things that have only been mentioned lightly, like compaction, which doesn't create a major visible effect, but can destroy the potential of the bottom for infauna. These are some of the things that I hope at the next meeting will be discussed even though they're almost invisible. Thank you.

DR. SISSENWINE: Thank you. Let me be provocative. Are we not studying the microfauna because it's not charismatic and doesn't get people's attention, or it's to hard to study, or it's not affected by fishing, or we just overlooked it? Any comments?

DR. THRUSH: Yeah. I would say that I'm pleased to hear you say that people need to study macrofauna and microfauna. But I would take heart in the fact that John Gauvin is a convert to the beauty of tube worms. You know, that's wonderful. And you're absolutely right, those animals play important roles in the ecosystem services. I alluded to that very briefly because of time constraints. But there's a paper just out in Annual Review of Ecology and Systematics that deals with those kinds of things. I'm one of the authors on that.

This is research for the future. It certainly hasn't been well developed yet, partly that's because the fisheries research and the fisheries impact research hasn't really integrated well with ecological research and that's something that needs to happen.

DR. SISSENWINE: Of course, it's clear that microfauna are important ecologically, but the issue is the susceptibility to impacts from fishing. I recall one of the Simon Jennings' papers actually addressed this specifically, and there wasn't much evidence of an impact. We do not necessarily know enough, but I think that is a consideration in terms of where the priorities lie.

DR. HIRSHFIELD: My name is Mike Hirshfield and I'm with Oceana. I'd like to make a pitch for charismatic megafauna. One of the things that I learned during this session was how sobering Simon Thrush's recovery graphs are. I was very impressed by them. And I was equally impressed with Mike Fogarty's stop light chart. With the predilection towards action, which I have, and in response to the question you asked, "what do we know enough about to act on now," and almost hearing a consensus (perhaps an unspoken consensus, but one that I think that this group ought to be able to speak), it strikes me that we ought to be talking seriously about a global ban on trawling in corals. I do not see where we can make the case that trawling in corals is a good thing. Thanks.

DR. SISSENWINE: Thank you. Anyone want to comment on that? John?

MR. GAUVIN: Well, my comment is that I think trawling avoids corals. I know in Alaska we do everything we can to avoid corals for natural reasons: they rip up our gear, they're generally associated with substrates that are very, very difficult to fish, if not impossible, and I think we somehow overlap with corals. We don't know whether that overlap and the amount of coral that is taken or harmed in that is substantive to the amount of coral we have, but certainly it's no one's desire to catch corals with a trawl.

I think you really ought to open up your thinking to, based on some of the video I saw last night, other gear and their effects on corals, because it was clear to me that while they may not have a crushing effect, if you knock it over and it's no longer able to serve its life function, it's dead. So, I mean, where sensitive substrates, like corals are, I'm not sure any gear is going to fish there without having an impact.

DR. SISSENWINE: Elliott?

DR. NORSE: I am glad to hear the newfound concern that seems to be essentially unanimous about trawling in places where there are corals. I would also add sponges and other large structureforming organisms. I'm sure John would agree with that. But I don't agree that they are not deliberately targeted. I have read in Canada and I have heard on the Gulf Coast of Mexico unsubstantiated allusions to what people in the trawling industry call "conditioning the bottom." Conditioning the bottom means removing obstructions from the bottom so as to make it easier to trawl, commonly, when a pair of trawlers pull a chain between them to get rid of "obstructions" on the seafloor. Of course, one person's obstruction is another person's coral, and that's a problem. That's one of the reasons why we find that corals' distributions are so peculiar. In some places, we see them in a certain habitat, but in other places, we don't. And that reason could be biological, but it could also be historical, as a result of human fishing activity.

So I think responsible fishermen will avoid corals, sponges, and other structure-forming seafloor organisms, and I think an increasing number of people in the general public and decision making communities are sensitized to that, but I wouldn't quite say that coral and sponge communities are always and have always been avoided.

DR. SISSENWINE: Okay. Food for thought on this particular issue. I think that the comment that came from the audience and the response are useful. They indicate that there is substantial sentiment for protecting coldwater corals and other large, three-dimensional biogenic habitats.

An issue, of course, is do we know where they are well enough to protect them? In some cases, maybe we do, and in some cases, we don't.

And then a related issue is, if we actually had a very complete map of where they are and where they might have been once and could potentially be restored, what magnitude of closures are we talking about? The more we map, the more evidence there is that these communities are widely distributed and possibly even common. Prior to fishing, they might have been even more common. When we thought they were rare, it was easy to argue to protect them all. However, is this the case if they are common? I don't expect an answer, but it's food for thought. Go ahead.

MS. LUTTENBERG: My name is Danielle Luttenberg and I'm with Environmental Defense in Boston. My question is about process. It's not the ecological process we're talking about, but rather the management or the policy process that comes out of this. As John Gauvin noted in his talk on Tuesday, and also as Allison reminded us this morning, the EFH provisions of the Sustainable Fisheries Act really do refer to commercial exploitation and how we protect the habitat of commercial species, not conservation for its own sake. And yet, as Clarence Pautzke noted just before in this conference and as is seen in councils and in lawsuits, there's a whole wide spectrum of values being represented when a group like this gets together. We're talking about conservation, biodiversity, improving fish stocks, corals, and so on, and as I said, and as the lawsuits indicate, there are other parties and other values that are coming up here that aren't just about fisheries exploitation. Elliott brought this up in his talk on the first day.

So we say we want to move forward, but the question is how do we move forward? Clarence Pautzke suggested that the council process is right now the most immediate that we have and so we have a forum. My question is, is the council process appropriate for this? Are the councils or is the council process really equipped to handle the interplay of all those different values of the conservationists who are showing up, of the members of the public, of the fishermen who are concerned about the industry, and if so, how do we guarantee the participation of all those different values and those different players in a process that really is right now set up to represent the commercial or recreational exploitation?

DR. SISSENWINE: Go ahead, Clarence.

DR. PAUTZKE: I would say you're going to find out in the next year and a half. The broad EIS is being done as far as I know for the council, and I'm starting to leave it more in my rearview mirror as I go to the research side of the house. But they have a committee together that's looking at various areas that can be designated as protective areas for corals, bunions, sea onions, those kinds of things that you're worried about protecting. Not necessarily to increase the fish stocks per se, because we think we got them under control pretty well up there, but I think there are benefits in having areas that are off limits. And I think that the council is trying to navigate its way through that.

Now, how do you become involved in that process? Well, part of it is, is you're part of a committee structure and the public process structure that is leading up to definition of the alternatives you're going to be looking at, which, as I understand it, is quite broad and would close a number of areas. John Gauvin has been involved with it. People from the Alaska Marine Conservation Council have been involved with it. I'm just not sure of all the groups represented. But you do have a very public process where you can work through it from a committee structure all the way up through your advisory panel and also your scientific and statistical committee through the council. And we're going to see over this next year and a half whether they come up with some areas that are protective for coral.

DR. SISSENWINE: Okay. I guess I'd also add as a comment that while the role of habitat within the Magnuson-Stevenson Act of the Sustainable Fishery Act is pretty clearly related to fish production, the actual implementation of that act requires basically compliance with NEPA, which broadens the context.

So how it all plays out, I think, is something that, as Clarence says, we'll learn over the next year or two, three. We're also inventing these things, I think all of us collectively, people in the agency, people on the councils and the public, as we go along. Your question, I think, is a very interesting one. Next.

MR. JOYCE: I'm John Joyce from the Marine and Fisheries Division of the Marine Institute in Dublin, Ireland, where we're facing now a total ban on cod fishing in the Irish Sea. And so at the risk of not being able to walk back to my hotel room because my foot is in my mouth, I want to ask what I see is the one big question, which is how do you mediate successfully between the diametrically opposing views of the fishermen who want to pursue their living on the one hand and the people who want to preserve the environment on the other so that both sides can feel heard and respected and, as one speaker has said, we can move forward together constructively? Thank you.

DR. SISSENWINE: Anyone want to take a shot at that? Elliott?

DR. NORSE: A friend who isn't here, Peter Auster, says, "No fish, no fishing." If the Irish Sea is closed to fishing, who benefits, the biodiversity fan who doesn't eat fish or the fisherman whose boat is tied up and not fishing? I think the answer is neither of the above. Forgive me, I read about the closure and that's a place none of us want to go.

Now, I haven't heard anybody say the reason that we've gotten there is because we have been too stringent and excessive in protecting the environment. The contrary argument, however—

MR. JOYCE: My question is, how do you get people to talk to each other?

DR. SISSENWINE: Well, I think that, in fact, this meeting in its structure was one example of the attempt to do that. And actually, I made a comment earlier that one hypothesis is that if we have more information and better communication, we will be able to gradually converge on consensus solutions.

The alternative hypothesis is even if both of those are perfect, we're going to still disagree on a lot. So we are testing these extreme hypotheses as we move forward on some of these very complex issues.

MR. GAUVIN: Mike, can I?

DR. SISSENWINE: Go ahead, John.

MR. GAUVIN: An aspect I think that would help any system that's faced with cutbacks and meeting objectives of rebuilding fisheries, I think the fishing industry will be a better partner in that in rebuilding and cutting back if you have a rights-based management system, which gives them an ability to live better under the smaller quotas, if that's what's involved, and certainly it gives an incentive to weather the down time because they can at least benefit from the savings or the yields that could occur if the fishery rebuilds. And I think that makes everyone a better citizen when you have that kind of system. So I recommend that that could play a role. I guess we could tallk about that now since there's no longer a ban on rights base management for at least now.

DR. SISSENWINE: Okay. Let's move on.

MS. KNIGHT: I'm Emily Knight. I'm from the University of Maine. I'm a graduate student with Les Watling, and actually my question is very close to the one that was just asked. I'm working with fishermen on my thesis and we're studying the effects of trawling and it's with a couple of draggermen, so it's studying the effects of what they do. And what I was noticing recently in the news, I come from the northeast perspective, and you probably heard about it on the news, the problems with the NMFS stock survey and this huge argument that's exploded in New England about whether or not the stock survey is accurate or not basically fishermen are claiming the cable attached to the trawl doors were askew, and so the survey would have caught less fish than a commercial boat. But then the scientist's point of view from NMFS is that they've been doing it this way for a long time and so the long-term statistical trend must be maintained and would still be precise if it's a declining stock. And you listen to them argue about it on the news and the thing that I started thinking to myself is that, in the end, it doesn't really matter who's right until who we're all doing this for believes in it, including the fishermen. Even when I take my videos and I show it to the fishermen, even if they see an impact, a lot of them still won't say "I don't believe trawling has an impact." And they don't say it because they're ignorant and they don't say it because they don't understand. They usually say it, in New England anyway, because they feel like if they admit anything that it's going to be taken away from them.

And so my question goes to that how do we better involve them in the process of the conclusions we come to so that they feel comfortable in it? Because it seems to me they don't feel comfortable in it because they're not directly involved in it. They feel they're sort of out of control of their destiny in a way. That's my question.

DR. SISSENWINE: Okay.

MR. STOLPE: If I may. Two of the fisheries I work for, Monk Fish and the Sea Scallop Fishery, are heavily involved in cooperative research with the NMFS. Both of these fisheries have done rather well because of the level of cooperative research that they've been involved in with NMFS. I hate to use the term "trawl gate" because it has some connotations that I don't think really apply to the situation. I don't think anybody in a responsible position is trying to imply that there was any intentional wrongdoing or anything like that. I think it's a situation that came about that wasn't handled sensitively by either side, but I think it's going to be worked out in the not too distant future.

DR. SISSENWINE: Well, the data will speak for itself, let's put it that way.

MS. KNIGHT: It doesn't even seem to me, in the argument, it doesn't really seem to matter to me who is right because I could probably take all the papers that every scientist in here has done on trawling impact and if a fisherman doesn't want to believe it, they're not going to believe it, and they're not doing that out of denial, even, you know; they have a different interest because they don't see it as in their interest to help science.

MR. STOLPE: I'd have to disagree with you there. The fishermen that I've dealt with that have been involved in the science tend to go along with it. And, of course—

MS. KNIGHT: And I agree. I want them to be involved more.

MR. STOLPE: Of course, the fishermen I'm familiar with that have been involved with the science have come out ahead because of the improvements that their participation contributed to the science. I don't know what the result would be if we set up a cooperative research program and the fishermen came out behind where they would have been without it. I don't know that I'd want to know because I would probably be held to be responsible for part of it. But until this situation with the ground cables came up, we were moving in a fairly positive direction with cooperative research. And I think getting the industry involved in doing the research is probably the best way to get the industry to accept the research. In the northeast, it hasn't just been limited to monk fish and scallops. It's also been going on for quite some time with surf clams and ocean quahogs. Also, though not at as high a level, with scup as well.

DR. SISSENWINE: There are various examples of this that are successful, mostly dealing with stock assessments, however, relatively little with habitat issues, although this should change in the future. I think there are lots of opportunities for cooperative research, including habitat studies, but there are bound to be some failures along with successes. This is a risk with all research, but with cooperative research there is an increasing dimension of differences in professional cultures and experiences of the people involved, and often a controversial, political backdrop. I think we just have to accept the reality that there are a lot of humans involved in all this activity.

DR. NORSE: I think the paradigm, and one of the talks the first day, John, it wasn't yours, was it Clarence's?, said that in 1976, when the Magnuson Act was originally passed, what we wanted to do is get the foreign fishermen out of our waters and rational, efficient, full utilization was the goal of fishery management. Well, lots happened in 26 years, and we feel differently about things now. So maybe we need to rephrase the question not how do we get the most out of the sea in terms of tons of fish, but how can we help people who fish to make a living put food on our tables in a way that doesn't degrade what we, as a society, value?

DR. SISSENWINE: John?

MR. GAUVIN: Well, I add to Elliott's last point that dozens of fishermen, they're out there, it's a tough business and it's tough to do what they value as an occupation. But, Emily, to your point, you said that fishermen say they doubt this and they don't think trawling has an effect, but they're doing the project with you. I mean, you're going to get out there. I, through some coercion, got my members to fund some research that you might have seen a poster on. I mean, they were certainly worried that, you know, bring in this friendly fire, you know. Didn't you pay for this research that just proved that trawling has this effect? I mean, they were willing to do that because they wanted to know the effect. Obviously they don't think they're having a nefarious effect because they're fishing the same areas every year, the old argument.

So I think fishermen are at the table, and I think they are willing to ask the questions. I think the buy-in is a good point, you need to have it, but I think the industry is quickly getting there. And I'm happy to see New England working with this, should say Les Watling (I may have said Elliot Norse but I meant Les Watling. Elliot is not involved with field research at all on the East Coast) Elliott Norse and you on these projects because I think they're there.

DR. SISSENWINE: Okay. Next person?

DR. KENCHINGTON: I am Trevor Kenchington, a consultant fishery scientist with Gadus Associates, from Nova Scotia. I work for a range of clients, including various fishing industry interests from Newfoundland south to Florida.

I want to throw out three lessons from this conference and see if there is any reaction from the panel. One should not be a new lesson, but I think, on the basis of the last few days, that many people still have not learned it. One came as a revelation to me in the last few days, based on what I have heard. One is something that I had hoped we might see discussed during the conference but I have not heard mentioned.

The first of those three has already been mentioned by a couple of people but it really has not received its due. It is that the effects of fishing gear on habitats that are obvious to the human eye may not be the ones that really matter. Hence, simply being able to say that neither the sorts of things we thought would happen nor those which we could easily have seen happen did happen does not mean that we have written off all problems. It is not that I expect that, for example, disturbance to biogeochemical processes in the seabed is a serious problem, but it might be and so might other unexpected effects of gear. We need to be able to wipe those off the slate, as well as the more obvious effects, before we are clear to draw conclusions about gear impacts being minor.

My second point: Studies of seabed impacts started with a first field experiment conducted by Michael Graham in the 1930s. It produced a truly lousy data set, but Graham nevertheless concluded that there is no problem. Other people from then up until the mid-1980s were gathering other lousy data sets and saying there is no problem. Through the 90s, with slightly better data sets, people were saying there is a massive problem. Now, based on what I have seen and heard over the last few days, I think we have at last begun to move to a more sophisticated level of understanding. We are now seeing studies of some gears in some habitats where the impacts look serious and some where they do not. Looking around at the posters and the presentations, by my own informal scoring system, I make it a dead heat between studies whose authors say, in effect, "we really didn't find anything" and studies whose authors say "we did." What is even more interesting is that the particular combinations of gear and habitats that seem to pose problems and those that do not were not what I would have anticipated. I would have expected that otter trawling on a fine sand bottom in deep water off Alaska would have had minimal impacts. Yet we heard, I think from Dr. Stone, that there are sea whips, there are things that live around the sea whips, and naturally an otter trawler makes a mess of sea whips when towing through them. On the other hand, there is a very interesting poster, presented, I think, by Ms. Allen of Louisiana State, concerning a fishery that uses toothed dredges to pick up oysters off an oyster reef, a biogenic structure that I would have expected to be exceptionally vulnerable to toothed gear, and yet they could find virtually no effect at all of the dredging. That raises the worrying possibility that not only do we need to focus on the impacts of certain gears on certain vulnerable kinds of sea bed, rather than on a broadbrush gear-impacts problem, but also that we may not be able to predict ahead of time which gears and seabeds we need to focus on.

The third lesson is the one that I had hoped there might be some discussion of, but there has not been. A lot of seabed-impacts research over the last 10 or 15 years, indeed running back further than that, has been an attempt to answer the question "Can we detect an impact of fishing gear on the sea bed?" Frankly, that is not very interesting. I can look at an otter trawler, at the gear on the deck of a trawler, and at a sea bed or a photograph of a seabed that is too deep to dive on, and I know that there is going to be an impact. Whether or not you can prove it scientifically is just a measure of whether or not you throw enough scientific resources at the problem. The question we should be asking is "How serious is that impact?" And that raises a whole other question of how great an impact matters—which is largely a policy matter, but it is one where scientists are going to have to advise the policymakers because, on such a matter, they are not capable understanding the implications of different management options without a lot of guidance. I think that this quantification of impacts is a whole additional step that, judging by the last few days, we are really not making yet. I think we need to move onto it.

I would be delighted to hear any comments on any of those three from the panel. Thank you.

DR. SISSENWINE: Anyone want to comment?

MR. GAUVIN: Trevor Kensington's points are well taken. The third point is probably a much better articulation of what I was trying to say, I think. That we detected an effect is not the key issue, but what is the meaning of the effect, and I think that's what we need to go on. I guess I'm wondering if other panelists or the public have any suggestions given the time frame this country is on with their EFH process because, boy, that's a key issue. My worry is that even though the unknowns that we can do things that we think are going to make it better may make it worse.

DR. SISSENWINE: Go ahead, Clarence.

DR. PAUTZKE: My answer relates to several different questions concerning which components of the ecosystem we should be looking at (i.e., how far we go down in the sediment to look at meiofauna versus, say, examining macrofauna on the surface), and I think that basically we've got a great demand for research on a multilayer ecosystem out there and relatively meager funding to do that research. Whether you are a scientist in the fisheries science centers or in academia, research has to be carefully prioritized. There is never going to be sufficient funding to span the whole range from studying impacts on meiofauna on a grain of sand to large scale impacts of bottom trawling on coral gardens. Research monies will flow where there is perceived to be the greatest interest. If everybody is interested in corals and sea onions and sponges, then probably you're going to see a lot of research on fisheries impacts on those areas. And managers still need regular funding to ensure the continuation of stock assessments for target species. We are always going to be squeezed for funding, and I do not see the funding environment getting any better in the near future.

DR. SISSENWINE: In response to the question about what level of change matters, I can think of three different contexts for an answer. One context is amenable to research, not easy research, but research I can sort of lay out in my head and figure out how to do it. Another requires very difficult research such that I am not convinced that anyone has a realistic plan for how to do it. And for the third context, research may be irrelevant.

The first would be, does it matter to the fish in terms of their productivity. And I think we know how to, although it's expensive, we know how to do various studies that could address this, such as short-term indicators of the growth rate of a fish in different habitat types. We can do biochemically, for example. Not easily, but it can be done.

The second one is, does it matter to the functional value of the ecosystem and all its goods and services, and that's a big, big issue and a big research problem.

The third one is, regardless of the other two, does it matter to the public for whatever reason? And that one may be informed by research, but ultimately public policy choices are going to be made by whatever processes there are that determine public policy, including politics and media campaigns, for better or for worse.

Go ahead.

MS. MANZANILLA: My name is Silvia Manzanilla. I work for the Ministry of Environment in Mexico City. And I congratulate you really for this very successful symposium. I do think that I didn't hear the role that benthic communities have and their links to so many other multiple parts of the ecosystem. The impact and its meaningfulness is not just restricted to the impact of the coral being torn apart; it goes beyond. There is almost no funding for research right now in my country, whereas in this country, you have a lot of experts dealing with so many parts of the food chain and the cascading impacts on the other elements of the ecosystem. Perhaps it would be useful to organize another symposium that could bring about this type of research, and we could start talking to experts that deal with all the parts of the components of the food chain.

On the other hand, I believe you mentioned that you wanted evidence, hard evidence of the impacts. You said "show me the tracks of the trawlers, show them to me." You know, it's true, sometimes we have to really take a picture of the tracks to show the damage, but I'm tired of dealing with that type of point of view. In my country, we do not have much of the baseline information on food chains, ecophysiology, or the relationships of different organisms in the ecosystem. We are starting to collect it in a few regions of our exclusive economic zone. We don't have the technology or the resources to prove the impact of the tracks, and this is consistent with what the trawlers want us to show them too. We have very limited capabilities of data analysis, even starting where you are maybe 50 years ago in many ways.

So in terms of this and putting yourself in our shoes in Mexico City and maybe in Mexico and other developing countries and nations, there is a prediction model, which states that in 2025, maybe in two decades, 75% of the population, of the human population in the developing world, will end up in the coast exerting pressure on the marine resources. In an underdeveloped country, we don't have data to prove or even sustain closure in critical areas. So planning for us and management is done over a great deal of uncertainty. Almost no data. We have to extrapolate whatever we see here reported, or the experiences of other countries, and try to start looking at analogies within our areas. The industry, believe me, sir, the industry in Mexico is really ill-informed and doesn't want to be informed. Maybe here they are approaching the research and they are really connecting with researchers. In my country, not only they don't want to connect with researchers, they mistrust all the research you all are doing also. They don't want to even hear it. That's one thing.

We are trying to implement conservation and regulations under social pressures of poor people migrating to the coastlines with really nothing to eat, specially the rural people that seek new economic alternatives. On the other hand, we're dealing with a very rich industry that is overfishing, has overcapitalized their fleets, and are not used to any restrictions. They are investing lots of money in their technology, and they are really depleting the resources. And not in the benefit of the most important part of a population that are barely making a living on marine resources.

Then we're dealing with lack of environmental awareness. Our level of environmental education is where you were maybe 30 years ago. And we're slowly, slowly advancing. Then we have little legislation dealing with the environment. Of course, the fishing industry and the fishing activities will never, have never acknowledged this legislation; they have chosen to ignore it. Environmental legislation is not there for them. And we're slowly, slowly starting to put little more limits, making ourselves heard, and it is not easy.

But then my question finally is, could you please tell me, being that in two decades we will have a high percentage of the human population exerting this type of pressure over marine resources (in our country it may be more than that), and taking into account the precautionary approach considering we have are dealing with such uncertainty, what would be the three main points you would recommend for immediate action in management dealing with a precautionary approach in fishery?

DR. SISSENWINE: As you were speaking, I was going to note how important Jake's talk was in the beginning. He would like to speak for himself, but first let me give the panel a chance

DR. PAUTZKE: Well, I do want to make one comment.

DR. SISSENWINE: Go ahead.

DR. PAUTZKE: Well, I didn't want to leave her with the impression I was demanding to see pictures of trawler tracks before any protective action should be taken. I was trying to make the point that those types of pictures can be very compelling to managers if you want regulatory change. Those are the appropriate types of evidence. They are much more compelling than the inflammatory pictures shown earlier that are not even on the topic of fish habitat, but just used to raise the level of emotion. Those types of inflammatory pictures are off issue and off message and rarely successful in achieving regulatory change of the type requested. On the other hand, evidence of trawl damage is compelling to decision makers and can lead to additional habitat protection.

DR. SISSENWINE: I guess I'd comment that I think this is, this particular statement is a very important one, and I think that we in this community, most of us, probably don't hear often enough and don't realize strongly enough how advantaged we are in terms of the processes we have and the science we have compared to the vast majority of places in the world. Compared to a lot of the places I'm involved in, Mexico looks pretty hightech, for example compared to situations in the coast of India and a lot of other places. So I do think this is a real critical issue to all of us that are concerned broadly about ecology and human well-being.

In terms of your question as to what are the priority areas that are needed, the one that emerges from every study group that I'm on, and that's a lot of them, like it is for Jake, is some sort of rights-based management regime. This doesn't mean individual transferable quotas necessarily. It means a rights-based regime that gives people some control over their destiny and deals with basically the folly of the alternative, which is a competitive-based allocation scheme.

So that would be my message. Unfortunately, in places where governing situations are weak at best, it's not an easy description to fill, but I haven't seen a group of experts at the international level come up with any suggestions that don't ultimately include some way of addressing the failure of the incentive systems unless there's a rights-based regime. Elliott wanted to comment.

DR. NORSE: Silvia's question really touches a nerve with me, and it tells me how important what happens here is. Forgive me, but I'm in the United States, and although we have many distinguished and important international participants in this, somehow my thinking is narrow enough so that I keep snapping back to the United States, and then it occurs to me, when I hear Silvia's comments, that it must happen here not only for the sake of our cod and our rockfishes and our corals and our meiofauna, but it also has to happen here because we have the richest scientific research infrastructure in the world. We have more toys, more scientists, more students. We have, whether I like it or not—as Clarence points out— a very elaborate management process. And if we can't do it here, that's sending a signal everywhere that it can't be done.

Now, if we can do it here, that doesn't mean that we should have Mexico and Guatemala and Brunei duplicate our system. They can't and shouldn't. But, nonetheless, if we fail to do it right, we're surely sending a signal.

DR. SISSENWINE: Okay. Clarence.

DR. PAUTZKE: I just want to say that within the spectrum of the population of fishermen that we deal with, you're always going to find those that are in complete denial that anything needs to be done to improve sustainability of the resource. They will maintain that position until there are no fish coming up in their net, and even then, they will come up with excuses why nothing needs to be done. Then there are the fishermen who want to find out what their fishing impacts are and are willing to advance solutions. Over time, we need to identify those fishermen who are looking for ways to improve, those people that want to listen, want to learn, and are analytical in their thinking. Those views need to prevail if we are going to have sustainable fisheries and protect our marine ecosystems.

DR. SISSENWINE: Okay. Thank you. Go ahead, Jake.

MR. RICE: Jake Rice, Director, Science Advisory Secretariat for Department of Fisheries and Oceans, Canada. In light of the time, I won't do anything more than encourage people who have made statements during this panel about "How do we get the engagement of citizens," that all the indications are we are moving in the right direction in response to Danielle, to John, and people like that. For people like Mike and I who spend a lot of our life at meetings(this is my 32nd week of meetings so far this year), everywhere one goes, the engagement and collaboration among resource users, conservationists, managers, and scientists is headed in the right direction. And there's another point that we can draw from that. There's a lot of science that still needs to be done. We've heard all about it. But there's no fact out there that anybody is going to find that will suddenly have everyone in this room and the constituencies that they speak for think exactly the same way about the issues. There is already absolutely no diversity of viewpoint about the direction in which we need to be going. It is only a matter of difference in views of about how fast, exactly where it should happen, and exactly how to distribute the costs that will have to be paid to move in that direction. That is where over and over again the issue comes down not to whose list of facts is longest, but how do you make the process work.

I can't say that I share Clarence's optimism that the council system is the best process by which to bring this much more diverse range of values to bear on decision making about the pace of which we make change in the direction of less impact of fisheries on ecosystems. My next point is not a question, (although I invite some comment on it), for those of you who care about this, make sure that you take time away from your science to participate in the advisory and management process in your own jurisdiction. Go to the United Nations Food and Agriculture Organization Web site in about another month or so where there will be an Annex to the guidelines for the Code of Conduct for Responsible Fishing, to do with bringing the ecosystem approach to fisheries into the Code.

It's informative to everyone that more than half of that whole Annex is on the process of developing management plans, not the products that you're trying to produce, but the process that will produce them. Whatever constituency you're part of, you should really focus on making sure that you're improving the process to get these viewpoints together in one room and speaking to each other. There are a huge number of people who can guarantee that you won't make any progress if they don't buy into the process you're part of.

There are guidelines that have been developed in Australia that are worth looking at. We have done formal experimental processes in Canada for doing it. Those products and that knowledge is out there of how to make the process work, and that's much more important than pursuing any single piece of research that I've heard in the four days I've been here.

DR. SISSENWINE: Thank you, Jake. Go ahead, please.

MR. VERSAGGI: My name is Sal Versaggi. I'm with Versaggi Shrimp Corporation, a local shrimping company here in Tampa, Florida. I'm a harvester, commercial. I come to these meetings to kind of educate myself, and they're very good. Not that I understand it all, I'm not a scientific person, but I think what we're doing is the right thing, and I think we're going down the right path, but there are a few things that I think are maybe a little bit biased. I see everything is directed towards the commercial sector of the industry. There's also a recreational sector. I heard very little about the recreational impacts on fisheries and what they do.

I picked up this little handout from Oceana, an activist group. They are probably our number one enemy. But, you know, they talk about 8,000 fish traps, but they don't say that we have a law now in the Gulf that, you know, these are going to be fazed out by the year 2007. There's a sunset program for that.

They go on to say there's 450,000 lobster traps. Well, there's also a plan for reduction in lobster traps. They don't tell you that. There's 1.3 million stone crab traps. There's a reduction plan for that also with the Gulf Council. That will be cut in half approximately. They talk about 3,000 to 4,000 shrimp boats, but they don't say anything about the 850,000 recreational boats that are going through estuaries, dealing out 25% of the hydrocarbons in the sea grasses and things like that.

When I come to a meeting like this and I see this kind of stuff as a handout on the table, it makes me very skeptical. And then we get into the credibility gaps. Well, this is part of the stuff that causes credibility gaps. I mean, I've got to be able to trust the people in the forums, and so on and so forth, that I'm going to.

You go to the Gulf Council meetings and all of a sudden there's a disconnect. You got all this great scientific evidence and then it gets very political because those are all politically appointed positions. And these people looking at it, and sometimes we've had a science, a hundred percent can be on their side, and they still go ahead and do what they want to do and ignore the science.

So, you know, with that kind of attitude, I don't see where we're going to get very far, and I don't see where the cooperation is going to come in.

This other thing about, you know, best available science. Sometimes the best available science is trash. Everybody admits it. But we're under a mandate from Congress, we're under a mandate from the Court, we're under a mandate for something else and time is our enemy. Time is your enemy, the scientists, time is our enemy, because we've got to satisfy the courthouse. Now we have regulations with litigation. I don't know how we're going to circumvent all these things and bring everything together.

I've worked with a program this past year with the Florida Marine Research Institute with electronic boxes and so on and so forth. I wouldn't do that with everybody, but I have trust in this particular person. And so many times, we have tried to work with government, and so many times, we've tried to work with the council, and so on and so forth, and anything that you give, it seems like they turn it around and use it against you. And it's always the commercial guys. The commercial guys are tired of being beat up. Thank you.

DR. SISSENWINE: Okay. I'd comment that Oceana is not a sponsor of this meeting. They're free to put out their material as anyone else is, and you judge it on its merit.

Anyone here want to comment more generally or respond to anything he said? Okay. Go ahead, you're next.

MR. RUBEC: My name is Peter Rubec. I'm a Research Scientist with the Florida Fish and Wildlife Conservation Commission, Florida Marine Research Institute (FMRI), situated in St. Petersburg, Florida. I have a varied background. I have worked as a stock assessment scientist in Canada, and I have worked in coastal fisheries in Texas. More recently, I've been involved in fish habitat suitability models (HSM) in estuaries at FMRI, and we're getting to the point where we think we can take habitat information and FMRI's fisheries independent monitoring (FIM) data and model it to predict the distributions and abundance of species (by life stages and seasons) like spotted seatrout, pinfish, and bay anchovy in Tampa Bay and in Charlotte Harbor. In addition, I've been working with Mr. Versaggi and a technical company situated in Tampa on development and evaluation of an electronic logbook (ELB). The idea there is that we're working with the fishing industry to gather not only catch and effort data, but associated environmental data. I see these things (HSM and ELB) coming together in terms of new models that can predict fish abundance in relationship to habitat from fisheries dependent data. The only way I see us getting to where we need to go in terms of identifying the areas that are being impacted by fishing is working with the fishing industry and by developing models that actually can assess where fishing occurs and can assess the impacts (of fishing on benthic habitats).

I am in the geographic information systems (GIS) group at FMRI. I agree there is a need to create marine protected areas (MPAs). The only thing wrong with the creation of MPAs is that they result in closing off areas to fishing. An MPA may protect biodiversity; it may protect corals. It's probably protecting juvenile groupers and other fish. So, MPAs may enhance fish production. In more open areas like sand bottoms, I haven't seen as much impact from trawling as I expected to see presented at this conference. For instance, one poster documented no negative impact of shrimp trawling on sand bottoms in the Gulf of Mexico.

But the only way we're going to answer those kinds of questions is to work with the fishing industry and develop new models that are like the HSM derived from FIM data that I'm working on now. The HSM are spatial models. They work within the GIS. They spatially predict animal abundance in relationship to habitat parameters. The next step in the models that I think is needed that doesn't exist now is to spatially predict fishing mortality.

I participated in another conference last week in Sarasota where fisheries ecosystem models were defined as being like ECOSIM, ECOSPACE, and multi-species virtual population analyses. These are multi-species models that involve food webs. The only thing wrong with those models is that they don't have environmental data and information on environmental change (e.g., climate change) or information on benthic habitats in the models. We need all of those things in the models, and I'm hoping that's the direction you'll go. I see this as a whole new type of model that needs to be developed.

As far as maybe the poorer countries that can't afford to sample a hundred species and look at their food web, GIS-based management strategies are still a good idea. They can be done quickly and easily with GIS. I'm advocating that the areas used by fisheries need to be identified. One example of this is through the use of vessel monitoring systems, like the study that we have recently done with the ELB (poster presented at the conference). The ELB study found that the pink shrimp fishery off the west coast of Florida was localized in two areas (east of the Dry Tortugas and between Tampa and the Florida Middlegrounds). A lot of fisheries are much more localized than people tend to assume. We need to start protecting areas necessary to sustain fisheries. We need to protect areas for the fisheries. The term for this is Territorial Use Rights in Fisheries. When the councils allocate and manage areas for specific fisheries; then maybe it will be possible to control the amount of fishing effort in those areas.

We can only get that kind of information from the fishery. We really need to understand where fishing effort is localized. Data supplied by the fishermen can lead to spatial comanagement of the fisheries. I haven't heard what I just stated presented at this conference. So, I'm saying it here now. Thank you.

DR. SISSENWINE: Okay. Thank you. Elliott, you wanted to comment on that?

DR. NORSE: Yes, thanks. I would agree with you that ocean zoning has the potential, at least, to be a way to spatially segregate people so that the endless competition over every square meter of seabed is reduced. There has to be room for people to fish. There has to be room for other concerns, maintaining biodiversity. There's that word again. Biodiversity.

What I feel sad about is that there hasn't been a whole lot of study of the effects of shrimp trawling in the Gulf of Mexico. Shrimp trawling has been happening for a long time, and we have a situation that may in some ways be comparable to the North Sea. The science developed in these places long after the fishery developed, and we didn't have anything like a pristine ecosystem. What was the bottom of the Gulf like? We saw the pictures of what the northwest shelf of Australia was like when there was no trawling and then trawling began, but we don't know much about what it was like in the Gulf of Mexico or North Sea before trawling began.

When I was doing my dissertation research, I would read studies on organisms that trawls caught in the Gulf of Mexico in the past. I don't think those organisms come up much anymore.

So to say that we don't see effects today may be because we have created a shifting baseline and we're looking at effects on a heavily impacted system from the agent of impact. We're decades too late to detect the major impacts.

DR. SISSENWINE: Okay. I think we'll take one last comment. Yes? Yes. And then we're going to wrap up.

DR. BARNES: I'm Peter Barnes, marine geologist with the U.S. Geological Survey. As a geologist, I look at the substrate as a support system for biology. This meeting is focused on the problems related to benthic fisheries and their management. However, the offshore realm is a public resource with many uses and resources. Focusing only on fisheries and their management might be shortsighted or too narrow. Today, there are other multiple managers in the offshore realm, Minerals Management Service (MMS), Department of Defense, NOAA Sanctuaries, the Coast Guard, the Corps of Engineers, and others. Onshore federal natural resources are also managed by many agencies—National Park Service (NPS), Bureau of Land Management (BLM), Fish and Wildlife Service (F&WS), Department of Agriculture. My question is twofold: (1) Are there analogies with onshore management of public resources that can effectively be used to manage offshore (and fishery) resources? (e.g., should we consider the F&WS model, BLM, MMS, or NPS model?) (2) Are there any onshore models that look like fishery management councils?

DR. SISSENWINE: Anyone?

DR. PAUTZKE: Is there an onshore one?

DR. SISSENWINE: Is there something onshore that manages one sector that has impacts that are overlapping with all sorts of others?

DR. BARNES: Public resource.

DR. SISSENWINE: Public. There are grazing boards that have some similarities to fishery management councils. Go ahead.

DR. PAUTZKE: Isn't another good example the Chesapeake? Isn't there some kind of authority down in the Chesapeake Bay area where they're trying to work together to protect the Chesapeake ecosystem and there's a number of federal and state and local organizations that are working through that to come up with some kind of a comprehensive plan.

DR. SISSENWINE: There are many examples where various sectors are brought together to communicate and coordinate decisions that impact an ecosystem. And this could be done to a greater extent within the council framework. There are relatively few examples where there's a single authority that actually can regulate all of the activities for a geographic area for an ecosystem. I think the Great Lakes Commission has considerable power in that regard. But there are relatively few of examples. I don't really know the terrestrial governing literature that well, and I think that's what we're talking about. But I think there's certainly a lot that all of us could learn from those experiences, some of which may be very applicable and some of which may not for a lot of reasons.

We've had an exciting symposium overall, and I think this final session has turned out to be one with more energy in it than I would have expected at this time of the day and at this time of the meeting, so I thank you all for that.

I make a few final observations, some of which are things we've already heard, some of which have only very, very modestly been touched on during the meeting. We've been talking about ecosystem alterations caused by fisheries. And I use the word "alteration" purposely because I think that we all agree that fisheries do alter ecosystems, but whether the alterations are positive or negative depends on the "eyes of the beholder." I think we also all acknowledge that we can't be the most successful species on earth without altering the ecosystem we're in, and that in fact all species in some ways contribute to alteration of ecosystems. Those species that are responsible for the biogenic communities are altering the habitat in a sense. So there's all sorts of alterations that go on. We tend to be using the terms of "adverse" or "damage" or things like that very often, and these are the ones that raise disagreements or emotions; the term "alteration" doesn't. And, of course, I think terms like "damage" or "adversely" generate reactions because we're not necessarily all coming from the same perspective. Some of us are coming from issues of aesthetics and human values and so on. Some of us are addressing issues of biodiversity. Some of us are addressing issues of fish productivity.

And so the degree to which you might label or put values on a particular alteration is always going to be a function of perspective or context.

It has been pointed out by some of the speakers, and I think it's worth keeping in mind, that the specific perspective built into the habitat provisions of the Sustainable Fisheries Act is with respect to fish productivity, but nevertheless, there are lots of other legitimate perspectives that need to be balanced in the equation as we move forward. And that's the challenge, to figure out how to do so because the values are so varied among us.

I think the same issue of perspectives applies to much of the debate about marine-protected areas, which seems to have gotten closely related to the subject of this meeting. When we discuss the usefulness of marine-protected areas, and frankly, I think at some points in the discussion, people are very clear about making the distinction about their effectiveness relative to objectives of biodiversity or other objectives and their effectiveness relative to fisheries. And at other times in the discussion, those distinctions seem to be very confused. So even during the course of a single talk, you're not quite sure whether you agree or disagree because perspectives change. I think perspectives are very important to keep in mind as we try to be successful in moving forward by better science and better communication.

Another topic I wanted to make a quick comment on is the importance of habitat characterization and habitat mapping. This is a really exciting area, and I think there is widespread agreement that we need to do more, we need more data. We have technology so we can get very high resolution data, and certainly we want high resolution data or comparable resolution data in terms of the geology and the biology so that we can match these things up and do things like ground truthing and really understand processes.

However, when I think about habitat mapping, I come back to another set of meetings I sit in, fishery management council meeting, and some of us sit in a number of them. People familiar with fisheries management recognize the mismatch between efforts to map with a resolution of meters or tens of meters, while the spatial scale of fisheries management is tens of kilometers at best. Nobody is going to manage habitat in a regulatory sense on the scales that we're all saying we want scientifically. So I think we have to think carefully about our calls for various high-resolution data and mapping and recognize the distinction between certain needs for industry purposes, so you can catch fish without damaging habitat, and other needs for matching up different types of data for research purposes, and still other very different needs if we're actually going to apply the data to management. I think it's important to keep those things in mind, because when we say we need a high resolution map of everything, we're probably dooming ourselves from the start because of the costs that might be involved.

Another, I thought, very important talk that was given here dealt with the issue of how we shape the future by shifting human values in ethics. And I think that a lot of the things that are going on around us in terms of how the science evolves and how it's used and so forth is clearly a reflection of those shifting values and ethics.

I guess the question I pose to people is how fast are they shifting? Are they shifting fast enough? Can they be shifted in just the right amount or is it inevitable that there's an overcompensation or a pendulum swinging too far effect as values change in the future? I think that's a challenge that society faces in dealing with this situation concerning fisheries and habitat.

A lot of the issues we're dealing with relate to the question of burden of proof. And we've heard many comments about the precautionary approach and how you apply the burden of proof. And, of course, this discussion connotes who does the burden of proof belong to, who has to prove things? What we don't hear discussed very often that I think is even more important is what is the standard of proof? Is the standard of proof the same as in a criminal trail, "beyond a shadow of a doubt," or is it "the preponderance of evidence" such as in a civil suit? I think that that particular aspect of the decision processes we're dealing with needs to be more clearly understood by the participants if we're going to be successful in moving forward with things like the precautionary approach and the debate about burden of proof.

Let me make one final comment about the subject matter we're dealing with, and that is we've been dealing with a very specific aspect of fisheries management, and that relates to how we might manage to address the issue of the alterations, sometimes damaging, that fisheries have on habitat. I don't think we should consider the topic in isolation from the bigger fisheries management picture. I say that for a number of reasons. I think that it's pretty clear that, generally, there's an acceptance with fishery management problems around the world that we need to deal with issues of too much fishing, and that certainly reducing the actual amount of fishing activity can only be beneficial for habitat.

We also agree generally that there's an issue of overcapitalization which is making fisheries economically marginal in many cases. I'd argue, then, that the overall solution to these problems is very interrelated because one of the things that could help us very much with making fisheries more responsive and more capable of dealing with some of these issues of habitat would be to have fisheries that were more profitable and could invest in the solutions and could be more rational about participation in the process to find solutions.

So I just urge people to look at the whole picture of what's necessary to deal with a greater rationalization of our fishery sector so that it can not only be successful in terms of producing fish, but can also be prudent in terms of its investments to take account of the externalities that might be associated with or imposed on other parts of the ecosystem.

My final comment is, yes, we do agree we make progress by reducing fishing. Yes, we agree we ought to be protecting certain obviously vulnerable habitats, three-dimensional, complex biogenic habitats. Yes, we all agree we need more research. Those are the general things. The devil is in the detail. I think it's time to step up and take on the devil on a lot of fronts.

I'll turn it over to Jim or the organizers who are really responsible for this meeting today to say the last words. I just offer my thanks and thanks to the entire panel.

DR. THOMAS: Peter Barnes and I want to say thank you to all of you, particularly to you, Mike, and the panel. We really have appreciated it. We think we have had a very, very successful symposium. Peter Barnes and I have truly enjoyed this meeting. We hope all of you have, too. Thank you all very much. Have a safe trip home.

(Proceedings were concluded at 4:12 p.m.)

* * * * * *

Index

A

abalone *Haliotis* spp., 60, *61*
Abra spp., 486, 658
abundance estimates, 675–677
Acanthogorgia armata, 372–373, 376–379. *see also* gorgonian corals
accounting for ecosystem goods and services, 47–49
acoustic mapping
    acoustic and geological habitat data, 669–671
    acoustic reflectance patterns, 211–212
    benthic habitat characteristics, 209–210
    NOAA hydrographic surveys, 201–202
    seabed classification and fish census data, 318–319
    seabed properties, 87
    subtidal oyster reefs, 154–158
Adams Point, Great Bay Estuary, 153–158
Adriatic Sea
    mud habitat composition, 549–550, 554–557
    sand habitat composition, 557–563
advanced notice of proposed rulemaking (ANPR), 21
adversity-disturbance domain, 142
adversity index, 149–150
Aeginina longicornis, 354
Aequipecten operclaris, 658
age-structured biological model, 746
Agenda 21, 41, *45*
Alaska
    5-year sweep rates, 684–685
    bottom trawling, 679–689
    essential fish habitat (EFH), 115, 698, 842–843
    fishing intensity distributions, 680–684
    living substrate, 289–298
    spatial and temporal distributions of bottom trawling, 684–685
    sweep rates modeling, 684
Alaskan Pollack fishery case study, 96–97
Alcyonidium digitatum, 658
Allee et al. (2000) classification, 185
American lobster. *see* northern lobster
American Oceans Campaign (AOC), 32–33
American plaice, 341, 394–397
    feeding behavior, 407–408
    trawling-induced fish diet changes, 392, 400–404
Ampharete, 658
Amphiura spp., 484–486
Anacapa Island, California, 161–164
Anapagurus brevicarpus, 563
anemones, 289–298
Anoplopoma fimbria. see sablefish
ANPR. *see* advanced notice of proposed rulemaking (ANPR)
anthropogenic effects
    Irish deepwater coral reefs, 823–824
    and natural disturbance in tilefish, 634–635
    vs. fishing effects, 14
antifouling paints on ships, 611
AOC. *see* American Oceans Campaign (AOC)
Arc/Info conversion of sonar images, 163
Arctic lyre crab. *see* toad crab
Arctic surfclam, 384
Arctica islandica. see ocean quahog
area closures
    fisheries management actions, 95
    Georges Bank, 345–366
    impact distribution of fishing activity, 83–84
Arnoglossus laterra. see scaldfish
Ascension Canyon, 314
ascidians, 289–298
Aspidosiphon müelleri, 563
Astarte borealis, 339
Astarte crebricostata, 523
Astarte elliptica, 339, 523
Asterias, 339
Asterias amurensis, 446, 455–456
Asterias rubens, 658, 659
Asterias vulgaris. see sea star
Atka mackerel, 291, 295–296, 682–634
Atlantic cod, 13, 25, 266–275, 371, 392, 394–400, 406–408, 610
Atlantic halibut, 371
Atlantic jackknife, 173–175
Atlantic (northeast), trawling impact on coral habitats, 807–814
Atlantic (northwest)
    bottom trawls, 786
    epifaunal communities, 786
    fish habitat conservation, 785–793
    habitat vulnerability and availability, 787–788
    snow crab resource and shrimp trawling, 590–591
    synthesis of vulnerability, risk, availability, 789–790
Atlantic red deepsea crab, 13
Atlantic rock crab, 173–175, 333, 339
Atlantic salmon, 25
Atlantic surfclam, 691–693
Atlantic (west), *Oculina* coral ecosystem, 795–803

B

backscatter data
    intensity map of Georges Bank, 143–145
    multibeam echosounder survey technology, 179–182
bacteria, southwestern Gulf of Mexico, 225–228
Baie des Chaleurs, Canada, 590
Balanus balanus, 523
Barents Sea Bear Island fishery protection zone
    community analyses, 522–523
    community composition, 526–527
    experimental area, 524–526
    faunal composition, 521–522
    otter trawling effects, 519–527
    species abundance and biomass, 523
barnacles, 289
basic bottom trawl survey, 37
basket stars, 415, 417
Bay of Fundy, eastern Canada, 598–599
Bear Island fishery protection zone, 519–527
bedrock, lake trout spawning habitat, 166
before-after-control-impact study, 597
Belgica mound, 821
benthic assemblages and habitat template, 150–151
benthic fisheries, Mexico, 60, *61*
benthic habitat management and social science, 713–722
benthic habitats
    acoustic diversity, 209–210
    assessment in Gulf of Alaska, 205
    and carry capacity, 769–782
    commercial otter trawling effects, 439–457
    composition changes, 481–482
    deep-sea trawling impact, 503–513
    effects of hydraulic clam dredges, 691–693
    fishing gear and fiber optics cable survey, 629–630
    Georges Bank, Canada, 141–152, 361, 363
    GIS technique, 201
    Grays Reef National Marine Sanctuary, 201
    ground-truthing with video mosaic images, 171–176
    local Fourier histogram (LFH) texture feature classification, 171–172
    and macrobenthic communities, 411–422
    mapping, 139, *140*
    multibeam technique, 201
    reducing/eliminating undesirable effects of fishing, 43–47
    seabed classification with multibeam sonar, 209
    side-scan sonar, 201
    towed gear effects, 74
    trawling impacts, 477–487
benthic impacts and spatial distribution, Florida Keys National Marine Sanctuary, 592–593

benthic incubation microcosms chambers (BMIC), 220–221
benthic invertebrates, 315, 429, 600–601
benthic megafauna recovery, Georges Bank, 325–341
benthic nutrient dynamics, 491–500
benthic perturbations, walrus foraging *vs.* trawling, 633
benthic structure-formers, loss of, 110–111
Bergen Ministerial Declaration, 42, *45*
Bering Sea benthic communities, 439–457
biodiversity conservation, 103–104
biodiversity index, 548
bioeconomic modeling and analysis, 47
biogenic structures abundances, 466, 467
biogeochemistry and bioturbation, 77–80
biological and socioeconomic implications of limited-access fishery management system, 833
biological diversity, 103–104
biological reference points and carrying capacity, 771–775
biological surveys, 842
biological traits analysis
    of benthic organisms, 480–481
    bottom trawling effects, 478–479
    composition changes in benthic community, 481–482
    species composition comparison, 482–487
"biology of scaling," 434
biomass, production, turnover time, 74–77
biomass and species abundance, 351–356
bioroughness, 79
bioturbation, 79
bioturbation and biogeochemistry, 77–80
bivalves, 289
black grouper, 318–319
blocks and block sizes, 680–688
blue cod and oyster dredging, 591
blue mussels populations, 610
blueline tilefish, 301, 308
BMIC. *see* benthic incubation microcosms chambers (BMIC)
bocaccio, 119–120, 668
body size
    effect of heavy trawling, 428–430
    trawling effects on red king crab, 432–435
boreal astarte, 339, 416
bottom discriminating sonar, 203
bottom fishing disturbance. *see also* bottom trawling; trawling and dredging
    beam trawling, 75
    biomass, production, and turnover time, 74–77
    bioturbation and biogeochemistry, 77–80
    food webs, 80
    macrofauna abundance, 73–77
    meiofauna abundance, 75–77

nutrient fluxes, 78
oxygen depletion from sediment resuspension, 78
patterns of trawling disturbance, 81–84
production fluctuations, 75
production to biomass ratios increases, 75
shifts in community structure, 75
shifts to small, free-living species, 75
surface sediment resuspension, 78
trophic structure effects, 80
turnover times, 75
bottom impacts of trawling, 60
bottom referencing underwater towed instrument vehicle (BRUTIV), 414
bottom trawl records, 680
bottom trawling. *see also* bottom fishing disturbance; trawling and dredging
biological traits analysis, 478–479
bottom trawl records, 680
and collapse of Alaska's fishery, 631
effects on burrowing cerinthid anemones, 603–604
effects on soft-sediment epibenthic communities, 461–473
fishing effort distributions, 680
haddock fisheries, 253–255
infauna identification, 37
prey taxa abundance, 471–473
seabed contact reduction, 837
spatial and temporal distributions, 679–689
brachiopods, 289
BRD. *see* bycatch reduction devices (BRD)
Bristol Bay, Alaska, 425–435, 631
brittle stars, 274, 329, 333, 354, 417–418, 523
brown rock shrimp, 27
brown shrimp, 571
Browns Bank scallop fishery case study, 96
BRUTIV. *see* bottom referencing underwater towed instrument vehicle (BRUTIV)
bryozoans, 289–298, 446
bunions, sea onions, corals, protective areas, 857
burrowing amphipods, 446
burrowing cerinthid anemones, 603–604
bycatch, 50–51, 60, 62
bycatch reduction devices (BRD), 277–278

C

cable lines, 204–205
Calcareous rhodophytes, 609
Campeche Bay and Tamaulipas shelf, 220
Canada's Oceans Act, *46*
Canadian Atlantic and Pacific trawling activity, 696–697
Canadian Environment Assessment Act, *46*
Canadian hydraulic clam dredge fishery, 384–386

Canadian legislation precautionary approach language, *6*
canary rockfish, 120, 668
Cancer borealis. see Jonah crab
Cancer irroratus, 339
Cancer magister. see Dungeness crab
Cancer pagurus, 660
Caranx trachurus. see horse mackerel
Caribbean Fishery Management Council (CFMC), 28–29
Caribbean spiny lobster. *see* spiny lobster
carnivorous polychaetes, 446
carrying capacity
and biological reference points, 771–775
fishing activity impact, 769–782
Cartagena Protocol on Biosafety, *46*
case studies
Alaskan Pollack fishery, 96–97
Browns Bank scallop fishery, 96
closed areas, 96
decision-making framework, 96–97
ecologically sustainable fisheries, 839
fishing effort controls, 96
gear modifications, 96–97
Georges Bank closed areas, 96
Heceta Bank, Oregon, 120–135
Monterey Bay, California, 671–675
Catham Rise, New Zealand, 593
Caulolatilus microps. see blueline tilefish
CBD. *see* Convention on Biological Diversity (CBD)
Cerastoderma edule, 610
CFMC. *see* Caribbean Fishery Management Council (CFMC)
CFP. *see* Common Fisheries Policy (CFP); European Union's Common Fisheries Policy (CFP)
Chaceon fenneri. see golden deepseas crab
Chaceon quinquedens. see Atlantic red deepsea crab
chaos and experience-based knowledge (EBK), 719
Chionoecetes opilio, 590–591
Chlamys islandica. see Iceland scallops
chronic bottom trawling and size structure benthic invertebrates, 425–435
chronic disturbance effects, 566–567
chronic effects of fishing disturbances, 462
chronic study design, 440–442
Ciliatocardium ciliatum, 523, 524
Citharichthys spp., 674
clam dredges on soft corals, 383–389
Clinocardium ciliatum. see hairy cockle
closed areas case study, 96
co-management, cooperation, community concepts, 718–719
coastal systems *vs.* deepwater trawl fisheries, 695–696
cobble habitat, 605
cockle populations, 610

Code of Conduct for Responsible Fishing, 46
coldwater coral reefs, 509, 512, 819–829
colonial hydroids community and life history divergence, 598–599
commercial fishery, Lake Michigan, 760
commercial fishing gears, 698–699, 841–842
commercial otter trawling. see also otter trawling
    chronic study design, 440–442
    ecological implications, 454–456
    effect on benthic communities in southeastern Bering Sea, 439–457
    essential fish habitat (EFH) management considerations, 456
    experimental study design, 442–443
    individual macrofauna taxa chronic study, 446
    individual macrofauna taxa experimental study, 448
    macrofauna community impact from, 451–454
    macrofauna community indices chronic study, 444–445
    macrofauna community indices experimental study, 447–448
    megafauna chronic study, 446
    megafauna experimental study, 448
    multivariate macrofauna assemblages chronic study, 445–446
    multivariate macrofauna assemblages experimental study, 448
    trawling catch composition experimental study, 448–450
    yellowfin sole, 440
    yellowfin sole diet composition experimental study, 450–451
    yellowfin sole sampling, 443
Common Fisheries Policy (CFP), 828–829
common property systems, 717–718
common seals, 611
Commons Petition (1376), 11
community, co-management, cooperation concepts, 718–719
community activity changes with time-lapse photographs, 245
community-based management system, 720
community characteristics changes, 510–512
community conservation, 720–721
community structure shifts, 75
comparative risk assessment, 94, 779–781
complexity and inadequate knowledge, 48
composition changes in benthic community, 481–482
concept of sustainability, complexity of, 44, 47
Connor et al. (2003) classification, 185
conservation, nonmarine vs. marine, 104–107
conservation harvesting technology development, 759
conservation paradigm, 785–792
conservation reference points, 53–54

conservation thinking
    biological diversity, 103–104
    dominionistic attitude, 102
    endangered species approach, 102–103
    habitat complexity and species diversity, 108
    place-based conservation, 102
    spatial complexity/diversity of ecosystems, 108
    species conservation, 102
    states of, 101–104
    terrestrial vs. freshwater realms, 107–108
    utilitarian attitude, 102, 103
continental shelf mudbelt habitat, 246
continental slope seascape, 699
Convention on Biological Diversity (CBD), 45, 651
cooperation, co-management, community concepts, 718–719
Copenhagen Declaration, 42
Coral Conservation Area, Northeastern Channel, Nova Scotia, 380
coral reef conservation, 819–829
coral reef habitats
    coral conservation, 819–829
    decline of, 838
    demersal trawling impact, 807–814
    impact from spiny lobster trap gear, 592
    Irish deepwater coral reefs, 823–824
    trap fishing, 599
"coral rubble facies," 812
corals, bunions, sea onions protectives areas, 857
corals, deepwater gorgonian. see gorgonian corals
Corbula gibba, 549
Corpus Christi Fish Pass, 572
Cowardin et al. (1979) classification, 185
cowcod, 120, 668
crab and shrimp trawling impacts, 630–631
crabs, 291, 296
Crangon crangon, 610
Crangon septemspinosa. see sand shrimp
Crassostrea virginica. see eastern oyster
crustaceans, 290
cultural systems, knowledge, identity, power, 714
cyclic behavior and solar activity, 610
Cyclocardia borealis, 339
Cyclocardia novangliae. see New England cyclocardia

D

daisy brittlestar, 339, 341, 398–399, 402
damage coefficient, 749–750
damage function, 749–757
darkblotched rockfish, 120, 668
Darwin Mounds, 807–814
database for benthic fisheries and tourist income, 760–761

DDT. *see* dichloro-diphenyl-trichloro-ethane (DDT)
"dead coral facies," 812
decision-making framework
    case studies, 96–97
    ecological risk assessment, 94–95
    establishment of areas closed to fishing, 95
    federal fishery management plans, 841
    fishing effort reductions, 95
    gear modifications, 95–96
    management actions, 95–96
    recommendations, 97–98
    research needs, 98
    risk evaluation in, 94
    risk management, 95
    trawling and dredging effects, 94–96
deep-sea benthic communities, effects of discards, 512–513
deep-sea trawling, benthic ecosystem impacts, 503–513
deep seabed trawling impacts
    changes in community characteristics, 510–512
    classification, 504–505
    direct effects, 505–508
    indirect effects, 510–513
    physical impacts on epifaunal/infaunal communities, 508–510
deepwater coral, 13
deepwater fishing impacts
    Irish deepwater fisheries, 820–821
    seabed impacts, 503–504
deepwater gorgonian corals. *see* gorgonian corals
deepwater habitats
    laser technology for investigating, 204
    line-scan imaging technology (LLS), 204
deepwater trawling impact
    coastal systems, 695
demersal fishes
    and benthic invertebrates, 315
    and otter trawling, 391–408
    and upper slope rockfish, 314
demersal trawling impact
    coral habitats, 807–814
    survey techniques, 808
denitrification rates, 79
density-dependent habitat selection, 748–749
density estimates, juvenile red snapper habitat, 279, 280–287
descriptive habitat names, 197–198
descriptive metrics, ecosystem-based management, 654–658, 662, 663
Desmopyllum cristagalli, 823
destructive fishing practices and marine ecosystem-based management paradigm, 101–112
DG Environment, 42

DG FIsheries, 42
DGPS. *see* differential global positioning system (DGPS)
Dichelopandalus leptocerus, 339
dichloro-diphenyl-trichloro-ethane (DDT), 611
diet comparisons, otter trawling and demersal fish feeding, 394–396
differential global positioning system (DGPS), 173
discards, 316, 512–513
DISCOL study, 512
disturbance, impact of, 215–218
disturbance ecology and seafloor communities, 640–641
disturbance index, 147, 149
diversity/spatial complexity of ecosystems, 108
dominionistic values, 102, 707–708
Dosidiscus gigas. see giant squid
Dreissena bugensis. see quagga
Dreissena polymorpha. see zebra mussel
drill cuttings dumping, 614
DRUMS. *see* dynamically responding underwater matrix sonar (DRUMS)
Dublin Bay prawn, 479–480
Ducks Unlimited, 102
Dungeness crab, 291
dynamic segmentation GIS, 131, 133–134
dynamically responding underwater matrix sonar (DRUMS), 414

E

EAARL. *see* Experimental Advanced Airborne Research Lidar (EAARL)
eastern oyster, 153–158
EBK. *see* experience-based knowledge (EBK)
Echinarachnius parma. see sand dollars
ecocertification, 51
ecological cost accounting systems, 44
ecological economic and social dimensions, managing, 47
ecological implications and commercial otter trawling, 454–456
ecological quality (EcoQ), 651
ecological risk assessment, 94–95
ecologically sustainable fisheries case study, 839
economic and social impacts, 714
economic effects of marine protected areas (MPA), 49
economics and spatial analysis, fishers' behavior, 713–714
EcoQ. *see* ecological quality (EcoQ)
ecosim model, 63–64
ecosystem alterations caused by fisheries, 865
ecosystem approach to fisheries, adopting, 42–44
"Ecosystem Approach to Fisheries Management," 74
ecosystem-based management
    analysis of similarity, 655

descriptive metrics, 654–658, 663
ecosystem consequences of bottom fishing, 84–87
epibenthos abundance data, 656–661
epifauna abundance data, 654
fisheries management in Mexico, 63–64
functional groups, 655–656
functioning of ecosystem, 655
indicator taxa, 655
infauna abundance data, 652–654, 656, 658–659
K-dominance, 655
marine benthos, 651–664
modeling, 85–86
multi-dimensional scaling ordination, 655
paradigm shift, 111–112
performance metrics, 655–656, 658–661, 662
trophic groups, 655
univariate metrics, 654–655
"ecosystem goods and services," 44
ecosystem goods and services, accounting for, 47–49
ecosystem objectives, 53
ecosystem resilience, 62
Ecotrust's Groundfish Restructuring (GFR) project, 735–737
ectoprocts, 295
edible oysters, 611
EE-Fisheye, 278
eel grass, 289
effort distributions, trawling disturbance, 83–84
effort model, 732–735
Eider ducks breeding success, 610
El Niño Southern Oscillation (ENSO) cycle, 610
electronic logbook, 320
ELMR. *see* Estuarine Living Marine Resources (ELMR) program reports
endangered species approach. conservation thinking, 102–103
England, northeastern coast, ecosystem-based management of marine benthos, 654–656
England, northeastern coast of
epifauna abundance data, 654
indicator taxa, 655
infauna abundance data, 652–654
performance metrics, 655–656
univariate metrics, 654–655
English Channel otter trawling, 316
Ensis directus. see Atlantic jackknife
ENSO cycle. *see* El Niño Southern Oscillation (ENSO) cycle
environmental and fishing disturbance indicators, 632–633
environmental awareness and conservation harvesting, 759
environmental management, legal frameworks, 714–715
environmental performance indicators, European marine fisheries, 840–841
EORR. *see* Experimental *Oculina* Research Reserve (EORR)
epibenthic communities and bottom trawling, 461–473
epibenthic sled, 414, 416–417, 521
epibenthos abundance data, 656–661
epibenthos assemblages, 834
epifauna abundance
assessment of, 464
deep seabed trawling impacts, 508–510
effects of bottom trawling, 464
northeastern coast of England, 654
northwest Atlantic continental shelf, 786
and species diversity, 464–469
Epinephelus drummondhayi. see speckled hind
Epinephelus flavolimbatus. see yellowedge grouper
Epinephelus morio. see red grouper
Epinephelus nigritus. see warsaw grouper
Erichthonius rubricornis, 354
Erichthonuis fasciatus, 397–404
Erimacrus isenbeckii. see hair crab
ESA. *see* U.S. Endangered Species Act (ESA)
essential fish habitat designation
brown rock shrimp, 27
Caribbean Fishery Management Council (CFMC), 28–29
Gulf of Mexico continental shelf, 315
Gulf of Mexico Fishery Management Council (GMFMC), 28
Mid-Atlantic Fishery Management Council (MAFMC), 26
New England Fishery Management Council (NEFMC), 24–25
North Pacific Fishery Management Council (NPMC), 31
Pacific Fishery Management Council (PFMC), 30–31
red drum, 27
Western Pacific Fishery Management Council, 29
essential fish habitats
criteria for designating, 92–93
defined, 12, 22–23, 770–771
fish abundance mean ranks, 268–271
fishing effects and legal requirement, 765–766
fishing effects in Alaska, 842–843
fishing ground locations, 268–271
geographical information systems (GIS), 698
groundfish surveys and fishers' knowledge, 265–275
identification of potential with fishers' knowledge, 267–268
identification of potential with national groundfish surveys, 266–267

identifying possible groundfish, 265–275
provisions, 22
secretarial guidelines, 21–24
esthetic values, 707
Estuarine Living Marine Resources (ELMR) program reports, 24–25
estuarine soft-bottom organisms, trawling impacts, 630–631
ethic toward the sea, 706–709
ethics, social and environmental, 714
ethics and the sea, 703–710
Ethusa mascarone, 549
Etrumeus, 59
Eualus pusiolus, 397–399, 402, 404
EUNIS (European Environment Agency 2004) classification, 185
European Community Satellite Vessel Monitoring System (VMS), 81–82
European marine fisheries, environmental performance indicators, 840–841
European (northern) continental margin, deep-sea trawling impact, 503–513
European Union's Common Fisheries Policy (CFP), 504
European VMS. *see* European Community Satellite Vessel Monitoring System (VMS)
European whiting fisheries, 266–275
events, defined, 216
evidence-based knowledge and social science, 719–720
evolutionary perspective, habitat mapping, 142
excavated seep pockmarks, 309, 310
Exclusive Economic Zone of Mexico, 59
Exclusive Economic Zone of the United States, 13
Executive Order 13158, 19–20
exergy index, 548–549
Exogone hebes, 354
experience-based knowledge (EBK) and chaos, 719
Experimental Advanced Airborne Research Lidar (EAARL), 202
experimental design
Barents Sea Bear Island fishery protection zone, 524–526
commercial otter trawling, 442–443
epibenthic sled sampling, 521
Grand Banks otter trawling experiment, 419–421
macrobenthic communities and benthic habitat, 412
trawling effects on macrobenthic and benthic habitat, 412
Experimental *Oculina* Research Reserve (EORR), 796
experimental trawling intensity, Grand Banks otter trawling, 418–419
exploitation, sustainability concept, 64–65
exposure assessment model, 94
externalities, unsustainability driver, *48*
extinction events vulnerability, 623–624

F

Farfantepenaeus aztecus. see brown shrimp
Farfantepenaeus duorarum. see pink shrimp
fast-track methods, trawl impacts, 203
fathometers, 11
FEAM. *see* Fisheries Economic Assessment model (FEAM)
fecundity, sustainability concept, 64–65
federal fishery management plans, 841
feeding and growth to maturity, 23
feeding behavior, 407–408
feeding changes, otter trawling and demersal fish feeding, 406–408
FIB index. *see* fishing-in-balance (FIB) index
FILLS. *see* fluorescence imaging laser line scan (FILLS) imagery
Filograna implexa, 334, 337
Final Rule, 12, 13
fine-scale groundfish distributions and large-scale seafloor maps, 667–677
Finn cycling index, 62
fish abundance mean ranks, 268–271
fish burrow pockmarks, 308, 310
fish census data and acoustic seabed classification, 318–319
fish diet changes, trawling-induced, 400–404
fish finders, 11
fish habitat conservation, protecting with Magnuson-Stevens Act, 19–39
fish habitat conservation paradigm, 785–793
fish habitat studies, high-resolution geological and biological data, 119–137
fish habitat suitability modeling, 864
fish-seagrass associations, 313
fisheries, rebuilding, 858
Fisheries Economic Assessment model (FEAM), 730, 732
fisheries management
in Mexico, 59–68
models in marine protected areas, 745–757
North Sea, 835–836
and scientific information, 851–852
and sustaining ecosystems, 854
fishers' behavior
cooperation, 859
knowledge and identification of potential essential fish habitats (EFH), 267–268
knowledge and surveys, 265–275
marine reserves, 761
spatial analysis and economics, 713–714
trawling, 858
Fishery Conservation Amendments of 1990, 20
Fishery Management Plan (FMP), 12, 13, 20, 119–120
fishery resource, defined, 20

fishing activities
    driving force of marine change, 611
    impact on benthic habitat and carrying capacity, 769–782
    indirect effects, 854–855
    living substrate vulnerability, 290
        potential threat to essential fish habitat, 12
fishing area sampling with GIS technique, 697
fishing capacity, impediments to progress, 43
"fishing down the food web," 60
fishing effects
    assessment of, 23
    essential fish habitat (EFH) in Alaska, 842–843
    gorgonian corals, 369–380
    Gulf of Alaska habitat, 247
    on habitat, 245–246
    initial effects in unfished areas, 74–76
    legal requirements and essential fish habitats (EFH), 765–766
    organic carbon content of sand habitats, 596–597
    seabed community diversity, 696
    seamount benthic habitat and fauna, 593
    studies needed on, 14
    tilefish habitat, 319
    vs. anthropogenic effects, 14
fishing effort, 752
    Alaska bottom trawling, 680
    Alaska haul records, 681
    case studies, 96
    comparisons inside and outside closed area, 359, 361
    gorgonian corals, 373, 376
    management actions, 95
    marine protected areas (MPA) model, 749–751
    model, 732–735
    risk management, 777
    sub-model OCEAN (ocean communities economics/ ecological/Equity analysis), 731
fishing gear
    effects on benthic habitats workshop, 833–834
    effects on habitat, 859
    and fiber optics cable, 629–630
    impact studies, 604
    zonal application of, 778
fishing ground locations
    groundfish essential fish habitats (EFH), 273–274
    identification of potential essential fish habitats (EFH), 268–271
fishing impact, 21
    balance between cumulative levels of fishing and natural disturbance, 81
    benthic ecosystem, 477–487
    and extirpations in North Sea benthos, 623–624

Irish deepwater coral reefs, 819–829, 828
Mexican ecosystems, 60–63
spatial and temporal distribution, 81
fishing-in-balance (FIB) index, 62–63
fishing intensity distributions, 680–684
fishing on benthic habitats, impediments to progress, 43–47
fishing pressure on Dublin Bay prawn, 479–480
fishing restrictions
    Atlantic surfclams, 26
    Caribbean Fishery Management Council (CFMC), 29
    Gulf of Mexico Fishery Management Council (GMFMC), 28
    Loligo pealei, 26
    Mid-Atlantic Fishery Management Council (MAFMC), 26
    New England Fishery Management Council (NEFMC), 25–26
    North Pacific Fishery Management Council (NPMC), 31–32
    Pacific Fishery Management Council (PFMC), 31
    south Atlantic Fishery Management Council (SAFMC), 27–28
fishing rights, 717–718
flatfishes, 291, 295, 296, 602–603
flathead sole, 683
fleet substitutions and gear modifications, 49–51
Florida, habitat classification, 207–208
Florida Keys National Marine Sanctuary
    acoustic seabed classification and fish census data, 318–319
    hook-and-line fisheries, 592–593
    monitoring changes, 840
    spiny lobster trap gear, 592
    spiny lobster traps and seagrass, 579–587
Flower Garden Banks pockmarks, 302–304, 309
fluorescence imaging laser line scan (FILLS) imagery, 211
flux rates of trawling impact, 498–500
FMP. *see* fishery management plan (FMP)
food webs and bottom fishing disturbance, 80
fourspot flounder, 354
Foveaux Strait, southern New Zealand, 595
Framework Convention on Climate Change, *45*
Framework for the Description, Identification, Conservation, and Enhancement of EFH, 21
free distribution theory, 838–839
frequency of occurrence, 291
freshwater *vs.* terrestrial realms, 107–108
Fundy Bay, eastern Canada colonial hydroids community, 598–599
fused ivory tree coral. *see* ivory tree coral

G

gadids, 291, 296
Gadus macrocephalus. see Pacific cod
Gadus morhua. see Atlantic cod
gag, 13, 799, 800
gas-escape pockmarks, 308–310
gear effects, application and scaling of, 14
gear modifications
  case studies, 96–97
  and fleet substitutions, 49–51
  management actions, 95–96
  reduced gear efficiency, 50–51
  risk management, 777–778
geographic information systems (GIS)
  assessing designs, 697
  assessing lobster trap designs, 697
  benthic habitat characterization, 201
  quantitative seafloor habitat, 206
geologic controls, usSEABED, 210–211
Georges Bank, 850
  backscatter intensity map, 143–145
  benthic assemblages and habitat template, 150–151
  benthic habitat characterization, 141–152
  benthic megafauna recovery, 325–341
  case studies, 96
  Closed Area I example, 601–602
  Closed Area II example, 589–590
  closed areas, 347–351, 361, 363
  comparisons inside and outside closed area
  fishing effects, 359, 361
  habitat, 356, 359
  macrofauna, 356
  nektonic macrofauna, 361, 363
geologic, oceanographic, benthic attributes, 346–350
  habitat disturbance mapping, 147–150
  multibeam bathymetric data, 142–143
  oceanographic information, 145, 147
  organic carbon content of sand habitats, 596–597
  sampling methods of closed areas, 351
  sand habitats, 600
  sea scallop fishing grounds, 597
  shallow sites and benthic megafauna recovery, 329–336
  species abundance and biomass, 351–356
  surficial sediment deposits, 142
  underwater photographs, 145
Gersemia rubiformis, 384, 386
giant scallops, 96, 598
giant squid, 59
gill netting, 50–51, 63
glacial deposits and lake trout spawning habitat, 166–167
glacial retreat, effects on seafloor, 236–237
Glacier Bay, Alaska halibut, 235–241
glass sponge, 510
global ban, trawling in corals, 856
GMFMC. *see* Gulf of Mexico Fishery Management Council (GMFMC)
golden deepseas crab, 27
gorgonian corals
  damaged corals, 373
  effects of fisheries damage, 379
  effects of fishing on, 369–380
  fishing effort distribution, 373, 376
  general distribution of, 372–373
  groundfish landings, 372
  habitat description, 373
  historic changes, 378–379
  mobile fishing gear impact, 596
  Northeastern Channel, Nova Scotia, 369–380
  recovery of damaged habitats, 379–380
  seabed inspection, 371
  significance of damage, 376–379
  survey sites, 371
  video analysis, 371–372
Gorgonocephalus lamarkii, 379
grab samples, sediment mapping with, 37
grain-size analysis, 188
Grand Banks otter trawling experiment
  effects on habitat, 416
  epibenthic sled effects on organisms, 416–417
  experimental design, 419–421
  experimental trawling intensity, 418–419
  impacted organisms, 419
  otter trawl catch, 415–416
  videograb effects on organisms, 417–418
gravel habitats, 325–341
Grays Reef National Marine Sanctuary, 201
Great Barrier Reef World Heritage Area, 839
Great Bay Estuary, 153–158
Great Lakes benthic habitat, 207
Greene et al. (1999) classification, 185
greenlings, 291
Greenpeace, 20
ground-truthing, 171–176, 671, 673
groundfish
  association with living substrate, 289–290
  biomass estimates, 120
  fine-scale distributions and large-scale seafloor maps, 667–677
  fishing ground locations, 273–274
  habitat, 668–671
  habitat management objectives, 115
  habitat preferences studies, 14–15
  identifying possible essential fish habitats (EFH), 265–275

interpreting acoustic and geological habitat data, 669–671
population abundance, 669–671
RACEBASE survey database, 290–291
stock assessment, 668–671
surveys and fishers' knowledge, 265–275
groundfish-habitat study, Heceta Bank, Oregon, 120–135
Gulf of Alaska
   assessing benthic habitat, 205
   chronic effects of fishing disturbances, 462
   fishing effects on habitat, 247
   gorgonian corals and sponges, 596
   otter trawling effects, 461–452
   soft-sediment epibenthic communities, 461–473
Gulf of California porpoise. *see* vaquita
Gulf of Maine, 208, 602
Gulf of Mexico
   bacteria, 225–228
   benthic biomass, 225
   environmental factors, 223–228
   essential fish habitat (EFH) identification, 315
   estimation of shrimp trawling effects, 315
   juvenile red snapper habitat, 313–314
   macrofauna, 228
   meiofauna, 228
   pockmarks, 301–310
   predator and prey dynamics, 320–321
   red snapper habitat, 277–278, 277–287
   seafloor pockmarks, 301–310
   sediment oxygen consumption and benthic biomass, 228
   sediment oxygen consumption and factor correlations, 225
   sediment oxygen consumption (SOC), 219–232, 222–223
   trawling activity impact, 230
Gulf of Mexico Fishery Management Council (GMFMC), 28

H

habitat
   classifying and mapping, 134
   defined, 141–142, 183–184
   fishing effects on, 245–246
   habitat regulations, 22–23
habitat areas of particular concern (HAPC), 12, 23
   golden deepseas crab, 27
   Gulf of Mexico Fishery Management Council (GMFMC), 28
   Mid-Atlantic Fishery Management Council (MAFMC), 26
   New England Fishery Management Council (NEFMC), 25
   North Pacific Fishery Management Council (NPMC), 31
   Pacific Fishery Management Council (PFMC), 31
   South Atlantic Fishery Management Council (SAFMC), 27
   spotted spiny lobster *Panulirus guttatus,* 27
   Western Pacific Fishery Management Council, 29
habitat characterization, 184, 866
   influence of events, 216
   and mapping., 13
   Oculina Banks Habitat Area of Particular Concern (OHAPC), 800
habitat classification, 184, 186–189
   estuarine and marine environments, 207–208
   fauna flora, 188–189
   Florida, 207–208
   grain size analysis, 188
   habitat association and usage, 189
   habitat recovery from disturbance, 189
   regional approach, 185–186
   seabed dynamics and currents, 186, 187
   seabed roughness, 188
   seabed texture, hardness, and layering in upper 5–10 cm, 186, 187
   topographical setting, 186
habitat classification schemes
   Allee et al. (2000) classification, 185
   Connor et al. (2003) classification, 185
   Cowardin et al. (1979) classification, 185
   EUNIS (European Environment Agency 2004) classification, 185
   Greene et al (1999) classification, 185
   habitat names, 189–190
   probable marine sublittoral habitat types, 190–191
   scale, 189
habitat complexity and species diversity, 108
habitat damage, 749
habitat descriptions, 194–197
habitat disturbance mapping, 147–149
habitat disturbance model, 642–646
habitat effect model, 685
habitat management framework, 639–646
habitat mapping, 184, 866. *see also* mapping
   evolutionary perspective, 142
   Georges Bank, Canada, 149–150
   managing ecosystem consequences of bottom fishing, 86–87
   trawling and dredging effects, 94
habitat-naming conventions, 190
habitat protection
   alternatives by North Pacific Fishery Management Council (NPMC), 33–36
   Magnuson-Stevens Act, 20–24
   provisions, 20

habitat recovery from disturbance, 189
habitat regulations, 21–24
habitat research costs, 36–37
habitat risk, 787–789
habitat template and benthic assemblages, 150–152
habitat template approach, 142
habitat vulnerability and availability paradigm, 787–788
haddock, 96, 266–275, 352, 371
    bathymetry, 257–261
    bottom trawl surveys, 253–255
    distributions, 255–257, 261–263
    feeding behavior, 407–408
    otter trawling and demersal fish feeding, 392, 394–400
    Scotian shelf, 251–263
    study methods, 253–255
    trawling-induced fish diet changes, 400–404
hair crab, 291
hairy cockle, 416
halfbanded rockfish, 674
halibut habitat, Glacier Bay, Alaska, 235–241
Haliotis sorenseni. see white abalone
Haliotis spp., 60, *61*
Halipteris willemoesi abundance, 466, 468–469, 471–473
Halodule wrightii, 580–581
HAPC. *see* habitat areas of particular concern (HAPC)
haul records, Alaska, 681
healthy ecosystems, defined, 12
Heceta Bank, Oregon
    case studies, 120–135
    classifying and mapping habitats, 134
    habitat-groundfish study, 120–135
    marine benthic invertebrates survey, 126–127
    remotely operated vehicle remotely operated platform for ocean science (ROV *ROPOS*), 123, 126–127
    sampling effort, 127
    seafloor mapping, 123, *124, 125*
    spatial analysis, 131, 133–134
    substratum (seafloor) classification, 127–131
    tools for direct observation and sampling, 123, 126
Helsinki Convention, *45*
Heraklion Bay, Cretan Sea, Eastern Mediterranean, 529–535
hermit crabs, 415, 446, 523
Heteromastus, 656
Hiatella sp., 523
high-resolution geological and biological data, fish habitat studies, 119–137
high-resolution habitat characterization, FILLS imagery, 211
high-resolution multibeam bathymetry and backscatter imagery, 134
Hippoglossoides elassodon. see flathead sole

Hippoglossoides platessoides. see American plaice
Hippoglossus hippoglossus. see Atlantic halibut
Hippoglossus stenolepis. see Pacific halibut
historic *vs.* contemporary communities, North Sea benthos, 622
Homarus americanus. see northern lobster
hook-and-line fisheries, 50–51, 592–593
Hoplostethus atlanticus. see orange roughy
horse mackerel, 610
horse mussel, 397–399, 402, 404
HR 39, 21
HR 1533, 20
Hudson, Canyon, New Jersey, tilefish habitat, 319
human-made influences, theoretical and methodological basis of estimations, 595–596
humanistic values, ethic toward the sea, 707–708
Hurricane Michelle, 582, 584
Hyalonema sp., 510
Hyas coarctatus. see toad crab
hydraulic dredges
    effects on benthic habitat, 691–693
    impact on biological and physical structures, 776
    northwest Atlantic continental shelf, 786
    soft coral *Gersemia rubiformis,* 383–389
hydrographic surveys, acoustic backscatter imagery, 201–202
hydroids, 289
hydrozoa, 446
hyperbenthic communities, 529–536
hypothetical habitat map, 172–173

I

ice gouging, 247
iceberg gouges
    effects on Pacific halibut community, 239–241
    and halibut habitat, 235–241
    mode of formation, 238–239
Iceland scallops, 415
ICES. *see* International Council for the Exploration of the Sea (ICES)
impact distribution of fishing activity, closed areas, 83–84
impediments to progress, 43–47
imposex, 611
inappropriate incentives, *48*
individual transferable quota (ITQ) management program, 692
infauna abundance data, 652–659
infaunal communities, 505–508, 510, 786
initial fishing effects in unfished areas, 74–76
inorganic nutrient releases from sediment, trawl ground rope disturbances, 542–543
International Council for the Exploration of the Sea (ICES) fisheries advice, 52–53

rectangles, 81
WGECO screening approach, 53–54
international instruments, ecosystem approach to fisheries management, 41–42
international instruments, precautionary approach language/wording, *45–46*
invertebrates, effects of chronic bottom trawling, 425–435
Irish deepwater coral reefs
    anthropogenic impacts, 823–824
    conservation initiatives, 828–829
    coral habitat, 823–824
    deepwater fishing impacts, 821
    Desmopyllum cristagalli, 823
    fishing impacts, 819–829, 828
    Lophelia pertusa, 823
    Madrepora oculata, 823
    management of, 824–828
    non-coral habitat areas, 824
    threats to, 828
Irish deepwater fisheries, 820–821
Irish exclusive fisheries zone, 828–829
Irish Sea, ban on cod fishing, 857
ITQ systems, 718
ivory tree coral, 795–803

J

Jeff's Reef, 796, 798, 800
Jonah crab, 173–175, 371
juvenile red snapper habitat, 277–287, 313–314
    seabed characterization, 279
    trawl sampling, 279–280

K

K-dominance, 655
kelp, 289
knowledge, identity, power cultural systems, 714
Kodiak Island, effects of bottom trawling, 462–463
Koge Stateholders Conference, 42
Kyoto Accords, 48
Kyoto Declaration, 42

L

Labidoplax digitata, 563
Laevicardium oblongum, 563
lake bed sediments, lake trout spawning habitat, 167–169
lake bed sediments and lake trout spawning habitat, 167–169
Lake Erie macrozoobenthos sampling, 206–207
Lake Michigan, 165–169, 760

lake trout spawning habitat
    bedrock, 166
    glacial deposits, 166–167
    ideal habitat, 168–169
    modern lake bed sediments, 167–169
    scanning hydrographic operational airborne lidar survey (SHOALS) program, 166
    substrate morphology and geology characterization, 165–169
Lanice spp., 486
large-scale ecosystem consequences, towed gear impacts, 74
large-scale seafloor maps and fine-scale groundfish distributions, 667–677
laser technology, 165–169, 204
lead-lines, 13
legal frameworks for environmental management, 714–715
legislation, *45–46*
Lepidopsetta bilineata. see rock sole
Lepidopsetta polyxystra. see northern rock sole
lesser weever, 610
LFH texture feature classification. *see* local Fourier histogram (LFH) texture feature classification
Limanda aspera. see yellowfin sole
Limanda ferruginea. see yellowfin flounder
limited-access fishery management system, 833
line-scan imaging technology (LLS), 204
lingcod, 120
little skate, 396
"live coral facies," 812
living substrate
    in Alaska, 289–298
    anemones, 292–295
    ascidians, 295
    association with crustaceans, 290
    association with groundfish, 289–290
    Atka mackerel, 295–296
    bryozoans, 295
    crabs, 291, 296
    ectoprocts, 295
    flatfish, 291
    frequency of occurrence in databases, 291
    gadids, 296
    greenlings, 291
    NMFS trawl surveys, 291
    observations, 292
    RACEBASE survey database, 290–291
    rockfish, 295–296
    sablefish, 291
    sea pens, 295
    sea whips, 295
    shortspine thornyhead *Sebastolobus alascanus,* 291
    skates, 291

sponges, 292, 295–296
  vulnerability to fishing activities, 290
LLS. *see* line-scan imaging technology (LLS)
lobster enclosure, mosaic image analysis, 174–175
lobster traps, assessing designs, 697
lobsters, 60
local Fourier histogram (LFH) texture feature classification, 171–172
Logachev mound, 821
Loligo pealei, 26
London Declaration, *45*
long-term, large scale biological surveys, 842
Lophelia pertusa, 509, 807–808, 814, 823, 827
Lopholatilus chamaeleonticeps. see tilefish
Louisiana oyster industry, 211–212
Lucinella divaricata, 563
Lumbriconereis impatiens, 549
Lutjanus campechanus. see red snapper

M

Maastricht Treaty on European Union, *45*
macro-epifauna and fish sensitivity, 318
macrobenthic communities and benthic habitat, effect of otter trawling, 411–422
macrofauna
  assemblages chronic study, 445–446
  bottom fishing disturbance, 73–77
  community indices study, 444–445, 447–448
  comparisons inside and outside closed area, 356
  impact from commercial otter trawling, 451–454
  southwestern Gulf of Mexico, 228
  taxa study, 446, 448
macroinfaunal recolonization dynamics and trawl fishing disturbance, 545–567
macrozoobenthos sampling, Lake Erie, 206–207
Madison-Swanson Fishery Reserve pockmarks, 304, 308
Madrepora oculata, 807, 814, 823
MAFMC. *see* Mid-Atlantic Fishery Management Council (MAFMC)
Magnuson-Stevens Fishery Act, 11, *22*
  evolution of habitat protection provisions, 20
  protecting fish habitat, 19–39
  Secretarial Guidelines on Essential Fish Habitat, 21–24
  social science in marine resource management, 716
  Sustainable Fisheries Act of 1996, 20–21
management actions
  decision-making framework, 95–96
  on effort distributions for trawling disturbance, 83–84
  evaluating tools on the axes of sustainability, 51–54
  listing objectives, 52–54
  studies needed, 14

mapping, 13. *see also* habitat mapping
  coldwater corals, 509, 512
  ecosystem consequences of bottom fishing, 86–87
  Lophelia pertusa, 509
  subtidal oyster reefs, 154
Margalef index, 548
Margarites sordidus, 416
marine and estuarine environments, habitat classification, 207–208
marine benthic invertebrates survey, Heceta Bank, Oregon, 126–127
marine ecosystem-based management paradigm, 101–112
marine fisheries and conservation, and social science, 715
marine protected areas (MPA), 19–20, 866
  analysis of stylized fisheries, 745–757
  bioeconomic models with fishing effort, 746–747
  economic effects of, 49
  effectiveness on benthic fauna, 601–602
  evaluating tools on the axes of sustainability, 49
  fish census data and acoustic seabed classification, 318–319
  fisheries management models, 745–757
  literature review, 746–747
  modified Leslie population matrix, 746
  National Ocean and Atmospheric Administration (NOAA), 20
  social effects of, 49, 715
  unsustainability of, 49
marine protected areas (MPA) model
  age-structured biological model, 746
  damage coefficient, 749–750
  density-dependent habitat selection, 748
  fishing effort, 749–751
  targeted population and economic system, 747–751
marine reserves
  fishermen behavior, 761
  risk management, 778–779
marine sublittoral habitats, 198
  classification of, 183–200
  classification structure, 192–194
  habitat classification schemes, 190–191
marine *vs.* nonmarine conservation, 104–107
maximum economic yield (MEY), 751
mean ranks
  distribution of fish abundance, 273–274
  of fish abundance, 268–271
mean trophic level (MTL) of catch, 62–63
megafauna chronic study, 446
megafauna experimental study, 448
megahabitats, 185
meiofauna, 75–77, 228
Merlangius merlangus. see European whiting
mesohabitats, 185
Mexican central Pacific shrimp fishing, 632–633

Mexico
- abalone *Haliotis* spp., 60
- clams, *60*
- current status of benthic fisheries, 60
- effects of gill netting, 63
- *Etrumeus,* 59
- fisheries management, 59–68
- giant squid, 59
- impact of fishing on ecosystems, 60–63
- lobster *Panulirus* spp., 60
- *Oligoplites,* 59
- *Opistonema,* 59
- oysters, *60*
- Pacific sardine, *59*
- red grouper, *59*
- *Scomber,* 59
- shrimp fishery, 60, 62–63
- status of benthic fisheries, *61*
- sustainability concept, 64–66
- yellowfin tuna, 59

MEY fisheries, 751
microcosm nutrient concentrations and fluxes, 493–498
Mid-Atlantic Fishery Management Council (MAFMC), 26
military activities on macrofauna, 615
mineralization shift, aerobic to anaerobic, 78
Mississippi and Alabama Pinnacles Area pockmarks, 304, 309
MLML and TGPI fisheries habitat mapping Web site, 180
MMPA. *see* U.S. Marine Mammal Protection Act (MMPA)
mobile bottom fishing methods, spatial extent and nature of, 695
mobile fishing gear impact, sponges and gorgonian corals, 596
modeling, ecosystem-based fisheries management, 85–86
modified Leslie population matrix, 746
Modiolus modiolus. see horse mussel
Monterey Bay, California case study, 671–675
Monterey Bay National Marine Sanctuary, 204, 246
moralistic values, 708
mosaic images, 171–176
MPA. *see* marine protected areas (MPA)
MTL of catch. *see* mean trophic level (MTL) of catch
mud habitats, Adriatic Sea, 545–567
"mud line," 506
mud organisms, study of, 855
"mud patch," 786
mullets, *61*
multi-dimensional scaling ordination, 655
multibeam technique, 13
- backscatter sampling, 180–182
- bathymetric data, Georges Bank, 142
- bathymetry and backscatter data, 302
- bathymetry and backscatter imagery, 134
- benthic habitat characterization, 201
- benthic habitat mapping, 209
- echosounder survey technology, 179–182
- seabed habitat mapping, 179–182
- sonar, 87, 134
- swath mapping sonars and geographic positioning systems, 123

multivariate macrofauna assemblages study, 445–446, 448
Mycteroperca bonaci. see black grouper
Mycteroperca microlepis. see gag
Mycteroperca phenax. see scamp
Mytilus edulis. see blue mussels

N

Nannie Island, Great Bay Estuary, 153–158
National Forest Management Act, 111
National Institute of Fisheries, Mexico, 63–64
National Marine Fisheries Service (NMFS), 19, 21–24
National Oceanic and Atmospheric Administration (NOAA), 13, 14–15, 20
National Research Council (NRC) study, 19, 92–98
national research plan, 849
National Wildlife Refuge system, 102
natural and anthropogenic disturbance assessment, side-scan sonar, 629
natural events and seafloor response, 215–218
natural mortality sustainability concept, 65–66
natural processes, impact of, 215–218
natural size variability, benthic invertebrates, 429–430
natural variability in marine ecosystems, 609–612
naturalistic values, ethic toward the sea, 708–709
nature reserve and shellfish fishing, 835
necessary essential fish habitat, defined, 22–23
NEFMC. *see* New England Fishery Management Council (NEFMC)
negativistic values, 709
nektonic macrofauna, Georges Bank closed areas, 361, 363
Nephrops norvegicus, 479–480
Nereis zonata, 333, 339
New England cyclocardia, 416
New England Fishery Management Council (NEFMC), 24–26
New York Bight, 315, 634–635
New Zealand, 591, 695–696
NMFS. *see* National Marine Fisheries Service (NMFS)
NOAA. *see* National Oceanic and Atmospheric Administration (NOAA)
Noctiluca scintillans, 610
non-coral habitat areas, Irish deepwater coral reefs, 824
nonfishing impacts on habitat, legislation, 20

nonmarine vs. marine conservation, 104–107
North Atlantic Oscillation, 610
North Pacific Fishery Management Council (NPFMC)
    alternatives for habitat protection, 33–36
    essential fish habitat designation, 31
    fishing restrictions, 31–32
    habitat areas of particular concern, 31
    habitat protection, 33–36
North Sea
    biological traits of, 477–487
    blue mussel populations, 610
    cockle populations, 610
    Eider duck breeding success, 610
    epibenthos assemblages, 834
    fisheries ecosystem plan, 835–836
    fishing impacts and extirpations, 623–624
    historic vs. contemporary communities, 622
    military activities on macrofauna on benthic habitat, 615
    plaice, 610
    pre-fishing state, 619–625
    reconstructing pre-fishing state, 620–622
    taxa inventories, 622–623
    trawling impact on benthic nutrient dynamics, 491–500
    vulnerability criteria and scoring, 624, 625
Northeastern Channel, Nova Scotia, 369–380
Northern Anacapa Island, California, 161–164
northern cutthroat eel, 513
northern horse mussel. *see* horse mussel
northern lobster, 173
northern pink shrimp, 590–591
northern rock sole, 290
Northern Shrimp Zone (NSZ), 571–576
Northria conchylega, 416
NRC. *see* National Research Council (NRC)
NSZ. *see* Northern Shrimp Zone (NSZ)
Nuculana minuta, 523
Nuculidae bivalves, 446
nutrient analysis, 78, 493–498
nutrient releases and sediment biogeochemistry, 539–543

O

objectives-based fisheries management, 52–54
Ocean Habitat Protection Act, 836–837
OCEAN (ocean communities economics/ecological/equity analysis)
    analysis sensitivity, 731–735
    data sources, 729–730
    effects of area closures, 739–742
    fishing effort sub-model, 731
    framework, 729
ocean quahog, 691–693

ocean zoning, 864
Oceana, 863
Oculina coral ecosystem, 795–803
Oculina Habitat Area of Particular Concern (OHAPC), 796
    causes of habitat decline, 801–803
    fish densities, 798
    fish populations, 798–799, 800
    habitat characterization, 800
    habitat conditions, 796–798
    habitat modules, 798, 799
    reef ball clusters, 798, 799, 800
Oculina varicosa. *see* ivory tree coral
OHAPC. *see Oculina* Habitat Area of Particular Concern (OHAPC)
Oligoplites, 59
Olympic Coast National Marine Sanctuary, 629–630
open-access equilibrium, 748
Ophiodon elongatus. *see* lingcod
Ophiopholis aculeata. *see* daisy brittlestar
Ophiothrix fragilis, 656–657
Ophiura spp., 486
Opistonema, 59
optical ground-truthing, 176–177
orange roughy, 509
organic carbon content of sand habitats, effects of fishing, 596–597
OSPAR Convention, 45
Ostrea edulis. *see* edible oysters
otter trawling
    effect on benthic nutrient releases and sediment biogeochemistry, 539–543
    effect on hyperbenthic communities, 529–535
    effects on benthic habitat and macrobenthic communities, 411–422
    effects on sub-Arctic benthic assemblage, 519–527
    English Channel, 316
    experimental design, 412
    fish landings, discards, benthic material, 316
    Grand Banks otter trawling experiment, 415–416
    ground ropes, 540–541
    Gulf of Alaska, 461–452
    impacts, 776
    inorganic nutrient releases from sediment, 542–543
    physical damage to benthic organisms
        Barents Sea Bear Island fishery protection zone, 523–524
    trawling-induced fish diet changes, 400–404
otter trawling and demersal fish feeding
    American plaice, 392
    Atlantic cod, 392, 396–400
    daisy brittlestar, 398
    diet comparisons, 394–396
    experimental site and design, 392–393

feeding changes, 406–408
haddock, 392, 396–400
horse mussel, 397
little skate, 396
sample processing, 393–394
stomach contents, 396–400
stomach samples, 394
trawl catch composition, 394–396
Western Bank, Nova Scotia, 391–408
winter flounder, 392, 396–400
yellowtail flounder, 392
overcapacity, 11, 44
overhead/ascendancy (O/A) ratio, 62
oxygen depletion from sediment resuspension, 78
oyster fishing, 591, 595
oyster habitats, 211–212, 629
oysters, 60

P

Pacific cod, 683
Pacific Fishery Management Council (PFMC), 31
    capacity estimates, 735–739
    essential fish habitat designation, 30–31
    Fishery Management Plan, 119–120
    fishing restrictions, 31
    habitat areas of particular concern, 31
    socioeconomic implications of marine resource management measures, 728
Pacific halibut, 235–241
    effects of iceberg gouging, 239–241
    field methods and observations, 237–238
    ice gouging effects on community structure, 247
    living substrate in Alaska, 290
Pacific ocean perch, 120, 290, 668
Pacific sardine, 59
Pamlico River Estuary, North Carolina, 632, 635–636
Pandalus borealis. see northern pink shrimp
Pandalus propinquus, 379
Panulirus argus. see spiny lobster
Panulirus guttatus. see spotted spiny lobster
paradigm of vulnerability, risk, availability, 785–793
Paragorgia arborea, 372–373, 376–379
Paralichthys dentatus. see summer flounder
Paralichthys oblongus. see fourspot flounder
Paralithodes camtschaticus. see red king crab
Parapercis colias. see blue cod
Parthenope massena, 549
patch model, 642–646
patchiness of effort, 81–83, 82
patterns of disturbance, 81
patterns of disturbance, 81–82
PCB. *see* poly-chlorine-biphenyls (PCB)
Pecten maximus, 659

Penaeus aztecus. see brown shrimp
Pentapora foliacea, 659
performance metrics, 655–656, 658–661, 662
Petromyzon marinus. see sea lamprey
PFMC. *see* Pacific Fishery Management Council (PFMC)
Phaxas, 658
Philine aperta, 549
Phoca vitulina. see common seals
Phocoena sinus. see vaquita
pink shrimp, 63–64
pipelines and cables impact on benthic habitat, 615
Piscataqua River Rocky Region Mosaic substrate transitions, 174
place-based conservation, 102
Placopecten magellanicus, 339–340
plaice, 610
Pleurogrammus monopterygius. see Atka mackerel
Pleurronectes platessa. see plaice
pockmarks
    blueline tilefish habitat, 301
    data collection methods, 302
    excavated seep, 309, 310
    fish burrow, 308, 310
    Flower Garden Banks, 302–304
    gas-escape, 308, 310
    Gulf of Mexico outer shelf, 301–310
    Madison-Swanson, 304
    Mississippi and Alabama Pinnacles Area, 304
    multibeam bathymetry and backscatter data, 302
    red grouper habitat, 301
    sediment samples, 302
    side-scan sonar data, 302
    Steamboat Lumps, 305–307
    Twin Ridges, 304–305
    video observations, 302
pollock, 21
pollution and eutrophication impact on benthic habitat, 615
poly-chlorine-biphenyls (PCB), 610
Porcupine Seabight, 821
pore water release, 491–492
Port Phillip Bay, Melbourne, Australia, 313
Potamilla neglecta, 329
potential groundfish essential fish habitats, fishers' questionnaires, 267–268, 271–273
potential marine protected areas, free distribution theory, 838–839
pre-fishing state, North Sea benthos, 619–625
precautionary approach language, 44–46
precautionary reference points, 53–54
predator and prey dynamics in deep-water reefs, 320–321
predictive models, 14
prey species, habitat regulations, 23–24
prey taxa abundance, 471–473

Primnoa resedaeformis, 370–373, 376–379
private property rights, 718
probable marine sublittoral habitat types, 190–191, 198
production, turnover time, and biomass, 74–77
production fluctuations, 75
production/respiration (P/R) ratio, 62
production to biomass ratios increases, 75
pseudo-side-scan sonar and multibeam bathymetry, 37
Pseudopleuronectes americanus. see winter flounder
Public Law
    101-627, 20
    104-297, 19, 21
Pycnopodia helianthoides, 455

Q

quadrat counts, subtidal oyster reefs, 153–158
quagga, 207
quantitative measure of acoustic diversity, 209–210
quantitative seafloor habitat, GIS terrain analysis, 206
quasi-privatization right, fishing rights, 718
queen conch, *61*

R

RACEBASE survey database, 290–291
radionuclides evaluation of sediment disturbance, 632
rapido trawl, 547
RBDI. *see* relative benthic fauna damage index (RBDI)
recolonization dynamics
    comparison of indicators, 565–566
    sand and mud habitats, 564–565
    studies, 545
    trawl fishing disturbance, 545–567
red drum, 27
red grouper
    excavated seep pockmarks, 309
    habitat changes, 245–246
    Mexico, *59, 61*
    and pockmarks, 301
red hake, 354
red herrings, 853
red king crab, 290, 428, 432–435, 631
red snapper habitat, *61*
    catch data, 278–279
    density estimates, 279
    density juvenile estimates, 280–287
    northern Gulf of Mexico continental shelf, 277–287
    seabed characterization, 279
    trawl sampling, 279–280
red urchin, *61*
redfish, 379
reef ball clusters, 798, 800

reef epibenthic communities protection, 838
Regional Fishery Management Council actions to protect habitat
    Caribbean Fishery Management Council (CFMC), 28–29
    Gulf of Mexico Fishery Management Council (GMFMC), 28
    Mid-Atlantic Fishery Management Council (MAFMC), 26
    New England Fishery Management Council (NEFMC), 24–26
    North Pacific Fishery Management Council (NPMC), 31–32
    Pacific Fishery Management Council (PFMC), 30–31
    South Atlantic Fishery Management Council (SAFMC), 26–28
    Western Pacific Fishery Management Council (WPFMC), 29–30
regulatory change, social and economic impacts, 714
relative benthic fauna damage index (RBDI), 612–616
remote sensing techniques evaluation, sediment disturbance, 632
remote species identification and quantification, fish stock assessments, 316–317
remotely operated vehicle remotely operated platform for ocean science (ROV *ROPOS*), 123, 126–127
remotely operated vehicle (ROV), 120, 203
resource rent, 748
restoration time, 612
Reykjavik Declaration, 42
Rio Declaration (Agenda 21), *45*
risk and uncertainty, 773–775
risk evaluation, 94, 775–776
risk management, 95, 777–779
rock sole, 683
Rockall Trough, 821
rockfish, 13, 119–120, 291, 295–296, 314
ROV *ROPOS. see* remotely operated vehicle remotely operated platform for ocean science (ROV *ROPOS*)
RoxAnn, 203

S

Sabellaria spinulosa, 75
sabellid worms, 75, 398
sablefish, 291
SAFMC. *see* South Atlantic Fishery Management Council (SAFMC)
Salvelinus namaycush. see lake trout
sand and gravel extraction impact on benthic habitat, 612–615
sand dollars, 417–418

sand habitats
    Adriatic Sea, 545–567
    habitat structure, 600
    organic carbon content, 596–597
    recolonization dynamics, 564–565
sand shrimp, 333
sanddabs, 674
Santa Maria Key's causeway, 634
Sardinops sagax. see Pacific sardine
scaldfish, 610
scale, 189
scales of disturbance, seafloor, 639–646
scales of fishing disturbance, 641–642
scallop dredging, 79, 590, 598, 786
scallop fishing effort data, 326–329
scallop habitat mapping, 208
scamp, 799, 800
scanning hydrographic operational airborne lidar survey (SHOALS) program, 166
Sciaenops ocellatus, 27
Scientific and Statistical Committee (SSC) Economics Subcommittee, 735–739
scientific study, ethic toward the sea, 706–707
Scientists' Statement on Protecting the World's Deep-sea Coral and Sponge Ecosystems, 111
Scomber, 59
Scorpaenida, 119–120
Scotian Shelf, 208, 251–263, 605
sea lamprey, 165
sea onions, corals, bunions, protective areas, 857
sea pens, 289–298
sea scallops, 346, 597
sea star, 333
sea urchins, 333, 415, 417
sea whips, 289–298, 464–468, 466, 471–473
seabed
    classification with multibeam sonar, 209
    community diversity, 696
    contact reduction by trawls, 837
    deepwater fishing, 503–504
    dynamics and currents, 186, 187
    gorgonian corals, 371
    habitat characterization data, 12
    habitat classification structure, 188
    habitat mapping, 179–182
    impacts research, 859–860
    mapping, 87
    probable marine sublittoral habitat types, 190–191
seafloor
    classification, 135, *136*
    disturbance ecology, 640–641
    GIS terrain analysis, 206
    habitat studies, 14–15
    habitats and communities, 640
    habitats effects from commercial fishing gears, 841–842
    mapping, 123, *124, 125*, 139, *140*, 141, 208
    pockmarks, 301–310
    response to natural events, 215–218
    spatial and temporal scales of disturbance, 639–646
    structures, 108–111
    structures and trawling and dredging impacts, 109–111
    time-lapse photographs, 245
seagrass, 313
    mapping with waveform-resolving lidar, 202
    spiny lobster traps, 579–587
    and trawling, 634
seamount benthic habitat and fauna, 593
Sebastes alutus. see Pacific ocean perch
Sebastes crameri. see darkblotched rockfish
Sebastes entomelas. see widow rockfish
Sebastes hopkinski. see squarespot rockfish
Sebastes levis. see cowcod
Sebastes paucispinis. see bocaccio
Sebastes pinniger. see canary rockfish
Sebastes ruberrimus. see yelloweye rockfish
Sebastes semicinctus. see halfbanded rockfish
Sebastolobus alascanus. see shortspine thornyhead
Secretary of Commerce, 12
sediment biogeochemistry, 79–80, 539–543
sediment characterization and trawling impact, 493
sediment disturbance, 632
sediment disturbance and scallop dredges, 79
sediment mapping, 37
sediment nutrient flux, 79
sediment oxygen consumption (SOC)
    benthic incubation microcosms chambers (BMIC), 220–221
    box cores sampling, 220–222
    measuring, 219–220
    southwestern Gulf of Mexico, 219–232, 225, 228
    Tamaulipas shelf and Campeche Bay, 220
sediment volume calculations, 309
sedimentology, assessing trawl impacts, 203
Senate bills, 20, 21
Senate Committee on Commerce, Science, and Transportation, 20
Serripes groenlandicus, 523, 524
sessile habitat-forming species reductions, 75
shallow sites, benthic megafauna recovery, 329–336
Shannon index, 548
SHOALS. *see* scanning hydrographic operational airborne lidar survey (SHOALS)
shortspine thornyhead, 291
shrimp fishery
    bycatch, 60, 62, 277–278
    catch and environmental data, 320

cessation in Corpus Christi Fish Pass, 572, 574–575
cessation in Texas benthic habitats, 571–577
effect on snow crab resource, 590–591
effects estimation, 315
effects on soft sediment habitats, 602
Mexico, 60, 62–63
shrimp and crab trawling impacts, estuarine soft-bottom organisms, 630–631
Sicyonia brevirostris. see brown rock shrimp
side-scan sonar, 13, 414
    acoustic reflectance patterns, 211–212
    assessing trawl impacts, 203
    benthic habitat characterization, 201
    natural and anthropogenic disturbance assessment, 629
    pockmarks, 302
    subtidal oyster reefs, 154, 156–157
    with video/grab sampling verification, 37
    white abalone habitat, 162
single-beam acoustic systems, 87
single-beam echo sounders, 13
Sipunculus nudus, 563
site-specific gear effects, application and scaling of, 14
skates, 291
small-scale analysis, subtidal fish guilds and habitats, 317
smooth bottom trawl gear, soft bottom habitat, 594
snapper-grouper spp., 27
snippet, 179–182
snow crabs, 415, 416–417, 590–591
SOC. *see* sediment oxygen consumption (SOC)
social and economic impact analysis (SIA), 716–717
social and economic impacts, regulatory change, 714
social and environmental ethics, 714
social and political pressures, 11
social effects of marine protected areas (MPA), 49
social impact analysis, 721
social impacts of marine protected areas (MPA), 715
social pressures and conservation, regulations, 861
social science
    assessment, 94
    in benthic habitat management, 713–722
    and evidence-based knowledge, 719–720
    Magnuson-Stevens Act, 716
    in marine resource management, 716–717
    underrepresentation in marine fisheries and conservation, 715
social sustainability, 47
socioeconomic implications
    in limited-access fishery management system, 833
    of marine resource management measures, 728
    spatial tools for assessing, 727–743
socioeconomic studies, 14
soft corals, 383–389, 415
soft-sediment habitats, 461–473, 602

solar activity and cyclic behavior in coastal seas, 610
Solenomilia variabilis, 509
Somateria mollissima. see Eider ducks
sonar. *see* side-scan sonar
South Atlantic Fishery Management Council (SAFMC), 27–28
Southern Shrimp Zone (SSZ), 571–577
spatial analysis
    fishers' behavior, 713–714
    Heceta Bank, Oregon, 131, 133–134
spatial and temporal distributions
    analyzing data by blocks, 687–688
    block sizes, 684–685
    bottom trawling in Alaska, 679–689
    habitat effect model, 685
    sweep rates modeling, 684, 685–688
    trawling activity, 696–697
spatial and temporal scales of disturbance, habitat management framework, 639–646
spatial complexity/diversity of ecosystems, 108
spatial distribution, 592–593, 698–699
spatial tools
    socioeconomic implications of management measures, 727–743
spawning, breeding, feeding, or growth to maturity, 23
Species at Risk Act, *46*
species composition comparison, biological traits analysis, 482–487
species composition shifts, 111
species conservation, 102
species distribution, 13
species diversity and epifauna abundance, 467–469
species diversity and habitat complexity, 108
speckled hind, 799, 800
spiny lobster, *61,* 579–587, 592
Spirontocaris spinus, 329, 523
Spisula solidissima. see Atlantic surfclam
sponges, 289–298, 596, 659
spotted spiny lobster, 27
squarespot rockfish, 674
SSZ. *see* Southern Shrimp Zone (SSZ)
static fishing gears impact, 776
Steamboat Lumps, 305–307, 308, 309
Stellwagen Bank National Marine Sanctuary Region, 603–604
STIL. *see* streak tube imaging lidar (STIL)
stock assessment, groundfish, 668–671
stock assessment paradigm, 106–107, 111
stock-recruitment relationship, 66
stomach contents and demersal fishes, 394, 396–400
Straddling Stocks Treaty, *46*
streak tube imaging lidar (STIL), 211
Strengthened Protection for Fisheries Habitat, 21
Strongylocentrotus droebachiensis, 339

Strongylocentrotus spp., 524
structural complexity of habitats, 197
sub-Arctic benthic assemblage, 519–527
submersible *Delta,* 123, 126
substrate, defined, 22
substrate morphology and geology. laser technology, 165–169
substratum (seafloor) classification, 127–131
subtidal fish guilds and habitats, 317
subtidal oyster reefs, 153–158
summer flounder *Paralichthys dentatus,* 26
surface sediment resuspension, 78
surficial sediment deposits, 142–143, 786
survey techniques
    groundfish population abundance, 669–671
    impact of demersal trawling, 808
suspensivores, 455
sustainability concept, 64–66
sustainable fisheries, 12
Sustainable Fisheries Act, 12, 20–21, 32–33, 92, 765–766
sweep rates modeling, 684, 685–688
symbolic issues, ethic toward the sea, 709
Synaphobranchus kaupi. see northern cutthroat eel
Syringammina fragilissima. see xenophyophore
Syringodium filiforme, 580–581, 580–587, 583–587

T

TAC. *see* total allowable catch (TAC)
Tamaulipas shelf and Campeche Bay, 220
target reference points, 54
targeted population and economic system, 747–751
taxa inventories, 622–623
TBT. *see* tributyltin (TBT)
Tellinidae bivalves, 446
temporal and spatial scales of disturbance, 639–646
temporal context and management choices, 43–44
terrestrial conservation, 104–107
terrestrial *vs.* freshwater realms, 107–108
Texas benthic habitats, shrimp trawling cessation, 571–577
textural analysis of sonar data, 162–163
Thalassia testudinum, 580–587
Thelepus cincinnatus, 329, 339, 398, 404
theoretical and methodological basis of estimations, human-made influences, 595–596
Theragra chalcogramma. see walleye pollock
Thracia spp., 523, 524
tilefish, 26, 308, 319, 634–635
time-lapse photographs, 245
TL. *see* trophic level (TL)
toad crab, 329, 339
tools on the axes of sustainability, 47–54

topographic position index (TPI), 130–131
total allowable catch (TAC), 106–107
totoaba, 63
tourist income and database for benthic fisheries, 760–761
towed trawl simulator sledge (TTSS), 541
TPI. *see* topographic position index (TPI)
Trachinus vipera. see lesser weever
Trachythyone elongata, 549
trans-Pacific cable lines, 204–205
"Transition to Sustainability," 728
trap-induced injuries and seagrass, 584–587
trawl catch and RoxAnn, 412–414
trawl catch composition, 394–396
trawl fishing disturbance and recolonization dynamics, 545–567
trawl marks analysis, 507–508
trawler tracks, 862
trawling
    deep seabed impacts, 510–512
    effect on body size, 428–430
    effects of, 11
    effects of chronic bottom, 425–435
    fishermen behavior, 858
    frequency and geographic distribution of, 93–94
    and seagrass, 634
    and wind disturbance, 635–636
trawling and dredging, 106–107
    assessing trawl impacts, 203
    fishing, 849–850
    framework for decision making, 94–96
    frequency and geographic distribution of trawl and dredge effort, 93–94
    gear-specific effects, 93
    habitat mapping, 94
    impacts, 786
    impacts on seafloor structures, 109–111
    National Research Council (NRC) study, 92–98
    review of, 93–94
    species composition shifts, 111
    trawl damage, 749
    trawl fishing disturbance, 545–567
trawling catch composition experimental study, 448–450
trawling disturbance
    aircraft records, 81
    altering sediment nutrient flux, 79
    benthic ecosystem, 477–487
    benthic nutrient dynamics, 491–500
    Darwin Mounds, 809–814
    epifaunal communities, 508–510
    experimental treatments and sampling procedure, 492–493
    flux rates, 498–500
    hyperbenthic communities, 533–536

infaunal communities, 510
management actions on effort distributions, 83–84
microcosm nutrient concentrations and fluxes, 493–498
nutrient analysis, 493
patchiness of effort, 81–83
patterns of, 81–84
pore water release, 491–492
and production, 82
quantifying, 81–82
sediment characterization, 493
trawling-induced fish diet changes, 400–404
tributyltin (TBT), 611, 615
trophic cascades, 855
trophic guilds, 566
trophic level (TL), 62–63
tropical storm Gabrielle, 582, 584
TTSS. *see* towed trawl simulator sledge
tube-building worm, 75
tube-dwelling amphipods, 446
tube-dwelling polychaetes, 446
tunas, 59
turnover time, biomass, and production, 74–77
Turritellidae, 446
Twin Ridges, 304–305, 308
"type conversion," 111

U

uncertainty and risk, 773–775
United Nations Fish Stock Agreement, 42
United States (northeast)
commercial fishing gears, 698–699
effects of fishing gear on benthic habitats workshop, 833–834
effects of hydraulic clam dredges, 691–693
univariate metrics, 654–655
unsustainability of fisheries, 47–49, 51
upper slope rockfish, 314
U.S. Continental Shelf, 13
U.S. Endangered Species Act (ESA), 102–103
U.S. Geological Survey (USGS), 12
U.S. Marine Mammal Protection Act (MMPA), 705–706
USGS. *see* U.S. Geological Survey (USGS)
usSEABED, 210–211
utilitarian attitude, 102
utilitarian values, 706
utilization, 786

V

vaquita, *63*
vessel size impact on benthic habitat, 44

video analysis
gorgonian corals, 371–372
ground-truthing, benthic habitat characteristics, 171–176
pockmarks, 302
soft corals abundance, 386–387
subtidal oyster reefs, 154–158
white abalone habitat, 162
video sledge, 203
videograb, 414, 417–418
vulnerability scoring, 622, 624, 625
vulnerability to fishing, 786

W

walleye pollock, 96–97, 683
walrus foraging *vs.* trawling, 633
warsaw grouper, 800
water column processes, 635–636
water-sediment interface, 855
waters, defined, 22
waved whelks, 339, 611
waveform-resolving lidar, 202
West Coast cooperative research program, 759
West Coast groundfish overfishing, 119–120
Western Bank, Nova Scotia, 391–408
Western Pacific Fishery Management Council, 29
whale fishery, changing ethics, 704–706
whelks, 415
Whidbey Passage study, 235–241
white abalone, 161–164, 163
widow rockfish, 120, 668
"wigs," 274
wind trawling disturbance, 635–636
winter flounder, 354, 392, 394–8, 404
feeding behavior, 407–408
trawling-induced fish diet changes, 400–404
winter skate, 354
World Summit on Sustainable Development, 41

X

xenophyophore, 812
"Xenophyophore facies," 811–812, 814

Y

yellow flounder, 394–397
yellowedge grouper, 308
yelloweye rockfish, 120, 668
yellowfin flounder, 96
yellowfin sole, 440, 443, 450–451, 455, 683

yellowfin tuna, 59
yellowtail flounder, 352, 354, 392, 400–404, 407–408

Z

zebra mussel communities, 206–207
zonal application of fishing gear, 778
Zostera marina, 609